HYDRODYNAMICS

BY

SIR HORACE LAMB, M.A., LL.D., Sc.D., F.R.S.

HONORARY FELLOW OF TRINITY COLLEGE, CAMBRIDGE; LATELY PROFESSOR OF MATHEMATICS IN THE VICTORIA UNIVERSITY OF MANCHESTER

SIXTH EDITION

DOVER PUBLICATIONS
NEW YORK

This Dover edition, first published in 1945, is an unabridged and unaltered republication of the 6th (1932) edition of the work originally published in 1879.

This edition is published by special arrangement with Cambridge University Press.

International Standard Book Number: 0-486-60256-7
Library of Congress Catalog Card Number: 46-1891

Manufactured in the United States of America
Dover Publications, Inc.
180 Varick Street
New York, N.Y. 10014

PREFACE

THIS may be regarded as the sixth edition of a *Treatise on the Mathematical Theory of the Motion of Fluids*, published in 1879. Subsequent editions, largely remodelled and extended, have appeared under the present title.

In this issue no change has been made in the general plan and arrangement, but the work has again been revised throughout, some important omissions have been made good, and much new matter has been introduced.

The subject has in recent years received considerable developments, in the theory of the tides for instance, and in various directions bearing on the problems of aeronautics, and it is interesting to note that the "classical" Hydrodynamics, often referred to with a shade of depreciation, is here found to have a widening field of practical applications. Owing to the elaborate nature of some of these researches it has not always been possible to fit an adequate account of them into the frame of this book, but attempts have occasionally been made to give some indication of the more important results, and of the methods employed.

As in previous editions, pains have been taken to make due acknowledgment of authorities in the footnotes, but it appears necessary to add that the original proofs have often been considerably modified in the text.

I have again to thank the staff of the University Press for much valued assistance during the printing.

HORACE LAMB

April 1932

CONTENTS

CHAPTER I

THE EQUATIONS OF MOTION

CHAPTER II

INTEGRATION OF THE EQUATIONS IN SPECIAL CASES

CHAPTER III

IRROTATIONAL MOTION

CHAPTER IV

MOTION OF A LIQUID IN TWO DIMENSIONS

CHAPTER V

IRROTATIONAL MOTION OF A LIQUID: PROBLEMS IN THREE DIMENSIONS

CHAPTER VI

ON THE MOTION OF SOLIDS THROUGH A LIQUID: DYNAMICAL THEORY

CHAPTER VII

VORTEX MOTION

CHAPTER VIII

TIDAL WAVES

CHAPTER IX

SURFACE WAVES

CHAPTER X

WAVES OF EXPANSION

363
195

CHAPTER XI

VISCOSITY

CHAPTER XII

ROTATING MASSES OF LIQUID

HYDRODYNAMICS

CHAPTER I

THE EQUATIONS OF MOTION

1. THE following investigations proceed on the assumption that the matter with which we deal may be treated as practically continuous and homogeneous in structure; *i.e.* we assume that the properties of the smallest portions into which we can conceive it to be divided are the same as those of the substance in bulk.

The fundamental property of a fluid is that it cannot be in equilibrium in a state of stress such that the mutual action between two adjacent parts is oblique to the common surface. This property is the basis of Hydrostatics, and is verified by the complete agreement of the deductions of that science with experiment. Very slight observation is enough, however, to convince us that oblique stresses may exist in fluids *in motion*. Let us suppose for instance that a vessel in the form of a circular cylinder, containing water (or other liquid), is made to rotate about its axis, which is vertical. If the angular velocity of the vessel be constant, the fluid is soon found to be rotating with the vessel as one solid body. If the vessel be now brought to rest, the motion of the fluid continues for some time, but gradually subsides, and at length ceases altogether; and it is found that during this process the portions of fluid which are further from the axis lag behind those which are nearer, and have their motion more rapidly checked. These phenomena point to the existence of mutual actions between contiguous elements which are partly tangential to the common surface. For if the mutual action were everywhere wholly normal, it is obvious that the moment of momentum, about the axis of the vessel, of any portion of fluid bounded by a surface of revolution about this axis, would be constant. We infer, moreover, that these tangential stresses are not called into play so long as the fluid moves as a solid body, but only whilst a change of shape of some portion of the mass is going on, and that their tendency is to oppose this change of shape.

2. It is usual, however, in the first instance to neglect the tangential stresses altogether. Their effect is in many practical cases small, and, independently of this, it is convenient to divide the not inconsiderable difficulties of our subject by investigating first the effects of purely normal stress. The further consideration of the laws of tangential stress is accordingly deferred till Chapter XI.

If the stress exerted across any small plane area situate at a point P of the fluid be wholly normal, its intensity (per unit area) is the same for all aspects of the plane. The following proof of this theorem is given here for purposes of reference. Through P draw three straight lines PA, PB, PC mutually at right angles, and let a plane whose direction-cosines relatively to these lines are l, m, n, passing infinitely close to P, meet them in A, B, C. Let p, p_1, p_2, p_3 denote the intensities of the stresses* across the faces ABC, PBC, PCA, PAB, respectively, of the tetrahedron $PABC$. If Δ be the area of the first-mentioned face, the areas of the others are, in order, $l\Delta$, $m\Delta$, $n\Delta$. Hence if we form the equation of motion of the tetrahedron parallel to PA we have $p_1 \,.\, l\Delta = pl \,.\, \Delta$, where we have omitted the terms which express the rate of change of momentum, and the component of the extraneous forces, because they are ultimately proportional to the mass of the tetrahedron, and therefore of the third order of small linear quantities, whilst the terms retained are of the second. We have then, ultimately, $p = p_1$, and similarly $p = p_2 = p_3$, which proves the theorem.

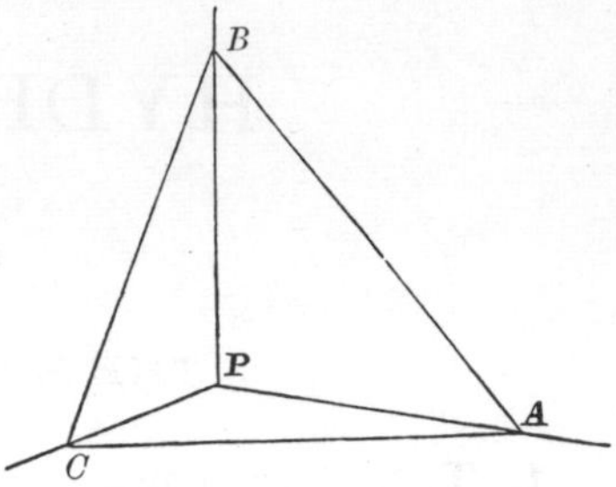

3. The equations of motion of a fluid have been obtained in two different forms, corresponding to the two ways in which the problem of determining the motion of a fluid mass, acted on by given forces and subject to given conditions, may be viewed. We may either regard as the object of our investigations a knowledge of the velocity, the pressure, and the density, at all points of space occupied by the fluid, for all instants; or we may seek to determine the history of every particle. The equations obtained on these two plans are conveniently designated, as by German mathematicians, the 'Eulerian' and the 'Lagrangian' forms of the hydrokinetic equations, although both forms are in reality due to Euler†.

The Eulerian Equations.

4. Let u, v, w be the components, parallel to the co-ordinate axes, of the velocity at the point (x, y, z) at the time t. These quantities are then functions of the independent variables x, y, z, t. For any particular value of t they define the motion at that instant at all points of space occupied by

* Reckoned positive when pressures, negative when tensions. Most fluids are, however, incapable under ordinary conditions of supporting more than an exceedingly slight degree of tension, so that p is nearly always positive.

† "Principes généraux du mouvement des fluides," *Hist. de l'Acad. de Berlin*, 1755.

"De principiis motus fluidorum," *Novi Comm. Acad. Petrop.* xiv. 1 (1759).

Lagrange gave three investigations of the equations of motion; first, incidentally, in

the fluid; whilst for particular values of x, y, z they give the history of what goes on at a particular place.

We shall suppose, for the most part, not only that u, v, w are finite and continuous functions of x, y, z, but that their space-derivatives of the first order ($\partial u/\partial x$, $\partial v/\partial x$, $\partial w/\partial x$, &c.) are everywhere finite*; we shall understand by the term 'continuous motion,' a motion subject to these restrictions. Cases of exception, if they present themselves, will require separate examination. In continuous motion, as thus defined, the relative velocity of any two neighbouring particles P, P' will always be infinitely small, so that the line PP' will always remain of the same order of magnitude. It follows that if we imagine a small closed surface to be drawn, surrounding P, and suppose it to move with the fluid, it will always enclose the same matter. And *any* surface whatever, which moves with the fluid, completely and permanently separates the matter on the two sides of it.

5. The values of u, v, w for successive values of t give as it were a series of pictures of consecutive stages of the motion, in which however there is no immediate means of tracing the identity of any one particle.

To calculate the rate at which any function $F(x, y, z, t)$ varies for a moving particle, we may remark that at the time $t+\delta t$ the particle which was originally in the position (x, y, z) is in the position $(x+u\delta t,\ y+v\delta t,\ z+w\delta t)$, so that the corresponding value of F is

$$F(x+u\delta t,\ y+v\delta t,\ z+w\delta t,\ t+\delta t)=F+u\delta t\frac{\partial F}{\partial x}+v\delta t\frac{\partial F}{\partial y}+w\delta t\frac{\partial F}{\partial z}+\delta t\frac{\partial F}{\partial t}.$$

If, after Stokes, we introduce the symbol D/Dt to denote a differentiation following the motion of the fluid, the new value of F is also expressed by $F+DF/Dt\,.\,\delta t$, whence

$$\frac{DF}{Dt}=\frac{\partial F}{\partial t}+u\frac{\partial F}{\partial x}+v\frac{\partial F}{\partial y}+w\frac{\partial F}{\partial z}. \qquad\qquad (1)$$

6. To form the dynamical equations, let p be the pressure, ρ the density, X, Y, Z the components of the extraneous forces per unit mass, at the point (x, y, z) at the time t. Let us take an element having its centre at (x, y, z), and its edges δx, δy, δz parallel to the rectangular co-ordinate axes. The rate at which the x-component of the momentum of this element is increasing is $\rho\,\delta x\,\delta y\,\delta z\,Du/Dt$; and this must be equal to the x-component of the forces

connection with the principle of Least Action, in the *Miscellanea Taurinensia*, ii. (1760) [*Oeuvres*, Paris, 1867–92, i.]; secondly in his "Mémoire sur la Théorie du Mouvement des Fluides," *Nouv. mém. de l'Acad. de Berlin*, 1781 [*Oeuvres*, iv.]; and thirdly in the *Mécanique Analytique*. In this last exposition he starts with the second form of the equations (Art. 14, below), but translates them at once into the 'Eulerian' notation.

* It is important to bear in mind, with a view to some later developments under the head of Vortex Motion, that these derivatives need not be assumed to be continuous.

acting on the element. Of these the extraneous forces give $\rho\,\delta x\,\delta y\,\delta z\,X$. The pressure on the yz-face which is nearest the origin will be ultimately

$$(p - \tfrac{1}{2}\partial p/\partial x\,.\,\delta x)\,\delta y\,\delta z\,*,$$

that on the opposite face

$$(p + \tfrac{1}{2}\partial p/\partial x\,.\,\delta x)\,\delta y\,\delta z.$$

The difference of these gives a resultant $-\partial p/\partial x\,.\,\delta x\,\delta y\,\delta z$ in the direction of x-positive. The pressures on the remaining faces are perpendicular to x. We have then

$$\rho\,\delta x\,\delta y\,\delta z\,\frac{Du}{Dt} = \rho\,\delta x\,\delta y\,\delta z\,X - \frac{\partial p}{\partial x}\,\delta x\,\delta y\,\delta z.$$

Substituting the value of Du/Dt from (1), and writing down the symmetrical equations, we have

$$\left.\begin{aligned}\frac{\partial u}{\partial t} + u\frac{\partial u}{\partial x} + v\frac{\partial u}{\partial y} + w\frac{\partial u}{\partial z} &= X - \frac{1}{\rho}\frac{\partial p}{\partial x},\\ \frac{\partial v}{\partial t} + u\frac{\partial v}{\partial x} + v\frac{\partial v}{\partial y} + w\frac{\partial v}{\partial z} &= Y - \frac{1}{\rho}\frac{\partial p}{\partial y},\\ \frac{\partial w}{\partial t} + u\frac{\partial w}{\partial x} + v\frac{\partial w}{\partial y} + w\frac{\partial w}{\partial z} &= Z - \frac{1}{\rho}\frac{\partial p}{\partial z}\end{aligned}\right\} \quad \text{..............(2)}$$

7. To these dynamical equations we must join, in the first place, a certain kinematical relation between u, v, w, ρ, obtained as follows.

If Q be the volume of a moving element, we have, on account of the constancy of mass,

$$\frac{D\,.\,\rho Q}{Dt} = 0,$$

or

$$\frac{1}{\rho}\frac{D\rho}{Dt} + \frac{1}{Q}\frac{DQ}{Dt} = 0. \quad \text{..............................(1)}$$

To calculate the value of $1/Q\,.\,DQ/Dt$, let the element in question be that which at time t fills the rectangular space $\delta x\,\delta y\,\delta z$ having one corner P at (x, y, z), and the edges PL, PM, PN (say) parallel to the co-ordinate axes. At time $t + \delta t$ the same element will form an oblique parallelepiped, and since the velocities of the particle L relative to the particle P are $\partial u/\partial x\,.\,\delta x$, $\partial v/\partial x\,.\,\delta x$, $\partial w/\partial x\,.\,\delta x$, the projections of the edge PL on the co-ordinate axes become, after the time δt,

$$\left(1 + \frac{\partial u}{\partial x}\,\delta t\right)\delta x, \qquad \frac{\partial v}{\partial x}\,\delta t\,.\,\delta x, \qquad \frac{\partial w}{\partial x}\,\delta t\,.\,\delta x,$$

respectively. To the first order in δt, the length of this edge is now

$$\left(1 + \frac{\partial u}{\partial x}\,\delta t\right)\delta x,$$

and similarly for the remaining edges. Since the angles of the parallelepiped

* It is easily seen, by Taylor's theorem, that the mean pressure over any face of the element $\delta x\,\delta y\,\delta z$ may be taken to be equal to the pressure at the centre of that face.

differ infinitely little from right angles, the volume is still given, to the first order in δt, by the product of the three edges, *i.e.* we have

$$Q+\frac{DQ}{Dt}\delta t=\left\{1+\left(\frac{\partial u}{\partial x}+\frac{\partial v}{\partial y}+\frac{\partial w}{\partial z}\right)\delta t\right\}\delta x\,\delta y\,\delta z,$$

or

$$\frac{1}{Q}\frac{DQ}{Dt}=\frac{\partial u}{\partial x}+\frac{\partial v}{\partial y}+\frac{\partial w}{\partial z}. \qquad \text{.......................(2)}$$

Hence (1) becomes

$$\frac{D\rho}{Dt}+\rho\left(\frac{\partial u}{\partial x}+\frac{\partial v}{\partial y}+\frac{\partial w}{\partial z}\right)=0. \qquad \text{....................(3)}$$

This is called the 'equation of continuity.'

The expression

$$\frac{\partial u}{\partial x}+\frac{\partial v}{\partial y}+\frac{\partial w}{\partial z}, \qquad \text{...........................(4)}$$

which, as we have seen, measures the rate of dilatation of the fluid at the point (x, y, z), is conveniently called the 'expansion' at that point. From a more general point of view the expression (4) is called the 'divergence' of the vector (u, v, w); it is often denoted briefly by

$$\operatorname{div}(u, v, w).$$

The preceding investigation is substantially that given by Euler*. Another, and now more usual, method of obtaining the equation of continuity is, instead of following the motion of a fluid element, to fix the attention on an element $\delta x\,\delta y\,\delta z$ of space, and to calculate the change produced in the included mass by the flux across the boundary. If the centre of the element be at (x, y, z), the amount of matter which per unit time enters it across the yz-face nearest the origin is

$$\left(\rho u-\tfrac{1}{2}\frac{\partial\,.\,\rho u}{\partial x}\delta x\right)\delta y\,\delta z,$$

and the amount which leaves it by the opposite face is

$$\left(\rho u+\tfrac{1}{2}\frac{\partial\,.\,\rho u}{\partial x}\delta x\right)\delta y\,\delta z.$$

The two faces together give a gain

$$-\frac{\partial\,.\,\rho u}{\partial x}\delta x\,\delta y\,\delta z,$$

per unit time. Calculating in the same way the effect of the flux across the remaining faces, we have for the total gain of mass, per unit time, in the space $\delta x\,\delta y\,\delta z$, the formula

$$-\left(\frac{\partial\,.\,\rho u}{\partial x}+\frac{\partial\,.\,\rho v}{\partial y}+\frac{\partial\,.\,\rho w}{\partial z}\right)\delta x\,\delta y\,\delta z.$$

Since the quantity of matter in any region can vary only in consequence of the flux across the boundary, this must be equal to

$$\frac{\partial}{\partial t}(\rho\,\delta x\,\delta y\,\delta z),$$

* *l.c. ante* p. 2.

whence we get the equation of continuity in the form

$$\frac{\partial \rho}{\partial t} + \frac{\partial . \rho u}{\partial x} + \frac{\partial . \rho v}{\partial y} + \frac{\partial . \rho w}{\partial z} = 0. \quad \text{.....................(5)}$$

8. It remains to put in evidence the physical properties of the fluid, so far as these affect the quantities which occur in our equations.

In an 'incompressible' fluid, or liquid, we have $D\rho/Dt = 0$, in which case the equation of continuity takes the simple form

$$\frac{\partial u}{\partial x} + \frac{\partial v}{\partial y} + \frac{\partial w}{\partial z} = 0. \quad \text{..............................(1)}$$

It is not assumed here that the fluid is of *uniform* density, though this is of course by far the most important case.

If we wish to take account of the slight compressibility of actual liquids, we shall have a relation of the form

$$p = \kappa (\rho - \rho_0)/\rho_0, \quad \text{..............................(2)}$$

or

$$\rho/\rho_0 = 1 + p/\kappa, \quad \text{..............................(3)}$$

where κ denotes what is called the 'elasticity of volume.'

In the case of a gas whose temperature is uniform and constant we have the 'isothermal' relation

$$p/p_0 = \rho/\rho_0, \quad \text{..............................(4)}$$

where p_0, ρ_0 are any pair of corresponding values for the temperature in question.

In most cases of motion of gases, however, the temperature is not constant, but rises and falls, for each element, as the gas is compressed or rarefied. When the changes are so rapid that we can ignore the gain or loss of heat by an element due to conduction and radiation, we have the 'adiabatic' relation

$$p/p_0 = (\rho/\rho_0)^\gamma, \quad \text{..............................(5)}$$

where p_0 and ρ_0 are any pair of corresponding values for the element considered. The constant γ is the ratio of the two specific heats of the gas; for atmospheric air, and some other gases, its value is about 1·408.

9. At the boundaries (if any) of the fluid, the equation of continuity is replaced by a special surface-condition. Thus at a *fixed* boundary, the velocity of the fluid perpendicular to the surface must be zero, *i.e.* if l, m, n be the direction-cosines of the normal,

$$lu + mv + nw = 0. \quad \text{..............................(1)}$$

Again at a surface of discontinuity, *i.e.* a surface at which the values of u, v, w change abruptly as we pass from one side to the other, we must have

$$l(u_1 - u_2) + m(v_1 - v_2) + n(w_1 - w_2) = 0, \quad \text{..............(2)}$$

where the suffixes are used to distinguish the values on the two sides. The same relation must hold at the common surface of a fluid and a moving solid.

The general surface-condition, of which these are particular cases, is that if $F(x, y, z, t) = 0$ be the equation of a bounding surface, we must have at every point of it

$$DF/Dt = 0. \qquad \text{(3)}$$

For the velocity relative to the surface of a particle lying in it must be wholly tangential (or zero), otherwise we should have a finite flow of fluid across it. It follows that the instantaneous rate of variation of F for a surface-particle must be zero.

A fuller proof, given by Lord Kelvin*, is as follows. To find the rate of motion ($\dot{\nu}$) of the surface $F(x, y, z, t)=0$, normal to itself, we write

$$F(x+l\dot{\nu}\,\delta t,\ \ y+m\dot{\nu}\,\delta t,\ \ z+n\dot{\nu}\,\delta t,\ \ t+\delta t)=0,$$

where l, m, n are the direction-cosines of the normal at (x, y, z). Hence

$$\dot{\nu}\left(l\frac{\partial F}{\partial x}+m\frac{\partial F}{\partial y}+n\frac{\partial F}{\partial z}\right)+\frac{\partial F}{\partial t}=0$$

Since
$$(l, m, n)=\left(\frac{\partial F}{\partial x},\ \frac{\partial F}{\partial y},\ \frac{\partial F}{\partial z}\right)\div R,$$

where
$$R=\left\{\left(\frac{\partial F}{\partial x}\right)^2+\left(\frac{\partial F}{\partial y}\right)^2+\left(\frac{\partial F}{\partial z}\right)^2\right\}^{\frac{1}{2}},$$

we have
$$\dot{\nu}=-\frac{1}{R}\frac{\partial F}{\partial t}. \qquad \text{(4)}$$

At every point of the surface we must have

$$\dot{\nu}=lu+mv+nw,$$

which leads, on substitution of the above values of l, m, n, to the equation (3).

The partial differential equation (3) is also satisfied by any surface moving with the fluid. This follows at once from the meaning of the operator D/Dt. A question arises as to whether the converse necessarily holds; *i.e.* whether a moving surface whose equation $F=0$ satisfies (3) will always consist of the same particles. Considering any such surface, let us fix our attention on a particle P situate on it at time t. The equation (3) expresses that the rate at which P is separating from the surface is at this instant zero: and it is easily seen that *if the motion be continuous* (according to the definition of Art. 4), the normal velocity, relative to the moving surface F, of a particle at an infinitesimal distance ζ from it is of the order ζ, viz. it is equal to $G\zeta$ where G is finite. Hence the equation of motion of the particle P relative to the surface may be written

$$D\zeta/Dt=G\zeta.$$

This shews that $\log\zeta$ increases at a finite rate, and since it is negative infinite to begin with (when $\zeta=0$), it remains so throughout, *i.e.* ζ remains zero for the particle P.

The same result follows from the nature of the solution of

$$\frac{\partial F}{\partial t}+u\frac{\partial F}{\partial x}+v\frac{\partial F}{\partial y}+w\frac{\partial F}{\partial z}=0, \qquad \text{(5)}$$

considered as a partial differential equation in F†. The subsidiary system of ordinary differential equations is

$$dt=\frac{dx}{u}=\frac{dy}{v}=\frac{dz}{w}, \qquad \text{(6)}$$

* (W. Thomson) "Notes on Hydrodynamics," *Camb. and Dub. Math. Journ.* Feb. 1848. [*Mathematical and Physical Papers*, Cambridge, 1882..., i. 83.]

† Lagrange, *Oeuvres*, iv. 706.

in which x, y, z are regarded as functions of the independent variable t. These are evidently the equations to find the paths of the particles, and their integrals may be supposed put in the forms

$$x=f_1(a, b, c, t), \quad y=f_2(a, b, c, t), \quad z=f_3(a, b, c, t), \quad \ldots\ldots\ldots\ldots(7)$$

where the arbitrary constants a, b, c are any three quantities serving to identify a particle; for instance they may be the initial co-ordinates. The general solution of (5) is then found by elimination of a, b, c between (7) and

$$F=\psi(a, b, c), \quad \ldots\ldots\ldots\ldots(8)$$

where ψ is an arbitrary function. This shews that a particle once in the surface $F=0$ remains in it throughout the motion.

Equation of Energy.

10. In most cases which we shall have occasion to consider the extraneous forces have a potential; viz. we have

$$X, Y, Z = -\frac{\partial\Omega}{\partial x}, \; -\frac{\partial\Omega}{\partial y}, \; -\frac{\partial\Omega}{\partial z}. \quad \ldots\ldots\ldots\ldots(1)$$

The physical meaning of Ω is that it denotes the potential energy, per unit mass, at the point (x, y, z), in respect of forces acting at a distance. It will be sufficient for the present to consider the case where the field of extraneous force is constant with respect to the time, *i.e.* $\partial\Omega/\partial t = 0$. If we now multiply the equations (2) of Art. 6 by u, v, w, in order, and add, we obtain a result which may be written

$$\tfrac{1}{2}\rho\frac{D}{Dt}(u^2+v^2+w^2)+\rho\frac{D\Omega}{Dt} = -\left(u\frac{\partial p}{\partial x}+v\frac{\partial p}{\partial y}+w\frac{\partial p}{\partial z}\right).$$

If we multiply this by $\delta x\,\delta y\,\delta z$, and integrate over any region, we find

$$\frac{D}{Dt}(T+V) = -\iiint\left(u\frac{\partial p}{\partial x}+v\frac{\partial p}{\partial y}+w\frac{\partial p}{\partial z}\right)dx\,dy\,dz, \quad \ldots\ldots\ldots(2)$$

where $\quad T=\tfrac{1}{2}\iiint\rho(u^2+v^2+w^2)\,dx\,dy\,dz, \qquad V=\iiint\Omega\rho\,dx\,dy\,dz, \quad \ldots\ldots(3)$

i.e. T and V denote the kinetic energy and the potential energy in relation to the field of extraneous force, of the fluid which at the moment occupies the region in question. The triple integral on the right-hand side of (2) may be transformed by a process which will often recur in our subject. Thus, by a partial integration,

$$\iiint u\frac{\partial p}{\partial x}\,dx\,dy\,dz = \iint[pu]\,dy\,dz - \iiint p\frac{\partial u}{\partial x}\,dx\,dy\,dz,$$

where $[pu]$ is used to indicate that the values of pu at the points where the boundary of the region is met by a line parallel to x are to be taken, with proper signs. If l, m, n be the direction-cosines of the *inwardly* directed normal to any element δS of this boundary, we have $\delta y\,\delta z = \pm\, l\,\delta S$, the signs alternating at the successive intersections referred to. We thus find that

$$\iint[pu]\,dy\,dz = -\iint pul\,dS,$$

where the integration extends over the whole bounding surface. Transforming the remaining terms in a similar manner, we obtain

$$\frac{D}{Dt}(T+V)=\iint p\,(lu+mv+nw)\,dS+\iiint p\left(\frac{\partial u}{\partial x}+\frac{\partial v}{\partial y}+\frac{\partial w}{\partial z}\right)dx\,dy\,dz. \quad \text{...(4)}$$

In the case of an incompressible fluid this reduces to the form

$$\frac{D}{Dt}(T+V)=\iint (lu+mv+nw)\,p dS. \quad \text{.................(5)}$$

Since $lu+mv+nw$ denotes the velocity of a fluid particle in the direction of the normal, the latter integral expresses the rate at which the pressures $p\delta S$ exerted from without on the various elements δS of the boundary are doing work. Hence the total increase of energy, kinetic and potential, of any portion of the liquid, is equal to the work done by the pressures on its surface.

In particular, if the fluid be bounded on all sides by fixed walls, we have

$$lu+mv+nw=0$$

over the boundary, and therefore

$$T+V=\text{const.} \quad \text{..............................(6)}$$

A similar interpretation can be given to the more general equation (4), provided p be a function of ρ only. If we write

$$E=-\int p d\left(\frac{1}{\rho}\right), \quad \text{............................(7)}$$

then E measures the work done by unit mass of the fluid against external pressure, as it passes, under the supposed relation between p and ρ, from its actual volume to some standard volume. For example, if the unit mass were enclosed in a cylinder with a sliding piston of area A, then when the piston is pushed outwards through a space δx, the work done is $pA\,.\,\delta x$, of which the factor $A\delta x$ denotes the increment of volume, *i.e.* of ρ^{-1}. In the case of the adiabatic relation we find

$$E=\frac{1}{\gamma-1}\left(\frac{p}{\rho}-\frac{p_0}{\rho_0}\right). \quad \text{.........................(8)}$$

We may call E the intrinsic energy of the fluid, per unit mass. Now, recalling the interpretation of the expression

$$\partial u/\partial x+\partial v/\partial y+\partial w/\partial z,$$

given in Art. 7, we see that the volume-integral in (4) measures the rate at which the various elements of the fluid are losing intrinsic energy by expansion; it is therefore equal to $-DW/Dt$,

where

$$W=\iiint E\rho\,dx\,dy\,dz. \quad \text{............................(9)}$$

Hence

$$\frac{D}{Dt}(T+V+W)=\iint p\,(lu+mv+nw)\,dS. \quad \text{...........(10)}$$

The total energy, which is now partly kinetic, partly potential in relation to a constant field of force, and partly intrinsic, is therefore increasing at a rate equal to that at which work is being done on the boundary by pressure from without.

On the isothermal hypothesis we should have

$$E = c^2 \log(\rho/\rho_0), \qquad \text{.............................(11)}$$

where $c^2 = p_0/\rho_0$. This measures the 'free energy' per unit mass. With this definition of E we have an equation of the same *form* as (10), although the meaning is different.

Transfer of Momentum.

10 a. If we fix our attention on the fluid which at the instant t occupies a certain region, the space which it occupies after a time δt will differ from the original region by the addition of a surface film of (positive or negative) thickness

$$(lu + mv + nw)\,\delta t,$$

where (l, m, n) is the direction of the outward normal to the surface. Hence it is easy to see that the rate, at time t, at which the momentum of this particular portion of fluid is increasing is equal to the rate of increase of the momentum contained in a *fixed* region having the same boundary, together with the flux of momentum outwards across the boundary.

In symbols, considering momentum parallel to Ox, we have

$$\begin{aligned}\iiint \frac{Du}{Dt}\rho\,dx\,dy\,dz &= \iiint \rho\left(\frac{\partial u}{\partial t} + u\frac{\partial u}{\partial x} + v\frac{\partial u}{\partial y} + w\frac{\partial u}{\partial z}\right) dx\,dy\,dz \\ &= \iiint \rho\frac{\partial u}{\partial t}\,dx\,dy\,dz + \iint \rho u\,(lu + mv + nw)\,dS \\ &\qquad - \iiint u\left(\frac{\partial(\rho u)}{\partial x} + \frac{\partial(\rho v)}{\partial y} + \frac{\partial(\rho w)}{\partial z}\right) dx\,dy\,dz \\ &= \frac{d}{dt}\iiint \rho u\,dx\,dy\,dz + \iint \rho u\,(lu + mv + nw)\,dS, \qquad \text{........(1)}\end{aligned}$$

by Art. 7 (5).

In steady motion (Art. 21) the first term on the right hand disappears, and the rate of increase of momentum of any portion of fluid is equal to the flux of momentum outwards across its boundary.

Conversely, if we apply the above principle to the fluid contained at any instant in a rectangular space $\delta x\,\delta y\,\delta z$, we reproduce the equation of motion (Art. 6).

Impulsive Generation of Motion.

11. If at any instant impulsive forces act bodily on the fluid, or if the boundary conditions suddenly change, a sudden alteration in the motion may take place. The latter case may arise, for instance, when a solid immersed in the fluid is suddenly set in motion.

Let ρ be the density, u, v, w the component velocities immediately before, u', v', w' those immediately after the impulse, X', Y', Z' the components of the extraneous impulsive forces per unit mass, ϖ the impulsive pressure, at the point (x, y, z). The change of momentum parallel to x of the element defined in Art. 6 is then $\rho\,\delta x\,\delta y\,\delta z\,(u'-u)$; the x-component of the extraneous impulsive forces is $\rho\,\delta x\,\delta y\,\delta z\,X'$, and the resultant impulsive pressure in the same direction is $-\partial\varpi/\partial x\,.\,\delta x\,\delta y\,\delta z$. Since an impulse is to be regarded as an infinitely great force acting for an infinitely short time (τ, say), the effects of all finite forces during this interval are to be neglected.

Hence, $$\rho\,\delta x\,\delta y\,\delta z\,(u'-u) = \rho\,\delta x\,\delta y\,\delta z\,X' - \frac{\partial\varpi}{\partial x}\,\delta x\,\delta y\,\delta z,$$

or
$$u'-u = X' - \frac{1}{\rho}\frac{\partial\varpi}{\partial x}.$$

Similarly,
$$\left.\begin{aligned} v'-v &= Y' - \frac{1}{\rho}\frac{\partial\varpi}{\partial y}, \\ w'-w &= Z' - \frac{1}{\rho}\frac{\partial\varpi}{\partial z} \end{aligned}\right\} \quad\text{.........................(1)}$$

These equations might also have been deduced from (2) of Art. 6, by multiplying the latter by δt, integrating between the limits 0 and τ, putting

$$X' = \int_0^\tau X\,dt, \quad Y' = \int_0^\tau Y\,dt, \quad Z' = \int_0^\tau Z\,dt, \quad \varpi = \int_0^\tau p\,dt,$$

and then making τ tend to the limit zero.

In a liquid an instantaneous change of motion can be produced by the action of impulsive pressures only, even when no impulsive forces act bodily on the mass. In this case we have $X', Y', Z' = 0$, so that

$$\left.\begin{aligned} u'-u &= -\frac{1}{\rho}\frac{\partial\varpi}{\partial x}, \\ v'-v &= -\frac{1}{\rho}\frac{\partial\varpi}{\partial y}, \\ w'-w &= -\frac{1}{\rho}\frac{\partial\varpi}{\partial z}. \end{aligned}\right\} \quad\text{.............................(2)}$$

If we differentiate these equations with respect to x, y, z, respectively, and add, and if we further suppose the density to be uniform, we find by Art. 8 (1) that

$$\frac{\partial^2\varpi}{\partial x^2} + \frac{\partial^2\varpi}{\partial y^2} + \frac{\partial^2\varpi}{\partial z^2} = 0.$$

The problem then, in any given case, is to determine a value of ϖ satisfying this equation and the proper boundary conditions*; the instantaneous change of motion is then given by (2).

* It will appear in Chapter III. that the value of ϖ is thus determinate, save as to an additive constant.

Equations referred to Moving Axes.

12. It is sometimes convenient in special problems to employ a system of rectangular axes which is itself in motion. The motion of this frame may be specified by the component velocities **u**, **v**, **w** of the origin, and the component rotations **p**, **q**, **r**, all referred to the instantaneous positions of the axes. If u, v, w be the component velocities of a fluid particle at (x, y, z), the rates of change of its co-ordinates relative to the moving frame will be

$$\frac{Dx}{Dt}=u-\mathbf{u}+\mathbf{r}y-\mathbf{q}z,\quad \frac{Dy}{Dt}=v-\mathbf{v}+\mathbf{p}z-\mathbf{r}x,\quad \frac{Dz}{Dt}=w-\mathbf{w}+\mathbf{q}x-\mathbf{p}y. \quad \text{...(1)}$$

After a time δt the velocities of the particle parallel to the new positions of the co-ordinate axes will have become

$$u+\left(\frac{\partial u}{\partial t}+\frac{\partial u}{\partial x}\frac{Dx}{Dt}+\frac{\partial u}{\partial y}\frac{Dy}{Dt}+\frac{\partial u}{\partial z}\frac{Dz}{Dt}\right)\delta t, \text{ \&c., \&c.} \quad \text{............(2)}$$

To find the component accelerations we must resolve these parallel to the original positions of the axes in the manner explained in books on Dynamics. In this way we obtain the expressions

$$\left.\begin{aligned} &\frac{\partial u}{\partial t}-\mathbf{r}v+\mathbf{q}w+\frac{\partial u}{\partial x}\frac{Dx}{Dt}+\frac{\partial u}{\partial y}\frac{Dy}{Dt}+\frac{\partial u}{\partial z}\frac{Dz}{Dt},\\ &\frac{\partial v}{\partial t}-\mathbf{p}w+\mathbf{r}u+\frac{\partial v}{\partial x}\frac{Dx}{Dt}+\frac{\partial v}{\partial y}\frac{Dy}{Dt}+\frac{\partial v}{\partial z}\frac{Dz}{Dt},\\ &\frac{\partial w}{\partial t}-\mathbf{q}u+\mathbf{p}v+\frac{\partial w}{\partial x}\frac{Dx}{Dt}+\frac{\partial w}{\partial y}\frac{Dy}{Dt}+\frac{\partial w}{\partial z}\frac{Dz}{Dt}. \end{aligned}\right\} \quad \text{...............(3)}$$

These will replace the expressions in the left-hand members of Art. 6 (2)*.

The general equation of continuity is

$$\frac{\partial\rho}{\partial t}+\frac{\partial}{\partial x}\left(\rho\frac{Dx}{Dt}\right)+\frac{\partial}{\partial y}\left(\rho\frac{Dy}{Dt}\right)+\frac{\partial}{\partial z}\left(\rho\frac{Dz}{Dt}\right)=0, \quad \text{............(4)}$$

reducing in the case of incompressibility to the form

$$\frac{\partial u}{\partial x}+\frac{\partial v}{\partial y}+\frac{\partial w}{\partial z}=0 \quad \text{..............................(5)}$$

as before.

The Lagrangian Equations.

13. Let a, b, c be the initial co-ordinates of any particle of fluid, x, y, z its co-ordinates at time t. We here consider x, y, z as functions of the independent variables a, b, c, t; their values in terms of these quantities give the whole history of every particle of the fluid. The velocities parallel to

* Greenhill, "On the General Motion of a Liquid Ellipsoid...," *Proc. Camb. Phil. Soc.* iv. 4 (1880).

the axes of co-ordinates of the particle (a, b, c) at time t are $\partial x/\partial t$, $\partial y/\partial t$, $\partial z/\partial t$, and the component accelerations in the same directions are $\partial^2 x/\partial t^2$, $\partial^2 y/\partial t^2$, $\partial^2 z/\partial t^2$. Let p be the pressure and ρ the density in the neighbourhood of this particle at time t; X, Y, Z the components of the extraneous forces per unit mass acting there. Considering the motion of the mass of fluid which at time t occupies the differential element of volume $\delta x \delta y \delta z$, we find, by the same reasoning as in Art. 6,

$$\frac{\partial^2 x}{\partial t^2} = X - \frac{1}{\rho}\frac{\partial p}{\partial x},$$

$$\frac{\partial^2 y}{\partial t^2} = Y - \frac{1}{\rho}\frac{\partial p}{\partial y},$$

$$\frac{\partial^2 z}{\partial t^2} = Z - \frac{1}{\rho}\frac{\partial p}{\partial z}.$$

These equations contain differential coefficients with respect to x, y, z, whereas our independent variables are a, b, c, t. To eliminate these differential coefficients, we multiply the above equations by $\partial x/\partial a$, $\partial y/\partial a$, $\partial z/\partial a$, respectively, and add; a second time by $\partial x/\partial b$, $\partial y/\partial b$, $\partial z/\partial b$, and add; and again a third time by $\partial x/\partial c$, $\partial y/\partial c$, $\partial z/\partial c$, and add. We thus get the three equations

$$\left(\frac{\partial^2 x}{\partial t^2} - X\right)\frac{\partial x}{\partial a} + \left(\frac{\partial^2 y}{\partial t^2} - Y\right)\frac{\partial y}{\partial a} + \left(\frac{\partial^2 z}{\partial t^2} - Z\right)\frac{\partial z}{\partial a} + \frac{1}{\rho}\frac{\partial p}{\partial a} = 0,$$

$$\left(\frac{\partial^2 x}{\partial t^2} - X\right)\frac{\partial x}{\partial b} + \left(\frac{\partial^2 y}{\partial t^2} - Y\right)\frac{\partial y}{\partial b} + \left(\frac{\partial^2 z}{\partial t^2} - Z\right)\frac{\partial z}{\partial b} + \frac{1}{\rho}\frac{\partial p}{\partial b} = 0,$$

$$\left(\frac{\partial^2 x}{\partial t^2} - X\right)\frac{\partial x}{\partial c} + \left(\frac{\partial^2 y}{\partial t^2} - Y\right)\frac{\partial y}{\partial c} + \left(\frac{\partial^2 z}{\partial t^2} - Z\right)\frac{\partial z}{\partial c} + \frac{1}{\rho}\frac{\partial p}{\partial c} = 0.$$

These are the 'Lagrangian' forms of the dynamical equations.

14. To find the form which the equation of continuity assumes in terms of our present variables, we consider the element of fluid which originally occupied a rectangular parallelepiped having its centre at the point (a, b, c), and its edges δa, δb, δc parallel to the axes. At the time t the same element forms an oblique parallelepiped. The centre now has for its co-ordinates x, y, z; and the projections of the edges on the co-ordinate axes are respectively

$$\frac{\partial x}{\partial a}\delta a, \quad \frac{\partial y}{\partial a}\delta a, \quad \frac{\partial z}{\partial a}\delta a;$$

$$\frac{\partial x}{\partial b}\delta b, \quad \frac{\partial y}{\partial b}\delta b, \quad \frac{\partial z}{\partial b}\delta b;$$

$$\frac{\partial x}{\partial c}\delta c, \quad \frac{\partial y}{\partial c}\delta c, \quad \frac{\partial z}{\partial c}\delta c.$$

The volume of the parallelepiped is therefore

$$\begin{vmatrix} \frac{\partial x}{\partial a}, & \frac{\partial y}{\partial a}, & \frac{\partial z}{\partial a} \\ \frac{\partial x}{\partial b}, & \frac{\partial y}{\partial b}, & \frac{\partial z}{\partial b} \\ \frac{\partial x}{\partial c}, & \frac{\partial y}{\partial c}, & \frac{\partial z}{\partial c} \end{vmatrix} \delta a\, \delta b\, \delta c,$$

or, as it is often written, $\dfrac{\partial (x, y, z)}{\partial (a, b, c)} \delta a\, \delta b\, \delta c.$

Hence, since the mass of the element is unchanged, we have

$$\rho \frac{\partial (x, y, z)}{\partial (a, b, c)} = \rho_0, \quad \text{..............................(1)}$$

where ρ_0 is the initial density at (a, b, c).

In the case of an incompressible fluid $\rho = \rho_0$, so that (1) becomes

$$\frac{\partial (x, y, z)}{\partial (a, b, c)} = 1. \quad \text{..............................(2)}$$

Weber's Transformation.

15. If as in Art. 10 the forces X, Y, Z have a potential Ω, the dynamical equations of Art. 13 may be written

$$\frac{\partial^2 x}{\partial t^2}\frac{\partial x}{\partial a} + \frac{\partial^2 y}{\partial t^2}\frac{\partial y}{\partial a} + \frac{\partial^2 z}{\partial t^2}\frac{\partial z}{\partial a} = -\frac{\partial \Omega}{\partial a} - \frac{1}{\rho}\frac{\partial p}{\partial a}, \text{ \&c., \&c.}$$

Let us integrate these equations with respect to t between the limits 0 and t. We remark that

$$\int_0^t \frac{\partial^2 x}{\partial t^2}\frac{\partial x}{\partial a} dt = \left[\frac{\partial x}{\partial t}\frac{\partial x}{\partial a}\right]_0^t - \int_0^t \frac{\partial x}{\partial t}\frac{\partial^2 x}{\partial a \partial t} dt$$
$$= \frac{\partial x}{\partial t}\frac{\partial x}{\partial a} - u_0 - \tfrac{1}{2}\frac{\partial}{\partial a}\int_0^t \left(\frac{\partial x}{\partial t}\right)^2 dt,$$

where u_0 is the initial value of the x-component of velocity of the particle (a, b, c). Hence if we write

$$\chi = \int_0^t \left[\int \frac{dp}{\rho} + \Omega - \tfrac{1}{2}\left\{\left(\frac{\partial x}{\partial t}\right)^2 + \left(\frac{\partial y}{\partial t}\right)^2 + \left(\frac{\partial z}{\partial t}\right)^2\right\}\right] dt, \quad \text{........(1)}$$

we find*

$$\left.\begin{aligned} \frac{\partial x}{\partial t}\frac{\partial x}{\partial a} + \frac{\partial y}{\partial t}\frac{\partial y}{\partial a} + \frac{\partial z}{\partial t}\frac{\partial z}{\partial a} - u_0 &= -\frac{\partial \chi}{\partial a}; \\ \frac{\partial x}{\partial t}\frac{\partial x}{\partial b} + \frac{\partial y}{\partial t}\frac{\partial y}{\partial b} + \frac{\partial z}{\partial t}\frac{\partial z}{\partial b} - v_0 &= -\frac{\partial \chi}{\partial b}; \\ \frac{\partial x}{\partial t}\frac{\partial x}{\partial c} + \frac{\partial y}{\partial t}\frac{\partial y}{\partial c} + \frac{\partial z}{\partial t}\frac{\partial z}{\partial c} - w_0 &= -\frac{\partial \chi}{\partial c}. \end{aligned}\right\} \quad \text{................(2)}$$

* H. Weber, "Ueber eine Transformation der hydrodynamischen Gleichungen," *Crelle*, lxviii. (1868). It is assumed in (1) that the density ρ, if not uniform, is a function of p only.

These three equations, together with

$$\frac{\partial \chi}{\partial t}=\int \frac{dp}{\rho}+\Omega-\tfrac{1}{2}\left\{\left(\frac{\partial x}{\partial t}\right)^2+\left(\frac{\partial y}{\partial t}\right)^2+\left(\frac{\partial z}{\partial t}\right)^2\right\}, \qquad \text{.........(3)}$$

and the equation of continuity, are the partial differential equations to be satisfied by the five unknown quantities x, y, z, p, χ; ρ being supposed already eliminated by means of one of the relations of Art. 8.

The initial conditions to be satisfied are

$$x=a, \quad y=b, \quad z=c, \quad \chi=0.$$

16. It is to be remarked that the quantities a, b, c need not be restricted to mean the initial co-ordinates of a particle; they may be any three quantities which serve to identify a particle, and which vary continuously from one particle to another. If we thus generalize the meanings of a, b, c, the form of the dynamical equations of Art. 13 is not altered; to find the form which the equation of continuity assumes, let x_0, y_0, z_0 now denote the initial co-ordinates of the particle to which a, b, c refer. The initial volume of the parallepiped, whose centre is at (x_0, y_0, z_0) and whose edges correspond to variations δa, δb, δc of the parameters a, b, c, is

$$\frac{\partial(x_0, y_0, z_0)}{\partial(a, b, c)}\delta a\,\delta b\,\delta c,$$

so that we have

$$\rho\frac{\partial(x, y, z)}{\partial(a, b, c)}=\rho_0\frac{\partial(x_0, y_0, z_0)}{\partial(a, b, c)}, \qquad \text{.....................(1)}$$

or, for an incompressible fluid,

$$\frac{\partial(x, y, z)}{\partial(a, b, c)}=\frac{\partial(x_0, y_0, z_0)}{\partial(a, b, c)}. \qquad \text{.....................(2)}$$

Equations in Polar Co-ordinates.

16 a. In the preceding investigations Cartesian co-ordinates have been employed, as is usually most consistent in the proof of general theorems. For special purposes polar co-ordinates are occasionally useful, and the appropriate formulae, on the 'Eulerian' plan, are accordingly given here for reference.

In *plane polars* we may use u and v to denote the radial and transversal velocities, respectively, at the point (r, θ) at time t. Since the radius vector of a particle is revolving at the rate v/r, the ordinary theory of rotating axes gives for the component accelerations:

$$\frac{Du}{Dt}-\frac{v}{r}\cdot v, \quad \frac{Dv}{Dt}+\frac{v}{r}\cdot u, \qquad \text{..................................(1)}$$

where, by the method of Art. 5,

$$\frac{D}{Dt}=\frac{\partial}{\partial t}+u\frac{\partial}{\partial r}+v\frac{\partial}{r\partial\theta}. \qquad \text{..................................(2)}$$

The 'expansion' (Δ) is found by calculating the rate of flux out of the quasi-rectangular element whose sides are δr and $r\delta\theta$; thus

$$\Delta=\frac{\partial u}{\partial r}+\frac{u}{r}+\frac{\partial v}{r\partial\theta}. \qquad \text{..................................(3)}$$

In *spherical polars* we denote the radial velocity at (r, θ, ϕ) by u, the velocity at right angles r in the plane of θ by v, and the velocity at right angles to the plane of θ by w. A triad of lines drawn from the origin parallel to these directors, when taken in this order, will, on the usual conventions, form a right-handed system. The changes in the angular co-ordinates of a particle in time δt are given by

$$r\,\delta\theta = v\,\delta t, \quad r\sin\theta\,\delta\phi = w\,\delta t.$$

This involves a rotation of the above system relative to its instantaneous position, with components

$$\cos\theta\,\delta\phi, \quad -\sin\theta\,\delta\phi, \quad \delta\theta.$$

Hence if $\mathbf{p}$, $\mathbf{q}$, $\mathbf{r}$ are the components of the instantaneous angular velocity of the system, we have

$$\mathbf{p} = \frac{w}{r}\cot\theta, \quad \mathbf{q} = -\frac{w}{r}, \quad \mathbf{r} = \frac{v}{r}. \qquad \text{(4)}$$

The required accelerations of the particle which is at (r, θ, ϕ) are therefore

$$\left.\begin{aligned} \frac{Du}{Dt} - \mathbf{r}v + \mathbf{q}w &= \frac{Du}{Dt} - \frac{v^2+w^2}{r}, \\ \frac{Dv}{Dt} - \mathbf{p}w + \mathbf{r}u &= \frac{Dv}{Dt} + \frac{uv}{r} - \frac{w^2}{r}\cot\theta, \\ \frac{Dw}{Dt} - \mathbf{q}u + \mathbf{p}v &= \frac{Dw}{Dt} + \frac{wu}{r} + \frac{vw}{r}\cot\theta, \end{aligned}\right\} \qquad \text{(5)}$$

where

$$\frac{D}{Dt} = \frac{\partial}{\partial t} + u\frac{\partial}{\partial r} + v\frac{\partial}{r\,\partial\theta} + w\frac{\partial}{r\sin\theta\,\partial\phi}. \qquad \text{(6)}$$

The expansion is found by calculating the flux out of the quasi-rectangular space whose edges are δr, $r\,\delta\theta$, $r\sin\theta\,\delta\phi$, and is

$$\Delta = \frac{\partial u}{\partial r} + 2\frac{u}{r} + \frac{\partial v}{r\,\partial\theta} + \frac{v}{r}\cot\theta + \frac{\partial w}{r\sin\theta\,\partial\phi}. \qquad \text{(7)}$$

CHAPTER II

INTEGRATION OF THE EQUATIONS IN SPECIAL CASES

17. IN a large and important class of cases the component velocities u, v, w can be expressed in terms of a single-valued function ϕ, as follows:

$$u, v, w = -\frac{\partial\phi}{\partial x}, \quad -\frac{\partial\phi}{\partial y}, \quad -\frac{\partial\phi}{\partial z}. \quad \text{..................(1)*}$$

Such a function is called a 'velocity-potential,' from its analogy with the potential function which occurs in the theories of Attractions, Electrostatics, &c. The general theory of the velocity-potential is reserved for the next chapter; but we give at once a proof of the following important theorem:

If a velocity potential exist, at any one instant, for any finite portion of a perfect fluid in motion under the action of forces which have a potential, then, provided the density of the fluid be either constant or a function of the pressure only, a velocity-potential exists for the same portion of the fluid at all instants before or after†.

In the equations of Art. 15, let the instant at which the velocity-potential ϕ_0 exists be taken as the origin of time; we have then

$$u_0 da + v_0 db + w_0 dc = -d\phi_0,$$

throughout the portion of the mass in question. Multiplying the equations (2) of Art. 15 in order by da, db, dc, and adding, we get

$$\frac{\partial x}{\partial t}dx + \frac{\partial y}{\partial t}dy + \frac{\partial z}{\partial t}dz - (u_0 da + v_0 db + w_0 dc) = -d\chi,$$

or, in the 'Eulerian' notation,

$$u dx + v dy + w dz = -d(\phi_0 + \chi) = -d\phi, \text{ say.}$$

Since the upper limit of t in Art. 15 (1) may be positive or negative, this proves the theorem.

It is to be particularly noticed that this continued existence of a velocity-potential is predicated, not of regions of space, but of portions of matter.

* The reasons for the introduction of the *minus* sign are stated in the Preface. The theory of 'cyclic' velocity-potentials is discussed later.

† Lagrange, "Mémoire sur la Théorie du Mouvement des Fluides," *Nouv. mém. de l'Acad. de Berlin*, 1781 [*Oeuvres*, iv. 714]. The argument is reproduced in the *Mécanique Analytique*.

Lagrange's statement and proof were alike imperfect; the first rigorous demonstration is due to Cauchy, "Mémoire sur la Théorie des Ondes," *Mém. de l'Acad. roy. des Sciences*, i. (1827) [*Oeuvres Complètes*, Paris, 1882..., 1re Série, i. 38]; the date of the memoir is 1815. Another proof is given by Stokes, *Camb. Trans.* viii. (1845) (see also *Math. and Phys. Papers*, Cambridge, 1880..., i. 106, 158, and ii. 36), together with an excellent historical and critical account of the whole matter.

A portion of matter for which a velocity-potential exists moves about and carries this property with it, but the part of space which it originally occupied may, in the course of time, come to be occupied by matter which did not originally possess the property, and which therefore cannot have acquired it.

The class of cases in which a single-valued velocity-potential exists includes all those where the motion has originated from rest under the action of forces of the kind here supposed; for then we have, initially,

$$u_0 da + v_0 db + w_0 dc = 0,$$

or

$$\phi_0 = \text{const.}$$

The restrictions under which the above theorem has been proved must be carefully remembered. It is assumed not only that the extraneous forces X, Y, Z, estimated at per unit mass, have a potential, but that the density ρ is either uniform or a function of p only. The latter condition is violated, for example, in the case of the convection currents generated by the unequal application of heat to a fluid; and again, in the wave-motion of a heterogeneous but incompressible fluid arranged originally in horizontal layers of equal density. Another case of exception is that of 'electro-magnetic rotations'; see Art. 29.

18. A comparison of the formulae (1) with the equations (2) of Art. 11 leads to a simple physical interpretation of ϕ.

Any actual state of motion of a liquid, for which a (single-valued) velocity-potential exists, could be produced instantaneously from rest by the application of a properly chosen system of impulsive pressures. This is evident from the equations cited, which shew, moreover, that $\phi = \varpi/\rho + \text{const.}$; so that $\varpi = \rho\phi + C$ gives the requisite system. In the same way $\varpi = -\rho\phi + C$ gives the system of impulsive pressures which would completely stop the motion*. The occurrence of an arbitrary constant in these expressions merely shews that a pressure uniform throughout a liquid mass produces no effect on the motion.

In the case of a gas, ϕ may be interpreted as the potential of the extraneous impulsive forces by which the actual motion at any instant could be produced instantaneously from rest.

A state of motion for which a velocity-potential does not exist cannot be generated or destroyed by the action of impulsive pressures, or of extraneous impulsive forces having a potential.

19. The existence of a velocity-potential indicates, besides, certain *kinematical* properties of the motion.

A 'line of motion' is defined to be a line drawn from point to point, so

* This interpretation was given by Cauchy, *loc. cit.*, and by Poisson, *Mém. de l'Acad. roy. des Sciences*, i. (1816).

that its direction is everywhere that of the motion of the fluid. The differential equations of the system of such lines are

$$\frac{dx}{u} = \frac{dy}{v} = \frac{dz}{w}. \quad \text{.............................(2)}$$

The relations (1) shew that when a velocity-potential exists the lines of motion are everywhere perpendicular to a system of surfaces, viz. the 'equipotential' surfaces $\phi = \text{const.}$

Again, if from the point (x, y, z) we draw a linear element δs in the direction (l, m, n), the velocity resolved in this direction is $lu + mv + nw$, or

$$-\frac{\partial\phi}{\partial x}\frac{dx}{ds} - \frac{\partial\phi}{\partial y}\frac{dy}{ds} - \frac{\partial\phi}{\partial z}\frac{dz}{ds}, \text{ which} = -\frac{\partial\phi}{\partial s}.$$

The velocity in any direction is therefore equal to the rate of decrease of ϕ in that direction.

Taking δs in the direction of the normal to the surface $\phi = \text{const.}$, we see that if a series of such surfaces be drawn corresponding to equidistant values of ϕ, the common difference being infinitely small, the velocity at any point will be inversely proportional to the distance between two consecutive surfaces in the neighbourhood of the point.

Hence, if any equipotential surface intersect itself, the velocity is zero at the intersection. The intersection of two *distinct* equipotential surfaces would imply an infinite velocity.

20. Under the circumstances stated in Art. 17, the equations of motion are at once integrable throughout that portion of the fluid mass for which a velocity-potential exists provided ρ is either constant, or a definite function of p. For in virtue of the relations

$$\partial v/\partial z = \partial w/\partial y, \quad \partial w/\partial x = \partial u/\partial z, \quad \partial u/\partial y = \partial v/\partial x,$$

which are implied in (1), the equations of Art. 6 may be written

$$-\frac{\partial^2\phi}{\partial x\partial t} + u\frac{\partial u}{\partial x} + v\frac{\partial v}{\partial x} + w\frac{\partial w}{\partial x} = -\frac{\partial\Omega}{\partial x} - \frac{1}{\rho}\frac{\partial p}{\partial x}, \text{ \&c., \&c.} \quad \text{.........(3)}$$

These have the integral

$$\frac{p}{\rho} + \tfrac{1}{2}q^2 + E = \frac{\partial\phi}{\partial t} + F(t). \quad \text{.........................(4)}$$

Here q denotes the resultant velocity $(u^2 + v^2 + w^2)^{\frac{1}{2}}$, $F(t)$ is an arbitrary function of t, and E is defined by Art. 10 (7), and has (in the case of a gas) the interpretation there given.

Our equations take a specially simple form in the case of an incompressible fluid; viz. we then have

$$\frac{p}{\rho} = \frac{\partial\phi}{\partial t} - \Omega - \tfrac{1}{2}q^2 + F(t), \quad \text{.......................(5)}$$

with the equation of continuity

$$\frac{\partial^2 \phi}{\partial x^2} + \frac{\partial^2 \phi}{\partial y^2} + \frac{\partial^2 \phi}{\partial z^2} = 0, \quad \text{.........................(6)}$$

which is the equivalent of Art. 1 (8). When, as in many cases which we shall have to consider, the boundary conditions are purely kinematical, the process of solution consists in finding a function which shall satisfy (5) and the prescribed surface-conditions. The pressure p is then given by (4), and is thus far indeterminate to the extent of an additive function of t. It becomes determinate when the value of p at some point of the fluid is given for all values of t. Since the term $F(t)$ is without influence on *resultant* pressures it is frequently omitted.

Suppose, for example, that we have a solid or solids moving through a liquid completely enclosed by fixed boundaries, and that it is possible (*e.g.* by means of a piston) to apply an arbitrary pressure at some point of the boundary. Whatever variations are made in the magnitude of the force applied to the piston, the motion of the fluid and of the solids will be absolutely unaffected, the pressure at all points instantaneously rising or falling by equal amounts. Physically, the origin of the paradox (such as it is) is that the fluid is treated as absolutely incompressible. In actual liquids changes of pressure are propagated with very great, but not infinite, velocity.

If the co-ordinate axes are in motion, the formula for the pressure is

$$\frac{p}{\rho} = \frac{\partial \phi}{\partial t} - \Omega - \tfrac{1}{2} q^2$$
$$- \mathbf{p}\left(y\frac{\partial \phi}{\partial z} - z\frac{\partial \phi}{\partial y}\right) - \mathbf{q}\left(z\frac{\partial \phi}{\partial x} - x\frac{\partial \phi}{\partial z}\right) - \mathbf{r}\left(x\frac{\partial \phi}{\partial y} - y\frac{\partial \phi}{\partial x}\right), \quad \text{......(7)}$$

where $$q^2 = (u - \mathbf{u})^2 + (v - \mathbf{v})^2 + (w - \mathbf{w})^2. \quad \text{.................(8)}$$

This easily follows from the formulae for the accelerations given in Art. 12 (3).

Steady Motion.

21. When at every point the velocity is constant in magnitude and direction, *i.e.* when

$$\frac{\partial u}{\partial t} = 0, \quad \frac{\partial v}{\partial t} = 0, \quad \frac{\partial w}{\partial t} = 0, \quad \text{.......................(1)}$$

everywhere, the motion is said to be 'steady.'

In steady motion the lines of motion coincide with the paths of the particles. For if P, Q be two consecutive points on a line of motion, a particle which is at any instant at P is moving in the direction of the tangent at P, and will, therefore, after an infinitely short time arrive at Q. The motion being steady, the lines of motion remain the same. Hence the direction of motion at Q is along the tangent to the same line of motion, *i.e.* the particle continues to describe the line, which is now appropriately called a 'stream-line.'

The stream-lines drawn through an infinitesimal contour define a tube, which may be called a 'stream-tube.'

In steady motion the equations (3) of Art. 20 give

$$\int \frac{dp}{\rho} = -\Omega - \tfrac{1}{2}q^2 + \text{constant.} \quad \text{.....................(2)}$$

The law of variation of pressure *along a stream-line* can however in this case be found without assuming the existence of a velocity-potential. For if δs denote an element of a stream-line, the acceleration in the direction of motion is $q\partial q/\partial s$, and we have

$$q\frac{\partial q}{\partial s} = -\frac{\partial \Omega}{\partial s} - \frac{1}{\rho}\frac{\partial p}{\partial s}, \quad \text{..............................(3)}$$

whence integrating along the stream-line,

$$\int \frac{dp}{\rho} = -\Omega - \tfrac{1}{2}q^2 + C. \quad \text{..........................(4)}$$

This is similar in form to (2), but is more general in that it does not assume the existence of a velocity-potential. It must however be carefully noticed that the 'constant' of equation (2) and the 'C' of equation (4) have different meanings, the former being an absolute constant, while the latter is constant along any particular stream-line, but may vary as we pass from one stream-line to another.

22. The theorem (4) stands in close relation to the principle of energy. If this be assumed independently, the formula may be deduced as follows*. Taking first the particular case of a liquid, consider the filament of fluid which at a given instant occupies a length AB of a stream-tube, the direction of motion being from A to B. Let p be the pressure, q the velocity, Ω the potential of the extraneous forces, σ the area of the cross-section, at A, and let the values of the same quantities at B be distinguished by accents. After a short interval of time the filament will occupy a length A_1B_1; let m be the mass included between the cross-sections at A and A_1, or B and B_1. Since the motion is steady, the gain of energy by the filament will be

$$m(\tfrac{1}{2}q'^2 + \Omega') - m(\tfrac{1}{2}q^2 + \Omega).$$

Again, the net work done on it is $pm/\rho - p'm/\rho$. Equating the increment of energy to the work done, we have

$$\frac{p}{\rho} + \tfrac{1}{2}q^2 + \Omega = \frac{p'}{\rho} + \tfrac{1}{2}q'^2 + \Omega',$$

or, using C in the same sense as before,

$$\frac{p}{\rho} = -\Omega - \tfrac{1}{2}q^2 + C, \quad \text{..........................(5)}$$

which is what the equation (4) becomes when ρ is constant.

* This is really a reversion to the methods of Daniel Bernoulli, *Hydrodynamica*, Argentorati, 1738.

To prove the corresponding formula for compressible fluids, we remark that the fluid crossing any section has now, in addition to its energies of motion and position, the energy ('intrinsic' or 'free' as the case may be)

$$-\int p\,d\left(\frac{1}{\rho}\right), \quad \text{or} \quad -\frac{p}{\rho}+\int\frac{dp}{\rho},$$

per unit mass. The addition of these terms in (5) gives the equation (4).

In the case of a gas subject to the adiabatic law

$$p/p_0=(\rho/\rho_0)^\gamma, \qquad \text{.................................(6)}$$

the equation (4) takes the form

$$\frac{\gamma}{\gamma-1}\frac{p}{\rho}=-\Omega-\tfrac{1}{2}q^2+C. \qquad \text{......................(7)}$$

23. The preceding equations shew that, in steady motion, and for points along any one stream-line*, the pressure is, *cæteris paribus*, greatest where the velocity is least, and *vice versâ*. This statement becomes evident when we reflect that a particle passing from a place of higher to one of lower pressure must have its motion accelerated, and *vice versâ* †.

It follows that in any case to which the equations of the last Article apply there is a limit which the velocity cannot exceed‡. For instance, let us suppose that we have a liquid flowing from a reservoir where the velocity may be neglected, and the pressure is p_0, and that we may neglect extraneous forces. We have then, in (5), $C=p_0/\rho$, and therefore

$$p=p_0-\tfrac{1}{2}\rho q^2. \qquad \text{............................(8)}$$

Now although it is found that a liquid from which all traces of air or other dissolved gas have been eliminated can sustain a negative pressure, or tension, of considerable magnitude§, this is not the case with fluids such as we find them under ordinary conditions. Practically, then, the equation (8) shews that q cannot exceed $(2p_0/\rho)^{\frac{1}{2}}$. This limiting velocity is that with which the fluid would escape from the reservoir into a vacuum. In the case of water at atmospheric pressure it is the velocity 'due to' the height of the water-barometer, or about 45 feet per second.

If in any case of fluid motion of which we have succeeded in obtaining the analytical expression, we suppose the motion to be gradually accelerated until the velocity at some point reaches the limit here indicated, a cavity will be formed there, and the conditions of the problem are more or less changed.

It will be shewn, in the next chapter (Art. 44), that in irrotational motion of a liquid, whether 'steady' or not, the place of least pressure is always at

* It will be shewn later that this restriction is unnecessary when a velocity-potential exists.

† Some interesting practical illustrations of this principle are given by Froude, *Nature*, xiii. 1875.

‡ Cf. Helmholtz, "Ueber discontinuirliche Flüssigkeitsbewegungen," *Berl. Monatsber.* April 1868; *Phil. Mag.* Nov. 1868 [*Wissenschaftliche Abhandlungen*, Leipzig, 1882–3, i. 146].

O. Reynolds, *Manch. Mem.* vi. (1877) [*Scientific Papers*, Cambridge, 1900... , i. 231].

some point of the boundary, provided the extraneous forces have a potential Ω satisfying the equation

$$\frac{\partial^2 \Omega}{\partial x^2} + \frac{\partial^2 \Omega}{\partial y^2} + \frac{\partial^2 \Omega}{\partial z^2} = 0.$$

This includes, of course, the case of gravity.

In the general case of a fluid in which p is a given function of ρ we have, putting $\Omega = 0$, $q_0 = 0$, in (4),

$$q^2 = 2\int_p^{p_0} \frac{dp}{\rho}. \qquad \text{.............(9)}$$

For a gas subject to the adiabatic law, this gives

$$q^2 = \frac{2\gamma}{\gamma - 1}\frac{p_0}{\rho_0}\left\{1 - \left(\frac{p}{p_0}\right)^{\frac{\gamma-1}{\gamma}}\right\} \qquad \text{.............(10)}$$

$$= \frac{2}{\gamma - 1}(c_0^2 - c^2), \qquad \text{.............(11)}$$

if $c, = (\gamma p/\rho)^{\frac{1}{2}}, = (dp/d\rho)^{\frac{1}{2}}$, denote the velocity of sound in the gas when at pressure p and density ρ, and c_0 the corresponding velocity for gas under the conditions which obtain in the reservoir. (See Chapter X.) Hence the limiting velocity is

$$\left(\frac{2}{\gamma - 1}\right)^{\frac{1}{2}} . c_0,$$

or $2{\cdot}214 c_0$, if $\gamma - 1{\cdot}408$.

24. We conclude this chapter with a few simple applications of the equations.

Flow of Liquids.

Let us take in the first instance the problem of the efflux of a liquid from a small orifice in the walls of a vessel which is kept filled up to a constant level, so that that motion may be regarded as steady.

The origin being taken in the upper surface, let the axis of z be vertical, and its positive direction downwards, so that $\Omega = -gz$. If we suppose the area of the upper surface large compared with that of the orifice, the velocity at the former may be neglected. Hence, determining the value of C in Art. 21 (4) so that $p = P$ (the atmospheric pressure) when $z = 0$, we have*

$$\frac{p}{\rho} = \frac{P}{\rho} + gz - \tfrac{1}{2}q^2. \qquad \text{.............(1)}$$

At the surface of the issuing jet we have $p = P$, and therefore

$$q^2 = 2gz, \qquad \text{.............(2)}$$

i.e. the velocity is that due to the depth below the upper surface. This is known as *Torricelli's Theorem*†.

* This result is due to D. Bernoulli, *l.c. ante* p. 21.

† "De motu gravium naturaliter accelerato," Firenze, 1643.

We cannot however at once apply this result to calculate the rate of efflux of the fluid, for two reasons. In the first place, the issuing fluid must be regarded as made up of a great number of elementary streams converging from all sides towards the orifice. Its motion is not, therefore, throughout the area of the orifice, everywhere perpendicular to this area, but becomes more and more oblique as we pass from the centre to the sides. Again, the converging motion of the elementary streams must make the pressure at the orifice somewhat greater in the interior of the jet than at the surface, where it is equal to the atmospheric pressure. The velocity, therefore, in the interior of the jet will be somewhat less than that given by (2).

Experiment shews however that the converging motion above spoken of ceases at a short distance beyond the orifice, and that (in the case of a circular orifice) the jet then becomes approximately cylindrical. The ratio of the area of the section S' of the jet at this point (called the 'vena contracta') to the area S of the orifice is called the 'coefficient of contraction.' If the orifice be simply a hole in a thin wall, this coefficient is found experimentally to be about ·62.

The paths of the particles at the vena contracta being nearly straight, there is little or no variation of pressure as we pass from the axis to the outer surface of the jet. We may therefore assume the velocity there to be uniform throughout the section, and to have the value given by (2), where z now denotes the depth of the vena contracta below the surface of the liquid in the vessel. The rate of efflux is therefore

$$(2gz)^{\frac{1}{2}} \cdot \rho S'. \qquad \text{(3)}$$

The calculation of the form of the issuing jet presents difficulties which have only been overcome in a few ideal cases of motion in two dimensions. (See Chapter IV.) It may however be shewn that the coefficient of contraction must, in general, lie betwen $\frac{1}{2}$ and 1. To put the argument in its simplest form, let us first take the case of liquid issuing from a vessel the pressure in which, at a distance from the orifice, exceeds that in the external space by the amount P, gravity being neglected. When the orifice is closed by a plate, the resultant pressure of the fluid on the containing vessel is of course *nil*. If when the plate is removed we assume (for the moment) that the pressure on the walls remains sensibly equal to P, there will be an unbalanced pressure PS acting on the vessel in the direction opposite to that of the jet, and tending to make it recoil. The equal and contrary reaction on the fluid produces in unit time the velocity q in the mass $\rho q S'$ flowing through the 'vena contracta,' whence

$$PS = \rho q^2 S'. \qquad \text{(4)}$$

The principle of energy gives, as in Art. 22,

$$P = \tfrac{1}{2}\rho q^2, \qquad \text{(5)}$$

so that, comparing, we have $S' = \frac{1}{2}S$. The formula (1) shews that the pressure on the walls, especially in the neighbourhood of the orifice, will in reality fall somewhat below the static pressure P, so that the left-hand side of (4) is an under-estimate. The ratio S'/S will therefore in general be $> \frac{1}{2}$.

In one particular case, viz. where a short cylindrical tube, projecting inwards, is attached to the orifice, the assumption above made is sufficiently exact, and the consequent value $\frac{1}{2}$ for the coefficient then agrees with experiment.

The reasoning is easily modified so as to take account of gravity (or other conservative forces). We have only to substitute for P the excess of the static pressure at the level of the orifice over the pressure outside. The difference of level between the orifice and the 'vena contracta' is here neglected*.

Another important application of Bernoulli's theorem is to the measurement of the velocity of a stream by means of a 'Pitot tube.' This consists of a fine tube open at one end, which points up-stream, and connected at the other end with a manometer. Along the stream-line which is in a line with the axis of the tube the velocity falls rapidly from q to 0, so that the manometer indicates the value of the 'total head' $p + \frac{1}{2}\rho q^2$ in the neighbourhood. A second manometer connected with a tube closed at the end, but with minute perforations in the wall, past which the stream glides, determines the value of the 'static pressure' p. The density ρ being known, a comparison of the readings gives the value of q. The two contrivances are often combined in one instrument. The method is extensively used in Aerodynamics, the compressibility of the air being found to have little effect up to speeds of the order of 200 ft. per sec.

Flow of a Gas.

24 a. The steady flow of a gas subject to the adiabatic law presents some features of interest.

Let σ be the cross-section at any point of a stream-tube, and δs an element of the length in the direction of flow. Omitting extraneous forces we have in place of Art. 23 (10)

$$q^2 - q_0^2 = \frac{2\gamma}{\gamma - 1}\frac{p_0}{\rho_0}\left\{1 - \left(\frac{p}{p_0}\right)^{\frac{\gamma-1}{\gamma}}\right\}, \quad \text{.................(1)}$$

* The above theory is due to Borda (*Mém. de l'Acad. des Sciences*, 1766), who also made experiments with the special form of mouth-piece referred to, and found $S/S' = 1{\cdot}942$. It was re-discovered by Hanlon, *Proc. Lond. Math. Soc.* iii. 4 (1869); the question is further elucidated in a note appended to this paper by Maxwell. See also Froude and J. Thomson, *Proc. Glasgow Phil. Soc.* x. (1876). It has been remarked by several writers that in the case of a diverging conical mouth-piece projecting inwards the section at the vena contracta may be less than half the area of the *internal* orifice.

where the zero suffix relates to some fixed section of the tube. If c be the velocity of sound corresponding to the local values of p and ρ this may be written

$$q^2 + \frac{2}{\gamma - 1} c^2 = q_0^2 + \frac{2}{\gamma - 1} c_0^2. \qquad (2)$$

Again, since the mass crossing any section in unit time is the same,

$$\rho q \sigma = \rho_0 q_0 \sigma_0. \qquad (3)$$

Hence

$$\frac{1}{\sigma}\frac{d\sigma}{ds} = -\frac{1}{q}\frac{dq}{ds} - \frac{1}{\rho}\frac{d\rho}{dp}\frac{dp}{ds}$$

$$= -\frac{1}{q}\frac{dq}{ds}\left(1 - \frac{q^2}{c^2}\right). \qquad (4)$$

It follows from (2) and (4) that in a *converging* tube q will increase and c diminish, or *vice versâ*, according as q is less or greater than c. For a diverging tube the statements must be reversed. Briefly, we may say that in a converging tube the stream velocity and the local velocity of sound continually approach one another, whilst in a diverging tube they separate more and more.

These results follow also from a graphical representation of the equations (2) and (3). Since c^2 is proportional to $\rho^{\gamma-1}$, the latter may be written

$$c^{\frac{2}{\gamma-1}} q\sigma = c_0^{\frac{2}{\gamma-1}} q_0 \sigma_0. \qquad (5)$$

If we take abscissae proportional to c and ordinates to q, the equation (2) represents an ellipse of invariable shape, drawn through the point (c_0, q_0). For any assigned value of σ/σ_0 the equation (5) represents a sort of hyperbolic curve. For a certain value (σ') of σ this will touch the ellipse, and we then have $q=c$.

The curves AA', BB', CC' in the annexed diagram correspond to the ratios

$$\frac{\sigma}{\sigma'} = 8, 4, 2,$$

respectively, whilst the point D corresponds to the minimum section σ'. For still smaller values of σ the intersections with the ellipse are imaginary, and steady adiabatic flow becomes impossible. The diagram shews that for any section greater than σ' there are *two* possible pairs of values of q and c, as has been remarked by Osborne Reynolds and others.

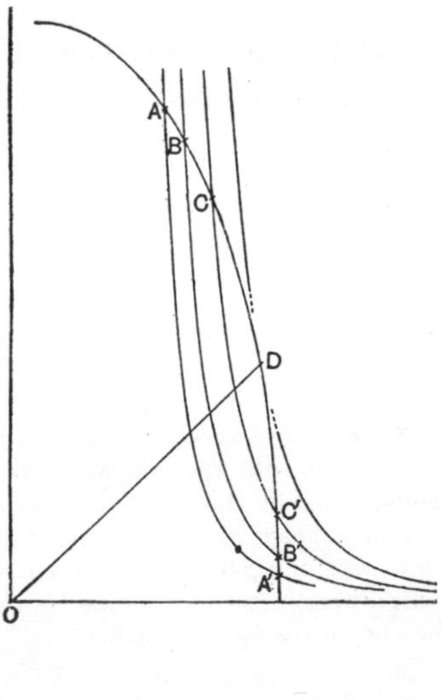

When q is less than c the representative point on the ellipse lies below OD. In a converging tube it assumes a sequence of positions such as A', B', C', the stream-velocity increasing, and the velocity of sound decreasing, as the critical section σ' is approached. When q is greater than c, on the other hand, the representative point lies above OD. In a converging tube we have a sequence such as A, B, C; the stream-velocity decreases, and the velocity of sound increases.

25. We consider more particularly the efflux of a gas, supposed to flow through a small orifice from a vessel in which the pressure is p_0 and density ρ_0 into a space where the pressure is p_1.

If the ratio p_0/p_1 of the pressure inside and outside the vessel do not exceed a certain limit, to be indicated presently, the flow will take place in much the same manner as in the case of a liquid, and the rate of discharge may be found by putting $p=p_1$ in Art. 23 (10), and multiplying the resulting value of q by the area σ_1 of the vena contracta. This gives for the rate of discharge of mass*

$$q_1\rho_1\sigma_1=\left(\frac{2}{\gamma-1}\right)^{\frac{1}{2}} c_0\rho_0\left\{\left(\frac{p_1}{p_0}\right)^{\frac{2}{\gamma}}-\left(\frac{p_1}{p_0}\right)^{\frac{\gamma+1}{\gamma}}\right\}^{\frac{1}{2}}.\sigma_1. \quad\text{.....................(6)}$$

It is plain however that there must be a limit to the applicability of this result; for otherwise we should be led to the paradoxical conclusion that when $p_1=0$, *i.e.* the discharge is into a vacuum, the flux of matter is *nil*. The elucidation of this point is due to Prof Osborne Reynolds†. It appears that $q\rho$ is a maximum, *i.e.* the section of an elementary stream is a minimum, when as appears from (4) the velocity of the stream is equal to the velocity of sound in gas of the pressure and density which prevail there. On the adiabatic hypothesis this gives, by Art. 23 (11),

$$\frac{c}{c_0}=\left(\frac{2}{\gamma+1}\right)^{\frac{1}{2}}, \quad\text{.......................................(7)}$$

and therefore
$$\frac{\rho}{\rho_0}=\left(\frac{2}{\gamma+1}\right)^{\frac{1}{\gamma-1}}, \qquad \frac{p}{p_0}=\left(\frac{2}{\gamma+1}\right)^{\frac{\gamma}{\gamma-1}}, \quad\text{........................(8)}$$

or, if $\gamma=1{\cdot}408$,
$$\rho={\cdot}634\rho_0, \quad p={\cdot}527p_0. \quad\text{...................................(9)}$$

If p_1 be less than this value, the stream after passing the point in question widens out again, until it is lost at a distance in the eddies due to viscosity. The minimum sections of the elementary streams will be situate in the neighbourhood of the orifice, and their sum S may be called the virtual area of the latter. The velocity of efflux, as found from (2), is

$$q={\cdot}911c_0.$$

The rate of discharge is then $=q\rho S$, where q and ρ have the values just found, and is therefore approximately independent of the external pressure p_1 so long as this falls below $527p_0$. The physical reason of this is (as pointed out by Reynolds) that, so long as the velocity at any point exceeds the velocity of sound under the conditions which obtain there, no change of pressure can be propagated backwards beyond this point so as to affect the motion higher up the stream ‡.

Some recent experiments of Stanton § confirm in all essentials the views of Reynolds, and clear up some apparent discrepancies.

Under similar circumstances as to pressure, the velocities of efflux of different gases are (so far as γ can be assumed to have the same value for each) proportional to the corresponding velocities of sound. Hence (as we shall see in Chapter x.) the velocity of efflux will vary inversely, and the rate of discharge of mass will vary directly, as the square root of the density‖.

* A result equivalent to this was given by Saint Venant and Wantzel, *Journ. de l'École Polyt.* xvi. 92 (1839), and was discussed by Stokes, *Brit. Ass. Reports for* 1846 [*Papers*, i. 176].

† "On the Flow of Gases," *Proc. Manch. Lit. and Phil. Soc.* Nov. 17, 1885; *Phil. Mag.* March 1886 [*Papers*, ii. 311]. A similar explanation was given by Hugoniot, *Comptes Rendus*, June 28, July 26, and Dec. 13, 1886.

‡ For a further discussion and references see Rayleigh, "On the Discharge of Gases under High Pressures," *Phil. Mag.* (6) xxxii. 177 (1916) [*Scientific Papers*, Cambridge, 1899–1920, vi. 407].

§ *Proc. Roy. Soc.* A, cxi. 306 (1926).

‖ Cf. Graham, *Phil. Trans.* 1846.

Rotating Liquid.

26. Let us next take the case of a mass of liquid rotating, under the action of gravity only, with constant and uniform angular velocity ω about the axis of z, supposed drawn vertically upwards.

By hypothesis, $$u, v, w = -\omega y, \omega x, 0,$$
$$X, Y, Z = 0, 0, -g.$$

The equation of continuity is satisfied identically, and the dynamical equations obviously are

$$-\omega^2 x = -\frac{1}{\rho}\frac{\partial p}{\partial x}, \quad -\omega^2 y = -\frac{1}{\rho}\frac{\partial p}{\partial y}, \quad 0 = -\frac{1}{\rho}\frac{\partial p}{\partial z} - g. \quad \text{......(1)}$$

These have the common integral

$$\frac{p}{\rho} = \tfrac{1}{2}\omega^2 (x^2 + y^2) - gz + \text{const.} \quad \text{........................(2)}$$

The free surface, $p = \text{const.}$, is therefore a paraboloid of revolution about the axis of z, having its concavity upwards, and its latus rectum $= 2g/\omega^2$.

Since $$\frac{\partial v}{\partial x} - \frac{\partial u}{\partial y} = 2\omega,$$

a velocity-potential does not exist. A motion of this kind could not therefore be generated in a 'perfect' fluid, *i.e.* in one unable to sustain tangential stress.

27. Instead of supposing the angular velocity ω to be uniform, let us suppose it to be a function of the distance r from the axis, and let us inquire what form must be assigned to this function in order that a velocity-potential may exist for the motion. We find

$$\frac{\partial v}{\partial x} - \frac{\partial u}{\partial y} = 2\omega + r\frac{d\omega}{dr},$$

and in order that this may vanish we must have $\omega r^2 = \mu$, a constant. The velocity at any point is then $= \mu/r$, so that the equation (2) of Art. 21 becomes

$$\frac{p}{\rho} = \text{const.} - \tfrac{1}{2}\frac{\mu^2}{r^2}, \quad \text{..............................(1)}$$

if no extraneous forces act. To find the value of ϕ we have, using polar co-ordinates,

$$\frac{\partial \phi}{\partial r} = 0, \quad \frac{\partial \phi}{r\partial \theta} = -\frac{\mu}{r},$$

whence $$\phi = -\mu\theta + \text{const.} = -\mu \tan^{-1}\frac{y}{x} + \text{const.} \quad \text{..............(2)}$$

We have here an instance of a 'cyclic' function. A function is said to be 'single-valued' throughout any region of space when we can assign to every point of that region a definite value of the function in such a way that these values shall form a continuous system. This is not possible with the function

(2); for the value of ϕ, if it vary continuously, changes by $-2\pi\mu$ as the point to which it refers describes a complete circuit round the origin. The general theory of cyclic velocity-potentials will be given in the next chapter.

If gravity act, and if the axis of z be drawn vertically upwards, we must add to (1) the term $-gz$. The form of the free surface is therefore that generated by the revolution of the hyperbolic curve $x^2z = \text{const.}$ about the axis of z.

By properly fitting together the two preceding solutions we obtain the case of Rankine's 'combined vortex.' Thus the motion being everywhere in coaxial circles, let us suppose the velocity to be equal to ωr from $r=0$ to $r=a$, and to $\omega a^2/r$ for $r>a$. The corresponding forms of the free surface are then given by

$$z=\frac{\omega^2}{2g}(r^2-a^2)+C,$$

and

$$z=\frac{\omega^2}{2g}\left(a^2-\frac{a^4}{r^2}\right)+C,$$

these being continuous with one another when $r=a$. The depth of the central depression below the general level of the surface is therefore ω^2a^2/g.

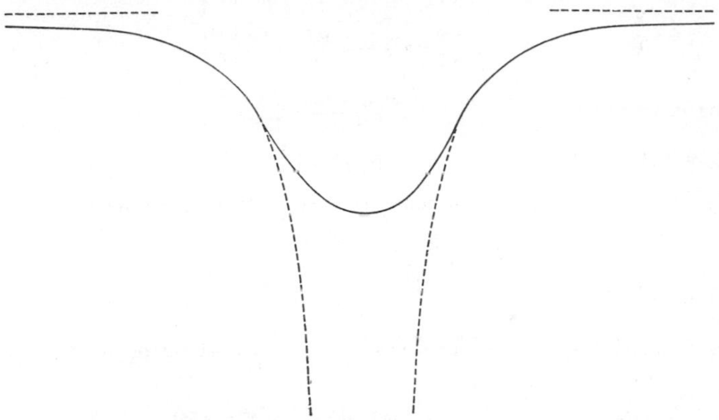

28. To illustrate, by way of contrast, the case of extraneous forces not having a potential, let us suppose that a mass of liquid filling a right circular cylinder moves from rest under the action of the forces

$$X=Ax+By, \quad Y=B'x+Cy, \quad Z=0,$$

the axis of z being that of the cylinder.

If we assume $u=-\omega y$, $v=\omega x$, $w=0$, where ω is a function of t only, these values satisfy the equation of continuity and the boundary conditions. The dynamical equations are evidently

$$\left.\begin{aligned} -y\frac{d\omega}{dt}-\omega^2x&=Ax+By-\frac{1}{\rho}\frac{\partial p}{\partial x}, \\ x\frac{d\omega}{dt}-\omega^2y&=B'x+Cy-\frac{1}{\rho}\frac{\partial p}{\partial x}. \end{aligned}\right\} \quad \text{.............................(1)}$$

Differentiating the first of these with respect to y, and the second with respect to x, and subtracting, we eliminate p, and find

$$\frac{d\omega}{dt}=\tfrac{1}{2}(B'-B). \quad \text{...(2)}$$

The fluid therefore rotates as a whole about the axis of z with constantly accelerated angular velocity, except in the particular case when $B=B'$. To find p, we substitute the value of $d\omega/dt$ in (1) and integrate; we thus get

$$\frac{p}{\rho}=\tfrac{1}{2}\omega^2(x^2+y^2)+\tfrac{1}{2}(Ax^2+2\beta xy+Cy^2)+\text{const.},$$

where $2\beta=B+B'$.

29. As a final example, we will take one suggested by the theory of 'electro-magnetic rotations.'

If an electric current be made to pass radially from an axial wire, through a conducting liquid, to the walls of a metallic containing cylinder, in a uniform magnetic field, the extraneous forces will be of the type*

$$X,\ Y,\ Z=-\frac{\mu y}{r^2},\ \frac{\mu x}{r^2},\ 0.$$

Assuming $u=-\omega y$, $v=\omega x$, $w=0$, where ω is a function of r and t only, we have

$$\left.\begin{aligned}-y\frac{\partial\omega}{\partial t}-\omega^2 x&=-\frac{\mu y}{r^2}-\frac{1}{\rho}\frac{\partial p}{\partial x},\\ x\frac{\partial\omega}{\partial t}-\omega^2 y&=\frac{\mu x}{r^2}-\frac{1}{\rho}\frac{\partial p}{\partial y}.\end{aligned}\right\}\quad\text{.................................(1)}$$

Eliminating p, we obtain $$2\frac{\partial\omega}{\partial t}+r\frac{\partial^2\omega}{\partial r\,\partial t}=0.$$

The solution of this is $$\omega=F(t)/r^2+f(r),$$

where F and f denote arbitrary functions. If $\omega=0$ when $t=0$, we have

$$F(0)/r^2+f(r)=0,$$

and therefore $$\omega=\frac{F(t)-F(0)}{r^2}=\frac{\lambda}{r^2},\quad\text{.......................................(2)}$$

where λ is a function of t which vanishes for $t=0$. Substituting in (1), and integrating, we find

$$\frac{p}{\rho}=\left(\mu-\frac{d\lambda}{dt}\right)\tan^{-1}\frac{y}{x}-\tfrac{1}{2}\omega^2r^2+\chi(t).$$

Since p is essentially a single-valued function, we must have $d\lambda/dt=\mu$, or $\lambda=\mu t$. Hence the fluid rotates with an angular velocity which varies inversely as the square of the distance from the axis, and increases constantly with the time.

If C denote the total flux of electricity outwards, per unit length of the axis, and γ the component of the magnetic force, parallel to the axis, we have $\mu=\gamma C/2\pi\rho$. The above case is specially simple, in that the forces X, Y, Z have a potential ($\Omega=-\mu\tan^{-1}y/x$), though a 'cyclic' one. As a rule, in electro-magnetic rotations, this is not the case.

CHAPTER III

IRROTATIONAL MOTION

30. THE present chapter is devoted mainly to an exposition of some general theorems relating to the kinds of motion already considered in Arts. 17–20; viz. those in which $u\,dx + v\,dy + w\,dz$ is an exact differential throughout a finite mass of fluid. It is convenient to begin with the following analysis, due to Stokes*, of the motion of a fluid element in the most general case.

The component velocities at the point (x, y, z) being u, v, w, the relative velocities at an infinitely near point $(x + \delta x, y + \delta y, z + \delta z)$ are

$$\left.\begin{aligned} \delta u &= \frac{\partial u}{\partial x}\delta x + \frac{\partial u}{\partial y}\delta y + \frac{\partial u}{\partial z}\delta z, \\ \delta v &= \frac{\partial v}{\partial x}\delta x + \frac{\partial v}{\partial y}\delta y + \frac{\partial v}{\partial z}\delta z, \\ \delta w &= \frac{\partial w}{\partial x}\delta x + \frac{\partial w}{\partial y}\delta y + \frac{\partial w}{\partial z}\delta z. \end{aligned}\right\} \quad \text{.....................(1)}$$

If we write

$$\left.\begin{aligned} a &= \frac{\partial u}{\partial x}, & b &= \frac{\partial v}{\partial y}, & c &= \frac{\partial w}{\partial z}, \\ f &= \frac{\partial w}{\partial y} + \frac{\partial v}{\partial z}, & g &= \frac{\partial u}{\partial z} + \frac{\partial w}{\partial x}, & h &= \frac{\partial v}{\partial x} + \frac{\partial u}{\partial y}, \\ \xi &= \frac{\partial w}{\partial y} - \frac{\partial v}{\partial z}, & \eta &= \frac{\partial u}{\partial z} - \frac{\partial w}{\partial x}, & \zeta &= \frac{\partial v}{\partial x} - \frac{\partial u}{\partial y}\dagger, \end{aligned}\right\} \quad \text{..........(2)}$$

equations (1) may be written

$$\left.\begin{aligned} \delta u &= \quad a\delta x + \tfrac{1}{2}h\delta y + \tfrac{1}{2}g\delta z + \tfrac{1}{2}(\eta\delta z - \zeta\delta y), \\ \delta v &= \tfrac{1}{2}h\delta x + \quad b\delta y + \tfrac{1}{2}f\delta z + \tfrac{1}{2}(\zeta\delta x - \xi\delta z), \\ \delta w &= \tfrac{1}{2}g\delta x + \tfrac{1}{2}f\delta y + \quad c\delta z + \tfrac{1}{2}(\xi\delta y - \eta\delta x). \end{aligned}\right\} \quad \text{..............(3)}$$

Hence the motion of a small element having the point (x, y, z) for its centre may be conceived as made up of three parts.

* "On the Theories of the Internal Friction of Fluids in Motion, &c." *Camb. Phil. Trans.* viii. (1845) [*Papers*, i, 80].

† There is here a deviation from the traditional convention. It has been customary to use symbols such as ξ, η, ζ (Helmholtz) or $\omega', \omega'', \omega'''$ (Stokes) to denote the component *rotations*

$$\frac{1}{2}\left(\frac{\partial w}{\partial y} - \frac{\partial v}{\partial z}\right), \ \frac{1}{2}\left(\frac{\partial u}{\partial z} - \frac{\partial w}{\partial x}\right), \ \frac{1}{2}\left(\frac{\partial v}{\partial x} - \frac{\partial u}{\partial y}\right)$$

of a fluid element. The fundamental kinematical theorem is however that of Art. 32 (3), and the definition of ξ, η, ζ adopted in the text avoids the intrusion of an unnecessary factor 2 (or $\frac{1}{2}$ as the case may be) in this and in a whole series of subsequent formulae relating to vortex motion. It also improves the electro-magnetic analogy of Art. 148.

The first part, whose components are u, v, w, is a motion of *translation* of the element as a whole.

The second part, expressed by the first three terms on the right-hand sides of the equations (3), is a motion such that, if δx, δy, δz be regarded as current co-ordinates, every point is moving in the direction of the normal to that quadric of the system

$$a(\delta x)^2 + b(\delta y)^2 + c(\delta z)^2 + f\delta y\,\delta z + g\,\delta z\,\delta x + h\,\delta x\,\delta y = \text{const.} \quad \ldots(4)$$

on which it lies. If we refer these quadrics to their principal axes, the corresponding parts of the velocities parallel to these axes will be

$$\delta u' = a'\delta x', \quad \delta v' = b'\delta y', \quad \delta w' = c'\delta z', \quad \ldots\ldots\ldots\ldots(5)$$

if

$$a'(\delta x')^2 + b'(\delta y')^2 + c'(\delta z')^2 = \text{const.}$$

is what (4) becomes by the transformation. The formulae (5) express that the length of every line in the element parallel to x' is being elongated at the (positive or negative) rate a', whilst lines parallel to y' and z' are being elongated in like manner at the rates b' and c' respectively. Such a motion is called one of *pure strain* and the principal axes of the quadrics (4) are called the axes of the strain.

The last two terms on the right-hand sides of the equations (3) express a *rotation* of the element as a whole about an instantaneous axis; the component angular velocities of the rotation being $\frac{1}{2}\xi$, $\frac{1}{2}\eta$, $\frac{1}{2}\zeta$*.

The vector whose components are ξ, η, ζ may conveniently be called the 'vorticity' of the medium at the point (x, y, z).

This analysis may be illustrated by the so-called 'laminar' motion of a liquid. Thus if

$$u = \mu y, \quad v = 0, \quad w = 0,$$

we have

$$a, b, c, f, g, \xi, \eta = 0, \quad h = \mu, \quad \zeta = -\mu.$$

If A represent a rectangular fluid element bounded by planes parallel to the co-ordinate planes, then B represents the change produced in this in a short time by the strain alone, and C that due to the strain *plus* the rotation.

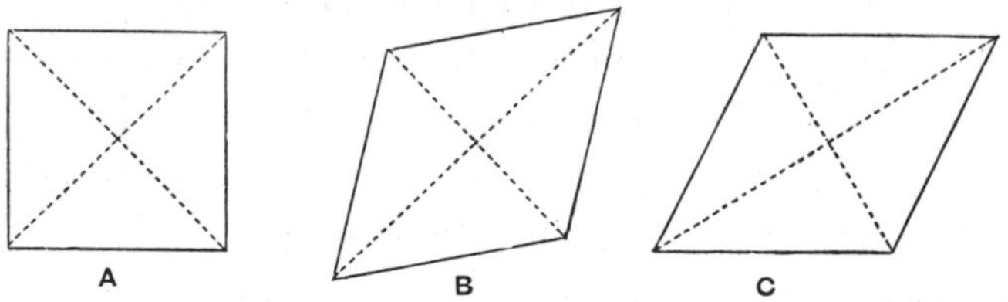

It is easily seen that the above resolution of the motion is unique. If we assume that the motion relative to the point (x, y, z) can be made up of a strain and a rotation in which the axes and coefficients of the strain and the axis and angular velocity of the rotation are arbitrary, then calculating the

* The quantities corresponding to $\frac{1}{2}\xi$, $\frac{1}{2}\eta$, $\frac{1}{2}\zeta$ in the theory of the infinitely small *displacements* of a continuous medium had been interpreted by Cauchy as expressing the 'mean rotations' of an element, *Exercices d'Analyse et de Physique*, ii. 302 (Paris, 1841).

relative velocities δu, δv, δw, we get expressions similar to those on the right-hand sides of (3), but with arbitrary values of $a, b, c, f, g, h, \xi, \eta, \zeta$. Equating coefficients of δx, δy, δz, however, we find that a, b, c, &c. must have respectively the same values as before. Hence the directions of the axes of the strain, the rates of extension or contraction along them, and the axis and the amount of the vorticity, at any point of the fluid, depend only on the state of relative motion at that point, and not on the position of the axis of reference.

When throughout a finite portion of a fluid mass we have ξ, η, ζ all zero, the relative motion of any element of that portion consists of a pure strain only, and is called 'irrotational.'

31. The value of the integral

$$\int (u\,dx + v\,dy + w\,dz),$$

or

$$\int \left(u \frac{dx}{ds} + v \frac{dy}{ds} + w \frac{dz}{ds} \right) ds,$$

taken along any line $ABCD$, is called* the 'flow' of the fluid from A to D along that line. We shall denote it for shortness by $I\,(ABCD)$.

If A and D coincide, so that the line forms a closed curve, or circuit, the value of the integral is called the 'circulation' in that circuit. We denote it by $I\,(ABCA)$. If in either case the integration be taken in the opposite direction, the signs of dx/ds, dy/ds, dz/ds will be reversed, so that we have

$$I\,(AD) = -I\,(DA), \quad \text{and} \quad I\,(ABCA) = -I\,(ACBA).$$

It is also plain that

$$I\,(ABCD) = I\,(AB) + I\,(BC) + I\,(CD).$$

Again, any surface may be divided, by a double series of lines crossing it, into infinitely small elements. The sum of the circulations round the boundaries of these elements, taken all in the same sense, is equal to the circulation round the original boundary of the surface (supposed for the moment to consist of a single closed curve). For, in the sum in question, the flow along each side common to two elements comes in twice, once for each element, but with opposite signs, and therefore disappears from the result. There remain then only the flows along those sides which are parts of the original boundary; whence the truth of the above statement.

From this it follows, by considerations of continuity, that the circulation round the boundary of any surface-element δS, having a given position and aspect, is ultimately proportional to the area of the element.

* Sir W. Thomson, "On Vortex Motion," *Edin. Trans.* xxv. (1869) [*Papers*, iv. 13].

If the element be a rectangle $\delta y \delta z$ having its centre at the point (x, y, z), then calculating the circulation round it in the direction shewn by the arrows in the annexed figure, we have

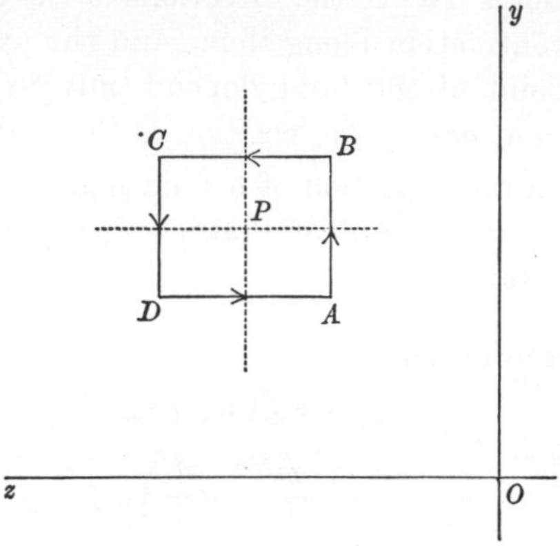

$$I(AB) = \{v - \tfrac{1}{2}(\partial v/\partial z)\,\delta z\}\,\delta y, \qquad I(BC) = \{w + \tfrac{1}{2}(\partial w/\partial y)\,\delta y\}\,\delta z,$$
$$I(CD) = -\{v + \tfrac{1}{2}(\partial v/\partial z)\,\delta z\}\,\delta y, \qquad I(DA) = -\{w - \tfrac{1}{2}(\partial w/\partial y)\,\delta y\}\,\delta z,$$

and therefore
$$I(ABCDA) = \left(\frac{\partial w}{\partial y} - \frac{\partial v}{\partial z}\right)\delta y o z.$$

In this way we infer that the circulations round the boundaries of any infinitely small areas δS_1, δS_2, δS_3, having their planes parallel to the co-ordinate planes, are

$$\xi\delta S_1, \quad \eta\delta S_2, \quad \zeta\delta S_3, \qquad \text{..............................(1)}$$

respectively.

Again, referring to the figure and the notation of Art. 2, we have

$$I(ABCA) = I(PBCP) + I(PCAP) + I(PABP)$$
$$= \xi \,.\, l\Delta + \eta \,.\, m\Delta + \zeta \,.\, n\Delta,$$

whence we infer that the circulation round the boundary of *any* infinitely small area δS is

$$(l\xi + m\eta + n\zeta)\,\delta S. \qquad \text{..............................(2)}$$

We have here an independent proof that the quantities ξ, η, ζ, as defined by Art. 30 (2), may be regarded as the components of a vector.

It will be observed that some convention is implied as to the relation between the sense in which the circulation round the boundary of δS is estimated, and the sense of the normal (l, m, n). In order to have a clear understanding on this point, we shall suppose in this book that the axes of co-ordinates form a *right-handed* system; thus if the axes of x and y point E. and N. respectively, that of z will point vertically upwards*. The sense in

* Maxwell, *Proc. Lond. Math. Soc.* (1) iii. 279, 280. Thus in the above diagram the axis of x is supposed drawn towards the reader.

which the circulation, as given by (2), is estimated is then related to the direction of the normal (l, m, n) in the manner typified by a right-handed screw*.

32. Expressing now that the circulation round the edge of any finite surface is equal to the sum of the circulations round the boundaries of the infinitely small elements into which the surface may be divided, we have, by (2),

$$\int (u\,dx + v\,dy + w\,dz) = \iint (l\xi + m\eta + n\zeta)\,dS, \quad \ldots\ldots\ldots\ldots\ldots(3)$$

or, substituting the values of ξ, η, ζ from Art. 30,

$$\int (u\,dx + v\,dy + w\,dz) = \iint \left\{ l\left(\frac{\partial w}{\partial y} - \frac{\partial v}{\partial z}\right) + m\left(\frac{\partial u}{\partial z} - \frac{\partial w}{\partial x}\right) + n\left(\frac{\partial v}{\partial x} - \frac{\partial v}{\partial y}\right)\right\} dS; \ldots(4)$$

where the single-integral is taken along the bounding curve, and the double-integral over the surface†. In these formulae the quantities l, m, n are the direction-cosines of the normal drawn always on one side of the surface, which we may term the positive side; the direction of integration in the first member is then that in which a man walking on the surface, on the positive side of it, and close to the edge, must proceed so as to have the surface always on his left hand.

The theorem (3) or (4) may evidently be extended to a surface whose boundary consists of two or more closed curves, provided the integration in the first member be taken round each of these in the proper direction, according to the rule just given. Thus, if the surface integral in (4) extend over the shaded portion of the annexed figure, the directions in which the circulations in the several parts of the boundary are to be taken are shewn by the arrows, the positive side of the surface being that which faces the reader.

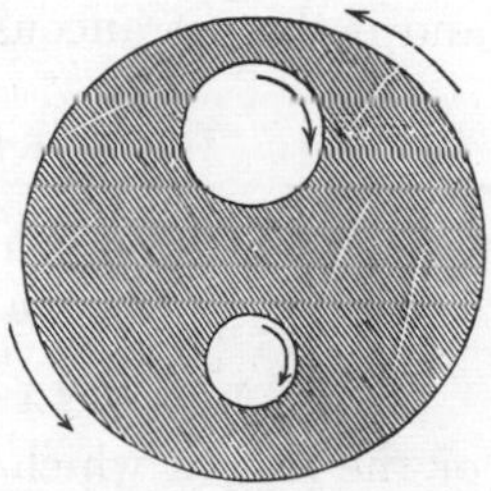

The value of the surface-integral taken over a *closed* surface is zero.

It should be noticed that (4) is a theorem of pure mathematics, and is true whatever functions u, v, w may be of x, y, z, provided only they be continuous and differentiable at all points of the surface‡.

33. The rest of this chapter is devoted to a study of the kinematical properties of irrotational motion in general, as defined by the equations

$$\xi, \eta, \zeta = 0, \quad \ldots\ldots\ldots\ldots\ldots\ldots\ldots\ldots\ldots\ldots\ldots(1)$$

* See Maxwell, *Electricity and Magnetism*, Oxford, 1873, Art. 23.

† This theorem is due to Stokes, *Smith's Prize Examination Papers for* 1854. The first published proof appears to have been given by Hankel, *Zur allgem. Theorie der Bewegung der Flüssigkeiten*, Göttingen, 1861. That given above is due to Lord Kelvin, *l.c. ante* p. 33. See also Thomson and Tait, *Natural Philosophy*, Art. 190 (*j*), and Maxwell, *Electricity and Magnetism*, Art. 24.

‡ It is not necessary that their differential coefficients should be continuous.

i.e. the circulation in every *infinitely small* circuit is assumed to be zero. The existence and properties of the velocity-potential in the various cases that may arise will appear as consequences of this definition.

The physical importance of the subject rests on the fact that if the motion of any portion of a fluid mass be irrotational at any one instant it will under certain very general conditions continue to be irrotational. Practically, as will be seen, this has already been established by Lagrange's theorem, proved in Art. 17, but the importance of the matter warrants a repetition of the investigation, in terms of the Eulerian notation, in the form given by Lord Kelvin*.

Consider first any terminated line AB drawn in the fluid, and suppose every point of this line to move always with the velocity of the fluid at that point. Let us calculate the rate at which the flow along this line, from A to B, is increasing. If $\delta x, \delta y, \delta z$ be the projections on the co-ordinate axes of an element of the line, we have

$$\frac{D}{Dt}(u\,\delta x) = \frac{Du}{Dt}\,\delta x + u\,\frac{D\,\delta x}{Dt}.$$

Now $D\delta x/Dt$, the rate at which δx is increasing in consequence of the motion of the fluid, is equal to the difference of the velocities parallel to x at the two ends of the element, *i.e.* to δu; and the value of Du/Dt is given by Art. 5. Hence, and by similar considerations, we find, if ρ be a function of p only, and if the extraneous forces X, Y, Z have a potential Ω,

$$\frac{D}{Dt}(u\,\delta x + v\,\delta y + w\,\delta z) = -\frac{\delta p}{\rho} - \delta\Omega + u\,\delta u + v\,\delta v + w\,\delta w.$$

Integrating along the line, from A to B, we get

$$\frac{D}{Dt}\int_A^B (u\,dx + v\,dy + w\,dz) = \left[-\int\frac{dp}{\rho} - \Omega + \tfrac{1}{2}q^2\right]_A^B, \quad \text{.........(2)}$$

or the rate at which the flow from A to B is increasing is equal to the excess of the value which $-\int dp/\rho - \Omega + \frac{1}{2}q^2$ has at B over that which it has at A. This theorem comprehends the whole of the dynamics of a perfect fluid. For instance, equations (2) of Art. 15 may be derived from it by taking as the line AB the infinitely short line whose projections were originally $\delta a, \delta b, \delta c$, and equating separately to zero the coefficients of these infinitesimals.

If Ω be single-valued, the expression within brackets on the right-hand side of (2) is a single-valued function of x, y, z. Hence if the integration on the left-hand side be taken round a closed curve, so that B coincides with A, we have

$$\frac{D}{Dt}\int (u\,dx + v\,dy + w\,dz) = 0, \quad \text{........................(3)}$$

or, the circulation in any circuit moving with the fluid does not alter with the time.

* *l.c. ante* p. 33.

It follows that if the motion of any portion of a fluid mass be initially irrotational it will always retain this property; for otherwise the circulation in every infinitely small circuit would not continue to be zero, as it is initially by virtue of Art. 32 (3).

34. Considering now any region occupied by irrotationally-moving fluid, we see from Art. 32 (3) that the circulation is zero in every circuit which can be filled up by a continuous surface lying wholly in the region, or which in other words is capable of being contracted to a point without passing out of the region. Such a circuit is said to be 'reducible.'

Again, let us consider two paths ACB, ADB, connecting two points A, B of the region, and such that either may by continuous variation be made to coincide with the other, without ever passing out of the region. Such paths are called 'mutually reconcileable.' Since the circuit $ACBDA$ is reducible, we have $I\,(ACBDA) = 0$, or since $I\,(BDA) = -I\,(ADB)$,

$$I\,(ACB) = I\,(ADB);$$

i.e. the flow is the same along any two reconcileable paths.

A region such that *all* paths joining any two points of it are mutually reconcileable is said to be 'simply-connected.' Such a region is that enclosed within a sphere, or that included between two concentric spheres. In what follows, as far as Art. 46, we contemplate only simply-connected regions.

35. The irrotational motion of a fluid within a simply-connected region is characterized by the existence of a single-valued velocity-potential. Let us denote by $-\phi$ the flow to a variable point P from some fixed point A, viz.

$$\phi = -\int_A^P (u\,dx + v\,dy + w\,dz). \qquad \text{......................(1)}$$

The value of ϕ has been shewn to be independent of the path along which the integration is effected, provided it lie wholly within the region. Hence ϕ is a single-valued function of the position of P; let us suppose it expressed in terms of the co-ordinates (x, y, z) of that point. By displacing P through an infinitely short space parallel to each of the axes of co-ordinates in succession, we find

$$u = -\frac{\partial\phi}{\partial x}, \quad v = -\frac{\partial\phi}{\partial y}, \quad w = -\frac{\partial\phi}{\partial z}, \qquad \text{....................(2)}$$

i.e. ϕ is a velocity-potential, according to the definition of Art. 17.

The substitution of any other point B for A, as the lower limit of the integral in (1), simply adds an arbitrary constant to the value of ϕ, viz. the flow from A to B. The original definition of ϕ in Art. 17, and the physical interpretation in Art. 18, alike leave the function indeterminate to the extent of an additive constant.

As we follow the course of any line of motion the value of ϕ continually decreases; hence in a simply-connected region the lines of motion cannot form closed curves.

36. The function ϕ with which we have here to do is, together with its first differential coefficients, by the nature of the case, finite, continuous, and single-valued at all points of the region considered. In the case of incompressible fluids, which we now proceed to consider more particularly, ϕ must also satisfy the equation of continuity, (6) of Art. 20, or as we shall in future write it, for shortness,

$$\nabla^2\phi = 0, \qquad\qquad (1)$$

at every point of the region. Hence ϕ is now subject to mathematical conditions identical with those satisfied by the potential of masses attracting or repelling according to the law of the inverse square of the distance, at all points external to such masses; so that many of the results proved in the theories of Attractions, Electrostatics, Magnetism, and the Steady Flow of Heat, have also a hydrodynamical application. We proceed to develop those which are most important from this point of view.

In any case of motion of an incompressible fluid the surface-integral of the normal velocity taken over any surface, open or closed, is conveniently called the 'flux' across the surface. It is of course equal to the volume of fluid crossing the surface per unit time.

When the motion is irrotational, the flux is given by

$$-\iint \frac{\partial\phi}{\partial n}\,dS,$$

where δS is an element of the surface, and δn an element of the normal to it, drawn in the proper direction. In any region occupied wholly by liquid, the total flux across the boundary is zero, *i.e.*

$$\iint \frac{\partial\phi}{\partial n}\,dS = 0, \qquad\qquad (2)$$

the element δn of the normal being drawn always on one side (say inwards), and the integration extending over the whole boundary. This may be regarded as a generalized form of the equation of continuity (1).

The lines of motion drawn through the various points of an infinitesimal circuit define a tube, which may be called a tube of flow. The product of the velocity (q) into the cross-section (σ, say) is the same at all points of such a tube.

We may, if we choose, regard the whole space occupied by the fluid as made up of tubes of flow, and suppose the size of the tubes so adjusted that the product $q\sigma$ is the same for each. The flux across any surface is then proportional to the number of tubes which cross it. If the surface be closed, the equation (2) expresses the fact that as many tubes cross the surface inwards as outwards. Hence a line of motion cannot begin or end at a point internal to the fluid.

37. The function ϕ cannot be a maximum or a minimum at a point in the interior of the fluid; for, if it were, we should have $\partial\phi/\partial n$ everywhere positive, or everywhere negative, over a small closed surface surrounding the point in question. Either of these suppositions is inconsistent with (2).

Further, the square of the velocity cannot be a *maximum* at a point in the interior of the fluid. For let the axis of x be taken parallel to the direction of the velocity at any point P. The equation (1), and therefore also the equation (2), is satisfied if we write $\partial\phi/\partial x$ for ϕ. The above argument then shews that $\partial\phi/\partial x$ cannot be a maximum or a minimum at P. Hence there must be points in the immediate neighbourhood of P at which $(\partial\phi/\partial x)^2$ and therefore *a fortiori*

$$\left(\frac{\partial\phi}{\partial x}\right)^2 + \left(\frac{\partial\phi}{\partial y}\right)^2 + \left(\frac{\partial\phi}{\partial z}\right)^2$$

is greater than the square of the velocity at P*.

On the other hand, the square of the velocity may be a *minimum* at some point of the fluid. The simplest case is that of a *zero* velocity; see, for example, the figure of Art. 69, below.

38. Let us apply (2) to the boundary of a finite spherical portion of the liquid. If r denote the distance of any point from the centre of the sphere, $\delta\varpi$ the elementary solid angle subtended at the centre by an element δS of the surface, we have

$$\partial\phi/\partial n = -\partial\phi/\partial r,$$

and $\delta S = r^2\delta\varpi$. Omitting the factor r^2, (2) becomes

$$\iint \frac{\partial\phi}{\partial r}\, d\varpi = 0,$$

or

$$\frac{\partial}{\partial r}\iint \phi\, d\varpi = 0. \qquad \text{.............................(3)}$$

Since $1/4\pi\,.\iint\phi\,d\varpi$ or $1/4\pi r^2\,.\iint\phi\,dS$ measures the mean value of ϕ over the surface of the sphere, (3) shews that this mean value is independent of the radius. It is therefore the same for any sphere, concentric with the former one, which can be made to coincide with it by gradual variation of the radius, without ever passing out of the region occupied by the irrotationally moving liquid. We may therefore suppose the sphere contracted to a point, and so obtain a simple proof of the theorem, first given by Gauss in his

* This theorem was enunciated, in another connection, by Lord Kelvin, *Phil. Mag.* Oct. 1850 [*Reprint of Papers on Electrostatics, &c.*, London, 1872, Art. 665]. The above demonstration is due to Kirchhoff, *Vorlesungen über mathematische Physik, Mechanik*, Leipzig, 1876. For another proof see Art. 44 below.

memoir* on the theory of Attractions, that the mean value of ϕ over any spherical surface throughout the interior of which (1) is satisfied, is equal to its value at the centre.

The theorem, proved in Art. 37, that ϕ cannot be a maximum or a minimum at a point in the interior of the fluid, is an obvious consequence of the above.

The above proof appears to be due, in principle, to Frost †. Another demonstration, somewhat different in form, was given by the late Lord Rayleigh ‡. The equation (1), being linear, will be satisfied by the arithmetic mean of any number of separate solutions $\phi_1, \phi_2, \phi_3, \ldots$. Let us suppose an infinite number of systems of rectangular axes to be arranged uniformly about any point P as origin, and let $\phi_1, \phi_2, \phi_3, \ldots$ be the velocity-potentials of motions which are the same with respect to these several systems as the original motion ϕ is with respect to the system x, y, z. In this case the arithmetic mean ($\bar{\phi}$, say) of the functions $\phi_1, \phi_2, \phi_3, \ldots$ will be a function of r, the distance from P, only. Expressing that in the motion (if any) represented by $\bar{\phi}$, the flux across any spherical surface which can be contracted to a point, without passing out of the region occupied by the fluid, would be zero, we have

$$4\pi r^2 . \frac{\partial \bar{\phi}}{\partial r} = 0,$$

or $\bar{\phi} = \text{const.}$

39. Again, let us suppose that the region occupied by the irrotationally moving fluid is 'periphractic,'§ *i.e.* that it is limited internally by one or more closed surfaces, and let us apply (2) to the space included between one (or more) of these internal boundaries, and a spherical surface completely enclosing it (or them) and lying wholly in the fluid. If M denote the total flux into this region, across the internal boundary, we find, with the same notation as before,

$$\iint \frac{\partial \phi}{\partial r} dS = -M,$$

the surface-integral extending over the sphere only. This may be written

$$\frac{1}{4\pi} \frac{\partial}{\partial r} \iint \phi \, d\varpi = -\frac{M}{4\pi r^2},$$

whence

$$\frac{1}{4\pi r^2} \iint \phi \, dS = \frac{1}{4\pi} \iint \phi \, d\varpi = \frac{M}{4\pi r} + C. \quad \ldots\ldots\ldots\ldots\ldots(4)$$

* "Allgemeine Lehrsätze, u.s.w.," *Resultate aus den Beobachtungen des magnetischen Vereins*, 1839 [*Werke*, Göttingen, 1870–80, v. 199].

† *Quarterly Journal of Mathematics*, xii. (1873).

‡ *Messenger of Mathematics*, vii. 69 (1878) [*Papers*, i. 347].

§ See Maxwell, *Electricity and Magnetism*, Arts. 18, 22. A region is said to be 'aperiphractic' when every closed surface drawn in it can be contracted to a point without passing out of the region.

That is, the mean value of ϕ over any spherical surface drawn under the above-mentioned conditions is equal to $M/4\pi r + C$, where r is the radius, M an absolute constant, and C a quantity which is independent of the radius but may vary with the position of the centre*.

If however the original region throughout which the irrotational motion holds be unlimited externally, and if the first derivative (and therefore all the higher derivatives) of ϕ vanish at infinity, then C is the same for *all* spherical surfaces enclosing the whole of the internal boundaries. For if such a sphere be displaced parallel to x†, without alteration of size, the rate at which C varies in consequence of this displacement is, by (4), equal to the mean value of $\partial\phi/\partial x$ over the surface. Since $\partial\phi/\partial x$ vanishes at infinity, we can by taking the sphere large enough make the latter mean value as small as we please. Hence C is not altered by a displacement of the centre of the sphere parallel to x. In the same way we see that C is not altered by a displacement parallel to y or z; *i.e.* it is absolutely constant.

If the internal boundaries of the region considered be such that the total flux across them is zero, *e.g.* if they be the surfaces of solids, or of portions of incompressible fluid whose motion is rotational, we have $M = 0$, so that the mean value of ϕ over *any* spherical surface enclosing them all is the same.

40. (α) If ϕ be constant over the boundary of any simply-connected region occupied by liquid moving irrotationally, it has the same constant value throughout the interior of that region. For if not constant it would necessarily have a maximum or a minimum value at some point of the region.

Otherwise: we have seen in Arts. 35, 36 that the lines of motion cannot begin or end at any point of the region, and that they cannot form closed curves lying wholly within it. They must therefore traverse the region, beginning and ending on its boundary. In our case however this is impossible, for such a line always proceeds from places where ϕ is greater to places where it is less. Hence there can be no motion, *i.e.*

$$\frac{\partial\phi}{\partial x},\quad \frac{\partial\phi}{\partial y},\quad \frac{\partial\phi}{\partial z} = 0,$$

and therefore ϕ is constant and equal to its value at the boundary.

(β) Again, if $\partial\phi/\partial n$ be zero at every point of the boundary of such a region as is above described, ϕ will be constant throughout the interior. For the condition $\partial\phi/\partial n = 0$ expresses that no lines of motion enter or leave the region, but that they are all contained within it. This is however, as we have seen, inconsistent with the other conditions which the lines must conform to. Hence, as before, there can be no motion, and ϕ is constant.

* It is understood, of course, that the spherical surfaces to which this statement applies are reconcileable with one another, in a sense analogous to that of Art. 34.

† Kirchhoff, *Mechanik*, p. 191.

This theorem may be otherwise stated as follows: no continuous irrotational motion of a liquid can take place in a simply-connected region bounded entirely by fixed rigid walls.

(γ) Again, let the boundary of the region considered consist partly of surfaces S over which ϕ has a given constant value, and partly of other surfaces Σ over which $\partial\phi/\partial n = 0$. By the previous argument, no lines of motion can pass from one point to another of S, and none can cross Σ. Hence no such lines exist; ϕ is therefore constant as before, and equal to its value at S.

It follows from these theorems that the irrotational motion of a liquid in a simply-connected region is determined when either the value of ϕ, or the value of the inward normal velocity $-\partial\phi/\partial n$, is prescribed at all points of the boundary, or (again) when the value of ϕ is given over part of the boundary, and the value of $-\partial\phi/\partial n$ over the remainder. For if ϕ_1, ϕ_2 be the velocity-potentials of two motions each of which satisfies the prescribed boundary-conditions, in any one of these cases, the function $\phi_1 - \phi_2$ satisfies the condition (α) or (β) or (γ) of the present Article, and must therefore be constant throughout the region.

41. A class of cases of great importance, but not strictly included in the scope of the foregoing theorems, occurs when the region occupied by the irrotationally moving liquid extends to infinity, but is bounded internally by one or more closed surfaces. We assume, for the present, that this region is simply-connected, and that ϕ is therefore single-valued.

If ϕ be constant over the internal boundary of the region, and tend everywhere to the same constant value at an infinite distance from the internal boundary, it is constant throughout the region. For otherwise ϕ would be a maximum or a minimum at some point within the region.

We infer, exactly as in Art. 40, that if ϕ be given arbitrarily over the internal boundary, and have a given constant value at infinity, its value is everywhere determinate.

Of more importance in our present subject is the theorem that, if the normal velocity be zero at every point of the internal boundary, and if the fluid be at rest at infinity, then ϕ is everywhere constant. We cannot however infer this at once from the proof of the corresponding theorem in Art. 40. It is true that we may suppose the region limited externally by an infinitely large surface at every point of which $\partial\phi/\partial n$ is infinitely small; but it is conceivable that the integral $\iint \partial\phi/\partial n \, . \, dS$, taken over a portion of this surface, might still be finite, in which case the investigation referred to would fail. We proceed therefore as follows.

Since the velocity tends to the limit zero at an infinite distance from the internal boundary (S, say), it must be possible to draw a closed surface Σ

completely enclosing S, beyond which the velocity is everywhere less than a certain value ϵ, which value may, by making Σ large enough, be made as small as we please. Now in any direction from S let us take a point P at such a distance beyond Σ that the solid angle which Σ subtends at it is infinitely small; and with P as centre let us describe two spheres, one just excluding, the other just including S. We shall prove that the mean value of ϕ over each of these spheres is, within an infinitely small amount, the same. For if Q, Q' be points of these spheres on a common radius PQQ', then if Q, Q' fall within Σ the corresponding values of ϕ may differ by a finite amount; but since the portion of either spherical surface which falls within Σ is an infinitely small fraction of the whole, no finite difference in the mean values can arise from this cause. On the other hand, when Q, Q' fall without Σ, the corresponding values of ϕ cannot differ by so much as $\epsilon \,.\, QQ'$, for ϵ is by definition a superior limit to the rate of variation of ϕ. Hence, the mean values of ϕ over the two spherical surfaces must differ by less than $\epsilon \,.\, QQ'$. Since QQ' is finite, whilst ϵ may by taking Σ large enough be made as small as we please, the difference of the mean values may, by taking P sufficiently distant, be made infinitely small.

Now we have seen in Arts. 38, 39 that the mean value of ϕ over the inner sphere is equal to its value at P, and that the mean value over the outer sphere is (since $M = 0$) equal to a constant quantity C. Hence, ultimately, the value of ϕ at infinity tends everywhere to the constant value C.

The same result holds even if the normal velocity be not zero over the internal boundary; for in the theorem of Art. 39 M is divided by r, which is in our case infinite.

It follows that if $\partial\phi/\partial n = 0$ at all points of the internal boundary, and if the fluid be at rest at infinity, it must be everywhere at rest. For no lines of motion can begin or end on the internal boundary. Hence such lines, if they existed, must come from an infinite distance, traverse the region occupied by the fluid, and pass off again to infinity; *i.e.* they must form infinitely long courses between places where ϕ has, with an infinitely small amount, the same value C, which is impossible.

The theorem that, if the fluid be at rest at infinity, the motion is determinate when the value of $-\partial\phi/\partial n$ is given over the internal boundary, follows by the same argument as in Art. 40.

Green's Theorem.

42. In treatises on Electrostatics, &c., many important properties of the potential are usually proved by means of a certain theorem due to Green. Of these the most important from our present point of view have already been given; but as the theorem in question leads, amongst other things, to a useful expression for the kinetic energy in any case of irrotational motion, some account of it will properly find a place here.

Let U, V, W be any three functions which are finite, single-valued and differentiable at all points of a connected region completely bounded by one or more closed surfaces S; let δS be an element of any one of these surfaces, and l, m, n the direction-cosines of the normals to it drawn inwards. We shall prove in the first place that

$$\iint (lU + mV + nW)\, dS = -\iiint \left(\frac{\partial U}{\partial x} + \frac{\partial V}{\partial y} + \frac{\partial W}{\partial z}\right) dx\,dy\,dz, \quad\text{......(1)}$$

where the triple-integral is taken throughout the region, and the double-integral over its boundary.

If we conceive a series of surfaces drawn so as to divide the region into any number of separate parts, the integral

$$\iint (lU + mV + nW)\, dS, \quad\text{..........................(2)}$$

taken over the original boundary, is equal to the sum of the similar integrals each taken over the whole boundary of one of these parts. For, for every element $\delta\sigma$ of a dividing surface, we have, in the integrals corresponding to the parts lying on the two sides of this surface, elements $(lU + mV + nW)\,\delta\sigma$, and $(l'U + m'V + n'W)\,\delta\sigma$, respectively. But the normals to which l, m, n and l', m', n' refer being drawn inwards in each case, we have $l' = -l$, $m' = -m$, $n' = -n$; so that, in forming the sum of the integrals spoken of, the elements due to the dividing surfaces disappear, and we have left only those due to the original boundary of the region.

Now let us suppose the dividing surfaces to consist of three systems of planes, drawn at infinitesimal intervals, parallel to yz, zx, xy, respectively. If x, y, z be the co-ordinates of the centre of one of the rectangular spaces thus formed, and δx, δy, δz the lengths of its edges, the part of the integral (2) due to the yz-face nearest the origin is

$$\left(U - \tfrac{1}{2}\frac{\partial U}{\partial x}\partial x\right)\delta y\,\delta z,$$

and that due to the opposite face is

$$-\left(U + \tfrac{1}{2}\frac{\partial U}{\partial x}\delta x\right)\delta y\,\delta z.$$

The sum of these is $-\partial U/\partial x \,.\, \delta x\,\delta y\,\delta z$. Calculating in the same way the parts of the integral due to the remaining pairs of faces, we get for the final result

$$-\left(\frac{\partial U}{\partial x} + \frac{\partial V}{\partial y} + \frac{\partial W}{\partial z}\right)\delta x\,\delta y\,\delta z.$$

Hence (1) simply expresses the fact that the surface-integral (2), taken over the boundary of the region, is equal to the sum of the similar integrals taken over the boundaries of the elementary spaces of which we have supposed it built up.

It is evident from (1), or it may be proved directly by transformation of co-ordinates, that if U, V, W be regarded as components of a vector, the expression

$$\frac{\partial U}{\partial x}+\frac{\partial V}{\partial y}+\frac{\partial W}{\partial z}$$

is a 'scalar' quantity, *i.e.* its value is unaffected by any such transformation. It is now usually called the 'divergence' of the vector-field at the point (x, y, z).

The interpretation of (1), when (U, V, W) is the velocity of a continuous substance, is obvious. In the particular case of irrotational motion we obtain

$$\iint \frac{\partial \phi}{\partial n}\, dS = -\iiint \nabla^2 \phi \, dx\, dy\, dz, \quad \text{......................(3)}$$

where δn denotes an element of the inwardly-directed normal to the surface S.

Again, if we put $U, V, W = \rho u, \rho v, \rho w$, respectively, we reproduce in substance the second investigation of Art. 7.

Another useful result is obtained by putting $U, V, W = u\phi, v\phi, w\phi$, respectively, where u, v, w satisfy the relation

$$\frac{\partial u}{\partial x}+\frac{\partial v}{\partial y}+\frac{\partial w}{\partial z}=0$$

throughout the region, and make

$$lu + mv + nw = 0$$

over the boundary. We find

$$\iiint \left(u\frac{\partial \phi}{\partial x}+v\frac{\partial \phi}{\partial y}+w\frac{\partial \phi}{\partial z}\right) dx\, dy\, dz = 0. \quad \text{..................(4)}$$

The function ϕ is here merely restricted to be finite, single-valued, and continuous, and to have its first differential coefficients finite, throughout the region.

43. Now let ϕ, ϕ' be any two functions which, together with their first and second derivatives, are finite and single-valued throughout the region considered; and let us put

$$U, V, W = \phi\frac{\partial \phi'}{\partial x},\ \phi\frac{\partial \phi'}{\partial y},\ \phi\frac{\partial \phi'}{\partial z},$$

respectively, so that $\qquad lU + mV + nW = \phi\dfrac{\partial \phi'}{\partial n}.$

Substituting in (1) we find

$$\iint \phi\frac{\partial \phi'}{\partial n}\, dS = -\iiint\left(\frac{\partial \phi}{\partial x}\frac{\partial \phi'}{\partial x}+\frac{\partial \phi}{\partial y}\frac{\partial \phi'}{\partial y}+\frac{\partial \phi}{\partial z}\frac{\partial \phi'}{\partial z}\right) dx\, dy\, dz$$
$$- \iiint \phi \nabla^2 \phi' \, dx\, dy\, dz. \quad \text{.........(5)}$$

By interchanging ϕ and ϕ' we obtain

$$\iint \phi' \frac{\partial \phi}{\partial n} dS = -\iiint \left(\frac{\partial \phi}{\partial x}\frac{\partial \phi'}{\partial x} + \frac{\partial \phi}{\partial y}\frac{\partial \phi'}{\partial y} + \frac{\partial \phi}{\partial z}\frac{\partial \phi'}{\partial z} \right) dx\,dy\,dz$$
$$- \iiint \phi' \nabla^2 \phi\, dx\,dy\,dz. \quad \text{.........(6)}$$

Equations (5) and (6) together constitute Green's theorem*

44. If ϕ, ϕ' be the velocity-potentials of two distinct modes of irrotational motion of a liquid, so that

$$\nabla^2 \phi = 0, \quad \nabla^2 \phi' = 0, \quad \text{.........................(1)}$$

we obtain

$$\iint \phi \frac{\partial \phi'}{\partial n} dS = \iint \phi' \frac{\partial \phi}{\partial n} dS. \quad \text{........................(2)}$$

If we recall the physical interpretation of the velocity-potential, given in Art. 18, then, regarding the motion as generated in each case impulsively from rest, we recognize this equation as a particular case of the dynamical theorem that

$$\Sigma\, p_r \dot{q}_r' = \Sigma\, p_r' \dot{q}_r,$$

where p_r, $\dot{q}_r$ and p_r', $\dot{q}_r'$ are generalized components of impulse and velocity, in any two possible motions of a system†.

Again, in Art. 43 (6) let $\phi' = \phi$, and let ϕ be the velocity-potential of a liquid. We obtain

$$\iiint \left\{ \left(\frac{\partial \phi}{\partial x}\right)^2 + \left(\frac{\partial \phi}{\partial y}\right)^2 + \left(\frac{\partial \phi}{\partial z}\right)^2 \right\} dx\,dy\,dz = -\iint \phi \frac{\partial \phi}{\partial n} dS. \quad \text{.........(3)}$$

To interpret this we multiply both sides by $\frac{1}{2}\rho$. Then on the right-hand side $-\partial\phi/\partial n$ denotes the normal velocity of the fluid inwards, whilst $\rho\phi$ is, by Art. 18, the impulsive pressure necessary to generate the motion. It is a proposition in Dynamics‡ that the work done by an impulse is measured by the product of the impulse into half the sum of the initial and final velocities, resolved in the direction of the impulse, of the point to which it is applied. Hence the right-hand side of (3), when modified as described, expresses the work done by the system of impulsive pressures which, applied to the surface S, would generate the actual motion; whilst the left-hand side gives the kinetic energy of this motion. The formula asserts that these two quantities are equal. Hence if T denote the total kinetic energy of the liquid, we have the very important formula

$$2T = -\rho \iint \phi \frac{\partial \phi}{\partial n} dS. \quad \text{.........................(4)}$$

If in (3), in place of ϕ, we write $\partial\phi/\partial x$, which will of course satisfy $\nabla^2 \partial\phi/\partial x = 0$, and apply the resulting theorem to the region included within a spherical surface of radius r

* G. Green, *Essay on Electricity and Magnetism*, Nottingham, 1828, Art. 3 [*Mathematical Papers* (ed. Ferrers), Cambridge, 1871, p. 3].

† Thomson and Tait, *Natural Philosophy*, Art. 313, equation (11).

‡ *Ibid.* Art. 308.

having any point (x, y, z) as centre, then with the same notation as in Art. 39, we have

$$\tfrac{1}{2}r^2\frac{\partial}{\partial r}\iiint u^2\,d\varpi = \iint u\frac{\partial u}{\partial r}\,dS = -\iint \frac{\partial\phi}{\partial x}\frac{\partial}{\partial n}\left(\frac{\partial\phi}{\partial x}\right)dS$$

$$= \iiint\left\{\left(\frac{\partial^2\phi}{\partial x^2}\right)^2 + \left(\frac{\partial^2\phi}{\partial x\,\partial y}\right)^2 + \left(\frac{\partial^2\phi}{\partial x\,\partial z}\right)^2\right\}dx\,dy\,dz.$$

Hence, writing $q^2 = u^2 + v^2 + w^2$,

$$\tfrac{1}{2}r^2\frac{\partial}{\partial r}\iint q^2\,d\varpi = \iiint\left\{\left(\frac{\partial^2\phi}{\partial x^2}\right)^2 + \left(\frac{\partial^2\phi}{\partial y^2}\right)^2 + \left(\frac{\partial^2\phi}{\partial z^2}\right)^2 + 2\left(\frac{\partial^2\phi}{\partial y\,\partial z}\right)^2 + 2\left(\frac{\partial^2\phi}{\partial z\,\partial x}\right)^2 + 2\left(\frac{\partial^2\phi}{\partial x\,\partial y}\right)^2\right\}dx\,dy\,dz. \quad\text{......(5)}$$

Since this latter expression is essentially positive, the mean value of q^2, taken over a sphere having any given point as centre, increases with the radius of the sphere. Hence q^2 cannot be a maximum at any point of the fluid, as was proved otherwise in Art. 37.

Moreover, recalling the formula for the pressure in any case of irrotational motion of a liquid, viz.

$$\frac{p}{\rho} = \frac{\partial\phi}{\partial t} - \Omega - \tfrac{1}{2}q^2 + F(t), \quad\text{......(6)}$$

we infer that, provided the potential Ω of the external forces satisfy the condition

$$\nabla^2\Omega = 0, \quad\text{......(7)}$$

the mean value of p over a sphere described with any point in the interior of the fluid as centre will diminish as the radius increases. The place of least pressure will therefore be somewhere on the boundary of the fluid. This has a bearing on the point discussed in Art. 23.

45. In this connection we may note a remarkable theorem discovered by Lord Kelvin*, and afterwards generalized by him into an universal property of dynamical systems started impulsively from rest under prescribed velocity-conditions†.

The irrotational motion of a liquid occupying a simply-connected region has less kinetic energy than any other motion consistent with the same normal motion of the boundary.

Let T be the kinetic energy of the irrotational motion to which the velocity-potential ϕ refers, and T_1 that of another motion given by

$$u = -\frac{\partial\phi}{\partial x} + u_0, \quad v = -\frac{\partial\phi}{\partial y} + v_0, \quad w = -\frac{\partial\phi}{\partial z} + w_0, \quad\text{......(8)}$$

where, in virtue of the equation of continuity, and the prescribed boundary-condition, we must have

$$\frac{\partial u_0}{\partial x} + \frac{\partial v_0}{\partial y} + \frac{\partial w_0}{\partial z} = 0$$

throughout the region, and $lu_0 + mv_0 + nw_0 = 0$ over the boundary. Further let us write

$$T_0 = \tfrac{1}{2}\rho\iiint(u_0^2 + v_0^2 + w_0^2)\,dx\,dy\,dz. \quad\text{......(9)}$$

* (W. Thomson) "On the Vis-Viva of a Liquid in Motion," *Camb. and Dub. Math. Journ.* 1849 [*Papers*, i. 107].

† Thomson and Tait, Art. 312.

We find $$T_1 = T + T_0 - \rho\iiint\left(u_0\frac{\partial\phi}{\partial x} + v_0\frac{\partial\phi}{\partial y} + w_0\frac{\partial\phi}{\partial z}\right)dx\,dy\,dz.$$

Since the last integral vanishes, by Art. 42 (4), we have

$$T_1 = T + T_0, \qquad\qquad (10)$$

which proves the theorem*.

46. We shall require to know, hereafter, the form assumed by the expression (4) for the kinetic energy when the fluid extends to infinity and is at rest there, being limited internally by one or more closed surfaces S. Let us suppose a large closed surface Σ described so as to enclose the whole of S. The energy of the fluid included between S and Σ is

$$-\tfrac{1}{2}\rho\iint\phi\frac{\partial\phi}{\partial n}\,dS - \tfrac{1}{2}\rho\iint\phi\frac{\partial\phi}{\partial n}\,d\Sigma, \qquad\qquad (11)$$

where the integration in the first term extends over S, that in the second over Σ. Since we have, by the equation of continuity,

$$\iint\frac{\partial\phi}{\partial n}\,dS + \iint\frac{\partial\phi}{\partial n}\,d\Sigma = 0,$$

the expression (11) may be written

$$-\tfrac{1}{2}\rho\iint(\phi - C)\frac{\partial\phi}{\partial n}\,dS - \tfrac{1}{2}\rho\iint(\phi - C)\frac{\partial\phi}{\partial n}\,d\Sigma, \qquad\qquad (12)$$

where C may be any constant, but is here supposed to be the constant value to which ϕ was shewn in Art. 39 to tend at an infinite distance from S. Now the whole region occupied by the fluid may be supposed made up of tubes of flow, each of which must pass either from one point of the internal boundary to another, or from that boundary to infinity. Hence the value of the integral

$$\iint\frac{\partial\phi}{\partial n}\,d\Sigma,$$

taken over any surface, open or closed, finite or infinite, drawn within the region, must be finite. Hence ultimately, when Σ is taken infinitely large and infinitely distant all round from S, the second term of (12) vanishes, and we have

$$2T = -\rho\iint(\phi - C)\frac{\partial\phi}{\partial n}\,dS, \qquad\qquad (13)$$

where the integration extends over the internal boundary only.

If the total flux across the internal boundary be zero, we have

$$\iint\frac{\partial\phi}{\partial n}\,dS = 0,$$

so that (13) may be written $$2T = -\rho\iint\phi\frac{\partial\phi}{\partial n}\,dS, \qquad\qquad (14)$$
simply.

* Some extensions of this result are discussed by Leathem, *Cambridge Tracts*, No. 1, 2nd ed. (1913). They supply further interesting illustrations of Kelvin's general dynamical principle.

On Multiply-connected Regions.

47. Before discussing the properties of irrotational motion in multiply-connected regions we must examine more in detail the nature and classification of such regions. In the following synopsis of this branch of the geometry of position we recapitulate for the sake of completeness one or two definitions already given.

We consider any connected region of space, enclosed by boundaries. A region is 'connected' when it is possible to pass from any one point of it to any other by an infinity of paths, each of which lies wholly in the region.

Any two such paths, or any two circuits, which can by continuous variation be made to coincide without ever passing out of the region, are said to be 'mutually reconcileable.' Any circuit which can be contracted to a point without passing out of the region is said to be 'reducible.' Two reconcileable paths, combined, form a reducible circuit. If two paths or two circuits be reconcileable, it must be possible to connect them by a continuous surface, which lies wholly within the region, and of which they form the complete boundary: and conversely.

It is further convenient to distinguish between 'simple' and 'multiple' irreducible circuits. A 'multiple' circuit is one which can by continuous variation be made to appear, in whole or in part, as the repetition of another circuit a certain number of times. A 'simple' circuit is one with which this is not possible.

A 'barrier,' or 'diaphragm,' is a surface drawn across the region, and limited by the line or lines in which it meets the boundary. Hence a barrier is necessarily a connected surface, and cannot consist of two or more detached portions.

A 'simply-connected' region is one such that all paths joining any two points of it are reconcileable, or such that all circuits drawn within it are reducible.

A 'doubly-connected' region is one such that two irreconcileable paths, and no more, can be drawn between any two points A, B of it; viz. any other path joining AB is reconcileable with one of these, or with a combination of the two taken each a certain number of times. In other words, the region is such that one (simple) irreducible circuit can be drawn in it, whilst all other circuits are either reconcileable with this (repeated, if necessary), or are reducible. As an example of a doubly-connected region we may take that enclosed by the surface of an anchor-ring, or that external to such a ring and extending to infinity.

Generally, a region such that n irreconcileable paths, and no more, can be drawn between any two points of it, or such that $n-1$ (simple) irreducible and irreconcileable circuits, and no more, can be drawn in it, is said to be 'n-ply-connected.'

The shaded portion of the figure on p. 35 is a triply-connected space of two dimensions.

It may be shewn that the above definition of an n-ply-connected space is self-consistent. In such simple cases as $n=2$, $n=3$, this is sufficiently evident without demonstration.

48. Let us suppose, now, that we have an n-ply-connected region, with $n-1$ simple independent irreducible circuits drawn in it. It is possible to draw a barrier meeting any one of these circuits in one point only, and not meeting any of the $n-2$ remaining circuits. A barrier drawn in this manner does not destroy the continuity of the region, for the interrupted circuit remains as a path leading round from one side to the other. The order of connection of the region is however diminished by unity; for every circuit drawn in the modified region must be reconcileable with one or more of the $n-2$ circuits not met by the barrier.

A second barrier, drawn in the same manner, will reduce the order of connection again by one, and so on; so that by drawing $n-1$ barriers we can reduce the region to a simply-connected one.

A simply-connected region is divided by a barrier into two separate parts; for otherwise it would be possible to pass from a point on one side of the barrier to an adjacent point on the other side by a path lying wholly within the region, which path would in the original region form an irreducible circuit.

Hence in an n-ply-connected region it is possible to draw $n-1$ barriers, and no more, without destroying the continuity of the region. This property is sometimes adopted as the definition of an n-ply-connected space.

Irrotational Motion in Multiply-connected Spaces.

49. The circulation is the same in any two reconcileable circuits $ABCA$, $A'B'C'A'$ drawn in a region occupied by fluid moving irrotationally. For the two circuits may be connected by a continuous surface lying wholly within the region; and if we apply the theorem of Art. 32 to this surface, we have, remembering the rule as to the direction of integration round the boundary.

$$I(ABCA)+I(A'C'B'A')=0,$$

or

$$I(ABCA)=I(A'B'C'A').$$

If a circuit $ABCA$ be reconcileable with two or more circuits $A'B'C'A'$, $A''B''C''A''$, &c., combined, we can connect all these circuits by a continuous surface which lies wholly within the region, and of which they form the complete boundary. Hence

$$I(ABCA)+I(A'C'B'A')+I(A''C''B''A'')+\&c.=0,$$

or

$$I(ABCA)=I(A'B'C'A')+I(A''B''C''A'')+\&c.;$$

i.e. the circulation in any circuit is equal to the sum of the circulations in the several members of any set of circuits with which it is reconcileable.

Let the order of connection of the region be $n+1$, so that n independent simple irreducible circuits $a_1, a_2, \ldots a_n$ can be drawn in it; and let the circulations in these be $\kappa_1, \kappa_2, \ldots \kappa_n$, respectively. The sign of any κ will of course depend on the direction of integration round the corresponding circuit; let the direction in which κ is estimated be called the positive direction in the circuit. The value of the circulation in any other circuit can now be found at once. For the given circuit is necessarily reconcileable with some combination of the circuits $a_1, a_2, \ldots a_n$; say with a_1 taken p_1 times, a_2 taken p_2 times and so on, where of course any p is negative when the corresponding circuit is taken in the negative direction. The required circulation then is

$$p_1\kappa_1 + p_2\kappa_2 + \ldots + p_n\kappa_n. \qquad \text{(1)}$$

Since any two paths joining two points A, B of the region together form a circuit, it follows that the values of the flow in the two paths differ by a quantity of the form (1), where, of course, in particular cases some or all of the p's may be zero.

50. Let us denote by $-\phi$ the flow to a variable point P from a fixed point A, viz.

$$\phi = -\int_A^P (u\,dx + v\,dy + w\,dz) \qquad \text{(2)}$$

So long as the path of integration from A to P is not specified, ϕ is indeterminate to the extent of a quantity of the form (1).

If however n barriers be drawn in the manner explained in Art. 48, so as to reduce the region to a simply-connected one, and if the path of integration in (2) be restricted to lie within the region as thus modified (*i.e.* it is not to cross any of the barriers), then ϕ becomes a single-valued function, as in Art. 35. It is continuous throughout the modified region, but its values at two adjacent points on opposite sides of a barrier differ by $\pm\kappa$. To derive the value of ϕ when the integration is taken along any path in the unmodified region we must *subtract* the quantity (1), where any p denotes the number of times this path crosses the corresponding barrier. A crossing in the positive direction of the circuits interrupted by the barrier is here counted as positive, a crossing in the opposite direction as negative.

By displacing P through an infinitely short space parallel to each co-ordinate axis in succession, we find

$$u, v, w = -\frac{\partial\phi}{\partial x},\ -\frac{\partial\phi}{\partial y},\ -\frac{\partial\phi}{\partial z};$$

so that ϕ satisfies the definition of a velocity-potential (Art. 17). It is now however a many-valued or cyclic function; *i.e.* it is not possible to assign to every point of the original region a unique and definite value of ϕ, such values forming a continuous system. On the contrary, whenever P describes an irreducible circuit, ϕ will not, in general, return to its original value, but will differ from it by a quantity of the form (1). The quantities $\kappa_1, \kappa_2, \dots \kappa_n$, which specify the amounts by which ϕ *decreases* as P describes the several independent circuits of the region, may be called the 'cyclic constants' of ϕ.

It is an immediate consequence of the 'circulation-theorem' of Art. 33 that under the conditions there presupposed the cyclic constants do not alter with the time. The necessity for these conditions is exemplified in the problem of Art. 29, where the potential of the extraneous forces is itself a cyclic function.

The foregoing theory may be illustrated by the case of Art. 27 (2), where the region (as limited by the exclusion of the origin, since the formula would give an infinite velocity there) is doubly-connected; for we can connect any two points A, B of it by two irreconcileable paths passing on opposite sides of the axis of z, *e.g.* ACB, ADB in the figure. The portion of the plane zx for which x is positive, may be taken as a barrier, and the region is thus made simply-connected. The circulation in any circuit meeting this barrier once only, *e.g.* in $ACBDA$, is

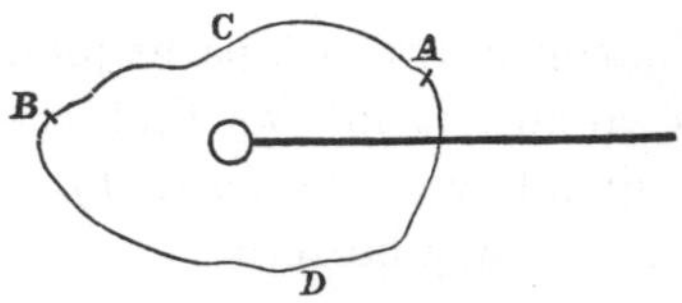

$$\int_0^{2\pi} \mu/r \,.\, r\,d\theta, \text{ or } 2\pi\mu.$$

That in any circuit not meeting the barrier is zero. In the modified region ϕ may be put equal to a single-valued function, viz. $-\mu\theta$, but its value on the positive side of the barrier is zero, that at an adjacent point on the negative is $-2\pi\mu$.

More complex illustrations of irrotational motion in multiply-connected spaces of two dimensions will present themselves in the next chapter.

51. Before proceeding further we may briefly indicate a somewhat different method of presenting the above theory.

Starting from the existence of a velocity-potential as the characteristic of the class of motions which we propose to study, and adopting the second definition of an $n+1$-ply-connected region, indicated in Art. 48, we remark that in a simply-connected region every equipotential surface must either be a closed surface, or else form a barrier dividing the region into two separate parts. Hence, supposing the whole system of such surfaces drawn, we see that if a closed curve cross any given equipotential surface once it must cross it again, and in the opposite direction. Hence, corresponding to any element of the curve, included between two consecutive equipotential surfaces, we have a second element such that the flow along it, being equal to the difference between the corresponding values of ϕ, is equal and opposite to that along the former; so that the circulation in the whole circuit is zero.

If however the region be multiply-connected, an equipotential surface may form a barrier without dividing it into two separate parts. Let as many such surfaces be drawn as is possible without destroying the continuity of the region. The number of these cannot, by definition, be greater than n. Every other equipotential surface which is not closed will be reconcileable (in an obvious sense) with one or more of these barriers. A curve drawn from one side of a barrier round to the other, without meeting any of the remaining barriers, will cross every equipotential surface reconcileable with the first barrier an odd number of times, and every other equipotential surface an even number of times. Hence the circulation in the circuit thus formed will not vanish, and ϕ will be a cyclic function.

In the method adopted above we have based the whole theory on the equations

$$\frac{\partial w}{\partial y} - \frac{\partial v}{\partial z} = 0, \quad \frac{\partial u}{\partial z} - \frac{\partial w}{\partial x} = 0, \quad \frac{\partial v}{\partial x} - \frac{\partial u}{\partial y} = 0, \quad \text{...........(3)}$$

and have deduced the existence and properties of the velocity-potential in the various cases as necessary consequences of these. In fact, Arts. 34, 35, and 49, 50 may be regarded as an inquiry into the nature of the solution of this system of differential equations, as depending on the character of the region through which they hold.

The integration of (3), when we have, on the right-hand side, instead of zero, known functions of x, y, z, will be treated in Chapter VII.

52. Proceeding now, as in Art. 36, to the particular case of an incompressible fluid, we remark that whether ϕ be cyclic or not, its first derivatives $\partial\phi/\partial x$, $\partial\phi/\partial y$, $\partial\phi/\partial z$, and therefore all the higher derivatives, are essentially single-valued functions, so that ϕ will still satisfy the equation of continuity

$$\nabla^2\phi = 0, \text{...................................(1)}$$

or the equivalent form

$$\iint \frac{\partial \phi}{\partial n} dS = 0, \text{...............................(2)}$$

where the surface-integration extends over the whole boundary of any portion of the fluid.

The theorem (α) of Art. 40, viz. that ϕ must be constant throughout the interior of any region at every point of which (1) is satisfied, if it be constant over the boundary, still holds when the region is multiply-connected. For ϕ, being constant over the boundary, is necessarily single-valued.

The remaining theorems of Art. 40, being based on the assumption that the stream-lines cannot form closed curves, will require modification. We must introduce the additional condition that the circulation is to be zero in each circuit of the region.

Removing this restriction, we have the theorem that the irrotational motion of a liquid occupying an n-ply-connected region is determinate when the normal velocity at every point of the boundary is prescribed, as well as the value of the circulation in each of the n independent and irreducible circuits which can be drawn in the region. For if ϕ_1, ϕ_2 be the (cyclic) velocity-potentials of two motions satisfying the above conditions, then

$\phi=\phi_1-\phi_2$ is a single-valued function which satisfies (1) at every point of the region, and makes $\partial\phi/\partial n=0$ at every point of the boundary. Hence, by Art. 40, ϕ is constant, and the motions determined by ϕ_1 and ϕ_2 are identical.

The theory of multiple connectivity seems to have been first developed by Riemann* for spaces of two dimensions, *à propos* of his researches on the theory of functions of a complex variable, in which connection also cyclic functions satisfying the equations

$$\frac{\partial^2\phi}{\partial x^2}+\frac{\partial^2\phi}{\partial y^2}=0$$

through multiply-connected regions, present themselves.

The bearing of the theory on Hydrodynamics and the existence in certain cases of many-valued velocity-potentials were first pointed out by von Helmholtz†. The subject of cyclic irrotational motion in multiply-connected regions was afterwards taken up and fully investigated by Lord Kelvin in the paper on vortex-motion already referred to‡.

Kelvin's Extension of Green's Theorem.

53. It was assumed in the proof of Green's theorem that ϕ and ϕ' were both single-valued functions. If either be a cyclic function, as may be the case when the region to which the integrations in Art. 43 refer is multiply-connected, the statement of the theorem must be modified. Let us suppose, for instance, that ϕ is cyclic; the surface-integral on the left-hand side of Art. 43 (5), and the second volume-integral on the right-hand side, are then indeterminate, on account of the indeterminateness in the value of ϕ itself. To remove this indeterminateness, let the barriers necessary to reduce the region to a simply-connected one be drawn, as explained in Art. 48. We may now suppose ϕ to be continuous and single-valued throughout the region thus modified; and the equation referred to will then hold, provided the two sides of each barrier be reckoned as part of the boundary of the region, and therefore included in the surface-integral on the left-hand side. Let $\delta\sigma_1$, be an element of one of the barriers, κ_1 the cyclic constant corresponding to that barrier, $\partial\phi'/\partial n$ the rate of variation of ϕ' in the positive direction of the normal to $\delta\sigma_1$. Since, in the parts of the surface-integral due to the two sides of $\delta\sigma_1$, $\partial\phi'/\partial n$ is to be taken with opposite signs, whilst the value of ϕ on the positive side exceeds that on the negative side by κ_1, we get finally for the element of the integral due to $\delta\sigma_1$, the value $\kappa_1\partial\phi'/\partial n\,.\,\delta\sigma_1$. Hence Art. 43 (5) becomes, in the altered circumstances,

$$\iint\phi\frac{\partial\phi'}{\partial n}\,dS+\kappa_1\iint\frac{\partial\phi'}{\partial n}\,d\sigma_1+\kappa_2\iint\frac{\partial\phi'}{\partial n}\,d\sigma_2+\ldots$$
$$=-\iiint\left(\frac{\partial\phi}{\partial x}\frac{\partial\phi'}{\partial x}+\frac{\partial\phi}{\partial y}\frac{\partial\phi'}{\partial y}+\frac{\partial\phi}{\partial z}\frac{\partial\phi'}{\partial z}\right)dx\,dy\,dz-\iiint\phi\nabla^2\phi'\,dx\,dy\,dz\,;\quad\ldots(1)$$

* *Grundlagen für eine allgemeine Theorie der Functionen einer veränderlichen complexen Grösse*, Göttingen, 1851 [*Mathematische Werke*, Leipzig, 1876, p. 3]. Also: "Lehrsätze aus der Analysis Situs," *Crelle*, liv. (1857) [*Werke*, p. 84]. † *Crelle*, lv. (1858).

‡ See also Kirchhoff, "Ueber die Kräfte welche zwei unendlich dünne starre Ringe in einer Flüssigkeit scheinbar auf einander ausüben können," *Crelle*, lxxi. (1869) [*Gesammelte Abhandlungen*, Leipzig, 1882, p. 404].

where the surface-integrations indicated on the left-hand side extend, the first over the original boundary of the region only, and the rest over the several barriers. The coefficient of any κ is evidently *minus* the total flux across the corresponding barrier, in a motion of which ϕ' is the velocity-potential. The values of ϕ in the first and last terms of the equation are to be assigned in the manner indicated in Art. 50.

If ϕ' also be a cyclic function, having the cyclic constants κ_1', κ_2', &c., then Art. 43 (6) becomes in the same way

$$\iint \phi' \frac{\partial \phi}{\partial n} dS + \kappa_1' \iint \frac{\partial \phi}{\partial n} d\sigma_1 + \kappa_2' \iint \frac{\partial \phi}{\partial n} d\sigma_2 + \ldots$$

$$= -\iiint \left(\frac{\partial \phi}{\partial x}\frac{\partial \phi'}{\partial x} + \frac{\partial \phi}{\partial y}\frac{\partial \phi'}{\partial y} + \frac{\partial \phi}{\partial z}\frac{\partial \phi'}{\partial z} \right) dx\,dy\,dz - \iiint \phi' \nabla^2 \phi \, dx\,dy\,dz. \quad \ldots(2)$$

Equations (1) and (2) together constitute Lord Kelvin's extension of Green's theorem.

54. If ϕ, ϕ' are both velocity-potentials of a liquid, we have

$$\nabla^2 \phi = 0, \quad \nabla^2 \phi' = 0, \quad \ldots\ldots\ldots\ldots(3)$$

and therefore

$$\iint \phi \frac{\partial \phi'}{\partial n} dS + \kappa_1 \iint \frac{\partial \phi'}{\partial n} d\sigma_1 + \kappa_2 \iint \frac{\partial \phi'}{\partial n} d\sigma_2 + \ldots$$

$$= \iint \phi' \frac{\partial \phi}{\partial n} dS + \kappa_1' \iint \frac{\partial \phi}{\partial n} d\sigma_1 + \kappa_2' \iint \frac{\partial \phi}{\partial n} d\sigma_2 + \ldots. \quad \ldots(4)$$

To obtain a physical interpretation of this theorem it is necessary to explain in the first place a method, imagined by Lord Kelvin, of generating any given cyclic irrotational motion of a liquid in a multiply-connected space.

Let us suppose the fluid to be enclosed in a perfectly smooth and flexible membrane occupying the position of the boundary. Further, let n barriers be drawn, as in Art. 48, so as to convert the region into a simply-connected one, and let their places be occupied by similar membranes, infinitely thin, and destitute of inertia. The fluid being initially at rest, let each element of the first-mentioned membrane be suddenly moved inwards with the given (positive or negative) normal velocity $-\partial\phi/\partial n$, whilst uniform impulsive pressures $\kappa_1\rho, \kappa_2\rho, \ldots \kappa_n\rho$ are simultaneously applied to the negative sides of the respective barrier-membranes. The motion generated will be characterized by the following properties. It will be irrotational, being generated from rest; the normal velocity at every point of the original boundary will have the prescribed value; the values of the impulsive pressure at two adjacent points on opposite sides of a membrane will differ by the corresponding value of $\kappa\rho$, and the values of the velocity-potential will therefore differ by the corresponding value of κ; finally, the motion on one side of a barrier will be continuous with that on the other. To prove the last statement we remark, first, that the velocities normal to the barrier at two adjacent points on opposite sides of it are the same, being each equal to the normal velocity of the adjacent portion of the membrane. Again, if P, Q be two consecutive points on a barrier, and if the corresponding values of ϕ be on the positive side ϕ_P, ϕ_Q, and on the negative side ϕ'_P, ϕ'_Q, we have

$$\phi_P - \phi'_P = \kappa = \phi_Q - \phi'_Q,$$

and therefore $$\phi_Q - \phi_P = \phi'_Q - \phi'_P,$$

i.e., if $PQ = \delta s$, $$\partial\phi/\partial s = \partial\phi'/\partial s.$$

Hence the tangential velocities at two adjacent points on opposite sides of the barrier also agree. If then we suppose the barrier-membranes to be liquefied immediately after the impulse, we obtain the irrotational motion in question.

The physical interpretation of (4), when multiplied by $-\rho$, now follows as in Art. 44. The values of $\rho\kappa$ are additional components of momentum, and those of $-\iint \partial\phi/\partial n \,.\, d\sigma$, the fluxes through the various apertures of the region, are the corresponding generalized velocities.

55. If in (2) we put $\phi' = \phi$, and suppose ϕ to be the velocity-potential of an incompressible fluid, we find

$$2T = \rho \iiint \left\{ \left(\frac{\partial\phi}{\partial x}\right)^2 + \left(\frac{\partial\phi}{\partial y}\right)^2 + \left(\frac{\partial\phi}{\partial z}\right)^2 \right\} dx\,dy\,dz$$

$$= -\rho \iint \phi \frac{\partial\phi}{\partial n}\, dS - \rho\kappa_1 \iint \frac{\partial\phi}{\partial n}\, d\sigma_1 - \rho\kappa_2 \iint \frac{\partial\phi}{\partial n}\, d\sigma_2 - \ldots \qquad \ldots\ldots(5)$$

The last member of this formula has a simple interpretation in terms of the artificial method of generating cyclic irrotational motion just explained. The first term has already been recognized as equal to twice the work done by the impulsive pressure $\rho\phi$ applied to every part of the original boundary of the fluid. Again, $\rho\kappa_1$ is the impulsive pressure applied, in the positive direction, to the infinitely thin massless membrane by which the place of the first barrier was supposed to be occupied; so that the expression

$$-\tfrac{1}{2} \iint \rho\kappa_1 \,.\, \frac{\partial\phi}{\partial n}\, d\sigma_1$$

denotes the work done by the impulsive forces applied to that membrane; and so on. Hence (5) expresses the fact that the energy of the motion is equal to the work done by the whole system of impulsive forces by which we may suppose it generated.

In applying (5) to the case where the fluid extends to infinity and is at rest there, we may replace the first term of the third member by

$$-\rho \iint (\phi - C) \frac{\partial\phi}{\partial n}\, dS, \qquad \ldots\ldots\ldots\ldots\ldots\ldots\ldots\ldots(6)$$

where the integration extends over the internal boundary only. The proof is the same as in Art 46. When the total flux across this boundary is zero, this reduces to

$$-\rho \iint \phi \frac{\partial \phi}{\partial n} dS. \quad \text{.............................(7)}$$

The minimum theorem of Lord Kelvin, given in Art. 45, may now be extended as follows:

The irrotational motion of a liquid in a multiply-connected region has less kinetic energy than any other motion consistent with the same normal motion of the boundary and the same value of the total flux through each of the several independent channels of the region.

The proof is left to the reader.

Sources and Sinks.

56. The analogy with the theories of Electrostatics, the Steady Flow of Heat, &c., may be carried further by means of the conception of sources and sinks.

A 'simple source' is a point from which fluid is imagined to flow out uniformly in all directions. If the total flux outwards across a small closed surface surrounding the point be m, then m is called the 'strength' of the source. A negative source is called a 'sink.' The continued existence of a source or a sink would postulate of course a continual creation or annihilation of fluid at the point in question.

The velocity-potential at any point P, due to a simple source, in a liquid at rest at infinity, is

$$\phi = m/4\pi r, \quad \text{................................(1)}$$

where r denotes the distance of P from the source. For this gives a radial flow from the point, and if $\delta S, = r^2 \delta\varpi$, be an element of a spherical surface having its centre at the source, we have

$$-\iint \frac{\partial \phi}{\partial r} dS = m,$$

a constant, so that the equation of continuity is satisfied, and the flux outwards has the value appropriate to the strength of the source.

A combination of two equal and opposite sources $\pm m'$, at a distance δs apart, where, in the limit, δs is taken to be infinitely small, and m' infinitely great, but so that the product $m'\delta s$ is finite and equal to μ (say), is called a 'double source' of strength μ, and the line δs, considered as drawn in the direction from $-m'$ to $+m'$, is called its axis.

To find the velocity-potential at any point (x, y, z) due to a double source

μ situate at (x', y', z'), and having its axis in the direction (l, m, n), we remark that, f being any continuous function,

$$f(x'+l\delta s, y'+m\delta s, z'+n\delta s)-f(x', y', z')$$
$$=\left(l\frac{\partial}{\partial x'}+m\frac{\partial}{\partial y'}+n\frac{\partial}{\partial z'}\right)f(x', y'\, z').\,\delta s,$$

ultimately. Hence, putting $f(x', y', z')=m'/4\pi r$, where

$$r=\{(x-x')^2+(y-y')^2+(z-z')^2\}^{\frac{1}{2}},$$

we find
$$\phi=\frac{\mu}{4\pi}\left(l\frac{\partial}{\partial x'}+m\frac{\partial}{\partial y'}+n\frac{\partial}{\partial z'}\right)\frac{1}{r}, \quad\text{.................(2)}$$
$$=-\frac{\mu}{4\pi}\left(l\frac{\partial}{\partial x}+m\frac{\partial}{\partial y}+n\frac{\partial}{\partial z}\right)\frac{1}{r}, \quad\text{.................(3)}$$
$$=\frac{\mu}{4\pi}\frac{\cos\vartheta}{r^2}, \quad\text{...(4)}$$

where, in the latter form, ϑ denotes the angle which the line r, considered as drawn from (x', y', z') to (x, y, z), makes with the axis (l, m, n).

We might proceed, in a similar manner (see Art. 82), to build up sources of higher degrees of complexity, but the above is sufficient for our immediate purpose.

Finally, we may imagine simple or double sources, instead of existing at isolated points, to be distributed continuously over lines, surfaces, or volumes.

57. We can now prove that any continuous acyclic irrotational motion of a liquid mass may be regarded as due to a distribution of simple and double sources over the boundary.

This depends on the theorem, proved in Art. 44, that if ϕ, ϕ' be any two single-valued functions which satisfy $\nabla^2\phi=0$, $\nabla^2\phi'=0$ throughout a given region, then

$$\iint\phi\frac{\partial\phi'}{\partial n}\,dS=\iint\phi'\frac{\partial\phi}{\partial n}\,dS, \quad\text{.......................(5)}$$

where the integration extends over the whole boundary. In the present application, we take ϕ to be the velocity-potential of the motion in question, and put $\phi'=1/r$, the reciprocal of the distance of any point of the fluid from a fixed point P.

We will first suppose that P is in the space occupied by the fluid. Since ϕ' then becomes infinite at P, it is necessary to exclude this point from the region to which the formula (5) applies; this may be done by describing a small spherical surface about P as centre. If we now suppose $\delta\Sigma$ to refer to this surface, and δS to the original boundary, the formula gives

$$\iint\phi\frac{\partial}{\partial n}\left(\frac{1}{r}\right)d\Sigma+\iint\phi\frac{\partial}{\partial n}\left(\frac{1}{r}\right)dS=\iint\frac{1}{r}\frac{\partial\phi}{\partial n}\,d\Sigma+\iint\frac{1}{r}\frac{\partial\phi}{\partial n}\,dS. \quad\text{......(6)}$$

At the surface Σ we have $\partial/\partial n\,(1/r) = -1/r^2$; hence if we put $\delta\Sigma = r^2 d\varpi$, and finally make $r \to 0$, the first integral on the left-hand becomes $= -4\pi\phi_P$, where ϕ_P denotes the value of ϕ at P, whilst the first integral on the right vanishes. Hence

$$\phi_P = -\frac{1}{4\pi}\iint \frac{1}{r}\frac{\partial\phi}{\partial n}\,dS + \frac{1}{4\pi}\iint \phi\,\frac{\partial}{\partial n}\left(\frac{1}{r}\right) dS. \quad \text{..............(7)}$$

This gives the value of ϕ at any point P of the fluid in terms of the values of ϕ and $\partial\phi/\partial n$ at the boundary. Comparing with the formulae (1) and (2) we see that the first term is the velocity-potential due to a surface distribution of simple sources, with a density $-\partial\phi/\partial n$ per unit area, whilst the second term is the velocity-potential of a distribution of double sources, with axes normal to the surface, the density being ϕ. It will appear from equation (10), below, that this is only one out of an infinite number of surface-distributions which will give the same value of ϕ throughout the interior.

When the fluid extends to infinity in every direction and is at rest there, the surface-integrals in (7) may, on a certain understanding, be taken to refer to the internal boundary alone. To see this, we may take as external boundary an infinite sphere having the point P as centre. The corresponding part of the first integral in (7) vanishes, whilst that of the second is equal to C, the constant value to which, as we have seen in Art. 41, ϕ tends at infinity. It is convenient, for facility of statement, to suppose $C = 0$; this is legitimate since we may always add an arbitrary constant to ϕ.

When the point P is external to the surface, ϕ' is finite throughout the original region, and the formula (5) gives at once

$$0 = -\frac{1}{4\pi}\iint \frac{1}{r}\frac{\partial\phi}{\partial n}\,dS + \frac{1}{4\pi}\iint \phi\,\frac{\partial}{\partial n}\left(\frac{1}{r}\right) dS, \quad \text{..............(8)}$$

where, again, in the case of a liquid extending to infinity, and at rest there, the terms due to the infinitely distant part of the boundary may be omitted.

58. The distribution expressed by (7) can, further, be replaced by one of simple sources only, or of double sources only, over the boundary.

Let ϕ be the velocity-potential of the fluid occupying a certain region, and let ϕ' now denote the velocity-potential of any possible acyclic irrotational motion through the rest of infinite space, with the condition that ϕ, or ϕ', as the case may be, vanishes at infinity. Then, if the point P be internal to the first region, and therefore external to the second, we have

$$\left.\begin{aligned} \phi_P &= -\frac{1}{4\pi}\iint \frac{1}{r}\frac{\partial\phi}{\partial n}\,dS + \frac{1}{4\pi}\iint \phi\,\frac{\partial}{\partial n}\left(\frac{1}{r}\right) dS, \\ 0 &= -\frac{1}{4\pi}\iint \frac{1}{r}\frac{\partial\phi'}{\partial n'}\,dS + \frac{1}{4\pi}\iint \phi'\,\frac{\partial}{\partial n'}\left(\frac{1}{r}\right) dS, \end{aligned}\right\} \quad \text{...........(9)}$$

where δn, $\delta n'$ denote elements of the normal to dS, drawn inwards to the first and second regions respectively, so that $\partial/\partial n' = -\partial/\partial n$. By addition, we have

$$\phi_P = -\frac{1}{4\pi}\iint \frac{1}{r}\left(\frac{\partial \phi}{\partial n} + \frac{\partial \phi'}{\partial n'}\right) dS + \frac{1}{4\pi}\iint (\phi - \phi') \frac{\partial}{\partial n}\left(\frac{1}{r}\right) dS. \quad \text{...(10)}$$

The function ϕ' will be determined by the surface-values of ϕ' or $\partial\phi'/\partial n'$, which are as yet at our disposal.

Let us in the first place make $\phi' = \phi$ at the surface. The tangential velocities on the two sides of the boundary are then continuous, but the normal velocities are discontinuous. To assist the ideas, we may imagine a liquid to fill infinite space, and to be divided into two portions by an infinitely thin vacuous sheet within which an impulsive pressure $\rho\phi$ is applied, so as to generate the given motion from rest. The last term of (10) disappears, so that

$$\phi_P = -\frac{1}{4\pi}\iint \frac{1}{r}\left(\frac{\partial \phi}{\partial n} + \frac{\partial \phi'}{\partial n'}\right) dS, \quad \text{.....................(11)}$$

that is, the motion (on either side) is that due to a surface-distribution of simple sources, of density

$$-\left(\frac{\partial \phi}{\partial n} + \frac{\partial \phi'}{\partial n'}\right)^*.$$

Secondly, we may suppose that $\partial\phi'/\partial n = \partial\phi/\partial n$ over the boundary. This gives continuous normal velocity, but discontinuous tangential velocity, over the original boundary. The motion may in this case be imagined to be generated by giving the prescribed normal velocity $-\partial\phi/\partial n$ to every point of an infinitely thin membrane coincident in position with the boundary. The first term of (10) now vanishes, and we have

$$\phi_P = \frac{1}{4\pi}\iint (\phi - \phi') \frac{\partial}{\partial n}\left(\frac{1}{r}\right) dS, \quad \text{.....................(12)}$$

shewing that the motion on either side may be conceived as due to a surface-distribution of double sources, with density

$$\phi - \phi'.$$

It may be shewn that the above representations of ϕ in terms of simple sources alone, or of double sources alone, are unique; whereas the representation of Art. 57 is indeterminate†.

It is obvious that *cyclic* irrotational motion of a liquid cannot be reproduced by any arrangement of simple sources. It is easily seen, however, that it may be represented by a certain distribution of double sources over the boundary, together with a uniform distribution of double sources over each of the barriers necessary to render the region occupied by the fluid simply-connected. In fact, with the same notation as in Art. 53, we find

$$\phi_P = \frac{1}{4\pi}\iint (\phi - \phi') \frac{\partial}{\partial n}\left(\frac{1}{r}\right) dS + \frac{\kappa_1}{4\pi}\iint \frac{\partial}{\partial n}\left(\frac{1}{r}\right) d\sigma_1 + \frac{\kappa_2}{4\pi}\iint \frac{\partial}{\partial n}\left(\frac{1}{r}\right) d\sigma_2 + \ldots, \quad \text{...(13)}$$

* This investigation was first given by Green, from the point of view of Electrostatics, *l.c. ante* p. 46.

† Cf. Larmor, "On the Mathematical Expression of the Principle of Huyghens," *Proc. Lond. Math. Soc.* (2) i. 1 (1903) [*Math. and Phys. Papers*, Cambridge, 1929, ii. 240].

where ϕ is the single-valued velocity-potential which obtains in the modified region, and ϕ' is the velocity-potential of the acyclic motion which is generated in the external space when the proper normal velocity $-\partial\phi/\partial n$ is given to each element δS of a membrane coincident in position with the original boundary.

Another mode of representing the irrotational motion of a liquid, whether cyclic or not, will present itself in the chapter on Vortex Motion.

We here close this account of the theory of irrotational motion. The mathematical reader will doubtless have noticed the absence of some important links in the chain of our propositions. For example, apart from physical considerations, no proof has been offered that a function ϕ exists which satisfies the conditions of Art. 36 throughout any given simply-connected region, and has arbitrarily prescribed values over the boundary. The formal proof of 'existence-theorems' of this kind is not attempted in the present treatise. For a review of the literature of this part of the subject the reader may consult the authors cited below*.

* H. Burkhardt and W. F. Meyer, "Potentialtheorie," and A. Sommerfeld, "Randwerthaufgaben in der Theorie d. part. Diff.-Gleichungen," *Encyc. d. math. Wiss.* ii. (1900).

CHAPTER IV

MOTION OF A LIQUID IN TWO DIMENSIONS

59. If the velocities u, v be functions of x, y only, while w is zero, the motion takes place in a series of planes parallel to xy, and is the same in each of these planes. The investigation of the motion of a liquid under these circumstances is characterized by certain analytical peculiarities; and the solutions of several problems of great interest are readily obtained.

Since the whole motion is known when we know that in the plane $z = 0$, we may confine our attention to that plane. When we speak of points and lines drawn in it, we shall understand them to represent respectively the straight lines parallel to the axis of z, and the cylindrical surfaces having their generating lines parallel to the axis of z, of which they are the traces.

By the flux across any curve we shall understand the volume of fluid which in unit time crosses that portion of the cylindrical surface, having the curve as base, which is included between the planes $z = 0$, $z = 1$.

Let A, P be any two points in the plane xy. The flux across any two lines joining AP is the same, provided they can be reconciled without passing out of the region occupied by the moving liquid; for otherwise the space included between these two lines would be gaining or losing matter. Hence if A be fixed, and P variable, the flux across any line AP is a function of the position of P. Let ψ be this function; more precisely, let ψ denote the flux across AP *from right to left*, as regards an observer placed on the curve, and looking along it from A in the direction of P. Analytically, if l, m be the direction-cosines of the normal (drawn to the left) to any element δs of the curve, we have

$$\psi = \int_A^P (lu + mv)\, ds. \qquad (1)$$

If the region occupied by the liquid be aperiphractic (see p. 40), ψ is necessarily a single-valued function, but in periphractic regions the value of ψ may depend on the nature of the path AP. For spaces of two dimensions, however, periphraxy and multiple-connectivity become the same thing, so that the properties of ψ, when it is a many-valued function, in relation to the nature of the region occupied by the moving liquid, may be inferred from Art. 50, where we have discussed the same question with regard to ϕ. The cyclic constants of ψ, when the region is periphractic, are the values of the flux across the closed curves forming the several parts of the internal boundary.

A change, say from A to B, of the point from which ψ is reckoned has merely the effect of adding a constant, viz. the flux across a line BA, to the value of ψ; so that we may, if we please, regard ψ as indeterminate to the extent of an additive constant.

If P move about in such a manner that the value of ψ does not alter, it will trace out a curve such that no fluid anywhere crosses it, *i.e.* a stream-line. Hence the curves $\psi=$ const. are the stream-lines, and ψ is called the 'stream-function.'

If P receive an infinitesimal displacement $PQ\,(=\delta y)$ parallel to y, the increment of ψ is the flux across PQ from right to left, *i.e.* $\delta\psi=-u\,.\,PQ$, or

$$u=-\frac{\partial\psi}{\partial y}. \qquad\qquad (2)$$

Again, displacing P parallel to x, we find in the same way

$$v=\frac{\partial\psi}{\partial x}. \qquad\qquad (3)$$

The existence of a function ψ related to u and v in this manner might also have been inferred from the form which the equation of continuity takes in this case, viz.

$$\frac{\partial u}{\partial x}+\frac{\partial v}{\partial y}=0, \qquad\qquad (4)$$

which is the analytical condition that $u\,dy-v\,dx$ should be an exact differential*.

The foregoing considerations apply whether the motion be rotational or irrotational. The formulae for the components of vorticity, given in Art. 30, become

$$\xi=0,\quad \eta=0,\quad \zeta=\frac{\partial^2\psi}{\partial x^2}+\frac{\partial^2\psi}{\partial y^2}; \qquad\qquad (5)$$

so that in irrotational motion we have

$$\frac{\partial^2\psi}{\partial x^2}+\frac{\partial^2\psi}{\partial y^2}=0. \qquad\qquad (6)$$

60. In what follows we confine ourselves to the case of irrotational motion, which is, as we have already seen, characterized by the existence, in addition, of a velocity-potential ϕ, connected with u, v by the relations

$$u=-\frac{\partial\phi}{\partial x},\quad v=-\frac{\partial\phi}{\partial y}, \qquad\qquad (1)$$

and, since we are considering the motion of incompressible fluids only, satisfying the equation of continuity

$$\frac{\partial^2\phi}{\partial x^2}+\frac{\partial^2\phi}{\partial y^2}=0. \qquad\qquad (2)$$

* The function ψ was introduced in this way by Lagrange, *Nouv. mém. de l'Acad. de Berlin*, 1781 [*Oeuvres*, iv. 720]. The kinematical interpretation is due to Rankine, "On Plane Water-Lines in Two Dimensions," *Phil. Trans.* 1864 [*Miscellaneous Scientific Papers*, London, 1881, p. 495].

The theory of the function ϕ, and the relation between its properties and the nature of the two-dimensional space through which the irrotational motion holds, may be readily inferred from the corresponding theorems in three dimensions proved in the last chapter. The alterations, whether of enunciation or of proof, which are requisite to adapt these to the case of two dimensions are for the most part purely verbal.

For instance, we have the theorem that the mean value of ϕ over the circumference of a circle is equal to its value at the centre, provided the circle can be contracted to a point, remaining always within the region occupied by the fluid.

Again, if this region extends to infinity, being bounded internally by one or more closed curves, and if the velocities tend to a zero limit at infinity, the value of ϕ tends there to a constant limit, provided the total flux across the internal boundaries is zero. This latter proviso is now essential.

The fundamental solution of the equation (2) has the form $\phi = C \log r$, where r denotes distance from a fixed point. This is the case of a two-dimensional source, for if we write

$$\phi = -\frac{m}{2\pi} \log r \quad \text{.............................(3)}$$

the flux outwards across a circle surrounding the point is

$$-\frac{\partial \phi}{\partial r} \, . \, 2\pi r = m. \quad \text{.............................(4)}$$

The constant m accordingly measures the output, or 'strength', of the source. We get essentially the same result if we imagine point sources of the type explained in Art. 56 to be distributed with uniform line-density m along its axis of z. The velocity in that case will be in the direction of r, and equal to $m/2\pi r$, consistently with (3). We have here the conception of a 'line-source' (in three dimensions).

For a double source, or 'doublet', as it is sometimes called, we have the formula

$$\phi = -\frac{\mu}{2\pi} \frac{\partial}{\partial s} (\log r), \text{.............................(5)}$$

where the symbol $\partial/\partial s$ indicates a space-differentiation in the direction of the axis of the source. If ϑ be the angle which direction of r increasing makes with this axis, we have $\delta r = - \delta s \cos \vartheta$, and therefore

$$\phi = \frac{\mu}{2\pi} \frac{\cos \vartheta}{r}. \quad \text{.............................(6)}$$

Again we might establish a system of formulae analogous to those of Art. 58. In particular, corresponding to Art. 58 (12), we have

$$\phi_P = -\frac{1}{2\pi} \int (\phi - \phi') \frac{\partial}{\partial n} (\log r) \, ds, \text{.....................(7)}$$

giving the value of ϕ in any region in terms of a distribution of double sources over the boundary. This will apply to the case of a fluid unlimited externally, provided the velocities tend to zero at infinity, and that the total flux outwards is zero. As in Art. 58 the function ϕ' refers to the space within the inner boundary, and is subject to the condition that $\partial\phi'/\partial n = \partial\phi/\partial n$ at this boundary. A deduction from this formula will be given presently (Art. 72 a).

60 a. The foregoing kinematical relations have exact analogies in the theory of electric conduction. In the case of a uniform plane sheet we have

$$\sigma f = -\frac{\partial V}{\partial x}, \quad \sigma g = -\frac{\partial V}{\partial y}, \quad \text{.......................(1)}$$

with

$$\frac{\partial f}{\partial x} + \frac{\partial g}{\partial y} = 0, \quad \text{..................................(2)}$$

where (f, g) is the current density, V is the electric potential, and σ is the specific resistance of the material. If we write

$$u = \sigma f, \quad v = \sigma g, \quad \phi = V, \quad \text{.......................(3)}$$

these become identical with the hydrodynamical relations. This has suggested a practical method of solution of two-dimensional hydrokinetic problems. The current sheet may consist of a thin layer of feebly conducting fluid (H_2SO_4) contained in a rectangular tank, two opposite walls of which are metallic and maintained at a constant difference of potential whilst the remaining walls (and the bottom) are insulators. The equipotential lines, to which the current lines are orthogonal, are easily traced electrically, and in this way practical solutions can be obtained of problems of flow of a stream past an obstacle (represented by a non-conducting disk in the electrical experiment) which are not easily treated by analysis*.

Again, instead of (3) we may put

$$u = -\sigma g, \quad v = \sigma f, \quad \psi = -V. \quad \text{.....................(4)}$$

The hydrodynamical relations are satisfied, but the stream-lines are now represented by the lines of equal electric potential, and can therefore be found directly. An obstacle has now to be represented by a disk whose conductivity so greatly exceeds that of the surrounding stratum that it may be regarded as practically perfect. This analogy has the further advantage that circulation can also be represented. For if (l, m) be the direction of the outward normal to the contour of the obstacle, the circulation is

$$\int (lv - mu)\, ds = \sigma \int (lf + mg)\, ds, \quad \text{.....................(5)}$$

* For experimental details reference should be made to E. F. Relf, *Phil. Mag.* (6) xlviii. (1924). As a test of the method the diagram on p. 86 *infra* was reproduced with remarkable accuracy. The circulation round a lamina was also determined and compared with theory.

and is therefore proportional to the total current outwards in the electric analogy. For this purpose the disk is connected with one terminal of a suitable battery, the other terminal being connected with one of the conducting walls of the tank.

61. The kinetic energy T of a portion of fluid bounded by a cylindrical surface whose generating lines are parallel to the axis of z, and by two planes perpendicular to the axis of z at unit distance apart, is given by the formula

$$2T = \rho \iint \left\{ \left(\frac{\partial \phi}{\partial x}\right)^2 + \left(\frac{\partial \phi}{\partial y}\right)^2 \right\} dx\,dy = -\rho \int \phi \frac{\partial \phi}{\partial n}\, ds, \quad \text{.........(1)}$$

where the surface-integral is taken over the portion of the plane xy cut off by the cylindrical surface, and the line-integral round the boundary of this portion. Since $\partial\phi/\partial n = -\partial\psi/\partial s$, the formula (1) may be written

$$2T = \rho \int \phi\, d\psi, \quad \text{...............................(2)}$$

the integration being carried in the positive direction round the boundary.

If we attempt by a process similar to that of Art. 46 to calculate the energy in the case where the region extends to infinity, we find that its value is infinite, except when the total flux outwards (M) is zero. For if we introduce a circle of great radius r as the external boundary of the portion of the plane xy considered, we find that the corresponding part of the integral on the right-hand side of (1) increases indefinitely with r. The only exception is when $M=0$, in which case we may suppose the line-integral in (1) to extend over the internal boundary only.

If the cylindrical part of the boundary consist of two or more separate portions one of which embraces all the rest, the enclosed region is multiply-connected, and the equation (1) needs a correction, which may be applied exactly as in Art. 55.

Conformal Transformations.

62. The functions ϕ and ψ are connected by the relations

$$\frac{\partial \phi}{\partial x} = \frac{\partial \psi}{\partial y}, \qquad \frac{\partial \phi}{\partial y} = -\frac{\partial \psi}{\partial x}. \quad \text{.......................(1)}$$

These conditions are fulfilled by equating $\phi + i\psi$, where i stands as usual for $\surd(-1)$, to any ordinary algebraic or transcendental function of $x + iy$, say

$$\phi + i\psi = f(x + iy). \quad \text{.............................(2)}$$

For then
$$\frac{\partial}{\partial y}(\phi + i\psi) = if'(x + iy) = i\frac{\partial}{\partial x}(\phi + i\psi), \quad \text{..............(3)}$$

whence, equating separately the real and the imaginary parts, we see that the equations (1) are satisfied.

Hence any assumption of the form (2) gives a possible case of irrotational motion. The curves $\phi = \text{const.}$ are the curves of equal velocity-potential, and the curves $\psi = \text{const.}$ are the stream-lines. Since, by (1),

$$\frac{\partial \phi}{\partial x}\frac{\partial \psi}{\partial x} + \frac{\partial \phi}{\partial y}\frac{\partial \psi}{\partial y} = 0,$$

we see that these two systems of curves cut one another at right angles, as already proved. Since the relations (1) are unaltered when we write $-\psi$ for ϕ, and ϕ for ψ, we may, if we choose, look upon the curves $\psi = \text{const.}$ as the equipotential curves, and the curves $\phi = \text{const.}$ as the stream-lines; so that every assumption of the kind indicated gives us *two* possible cases of irrotational motion.

For shortness, we shall through the rest of this chapter follow the usual notation of the Theory of Functions, and write

$$z = x + iy, \quad \text{..............................(4)}$$

$$w = \phi + i\psi. \quad \text{..............................(5)}$$

From a modern point of view, the fundamental property of a *function* of a complex variable is that it has a definite differential coefficient with respect to that variable*. If ϕ, ψ denote any functions whatever of x and y, then corresponding to every value of $x + iy$ there must be one or more definite values of $\phi + i\psi$; but the ratio of the differential of this function to that of $x + iy$, viz.

$$\frac{\delta\phi + i\delta\psi}{\delta x + i\delta y}, \quad \text{or} \quad \frac{\left(\frac{\partial \phi}{\partial x} + i\frac{\partial \psi}{\partial x}\right)\delta x + \left(\frac{\partial \phi}{\partial y} + i\frac{\partial \psi}{\partial y}\right)\delta y}{\delta x + i\delta y},$$

depends in general on the ratio $\delta x : \delta y$. The condition that it should be the same for all values of the latter ratio is

$$\frac{\partial \phi}{\partial y} + i\frac{\partial \psi}{\partial y} = i\left(\frac{\partial \phi}{\partial x} + i\frac{\partial \psi}{\partial x}\right), \quad \text{......................(6)}$$

which is equivalent to (1) above. This property was adopted by Riemann as the definition of a function of the complex variable $x + iy$; viz. such a function must have, for every assigned value of the variable, not only a definite value or system of values, but also for each of these values a definite differential coefficient. The advantage of this definition is that it is quite independent of the existence of an analytical expression for the function.

If the complex quantities z and w be represented geometrically after the manner of Argand and Gauss, the differential coefficient dw/dz may be interpreted as the operator which transforms an infinitesimal vector δz into the corresponding vector δw. It follows then, from the above property, that corresponding figures in the planes of z and w are similar in their infinitely small parts.

* See, for example, Forsyth, *Theory of Functions*, 3rd ed., Cambridge, 1918, cc. i., ii.

For instance, in the plane of w the straight lines $\phi = \text{const.}$, $\psi = \text{const.}$, where the constants have assigned to them a series of values in arithmetical progression, the common difference being infinitesimal and the same in each case, form two systems of straight lines at right angles, dividing the plane into infinitely small squares. Hence in the plane xy the corresponding curves $\phi = \text{const.}$, $\psi = \text{const.}$, the values of the constants being assigned as before, cut one another at right angles (as has already been proved otherwise) and divide the plane into infinitely small squares.

Conversely, if ϕ, ψ be any two functions of x, y such that the curves $\phi = m\epsilon$, $\psi = n\epsilon$, where ϵ is infinitesimal, and m, n are any integers, divide the plane xy into elementary squares, it is evident geometrically that

$$\frac{\partial x}{\partial \phi} = \pm \frac{\partial y}{\partial \psi}, \qquad \frac{\partial x}{\partial \psi} = \mp \frac{\partial y}{\partial \phi}.$$

If we take the upper signs, these are the conditions that $x + iy$ should be a function of $\phi + i\psi$. The case of the lower signs is reduced to this by reversing the sign of ψ. Hence the equation (2) contains the *complete* solution of the problem of conformal representation of one plane on another*.

The similarity of corresponding infinitely small portions of the planes w and z breaks down at points where the differential coefficient dw/dz is zero or infinite. Since

$$\frac{dw}{dz} = \frac{\partial \phi}{\partial x} + i \frac{\partial \psi}{\partial x}, \quad \text{..............................(7)}$$

the corresponding value of the velocity, in the hydrodynamical application, is zero or infinite.

In all physical applications, w must be a single-valued, or at most a cyclic function of z in the sense of Art. 50, throughout the region with which we are concerned. Hence in the case of a 'multiform' function, this region must be confined to a single sheet of the corresponding Riemann's surface, and 'branch-points' therefore must not occur in its interior.

63. We can now proceed to some applications of the foregoing method.

First let us assume $\qquad w = Az^n,$

A being real. Introducing polar co-ordinates, r, θ, we have

$$\left.\begin{aligned} \phi &= Ar^n \cos n\theta, \\ \psi &= Ar^n \sin n\theta. \end{aligned}\right\} \quad \text{..............................(1)}$$

The following cases may be noticed.

1°. If $n = 1$, the stream-lines are a system of straight lines parallel to x, and the equipotential curves are a similar system parallel to y. In this case *any* corresponding figures in the planes of w and z are similar, whether they be finite or infinitesimal.

* Lagrange, "Sur la construction des cartes géographiques," *Nouv. mém. de l'Acad. de Berlin*, 1779 [*Oeuvres*, iv. 636]. For the further history of the problem, see Forsyth, *Theory of Functions*, c. xix.

2°. If $n = 2$, the curves $\phi = \text{const.}$ are a system of rectangular hyperbolas having the axes of co-ordinates as their principal axes, and the curves $\psi = \text{const.}$ are a similar system, having the co-ordinate axes as asymptotes. The lines $\theta = 0$, $\theta = \frac{1}{2}\pi$ are parts of the same stream-line $\psi = 0$, so that we may take the positive parts of the axes of x, y as fixed boundaries, and thus obtain the case of a fluid in motion in the angle between two perpendicular walls.

3°. If $n = -1$, we get two systems of circles touching the axes of co-ordinates at the origin. Since now $\phi = A/r \,.\, \cos\theta$, the velocity at the origin is infinite; we must therefore suppose the region to which our formulae apply to be limited internally by a closed curve.

4°. If $n = -2$, each system of curves is composed of a double system of lemniscates. The axes of the system $\phi = \text{const.}$ coincide with x or y; those of the system $\psi = \text{const.}$ bisect the angles between these axes.

5°. By properly choosing the value of n we get a case of irrotational motion in which the boundary is composed of two rigid walls inclined at any angle α. The equation of the stream-lines being

$$r^n \sin n\theta = \text{const.}, \qquad \text{.........................(2)}$$

we see that the lines $\theta = 0$, $\theta = \pi/n$ are parts of the same stream-line. Hence if we put $n = \pi/\alpha$, we obtain the required solution in the form

$$\phi = A r^{\frac{\pi}{\alpha}} \cos\frac{\pi\theta}{\alpha}, \qquad \psi = A r^{\frac{\pi}{\alpha}} \sin\frac{\pi\theta}{\alpha}. \qquad \text{.................(3)}$$

The component velocities along and perpendicular to r are

$$-A\frac{\pi}{\alpha} r^{\frac{\pi}{\alpha}-1} \cos\frac{\pi\theta}{\alpha}, \text{ and } A\frac{\pi}{\alpha} r^{\frac{\pi}{\alpha}-1} \sin\frac{\pi\theta}{\alpha}, \qquad \text{..............(4)}$$

and are therefore zero, finite, or infinite at the origin, according as α is less than, equal to, or greater than π.

64. We take some examples of cyclic functions.

1°. The assumption $\qquad w = -\mu \log z, \qquad$ ....(1)

where μ is real, gives $\qquad \phi = -\mu \log r, \quad \psi = -\mu\theta. \qquad$ ....(2)

The velocity at a distance r from the origin is μ/r; this point must therefore be isolated by drawing a closed curve round it.

If we take the radii $\theta = \text{const.}$ as the stream-lines we get the case of a (two-dimensional) source, of strength $2\pi\mu$, at the origin. (See Art. 60.)

If the circles $r = \text{const.}$ be taken as stream-lines we have the case of Art. 27; the motion is now cyclic, the circulation in any circuit embracing the origin being $2\pi\mu$.

2°. Let us take $$w = -\mu \log \frac{z-a}{z+a}. \qquad\qquad (3)$$

If we denote by r_1, r_2 the radii drawn to any point in the plane xy from the points $(\pm a, 0)$, and by θ_1, θ_2 the angles which these radii make with the positive direction of the axis of x, we have

$$z - a = r_1 e^{i\theta_1}, \quad z + a = r_2 e^{i\theta_2},$$

whence $$\phi = -\mu \log r_1/r_2, \quad \psi = -\mu(\theta_1 - \theta_2). \qquad\qquad (4)$$

The curves $\phi = \text{const.}$, $\psi = \text{const.}$ form two orthogonal systems of 'coaxal' circles.

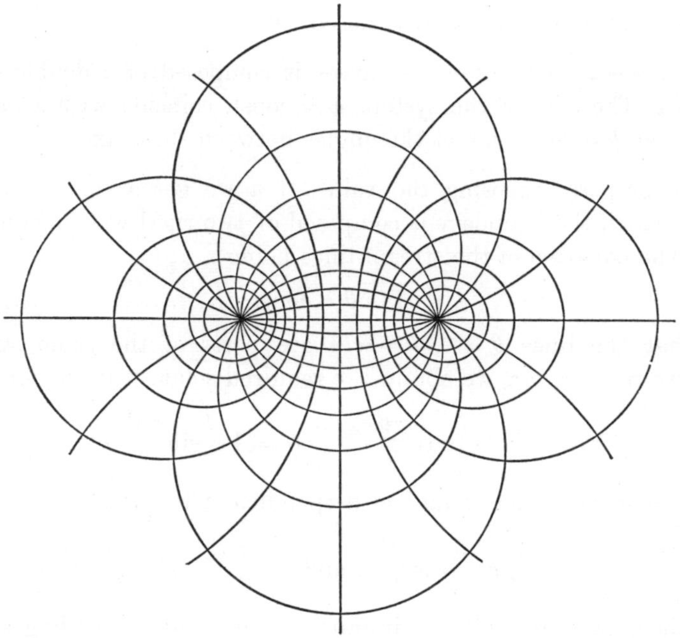

Either of these systems may be taken as the equipotential curves, and the other system will then form the stream-lines. In either case the velocity at the points $(\pm a, 0)$ will be infinite. If these points be accordingly isolated by drawing closed curves round them, the rest of the plane xy becomes a triply-connected region.

If the circles $\theta_1 - \theta_2 = \text{const.}$ be taken as the stream-lines we have the case of a source and a sink, of equal intensities, situate at the points $(\pm a, 0)$. If a is diminished indefinitely, whilst μa remains finite, we reproduce the assumption of Art. 60 (5), which corresponds to the case of a double line-source at the origin. The lines of motion are shewn (in part) on p. 76.

If, on the other hand, we take the circles $r_1/r_2 = \text{const.}$ as the stream-lines we get a case of cyclic motion, viz. the circulation in any circuit embracing the first (only) of the above points is $2\pi\mu$, that in a circuit embracing the second is $-2\pi\mu$; whilst that in a circuit embracing both is zero. This example will have additional interest for us when in Chapter VII. we come to treat of 'Rectilinear Vortices.'

3°. By a simple combination of sources we can represent the flow past a circular barrier due to a source at a given external point P.

Let Q be the inverse point of P with respect to the circle, and imagine equal sources μ at

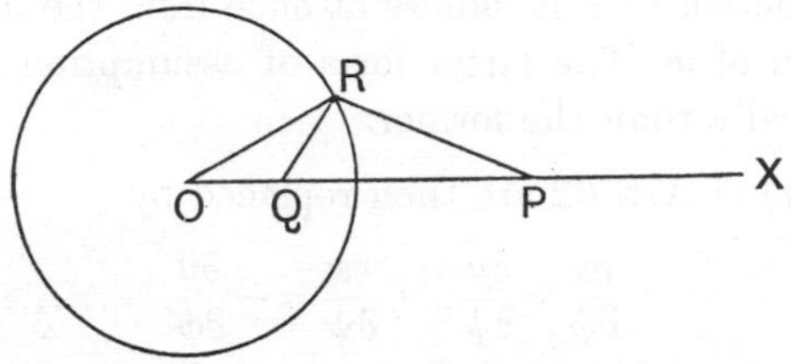

P and Q, and a sink $-\mu$ at the centre O. Then, referring to (2) above, the value of ψ at a point R on the circumference is

$$\psi = -\mu(RPX + RQX - ROX) = -\mu(RPX + ORQ) = -\mu(RPX + RPO) = -\pi\mu,$$

a constant over the circle*.

4°. The potential- and stream-functions due to a row of equal and equidistant sources at the points (0, 0), (0, $\pm a$), (0, $\pm 2a$), ... are given by the formula

$$w \propto \log z + \log(z - ia) + \log(z - ia) + \log(z - 2ia) + \log(z + 2ia) + \ldots, \quad \text{.........(5)}$$

or, say,

$$w = C \log \sinh \frac{\pi z}{a}, \quad \text{.........(6)}$$

where C is real. This makes

$$\phi = \frac{1}{2} C \log \frac{1}{2}\left(\cosh \frac{2\pi x}{a} - \cos \frac{2\pi y}{a}\right), \quad \psi = C \tan^{-1}\left\{\frac{\tan(\pi y/a)}{\tanh(\pi x/a)}\right\}, \quad \text{.........(7)}$$

in agreement with a result given by Maxwell†. The formulae apply also to the case of a source midway between two fixed boundaries $y = \pm\frac{1}{2}a$.

The case of a row of *double* sources having their axes parallel to x is obtained by differentiating (6) with respect to z. Omitting a factor we have

$$w = C \coth \frac{\pi z}{a}, \quad \text{.........(8)}$$

or

$$\phi = \frac{C \sinh(2\pi x/a)}{\cosh(2\pi x/a) - \cos(2\pi y/a)}, \quad \psi = -\frac{C \sin(2\pi y/a)}{\cosh(2\pi x/a) - \cos(2\pi y/a)}. \quad \text{.........(9)}$$

Superposing a uniform motion parallel to x negative, we have

$$w = z + C \coth \frac{\pi z}{a}, \quad \text{.........(10)}$$

or

$$\phi = x + \frac{C \sinh(2\pi x/a)}{\cosh(2\pi x/a) - \cos(2\pi y/a)}, \quad \psi = y - \frac{C \sin(2\pi y/a)}{\cosh(2\pi x/a) - \cos(2\pi y/a)}. \quad \text{...(11)}$$

The stream-line $\psi = 0$ now consists in part of the line $y = 0$, and in part of an oval curve whose semi-diameters parallel to x and y are given by the equations

$$\sinh^2 \frac{\pi x}{a} = \frac{\pi C}{a}, \quad y \tan \frac{\pi y}{a} = C. \quad \text{.........(12)}$$

* Kirchhoff, *Pogg. Ann.*, lxiv. (1845) [*Ges. Abh.* 1].

† *Electricity and Magnetism*, Art. 203.

If we put $$C=\pi b^2/a, \qquad \text{.....(13)}$$
where b is small compared with a*, these semi-diameters are each equal to b, approximately. We thus obtain the potential- and stream-functions for a liquid flowing through a grating of parallel cylindrical bars of small circular section. The second of equations (11) becomes in fact, for small values of x, y,

$$\psi=y\left(1-\frac{b^2}{x^2+y^2}\right). \qquad \text{.....(14)}$$

65. If w be a function of z, it follows at once from the definition of Art. 62 that z is a function of w. The latter form of assumption is sometimes more convenient analytically than the former.

The relations (1) of Art. 62 are then replaced by

$$\frac{\partial x}{\partial \phi}=\frac{\partial y}{\partial \psi}, \qquad \frac{\partial x}{\partial \psi}=-\frac{\partial y}{\partial \phi}. \qquad \text{.....(1)}$$

Also since $$\frac{dw}{dz}=\frac{\partial \phi}{\partial x}+i\frac{\partial \psi}{\partial x}=-u+iv,$$

we have $$-\frac{dz}{dw}=\frac{1}{u-iv}=\frac{1}{q}\left(\frac{u}{q}+i\frac{v}{q}\right),$$

where q is the resultant velocity at (x, y). Hence if we write

$$\zeta=-\frac{dz}{dw}, \qquad \text{.....(2)}$$

and imagine the properties of the function ζ to be exhibited graphically in the manner already explained, the vector drawn from the origin to any point in the plane of ζ will agree in direction with, and be in magnitude the reciprocal of, the velocity at the corresponding point of the plane of z.

Again, since $1/q$ is the modulus of dz/dw, *i.e.* of $\partial x/\partial\phi + i\partial y/\partial\phi$, we have

$$\frac{1}{q^2}=\left(\frac{\partial x}{\partial \phi}\right)^2+\left(\frac{\partial y}{\partial \phi}\right)^2, \qquad \text{.....(3)}$$

which may, by (1), be put into the equivalent forms

$$\frac{1}{q^2}=\left(\frac{\partial x}{\partial \phi}\right)^2+\left(\frac{\partial x}{\partial \psi}\right)^2=\left(\frac{\partial y}{\partial \psi}\right)^2+\left(\frac{\partial y}{\partial \phi}\right)^2=\left(\frac{\partial y}{\partial \psi}\right)^2+\left(\frac{\partial x}{\partial \psi}\right)^2=\frac{\partial x}{\partial \phi}\frac{\partial y}{\partial \psi}-\frac{\partial x}{\partial \psi}\frac{\partial y}{\partial \phi}. \qquad \text{.....(4)}$$

The last formula, viz. $$\frac{1}{q^2}=\frac{\partial(x, y)}{\partial(\phi, \psi)}, \qquad \text{.....(5)}$$

expresses the fact that corresponding elementary areas in the planes of z and w are in the ratio of the square of the modulus of dz/dw to unity.

* The approximately circular form holds however for a considerable range of values of C. Thus if we put $C=\frac{1}{4}a$, we find from (12)

$$x/a=\cdot 254, \qquad y/a=\cdot 250.$$

The two diameters are very nearly equal, although the breadth of the oval is half the interval between the stream-lines $y=\pm\frac{1}{2}a$.

66. The following examples of this procedure are important.

1°. Assume $$z = c \cosh w, \quad \text{.................................(1)}$$

or $$\left.\begin{aligned} x &= c \cosh \phi \cos \psi, \\ y &= c \sinh \phi \sin \psi. \end{aligned}\right\} \quad \text{..........................(2)}$$

The curves $\phi = \text{const.}$ are the ellipses

$$\frac{x^2}{c^2 \cosh^2 \phi} + \frac{y^2}{c^2 \sinh^2 \phi} = 1, \quad \text{....................(3)}$$

and the curves $\psi = \text{const.}$ are the hyperbolas

$$\frac{x^2}{c^2 \cos^2 \psi} - \frac{y^2}{c^2 \sin^2 \psi} = 1, \quad \text{....................(4)}$$

these conics having the common foci ($\pm c$, 0). The two systems of curves are shewn below.

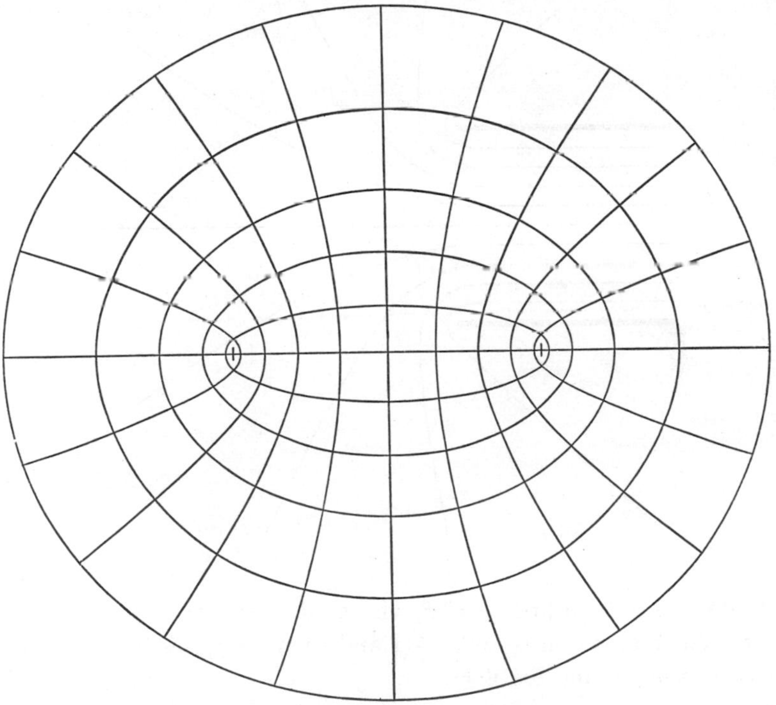

Since at the foci we have $\phi = 0$, $\psi = n\pi$, n being some integer, we see by (2) of the preceding Art. that the velocity there is infinite. If the hyperbolas be taken as the stream-lines, the portions of the axis of x which lie outside the points ($\pm c$, 0) may be taken as rigid boundaries. We obtain in this manner the case of a liquid flowing from one side to the other of a thin plane partition, through an aperture of breadth $2c$.

If the ellipses be taken as the stream-lines we get the case of a liquid circulating round an elliptic cylinder, or, as an extreme case, round a lamina whose section is the line joining the foci ($\pm c$, 0).

At an infinite distance from the origin ϕ is infinite, of the order $\log r$, where r is the radius vector; and the velocity is infinitely small of the order $1/r$.

2°. Let $$z = w + e^w, \qquad\text{.................................(5)}$$

or $$x = \phi + e^\phi \cos\psi, \quad y = \psi + e^\phi \sin\psi. \qquad\text{..................(6)}$$

The stream-line $\psi = 0$ coincides with the axis of x. Again, the portion of the line $y = \pi$ between $x = -\infty$ and $x = -1$, considered as a line bent back on

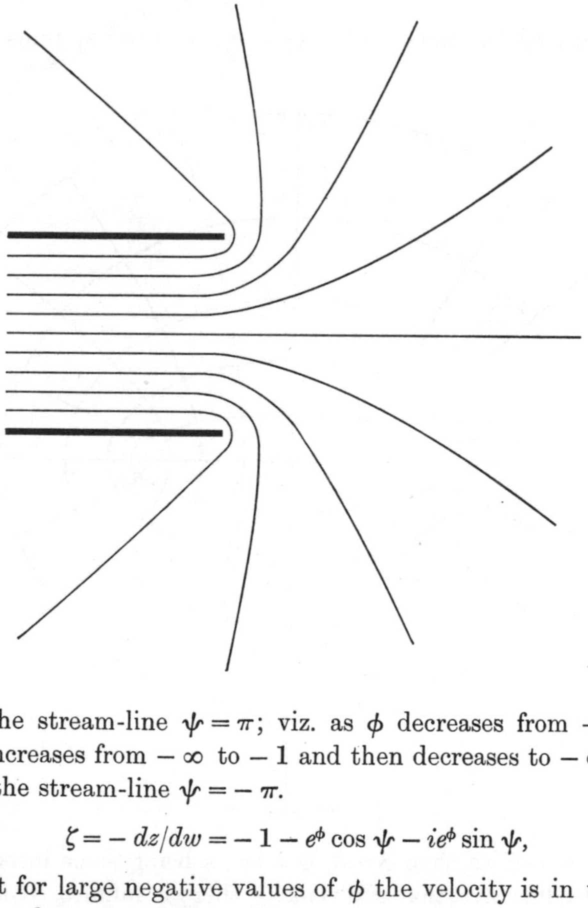

itself, forms the stream-line $\psi = \pi$; viz. as ϕ decreases from $+\infty$ through 0 to $-\infty$, x increases from $-\infty$ to -1 and then decreases to $-\infty$ again. Similarly for the stream-line $\psi = -\pi$.

Since $$\zeta = -dz/dw = -1 - e^\phi \cos\psi - ie^\phi \sin\psi,$$

it appears that for large negative values of ϕ the velocity is in the direction of x-negative, and equal to unity, whilst for large positive values it is zero.

The above formulae therefore express the motion of a liquid flowing into a canal bounded by two thin parallel walls from an open space. At the ends of the walls we have $\phi = 0$, $\psi = \pm\pi$, and therefore $\zeta = 0$, *i.e.* the velocity is infinite. The forms of the stream-lines, drawn, as in all similar cases in this chapter, for equidistant values of ψ, are shewn in the figure on p. 74*.

If the walls instead of being parallel make angles $\pm\beta$ with the line of symmetry, the appropriate formula is

$$z=\frac{1-n}{n}(1-e^{-nw})+e^{(1-n)w}, \quad\text{.................................(7)}$$

where $n=\beta/\pi$. The stream-lines $\psi=\pm\pi$ follow the course of the walls†. This agrees with (5) when n tends to the limit 0, whilst if $n=\frac{1}{2}$ we have virtually the case shewn on p. 73.

If we change the sign of w in (5) the direction of flow is reversed. If we further superpose a uniform stream in the negative direction of x, by writing $w-z$ for w, we obtain‡

$$w=e^{z-w}, \text{ or } z=w+\log w. \quad\text{.................................(8)}$$

The velocity between the walls at a great distance to the left is now annulled, and we have an idealized representation of a Pitot tube (Art. 24). The stream-lines can be plotted from the formulae

$$x=\phi+\tfrac{1}{2}\log(\phi^2+\psi^2), \quad y=\psi+\tan^{-1}(\psi/\phi). \quad\text{........................(9)}$$

67. It is known that a function $f(z)$ which is finite, continuous, and single-valued, and has its first derivative finite, at all points of the space included between two concentric circles about the origin, can be expanded in the form

$$f(z)=A_0+A_1z+A_2z^2+\ldots+B_1z^{-1}+B_2z^{-2}+\ldots. \quad\text{.........(1)}$$

If the above conditions be satisfied at all points within a circle having the origin as centre, we retain only the ascending series; if at all points without such a circle, the descending series, with the addition of the constant A_0, is sufficient. If the conditions be fulfilled for all points of the plane xy without exception, $f(z)$ can be no other than a constant A_0.

Putting $f(z)=\phi+i\psi$, introducing polar co-ordinates, and writing the complex constants A_n, B_n in the forms P_n+iQ_n, R_n+iS_n, respectively, we obtain—

$$\left.\begin{aligned}\phi&=P_0+\Sigma_1^\infty r^n(P_n\cos n\theta-Q_n\sin n\theta)+\Sigma_1^\infty r^{-n}(R_n\cos n\theta+S_n\sin n\theta),\\ \psi&=Q_0+\Sigma_1^\infty r^n(Q_n\cos n\theta-P_n\sin n\theta)+\Sigma_1^\infty r^{-n}(S_n\cos n\theta-R_n\sin n\theta).\end{aligned}\right\}\ldots(2)$$

These formulae are convenient in treating problems where we have the value of ϕ, or of $\partial\phi/\partial n$, given over concentric circular boundaries. This value may be expanded for each boundary in a series of sines and cosines of multiples of θ, by Fourier's theorem. The series thus found must be equivalent to those obtained from (2); whence, equating separately coefficients of $\sin n\theta$ and $\cos n\theta$, we obtain equations to determine P_n, Q_n, R_n, S_n.

* This example was given by Helmholtz, *Berl. Monatsber.* April 23, 1868 [*Phil. Mag.* Nov. 1868; *Wiss. Abh.* i. 154].

† R. A. Harris, "On Two-Dimensional Fluid Motion through Spouts composed of two Plane Walls," *Ann. of Math.* (2), ii. (1901). A diagram is given for the case of $\beta=\frac{1}{4}\pi$.

‡ Rayleigh, *Proc. Roy. Soc.* A, xci. 503 (1915) [*Papers*, vi. 329], where a few of the stream-lines are traced.

68. As a simple example let us take the case of an infinitely long circular cylinder of radius a moving with velocity U perpendicular to its length, in an infinite mass of liquid which is at rest at infinity.

Let the origin be taken in the axis of the cylinder, and the axes of x, y in a plane perpendicular to its length. Further let the axis of x be in the direction of the velocity U. The motion, supposed originated from rest, will necessarily be irrotational, and ϕ will be single-valued. Also, since $\int \partial\phi/\partial n . ds$, taken round the section of the cylinder, is zero, ψ is also single-valued (Art. 59), so that the formulae (2) apply. Moreover, since $\partial\phi/\partial n$ is given at every point of the internal boundary of the fluid, viz.

$$-\frac{\partial\phi}{\partial r} = U\cos\theta, \text{ for } r = a, \quad \text{.......................(3)}$$

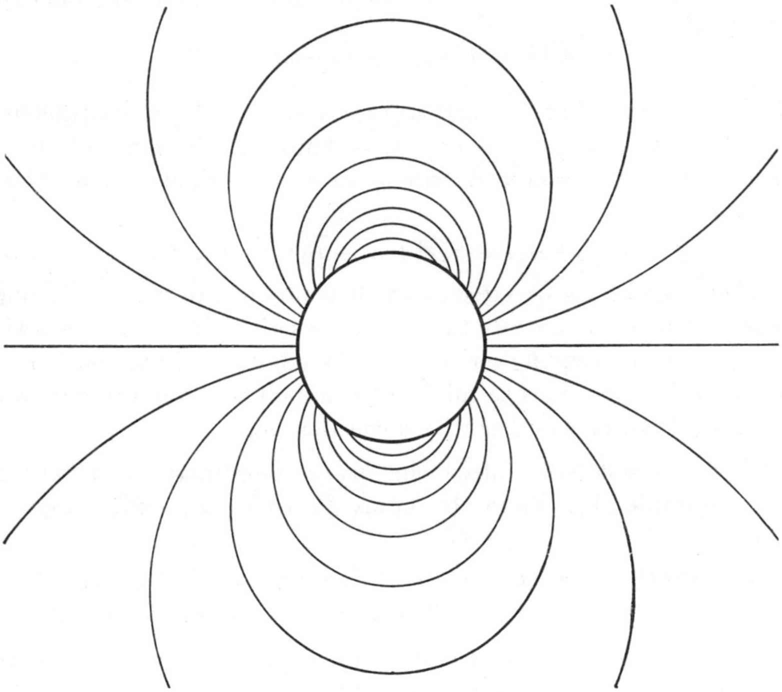

and since the fluid is at rest at infinity, the problem is determinate, by Art. 41. These conditions give $P_n = 0$, $Q_n = 0$, and

$$U\cos\theta = \Sigma_1^\infty \, na^{-n-1}(R_n \cos n\theta + S_n \sin n\theta),$$

which can only be satisfied by making $R_1 = Ua^3$, and all the other coefficients zero. The complete solution is therefore

$$\phi = \frac{Ua^2}{r}\cos\theta, \quad \psi = -\frac{Ua^2}{r}\sin\theta. \quad \text{.................(4)}$$

The stream-lines $\psi = \text{const.}$ are circles, as shewn above. Comparing with Art. 60 (6) we see that the effect is that of a double source at the origin.

The kinetic energy of the liquid is given by the formula (2) of Art. 61, viz.

$$2T = \rho \int \phi \, d\psi = \rho U^2 a^2 \int_0^{2\pi} \cos^2\theta \, d\theta = M' U^2, \quad \text{...........(5)}$$

if M', $= \pi a^2 \rho$, be the mass of fluid displaced by unit length of the cylinder. This result shews that the whole effect of the presence of the fluid may be represented by an addition M' to the inertia per unit length of the cylinder. Thus, in the case of rectilinear motion, if we have an extraneous force X per unit length acting on the cylinder, the equation of energy gives

$$\frac{d}{dt}(\tfrac{1}{2} M U^2 + \tfrac{1}{2} M' U^2) = XU,$$

or

$$(M + M') \frac{dU}{dt} = X, \quad \text{........................(6)}$$

where M represents the mass of the cylinder itself.

Writing this in the form

$$M \frac{dU}{dt} = X - M' \frac{dU}{dt},$$

we learn that the pressure of the fluid is equivalent to a force $-M'dU/dt$ per unit length in the direction of motion. This vanishes when U is constant.

The above result can be verified by direct calculation. By Art. 20 (7), (8) the pressure is given by the formula

$$\frac{p}{\rho} = \frac{\partial\phi}{\partial t} - \tfrac{1}{2} q^2 + F(t), \quad \text{......................................(7)}$$

provided q denotes the velocity of the fluid relative to the axis of the moving cylinder. The term due to the extraneous forces (if any) acting on the fluid has been omitted; the effect of these would be given by the rules of Hydrostatics. We have, for $r = a$,

$$\frac{\partial\phi}{\partial t} = a \frac{dU}{dt} \cos\theta, \quad q^2 = 4U^2 \sin^2\theta, \quad \text{.............................(8)}$$

whence

$$p = \rho \left(a \frac{dU}{dt} \cos\theta - 2U^2 \sin^2\theta + F(t) \right). \quad \text{........................(9)}$$

The resultant force on unit length of the cylinder is evidently parallel to the initial line $\theta = 0$; to find its amount we multiply by $-a d\theta \,.\, \cos\theta$ and integrate with respect to θ between the limits 0 and 2π. The result is $-M'dU/dt$, as before.

If in the above example we impress on the fluid and the cylinder a velocity $-U$ we have the case of a current flowing with the general velocity U past a fixed circular cylinder. Adding to ϕ and ψ the terms $Ur\cos\theta$ and $Ur\sin\theta$, respectively, we get

$$\phi = U\left(r + \frac{a^2}{r}\right)\cos\theta, \quad \psi = U\left(r - \frac{a^2}{r}\right)\sin\theta. \quad \text{........(10)}$$

The stream-lines are shewn on the next page.

If no extraneous forces act, and if U be constant, the resultant force on the cylinder is zero. Cf. Art. 92.

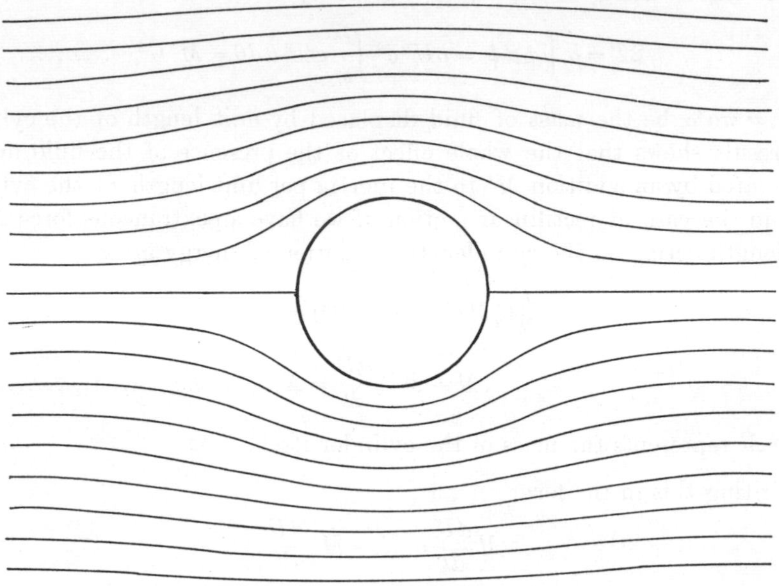

69. To render the formula (1) of Art. 67 capable of representing *any* case of continuous irrotational motion in the space between two concentric circles, we must add to the right-hand side the term

$$A \log z. \quad \text{..........................} (1)$$

If $A = P + iQ$, the corresponding terms in ϕ, ψ are

$$P \log r - Q\theta, \quad P\theta + Q \log r, \quad \text{..................} (2)$$

respectively. The meaning of these terms is evident; thus $2\pi P$, the cyclic constant of ψ, is the flux across the inner (or outer) circle; and $2\pi Q$, the cyclic constant of ϕ, is the circulation in any circuit embracing the origin.

For example, returning to the problem of the last Art., let us suppose that in addition to the motion produced by the cylinder we have an independent circulation round it, the cyclic constant being κ. The boundary-condition is then satisfied by

$$\phi = U \frac{a^2}{r} \cos\theta - \frac{\kappa}{2\pi}\theta. \quad \text{......................} (3)$$

The effect of the cyclic motion, superposed on that due to the cylinder, will be to augment the velocity on one side, and to diminish (and, it may be, to reverse) it on the other. Hence when the cylinder moves in a straight line with constant velocity, there will be a diminished pressure on one side, and an increased pressure on the other, so that a constraining force must be applied at right angles to the direction of motion.

The figure shews the lines of flow. At a distance from the origin they approximate to the form of concentric circles, the disturbance due to the cylinder becoming small in comparison with the cyclic motion. When, as in the case represented, $U > \kappa/2\pi a$, there is a point of zero velocity in the fluid. The stream-line system has the same configuration in all cases, the only effect of a change in the value of U being to alter the radius of the cylinder on the scale of the diagram.

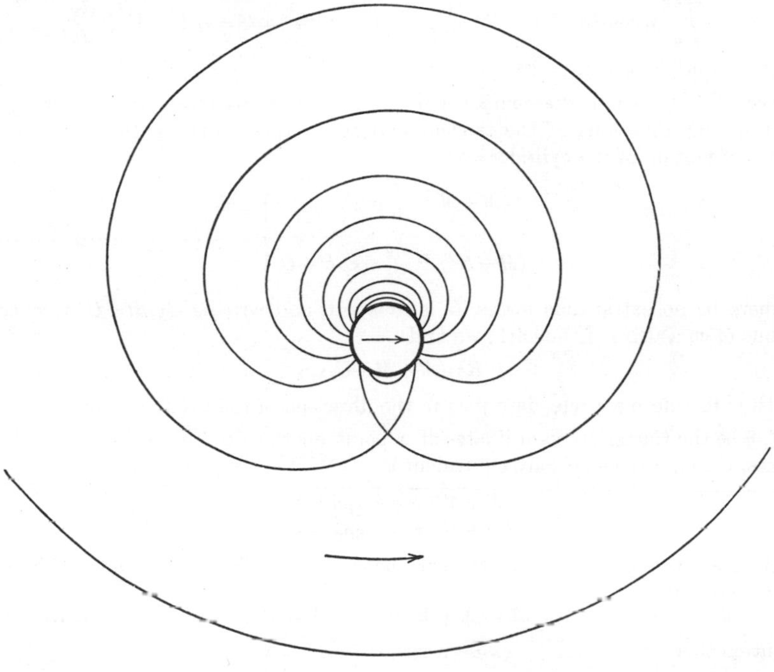

When the problem is reduced to one of steady motion we have in place of (3)

$$\phi = U\left(r + \frac{a^2}{r}\right)\cos\theta - \frac{\kappa}{2\pi}\theta, \quad \text{.....................(4)}$$

whence

$$\frac{p}{\rho} = \text{const.} - \tfrac{1}{2}q^2$$
$$= \text{const.} - \tfrac{1}{2}\left(2U\sin\theta + \frac{\kappa}{2\pi a}\right)^2, \quad \text{.....................(5)}$$

for $r = a$. The resultant pressure on the cylinder is therefore

$$-\int_0^{2\pi} p\sin\theta\, a\, d\theta = +\kappa\rho U, \quad \text{.....................(6)}$$

at right angles to the general direction of the stream. This result is independent of the radius of the cylinder. It will be shewn later that it holds for any form of section*.

To calculate the effect of the fluid pressures on the cylinder when moving in any manner we may conveniently adopt moving axes, the origin being taken at the centre, and the axis of x in the direction of the velocity U. If χ be the angle which this makes with a fixed direction, the equation (6) of Art. 20 gives

$$\frac{p}{\rho} = \frac{\partial\phi}{\partial t} - \tfrac{1}{2}q^2 - \frac{d\chi}{dt}\frac{\partial\phi}{\partial\theta}, \quad \text{.....................(7)}$$

* This important theorem is due to Kutta and Joukowski; see Kutta, *Sitzb. d. k. bayr. Akad. d. Wiss.* 1910. Proofs are given later (Arts. 72 b, 372).

where q now denotes fluid velocity relative to the origin, to be calculated from the relative velocity-potential $\phi + Ur\cos\theta$, ϕ being given by (3). We find, for $r=a$,

$$\frac{p}{\rho}=a\frac{dU}{dt}\cos\theta-\tfrac{1}{2}\left(2U\sin\theta+\frac{\kappa}{2\pi a}\right)^2+aU\frac{d\chi}{dt}\sin\theta+\frac{\kappa}{2\pi}\frac{d\chi}{dt}. \qquad (8)$$

The resultant pressures parallel to x and y are therefore

$$-\int_0^{2\pi} p\cos\theta\, a\,d\theta=-M'\frac{dU}{dt},\quad -\int_0^{2\pi} p\sin\theta\, a\,d\theta=\kappa\rho U-M'U\frac{d\chi}{dt}, \qquad (9)$$

where $M'=\pi\rho a^2$ as before.

Hence, if P, Q denote the components of the extraneous forces, if any, acting on the cylinder in the directions of the tangent and the normal to the path, respectively, the equations of motion of the cylinder are

$$\left.\begin{aligned}(M+M')\frac{dU}{dt}&=P,\\ (M+M')\,U\frac{d\chi}{dt}&=\kappa\rho U+Q.\end{aligned}\right\} \qquad (10)$$

If there be no extraneous forces, U is constant, and writing $d\chi/dt=U/R$, where R is the radius of curvature of the path, we find

$$R=(M+M')\,U/\kappa\rho. \qquad (11)$$

The path is therefore a circle, described in the direction of the cyclic motion*.

If ξ, η be the Cartesian co-ordinates of a point on the axis of the cylinder relative to fixed axes, the equations (10) are equivalent to

$$\left.\begin{aligned}(M+M')\,\ddot{\xi}&=-\kappa\rho\dot{\eta}+X,\\ (M+M')\,\ddot{\eta}&=\kappa\rho\dot{\xi}+Y,\end{aligned}\right\} \qquad (12)$$

where X, Y are the components of the extraneous forces. To find the effect of a constant force, we may put

$$X=(M+M')\,g',\qquad Y=0. \qquad (13)$$

The solution then is

$$\left.\begin{aligned}\xi&=\alpha+c\cos(nt+\epsilon),\\ \eta&=\beta+\frac{g'}{n}t+c\sin(nt+\epsilon),\end{aligned}\right\} \qquad (14)$$

where α, β, c, ϵ are arbitrary constants, and

$$n=\kappa\rho/(M+M'). \qquad (15)$$

This shews that the path is a trochoid, described with a mean velocity g'/n perpendicular to x†. It is remarkable that the cylinder has on the whole no progressive motion in the direction of the extraneous force. In the particular case $c=0$ its path is a straight line perpendicular to the force. The problem is an illustration of the theory of 'gyrostatic systems,' to be referred to in Chapter VI.

70. The formula (1) of Art. 67, as amended by the addition of the term $A\log z$, may readily be generalized so as to apply to any case of irrotational motion in a region with circular boundaries, one of which encloses all the rest. In fact, for each internal boundary we have a series of the form

$$A\log(z-c)+\frac{A_1}{z-c}+\frac{A_2}{(z-c)^2}+\dots, \qquad (1)$$

* Rayleigh, "On the Irregular Flight of a Tennis Ball," *Mess. of Math.* vii. (1878) [*Papers*, i. 344]; Greenhill, *Mess. of Math.* ix. 113 (1880).

† Greenhill, *l.c.*

where $c, = a+ib$ say, refers to the centre, and the coefficients $A, A_1, A_2, \ldots$ are in general complex quantities. The difficulty however of determining these coefficients so as to satisfy given boundary conditions is now so great as to render this method of very limited application.

Indeed the determination of the irrotational motion of a liquid subject to given boundary conditions is a problem whose exact solution can be effected by direct processes in only a limited number of cases. When the boundaries consist of fixed *straight* walls, a method of transformation devised by Schwarz* and Christoffel†, to be explained in Art. 73, is available. Most of the problems however whose solution is known have been obtained by an inverse method, viz. we take some known form of ϕ or ψ and inquire what boundary conditions it can be made to satisfy. Some simple examples of this procedure have already been given in Arts. 63, 64.

If we take a known problem of flow with given fixed boundaries, where $w=f(z)$, say, and apply a conformal transformation $z=\chi(z')$, the transformed boundaries in the plane of z' will still be stream-lines, and in this way we derive the solution of a new problem. It is sometimes advantageous to effect the transformation in two or more successive steps.

A problem which has led to important transformations in this way is that of the flow past a fixed circular cylinder. It is easily seen from Arts. 68, 69 that the general solution of this is

$$w = U\left(z+\frac{a^2}{z}\right) - iV\left(z-\frac{a^2}{z}\right) + \frac{i\kappa}{2\pi}\log\frac{z}{a}, \quad\text{..........................(2)}$$

where $-U$, $-V$ are the component velocities at infinity, and κ is the circulation. The procedure followed is to write

$$z = t + c, \quad\text{..(3)}$$

where t is an intermediate complex variable and $|c| < a$, and finally

$$z' = t + \frac{b^2}{t}. \quad\text{..(4)}$$

It is obvious that the infinitely distant regions of the planes z and z' will be identical, and the general direction of the stream, and the value of the circulation, therefore the same. The constants c and b are adjusted so that the points $t = \pm b$ in the plane of t may correspond to two arbitrary points A, B in the plane of z.

For instance, let AB be a chord of the circle $r=a$, parallel to Ox and subtending an angle 2β at the centre O. Referring to the figure on the next page we find

$$c = -ia\cos\beta, \quad b = a\sin\beta. \quad\text{................................(5)}$$

* "Ueber einige Abbildungsaufgaben," *Crelle*, lxx. [*Gesammelte Abhandlungen*, Berlin, 1890, ii. 65].

† "Sul problema delle temperature stazionarie e la rappresentazione di una data superficie," *Ann. di. Mat.* (2) i. 89. See also Kirchhoff, "Zur Theorie des Condensators," *Berl. Monatsber.* 1877 [*Ges. Abh.* 101]. Many of the solutions which can thus be obtained have interesting applications in Electrostatics, Heat-Conduction, &c. See, for example, J. J. Thomson, *Recent Researches in Electricity and Magnetism*, Oxford, 1893.

Then if P be any other point in the plane of z we have

$$z=\overline{OP},\quad t=\overline{CP}. \qquad\text{(6)}$$

It follows from (4) that

$$\frac{z'-2b}{z'+2b}=\left(\frac{t-b}{t+b}\right)^2. \qquad\text{(7)}$$

Writing for a moment

$$t-b=r_1e^{i\theta_1},\quad t+b=r_2e^{i\theta_2},\quad z'-2b=r_1'e^{i\theta_1'},\quad z'+2b=r_2'e^{i\theta_2'}, \qquad\text{(8)}$$

we have

$$\theta_1'-\theta_2'=2(\theta_1-\theta_2). \qquad\text{(9)}$$

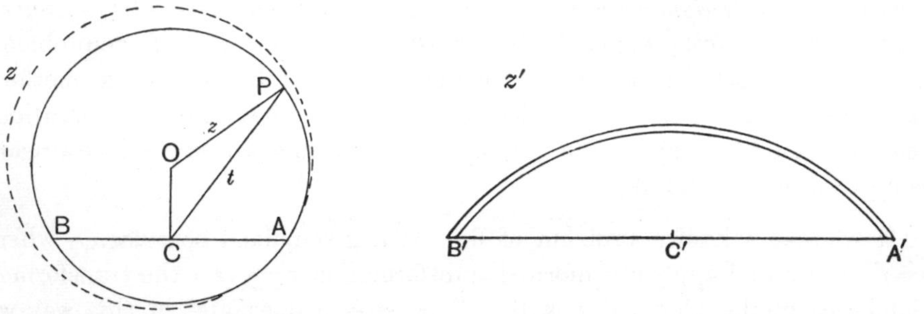

Now let P describe the circle in the plane of z, in the positive direction, starting from A. The corresponding point P' in the plane of z' will, by (9), move so that the angle $A'P'B'$ is constant and equal to 2β, the path therefore being an arc of a circle. As P passes B, θ_2 increases by π; hence in order that the equation (9) may subsist, θ_2' must increase by 2π. Hence as P completes its circle, P' moves back again along the arc $B'A'$. We thus obtain the case of a stream flowing in an arbitrary direction and with arbitrary circulation past a cylindrical lamina whose section is an arc of a circle*.

Since

$$\frac{dw}{dz'}=\frac{dw}{dz}\Big/\left(1-\frac{b^2}{t^2}\right), \qquad\text{(10)}$$

the velocity at the edges A', B' will be infinite. It can be made finite, however, at *one* edge, say B', by a suitable determination of the circulation, viz.

$$\kappa=4\pi a(U\cos\beta-V\sin\beta). \qquad\text{(11)}$$

The flow at B' is then given by

$$u-iv=(U\sin\beta+V\cos\beta)\sin\beta e^{2i\beta}, \qquad\text{(12)}$$

and is of course tangential to the arc. If the general velocity of the stream is W, at an inclination α to $B'A'$, we have

$$U=-W\cos\alpha,\quad V=-W\sin\alpha. \qquad\text{(13)}$$

Also, if R is the radius of the arc,

$$a\sin\beta=R\sin 2\beta. \qquad\text{(14)}$$

The 'lift,' therefore, at right angles to the stream, as given by Art. 72 b, is

$$4\pi\rho W^2R\frac{\sin 2\beta}{\sin\beta}\cos(\alpha+\beta). \qquad\text{(15)}$$

If instead of the circle $r=a$ in the figure we take as the circle to be transformed a circle touching it at A, and just *including* B, we get the profile of a Joukowsky aerofoil, of

* Kutta, *l.c. ante* p. 79. Some related problems are discussed by Blasius, *Zeitschr. f. Math. u. Phys.* lix. 225 (1911).

which the circular arc is, as it were, the skeleton*. This has a cusp at the point corresponding to A, and so involves an infinite velocity at this point (only). This singularity may be avoided by giving a suitable value to κ.

A simple method of obtaining solutions in two important cases of two-dimensioned motion is explained in the following Arts.

71. Case I. The boundary of the fluid consists of a rigid cylindrical surface which is in motion with velocity U in a direction perpendicular to the length.

Let us take as axis of x the direction of this velocity U, and let δs be an element of the section of the surface by the plane xy.

Then at all points of this section the velocity of the fluid in the direction of the normal, which is denoted by $\partial\psi/\partial s$, must be equal to the velocity of the boundary normal to itself, or $-U dy/ds$. Integrating along the section, we have

$$\psi = -Uy + \text{const.} \qquad \text{.............................(1)}$$

If we take any admissible form of ψ, this equation defines a system of curves each of which would by its motion parallel to x give rise to the stream-lines $\psi = \text{const.}$†. We give a few examples.

1°. If we choose for ψ the form $-Uy$, (1) is satisfied identically for all forms of the boundary. Hence the fluid contained within a cylinder of any shape which has a motion of translation only may move as a solid body. If, further, the cylindrical space occupied by the fluid be simply-connected, this is the only kind of irrotational motion possible. This is otherwise evident from Art. 40; for the motion of the fluid and the solid as one mass evidently satisfies all the conditions, and is therefore the only solution which the problem admits of.

2°. Let $\psi = A/r \,.\, \sin\theta$; then (1) becomes

$$\frac{A}{r}\sin\theta = -Ur\sin\theta = \text{const.} \qquad \text{.....................(2)}$$

In this system of curves is included a circle of radius a, provided $A/a = -Ua$. Hence the motion produced in an infinite mass of liquid by a circular cylinder moving through it with velocity U perpendicular to its length, is given by

$$\psi = -\frac{Ua^2}{r}\sin\theta, \qquad \text{.............................(3)}$$

which agrees with Art. 68.

* For further developments, and modifications of the method, reference may be made to Glauert, *Aerofoil and Airscrew Theory*, Cambridge, 1926.

† Cf. Rankine, *l.c. ante* p. 63, where the method is applied to obtain curves resembling the lines of ships.

3°. Let us introduce the elliptic co-ordinates ξ, η, connected with x, y by the relation

$$x + iy = c \cosh(\xi + i\eta), \qquad (4)$$

or

$$\left.\begin{aligned} x &= c \cosh \xi \cos \eta, \\ y &= c \sinh \xi \sin \eta, \end{aligned}\right\} \qquad (5)$$

(cf. Art. 66), where ξ may be supposed to range from 0 to ∞, and η from 0 to 2π. If we now put

$$\phi + i\psi = Ce^{-(\xi+i\eta)}, \qquad (6)$$

where C is some real constant, we have

$$\psi = -Ce^{-\xi} \sin \eta, \qquad (7)$$

so that (1) becomes $Ce^{-\xi} \sin \eta = Uc \sinh \xi \sin \eta + \text{const.}$

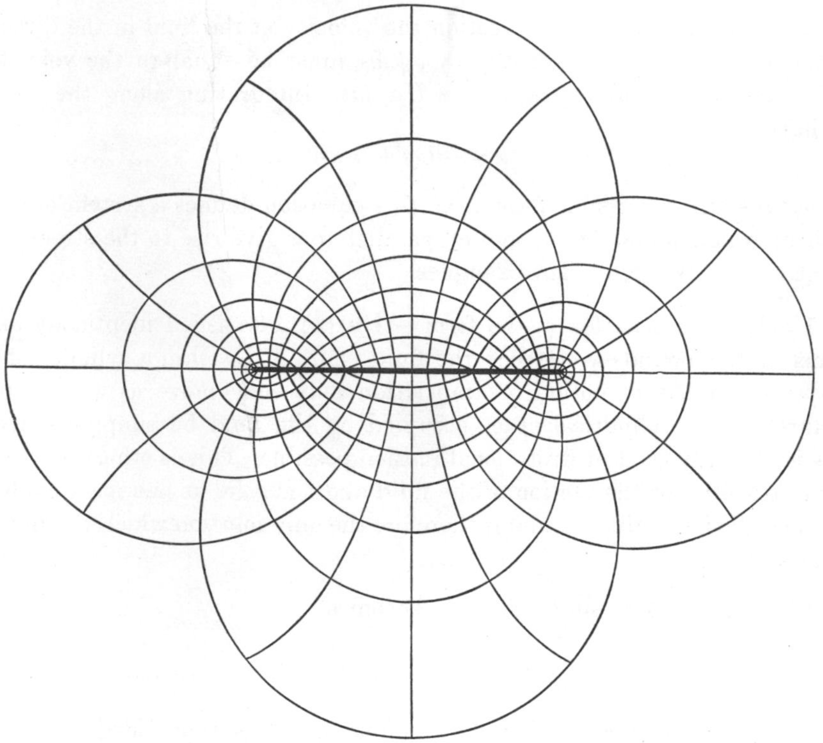

In this system of curves is included the ellipse whose parameter ξ_0 is determined by

$$Ce^{-\xi_0} = Uc \sinh \xi_0.$$

If a, b be the semi-axes of the ellipse we have

$$a = c \cosh \xi_0, \quad b = c \sinh \xi_0,$$

so that

$$C = \frac{Ubc}{a-b} = Ub\left(\frac{a+b}{a-b}\right)^{\frac{1}{2}}.$$

Hence the formula

$$\psi = -Ub\left(\frac{a+b}{a-b}\right)^{\frac{1}{2}} e^{-\xi} \sin \eta \qquad (8)$$

gives the motion produced in an infinite mass of liquid by an elliptic cylinder of semi-axes a, b, moving parallel to the greater axis with velocity U.

That the above formulae make the velocity zero at infinity appears from the consideration that, when ξ is large, δx and δy are of the same order as $e^{\xi}\delta\xi$ and $e^{\xi}\delta\eta$, so that $\partial\psi/\partial x$, $\partial\psi/\partial y$ are of the order $e^{-2\xi}$ or $1/r^2$, ultimately, where r denotes the distance of any point from the axis of the cylinder. At infinity ψ tends to the form $A \sin\theta/r$ as in the case of a double source.

If the motion of the cylinder were parallel to the minor axis, the formula would be

$$\psi = Va\left(\frac{a+b}{a-b}\right)^{\frac{1}{2}} e^{-\xi}\cos\eta. \quad\text{.......................(9)}$$

The stream-lines are in each case the same for all confocal elliptic forms of the cylinder, so that the formulae hold even when the section reduces to the straight line joining the foci. In this case (9) becomes

$$\psi = Vce^{-\xi}\cos\eta, \quad\text{............................(10)}$$

which would give the motion produced by an infinitely long lamina of breadth $2c$ moving 'broadside on' in an infinite mass of liquid. Since however this solution makes the velocity infinite at the edges, it is subject to the practical limitation already indicated in several instances*.

The kinetic energy of the fluid is given by

$$2T = \rho\int\phi\, d\psi = \rho C^2 e^{-2\xi_0}\int_0^{2\pi}\cos^2\eta\, d\eta$$
$$= \pi\rho b^2 U^2, \quad\text{............................(11)}$$

where b is the half-breadth of the cylinder perpendicular to the direction of motion.

Where there is circulation κ round the cylinder we have merely to add to the above values of ψ a term $\kappa\xi/2\pi$. In the case of the lamina the value of κ may be adjusted so as to make the velocity finite at one edge, but not at both.

If the units of length and time be properly chosen we may write for (4) and (6)

$$x + iy = \cosh(\xi + i\eta), \qquad \phi + i\psi = e^{-(\xi+i\eta)},$$

whence

$$x = \phi\left(1 + \frac{1}{\phi^2+\psi^2}\right), \qquad y = \psi\left(1 - \frac{1}{\phi^2+\psi^2}\right).$$

These formulae are convenient for tracing the curves ϕ=const., ψ=const., which are figured on the preceding page.

By superposition of the results (8) and (9) we obtain, for the case of an elliptic cylinder having a velocity of translation whose components are U, V,

$$\psi = -\left(\frac{a+b}{a-b}\right)^{\frac{1}{2}} e^{-\xi}(Ub\sin\eta - Va\cos\eta). \quad\text{.......................(12)}$$

To find the motion relative to the cylinder we must add to this the expression

$$Uy - Vx = c(U\sinh\xi\sin\eta - V\cosh\xi\cos\eta). \quad\text{.....................(13)}$$

* This investigation was given in the *Quart. Journ. of Math.* xiv. (1875). Results equivalent to (8), (9) had however been obtained, in a different manner, by Beltrami, "Sui principii fondamentali dell' idrodinamica razionale," *Mem. dell' Accad. delle Scienze di Bologna*, 1873, p. 394. [*Opere matematiche*, Milano, 1904, ii. 202.]

For example, the stream-function for a current impinging at an angle of 45° on a plane lamina whose edges are at $x = \pm c$ is

$$\psi = -\frac{1}{\sqrt{2}} q_0 c \sinh \xi (\cos \eta - \sin \eta), \qquad \text{.........................(14)}$$

where q_0 is the velocity at infinity. This immediately verifies, for it makes $\psi = 0$ for $\xi = 0$, and gives

$$\psi = -\frac{q_0}{\sqrt{2}}(x - y)$$

for $\xi = \infty$. The stream-lines for this case (turned through 45° for convenience) are shewn below. They will serve to illustrate some results to be obtained later in Chapter VI.

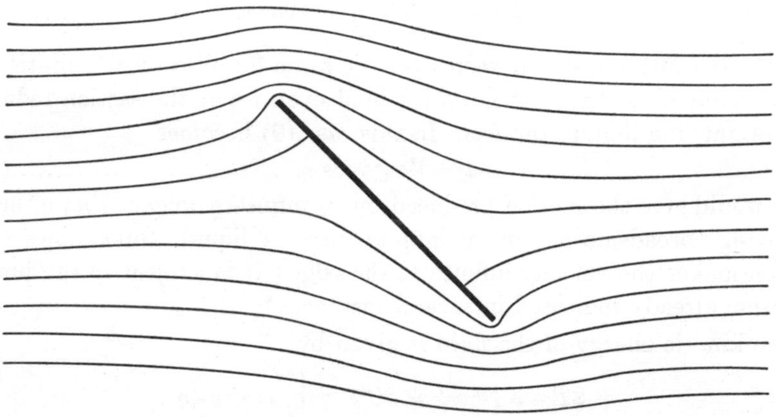

If we trace the course of the stream-line $\psi = 0$ from $\phi = +\infty$ to $\phi = -\infty$, we find that it consists in the first place of the hyperbolic arc $\eta = \frac{1}{4}\pi$, meeting the lamina at right angles; it then divides into two portions, following the faces of the lamina, which finally re-unite and are continued as the hyperbolic arc $\eta = \frac{5}{4}\pi$. The points where the hyperbolic arcs abut on the lamina are points of zero velocity, and therefore of maximum pressure*. It is plain that the fluid pressures on the lamina are equivalent to a couple tending to set it broadside on to the stream; and it is easily found that the moment of this couple, per unit length, is $\frac{1}{2}\pi\rho q_0^2 c^2$†. Compare Art. 124.

72. CASE II. The boundary of the fluid consists of a rigid cylindrical surface rotating with angular velocity ω about an axis parallel to its length.

Taking the origin in the axis of rotation, and the axes of x, y in a perpendicular plane, then, with the same notation as before, $\partial\psi/\partial s$ will be equal to the normal component of the velocity of the boundary, or

$$\frac{\partial \psi}{\partial s} = \omega r \frac{dr}{ds},$$

* Prof. Hele Shaw has made a number of beautiful experimental verifications of the forms of the stream-lines in cases of steady irrotational motion in two dimensions, including those figured on p. 78 and on this page; see *Trans. Inst. Nav. Arch.* xl. (1898). The theory of his method will find a place in Chapter XI.

† When the general direction of the stream makes an angle α with the lamina the couple is $\frac{1}{2}\pi\rho q_0^2 c^2 \sin 2\alpha$. Cisotti, *Ann. di. mat.* (3), xix. 83 (1912).

if r denote the radius vector from the origin. Integrating we have, at all points of the boundary,

$$\psi = \tfrac{1}{2}\omega r^2 + \text{const.} \qquad \text{.............................(1)}$$

If we assume any possible form of ψ, this will give us the equation of a series of curves, each of which would, by rotating round the origin, produce the system of stream-lines determined by ψ.

As examples we may take the following:

1°. If we assume $\quad \psi = Ar^2 \cos 2\theta = A(x^2 - y^2), \qquad \text{....................(2)}$

the equation (1) becomes

$$(\tfrac{1}{2}\omega - A)\,x^2 + (\tfrac{1}{2}\omega + A)\,y^2 = C,$$

which, for any given value of A, represents a system of similar conics. That this system may include the ellipse

$$\frac{x^2}{a^2} + \frac{y^2}{b^2} = 1,$$

we must have

$$(\tfrac{1}{2}\omega - A)\,a^2 = (\tfrac{1}{2}\omega + A)\,b^2,$$

or

$$A = \tfrac{1}{2}\omega \,.\, \frac{a^2 - b^2}{a^2 + b^2}.$$

Hence the formula

$$\psi = \tfrac{1}{2}\omega \,.\, \frac{a^2 - b^2}{a^2 + b^2}(x^2 - y^2) \qquad \text{.......................(3)}$$

gives the motion of a liquid contained within a hollow cylinder whose section is an ellipse with semi-axes a, b, produced by the rotation of the cylinder about its longitudinal axis with angular velocity ω. The arrangement of the stream-lines $\psi = \text{const.}$ is shewn on the next page.

The corresponding formula for ϕ is

$$\phi = -\,\omega \,.\, \frac{a^2 - b^2}{a^2 + b^2} \,.\, xy. \qquad \text{..........................(4)}$$

The kinetic energy of the fluid, per unit length of the cylinder, is given by

$$2T = \rho \iint \left\{ \left(\frac{\partial \phi}{\partial x}\right)^2 + \left(\frac{\partial \phi}{\partial y}\right)^2 \right\} dx\,dy = \tfrac{1}{4}\,\frac{(a^2 - b^2)^2}{a^2 + b^2}\,\omega^2 \times \pi\rho ab. \qquad \text{......(5)}$$

This is less than if the fluid were to rotate with the boundary, as one rigid mass, in the ratio of

$$\left(\frac{a^2 - b^2}{a^2 + b^2}\right)^2$$

to unity. We have here an illustration of Lord Kelvin's minimum theorem, proved in Art. 45.

2°. With the same notation of elliptic co-ordinates as in Art. 71, 3°, let us assume

$$\phi + i\psi = Cie^{-2(\xi + i\eta)}. \qquad \text{..........................(6)}$$

Since

$$x^2 + y^2 = \tfrac{1}{2}c^2(\cosh 2\xi + \cos 2\eta),$$

the equation (1) becomes

$$Ce^{-2\xi}\cos 2\eta - \tfrac{1}{4}\omega c^2(\cosh 2\xi + \cos 2\eta) = \text{const.}$$

This system of curves includes the ellipse whose parameter is ξ_0, provided

$$Ce^{-2\xi_0} - \tfrac{1}{4}\omega c^2 = 0,$$

or, using the values of a, b already given,

$$C = \tfrac{1}{4}\omega (a+b)^2,$$

so that

$$\left.\begin{aligned} \psi &= \tfrac{1}{4}\omega (a+b)^2 e^{-2\xi} \cos 2\eta, \\ \phi &= \tfrac{1}{4}\omega (a+b)^2 e^{-2\xi} \sin 2\eta. \end{aligned}\right\} \quad \text{.......................(7)}$$

At a great distance from the origin the velocity is of the order $1/r^3$.

The above formulae therefore give the motion of an infinite mass of liquid, otherwise at rest, produced by the rotation of an elliptic cylinder about its axis with angular velocity ω*. The diagram shews the stream-lines both inside and outside a rigid elliptical cylindrical case rotating about its axis.

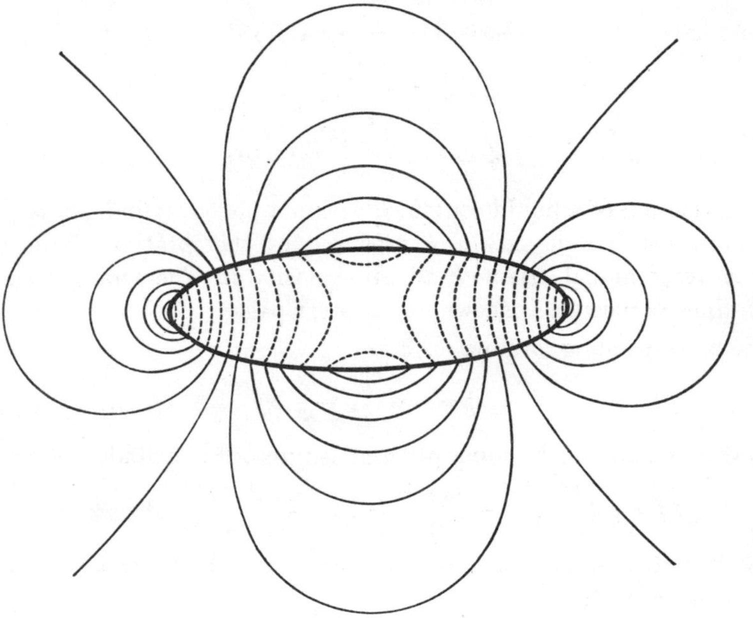

The kinetic energy of the external fluid is given by

$$2T = \tfrac{1}{8}\pi\rho c^4 \,.\, \omega^2. \quad \text{..............................(8)}$$

It is remarkable that this is the same for all confocal elliptic forms of the section of the cylinder.

Combining these results with those of Arts. 66, 71 we find that if an elliptic cylinder be moving with velocities U, V parallel to the principal axes of its cross-section, and rotating with angular velocity ω, and if (further) the

* *Quart. Journ. Math.* xiv. (1875); see also Beltrami, *l.c. ante* p. 85.

fluid be circulating irrotationally round it, the cyclic constant being κ, then the stream-function relative to the aforesaid axes is

$$\psi = -\sqrt{\left(\frac{a+b}{a-b}\right)}\, e^{-\xi}\,(Ub \sin\eta - Va\cos\eta) + \tfrac{1}{4}\omega\,(a+b)^2 e^{-2\xi}\cos 2\eta + \frac{\kappa}{2\pi}\,\xi. \qquad \text{......(9)}$$

The *paths* followed by the particles of fluid in several of the preceding cases, as distinguished from the stream-lines, have been studied by Prof. W. B. Morton *; they are very remarkable. The particular case of the circular cylinder (Art. 68) was examined by Maxwell †.

3°. Let us assume $\quad \psi = Ar^3 \cos 3\theta = A(x^3 - 3xy^2).$

The equation (1) of the boundary then becomes

$$A\,(x^3 - 3xy^2) - \tfrac{1}{2}\omega\,(x^2+y^2) = C. \qquad \text{..............................(10)}$$

We may choose the constants so that the straight line $x=a$ shall form part of the boundary. The conditions for this are

$$Aa^3 - \tfrac{1}{2}\omega a^2 = C, \qquad 3Aa + \tfrac{1}{2}\omega = 0.$$

Substituting in (10) the values of A, C hence derived, we have

$$x^3 - a^3 - 3xy^2 + 3a\,(x^2 - a^2 + y^2) = 0.$$

Dividing out by $x-a$, we get $\quad x^2 + 4ax + 4a^2 - 3y^2,$

or $\quad x + 2a = \pm\sqrt{3}\,.\,y.$

The rest of the boundary consists therefore of two straight lines passing through the point $(-2a, 0)$, and inclined at angles of 30° to the axis of x.

We have thus obtained the formulae for the motion of the fluid contained within a vessel in the form of an equilateral prism, when the latter is rotating with angular velocity ω about an axis parallel to its length and passing through the centre of its section; viz. we have

$$\psi = -\tfrac{1}{6}\frac{\omega}{a} r^3 \cos 3\theta, \qquad \phi = \tfrac{1}{6}\frac{\omega}{a} r^3 \sin 3\theta, \qquad \text{........................(11)}$$

where $2\sqrt{3}a$ is the length of a side of the prism ‡.

4°. In the case of a liquid contained in a rotating cylinder whose section is a circular sector of radius a and angle 2α, the axis of rotation passing through the centre, we may assume

$$\psi = \tfrac{1}{2}\omega r^2 \frac{\cos 2\theta}{\cos 2\alpha} + \Sigma A_{2n+1}\left(\frac{r}{a}\right)^{(2n+1)\pi/2\alpha} \cos\,(2n+1)\frac{\pi\theta}{2\alpha}, \qquad \text{..............(12)}$$

the middle radius being taken as initial line. For this makes $\psi = \tfrac{1}{2}\omega r^2$ for $\theta = \pm\alpha$, and the constants A_{2n+1} can be determined by Fourier's method so as to make $\psi = \tfrac{1}{2}\omega a^2$ for $r=a$. We find

$$A_{2n+1} = (-)^{n+1}\,\omega a^2 \left\{\frac{1}{(2n+1)\pi - 4\alpha} - \frac{2}{(2n+1)\pi} + \frac{1}{(2n+1)\pi + 4\alpha}\right\}. \qquad \text{......(13)}$$

The conjugate expression for ϕ is

$$\phi = -\tfrac{1}{2}\omega r^2 \frac{\sin 2\theta}{\cos 2\alpha} - \Sigma A_{2n+1}\left(\frac{r}{a}\right)^{(2n+1)\pi/2\alpha} \sin\,(2n+1)\frac{\pi\theta}{2\alpha}. \qquad \text{............(14)}$$

* *Proc. Roy. Soc.* A, lxxxix. 106 (1913).

† *Proc. Lond. Math. Soc.* iii. 82 (1870) [*Papers*, ii. 208].

‡ The problem of fluid motion in a rotating cylindrical case is to a certain extent mathematically identical with that of the torsion of a uniform rod or bar. The examples numbered '1°' and '3°' are mere adaptations of two of de Saint-Venant's solutions of the latter problem. See Thomson and Tait, Art. 704 *et seq.*

The kinetic energy is given by

$$2T = -\rho \int \phi \frac{\partial \phi}{\partial n} ds = -2\rho\omega \int_0^a \phi_\alpha r\, dr, \qquad (15)$$

where ϕ_α denotes the value of ϕ for $\theta = \alpha$, the value of $\partial\phi/\partial n$ being zero over the circular part of the boundary*.

The case of the semicircle $\alpha = \frac{1}{2}\pi$ will be of use to us later. We then have

$$A_{2n+1} = (-)^{n+1} \frac{\omega a^2}{\pi} \left\{ \frac{1}{2n-1} - \frac{2}{2n+1} + \frac{1}{2n+3} \right\}; \qquad (16)$$

and therefore

$$\int_0^a \phi_\alpha r\, dr = \frac{\omega a^4}{\pi} \Sigma \frac{1}{2n+3} \left\{ \frac{1}{2n-1} - \frac{2}{2n+1} + \frac{1}{2n+3} \right\} = -\frac{\omega a^4}{\pi} \left(2 - \frac{\pi^2}{8} \right).$$

Hence †

$$2T = \tfrac{1}{2}\pi\rho\omega^2 a^4 \left(\frac{8}{\pi^2} - \tfrac{1}{2} \right) = \cdot 3106 a^2 \times \tfrac{1}{2}\pi\rho\omega^2 a^2. \qquad (17)$$

This is less than if the fluid were solidified, in the ratio of 6212 to 1. Cf. Art. 45.

72 a. We have seen in several instances that when a cylinder has a motion of translation though an infinite fluid the effect at a great distance is that of a double source. A general formula for this can be given in terms of certain constants which occur in the expression for the kinetic energy of the fluid‡.

If we write

$$\phi = U\phi_1 + V\phi_2, \qquad (1)$$

where (U, V) is the velocity of the cylinder, the functions ϕ_1, ϕ_2 are determined by the conditions that $\nabla^2\phi_1 = 0$, $\nabla^2\phi_2 = 0$ throughout the external space, that their derivatives vanish at infinity, and that at the contour of the cylinder

$$-\frac{\partial \phi_1}{\partial n} = l, \quad -\frac{\partial \phi_2}{\partial n} = m, \qquad (2)$$

where (l, m) is the direction of the outward normal. Hence the energy of the fluid is given by

$$\frac{2T}{\rho} = -\int \phi \frac{\partial \phi}{\partial n} ds = AU^2 + 2HUV + BV^2, \qquad (3)$$

where

$$\left.\begin{aligned} A &= -\int \phi_1 \frac{\partial \phi_1}{\partial n} ds = \int l\phi_1 ds, \\ B &= -\int \phi_2 \frac{\partial \phi_2}{\partial n} ds = \int m\phi_2 ds, \\ H &= -\int \phi_1 \frac{\partial \phi_2}{\partial n} ds = -\int \phi_2 \frac{\partial \phi_1}{\partial n} ds = \int m\phi_1 ds = \int l\phi_2 ds. \end{aligned}\right\} \qquad (4)$$

The two forms of H are equal by the two-dimensional form of Green's Theorem. Cf. Art. 121, where the general three-dimensional case is discussed.

Referring to Art. 60 (7), suppose that a cylinder of any form of section is moving with unit velocity parallel to the axis of x. Taking an origin within the contour, and writing

$$\begin{aligned} r^2 &= (x_0 - x)^2 + (y_0 - y)^2 \\ &= r_0^2 - 2(xx_0 + yy_0) + \ldots, \end{aligned} \qquad (5)$$

* This problem was first solved by Stokes, "On the Critical Values of the Sums of Periodic Series," *Camb. Trans.* viii. (1847) [*Papers*, i. 305]. See also Hicks, *Mess. of Math.* viii. 42 (1878); Greenhill, *ibid.* viii. 89, and x. 83.

† Greenhill, *l.c.*

‡ Cf. *Proc. Roy. Soc.* A, cxi. 14 (1926) and Art. 300 *infra*.

where (x_0, y_0) is a distant point at which the value of ϕ is required, and (x, y) a point of the contour, we have

$$\log r = \log r_0 - \frac{xx_0 + yy_0}{r_0^2} + \ldots, \quad \text{.....(6)}$$

and

$$\frac{\partial}{\partial n}(\log r) = -\frac{lx_0 + my_0}{r_0^2},$$

approximately. Writing

$$\phi = \phi_1, \quad \phi' = -x \quad \text{.....(7)}$$

in the formula referred to, we find

$$2\pi\phi_P = \frac{(A+Q)\,x_0 + Hy_0}{r_0^2}, \quad \text{.....(8)}$$

where A and H are defined by (4), and

$$Q = \int lx\,ds, \quad \text{.....(9)}$$

i.e. Q denotes the sectional area of the cylinder.

The flow at a great distance is accordingly that due to a double source, but the axis of the source does not in general coincide with the direction of motion of the cylinder.

The generalization of (8) is obvious. When the cylinder has a velocity (U, V) we have

$$2\pi r_0^2 \phi_P = \{(A+Q)\,U + HV\}\,x_0 + \{HU + (B+Q)\,V\}\,y_0. \quad \text{.....(10)}$$

In terms of the complex variables w, z, this may be written

$$w = (\alpha + i\beta)/z_0, \quad \text{.....(11)}$$

with

$$2\pi\alpha = (A+Q)\,U + HV, \quad 2\pi\beta = HU + (B+Q)\,V. \quad \text{.....(12)}$$

For an elliptic cross-section we have, by comparison of Art. 71 (11) with (3) above, $A = \pi b^2$, $B = \pi a^2$, whilst $Q = \pi ab$. Hence

$$\phi_P = (a+b)\,(bUx_0 + aVy_0)/2r_0^3. \quad \text{.....(13)}$$

72 b. The hydrodynamic forces on a fixed cylinder due to the steady irrotational motion of a surrounding fluid have already been calculated in one or two cases. A general method, available whenever the form of w, $= \phi + i\psi$, for the fluid motion is known, has been given by Blasius*.

The pressures on the contour may be reduced to a force (X, Y) at the origin, and a couple N. If θ be the angle which the velocity q makes with the axis of x, we have

$$Y + iX = -\tfrac{1}{2}\rho\int q^2(\cos\theta - i\sin\theta)\,ds, \quad \text{.....(1)}$$

where the integral is taken round the contour of the cylinder.

This may be written

$$Y + iX = -\tfrac{1}{2}\rho\int (qe^{-i\theta})^2\,e^{i\theta}\,ds = -\tfrac{1}{2}\rho\int\left(\frac{dw}{dz}\right)^2 dz, \quad \text{.....(2)}$$

This gives X and Y.

Again, if ϑ be the angle which an element δs of the contour makes with the radius vector (produced),

$$N = \int pr\cos\vartheta\,ds = \int pr\,dr = -\tfrac{1}{2}\rho\int(u^2+v^2)(x\,dx + y\,dy), \quad \text{.....(3)}$$

* "Funktiontheoretische Methoden in der Hydrodynamik," *Zeitschr. f. Math. u. Phys.* lviii. (1910).

Now along a stream-line we have $v\,dx = u\,dy$, whence

$$(u - iv)^2 (dx + i\,dy) = (u^2 + v^2)(dx - i\,dy)$$

and

$$(u - iv)^2 (x + iy)(dx + i\,dy) = (u^2 + v^2)\{x\,dx + y\,dy + i(y\,dx - x\,dy)\}.$$

Hence N is given by the *real part* of the integral

$$-\tfrac{1}{2}\rho \int \left(\frac{dw}{dz}\right)^2 z\,dz. \quad \text{.....(4)}$$

In the case of a cylinder immersed in a uniform stream, with circulation, the value of w at a great distance tends to the form

$$w = A + Bz + C \log z. \quad \text{.....(5)}$$

Since in (4) there are no singularities of the integrand in the space occupied by the fluid, the integral may be replaced by that round an infinite enclosing contour. On this understanding

$$\int \left(\frac{dw}{dz}\right)^2 dz = \int \left(B^2 + \frac{2BC}{z} + \frac{C^2}{z^2}\right) dz = 2BC \int \frac{dz}{z} = 4\pi i BC. \quad \text{.....(6)}$$

If the stream at infinity is (U, V), and if κ denote the circulation, we have

$$B = -(U - iV), \quad C = -i\kappa/2\pi. \quad \text{.....(7)}$$

Hence
$$X = \kappa\rho V, \quad Y = -\kappa\rho U \quad \text{.....(8)}$$

which is the generalization of the result obtained in Art. 69 for the particular case of a circular section.

For the calculation of the moment N the expression in (5) must be carried a stage further. Writing

$$w = A + Bz + C \log z + \frac{D}{z}, \quad \text{.....(9)}$$

we have

$$\left(\frac{dw}{dz}\right)^2 = B^2 + \frac{2BC}{z} + \frac{C^2 - 2BD}{z^2} + \ldots. \quad \text{.....(10)}$$

Omitting all the terms which disappear in the case of an infinite contour we have

$$\int \left(\frac{dw}{dz}\right)^2 z\,dz = 2\pi i (C^2 - 2BD). \quad \text{.....(11)}$$

Substituting the values of B and C from (7), writing $D = \alpha + i\beta$, and taking the real part, we find

$$N = 2\pi\rho(\beta U - \alpha V). \quad \text{.....(12)}$$

If by the superposition of a general velocity $(-U, -V)$ the fluid were reduced to rest at infinity, the term D/z in (9) would be due to a translation of the cylinder with this velocity. Hence the values of α, β are as given in Art. 72, except that the signs are reversed. Hence (12) gives

$$N = \rho\{(A - B)\,UV - H(U^2 - V^2)\}. \quad \text{.....(13)}$$

Thus for an elliptic section referred to its principal axes

$$N = -\pi\rho(a^2 - b^2)\,UV. \quad \text{.....(14)}$$

As a further application of Blasius' formula we may calculate the force on a fixed cylinder due to an external source.

We write
$$w = -\mu \log (z-c) + f(z), \qquad \text{.............................(15)}$$
where the first term represents the source at $z=c$, say, and $f(z)$ its image in the cylinder, *i.e.* $f(z)$ is the addition necessary to annul the normal velocity at the contour, due to the source. Hence
$$\frac{dw}{dz} = -\frac{\mu}{z-c} + f'(z). \qquad \text{.............................(16)}$$

The contour integral in (2) is now equal to the integral round an infinite contour *minus* the integral (in the positive sense) round the singularity at $z=c$. The infinite contour gives a zero result. In the neighbourhood of the singularity the only part of $(dw/dz)^2$ which need be taken into account is that containing the *first* power of $z-c$ in the denominator, viz.
$$-\frac{2\mu f'(c)}{z-c},$$
ultimately. Hence
$$Y + iX = -2\pi i \mu \rho f'(c). \qquad \text{.............................(17)}$$

The form of $f(z)$ for the case of a *circular* cylinder is already known from Art. 64, 3°. The source being supposed on the axis of x, so that c is real, we have
$$f(z) = -\mu \log (z - a^2/c) + \mu \log z, \qquad \text{.............................(18)}$$
$$f'(c) = -\frac{\mu a^2}{c(c^2-a^2)}, \qquad \text{.............................(19)}$$
$$X = \frac{2\pi\mu^2 a^2}{c(c^2-a^2)}, \quad Y = 0. \qquad \text{.............................(20)*}$$

In the general case, an approximation to the asymptotic form which $f(z)$ assumes, when the distance of the source is great compared with the dimensions of the cross-section, is obtained if we suppose it to represent the effect of a translation of the cylinder with a velocity equal and opposite to that which the source would produce in the neighbourhood, if the cylinder were absent. Thus, the source being still assumed to be on the axis of x, we have from Art. 72 a
$$f(z) = \frac{(A+Q+iH)\,U}{2\pi z}, \qquad \text{.............................(21)}$$
where $U = \mu/c$. Hence
$$f'(c) = -\frac{(A+Q+iH)\,\mu}{2\pi c^3}, \qquad \text{.............................(22)}$$
and therefore
$$X = \frac{(A+Q)\,\mu^2\rho}{c^3}, \quad Y = -\frac{H\mu^2\rho}{c^3}. \qquad \text{.............................(23)}$$

If $f(=\mu^2/c^3)$ is the acceleration at the position of the origin, in the undisturbed stream, these results may be written
$$X = \rho(A+Q)f, \quad Y = -\rho H f. \qquad \text{.............................(24)}$$

For a circular section $A = \pi a^2$, $H=0$, $Q = \pi a^2$, and the formula (20) is verified, if we neglect terms of the order a^2/c^2.

A number of elegant applications of Blasius' method, relating to the mutual action of circular cylinders, with circulation, have been made by Cisotti†. One of his results may be quoted. A cylinder of radius b is fixed excentrically within a cylindrical tunnel of radius a, and the intervening space is occupied by fluid having a circulation κ. The resultant force on the cylinder is towards the nearest part of the tunnel wall, and has the value
$$\kappa^2 d^2 \div 2\pi \sqrt{\{(a+b+d)(a+b-d)(a-b+d)(a-b-d)\}},$$
where d is the distance between the axes.

* The result is due to Prof. G. I. Taylor. † *Rend. d. r. Accad. d. Lincei* (6) i. (1925–6).

Free Stream-Lines.

73. The first solution of a problem of two-dimensional motion in which the fluid is bounded partly by fixed plane walls and partly by surfaces of constant pressure, was given by Helmholtz*. Kirchhoff† and others have since elaborated a general method of dealing with such questions. If the surfaces of constant pressure be regarded as free, we have a theory of jets, which furnishes some interesting results in illustration of Art. 24. Again since the space beyond these surfaces may be filled with liquid at rest, without altering the conditions of the problem, we obtain also a number of cases of 'discontinuous motion,' which are mathematically possible with perfect fluids, but whose practical significance is more open to question. We shall return to this point at a later stage (Chap. XI.); in the meantime we shall speak of the surfaces of constant pressure as 'free.' Extraneous forces, such as gravity, being neglected, the velocity must be constant along any such surface, by Art. 21 (2).

The method in question is based on the properties of the function ζ introduced in Art. 65. The moving fluid is supposed bounded by stream-lines $\psi = \text{const.}$, which consist partly of straight walls, and partly of lines along which the resultant velocity (q) is constant. For convenience, we may in the first instance suppose the units of length and time to be so adjusted that this constant velocity is equal to unity. Then in the plane of the function ζ the lines for which $q = 1$ are represented by arcs of a circle of unit radius, having the origin as centre, and the straight walls (since the direction of the flow along each is constant) by radial lines drawn outwards from the circumference. The points where these lines meet the circle correspond to the points where the bounding stream-lines change their character.

Consider, next, the function $\log \zeta$. In the plane of this function the circular arcs for which $q = 1$ become transformed into portions of the imaginary axis, and the radial lines into lines parallel to the real axis, since if $\zeta = q^{-1} e^{i\theta}$ we have

$$\log \zeta = \log \frac{1}{q} + i\theta. \qquad \text{.............................(1)}$$

It remains, then, to determine a relation of the form‡

$$\log \zeta = f(w), \qquad \text{................................(2)}$$

where $w = \phi + i\psi$, as usual, such that the rectilinear boundaries in the plane of $\log \zeta$ shall correspond to straight lines $\psi = \text{const.}$ in the plane of w. There are further conditions of correspondence between special points, one on the boundary, and one in the interior, of each region, which render the problem determinate.

* *Loc. cit. ante* p. 75.

† "Zur Theorie freier Flüssigkeitsstrahlen," *Crelle*, lxx. (1869) [*Ges. Abh.* p. 416]. See also his *Mechanik*, cc. xxi., xxii.

‡ The use of log ζ, in place of ζ, is due to Planck, *Wied. Ann.* xxi. (1884).

When the correspondence between the planes of ζ and w has been established, the connection between z and w is to be found, by integration, from the relation

$$\frac{dz}{dw} = -\zeta. \qquad \text{.................................(3)}$$

The arbitrary constant which appears in the result is due to the arbitrary position of the origin in the plane of z.

The problem is thus reduced to one of conformal representation between two areas bounded by straight lines*. This is resolved by the method of Schwarz and Christoffel, already referred to †, in which each area is represented in turn on a half-plane. Let $Z\,(=X+iY)$ and t be two complex variables connected by the relation

$$\frac{dZ}{dt} = A\,(a-t)^{-\alpha/\pi}\,(b-t)^{-\beta/\pi}\,(c-t)^{-\gamma/\pi}\,\ldots, \qquad \text{...........(4)}$$

where $a, b, c, \ldots$ are real quantities in ascending order of magnitude, whilst $\alpha, \beta, \gamma, \ldots$ are angles (not necessarily all positive) such that

$$\alpha+\beta+\gamma+\ldots = 2\pi; \qquad \text{.........................(5)}$$

and consider the line made up of portions of the real axis of t with small semi-circular indentations (on the upper side) about the points $a, b, c, \ldots$. If a point describe this line from $t=-\infty$ to $t=+\infty$, the modulus only of the expression in (4) will vary so long as a straight portion is being described, whilst the effect of the clockwise description of the semi-circular portions is to introduce factors $e^{i\alpha}, e^{i\beta}, e^{i\gamma}, \ldots$ in succession. Hence, regarding dZ/dt as an operator which converts δt into δZ, we see that the upper half of the plane of t is conformably represented on the area of a closed polygon whose *exterior* angles are $\alpha, \beta, \gamma, \ldots$, by the formula

$$Z = A\int (a-t)^{-\alpha/\pi}\,(b-t)^{-\beta/\pi}\,(c-t)^{-\gamma/\pi}\ldots dt + B, \qquad \text{........(6)}$$

provided the path of integration in the t-plane lies wholly within the region above delimited. When $a, b, c, \ldots, \alpha, \beta, \gamma, \ldots$ are given, the polygon is completely determinate as to shape; the complex constants A, B only affect its scale and orientation, and its position, respectively.

As already indicated, we are specially concerned with the conformal representation of *rectangular* areas. If $\alpha=\beta=\gamma=\delta=\frac{1}{2}\pi$, the formula (6) becomes

$$Z = A\int \frac{dt}{\sqrt{\{(a-t)\,(b-t)\,(c-t)\,(d-t)\}}} + B. \qquad \text{.............(7)}$$

It is easily seen that the rectangle is finite in all its dimensions unless *two* at least of the points a, b, c, d are at infinity. The excepted case is the one

* See Forsyth, *Theory of Functions*, c. xx.

† See the footnotes on p. 81 *ante*.

specially important to us; the two finite points may then conveniently be taken to be $t = \pm 1$, so that

$$Z = A\int \frac{dt}{\sqrt{(t^2-1)}} + B$$

$$= A \cosh^{-1} t + B. \quad \text{.............................(8)}$$

In particular, the assumption

$$t = \cosh\frac{Z}{k}, \quad \text{..............................(9)}$$

where k is real, transforms the space bounded by the positive halves of the lines $Y=0$, $Y=\pi k$, and the intervening portion of the axis of Y, into the upper half of the plane t. Cf. Art. 66, 1°.

Again, if the two finite points coincide, say at the origin of t, we have

$$Z = A\int\frac{dt}{t} + B = A\log t + B. \quad \text{....................(10)}$$

This transforms the upper half of the t-plane into a strip bounded by two parallel straight lines. For example, if

$$t = e^{Z/k}, \quad \text{..................................(11)}$$

where k is real, these may be the lines $Y=0$, $Y=\pi k$.

74. As a first application of the method in question, we may take the case of a fluid escaping from a large vessel by a straight canal projecting inwards*. This is the two-dimensional form of Borda's mouthpiece, referred to in Art. 24.

The boundaries of corresponding areas in the planes of ζ, $\log\zeta$, and w, respectively, are easily traced, and are shewn in the figures†. It remains to connect the areas in the planes of $\log\zeta$ and w each with the upper half-plane of an intermediate variable t. It appears from equations (8) and (10) of the preceding Art. that this is accomplished by the substitutions

$$\log\zeta = A\cosh^{-1} t + B, \quad w = C\log t + D. \quad \text{..............(1)}$$

We have here made the corners A, A' in the plane of $\log\zeta$ correspond to $t = \pm 1$, and we have also assumed that $t=0$ corresponds to $w = -\infty$, as is evident on inspection of the figures. To specify more precisely the values of the cyclic functions $\cosh^{-1} t$ and $\log t$ we will assume that they both vanish at $t=1$, and that their values at other points in the positive half-plane are determined by considerations of continuity. It follows that when $t=-1$ the value of each function will be $i\pi$. At the points A', A in the plane of $\log\zeta$,

* This problem was first solved by Helmholtz, *l.c. ante* p. 75.

† The heavy lines correspond to rigid boundaries, and the fine continuous lines to free surfaces. Corresponding points in the various figures are indicated by the same letters.

we have, on the simplest convention, $\log\zeta = 0$ and $2i\pi$, respectively; whence, towards determining the constants in (1), we have

$$0 = B, \quad 2i\pi = i\pi A + B,$$

so that
$$\log\zeta = 2\cosh^{-1} t. \quad \text{.............................}(2)$$

Again, in the plane of w we take the line II' as the line $\psi = 0$; and if the final breadth of the issuing jet be $2b$, the bounding stream-lines will be $\psi = \pm b$. We may further suppose that $\phi = 0$ is the equipotential curve passing through A and A'. Hence, from (1),

$$ib = i\pi C + D, \quad -ib = D,$$

so that
$$w = \frac{2b}{\pi}\log t - ib. \quad \text{.............................}(3)$$

It is easy to eliminate t between (2) and (3), and thence to find the relation

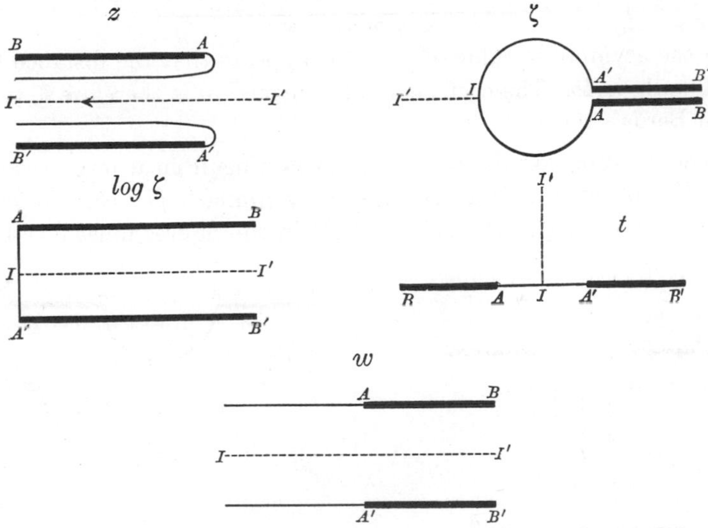

between z and w by integration, but the formulae are perhaps more convenient in their present shape.

The course of either free stream-line, say $A'I$, from its origin at A', is now easily traced. For points of this line t is real and ranges from 1 to 0; we have, moreover, from (2), $i\theta = 2\cosh^{-1} t$, or $t = \cos\frac{1}{2}\theta$. Hence, also, from (3),

$$\phi = \frac{2b}{\pi}\log\cos\tfrac{1}{2}\theta. \quad \text{.............................}(4)$$

Since, along this line, we have $d\phi/ds = -q = -1$, we may put $\phi = -s$, where the arc s is measured from A'. The intrinsic equation of the curve is therefore

$$s = \frac{2b}{\pi}\log\sec\tfrac{1}{2}\theta. \quad \text{.............................}(5)$$

From this we deduce in the ordinary way

$$x = \frac{2b}{\pi}(\sin^2 \tfrac{1}{2}\theta - \log \sec \tfrac{1}{2}\theta), \quad y = \frac{b}{\pi}(\theta - \sin\theta), \quad \text{............(6)}$$

if the origin be at A'. By giving θ a series of values ranging from 0 to π, the curve is easily plotted*.

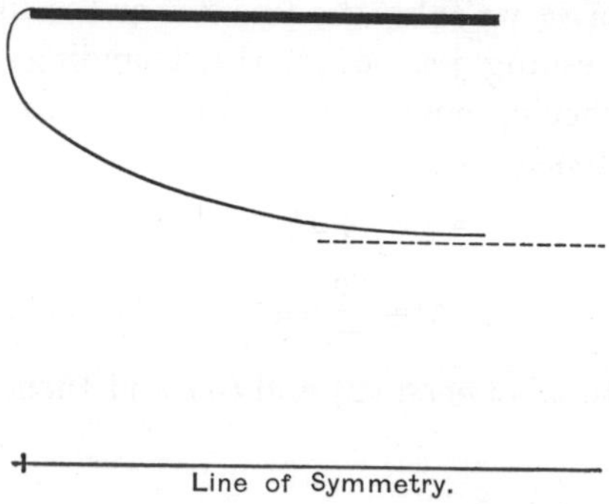

Since the asymptotic value of y is b, it appears that the distance between the fixed walls is $4b$. The coefficient of contraction is therefore $\frac{1}{2}$, in accordance with Borda's theory.

75. The solution for the case of fluid issuing from a large vessel by an aperture in a plane wall is analytically very similar. The chief difference is that the values of $\log \zeta$ at the points A, A' in the figures must now be taken

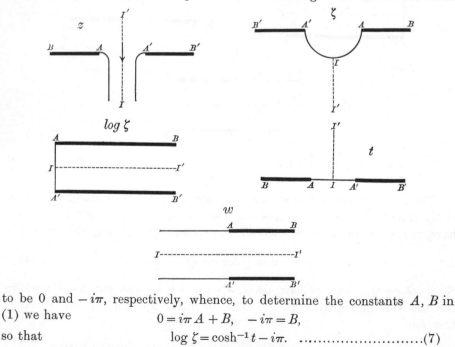

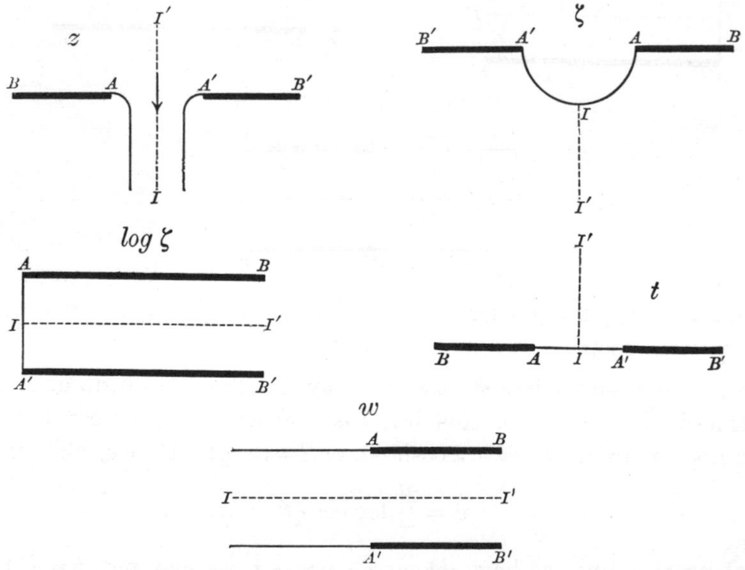

to be 0 and $-i\pi$, respectively, whence, to determine the constants A, B in (1) we have

$$0 = i\pi A + B, \quad -i\pi = B,$$

so that

$$\log \zeta = \cosh^{-1} t - i\pi. \quad \text{..........................(7)}$$

* To correspond exactly with p. 97 the figure should be turned through 180°.

The relation between w and t is exactly as before, viz.

$$w = \frac{2b}{\pi} \log t - ib, \quad\text{..............................(8)}$$

where $2b$ is the final breadth of the stream, between the free boundaries.

For the stream-line AI, t is real, and ranges from -1 to 0. Since, also, $i\theta = \cosh^{-1} t - i\pi$ we may put $t = \cos(\theta + \pi)$, where θ varies from 0 to $-\frac{1}{2}\pi$. Hence, from (8), with $\phi = -s$, we have, for the intrinsic equation of the stream-line,

$$s = \frac{2b}{\pi} \log(-\sec\theta). \quad\text{..........................(9)}$$

From this we find

$$x = \frac{4b}{\pi} \sin^2 \tfrac{1}{2}\theta, \quad y = \frac{2b}{\pi} \{\log \tan(\tfrac{1}{4}\pi + \tfrac{1}{2}\theta) - \sin\theta\}, \quad\text{.........(10)}$$

if the point A in the plane of z be taken as origin*. The curve is shewn (in an altered position) below.

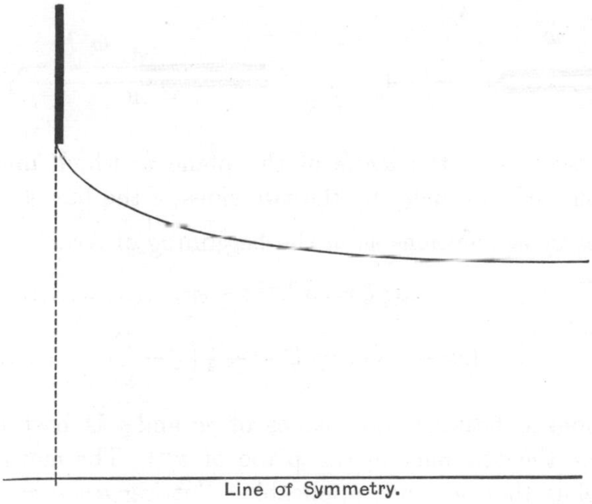

The asymptotic value of x, corresponding to $\theta = -\frac{1}{2}\pi$, is $2b/\pi$, the half width of the aperture is therefore $(\pi + 2)b/\pi$, and the coefficient of contraction is

$$\pi/(\pi + 2) = \cdot 611.$$

76. In the next example a stream of infinite breadth is supposed to impinge directly on a fixed plane lamina, and thence to divide into two portions bounded internally by free surfaces.

The middle stream-line, after meeting the lamina at right angles, branches off into two parts, which follow the lamina to the edges, and thence form the

* This example was given by Kirchhoff (*l.c.*), and discussed more fully by Rayleigh, "Notes on Hydrodynamics," *Phil. Mag.* Dec. 1876 [*Papers*, i. 297].

free boundaries. Let this be the line $\psi = 0$, and let us further suppose that at the point of divergence we have $\phi = 0$. The forms of the boundaries in the various planes are shewn in the figures. The region occupied by the moving

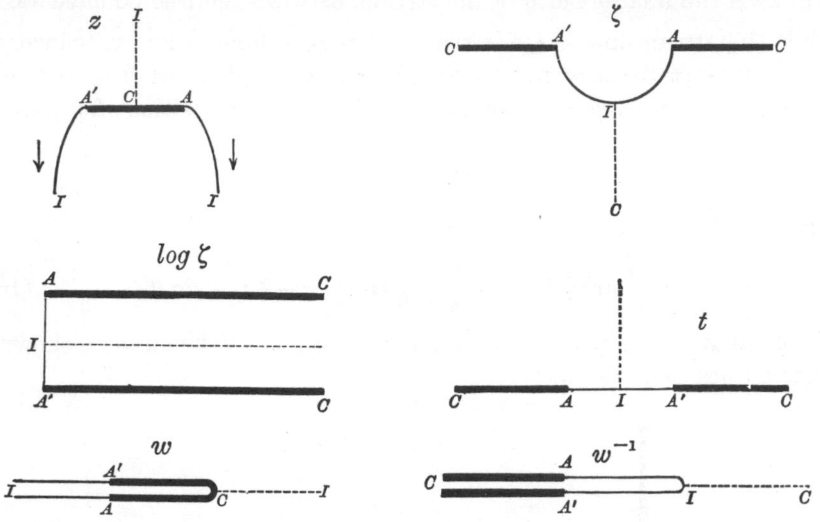

fluid now corresponds to the *whole* of the plane w, which must however be regarded as bounded internally by the two sides of the line $\psi = 0$, $\phi < 0$.

With the same conventions as in the beginning of Art. 75, we have

$$\log \zeta = \cosh^{-1} t - i\pi, \quad \text{.........................(1)}$$

or

$$t = -\cosh(\log \zeta) = -\tfrac{1}{2}\left(\zeta - \frac{1}{\zeta}\right). \quad \text{.................(2)}$$

The correspondence between the planes of w and t is best established by considering first the boundary in the plane of w^{-1}. The method of Schwarz and Christoffel is then at once applicable. Putting $\alpha = -\pi$, $\beta = \gamma = \ldots = 0$, in Art. 73 (4), we have

$$\frac{dw^{-1}}{dt} = At, \qquad w^{-1} = \tfrac{1}{2}At^2 + B. \quad \text{....................(3)}$$

At I we have $t = 0$, $w^{-1} = 0$, so that $B = 0$, or (say)

$$w = -\frac{C}{t^2}. \quad \text{...................................(4)}$$

To connect C (which is easily seen to be real) with the breadth (l) of the lamina, we notice that along CA we have $\zeta = q^{-1}$, and therefore, from (2),

$$t = -\tfrac{1}{2}\left(\frac{1}{q} + q\right), \qquad q = -t - \sqrt{(t^2 - 1)}, \quad \text{................(5)}$$

the sign of the radical being determined so as to make $q=0$ for $t=-\infty$. Also, $dx/d\phi=-1/q$. Hence, integrating along CA in the first figure we have

$$l=2\int_{-\infty}^{-1}\frac{dx}{d\phi}\frac{d\phi}{dt}\,dt=-4C\int_{-\infty}^{-1}\frac{dt}{qt^3}=-4C\int_{-\infty}^{-1}\{-t+\surd(t^2-1)\}\frac{dt}{t^3},\ \ldots(6)$$

whence

$$C=\frac{l}{\pi+4}.\ \ldots\ldots(7)$$

Along the free boundary AI, we have $\log\zeta=i\theta$, and therefore, from (2) and (4),

$$t=-\cos\theta,\qquad \phi=-C\sec^2\theta.\ \ldots\ldots(8)$$

The intrinsic equation of the curve is therefore

$$s=\frac{l}{\pi+4}\sec^2\theta,\ \ldots\ldots(9)$$

where θ ranges from 0 to $-\frac{1}{2}\pi$. This leads to

$$\left.\begin{aligned} x&=\frac{2l}{\pi+4}(\sec\theta+\tfrac{1}{4}\pi),\\ y&=\frac{l}{\pi+4}\{\sec\theta\tan\theta-\log\tan(\tfrac{1}{4}\pi+\tfrac{1}{2}\theta)\},\end{aligned}\right\}\ \ldots\ldots(10)$$

the origin being at the centre of the lamina.

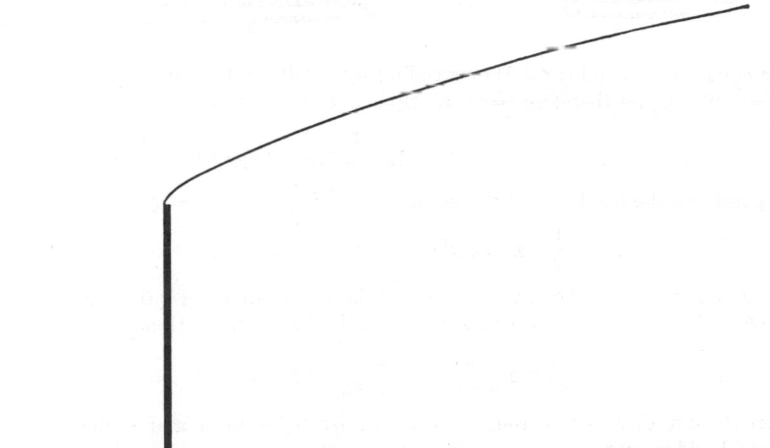

The excess of pressure on the anterior face of the lamina is, by Art. 23 (7), equal to $\frac{1}{2}\rho(1-q^2)$. Hence the resultant force on the lamina is

$$\rho\int_{-\infty}^{-1}(1-q^2)\frac{dx}{dt}\,dt=-2\rho C\int_{-\infty}^{-1}\left(\frac{1}{q}-q\right)\frac{dt}{t^3}=-4\rho C\int_{-\infty}^{-1}\surd(t^2-1)\frac{dt}{t^3}=\pi\rho C.$$
$$\ldots\ldots(11)$$

It is evident from Art. 23 (7), and from the obvious geometrical similarity of the motion in all cases, that the resultant pressure (P_0, say) will vary as the square of the general velocity of the stream. We thus find, for an arbitrary velocity q_0*,

$$P_0 = \frac{\pi}{\pi+4}\rho q_0^2 \,.\, l = \cdot 440 \rho q_0^2 \,.\, l. \quad \text{.....................(12)}$$

77. If the stream be oblique to the lamina, making an angle α, say, with its plane, the problem is modified in the manner shewn in the figures.

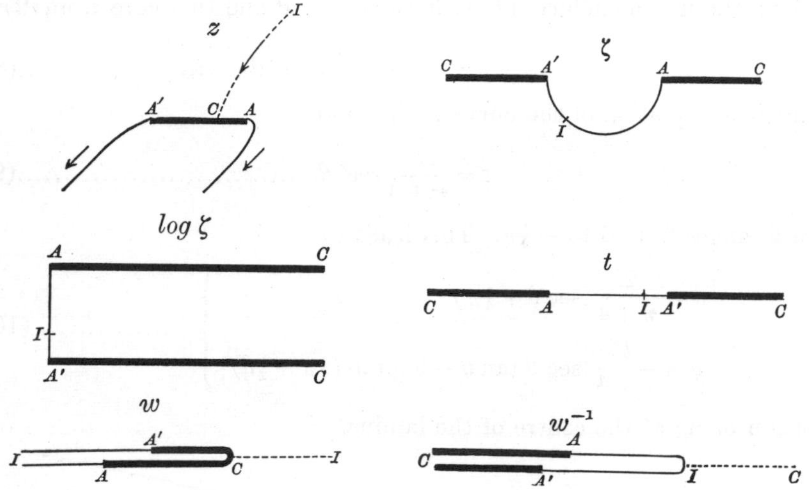

The equations (1) and (2) of the preceding Art. still apply; but at the point I we now have $\zeta = e^{-i(\pi-\alpha)}$, and therefore $t = \cos\alpha$. Hence, in place of (4)†,

$$w = -\frac{C}{(t-\cos\alpha)^2}. \quad \text{.....................................(13)}$$

At points on the front face of the lamina, we have, since $q^{-1} = |\zeta|$,

$$\frac{1}{q} = \pm t + \surd(t^2-1), \quad q = \pm t - \surd(t^2-1), \quad \text{.........................(14)}$$

where the upper or the lower signs are to be taken according as $t \gtrless 0$, *i.e.* according as the point referred to lies to the left or right of C in the first figure. Hence

$$\frac{dx}{dt} = \pm\frac{1}{q}\frac{d\phi}{dt} = \frac{2C}{(t-\cos\alpha)^3}\{t \pm \surd(t^2-1)\}. \quad \text{.........................(15)}$$

Between A' and C, t varies from 1 to ∞, whilst between A and C the range is from $-\infty$ to -1. If we put

$$t = \frac{1-\cos\alpha\cos\omega}{\cos\alpha - \cos\omega},$$

the corresponding ranges of ω will be from π to α, and from α to 0, respectively; and we find

$$\frac{dt}{(t-\cos\alpha)^3} = -\frac{\cos\alpha - \cos\omega}{\sin^4\alpha}\sin\omega\, d\omega, \quad \pm\surd(t^2-1) = \frac{\sin\alpha\sin\omega}{\cos\alpha-\cos\omega}.$$

* Kirchhoff, *l.c. ante* p. 94; Rayleigh, "On the Resistance of Fluids," *Phil. Mag.* Dec. 1876 [*Papers*, i. 287].

† The solution up to this point was given by Kirchhoff (*Crelle, l.c.*); the subsequent discussion is taken, with merely analytical modifications, from the paper by Rayleigh.

Hence
$$\frac{dx}{d\omega} = -\frac{2C}{\sin^4\alpha}(1-\cos\alpha\cos\omega+\sin\alpha\sin\omega)\sin\omega, \quad \text{.................(16)}$$
and therefore
$$x = \frac{C}{\sin^4\alpha}\{2\cos\omega+\cos\alpha\sin^2\omega+\sin\alpha\sin\omega\cos\omega+(\tfrac{1}{2}\pi-\omega)\sin\alpha\}, \quad \text{......(17)}$$
where the origin has been adjusted so that x shall have equal and opposite values when $\omega=0$ and $\omega=\pi$, respectively; *i.e.* it has been taken at the centre of the lamina. Hence, in terms of C, the whole breadth is
$$l = \frac{4+\pi\sin\alpha}{\sin^4\alpha}\,.\,C. \quad \text{.................................(18)}$$

The distance, from the centre, of the point ($\omega=\alpha$) at which the stream divides is
$$x = \frac{2\cos\alpha\,(1+\sin^2\alpha)+(\tfrac{1}{2}\pi-\alpha)\sin\alpha}{4+\pi\sin\alpha}\,.\,l. \quad \text{.......................(19)}$$

To find the total pressure on the front face, we have
$$\tfrac{1}{2}\rho\,(1-q^2)\,dx = \pm\tfrac{1}{2}\rho\left(\frac{1}{q}-q\right)\frac{d\phi}{dt}\,dt = \pm 2\rho C\sqrt{(t^2-1)}\,\frac{dt}{(t-\cos\alpha)^3}$$
$$= -\frac{2\rho C}{\sin^3\alpha}\,.\,\sin^2\omega\,d\omega. \quad \text{...(20)}$$

Integrated between the limits π and 0, this gives $\pi\rho C/\sin^3\alpha$. Hence, in terms of l, and of an arbitrary velocity q_0 of the stream, we find
$$P_0 - \frac{\pi\sin\alpha}{4+\pi\sin\alpha}\,.\,\rho q_0^{\,2}\,.\,l. \quad \text{.................................(21)}$$

To find the centre of pressure, we take moments about the centre of the lamina. Thus
$$\tfrac{1}{2}\rho\int(1-q^2)\,x\,dx = -\frac{2\rho C}{\sin^3\alpha}\,.\int_\pi^0 x\sin^2\omega\,d\omega$$
$$= \frac{\pi\rho C}{\sin^3\alpha}\times\tfrac{3}{4}\,\frac{C\cos\alpha}{\sin^4\alpha}, \quad \text{..............................(22)}$$
on substituting the value of x from (17). The first factor represents the total pressure; the abscissa $\bar{x}$ of the centre of pressure is therefore given by the second, or, in terms of the breadth,
$$\bar{x} = \tfrac{3}{4}\,\frac{\cos\alpha}{4+\pi\sin\alpha}\,.\,l. \quad \text{...(23)}$$

In the following table, derived from Rayleigh's paper, the column I gives the excess of pressure on the anterior face, in terms of its value when $\alpha=90°$; whilst columns II and III give respectively the distances of the centre of pressure, and of the point where the stream divides, from the centre of the lamina, expressed as fractions of the total breadth*.

α	I	II	III
90°	1·000	·000	·000
70°	·965	·037	·232
50°	·854	·075	·402
30°	·641	·117	·483
20°	·481	·139	·496
10°	·273	·163	·500

* For a comparison with experimental results see Rayleigh, *l.c.* and *Nature*, xlv. (1891) [*Papers*, iii. 491].

78. An interesting variation of the problem of Art. 76 has been discussed by Bobyleff*. A stream is supposed to impinge symmetrically on a bent lamina whose section consists of two equal straight lines forming an angle.

If 2α be the angle, measured on the down-stream side, the boundaries in the plane of ζ can be transformed, so as to have the same shape as in the Art. cited, by the assumption

$$\zeta = A\zeta'^n,$$

provided A and n be determined so as to make $\zeta'=1$ when $\zeta=e^{-i(\frac{1}{2}\pi-\alpha)}$, and $\zeta'=e^{-i\pi}$ when $\zeta=e^{-i(\frac{1}{2}\pi+\alpha)}$. This gives

$$A=e^{-i(\frac{1}{2}\pi-\alpha)}, \qquad n=2\alpha/\pi.$$

On the right-hand half of the lamina, t will be negative as before, and since $q^{-1}=|\zeta|$,

$$\frac{1}{q}=\{-t+\surd(t^2-1)\}^n, \quad q=\{-t-\surd(t^2-1)\}^n. \quad \text{.....................(24)}$$

Hence

$$\int_{-\infty}^{-1}\frac{1}{q}\frac{d\phi}{dt}dt=2C\int_{-\infty}^{-1}\{-t+\surd(t^2-1)\}^n\frac{dt}{t^3}=-C-nC\int_{-\infty}^{-1}\{-t+\surd(t^2-1)\}^n\frac{dt}{t^2\surd(t^2-1)},$$

$$\int_{-\infty}^{-1}q\frac{d\phi}{dt}dt=2C\int_{-\infty}^{-1}\{-t-\surd(t^2-1)\}^n\frac{dt}{t^3}=-C+nC\int_{-\infty}^{-1}\{-t-\surd(t^2-1)\}^n\frac{dt}{t^2\surd(t^2-1)}.$$

These can be reduced to known forms by the substitution

$$t=-\tfrac{1}{2}\left(\frac{1}{\surd\omega}+\surd\omega\right),$$

where ω ranges from 0 to 1. We thus find

$$\frac{1}{C}\int_{-\infty}^{-1}\frac{1}{q}\frac{d\phi}{dt}dt=-1-2n\int_0^1\frac{\omega^{-\frac{1}{2}n}}{(1+\omega)^2}d\omega=-1-n-n^2\int_0^1\frac{\omega^{-\frac{1}{2}n}}{1+\omega}d\omega, \quad \text{......(25)}$$

$$\frac{1}{C}\int_{-\infty}^{-1}q\frac{d\phi}{dt}dt=-1+2n\int_0^1\frac{\omega^{\frac{1}{2}n}}{(1+\omega)^2}d\omega=-1-n+n^2\int_0^1\frac{\omega^{\frac{1}{2}n-1}}{1+\omega}d\omega. \quad \text{......(26)}$$

We have here used the formulae

$$\int_0^1\frac{\omega^{-k}}{(1+\omega)^2}d\omega=\tfrac{1}{2}+k\int_0^1\frac{\omega^{-k}}{1+\omega}d\omega,$$

$$\int_0^1\frac{\omega^{k}}{(1+\omega)^2}d\omega=-\tfrac{1}{2}+k\int_0^1\frac{\omega^{k-1}}{1+\omega}d\omega,$$

where $1>k>0$.

Since, along the stream-line, $ds/d\phi=-1/q$, we have from (25), if b denote the half-breadth of the lamina,

$$b=C\left\{1+\frac{2\alpha}{\pi}+\frac{4\alpha^2}{\pi^2}\int_0^1\frac{\omega^{-\alpha/\pi}}{1+\omega}d\omega\right\}. \quad \text{..........................(27)}$$

The definite integral which occurs in this expression can be calculated from the formula

$$\int_0^1\frac{\omega^{-k}}{1+\omega}d\omega=\frac{1}{(1-k)(2-k)}+\tfrac{1}{2}\Psi(1-\tfrac{1}{2}k)-\tfrac{1}{2}\Psi(\tfrac{1}{2}-\tfrac{1}{2}k), \quad \text{...............(28)}$$

where $\Psi(m)$, $=d/dm\,.\log\Pi(m)$, is the function introduced and tabulated by Gauss†.

The normal pressure on either half is, by the method of Art. 76,

$$=-\tfrac{1}{2}\rho\int_{-\infty}^{-1}\left(\frac{1}{q}-q\right)\frac{d\phi}{dt}dt=\tfrac{1}{2}n^2C\rho\int_0^\infty\frac{\omega^{-\frac{1}{2}n}}{1+\omega}d\omega=\tfrac{1}{2}n^2C\rho\,.\frac{\pi}{\sin\frac{1}{2}n\pi}=\rho C\,.\frac{2\alpha^2}{\pi\sin\alpha}. \quad \text{...(29)}$$

* *Journal of the Russian Physico-Chemical Society*, xiii. (1881) [Wiedemann's *Beiblätter*, vi. 163]. The problem appears, however, to have been previously discussed in a similar manner by M. Réthy, *Klausenburger Berichte*, 1879. It is generalized by Bryan and Jones, *Proc. Roy. Soc.* A, xci. 354 (1915).

† "Disquisitiones generales circa seriem infinitam...," *Werke*, Göttingen, 1870..., iii. 161.

The resultant pressure in the direction of the stream is therefore

$$=\frac{4\alpha^2}{\pi}\rho C. \qquad \text{.......................................(30)}$$

Hence, for any arbitrary velocity q_0 of the stream, the resultant pressure is

$$P=\frac{4\alpha^2}{\pi L}\cdot\rho q_0{}^2 b, \qquad \text{.......................................(31)}$$

where L stands for the numerical factor in (27).

For $\alpha=\frac{1}{2}\pi$, we have $L=2+\frac{1}{2}\pi$, leading to the same result as in Art. 76 (12).

In the following table, taken (with a slight modification) from Bobyleff's paper, the second column gives the ratio P/P_0 of the resultant pressure to that experienced by a plane strip of the same area. This ratio is a maximum when $\alpha=100°$, about, the lamina being then concave on the up-stream side. In the third column the ratio of P to the distance ($2b\sin\alpha$) between the edges of the lamina is compared with $\frac{1}{2}\rho q_0{}^2$. For values of α nearly equal to 180°, this ratio tends to the value unity, as we should expect, since the fluid within the acute angle is then nearly at rest, and the pressure-excess therefore practically equal to $\frac{1}{2}\rho q_0{}^2$. The last column gives the ratio of the resultant pressure to that experienced by a plane strip of breadth $2b\sin\alpha$, as calculated from (12).

α	P/P_0	$P/\rho q_0{}^2 b\sin\alpha$	$P/P_0\sin\alpha$
10°	·039	·199	·227
20°	·140	·359	·409
30°	·278	·489	·555
40°	·433	·593	·674
45°	·512	·637	·724
50°	·589	·677	·769
60°	·733	·745	·846
70°	·854	·800	·909
80°	·945	·844	·959
90°	1·000	·879	1·000
100°	1·016	·907	1·031
110°	·995	·931	1·059
120°	·935	·950	1·079
130°	·840	·964	1·096
135°	·780	·970	1·103
140°	·713	·975	1·109
150°	·559	·984	1·119
160°	·385	·990	1·126
170°	·197	·996	1·132

Discontinuous Motions.

79. It must suffice to have given a few of the more important examples of steady motion with a free surface, treated by what is perhaps the most systematic method. Considerable additions to the subject have been made by Michell*, Love†, and other writers‡. It remains to say something of the physical

* "On the Theory of Free Stream-lines," *Phil. Trans.* A, clxxxi. (1890).

† "On the Theory of Discontinuous Fluid Motions in Two Dimensions," *Proc. Camb. Phil. Soc.* vii. (1891).

‡ For references see Love, *Encycl. d. math. Wiss.* iv. (3), 97.... A very complete account of the more important known solutions, with fresh additions and developments, is given by Greenhill,

considerations which led in the first instance to the investigation of such problems.

We have, in the preceding pages, had several instances of the flow of a liquid round a sharp projecting edge, and it appeared in each case that the velocity there was infinite. This is indeed a necessary consequence of the assumed irrotational character of the motion, whether the fluid be incompressible or not, as may be seen by considering the configuration of the equipotential surfaces (which meet the boundary at right angles) in the immediate neighbourhood.

The occurrence of infinite values of the velocity may be afforded by supposing the edge to be slightly rounded, but even then the velocity near the edge will much exceed that which obtains at a distance great in comparison with the radius of curvature.

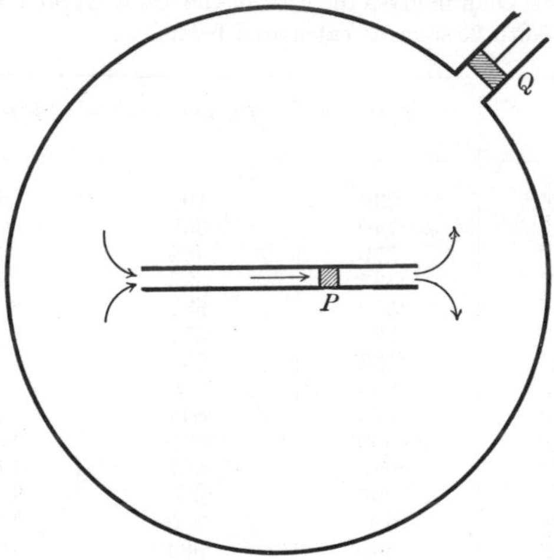

In order that the motion of a fluid may conform to such conditions, it is necessary that the pressure at a distance should greatly exceed that at the edge. This excess of pressure is demanded by the *inertia* of the fluid, which cannot be guided round a sharp curve, in opposition to centrifugal force, except by a distribution of pressure increasing with a very rapid gradient outwards.

Report on the Theory of a Stream-line past a Plane Barrier, published by the Advisory Committee for Aeronautics, 1910.

The extension to the case of *curved* rigid boundaries is discussed in a general manner in various papers by Levi-Civita and Cisotti. For these, reference may be made to the *Rend. d. Circolo Mat. di Palermo*, xxiii. xxv. xxvi. xxviii. and the *Rend. d. r. Accad. d. Lincei*, xx. xxi.; the working out of particular cases naturally presents great difficulties. The matter was treated later by Leathem, *Phil. Trans.* A, ccxx. 439 (1915) and H. Levy, *Proc. Roy. Soc.* A, xcii. 107 (1915). The theory of mutually impinging jets is treated very fully by Cisotti, "Vene confluenti," *Ann. di mat.* (3) xxiii. 285 (1914).

Hence, unless the pressure at a distance be very great, the maintenance of the motion in question would require a negative pressure at the corner, such as fluids under ordinary conditions are unable to sustain.

To put the matter in as definite a form as possible, let us imagine the following case. Let us suppose that a straight tube, whose length is large compared with the diameter, is fixed in the middle of a large closed vessel filled with frictionless liquid, and that this tube contains, at a distance from the ends, a sliding plug, or piston, P, which can be moved in any required manner by extraneous forces applied to it. The thickness of the walls of the tube is supposed to be small in comparison with the diameter; and the edges, at the two ends, to be rounded off, so that there are no sharp angles. Let us further suppose that at some point of the walls of the vessel there is a lateral tube, with a piston P, by means of which the pressure in the interior can be adjusted at will.

Everything being at rest to begin with, let a slowly increasing velocity be communicated to the plug P, so that (for simplicity) the motion at any instant may be regarded as approximately steady. At first, provided a sufficient force be applied to Q, a continuous motion of the kind indicated in the diagram on p. 74 will be produced in the fluid, there being in fact only one type of motion consistent with the conditions of the question. As the acceleration of the piston P proceeds, the pressure on Q may become enormous, even with very moderate velocities of P, and if Q be allowed to yield, an annular cavity will be formed at each end of the tube.

It is not easy to make out the further course of the motion in such a case from a theoretical standpoint, even in the case of a 'perfect' fluid. In actual liquids the problem is modified by viscosity, which prevents any slipping of the fluid immediately in contact with the tube, and must further exercise a considerable influence on such rapid different motions of the fluid as are here in question.

As a matter of observation, the motions of fluids are often found to differ widely, under the circumstances supposed in each case, from the types represented on such diagrams as those of pp. 73, 74, 84, 86. In such a case as we have just described, the fluid issuing from the mouth of the tube does not immediately spread out in all directions, but forms, at all events for some distance, a more or less compact stream, bounded on all sides by fluid nearly at rest. A familiar instance is the smoke-laden stream of gas issuing from a chimney. In all such cases, however, the motion in the immediate neighbourhood of the boundary of the stream is found to be wildly irregular*.

It was the endeavour to construct types of steady motion of a frictionless

* Certain experiments would indicate that jets may be formed *before* the 'limiting velocity' of Helmholtz is reached, and that viscosity plays an essential part in the process. Smoluchowski, "Sur la formation des veines d'efflux dans les liquides," *Bull. de l'Acad. de Cracovie*, 1904.

liquid, in two dimensions, which should resemble more closely what is observed in such cases as we have referred to, that led Helmholtz* and Kirchhoff* to investigate the theory of free stream-lines. It is obvious that we may imagine the space beyond a free boundary to be occupied, if we choose, by liquid of the same density at rest, since the condition of constant pressure along the stream-line is not thereby affected. In this way the problems of Arts. 76, 77, for example, give us a theory of the pressure exerted on a fixed lamina by a stream flowing past it, or (what comes to the same thing) the resistance experienced by a lamina when made to move with constant velocity through a liquid which would otherwise be at rest.

The question as to the practical validity of this theory will be referred to later in connection with some related problems (Chapter XI.).

Flow in a Curved Stratum.

80. The theory developed in Arts. 59, 60 may be readily extended to the two-dimensional motion of a *curved* stratum of liquid, whose thickness is small compared with the radii of curvature. This question has been discussed from the point of view of electric conduction, by Boltzmann†, Kirchhoff‡, Töpler§, and others.

As in Art. 59, we take a fixed point A, and a variable point P, on the surface defining the form of the stratum, and denote by ψ the flux across any curve AP drawn on this surface. Then ψ is a function of the position of P, and by displacing P in any direction through a small distance δs, we find that the flux across the element δs is given by $\partial\psi/\partial s \,.\, \delta s$. The *velocity* perpendicular to this element will be $\delta\psi/h\,\delta s$, where h is the thickness of the stratum, not assumed as yet to be uniform.

If, further, the motion be irrotational, we shall have in addition a velocity-potential ϕ, and the equipotential curves $\phi=$const. will cut the stream-lines $\psi=$const. at right angles.

In the case of *uniform* thickness, to which we now proceed, it is convenient to write ψ for ψ/h, so that the velocity perpendicular to an element δs is now given indifferently by $\partial\psi/\partial s$ and $\partial\phi/\partial n$, δn being an element drawn at right angles to δs in the proper direction. The further relations are then exactly as in the plane problem; in particular the curves $\phi=$const., $\psi=$const., drawn for a series of values in arithmetic progression, the common difference being infinitely small and the same in each case, will divide the surface into elementary squares. For, by the orthogonal property, the elementary spaces in question are rectangles, and if δs_1, δs_2 be elements of a stream-line and an equipotential line, respectively, forming the sides of one of these rectangles, we have $\partial\psi/\partial s_2=\partial\phi/\partial s_1$, whence $\delta s_1=\delta s_2$, since by construction $\delta\psi=\delta\phi$.

Any problem of irrotational motion in a curved stratum (of uniform thickness) is therefore reduced by orthomorphic projection to the corresponding problem *in plano*. Thus for a spherical surface we may use, among an infinity of other methods, that of stereographic projection. As a simple example of this, we may take the case of a stratum

* *ll. c. ante* pp. 75, 94.

† *Wiener Sitzungsberichte*, lii. 214 (1865) [*Wissenschaftliche Abhandlungen*, Leipzig, 1909, i. 1].

‡ *Berl. Monatsber.* July 19, 1875 [*Ges. Abh.* i. 56].

§ *Pogg. Ann.* clx. 375 (1877).

of uniform depth covering the surface of a sphere with the exception of two circular islands (which may be of any size and in any relative position). It is evident that the only (two-dimensional) irrotational motion which can take place in the doubly-connected space occupied by the fluid is one in which the fluid circulates in opposite directions round the two islands, the cyclic constants being equal in magnitude. Since circles project into circles, the plane problem is that solved in Art. 64, 2°, viz. the stream-lines are a system of coaxal circles with real 'limiting points' (A, B, say), and the equipotential lines are the orthogonal system passing through A, B. Returning to the sphere, it follows from well-known theorems of stereographic projection that the stream-lines (including the contours of the two islands) are the circles in which the surface is cut by a system of planes passing through a fixed line, viz. the intersection of the tangent planes at the points corresponding to A and B, whilst the equipotential lines are the circles in which the sphere is cut by planes passing through these points*.

In any case of transformation by orthomorphic projection, whether the motion be irrotational or not, the velocity ($\partial\psi/\partial n$) is transformed in the inverse ratio of a linear element, and therefore the kinetic energies of the portions of the fluid occupying corresponding areas are equal (provided, of course, the density and the thickness be the same). In the same way the circulation ($\int \partial\psi/\partial n \,.\, ds$) in any circuit is unaltered by projection.

* This example was given by Kirchhoff, in the electrical interpretation, the problem considered being the distribution of current in a uniform spherical conducting sheet, the electrodes being situate at any two points A, B of the surface.

CHAPTER V

IRROTATIONAL MOTION OF A LIQUID: PROBLEMS IN THREE DIMENSIONS

81. Of the methods available for obtaining solutions of the equation

$$\nabla^2\phi = 0 \quad \text{.................................(1)}$$

in three dimensions, the most important is that of Spherical Harmonics. This is especially suitable when the boundary conditions have relation to spherical or nearly spherical surfaces.

For a full account of this method we must refer to the special treatises*, but as the subject is very extensive, and has been treated from different points of view, it may be worth while to give a slight sketch, without formal proofs, or with mere indications of proofs, of such parts of it as are most important for our present purpose.

It is easily seen that since the operator ∇^2 is homogeneous with respect to x, y, z, the part of ϕ which is of any specified algebraic degree must satisfy (1) separately. Any such homogeneous solution of (1) is called a 'spherical solid harmonic' of the algebraic degree in question. If ϕ_n be a spherical solid harmonic of degree n, then if we write

$$\phi_n = r^n S_n, \quad \text{.................................(2)}$$

S_n will be a function of the direction (only) in which the point (x, y, z) lies with respect to the origin; in other words, a function of the position of the point in which the radius vector meets a unit sphere described with the origin as centre. It is therefore called a 'spherical surface harmonic' of order n†.

To any solid harmonic ϕ_n of degree n corresponds another of degree $-n-1$, obtained by division by r^{2n+1}; *i.e.* $\phi = r^{-2n-1}\phi_n$ is also a solution of (1). Thus, corresponding to any spherical surface-harmonic S_n, we have the two spherical solid harmonics $r^n S_n$ and $r^{-n-1}S_n$.

82. The most important case is when n is integral, and when the surface-harmonic S_n is further restricted to be finite over the unit sphere. In the

* Todhunter, *Functions of Laplace, Lamé, and Bessel*, Cambridge, 1875. Ferrers, *Spherical Harmonics*, Cambridge, 1877. Heine, *Handbuch der Kugelfunctionen*, 2nd ed., Berlin, 1878. Thomson and Tait, *Natural Philosophy*, 2nd ed., Cambridge, 1879, i. 171–218. Byerly, *Fourier's Series and Spherical, Cylindrical, and Ellipsoidal Harmonics*, Boston, U.S.A. 1893. Whittaker and Watson, *Modern Analysis*, 3rd ed., Cambridge, 1920.

For the history of the subject see Todhunter, *History of the Theories of Attraction, &c.*, Cambridge, 1873, ii. Also Wangerin, "Theorie d. Kugelfunktionen, u.s.w.," *Encycl. d. math. Wiss.* ii. (1) (1904).

† The symmetrical treatment of spherical solid harmonics in terms of Cartesian co-ordinates was introduced by Clebsch, in a much neglected paper, *Crelle*, lxi. 195 (1863). It was adopted independently by Thomson and Tait as the basis of their exposition.

form in which the theory (for this case) is presented by Thomson and Tait, and by Maxwell *, the primary solution of (1) is

$$\phi_{-1} = A/r. \quad \text{.................................(3)}$$

This represents as we have seen (Art. 56) the velocity potential due to a point-source at the origin. Since (1) is still satisfied when ϕ is differentiated with respect to x, y, or z, we derive a solution

$$\phi_{-2} = A\left(l\frac{\partial}{\partial x} + m\frac{\partial}{\partial y} + n\frac{\partial}{\partial z}\right)\frac{1}{r}. \quad \text{.....................(4)}$$

This is the velocity-potential of a double source at the origin, having its axis in the direction (l, m, n); see Art. 56 (3). The process can be continued, and the general type of spherical solid harmonic obtainable in this way is

$$\phi_{-n-1} = A\frac{\partial^n}{\partial h_1 \partial h_2 \ldots \partial h_n}\frac{1}{r}, \quad \text{........................(5)}$$

where
$$\frac{\partial}{\partial h_s} = l_s\frac{\partial}{\partial x} + m_s\frac{\partial}{\partial y} + n_s\frac{\partial}{\partial z},$$

l_s, m_s, n_s being arbitrary direction-cosines.

This may be regarded as the velocity-potential of a certain configuration of simple sources about the origin, the dimensions of this system being small compared with r. To construct this system we premise that from any given system of sources we may derive a system of higher order by first displacing it through a space $\frac{1}{2}h_s$ in the direction (l_s, m_s, n_s), and then superposing the *reversed* system, supposed displaced from its original position through a space $\frac{1}{2}h_s$ in the opposite direction. Thus, beginning with the case of a simple source O at the origin, a first application of the above process gives us two sources O_+, O_- equidistant from the origin, in opposite directions. The same process applied to the system O_+, O_- gives us four sources O_{++}, O_{-+}, O_{+-}, O_{--} at the corners of a parallelogram. The next step gives us eight sources at the corners of a parallelepiped, and so on. The velocity-potential, at a great distance, due to an arrangement of 2^n sources obtained in this way, will be given by (5), if $4\pi A = m'h_1h_2 \ldots h_n$, m' being the strength of the original source at O. The formula becomes exact, for all distances r, when h_1, h_2, ... h_n are diminished, and m' increased, indefinitely, but so that A is finite.

The surface-harmonic corresponding to (5) is given by

$$S_n = Ar^{n+1}\frac{\partial}{\partial h_1 \partial h_2 \ldots \partial h_n}\frac{1}{r}, \quad \text{........................(6)}$$

and the complementary solid harmonic by

$$\phi_n = r^n S_n = r^{2n+1}\phi_{-n-1}. \quad \text{........................(7)}$$

* *Electricity and Magnetism*, c. ix.

By the method of 'inversion*,' applied to the above configuration of sources, it may be shewn that the solid harmonic (7) of positive degree n may be regarded as the velocity-potential due to a certain arrangement of 2^n simple sources at infinity.

The lines drawn from the origin in the various directions (l_s, m_s, n_s) are called the 'axes' of the solid harmonic (5) or (7), and the points in which these lines meet the unit sphere are called the 'poles' of the surface-harmonic S_n. The formula (5) involves $2n+1$ arbitrary constants, viz. the angular co-ordinates (two for each) of the n poles, and the factor A. It can be shewn that this expression is equivalent to the most general form of spherical surface-harmonic which is of integral order n and finite over the unit sphere†.

83. In the original investigation of Laplace‡, the equation $\nabla^2\phi = 0$ is first expressed in terms of spherical polar co-ordinates, r, θ, ω, where

$$x = r\cos\theta, \qquad y = r\sin\theta\cos\omega, \qquad z = r\sin\theta\sin\omega.$$

The simplest way of effecting the transformation is to apply the theorem of Art. 36 (2) to the surface of a volume-element $r\delta\theta\,.\,r\sin\theta\delta\omega\,.\,\delta r$. Thus the difference of flux across the two faces perpendicular to r is

$$\frac{\partial}{\partial r}\left(\frac{\partial\phi}{\partial r}\,.\,r\delta\theta\,.\,r\sin\theta\delta\omega\right)\delta r.$$

Similarly for the two faces perpendicular to the meridian ($\omega = \text{const.}$) we find

$$\frac{\partial}{\partial\theta}\left(\frac{\partial\phi}{r\partial\theta}\,.\,r\sin\theta\delta\omega\,.\,\delta r\right)\delta\theta,$$

and for the two faces perpendicular to a parallel of latitude ($\theta = \text{const.}$)

$$\frac{\partial}{\partial\omega}\left(\frac{\partial\phi}{r\sin\theta\partial\omega}\,.\,r\delta\theta\,.\,\delta r\right)\delta\omega.$$

Hence, by addition,

$$\sin\theta\frac{\partial}{\partial r}\left(r^2\frac{\partial\phi}{\partial r}\right) + \frac{\partial}{\partial\theta}\left(\sin\theta\frac{\partial\phi}{\partial\theta}\right) + \frac{1}{\sin\theta}\frac{\partial^2\phi}{\partial\omega^2} = 0. \qquad \text{............(1)}$$

This might of course have been derived from Art. 81 (1) by the usual method of change of independent variables.

If we now assume that ϕ is homogeneous, of degree n, and put

$$\phi = r^n S_n,$$

we obtain
$$\frac{1}{\sin\theta}\frac{\partial}{\partial\theta}\left(\sin\theta\frac{\partial S_n}{\partial\theta}\right) + \frac{1}{\sin^2\theta}\frac{\partial^2 S_n}{\partial\omega^2} + n(n+1)S_n = 0, \qquad \text{........(2)}$$
which is the general differential equation of spherical surface-harmonics.

* Explained by Thomson and Tait, *Natural Philosophy*, Art. 515.

† Sylvester, *Phil. Mag.* (5), ii. 291 (1876) [*Mathematical Papers*, Cambridge, 1904..., iii. 37].

‡ "Théorie de l'attraction des sphéroides et de la figure des planètes," *Mém. de l'Acad. roy. des Sciences*, 1782 [*Oeuvres Complètes*, Paris, 1878..., x. 341]; *Mécanique Céleste*, Livre 2me, c. ii.

Since the product $n(n+1)$ is unchanged in value when we write $-n-1$ for n, it appears that

$$\phi = r^{-n-1} S_n$$

will also be a solution of (1), as already stated (Art. 81).

84. In the case of symmetry about the axis of x, the term $\partial^2 S_n/\partial\omega^2$ disappears, and putting $\cos\theta = \mu$ we get

$$\frac{d}{d\mu}\left\{(1-\mu^2)\frac{dS_n}{d\mu}\right\} + n(n+1)S_n = 0, \quad \text{.................(1)}$$

the differential equation of spherical 'zonal' harmonics*. This equation, containing only terms of two different dimensions in μ, is adapted for integration by series. We thus obtain

$$S_n = A\left\{1 - \frac{n(n+1)}{1.2}\mu^2 + \frac{(n-2)n(n+1)(n+3)}{1.2.3.4}\mu^4 - \ldots\right\}$$
$$+ B\left\{\mu - \frac{(n-1)(n+2)}{1.2.3}\mu^3 + \frac{(n-3)(n-1)(n+2)(n+4)}{1.2.3.4.5}\mu^5 - \ldots\right\}. \quad \text{......(2)}$$

The series which here present themselves are of the kind called 'hypergeometric'; viz. if we write, after Gauss†,

$$F(\alpha,\beta,\gamma,x) = 1 + \frac{\alpha.\beta}{1.\gamma}x + \frac{\alpha.\alpha+1.\beta.\beta+1}{1.2.\gamma.\gamma+1}x^2$$
$$+ \frac{\alpha.\alpha+1.\alpha+2.\beta.\beta+1.\beta+2}{1.2.3.\gamma.\gamma+1.\gamma+2}x^3 + \ldots, \quad \text{.........(3)}$$

we have

$$S_n = AF(-\tfrac{1}{2}n, \tfrac{1}{2}+\tfrac{1}{2}n, \tfrac{1}{2}, \mu^2) + B\mu F(\tfrac{1}{2}-\tfrac{1}{2}n, 1+\tfrac{1}{2}n, \tfrac{3}{2}, \mu^2). \quad \text{......(4)}$$

The series (3) is of course essentially convergent when x lies between 0 and 1; but when $x=1$ it is convergent if, and only if,

$$\gamma - \alpha - \beta > 0.$$

In this case we have
$$F(\alpha,\beta,\gamma,1) = \frac{\Pi(\gamma-1).\Pi(\gamma-\alpha-\beta-1)}{\Pi(\gamma-\alpha-1).\Pi(\gamma-\beta-1)}, \quad \text{.......................(5)}$$

where $\Pi(m)$ is in Gauss's notation the equivalent of Euler's $\Gamma(m+1)$.

The degree of divergence of the series (3) when

$$\gamma - \alpha - \beta < 0,$$

as x approaches the value 1, is given by the theorem‡

$$F(\alpha,\beta,\gamma,x) = (1-x)^{\gamma-\alpha-\beta} F(\gamma-\alpha, \gamma-\beta, \gamma, x). \quad \text{.....................(6)}$$

Since the latter series will now be convergent when $x=1$, we see that $F(\alpha,\beta,\gamma,x)$ becomes divergent as $(1-x)^{\gamma-\alpha-\beta}$; more precisely, for values of x infinitely nearly equal to unity, we have

$$F(\alpha,\beta,\gamma,x) = \frac{\Pi(\gamma-1).\Pi(\alpha+\beta-\gamma-1)}{\Pi(\alpha-1).\Pi(\beta-1)}(1-x)^{\gamma-\alpha-\beta}, \quad \text{..............(7)}$$

ultimately.

* So called by Thomson and Tait, because the nodal lines ($S_n=0$) divide the unit sphere into parallel belts.

† *l.c. ante* p. 104.

‡ Forsyth, *Differential Equations*, 3rd ed., London, 1903, c. vi.

For the critical case where $\gamma-\alpha-\beta=0$,

we may have recourse to the formula

$$\frac{d}{dx}F(\alpha,\beta,\gamma,x)=\frac{\alpha\beta}{\gamma}F(\alpha+1,\beta+1,\gamma+1,x), \quad \ldots\ldots\ldots\ldots(8)$$

which, with (6), gives in the case supposed

$$\frac{d}{dx}F(\alpha,\beta,\gamma,x)=\frac{\alpha\beta}{\gamma}(1-x)^{-1}.F(\gamma-\alpha,\gamma-\beta,\gamma+1,x)$$

$$=\frac{\alpha\beta}{\gamma}(1-x)^{-1}.F(\alpha,\beta,\alpha+\beta+1,x). \quad \ldots\ldots\ldots\ldots(9)$$

The last factor is now convergent when $x=1$, so that $F(\alpha,\beta,\gamma,x)$ is ultimately divergent as $\log(1-x)$. More precisely we have, for values of x near this limit,

$$F(\alpha,\beta,\alpha+\beta,x)=\frac{\Pi(\alpha+\beta-1)}{\Pi(\alpha-1).\Pi(\beta-1)}\log\frac{1}{1-x}. \quad \ldots\ldots\ldots\ldots(10)$$

85. Of the two series which occur in the general expression (Art. 84 (2)) of a zonal harmonic, the former terminates when n is an even, and the latter when n is an odd integer. For other values of n both series are essentially convergent for values of μ between ± 1, but since in each case we have $\gamma-\alpha-\beta=0$, they diverge at the limits $\mu=\pm 1$, becoming infinite as $\log(1-\mu^2)$.

It follows that the terminating series corresponding to integral values of n are the only zonal surface-harmonics which are finite over the unit sphere. If we reverse the series we find that both these cases (n even, and n odd) are included in the formula*

$$P_n(\mu)=\frac{1.3.5\ldots(2n-1)}{1.2.3\ldots n}\left\{\mu^n-\frac{n(n-1)}{2(2n-1)}\mu^{n-2}+\frac{n(n-1)(n-2)(n-3)}{2.4.(2n-1)(2n-3)}\mu^{n-4}-\ldots\right\}, \quad \ldots\ldots(1)$$

where the constant factor has been adjusted so as to make $P_n(\mu)=1$ for $\mu=1$†. The formula may also be written

$$P_n(\mu)=\frac{1}{2^n.n!}\frac{d^n}{d\mu^n}(\mu^2-1)^n. \quad \ldots\ldots\ldots\ldots(2)$$

The series (1) may otherwise be obtained by development of Art. 82 (6), which in the case of the zonal harmonic assumes the form

$$S_n=Ar^{n+1}\frac{\partial^n}{\partial x^n}\frac{1}{r}. \quad \ldots\ldots\ldots\ldots(3)$$

* For n even this corresponds to $A=(-)^{\frac{1}{2}n}\dfrac{1.3.5\ldots(n-1)}{2.4\ldots n}$, $B=0$; whilst for n odd we have $A=0$, $B=(-)^{\frac{1}{2}(n-1)}\dfrac{3.5\ldots n}{2.4\ldots(n-1)}$. See Heine, i. 12, 147.

† Tables of $P_1, P_2, \ldots P_7$ were calculated by Glaisher, for values of μ at intervals of ·01, *Brit. Ass. Report*, 1879, and are reprinted by Dale, *Five-Figure Tables...*, London, 1903. A table of the same functions for every degree of the quadrant, calculated under the direction of Prof. Perry, was published in the *Phil. Mag.* for Dec. 1891. Both tables are reproduced in Byerly's treatise, also by Jahnke and Emde, *Funktionentafeln*, Leipzig, 1909. The values of the first 20 zonal harmonics, at intervals of 5°, have been calculated by Prof. A. Lodge, *Phil. Trans.* A. cciii. (1904).

As particular cases of (2) we have

$$P_0(\mu)=1,\quad P_1(\mu)=\mu,\quad P_2(\mu)=\tfrac{1}{2}(3\mu^2-1),\quad P_3(\mu)=\tfrac{1}{2}(5\mu^3-3\mu).$$

Expansions of P_n in terms of other functions of θ as independent variables, in places of μ, have been obtained by various writers. For example, we have

$$P_n(\cos\theta)=1-\frac{n(n+1)}{1^2}\sin^2\tfrac{1}{2}\theta+\frac{(n-1)n(n+1)(n+2)}{1^2.2^2}\sin^4\tfrac{1}{2}\theta-\ldots.\ \ldots(4)$$

This may be deduced from (2)*, or it may be obtained independently by putting $\mu=1-2z$ in Art. 84 (1), and integrating by a series.

The function $P_n(\mu)$ was first introduced into analysis by Legendre† as the coefficient of h^n in the expansion of

$$(1-2\mu h+h^2)^{-\frac{1}{2}}.$$

The connection of this with our present point of view is that if ϕ be the velocity-potential of a unit source on the axis of x at a distance c from the origin, we have, on Legendre's definition, for values of r less than c,

$$4\pi\phi=(c^2-2\mu cr+r^2)^{-\frac{1}{2}}$$

$$=\frac{1}{c}+P_1\frac{r}{c^2}+P_2\frac{r^2}{c^3}+\ldots.\ \ldots\ldots(5)$$

Each term in this expansion must separately satisfy $\nabla^2\phi=0$, and therefore the coefficient P_n must be a solution of Art. 84 (1). Since P_n, as thus defined, is obviously finite for all values of μ, and becomes equal to unity for $\mu=1$, it must be identical with (1).

For values of r greater than c, the corresponding expansion is

$$4\pi\phi=\frac{1}{r}+P_1\frac{c}{r^2}+P_2\frac{c^2}{r^3}+\ldots.\ \ldots\ldots(6)$$

We can hence deduce expressions, which will be useful to us later, Art. 98, for the velocity-potential due to a *double-source* of unit strength, situate on the axis of x at a distance c from the origin, and having its axis pointing *from* the origin. This is evidently equal to $\partial\phi/\partial c$, where ϕ has either of the above forms; so that the required potential is, for $r<c$,

$$-\frac{1}{4\pi}\left(\frac{1}{c^2}+2P_1\frac{r}{c^3}+3P_2\frac{r^2}{c^4}-\ldots\right),\ \ldots\ldots(7)$$

and for $r>c$,

$$\frac{1}{4\pi}\left(P_1\frac{1}{r^2}+2P_2\frac{c}{r^3}+\ldots\right).\ \ldots\ldots(8)$$

The remaining solution of Art. 84 (1), in the case of n integral, can be put into the more compact form‡

$$Q_n(\mu)=\tfrac{1}{2}P_n(\mu)\log\frac{1+\mu}{1-\mu}-Z_n,\ \ldots\ldots(9)$$

where

$$Z_n=\frac{2n-1}{1.n}P_{n-1}+\frac{2n-5}{3(n-1)}P_{n-3}+\ldots.\ \ldots\ldots(10)$$

* Murphy, *Elementary Principles of the Theories of Electricity, &c.*, Cambridge, 1833, p. 7. [Thomson and Tait, Art. 782.]

† "Sur l'attraction des sphéroides homogènes," *Mém. des Savans Étrangers*, x. (1785).

‡ This is equivalent to Art. 84 (4) with, for n even, $A=0$, $B=(-)^{\frac{1}{2}n}\dfrac{2.4\ldots n}{1.4\ldots(n-1)}$; whilst for n odd we have $A=(-)^{\frac{1}{2}(n+1)}\dfrac{2.4\ldots(n-1)}{3.5\ldots n}$, $B=0$. See Heine, i. 141, 147.

This function $Q_n(\mu)$ is sometimes called the zonal harmonic 'of the second kind.'

Thus

$$Q_0(\mu)=\tfrac{1}{2}\log\frac{1+\mu}{1-\mu},\qquad Q_2(\mu)=\tfrac{1}{4}(3\mu^2-1)\log\frac{1+\mu}{1-\mu}-\tfrac{3}{2}\mu,$$

$$Q_1(\mu)=\tfrac{1}{2}\mu\log\frac{1+\mu}{1-\mu}-1,\qquad Q_3(\mu)=\tfrac{1}{4}(5\mu^2-3\mu)\log\frac{1+\mu}{1-\mu}-\tfrac{5}{2}\mu^2+\tfrac{2}{3}.$$

86. When we abandon the restriction as to symmetry about the axis of x, we may suppose S_n, if a finite and single-valued function of ω, to be expanded in a series of terms varying as $\cos s\omega$ and $\sin s\omega$ respectively. If this expansion is to apply to the whole sphere (*i.e.* from $\omega=0$ to $\omega=2\pi$), we may further (by Fourier's theorem) suppose the values of s to be integral. The differential equation satisfied by any such term is

$$\frac{d}{d\mu}\left\{(1-\mu^2)\frac{dS_n}{d\mu}\right\}+\left\{n(n+1)-\frac{s^2}{1-\mu^2}\right\}S_n=0. \quad\text{.........} \quad (1)$$

If we put $$S_n=(1-\mu^2)^{\frac{1}{2}s}v,$$

this takes the form

$$(1-\mu^2)\frac{d^2v}{d\mu^2}-2(s+1)\mu\frac{dv}{d\mu}+(n-s)(n+s+1)v=0,$$

which is suitable for integration by series. We thus obtain

$$S_n=A(1-\mu^2)^{\frac{1}{2}s}\left\{1-\frac{(n-s)(n+s+1)}{1.2}\mu^2+\frac{(n-s-2)(n-s)(n+s+1)(n+s+3)}{1.2.3.4}\mu^4-\ldots\right\}$$
$$+B(1-\mu^2)^{\frac{1}{2}s}\left\{\mu-\frac{(n-s-1)(n+s+2)}{1.2.3}\mu^3+\frac{(n-s-3)(n-s-1)(n+s+2)(n+s+4)}{1.2.3.4.5}\mu^5-\ldots\right\}, \quad\ldots(2)$$

the factor $\cos s\omega$ or $\sin s\omega$ being for the moment omitted. In the hypergeometric notation this may be written

$$S_n=(1-\mu^2)^{\frac{1}{2}s}\{AF(\tfrac{1}{2}s-\tfrac{1}{2}n,\ \tfrac{1}{2}+\tfrac{1}{2}s+\tfrac{1}{2}n,\ \tfrac{1}{2},\ \mu^2)$$
$$+B\mu F(\tfrac{1}{2}+\tfrac{1}{2}s-\tfrac{1}{2}n,\ 1+\tfrac{1}{2}s+\tfrac{1}{2}n,\ \tfrac{3}{2},\ \mu^2)\}. \quad\ldots(3)$$

These expressions converge when $\mu^2<1$, but since in each case we have

$$\gamma-\alpha-\beta=-s,$$

the series become infinite as $(1-\mu^2)^{-s}$ at the limits $\mu=\pm 1$, unless they terminate*. The former series terminates when $n-s$ is an even, and the

* Rayleigh, *Theory of Sound*, London, 1877, Art. 338.

latter when it is an odd integer. By reversing the series we can express both these finite solutions by the single formula*

$$P_n^s(\mu) = \frac{(2n)!}{2^n(n-s)!\,n!}(1-\mu^2)^{\frac{1}{2}s}\left\{\mu^{n-s} - \frac{(n-s)(n-s-1)}{2\,.\,(2n-1)}\mu^{n-s-2}\right.$$
$$\left.+\frac{(n-s)(n-s-1)(n-s-2)(n-s-3)}{2\,.\,4\,.\,(2n-1)(2n-3)}\mu^{n-s-4} - \ldots\right\} \quad \ldots\ldots(4)$$

On comparison with Art. 85 (1) we find that

$$P_n^s(\mu) = (1-\mu^2)^{\frac{1}{2}s}\frac{d^s P_n(\mu)}{d\mu^s}. \quad \ldots\ldots\ldots\ldots\ldots\ldots(5)$$

That this is a solution of (1) may of course be verified independently.

In terms of $\sin\frac{1}{2}\theta$, we have

$$P_n^s(\cos\theta) = \frac{(n+s)!}{2^s(n-s)!\,s!}\sin^s\theta\left\{1 - \frac{(n-s)(n+s+1)}{1\,.\,(s+1)}\sin^2\tfrac{1}{2}\theta\right.$$
$$\left.+\frac{(n-s-1)(n-s)(n+s+1)(n+s+2)}{1\,.\,2\,.\,(s+1)(s+2)}\sin^4\tfrac{1}{2}\theta - \ldots\right\} \quad \ldots\ldots(6)$$

This corresponds to Art. 85 (4), from which it can easily be derived.

Collecting our results we learn that a surface-harmonic which is finite over the unit sphere is necessarily of integral order, and is further expressible, if n denote the order, in the form

$$S_n = A_0 P_n(\mu) + \sum_{s=1}^{s=n}(A_s\cos s\omega + B_s\sin s\omega)\,P_n^s(\mu), \quad \ldots\ldots\ldots(7)$$

containing $2n+1$ arbitrary constants. The terms of this involving ω are called 'tesseral' harmonics, with the exception of the last two, which are given by the formula

$$(1-\mu^2)^{\frac{1}{2}n}(A_n\cos n\omega + B_n\sin n\omega),$$

and are called 'sectorial' harmonics†; the names being suggested by the forms of the compartments into which the unit sphere is divided by the nodal lines $S_n = 0$.

The formula for the tesseral harmonic of rank s may be obtained otherwise from the general expression (6) of Art. 82 by making $n-s$ out of the n poles of the harmonic coincide at the point $\theta = 0$ of the sphere, and distributing the remaining s poles evenly round the equatorial circle $\theta = \frac{1}{2}\pi$.

The remaining solution of (1), in the case of n integral, may be put in the form

$$S_n = (A_s\cos s\omega + B_s\sin s\omega)\,Q_n^s(\mu), \quad \ldots\ldots\ldots\ldots\ldots\ldots(8)$$

* There are great varieties of notation in connection with these 'associated functions,' as they have been called. That chosen in the text was proposed by F. Neumann; and is adopted by Whittaker and Watson, p. 323.

† The prefix 'spherical' is implied; it is often omitted for brevity.

where*

$$Q_n^s(\mu) = (1-\mu^2)^{\frac{1}{2}s}\frac{d^s Q_n(\mu)}{d\mu^s}. \quad \text{.......................(9)}$$

This is sometimes called a tesseral harmonic 'of the second kind.'

87. Two surface-harmonics S, S' are said to be 'conjugate,' or 'orthogonal,' when

$$\iint SS'\, d\varpi = 0, \quad \text{.................................(1)}$$

where $\delta\varpi$ is an element of surface of the unit sphere, and the integration extends over this sphere.

It may be shewn that any two surface-harmonics, of different orders, which are finite over the unit sphere, are orthogonal, and also that the $2n+1$ harmonics of any given order n, of the zonal, tesseral, and sectorial types specified in Arts. 85, 86, are all mutually orthogonal. It will appear, later, that the orthogonal property is of great importance in the physical applications of the subject.

Since $\delta\varpi = \sin\theta\,\delta\theta\,\delta\omega = -\,\delta\mu\,\delta\omega$, we have, as particular cases of this theorem,

$$\int_{-1}^{1} P_m(\mu)\, d\mu = 0, \quad \text{......................(2)}$$

$$\int_{-1}^{1} P_m(\mu) \,.\, P_n(\mu)\, d\mu = 0, \quad \text{.......... ....(3)}$$

and

$$\int_{-1}^{1} P_m^s(\mu) \,.\, P_n^s(\mu)\, d\mu = 0, \quad \text{......................(4)}$$

provided m, n are unequal.

For $m=n$, it may be shewn† that

$$\int_{-1}^{1} \{P_n(\mu)\}^2\, d\mu = \frac{2}{2n+1}, \quad \text{........................(5)}$$

$$\int_{-1}^{1} \{P_n^s(\mu)\}^2\, d\mu = \frac{(n+s)!}{(n-s)!}\frac{2}{2n+1}. \quad \text{.................(6)}$$

88. We may also quote the theorem that any arbitrary function $f(\mu, \omega)$ of the position of a point on the unit sphere can be expanded in a series of surface-harmonics, obtained by giving n all integral values from 0 to ∞, in Art. 86 (7). The formulae (5) and (6) are useful in determining the coefficients in this expansion.

Thus, in the case of symmetry about an axis, the theorem takes the form

$$f(\mu) = C_0 + C_1 P_1(\mu) + C_2 P_2(\mu) + \ldots + C_n P_n(\mu) + \ldots. \quad \text{......(7)}$$

If we multiply both sides by $P_n(\mu)\, d\mu$, and integrate between the limits ± 1, we find

$$C_0 = \tfrac{1}{2}\int_{-1}^{1} f(\mu)\, d\mu, \quad \text{............................(8)}$$

* A table of the functions $Q_n(\mu)$, $Q_n^s(\mu)$, for various values of n and s, is given by Bryan, *Proc. Camb. Phil. Soc.* vi. 297.

† Ferrers, p. 86; Whittaker and Watson, pp. 306, 325.

and, generally,

$$C_n = \frac{2n+1}{2} \int_{-1}^{1} f(\mu) P_n(\mu) \, d\mu. \quad \text{.....................(9)}$$

For the analytical proof of the theorem recourse must be had to the special treatises*; the physical grounds for assuming the possibility of this and other similar expansions will appear, incidentally, in connection with various problems.

89. Solutions of the equation $\nabla^2\phi = 0$ may also be obtained by the usual method of treating linear equations with constant coefficients†. Thus, the equation is satisfied by

$$\phi = e^{\alpha x + \beta y + \gamma z},$$

or, more generally, by $\qquad \phi = f(\alpha x + \beta y + \gamma z), \quad \text{..........................(1)}$

provided $\qquad \alpha^2 + \beta^2 + \gamma^2 = 0. \quad \text{..............................(2)}$

For example, we may put

$$\alpha, \beta, \gamma = 1, \quad i \cos \vartheta, \quad i \sin \vartheta, \quad \text{.....................(3)}$$

or, again, $\qquad \alpha, \beta, \gamma = 1, \quad i \cosh u, \quad \sinh u. \quad \text{.....................(4)}$

It may be shewn‡ that the most general solution possible can be obtained by superposition of solutions of the type (1).

Using (3), and introducing the cylindrical co-ordinates x, ϖ, ω, where

$$y = \varpi \cos \omega, \qquad z = \varpi \sin \omega, \quad \text{.....................(5)}$$

we build up a solution symmetrical about the axis of x if we take

$$\phi = \frac{1}{2\pi} \int_0^{2\pi} f\{x + i\varpi \cos(\vartheta - \omega)\} \, d\vartheta.$$

For, since the integration extends over a whole circumference, it is immaterial where the origin of ϑ is placed, and the formula may therefore be written§

$$\phi = \frac{1}{2\pi} \int_0^{2\pi} f(x + i\varpi \cos \vartheta) \, d\vartheta = \frac{1}{\pi} \int_0^{\pi} f(x + i\varpi \cos \vartheta) \, d\vartheta. \quad \text{.......(6)}$$

This is remarkable as giving a value of ϕ, symmetrical about the axis of x, in terms of its values $f(x)$ at points of this axis. It may be shewn, by means of the theorem of Art. 38, that the form of ϕ is in such a case completely determined by the values over any finite length of the axis‖.

As particular cases of (6) we have the functions

$$\frac{1}{\pi} \int_0^{\pi} (x + i\varpi \cos \vartheta)^n \, d\vartheta, \qquad \frac{1}{\pi} \int_0^{\pi} (x + i\varpi \cos \vartheta)^{-n-1} \, d\vartheta,$$

* For an account of the more recent investigations of the question, see Wangerin, *l.c.*

† Forsyth, *Differential Equations*, p. 444.

‡ Whittaker, *Month. Not. R. Ast. Soc.* lxii. (1902).

§ Whittaker and Watson, *Modern Analysis*, c. xviii.

‖ Thomson and Tait, Art. 498.

where n will be supposed to be integral. Since these are solid harmonics finite over the unit sphere, and since, for $\varpi = 0$, they reduce to r^n and r^{-n-1}, they must be equivalent to $P_n(\mu) r^n$, and $P_n(\mu) r^{-n-1}$, respectively. We thus obtain the forms

$$P_n(\mu) = \frac{1}{\pi}\int_0^\pi \{\mu + i\sqrt{(1-\mu^2)}\cos\vartheta\}^n d\vartheta, \qquad (7)$$

$$P_n(\mu) = \frac{1}{\pi}\int_0^\pi \frac{d\vartheta}{\{\mu + i\sqrt{(1-\mu^2)}\cos\vartheta\}^{n+1}}, \qquad (8)$$

due originally to Laplace* and Jacobi†, respectively.

90. As a first application of the foregoing theory let us suppose that an arbitrary distribution of impulsive pressure is applied to the surface of a spherical mass of fluid initially at rest. This is equivalent to prescribing an arbitrary value of ϕ over the surface; the value of ϕ in the interior is thence determinate, by Art. 40. To find it, we may suppose the given surface-value to be expanded, in accordance with the theorem quoted in Art. 88, in a series of surface-harmonics of integral order, thus

$$\phi = S_0 + S_1 + S_2 + \ldots + S_n + \ldots. \qquad (1)$$

The required value is then

$$\phi = S_0 + \frac{r}{a} S_1 + \frac{r^2}{a^2} S_2 + \ldots + \frac{r^n}{a^n} S_n + \ldots, \qquad (2)$$

for this satisfies $\nabla^2\phi = 0$, and assumes the prescribed form (1) when $r = a$, the radius of the sphere.

The corresponding solution for the case of a prescribed value of ϕ over the surface of a spherical cavity in an infinite mass of liquid initially at rest is evidently

$$\phi = \frac{a}{r} S_0 + \frac{a^2}{r^2} S_1 + \frac{a^3}{r^3} S_2 + \ldots + \frac{a^{n+1}}{r^{n+1}} S_n + \ldots. \qquad (3)$$

Combining these two results we get the case of an infinite mass of fluid whose continuity is interrupted by an infinitely thin vacuous stratum, of spherical form, within which an arbitrary impulsive pressure is applied. The values (2) and (3) of ϕ are of course continuous at the stratum, but the values of the normal velocity are discontinuous, viz. we have, for the internal fluid,

$$\frac{\partial\phi}{\partial r} = \Sigma n \frac{S_n}{a},$$

and for the external fluid

$$\frac{\partial\phi}{\partial r} = -\Sigma(n+1)\frac{S_n}{a}.$$

* *Méc. Cél.* Livre 11me, c. ii.

† *Crelle*, xxvi. (1843) [*Gesammelte Werke*, Berlin, 1881..., vi. 148].

The motion, whether internal or external, is therefore that due to a distribution of simple sources with surface-density

$$\Sigma(2n+1)\frac{S_n}{a} \quad\ldots\ldots\ldots\ldots\ldots\ldots(4)$$

over the sphere; see Art. 58.

91. Let us next suppose that, instead of the impulsive pressure, it is the normal velocity which is prescribed over the spherical surface; thus

$$\frac{\partial\phi}{\partial r} = S_1 + S_2 + \ldots + S_n + \ldots, \quad\ldots\ldots\ldots\ldots(1)$$

the term of zero order being necessarily absent, since we must have

$$\iint \frac{\partial\phi}{\partial r}\, d\varpi = 0, \quad\ldots\ldots\ldots\ldots\ldots\ldots(2)$$

on account of the constancy of volume of the included mass.

The value of ϕ for the internal space is of the form

$$\phi = A_1 r S_1 + A_2 r^2 S_2 + \ldots + A_n r^n S_n + \ldots, \quad\ldots\ldots\ldots(3)$$

for this is finite and continuous, and satisfies $\nabla^2\phi = 0$, and the constants can be determined so as to make $\partial\phi/\partial r$ assume the given surface-value (1); viz. we have $nA_n a^{n-1} = 1$. The required solution is therefore

$$\phi = a\Sigma\frac{1}{n}\frac{r^n}{a^n}S_n. \quad\ldots\ldots\ldots\ldots\ldots\ldots(4)$$

The corresponding solution for the external space is found in like manner to be

$$\phi = -a\Sigma\frac{1}{n+1}\frac{a^{n+1}}{r^{n+1}}S_n. \quad\ldots\ldots\ldots\ldots\ldots(5)$$

The two solutions, taken together, give the motion produced in an infinite mass of liquid which is divided into two portions by a thin spherical membrane, when a prescribed normal velocity is given to every point of the membrane, subject to the condition (2).

The value of ϕ changes from $a\Sigma S_n/n$ to $-a\Sigma S_n/(n+1)$, as we cross the membrane, so that the *tangential* velocity is now discontinuous. The motion, whether inside or outside, is that due to a *double-sheet* of density

$$-a\Sigma\frac{2n+1}{n(n+1)}S_n; \quad\ldots\ldots\ldots\ldots\ldots\ldots(6)$$

see Art. 58.

The kinetic energy of the internal fluid is given by the formula (4) of Art. 44, viz.

$$2T = \rho\iint\phi\frac{\partial\phi}{\partial r}\,dS = \rho a^3\,\Sigma\frac{1}{n}\iint S_n^2\,d\varpi, \quad\ldots\ldots\ldots\ldots(7)$$

the parts of the integral which involve products of surface-harmonics of different orders disappearing in virtue of the orthogonal property of Art. 87.

For the external fluid we have

$$2T = -\rho \iint \phi \frac{\partial \phi}{\partial r} dS = \rho a^3 \Sigma \frac{1}{n+1} \iint S_n^2 d\varpi. \quad \text{..............(8)}$$

91 a. The harmonic of zero order lends itself at once to the discussion of the two mathematically cognate problems of the collapse of a spherical bubble in water, and the expansion of a spherical cavity due to the pressure of an included gas, as in the case of a submarine mine.

In the former problem*, if R_0 be the initial radius of the bubble, and R its value at time t, we have

$$\phi = \frac{R^2 \dot{R}}{r}, \quad \text{..(1)}$$

since this makes $-\partial\phi/\partial r = \dot{R}$, for $r = R$. Hence, putting $\Omega = 0$ in Art. 22 (5), we have

$$\frac{p - p_0}{\rho} = \frac{R^2 \ddot{R} + 2R\dot{R}^2}{r} - \frac{R^4 \dot{R}^2}{2r^4}, \quad \text{...............................(2)}$$

if p_0 be the pressure at $r = \infty$. Hence, putting $r = R$ and neglecting the internal pressure

$$R\ddot{R} + \tfrac{3}{2}\dot{R}^2 = -\frac{p_0}{\rho}, \quad \text{.................................(3)}$$

the integral of which is

$$R^3 \dot{R}^2 = \tfrac{2}{3}\frac{p_0}{\rho}(R_0^3 - R^3). \quad \text{..(4)}$$

This cannot easily be integrated further, but the time (t_1) of total collapse can be found; thus, putting $R = R_0 x^{\frac{1}{3}}$,

$$t_1 = R_0 \sqrt{\left(\frac{\rho}{6p_0}\right)} \int_0^1 x^{-\frac{1}{6}}(1-x)^{-\frac{1}{2}} dx = R_0 \sqrt{\left(\frac{\rho}{6p_0}\right)} \frac{\Gamma(\frac{5}{6})\Gamma(\frac{1}{2})}{\Gamma(\frac{4}{3})} = \cdot 915 R_0 \sqrt{(\rho/p_0)}. \quad \text{...(5)}$$

Thus if $\rho = 1$, $R_0 = 1$ cm., and $p_0 = 10^6$ C.G.S. (1 atmosphere), $t_1 = \cdot 000915$ sec.

The kinetic energy at any instant is

$$2\pi\rho R^3 \dot{R}^2 = \tfrac{4}{3}\pi p_0 (R_0^3 - R^3), \quad \text{................................(6)}$$

as is indeed obvious from a consideration of the work done at a distance on the fluid. When the collapse occurs, the energy destroyed, or rather converted into other forms, is $\frac{4}{3}\pi p_0 R_0^3$. If $R_0 = 1$, $p_0 = 10^6$, this is $4\cdot 18 \times 10^6$ ergs, or about $\cdot 308$ of a ft.-lb.

The equations (1) and (2) are applicable also to the problem of the expanding cavity, but we now neglect the pressure p_0 at a distance. If p_1 be the initial pressure in the cavity, when $R = R_0$, and $\dot{R} = 0$, the internal pressure at time t is given by

$$\frac{p}{p_1} = \left(\frac{R_0}{R}\right)^{3\gamma}, \quad \text{...(7)}$$

if we assume the adiabatic law of expansion. Hence

$$R\ddot{R} + \tfrac{3}{2}\dot{R}^2 = c_0^2 \left(\frac{R_0}{R}\right)^{3\gamma}, \quad \text{.................................(8)}$$

where

$$c_0 = \sqrt{(p_1/\rho)}. \quad \text{..(9)}$$

This quantity c_0 is of the nature of a velocity, and determines the rapidity with which changes take place. The integral of (8) is

$$\frac{\dot{R}^2}{c_0^2} = \frac{2}{3(\gamma - 1)} \left\{ \left(\frac{R_0}{R}\right)^3 - \left(\frac{R_0}{R}\right)^{3\gamma} \right\}. \quad \text{.........................(10)}$$

* Besant, *Hydrostatics and Hydrodynamics*, Cambridge, 1859; Rayleigh, *Phil. Mag.* xxxiv. 94 (1917) [*Papers*, vi. 504].

It appears from (8) that the initial acceleration ($\ddot{R}$) in the radius is c_0^2/R_0, whatever the law of expansion. For (8) and (10) we find that the maximum of $\dot{R}$ occurs when

$$(R/R_0)^{3\gamma-3}=\gamma, \qquad (11)$$

and is given by

$$\frac{\dot{R}^2}{c_0^2}=\frac{2}{3\gamma^{\gamma(\gamma-1)}}. \qquad (12)$$

The solution is not easily completed except in the special case of $\gamma=\frac{4}{3}$. Writing

$$R/R_0=1+z, \qquad (13)$$

we have then

$$(1+z)^2\frac{dz}{dt}=\frac{c_0}{R}\sqrt{(2z)}, \qquad (14)$$

whence

$$c_0 t/R_0=\sqrt{(2z)}\,(1+\tfrac{2}{3}z+\tfrac{1}{5}z^2). \qquad (15)$$

As a concrete illustration, suppose the initial diameter of the cavity to be 1 metre, and the initial pressure p_1 to be 1000 atmospheres, which makes $c_0=3{\cdot}16\times10^4$ cm./sec. It is then found that the radius of the cavity is doubled in $\frac{1}{250}$ of a second, and multiplied five-fold in about $\frac{1}{30}$ sec. The initial acceleration of the radius is $2{\cdot}00\times10^7$ cm./sec.^{2}, shewing that the neglect of gravity in the early stages of the motion is amply justified. The maximum of $\dot{R}$ occurs when $R/R_0=\frac{4}{3}$, $t={\cdot}0016$ sec., and is about 145 metres per second, or about one-tenth of the velocity of sound in water. With initial pressures of the order of 10,000 atmospheres or more, we should have velocities comparable with the velocity of sound, and the effect of compressibility would be no longer negligible*.

92. The harmonic of the first order is involved in the problem of the motion of a solid sphere in an infinite mass of liquid which is at rest at infinity. If we take the origin at the centre of the sphere, and the axis of x in the direction of motion, the normal velocity at the surface is Ux/r, $=U\cos\theta$, where U is the velocity of the centre. Hence the conditions to determine ϕ are (1°) that we must have $\nabla^2\phi=0$ everywhere, (2°) that the space-derivatives of ϕ must vanish at infinity, and (3°) that at the surface of the sphere ($r=a$) we must have

$$-\frac{\partial\phi}{\partial r}=U\cos\theta. \qquad (1)$$

The form of this suggests at once the zonal harmonic of the first order; we therefore assume

$$\phi=A\frac{\partial}{\partial x}\frac{1}{r}=-A\frac{\cos\theta}{r^2}.$$

The condition (1) gives $-2A/a^3=U$, so that the required solution is†

$$\phi=\tfrac{1}{2}U\frac{a^3}{r^2}\cos\theta. \qquad (2)$$

It appears on comparison with Art. 56 (4) that the motion of the fluid is the same as would be produced by a *double-source* of strength $2\pi Ua^3$, situate at the centre of the sphere. For the forms of the lines of motion see p. 128.

* This discussion is taken from a paper "The early stages of a submarine explosion," *Phil. Mag.* xlv. 257 (1923).

† Stokes, "On some cases of Fluid Motion," *Camb. Trans.* viii. (1843) [*Papers*, i. 17]. Dirichlet, "Ueber die Bewegung eines festen Körpers in einem incompressibeln flüssigen Medium," *Berl. Monatsber.* 1852 [*Werke*, Berlin, 1889–97, ii. 115].

To find the energy of the fluid motion we have

$$2T = -\rho \iint \phi \frac{\partial \phi}{\partial r} dS = \tfrac{1}{2} \rho a U^2 \int_0^\pi \cos^2\theta \,.\, 2\pi a \sin\theta \,.\, a\, d\theta$$
$$= \tfrac{2}{3} \pi \rho a^3 U^2 = M' U^2, \quad \text{.......................(3)}$$

if $M' = \frac{2}{3}\pi\rho a^3$. It appears, exactly as in Art. 68, that the effect of the fluid pressure is equivalent simply to an addition to the *inertia* of the solid, the increment being now *half* the mass of the fluid displaced*.

Thus in the case of rectilinear motion of the sphere, if no external forces act on the fluid, the resultant pressure is equivalent to a force

$$-M' \frac{dU}{dt}, \quad \text{................................(4)}$$

in the direction of motion, vanishing when U is constant. Hence if the sphere be set in motion and left to itself, it will continue to move in a straight line with constant velocity.

The behaviour of a solid projected in an actual fluid is of course quite different; a continual application of force is necessary to maintain the motion, and if this be not supplied the solid is gradually brought to rest. It must be remembered, however, in making this comparison, that in a 'perfect' fluid there is no dissipation of energy, and that if, further, the fluid be incompressible, the solid cannot lose its kinetic energy by transfer to the fluid, since, as we have seen in Chapter III., the motion of the fluid is entirely determined by that of the solid, and therefore ceases with it.

If we wish to verify the preceding result by direct calculation from the formula

$$\frac{p}{\rho} = \frac{\partial \phi}{\partial t} - \tfrac{1}{2} q^2 + F(t), \quad \text{...................................(5)}$$

we must remember that the origin is in motion, and that the values of r and θ for a fixed point of space are therefore increasing at the rates $-U\cos\theta$, and $(U\sin\theta)/r$, respectively; or we may appeal to Art. 20 (6). In either way we find

$$\frac{p}{\rho} = \tfrac{1}{2} a \frac{dU}{dt} \cos\theta + \tfrac{9}{16} U^2 \cos 2\theta - \tfrac{1}{16} U^2 + F(t). \quad \text{.....................(6)}$$

The last three terms are the same for surface-elements in the positions θ and $\pi - \theta$; so that, when U is constant, the pressures on the various elements of the anterior half of the sphere are balanced by equal pressures on the corresponding elements of the posterior half. But when the motion of the sphere is being accelerated there is an excess of pressure on the anterior, and a defect on the posterior half. The reverse holds when the motion is being retarded. The resultant effect in the direction of motion is

$$-\int_0^\pi 2\pi a \sin\theta \,.\, a\, d\theta \,.\, p\cos\theta = -\tfrac{2}{3}\pi\rho a^3 \frac{dU}{dt},$$

as before.

93. The same method can be applied to find the motion produced in a liquid contained between a solid sphere and a fixed concentric spherical boundary, when the sphere is moving with given velocity U.

* Stokes, *l.c.* The result had been obtained otherwise, on the hypothesis of infinitely small motion, by Green, "On the Vibration of Pendulums in Fluid Media," *Edin. Trans.* 1833 [*Papers*, p. 315].

The centre of the sphere being taken as origin, it is evident, since the space occupied by the fluid is limited both externally and internally, that solid harmonics of both positive and negative degrees are admissible; they are in fact required, in order to satisfy the boundary conditions, which are

$$-\partial\phi/\partial r = U\cos\theta,$$

for $r=a$, the radius of the spheres, and

$$\partial\phi/\partial r = 0,$$

for $r=b$, the radius of the external boundary, the axis of x being as before in the direction of motion.

We therefore assume
$$\phi = \left(Ar + \frac{B}{r^2}\right)\cos\theta, \qquad\qquad (1)$$

and the conditions in question give

$$A - \frac{2B}{a^3} = -U, \qquad A - \frac{2B}{b^3} = 0,$$

whence
$$A = \frac{a^3}{b^3 - a^3}U, \qquad B = \frac{1}{2}\frac{a^3 b^3}{b^3 - a^3}U. \qquad\qquad (2)$$

The kinetic energy of the fluid motion is given by

$$2T = -\rho\iint \phi\frac{\partial\phi}{\partial r}\,dS,$$

the integration extending over the inner spherical surface, since at the outer we have $\partial\phi/\partial r = 0$. We thus find

$$2T = \tfrac{2}{3}\pi\frac{b^3 + 2a^3}{b^3 - a^3}\rho a^3 U^2. \qquad\qquad (3)$$

It appears that the effective addition to the inertia of the sphere is now*

$$\tfrac{2}{3}\pi\frac{b^3 + 2a^3}{b^3 - a^3}\rho a^3. \qquad\qquad (4)$$

As b diminishes from ∞ to a, this increases continually from $\frac{2}{3}\pi\rho a^3$ to ∞, in accordance with Lord Kelvin's minimum theorem (Art. 45). In other words, the introduction of a rigid spherical partition in the problem of Art. 92 acts as a constraint increasing the kinetic energy for any given velocity of the sphere, and so virtually increasing the inertia of the system.

94. In all cases where the motion of a liquid takes place in a series of planes passing through a common line, and is the same in each such plane, there exists a stream-function analogous in some of its properties to the two-dimensional stream-function of the last Chapter. If in any plane through the axis of symmetry we take two points A and P, of which A is arbitrary, but fixed, while P is variable, then considering the annular surface generated by any line AP, it is plain that the flux across this surface is a function of the position of P. Denoting this function by $2\pi\psi$, and taking the axis of x to coincide with that of symmetry, we may say that ψ is a function of x and ϖ, where x is the abscissa of P, and ϖ, $= (y^2 + z^2)^{\frac{1}{2}}$, is its distance from the axis. The curves $\psi = \text{const.}$ are evidently stream-lines.

If P' be a point infinitely near to P in a meridian plane, it follows from the above definition that the velocity normal to PP' is equal to

$$\frac{2\pi\delta\psi}{2\pi\varpi\,.\,PP'},$$

* Stokes, *l.c. ante* p. 123.

whence, taking PP' parallel first to ϖ and then to x,

$$u=-\frac{1}{\varpi}\frac{\partial\psi}{\partial\varpi}, \quad v=\frac{1}{\varpi}\frac{\partial\psi}{\partial x}, \quad \text{......................(1)}$$

where u and v are the components of fluid velocity in the directions of x and ϖ respectively, the convention as to sign being similar to that of Art. 59.

These kinematical relations may also be inferred from the form which the equation of continuity takes under the present circumstances. If we express that the total flux into the annular space generated by the revolution of an elementary rectangle $\delta x\,\delta\varpi$ is zero, we find

$$\frac{\partial}{\partial x}(u\,.\,2\pi\varpi\,\delta\varpi)\,\delta x+\frac{\partial}{\partial\varpi}(v\,.\,2\pi\varpi\,\delta x)\,\delta\varpi=0,$$

or

$$\frac{\partial}{\partial x}(\varpi u)+\frac{\partial}{\partial\varpi}(\varpi v)=0, \quad \text{..........................(2)}$$

which shews that

$$\varpi v\,.\,dx-\varpi u\,.\,d\varpi$$

is an exact differential. Denoting this by $d\psi$ we obtain the relations (1)*.

So far the motion has not been assumed to be irrotational; the condition that it should be so is

$$\frac{\partial v}{\partial x}-\frac{\partial u}{\partial\varpi}=0,$$

which leads to

$$\frac{\partial^2\psi}{\partial x^2}+\frac{\partial^2\psi}{\partial\varpi^2}-\frac{1}{\varpi}\frac{\partial\psi}{\partial\varpi}=0. \quad \text{..........................(3)}$$

The differential equation of ϕ is obtained by writing

$$u=-\frac{\partial\phi}{\partial x}, \quad v=-\frac{\partial\phi}{\partial\varpi}$$

in (2), viz. it is

$$\frac{\partial^2\phi}{\partial x^2}+\frac{\partial^2\phi}{\partial\varpi^2}+\frac{1}{\varpi}\frac{\partial\phi}{\partial\varpi}=0. \quad \text{..........................(4)}$$

It appears that the functions ϕ and ψ are not now (as they were in Art. 62) interchangeable. They are, indeed, of different dimensions.

The kinetic energy of the liquid contained in any region bounded by surfaces of revolution about the axis is given by

$$\begin{aligned}2T&=-\rho\iint\phi\frac{\partial\phi}{\partial n}\,dS\\&=\rho\int\phi\frac{\partial\psi}{\varpi\partial s}\,.\,2\pi\varpi ds\\&=2\pi\rho\int\phi\,d\psi, \quad \text{..............................(5)}\end{aligned}$$

* The stream-function for the case of symmetry about an axis was introduced in this manner by Stokes, "On the Steady Motion of Incompressible Fluids," *Camb. Trans.* vii. (1842) [*Papers*, i. 1]. Its analytical theory has been treated very fully by Sampson, "On Stokes' Current-Function," *Phil. Trans.* A, clxxxii. (1891).

δs denoting an element of the meridian section of the bounding surfaces, and the integration extending round the various parts of this section, in the proper directions. Compare Art. 61 (2).

95. In the case of a point-source at the origin whose velocity-potential is

$$\phi = \frac{1}{r}, \qquad \text{.......................(1)}$$

the flux through any closed curve is numerically equal to the solid angle which the curve subtends at the origin. Hence for a circle with Ox as axis, whose radius subtends an angle θ at O, we have, attending to the sign,

$$2\pi\psi = -2\pi(1-\cos\theta).$$

Omitting the constant term we have

$$\psi = \frac{x}{r} = \frac{\partial r}{\partial x}. \qquad \text{.......................(2)}$$

The solutions corresponding to any number of simple sources situate at various points of the axis of x may evidently be superposed; thus for the double-source

$$\phi = -\frac{\partial}{\partial x}\frac{1}{r} = \frac{\cos\theta}{r^2}, \qquad \text{.......................(3)}$$

we have

$$\psi = -\frac{\partial^2 r}{\partial x^2} = -\frac{\varpi^2}{r^3} = -\frac{\sin^2\theta}{r}. \qquad \text{.......................(4)}$$

And, generally, to the zonal solid harmonic of degree $-n-1$, viz. to

$$\phi = A\frac{\partial^n}{\partial x^n}\frac{1}{r}, \qquad \text{.......................(5)}$$

corresponds*

$$\psi = A\frac{\partial^{n+1} r}{\partial x^{n+1}}. \qquad \text{.......................(6)}$$

A more general formula, applicable to harmonics of any degree, fractional or not, may be obtained as follows. Using spherical polar co-ordinates r, θ, the component velocities along r, and perpendicular to r in the plane of the meridian, are found by making the linear element PP' of Art. 94 coincide successively with $r\,\delta\theta$ and δr, respectively, viz. they are

$$-\frac{1}{r\sin\theta}\frac{\partial\psi}{r\partial\theta}, \quad \frac{1}{r\sin\theta}\frac{\partial\psi}{\partial r}. \qquad \text{.......................(7)}$$

Hence in the case of irrotational motion we have

$$\frac{\partial\psi}{\sin\theta\,\partial\theta} = r^2\frac{\partial\phi}{\partial r}, \quad \frac{\partial\psi}{\partial r} = -\sin\theta\frac{\partial\phi}{\partial\theta}. \qquad \text{.......................(8)}$$

Thus if

$$\phi = r^n S_n, \qquad \text{.......................(9)}$$

where S_n is a zonal harmonic of order n, we have, putting $\mu = \cos\theta$,

$$\frac{\partial\psi}{\partial\mu} = -nr^{n+1}S_n, \quad \frac{\partial\psi}{\partial r} = r^n(1-\mu^2)\frac{dS_n}{d\mu}.$$

* Stefan, "Ueber die Kraftlinien eines um eine Axe symmetrischen Feldes," *Wied. Ann.* xvii. (1882).

The latter equation gives

$$\psi = \frac{1}{n+1} r^{n+1} (1-\mu^2) \frac{dS_n}{d\mu}, \quad \text{.....................(10)}$$

which must necessarily also satisfy the former; this is readily verified by means of Art. 84 (1).

Thus in the case of the zonal harmonic P_n, we have as corresponding values

$$\phi = r^n P_n(\mu), \quad \psi = \frac{1}{n+1} r^{n+1} (1-\mu^2) \frac{dP_n}{d\mu}, \quad \text{............(11)}$$

and

$$\phi = r^{-n-1} P_n(\mu), \quad \psi = -\frac{1}{n} r^{-n} (1-\mu^2) \frac{dP_n}{d\mu}, \quad \text{............(12)}$$

of which the latter must be equivalent to (5) and (6). The same relations hold of course with regard to the zonal harmonic of the second kind, Q_n.

96. We saw in Art. 92 that the motion produced by a solid sphere in an infinite mass of liquid may be regarded as due to a double-source at the

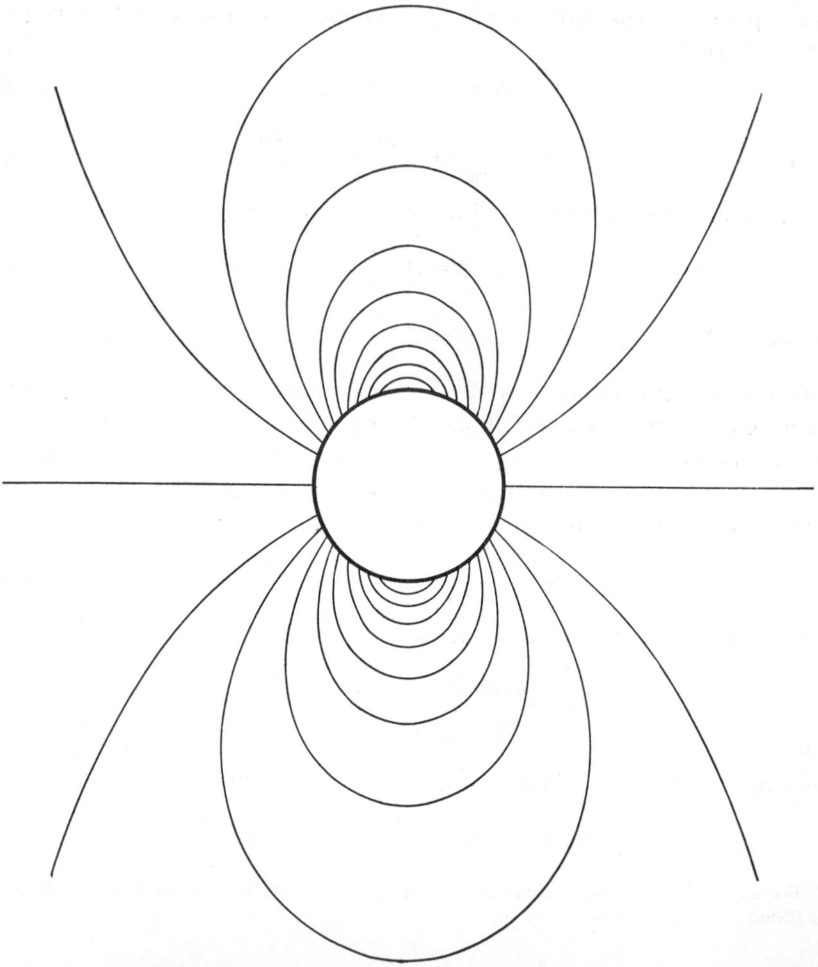

centre. Comparing the formulae there given with Art. 95 (4), it appears that the stream-function due to the sphere is

$$\psi = -\tfrac{1}{2}U\frac{a^3}{r}\sin^2\theta. \qquad\qquad (1)$$

The forms of the lines of motion corresponding to a number of equidistant values of ψ are shewn on the opposite page. The stream-lines *relative to the sphere* are figured in a diagram near the end of Chapter VII.

Again, the stream-function due to two double-sources having their axes oppositely directed along the axis of x will be of the form

$$\psi = \frac{A\varpi^2}{r_1^3} - \frac{B\varpi^2}{r_2^3}, \qquad\qquad (2)$$

where r_1, r_2 denote the distances of any point from the positions P and Q, say, of the two sources. At the stream-surface $\psi=0$ we have

$$r_1/r_2=(A/B)^{\frac{1}{3}},$$

i.e. the surface is a sphere in relation to which P and Q are inverse points. If O be the centre of this sphere, and a its radius, we find

$$A/B=OP^3/a^3=a^3/OQ^3. \qquad\qquad (3)$$

This sphere may be taken as a fixed boundary to the fluid on either side, and we thus obtain the motion due to a double-source (or say to an infinitely small sphere moving along Ox) in presence of a fixed spherical boundary. The disturbance of the stream-lines by the fixed sphere is that due to a double-source of the opposite sign placed at the 'inverse' point, the ratio of the strengths being given by (3)*. This fictitious double-source may be called the 'image' of the original one.

There is also a simple construction for the image of a point-source in a fixed sphere. The image of a source m at P will consist of a source $m\,.\,OQ/a$ at the inverse point Q, together with a line of sinks extending with uniform line-density $-m/a$ from P to the centre O†.

This might be deduced by integration from the preceding result, but a direct verification is simpler. It follows at once from Art. 95 (2) that the stream-function due to a line of sources of density m would be

$$\psi=m(r-r'), \qquad\qquad (4)$$

where r, r' are the distances of the two ends of the line from the point considered. Hence the arrangement of sources just described will give, at any point R on the sphere,

$$\psi=-m\,.\cos RPO-m\,.\frac{OQ}{a}\cos OQR-\frac{m}{a}(OR-QR). \qquad\qquad (5)$$

Since $\qquad QR=OR\cos ORQ+OQ\cos OQR$, and $RPO=ORQ$,

this reduces to $\psi=-m$, a constant over the sphere.

For the calculation of the force on the sphere we have recourse to zonal harmonics. Referred to O as origin the velocity-potential of the original source, in the neighbourhood of the sphere, is given by

$$\phi/m=\frac{1}{c}+\frac{r\cos\theta}{c^2}+\frac{r^2(3\cos^2\theta-1)}{2c^3}+\ldots \qquad\qquad (6)$$

* This result was given by Stokes, "On the Resistance of a Fluid to two Oscillating Spheres," *Brit. Ass. Report*, 1847 [*Papers*, i. 230].

† Hicks, *l.c. infra*, p. 134. See the diagram on p. 71 *ante*.

The motion reflected from the sphere will be given by

$$\phi'/m = \frac{a^3 \cos\theta}{2c^2 r^2} + \frac{a^5(3\cos^2\theta - 1)}{3c^3 r^3} + \ldots, \quad \text{.........................(7)}$$

since this makes $\partial/\partial r\,(\phi+\phi')=0$, for $r=a$. The velocity at the surface will therefore be

$$q = -\frac{\partial}{a\partial\theta}(\phi+\phi') = \frac{3m}{2c^2}\sin\theta + \frac{5ma}{3c^3}\sin\theta\cos\theta + \ldots. \quad \text{.................(8)}$$

For an approximate result we may stop the expansion at this point. The resultant force towards P is then

$$X = -\int_0^\pi p\cos\theta \,.\, 2\pi a^2 \sin\theta\, d\theta = \pi\rho a^2 \int_0^\pi q^2 \sin\theta\cos\theta\, d\theta = \frac{4\pi\rho a^3 m^2}{c^5}. \quad \text{......(9)}$$

If f be the acceleration at O when the sphere is absent, viz. $f=2m^2/c^5$, we have

$$X = 2\pi\rho a^3 f. \quad \text{....................................(10)*.}$$

97. Rankine† employed a method similar to that of Art. 71 to discover forms of solids of revolution which will by motion parallel to their axes generate in a surrounding liquid any given type of irrotational motion symmetrical about an axis.

The velocity of the solid being U, and δs denoting an element of the meridian, the normal velocity at any point of the surface is $U\partial\varpi/\partial s$, and that of the fluid in contact is given by $-\partial\psi/\varpi\partial s$. Equating these and integrating along the meridian, we have

$$\psi = -\tfrac{1}{2}U\varpi^2 + \text{const.} \quad \text{..........................(1)}$$

If in this we substitute the value of ψ due to any distribution of sources along the axis of symmetry, we obtain the equation of a family of stream-lines. If the sum of the strengths is zero, one of these lines will serve as the profile of a finite solid of revolution past which the flow takes place.

In this way we may readily verify the solution already obtained for the sphere; thus, assuming

$$\psi = A\varpi^2/r^3, \quad \text{..............................(2)}$$

we find that (1) is satisfied for $r=a$, provided

$$A = -\tfrac{1}{2}Ua^3, \quad \text{..............................(3)}$$

which agrees with Art. 96 (1).

By a continuous distribution of sources and sinks along the axis it has been found possible to imitate forms which have empirically been found advantageous for the profiles of air-ships. The fluid pressures can in such cases be calculated, and the results compared with experiment.

98. The motion of a liquid bounded by *two* spherical surfaces can be found by successive approximations in certain cases. For two solid spheres moving in the line of centres the solution is greatly facilitated by the result given at the end of Art. 96, as to the 'image' of a double-source in a fixed sphere.

* Prof. G. I. Taylor, *Aeronautical Research Committee, R. & M.* 1166 (1928).

† "On the Mathematical Theory of Stream Lines, especially those with Four Foci and upwards," *Phil. Trans.* 1871, p. 267 (not included in the collection referred to on p. 63 *ante*).

Let a, b be the radii, and c the distance between the centres A, B. Let U be the velocity of A towards B, U' that of B towards A. Also, P being any point, let $AP=r$, $BP=r'$, $PAB=\theta$, $PBA=\theta'$. The velocity-potential will be of the form

$$U\phi+U'\phi', \qquad \text{.......(1)}$$

where the functions ϕ and ϕ' are to be determined by the conditions that

$$\nabla^2\phi=0, \qquad \nabla^2\phi'=0, \qquad \text{.......(2)}$$

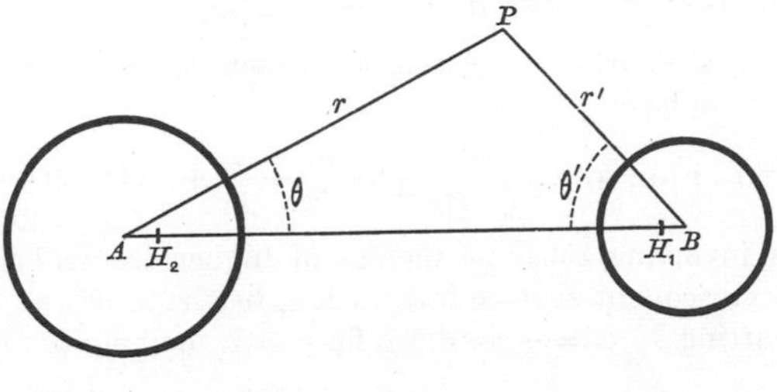

throughout the fluid, that their space-derivatives vanish at infinity, and that

$$\frac{\partial\phi}{\partial r}=-\cos\theta, \qquad \frac{\partial\phi'}{\partial r}=0, \qquad \text{.......(3)}$$

over the surface of A, whilst

$$\frac{\partial\phi}{\partial r'}=0, \qquad \frac{\partial\phi'}{\partial r'}=-\cos\theta', \qquad \text{.......(4)}$$

over the surface of B. It is evident that ϕ is the value of the velocity-potential when A moves with unit velocity towards B, while B is at rest; and similarly for ϕ'.

To find ϕ, we remark that if B were absent the motion of the fluid would be that due to a certain double-source at A having its axis in the direction AB. The theorem of Art. 96 shews that we may satisfy the condition of zero normal velocity over the surface of B by introducing a double-source, viz. the 'image' of that at A in the sphere B. This image is at H_1, the inverse point of A with respect to the sphere B; its axis coincides with AB, and its strength is $-\mu_0 b^3/c^3$, where μ_0 is the strength of the original source at A, viz.

$$\mu_0=2\pi a^3.$$

The resultant motion due to the two sources at A and H_1 will however violate the condition to be satisfied at the surface of the sphere A, and in order to neutralize the normal velocity at this surface, due to H_1, we must superpose a double-source at H_2, the image of H_1 in the sphere A. This will introduce a normal velocity at the surface of B, which may again be neutralized by adding the image of H_2 in B, and so on. If $\mu_1, \mu_2, \mu_3, \ldots$ be the strengths of the successive images, and $f_1, f_2, f_3, \ldots$ their distances from A, we have

$$\left.\begin{aligned}
f_1&=c-\frac{b^2}{c}, & f_2&=\frac{a^2}{f_1}, & \frac{\mu_1}{\mu_0}&=-\frac{b^3}{c^3}, & \frac{\mu_2}{\mu_1}&=-\frac{a^3}{f_1^3},\\
f_3&=c-\frac{b^2}{c-f_2}, & f_4&=\frac{a^2}{f_3}, & \frac{\mu_3}{\mu_2}&=-\frac{b^3}{(c-f_2)^3}, & \frac{\mu_4}{\mu_3}&=-\frac{a^3}{f_3^3},\\
f_5&=c-\frac{b^2}{c-f_4}, & f_6&=\frac{a^2}{f_5}, & \frac{\mu_5}{\mu_4}&=-\frac{b^3}{(c-f_4)^3}, & \frac{\mu_6}{\mu_5}&=-\frac{a^3}{f_5^3},
\end{aligned}\right\} \qquad \text{.......(5)}$$

and so on, the laws of formation being obvious. The images continually diminish in intensity, and this very rapidly if the radius of either sphere is small compared with the shortest distance between the two surfaces.

The formula for the kinetic energy is

$$2T=-\rho\iint(U\phi+U'\phi')\left(U\frac{\partial\phi}{\partial n}+U'\frac{\partial\phi'}{\partial n}\right)dS=LU^2+2MUU'+NU'^2, \quad \text{......(6)}$$

provided

$$L=-\rho\iint\phi\frac{\partial\phi}{\partial n}dS_A,\quad M=-\rho\iint\phi\frac{\partial\phi'}{\partial n}dS_B=-\rho\iint\phi'\frac{\partial\phi}{\partial n}dS_A,\quad N=-\rho\iint\phi'\frac{\partial\phi'}{\partial n}dS_B, \quad \text{..(7)}$$

where the suffixes indicate over which sphere the integration is to be effected. The equality of the two forms of M follows from Green's Theorem (Art. 44).

The value of ϕ near the surface of A can be written down at once from the results (7) and (8) of Art. 85, viz. we have

$$4\pi\phi=(\mu_0+\mu_2+\mu_4+\ldots)\frac{\cos\theta}{r^2}-2\left(\frac{\mu_1}{f_1^3}+\frac{\mu_3}{f_3^3}+\ldots\right)r\cos\theta+\&c., \quad \text{............(8)}$$

the remaining terms, involving zonal harmonics of higher orders, being omitted, as they will disappear in the subsequent surface-integration, in virtue of the orthogonal property of Art. 87. Hence, putting $\partial\phi/\partial n=-\cos\theta$, we find with the help of (5)

$$L=\tfrac{1}{3}\rho\,(\mu_0+3\mu_2+3\mu_4+\ldots)=\tfrac{2}{3}\pi\rho a^3\left(1+3\frac{a^3b^3}{c^3f_1^3}+3\frac{a^6b^6}{c^3f_1^3(c-f_2)^3f_3^3}+\ldots\right). \quad \text{......(9)}$$

It appears that the inertia of the sphere A is in all cases increased by the presence of a fixed sphere B. Compare Art. 93.

The value of N may be written down from symmetry, viz. it is

$$N=\tfrac{2}{3}\pi\rho b^3\left(1+3\frac{a^3b^3}{c^3f_1'^3}+3\frac{a^6b^6}{c^3f_1'^3(c-f_2')^3f_3'^3}+\ldots\right), \quad \text{..................(10)}$$

where

$$\left.\begin{aligned} f_1'&=c-\frac{a^2}{c}, & f_2'&=\frac{b^2}{f_1'},\\ f_3'&=c-\frac{a^2}{c-f_2'}, & f_4'&=\frac{b^2}{f_3'},\\ f_5'&=c-\frac{a^2}{c-f_4'}, & f_6'&=\frac{b^2}{f_5'},\end{aligned}\right\} \quad \text{..............................(11)}$$

and so on.

To calculate M we require the value of ϕ' near the surface of the sphere A; this is due to double-sources $\mu_0', \mu_1', \mu_2', \mu_3', \ldots$ at distances $c, c-f_1', c-f_2', c-f_3', \ldots$ from A, where $\mu_0'=-2\pi b^3$, and

$$\left.\begin{aligned} \frac{\mu_1'}{\mu_0'}&=-\frac{a^3}{c^3}, & \frac{\mu_2'}{\mu_1'}&=-\frac{b^3}{f_1'^3},\\ \frac{\mu_3'}{\mu_2'}&=-\frac{a^3}{(c-f_2')^3}, & \frac{\mu_4'}{\mu_3'}&=-\frac{b^3}{f_3'^3},\\ \frac{\mu_5'}{\mu_4'}&=-\frac{a^3}{(c-f_4')^3}, & \frac{\mu_6'}{\mu_5'}&=-\frac{b^3}{f_5'^3},\end{aligned}\right\} \quad \text{..........................(12)}$$

and so on. This gives, for points near the surface of A,

$$4\pi\phi'=(\mu_1'+\mu_3'+\mu_5'+\ldots)\frac{\cos\theta}{r^2}-2\left(\frac{\mu_0'}{c^3}+\frac{\mu_2'}{(c-f_2')^3}+\frac{\mu_4'}{(c-f_4')^3}+\ldots\right)r\cos\theta+\&c. \quad \text{......(13)}$$

Hence

$$M=-\rho\iint\phi'\frac{\partial\phi}{\partial n}dS_A=\rho\,(\mu_1'+\mu_3'+\mu_5'+\ldots)$$

$$=2\pi\rho\frac{a^3b^3}{c^3}\left\{1+\frac{a^3b^3}{f_1'^3(c-f_2')^3}+\frac{a^6b^6}{f_1'^3f_3'^3(c-f_2')^3(c-f_4')^3}+\ldots\right\}. \quad \text{.........(14)}$$

When the ratios a/c and b/c are both small we have

$$L=\tfrac{2}{3}\pi\rho a^3\left(1+3\frac{a^3b^3}{c^6}\right),\quad M=2\pi\rho\frac{a^3b^3}{c^3},\quad N=\tfrac{2}{3}\pi\rho b^3\left(1+3\frac{a^3b^3}{c^6}\right),\ \ldots\ldots(15)$$

approximately*.

If in the preceding results we put $b=a$, $U'=U$, the plane bisecting AB at right angles will be a plane of symmetry, and may therefore be taken as a fixed boundary to the fluid on either side. Hence, putting $c=2h$, we find, for the kinetic energy of the liquid when a sphere is in motion perpendicular to a rigid plane boundary, at a distance h from it,

$$2T=\tfrac{2}{3}\pi\rho a^3\left(1+\tfrac{3}{8}\frac{a^3}{h^3}+\ldots\right)U^2,\ \ldots\ldots(16)$$

a result due to Stokes.

99. When the spheres are moving at right angles to the line of centres the problem is more difficult; we shall therefore content ourselves with the first steps in the approximation, referring, for a more complete treatment, to the papers cited on p. 134.

Let the spheres be moving with velocities V, V' in parallel directions at right angles to A, B, and let r, θ, ω and r', θ', ω' be two systems of spherical polar co-ordinates having their origins at A and B respectively, and their polar axes in the directions of the velocities V, V'. The velocity-potential will be of the form

$$V\phi+V'\phi',$$

with the surface-conditions

$$\frac{\partial\phi}{\partial r}=-\cos\theta,\quad \frac{\partial\phi'}{\partial r}=0,\quad \text{for } r=a,\ \ldots\ldots(1)$$

and

$$\frac{\partial\phi}{\partial r'}=0,\quad \frac{\partial\phi'}{\partial r'}=-\cos\theta',\quad \text{for } r'=b.\ \ldots\ldots(2)$$

If the sphere B were absent the velocity-potential due to unit velocity of A would be

$$\tfrac{1}{2}\frac{a^3}{r^2}\cos\theta.$$

Since $r\cos\theta=r'\cos\theta'$, the value of this in the neighbourhood of B will be

$$\tfrac{1}{2}\frac{a^3}{c^3}r'\cos\theta',$$

approximately. The normal velocity at the surface of B, due to this, will be cancelled by the addition of the term

$$\tfrac{1}{4}\frac{a^3b^3}{c^3}\frac{\cos\theta'}{r'^2},$$

which in the neighbourhood of A becomes equal to

$$\tfrac{1}{4}\frac{a^3b^3}{c^6}r\cos\theta,$$

nearly. To rectify the normal velocity at the surface of A, we add the term

$$\tfrac{1}{8}\frac{a^6b^3}{c^6}\frac{\cos\theta}{r^2}.$$

Stopping at this point, and collecting our results, we have, over the surface of A,

$$\phi=\tfrac{1}{2}a\left(1+\tfrac{3}{4}\frac{a^3b^3}{c^6}\right)\cos\theta,\ \ldots\ldots(3)$$

and at the surface of B,

$$\phi=\tfrac{3}{4}b\,.\,\frac{a^3}{c^3}\cos\theta'.\ \ldots\ldots(4)$$

* To this degree of approximation the results may be more easily obtained without the use of 'images,' the procedure being similar to that of the next Art.

Hence if we denote by P, Q, R the coefficients in the expression for the kinetic energy, viz.

$$2T = PV^2 + 2QVV' + RV'^2, \quad \text{.............................(5)}$$

we have

$$\left.\begin{aligned} P &= -\rho\iint \phi\frac{\partial\phi}{\partial n}\, dS_A = \tfrac{2}{3}\pi\rho a^3\left(1+\tfrac{3}{4}\frac{a^3b^3}{c^6}\right),\\ Q &= -\rho\iint \phi\frac{\partial\phi'}{\partial n}\, dS_B = \pi\rho\,\frac{a^3b^3}{c^3},\\ R &= -\rho\iint \phi'\frac{\partial\phi'}{\partial n}\, dS_B = \tfrac{2}{3}\pi\rho b^3\left(1+\tfrac{3}{4}\frac{a^3b^3}{c^6}\right). \end{aligned}\right\} \quad \text{.....................(6)}$$

The case of a sphere moving parallel to a fixed plane boundary, at a distance h, is obtained by putting $b = a$, $V = V'$, $c = 2h$, and halving the consequent value of T; thus

$$2T = \tfrac{2}{3}\pi\rho a^3\left(1+\tfrac{3}{16}\frac{a^3}{h^3}\right)V^2. \quad \text{..............................(7)}$$

This result, which was also given by Stokes, may be compared with that of Art. 98 (16)*.

Cylindrical Harmonics.

100. In terms of the cylindrical co-ordinates x, ϖ, ω introduced in Art. 89, the equation $\nabla^2\phi = 0$ takes the form

$$\frac{\partial^2\phi}{\partial x^2} + \frac{\partial^2\phi}{\partial\varpi^2} + \frac{1}{\varpi}\frac{\partial\phi}{\partial\varpi} + \frac{1}{\varpi^2}\frac{\partial^2\phi}{\partial\omega^2} = 0. \quad \text{...................(1)}$$

This may be obtained by direct transformation, or more simply by expressing that the total flux across the boundary of an element $\delta x\,.\,\delta\varpi\,.\,\varpi\delta\omega$ is zero, after the manner of Art. 83.

In the case of symmetry about the axis of x, the equation reduces to the form (4) of Art. 94. A particular solution is then $\phi = e^{\pm kx}\chi(\varpi)$, provided

$$\chi''(\varpi) + \frac{1}{\varpi}\chi'(\varpi) + k^2\chi(\varpi) = 0. \quad \text{...................(2)}$$

This is the differential equation of 'Bessel's Functions' of zero order. Its complete primitive consists, of course, of the sum of two definite functions of ϖ, each multiplied by an arbitrary constant. That solution which is finite for $\varpi = 0$ is easily found in the form of an ascending series; it is usually denoted by $CJ_0(k\varpi)$, where

$$J_0(\zeta) = 1 - \frac{\zeta^2}{2^2} + \frac{\zeta^2}{2^2.4^2} - \dots \quad \text{...................(3)}$$

* For a fuller analytical treatment of the problem of the motion of two spheres we refer to the following papers: W. M. Hicks, "On the Motion of two Spheres in a Fluid," *Phil. Trans.* 1880, p. 455; R. A. Herman, "On the Motion of two Spheres in Fluid," *Quart. Journ. Math.* xxii. (1887); Basset, "On the Motion of Two Spheres in a Liquid, &c." *Proc. Lond. Math. Soc.* xviii. 369 (1887). See also C. Neumann, *Hydrodynamische Untersuchungen*, Leipzig, 1883; Basset, *Hydrodynamics*, Cambridge, 1888. The mutual influence of 'pulsating' spheres, *i.e.* of spheres which periodically change their *volume*, has been studied by C. A. Bjerknes, with a view to a mechanical illustration of electric and other forces. A full account of these researches is given by his son Prof. V. Bjerknes in *Vorlesungen über hydrodynamische Fernkräfte*, Leipzig, 1900–1902. The question is also treated by Hicks, *Camb. Proc.* iii. 276 (1879), iv. 29 (1880), and by Voigt, *Gött. Nachr.* 1891, p. 37.

We have thus obtained solutions of $\nabla^2\phi = 0$ of the types*

$$\phi = e^{\pm kx} J_0(k\varpi). \quad\text{.............................(4)}$$

It is easily seen from Art. 94 (1) that the corresponding value of the stream-function is

$$\psi = \mp\, \varpi e^{\pm kx} J_0'(k\varpi). \quad\text{.........................(5)}$$

The formula (4) may be recognized as a particular case of Art. 89 (6); viz. it is equivalent to

$$\phi = \frac{1}{\pi}\int_0^\pi e^{\pm k(x+i\varpi\cos\vartheta)}\, d\vartheta, \quad\text{......................(6)}$$

since

$$J_0(\zeta) = \frac{1}{\pi}\int_0^\pi \cos(\zeta\cos\vartheta)\, d\vartheta = \frac{1}{\pi}\int_0^\pi e^{i\zeta\cos\vartheta}\, d\vartheta, \quad\text{...........(7)}$$

as may be verified by developing the cosine, and integrating term by term.

Again, (4) may also be identified as the limiting form assumed by a spherical solid zonal harmonic when the order (n) is made infinite, provided that at the same time the distance of the origin from the point considered be made infinitely great, the two infinities being subject to a certain relation†.

Thus we may take

$$\phi = \frac{r^n}{a^n} P_n(\mu) = \left(1 + \frac{x}{a}\right)^n \chi_n(\varpi), \quad\text{...................(8)}$$

where we have temporarily changed the meanings of x and ϖ, viz.

$$r = a + x, \qquad \varpi = 2a\sin\tfrac{1}{2}\theta,$$

whilst

$$\chi_n(\varpi) = 1 - \frac{n(n+1)}{2^2}\frac{\varpi^2}{a^2} + \frac{(n-1)n(n+1)(n+2)}{2^2 . 4^2}\frac{\varpi^4}{a^4} - \ldots; \quad\text{......(9)}$$

see Art. 85 (4). If we now put $k = n/a$, and suppose a and n to become infinite, whilst k remains finite, the symbols x and ϖ will regain their former meanings, and we reproduce the formula (4) with the upper sign in the exponential. The lower sign is obtained if we start with

$$\phi = \frac{a^{n+1}}{r^{n+1}} P_n(\mu).$$

The same procedure leads to an expression of an arbitrary function of ϖ in terms of the Bessel's Function of zero order‡. According to Art. 88, an arbitrary function of latitude on the surface of a sphere can be expanded in spherical zonal harmonics, thus

$$F(\mu) = \Sigma(n+\tfrac{1}{2}) P_n(\mu) \int_{-1}^{1} F(\mu') P_n(\mu')\, d\mu'. \quad\text{...........(10)}$$

* Except as to notation these solutions are to be found in Poisson, *l.c. ante* p. 18.

† This process was indicated, without the restriction to symmetry, by Thomson and Tait, Art. 783 (1867).

‡ The procedure appears to be due substantially to C. Neumann (1862).

If we denote by ϖ the length of the chord drawn to the variable point from the pole ($\theta = 0$) of the sphere, we have

$$\varpi = 2a \sin \tfrac{1}{2}\theta, \qquad \varpi\,\delta\varpi = -a^2 \delta\mu,$$

where a is the radius, so that the formula may be written

$$f(\varpi) = \frac{1}{a^2} \Sigma (n + \tfrac{1}{2}) H_n(\varpi) \int_0^{2a} f(\varpi') H_n(\varpi') \varpi' d\varpi'. \quad \ldots\ldots(11)$$

If we now put
$$k = \frac{n}{a}, \qquad \delta k = \frac{1}{a},$$

and finally make a infinite, we are led to the important theorem*:

$$f(\varpi) = \int_0^\infty J_0(k\varpi)\, k\, dk \int_0^\infty f(\varpi') J_0(k\varpi') \varpi' d\varpi'. \quad \ldots\ldots\ldots\ldots(12)$$

101. If in (1) we suppose ϕ to be expanded in a series of terms varying as $\cos s\omega$ or $\sin s\omega$, each such term will be subject to an equation of the form

$$\frac{\partial^2 \phi}{\partial x^2} + \frac{\partial^2 \phi}{\partial \varpi^2} + \frac{1}{\varpi}\frac{\partial \phi}{\partial \varpi} - \frac{s^2}{\varpi^2}\phi = 0. \quad \ldots\ldots\ldots\ldots\ldots\ldots\ldots(13)$$

This will be satisfied by $\phi = e^{\pm kx} \chi(\varpi)$, provided

$$\chi''(\varpi) + \frac{1}{\varpi}\chi'(\varpi) + \left(k^2 - \frac{s^2}{\varpi^2}\right)\chi(\varpi) = 0, \quad \ldots\ldots\ldots\ldots\ldots(14)$$

which is the differential equation of Bessel's Functions of order s†. The solution which is finite for $\varpi = 0$ may be written $\chi(\varpi) = C J_s(k\varpi)$, where

$$J_s(\zeta) = \frac{\zeta^s}{2^s . \Pi(s)} \left\{1 - \frac{\zeta^2}{2(2s+2)} + \frac{\zeta^4}{2 . 4 (2s+2)(2s+4)} - \ldots\right\}. \quad \ldots(15)$$

The complete solution of (14) involves, in addition, a Bessel's Function 'of the second kind' with whose form we shall be concerned at a later period in our subject‡.

We have thus obtained solutions of the equation $\nabla^2 \phi = 0$, of the types

$$\phi = e^{\pm kx} J_s(k\varpi) \left.\begin{matrix}\cos \\ \sin\end{matrix}\right\} s\omega. \quad \ldots\ldots\ldots\ldots\ldots\ldots\ldots\ldots(16)$$

* For more rigorous proofs, and for the history of the theorem, see Watson, *l.c. infra*.

† Forsyth, Art. 100; Whittaker and Watson, c. xvii.

‡ For the further theory of the Bessel's Functions of both kinds recourse may be had to Gray and Mathews, *Treatise on Bessel Functions*, 2nd ed., London, 1922, and to G. N. Watson, *Theory of Bessel Functions*, Cambridge, 1923, where ample references are given to previous writers. An account of the subject, from the physical point of view, will be found in Rayleigh's *Theory of Sound*, cc. ix., xviii., with many important applications.

Numerical tables of the functions $J_s(\zeta)$ have been constructed by Bessel and Hansen, and more recently by Meissel (*Berl. Abh.* 1888). These are reproduced by Gray and Mathews, and, with valuable extensions, in Watson's treatise. Abridged tables are included in the collections of Dale and of Jahnke and Emde referred to on p. 114.

These may also be obtained as limiting forms of the spherical solid harmonics

$$\frac{r^n}{a^n} P_n^s(\mu) \left.\begin{matrix}\cos\\ \sin\end{matrix}\right\} s\omega, \qquad \frac{a^{n+1}}{r^{n+1}} P_n^s(\mu) \left.\begin{matrix}\cos\\ \sin\end{matrix}\right\} s\omega,$$

with the help of the expansion (6) of Art. 86*.

102. The formula (12) of Art. 100 enables us to write down expressions, which are sometimes convenient, for the value of ϕ on one side of an infinite plane ($x = 0$) in terms of the values of ϕ or $\partial\phi/\partial n$ at points of this plane, in the case of symmetry about an axis (Ox) normal to the plane†. Thus if

$$\phi = F(\varpi), \quad \text{for } x = 0, \qquad \text{.......................(1)}$$

we have, on the side $x > 0$,

$$\phi = \int_0^\infty e^{-kx} J_0(k\varpi)\, k\, dk \int_0^\infty F(\varpi') J_0(k\varpi')\, \varpi' d\varpi'. \qquad \text{........(2)}$$

Again, if

$$-\frac{\partial\phi}{\partial x} = f(\varpi), \quad \text{for } x = 0, \qquad \text{......................(3)}$$

we have

$$\phi = \int_0^\infty e^{-kx} J_0(k\varpi)\, dk \int_0^\infty f(\varpi') J_0(k\varpi')\, \varpi' d\varpi'. \qquad \text{...........(4)}$$

The exponentials have been chosen so as to vanish for $x = \infty$.

Another solution of these problems has already been given in Art. 58, from equations (12) and (11) of which we derive

$$\phi = \frac{1}{2\pi} \iint \phi \frac{\partial}{\partial x}\left(\frac{1}{r}\right) dS, \qquad \text{.........................(5)}$$

and

$$\phi = -\frac{1}{2\pi} \iint \frac{\partial\phi}{\partial x} \frac{dS}{r}, \qquad \text{.............................(6)}$$

respectively, where r denotes distance from the element δS of the plane to the point at which the value of ϕ is required.

We proceed to a few applications of the general formulae (2) and (4).

1°. If, in (4), we assume $f(\varpi)$ to vanish for all but infinitesimal values of ϖ, and to become infinite for these in such a way that

$$\int_0^\infty f(\varpi)\, 2\pi\varpi\, d\varpi = \tfrac{1}{2}.$$

we obtain

$$4\pi\phi = \int_0^\infty e^{-kx} J_0(k\varpi)\, dk, \qquad \text{..................................(7)}$$

and therefore, since $J_0' = -J_1$,

$$4\pi\psi = -\varpi \int_0^\infty e^{-kx} J_1(k\varpi)\, dk, \qquad \text{............................(8)}$$

by Art. 100 (5).

* The connection between spherical surface-harmonics and Bessel's Functions was noticed by Mehler, "Ueber die Vertheilung d. statischen Elektricität in einem v. zwei Kugelkalotten begrenzten Körper," *Crelle*, lxviii. (1868). It was investigated independently by Rayleigh, "On the Relation between the Functions of Laplace and Bessel," *Proc. Lond. Math. Soc.* ix. 61 (1878) [*Papers*, i. 338]; see also *Theory of Sound*, Arts. 336, 338.

There are also methods of deducing Bessel's Functions 'of the second kind' as limiting forms of the spherical harmonics $Q_n(\mu)$, $Q_n^s(\mu) \left.\begin{matrix}\cos\\ \sin\end{matrix}\right\} s\omega$; for these see Heine, i. 184, 232.

† The method may be extended so as to be free from this restriction.

By comparison with the primitive expressions for a point-source at the origin (Art. 95), we infer that

$$\int_0^\infty e^{-kx} J_0(k\varpi)\,dk = \frac{1}{r}, \qquad \int_0^\infty e^{-kx} J_1(k\varpi)\,dk = \frac{\varpi}{r(r+x)}, \quad\ldots\ldots(9)$$

where $r=\surd(x^2+\varpi^2)$; these are in fact known results*.

2°. Let us next suppose that sources are distributed with uniform density over the plane area contained by the circle $\varpi=a$, $x=0$. Using the series for J_0, J_1, or otherwise, we find

$$\int_0^a J_0(k\varpi)\,\varpi\,d\varpi = \frac{a}{k} J_1(ka). \quad\ldots\ldots(10)$$

Hence†

$$\phi = \frac{1}{\pi a}\int_0^\infty e^{-kx} J_0(k\varpi) J_1(ka)\frac{dk}{k}, \qquad \psi = -\frac{\varpi}{\pi a}\int_0^\infty e^{-kx} J_1(k\varpi) J_1(ka)\frac{dk}{k}, \quad\ldots\ldots(11)$$

where the constant factor has been chosen so as to make the total flux through the circle equal to unity.

3°. Again, if the density of the sources, within the same circle, vary as $1/\surd(a^2-\varpi^2)$, we have to deal with the integral‡

$$\int_0^a J_0(k\varpi)\frac{\varpi\,d\varpi}{\surd(a^2-\varpi^2)} = a\int_0^{\frac{1}{2}\pi} J_0(ka\sin\vartheta)\sin\vartheta\,d\vartheta = \frac{\sin ka}{k}, \quad\ldots\ldots(12)$$

where the evaluation is effected by substituting the series form of J_0, and treating each term separately. Hence

$$\phi = \frac{1}{2\pi a}\int_0^\infty e^{-kx} J_0(k\varpi)\sin ka\frac{dk}{k}, \qquad \psi = -\frac{\varpi}{2\pi a}\int_0^\infty e^{-kx} J_1(k\varpi)\sin ka\frac{dk}{k}, \quad\ldots\ldots(13)$$

if the constant factor be determined by the same condition as before§.

It is a known theorem of Electrostatics that the assumed law of density makes ϕ constant over the circular area. It may be shewn independently that

$$\left.\begin{aligned} &\int_0^\infty J_0(k\varpi)\sin ka\frac{dk}{k} = \tfrac{1}{2}\pi, \text{ or } \sin^{-1}\frac{a}{\varpi}, \\ &\int_0^\infty J_1(k\varpi)\sin ka\frac{dk}{k} = \frac{a-\surd(a^2-\varpi^2)}{\varpi}, \text{ or } \frac{a}{\varpi}, \end{aligned}\right\} \quad\ldots\ldots(14)$$

according as $\varpi \lesseqgtr a$||. The formulae (13) therefore express the flow of a liquid through a circular aperture in a thin plane rigid wall. Another solution will be obtained in Art. 108. The corresponding problem in two dimensions was solved in Art. 66, 1°.

4°. Let us next suppose that when $x=0$, we have $\phi=C\surd(a^2-\varpi^2)$ for $\varpi<a$, and $\phi=0$ for $\varpi>a$. We find

$$\int_0^a J_0(k\varpi)\surd(a^2-\varpi^2)\,\varpi\,d\varpi = a^3\int_0^{\frac{1}{2}\pi} J_0(ka\sin\vartheta)\sin\vartheta\cos^2\vartheta\,d\vartheta = a^3\psi_1(ka), \quad\ldots(15)$$

provided

$$\psi_1(\zeta) = \tfrac{1}{3}\left(1-\frac{\zeta^2}{2.5}+\frac{\zeta^4}{2.4.5.7}-\ldots\right) = -\frac{d}{\zeta d\zeta}\frac{\sin\zeta}{\zeta}. \quad\ldots\ldots(16)$$

Hence, by (2),

$$\phi = -C\int_0^\infty e^{-kx} J_0(k\varpi)\frac{d}{dk}\left(\frac{\sin ka}{k}\right)dk. \quad\ldots\ldots(17)$$

* The former is due to Lipschitz, *Crelle*, lvi. 189 (1859); see Watson, p. 384. The latter follows by differentiation with respect to ϖ and integration with respect to x.

† Cf. H. Weber, *Crelle*, lxxv. 88; Heine, ii. 180.

‡ The formula (12) has been given by various writers; see Rayleigh, *Papers*, iii. 98; Hobson, *Proc. Lond. Math. Soc.* xxv. 71 (1893).

§ Cf. H. Weber, *Crelle*, lxxv. (1873); Heine, ii. 192.

|| H. Weber, *Crelle*, lxxv.; Watson, p. 405. See also *Proc. Lond. Math. Soc.* xxxiv. 282.

This gives, for $x=0$,

$$-\left(\frac{\partial\phi}{\partial x}\right)_0 = C\int_0^\infty J_0(k\varpi)\sin ka\frac{dk}{k} + C\varpi\int_0^\infty J_0'(k\varpi)\sin ka\,dk, \quad\text{.........(18)}$$

after a partial integration. The value of the former integral is given in (14), and that of the latter can be deduced from it by differentiation with respect to ϖ. Hence

$$-\left(\frac{\partial\phi}{\partial x}\right)_0 = \tfrac{1}{2}\pi C, \text{ or } C\left(\sin^{-1}\frac{a}{\varpi} - \frac{a}{\sqrt{(\varpi^2-a^2)}}\right), \quad\text{.........(19)}$$

according as $\varpi \lesseqgtr a$. It follows that if $C=2/\pi\,.\,U$, the formula (17) will relate to the motion of a thin circular disk with velocity U normal to its plane, in an infinite mass of liquid. The expression for the kinetic energy is

$$2T = -\rho\iint\phi\frac{\partial\phi}{\partial n}dS = \pi\rho C^2\int_0^a\sqrt{(a^2-\varpi^2)}\,2\pi\varpi\,d\varpi = \tfrac{2}{3}\pi^2\rho a^3C^2,$$

or

$$2T = \tfrac{8}{3}\rho a^3U^2. \quad\text{.........(20)}$$

The effective addition to the inertia of the disk is therefore $2/\pi$ (= ·6366) times the mass of a spherical portion of the fluid, of the same radius. For another investigation of this question, see Art. 108.

Ellipsoidal Harmonics.

103. The method of Spherical Harmonics can also be adapted to the solution of the equation

$$\nabla^2\phi = 0, \quad\text{.........(1)}$$

under boundary-conditions having relation to ellipsoids of revolution*.

Beginning with the case where the ellipsoids are *prolate*, we write

$$\left.\begin{aligned} x = k\cos\theta\cosh\eta = k\mu\zeta, \qquad y = \varpi\cos\omega, \qquad z = \varpi\sin\omega, \\ \text{where}\qquad \varpi = k\sin\theta\sinh\eta = k(1-\mu^2)^{\frac{1}{2}}(\zeta^2-1)^{\frac{1}{2}}. \end{aligned}\right\}\ldots \quad (2)$$

The surfaces $\zeta=$ const., $\mu=$ const. are confocal ellipsoids and hyperboloids of two sheets, respectively, the common foci being the points $(\pm k, 0, 0)$. The value of ζ may range from 1 to ∞, whilst μ lies between ± 1. The co-ordinates μ, ζ, ω form an orthogonal system, and the values of the linear elements δs_μ, δs_ζ, δs_ω described by the point (x, y, z) when μ, ζ, ω separately vary are

$$\delta s_\mu = k\left(\frac{\zeta^2-\mu^2}{1-\mu^2}\right)^{\frac{1}{2}}\delta\mu, \quad \delta s_\zeta = k\left(\frac{\zeta^2-\mu^2}{\zeta^2-1}\right)^{\frac{1}{2}}\delta\zeta, \quad \delta s_\omega = k(1-\mu^2)^{\frac{1}{2}}(\zeta^2-1)^{\frac{1}{2}}\delta\omega. \quad\text{......(3)}$$

To express (1) in terms of our new variables we equate to zero the total flux across the walls of a volume element $\delta s_\mu\delta s_\zeta\delta s_\omega$, and obtain

$$\frac{\partial}{\partial\mu}\left(\frac{\partial\phi}{\partial s_\mu}\delta s_\zeta\delta s_\omega\right)\delta\mu + \frac{\partial}{\partial\zeta}\left(\frac{\partial\phi}{\partial s_\zeta}\delta s_\mu\delta s_\omega\right)\delta\zeta + \frac{\partial}{\partial\omega}\left(\frac{\partial\phi}{\partial s_\omega}\delta s_\mu\delta s_\zeta\right)\delta\omega = 0,$$

or, on substitution from (3),

$$\frac{\partial}{\partial\mu}\left\{(1-\mu^2)\frac{\partial\phi}{\partial\mu}\right\} + \frac{\partial}{\partial\zeta}\left\{(\zeta^2-1)\frac{\partial\phi}{\partial\zeta}\right\} + \frac{\zeta^2-\mu^2}{(1-\mu^2)(\zeta^2-1)}\frac{\partial^2\phi}{\partial\omega^2} = 0.$$

* Heine, "Ueber einige Aufgaben, welche auf partielle Differentialgleichungen führen," *Crelle*, xxvi. 185 (1843), and *Kugelfunctionen*, ii. Art. 38. See also Ferrers, c. vi.

This may also be written

$$\frac{\partial}{\partial\mu}\left\{(1-\mu^2)\frac{\partial\phi}{\partial\mu}\right\}+\frac{1}{1-\mu^2}\frac{\partial^2\phi}{\partial\omega^2}=\frac{\partial}{\partial\zeta}\left\{(1-\zeta^2)\frac{\partial\phi}{\partial\zeta}\right\}+\frac{1}{1-\zeta^2}\frac{\partial^2\phi}{\partial\omega^2}. \quad\text{.....(4)}$$

104. If ϕ be a finite function of μ and ω, from $\mu=-1$ to $\mu=+1$ and from $\omega=0$ to $\omega=2\pi$, it may be expanded in a series of surface harmonics of integral orders, of the types given by Art. 86 (7), where the coefficients are functions of ζ; and it appears on substitution in (4) that each term of the expansion must satisfy the equation separately. Taking first the case of the zonal harmonic, we write

$$\phi=P_n(\mu).Z, \quad\text{.....(5)}$$

and on substitution we find, in virtue of Art. 84 (1),

$$\frac{d}{d\zeta}\left\{(1-\zeta^2)\frac{dZ}{d\zeta}\right\}+n(n+1)Z=0, \quad\text{.....(6)}$$

which is of the same form as the equation referred to. We thus obtain the solutions

$$\phi=P_n(\mu).P_n(\zeta), \quad\text{.....(7)}$$

and

$$\phi=P_n(\mu).Q_n(\zeta), \quad\text{.....(8)}$$

where

$$Q_n(\zeta)=P_n(\zeta)\int_\zeta^\infty\frac{d\zeta}{\{P_n(\zeta)\}^2(\zeta^2-1)},$$

$$=\tfrac{1}{2}P_n(\zeta)\log\frac{\zeta+1}{\zeta-1}-\frac{2n-1}{1.n}P_{n-1}(\zeta)-\frac{2n-5}{3(n-1)}P_{n-3}(\zeta)-\ldots,$$

$$=\frac{n!}{1.3\ldots(2n+1)}\left\{\zeta^{-n-1}+\frac{(n+1)(n+2)}{2(2n+3)}\zeta^{-n-3}+\frac{(n+1)(n+2)(n+3)(n+4)}{2.4(2n+3)(2n+5)}\zeta^{-n-5}+\ldots\right\}. \quad\text{...(9)*}$$

The solution (7) is finite when $\zeta=1$, and is therefore adapted to the space *within* an ellipsoid of revolution; while (8) is infinite for $\zeta=1$, but vanishes for $\zeta=\infty$, and is therefore appropriate to the external region. As particular cases of the formula (9) we note

$$Q_0(\zeta)=\tfrac{1}{2}\log\frac{\zeta+1}{\zeta-1}, \qquad Q_1(\zeta)=\tfrac{1}{2}\zeta\log\frac{\zeta+1}{\zeta-1}-1,$$

$$Q_2(\zeta)=\tfrac{1}{4}(3\zeta^2-1)\log\frac{\zeta+1}{\zeta-1}-\tfrac{3}{2}\zeta.$$

The definite-integral form of Q_n shews that

$$P_n(\zeta)\frac{dQ_n(\zeta)}{d\zeta}-\frac{dP_n(\zeta)}{d\zeta}Q_n(\zeta)=-\frac{1}{\zeta^2-1}. \quad\text{.....(10)}$$

The expressions for the stream-function corresponding to (7) and (8) are readily found; thus, from the definition of Art. 94,

$$\frac{\partial\phi}{\partial s_\zeta}=-\frac{1}{\varpi}\frac{\partial\psi}{\partial s_\mu}, \qquad \frac{\partial\phi}{\partial s_\mu}=\frac{1}{\varpi}\frac{\partial\psi}{\partial s_\zeta}, \quad\text{.....(11)}$$

* Ferrers, c. v.; Todhunter, c. vi.; Forsyth, Arts. 96–99.

whence $$\frac{\partial\psi}{\partial\mu} = -k(\zeta^2-1)\frac{\partial\phi}{\partial\zeta},\quad \frac{\partial\psi}{\partial\zeta} = k(1-\mu^2)\frac{\partial\phi}{\partial\mu}. \quad\text{..............(12)}$$

Thus, in the case of (7), we have

$$\frac{\partial\psi}{\partial\mu} = -k(\zeta^2-1)\frac{dP_n(\zeta)}{d\zeta}\,.\,P_n(\mu)$$

$$= \frac{k}{n(n+1)}(\zeta^2-1)\frac{dP_n(\zeta)}{d\zeta}\,.\,\frac{d}{d\mu}\left\{(1-\mu^2)\frac{dP_n(\mu)}{d\mu}\right\},$$

whence $$\psi = \frac{k}{n(n+1)}(1-\mu^2)\frac{dP_n(\mu)}{d\mu}\,.\,(\zeta^2-1)\frac{dP_n(\zeta)}{d\zeta}. \quad\text{.........(13)}$$

The same result will follow of course from the second of equations (12).

In the same way, the stream-function corresponding to (8) is

$$\psi = \frac{k}{n(n+1)}(1-\mu^2)\frac{dP_n(\mu)}{d\mu}\,.\,(\zeta^2-1)\frac{dQ_n(\zeta)}{d\zeta}. \quad\text{.........(14)}$$

105. We can apply this to the case of an ovary ellipsoid moving parallel to its axis in an infinite mass of liquid. The elliptic co-ordinates must be chosen so that the ellipsoid in question is a member of the confocal family, say that for which $\zeta = \zeta_0$. Comparing with Art. 103 (2) we see that if a, c be the polar and equatorial radii, and e the eccentricity of the meridian section, we must have

$$k = ae, \qquad \zeta_0 = 1/e, \qquad k(\zeta_0^2-1)^{\frac{1}{2}} = c.$$

The surface-condition is given by Art. 97 (1), viz. we must have

$$\psi = -\tfrac{1}{2}Uk^2(1-\mu^2)(\zeta^2-1) + \text{const.}, \quad\text{..................(1)}$$

for $\zeta = \zeta_0$. Hence putting $n = 1$ in Art. 104 (14), and introducing an arbitrary multiplier A, we have

$$\psi = \tfrac{1}{2}Ak(1-\mu^2)(\zeta^2-1)\left\{\tfrac{1}{2}\log\frac{\zeta+1}{\zeta-1} - \frac{\zeta}{\zeta^2-1}\right\}, \quad\text{.........(2)}$$

with the condition

$$A = Uk \div \left\{\frac{\zeta_0}{\zeta_0^2-1} - \tfrac{1}{2}\log\frac{\zeta_0+1}{\zeta_0-1}\right\} = Ua \div \left\{\frac{1}{1-e^2} - \frac{1}{2e}\log\frac{1+e}{1-e}\right\}. \quad\text{...(3)}$$

The corresponding formula for the velocity-potential is

$$\phi = A\mu\left\{\tfrac{1}{2}\zeta\log\frac{\zeta+1}{\zeta-1} - 1\right\}. \quad\text{.....................(4)}$$

The kinetic energy, and thence the inertia-coefficient due to the fluid, may be readily calculated by the formula (5) of Art. 94.

106. Leaving the case of symmetry, the solutions of $\nabla^2\phi = 0$ when ϕ is a tesseral or sectorial harmonic in μ and ω are found by a similar method to be of the types

$$\phi = P_n^s(\mu)\,.\,P_n^s(\zeta)\begin{matrix}\cos\\ \sin\end{matrix}\Big\}\,s\omega, \quad\text{........................(1)}$$

$$\phi = P_n^s(\mu)\,.\,Q_n^s(\zeta)\begin{matrix}\cos\\ \sin\end{matrix}\Big\}\,s\omega, \quad\text{........................(2)}$$

where, as in Art. 86, $$P_n^s(\mu) = (1-\mu^2)^{\frac{1}{2}s} \frac{d^s P_n(\mu)}{d\mu^s}, \quad \text{.....................(3)}$$

whilst (to avoid imaginaries) we write

$$P_n^s(\zeta) = (\zeta^2 - 1)^{\frac{1}{2}s} \frac{d^s P_n(\zeta)}{d\zeta^s}, \quad \text{.....................(4)}$$

and $$Q_n^s(\zeta) = (\zeta^2 - 1)^{\frac{1}{2}s} \frac{d^s Q_n(\zeta)}{d\zeta^s}. \quad \text{.....................(5)}$$

It may be shewn that

$$Q_n^s(\zeta) = (-)^s \frac{(n+s)!}{(n-s)!} P_n^s(\zeta) . \int_\zeta^\infty \frac{d\zeta}{\{P_n^s(\zeta)\}^2 . (\zeta^2 - 1)}, \quad \text{.........(6)}$$

whence $$P_n^s(\zeta) \frac{dQ_n^s(\zeta)}{d\zeta} - \frac{dP_n^s(\zeta)}{d\zeta} Q_n^s(\zeta) = (-)^{s+1} \frac{(n+s)!}{(n-s)!} \frac{1}{\zeta^2 - 1}. \quad \text{.....(7)}$$

As examples we may take the case of an ovary ellipsoid moving parallel to an *equatorial* axis, say that of y, or rotating about this axis.

1°. In the former case, the surface-condition is

$$\frac{\partial \phi}{\partial \zeta} = -V \frac{\partial y}{\partial \zeta},$$

for $\zeta = \zeta_0$, where V is the velocity of translation, or

$$\frac{\partial \phi}{\partial \zeta} = -V . \frac{k\zeta_0}{(\zeta_0^2 - 1)^{\frac{1}{2}}} . (1 - \mu^2)^{\frac{1}{2}} \cos \omega. \quad \text{..................(8)}$$

This is satisfied by putting $n = 1$, $s = 1$, in (2), viz.

$$\phi = A(1-\mu^2)^{\frac{1}{2}} (\zeta^2 - 1)^{\frac{1}{2}} . \left\{\tfrac{1}{2} \log \frac{\zeta + 1}{\zeta - 1} - \frac{\zeta}{\zeta^2 - 1}\right\} \cos \omega, \quad \text{.........(9)}$$

the constant A being given by

$$A \left\{\tfrac{1}{2} \log \frac{\zeta_0 + 1}{\zeta_0 - 1} - \frac{\zeta_0^2 - 2}{\zeta_0(\zeta_0^2 - 1)}\right\} = -kV. \quad \text{..............(10)}$$

2°. In the case of rotation about Oy, if Ω_y be the angular velocity, we must have

$$\frac{\partial \phi}{\partial \zeta} = -\Omega_y \left(z \frac{\partial x}{\partial \zeta} - x \frac{\partial z}{\partial \zeta}\right),$$

for $\zeta = \zeta_0$ or $$\frac{\partial \phi}{\partial \zeta} = k^2 \Omega_y . \frac{1}{(\zeta_0^2 - 1)^{\frac{1}{2}}} . \mu (1 - \mu^2)^{\frac{1}{2}} \sin \omega. \quad \text{...............(11)}$$

Putting $n = 2$, $s = 1$, in the formula (2) we find

$$\phi = A\mu(1-\mu^2)^{\frac{1}{2}} (\zeta^2 - 1)^{\frac{1}{2}} \left\{\tfrac{3}{2} \zeta \log \frac{\zeta + 1}{\zeta - 1} - 3 - \frac{1}{\zeta^2 - 1}\right\} \sin \omega, \quad \text{...(12)}$$

A being determined by comparison with (11).

107. When the ellipsoid is of the *oblate* or 'planetary' form, the appropriate co-ordinates are given by

$$\left.\begin{aligned} x = k \cos \theta \sinh \eta = k\mu\zeta, \quad y = \varpi \cos \omega, \quad z = \varpi \sin \omega, \\ \text{where} \quad \varpi = k \sin \theta \cosh \eta = k(1-\mu^2)^{\frac{1}{2}} (\zeta^2 + 1)^{\frac{1}{2}}. \end{aligned}\right\} \quad \text{......(1)}$$

Here ζ may range from 0 to ∞ (or, in some applications, from $-\infty$ through 0 to $+\infty$), whilst μ lies between ± 1. The quadrics $\zeta=$ const., $\mu=$ const. are planetary ellipsoids and hyperboloids of revolution of one sheet, all having the common focal circle $x=0$, $\varpi=k$. As limiting forms we have the ellipsoid $\zeta=0$, which coincides with the portion of the plane $x=0$ for which $\varpi<k$, and the hyperboloid $\mu=0$ coinciding with the remaining portion of this plane.

With the same notation as before we find

$$\delta s_\mu = k\left(\frac{\zeta^2+\mu^2}{1-\mu^2}\right)^{\frac{1}{2}}\delta\mu,\quad \delta s_\zeta = k\left(\frac{\zeta^2+\mu^2}{\zeta^2+1}\right)^{\frac{1}{2}}\delta\zeta,\quad \delta s_\omega = k(1-\mu^2)^{\frac{1}{2}}(\zeta^2+1)^{\frac{1}{2}}\delta\omega; \quad \text{......(2)}$$

and the equation of continuity becomes

$$\frac{\partial}{\partial\mu}\left\{(1-\mu^2)\frac{\partial\phi}{\partial\mu}\right\}+\frac{\partial}{\partial\zeta}\left\{(\zeta^2+1)\frac{\partial\phi}{\partial\zeta}\right\}+\frac{\zeta^2+\mu^2}{(1-\mu^2)(\zeta^2+1)}\frac{\partial^2\phi}{\partial\omega^2}=0,$$

or

$$\frac{\partial}{\partial\mu}\left\{(1-\mu^2)\frac{\partial\phi}{\partial\mu}\right\}+\frac{1}{1-\mu^2}\frac{\partial^2\phi}{\partial\omega^2}=-\frac{\partial}{\partial\zeta}\left\{(\zeta^2+1)\frac{\partial\phi}{\partial\zeta}\right\}+\frac{1}{\zeta^2+1}\frac{\partial^2\phi}{\partial\omega^2}. \quad \text{...(3)}$$

This is of the same form as Art. 103 (4), with $i\zeta$ in place of ζ, and the like correspondence will run through the subsequent formulae.

In the case of symmetry about the axis we have the solutions

$$\phi = P_n(\mu)\,.\,p_n(\zeta), \quad \text{......(4)}$$

and

$$\phi = P_n(\mu)\,.\,q_n(\zeta), \quad \text{......(5)}$$

where

$$p_n(\zeta)=\frac{1.3.5\ldots(2n-1)}{n!}\left\{\zeta^n+\frac{n(n-1)}{2(2n-1)}\zeta^{n-2}+\frac{n(n-1)(n-2)(n-3)}{2.4(2n-1)(2n-3)}\zeta^{n-4}+\ldots\right\}, \quad \text{...(6)}$$

and

$$q_n(\zeta)=p_n(\zeta)\int_\zeta^\infty\frac{d\zeta}{\{p_n(\zeta)\}^2(\zeta^2+1)},$$

$$=(-)^n\left\{p_n(\zeta)\cot^{-1}\zeta-\frac{2n-1}{1.n}p_{n-1}(\zeta)+\frac{2n-5}{3(n-1)}p_{n-3}(\zeta)-\ldots\right\},$$

$$=\frac{n!}{1.3.5\ldots(2n+1)}\left\{\zeta^{-n-1}-\frac{(n+1)(n+2)}{2(2n+3)}\zeta^{-n-3}+\frac{(n+1)(n+2)(n+3)(n+4)}{2.4(2n+3)(2n+5)}\zeta^{-n-5}-\ldots\right\}, \quad \text{......(7)}$$

the latter expansion being however convergent only when $\zeta>1$*. As before, the solution (4) is appropriate to the region included within an ellipsoid of the family $\zeta=$ const., and (5) to the external space.

We note that

$$p_n(\zeta)\frac{dq_n(\zeta)}{d\zeta}-\frac{dp_n(\zeta)}{d\zeta}q_n(\zeta)=-\frac{1}{\zeta^2+1}. \quad \text{......(8)}$$

As particular cases of the formula (7) we have

$$q_0(\zeta)=\cot^{-1}\zeta,\quad q_1(\zeta)=1-\zeta\cot^{-1}\zeta,$$
$$q_2(\zeta)=\tfrac{1}{2}(3\zeta^2+1)\cot^{-1}\zeta-\tfrac{3}{2}\zeta.$$

* The reader may easily adapt the demonstrations referred to in Art. 104 to the present case.

The formulae for the stream-function corresponding to (4) and (5) are

$$\psi = \frac{k}{n(n+1)}(1-\mu^2)\frac{dP_n(\mu)}{d\mu}\,.\,(\zeta^2+1)\frac{dp_n(\zeta)}{d\zeta}, \quad\text{.........(9)}$$

and

$$\psi = \frac{k}{n(n+1)}(1-\mu^2)\frac{dP_n(\mu)}{d\mu}\,.\,(\zeta^2+1)\frac{dq_n(\zeta)}{d\zeta}, \quad\text{........(10)}$$

108. 1°. The simplest case of Art. 107 (5) is when $n=0$, viz.

$$\phi = A\cot^{-1}\zeta, \quad\text{..............................(1)}$$

where ζ is supposed to range from $-\infty$ to $+\infty$. The formula (10) of the last Art. then assumes an indeterminate form, but we find by the method of Art. 104,

$$\psi = Ak\mu, \quad\text{..................................(2)}$$

where μ ranges from 0 to 1. This solution represents the flow of a liquid through a circular aperture in an infinite plane wall, viz. the aperture is the portion of the plane yz for which $\varpi < k$. The velocity at any point of the aperture ($\zeta=0$) is

$$u = -\frac{1}{\varpi}\frac{\partial\psi}{\partial\omega} = \frac{A}{(k^2-\varpi^2)^{\frac{1}{2}}},$$

since, when $x=0$, $k\mu=(k^2-\varpi^2)^{\frac{1}{2}}$. The velocity is therefore infinite at the edge. Compare Art. 102, 3°.

2°. Again, the motion due to a planetary ellipsoid ($\zeta=\zeta_0$) moving with velocity U parallel to its axis in an infinite mass of liquid is given by

$$\phi = A\mu(1-\zeta\cot^{-1}\zeta), \quad \psi = \tfrac{1}{2}Ak(1-\mu^2)(\zeta^2+1)\left\{\frac{\zeta}{\zeta^2+1}-\cot^{-1}\zeta\right\}, \quad\text{...(3)}$$

where

$$A = -kU \div \left\{\frac{\zeta}{\zeta_0^2+1}-\cot^{-1}\zeta_0\right\}.$$

Denoting the polar and equatorial radii by a and c, and the eccentricity of the meridian section by e, we have

$$a = k\zeta_0, \quad c = k(\zeta_0^2+1)^{\frac{1}{2}}, \quad e = (\zeta_0^2+1)^{-\frac{1}{2}}.$$

In terms of these quantities

$$A = -Uc \div \left\{(1-e^2)^{\frac{1}{2}} - \frac{1}{e}\sin^{-1}e\right\}. \quad\text{...................(4)}$$

The forms of the lines of motion, for equidistant values of ψ, are shewn on the next page. Cf. Art. 71, 3°.

The most interesting case is that of the circular disk, for which $e=1$, and $A = 2Uc/\pi$. The value of ϕ given in (3) becomes equal to $\pm A\mu$, or $\pm A(1-\varpi^2/c^2)^{\frac{1}{2}}$, for the two sides of the disk, and the normal velocity to $\pm U$. Hence the formula (4) of Art. 44 gives

$$2T = \tfrac{8}{3}\rho c^3 U^2, \quad\text{..............................(5)}$$

as in Art. 102 (20).

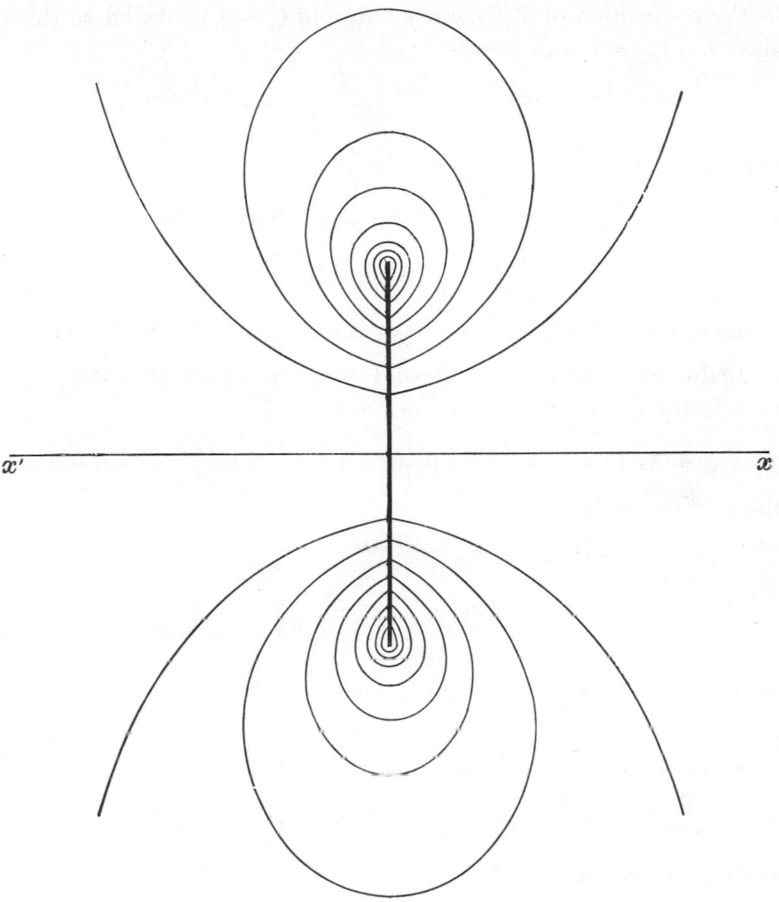

109. The solutions of the equation Art. 107 (3) in tesseral harmonics are

$$\phi = P_n^s(\mu) \,.\, p_n^s(\zeta) \,.\, \left.\begin{matrix}\cos\\ \sin\end{matrix}\right\} s\omega, \qquad\qquad (1)$$

and

$$\phi = P_n^s(\mu) \,.\, q_n^s(\zeta) \,.\, \left.\begin{matrix}\cos\\ \sin\end{matrix}\right\} s\omega, \qquad\qquad (2)$$

where

$$p_n^s(\zeta) = (\zeta^2+1)^{\frac{1}{2}s} \frac{d^s p_n(\zeta)}{d\zeta^s}, \qquad\qquad (3)$$

and

$$q_n^s(\zeta) = (\zeta^2+1)^{\frac{1}{2}s} \frac{d^s q_n(\zeta)}{d\zeta^s},$$

$$= (-)^s \frac{(n+s)!}{(n-s)!} \,.\, p_n^s(\zeta) \,.\, \int_\zeta^\infty \frac{d\zeta}{\{p_n^s(\zeta)\}^2(\zeta^2+1)}. \qquad\qquad (4)$$

These functions possess the property

$$p_n^s(\zeta) \frac{dq_n^s(\zeta)}{d\zeta} - \frac{dp_n^s(\zeta)}{d\zeta} q_n^s(\zeta) = (-)^{s+1} \frac{(n+s)!}{(n-s)!} \frac{1}{\zeta^2+1}. \qquad\qquad (5)$$

We may apply these results as in Art. 108.

1°. For the motion of a planetary ellipsoid ($\zeta = \zeta_0$) parallel to the axis of y we have $n = 1$, $s = 1$, and thence

$$\phi = A(1-\mu^2)^{\frac{1}{2}}(\zeta^2+1)^{\frac{1}{2}}\left\{\frac{\zeta}{\zeta^2+1} - \cot^{-1}\zeta\right\}\cos\omega, \qquad (6)$$

with the condition

$$\frac{\partial\phi}{\partial\zeta} = -V\frac{\partial y}{\partial\zeta},$$

for $\zeta = \zeta_0$, V denoting the velocity of the solid. This gives

$$A\left\{\frac{\zeta_0^2+2}{\zeta_0(\zeta_0^2+1)} - \cot^{-1}\zeta_0\right\} = -kV. \qquad (7)$$

In the case of the disk ($\zeta_0 = 0$), we have $A = 0$, as we should expect.

2°. Again, for a planetary ellipsoid rotating about the axis of y with angular velocity Ω_y, we have, putting $n = 2$, $s = 1$,

$$\phi = A\mu(1-\mu^2)^{\frac{1}{2}}(\zeta^2+1)^{\frac{1}{2}}\left\{3\zeta\cot^{-1}\zeta - 3 + \frac{1}{\zeta^2+1}\right\}\sin\omega, \qquad (8)$$

with the surface-condition

$$\frac{\partial\phi}{\partial\zeta} = -\Omega_y\left(z\frac{\partial x}{\partial\zeta} - x\frac{\partial z}{\partial\zeta}\right)$$

$$= -\frac{k^2\Omega_y}{(\zeta_0^2+1)^{\frac{1}{2}}}\cdot\mu(1-\mu^2)^{\frac{1}{2}}\sin\omega. \qquad (9)$$

For the circular disk ($\zeta_0 = 0$) this gives

$$\tfrac{3}{2}\pi A = -k^2\Omega_y. \qquad (10)$$

At the two surfaces of the disk we have

$$\phi = \mp 2A\mu(1-\mu^2)^{\frac{1}{2}}\sin\omega, \quad \frac{\partial\phi}{\partial n} = \mp k\Omega_y(1-\mu^2)^{\frac{1}{2}}\sin\omega,$$

and substituting in the formula

$$2T = -\rho\iint\phi\frac{\partial\phi}{\partial n}\,\varpi\, d\varpi\, d\omega,$$

we obtain

$$2T = \tfrac{16}{45}\rho c^5\,.\,\Omega_y^2. \qquad (11)^*$$

110. In questions relating to ellipsoids with three unequal axes we may employ the more general type of Ellipsoidal Harmonics, usually known by the name of 'Lamé's Functions†.' Without attempting a formal account of these functions, we will investigate some solutions of the equation

$$\nabla^2\phi = 0, \qquad (1)$$

in ellipsoidal co-ordinates, which are analogous to spherical harmonics of the first and second orders, with a view to their hydrodynamical applications.

* For further solutions in terms of the present co-ordinates see Nicholson, *Phil. Trans.* A, ccxxiv. 49 (1924).

† See, for example, Ferrers, *Spherical Harmonics*, c. vi.; W. D. Niven, *Phil. Trans.* A, clxxxii. 182 (1891) and *Proc. Roy. Soc.* A, lxxix. 458 (1906); Poincaré, *Figures d'Équilibre d'une Masse Fluide*, Paris, 1902, c. vi.; Darwin, *Phil. Trans.* A, cxcvii. 461 (1901) [*Scientific Papers*, Cambridge, 1907–11, iii. 186]; Whittaker and Watson, c. xxiii. An outline of the theory is given by Wangerin, *l.c. ante* p. 110.

It is convenient to prefix an investigation of the motion of a liquid contained in an ellipsoidal envelope, which can be treated at once by Cartesian methods.

Thus, when the envelope is in motion parallel to the axis of x with velocity U, the enclosed fluid moves as a solid, and the velocity-potential is simply $\phi = -Ux$.

Next let us suppose that the envelope is rotating about a principal axis (say that of x) with angular velocity Ω_x. The equation of the surface being

$$\frac{x^2}{a^2}+\frac{y^2}{b^2}+\frac{z^2}{c^2}=1, \quad\text{..........................(2)}$$

the surface-condition is

$$-\frac{x}{a^2}\frac{\partial\phi}{\partial x}-\frac{y}{b^2}\frac{\partial\phi}{\partial y}-\frac{z}{c^2}\frac{\partial\phi}{\partial z}=-\frac{y}{b^2}\Omega_x z+\frac{z}{c^2}\Omega_x y.$$

We therefore assume $\phi = Ayz$, which is evidently a solution of (1), and obtain, on determining the constant by the condition just written,

$$\phi=-\frac{b^2-c^2}{b^2+c^2}\Omega_x\,.\,yz.$$

Hence, if the centre be moving with a velocity whose components are U, V, W and if Ω_x, Ω_y, Ω_z be the angular velocities about the principal axes, we have by superposition*

$$\phi=-Ux-Vy-Wz-\frac{b^2-c^2}{b^2+c^2}\Omega_x yz-\frac{c^2-a^2}{c^2+a^2}\Omega_y zx-\frac{a^2-b^2}{a^2+b^2}\Omega_z xy. \quad\text{...(3)}$$

We may also include the case where the envelope is changing its form, but so as to remain ellipsoidal. If in (2) the lengths (only) of the axes are changing at the rates $\dot{a}$, $\dot{b}$, $\dot{c}$, respectively, the general boundary-condition, Art. 9 (3), becomes

$$\frac{x^2}{a^3}\dot{a}+\frac{y^2}{b^3}\dot{b}+\frac{z^2}{c^3}\dot{c}+\frac{x}{a^2}\frac{\partial\phi}{\partial x}+\frac{y}{b^2}\frac{\partial\phi}{\partial y}+\frac{z}{c^2}\frac{\partial\phi}{\partial z}=0, \quad\text{.............(4)}$$

which is satisfied† by

$$\phi=-\tfrac{1}{2}\left(\frac{\dot{a}}{a}x^2+\frac{\dot{b}}{b}y^2+\frac{\dot{c}}{c}z^2\right). \quad\text{........................(5)}$$

The equation (1) requires that

$$\frac{\dot{a}}{a}+\frac{\dot{b}}{b}+\frac{\dot{c}}{c}=0, \quad\text{..............................(6)}$$

which is in fact the condition which must be satisfied by the changing ellipsoidal surface in order that the enclosed volume ($\frac{4}{3}\pi abc$) may be constant.

* This result appears to have been published independently by Beltrami, Bjerknes, and Maxwell, in 1873. See Hicks, "Report on Recent Progress in Hydrodynamics," *Brit. Ass. Rep.* 1882, and Kelvin's *Papers*, iv. 197 (footnote).

† C. A. Bjerknes, "Verallgemeinerung des Problems von den Bewegungen, welche in einer ruhenden unelastischen Flüssigkeit die Bewegung eines Ellipsoids hervorbringt," *Göttinger Nachrichten*, 1873, pp. 448, 829.

111. The solutions of the corresponding problems for an infinite mass of fluid bounded *internally* by an ellipsoid involve the use of a special system of orthogonal curvilinear co-ordinates.

If x, y, z be functions of three parameters λ, μ, ν, such that the surfaces

$$\lambda = \text{const.}, \quad \mu = \text{const.}, \quad \nu = \text{const.} \quad \text{.................(1)}$$

are mutually orthogonal at their intersections, and if we write

$$\left.\begin{aligned} \frac{1}{h_1^2} &= \left(\frac{\partial x}{\partial \lambda}\right)^2 + \left(\frac{\partial y}{\partial \lambda}\right)^2 + \left(\frac{\partial z}{\partial \lambda}\right)^2, \\ \frac{1}{h_2^2} &= \left(\frac{\partial x}{\partial \mu}\right)^2 + \left(\frac{\partial y}{\partial \mu}\right)^2 + \left(\frac{\partial z}{\partial \mu}\right)^2, \\ \frac{1}{h_3^2} &= \left(\frac{\partial x}{\partial \nu}\right)^2 + \left(\frac{\partial y}{\partial \nu}\right)^2 + \left(\frac{\partial z}{\partial \nu}\right)^2, \end{aligned}\right\} \quad \text{....................(2)}$$

the direction-cosines of the normals to the three surfaces which pass through (x, y, z) will be

$$\left(h_1 \frac{\partial x}{\partial \lambda}, h_1 \frac{\partial y}{\partial \lambda}, h_1 \frac{\partial z}{\partial \lambda}\right), \quad \left(h_2 \frac{\partial x}{\partial \mu}, h_2 \frac{\partial y}{\partial \mu}, h_2 \frac{\partial z}{\partial \mu}\right), \quad \left(h_3 \frac{\partial x}{\partial \nu}, h_3 \frac{\partial y}{\partial \nu}, h_3 \frac{\partial z}{\partial \nu}\right), \ldots(3)$$

respectively. It follows that the lengths of linear elements drawn in the directions of these normals will be

$$\delta\lambda/h_1, \qquad \delta\mu/h_2, \qquad \delta\nu/h_3.$$

Hence if ϕ be the velocity-potential of a fluid motion, the total flux into the rectangular space included between the six surfaces $\lambda \pm \frac{1}{2}\delta\lambda$, $\mu \pm \frac{1}{2}\delta\mu$, $\nu \pm \frac{1}{2}\delta\nu$ will be

$$\frac{\partial}{\partial \lambda}\left(h_1 \frac{\partial \phi}{\partial \lambda} \cdot \frac{\delta\mu}{h_2} \cdot \frac{\delta\nu}{h_3}\right)\delta\lambda + \frac{\partial}{\partial \mu}\left(h_2 \frac{\partial \phi}{\partial \mu} \cdot \frac{\delta\nu}{h_3} \cdot \frac{\delta\lambda}{h_1}\right)\delta\mu + \frac{\partial}{\partial \nu}\left(h_3 \frac{\partial \phi}{\partial \nu} \cdot \frac{\delta\lambda}{h_1} \cdot \frac{\delta\mu}{h_2}\right)\delta\nu.$$

It appears from Art. 42 (3) that the same flux is expressed by $\nabla^2\phi$ multiplied by the volume of the space, *i.e.* by $\delta\lambda\, \delta\mu\, \delta\nu / h_1 h_2 h_3$. Hence*

$$\nabla^2\phi = h_1 h_2 h_3 \left\{\frac{\partial}{\partial \lambda}\left(\frac{h_1}{h_2 h_3} \frac{\partial \phi}{\partial \lambda}\right) + \frac{\partial}{\partial \mu}\left(\frac{h_2}{h_3 h_1} \frac{\partial \phi}{\partial \mu}\right) + \frac{\partial}{\partial \nu}\left(\frac{h_3}{h_1 h_2} \frac{\partial \phi}{\partial \nu}\right)\right\}. \quad \ldots(4)$$

Equating this to zero, we obtain the general equation of continuity in orthogonal co-ordinates, of which particular cases have already been investigated in Arts. 83, 103, 107.

The theory of triple orthogonal systems of surfaces is very attractive mathematically, and abounds in interesting and elegant formulae. We may note that if λ, μ, ν be regarded as functions of x, y, z, the direction-cosines

* The above method was given in a paper by W. Thomson, "On the Equations of Motion of Heat referred to Curvilinear Co-ordinates," *Camb. Math. Journ.* iv. 179 (1843) [*Papers*, i. 25]. Reference may also be made to Jacobi, "Ueber eine particuläre Lösung der partiellen Differentialgleichung......," *Crelle*, xxxvi. 113 (1847) [*Werke*, ii. 198].

The transformation of $\nabla^2\phi$ to general orthogonal co-ordinates was first effected by Lamé, "Sur les lois de l'équilibre du fluide éthéré," *Journ. de l'École Polyt.* xiv. 191 (1834). See also his *Leçons sur les Coordonnées Curvilignes*, Paris, 1859, p. 22.

of the three line-elements above considered can also be expressed in the forms

$$\left(\frac{1}{h_1}\frac{\partial\lambda}{\partial x}, \frac{1}{h_1}\frac{\partial\lambda}{\partial y}, \frac{1}{h_1}\frac{\partial\lambda}{\partial z}\right), \left(\frac{1}{h_2}\frac{\partial\mu}{\partial x}, \frac{1}{h_2}\frac{\partial\mu}{\partial y}, \frac{1}{h_2}\frac{\partial\mu}{\partial z}\right), \left(\frac{1}{h_3}\frac{\partial\nu}{\partial x}, \frac{1}{h_3}\frac{\partial\nu}{\partial y}, \frac{1}{h_3}\frac{\partial\nu}{\partial z}\right), \quad \text{.........(5)}$$

from which, and from (3), various interesting relations can be inferred. The formulae already given are, however, sufficient for our present purpose.

112. In the applications to which we now proceed the triple orthogonal system consists of the confocal quadrics

$$\frac{x^2}{a^2+\theta}+\frac{y^2}{b^2+\theta}+\frac{z^2}{c^2+\theta}-1=0, \quad \text{....................(1)}$$

whose properties are explained in books on Solid Geometry. Through any given point (x, y, z) there pass three surfaces of the system, corresponding to the three roots of (1), considered as a cubic in θ. If (as we shall for the most part suppose) $a > b > c$, one of these roots (λ, say) will lie between ∞ and $-c^2$, another (μ) between $-c^2$ and $-b^2$, and the third (ν) between $-b^2$ and $-a^2$. The surfaces λ, μ, ν are therefore ellipsoids, hyperboloids of one sheet, and hyperboloids of two sheets, respectively.

It follows immediately from this definition of λ, μ, ν, that

$$\frac{x^2}{a^2+\theta}+\frac{y^2}{b^2+\theta}+\frac{z^2}{c^2+\theta}-1=\frac{(\lambda-\theta)(\mu-\theta)(\nu-\theta)}{(a^2+\theta)(b^2+\theta)(c^2+\theta)}, \quad \text{.........(2)}$$

identically, for all values of θ. Hence multiplying by $a^2+\theta$, and afterwards putting $\theta=-a^2$, we obtain the first of the following equations:

$$\left.\begin{aligned} x^2 &= \frac{(a^2+\lambda)(a^2+\mu)(a^2+\nu)}{(a^2-b^2)(a^2-c^2)}, \\ y^2 &= \frac{(b^2+\lambda)(b^2+\mu)(b^2+\nu)}{(b^2-c^2)(b^2-a^2)}, \\ z^2 &= \frac{(c^2+\lambda)(c^2+\mu)(c^2+\nu)}{(c^2-a^2)(c^2-b^2)}. \end{aligned}\right\} \quad \text{....................(3)}$$

These give $$\frac{\partial x}{\partial\lambda}=\tfrac{1}{2}\frac{x}{a^2+\lambda}, \quad \frac{\partial y}{\partial\lambda}=\tfrac{1}{2}\frac{y}{b^2+\lambda}, \quad \frac{\partial z}{\partial\lambda}=\tfrac{1}{2}\frac{z}{c^2+\lambda}, \quad \text{...........(4)}$$

and thence, in the notation of Art. 111 (2),

$$\frac{1}{h_1^2}=\tfrac{1}{4}\left\{\frac{x^2}{(a^2+\lambda)^2}+\frac{y^2}{(b^2+\lambda)^2}+\frac{z^2}{(c^2+\lambda)^2}\right\}. \quad \text{..............(5)}$$

If we differentiate (2) with respect to θ and afterwards put $\theta=\lambda$, we deduce the first of the following three relations:

$$\left.\begin{aligned} h_1^2 &= 4\,\frac{(a^2+\lambda)(b^2+\lambda)(c^2+\lambda)}{(\lambda-\mu)(\lambda-\nu)}, \\ h_2^2 &= 4\,\frac{(a^2+\mu)(b^2+\mu)(c^2+\mu)}{(\mu-\nu)(\mu-\lambda)}, \\ h_3^2 &= 4\,\frac{(a^2+\nu)(b^2+\nu)(c^2+\nu)}{(\nu-\lambda)(\nu-\mu)}. \end{aligned}\right\} \quad \text{....................(6)}$$

The remaining relations of the sets (3) and (6) have been written down from symmetry*.

Substituting in Art. 111 (4), we find†

$$\nabla^2\phi = -\frac{4}{(\mu-\nu)(\nu-\lambda)(\lambda-\mu)}\left[(\mu-\nu)\left\{(a^2+\lambda)^{\frac{1}{2}}(b^2+\lambda)^{\frac{1}{2}}(c^2+\lambda)^{\frac{1}{2}}\frac{\partial}{\partial\lambda}\right\}^2 + (\nu-\lambda)\left\{(a^2+\mu)^{\frac{1}{2}}(b^2+\mu)^{\frac{1}{2}}(c^2+\mu)^{\frac{1}{2}}\frac{\partial}{\partial\mu}\right\}^2 + (\lambda-\mu)\left\{(a^2+\nu)^{\frac{1}{2}}(b^2+\nu)^{\frac{1}{2}}(c^2+\nu)^{\frac{1}{2}}\frac{\partial}{\partial\nu}\right\}^2\right]\phi. \quad \text{......(7)}$$

113. The particular solutions of the transformed equation $\nabla^2\phi=0$ which first present themselves are those in which ϕ is a function of one (only) of the variables λ, μ, ν. Thus ϕ may be a function of λ alone, provided

$$(a^2+\lambda)^{\frac{1}{2}}(b^2+\lambda)^{\frac{1}{2}}(c^2+\lambda)^{\frac{1}{2}}\frac{d\phi}{d\lambda} = \text{const.},$$

whence

$$\phi = C\int_\lambda^\infty \frac{d\lambda}{\Delta}, \quad \text{..............................(1)}$$

if

$$\Delta = \{(a^2+\lambda)(b^2+\lambda)(c^2+\lambda)\}^{\frac{1}{2}}, \quad \text{....................(2)}$$

the additive constant which attaches to ϕ being chosen so as to make ϕ vanish for $\lambda=\infty$.

In this solution, which corresponds to $\phi=A/r$ in spherical harmonics, the equipotential surfaces are the confocal ellipsoids, and the motion in the space external to any one of these (say that for which $\lambda=0$) is that due to a certain arrangement of simple sources over it. The velocity at any point is given by the formula

$$-h_1\frac{d\phi}{d\lambda} = C\frac{h_1}{\Delta}. \quad \text{............................(3)}$$

At a great distance from the origin the ellipsoids λ become spheres of radius $\lambda^{\frac{1}{2}}$, and the velocity is therefore ultimately equal to $2C/r^2$, where r denotes the distance from the origin. Over any particular equipotential surface λ, the velocity varies as the perpendicular from the centre on the tangent plane.

To find the distribution of sources over the surface $\lambda=0$ which would produce the actual motion in the external space, we substitute for ϕ the value (1), in the formula (11) of Art. 58, and for ϕ' (which refers to the internal space) the constant value

$$\phi' = C\int_0^\infty \frac{d\lambda}{\Delta}. \quad \text{..............................(4)}$$

* It will be noticed that h_1, h_2, h_3 are double the perpendiculars from the origin on the tangent planes to the three quadrics λ, μ, ν.

† Cf. Lamé, "Sur les surfaces isothermes dans les corps solides homogènes en équilibre de empérature," *Liouville*, ii. 147 (1837).

The formula referred to then gives, for the surface-density of the required distribution,

$$\frac{C}{abc} \cdot h_1. \qquad\qquad (5)$$

The solution (1) may also be interpreted as representing the motion due to a change in the dimensions of the ellipsoid, such that the surface remains similar to itself, and retains the directions of its principal axes unchanged. If we put

$$\dot{a}/a = \dot{b}/b = \dot{c}/c, \quad = k, \text{ say},$$

the surface-condition Art. 110 (4) becomes

$$-\partial\phi/\partial n = \tfrac{1}{2}kh_1,$$

which is identical with (3), if we put $\lambda = 0$, $C = \frac{1}{2}kabc$.

A particular case of (5) is where the sources are distributed over the *elliptic disk* for which $\lambda = -c^2$, and therefore $z^2 = 0$. This is important in Electrostatics, but a more interesting application from the present point of view is to the flow through an *elliptic aperture*, viz. if the plane xy be occupied by a thin rigid partition with the exception of the part included by the ellipse

$$\frac{x^2}{a^2} + \frac{y^2}{b^2} = 1, \qquad z = 0,$$

we assume, putting $c = 0$ in previous formulae,

$$\phi = \mp A \int_0^\lambda \frac{d\lambda}{(a^2+\lambda)^{\frac{1}{2}}(b^2+\lambda)^{\frac{1}{2}}\lambda^{\frac{1}{2}}}, \qquad\qquad (6)$$

where the upper limit is the positive root of

$$\frac{x^2}{a^2+\lambda} + \frac{y^2}{b^2+\lambda} + \frac{z^2}{\lambda} = 1, \qquad\qquad (7)$$

and the negative or the positive sign is to be taken according as the point for which ϕ is required lies on the positive or the negative side of the plane xy. The two values of ϕ are continuous at the aperture, where $\lambda = 0$. As before, the velocity at a great distance is equal to $2A/r^2$, and the total flux through the area $2\pi r^2$ is therefore $4\pi A$. The total range of ϕ from $\lambda = -\infty$ to $\lambda = +\infty$ is

$$2A \int_0^\infty \frac{d\lambda}{(a^2+\lambda)^{\frac{1}{2}}(b^2+\lambda)^{\frac{1}{2}}\lambda^{\frac{1}{2}}} = 4A \int_0^{\frac{1}{2}\pi} \frac{d\theta}{\sqrt{(a^2\sin^2\theta + b^2\cos^2\theta)}}$$

The 'conductivity,' therefore, of the aperture (to borrow a term from electricity) is

$$\pi \div \int_0^{\frac{1}{2}\pi} \frac{d\theta}{\sqrt{(a^2\sin^2\theta + b^2\cos^2\theta)}}. \qquad\qquad (8)$$

For a circular aperture this $= 2a$.

For points in the aperture the velocity may be found immediately from (6) and (7); thus we may put

$$\delta z = \pm \lambda^{\frac{1}{2}}\left(1 - \frac{x^2}{a^2} - \frac{y^2}{b^2}\right)^{\frac{1}{2}}, \qquad \delta\phi = \mp \frac{2A\lambda^{\frac{1}{2}}}{ab},$$

approximately, since λ is small, whence

$$-\frac{\partial\phi}{\partial z}=\frac{2A}{ab}\cdot\left(1-\frac{x^2}{a^2}-\frac{y^2}{b^2}\right)^{-\frac{1}{2}} \quad\text{.....................(9)}$$

This becomes infinite, as we should expect, at the edge. The particular case of a *circular* aperture has already been solved otherwise in Arts. 102, 108.

114. We proceed to investigate the solution of $\nabla^2\phi=0$, finite at infinity, which corresponds, for the space external to the ellipsoid, to the solution $\phi=x$ for the internal space. Following the analogy of spherical harmonics we may assume for trial

$$\phi=x\chi, \quad\text{.....................................(1)}$$

which gives

$$\nabla^2\chi+\frac{2}{x}\frac{\partial\chi}{\partial x}=0, \quad\text{..............................(2)}$$

and inquire whether this can be satisfied by making χ equal to some function of λ only. On this supposition we shall have, by Art. 111,

$$\frac{\partial\chi}{\partial x}=h_1\frac{d\chi}{d\lambda}\cdot h_1\frac{\partial x}{\partial\lambda},$$

and therefore, by Art. 112 (4), (6),

$$\frac{2}{x}\frac{\partial\chi}{\partial x}=4\frac{(b^2+\lambda)(c^2+\lambda)}{(\lambda-\mu)(\lambda-\nu)}\frac{d\chi}{d\lambda}.$$

On substituting the value of $\nabla^2\chi$ in terms of λ, the equation (2) becomes

$$\left\{(a^2+\lambda)^{\frac{1}{2}}(b^2+\lambda)^{\frac{1}{2}}(c^2+\lambda)^{\frac{1}{2}}\frac{d}{d\lambda}\right\}^2\chi=-(b^2+\lambda)(c^2+\lambda)\frac{d\chi}{d\lambda},$$

which may be written

$$\frac{d}{d\lambda}\log\left\{(a^2+\lambda)^{\frac{1}{2}}(b^2+\lambda)^{\frac{1}{2}}(c^2+\lambda)^{\frac{1}{2}}\frac{d\chi}{d\lambda}\right\}=-\frac{1}{a^2+\lambda}.$$

Hence

$$\chi=C\int_\lambda^\infty\frac{d\lambda}{(a^2+\lambda)^{\frac{3}{2}}(b^2+\lambda)^{\frac{1}{2}}(c^2+\lambda)^{\frac{1}{2}}}, \quad\text{.................(3)}$$

the arbitrary constant which presents itself in the second integration being chosen as before so as to make χ vanish at infinity.

The solution contained in (1) and (3) enables us to find the motion of a liquid, at rest at infinity, produced by the translation of a solid ellipsoid through it, parallel to a principal axis. The notation being as before, and the ellipsoid

$$\frac{x^2}{a^2}+\frac{y^2}{b^2}+\frac{z^2}{c^2}=1 \quad\text{............................(4)}$$

being supposed in motion parallel to x with velocity U, the surface-condition is

$$\frac{\partial\phi}{\partial\lambda}=-U\frac{\partial x}{\partial\lambda}, \quad\text{for } \lambda=0. \quad\text{.......................(5)}$$

Let us write, for shortness,

$$\alpha_0 = abc\int_0^\infty \frac{d\lambda}{(a^2+\lambda)\,\Delta}, \quad \beta_0 = abc\int_0^\infty \frac{d\lambda}{(b^2+\lambda)\,\Delta}, \quad \gamma_0 = abc\int_0^\infty \frac{d\lambda}{(c^2+\lambda)\,\Delta}, \qquad \text{......(6)}$$

where

$$\Delta = \{(a^2+\lambda)\,(b^2+\lambda)\,(c^2+\lambda)\}^{\frac{1}{2}}. \qquad \text{....................(7)}$$

It will be noticed that these quantities α_0, β_0, γ_0 are purely numerical. The conditions of our problem are satisfied by

$$\phi = Cx\int_\lambda^\infty \frac{d\lambda}{(a^2+\lambda)\,\Delta}, \qquad \text{..........................(8)}$$

provided

$$C = \frac{abc}{2-\alpha_0}\,U. \qquad \text{..................................(9)}$$

The corresponding solution when the ellipsoid moves parallel to y or z can be written down from symmetry, and by superposition we derive the case where the ellipsoid has any motion of translation whatever*.

At a great distance from the origin, the formula (8) becomes equivalent to

$$\phi = \tfrac{2}{3}\,C\,\frac{x}{r^3}, \qquad \text{..............(10)}$$

which is the velocity-potential of a double source at the origin, of strength $\frac{8}{3}\pi C$, or

$$\tfrac{8}{3}\,\frac{\pi}{2-\alpha_0}\,abcU,$$

compare Art. 92.

The kinetic energy of the fluid is given by

$$2T = -\rho\iint\phi\,\frac{\partial\phi}{\partial n}\,dS = \frac{\alpha_0}{2-\alpha_0}\,.\,\rho U^2\,.\iint xl\,dS,$$

where l is the cosine of the angle which the normal to the surface makes with the axis of x. Since the latter integral is equal to the volume of the ellipsoid, we have

$$2T = \frac{\alpha_0}{2-\alpha_0}\,.\,\tfrac{4}{3}\pi\,abc\rho\,.\,U^2. \qquad \text{........................(11)}$$

The inertia-coefficient is therefore equal to the fraction

$$k = \frac{\alpha_0}{2-\alpha_0} \qquad \text{..............................(12)}$$

of the mass displaced by the solid. For the case of the sphere ($a=b=c$) we find $\alpha_0=\frac{2}{3}$, $k=\frac{1}{2}$, in agreement with Art. 92. If we put $a=b$, we get the case of an ellipsoid of revolution.

* This problem was first solved by Green, "Researches on the Vibration of Pendulums in Fluid Media," *Trans. R. S. Edin.* xiii. 54 (1833) [*Papers*, p. 315]. The investigation is much shortened if we assume at once from the Theory of Attractions that (8) is a solution of $\nabla^2\phi=0$, being in fact (except for a constant factor) the x-component of the attraction of a homogeneous ellipsoid at an external point.

For the prolate ellipsoid ($b=c$, $a>b$) we find

$$\alpha_0=\frac{2(1-e^2)}{e^3}\left(\tfrac{1}{2}\log\frac{1+e}{1-e}-e\right), \qquad \text{.............................(13)}$$

$$\beta_0=\gamma_0=\frac{1}{e^2}-\frac{1-e^2}{2e^3}\log\frac{1+e}{1-e}, \qquad \text{.............................(14)}$$

where e is the eccentricity of the meridian section. The formulae for an oblate ellipsoid are given in Art. 373. The values of k for a prolate ellipsoid moving respectively 'end-on' and 'broadside on,' viz.

$$k_1=\frac{\alpha_0}{2-\alpha_0}, \qquad k_2=\frac{\beta_0}{2-\beta_0}, \qquad \text{.............................(15)}$$

are tabulated on the opposite page for a series of values of the ratio a/b.

For an elliptic disk ($a\to 0$) the formula (11) becomes nugatory, since $\alpha_0\to 2$. A separate calculation, starting from (1) and (3), leads to the result

$$2T=\tfrac{4}{3}\pi\rho b^2c^2U^2\div\int_0^{\frac{1}{2}\pi}\sqrt{(b^2\sin^2\theta+c^2\cos^2\theta)}\,d\theta. \qquad \text{.................(16)}$$

For $b=c$ this reproduces the result (20) of Art. 102.

115. We next inquire whether the equation $\nabla^2\phi=0$ can be satisfied by

$$\phi=yz\chi, \qquad \text{.................................(1)}$$

where χ is a function of λ only. This requires

$$\nabla^2\chi+\frac{2}{y}\frac{\partial\chi}{\partial y}+\frac{2}{z}\frac{\partial\chi}{\partial z}=0. \qquad \text{.........................(2)}$$

Now, from Art. 112 (4), (6),

$$\frac{2}{y}\frac{\partial\chi}{\partial y}+\frac{2}{z}\frac{\partial\chi}{\partial z}=2h_1^2\left(\frac{1}{y}\frac{\partial y}{\partial\lambda}+\frac{1}{z}\frac{\partial z}{\partial\lambda}\right)\frac{d\chi}{d\lambda}$$

$$=4\frac{(a^2+\lambda)(b^2+\lambda)(c^2+\lambda)}{(\lambda-\mu)(\lambda-\nu)}\left(\frac{1}{b^2+\lambda}+\frac{1}{c^2+\lambda}\right)\frac{d\chi}{d\lambda}.$$

On substitution in (2) we find, by Art. 112 (7),

$$\frac{d}{d\lambda}\log\left\{(a^2+\lambda)^{\frac{1}{2}}(b^2+\lambda)^{\frac{3}{2}}(c^2+\lambda)^{\frac{3}{2}}\frac{d\chi}{d\lambda}\right\}=-\frac{1}{b^2+\lambda}-\frac{1}{c^2+\lambda},$$

whence

$$\chi=C\int_\lambda^\infty\frac{d\lambda}{(b^2+\lambda)(c^2+\lambda)\Delta}, \qquad \text{.......................(3)}$$

the second constant of integration being chosen as before.

For a rigid ellipsoid rotating about the axis of x with angular velocity Ω_x, the surface-condition is

$$\frac{\partial\phi}{\partial\lambda}=\Omega_x\left(z\frac{\partial y}{\partial\lambda}-y\frac{\partial z}{\partial\lambda}\right), \qquad \text{.........................(4)}$$

for $\lambda=0$. Assuming*

$$\phi=Cyz\int_\lambda^\infty\frac{d\lambda}{(b^2+\lambda)(c^2+\lambda)\Delta}, \qquad \text{.....................(5)}$$

* The expression (5) differs only by a factor from

$$y\frac{\partial\Phi}{\partial z}-z\frac{\partial\Phi}{\partial y},$$

where Φ is the gravitation potential of a uniform solid ellipsoid at an external point (x, y, z). Since $\nabla^2\Phi=0$ it easily follows that the above is also a solution of the equation $\nabla^2\phi=0$.

we find that the surface-condition (4) is satisfied, provided

$$-\frac{C}{ab^3c^3}+\tfrac{1}{2}C\left(\frac{1}{b^2}+\frac{1}{c^2}\right)\frac{\gamma_0-\beta_0}{abc\,(b^2-c^2)}=\tfrac{1}{2}\Omega_x\left(\frac{1}{b^2}-\frac{1}{c^2}\right),$$

or

$$C=\frac{(b^2-c^2)^2}{2\,(b^2-c^2)+(b^2+c^2)\,(\beta_0-\gamma_0)}\,abc\,\Omega_x. \quad\text{..............(6)}$$

The formulae for the cases of rotation about y or z can be written down from symmetry*.

The formula for the kinetic energy is

$$2T=-\rho\iint\phi\frac{\partial\phi}{\partial n}\,dS$$

$$=\rho C\Omega_x{}^2.\int_0^\infty\frac{d\lambda}{(a^2+\lambda)^{\frac{1}{2}}\,(b^2+\lambda)^{\frac{3}{2}}\,(c^2+\lambda)^{\frac{3}{2}}}.\iint(ny-mz)\,yz\,dS,$$

if (l, m, n) denote the direction-cosines of the normal to the ellipsoid. The latter integral

$$=\iiint(y^2-z^2)\,dx\,dy\,dz=\tfrac{1}{5}(b^2-c^2)\,.\,\tfrac{4}{3}\pi abc.$$

Hence we find

$$2T=\tfrac{1}{5}\,.\,\frac{(b^2-c^2)^2\,(\gamma_0-\beta_0)}{2\,(b^2-c^2)+(b^2+c^2)\,(\beta_0-\gamma_0)}\,.\,\tfrac{4}{3}\pi abc\rho\,.\,\Omega_x{}^2. \quad\text{.........(7)}$$

For a prolate ellipsoid ($b=c$, $a>b$) rotating about an equatorial diameter, the ratio of the inertia coefficient to the moment of inertia, about the same diameter, of the mass of fluid displaced is found to be

$$k'=\frac{e^4\,(\beta_0-\alpha_0)}{(2-e^2)\,\{2e^2-(2-e^2)\,(\beta_0-\alpha_0)\}}. \quad\text{..............................(8)}$$

The values of k_1, k_2 (defined in Art. 114), and k' are shewn in the accompanying table.

a/b	k_1	k_2	k'
1	0·5	0·5	0
1·50	0·305	0·621	0·094
2·00	0·209	0·702	0·240
2·51	0·156	0·763	0·367
2·99	0·122	0·803	0·465
3·99	0·082	0·860	0·608
4·99	0·059	0·895	0·701
6·01	0·045	0·918	0·764
6·97	0·036	0·933	0·805
8·01	0·029	0·945	0·840
9·02	0·024	0·954	0·865
9·97	0·021	0·960	0·883
∞	0	1	1

The two remaining types of ellipsoidal harmonic of the second order, finite at the origin, are given by the expression

$$\frac{x^2}{a^2+\theta}+\frac{y^2}{b^2+\theta}+\frac{z^2}{c^2+\theta}-1, \quad\text{..................................(9)}$$

* The solution contained in (5) and (6) is due to Clebsch, "Ueber die Bewegung eines Ellipsoides in einer tropfbaren Flüssigkeit," *Crelle*, lii. 103, liii. 287 (1854–6).

where θ is either root of $$\frac{1}{a^2+\theta}+\frac{1}{b^2+\theta}+\frac{1}{c^2+\theta}=0, \qquad\text{.................................(10)}$$

this being the condition that (9) should satisfy $\nabla^2\phi=0$.

The method of obtaining the corresponding solutions for the external space is explained in the treatise of Ferrers. These solutions would enable us to express the motion produced in a surrounding liquid by variations in the lengths of the axes of an ellipsoid, subject to the condition of no variation of volume:

$$\dot{a}/a+\dot{b}/b+\dot{c}/c=0. \qquad\text{.....................................(11)}$$

We have already found in Art. 113, the solution for the case where the ellipsoid expands (or contracts) remaining similar to itself; so that by superposition we could obtain the case of an internal boundary changing its position and dimensions in any manner whatever, subject only to the condition of remaining ellipsoidal. This extension of the results arrived at by Green and Clebsch was first treated, though in a different manner from that here indicated, by Bjerknes*.

116. The investigations of this chapter have related almost entirely to the case of spherical or ellipsoidal boundaries. It will be understood that solutions of the equation $\nabla^2\phi=0$ can be carried out, on lines more or less similar, which are appropriate to other forms of boundary. The surface which comes next in interest, from the point of view of the present subject, is that of the anchor-ring or 'torus'; this case has been very ably treated, by distinct methods, by Hicks, and Dyson†. We may also refer to the analytically remarkable problem of the spherical bowl, which has been investigated by Basset‡.

APPENDIX TO CHAPTER V

THE HYDRODYNAMICAL EQUATIONS REFERRED TO GENERAL ORTHOGONAL CO-ORDINATES

We follow the notation of Art. 111, with this modification that differentiations of x, y, z with respect to the independent variables λ, μ, ν are indicated by the suffixes 1, 2, 3, respectively. Thus the direction-cosines of the normal to the surface $\lambda=$const. are

$$(h_1x_1,\ h_1y_1,\ h_1z_1),$$

and so on.

If u, v, w be the component velocities along the three normals, the total flux out of the quasi-rectangular region whose edges are $\delta\lambda/h_1$, $\delta\mu/h_2$, $\delta\nu/h_3$ will be

$$\frac{\partial}{\partial\lambda}\left(\frac{u\,\delta\mu\,\delta\nu}{h_2h_3}\right)\delta\lambda+\frac{\partial}{\partial\mu}\left(\frac{v\,\delta\nu\,\delta\lambda}{h_3h_1}\right)\delta\mu+\frac{\partial}{\partial\nu}\left(\frac{w\,\delta\lambda\,\delta\mu}{h_1h_2}\right)\delta\nu,$$

whence the expression for the expansion, viz.

$$\Delta=h_1h_2h_3\left\{\frac{\partial}{\partial\lambda}\left(\frac{u}{h_2h_3}\right)+\frac{\partial}{\partial\mu}\left(\frac{v}{h_3h_1}\right)+\frac{\partial}{\partial\nu}\left(\frac{w}{h_1h_2}\right)\right\}; \qquad\text{.....................(1)}$$

cf. Art. 111 (4).

* *l.c. ante* p. 147.

† Hicks, "On Toroidal Functions," *Phil. Trans.* clxxii. 609 (1881); Dyson, "On the Potential of an Anchor-Ring," *Phil. Trans.* clxxxiv. 43 (1892); see also C. Neumann, *l.c. ante* p. 134.

‡ "On the Potential of an Electrified Spherical Bowl, &c.," *Proc. Lond. Math. Soc.* (1) xv. 286 (1885); *Hydrodynamics*, i. 149.

The circulation round a rectangular circuit on the surface $\lambda=$const., whose sides are $\delta\mu/h_2$, $\delta\nu/h_3$, is

$$\frac{\partial}{\partial\mu}\left(\frac{w\,\delta\nu}{h_3}\right)\delta\mu-\frac{\partial}{\partial\nu}\left(\frac{v\,\delta\mu}{h_2}\right)\delta\nu. \quad\text{.............................(2)}$$

Dividing by the area of the circuit we get the first of the following formulae for the components of vorticity about the three normals:

$$\left.\begin{aligned}\xi&=h_2h_3\left\{\frac{\partial}{\partial\mu}\left(\frac{w}{h_3}\right)-\frac{\partial}{\partial\nu}\left(\frac{v}{h_2}\right)\right\},\\ \eta&=h_3h_1\left\{\frac{\partial}{\partial\nu}\left(\frac{u}{h_1}\right)-\frac{\partial}{\partial\lambda}\left(\frac{w}{h_3}\right)\right\},\\ \zeta&=h_1h_2\left\{\frac{\partial}{\partial\lambda}\left(\frac{v}{h_2}\right)-\frac{\partial}{\partial\mu}\left(\frac{u}{h_1}\right)\right\}.\end{aligned}\right\} \quad\text{.............................(3)}$$

To find expressions for the component accelerations, we note that in a time δt a particle changes its parameters from (λ, μ, ν) to $(\lambda+\delta\lambda, \mu+\delta\mu, \nu+\delta\nu)$, where

$$\delta\lambda/h_1=u\,\delta t,\quad \delta\mu/h_2=v\,\delta t,\quad \delta\nu/h_3=w\,\delta t.$$

The component velocities therefore become

$$u+\left(\frac{\partial u}{\partial t}+h_1u\frac{\partial u}{\partial\lambda}+h_2v\frac{\partial u}{\partial\mu}+h_3w\frac{\partial u}{\partial\nu}\right)\delta t,\ \&\text{c.},\ \&\text{c.}, \quad\text{....................(4)}$$

and we have to resolve these along the original directions of u, v, w. Now after a time δt the direction-cosines of the new direction of v become

$$h_2x_2+\frac{\partial}{\partial\lambda}(h_2x_2)\,h_1u\,\delta t+\frac{\partial}{\partial\mu}(h_2x_2)\,h_2v\,\delta t+\frac{\partial}{\partial\nu}(h_2x_2)\,h_3w\,\delta t,\ \&\text{c.},\ \&\text{c.},$$

where in the two expressions not written out the derivatives of x are to be replaced by those of y and z, respectively. Hence the cosine of the angle between the new direction of v and the original direction of u, viz. (h_1x_1, h_1y_1, h_1z_1), is

$$\{(x_1x_{12}+y_1y_{12}+z_1z_{12})\,h_1u+(x_1x_{22}+y_1y_{22}+z_1z_{22})\,h_2v+(x_1x_{23}+y_1y_{23}+z_1z_{23})\,h_3w\}\,h_1h_2\,\delta t. \quad\text{.........(5)}$$

Certain terms have been omitted from this expression in virtue of the relation

$$x_1x_2+y_1y_2+z_1z_2=0, \quad\text{...................................(6)}$$

which follows from the orthogonal property. Again, differentiating (6) with respect to ν and comparing with similar results we infer that

$$x_1x_{23}+y_1y_{23}+z_1z_{23}=0. \quad\text{...............................(7)*}$$

Also, differentiation of the identity

$$x_1^2+y_1^2+z_1^2=\frac{1}{h_1^2} \quad\text{...................................(8)}$$

with respect to μ gives

$$x_1x_{12}+y_1y_{12}+z_1z_{12}=\frac{1}{h_1}\frac{\partial}{\partial\mu}\left(\frac{1}{h_1}\right). \quad\text{............................(9)}$$

Again,

$$x_1x_{22}+y_1y_{22}+z_1z_{22}=\frac{\partial}{\partial\mu}(x_1x_2+y_1y_2+z_1z_2)-(x_2x_{12}+y_1y_{12}+z_1z_{12})=-\frac{1}{h_2}\frac{\partial}{\partial\lambda}\left(\frac{1}{h_2}\right). \quad\text{...(10)}$$

The expression (5) thus reduces to

$$\left\{u\frac{\partial}{\partial\mu}\left(\frac{1}{h_1}\right)-v\frac{\partial}{\partial\lambda}\left(\frac{1}{h_2}\right)\right\}h_1h_2\,\delta t. \quad\text{............................(11)}$$

* Forsyth, *Differential Geometry*, Cambridge (1912), p. 412.

In the same way the cosine of the angle between the new direction of w and the original direction of u is

$$\left\{u\frac{\partial}{\partial\nu}\left(\frac{1}{h_1}\right)-w\frac{\partial}{\partial\lambda}\left(\frac{1}{h_3}\right)\right\}h_1h_3\delta t. \qquad (12)$$

The acceleration in the original direction of u is thus found to be

$$\frac{\partial u}{\partial t}+h_1u\frac{\partial u}{\partial\lambda}+h_2v\frac{\partial u}{\partial\mu}+h_3w\frac{\partial u}{\partial\nu}$$
$$+h_1h_2v\left\{u\frac{\partial}{\partial\mu}\left(\frac{1}{h_1}\right)-v\frac{\partial}{\partial\lambda}\left(\frac{1}{h_2}\right)\right\}$$
$$+h_1h_3w\left\{u\frac{\partial}{\partial\nu}\left(\frac{1}{h_1}\right)-w\frac{\partial}{\partial\lambda}\left(\frac{1}{h_3}\right)\right\}, \qquad (13)^*$$

or, more symmetrically,

$$\frac{\partial u}{\partial t}+h_1u\frac{\partial u}{\partial\lambda}+h_2v\frac{\partial u}{\partial\mu}+h_3w\frac{\partial u}{\partial\nu}$$
$$+h_1u\left\{h_1u\frac{\partial}{\partial\lambda}\left(\frac{1}{h_1}\right)+h_2v\frac{\partial}{\partial\mu}\left(\frac{1}{h_1}\right)+h_3w\frac{\partial}{\partial\nu}\left(\frac{1}{h_1}\right)\right\}$$
$$-h_1\left\{h_1u^2\frac{\partial}{\partial\lambda}\left(\frac{1}{h_1}\right)+h_2v^2\frac{\partial}{\partial\lambda}\left(\frac{1}{h_2}\right)+h_3w^2\frac{\partial}{\partial\lambda}\left(\frac{1}{h_3}\right)\right\} \qquad (14)$$

The expressions for the acceleration in the direction of v and w follow by symmetry.

For example, in *cylindrical* co-ordinates we have

$$x=r\cos\theta, \quad y=r\sin\theta, \quad z=z.$$

Putting $\lambda=r, \quad \mu=\theta, \quad \nu=z,$

we have $h_1=1, \quad h_2=1/r, \quad h_3=1.$

The expansion is accordingly

$$\Delta=\frac{\partial u}{\partial r}+\frac{u}{r}+\frac{\partial v}{r\partial\theta}+\frac{\partial w}{\partial z}, \qquad (15)$$

and the components of vorticity are

$$\xi=\frac{\partial w}{r\partial\theta}-\frac{\partial v}{\partial z}, \quad \eta=\frac{\partial u}{\partial z}-\frac{\partial w}{\partial r}, \quad \zeta=\frac{\partial v}{\partial r}+\frac{v}{r}-\frac{\partial u}{r\partial\theta}. \qquad (16)$$

The component accelerations are

$$\left.\begin{aligned}&\frac{\partial u}{\partial t}+u\frac{\partial u}{\partial r}+v\frac{\partial u}{r\partial\theta}-\frac{v^2}{r}+w\frac{\partial u}{\partial z},\\&\frac{\partial v}{\partial t}+u\frac{\partial v}{\partial r}+v\frac{\partial v}{r\partial\theta}+\frac{uv}{r}+w\frac{\partial v}{\partial z},\\&\frac{\partial w}{\partial t}+u\frac{\partial w}{\partial r}+v\frac{\partial w}{r\partial\theta}+w\frac{\partial w}{\partial z}.\end{aligned}\right\} \qquad (17)$$

If in this formula we put $w=0$ we get the results for plane polar co-ordinates (Art. 16a).

In *spherical polars*

$$x=r\sin\theta\cos\omega, \quad y=r\sin\theta\sin\omega, \quad z=r\cos\theta.$$

Putting $\lambda=r, \quad \mu=\theta, \quad \nu=\omega,$

we have $h_1=1, \quad h_2=1/r, \quad h_3=1/r\sin\theta.$

* G. B. Jeffery, *Phil. Mag.* (6) xxix. 445 (1915).

Hence
$$\Delta=\frac{\partial u}{\partial t}+2\frac{u}{r}+\frac{\partial v}{r\partial\theta}+\frac{v}{r}\cot\theta+\frac{1}{r\sin\theta}\frac{\partial w}{\partial\omega}, \qquad (18)$$

$$\left.\begin{aligned}\xi&=\frac{\partial w}{r\partial\theta}-\frac{\partial v}{r\sin\theta\,\partial\omega}-\frac{w}{r}\cot\theta,\\ \eta&=\frac{\partial u}{r\sin\theta\,\partial\omega}-\frac{\partial w}{\partial r}-\frac{w}{r},\\ \zeta&=\frac{\partial v}{\partial r}+\frac{v}{r}-\frac{\partial u}{r\partial\theta}.\end{aligned}\right\} \qquad (19)$$

The component accelerations are

$$\left.\begin{aligned}&\frac{\partial u}{\partial t}+u\frac{\partial u}{\partial r}+v\frac{\partial u}{r\partial\theta}+w\frac{\partial u}{r\sin\theta\,d\omega}-\frac{v^2+w^2}{r},\\ &\frac{\partial v}{\partial t}+u\frac{\partial v}{\partial r}+v\frac{\partial v}{r\partial\theta}+w\frac{\partial v}{r\sin\theta\,d\omega}+\frac{uv}{r}-\frac{w^2}{r}\cot\theta,\\ &\frac{\partial w}{\partial t}+u\frac{\partial w}{\partial r}+v\frac{\partial w}{r\partial\theta}+w\frac{\partial w}{r\sin\theta\,d\omega}+\frac{uw}{r}+\frac{vw}{r}\cot\theta;\end{aligned}\right\} \qquad (20)$$

cf. Art. 16 a.

CHAPTER VI

ON THE MOTION OF SOLIDS THROUGH A LIQUID: DYNAMICAL THEORY

117. IN this chapter it is proposed to study the very interesting dynamical problem furnished by the motion of one or more solids in a frictionless liquid. The development of this subject is due mainly to Thomson and Tait* and to Kirchhoff†. The cardinal feature of the methods followed by these writers consists in this, that the solids and the fluid are treated as forming together one dynamical system, and thus the troublesome calculation of the effect of the fluid pressures on the surfaces of the solids is avoided.

To begin with the case of a single solid moving through an infinite mass of liquid, we will suppose in the first instance that the motion of the fluid is entirely due to that of the solid, and is therefore irrotational and acyclic. Some special cases of this problem have been treated incidentally in the foregoing pages, and it appeared that the whole effect of the fluid might be represented by an addition to the *inertia* of the solid. The same result will be found to hold in general, provided we use the term 'inertia' in a somewhat extended sense.

Under the circumstances supposed, the motion of the fluid is characterized by the existence of a single-valued velocity-potential ϕ which, besides satisfying the equation of continuity

$$\nabla^2\phi = 0, \quad \text{.................................(1)}$$

fulfils the following conditions: (1°) the value of $-\partial\phi/\partial n$, where δn denotes as usual an element of the normal at any point of the surface of the solid, drawn on the side of the fluid, must be equal to the velocity of the surface at that point normal to itself, and (2°) the differential coefficients $\partial\phi/\partial x$, $\partial\phi/\partial y$, $\partial\phi/\partial z$ must vanish at an infinite distance, in every direction, from the solid. The latter condition is rendered necessary by the consideration that a finite velocity at infinity would imply an infinite kinetic energy, which could not be generated by finite forces acting for a finite time on the solid. It is also the condition to which we are led by supposing the fluid to be enclosed within a fixed vessel infinitely large and infinitely distant, all round, from the moving body. For on this supposition the space occupied by the fluid may be conceived as made up of tubes of flow which begin and end on

* *Natural Philosophy*, Art. 320. Subsequent investigations by Lord Kelvin will be referred to later.

† "Ueber die Bewegung eines Rotationskörpers in einer Flüssigkeit," *Crelle*, lxxi. 237 (1869) [*Ges. Abh.* p. 376]; *Mechanik*, c. xix.

the surface of the solid, so that the total flux across any area, finite or infinite, drawn in the fluid must be finite, and therefore the velocity at infinity zero.

It has been shewn in Art. 41 that under the above conditions the motion of the fluid is determinate.

118. In the further study of the problem it is convenient to follow the method introduced by Euler in the dynamics of rigid bodies, and to adopt a system of rectangular axes Ox, Oy, Oz fixed in the body, and moving with it. If the motion of the body at any instant be defined by the angular velocities p, q, r about, and the translational velocities u, v, w of the origin parallel to, the instantaneous positions of these axes*, we may write, after Kirchhoff,

$$\phi = u\phi_1 + v\phi_2 + w\phi_3 + p\chi_1 + q\chi_2 + r\chi_3, \quad \text{..............(2)}$$

where, as will appear immediately, $\phi_1, \phi_2, \phi_3, \chi_1, \chi_2, \chi_3$ are certain functions of x, y, z determined solely by the configuration of the surface of the solid, relative to the co-ordinate axes. In fact, if l, m, n denote the direction-cosines of the normal, drawn towards the fluid, at any point of this surface, the kinematical surface-condition is

$$-\frac{\partial\phi}{\partial n} = l(u + qz - ry) + m(v + rx - pz) + n(w + py - qx),$$

whence, substituting the value (2) of ϕ, we find

$$\left.\begin{aligned} -\frac{\partial\phi_1}{\partial n} &= l, & -\frac{\partial\phi_2}{\partial n} &= m, & -\frac{\partial\phi_3}{\partial n} &= n, \\ -\frac{\partial\chi_1}{\partial n} &= ny - mz, & -\frac{\partial\chi_2}{\partial n} &= lz - nx, & -\frac{\partial\chi_3}{\partial n} &= mx - ly. \end{aligned}\right\} \quad \text{......(3)}$$

Since these functions must also satisfy (1), and have their derivatives zero at infinity, they are completely determinate, by Art. 41†.

119. Now whatever the motion of the solid and fluid at any instant, it might have been generated instantaneously from rest by a properly adjusted impulsive 'wrench' applied to the solid. This wrench is in fact that which would be required to counteract the impulsive pressures $\rho\phi$ on the surface, and, in addition, to generate the actual momentum of the solid. It is called by Lord Kelvin the 'impulse' of the system at the moment under consideration. It is to be noted that the impulse, as thus defined, cannot be asserted to be equivalent to the total momentum of the system, which is indeed in the present problem indeterminate‡. We proceed to shew however that the impulse varies, in consequence of extraneous forces acting on the solid, in exactly the same way as the momentum of a finite dynamical system.

* The symbols u, v, w, p, q, r are not at present required in their former meanings.

† For the particular case of an ellipsoidal surface, their values may be written down from the results of Arts. 114, 115.

‡ That is, the attempt to calculate it leads to 'improper' or 'indeterminate' integrals.

Let us in the first instance consider any actual motion of a solid, from time t_0 to time t_1, under any given forces applied to it, in a *finite* mass of liquid enclosed by a fixed envelope of any form. Let us imagine the motion to have been generated from rest, previously to the time t_0, by forces (whether continuous or impulsive) applied to the solid, and to be arrested, in like manner, by forces applied to the solid after the time t_1. Since the momentum of the system is null both at the beginning and at the end of this process, the time-integrals of the forces applied to the solid, together with the time-integral of the pressures exerted on the fluid by the envelope, must form an equilibrating system. The effect of these latter pressures may be calculated, by Art. 20, from the formula

$$\frac{p}{\rho} = \frac{\partial \phi}{\partial t} - \tfrac{1}{2} q^2 + F(t). \qquad \text{.........................(1)}$$

A pressure uniform over the envelope has no resultant effect; hence, since ϕ is constant at the beginning and end, the only effective part of the integral pressure $\int p\, dt$ is given by the term

$$-\tfrac{1}{2}\rho \int q^2 dt. \qquad \text{...............................(2)}$$

Let us now revert to the original form of our problem, and suppose the containing envelope to be infinitely large, and infinitely distant in every direction from the moving solid. It is easily seen by considering the arrangement of the tubes of flow (Art. 36) that the fluid velocity q at a great distance r from an origin in the neighbourhood of the solid will ultimately be, at most*, of the order $1/r^2$, and the integral pressure (2) therefore of the order $1/r^4$. Since the surface-elements of the envelope are of the order $r^2 \delta\varpi$, where $\delta\varpi$ is an elementary solid angle, the force- and couple-resultants of the integral pressure (2) will now both be null. The same statement therefore holds with regard to the time-integral of the forces applied to the solid.

If we imagine the motion to have been started *instantaneously* at time t_0, and to be arrested instantaneously at time t_1, the result at which we have arrived may be stated as follows:

The 'impulse' of the motion (in Lord Kelvin's sense) at time t_1 differs from the 'impulse' at time t_0 by the time-integral of the extraneous forces acting on the solid during the interval $t_1 - t_0$†.

It will be noticed that the above reasoning is substantially unaltered when the single solid is replaced by a group of solids, which may moreover be flexible instead of rigid, and even when these solids are replaced by masses of liquid which are moving rotationally.

120. To express the above result analytically, let $\xi, \eta, \zeta, \lambda, \mu, \nu$ be the components of the force- and couple-constituents of the impulse; and let

* It is really of the order $1/r^3$ when, as in the case considered, the total flux outwards is zero.

† Sir W. Thomson, *l.c. ante* p. 33. The form of the argument given above was kindly suggested to the author by Sir J. Larmor.

X, Y, Z, L, M, N designate in the same manner the system of extraneous forces. The whole variation of $\xi, \eta, \zeta, \lambda, \mu, \nu$, due partly to the motion of the axes to which these quantities are referred, and partly to the action of the extraneous forces, is then given by the formulae*

$$\left.\begin{aligned}\frac{d\xi}{dt} &= r\eta - q\zeta + X, & \frac{d\lambda}{dt} &= w\eta - v\zeta + r\mu - q\nu + L,\\ \frac{d\eta}{dt} &= p\zeta - r\xi + Y, & \frac{d\mu}{dt} &= u\zeta - w\xi + p\nu - r\lambda + M,\\ \frac{d\zeta}{dt} &= q\xi - p\eta + Z, & \frac{d\nu}{dt} &= v\xi - u\eta + q\lambda - p\mu + N.\end{aligned}\right\} \quad \text{.........(1)}$$

For at time $t + \delta t$ the moving axes make with their positions at time t angles whose cosines are

$$(1, r\,\delta t, -q\,\delta t), \quad (-r\,\delta t, 1, p\,\delta t), \quad (q\,\delta t, -p\,\delta t, 1),$$

respectively. Hence, resolving parallel to the new position of the axis of x,

$$\xi + \delta\xi = \xi + \eta\,.\,r\delta t - \zeta\,.\,q\delta t + X\,\delta t.$$

Again, taking moments about the new position of Ox, and remembering that O has been displaced through spaces $u\,\delta t$, $v\,\delta t$, $w\,\delta t$ parallel to the axes, we find

$$\lambda + \delta\lambda = \lambda + \eta\,.\,w\,\delta t - \zeta\,.\,v\,\delta t + \mu\,.\,r\,\delta t - \nu\,.\,q\,\delta t + L\,\delta t.$$

These, with the similar results which can be written down from symmetry, give the equations (1).

When no extraneous forces act, we verify at once that these equations have the integrals

$$\xi^2 + \eta^2 + \zeta^2 = \text{const.}, \qquad \lambda\xi + \mu\eta + \nu\zeta = \text{const.}, \quad \text{...........(2)}$$

which express that the magnitudes of the force- and couple-resultants of the impulse are constant.

121. It remains to express $\xi, \eta, \zeta, \lambda, \mu, \nu$ in terms of u, v, w, p, q, r. In the first place let $\mathbf{T}$ denote the kinetic energy of the *fluid*, so that

$$2\mathbf{T} = -\rho \iint \phi \frac{\partial \phi}{\partial n}\, dS, \quad \text{..........................(1)}$$

where the integration extends over the surface of the moving solid. Substituting the value of ϕ from Art. 118 (2), we get

$$\begin{aligned}2\mathbf{T} = \mathbf{A}u^2 + \mathbf{B}v^2 + \mathbf{C}w^2 + 2\mathbf{A}'vw + 2\mathbf{B}'wu + 2\mathbf{C}'uv\\ + \mathbf{P}p^2 + \mathbf{Q}q^2 + \mathbf{R}r^2 + 2\mathbf{P}'qr + 2\mathbf{Q}'rp + 2\mathbf{R}'pq\\ + 2p(\mathbf{F}u + \mathbf{G}v + \mathbf{H}w) + 2q(\mathbf{F}'u + \mathbf{G}'v + \mathbf{H}'w) + 2r(\mathbf{F}''u + \mathbf{G}''v + \mathbf{H}''w),\end{aligned} \quad \text{......(2)}$$

where the twenty-one coefficients **A**, **B**, **C**, &c. are certain constants

* Cf. Hayward, "On a Direct Method of Estimating Velocities, Accelerations, and all similar Quantities, with respect to Axes moveable in any manner in space," *Camb. Trans.* x. 1 (1856).

determined by the form and position of the surface relative to the co-ordinate axes. Thus, for example,

$$\left.\begin{aligned}\mathbf{A} &= -\rho\iint \phi_1\frac{\partial\phi_1}{\partial n}\,dS = \rho\iint \phi_1 l\,dS,\\ \mathbf{A}' &= -\tfrac{1}{2}\rho\iint\left(\phi_2\frac{\partial\phi_3}{\partial n}+\phi_3\frac{\partial\phi_2}{\partial n}\right)dS\\ &= -\rho\iint\phi_2\frac{\partial\phi_3}{\partial n}\,dS = -\rho\iint\phi_3\frac{\partial\phi_2}{\partial n}\,dS\\ &= \rho\iint\phi_2 n\,dS = \rho\iint\phi_3 m\,dS,\\ \mathbf{P} &= -\rho\iint\chi_1\frac{\partial\chi_1}{\partial n}\,dS = \rho\iint\chi_1(ny-mz)\,dS,\end{aligned}\right\}\quad\ldots\ldots\ldots\ldots(3)$$

the transformations depending on Art. 118 (3) and on a particular case of Green's Theorem (Art. 44 (2)). These expressions for the coefficients were given by Kirchhoff.

The actual values of the coefficients in the expression for $2\mathbf{T}$ have been found in the preceding chapter for the case of the ellipsoid, viz. we have from Arts. 114, 115

$$\mathbf{A}=\frac{\alpha_0}{2-\alpha_0}\cdot\tfrac{4}{3}\pi\rho abc,\qquad \mathbf{P}=\tfrac{1}{5}\frac{(b^2-c^2)^2(\gamma_0-\beta_0)}{2(b^2-c^2)+(b^2+c^2)(\beta_0-\gamma_0)}\cdot\tfrac{4}{3}\pi\rho abc,\quad\ldots\ldots\ldots(4)$$

with similar expressions for **B**, **C**, **Q**, **R**. The remaining coefficients, as will appear presently, in this case all vanish. We note that

$$\mathbf{A}-\mathbf{B}=\frac{2(\alpha_0-\beta_0)}{(2-\alpha_0)(2-\beta_0)}\cdot\tfrac{4}{3}\pi\rho abc,\quad\ldots\ldots\ldots\ldots(5)$$

so that if $a>b>c$, then $\mathbf{A}<\mathbf{B}<\mathbf{C}$, as might have been anticipated.

The formulae for an ellipsoid of revolution may be deduced by putting $b=c$; they may also be obtained independently by the method of Arts. 104–109. Thus for a circular disk ($a=0$, $b=c$) we have

$$\mathbf{A},\ \mathbf{B},\ \mathbf{C}=\tfrac{8}{3}\rho c^3,\ 0,\ 0;\qquad \mathbf{P},\ \mathbf{Q},\ \mathbf{R}=0,\ \tfrac{16}{45}\rho c^5,\ \tfrac{16}{45}\rho c^5.\quad\ldots\ldots\ldots(6)$$

121 a. When the motion of the solid is one of pure *translation* the formula for the kinetic energy of the fluid reduces to

$$2\mathbf{T}=\mathbf{A}u^2+\mathbf{B}v^2+\mathbf{C}w^2+2\mathbf{A}'vw+2\mathbf{B}'wu+2\mathbf{C}'uv.\quad\ldots\ldots\ldots(1)$$

We can now shew that the effect at a great distance is in all cases that of a suitable double source, and that the character of this source is completely defined by the coefficients in (1).

For this we have recourse to the formula (12) of Art. 58, viz.

$$4\pi\,\phi_P=\iint(\phi-\phi')\frac{\partial}{\partial n}\frac{1}{r}\,dS.\quad\ldots\ldots\ldots\ldots(2)$$

We may regard the boundary of the solid as a thin rigid shell, with fluid also in its interior, and assume the potentials ϕ and ϕ' to refer to the external and internal regions, respectively. Let (x_1, y_1, z_1) be the co-ordinates of the point P, which we suppose to be at a distance great compared with the dimensions of the solid, and (x, y, z) those of a surface-element δS. Then, writing

$$r_1 = \sqrt{(x_1^2 + y_1^2 + z_1^2)}, \quad r = \sqrt{\{(x_1 - x)^2 + (y_1 - y)^2 + (z_1 - z)^2\}},$$

we have, approximately,

$$\frac{1}{r} = \frac{1}{r_1} + \frac{xx_1 + yy_1 + zz_1}{r_1^3}, \quad \frac{\partial}{\partial n}\frac{1}{r} = \frac{lx_1 + my_1 + nz_1}{r_1^3}.$$

Suppose, now, that the shell is moving with unit velocity parallel to x, without rotation. Writing

$$\phi = \phi_1, \quad \phi' = -x, \qquad \text{.........................(3)}$$

we have

$$\iint \phi \frac{\partial}{\partial n}\frac{1}{r}\, dS = \frac{\mathbf{A}x_1 + \mathbf{C}'y_1 + \mathbf{B}'z_1}{\rho r_1^3}, \qquad \text{.........................(4)}$$

and

$$\iint \phi' \frac{\partial}{\partial n}\frac{1}{r}\, dS = -\frac{Qx_1}{r_1^3}, \qquad \text{.........................(5)}$$

where Q denotes the volume of the solid. We have, in fact,

$$\iint xl\, dS = Q, \quad \iint xm\, dS = 0, \quad \iint xn\, dS = 0. \qquad \text{.............(6)}$$

Hence

$$4\pi\, \phi_P = \frac{(\mathbf{A} + \rho Q)\, x_1 + \mathbf{C}'y_1 + \mathbf{B}'z_1}{\rho r_1^3}. \qquad \text{.................(7)*}$$

The effect at a distance is therefore that of a double source, but the axis of the source does not necessarily coincide with the direction of translation. If, however, the solid is moving parallel to an axis of permanent translation (Art. 124), the coefficients $\mathbf{C}'$ and $\mathbf{B}'$ vanish, and

$$4\pi\, \phi_P = \frac{(\mathbf{A} + \rho Q)\, x_1}{\rho r_1^3}. \qquad \text{.........................(8)}$$

For example, in the case of the sphere we have $\mathbf{A} = \frac{2}{3}\pi\rho a^3$, $Q = \frac{4}{3}\pi a^3$, and

$$\phi_P = \frac{a^3 x_1}{2r_1^3}, \qquad \text{...............................(9)}$$

as in Art. 92.

When the velocity (u, v, w) of the solid is general, the formula (7) is replaced by

$$\begin{aligned} 4\pi r_1^3 \rho \phi_P = {} & (\mathbf{A}u + \mathbf{C}'v + \mathbf{B}'w)\, x_1 \\ & + (\mathbf{C}'u + \mathbf{B}v + \mathbf{A}'w)\, y_1 + (\mathbf{B}'u + \mathbf{A}'v + \mathbf{C}w)\, z_1 \\ & + \rho Q\, (ux_1 + vy_1 + wz_1). \qquad \text{...............................(10)} \end{aligned}$$

Conversely a knowledge of the form of the velocity-potential at infinity due to a 'permanent' translation leads to a knowledge of the corresponding inertia-coefficient.

* From a paper "On Wave Resistance," *Proc. Roy. Soc.* cxi. 15 (1926).

For instance, in the Rankine ovals referred to in Art. 97 we have a distribution of sources along the axis of x, subject to the condition that the total 'strength' of these sources is zero. If the line-density of this distribution be m, we have

$$\phi = \int \frac{m\,d\xi}{\sqrt{\{(x_1-\xi)^2+y_1^2+z_1^2\}}} = \int \left(\frac{1}{r_1}+\frac{\xi x_1}{r_1^3}+\ldots\right) m\,d\xi,$$

or

$$\phi = \frac{x_1}{r_1^3}\int m\xi\,d\xi + \ldots, \quad\ldots\ldots(11)$$

since $\int m\,d\xi = 0$. Hence

$$\mathbf{A}/\rho + Q = 4\pi\int m\xi\,d\xi. \quad\ldots\ldots(12)^*$$

122. The kinetic energy, $\mathbf{T}_1$ say, of the solid alone is given by an expression of the form

$$\begin{aligned}2\mathbf{T}_1 = \mathbf{m}\,(u^2+v^2+w^2) \\ +\mathbf{P}_1p^2+\mathbf{Q}_1q^2+\mathbf{R}_1r^2+2\mathbf{P}_1'qr+2\mathbf{Q}_1'rp+2\mathbf{R}_1'pq \\ +2\mathbf{m}\,\{\alpha(vr-wq)+\beta(wp-ur)+\gamma(uq-vp)\}. \quad\ldots\ldots(1)\end{aligned}$$

Hence the total energy $\mathbf{T}+\mathbf{T}_1$, of the system, which we shall denote by T, is given by an expression of the same general form as in Art. 121, say

$$\begin{aligned}2T = Au^2+Bv^2+Cw^2+2A'vw+2B'wu+2C'uv \\ +Pp^2+Qq^2+Rr^2+2P'qr+2Q'rp+2R'pq \\ +2p(Fu+Gv+Hw)+2q(F'u+G'v+H'w)+2r(F''u+G''v+H''w). \\ \ldots\ldots(2)\end{aligned}$$

The values of the several components of the impulse in terms of the velocities u, v, w, p, q, r can now be found by a well-known dynamical method†. Let a system of indefinitely great forces (X, Y, Z, L, M, N) act for an indefinitely short time τ on the solid, so as to change the impulse from $(\xi, \eta, \zeta, \lambda, \mu, \nu)$ to $(\xi+\delta\xi,\ \eta+\delta\eta,\ \zeta+\delta\zeta,\ \lambda+\delta\lambda,\ \mu+\delta\mu,\ \nu+\delta\nu)$. The work done by the force X, viz.

$$\int_0^\tau Xu\,dt,$$

lies between

$$u_1\int_0^\tau X\,dt \quad\text{and}\quad u_2\int_0^\tau X\,dt,$$

where u_1 and u_2 are the greatest and least values of u during the time τ, *i.e.* it lies between $u_1\delta\xi$ and $u_2\delta\xi$. If we now introduce the supposition that $\delta\xi, \delta\eta, \delta\zeta, \delta\lambda, \delta\mu, \delta\nu$ are infinitely small, u_1 and u_2 are each equal to u, and the work done is $u\delta\xi$. In the same way we may calculate the work done by the remaining forces and couples. The total result must be equal to the increment of the kinetic energy, whence

$$\begin{aligned}u\delta\xi+v\delta\eta+w\delta\zeta+p\delta\lambda+q\delta\mu+r\delta\nu \\ =\delta T = \frac{\partial T}{\partial u}\delta u+\frac{\partial T}{\partial v}\delta v+\frac{\partial T}{\partial w}\delta w+\frac{\partial T}{\partial p}\delta p+\frac{\partial T}{\partial q}\delta q+\frac{\partial T}{\partial r}\delta r. \quad\ldots(3)\end{aligned}$$

* G. I. Taylor, *Proc. Roy. Soc.* cxx. 13 (1928).

† See Thomson and Tait, Art. 313, or Maxwell, *Electricity and Magnetism*, Part IV. c. v.

Now if the velocities be all altered in any given ratio, the impulses will be altered in the same ratio. If then we take

$$\frac{\delta u}{u}=\frac{\delta v}{v}=\frac{\delta w}{w}=\frac{\delta p}{p}=\frac{\delta q}{q}=\frac{\delta r}{r}=k,$$

it will follow that $$\frac{\delta \xi}{\xi}=\frac{\delta \eta}{\eta}=\frac{\delta \zeta}{\zeta}=\frac{\delta \lambda}{\lambda}=\frac{\delta \mu}{\mu}=\frac{\delta \nu}{\nu}=k.$$

Substituting in (3), we find

$$u\xi+v\eta+w\zeta+p\lambda+q\mu+r\nu$$
$$=u\frac{\partial T}{\partial u}+v\frac{\partial T}{\partial v}+w\frac{\partial T}{\partial w}+p\frac{\partial T}{\partial p}+q\frac{\partial T}{\partial q}+r\frac{\partial T}{\partial r}=2T, \quad \ldots(4)$$

since T is a homogeneous quadratic function. Now performing the arbitrary variation δ on the first and last members of (4), and omitting terms which cancel by (3), we find

$$\xi\delta u+\eta\delta v+\zeta\delta w+\lambda\delta p+\mu\delta q+\nu\delta r=\delta T.$$

Since the variations δu, δv, δw, δp, δq, δr are all independent, this gives the required formulae

$$\xi,\ \eta,\ \zeta=\frac{\partial T}{\partial u},\ \frac{\partial T}{\partial v},\ \frac{\partial T}{\partial w}, \qquad \lambda,\ \mu,\ \nu=\frac{\partial T}{\partial p},\ \frac{\partial T}{\partial q},\ \frac{\partial T}{\partial r}. \quad \ldots\ldots\ldots(5)$$

It may be noted that since $\xi, \eta, \zeta, \ldots$ are linear functions of $u, v, w, \ldots$, the latter quantities may also be expressed as linear functions of the former, so that T may be regarded as a homogeneous quadratic function of $\xi, \eta, \zeta, \lambda, \mu, \nu$. When expressed in this manner we may denote it by T'. The equation (3) then gives at once

$$u\,\delta\xi+v\,\delta\eta+w\,\delta\zeta+p\,\delta\lambda+q\,\delta\mu+r\,\delta\nu$$
$$=\frac{\partial T'}{\partial \xi}\,\delta\xi+\frac{\partial T'}{\partial \eta}\,\delta\eta+\frac{\partial T'}{\partial \zeta}\,\delta\zeta+\frac{\partial T'}{\partial \lambda}\,\delta\lambda+\frac{\partial T'}{\partial \mu}\,\delta\mu+\frac{\partial T'}{\partial \nu}\,\delta\nu,$$

whence $$u,\ v,\ w=\frac{\partial T'}{\partial \xi},\ \frac{\partial T'}{\partial \eta},\ \frac{\partial T'}{\partial \zeta}, \qquad p,\ q,\ r=\frac{\partial T'}{\partial \lambda},\ \frac{\partial T'}{\partial \mu},\ \frac{\partial T'}{\partial \nu}. \quad \ldots\ldots(6)$$

These formulae are in a sense reciprocal to (5).

We can utilize this last result to obtain, when no extraneous forces act, another integral of the equations of motion, in addition to those found in Art. 120. Thus

$$\frac{dT}{dt}=\frac{\partial T'}{\partial \xi}\frac{d\xi}{dt}+\ldots+\ldots+\frac{\partial T'}{\partial \lambda}\frac{d\lambda}{dt}+\ldots+\ldots$$
$$=u\frac{d\xi}{dt}+\ldots+\ldots+p\frac{d\lambda}{dt}+\ldots+\ldots,$$

which vanishes identically, by Art. 120 (1). Hence we have the equation of energy

$$T=\text{const.} \quad \ldots\ldots\ldots\ldots\ldots\ldots\ldots\ldots\ldots\ldots\ldots(7)$$

123. If in the formulae (5) we put, in the notation of Art. 121,

$$T = \mathbf{T} + \mathbf{T}_1,$$

it is known from the Dynamics of rigid bodies that the terms in $\mathbf{T}_1$ represent the linear and angular momentum of the solid by itself. Hence the remaining terms, involving $\mathbf{T}$, must represent the system of impulsive pressures exerted by the surface of the solid on the fluid, in the supposed instantaneous generation of the motion from rest.

This is easily verified. For example, the x-component of the above system of impulsive pressures is

$$\iint \rho\phi l dS = -\rho \iint \phi \frac{\partial \phi_1}{\partial n} dS$$
$$= \mathbf{A}u + \mathbf{C}'v + \mathbf{B}'w + \mathbf{F}p + \mathbf{F}'q + \mathbf{F}''r = \frac{\partial \mathbf{T}}{\partial u}, \quad \text{.........(8)}$$

by the formulae of Arts. 118, 121. In the same way, the moment of the impulsive pressures about Ox is

$$\iint \rho\phi (ny - mz)\, dS = -\rho \iint \phi \frac{\partial \chi_1}{\partial n} dS$$
$$= \mathbf{F}u + \mathbf{G}v + \mathbf{H}w + \mathbf{P}p + \mathbf{R}'q + \mathbf{Q}'r = \frac{\partial \mathbf{T}}{\partial p}. \quad \text{......(9)}$$

124. The equations of motion may now be written*

$$\left.\begin{aligned}
\frac{d}{dt}\frac{\partial T}{\partial u} &= r\frac{\partial T}{\partial v} - q\frac{\partial T}{\partial w} + X,\\
\frac{d}{dt}\frac{\partial T}{\partial v} &= p\frac{\partial T}{\partial w} - r\frac{\partial T}{\partial u} + Y,\\
\frac{d}{dt}\frac{\partial T}{\partial w} &= q\frac{\partial T}{\partial u} - p\frac{\partial T}{\partial v} + Z,\\
\frac{d}{dt}\frac{\partial T}{\partial p} &= w\frac{\partial T}{\partial v} - v\frac{\partial T}{\partial w} + r\frac{\partial T}{\partial q} - q\frac{\partial T}{\partial r} + L,\\
\frac{d}{dt}\frac{\partial T}{\partial q} &= u\frac{\partial T}{\partial w} - w\frac{\partial T}{\partial u} + p\frac{\partial T}{\partial r} - r\frac{\partial T}{\partial p} + M,\\
\frac{d}{dt}\frac{\partial T}{\partial r} &= v\frac{\partial T}{\partial u} - u\frac{\partial T}{\partial v} + q\frac{\partial T}{\partial p} - p\frac{\partial T}{\partial q} + N.
\end{aligned}\right\} \quad \text{..............(1)}$$

If in these we write $T = \mathbf{T} + \mathbf{T}_1$, and isolate the terms due to $\mathbf{T}$, we obtain expressions for the forces exerted on the moving solid by the pressure of the surrounding fluid; thus the total component ($\mathbf{X}$, say) of the fluid pressure parallel to x is

$$\mathbf{X} = -\frac{d}{dt}\frac{\partial \mathbf{T}}{\partial u} + r\frac{\partial \mathbf{T}}{\partial v} - q\frac{\partial \mathbf{T}}{\partial w}, \quad \text{.......................(2)}$$

* See Kirchhoff, *l.c. ante* p. 160; also Sir W. Thomson, "Hydrokinetic Solutions and Observations," *Phil. Mag.* (5) xlii. 362 (1871) [reprinted in *Baltimore Lectures*, Cambridge, 1904, p. 584].

and the moment (**L**) of the same pressures about x is*

$$\mathbf{L}=-\frac{d}{dt}\frac{\partial \mathbf{T}}{\partial p}+w\frac{\partial \mathbf{T}}{\partial v}-v\frac{\partial \mathbf{T}}{\partial w}+r\frac{\partial \mathbf{T}}{\partial q}-q\frac{\partial \mathbf{T}}{\partial r}. \quad \text{..............(3)}$$

For example, if the solid be constrained to move with a constant velocity (u, v, w), without rotation, we have

$$\left.\begin{array}{c}\mathbf{X},\ \mathbf{Y},\ \mathbf{Z}=0, \\ \mathbf{L},\ \mathbf{M},\ \mathbf{N}=w\dfrac{\partial \mathbf{T}}{\partial v}-v\dfrac{\partial \mathbf{T}}{\partial w},\ u\dfrac{\partial \mathbf{T}}{\partial w}-w\dfrac{\partial \mathbf{T}}{\partial u},\ v\dfrac{\partial \mathbf{T}}{\partial u}-u\dfrac{\partial \mathbf{T}}{\partial v},\end{array}\right\} \text{......(4)}$$

where $$2\mathbf{T}=\mathbf{A}u^2+\mathbf{B}v^2+\mathbf{C}w^2+2\mathbf{A}'vw+2\mathbf{B}'wu+2\mathbf{C}'uv.$$

The fluid pressures thus reduce to a couple, which moreover vanishes if

$$\frac{\partial \mathbf{T}}{\partial u}:u=\frac{\partial \mathbf{T}}{\partial v}:v=\frac{\partial \mathbf{T}}{\partial w}:w,$$

i.e. provided the velocity (u, v, w) be in the direction of one of the principal axes of the ellipsoid

$$\mathbf{A}x^2+\mathbf{B}y^2+\mathbf{C}z^2+2\mathbf{A}'yz+2\mathbf{B}'zx+2\mathbf{C}'xy=\text{const.} \quad \text{.........(5)}$$

Hence, as was first pointed out by Kirchhoff, there are, for any solid, three mutually perpendicular directions of permanent translation; that is to say, if the solid be set in motion parallel to one of these, without rotation, and left to itself, it will continue to move in this manner. It is evident that these directions are determined solely by the configuration of the surface of the body. It must be observed however that the impulse necessary to produce one of these permanent translations does not in general reduce to a single force; thus if the axes of co-ordinates be chosen, for simplicity, parallel to the three directions in question, so that $A', B', C'=0$, we have, corresponding to the motion u alone,

$$\xi,\ \eta,\ \zeta=Au,\ 0,\ 0; \qquad \lambda,\ \mu,\ \nu=Fu,\ F'u,\ F''u,$$

so that the impulse consists of a wrench of pitch F/A.

With the same choice of axes, the components of the couple which is the equivalent of the fluid pressures on the solid, in the case of any uniform translation (u, v, w), are

$$\mathbf{L},\ \mathbf{M},\ \mathbf{N}=(\mathbf{B}-\mathbf{C})\,vw,\ (\mathbf{C}-\mathbf{A})\,wu,\ (\mathbf{A}-\mathbf{B})\,uv. \quad \text{...........(6)}$$

Hence if in the ellipsoid

$$\mathbf{A}x^2+\mathbf{B}y^2+\mathbf{C}z^2=\text{const.}, \quad \text{........................(7)}$$

we draw a radius vector r in the direction of the velocity (u, v, w) and erect the perpendicular h from the centre on the tangent plane at the extremity of r, the plane of the couple is that of h and r, its magnitude is proportional to $\sin(h, r)/h$, and its tendency is to turn the solid in the direction from h to r.

* The forms of these expressions being known, it is not difficult to verify them by direct calculation from the pressure-equation, Art. 20 (5). See a paper "On the Forces experienced by a Solid moving through a Liquid," *Quart. Journ. Math.* xix. 66 (1883).

Thus if the direction of (u, v, w) differs but slightly from that of the axis of x, the tendency of the couple is to diminish the deviation when **A** is the greatest, and to increase it when **A** is the least, of the three quantities **A**, **B**, **C**, whilst if **A** is intermediate to **B** and **C** the tendency depends on the position of r relative to the circular sections of the above ellipsoid. It appears then that of the three permanent translations one only is thoroughly stable, viz. that corresponding to the greatest of the three coefficients **A**, **B**, **C**. For example, the only stable direction of translation of an ellipsoid is that of its *least* axis; see Art. 121*.

125. The above, although the simplest, are not the only steady motions of which the body is capable, under the action of no extraneous forces. The instantaneous motion of the body at any instant consists, by a well-known theorem of Kinematics, of a twist about a certain screw; and the condition that this motion should be permanent is that it should not affect the configuration of the impulse (which is fixed in space) relatively to the body. This requires that the axes of the screw and of the corresponding impulsive wrench should coincide. Since the general equations of a straight line involve four independent constants, this gives four linear relations to be satisfied by the five ratios $u:v:w:p:q:r$. There exists then for every body, under the circumstances here considered, a singly-infinite system of possible steady motions.

The steady motions next in importance to the three permanent translations are those in which the impulse reduces to a *couple*. The equations (1) of Art. 120 shew that we may have $\xi, \eta, \zeta = 0$, and λ, μ, ν constant, provided

$$\lambda/p = \mu/q = \nu/r, \quad = k, \text{ say.} \qquad \text{(1)}$$

If the axes of co-ordinates have the special directions referred to in the preceding Art., the conditions $\xi, \eta, \zeta = 0$ give us at once u, v, w in terms of p, q, r, viz.

$$u = -\frac{Fp + F'q + F''r}{A}, \qquad v = -\frac{Gp + G'q + G''r}{B}, \qquad w = -\frac{Hp + H'q + H''r}{C}. \qquad \text{(2)}$$

Substituting these values in the expressions for λ, μ, ν obtained from Art. 122 (5), we find

$$\lambda, \mu, \nu = \frac{\partial \Theta}{\partial p}, \frac{\partial \Theta}{\partial q}, \frac{\partial \Theta}{\partial r}, \qquad \text{(3)}$$

provided

$$2\Theta(p, q, r) = \mathfrak{P}p^2 + \mathfrak{Q}q^2 + \mathfrak{R}r^2 + 2\mathfrak{P}'qr + 2\mathfrak{Q}'rp + 2\mathfrak{R}'pq, \qquad \text{(4)}$$

the coefficients in this expression being determined by formulae of the types

$$\mathfrak{P} = P - \frac{F^2}{A} - \frac{G^2}{B} - \frac{H^2}{C}, \quad \mathfrak{P}' = P' - \frac{F'F''}{A} - \frac{G'G''}{B} - \frac{H'H''}{C}. \qquad \text{(5)}$$

These formulae hold for any case in which the force-constituent of the impulse is zero. Introducing the conditions (1) of steady motion, the ratios $p:q:r$ are to be determined from the three equations

$$\left.\begin{aligned} \mathfrak{P}p + \mathfrak{R}'q + \mathfrak{Q}'r &= kp, \\ \mathfrak{R}'p + \mathfrak{Q}q + \mathfrak{P}'r &= kq, \\ \mathfrak{Q}'p + \mathfrak{P}'q + \mathfrak{R}r &= kr. \end{aligned}\right\} \qquad \text{(6)}$$

* The physical cause of this tendency of an elongated body to set itself broadside-on to the relative motion is clearly indicated in the diagram on p. 86. A number of interesting practical illustrations are given by Thomson and Tait, Art. 325.

The form of these shews that the line whose direction-ratios are $p:q:r$ must be parallel to one of the principal axes of the ellipsoid

$$\Theta(x, y, z) = \text{const.} \qquad\qquad (7)$$

There are therefore three permanent screw-motions such that the corresponding impulsive wrench in each case reduces to a couple only. The axes of these three screws are mutually at right angles, but do not in general intersect.

It may now be shewn that in all cases where the impulse reduces to a couple only, the motion can be completely determined. It is convenient, retaining the same directions of the axes, to change the origin. Now the origin may be transferred to any point (x, y, z) by writing

$$u+ry-qz, \qquad v+pz-rx, \qquad w+qx-py,$$

for u, v, w respectively. The coefficient of $2vr$ in the expression for the kinetic energy, Art. 122 (2), becomes $-Bx+G''$, that of $2wq$ becomes $Cx+H'$, and so on. Hence if we take

$$x=\tfrac{1}{2}\left(\frac{G''}{B}-\frac{H'}{C}\right), \quad y=\tfrac{1}{2}\left(\frac{H}{C}-\frac{F''}{A}\right), \quad z=\tfrac{1}{2}\left(\frac{F'}{A}-\frac{G}{B}\right), \qquad (8)$$

the coefficients in the transformed expression for $2T$ will satisfy the relations

$$\frac{G''}{B}=\frac{H'}{C}, \quad \frac{H}{C}=\frac{F''}{A}, \quad \frac{F'}{A}=\frac{G}{B}. \qquad (9)$$

If we denote the values of these pairs of equal quantities by α, β, γ respectively, the formulae (2) may be written

$$u=-\frac{\partial\Psi}{\partial p}, \quad v=-\frac{\partial\Psi}{\partial q}, \quad w=-\frac{\partial\Psi}{\partial r}, \qquad (10)$$

where
$$2\Psi(p, q, r)=\frac{F}{A}p^2+\frac{G'}{B}q^2+\frac{H''}{C}r^2+2\alpha qr+2\beta rp+2\gamma pq. \qquad (11)$$

The motion of the body at any instant may be conceived as made up of two parts; viz. a motion of translation equal to that of the origin, and one of rotation about an instantaneous axis passing through the origin. Since $\xi, \eta, \zeta=0$ the latter part is to be determined by the equations

$$\frac{d\lambda}{dt}=r\mu-q\nu, \qquad \frac{d\mu}{dt}=p\nu-r\lambda, \qquad \frac{d\nu}{dt}=q\lambda-p\mu,$$

which express that the vector (λ, μ, ν) is constant in magnitude and has a fixed direction in space. Substituting from (3),

$$\left.\begin{aligned} \frac{d}{dt}\frac{\partial\Theta}{\partial p}&=r\frac{\partial\Theta}{\partial q}-q\frac{\partial\Theta}{\partial r}, \\ \frac{d}{dt}\frac{\partial\Theta}{\partial q}&=p\frac{\partial\Theta}{\partial r}-r\frac{\partial\Theta}{\partial p}, \\ \frac{d}{dt}\frac{\partial\Theta}{\partial r}&=q\frac{\partial\Theta}{\partial p}-p\frac{\partial\Theta}{\partial q}. \end{aligned}\right\} \qquad (12)$$

These are identical in form with the equations of motion of a rigid body about a fixed point, so that we may make use of Poinsot's well-known solution of the latter problem. The angular motion of the body is obtained by making the ellipsoid (7), which is fixed in the body, roll on a plane

$$\lambda x+\mu y+\nu z=\text{const.},$$

which is fixed in space, with an angular velocity proportional to the length OI of the radius vector drawn from the origin to the point of contact I. The representation of the actual motion is then completed by impressing on the whole system of rolling ellipsoid and plane a velocity of translation whose components are given by (10). This velocity is in the direction of the normal OM to the tangent plane of the quadric

$$\Psi(x, y, z) = -\epsilon^3, \quad \text{.......................................(13)}$$

at the point P where OI meets it, and is equal to

$$\frac{\epsilon^3}{OP \, . \, OM} \times \text{angular velocity of body.} \quad \text{...........................(14)}$$

When OI does not meet the quadric (13), but the conjugate quadric obtained by changing the sign of ϵ, the sense of the velocity (14) is reversed*.

126. The problem of the integration of the equations of motion of a solid in the general case has engaged the attention of several mathematicians, but, as might be anticipated from the complexity of the question, the physical meaning of the results is not easily grasped†.

In what follows we shall in the first place inquire what simplifications occur in the formula for the kinetic energy, for special classes of solids, and then proceed to investigate one or two particular problems of considerable interest which can be treated without difficult mathematics.

The general expression for the kinetic energy contains, as we have seen, twenty-one coefficients, but by the choice of special directions for the co-ordinate axes, and a special origin, these can be reduced to fifteen‡.

The most symmetrical way of writing the general expression is

$$\begin{aligned} 2T = {} & Au^2 + Bv^2 + Cw^2 + 2A'vw + 2B'wu + 2C'uv \\ & + Pp^2 + Qq^2 + Rr^2 + 2P'qr + 2Q'rp + 2R'pq \\ & + 2Lup + 2Mvq + 2Nwr \\ & + 2F(vr + wq) + 2G(wp + ur) + 2H(uq + vp) \\ & + 2F'(vr - wq) + 2G'(wp - ur) + 2H'(uq - vp). \quad \text{.........(1)} \end{aligned}$$

It has been seen that we may choose the directions of the axes so that $A', B', C' = 0$, and it may easily be verified that by displacing the origin we can further make $F', G', H' = 0$. We shall henceforward suppose these simplifications to have been made.

1°. If the solid has a plane of symmetry, it is evident from the configuration of the relative stream-lines that a translation normal to this plane must be one of the permanent translations of Art. 124. If we take this plane as that of xy, it is further evident that the energy of the motion must be unaltered if we reverse the signs of w, p, q. This requires that P', Q', L, M, N, H should vanish. The three screws of Art. 125 are now pure rotations, but their axes do not in general intersect.

* The substance of this Art. is taken from a paper, "On the Free Motion of a Solid through an Infinite Mass of Liquid," *Proc. Lond. Math. Soc.* viii. 273 (1877). Similar results were obtained independently by Craig, "The Motion of a Solid in a Fluid," *Amer. Journ. of Math.* ii. 162 (1879).

† For references see Wien, *Lehrbuch d. Hydrodynamik*, Leipzig, 1900, p. 164.

‡ Cf. Clebsch, "Ueber die Bewegung eines Körpers in einer Flüssigkeit," *Math. Ann.* iii. 238 (1870). This paper deals with the 'reciprocal' form of the dynamical equations, obtained by substituting from Art. 122 (6) in Art. 120 (1).

2°. If the body has a second plane of symmetry, at right angles to the former one, we may take this as the plane xz. We find that in this case R' and G must also vanish, so that

$$2T = Au^2 + Bv^2 + Cw^2 + Pp^2 + Qq^2 + Rr^2 + 2F(vr + wq). \quad \text{......(2)}$$

The axis of x is the axis of one of the permanent rotations, and those of the other two intersect it at right angles, though not necessarily in the same point.

3°. If the body has a third plane of symmetry, say that of yz, at right angles to the two former ones, we have

$$2T = Au^2 + Bv^2 + Cw^2 + Pp^2 + Qq^2 + Rr^2. \quad \text{..............(3)}$$

4°. Returning to (2°), we note that in the case of a solid *of revolution* about Ox, the expression for $2T$ must be unaltered when we write $v, q, -w, -r$ for w, r, v, q, respectively, since this is equivalent to rotating the axes of y, z through a right angle. Hence $B = C$, $Q = R$, $F = 0$; and therefore

$$2T = Au^2 + B(v^2 + w^2) + Pp^2 + Q(q^2 + r^2). \quad \text{...........(4)*}$$

The same reduction obtains in some other cases, for example when the solid is a right prism whose section is any regular polygon†. This is seen at once from the consideration that, the axis of x coinciding with the axis of the prism, it is impossible to assign any uniquely symmetrical directions to the axes of y and z.

5°. If, in the last case, the form of the solid be similarly related to each of the co-ordinate planes (for example a sphere, or a cube), the expression (3) takes the form

$$2T = A(u^2 + v^2 + w^2) + P(p^2 + q^2 + r^2). \quad \text{..............(5)}$$

This again may be extended, for a like reason, to other cases, for example any regular polyhedron. Such a body is practically for the present purpose 'isotropic,' and its motion will be exactly that of a sphere under similar conditions.

6°. We may next consider another class of cases. Let us suppose that the body has a sort of skew symmetry about a certain axis (say that of x), viz. that it is identical with itself turned through two right angles about this axis, but has not necessarily a plane of symmetry‡. The expression for $2T$ must be unaltered when we change the signs of v, w, q, r, so that the coefficients Q', R', G, H must all vanish. We have then

$$2T = Au^2 + Bv^2 + Cw^2 + Pp^2 + Qq^2 + Rr^2 + 2P'qr$$
$$+ 2Lup + 2Mvq + 2Nwr + 2F(vr + wq). \quad \text{.........(6)}$$

* For the solution of the equations of motion in this case see Greenhill, "The Motion of a Solid in Infinite Liquid under no Forces," *Amer. Journ. of Math.* xx. 1 (1897).

† See Larmor, "On Hydrokinetic Symmetry," *Quart. Journ. Math.* xx. 261 (1884). [*Papers*, i. 77.]

‡ A two-bladed screw-propeller of a ship is an example of a body of this kind.

The axis of x is one of the directions of permanent translation; and is also the axis of one of the three screws of Art. 125, the pitch being $-L/A$. The axes of the two remaining screws intersect it at right angles, but not in general in the same point.

7°. If, further, the body be identical with itself turned through *one* right angle about the above axis, the expression (6) must be unaltered when $v, q, -w, -r$ are written for w, r, v, q, respectively. This requires that $B=C, Q=R, P'=0, M=N, F=0$. Hence*

$$2T = Au^2 + B(v^2+w^2) + Pp^2 + Q(q^2+r^2) + 2Lup + 2M(vq+wr). \quad \ldots(7)$$

The form of this expression is unaltered when the axes of y, z are turned in their own plane through any angle. The body is therefore said to possess helicoidal symmetry about the axis of x.

8°. If the body possess the same properties of skew symmetry about an axis intersecting the former one at right angles, we must evidently have

$$2T = A(u^2+v^2+w^2) + P(p^2+q^2+r^2) + 2L(pu+qv+rw). \quad \ldots(8)$$

Any direction is now one of permanent translation, and any line drawn through the origin is the axis of a screw of the kind considered in Art. 125, of pitch $-L/A$. The form of (8) is unaltered by any change in the directions of the axes of co-ordinates. The solid is therefore in this case said to be 'helicoidally isotropic.'

127. For the case of a solid of revolution, or of any other form to which the formula

$$2T = Au^2 + B(v^2+w^2) + Pp^2 + Q(q^2+r^2) \quad \ldots\ldots\ldots\ldots\ldots(1)$$

applies, the complete integration of the equations of motion was effected by Kirchhoff† in terms of elliptic functions.

The particular case where the solid moves without rotation about its axis, and with this axis always in one plane, admits of very simple treatment‡, and the results are very interesting.

If the fixed plane in question be that of xy we have $p, q, w=0$, so that the equations of motion, Art. 124 (1), reduce to

$$\left.\begin{aligned} A\frac{du}{dt} = rBv, \qquad B\frac{dv}{dt} = -rAu, \\ Q\frac{dr}{dt} = (A-B)\,uv. \end{aligned}\right\} \quad \ldots\ldots\ldots\ldots\ldots\ldots\ldots\ldots(2)$$

Let $\mathbf{x}$, $\mathbf{y}$ be the co-ordinates of the moving origin relative to fixed axes in the plane (xy) in which the axis of the solid moves, the axis of $\mathbf{x}$ coinciding with the line of the

* This result admits of the same kind of generalization as (4), *e.g.* it applies to a body shaped like a screw-propeller with *three* symmetrically-disposed blades. The integration of the equations of motion is discussed by Greenhill, "The Motion of a Solid in Infinite Liquid," *Amer. Journ. of Math.* xxviii. 71 (1906).

† *l.c. ante* p. 160.

‡ See Thomson and Tait, Art. 322; Greenhill, "On the Motion of a Cylinder through a Frictionless Liquid under no Forces," *Mess. of Math.* ix. 117 (1880).

resultant impulse (I, say) of the motion; and let θ be the angle which the line Ox (fixed in the solid) makes with $\mathbf{x}$. We have then

$$Au = I\cos\theta, \qquad Bv = -I\sin\theta, \qquad r = \dot\theta.$$

The first two of equations (2) merely express the fixity of the direction of the impulse in space; the third gives

$$Q\ddot\theta + \frac{A-B}{AB}I^2\sin\theta\cos\theta = 0. \quad \text{..........(3)}$$

We may suppose, without loss of generality, that $A > B$. If we write $2\theta = \vartheta$, (3) becomes

$$\ddot\vartheta + \frac{(A-B)I^2}{ABQ}\sin\vartheta = 0, \quad \text{..........(4)}$$

which is the equation of motion of the common pendulum. Hence the angular motion of the body is that of a 'quadrantal pendulum,' *i.e.* a body whose motion follows the same law in regard to a quadrant as the ordinary pendulum does in regard to a half-circumference. When θ has been determined from (3) and the initial conditions, $\mathbf{x}$, $\mathbf{y}$ are to be found from the equations

$$\left.\begin{aligned} \dot{\mathbf{x}} &= u\cos\theta - v\sin\theta = \frac{I}{A}\cos^2\theta + \frac{I}{B}\sin^2\theta, \\ \dot{\mathbf{y}} &= u\sin\theta + v\cos\theta = \left(\frac{I}{A} - \frac{I}{B}\right)\sin\theta\cos\theta = \frac{Q}{I}\ddot\theta, \end{aligned}\right\} \quad \text{..........(5)}$$

the latter of which gives

$$\mathbf{y} = \frac{Q}{I}\dot\theta, \quad \text{..........(6)}$$

as is otherwise obvious, the additive constant being zero since the axis of $\mathbf{x}$ is taken to be coincident with, and not merely parallel to, the line of the impulse I.

Let us first suppose that the body makes complete revolutions, in which case the first integral of (3) is of the form

$$\dot\theta^2 = \omega^2(1 - k^2\sin^2\theta), \quad \text{..........(7)}$$

where

$$k^2 = \frac{A-B}{ABQ}\cdot\frac{I^2}{\omega^2}. \quad \text{..........(8)}$$

Hence, reckoning t from the position $\theta = 0$, we have

$$\omega t = \int_0^\theta \frac{d\theta}{(1 - k^2\sin^2\theta)^{\frac{1}{2}}} = F(k, \theta), \quad \text{..........(9)}$$

in the usual notation of elliptic integrals. If we eliminate t between (5) and (7), and then integrate with respect to θ, we find

$$\left.\begin{aligned} \mathbf{x} &= \left(\frac{I}{A\omega} + \frac{Q\omega}{I}\right)F(k, \theta) - \frac{Q\omega}{I}E(k, \theta), \\ \mathbf{y} &= \frac{Q}{I}\dot\theta = \frac{Q\omega}{I}(1 - k^2\sin^2\theta)^{\frac{1}{2}}, \end{aligned}\right\} \quad \text{..........(10)}$$

the origin of $\mathbf{x}$ being taken to correspond to the position $\theta = 0$. The path can then be traced, in any particular case, by means of Legendre's Tables. See the curve marked I on the next page.

If, on the other hand, the solid does not make a complete revolution, but oscillates through an angle α on each side of the position $\theta = 0$, the proper form of the first integral of (3) is

$$\dot\theta^2 = \omega^2\left(1 - \frac{\sin^2\theta}{\sin^2\alpha}\right), \quad \text{..........(11)}$$

where

$$\sin^2\alpha = \frac{ABQ}{A-B}\cdot\frac{\omega^2}{I^2}. \quad \text{..........(12)}$$

If we put
$$\sin\theta = \sin\alpha\sin\psi,$$
this gives
$$\dot{\psi}^2 = \frac{\omega^2}{\sin^2\alpha}(1-\sin^2\alpha\sin^2\psi),$$
whence
$$\frac{\omega t}{\sin\alpha} = F(\sin\alpha, \psi). \quad\text{.....................................(13)}$$

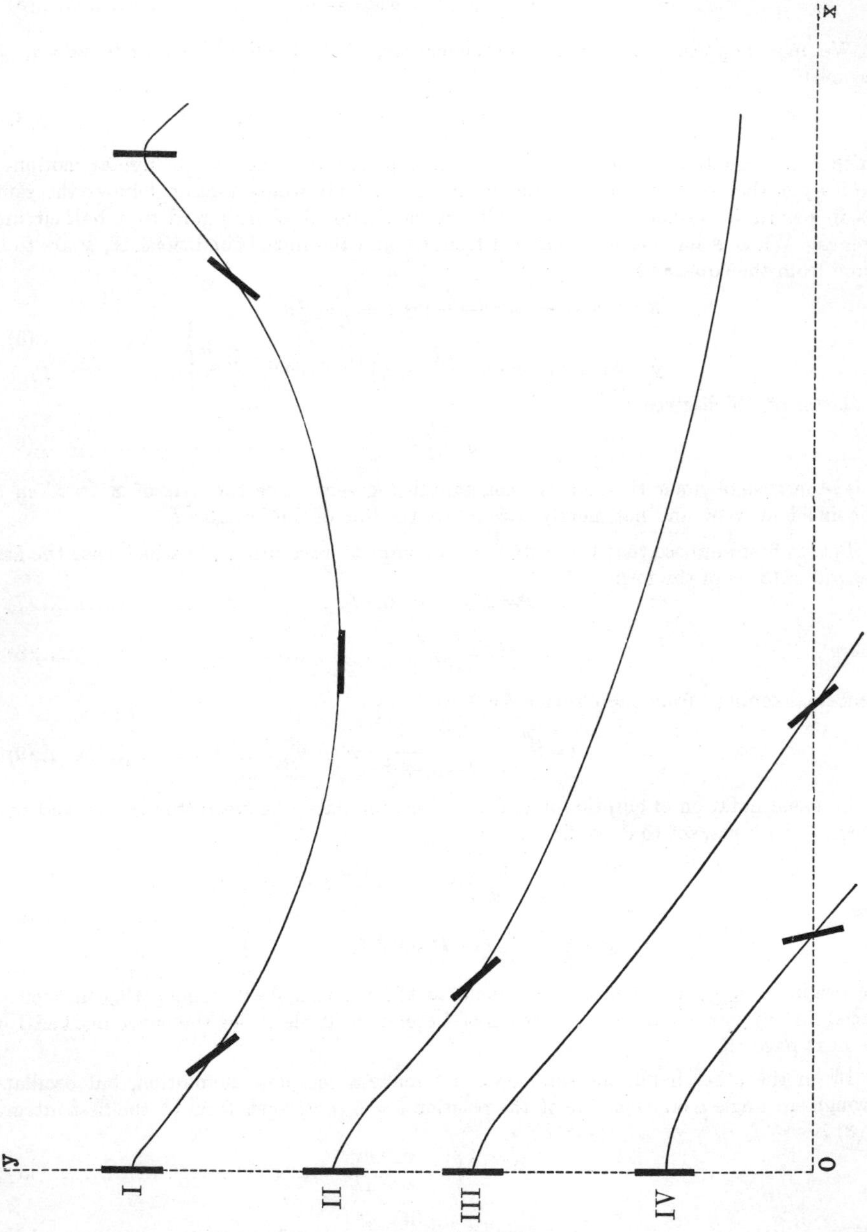

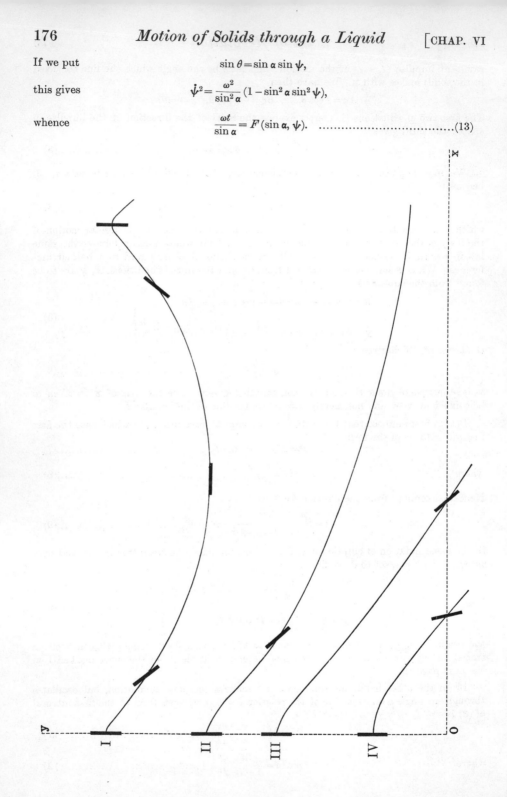

Transforming to ψ as independent variable, in (5), and integrating, we find

$$\left.\begin{aligned}\mathbf{x}&=\frac{I}{B\omega}\sin\alpha\,.\,F(\sin\alpha,\psi)-\frac{Q\omega}{I}\operatorname{cosec}\alpha\,.\,E(\sin\alpha,\psi),\\ \mathbf{y}&=\frac{Q\omega}{I}\cos\psi.\end{aligned}\right\}\quad\ldots\ldots\ldots\ldots(14)$$

The path of the point O is now a sinuous curve crossing the line of the impulse at intervals of time equal to a half-period of the angular motion. This is illustrated by the curves III and IV of the figure.

There remains a critical case between the two preceding, where the solid just makes a half-revolution, θ having as asymptotic limits the two values $\pm\frac{1}{2}\pi$. This case may be obtained by putting $k=1$ in (7), or $\alpha=\frac{1}{2}\pi$ in (11); and we find

$$\dot{\theta}=\omega\cos\theta,\quad\ldots\ldots\ldots\ldots(15)$$

$$\omega t=\log\tan\left(\tfrac{1}{4}\pi+\tfrac{1}{2}\theta\right),\quad\ldots\ldots\ldots\ldots(16)$$

$$\left.\begin{aligned}\mathbf{x}&=\frac{I}{B\omega}\log\tan\left(\tfrac{1}{4}\pi+\tfrac{1}{2}\theta\right)-\frac{Q\omega}{I}\sin\theta,\\ \mathbf{y}&=\frac{Q\omega}{I}\cos\theta.\end{aligned}\right\}\quad\ldots\ldots\ldots\ldots(17)$$

See the curve II of the figure*.

It is to be observed that the above investigation is not restricted to the case of a solid of revolution; it applies equally well to a body with two perpendicular planes of symmetry, moving parallel to one of these planes, provided the origin be properly chosen. If the plane in question be that of xy, then on transferring the origin to the point $(F/B, 0, 0)$ the last term in the formula (2) of Art. 126 disappears, and the equations of motion take the form (2) above. On the other hand, if the motion be parallel to zx we must transfer the origin to the point $(-F/C, 0, 0)$.

The results of this Article, with the accompanying diagram, serve to exemplify the statements made near the end of Art. 124. Thus the curve IV illustrates, with exaggerated amplitude, the case of a slightly disturbed *stable* steady motion parallel to an axis of permanent translation. The case of a slightly disturbed *unstable* steady motion would be represented by a curve contiguous to II, on one side or the other, according to the nature of the disturbance.

128. The mere question of the stability of the motion of a body parallel to an axis of symmetry may of course be treated more simply by approximate methods. Thus, in the case of a body with three planes of symmetry, as in Art. 126, 3°, slightly disturbed from a state of steady motion parallel to x, we find, writing $u=u_0+u'$, and assuming u', v, w, p, q, r to be all small,

$$\left.\begin{aligned}A\frac{du'}{dt}&=0,\quad B\frac{dv}{dt}=-Au_0r,\quad C\frac{dw}{dt}=Au_0q,\\ P\frac{dp}{dt}&=0,\quad Q\frac{dq}{dt}=(C-A)\,u_0w,\quad R\frac{dr}{dt}=(A-B)\,u_0v.\end{aligned}\right\}\quad\ldots\ldots(1)$$

* In order to bring out the peculiar features of the motion, the curves have been drawn for the somewhat extreme case of $A=5B$. In the case of an infinitely thin disk, without inertia of its own, we should have $A/B=\infty$; the curves would then have *cusps* where they meet the axis of $\mathbf{y}$. It appears from (5) that $\dot{\mathbf{x}}$ has always the same sign, so that *loops* cannot occur in any case.

In the various cases figured the body is projected always with the same impulse, but with different degrees of rotation. In the curve I, the maximum angular velocity is $\sqrt{2}$ times what it is in the critical case II; whilst the curves III and IV represent oscillations of amplitude 45° and 18° respectively.

Hence
$$B\frac{d^2v}{dt^2}+\frac{A(A-B)}{R}u_0^2v=0,$$

with a similar equation for r, and

$$C\frac{d^2w}{dt^2}+\frac{A(A-C)}{Q}u_0^2w=0, \qquad\qquad (2)$$

with a similar equation for q. The motion is therefore stable only when A is the greatest of the three quantities A, B, C.

It is evident from ordinary Dynamics that the stability of a body moving parallel to an axis of symmetry will be increased, or its instability (as the case may be) will be diminished, by communicating to it a rotation about this axis. This question has been examined by Greenhill*.

Thus, in the case of a solid of revolution slightly disturbed from a state of motion in which u and p are constant and the remaining velocities are zero, if we neglect squares and products of small quantities the first and fourth of equations (1) of Art. 124 give

$$du/dt=0, \quad dp/dt=0,$$

whence
$$u=u_0, \qquad p=p_0, \qquad\qquad (3)$$

say, where u_0, p_0 are constants. The remaining equations then take, on substitution from Art. 126 (3), the forms

$$B\left(\frac{dv}{dt}-p_0w\right)=-Au_0r, \qquad B\left(\frac{dw}{dt}+p_0v\right)=Au_0q, \qquad\qquad (4)$$

$$Q\frac{dq}{dt}+(P-Q)p_0r=-(A-B)u_0w, \qquad Q\frac{dr}{dt}-(P-Q)p_0q=(A-B)u_0v. \qquad (5)$$

If we assume that v, w, q, r vary as $e^{i\sigma t}$, and eliminate their ratios, we find

$$Q\sigma^2\pm(P-2Q)p_0\sigma-\left\{(P-Q)p_0^2+\frac{A}{B}(A-B)u_0^2\right\}=0. \qquad\qquad (6)$$

The condition that the roots of this should be real is that

$$P^2p_0^2+4\frac{A}{B}(A-B)Qu_0^2$$

should be positive. This is always satisfied when $A>B$, and can be satisfied in any case by giving a sufficiently great value to p_0.

This example illustrates the steadiness of flight which is given to an elongated projectile by rifling.

129. In the investigation of Art. 125 the term 'steady' was used to characterize modes of motion in which the 'instantaneous screw' preserved a constant relation to the moving solid. In the case of a solid of revolution, however, we may conveniently use the term in a somewhat wider sense, extending it to motions in which the vectors representing the velocities of translation and rotation are of constant magnitude, and make constant angles with the axis of symmetry and with each other, although their relation to points of the solid not on the axis may continually vary.

* "Fluid Motion between Confocal Elliptic Cylinders, &c." *Quart. Journ. Math.* xvi. 227 (1879).

The conditions to be satisfied in this case are most easily obtained from the equations of motion of Art. 124, which become, on substitution from Art. 126 (4),

$$\left.\begin{aligned} A\frac{du}{dt}&=B(rv-qw), & P\frac{dp}{dt}&=0, \\ B\frac{dv}{dt}&=Bpw-Aru, & Q\frac{dq}{dt}&=-(A-B)\,uw-(P-Q)\,pr, \\ B\frac{dw}{dt}&=Aqu-Bpv, & Q\frac{dr}{dt}&=(A-B)\,uv+(P-Q)\,pq. \end{aligned}\right\} \quad\text{..........(1)}$$

It appears that p is in any case constant, and that q^2+r^2 will also be constant provided

$$v/q=wr, \quad =k, \text{ say.} \quad\text{..........(2)}$$

This makes $du/dt=0$, and $v^2+w^2=\text{const.}$ It follows that k will also be constant; and it only remains to satisfy the equations

$$kB\frac{dq}{dt}=(kBp-Au)\,r, \qquad Q\frac{dq}{dt}=-\{(A-B)\,ku+(P-Q)\,p\}\,r.$$

These will be consistent provided

$$kB\{(A-B)\,ku+(P-Q)\,p\}+Q\,(kBp-Au)=0,$$

whence

$$\frac{u}{p}=\frac{kBP}{AQ-k^2B\,(A-B)}. \quad\text{..........(3)}$$

Hence by variation of k we obtain an infinite number of possible modes of steady motion, of the kind above defined. In each of these the instantaneous axis of rotation and the direction of translation of the origin are in one plane with the axis of the solid. It is easily seen that the origin describes a helix about the line of the impulse.

These results are due to Kirchhoff.

130. The only case of a body possessing *helicoidal* property, where simple results can be obtained, is that of the 'isotropic helicoid' defined by Art. 126 (8).

Let O be the centre of the body, and let us take as axes of co-ordinates at any instant a line Ox parallel to the axis of the impulse, a line Oy drawn outwards from this axis, and a line Oz perpendicular to the plane of the two former. If I and K denote the force- and couple-constituents of the impulse, we have

$$\left.\begin{aligned} Au+Lp&=\xi=I, & Av+Lq&=\eta=0, & Aw+Lr&=\zeta=0, \\ Pp+Lu&=\lambda=K, & Pq+Lv&=\mu=0, & Pr+Lw&=\nu=I\varpi, \end{aligned}\right\} \quad\text{..........(1)}$$

where ϖ denotes the distance of O from the axis of the impulse.

Since $AP-L^2\neq 0$, the second and fifth of these equations shew that $v=0$, $q=0$. Hence ϖ is constant throughout the motion, and the remaining quantities are also constant; in particular

$$u=\frac{PI-LK}{AP-L^2}, \qquad w=-\frac{LI\varpi}{AP-L^2}. \quad\text{..........(2)}$$

The origin O therefore describes a helix about the axis of the impulse, of pitch

$$\frac{K}{I}-\frac{P}{L}.$$

This example is due to Kelvin*.

* *l.c. ante* p. 168. It is there pointed out that a solid of the kind here in question may be constructed by attaching vanes to a sphere, at the middle points of twelve quadrantal arcs drawn so as to divide the surface into octants. The vanes are to be perpendicular to the surface, and are to be inclined at angles of 45° to the respective arcs. Larmor (*l.c. ante* p. 173) gives another example. "If...we take a regular tetrahedron (or other regular solid), and replace the edges by skew bevel faces placed in such wise that when looked at from any corner they all slope the same way, we have an example of an isotropic helicoid."

For some further investigations in the present connection see a paper by Miss Fawcett, "On the Motion of Solids in a Liquid," *Quart. Journ. Math.* xxvi. 231 (1893).

131. Before leaving this part of the subject we remark that the preceding theory applies, with obvious modifications, to the acyclic motion of a liquid occupying a cavity in a moving solid. If the origin be taken at the centre of inertia of the liquid, the formula for the kinetic energy of the fluid motion is of the type

$$2\mathbf{T} = \mathbf{m}(u^2 + v^2 + w^2) + \mathbf{P}p^2 + \mathbf{Q}q^2 + \mathbf{R}r^2 + 2\mathbf{P}'qr + 2\mathbf{Q}'rp + 2\mathbf{R}'pq. \quad \ldots(1)$$

For the kinetic energy is equal to that of the whole fluid mass ($\mathbf{m}$), supposed concentrated at its centre of inertia and moving with this point, together with the kinetic energy of the motion relative to the centre of inertia. The latter part of the energy is easily proved by the method of Arts. 118, 121 to be a homogeneous quadratic function of p, q, r.

Hence the fluid may be replaced by a solid of the same mass, having the same centre of inertia, provided the principal axes and moments of inertia be properly assigned.

The values of the coefficients in (1), for the case of an ellipsoidal cavity, may be calculated from Art. 110. Thus, if the axes of x, y, z coincide with the principal axes of the ellipsoid, we find

$$\mathbf{P}, \mathbf{Q}, \mathbf{R} = \tfrac{1}{5}\mathbf{m}\frac{(b^2-c^2)^2}{b^2+c^2}, \ \tfrac{1}{5}\mathbf{m}\frac{(c^2-a^2)^2}{c^2+a^2}, \ \tfrac{1}{5}\mathbf{m}\frac{(a^2-b^2)^2}{a^2+b^2}; \qquad \mathbf{P}', \mathbf{Q}', \mathbf{R}' = 0.$$

Case of a Perforated Solid.

132. If the moving solid have one or more apertures or perforations, so that the space external to it is multiply-connected, the fluid may have a motion independent of that of the solid, viz. a cyclic motion in which the circulations in the several irreducible circuits which can be drawn through the apertures may have any given constant values. We will briefly indicate how the foregoing methods may be adapted to this case.

Let κ, κ', κ'', ... be the circulations in the various circuits, and let $\delta\sigma$, $\delta\sigma'$, $\delta\sigma''$, ... be elements of the corresponding barriers, drawn as in Art. 48. Further, let l, m, n denote the direction-cosines of the normal, drawn towards the fluid at any point of the surface of the solid, or drawn on the positive side at any point of a barrier. The velocity-potential is then of the form

$$\phi + \phi_0,$$

where

$$\left.\begin{aligned} \phi &= u\phi_1 + v\phi_2 + w\phi_3 + p\chi_1 + q\chi_2 + r\chi_3, \\ \phi_0 &= \kappa\omega + \kappa'\omega' + \kappa''\omega'' + \ldots. \end{aligned}\right\} \quad \ldots\ldots\ldots\ldots\ldots(1)$$

The functions ϕ_1, ϕ_2, ϕ_3, χ_1, χ_2, χ_3 are determined by the same conditions as in Art. 118. To determine ω, we have the conditions: (1°) that it must satisfy $\nabla^2\omega = 0$ at all points of the fluid; (2°) that its derivatives must vanish at infinity; (3°) that $\partial\omega/\partial n$ must $= 0$ at the surface of the solid; and (4°) that ω must be a cyclic function, diminishing by unity whenever the point to which it refers completes a circuit cutting the first barrier once (only) in the positive direction, and recovering its original value whenever the point completes a circuit not cutting this barrier. It appears from Art. 52 that these conditions determine ω save as to an additive constant. In like manner the remaining functions ω', ω'', ... are determined.

By the formula (5) of Art. 55, twice the kinetic energy of the fluid is equal to

$$-\rho\iint(\phi+\phi_0)\frac{\partial}{\partial n}(\phi+\phi_0)\,dS$$
$$-\rho\kappa\iint\frac{\partial}{\partial n}(\phi+\phi_0)\,d\sigma-\rho\kappa'\iint\frac{\partial}{\partial n}(\phi+\phi_0)\,d\sigma'-\dots \qquad (2)$$

Since the cyclic constants of ϕ are zero, and since $\partial\phi_0/\partial n$ vanishes at the surface of the solid, we have, by Art. 54 (4),

$$\iint\phi_0\frac{\partial\phi}{\partial n}\,dS+\kappa\iint\frac{\partial\phi}{\partial n}\,d\sigma+\kappa'\iint\frac{\partial\phi}{\partial n}\,d\sigma'+\dots=\iint\phi\frac{\partial\phi_0}{\partial n}\,dS=0.$$

Hence (2) reduces to

$$-\rho\iint\phi\frac{\partial\phi}{\partial n}\,dS-\rho\kappa\iint\frac{\partial\phi_0}{\partial n}\,d\sigma-\rho\kappa'\iint\frac{\partial\phi_0}{\partial n}\,d\sigma'-\dots \qquad (3)$$

Substituting the values of ϕ, ϕ_0 from (1) we find that the kinetic energy of the fluid is equal to

$$\mathbf{T}+K, \qquad (4)$$

where $\mathbf{T}$ is a homogeneous quadratic function of u, v, w, p, q, r, of the form defined by Art. 121 (2) (3), and

$$2K=(\kappa,\kappa)\kappa^2+(\kappa',\kappa')\kappa'^2+\dots+2(\kappa,\kappa')\kappa\kappa'+\dots, \qquad (5)$$

where, for example,

$$\left.\begin{aligned}(\kappa,\kappa)&=-\rho\iint\frac{\partial\omega}{\partial n}\,d\sigma,\\ (\kappa,\kappa')&=-\tfrac{1}{2}\rho\iint\frac{\partial\omega'}{\partial n}\,d\sigma-\tfrac{1}{2}\rho\iint\frac{\partial\omega}{\partial n}\,d\sigma'\\ &=-\rho\iint\frac{\partial\omega'}{\partial n}\,d\sigma=-\rho\iint\frac{\partial\omega}{\partial n}\,d\sigma'.\end{aligned}\right\} \qquad (6)$$

The identity of the different forms of (κ, κ') follows from Art. 54 (4).

Hence the total energy of fluid and solid is given by

$$T=\mathfrak{T}+K, \qquad (7)$$

where $\mathfrak{T}$ is a homogeneous quadratic function of u, v, w, p, q, r of the same form as Art. 121 (8), and K is defined by (5) and (6) above.

133. The 'impulse' of the motion now consists partly of impulsive forces applied to the solid, and partly of impulsive pressures $\rho\kappa, \rho\kappa', \rho\kappa''$, ... applied uniformly (as explained in Art. 54) over the several membranes which are supposed for a moment to occupy the positions of the barriers. Let us denote by $\xi_1, \eta_1, \zeta_1, \lambda_1, \mu_1, \nu_1$ the components of the extraneous impulse applied to the solid. Expressing that the x-component of the momentum of the solid is equal to the similar component of the total impulse acting on it, we have

$$\frac{\partial \mathbf{T}_1}{\partial u} = \xi_1 - \rho \iint (\phi + \phi_0)\, l\, dS$$
$$= \xi_1 + \rho \iint (u\phi_1 + \ldots + p\chi_1 + \ldots + \kappa\omega + \ldots) \frac{\partial \phi_1}{\partial n} dS$$
$$= \xi_1 - \frac{\partial \mathbf{T}}{\partial u} + \rho\kappa \iint \omega \frac{\partial \phi_1}{\partial n} dS + \rho\kappa' \iint \omega' \frac{\partial \phi_1}{\partial n} dS + \ldots, \quad \ldots\ldots(1)$$

where, as before, $\mathbf{T}_1$ denotes the kinetic energy of the solid, and $\mathbf{T}$ that part of the energy of the fluid which is independent of the cyclic motion. Again, considering the angular momentum of the solid about the axis of x,

$$\frac{\partial \mathbf{T}_1}{\partial p} = \lambda_1 - \rho \iint (\phi + \phi_0)(ny - mz)\, dS$$
$$= \lambda_1 + \rho \iint (u\phi_1 + \ldots + p\chi_1 + \ldots + \kappa\omega + \ldots) \frac{\partial \chi_1}{\partial n} dS$$
$$= \lambda_1 - \frac{\partial \mathbf{T}}{\partial p} + \rho\kappa \iint \omega \frac{\partial \chi_1}{\partial n} dS + \rho\kappa' \iint \omega' \frac{\partial \chi_1}{\partial n} dS + \ldots. \quad \ldots\ldots\ldots(2)$$

Hence, since $\mathfrak{T} = \mathbf{T} + \mathbf{T}_1$, we have

$$\left.\begin{aligned} \xi_1 &= \frac{\partial \mathfrak{T}}{\partial u} - \rho\kappa \iint \omega \frac{\partial \phi_1}{\partial n} dS - \rho\kappa' \iint \omega' \frac{\partial \phi_1}{\partial n} dS - \ldots, \\ \lambda_1 &= \frac{\partial \mathfrak{T}}{\partial p} - \rho\kappa \iint \omega \frac{\partial \chi_1}{\partial n} dS - \rho\kappa' \iint \omega' \frac{\partial \chi_1}{\partial n} dS - \ldots, \end{aligned}\right\} \quad \ldots\ldots\ldots(3)$$

By virtue of Lord Kelvin's extension of Green's Theorem, already referred to, these may be written in the alternative forms

$$\left.\begin{aligned} \xi_1 &= \frac{\partial \mathfrak{T}}{\partial u} + \rho\kappa \iint \frac{\partial \phi_1}{\partial n} d\sigma + \rho\kappa' \iint \frac{\partial \phi_1}{\partial n} d\sigma' + \ldots, \\ \lambda_1 &= \frac{\partial \mathfrak{T}}{\partial p} + \rho\kappa \iint \frac{\partial \chi_1}{\partial n} d\sigma + \rho\kappa' \iint \frac{\partial \chi_1}{\partial n} d\sigma' + \ldots. \end{aligned}\right\} \quad \ldots\ldots\ldots\ldots(4)$$

Adding to these the terms due to the impulsive pressures applied to the barriers, we have, finally, for the components of the total impulse of the motion*,

$$\left.\begin{aligned} \xi, \eta, \zeta &= \frac{\partial \mathfrak{T}}{\partial u} + \xi_0, \quad \frac{\partial \mathfrak{T}}{\partial v} + \eta_0, \quad \frac{\partial \mathfrak{T}}{\partial w} + \zeta_0, \\ \lambda, \mu, \nu &= \frac{\partial \mathfrak{T}}{\partial p} + \lambda_0, \quad \frac{\partial \mathfrak{T}}{\partial q} + \mu_0, \quad \frac{\partial \mathfrak{T}}{\partial r} + \nu_0, \end{aligned}\right\} \quad \ldots\ldots\ldots\ldots\ldots(5)$$

where, for example,

$$\left.\begin{aligned} \xi_0 &= \rho\kappa \iint \left(l + \frac{\partial \phi_1}{\partial n}\right) d\sigma + \rho\kappa' \iint \left(l + \frac{\partial \phi_1}{\partial n}\right) d\sigma' + \ldots, \\ \lambda_0 &= \rho\kappa \iint \left(ny - mz + \frac{\partial \chi_1}{\partial n}\right) d\sigma + \rho\kappa' \iint \left(ny - mz + \frac{\partial \chi_1}{\partial n}\right) d\sigma' + \ldots. \end{aligned}\right\} \ldots(6)$$

* Cf. Sir W. Thomson, *l.c. ante* p. 168.

It is evident that the constants $\xi_0, \eta_0, \zeta_0, \lambda_0, \mu_0, \nu_0$ are the components of the impulse of the cyclic fluid motion which would remain if the solid were, by forces applied to it alone, brought to rest.

By the argument of Art. 119, the total impulse is subject to the same laws as the momentum of a finite dynamical system. Hence the equations of motion of the solid are obtained by substituting from (5) in the equations (1) of Art. 120 *.

134. As a simple example we may take the case of an annular solid of revolution.

If the axis of x coincide with that of the ring, we see by reasoning of the same kind as in Art. 126, 4° that if the situation of the origin on this axis be properly chosen we may write

$$2T = Au^2 + B(v^2+w^2) + Pp^2 + Q(q^2+r^2) + (\kappa, \kappa)\,\kappa^2. \quad \text{.................(1)}$$

Hence

$$\xi, \eta, \zeta = Au + \xi_0,\ Bv,\ Bw; \qquad \lambda, \mu, \nu = Pp,\ Qq,\ Qr. \quad \text{.................(2)}$$

Substituting in the equations of Art. 120, we find $dp/dt = 0$, or $p = \text{const.}$, as is otherwise obvious. Let us suppose that the ring is slightly disturbed from a state of motion in which v, w, p, q, r are zero, *i.e.* a steady motion parallel to the axis. In the beginning of the disturbed motion v, w, p, q, r will be small quantities whose products we may neglect. The first of the equations referred to then gives $du/dt = 0$, or $u = \text{const.}$, and the remaining equations become

$$\left.\begin{aligned} B\frac{dv}{dt} &= -(Au+\xi_0)\,r, & Q\frac{dq}{dt} &= -\{(A-B)\,u + \xi_0\}\,w, \\ B\frac{dw}{dt} &= (Au+\xi_0)\,q, & Q\frac{dr}{dt} &= \{(A-B)\,u + \xi_0\}\,v. \end{aligned}\right\} \quad \text{..............(3)}$$

Eliminating r, we find

$$BQ\frac{d^2v}{dt^2} = -(Au+\xi_0)\{(A-B)\,u+\xi_0\}\,v. \quad \text{.........................(4)}$$

Exactly the same equation is satisfied by w. It is therefore necessary and sufficient for stability that the coefficient of v on the right-hand side of (4) should be negative; and the time of a small oscillation, when this condition is satisfied, is †

$$2\pi\left[\frac{BQ}{(Au+\xi_0)\{(A-B)\,u+\xi_0\}}\right]^{\frac{1}{2}}. \quad \text{.............................(5)}$$

We may also notice another case of steady motion of the ring, viz. where the impulse reduces to a couple about a diameter. It is easily seen that the equations of motion are satisfied by $\xi, \eta, \zeta, \lambda, \mu = 0$, and ν constant; in which case

$$u = -\xi_0/A, \quad r = \text{const.}$$

The ring then rotates about an axis in the plane yz parallel to that of z, at a distance u/r from it ‡.

* This conclusion may be verified by direct calculation from the pressure-formula of Art. 20; see Bryan, "Hydrodynamical Proof of the Equations of Motion of a Perforated Solid, ...," *Phil. Mag.* (5) xxxv. 338 (1893).

† Sir W. Thomson, *l.c. ante* p. 168.

‡ For further investigations on this subject we refer to papers by Basset, "On the Motion of a Ring in an Infinite Liquid," *Proc. Camb. Phil. Soc.* vi. 47 (1887), and Miss Fawcett, *l.c. ante* p. 179.

The Forces on a Cylinder moving in Two Dimensions.

134 a. The two-dimensional problem of the motion of a cylindrical body, especially when there is circulation round it, is most simply treated by direct calculation of the pressures on the surface*. We assume as usual that the fluid is at rest at infinity.

Taking axes fixed in a cross-section, we denote by $(\mathbf{u}, \mathbf{v})$ the velocity of the origin, and by $\mathbf{r}$ the angular velocity, the symbols u, v being now required in their original sense as component velocities of the fluid. The pressure-equation is then

$$\frac{p}{\rho} = \frac{\partial \phi}{\partial t} - (\mathbf{u} - \mathbf{r}y)\frac{\partial \phi}{\partial x} - (\mathbf{v} + \mathbf{r}x)\frac{\partial \phi}{\partial y} - \tfrac{1}{2}q^2 + \text{const.}, \quad \text{.........(1)}$$

where $q^2 = u^2 + v^2$. The force $(\mathbf{X}, \mathbf{Y})$ and couple $(\mathbf{N})$ to which the pressures on the surface reduce are

$$\mathbf{X} = -\int pl\,ds, \quad \mathbf{Y} = -\int pm\,ds, \quad \mathbf{N} = -\int p\,(mx - ly)\,ds, \quad \text{......(2)}$$

where l, m are the direction cosines of the normal drawn outwards from an element δs of the contour, and the integration is taken round the perimeter. Now

$$\left.\begin{aligned} \tfrac{1}{2}\int q^2 l\,ds &= -\iint\left(u\frac{\partial u}{\partial x} + v\frac{\partial u}{\partial y}\right)dx\,dy = \int (lu + mv)\,u\,ds, \\ \tfrac{1}{2}\int q^2 m\,ds &= -\iint\left(u\frac{\partial v}{\partial x} + v\frac{\partial v}{\partial y}\right)dx\,dy = \int (lu + mv)\,v\,ds \end{aligned}\right\} \quad \text{......(3)}$$

in virtue of the relations

$$\partial v/\partial x = \partial u/\partial y, \qquad \partial u/\partial x + \partial v/\partial y = 0.$$

We have here omitted the various line-integrals taken over an infinite enclosing boundary, since at a great distance r the velocity is at most of the order $1/r$, whilst δs is of the order $r\,\delta\theta$. At the surface of the cylinder we have

$$lu + mv = l\,(\mathbf{u} - \mathbf{r}y) + m\,(\mathbf{v} + \mathbf{r}x). \quad \text{....................(4)}$$

Hence substituting from (1) in (2) we find

$$\begin{aligned} \frac{\mathbf{X}}{\rho} &= -\int \frac{\partial \phi}{\partial t}\,l\,ds + \int (mu - lv)\,(\mathbf{v} + \mathbf{r}x)\,ds \\ &= -\int \frac{\partial \phi}{\partial t}\,l\,ds + \int (\mathbf{v} + \mathbf{r}x)\frac{\partial \phi}{\partial s}\,ds, \quad \text{......................(5)} \end{aligned}$$

and similarly

$$\frac{\mathbf{Y}}{\rho} = -\int \frac{\partial \phi}{\partial t}\,m\,ds - \int (\mathbf{u} - \mathbf{r}y)\frac{\partial \phi}{\partial s}\,ds. \quad \text{......................(6)}$$

Again we find,

$$\tfrac{1}{2}\int q^2\,(mx - ly)\,ds = \int (lu + mv)\,(xv - yu)\,ds. \quad \text{..............(7)}$$

* *Aeronautical Research Committee, R. and M.* 1218 (1929). For another treatment see Glauert, *R. and M.* 1215 (1929).

Here also, the line-integrals round an infinitely remote boundary are omitted, since we may suppose that at this boundary $l/x = m/y$, and that $lu + mv$ is of the order $1/r^2$. The formula (2) for **N** thus becomes

$$\frac{\mathbf{N}}{\rho} = -\int \frac{\partial \phi}{\partial t}(mx - ly)\,ds + \int (\mathbf{u}x + \mathbf{v}y)(lv - mu)\,ds$$

$$= -\int \frac{\partial \phi}{\partial t}(mx - ly)\,ds - \int (\mathbf{u}x + \mathbf{v}y)\frac{\partial \phi}{\partial s}\,ds \quad \text{.................(8)}$$

We now write, in analogy with Arts. 118, 132,

$$\phi = \mathbf{u}\phi_1 + \mathbf{v}\phi_2 + \mathbf{r}\chi + \phi_0, \quad \text{.........................(9)}$$

where ϕ_0 represents the circulatory motion which would persist if the cylinder were brought to rest. It is therefore a cyclic function with, say, the cyclic constant κ. Comparing with (4) we have, at the surface of the cylinder,

$$\frac{\partial \phi_1}{\partial n} = -l, \quad \frac{\partial \phi_2}{\partial n} = -m, \quad \frac{\partial \chi}{\partial n} = -(mx - ly), \quad \frac{\partial \phi_0}{\partial n} = 0. \quad \text{......(10)}$$

In the absence of circulation the energy of the fluid would be

$$\mathbf{T} = -\tfrac{1}{2}\rho\int (\phi - \phi_0)\frac{\partial \phi}{\partial n}\,ds. \quad \text{.......................(11)}$$

Substituting from (9) and (10) this gives

$$2\mathbf{T} = \mathbf{A}\mathbf{u}^2 + 2\mathbf{H}\mathbf{u}\mathbf{v} + \mathbf{B}\mathbf{v}^2 + \mathbf{R}\mathbf{r}^2 + 2(\mathbf{L}\mathbf{u} + \mathbf{M}\mathbf{v})\,\mathbf{r}, \quad \text{........(12)}$$

where

$$\left.\begin{aligned}
&\mathbf{A} = \rho\int l\phi_1\,ds, \quad \mathbf{H} = \rho\int l\phi_2\,ds = \rho\int m\phi_1\,ds, \quad \mathbf{B} = \rho\int m\phi_2\,ds \\
&\mathbf{R} = \rho\int (mx - ly)\,\chi\,ds, \\
&\mathbf{L} = \rho\int l\chi\,ds = \rho\int (mx - ly)\,\phi_1\,ds, \quad \mathbf{M} = \rho\int m\chi\,ds = \rho\int (mx - ly)\,\phi_2\,ds.
\end{aligned}\right\}$$

......(13)

The leading terms in (5), (6), and (8) now take the forms

$$\left.\begin{aligned}
-\frac{d}{dt}(\mathbf{A}\mathbf{u} + \mathbf{H}\mathbf{v} + \mathbf{L}\mathbf{r}) &= -\frac{d}{dt}\frac{\partial \mathbf{T}}{\partial \mathbf{u}}, \\
-\frac{d}{dt}(\mathbf{H}\mathbf{u} + \mathbf{B}\mathbf{v} + \mathbf{M}\mathbf{r}) &= -\frac{d}{dt}\frac{\partial \mathbf{T}}{\partial \mathbf{v}}, \\
-\frac{d}{dt}(\mathbf{R}\mathbf{r} + \mathbf{L}\mathbf{u} + \mathbf{M}\mathbf{v}) &= -\frac{d}{dt}\frac{\partial \mathbf{T}}{\partial \mathbf{r}}.
\end{aligned}\right\} \quad \text{.................(14)}$$

Again, we have

$$\left.\begin{aligned}
\rho\int x\frac{\partial(\phi - \phi_0)}{ds}\,ds &= \rho\int m(\phi - \phi_0)\,ds = \mathbf{H}\mathbf{u} + \mathbf{B}\mathbf{v} + \mathbf{M}\mathbf{r} = \frac{\partial \mathbf{T}}{\partial \mathbf{v}}, \\
\rho\int y\frac{\partial(\phi - \phi_0)}{\partial s}\,ds &= -\rho\int l(\phi - \phi_0)\,ds = -(\mathbf{A}\mathbf{u} + \mathbf{H}\mathbf{v} + \mathbf{L}\mathbf{r}) = -\frac{\partial \mathbf{T}}{\partial \mathbf{u}}.
\end{aligned}\right\}$$

......(15)

Hence if we write

$$\int x\frac{\partial\phi_0}{\partial s}\,ds=\alpha,\quad \int y\frac{\partial\phi_0}{\partial s}\,ds=\beta, \quad\text{.....................(16)}$$

the expressions for the forces become

$$\left.\begin{aligned} \mathbf{X} &= -\frac{d}{dt}\frac{\partial\mathbf{T}}{\partial\mathbf{u}}+\mathbf{r}\frac{\partial\mathbf{T}}{\partial\mathbf{v}}-\kappa\rho\mathbf{v}+\rho\alpha\mathbf{r},\\ \mathbf{Y} &= -\frac{d}{dt}\frac{\partial\mathbf{T}}{\partial\mathbf{v}}-\mathbf{r}\frac{\partial\mathbf{T}}{\partial\mathbf{u}}+\kappa\rho\mathbf{u}+\rho\beta\mathbf{r},\\ \mathbf{N} &= -\frac{d}{dt}\frac{\partial\mathbf{T}}{\partial\mathbf{r}}+\mathbf{v}\frac{\partial\mathbf{T}}{\partial\mathbf{u}}-\mathbf{u}\frac{\partial\mathbf{T}}{\partial\mathbf{v}}-\rho\,(\alpha\mathbf{u}+\beta\mathbf{v}). \end{aligned}\right\} \quad\text{............(17)}$$

By turning the co-ordinate axes through a suitable angle, the coefficient **H** can be made to vanish. And by a suitable choice of origin we may also annul the coefficients **L**, **M**, or alternatively we may make $\alpha=0$, $\beta=0$. But these two determinations are *in general* incompatible, and neither of these special origins can be assumed to coincide with the mean centre of the area of the section.

The most interesting case, however, is where the section is symmetrical with respect to each of two perpendicular axes. If these are taken as axes of co-ordinates we have

$$\mathbf{H}=0,\quad \mathbf{L}=0,\quad \mathbf{M}=0,\quad \alpha=0,\quad \beta=0, \quad\text{..............(18)}$$

and the formulae (17) reduce to

$$\left.\begin{aligned} \mathbf{X} &= -\mathbf{A}\frac{d\mathbf{u}}{dt}+\mathbf{Brv}-\kappa\mathbf{v},\\ \mathbf{Y} &= -\mathbf{B}\frac{d\mathbf{v}}{dt}-\mathbf{Aru}+\kappa\mathbf{u},\\ \mathbf{N} &= -\mathbf{R}\frac{d\mathbf{r}}{dt}-(\mathbf{A}-\mathbf{B})\,\mathbf{uv}. \end{aligned}\right\} \quad\text{.....................(19)}$$

To form the equation of motion in this case we have only to modify the inertia coefficients, as in Art. 122. If the distribution of mass is also symmetrical, we write

$$A=\mathbf{A}+M,\quad B=\mathbf{B}+M,\quad R=\mathbf{R}+L, \quad\text{..............(20)}$$

where M represents the mass of the cylinder itself, and L its moment of inertia. Then

$$\left.\begin{aligned} A\frac{d\mathbf{u}}{dt}-B\mathbf{rv}+\kappa\rho\mathbf{v} &= X,\\ B\frac{d\mathbf{v}}{dt}+A\mathbf{ru}-\kappa\rho\mathbf{u} &= Y,\\ L\frac{d\mathbf{r}}{dt}-(A-B)\,\mathbf{uv} &= N, \end{aligned}\right\} \quad\text{.......................(21)}$$

where X, Y, N represent the effect of extraneous forces. When these are absent, and the circulation zero, the solution is as in Art. 127.

In the case of a circular section there is no point in supposing the co-ordinate axes to rotate. Putting $A = B$, $\mathbf{r} = 0$, we have

$$A\frac{d\mathbf{u}}{dt} + \kappa\rho\mathbf{u} = X, \quad A\frac{d\mathbf{v}}{dt} - \kappa\rho\mathbf{u} = Y, \quad \text{.................}(22)$$

as in Art. 69.

If the section is symmetrical with respect to one axis only, say that of x, we have $\mathbf{H} = 0$, $\mathbf{L} = 0$, $B = 0$. By a displacement of the origin along the axis of symmetry we can make $\mathbf{M} = 0$, but α will not in general vanish simultaneously. If there is no circulation the new origin corresponds to the 'centre of reaction' of Thomson and Tait*.

Equations of Motion in Generalized Co-ordinates.

135. When we have more than one moving solid, or when the fluid is bounded, wholly or in part, by fixed walls, we may have recourse to Lagrange's method of 'generalized co-ordinates.' This was first applied to hydrodynamical problems by Thomson and Tait†.

The systems ordinarily contemplated in Analytical Dynamics are of finite freedom; *i.e.* the position of every particle is completely determined when we know the values of a finite number of independent variables or 'generalized co-ordinates' $q_1, q_2, \ldots q_n$. The kinetic energy T can then be expressed as a quadratic function of the 'generalized velocity components' $\dot{q}_1, \dot{q}_2, \ldots \dot{q}_n$.

In the Hamiltonian method the actual motion of the system between any two instants t_0, t_1 is compared with a slightly varied motion. If ξ, η, ζ be the Cartesian co-ordinates of any particle m, and X, Y, Z the components of the total force acting on it, it is proved that

$$\int_{t_0}^{t_1} \{\Delta T + \Sigma (X\Delta\xi + Y\Delta\eta + Z\Delta\zeta)\}\, dt = 0, \quad \text{..............}(1)$$

provided the varied motion be such that

$$\Big[\Sigma m (\dot{\xi}\Delta\xi + \dot{\eta}\Delta\eta + \dot{\zeta}\Delta\zeta)\Big]_{t_0}^{t_1} = 0. \quad \text{....................}(2)$$

The summation Σ is understood to include all the particles of the system. The varied motion is usually supposed to be adjusted so that the initial and final positions of each particle shall be respectively the same as in the actual motion. The quantities $\Delta\zeta$, $\Delta\eta$, $\Delta\zeta$ then vanish at each limit of integration, and the condition (2) is fulfilled.

For a conservative system free from extraneous force (1) takes the form

$$\Delta \int_{t_0}^{t_1} (T - V)\, dt = 0. \quad \text{.........................}(3)$$

* *Natural Philosophy*, Art. 321.

† *Ibid.* Art. 331.

In words, if the actual motion of the system between any two configurations through which it passes be compared with *any* slightly varied motion, between the same configurations, which the system is (by the application of suitable forces) made to execute *in the same time*, the time-integral of the 'kinetic potential'* $V-T$ is stationary.

In terms of generalized co-ordinates, the equation (1) takes the form

$$\int_{t_0}^{t_1} (\Delta T + Q_1 \Delta q_1 + Q_2 \Delta q_2 + \ldots + Q_n \Delta q_n)\, dt = 0, \quad \ldots\ldots\ldots\ldots(4)$$

from which Lagrange's equations

$$\frac{d}{dt}\frac{\partial T}{\partial \dot{q}_r} - \frac{\partial T}{\partial q_r} = Q_r \quad \ldots\ldots\ldots\ldots\ldots\ldots\ldots\ldots\ldots\ldots(5)$$

can be deduced by a known process.

136. Proceeding now to the hydrodynamical problem, let $q_1, q_2, \ldots q_n$ be a system of generalized co-ordinates which serve to specify the configuration of the solids. We will suppose, for the present, that the motion of the fluid is entirely due to that of the solids, and is therefore irrotational and acyclic.

In this case the velocity-potential at any instant will be of the form

$$\phi = \dot{q}_1 \phi_1 + \dot{q}_2 \phi_2 + \ldots + \dot{q}_n \phi_n, \quad \ldots\ldots\ldots\ldots\ldots\ldots\ldots(1)$$

where $\phi_1, \phi_2, \ldots$ are determined in a manner analogous to that of Art. 118. The formula for the kinetic energy of the fluid is then

$$2\mathbf{T} = -\rho \iint \phi \frac{\partial \phi}{\partial n}\, dS = \mathbf{A}_{11} \dot{q}_1^2 + \mathbf{A}_{22} \dot{q}_2^2 + \ldots + 2\mathbf{A}_{12} \dot{q}_1 \dot{q}_2 + \ldots, \quad \ldots\ldots(2)$$

where

$$\mathbf{A}_{rr} = -\rho \iint \phi_r \frac{\partial \phi_r}{\partial n}\, dS, \qquad \mathbf{A}_{rs} = -\rho \iint \phi_r \frac{\partial \phi_s}{\partial n}\, dS = -\rho \iint \phi_s \frac{\partial \phi_r}{\partial n}\, dS, \quad \ldots(3)$$

the integrations extending over the instantaneous positions of the bounding surfaces of the fluid. The identity of the two forms of $\mathbf{A}_{rs}$ follows from Green's Theorem. The coefficients $\mathbf{A}_{rr}$, $\mathbf{A}_{rs}$ will in general be functions of the co-ordinates $q_1, q_2, \ldots q_n$.

If we add to (2) twice the kinetic energy, $\mathbf{T}_1$, of the solids themselves, we get an expression of the same form, with altered coefficients, say

$$2T = A_{11} \dot{q}_1^2 + A_{22} \dot{q}_2^2 + \ldots + 2A_{12} \dot{q}_1 \dot{q}_2 + \ldots. \quad \ldots\ldots\ldots\ldots\ldots(4)$$

It remains to shew that, although our system is one of infinite freedom, the equations of motion of the solids can, under the circumstances presupposed, be obtained by substituting this value of T in the Lagrangian equations, Art. 135 (5). We are not at liberty to assume this without further examination, for the positions of the various particles of the fluid are

* The name was introduced by Helmholtz, "Die physikalische Bedeutung des Princips der kleinsten Wirkung," *Crelle*, c. 137, 213 (1886) [*Wiss. Abh.* iii. 203].

not determined by the instantaneous values $q_1, q_2, \ldots q_n$ of the co-ordinates of the solids. For instance, if the solids, after performing various evolutions, return each to its original position, the individual particles of the fluid will in general be found to be finitely displaced*.

Going back to the general formula (1) of Art. 135, let us suppose that in the varied motion, to which the symbol Δ refers, the solids undergo no change of size or shape, and that the fluid remains incompressible, and has, at the boundaries, the same displacement in the direction of the normal as the solids with which it is in contact. It is known that under these conditions the terms due to the internal reactions of the solids will disappear from the sum

$$\Sigma(X\Delta\xi + Y\Delta\eta + Z\Delta\zeta).$$

The terms due to the mutual pressures of the fluid elements are equivalent to

$$-\iiint\left(\frac{\partial p}{\partial x}\Delta\xi + \frac{\partial p}{\partial y}\Delta\eta + \frac{\partial p}{\partial z}\Delta\zeta\right)dx\,dy\,dz,$$

or

$$\iint p\,(l\Delta\xi + m\Delta\eta + n\Delta\zeta)\,dS + \iiint p\left(\frac{\partial\Delta\xi}{\partial x} + \frac{\partial\Delta\eta}{\partial y} + \frac{\partial\Delta\zeta}{\partial z}\right)dx\,dy\,dz,$$

where the former integral extends over the bounding surfaces, and l, m, n denote the direction-cosines of the normal, drawn towards the fluid. The volume-integral vanishes by the condition of incompressibility

$$\frac{\partial\Delta\xi}{\partial x} + \frac{\partial\Delta\eta}{\partial y} + \frac{\partial\Delta\zeta}{\partial z} = 0. \qquad\qquad (5)$$

The surface-integral vanishes at a fixed boundary, where

$$l\Delta\xi + m\Delta\eta + n\Delta\zeta = 0;$$

and in the case of a moving solid it is cancelled by the terms due to the pressure exerted by the fluid on the solid. Hence the symbols X, Y, Z may be taken to refer only to the remaining forces acting on the system, and we may write

$$\Sigma(X\Delta\xi + Y\Delta\eta + Z\Delta\zeta) = Q_1\Delta q_1 + Q_2\Delta q_2 + \ldots + Q_n\Delta q_n, \qquad (6)$$

where $Q_1, Q_2, \ldots Q_n$ are generalized components of force.

The varied motion of the fluid has still a high degree of generality. We will now further limit it by supposing that while the solids are, by suitable forces applied to them, made to execute an arbitrary motion, the fluid is left to take its own course in consequence of this. The varied motion of the fluid may accordingly be taken to be irrotational, in which case the varied kinetic energy $T + \Delta T$ of the system will be the same function of the varied co-ordinates $q_r + \Delta q_r$, and the varied velocities $\dot{q}_r + \Delta\dot{q}_r$, that the actual energy T is of q_r and $\dot{q}_r$.

* As a simple example, take the case of a circular disk which is made to move, without rotation, so that its centre describes a rectangle two of whose sides are normal to its plane; and examine the displacements of a particle initially in contact with the disk at its centre.

Again, considering the particles of the fluid alone, we shall have, on the same supposition,

$$\Sigma m(\dot{\xi}\Delta\xi+\dot{\eta}\Delta\eta+\dot{\zeta}\Delta\zeta)=-\rho\iiint\left(\frac{\partial\phi}{\partial x}\Delta x+\frac{\partial\phi}{\partial y}\Delta y+\frac{\partial\phi}{\partial z}\Delta z\right)dxdydz$$

$$=\rho\iint\phi\,(l\Delta\xi+m\Delta\eta+n\Delta\zeta)\,dS,$$

where use has again been made of the condition (5) of incompressibility. By the kinematical condition to be satisfied at the boundaries, we have

$$l\Delta\xi+m\Delta\eta+n\Delta\zeta=-\frac{\partial\phi_1}{\partial n}\Delta q_1-\frac{\partial\phi_2}{\partial n}\Delta q_2-\ldots-\frac{\partial\phi_n}{\partial n}\Delta q_n,$$

and therefore

$$\Sigma m(\dot{\xi}\Delta\xi+\dot{\eta}\Delta\eta+\dot{\zeta}\Delta\zeta)=-\rho\iint\phi\left(\frac{\partial\phi_1}{\partial n}\Delta q_1+\frac{\partial\phi_2}{\partial n}\Delta q_2+\ldots+\frac{\partial\phi_n}{\partial n}\Delta q_n\right)dS$$

$$=(\mathbf{A}_{11}\dot{q}_1+\mathbf{A}_{12}\dot{q}_2+\ldots+\mathbf{A}_{1n}\dot{q}_n)\Delta q_1+(\mathbf{A}_{21}\dot{q}_1+\mathbf{A}_{22}\dot{q}_2+\ldots+\mathbf{A}_{2n}\dot{q}_n)\Delta q_2$$
$$+\ldots+(\mathbf{A}_{n1}\dot{q}_1+\mathbf{A}_{n2}\dot{q}_2+\ldots+\mathbf{A}_{nn}\dot{q}_n)\Delta q_n$$

$$=\frac{\partial\mathbf{T}}{\partial\dot{q}_1}\Delta q_1+\frac{\partial\mathbf{T}}{\partial\dot{q}_2}\Delta q_2+\ldots+\frac{\partial\mathbf{T}}{\partial\dot{q}_n}\Delta q_n,\quad\ldots\ldots\ldots\ldots(7)$$

by (1), (2), (3) above. If we add the terms due to the solids, we find that the condition (2) of Art. 135 still holds; and the deduction of Lagrange's equations

$$\frac{d}{dt}\frac{\partial T}{\partial\dot{q}_r}-\frac{\partial T}{\partial q_r}=Q_r\quad\ldots\ldots\ldots\ldots(8)$$

then proceeds in the usual manner.

137. As a first application of the foregoing theory we may take an example given by Thomson and Tait*, where a sphere is supposed to move in a liquid which is limited only by an infinite plane wall.

Taking, for simplicity, the case where the centre moves in a plane perpendicular to that of the wall, let us specify its position at time t by rectangular co-ordinates x, y in this plane, of which y denotes distance from the wall. We have

$$2T=A\dot{x}^2+B\dot{y}^2,\quad\ldots\ldots\ldots\ldots(1)$$

where A and B are functions of y only, it being plain that the term $\dot{x}\dot{y}$ cannot occur, since the energy must remain unaltered when the sign of $\dot{x}$ is reversed. The values of A, B can be written down from the results of Arts. 98, 99, viz. if m denote the mass of the sphere, and a its radius, we have

$$A=m+\tfrac{2}{3}\pi\rho a^3\left(1+\tfrac{3}{16}\frac{a^3}{y^3}\right),\qquad B=m+\tfrac{2}{3}\pi\rho a^3\left(1+\tfrac{3}{8}\frac{a^3}{y^3}\right),\quad\ldots\ldots\ldots\ldots(2)$$

approximately, if y be great in comparison with a.

The equations of motion give

$$\frac{d}{dt}(A\dot{x})=X,\qquad\frac{d}{dt}(B\dot{y})-\tfrac{1}{2}\left(\frac{dA}{dy}\dot{x}^2+\frac{dB}{dy}\dot{y}^2\right)=Y,\quad\ldots\ldots\ldots\ldots(3)$$

where X, Y are the components of extraneous force, supposed to act on the sphere in a line through the centre.

* *Natural Philosophy*, Art. 321.

If there be no extraneous force, and if the sphere be projected in a direction normal to the wall, we have $\dot{x}=0$, and

$$B\dot{y}^2=\text{const.} \qquad \text{.........(4)}$$

Since B diminishes as y increases, the sphere experiences an acceleration *from* the wall.

Again, if the sphere be constrained to move in a line parallel to the wall, we have $\dot{y}=0$, and the necessary constraining force is

$$Y=-\tfrac{1}{2}\frac{dA}{dy}\dot{x}^2. \qquad \text{.........(5)}$$

Since dA/dy is negative, the sphere appears to be *attracted* by the wall. The reason of this is easily seen by reducing the problem to one of steady motion. The fluid velocity will evidently be greater, and the pressure therefore less, on the side of the sphere next the wall than on the further side; see Art. 23.

The above investigation will also apply to the case of two spheres projected in an unlimited mass of fluid, in such a way that the plane $y=0$ is a plane of symmetry in all respects.

138. Let us next take the case of two spheres moving in the line of centres.

The kinematical part of this problem has been treated in Art. 98. If we now denote by x, y the distances of the centres of the spheres A, B from some fixed origin O in the line joining them, we have

$$2T=L\dot{x}^2-2M\dot{x}\dot{y}+N\dot{y}^2, \qquad \text{.........(1)}$$

where the coefficients L, M, N are functions of $y-x$, or c, the distance between the centres. Hence the equations of motion are

$$\left.\begin{aligned}\frac{d}{dt}(L\dot{x}-M\dot{y})+\tfrac{1}{2}\left(\frac{dL}{dc}\dot{x}^2-2\frac{dM}{dc}\dot{x}\dot{y}+\frac{dN}{dc}\dot{y}^2\right)&=X,\\ \frac{d}{dt}(-M\dot{x}+N\dot{y})-\tfrac{1}{2}\left(\frac{dL}{dc}\dot{x}^2-2\frac{dM}{dc}\dot{x}\dot{y}+\frac{dN}{dc}\dot{y}^2\right)&=Y,\end{aligned}\right\} \qquad \text{.........(2)}$$

where X, Y are the forces acting on the spheres along the line of centres. If the radii a, b are both small compared with c, we have, by Art. 98 (15), keeping only the most important terms,

$$L=m+\tfrac{2}{3}\pi\rho a^3, \quad M=2\pi\rho\frac{a^3b^3}{c^3}, \quad N=m'+\tfrac{2}{3}\pi\rho b^3, \qquad \text{.........(3)}$$

approximately, where m, m' are the masses of the two spheres. Hence to this order of approximation

$$\frac{dL}{dc}=0, \quad \frac{dM}{dc}=-6\pi\rho\frac{a^3b^3}{c^4}, \quad \frac{dN}{dc}=0.$$

If each sphere be constrained to move with constant velocity, the force which must be applied to A to maintain its motion is

$$X=-\frac{dM}{dc}\dot{y}(\dot{y}-\dot{x})-\frac{dM}{dc}\dot{x}\dot{y}=6\pi\rho\frac{a^3b^3}{c^4}\dot{y}^2. \qquad \text{.........(4)}$$

This tends towards B, and depends only on the velocity of B. The spheres therefore appear to repel one another; and it is to be noticed that the apparent forces are not equal and opposite unless $\dot{x}=\pm\dot{y}$.

Again, if each sphere make small periodic oscillations about a mean position, the period being the same for each, the mean values of the first terms in (2) will be zero, and the spheres therefore will appear to act on one another with forces equal to

$$6\pi\rho\frac{a^3b^3}{c^4}[\dot{x}\dot{y}], \qquad \text{.........(5)}$$

where $[\dot{x}\dot{y}]$ denotes the mean value of $\dot{x}\dot{y}$. If $\dot{x}$, $\dot{y}$ differ in phase by less than a quarter-period, this force is one of repulsion, if by more than a quarter-period it is one of attraction.

Next, let B perform small periodic oscillations, while A is held at rest. The mean force which must be applied to A to prevent it from moving is

$$X = \tfrac{1}{2}\frac{dN}{dc}[\dot{y}^2], \qquad\qquad (6)$$

where $[\dot{y}^2]$ denotes the mean square of the velocity of B. To the above order of approximation dN/dc is zero; on reference to Art. 98 we find that the most important term in it is $-12\pi\rho a^3 b^6/c^7$, so that the force exerted on A is attractive, and equal to

$$6\pi\rho\frac{a^3 b^6}{c^7}[\dot{y}^2]. \qquad\qquad (7)$$

This result comes under a general principle enunciated by Kelvin. If we have two bodies immersed in a fluid, one of which (A) performs small vibrations while the other (B) is held at rest, the fluid velocity at the surface of B will on the whole be greater on the side nearer A than on that which is more remote. Hence the average pressure on the former side will be less than that on the latter, so that B will experience on the whole an attraction towards A. As practical illustrations of this principle we may cite the apparent attraction of a delicately-suspended card by a vibrating tuning-fork, and other similar phenomena studied experimentally by Guthrie* and explained in the above manner by Kelvin†.

Modification of Lagrange's Equations in the case of Cyclic Motion.

139. We return to the investigation of Art. 136, with the view of adapting it to the case where the fluid has cyclic irrotational motion through channels in the moving solids, or (it may be) in an enclosing vessel, independently of the motion due to the solids themselves.

Let us imagine barrier-surfaces to be drawn across the several apertures. In the case of channels in a containing vessel we shall suppose these ideal surfaces to be fixed in space, and in the case of channels in a moving solid we shall suppose them to be fixed relatively to the solid. Let $\dot{\chi}, \dot{\chi}', \dot{\chi}'', \ldots$ be the fluxes at time t across, and *relative to*, the several barriers; and let $\chi, \chi', \chi'', \ldots$ be the time-integrals of these fluxes, reckoned from some arbitrary epoch, these quantities determining (therefore) the volumes of fluid which have up to the time t crossed the respective barriers. It will appear that the analogy with a dynamical system of finite freedom is still conserved, provided the quantities $\chi, \chi', \chi'', \ldots$ be reckoned as generalized co-ordinates of the system, in addition to those $(q_1, q_2, \ldots q_n)$ which specify the positions of the moving solids. It is obvious already that the absolute values of $\chi, \chi', \chi'', \ldots$ will not enter into the expression for the kinetic energy, but only their rates of variation.

In the first place, we may shew that the motion of the fluid, in any given configuration of the solids, is completely determined by the instantaneous

* "On Approach caused by Vibration," *Phil. Mag.* (4) xl. 345 (1870).

† *Reprint of Papers on Electrostatics, &c.* Art. 741. For references to further investigations, both experimental and theoretical, by C. A. Bjerknes and others, on the mutual influence of oscillating spheres in a fluid, see Hicks, "Report on Recent Researches in Hydrodynamics," *Brit. Ass. Rep.* 1882, pp. 52...; Love, *Encycl. d. math. Wiss.* iv. (3), pp. 111, 112.

values of $\dot{q}_1, \dot{q}_2, \ldots \dot{q}_n, \dot{\chi}, \dot{\chi}', \dot{\chi}'', \ldots$. For if there were two modes of irrotational motion consistent with these values, then, in the motion which is the *difference* of these, the boundaries of the fluid would be at rest, and the flux across each barrier would be zero. The formula (5) of Art. 55 shews that under these conditions the kinetic energy would vanish.

It follows that the velocity-potential can be expressed in the form

$$\phi = \dot{q}_1\phi_1 + \dot{q}_2\phi_2 + \ldots + \dot{q}_n\phi_n + \dot{\chi}\Omega + \dot{\chi}'\Omega' + \ldots \quad \text{.........(1)}$$

Here ϕ_r is the velocity-potential of a motion in which q_r alone varies and the flux across each barrier is accordingly zero. Again Ω is the velocity-potential of a motion in which the solids are all at rest, whilst the flux through the first aperture is unity, and that through every other aperture is zero. It is to be observed that $\phi_1, \phi_2, \ldots \phi_n, \Omega, \Omega', \ldots$ are in general all of them cyclic functions, which may however be treated as single-valued, on the conventions of Art. 50.

The kinetic energy of the fluid is given by the expression

$$2\mathbf{T} = \rho \iiint \left\{ \left(\frac{\partial\phi}{\partial x}\right)^2 + \left(\frac{\partial\phi}{\partial y}\right)^2 + \left(\frac{\partial\phi}{\partial z}\right)^2 \right\} dx\,dy\,dz, \quad \text{.............(2)}$$

where the integral is taken over the region occupied by the fluid at the instant under consideration. If we substitute from (1) we obtain $\mathbf{T}$ as a homogeneous quadratic function of $\dot{q}_1, \dot{q}_2, \ldots \dot{q}_n, \dot{\chi}, \dot{\chi}', \dot{\chi}'', \ldots$ with coefficients which depend on the instantaneous configuration of the solids, and are therefore functions of $q_1, q_2, \ldots q_n$ only. Moreover, we find, by Art. 53 (1),

$$\frac{\partial\mathbf{T}}{\partial\dot{\chi}} = \rho \iiint \left\{ \frac{\partial\phi}{\partial x}\frac{\partial\Omega}{\partial x} + \frac{\partial\phi}{\partial y}\frac{\partial\Omega}{\partial y} + \frac{\partial\phi}{\partial z}\frac{\partial\Omega}{\partial z} \right\} dx\,dy\,dz$$

$$= -\rho \iint \phi \frac{\partial\Omega}{\partial n}\,dS - \rho\kappa \iint \frac{\partial\Omega}{\partial n}\,d\sigma - \rho\kappa' \iint \frac{\partial\Omega}{\partial n}\,d\sigma' - \ldots,$$

where $\kappa, \kappa', \ldots$ are the cyclic constants of ϕ, and the first surface-integral is to be taken over the surfaces of the solids, and the remaining ones over the several barriers. By the conditions which determine Ω, this reduces to the first equation of the system:

$$\frac{\partial\mathbf{T}}{\partial\dot{\chi}} = \rho\kappa, \quad \frac{\partial\mathbf{T}}{\partial\dot{\chi}'} = \rho\kappa', \ldots \quad \text{.......................(3)}$$

These shew that $\rho\kappa, \rho\kappa', \ldots$ are to be regarded as the generalized components of momentum corresponding to the velocity-components $\dot{\chi}, \dot{\chi}', \ldots$, respectively.

We have recourse to the general Hamiltonian formula (1) of Art. 135. We will suppose that the varied motion of the solids is subject only to the condition that the initial and final configurations are to be the same as in the actual motion; also that the *initial* position of each particle of the fluid is the same in the two motions. The expression

$$\Sigma m(\dot{\xi}\Delta\xi + \dot{\eta}\Delta\eta + \dot{\zeta}\Delta\zeta) \quad \text{.........................(4)}$$

will accordingly vanish at time t_0, but not in general at time t_1, in the absence of further restrictions.

We will now suppose that the varied motion of the fluid is irrotational, and accordingly determined by the instantaneous values of the varied generalized co-ordinates and velocities. Considering the particles of the fluid alone, we have

$$\Sigma m\,(\dot\xi\Delta\xi+\dot\eta\Delta\eta+\dot\zeta\Delta\zeta)=-\rho\iiint\left(\frac{\partial\phi}{\partial x}\Delta\xi+\frac{\partial\phi}{\partial y}\Delta\eta+\frac{\partial\phi}{\partial z}\Delta\zeta\right)dx\,dy\,dz$$
$$=\rho\iint\phi\,(l\Delta\xi+m\Delta\eta+n\Delta\zeta)\,dS+\rho\kappa\iint(l\Delta\xi+m\Delta\eta+n\Delta\zeta)\,d\sigma$$
$$+\rho\kappa'\iint(l\Delta\xi+m\Delta\eta+n\Delta\zeta)\,d\sigma'+\ldots,\quad\ldots\ldots(5)$$

where l, m, n are the direction-cosines of the normal to an element of the bounding surface, drawn towards the fluid, or (as the case may be) of the normal to an element of a barrier, drawn in the direction in which the corresponding circulation is estimated.

At time t_1 we shall have

$$l\Delta\xi+m\Delta\eta+n\Delta\zeta=0$$

at the surface of the solids, as well as at the fixed boundaries. Again, if AB represent one of the barriers in its position at time t_1, and if $A'B'$ represent the locus at the same instant, in the varied motion, of those particles which in the actual motion occupy the position AB, the volume included between AB and $A'B'$ will be equal to the corresponding $\Delta\chi$, whence

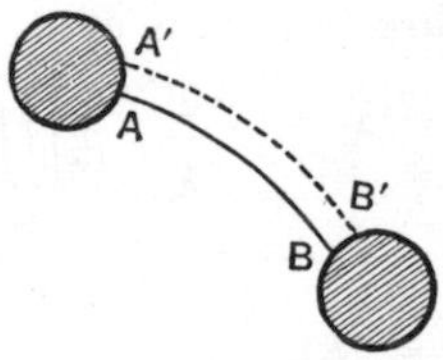

$$\left.\begin{aligned}&\iint(l\Delta\xi+m\Delta\eta+n\Delta\zeta)\,d\sigma=\Delta\chi,\\&\iint(l\Delta\xi+m\Delta\eta+n\Delta\zeta)\,d\sigma'=\Delta\chi',\\&\ldots\ldots\ldots\ldots\ldots\ldots\end{aligned}\right\}\quad\ldots\ldots\ldots\ldots(6)$$

The varied circulations are, from instant to instant, still at our disposal. We may suppose them to be so adjusted as to make $\Delta\chi$, $\Delta\chi'$, ... vanish at time t_1. The expression (4) will accordingly vanish, and if we further suppose that the external forces do on the whole no work when the boundary of the fluid is at rest, whatever relative displacements be given to the parts of the fluid, we have

$$\int_{t_0}^{t_1}\{\Delta T+Q_1\Delta q_1+Q_2\Delta q_2+\ldots+Q_n\Delta q_n\}\,dt=0,\quad\ldots\ldots\ldots(7)$$

s before.

By a partial integration, and remembering that by hypothesis

$$\Delta q_1,\ \Delta q_2,\ \ldots\ \Delta q_n,\ \Delta\chi,\ \Delta\chi',\ \ldots$$

vanish at the limits t_0, t_1, but are otherwise independent, we obtain n equations of the type

$$\frac{d}{dt}\frac{\partial T}{\partial \dot{q}_r} - \frac{\partial T}{\partial q_r} = Q_r, \qquad \text{(8)}$$

together with

$$\frac{d}{dt}\frac{\partial T}{\partial \dot{\chi}} = 0, \quad \frac{d}{dt}\frac{\partial T}{\partial \dot{\chi}'} = 0, \ldots \qquad \text{(9)}$$

140. Equations of the type (8) and (9) present themselves in various problems of ordinary Dynamics, *e.g.* in questions relating to gyrostats, where the co-ordinates $\chi, \chi', \ldots$, whose absolute values do not affect the kinetic or the potential energy of the system, are the angular co-ordinates of the gyrostats relative to their frames. The general theory of such systems has been treated by Routh*, Thomson and Tait†, and other writers.

We have seen that

$$\frac{\partial T}{\partial \dot{\chi}} = \rho\kappa, \quad \frac{\partial T}{\partial \dot{\chi}'} = \rho\kappa', \ldots, \qquad \text{(10)}$$

and the integration of (9) shews that the quantities $\kappa, \kappa', \ldots$ are constants with regard to the time, as is otherwise known (Art. 50). Let us write

$$R = T - \rho\kappa\dot{\chi} - \rho\kappa'\dot{\chi}' - \ldots. \qquad \text{(11)}$$

The equations (10), when written in full, determine $\dot{\chi}, \dot{\chi}', \ldots$ as linear functions of $\kappa, \kappa', \ldots$ and $\dot{q}_1, \dot{q}_2, \ldots \dot{q}_n$; and by substitution in (11) we can express R as a homogeneous quadratic function of the same quantities, with coefficients which of course in general involve the co-ordinates $q_1, q_2, \ldots q_n$. On this supposition we have, performing an arbitrary variation δ on both sides of (11), and omitting terms which cancel by (10),

$$\frac{\partial R}{\partial \dot{q}_1}\delta\dot{q}_1 + \ldots + \frac{\partial R}{\partial q_1}\delta q_1 + \ldots + \frac{\partial R}{\partial \kappa}\delta\kappa + \ldots$$
$$= \frac{\partial T}{\partial \dot{q}_1}\delta\dot{q}_1 + \ldots + \frac{\partial T}{\partial q_1}\delta q_1 + \ldots - \rho\dot{\chi}\delta\kappa - \ldots, \qquad \text{(12)}$$

where, for brevity, only one term of each kind is exhibited. Hence we obtain $2n$ equations of the types

$$\frac{\partial R}{\partial \dot{q}_r} = \frac{\partial T}{\partial \dot{q}_r}, \quad \frac{\partial R}{\partial q_r} = \frac{\partial T}{\partial q_r}, \qquad \text{(13)}$$

together with

$$\frac{\partial R}{\partial \kappa} = -\rho\dot{\chi}, \quad \frac{\partial R}{\partial \kappa'} = -\rho\dot{\chi}', \ldots \qquad \text{(14)}$$

Hence the equations (8) may be written

$$\frac{d}{dt}\frac{\partial R}{\partial \dot{q}_r} - \frac{\partial R}{\partial q_r} = Q_r, \qquad \text{(15)}$$

* *On the Stability of a Given State of Motion* (Adams Prize Essay), London, 1877; *Advanced Rigid Dynamics*, 6th ed., London, 1905.

† *Natural Philosophy*, 2nd ed., Art. 319 (1879). See also Helmholtz, "Principien der Statik monocyclischer Systeme," *Crelle*, xcvii. (1884) [*Wiss. Abh.* iii. 179]; Larmor, "On the Direct Application of the Principle of Least Action to the Dynamics of Solid and Fluid Systems," *Proc. Lond. Math. Soc.* (1) xv. (1884) [*Papers*, i. 31]; Basset, *Proc. Camb. Phil. Soc.* vi. 117 (1889).

where the velocities $\dot{\chi}, \dot{\chi}', \ldots$ corresponding to the 'ignored' co-ordinates $\chi, \chi', \ldots$ have now been eliminated*.

141. In order to shew more explicitly the nature of the modification introduced by the cyclic motions into the dynamical equations, we proceed as follows.

If we substitute in (11) from (14), we obtain

$$T = R - \left(\kappa \frac{\partial R}{\partial \kappa} + \kappa' \frac{\partial R}{\partial \kappa'} + \ldots\right). \quad \text{.....................(16)}$$

Now, remembering the composition of R, we may write for a moment

$$R = R_{2,0} + R_{1,1} + R_{0,2}, \quad \text{.......................(17)}$$

where $R_{2,0}$ is a homogeneous quadratic function of $\dot{q}_1, \dot{q}_2, \ldots \dot{q}_n$, $R_{0,2}$ is a homogeneous quadratic function of $\kappa, \kappa', \ldots$, and $R_{1,1}$ is bilinear in these two sets of variables. Hence (16) takes the form

$$T = R_{2,0} - R_{0,2}, \quad \text{.............................(18)}$$

or, as we shall henceforth write it,

$$T = \mathfrak{T} + K, \quad \text{.............................(19)}$$

where $\mathfrak{T}$ and K are homogeneous quadratic functions of $\dot{q}_1, \dot{q}_2, \ldots \dot{q}_n$, and of $\kappa, \kappa', \ldots$, respectively. It follows also from (17) that

$$R = \mathfrak{T} - K - \beta_1 \dot{q}_1 - \beta_2 \dot{q}_2 - \ldots - \beta_n \dot{q}_n, \quad \text{..............(20)}$$

where $\beta_1, \beta_2, \ldots$ are linear functions of $\kappa, \kappa', \ldots$, say

$$\left.\begin{aligned} \beta_1 &= \alpha_1 \kappa + \alpha_1' \kappa' + \ldots, \\ \beta_2 &= \alpha_1 \kappa + \alpha_2' \kappa' + \ldots, \\ &\ldots\ldots\ldots\ldots\ldots\ldots \\ \beta_n &= \alpha_n \kappa + \alpha_n' \kappa' + \ldots. \end{aligned}\right\} \quad \text{.......................(21)}$$

The meaning of the coefficients α (in the hydrodynamical problem) appears from (14) and (20). We find

$$\left.\begin{aligned} \rho \dot{\chi} &= \frac{\partial K}{\partial \kappa} + \alpha_1 \dot{q}_1 + \alpha_2 \dot{q}_2 + \ldots + \alpha_n \dot{q}_n, \\ \rho \dot{\chi}' &= \frac{\partial K}{\partial \kappa'} + \alpha_1' \dot{q}_1 + \alpha_2' \dot{q}_2 + \ldots + \alpha_n' \dot{q}_n, \\ &\ldots\ldots\ldots\ldots\ldots\ldots\ldots\ldots\ldots\ldots \end{aligned}\right\} \quad \text{..............(22)}$$

which shew that α_r is the contribution to the flux of matter across the first barrier due to unit rate of variation of the co-ordinate q_r, and so on.

If we now substitute from (20) in the equations (15) we obtain the general equations of motion of a 'gyrostatic system,' in the form†

* This investigation is due to Routh, *l.c.*; cf. Whittaker, *Analytical Dynamics*, Art. 38.

† These equations were first given in a paper by Sir W. Thomson, "On the Motion of Rigid Solids in a Liquid circulating irrotationally through perforations in them or in a Fixed Solid," *Phil. Mag.* (4) xlv. 332 (1873) [*Papers*, iv. 101]. See also C. Neumann, *Hydrodynamische Untersuchungen* (1883).

$$\left.\begin{aligned}
&\frac{d}{dt}\frac{\partial \mathfrak{T}}{\partial \dot{q}_1}-\frac{\partial \mathfrak{T}}{\partial q_1} \qquad\qquad\qquad + (1, 2)\, \dot{q}_2 + (1, 3)\, \dot{q}_3 + \ldots + (1, n)\, \dot{q}_n + \frac{\partial K}{\partial q_1} = Q_1,\\
&\frac{d}{dt}\frac{\partial \mathfrak{T}}{\partial \dot{q}_2}-\frac{\partial \mathfrak{T}}{\partial q_2} + (2, 1)\, \dot{q}_1 \qquad\qquad + (2, 3)\, \dot{q}_3 + \ldots + (2, n)\, \dot{q}_n + \frac{\partial K}{\partial q_2} = Q_2,\\
&\cdots\cdots\cdots\cdots\cdots\cdots\cdots\cdots\cdots\cdots\cdots\cdots\cdots\cdots\\
&\frac{d}{dt}\frac{\partial \mathfrak{T}}{\partial \dot{q}_n}-\frac{\partial \mathfrak{T}}{\partial q_n} + (n, 1)\, \dot{q}_1 + (n, 2)\, \dot{q}_2 + (n, 3)\, \dot{q}_3 + \ldots \qquad\qquad + \frac{\partial K}{\partial q_n} = Q_n,
\end{aligned}\right\} \quad \text{......(23)}$$

where

$$(r, s) = \frac{\partial \beta_s}{\partial q_r} - \frac{\partial \beta_r}{\partial q_s}. \quad \text{......(24)}$$

It is important to notice that $(r, s) = -(s, r)$, and $(r, r) = 0$.

If in the equations of motion of a fully-specified system of finite freedom (Art. 135 (4)) we reverse the sign of the time-element δt, the equations are unaltered. The motion is therefore reversible; that is to say, if as the system is passing through any assigned configuration the velocities $\dot{q}_1, \dot{q}_2, \ldots \dot{q}_n$ be all reversed, it will (if the forces be always the same in the same configuration) retrace its former path. It is important to observe that this statement does not in general hold of a gyrostatic system; thus, the terms in (23) which are linear in $\dot{q}_1, \dot{q}_2, \ldots \dot{q}_n$ change sign with δt, whilst the others do not. Hence, in the present application, the motion of the solids is not reversible, unless indeed we imagine the circulations $\kappa, \kappa', \ldots$ to be reversed simultaneously with the velocities $\dot{q}_1, \dot{q}_2, \ldots \dot{q}_n$*.

If we multiply the equations (23) by $\dot{q}_1, \dot{q}_2, \ldots \dot{q}_n$ in order, and add, we find, by an easy adaptation of the usual process,

$$\frac{d}{dt}(\mathfrak{T} + K) = Q_1\dot{q}_1 + Q_2\dot{q}_2 + \ldots + Q_n\dot{q}_n, \quad \text{......(25)}$$

or, if the system be conservative,

$$\mathfrak{T} + K + V = \text{const.} \quad \text{......(26)}$$

142. The results of Art. 141 may be applied to find the conditions of equilibrium of a system of solids surrounded by a liquid in cyclic motion. This problem of 'Kineto-Statics,' as it may be termed, is however more naturally treated by a simpler process.

The value of ϕ under the present circumstances can be expressed in the alternative forms

$$\phi = \dot{\chi}\Omega + \dot{\chi}'\Omega' + \ldots, \quad \text{......(1)}$$

$$\phi = \kappa\omega + \kappa'\omega' + \ldots; \quad \text{......(2)}$$

and the kinetic energy can accordingly be obtained as a homogeneous quadratic function either of $\dot{\chi}, \dot{\chi}', \ldots$, or of $\kappa, \kappa', \ldots$, with coefficients which are in each case functions of the co-ordinates $q_1, q_2, \ldots q_n$ which specify the

* Just as the motion of the axis of a top cannot be reversed unless we reverse the spin.

configuration of the solids. These two expressions for the energy may be distinguished by the symbols T_0 and K, respectively. Again, by Art. 55 (5) we have a third formula

$$2T = \rho\kappa\dot{\chi} + \rho\kappa'\dot{\chi}' + \dots \qquad (3)$$

The investigation at the beginning of Art. 139, shortened by the omission of the terms involving $\dot{q}_1, \dot{q}_2, \dots \dot{q}_n$, shews that

$$\rho\kappa = \frac{\partial T_0}{\partial \dot{\chi}}, \quad \rho\kappa' = \frac{\partial T_0}{\partial \dot{\chi}'}, \dots \qquad (4)$$

Again, the explicit formula for K is

$$2K = -\rho\kappa \iint \frac{\partial\phi}{\partial n}\, d\sigma - \rho\kappa' \iint \frac{\partial\phi}{\partial n}\, d\sigma' - \dots$$

$$= (\kappa, \kappa)\,\kappa^2 + (\kappa', \kappa')\,\kappa'^2 + \dots + 2\,(\kappa, \kappa')\,\kappa\kappa' + \dots, \qquad (5)$$

where

$$(\kappa, \kappa) = -\rho \iint \frac{\partial\omega}{\partial n}\, d\sigma, \quad (\kappa, \kappa') = -\rho \iint \frac{\partial\omega'}{\partial n}\, d\sigma = -\rho \iint \frac{\partial\omega}{\partial n}\, d\sigma', \dots (6)$$

and so on. Hence

$$\frac{\partial K}{\partial \kappa} = (\kappa, \kappa)\,\kappa + (\kappa, \kappa')\,\kappa' + \dots = -\rho \iint \frac{\partial\phi}{\partial n}\, d\sigma.$$

We thus obtain

$$\rho\dot{\chi} = \frac{\partial K}{\partial \kappa}, \quad \rho\dot{\chi}' = \frac{\partial K}{\partial \kappa'}, \dots \qquad (7)$$

Again, writing $T_0 + K$ for $2T$ in (3), and performing a total variation δ on both sides of the resulting identity, we find, on omitting terms which cancel in virtue of (4) and (7)*,

$$\frac{\partial T_0}{\partial q_r} + \frac{\partial K}{\partial q_r} = 0. \qquad (8)$$

This completes the requisite analytical formulae†.

If we now imagine the solids to be guided from rest in the configuration $(q_1, q_2, \dots q_n)$ to rest in an adjacent configuration

$$(q_1 + \Delta q_1,\ q_2 + \Delta q_2,\ \dots\ q_n + \Delta q_n),$$

the work required is $\quad Q_1 \Delta q_1 + Q_2 \Delta q_2 + \dots + Q_n \Delta q_n,$

where $Q_1, Q_2, \dots Q_n$ are the components of extraneous force which have to be applied to neutralize the pressures of the fluid on the solids. This must be equal to the increment ΔK of the kinetic energy, calculated on the supposition that the circulations $\kappa, \kappa', \dots$ are constant. Hence

$$Q_r = \frac{\partial K}{\partial q_r}. \qquad (9)$$

* It would be sufficient to assume *either* (4) *or* (7); the process then leads to an independent proof of the other set of formulae.

† It may be noted that the function R of Art. 140 now reduces to $-K$.

The forces representing the pressures of the fluid on the solids (when these are held at rest) are obtained by reversing the signs, viz. they are given by

$$Q_r' = -\frac{\partial K}{\partial q_r}; \quad \text{.............................(10)}$$

the solids therefore tend to move so that the kinetic energy of the cyclic motion diminishes.

In virtue of (8) we have, also,

$$Q_r' = \frac{\partial T_0}{\partial q_r}. \quad \text{.............................(11)}$$

143. The formula (19) of Art. 141 may be applied to find approximate expressions for the forces on a solid immersed in a non-uniform stream*.

Suppose we have a solid maintained at rest in a cyclic region in which a fluid is circulating irrotationally, and let K be the energy of the *fluid*, which will of course vary with the position of the solid. We will suppose the dimensions of the latter to be so small compared with the distances from the walls of the region that its position may be sufficiently given by point-co-ordinates (x, y, z). We have, then, for the components of the force exerted on it by the pressures of the fluid,

$$\mathbf{X} = -\frac{\partial K}{\partial x}, \quad \mathbf{Y} = -\frac{\partial K}{\partial y}, \quad \mathbf{Z} = -\frac{\partial K}{\partial z}. \quad \text{..................(1)}$$

It remains to find, approximately, the form of this function K of x, y, z. Let (u, v, w) be the velocity which the fluid would have at (x, y, z) if the solid were absent. If the solid were made to move with this velocity, and were of the same density as the surrounding fluid, the energy would be approximately the same as if the whole were fluid. It follows from Art. 141 (19) that in this case the energy of the fluid would be $\mathfrak{T} + K$, where

$$2\mathfrak{T} = \mathbf{A}u^2 + \mathbf{B}v^2 + \mathbf{C}w^2 + 2\mathbf{A}'vw + 2\mathbf{B}'wu + 2\mathbf{C}'uv, \quad \text{........(2)}$$

by Art. 124, and that of the solid would be

$$\tfrac{1}{2}\rho Q\,(u^2 + v^2 + w^2), \quad \text{.............................(3)}$$

where Q is the volume displaced. The expression

$$\mathfrak{T} + \tfrac{1}{2}\rho Q\,(u^2 + v^2 + w^2) + K \quad \text{.......................(4)}$$

has therefore a constant value, viz. that of the energy of a fluid filling the region, and having the given circulations. This determines the form of K.

Hence

$$\left.\begin{aligned} \mathbf{X} &= \frac{\partial \mathfrak{T}}{\partial x} + \tfrac{1}{2}\rho Q \frac{\partial}{\partial x}(u^2 + v^2 + w^2), \\ \mathbf{Y} &= \frac{\partial \mathfrak{T}}{\partial y} + \tfrac{1}{2}\rho Q \frac{\partial}{\partial y}(u^2 + v^2 + w^2), \\ \mathbf{Z} &= \frac{\partial \mathfrak{T}}{\partial z} + \tfrac{1}{2}\rho Q \frac{\partial}{\partial z}(u^2 + v^2 + w^2). \end{aligned}\right\} \quad \text{..................(5)}$$

* G. I. Taylor, "The Forces on a Body placed in a Curved or Converging Stream of Fluid," *Proc. Roy. Soc.* cxx. 260 (1928).

Since the forces on the solid must depend only on the motion of the fluid in the immediate neighbourhood, these expressions are general, and independent of the special conception employed in their derivation.

If the direction of the undisturbed stream, near the solid, be taken as the axis of x, the results simplify. Putting $v=0$, $w=0$, we have

$$\left.\begin{aligned}\mathbf{X}&=\left\{(\mathbf{A}+\rho Q)\frac{\partial u}{\partial x}+\mathbf{B}'\frac{\partial w}{\partial x}+\mathbf{C}'\frac{\partial v}{\partial x}\right\}u,\\ \mathbf{Y}&=\left\{(\mathbf{A}+\rho Q)\frac{\partial u}{\partial y}+\mathbf{B}'\frac{\partial w}{\partial y}+\mathbf{C}'\frac{\partial v}{\partial y}\right\}u,\\ \mathbf{Z}&=\left\{(\mathbf{A}+\rho Q)\frac{\partial u}{\partial z}+\mathbf{B}'\frac{\partial w}{\partial z}+\mathbf{C}'\frac{\partial v}{\partial z}\right\}u.\end{aligned}\right\} \qquad (6)$$

If, further, the stream is symmetrical with respect to the planes $y=0$, $z=0$ we have $\partial u/\partial y=0$, $\partial u/\partial z=0$, and therefore also $\partial v/\partial x=0$, $\partial w/\partial x=0$, on account of the assumed irrotational character. The symmetry also requires $\partial w/\partial y=\partial v/\partial z=0$. Hence

$$\left.\begin{aligned}\mathbf{X}&=(\mathbf{A}+\rho Q)\,u\frac{\partial u}{\partial x},\\ \mathbf{Y}&=\mathbf{C}'u\frac{\partial v}{\partial y},\\ \mathbf{Z}&=\mathbf{B}'u\frac{\partial w}{\partial z}.\end{aligned}\right\} \qquad (7)$$

First suppose that one of the axes of permanent translation (Art. 124) coincides with the direction of the stream. Then $\mathbf{C}'=0$, $\mathbf{B}'=0$, and

$$\mathbf{X}=(\mathbf{A}+\rho Q)f,\qquad \mathbf{Y}=0,\qquad \mathbf{Z}=0, \qquad (8)$$

where f is the acceleration in the undisturbed stream. Thus if the solid is spherical, $\mathbf{A}=\frac{2}{3}\pi\rho a^3$, $Q=\frac{4}{3}\pi a^3$, $\mathbf{X}=2\pi\rho a^3 f$. For a circular cylinder, reckoning per unit length,

$$\mathbf{A}=\pi\rho a^2,\ Q=\pi\rho a^2,\ \mathbf{X}=2\pi\rho a^2 f.$$

Next suppose merely that two of the axes of permanent translation lie in a plane with the direction of the stream. If the plane in question be that of xy we have $\mathbf{A}'=0$, $\mathbf{B}'=0$. If the stream is symmetrical about the axis of x, we have, further,

$$\frac{\partial v}{\partial y}=\frac{\partial w}{\partial z}=-\frac{1}{2}\frac{\partial u}{\partial x},$$

and the forces reduce to

$$\mathbf{X}=(\mathbf{A}+\rho Q)f,\qquad \mathbf{Y}=\tfrac{1}{2}\mathbf{C}'f,\qquad \mathbf{Z}=0. \qquad (9)$$

In the case of a circular disk,

$$\mathbf{A}=\tfrac{8}{3}\rho a^3\cos^2\alpha,\qquad \mathbf{C}'=-\tfrac{8}{3}\rho a^3\sin\alpha\cos\alpha,\qquad Q=0,$$

where α is the angle which the stream makes with the axis of symmetry. In the two-dimensional case of the elliptic cylinder,

$$\mathbf{A}=\pi\rho\,(b^2\cos^2\alpha+a^2\sin^2\alpha),\qquad \mathbf{C}'=\pi\rho\,(a^2-b^2)\sin\alpha\cos\alpha,\qquad Q=\pi ab,$$

where α is now the inclination of the stream to the major axis*.

The above theory has an interest in connection with the 'pressure-drop' in a wind-channel, as used for measuring the drag of aircraft models. The stream of air converges slightly towards the fan at the forward end of the tunnel, and the increase of velocity implies a fall of pressure. We have then

$$\rho f=-\frac{\partial p}{\partial x}. \qquad (10)$$

* These particular cases have been verified by direct calculation of the effect of the fluid pressures: *Aeronautical Research Committee*, R. and M. 1164 (1928).

The preceding formulae shew that it would be incorrect to calculate the value of $\mathbf{X}$ from the observed pressure-gradient as if it were a statical question, in which case we should have $\mathbf{X} = \rho Q f$ simply*.

Some further interesting examples of Kineto-Statics (not reproduced in the present edition) have been discussed by Sir W. Thomson†, Kirchhoff‡, and Boltzmann§.

144. We here take leave of this branch of our subject. To avoid, as far as may be, the suspicion of vagueness which sometimes attaches to the use of 'generalized co-ordinates,' an attempt has been made in this Chapter to put the question on as definite a basis as possible, even at the expense of some degree of prolixity in the methods.

To some writers‖ the matter has presented itself as a much simpler one. The problems are brought at one stroke under the sway of the ordinary formulae of Dynamics by the imagined introduction of an infinite number of 'ignored co-ordinates,' which would specify the configuration of the various particles of the fluid. The corresponding components of momentum are assumed all to vanish, with the exception (in the case of a cyclic region) of those which are represented by the circulations through the several apertures.

From a physical point of view it is difficult to refuse assent to such a generalization, especially when it has formed the starting-point of all the development of this part of the subject; but it is at least legitimate, and from the hydrodynamical standpoint even desirable, that it should be verified *à posteriori* by independent, if more pedestrian, methods.

Whichever procedure be accepted, the result is that the systems contemplated in this Chapter are found to comport themselves (so far as the 'palpable' co-ordinates $q_1, q_2, \ldots q_n$ are concerned) exactly like ordinary systems of finite freedom. The further development of the general theory belongs to Analytical Dynamics, and must accordingly be sought for in books and memoirs devoted to that subject. It may be worth while, however, to remark that the hydrodynamical systems afford extremely interesting and beautiful illustrations of the Principle of Least Action, the Reciprocal Theorems of Helmholtz, and other general dynamical theories.

* G. I. Taylor, *l.c.*

† "On the Forces experienced by Solids immersed in a Moving Liquid," *Proc. R. S. Edin.* 1870 [*Reprint*, Art. xli.].

‡ *l.c. ante* p. 54.

§ "Ueber die Druckkräfte welche auf Ringe wirksam sind die in eine bewegte Flüssigkeit tauchen," *Crelle*, lxxiii. (1871) [*Wiss. Abh.* i. 200].

‖ See Thomson and Tait, and Larmor, *ll. cit. ante* p. 195.

CHAPTER VII

VORTEX MOTION

145. OUR investigations have thus far been confined for the most part to the case of irrotational motion. We now proceed to the study of rotational or 'vortex' motion. This subject was first investigated by Helmholtz*; other and simpler proofs of some of his theorems were afterwards given by Kelvin in the paper on vortex motion already cited in Chapter III.

We shall, throughout this Chapter, use the symbols ξ, η, ζ to denote, as in Chapter III., the components of vorticity, viz.

$$\xi = \frac{\partial w}{\partial y} - \frac{\partial v}{\partial z}, \qquad \eta = \frac{\partial u}{\partial z} - \frac{\partial w}{\partial x}, \qquad \zeta = \frac{\partial v}{\partial x} - \frac{\partial u}{\partial y}. \quad \text{...........(1)}$$

A line drawn from point to point so that its direction is everywhere that of the instantaneous axis of rotation of the fluid is called a 'vortex-line.' The differential equations of the system of vortex-lines are

$$\frac{dx}{\xi} = \frac{dy}{\eta} = \frac{dz}{\zeta}. \quad \text{..............................(2)}$$

If through every point of a small closed curve we draw the corresponding vortex-line, we mark out a tube, which we call a 'vortex-tube.' The fluid contained within such a tube constitutes what is called a 'vortex-filament,' or simply a 'vortex.'

Let ABC, $A'B'C'$ be any two circuits drawn on the surface of a vortex-tube and embracing it, and let AA' be a connecting line also drawn on the surface. Let us apply the theorem of Art. 32 to the circuit $ABCAA'C'B'A'A$ and the part of the surface of the tube bounded by it. Since

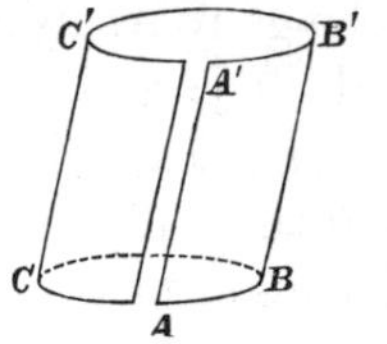

$$l\xi + m\eta + n\zeta = 0$$

at every point of this surface, the line-integral

$$\int (u\,dx + v\,dy + w\,dz),$$

taken round the circuit, must vanish; *i.e.* in the notation of Art. 31

$$I(ABCA) + I(AA') + I(A'C'B'A') + I(A'A) = 0,$$

which reduces to $$I(ABCA) = I(A'B'C'A').$$

Hence the circulation is the same in all circuits embracing the same vortex-tube.

* "Ueber Integrale der hydrodynamischen Gleichungen welche den Wirbelbewegungen entsprechen," *Crelle*, lv. (1858) [*Wiss. Abh.* i. 101].

Again, it appears from Art. 31 that the circulation round the boundary of any cross-section of the tube, made normal to its length, is $\omega\sigma$, where $\omega, = (\xi^2+\eta^2+\zeta^2)^{\frac{1}{2}}$, is the resultant vorticity of the fluid, and σ the infinitely small area of the section.

Combining these results we see that the product of the vorticity into the cross-section is the same at all points of a vortex. This product is conveniently taken as a measure of the 'strength' of the vortex*.

The foregoing proof is due to Kelvin; the theorem itself was first given by Helmholtz, as a deduction from the relation

$$\frac{\partial \xi}{\partial x}+\frac{\partial \eta}{\partial y}+\frac{\partial \zeta}{\partial z}=0, \qquad (3)$$

which follows at once from the values of ξ, η, ζ given by (1). In fact writing, in Art. 42 (1), ξ, η, ζ for U, V, W, respectively, we find

$$\iint (l\xi+m\eta+n\zeta)\,dS=0, \qquad (4)$$

where the integration extends over any closed surface lying wholly in the fluid. Applying this to the closed surface formed by two cross-sections of a vortex-tube and the part of the walls intercepted between them, we find $\omega_1\sigma_1=\omega_2\sigma_2$, where ω_1, ω_2 denote the vorticities at the sections σ_1, σ_2, respectively.

Kelvin's proof shews that the theorem is true even when ξ, η, ζ are discontinuous (in which case there may be an abrupt bend at some point of a vortex), provided only that u, v, w are continuous.

An important consequence of the above theorem is that a vortex-line cannot begin or end at any point in the interior of the fluid. Any vortex-lines which exist must either form closed curves, or else traverse the fluid, beginning and ending on its boundaries. Compare Art. 36.

The theorem of Art. 32 (3) may now be enunciated as follows: The circulation in any circuit is equal to the sum of the strengths of all the vortices which it embraces.

146. It was proved in Art. 33 that in a perfect fluid whose density is either uniform or a function of the pressure only, and which is subject to forces having a single-valued potential, the circulation in any circuit moving with the fluid is constant.

Applying this theorem to a circuit embracing a vortex-tube we find that the strength of any vortex is constant.

If we take at any instant a surface composed wholly of vortex-lines, the circulation in any circuit drawn on it is zero, by Art. 32, for we have $l\xi+m\eta+n\zeta=0$ at every point of the surface. The preceding Art. shews that if the surface be now supposed to move with the fluid, the circulation will always be zero in any circuit drawn on it, and therefore the surface will

* The *circulation* round a vortex being the most natural measure of its intensity.

always consist of vortex-lines. Again, considering two such surfaces, it is plain that their intersection must always be a vortex-line, whence we derive the theorem that the vortex-lines move with the fluid.

This remarkable theorem was first given by Helmholtz for the case of incompressibility; the preceding proof, by Kelvin, shews that it holds for all fluids subject to the conditions above stated.

The theorem that the circulation in any circuit moving with the fluid is invariable constitutes the sole and sufficient appeal to Dynamics which it is necessary to make in the investigations of this Chapter. It is based on the hypothesis of a continuous distribution of pressure, and (conversely) implies this. For if in any problem we have discovered functions u, v, w of x, y, z, t which satisfy the kinematical conditions, then, if this solution is to be also *dynamically* possible, the relation between the pressures about two moving particles A, B must be given by the formula (2) of Art. 33, viz.

$$\left[\int \frac{dp}{\rho} + \Omega - \tfrac{1}{2}q^2\right]_A^B = -\frac{D}{Dt}\int_A^B (u\,dx + v\,dy + w\,dz). \quad \text{..........(1)}$$

It is therefore necessary and sufficient that the expression on the right-hand side should be the same for all paths of integration (moving with the fluid) which can be drawn from A to B. This is secured if, and only if, the assumed values of u, v, w make the vortex-lines move with the fluid, and also make the strength of every vortex constant with respect to the time.

It is easily seen that the argument is in no way impaired if the assumed values of u, v, w make ξ, η, ζ discontinuous at certain surfaces, provided only that u, v, w are themselves everywhere continuous.

On account of their historical interest, one or two independent proofs of the preceding theorems may be briefly indicated, and their mutual relations pointed out.

Of these proofs, perhaps the most conclusive is based upon a slight generalization of some equations given originally by Cauchy in the introduction to his great memoir on Waves*, and employed by him to demonstrate Lagrange's velocity-potential theorem.

The equations (2) of Art. 15 yield, on elimination of the function χ by cross-differentiation,

$$\frac{\partial u}{\partial b}\frac{\partial x}{\partial c} - \frac{\partial u}{\partial c}\frac{\partial x}{\partial b} + \frac{\partial v}{\partial b}\frac{\partial y}{\partial c} - \frac{\partial v}{\partial c}\frac{\partial y}{\partial b} + \frac{\partial w}{\partial b}\frac{\partial z}{\partial c} - \frac{\partial w}{\partial c}\frac{\partial z}{\partial b} = \frac{\partial w_0}{\partial b} - \frac{\partial v_0}{\partial c}$$

(where u, v, w have been written in place of $\partial x/\partial t$, $\partial y/\partial t$, $\partial z/\partial t$, respectively), with two symmetrical equations. If in these equations we replace the differential coefficients of u, v, w with respect to a, b, c, by their values in terms of differential coefficients of the same quantities with respect to x, y, z, we obtain

$$\left.\begin{aligned}
\xi\frac{\partial(y,z)}{\partial(b,c)} + \eta\frac{\partial(z,x)}{\partial(b,c)} + \zeta\frac{\partial(x,y)}{\partial(b,c)} &= \xi_0,\\
\xi\frac{\partial(y,z)}{\partial(c,a)} + \eta\frac{\partial(z,x)}{\partial(c,a)} + \zeta\frac{\partial(x,y)}{\partial(c,a)} &= \eta_0,\\
\xi\frac{\partial(y,z)}{\partial(a,b)} + \eta\frac{\partial(z,x)}{\partial(a,b)} + \zeta\frac{\partial(x,y)}{\partial(a,b)} &= \zeta_0.
\end{aligned}\right\} \quad \text{..........................(2)}$$

* *l.c. ante* p. 17.

If we multiply these by $\partial x/\partial a$, $\partial x/\partial b$, $\partial x/\partial c$, in order, and add, then, taking account of the Lagrangian equation of continuity (Art. 14 (1)) we deduce the first of the following three symmetrical equations:

$$\left.\begin{aligned} \frac{\xi}{\rho} &= \frac{\xi_0}{\rho_0}\frac{\partial x}{\partial a} + \frac{\eta_0}{\rho_0}\frac{\partial x}{\partial b} + \frac{\zeta_0}{\rho_0}\frac{\partial x}{\partial c}, \\ \frac{\eta}{\rho} &= \frac{\xi_0}{\rho_0}\frac{\partial y}{\partial a} + \frac{\eta_0}{\rho_0}\frac{\partial y}{\partial b} + \frac{\zeta_0}{\rho_0}\frac{\partial y}{\partial c}, \\ \frac{\zeta}{\rho} &= \frac{\xi_0}{\rho_0}\frac{\partial z}{\partial a} + \frac{\eta_0}{\rho_0}\frac{\partial z}{\partial b} + \frac{\zeta_0}{\rho_0}\frac{\partial z}{\partial c}. \end{aligned}\right\} \quad \text{.................................(3)}$$

In the particular case of an incompressible fluid ($\rho=\rho_0$) these differ only in the use of the notation ξ, η, ζ from the equations given by Cauchy. They shew at once that if the initial values ξ_0, η_0, ζ_0 of the component vorticities vanish for any particle of the fluid, then ξ, η, ζ are always zero for that particle. This constitutes in fact Cauchy's proof of Lagrange's theorem.

To interpret (3) in the general case, let us take at time $t=0$ a linear element coincident with a vortex-line, say

$$\delta a,\ \delta b,\ \delta c = \epsilon\frac{\xi_0}{\rho_0},\ \epsilon\frac{\eta_0}{\rho_0},\ \epsilon\frac{\zeta_0}{\rho_0},$$

where ϵ is infinitesimal. If we suppose this element to move with the fluid the equations (3) shew that its projections on the co-ordinate axes at any other time will be given by

$$\delta x,\ \delta y,\ \delta z = \epsilon\frac{\xi}{\rho},\ \epsilon\frac{\eta}{\rho},\ \epsilon\frac{\zeta}{\rho},$$

i.e. the element will still form part of a vortex line, and its length (δs, say) will vary as ω/ρ, where ω is the resultant vorticity. But if σ be the cross-section of a vortex-filament having δs as axis, the product $\rho\sigma\delta s$ is constant with regard to the time. Hence the strength $\omega\sigma$ of the vortex is constant*.

The proof given originally by Helmholtz depends on a system of three equations which, when generalized so as to apply to any fluid in which ρ is a function of p only, become†

$$\left.\begin{aligned} \frac{D}{Dt}\left(\frac{\xi}{\rho}\right) &= \frac{\xi}{\rho}\frac{\partial u}{\partial x} + \frac{\eta}{\rho}\frac{\partial u}{\partial y} + \frac{\zeta}{\rho}\frac{\partial u}{\partial z}, \\ \frac{D}{Dt}\left(\frac{\eta}{\rho}\right) &= \frac{\xi}{\rho}\frac{\partial v}{\partial x} + \frac{\eta}{\rho}\frac{\partial v}{\partial y} + \frac{\zeta}{\rho}\frac{\partial v}{\partial z}, \\ \frac{D}{Dt}\left(\frac{\zeta}{\rho}\right) &= \frac{\xi}{\rho}\frac{\partial w}{\partial x} + \frac{\eta}{\rho}\frac{\partial w}{\partial y} + \frac{\zeta}{\rho}\frac{\partial w}{\partial z}. \end{aligned}\right\} \quad \text{.................................(4)}$$

These may be obtained as follows. The dynamical equations of Art. 6 may be written, when a force-potential Ω exists, in the forms

$$\left.\begin{aligned} \frac{\partial u}{\partial t} - v\zeta + w\eta &= -\frac{\partial \chi'}{\partial x}, \\ \frac{\partial v}{\partial t} - w\xi + u\zeta &= -\frac{\partial \chi'}{\partial y}, \\ \frac{\partial w}{\partial t} - u\eta + v\xi &= -\frac{\partial \chi'}{\partial z}, \end{aligned}\right\} \quad \text{.................................(5)}$$

provided

$$\chi' = \int\frac{dp}{\rho} + \tfrac{1}{2}q^2 + \Omega, \quad \text{.......................................(6)}$$

* See Nanson, *Mess. of Math.* iii. 120 (1874); Kirchhoff, *Mechanik*, c. xv. (1876); Stokes, *Papers*, ii. 47 (1883).

† Nanson, *l.c.*

where $q^2=u^2+v^2+w^2$. From the second and third of these we obtain, eliminating χ' by cross-differentiation,

$$\frac{\partial\xi}{\partial t}+v\frac{\partial\xi}{\partial y}+w\frac{\partial\xi}{\partial z}-u\left(\frac{\partial\eta}{\partial y}+\frac{\partial\zeta}{\partial z}\right)=\eta\frac{\partial u}{\partial y}+\zeta\frac{\partial u}{\partial z}-\xi\left(\frac{\partial v}{\partial y}+\frac{\partial w}{\partial z}\right).$$

Remembering the relation $$\frac{\partial\xi}{\partial x}+\frac{\partial\eta}{\partial y}+\frac{\partial\zeta}{\partial z}=0, \qquad\qquad (7)$$

and the equation of continuity

$$\frac{D\rho}{Dt}+\rho\left(\frac{\partial u}{\partial x}+\frac{\partial v}{\partial y}+\frac{\partial w}{\partial z}\right)=0, \qquad\qquad (8)$$

we easily deduce the first of equations (4).

To interpret these equations we take, at time t, a linear element whose projections on the co-ordinate axes are

$$\delta x,\ \delta y,\ \delta z=\epsilon\frac{\xi}{\rho},\ \epsilon\frac{\eta}{\rho},\ \epsilon\frac{\zeta}{\rho}, \qquad\qquad (9)$$

where ϵ is infinitesimal. If this element be supposed to move with the fluid, the rate at which δx is increasing is equal to the difference of the values of u at the two ends, whence

$$\frac{D\delta x}{Dt}=\epsilon\frac{\xi}{\rho}\frac{\partial u}{\partial x}+\epsilon\frac{\eta}{\rho}\frac{\partial u}{\partial y}+\epsilon\frac{\zeta}{\rho}\frac{\partial u}{\partial z}.$$

It follows, by (4), that

$$\frac{D}{Dt}\left(\delta x-\epsilon\frac{\xi}{\rho}\right)=0,\qquad \frac{D}{Dt}\left(\delta y-\epsilon\frac{\eta}{\rho}\right)=0,\qquad \frac{D}{Dt}\left(\delta z-\epsilon\frac{\zeta}{\rho}\right)=0. \qquad\qquad (10)$$

Helmholtz concludes that if the relations (9) hold at time t, they will hold at time $t+\delta t$, and so on, continually. The inference is, however, not quite rigorous; it is in fact open to the criticisms which Stokes* directed against various defective proofs of Lagrange's velocity-potential theorem†.

By way of establishing a connection with Kelvin's investigation we may notice that the equations (2) express that the circulation is constant in each of three infinitely small circuits initially perpendicular, respectively, to the three co-ordinate axes. Taking, for example, the circuit which initially bounded the rectangle $\delta b\,\delta c$, and denoting by A, B, C the areas of its projections at time t on the co-ordinate planes, we have

$$A=\frac{\partial(y,z)}{\partial(b,c)}\delta b\,\delta c,\qquad B=\frac{\partial(z,x)}{\partial(b,c)}\delta b\,\delta c,\qquad C=\frac{\partial(x,y)}{\partial(b,c)}\delta b\,\delta c,$$

so that the first of the equations referred to is equivalent‡ to

$$\xi A+\eta B+\zeta C=\xi_0\,\delta b\,\delta c. \qquad\qquad (11)$$

As an application of the equations (4) we may consider the motion of a liquid of uniform vorticity contained in a fixed ellipsoidal vessel§. The formulae

$$u=qz-ry,\qquad v=rx-pz,\qquad w=py-qx \qquad\qquad (12)$$

* *l.c. ante* p. 17.

† It may be mentioned that, in the case of an incompressible fluid, equations somewhat similar to (4) had been established by Lagrange, *Miscell. Taur.* ii. (1760) [*Oeuvres*, i. 442]. The author is indebted for this reference, and for the above remark on Helmholtz' investigation, to Sir J. Larmor. Equations equivalent to those given by Lagrange were obtained independently by Stokes, *l.c.*, and made the basis of a rigorous proof of the velocity-potential theorem.

‡ Nanson, *Mess. of Math.* vii. 182 (1878). A similar interpretation of Helmholtz' equations was given by the author of this work in the *Mess. of Math.* vii. 41 (1877).

Finally it may be noted that another proof of Lagrange's theorem, based on elementary dynamical principles, without special reference to the hydrokinetic equations, was indicated by Stokes, *Camb. Trans.* viii. [*Papers*, i. 113], and carried out by Kelvin in his paper on Vortex Motion.

§ Cf. Voigt, "Beiträge zur Hydrodynamik," *Gött. Nachr.* 1891, p. 71; Tedone, *Nuovo Cimento*, xxxiii. (1893). The artifice in the text is taken from Poincaré, "Sur la précession des corps déformables," *Bull. Astr.* 1910.

obviously represent a uniform rotation of the fluid as a solid within a spherical boundary. Transforming the co-ordinates and the corresponding velocities by homogeneous strain we obtain the formulae

$$\frac{u}{a}=\frac{qz}{c}-\frac{ry}{b},\quad \frac{v}{b}=\frac{rx}{a}-\frac{pz}{c},\quad \frac{w}{c}=\frac{py}{b}-\frac{qx}{a}, \qquad\ldots\ldots(13)$$

as representing a certain motion within a fixed ellipsoidal boundary

$$\frac{x^2}{a^2}+\frac{y^2}{b^2}+\frac{z^2}{c^2}=1. \qquad\ldots\ldots(14)$$

These make $$\xi=\left(\frac{b}{c}+\frac{c}{b}\right)p,\quad \eta=\left(\frac{c}{a}+\frac{a}{c}\right)q,\quad \zeta=\left(\frac{a}{b}+\frac{b}{a}\right)r. \qquad\ldots\ldots(15)$$

Substituting in (4) we obtain

$$(b^2+c^2)\frac{dp}{dt}=(b^2-c^2)\,qr, \qquad\ldots\ldots(16)$$

which may be written

$$a^2(b^2+c^2)\frac{dp}{dt}=\{b^2(c^2+a^2)-c^2(a^2+b^2)\}\,qr, \qquad\ldots\ldots(17)$$

with two similar equations. We have here an identity as to form with Euler's equations of free motion of a solid about a fixed point. We easily deduce the integrals

$$\frac{\xi^2}{a^2}+\frac{\eta^2}{b^2}+\frac{\zeta^2}{c^2}=\text{const.}, \qquad\ldots\ldots(18)$$

and $$\frac{b^2c^2\xi^2}{b^2+c^2}+\frac{c^2a^2\eta^2}{c^2+a^2}+\frac{a^2b^2\zeta^2}{a^2+b^2}=\text{const.}, \qquad\ldots\ldots(19)$$

the former of which is a verification of one of Helmholtz' theorems, whilst the latter follows from the constancy of the energy.

147. It is easily seen by the same kind of argument as in Art. 41 that no continuous irrotational motion is possible in an incompressible fluid filling infinite space, and subject to the condition that the velocity vanishes at infinity. This leads at once to the following theorem:

The motion of a fluid which fills infinite space, and is at rest at infinity, is determinate when we know the values of the expansion (θ, say) and of the component vorticities ξ, η, ζ, at all points of the region.

For, if possible, let there be two sets of values, u_1, v_1, w_1, and u_2, v_2, w_2, of the component velocities, each satisfying the equations

$$\frac{\partial u}{\partial x}+\frac{\partial v}{\partial y}+\frac{\partial w}{\partial z}=\theta, \qquad\ldots\ldots(1)$$

$$\frac{\partial w}{\partial y}-\frac{\partial v}{\partial z}=\xi,\qquad \frac{\partial u}{\partial z}-\frac{\partial w}{\partial x}=\eta,\qquad \frac{\partial v}{\partial x}-\frac{\partial u}{\partial y}=\zeta, \qquad\ldots\ldots(2)$$

throughout infinite space, and vanishing at infinity. The quantities

$$u'=u_1-u_2,\qquad v'=v_1-v_2,\qquad w'=w_1-w_2$$

will satisfy (1) and (2) with $\theta, \xi, \eta, \zeta=0$, and will vanish at infinity. Hence, in virtue of the result above stated, they will everywhere vanish, and there is only one possible motion satisfying the given conditions.

In the same way we can shew that the motion of a fluid occupying any *limited* simply-connected region is determinate when we know the values of the expansion, and of the component vorticities, at every point of the region, and the value of the normal velocity at every point of the boundary. In the case of an n-ply-connected region we must add to the above data the values of the circulations in n several independent circuits of the region.

148. If, in the case of infinite space, the quantities θ, ξ, η, ζ all vanish beyond some finite distance of the origin, the complete determination of u, v, w in terms of them can be effected as follows*.

The component velocities due to the *expansion* can be written down at once from Art. 56 (1), it being evident that the expansion θ' in an element $\delta x' \delta y' \delta z'$ is equivalent to a simple source of strength $\theta' \delta x' \delta y' \delta z'$. We thus obtain

$$u = -\frac{\partial \Phi}{\partial x}, \qquad v = -\frac{\partial \Phi}{\partial y}, \qquad w = -\frac{\partial \Phi}{\partial z}, \quad \ldots\ldots\ldots\ldots\ldots(1)$$

where

$$\Phi = \frac{1}{4\pi}\iiint \frac{\theta'}{r}\, dx'\, dy'\, dz', \quad \ldots\ldots\ldots\ldots\ldots\ldots\ldots\ldots\ldots(2)$$

r denoting the distance between the point (x', y', z') at which the volume-element of the integral is situate and the point (x, y, z) at which the values of u, v, w are required, viz.

$$r = \{(x - x')^2 + (y - y')^2 + (z - z')^2\}^{\frac{1}{2}}.$$

The integration includes all parts of space at which θ' differs from zero.

To find the velocities due to the *vortices*, we note that when there is no expansion, the flux across any two open surfaces bounded by the same curve as edge will be the same, and will therefore be determined solely by the configuration of the edge. This suggests that the flux through any closed curve may be expressed as a line-integral taken round the curve, say

$$\int (F dx + G dy + H dz). \quad \ldots\ldots\ldots\ldots\ldots\ldots\ldots\ldots\ldots(3)$$

On this hypothesis we should have, by the method of Art. 31,

$$u = \frac{\partial H}{\partial y} - \frac{\partial G}{\partial z}, \qquad v = \frac{\partial F}{\partial z} - \frac{\partial H}{\partial x}, \qquad w = \frac{\partial G}{\partial x} - \frac{\partial F}{\partial y}. \quad \ldots\ldots\ldots(4)$$

It is necessary and (as we have seen) sufficient that the functions F, G, H should satisfy

$$\frac{\partial w}{\partial y} - \frac{\partial v}{\partial z} = \frac{\partial}{\partial x}\left(\frac{\partial F}{\partial x} + \frac{\partial G}{\partial y} + \frac{\partial H}{\partial z}\right) - \nabla^2 F,$$

together with two similar equations. They will in any case be indeterminate to the extent of three additive functions of the forms $\partial\chi/\partial x$, $\partial\chi/\partial y$, $\partial\chi/\partial z$, respectively, and we may, if we please, suppose χ to be chosen so that

$$\frac{\partial F}{\partial x} + \frac{\partial G}{\partial y} + \frac{\partial H}{\partial z} = 0, \quad \ldots\ldots\ldots\ldots\ldots\ldots\ldots\ldots\ldots(5)$$

* The investigation which follows is substantially that given by Helmholtz. The kinematical problem in question was first solved, in a slightly different manner, by Stokes, "On the Dynamical Theory of Diffraction," *Camb. Trans.* ix. (1849) [*Papers*, ii. 254...].

in which case $\nabla^2 F = -\xi, \quad \nabla^2 G = -\eta, \quad \nabla^2 H = -\zeta.$ ....(6)

Particular solutions of these equations are obtained by equating F, G, H to the potentials of distributions of matter whose volume-densities are $\xi/4\pi$, $\eta/4\pi$, $\zeta/4\pi$, respectively; thus

$$F = \frac{1}{4\pi}\iiint \frac{\xi'}{r}\,dx'dy'dz', \quad G = \frac{1}{4\pi}\iiint \frac{\eta'}{r}\,dx'dy'dz', \quad H = \frac{1}{4\pi}\iiint \frac{\zeta'}{r}\,dx'dy'dz', \quad \text{......(7)}$$

where the accents attached to ξ, η, ζ are used to distinguish the values of these quantities at the point (x', y', z'). The integrations are to include, of course, all places where ξ, η, ζ differ from zero. It remains to shew that these values of F, G, H do in fact satisfy (5). Since $\partial/\partial x \,.\, r^{-1} = -\partial/\partial x' \,.\, r^{-1}$, the formulae (7) make

$$\frac{\partial F}{\partial x} + \frac{\partial G}{\partial y} + \frac{\partial H}{\partial z} = -\frac{1}{4\pi}\iiint\left(\xi'\frac{\partial}{\partial x'}\frac{1}{r} + \eta'\frac{\partial}{\partial y'}\frac{1}{r} + \zeta'\frac{\partial}{\partial z'}\frac{1}{r}\right)dx'dy'dz'.$$

The right-hand member vanishes, by a generalization of the theorem of Art. 42 (4)*, since

$$\frac{\partial \xi}{\partial x} + \frac{\partial \eta}{\partial y} + \frac{\partial \zeta}{\partial z} = 0$$

everywhere, whilst $l\xi + m\eta + n\zeta = 0$

at the surfaces of the vortices (where ξ, η, ζ may be discontinuous), and ξ, η, ζ vanish at infinity.

The complete solution of our problem is obtained by superposition of the results contained in (1) and (4), viz.

$$\left.\begin{aligned} u &= -\frac{\partial \Phi}{\partial x} + \frac{\partial H}{\partial y} - \frac{\partial G}{\partial z}, \\ v &= -\frac{\partial \Phi}{\partial y} + \frac{\partial F}{\partial z} - \frac{\partial H}{\partial x}, \\ w &= -\frac{\partial \Phi}{\partial z} + \frac{\partial G}{\partial x} - \frac{\partial F}{\partial y}, \end{aligned}\right\} \quad \text{..........................(8)}$$

where Φ, F, G, H have the values given in (2) and (7).

It may be added that the proviso that θ, ξ, η, ζ should vanish beyond a certain distance from the origin is not absolutely essential. It is sufficient if the data be such that the integrals in (2) and (7), when taken over infinite space, are convergent. This will certainly be the case if θ, ξ, η, ζ are ultimately of the order R^{-n}, where R denotes distance from the origin, and $n > 3$†.

When the region occupied by the fluid is not unlimited, but is bounded (in whole or in part) by surfaces at which the normal velocity is given, and when further (in the case of an n-ply connected region) the value of the circulation in each of n independent circuits is prescribed, the problem may

* The singularity which occurs at the point $r = 0$ is assumed to be treated here and elsewhere as in the theory of Attractions. The result is not affected.

† Cf. Leathem, *Cambridge Tracts*, No. 1 (2nd ed.), p. 44.

by a similar analysis be reduced to one of irrotational motion, of the kind considered in Chapter III., and there proved to be determinate. This may be left to the reader, with the remark that if the vortices traverse the region, beginning and ending on the boundary, it is convenient to imagine them continued beyond it, or along the boundary, in such a manner that they form re-entrant filaments, and to make the integrals (7) refer to the complete system of vortices thus obtained. On this understanding the condition (5) will still be satisfied.

There is an exact correspondence between the analytical relations above developed and certain formulae in Electro-magnetism. If, in the equations (1) and (2) of Art. 147, we write

$$\alpha,\ \beta,\ \gamma,\ \rho,\ u,\ v,\ w,\ \rho$$

for

$$u,\ v,\ w,\ \theta,\ \xi,\ \eta,\ \zeta,\ \theta,$$

respectively, we obtain

$$\left.\begin{aligned}\frac{\partial\alpha}{\partial x}+\frac{\partial\beta}{\partial y}+\frac{\partial\gamma}{\partial z}=\rho,\\ \frac{\partial\gamma}{\partial y}-\frac{\partial\beta}{\partial z}=u,\qquad \frac{\partial\alpha}{\partial z}-\frac{\partial\gamma}{\partial x}=v,\qquad \frac{\partial\beta}{\partial x}-\frac{\partial\alpha}{\partial y}=w,\end{aligned}\right\}\ \ldots\ldots\ldots\ldots\ldots\ldots(9)$$

which are the fundamental relations of the theory referred to; viz. α, β, γ are the components of magnetic force, u, v, w those of electric current, and ρ is the volume-density of the imaginary magnetic matter by which any magnetization present in the field may be represented*. Hence, the vortex-filaments correspond to electric circuits, the strengths of the vortices to the strengths of the currents in these circuits, sources and sinks to positive and negative magnetic poles, and, finally, fluid velocity to magnetic force†.

The analogy will of course extend to all results deduced from the fundamental relations; thus, in equations (8), Φ corresponds to the magnetic potential and F, G, H to the components of 'electro-magnetic momentum.'

149. To interpret the result contained in Art. 148 (8), we may calculate the values of u, v, w due to an isolated re-entrant vortex-filament situate in an infinite mass of incompressible fluid which is at rest at infinity.

Since $\theta=0$, we shall have $\Phi=0$. Again, to calculate the values of F, G, H, we may replace the volume-element $\delta x'\delta y'\delta z'$ by $\sigma'\delta s'$, where $\delta s'$ is an element of the length of the filament, and σ' its cross-section. Also

$$\xi'=\omega'\frac{dx'}{ds'},\qquad \eta'=\omega'\frac{dy'}{ds'},\qquad \zeta'=\omega'\frac{dz'}{ds'},$$

where ω' is the vorticity. Hence the formulae (7) of Art. 148 become

$$F=\frac{\kappa}{4\pi}\int\frac{dx'}{r},\qquad G=\frac{\kappa}{4\pi}\int\frac{dy'}{r},\qquad H=\frac{\kappa}{4\pi}\int\frac{dz'}{r},\ \ldots\ldots\ldots(1)$$

where κ, $=\omega'\sigma'$, measures the strength of the vortex, and the integrals are to be taken along the whole length of the filament.

* Cf. Maxwell, *Electricity and Magnetism*, Art. 607. The analogy has been improved by the adoption of the 'rational' system of electrical units advocated by Heaviside, *Electrical Papers*, London, 1892, i. 199.

† This analogy was first pointed out by Helmholtz; it has been extensively utilized by Kelvin in his papers on *Electrostatics and Magnetism*.

Hence, by Art. 148 (4), we have

$$u=\frac{\kappa}{4\pi}\int\left(\frac{\partial}{\partial y}\frac{1}{r}\,.\,dz'-\frac{\partial}{\partial z}\frac{1}{r}\,.\,dy'\right),$$

with similar results for v, w. We thus find*

$$\left.\begin{aligned} u&=\frac{\kappa}{4\pi}\int\left(\frac{dy'}{ds'}\frac{z-z'}{r}-\frac{dz'}{ds'}\frac{y-y'}{r}\right)\frac{ds'}{r^2},\\ v&=\frac{\kappa}{4\pi}\int\left(\frac{dz'}{ds'}\frac{x-x'}{r}-\frac{dx'}{ds'}\frac{z-z'}{r}\right)\frac{ds'}{r^2},\\ w&=\frac{\kappa}{4\pi}\int\left(\frac{dx'}{ds'}\frac{y-y'}{r}-\frac{dy'}{ds'}\frac{x-x'}{r}\right)\frac{ds'}{r^2}.\end{aligned}\right\}\quad\ldots\ldots\ldots\ldots(2)$$

If δu, δv, δw denote the parts of these expressions which involve the element $\delta s'$ of the filament, it appears that the resultant of δu, δv, δw is perpendicular to the plane containing the direction of the vortex-line at (x', y', z') and the line r, and that its sense is that in which the point (x, y, z) would be carried if it were attached to a rigid body rotating with the fluid element at (x', y', z'). For the magnitude of the resultant we have

$$\{(\delta u)^2+(\delta v)^2+(\delta w)^2\}^{\frac{1}{2}}=\frac{\kappa}{4\pi}\frac{\sin\chi\,\delta s'}{r^2},\quad\ldots\ldots\ldots\ldots(3)$$

where χ is the angle which r makes with the vortex-line at (x', y', z').

With the change of symbols indicated in the preceding Art. this result becomes identical with the law of action of an electric current on a magnetic pole†.

Velocity-Potential due to a Vortex.

150. At points external to the vortices there exists a velocity-potential, whose value may be obtained as follows. Taking for shortness the case of a single re-entrant vortex, we have, from the preceding Art., in the case of an incompressible fluid,

$$u=\frac{\kappa}{4\pi}\int\left(\frac{\partial}{\partial z'}\frac{1}{r}\,.\,dy'-\frac{\partial}{\partial y'}\frac{1}{r}\,.\,dz'\right).\quad\ldots\ldots\ldots\ldots(1)$$

By Stokes' Theorem (Art. 32 (4)) we can replace a line-integral extending round a closed curve by a surface-integral taken over any surface bounded by that curve; viz. we have, with a slight change of notation,

$$\int(P\,dx'+Q\,dy'+R\,dz')=\iint\left\{l\left(\frac{\partial R}{\partial y'}-\frac{\partial Q}{\partial z'}\right)+m\left(\frac{\partial P}{\partial z'}-\frac{\partial R}{\partial x'}\right)+n\left(\frac{\partial Q}{\partial x'}-\frac{\partial P}{\partial y'}\right)\right\}dS'.$$

If we put $\quad P=0,\qquad Q=\dfrac{\partial}{\partial z'}\dfrac{1}{r},\qquad R=-\dfrac{\partial}{\partial y'}\dfrac{1}{r},$

we find

$$\frac{\partial R}{\partial y'}-\frac{\partial Q}{\partial z'}=\frac{\partial^2}{\partial x'^2}\frac{1}{r'},\qquad \frac{\partial P}{\partial z'}-\frac{\partial R}{\partial x'}=\frac{\partial^2}{\partial x'\partial y'}\frac{1}{r'},\qquad \frac{\partial Q}{\partial x'}-\frac{\partial P}{\partial y'}=\frac{\partial^2}{\partial x'\partial z'}\frac{1}{r'},$$

* These are equivalent to the forms obtained by Stokes, *l.c. ante* p. 208.

† Ampère, *Théorie mathématique des phénomènes électro-dynamiques*, Paris, 1826.

so that (1) may be written

$$u = \frac{\kappa}{4\pi}\iint\left(l\frac{\partial}{\partial x'} + m\frac{\partial}{\partial y'} + n\frac{\partial}{\partial z'}\right)\frac{d}{dx'}\frac{1}{r}\,dS'.$$

Hence, and by similar reasoning, we have, since $\partial/\partial x' . r^{-1} = -\partial/\partial x . r^{-1}$,

$$u = -\frac{\partial\phi}{\partial x}, \quad v = -\frac{\partial\phi}{\partial y}, \quad w = -\frac{\partial\phi}{\partial z}, \quad \text{.................(2)}$$

where

$$\phi = \frac{\kappa}{4\pi}\iint\left(l\frac{\partial}{\partial x'} + m\frac{\partial}{\partial y'} + n\frac{\partial}{\partial z'}\right)\frac{1}{r}\,dS'. \quad \text{.................(3)}$$

Here l, m, n denote the direction-cosines of the normal to the element $\delta S'$ of a surface bounded by the vortex-filament.

The formula (3) may be otherwise written

$$\phi = \frac{\kappa}{4\pi}\iint\frac{\cos\vartheta}{r^2}\,dS', \quad \text{..........................(4)}$$

where ϑ denotes the angle between r and the normal (l, m, n). Since $\cos\vartheta\, dS'/r^2$ measures the elementary solid angle subtended by $\delta S'$ at (x, y, z), we see that the velocity-potential at any point, due to a single re-entrant vortex, is equal to the product of $\kappa/4\pi$ into the solid angle which a surface bounded by the vortex subtends at that point.

Since this solid angle changes by 4π when the point in question describes a circuit embracing the vortex, we verify that the value of ϕ given by (4) is cyclic, the cyclic constant being κ. Cf. Art. 145.

It may be noticed that the expression in (4) is equal to the flux (in the negative direction) through the aperture of the vortex, due to a point-source of strength κ at the point (x, y, z).

Comparing (4) with Art. 56 (4) we see that a vortex is, in a sense, equivalent to a uniform distribution of double sources over any surface bounded by it. The axes of the double sources must be supposed to be everywhere normal to the surface, and the density of the distribution to be equal to the strength of the vortex. It is here assumed that the relation between the positive direction of the normal and the positive direction of the axis of the vortex-filament is of the 'right-handed' type. See Art. 31.

Conversely, it may be shewn that any distribution of double sources over a *closed* surface, the axes being directed along the normals, may be replaced by a system of closed vortex-filaments lying in the surface*. The same thing will appear independently from the investigation of the next Art.

Vortex-Sheets.

151. We have so far assumed u, v, w to be continuous. We may now shew how cases where surfaces of discontinuity present themselves may be brought within the scope of our theorems.

* Cf. Maxwell, *Electricity and Magnetism*, Arts. 485, 652.

The case of a discontinuity in the *normal* velocity alone has already been treated in Art. 58. If u, v, w denote the component velocities on one side, and u', v', w' those on the other, it was found that the circumstances could be represented by imagining a distribution of simple sources, with surface-density

$$l(u'-u)+m(v'-v)+n(w'-w),$$

where l, m, n denote the direction-cosines of the normal drawn towards the side to which the accents refer.

Let us next consider the case where the *tangential* velocity (only) is discontinuous, so that

$$l(u'-u)+m(v'-v)+n(w'-w)=0. \quad\text{.................}(1)$$

We will suppose that the lines of *relative* motion, which are defined by the differential equations

$$\frac{dx}{u'-u}=\frac{dy}{v'-v}=\frac{dz}{w'-w}, \quad\text{.........................}(2)$$

are traced on the surface, and that the system of orthogonal trajectories to these lines is also drawn. Let PQ, $P'Q'$ be linear elements drawn close to the surface, on the two sides, parallel to a line of the system (2), and let PP' and QQ' be normal to the surface and infinitely small in comparison with PQ or $P'Q'$. The circulation in the circuit $P'Q'QP$ will then be equal to $(q'-q)\,PQ$, where q, q' denote the absolute velocities on the two sides. This is the same as if the position of the surface were occupied by an infinitely thin stratum of vortices, the orthogonal trajectories above-mentioned being the vortex-lines, and the vorticity ω and the (variable) thickness δn of the stratum being connected by the relation

$$\omega\,\delta n=q'-q. \quad\text{...............................}(3)$$

The same result follows from a consideration of the discontinuities which occur in the values of u, v, w as determined by the formulae (4) and (7) of Art. 148, when we apply these to the case of a stratum of thickness δn within which ξ, η, ζ are infinite, but so that $\xi\delta n, \eta\delta n, \zeta\delta n$ are finite*.

It was shewn in Arts. 147, 148 that any continuous motion of a fluid filling infinite space, and at rest at infinity, may be regarded as due to a suitable arrangement of sources and vortices distributed with finite density. We have now seen how by considerations of continuity we can pass to the case where the sources and vortices are distributed with infinite volume-density, but infinite surface-density, over surfaces. In particular, we may take the case where the infinite fluid in question is incompressible, and is divided into two portions by a closed surface over which the normal velocity is continuous, but the tangential velocity discontinuous, as in Art. 58 (12). This is

* Helmholtz, *l.c. ante* p. 202.

equivalent to a vortex-sheet; and we infer that every continuous irrotational motion, whether cyclic or not, of an incompressible substance occupying any region whatever, may be regarded as due to a certain distribution of vortices over the boundaries which separate it from the rest of infinite space. In the case of a region extending to infinity, the distribution is confined to the *finite* portion of the boundary, provided the fluid be at rest at infinity.

This theorem is complementary to the results obtained in Art. 58.

The foregoing conclusions may be illustrated by means of the results of Art. 91. Thus when a normal velocity S_n was prescribed over the sphere $r=a$, the values of the velocity-potential for the internal and external space were found to be

$$\phi=\frac{a}{n}\left(\frac{r}{a}\right)^n S_n, \text{ and } \phi=-\frac{a}{n+1}\left(\frac{a}{r}\right)^{n+1} S_n,$$

respectively. Hence if $\delta\epsilon$ be the angle which a linear element drawn on the surface subtends at the centre, the relative velocity estimated in the direction of this element will be

$$\frac{2n+1}{n(n+1)}\frac{\partial S_n}{\partial \epsilon}.$$

The resultant relative velocity is therefore tangential to the surface, and perpendicular to the contour lines (S_n=const.) of the surface-harmonic S_n, which are therefore the vortex-lines.

For example, if we have a thin spherical shell filled with and surrounded by liquid, moving as in Art. 92 parallel to the axis of x, the motion of the fluid, whether internal or external, will be that due to a system of vortices arranged in parallel circles on the sphere; the strength of an elementary vortex being proportional to the projection, on the axis of x, of the breadth of the corresponding zone of the surface*.

Impulse and Energy of a Vortex-System.

152. The following investigations relate to the case of a vortex-system of finite dimensions in an incompressible fluid which fills infinite space and is at rest at infinity.

The problem of finding a distribution of impulsive force (X', Y', Z') per unit mass which would generate the actual motion (u, v, w) instantaneously from rest is to some extent indeterminate, but a sufficient solution for our purpose may be obtained as follows.

We imagine a simply-connected surface S to be drawn enclosing all the vortices. We denote by ϕ the single-valued velocity-potential which obtains outside S, and by ϕ_1 that solution of $\nabla^2\phi=0$ which is finite throughout the interior of S, and is continuous with ϕ at this surface. In other words, ϕ_1 is the velocity-potential of the motion which would be produced within S by the application of impulsive pressures $\rho\phi$ over the surface. If we now assume

$$X'=u+\frac{\partial\phi_1}{\partial x}, \qquad Y'=v+\frac{\partial\phi_1}{\partial y}, \qquad Z'=w+\frac{\partial\phi_1}{\partial z} \quad\text{...........(1)}$$

* The same statements hold also for an ellipsoidal shell moving parallel to one of its principal axes. See Art. 114.

at internal points, and

$$X'=0, \qquad Y'=0, \qquad Z'=0 \quad \text{....................(2)}$$

at external points, it is evident on reference to Art. 11 that these forces would in fact generate the actual motion instantaneously from rest, the distribution of impulsive pressure being given by $\rho\phi$ at external, and $\rho\phi_1$ at internal, points. The forces are discontinuous at the surface, but the discontinuity is only in the normal component, the tangential components vanishing just inside and just outside owing to the continuity of ϕ with ϕ_1. Hence if (l, m, n) be the direction-cosines of the inward normal, we should have

$$mZ'-nY'=0, \quad nX'-lZ'=0, \quad lY'-mX'=0, \quad \text{............(3)}$$

at points just inside the surface.

Now if we integrate over the volume enclosed by S we have

$$\iiint(y\zeta-z\eta)\,dx\,dy\,dz=\iiint\left\{y\left(\frac{\partial v}{\partial x}-\frac{\partial u}{\partial y}\right)-z\left(\frac{\partial u}{\partial z}-\frac{\partial w}{\partial x}\right)\right\}dx\,dy\,dz$$

$$=\iiint\left\{y\left(\frac{\partial Y'}{\partial x}-\frac{\partial X'}{\partial y}\right)-z\left(\frac{\partial X'}{\partial z}-\frac{\partial Z'}{\partial x}\right)\right\}dx\,dy\,dz$$

$$=-\iint\{y(lY'-mX')-z(nX'-lZ')\}\,dS+2\iiint X'\,dx\,dy\,dz, \quad \text{......(4)}$$

where the surface-integral vanishes in virtue of (3).

Again

$$-\iiint(y^2+z^2)\,\xi\,dx\,dy\,dz=-\iiint(y^2+z^2)\left(\frac{\partial w}{\partial y}-\frac{\partial v}{\partial z}\right)dx\,dy\,dz$$

$$=-\iiint(y^2+z^2)\left(\frac{\partial Z'}{\partial y}-\frac{\partial Y'}{\partial z}\right)dx\,dy\,dz$$

$$=\iint(y^2+z^2)(mZ'-nY')\,dS+2\iiint(yZ'-zY')\,dx\,dy\,dz, \quad \text{........(5)}$$

where the surface-integral vanishes as before.

We thus obtain for the force- and couple-resultants of the impulse of the vortex-system the expressions

$$\left.\begin{aligned} P&=\tfrac{1}{2}\rho\iiint(y\zeta-z\eta)\,dx\,dy\,dz, & L&=-\tfrac{1}{2}\rho\iiint(y^2+z^2)\,\xi\,dx\,dy\,dz,\\ Q&=\tfrac{1}{2}\rho\iiint(z\xi-x\zeta)\,dx\,dy\,dz, & M&=-\tfrac{1}{2}\rho\iiint(z^2+x^2)\,\eta\,dx\,dy\,dz,\\ R&=\tfrac{1}{2}\rho\iiint(x\eta-y\xi)\,dx\,dy\,dz, & N&=-\tfrac{1}{2}\rho\iiint(x^2+y^2)\,\zeta\,dx\,dy\,dz.\end{aligned}\right\} \text{...(6)}$$

To apply these to the case of a single re-entrant vortex-filament of infinitely small section σ, we replace the volume element by $\sigma\delta s$, and write

$$\xi=\omega\frac{dx}{ds}, \qquad \eta=\omega\frac{dy}{ds}, \qquad \zeta=\omega\frac{dz}{ds}. \quad \text{................(7)}$$

Hence

$$P=\tfrac{1}{2}\rho\omega\sigma\int(y\,dz-z\,dy)=\kappa\rho\iint l'\,dS', \quad \text{.................(8)}$$

$$L=-\tfrac{1}{2}\rho\omega\sigma\int(y^2+z^2)\,dx=-\kappa\rho\iint(m'z-n'y)\,dS', \text{...........(9)}$$

with similar formulae. The line-integrals are supposed to be taken along the filament, and the surface-integrals over a barrier bounded by it, and l', m', n' are the direction-cosines of the normal to an element $\delta S'$ of the barrier. The identities of the different forms follow from Stokes' Theorem. We have also written κ for $\omega\sigma$, *i.e.* κ is the circulation round the filament*.

The whole investigation has reference of course to the instantaneous state of the system, but it may be recalled that, when no extraneous forces act, the impulse is, by the argument of Art. 119, constant in every respect.

153. Let us next consider the *energy* of the vortex-system. It is easily proved that under the circumstances presupposed, and in the absence of extraneous forces, this energy will be constant. For if T be the energy of the fluid bounded by any closed surface S, we have, putting $V=0$ in Art. 10 (5),

$$\frac{DT}{Dt}=\iint(lu+mv+nw)\,p\,dS. \quad\text{.......................(1)}$$

If the surface S enclose all the vortices, we may put

$$\frac{p}{\rho}=\frac{\partial\phi}{\partial t}-\tfrac{1}{2}q^2+F(t), \quad\text{..........................(2)}$$

and it easily follows from Art. 150 (4) that at a great distance R from the vortices p will be finite, and $lu+mv+nw$ of the order R^{-3}, whilst when the surface S is taken wholly at infinity, the elements δS vary as R^2. Hence, ultimately, the right-hand side of (1) vanishes, and we have

$$T=\text{const.} \quad\text{................................(3)}$$

We proceed to investigate one or two important kinematical expressions for T, still confining ourselves, for simplicity, to the case where the fluid (supposed incompressible) extends to infinity, and is at rest there, all the vortices being within a finite distance of the origin.

The first of these expressions is indicated by the electro-magnetic analogy pointed out in Art. 148. Since $\theta=0$, and therefore $\Phi=0$, we have

$$\begin{aligned}2T&=\rho\iiint(u^2+v^2+w^2)\,dx\,dy\,dz\\&=\rho\iiint\left\{u\left(\frac{\partial H}{\partial y}-\frac{\partial G}{\partial z}\right)+v\left(\frac{\partial F}{\partial z}-\frac{\partial H}{\partial x}\right)+w\left(\frac{\partial G}{\partial x}-\frac{\partial F}{\partial y}\right)\right\}dx\,dy\,dz,\end{aligned}$$

by Art. 148 (4). The last member may be replaced by the sum of a surface-integral

$$\rho\iint\{F(mw-nv)+G(nu-lw)+H(lv-mu)\}\,dS,$$

and a volume-integral

$$\rho\iiint\left\{F\left(\frac{\partial w}{\partial y}-\frac{\partial v}{\partial z}\right)+G\left(\frac{\partial u}{\partial z}-\frac{\partial w}{\partial x}\right)+H\left(\frac{\partial v}{\partial x}-\frac{\partial u}{\partial y}\right)\right\}dx\,dy\,dz.$$

* The expressions (8) and (9) were obtained by elementary reasoning by J. J. Thomson, *On the Motion of Vortex Rings* (Adams Prize Essay), London, 1883, pp. 5, 6, and the formulae (6) deduced from them, with, however, the opposite signs in the case of L, M, N. The correction is due to Mr Welsh.

An interesting test of the formulae as they now stand is afforded by the case of a spherical mass rotating as if solid and surrounded by fluid at rest, provided we take into account the spherical vortex-sheet which represents the discontinuity of velocity.

At points of the infinitely distant boundary, F, G, H are ultimately of the order R^{-2}. and u, v, w of the order R^{-3}, so that the surface-integral vanishes, and we have

$$T=\tfrac{1}{2}\rho\iiint(F\xi+G\eta+H\zeta)\,dx\,dy\,dz, \quad \text{..................(4)}$$

or, substituting the values of F, G, H from Art. 148 (7),

$$T=\frac{\rho}{8\pi}\iiiiiint\frac{\xi\xi'+\eta\eta'+\zeta\zeta'}{r}\,dx\,dy\,dz\,dx'\,dy'\,dz', \quad \text{...........(5)}$$

where each volume-integration extends over the whole space occupied by the vortices.

A slightly different form may be given to this expression as follows. Regarding the vortex-system as made up of filaments, let δs, $\delta s'$ be elements of length of any two filaments, σ, σ' the corresponding cross-sections, and ω, ω' the corresponding vorticities. The elements of volume may be taken to be $\sigma\delta s$ and $\sigma'\delta s'$, respectively, so that the expression following the integral signs in (5) is equivalent to

$$\frac{\cos\epsilon}{r}\,.\,\omega\sigma\delta s\,.\,\omega'\sigma'\delta s',$$

where ϵ is the angle between δs and $\delta s'$. If we put $\omega\sigma=\kappa$, $\omega'\sigma'=\kappa'$, we have

$$T=\frac{\rho}{4\pi}\,\Sigma\kappa\kappa'\iint\frac{\cos\epsilon}{r}\,ds\,ds', \quad \text{......................(6)}$$

where the double integral is to be taken along the axes of the filaments, and the summation Σ includes (once only) every pair of filaments which are present.

The factor of ρ in (6) is identical with the expression for the energy of a system of electric currents flowing along conductors coincident in position with the vortex-filaments, with strengths $\kappa, \kappa', \ldots$ respectively*. The above investigation is in fact merely an inversion of the argument given in treatises on Electro-magnetism, whereby it is proved that

$$\frac{1}{4\pi}\,\Sigma ii'\iint\frac{\cos\epsilon}{r}\,ds\,ds'=\tfrac{1}{2}\iiint(\alpha^2+\beta^2+\gamma^2)\,dx\,dy\,dz,$$

where i, i' denote the strengths of the currents in the linear conductors whose elements are denoted by δs, $\delta s'$, and α, β, γ are the components of magnetic force at any point of the field.

The theorem of this Art. is purely kinematical, and rests solely on the assumption that the functions u, v, w satisfy the equation of continuity,

$$\frac{\partial u}{\partial x}+\frac{\partial v}{\partial y}+\frac{\partial w}{\partial z}=0,$$

throughout infinite space, and vanish at infinity. It can therefore by an easy generalization be extended to a case considered in Art. 144, where a liquid is supposed to circulate irrotationally through apertures in fixed solids, the values of u, v, w being now taken to be zero at all points of space not occupied by the fluid. The investigation of Art. 151 shews that the distribution of velocity thus obtained may be regarded as due to a system of vortex-sheets coincident with the bounding surfaces. The energy of this system will be given by an obvious adaptation of the formula (6) above, and will therefore be proportional to that of the corresponding system of electric current-sheets. This proves a statement made by anticipation in Art. 144.

* The 'rational' system of electrical units being understood; see *ante* p. 210.

Under the circumstances stated at the beginning of Art. 152, we have another useful expression for T; viz.

$$T = \rho \iiint \{u(y\zeta - z\eta) + v(z\xi - x\zeta) + w(x\eta - y\xi)\}\, dx\,dy\,dz^{*}. \quad \ldots\ldots(7)$$

To verify this, we take the right-hand member, and transform it by the process already so often employed, omitting the surface-integrals for the same reason as in the preceding Art. The first of the three terms gives

$$\rho \iiint u \left\{ y \left(\frac{\partial v}{\partial x} - \frac{\partial u}{\partial y} \right) - z \left(\frac{\partial u}{\partial z} - \frac{\partial w}{\partial x} \right) \right\} dx\,dy\,dz$$

$$= -\rho \iiint \left\{ (vy + wz) \frac{\partial u}{\partial x} - u^2 \right\} dx\,dy\,dz.$$

Transforming the remaining terms in the same way, adding, and making use of the equation of continuity, we obtain

$$\rho \iiint \left(u^2 + v^2 + w^2 + xu \frac{\partial u}{\partial x} + yv \frac{\partial v}{\partial y} + zw \frac{\partial w}{\partial z} \right) dx\,dy\,dz,$$

or, finally, on again transforming the last three terms,

$$\tfrac{1}{2}\rho \iiint (u^2 + v^2 + w^2)\, dx\,dy\,dz.$$

In the case of a finite region the surface-integrals must be retained†. This involves the addition to the right-hand side of (7) of the term

$$\rho \iint \{(lu + mv + nw)(xu + yv + zw) - \tfrac{1}{2}(lx + my + nz) q^2\}\, dS, \quad \ldots\ldots(8)$$

where $q^2 = u^2 + v^2 + w^2$. This simplifies in the case of a *fixed* boundary.

The value of the expression (7) must be unaltered by any displacement of the origin of co-ordinates. Hence we must have

$$\iiint (v\zeta - w\eta)\, dx\,dy\,dz = 0, \quad \iiint (w\xi - u\zeta)\, dx\,dy\,dz = 0, \quad \iiint (u\eta - v\xi)\, dx\,dy\,dz = 0. \quad \ldots\ldots(9)$$

These equations, which may easily be verified by partial integration, follow also from the consideration that if there are no extraneous forces the components of the impulse parallel to the co-ordinate axes must be constant. Thus, taking first the case of a fluid enclosed in a fixed envelope of finite size, we have, in the notation of Art. 152,

$$P = \rho \iiint u\, dx\,dy\,dz - \rho \iint l\phi\, dS, \quad \ldots\ldots(10)$$

if ϕ denote the velocity-potential near the envelope, where the motion is irrotational.

Hence

$$\frac{dP}{dt} = \rho \iiint \frac{\partial u}{\partial t}\, dx\,dy\,dz - \rho \iint l \frac{\partial \phi}{\partial t}\, dS$$

$$= -\rho \iiint \frac{\partial \chi'}{\partial x}\, dx\,dy\,dz + \rho \iiint (v\zeta - w\eta)\, dx\,dy\,dz - \rho \iint l \frac{\partial \phi}{\partial t}\, dS, \quad \ldots\ldots(11)$$

by Art. 146 (5). The first and third terms of this cancel, since at the envelope we have $\chi' = \partial\phi/\partial t$, by Art. 20 (4) and Art. 146 (6). Hence for any re-entrant system of vortices enclosed in a fixed vessel, we have

$$\frac{dP}{dt} = \rho \iiint (v\zeta - w\eta)\, dx\,dy\,dz, \quad \ldots\ldots(12)$$

with two similar equations. It has been proved in Art. 119 that if the containing vessel be infinitely large, and infinitely distant from the vortices, P is constant. This gives the first of equations (9).

* *Motion of Fluids*, Art. 136 (1879).

† J. J. Thomson, *l.c. ante* p. 216.

Conversely from (9), established otherwise, we could infer the constancy of the components P, Q, R of the impulse*.

Rectilinear Vortices.

154. When the motion is in two dimensions x, y we have $w = 0$, whilst u, v are functions of x, y, only. Hence $\xi = 0$, $\eta = 0$, so that the vortex-lines are straight lines parallel to z. The theory then takes a very simple form.

The formulae (8) of Art. 148 are now replaced by

$$u = -\frac{\partial\phi}{\partial x} - \frac{\partial\psi}{\partial y}, \quad v = -\frac{\partial\phi}{\partial y} + \frac{\partial\psi}{\partial x}, \quad \text{.................(1)}$$

the functions ϕ, ψ being subject to the equations

$$\nabla_1^2\phi = -\theta, \quad \nabla_1^2\psi = \zeta, \quad \text{.........................(2)}$$

where $$\nabla_1^2 = \frac{\partial^2}{\partial x^2} + \frac{\partial^2}{\partial y^2},$$

and to the proper boundary-conditions.

In the case of an incompressible fluid, to which we will now confine ourselves, we have

$$u = -\frac{\partial\psi}{\partial y}, \quad v = \frac{\partial\psi}{\partial x}, \quad \text{..........................(3)}$$

where ψ is the stream-function of Art. 59. It is known from the Theory of Attractions that the solution of

$$\nabla_1^2\psi = \zeta, \quad \text{.................................(4)}$$

ζ being a given function of x, y, is

$$\psi = \frac{1}{2\pi}\iint \zeta' \log r\, dx'dy' + \psi_0, \quad \text{..... ....(5)}$$

where ζ' denotes the value of ζ at the point (x', y'), and r stands for

$$\{(x-x')^2 + (y-y')^2\}^{\frac{1}{2}}.$$

The 'complementary function' ψ_0 may be any solution of

$$\nabla_1^2\psi_0 = 0; \quad \text{...............................(6)}$$

it enables us to satisfy the boundary-conditions.

In the case of an unlimited mass of liquid, at rest at infinity, ψ_0 is constant. The formulae (3) and (5) then give

$$u = -\frac{1}{2\pi}\iint \zeta' \frac{y-y'}{r^2}\, dx'dy', \quad v = \frac{1}{2\pi}\iint \zeta' \frac{x-x'}{r^2}\, dx'dy'. \quad \text{......(7)}$$

Hence a vortex-filament whose co-ordinates are x', y' and whose strength is κ contributes to the motion at (x, y) a velocity whose components are

$$-\frac{\kappa}{2\pi}\cdot\frac{y-y'}{r^2}, \text{ and } \frac{\kappa}{2\pi}\cdot\frac{x-x'}{r^2}.$$

This velocity is perpendicular to the line joining the points (x, y), (x', y'), and its amount is $\kappa/2\pi r$.

* J. J. Thomson, *l.c.*

Let us calculate the integrals $\iint u\zeta dxdy$, and $\iint v\zeta dxdy$, where the integrations include all portions of the plane xy for which ζ does not vanish. We have

$$\iint u\zeta dxdy = -\frac{1}{2\pi}\iiiint \zeta\zeta' \frac{y-y'}{r^2}\, dxdydx'dy',$$

where each double integration includes the sections of all the vortices. Now, corresponding to any term

$$\zeta\zeta' \frac{y-y'}{r^2}\, dxdydx'dy'$$

of this result, we have another

$$\zeta\zeta' \frac{y'-y}{r^2}\, dxdydx'dy',$$

and these two neutralize each other. Hence, and by similar reasoning,

$$\iint u\zeta dxdy = 0, \qquad \iint v\zeta dxdy = 0. \qquad \text{.....................(8)}$$

If as before we denote the strength of a vortex by κ, these results may be written

$$\Sigma\kappa u = 0, \qquad \Sigma\kappa v = 0. \qquad \text{.........................(9)}$$

Since the strength of each vortex is constant with regard to the time, the equations (9) express that the point whose co-ordinates are

$$\bar{x} = \frac{\Sigma\kappa x}{\Sigma\kappa}, \qquad \bar{y} = \frac{\Sigma\kappa y}{\Sigma\kappa} \qquad \text{.......................(10)}$$

is fixed throughout the motion.

This point, which coincides with the centre of inertia of a film of matter distributed over the plane xy with the surface-density ζ, may be called the 'centre' of the system of vortices, and the straight line parallel to z of which it is the projection may be called the 'axis' of the system. If $\Sigma\kappa = 0$, the centre is at infinity, or else indeterminate.

155. Some interesting examples are furnished by the case of one or more isolated vortices of infinitely small section. Thus:

1°. Let us suppose that we have only one vortex-filament present, and that the vorticity ζ has the same sign throughout its infinitely small section. Its centre, as just defined, will lie either within the substance of the filament, or infinitely close to it. Since this centre remains at rest, the filament as a whole will be stationary, though its parts may experience relative motions, and its centre will not necessarily lie always in the same element of fluid. Any particle at a finite distance r from the centre of the filament will describe a circle about the latter as axis, with constant velocity $\kappa/2\pi r$. The region external to the vortex is doubly-connected; and the circulation in any (simple) circuit embracing it is of course κ. The irrotational motion of the surrounding fluid is the same as in Art. 27 (2).

2°. Next suppose that we have two vortices, of strengths κ_1, κ_2, respectively. Let A, B be their centres, O the centre of the system. The motion of each filament as a whole is entirely due to the other, and is therefore always perpendicular to AB. Hence the two filaments remain always at the same distance from one another, and rotate with constant angular velocity about O, which is fixed. This angular velocity is easily found; we have only to divide the velocity of A (say), viz. $\kappa_2/(2\pi \, . \, AB)$, by the distance AO, where

$$AO = \frac{\kappa_2}{\kappa_1 + \kappa_2} AB,$$

and so obtain
$$\frac{\kappa_1 + \kappa_2}{2\pi \, . \, AB^2}.$$

If κ_1, κ_2 be of the same sign, *i.e.* if the directions of rotation in the two vortices be the same, O lies between A and B; but if the rotations be of opposite signs, O lies in AB, or BA, produced.

If $\kappa_1 = -\kappa_2$, O is at infinity; but it is easily seen that A, B move with equal velocities $\kappa_1/(2\pi \, . \, AB)$ at right angles to AB, which remains fixed in direction. Such a combination of two equal and opposite vortices may be called a 'vortex-pair.' It is the two-dimensional analogue of a circular vortex-ring (Art. 160), and exhibits many of the properties of the latter.

The stream-lines of a vortex-pair form a system of coaxal circles, as shewn on p. 67, the vortices being at the limiting points $(\pm a, 0)$. To find the *relative*

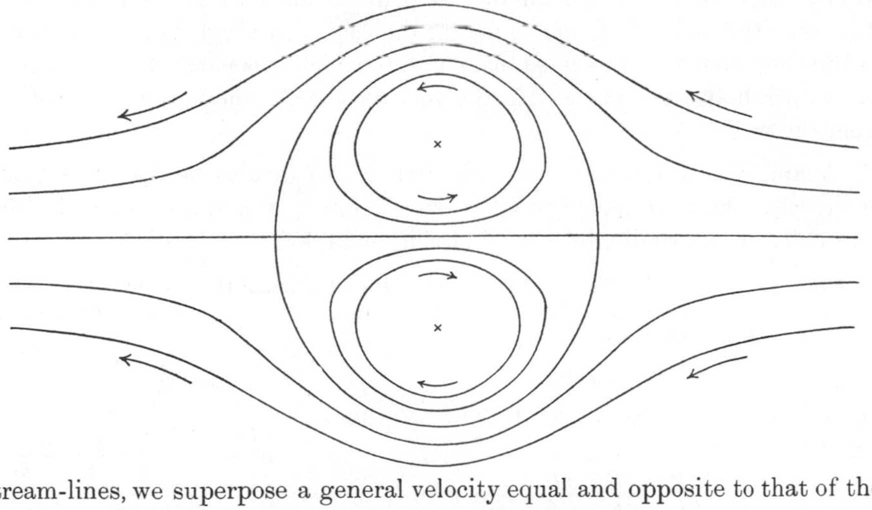

stream-lines, we superpose a general velocity equal and opposite to that of the vortices, and obtain, for the relative stream-function,

$$\psi = \frac{\kappa_1}{2\pi}\left(\frac{x}{2a} + \log \frac{r_1}{r_2}\right), \quad \text{..........................(1)}$$

in the notation of Art. 64, 2°. The figure (which is turned through 90° for convenience) shews a few of the lines. The line $\psi = 0$ consists partly of the axis of y, and partly of an oval surrounding both vortices.

It is plain that the particular portion of fluid enclosed within this oval accompanies the vortex-pair in its career, the motion at external points being exactly that which would be produced by a rigid cylinder having the same boundary; cf. Art. 71. The semi-axes of the oval are $2{\cdot}09\,a$ and $1{\cdot}73\,a$, approximately*.

A difficulty is sometimes felt, in this as in the analogous instance of a vortex-ring, in understanding why the vortices should not be stationary. If in the figure on p. 70 the filaments were replaced by solid cylinders of small circular section, the latter might indeed remain at rest, provided they were rigidly connected by some contrivance which did not interfere with the motion of the fluid; but in the absence of such a connection they would in the first instance be attracted towards one another, on the principle explained in Art. 23. This attraction is however neutralized if we superpose a general velocity V of suitable amount in the direction opposite to the cyclic motion half-way between the cylinders. To find V, we remark that the fluid velocities at the two points $(a \pm c, 0)$, where c is small, will be approximately equal in absolute magnitude, provided

$$V + \frac{\kappa}{2\pi c} - \frac{\kappa}{4\pi a} = \frac{\kappa}{2\pi c} + \frac{\kappa}{4\pi a} - V,$$

where κ is the circulation. Hence

$$V = \frac{\kappa}{4\pi a},$$

which is exactly the velocity of translation of the vortex-pair, in the original form of the problem†.

Since the velocity of the fluid at all points of the plane of symmetry is wholly tangential, we may suppose this plane to form a rigid boundary of the fluid on either side of it, and so obtain the case of a single rectilinear vortex in the neighbourhood of a fixed plane wall to which it is parallel. The filament moves parallel to the plane with the velocity $\kappa/4\pi h$, where h is the distance from the wall.

Again, since the stream-lines are circles, we can also derive the solution of the case where we have a single vortex-filament in a space bounded, either internally or externally, by a fixed circular cylinder.

Thus, in the figure, let EPD be the section of the cylinder, A the position of the vortex (supposed in this case external), and let B be the 'image' of A with respect to the circle EPD, viz. C being the centre, let

$$CB \,.\, CA = c^2,$$

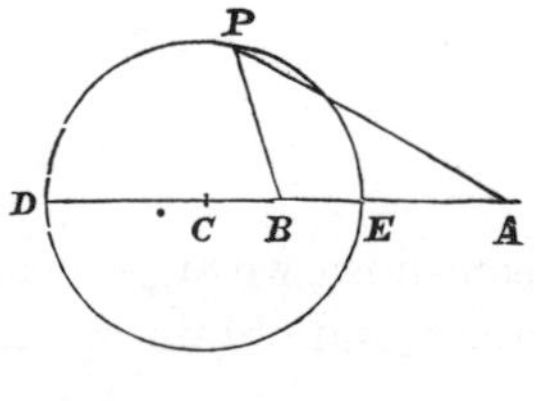

where c is the radius of the circle. If P be any point on the circle, we have

$$\frac{AP}{BP} = \frac{AE}{BE} = \frac{AD}{BD} = \text{const.};$$

so that the circle occupies the position of a stream-line due to a vortex-pair at A, B. Since the motion of the vortex A would be perpendicular to AB,

* Cf. Sir W. Thomson, "On Vortex Atoms," *Phil. Mag.* (4), xxxiv. 20 (1867) [*Papers*, iv. 1]; and Riecke, *Gött. Nachr.* 1888, where paths of fluid particles are also delineated.

† A more exact investigation is given by Hicks, "On the Condition of Steady Motion of Two Cylinders in a Fluid," *Quart. Journ. Math.* xvii. 194 (1881).

it is plain that all the conditions of the problem will be satisfied if we suppose A to describe a circle about the axis of the cylinder with the constant velocity

$$-\frac{\kappa}{2\pi . AB} = -\frac{\kappa . CA}{2\pi (CA^2 - c^2)},$$

where κ denotes the strength of A.

In the same way a single vortex of strength κ, situated inside a fixed circular cylinder, say at B, would describe a circle with constant velocity

$$\frac{\kappa . CB}{2\pi (c^2 - CB^2)}.$$

It is to be noticed, however*, that in the case of the external vortex the motion is not completely determinate unless, in addition to the strength κ, the value of the circulation in a circuit embracing the cylinder (but not the vortex) is prescribed. In the above solution, this circulation is that due to the vortex-image at B and is $-\kappa$. This may be annulled by the superposition of an additional vortex $+\kappa$ at C, in which case we have, for the velocity of A,

$$-\frac{\kappa . CA}{2\pi (CA^2 - c^2)} + \frac{\kappa}{2\pi . CA} = -\frac{\kappa c^2}{2\pi . CA (CA^2 - c^2)}.$$

For a prescribed circulation κ' we must add to this the term $\kappa'/2\pi . CA$.

L. Föppl †, using the method of images, has investigated the case of a cylinder advancing through fluid with velocity U, and followed by a vortex-pair symmetrically situated with respect to the line of advance of the centre. It appears that the vortices can maintain their position relative to the cylinder provided they lie on the curve

$$2ry = r^2 - a^2,$$

and that the strengths of the vortices corresponding to a given position on this curve are

$$\pm 2Uy\left(1 - \frac{a^4}{r^4}\right).$$

He finds, however, that the arrangement is unstable for anti-symmetrical disturbances.

Some paths of vortices in a stream past a cylindrical obstacle (with circulation) have been traced by Walton ‡. The path of a vortex in a semicircular region is investigated by K. De § by Routh's method referred to on p. 224.

3°. If we have four parallel rectilinear vortices whose centres form a rectangle $ABB'A'$, the strengths being κ for the vortices A', B, and $-\kappa$ for

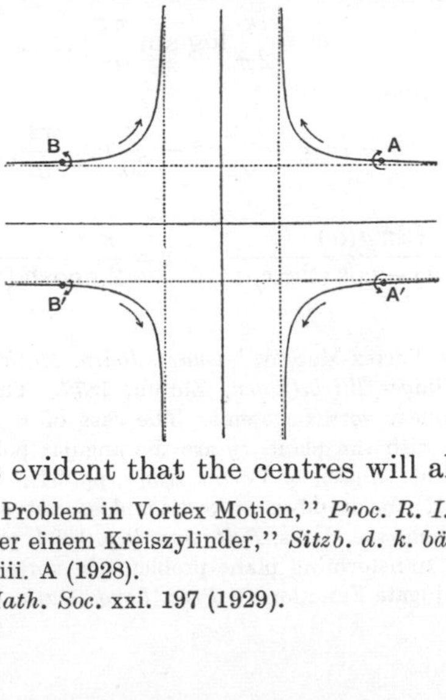

the vortices A, B', it is evident that the centres will always form a rectangle

* F. A. Tarleton, "On a Problem in Vortex Motion," *Proc. R. I. A.* December 12, 1892.

† "Wirbelbewegung hinter einem Kreiszylinder," *Sitzb. d. k. bayr. Akad. d. Wiss.* 1913.

‡ *Proc. R. I. Acad.* xxxviii. A (1928).

§ *Bull. of the Calcutta Math. Soc.* xxi. 197 (1929).

Further, the various rotations having the directions indicated in the figure, we see that the effect of the presence of the pair A, A' on B, B' is to separate them, and at the same time to diminish their velocity perpendicular to the line joining them. The planes which bisect AB, AA' at right angles may (either or both) be taken as fixed rigid boundaries. We thus get the case where a pair of vortices, of equal and opposite strengths, move towards (or from) a plane wall, or where a single vortex moves in the angle between two perpendicular walls.

If x, y be the co-ordinates of the vortex A relative to the planes of symmetry, we readily find

$$\dot{x} = -\frac{\kappa}{4\pi} \cdot \frac{x^2}{yr^2}, \qquad \dot{y} = \frac{\kappa}{4\pi} \cdot \frac{y^2}{xr^2}, \quad \text{.................................(2)}$$

where $r^2 = x^2 + y^2$. By division we obtain the differential equation of the path, viz.

$$\frac{dx}{x^3} + \frac{dy}{y^3} = 0,$$

whence

$$a^2 (x^2 + y^2) = 4x^2 y^2,$$

a being an arbitrary constant, or, transforming to polar co-ordinates,

$$r = \frac{a}{\sin 2\theta}. \quad \text{..(3)}$$

Also since

$$x\dot{y} - y\dot{x} = \frac{\kappa}{4\pi},$$

the vortex moves as if under a centre of force at the origin. This force is repulsive, and its law is that of the inverse cube*.

156. If we write, as in Chapter IV.,

$$z = x + iy, \qquad w = \phi + i\psi, \quad \text{.........................(1)}$$

the potential- and stream-functions due to an infinite row of equidistant vortices, each of strength κ, whose co-ordinates are

$$(0, 0), \quad (\pm a, 0), \quad (\pm 2a, 0), \ldots,$$

will be given by the formula

$$w = \frac{i\kappa}{2\pi} \log \sin \frac{\pi z}{a}; \quad \text{..............................(2)}$$

cf. Art. 64, 4°. This makes

$$u - iv = -\frac{dw}{dz} = -\frac{i\kappa}{2a} \cot \frac{\pi z}{a}, \quad \text{........................(3)}$$

whence

$$u = -\frac{\kappa}{2a} \frac{\sinh (2\pi y/a)}{\cosh (2\pi y/a) - \cos (2\pi x/a)}, \quad v = \frac{\kappa}{2a} \frac{\sin (2\pi x/a)}{\cosh (2\pi y/a) - \cos (2\pi x/a)}. \quad \text{......(4)}$$

* Greenhill, "On Plane Vortex-Motion," *Quart. Journ. Math.* xv. 10 (1878); Gröbli, *Die Bewegung paralleler geradliniger Wirbelfäden*, Zürich, 1877. These papers contain other interesting examples of rectilinear vortex-systems. The case of a system of equal and parallel vortices whose intersections with the plane xy are the angular points of a regular polygon was treated by J. J. Thomson in his *Motion of Vortex Rings*, pp. 94.... He finds that the configuration is stable if, and only if, the number of vortices does not exceed six. For some further references as to special problems see Hicks, *Brit. Ass. Rep.* 1882, pp. 41...; Love, *l.c. ante* p. 192.

An ingenious method of transforming plane problems in vortex-motion was given by Routh, "Some Applications of Conjugate Functions," *Proc. Lond. Math. Soc.* xii. 73 (1881).

These expressions make $u = \mp \frac{1}{2}\kappa/a$, $v = 0$, for $y = \pm \infty$; the row of vortices is in fact, as regards distant points, equivalent to a vortex-sheet of uniform strength κ/a (Art. 151).

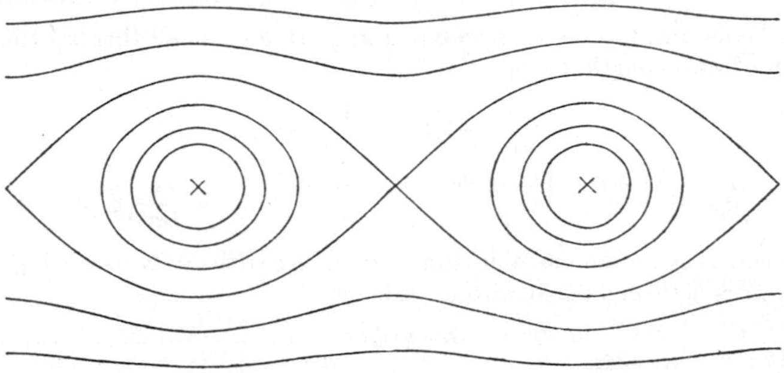

The diagram shews the arrangement of the stream-lines.

It follows easily that if there are two parallel rows of equidistant vortices, symmetrical with respect to the plane $y = 0$, the strengths being κ for the upper and $-\kappa$ for the lower row, as indicated on the next page, the whole system will advance with a uniform velocity

$$U = \frac{\kappa}{2a} \coth \frac{\pi b}{a}, \quad \text{......(5)}$$

where b is the distance between the two rows. The mean velocity in the plane of symmetry is κ/a. The velocity at a distance outside the two rows tends to the limit 0.

If the arrangement be modified so that each vortex in one row is opposite the centre of the interval between two consecutive vortices in the other row, as shewn on p. 228, the general velocity of advance is

$$V = \frac{\kappa}{2a} \tanh \frac{\pi b}{a}. \quad \text{......(6)}$$

The mean velocity in the medial plane is again κ/a.

The stability of these various arrangements has been discussed by von Kármán*. Taking first the case of the *single* row, let us suppose the vortex whose undisturbed co-ordinates are $(ma, 0)$ to be displaced to the point $(ma + x_m, y_m)$. The formulae of Art. 154 give, for the motion of the vortex initially at the origin,

$$\frac{dx_0}{dt} = -\frac{\kappa}{2\pi} \sum_m \frac{y_0 - y_m}{r_m^2}, \quad \frac{dy_0}{dt} = \frac{\kappa}{2\pi} \sum_m \frac{x_0 - x_m - ma}{r_m^2}, \quad \text{......(7)}$$

where

$$r_m^2 = (x_0 - x_m - ma)^2 + (y_0 - y_m)^2, \quad \text{......(8)}$$

and the summation with respect to m includes all positive and negative integral values, zero being of course excluded. If we neglect terms of the second order in the displacements, we find

$$\frac{dx_0}{dt} = -\frac{\kappa}{2\pi a^2} \sum_m \frac{y_0 - y_m}{m^2}, \quad \frac{dy_0}{dt} = -\frac{\kappa}{2\pi a^2} \sum_m \frac{1}{m} - \frac{\kappa}{2\pi a^2} \sum_m \frac{x_0 - x_m}{m^2} \quad \text{......(9)}$$

* "Flüssigkeits- u. Luftwiderstand," *Phys. Zeitschr.* xiii. 49 (1911); also *Gött. Nachr.* 1912, p. 547. The investigation is only given in outline in these papers; I have supplied various steps.

The first term in the value of dy_0/dt is to be omitted as being independent of the disturbance*.

Consider now a disturbance of the type

$$x_m = \alpha e^{im\phi}, \quad y_m = \beta e^{im\phi}, \qquad \text{.................................(10)}$$

where ϕ may be assumed to lie between 0 and 2π. If ϕ be small this has the character of an undulation of wave-length $2\pi a/\phi$. We find

$$\frac{d\alpha}{dt} = -\lambda\beta, \quad \frac{d\beta}{dt} = -\lambda\alpha, \qquad \text{.................................(11)}$$

where $$\lambda = \frac{\kappa}{\pi a^2}\left(\frac{1-\cos\phi}{1^2} + \frac{1-\cos 2\phi}{2^2} + \frac{1-\cos 3\phi}{3^2} + \ldots\right) = \frac{\kappa}{4\pi a^2}\phi(2\pi - \phi). \qquad \text{......(12)}$$

The arrangement is therefore unstable, the disturbance ultimately increasing as $e^{\lambda t}$. When the wave-length is large compared with a we have

$$\lambda = \tfrac{1}{2}\kappa\phi/a^2, \qquad \text{.......................................(13)}$$

approximately; cf. Art. 234.

Proceeding next to the case of the *symmetrical* double row, the positions at time t of vortices in the upper and lower rows may be taken to be

$$(ma + Ut + x_m,\ \tfrac{1}{2}b + y_m), \quad \text{and} \quad (na + Ut + x_n',\ -\tfrac{1}{2}b + y_n'),$$

respectively, where U denotes the general velocity of advance of the system, and the origin is in the plane of symmetry.

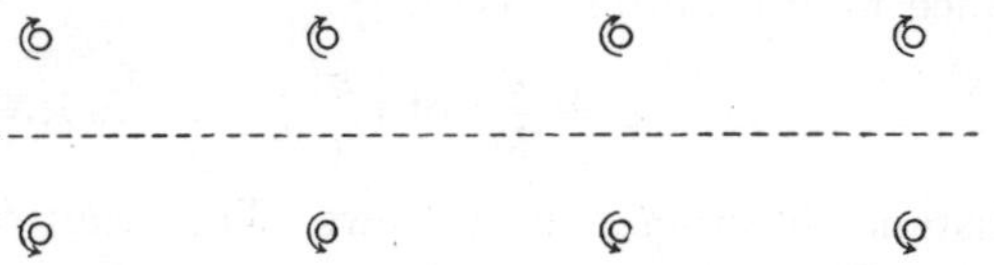

The component velocities of a vortex in the *upper* row, *e.g.* that for which $m=0$, due to the remaining vortices of the same row, will be given as before by (9), where the sum Σm^{-1} may be omitted. The components due to the vortex n of the lower row will be

$$\frac{\kappa}{2\pi}\frac{b + y_0 - y_n'}{r_n^2}, \quad -\frac{\kappa}{2\pi}\frac{x_0 - x_n' - na}{r_n^2},$$

where $$r_n^2 = (x_0 - x_n' - na)^2 + (y_0 - y_n' + b)^2.$$

If we neglect terms of the second order in the disturbance we find, after a little reduction,

$$\frac{2\pi}{\kappa}\left(\frac{dx_0}{dt} + U\right) = -\sum_m \frac{y_0 - y_m}{m^2a^2} + \sum_n \frac{b}{n^2a^2 + b^2} + \sum_n \frac{n^2a^2 - b^2}{(n^2a^2+b^2)^2}(y_0 - y_n')$$
$$+ \sum_n \frac{2nab}{(n^2a^2+b^2)^2}(x_0 - x_n'), \qquad \text{.............................(14)}$$

$$\frac{2\pi}{\kappa}\frac{dy_0}{dt} = -\sum_m \frac{x_0 - x_m}{m^2a^2} + \sum_n \frac{n^2a^2 - b^2}{(n^2a^2+b^2)^2}(x_0 - x_n')$$
$$- \sum_n \frac{2nab}{(n^2a^2+b^2)^2}(y_0 - y_n'), \qquad \text{.............................(15)}$$

where the summations with respect to n go from $-\infty$ to $+\infty$, including zero. The terms in (14) independent of the disturbance will cancel, since, by (5),

$$U = \frac{\kappa}{2a}\coth\frac{\pi b}{a} = \frac{\kappa}{2\pi}\sum_n \frac{b}{n^2a^2 + b^2}.$$

* In the summations the vortices are to be taken in pairs equidistant from the origin; otherwise the result would be indeterminate. The investigation may be regarded as applying to the central portions of a long, but not infinitely long, row; the term referred to is then negligible.

If we now put

$$x_m = \alpha e^{im\phi}, \quad y_m = \beta e^{im\phi}, \quad x_n' = \alpha' e^{in\phi}, \quad y_n' = \beta' e^{in\phi}, \quad \text{.................(16)}$$

where $0 < \phi < 2\pi$, the equations take the form

$$\left.\begin{aligned} \frac{2\pi a^2}{\kappa}\frac{d\alpha}{dt} &= \qquad\; - A\beta - B\alpha' - C\beta', \\ \frac{2\pi a^2}{\kappa}\frac{d\beta}{dt} &= -A\alpha \qquad\; - C\alpha' + B\beta'. \end{aligned}\right\} \quad \text{...........................(17)}$$

If we write, for shortness,

$$k = b/a, \quad \text{..(18)}$$

the values of the coefficients are*

$$A = \sum_m \frac{1 - e^{im\phi}}{m^2} - \sum_n \frac{n^2 - k^2}{(n^2 + k^2)^2} = \tfrac{1}{2}\phi(2\pi - \phi) + \frac{\pi^2}{\sinh^2 k\pi}, \quad \text{.................(19)}$$

$$B = \sum_n \frac{2nke^{in\phi}}{(n^2 + k^2)^2} = i\left\{\frac{\pi\phi\cosh k(\pi - \phi)}{\sinh k\pi} - \frac{\pi^2 \sinh k\phi}{\sinh^2 k\pi}\right\}, \quad \text{.................(20)}$$

$$C = \sum_n \frac{(n^2 - k^2)e^{in\phi}}{(n^2 + k^2)^2} = -\frac{\pi^2\cosh k\phi}{\sinh^2 k\pi} - \frac{\pi\phi\sinh k(\pi - \phi)}{\sinh k\pi}. \quad \text{...............(21)}$$

To deduce the equations relating to the *lower* row we have merely to reverse the signs of κ and b, and to interchange accented and unaccented letters. Hence

$$\left.\begin{aligned} \frac{2\pi a^2}{\kappa}\frac{d\alpha'}{dt} &= \qquad A\beta' - B\alpha + C\beta, \\ \frac{2\pi a^2}{\kappa}\frac{d\beta'}{dt} &= A\alpha' \qquad + C\alpha + B\beta. \end{aligned}\right\} \quad \text{..............................(22)}$$

The formulae (17) and (22) are the equations of motion of the vortex-system in what may be called a normal mode of the disturbance.

The solutions are of two types. In the first type we have

$$\alpha = \alpha', \quad \beta = -\beta', \quad \text{.......................................(23)}$$

and therefore

$$\left.\begin{aligned} \frac{2\pi a^2}{\kappa}\frac{d\alpha}{dt} &= -B\alpha - (A - C)\beta, \\ \frac{2\pi a^2}{\kappa}\frac{d\beta}{dt} &= -(A + C)\alpha - B\beta. \end{aligned}\right\} \quad \text{..............................(24)}$$

The solution involves exponentials $e^{\lambda t}$, the values of λ being given by

$$\frac{2\pi a^2}{\kappa}\lambda = -B \pm \surd(A^2 - C^2). \quad \text{...............................(25)}$$

In the second type we have

$$\alpha = -\alpha', \quad \beta = \beta', \quad \text{.......................................(26)}$$

and therefore

$$\left.\begin{aligned} \frac{2\pi a^2}{\kappa}\frac{d\alpha}{dt} &= B\alpha - (A + C)\beta, \\ \frac{2\pi a^2}{\kappa}\frac{d\beta}{dt} &= -(A - C)\alpha + B\beta. \end{aligned}\right\} \quad \text{..............................(27)}$$

The corresponding values of λ are given by

$$\frac{2\pi a^2}{\kappa}\lambda = B \pm \surd(A^2 - C^2). \quad \text{.................................(28)}$$

* The summations with respect to n can be derived from the Fourier expansion

$$\frac{\cosh k(\pi - \phi)}{\sinh k\pi} = \frac{1}{\pi}\left\{\frac{1}{k} + \frac{2k\cos\phi}{1^2 + k^2} + \frac{2k\cos 2\phi}{2^2 + k^2} + \dots\right\}.$$

Since B is a pure imaginary, whilst A and C are real, it is necessary for stability in each case that A^2 should not exceed C^2 for admissible values of ϕ. Now when $\phi=\pi$ we find

$$A+C=\tfrac{1}{2}\pi^2\tanh^2\tfrac{1}{2}k\pi,\quad A-C=\tfrac{1}{2}\pi^2\coth^2\tfrac{1}{2}k\pi, \quad\text{.....................(29)}$$

so that A^2-C^2 is positive. We conclude that both types are unstable.

Passing to the *unsymmetrical* case, we denote the positions of the displaced vortices by

$$(ma+Vt+x_m,\quad \tfrac{1}{2}b+y_m),\ \text{and}\ ((n+\tfrac{1}{2})a+Vt+x_n',\quad -\tfrac{1}{2}b+y_n'),$$

where V is given by (6). The requisite formulae are obtained by writing $n+\frac{1}{2}$ for n in preceding results.

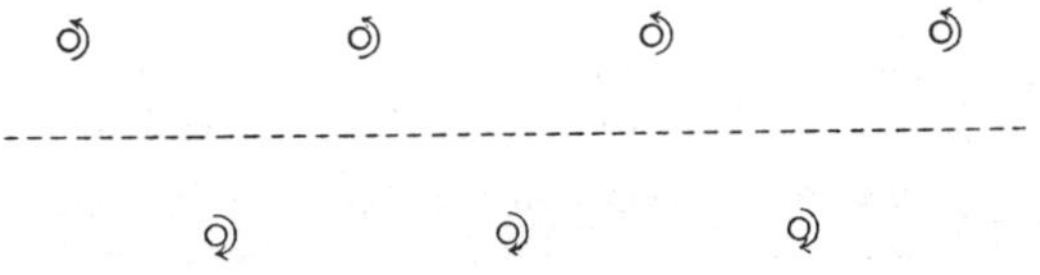

The equations (17) and (22) will accordingly apply, provided*

$$A=\sum_m\frac{1-e^{im\phi}}{m^2}-\sum_n\frac{(n+\frac{1}{2})^2-k^2}{\{(n+\frac{1}{2})^2+k^2\}^2}=\tfrac{1}{2}\phi(2\pi-\phi)-\frac{\pi^2}{\cosh^2 k\pi}, \quad\text{...............(30)}$$

$$B=\sum_n\frac{(2n+1)ke^{i(n+\frac{1}{2})\phi}}{\{(n+\frac{1}{2})^2+k^2\}^2}=i\left\{\frac{\pi\phi\sinh k(\pi-\phi)}{\cosh k\pi}+\frac{\pi^2\sinh k\phi}{\cosh^2 k\pi}\right\}, \quad\text{.........(31)}$$

$$C=\sum_n\frac{\{(n+\frac{1}{2})^2-k^2\}e^{i(n+\frac{1}{2})\phi}}{\{(n+\frac{1}{2})^2+k^2\}^2}=\frac{\pi^2\cosh k\phi}{\cosh^2 k\pi}-\frac{\pi\phi\cosh k(\pi-\phi)}{\cosh k\pi}. \quad\text{.........(32)}$$

These values of A, B, C are to be substituted in (25) and (28). As in the former case it is necessary for stability that A^2 should not be greater than C^2. Now when $\phi=\pi$, $C=0$; hence A must also vanish, or

$$\cosh^2 k\pi=2,\quad k\pi=\cdot 8814,\quad b/a=k=\cdot 281. \quad\text{.........................(33)}$$

The configuration is therefore unstable unless the ratio of the interval between the two rows to the distance between consecutive vortices has precisely this value.

To determine whether the arrangement is stable, under the above condition, for all values of ϕ from 0 to 2π, let us write for a moment $k(\pi-\phi)=x$, $k\pi=\mu$, so that

$$k^2A=-\tfrac{1}{2}x^2,\quad k^2C=\tfrac{1}{2}(\mu x\cosh\mu x\cosh x-\mu^2\sinh\mu\sinh x), \quad\text{............(34)}$$

where x may range between $\pm\mu$. Since A is an even and C an odd function of x, it is sufficient for comparison of absolute values to suppose x positive. Hence, writing

$$y=\mu\cosh\mu\cosh x-\mu^2\sinh\mu\frac{\sinh x}{x}-x, \quad\text{.........................(35)}$$

we have to ascertain whether this is positive for $0<x<\mu$. Since $\mu=\cdot 8814$, $\cosh\mu=\sqrt{2}$, $\sinh\mu=1$, y is positive for $x=0$, and it evidently vanishes for $x=\mu$. Again

$$\frac{dy}{dx}=\mu\cosh\mu\sinh x+\mu^2\sinh\mu\frac{\sinh x}{x^2}-\mu^2\sinh\mu\frac{\cosh x}{x}-1, \quad\text{............(36)}$$

which is equal to -1 for $x=0$, and vanishes for $x=\mu$. Finally,

$$\frac{d^2y}{dx^2}=\mu\cosh\mu\cosh x-\mu^2\sinh\mu\frac{\sinh x}{x}+2\mu^2\sinh\mu\frac{\cosh x}{x^2}-2\mu^2\sinh\mu\frac{\sinh x}{x^3}, \quad\text{...(37)}$$

* The summations with respect to n can be derived from the expansion

$$\frac{\sinh k(\pi-\phi)}{\cosh k\pi}=\frac{2}{\pi}\left\{\frac{k\cos\frac{1}{2}\phi}{(\frac{1}{2})^2+k^2}+\frac{k\cos\frac{3}{2}\phi}{(\frac{3}{2})^2+k^2}+\dots\right\}.$$

which is easily seen to be positive for all values of x, since $(\tanh x)/x<1$. Hence as x increases from 0 to μ, dy/dx is steadily increasing from -1 to 0, and is therefore negative. Hence y steadily diminishes from its initial positive value to zero, and is therefore positive.

We conclude that the configuration is definitely stable* except for $x=\pm\mu$, when $\phi=0$ or 2π, in which cases $B=0$, by (31), and therefore $\lambda=0$. Since the disturbed particles are then all in the same phase, the reason why the period of disturbance should be infinite is easily perceived.

This unsymmetrical configuration is of special interest because it is exemplified in the trail of vortices which is often observed in the wake of a cylindrical body advancing through a fluid. This has suggested further researches.

The effect of lateral rigid boundaries equidistant from the medial line on the stability of the configuration has been discussed by Rosenhead†. He finds that as the ratio a/h of the interval a between successive vortices in the same row to the distance h between the walls increases from zero to ·815 the unsymmetrical arrangement is stable only for a definite value of b/a, which decreases continuously from ·281 to ·256. But when $a/h>$ ·815 there is stability for a certain range of values of b/a. And when $a/h>1{\cdot}419$ the configuration is stable for all values of b/a.

The symmetrical configuration, on the other hand, is always unstable.

157. When, as in the case of a vortex-pair, or a system of vortex-pairs, the algebraic sum of the strengths of all the vortices is zero, we may work out a theory of the 'impulse,' in two dimensions, analogous to that given in Arts. 119, 152 for the case of a finite vortex-system. The detailed examination of this must be left to the reader. If P, Q denote the components of the impulse parallel to x and y, and N its moment about Oz, all reckoned per unit depth of the fluid parallel to z, it will be found that

$$\left.\begin{aligned} P=\rho\iint y\zeta\,dx\,dy, \quad Q=-\rho\iint x\zeta\,dx\,dy,\\ N=-\tfrac{1}{2}\rho\iint(x^2+y^2)\,\zeta\,dx\,dy. \end{aligned}\right\} \qquad\text{(1)}$$

For instance, in the case of a single vortex-pair, the strengths of the two vortices being $\pm\kappa$, and their distance apart c, the impulse is $\rho\kappa c$, in a line bisecting c at right angles.

The constancy of the impulse gives

$$\left.\begin{aligned} \Sigma\kappa x=\text{const.}, \quad \Sigma\kappa y=\text{const.},\\ \Sigma\kappa\,(x^2+y^2)=\text{const.} \end{aligned}\right\} \qquad\text{(2)}$$

It may also be shewn that the energy of the motion in the present case is given by

$$T=-\tfrac{1}{2}\rho\iint\psi\zeta\,dx\,dy=-\tfrac{1}{2}\rho\Sigma\kappa\psi. \qquad\text{(3)}$$

When $\Sigma\kappa$ is not zero, the energy and the moment of the impulse are both infinite, as may be easily verified in the case of a single rectilinear vortex.

* This is stated without proof by Kármán.

† *Phil. Trans.* A, ccviii. 275 (1929). See also Glauert, *Proc. Roy. Soc.* A, cxx. 34 (1928).

The theory of a system of isolated rectilinear vortices has been put in a very elegant form by Kirchhoff*.

Denoting the positions of the centres of the respective vortices by (x_1, y_1), (x_2, y_2), ... and their strengths by $\kappa_1, \kappa_2, \ldots$, it is evident from Art. 154 that we may write

$$\left.\begin{aligned} \kappa_1 \frac{dx_1}{dt} &= -\frac{\partial W}{\partial y_1}, & \kappa_1 \frac{dy_1}{dt} &= \frac{\partial W}{\partial x_1}, \\ \kappa_2 \frac{dx_2}{dt} &= -\frac{\partial W}{\partial y_2}, & \kappa_2 \frac{dy_2}{dt} &= \frac{\partial W}{\partial x_2}, \\ &\ldots\ldots\ldots, \end{aligned}\right\} \quad \text{.........(4)}$$

where

$$W = \frac{1}{2\pi} \Sigma \kappa_1 \kappa_2 \log r_{12}, \quad \text{.........(5)}$$

if r_{12} denote the distance between the vortices κ_1, κ_2.

Since W depends only on the *relative* configuration of the vortices, its value is unaltered when $x_1, x_2, \ldots$ are increased by the same amount, whence $\Sigma \partial W/\partial x_1 = 0$, and, in the same way, $\Sigma \partial W/\partial y_1 = 0$. This gives the first two of equations (2), but the proof is not now limited to the case of $\Sigma\kappa = 0$. The argument is in fact substantially the same as in Art. 154. Again, we obtain from (4)

$$\Sigma\kappa \left(x \frac{dx}{dt} + y \frac{dy}{dt} \right) = -\Sigma \left(x \frac{\partial W}{\partial y} - y \frac{\partial W}{\partial x} \right),$$

or if we introduce polar co-ordinates (r_1, θ_1), (r_2, θ_2), ... for the several vortices,

$$\Sigma\kappa r \frac{dr}{dt} = -\Sigma \frac{\partial W}{\partial \theta}. \quad \text{.........(6)}$$

Since W is unaltered by a rotation of the axes of co-ordinates in their own plane about the origin, we have $\Sigma \partial W/\partial \theta = 0$, whence

$$\Sigma\kappa r^2 = \text{const.}, \quad \text{.........(7)}$$

which agrees with the third of equations (2), but is free from the restriction there implied.

An additional integral of (4) is obtained as follows. We have

$$\Sigma\kappa \left(x \frac{dy}{dt} - y \frac{dx}{dt} \right) = \Sigma \left(x \frac{\partial W}{\partial x} + y \frac{\partial W}{\partial y} \right),$$

or

$$\Sigma\kappa r^2 \frac{d\theta}{dt} = \Sigma r \frac{\partial W}{\partial r}. \quad \text{.........(8)}$$

If every r be increased in the ratio $1+\epsilon$, where ϵ is infinitesimal, the increment of W is equal to $\Sigma \epsilon r . \partial W/\partial r$. But since the new configuration of the vortex-system is geometrically similar to the former one, the mutual distances r_{12} are altered in the same ratio $1+\epsilon$, and therefore, from (5), the increment of W is $\epsilon/2\pi . \Sigma\kappa_1\kappa_2$. Hence (8) may be written in the form

$$\Sigma\kappa r^2 \frac{d\theta}{dt} = \frac{1}{2\pi} \Sigma \kappa_1 \kappa_2. \quad \text{.........(9)}$$

158. The preceding results are independent of the form of the sections of the vortices, so long as the dimensions of these sections are small compared with the mutual distances of the vortices themselves. The simplest case is when the sections are circular, and it is of interest to inquire whether this form is stable. This question has been examined by Kelvin†.

* *Mechanik*, c. xx.

† Sir W. Thomson, "On the Vibrations of a Columnar Vortex," *Phil. Mag.* (5), x. 155 (1880) [*Papers*, iv. 152].

When the disturbance is in two dimensions only, the calculations are very simple. Let us suppose, as in Art. 27, that the space within a circle $r=a$, having the centre as origin, is occupied by fluid having a uniform vorticity ω, and that this is surrounded by fluid moving irrotationally. If the motion be continuous at this circle we have, for $r<a$,

$$\psi = -\tfrac{1}{4}\omega(a^2-r^2), \quad\text{.........(1)}$$

while for $r>a$,

$$\psi = -\tfrac{1}{2}\omega a^2 \log a/r. \quad\text{.........(2)}$$

To examine the effect of a slight irrotational disturbance, we assume, for $r<a$,

$$\left.\begin{aligned} \psi &= -\tfrac{1}{4}\omega(a^2-r^2) + A\frac{r^s}{a^s}\cos(s\theta-\sigma t), \\ \text{and, for } r>a, \quad \psi &= -\tfrac{1}{2}\omega a^2 \log\frac{a}{r} + A\frac{a^s}{r^s}\cos(s\theta-\sigma t), \end{aligned}\right\} \quad\text{.........(3)}$$

where s is integral, and σ is to be determined. The constant A must have the same value in these two expressions, since the radial component of the velocity, $-\partial\psi/r\partial\theta$, must be continuous at the boundary of the vortex, for which $r=a$, approximately. Assuming for the equation to this boundary

$$r = a + \alpha\cos(s\theta - \sigma t), \quad\text{.........(4)}$$

we have still to express that the transverse component $(\partial\psi/\partial r)$ of the velocity is continuous. This gives

$$\tfrac{1}{2}\omega r + s\frac{A}{a}\cos(s\theta-\sigma t) = \frac{\tfrac{1}{2}\omega a^2}{r} - s\frac{A}{a}\cos(s\theta-\sigma t).$$

Substituting from (4), and neglecting the square of α, we find

$$\omega\alpha = -2sA/a. \quad\text{.........(5)}$$

So far the work is purely kinematical; the dynamical theorem that the vortex-lines move with the fluid shews that the normal velocity of a particle on the boundary must be equal to that of the boundary itself. This condition gives

$$\frac{\partial r}{\partial t} = -\frac{\partial\psi}{r\partial\theta} - \frac{\partial\psi}{\partial r}\frac{\partial r}{r\partial\theta},$$

where r has the value (4), or

$$\sigma\alpha = s\frac{A}{a} + \tfrac{1}{2}\omega a \cdot \frac{s\alpha}{a}. \quad\text{.........(6)}$$

Eliminating the ratio A/α between (5) and (6) we find

$$\sigma = \tfrac{1}{2}(s-1)\omega. \quad\text{.........(7)}$$

Hence the disturbance represented by the plane harmonics in (3) consists of a system of corrugations travelling round the circumference of the vortex with an angular velocity

$$\sigma/s = (s-1)/s \cdot \tfrac{1}{2}\omega. \quad\text{.........(8)}$$

This is the angular velocity in space; relative to the rotating fluid the angular velocity is

$$\sigma/s - \tfrac{1}{2}\omega = -\tfrac{1}{2}\omega/s, \quad\text{.........(9)}$$

the direction being opposite to that of the rotation. When $s=2$, the disturbed section is an ellipse which rotates about its centre with angular velocity $\tfrac{1}{4}\omega$.

The three-dimensional oscillations of an isolated columnar vortex-filament have also been discussed by Kelvin in the paper cited. The columnar form is found to be stable for disturbances of a general character.

In a recent paper Rosenhead* has examined the stability of the Kármán unsymmetrical arrangement when the cross-sections are of *finite* area. The conclusion is that there is stability for strictly two-dimensional disturbances, but instability for sinusoidal longitudinal deformations, whose wave-length bears less than a certain ratio to the diameter.

* *Proc. Roy. Soc.* A, cxxvii. 590 (1930).

159. The particular case of a two dimensional elliptic disturbance can be solved without approximation as follows*.

Let us suppose that the space within the ellipse

$$\frac{x^2}{a^2}+\frac{y^2}{b^2}=1 \qquad\qquad (1)$$

is occupied by liquid having a uniform vorticity ω, whilst the surrounding fluid is moving irrotationally. It will appear that the conditions of the problem can all be satisfied if we imagine the elliptic boundary to rotate, without change of shape, with a constant angular velocity (n, say), to be determined.

The formula for the external space can be at once written down from Art. 72, 4°; viz. we have

$$\psi=\tfrac{1}{4}n(a+b)^2e^{-2\xi}\cos 2\eta+\tfrac{1}{2}\omega ab\xi, \qquad\qquad (2)$$

where ξ, η now denote the elliptic co-ordinates of Art. 71, 3°, and the cyclic constant κ has been put $=\pi ab\omega$.

The value of ψ for the internal space has to satisfy

$$\frac{\partial^2\psi}{\partial x^2}+\frac{\partial^2\psi}{\partial y^2}=\omega, \qquad\qquad (3)$$

with the boundary-condition $\quad \dfrac{ux}{a^2}+\dfrac{vy}{b^2}=-ny\,.\,\dfrac{x}{a^2}+nx\,.\,\dfrac{y}{b^2}. \qquad\qquad (4)$

These conditions are both fulfilled by

$$\psi=\tfrac{1}{2}\omega(Ax^2+By^2), \qquad\qquad (5)$$

provided $\qquad A+B=1, \qquad Aa^2-Bb^2=\dfrac{n}{\omega}(a^2-b^2). \qquad\qquad (6)$

It remains to express that there is no tangential slipping at the boundary of the vortex; *i.e.* that the values of $\partial\psi/\partial\xi$ obtained from (2) and (5) there coincide. Putting $x=c\cosh\xi\cos\eta$, $y=c\sinh\xi\sin\eta$, where $c=\surd(a^2-b^2)$, differentiating, and equating coefficients of $\cos 2\eta$, we obtain the additional condition

$$-\tfrac{1}{2}n(a+b)^2e^{-2\xi}=\tfrac{1}{2}\omega c^2(A-B)\cosh\xi\sinh\xi,$$

where ξ is the parameter of the ellipse (1). This is equivalent to

$$A-B=-\frac{n}{\omega}\cdot\frac{a^2-b^2}{ab}, \qquad\qquad (7)$$

since, at points of the ellipse, $\cosh\xi=a/c$, $\sinh\xi=b/c$.

Combined with (6) this gives $\qquad Aa=Bb=\dfrac{ab}{a+b}, \qquad\qquad (8)$

and $\qquad n=\dfrac{ab}{(a+b)^2}\omega. \qquad\qquad (9)$

When $a=b$, this agrees with our former approximate result.

The component velocities $\dot{x}$, $\dot{y}$ of a particle of the vortex relative to the principal axes of the ellipse are given by

$$\dot{x}=-\frac{\partial\psi}{\partial y}+ny, \qquad \dot{y}=\frac{\partial\psi}{\partial x}-nx,$$

whence we find $\qquad \dfrac{\dot{x}}{a}=-n\dfrac{y}{b}, \qquad \dfrac{\dot{y}}{b}=n\dfrac{x}{a}. \qquad\qquad (10)$

* Kirchhoff, *Mechanik*, c. xx.; Basset, *Hydrodynamics*, ii. 41.

Integrating, we find $x = ka\cos(nt+\epsilon)$, $y = kb\sin(nt+\epsilon)$, ...(11)

where k, ϵ are arbitrary constants, so that the *relative* paths of the particles are ellipses similar to the boundary of the vortex, described according to the harmonic law. If x', y' be the co-ordinates relative to axes fixed in space, we find

$$\left.\begin{aligned} x' &= x\cos nt - y\sin nt = \tfrac{1}{2}k(a+b)\cos(2nt+\epsilon) + \tfrac{1}{2}k(a-b)\cos\epsilon, \\ y' &= x\sin nt + y\cos nt = \tfrac{1}{2}k(a+b)\sin(2nt+\epsilon) - \tfrac{1}{2}k(a-b)\sin\epsilon. \end{aligned}\right\} \quad \text{.........(12)}$$

The absolute paths are therefore circles described with angular velocity $2n$*.

159 a. The motion of a solid in a liquid endowed with vorticity is a problem of considerable interest, but is unfortunately not very tractable. The only exception is when the motion is two-dimensional, and the vorticity uniform.

Let x_0, y_0 be the co-ordinates, relative to fixed axes, of a point C of the (cylindrical) solid; let x, y be the co-ordinates of any point of the fluid relative to parallel axes through C, and let (u, v) be the velocity relative to C. We have then

$$\left.\begin{aligned} \frac{\partial u}{\partial t} + \ddot{x}_0 + u\frac{\partial u}{\partial x} + v\frac{\partial v}{\partial x} - \zeta v &= -\frac{1}{\rho}\frac{\partial p}{\partial x}, \\ \frac{\partial v}{\partial t} + \ddot{y}_0 + u\frac{\partial u}{\partial y} + v\frac{\partial v}{\partial y} + \zeta u &= -\frac{1}{\rho}\frac{\partial p}{\partial y}; \end{aligned}\right\} \quad \text{..........................(1)}$$

cf. Arts. 12 (3) and 146 (5). Since

$$u = -\frac{\partial \psi}{\partial y}, \qquad v = \frac{\partial \psi}{\partial x}, \quad \text{.....................................(2)}$$

and ζ is constant, it appears that $\partial u/\partial t$ and $\partial v/\partial t$ are the derivatives with respect to x and y, respectively, of a certain function of x, y, t. Denoting this function by $-\partial\phi/\partial t$, we have

$$\frac{\partial}{\partial x}\left(\frac{\partial \phi}{\partial t}\right) = -\frac{\partial u}{\partial t} = \frac{\partial}{\partial y}\left(\frac{\partial \psi}{\partial t}\right), \quad \frac{\partial}{\partial y}\left(\frac{\partial \phi}{\partial t}\right) = -\frac{\partial v}{\partial t} = -\frac{\partial}{\partial x}\left(\frac{\partial \psi}{\partial t}\right), \quad \text{...........(3)}$$

which are the conditions that $\dfrac{d}{dt}(\phi + i\psi)$

should be a function of the complex variable $x+iy$. This consideration determines $\partial\phi/\partial t$ when the form of ψ is known†.

The equations (1) now give

$$\frac{p}{\rho} = \frac{\partial \phi}{\partial t} - (\ddot{x}_0 x + \ddot{y}_0 y) - \tfrac{1}{2}q^2 + \zeta\psi, \quad \text{.............................(4)}$$

where $$q^2 = u^2 + v^2. \quad \text{..(5)}$$

We proceed to apply these results to some cases of motion of a *circular* cylinder. The point C is naturally taken on its axis.

Let us suppose in the first instance that the undisturbed motion of the fluid consists of a uniform rotation ω about the origin, so that $\zeta = 2\omega$. The stream-function for the motion relative to a moving point (x_0, y_0) is then

$$\begin{aligned} \psi_0 &= \tfrac{1}{2}\omega\{(x_0+x)^2 + (y_0+y)^2\} + \dot{x}_0 y - \dot{y}_0 x \\ &= \tfrac{1}{2}\omega r^2 + \omega r(x_0\cos\theta + y_0\sin\theta) + \tfrac{1}{2}\omega(x_0^2 + y_0^2) + r(\dot{x}_0\sin\theta - \dot{y}_0\cos\theta), \quad \text{......(6)} \end{aligned}$$

* For further researches in this connection see Hill, "On the Motion of Fluid part of which is moving rotationally and part irrotationally," *Phil. Trans.* 1884; Love, "On the Stability of certain Vortex Motions," *Proc. Lond. Math. Soc.* (1) xxv. 18 (1893).

† Cf. Proudman, "On the Motion of Solids in a Liquid possessing Vorticity," *Proc. Roy. Soc.* A, xcii. 408 (1916).

where we have introduced polar co-ordinates relative to C. The relative stream-function for the disturbed motion will be

$$\psi=\tfrac{1}{2}\omega r^2+\omega\left(r-\frac{a^2}{r}\right)(x_0\cos\theta+y_0\sin\theta)+\tfrac{1}{2}\omega(x_0^2+y_0^2)+\left(r-\frac{a^2}{r}\right)(\dot{x}_0\sin\theta-\dot{y}_0\cos\theta). \quad \text{......(7)}$$

For this satisfies $\nabla_1^2\psi=2\omega$; it makes $\psi=\text{const.}$ for $r=a$; and it agrees with (6) for $r=\infty$.

Hence
$$\frac{\partial\psi}{\partial t}=\omega\left(r-\frac{a^2}{r}\right)(\dot{x}_0\cos\theta+\dot{y}_0\sin\theta)+\left(r-\frac{a^2}{r}\right)(\ddot{x}_0\sin\theta-\ddot{y}_0\cos\theta), \quad \text{............(8)}$$

and therefore

$$\frac{\partial\phi}{\partial t}=-\omega\left(r+\frac{a^2}{r}\right)(\dot{x}_0\sin\theta-\dot{y}_0\cos\theta)+\left(r+\frac{a^2}{r}\right)(\ddot{x}_0\cos\theta+\ddot{y}_0\sin\theta), \quad \text{.........(9)}$$

terms independent of r and θ being omitted. Again we have, for $r=a$,

$$\frac{\partial\psi}{r\partial\theta}=0, \quad \frac{\partial\psi}{\partial r}=\omega a+2\omega(x_0\cos\theta+y_0\sin\theta)+2(\dot{x}_0\sin\theta-\dot{y}_0\cos\theta),$$

and therefore

$$\tfrac{1}{2}q^2=2\omega^2a(x_0\cos\theta+y_0\sin\theta)+2\omega a(\dot{x}_0\sin\theta-\dot{y}_0\cos\theta)+\text{etc.}, \quad \text{............(10)}$$

where terms are omitted which will contribute nothing to the resultant force on the cylinder. Substituting in (4) we find, for $r=a$,

$$\frac{p}{\rho}=a(\ddot{x}_0\cos\theta+\ddot{y}_0\sin\theta)-4\omega a(\dot{x}_0\sin\theta-\dot{y}_0\cos\theta)-2\omega^2a(x_0\cos\theta+y_0\sin\theta)+\text{etc.} \quad \text{...(11)}$$

The component forces on the cylinder, due to fluid pressure, are therefore*

$$\left.\begin{aligned}-\int_0^{2\pi}p\cos\theta\,a\,d\theta&=-M'(\ddot{x}_0+4\omega\dot{y}_0-2\omega^2x_0),\\-\int_0^{2\pi}p\sin\theta\,a\,d\theta&=-M'(\ddot{y}_0-4\omega\dot{x}_0-2\omega^2y_0),\end{aligned}\right\} \quad \text{.....................(12)}$$

where $M'=\pi\rho a^2$. Hence if M be the mass per unit length of the cylinder itself, the equations of motion are

$$\left.\begin{aligned}\mu\ddot{x}+4\omega\dot{y}-2\omega^2x&=X/M',\\\mu\ddot{y}-4\omega\dot{x}-2\omega^2y&=Y/M',\end{aligned}\right\} \quad \text{.................................(13)}$$

where $\mu=1+M/M'$, and the zero suffixes have been omitted as no longer necessary. If we write $z=x+iy$, these equations are equivalent to

$$\mu\ddot{z}-4i\omega\dot{z}-2\omega^2z=(X+iY)/M'. \quad \text{..............................(14)}$$

To ascertain the free motion, when $X=0$, $Y=0$, we assume that $z\propto e^{im\omega t}$, and find

$$\mu m^2-4m+2=0. \quad \text{......................................(15)}$$

If $\mu<2$, *i.e.* if the mass of the cylinder is less than that of the fluid which it displaces, the values of m are real, and the solution has the form

$$z=Ae^{im_1\omega t}+Be^{im_2\omega t}, \quad \text{...................................(16)}$$

where m_1, m_2 are positive. This represents motion in a 'direct' epicyclic. As special cases circular paths are possible, and are stable. If on the other hand $\mu>2$, the values of m are complex, and the solution takes the form

$$z=(Ae^{\alpha t}+Be^{-\alpha t})e^{i\beta t}, \quad \text{...................................(17)}$$

the ultimate path being an equiangular spiral. If $\mu=2$, we have $(m-1)^2=0$, and

$$z=(A+Bt)e^{i\omega t}. \quad \text{.......................................(18)}$$

* Cf. G. I. Taylor, "Motion of Solids in Fluids when the Flow is not Irrotational," *Proc. Roy. Soc.* A, xciii. 99 (1916).

Hence, although it is possible as we should expect for a cylinder having the same mean density as the fluid to revolve with the latter in a circular path, this motion is unstable.

If there is a radial force whose direction revolves with the fluid, say

$$X+iY=Re^{i\omega t}, \qquad (19)$$

the equation (14) is satisfied, when $\mu=2$, by

$$z=re^{i\omega t}, \qquad (20)$$

provided

$$\ddot{r}=\tfrac{1}{2}R/M'. \qquad (21)$$

The cylinder can therefore move, relatively to the rotating fluid, along a radius*, but this motion, again, must be classed as unstable†.

Let us next suppose that the fluid when undisturbed is in laminar motion parallel to Ox, with constant vorticity 2ω, the stream-function being

$$\psi_0=\omega\,(y_0+y)^2=\tfrac{1}{2}\omega r^2\,(1-\cos 2\theta)+2\omega y_0\, r\sin\theta+\omega y_0^2. \qquad (22)$$

In the disturbed motion relative to the cylinder

$$\psi=\tfrac{1}{2}\omega r^2-\tfrac{1}{2}\omega\left(r^2-\frac{a^4}{r^2}\right)\cos 2\theta+2\omega y_0\left(r-\frac{a^2}{r}\right)\sin\theta+\omega y_0^2+\left(r-\frac{a^2}{r}\right)(\dot{x}_0\sin\theta-\dot{y}_0\cos\theta). \qquad (23)$$

Hence

$$\frac{\partial\psi}{\partial t}=2\omega\dot{y}_0\left(r-\frac{a^2}{r}\right)\sin\theta+\left(r-\frac{a^2}{r}\right)(\ddot{x}_0\sin\theta-\ddot{y}_0\cos\theta), \qquad (24)$$

the terms independent of r and θ being omitted. We write therefore

$$\frac{\partial\phi}{\partial t}=2\omega\dot{y}_0\left(r+\frac{a^2}{r}\right)\cos\theta+\left(r+\frac{a^2}{r}\right)(\ddot{x}_0\cos\theta+\ddot{y}_0\sin\theta). \qquad (25)$$

For $r=a$ we have from (23)

$$\frac{\partial\psi}{r\partial\theta}=0,\quad \frac{\partial\psi}{\partial r}=-\omega a+4\omega a\sin^2\theta+4\omega y_0\sin\theta+2\,(\dot{x}_0\sin\theta-\dot{y}_0\cos\theta), \qquad (26)$$

and therefore

$$\tfrac{1}{2}q^2=-4\omega^2 a y_0\sin\theta-2\omega a\,(\dot{x}_0\sin\theta-\dot{y}_0\cos\theta)+16\omega^2 a y_0\sin^3\theta$$
$$+8\omega a y_0\,(\dot{x}_0\sin^2\theta-\dot{y}_0\sin^2\theta\cos\theta)+\text{etc.}, \qquad (27)$$

those terms only being retained which will contribute to the resultant force on the cylinder. Substituting in (4) we find, for $r=a$,

$$\frac{p}{\rho}=a\,(\ddot{x}_0\cos\theta+\ddot{y}_0\sin\theta)+2\omega a\dot{x}_0\,(\sin\theta-4\sin^3\theta)+2\omega a\dot{y}_0\,(\cos\theta+4\sin^2\theta\cos\theta)$$
$$+4\omega^2 a y_0\,(\sin\theta-4\sin^3\theta)+\text{etc.} \qquad (28)$$

Hence‡

$$\left.\begin{aligned}-\int_0^{2\pi}p\cos\theta\, a\, d\theta&=-M'(\ddot{x}_0+4\omega\dot{y}_0),\\ -\int_0^{2\pi}p\sin\theta\, a\, d\theta&=-M'(\ddot{y}_0-4\omega\dot{x}_0-8\omega^2 y_0).\end{aligned}\right\} \qquad (29)$$

The equations of motion of the cylinder are therefore, omitting the suffixes,

$$\left.\begin{aligned}\mu\ddot{x}+4\omega\dot{y}&=X/M',\\ \mu\ddot{y}-4\omega\dot{x}-8\omega^2 y&=Y/M'.\end{aligned}\right\} \qquad (30)$$

We notice that the cylinder can remain at relative rest subject to a force

$$Y=-8\omega^2 M'y=4\omega M'U=2\kappa\rho U, \qquad (31)$$

* Cf. Taylor, *l.c.*

† Some cases of motion of a *sphere* in rotating fluid have been studied by Proudman, *l.c.*; S. F. Grace, *Proc. Roy. Soc.* A, cii. 89 (1922); and Taylor, *Proc. Roy. Soc.* **A**, cii. 180 (1922).

‡ Cf. Taylor, *l.c.*

where $U(=-2\omega y)$ is the velocity of the undisturbed stream at the level of the centre, and $\kappa\ (=2\pi a^2\omega)$ is the circulation immediately round the cylinder. This result may be contrasted with Art. 69 (6).

It is easily found from (30) that, if $\mu<2$, the path when there are no extraneous forces is a trochoid whose general direction of advance is parallel to the stream.

160. It was pointed out in Art. 80 that the motion of an incompressible fluid in a curved stratum of small and uniform thickness is completely defined by a stream-function ψ, so that any kinematical problem of this kind may be transformed by projection into one relating to a plane stratum. If, further, the projection be 'orthomorphic,' the kinetic energy of corresponding portions of liquid, and the circulations in corresponding circuits, are the same in the two motions. The latter statement shews that vortices transform into vortices of equal strengths. It follows at once from Art. 145 that in the case of a *closed* simply-connected surface the algebraic sum of the strengths of all the vortices present is zero.

We may apply this to motion in a spherical stratum. The simplest case is that of a pair of isolated vortices situated at antipodal points; the stream-lines are then parallel small circles, the velocity varying inversely as the radius of the circle. For a vortex-pair situate at *any* two points A, B, the stream-lines are coaxal circles as in Art. 80. It is easily found by the method of stereographic projection that the velocity at any point P is the resultant of two velocities $\kappa/2\pi a \,.\, \cot\frac{1}{2}\theta_1$ and $\kappa/2\pi a \,.\, \cot\frac{1}{2}\theta_2$, perpendicular respectively to the great-circle arcs AP, BP, where θ_1, θ_2 denote the lengths of these arcs, a the radius of the sphere, and $\pm\kappa$ the strengths of the vortices. The centre* (see Art. 154) of either vortex moves perpendicular to AB with a velocity $\kappa/2\pi a \,.\, \cot\frac{1}{2}AB$. The two vortices therefore describe parallel and equal small circles, remaining at a constant distance from each other.

Circular Vortices.

161. Let us next take the case where all the vortices present in the liquid (supposed unlimited as before) are circular, having the axis of x as a common axis. Let ϖ denote the distance of any point P from this axis, v the velocity in the direction of ϖ, and ω the resultant vorticity at P. It is evident that u, v, ω are functions of x, ϖ only.

Under these circumstances there exists a stream-function ψ, defined as in Art. 94, viz. we have

$$u=-\frac{1}{\varpi}\frac{\partial\psi}{\partial\varpi},\qquad v=\frac{1}{\varpi}\frac{\partial\psi}{\partial x},\qquad\text{.......................(1)}$$

whence

$$\omega=\frac{\partial v}{\partial x}-\frac{\partial u}{\partial\varpi}=\frac{1}{\varpi}\left(\frac{\partial^2\psi}{\partial x^2}+\frac{\partial^2\psi}{\partial\varpi^2}-\frac{1}{\varpi}\frac{\partial\psi}{\partial\varpi}\right).\qquad\text{..............(2)}$$

It is easily seen from the expressions (7) of Art. 148 that the vector (F, G, H) will under the present conditions be everywhere perpendicular to

* To prevent possible misconception it may be remarked that the centres of corresponding vortices are not necessarily corresponding points. The paths of these centres are therefore not in general projective.

the axis of x and the radius ϖ. If we denote its magnitude by S, the flux through the circle (x, ϖ) will be $2\pi\varpi S$, whence

$$\psi = -\varpi S. \quad \text{.........(3)}$$

To find the value of ψ at (x, ϖ) due to a single vortex-filament of circulation κ, whose co-ordinates are x', ϖ', we note that the element which makes an angle θ with the direction of S may be denoted by $\varpi'\delta\theta$, and therefore by Art. 149 (1)

$$\psi = -\varpi S = -\frac{\kappa\varpi\varpi'}{4\pi}\int_0^{2\pi}\frac{\cos\theta}{r}\,d\theta, \quad \text{.........(4)}$$

where $$r = \{(x-x')^2 + \varpi^2 + \varpi'^2 - 2\varpi\varpi'\cos\theta\}^{\frac{1}{2}}. \quad \text{.........(5)}$$

If we denote by r_1, r_2 the least and greatest distances, respectively, of the point P from the vortex, viz.

$$r_1^2 = (x-x')^2 + (\varpi-\varpi')^2, \quad r_2^2 = (x-x')^2 + (\varpi+\varpi')^2, \quad \text{......(6)}$$

we have $$r^2 = r_1^2\cos^2\tfrac{1}{2}\theta + r_2^2\sin^2\tfrac{1}{2}\theta, \quad 4\varpi\varpi'\cos\theta = r_1^2 + r_2^2 - 2r^2, \quad \text{......(7)}$$

and therefore

$$\psi = -\frac{\kappa}{8\pi}\left[(r_1^2 + r_2^2)\int_0^{\pi}\frac{d\theta}{\sqrt{(r_1^2\cos^2\frac{1}{2}\theta + r_2^2\sin^2\frac{1}{2}\theta)}} - 2\int_0^{\pi}\sqrt{(r_1^2\cos^2\tfrac{1}{2}\theta + r_2^2\sin^2\tfrac{1}{2}\theta)}\,d\theta\right]. \quad \text{......(8)}$$

The integrals are of the types met with in the theory of the 'arithmetico-geometrical mean.'* In the ordinary, less symmetrical, notation of 'complete' elliptic integrals we have

$$\psi = -\frac{\kappa}{2\pi}(\varpi\varpi')^{\frac{1}{2}}\left\{\left(\frac{2}{k} - k\right)F_1(k) - \frac{2}{k}E_1(k)\right\}, \quad \text{.........(9)}$$

provided $$k^2 = 1 - \frac{r_1^2}{r_2^2} = \frac{4\varpi\varpi'}{(x-x')^2 + (\varpi+\varpi')^2}. \quad \text{.........(10)}$$

The value of ψ at any assigned point can therefore be computed with the help of Legendre's tables.

A neater expression may be obtained by means of 'Landen's transformation,'† viz.

$$\psi = -\frac{\kappa}{2\pi}(r_1 + r_2)\{F_1(\lambda) - E_1(\lambda)\}, \quad \text{.........(11)}$$

provided $$\lambda = \frac{r_2 - r_1}{r_2 + r_1}. \quad \text{.........(12)}$$

The forms of the stream-lines corresponding to equidistant values of ψ are shewn on the next page. They are traced by a method devised by Maxwell, to whom the formula (11) is also due‡.

* See Cayley, *Elliptic Functions*, Cambridge, 1876, c. xiii.

† See Cayley, *l.c.*

‡ *Electricity and Magnetism*, Arts. 704, 705. See also Minchin, *Phil. Mag.* (5), xxxv. (1893); Nagaoka, *Phil. Mag.* (6), vi. (1903).

Expressions for the velocity-potential and the stream-function can also be obtained in the form of definite integrals involving Bessel's Functions.

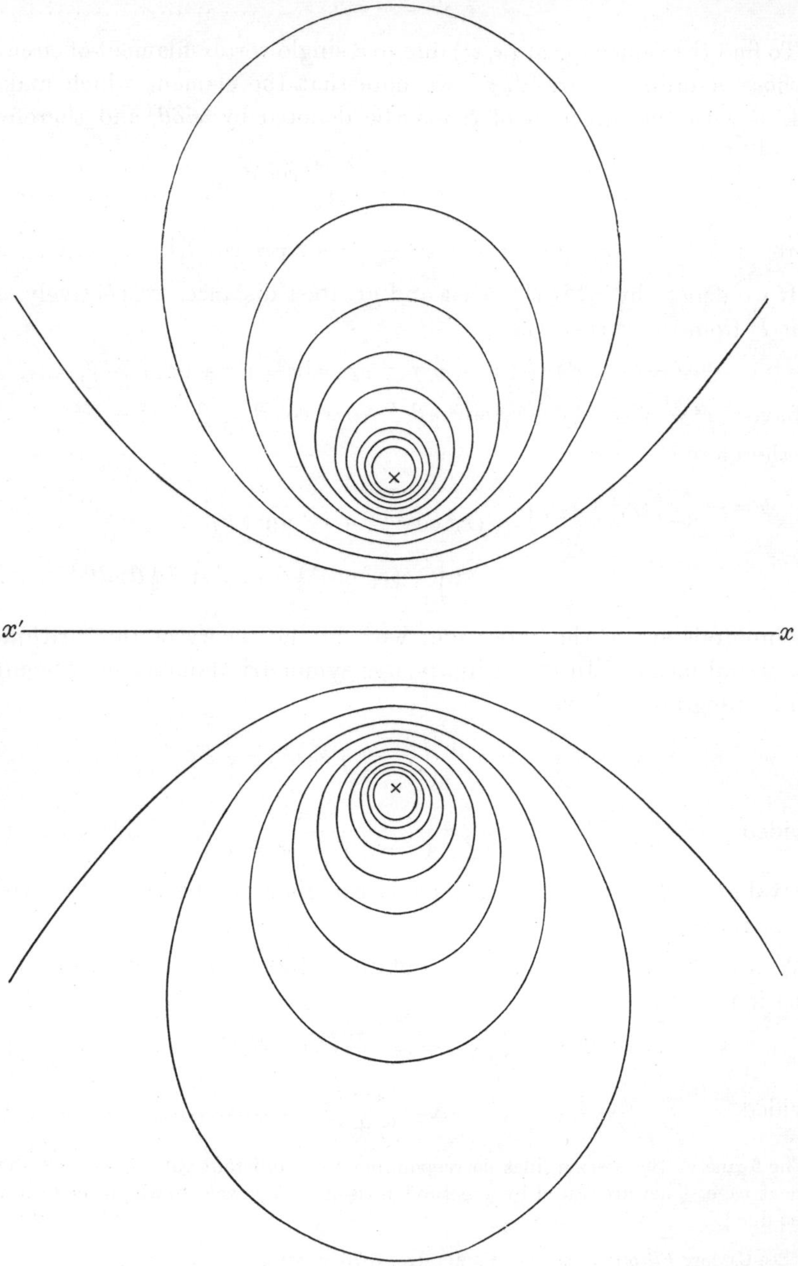

Thus, supposing the vortex to occupy the position of the circle $x=0$, $\varpi=a$, it is evident that the portions of the positive side of the plane $x=0$ which lie within and without this circle constitute two distinct equipotential surfaces. Hence, assuming that we have $\phi = \frac{1}{2}\kappa$ for $x=0$, $\varpi < a$, and $\phi = 0$ for $x=0$, $\varpi > a$, we obtain from Art. 102 (2)

$$\phi = \tfrac{1}{2}\kappa a \int_0^\infty e^{-kx} J_0(k\varpi) J_1(ka)\, dk, \quad\text{.............................(13)}$$

and therefore, in accordance with Art. 100 (5),

$$\psi = -\tfrac{1}{2}\kappa a\varpi \int_0^\infty e^{-kx} J_1(k\varpi) J_1(ka)\, dk. \quad\text{.......................(14)}$$

These formulae relate of course to the region $x > 0$*.

It was shewn in Art. 150 that the value of ϕ is that due to a system of double sources distributed with uniform density κ over the interior of the circle. The values of ϕ and ψ for a uniform distribution of *simple* sources over the same area have been given in Art. 102 (11). The above formulae (13) and (14) can thence be derived by differentiating with respect to x, and adjusting the constant factor†.

162. The energy of any system of circular vortices having the axis of x as a common axis, is

$$T = \pi\rho \iint (u^2 + v^2)\, \varpi\, dx\, d\varpi = \pi\rho \iint \left(v \frac{\partial\psi}{\partial x} - u \frac{\partial\psi}{\partial x}\right) dx\, d\varpi$$

$$= -\pi\rho \iint \psi\omega\, dx\, d\varpi \qquad = -\pi\rho\Sigma\kappa\psi, \quad\text{.......................(1)}$$

by a partial integration, the integrated terms vanishing at the limits. We have here used κ to denote the strength $\omega\,\delta x\,\delta\varpi$ of an elementary vortex-filament.

Again the formula (7) of Art. 153 becomes‡

$$T = 2\pi\rho \iint (\varpi u - xv)\, \varpi\omega\, dx\, dy = 2\pi\rho\Sigma\kappa\varpi\,(\varpi u - xv). \quad\text{.........(2)}$$

The impulse of the system obviously reduces to a force along Ox. By Art. 152 (6),

$$P = \tfrac{1}{2}\rho \iint (y\zeta - z\eta)\, dx\, dy\, dz = \pi\rho \iint \varpi^2\omega\, dx\, d\varpi = \pi\rho\Sigma\kappa\varpi^2. \quad\text{......(3)}$$

If we introduce the symbols x_0, ϖ_0 defined by the equations

$$x_0 = \frac{\Sigma\kappa\varpi^2 x}{\Sigma\kappa\varpi^2}, \quad \varpi_0{}^2 = \frac{\Sigma\kappa\varpi^2}{\Sigma\kappa}, \quad\text{....................(4)}$$

these determine a circle whose position evidently depends on the strengths and the configuration of the vortices, and not on the position of the origin on the axis of symmetry. It may be called the 'circular axis' of the whole system of vortex-rings.

* The formula for ψ occurs in Basset, *Hydrodynamics*, ii. 93. See also Nagaoka, *l.c.*

† Other expressions for ϕ and ψ can be obtained in terms of zonal spherical harmonics. Thus the value of ϕ is given in Thomson and Tait, Art. 546; and that of ψ can be deduced by the formulae (11), (12) of Art. 95 *ante*. The elliptic-integral forms are however the most useful for purposes of interpretation.

‡ At any point in the plane $z=0$ we have $y=\varpi$, $\xi=0$, $\eta=0$, $\zeta=\frac{1}{2}\omega$, $v=v$; the rest follows by symmetry.

Since κ is constant for each vortex, the constancy of the impulse shews, by (3) and (4), that the circular axis remains constant in radius. To find its motion parallel to x, we have, from (4),

$$\Sigma\kappa\,.\,\varpi_0{}^2\,.\,\frac{dx_0}{dt}=\Sigma\kappa\varpi^2\frac{dx}{dt}+2\Sigma\kappa\varpi x\frac{d\varpi}{dt}=\Sigma\kappa\varpi\,(\varpi u+2xv).\quad\ldots\ldots(5)$$

With the help of (2) this can be put in the form

$$\Sigma\kappa\,.\,\varpi_0{}^2\,.\,\frac{dx_0}{dt}=\frac{T}{2\pi\rho}+3\Sigma\kappa\,(x-x_0)\,\varpi v,\quad\ldots\ldots\ldots\ldots\ldots\ldots(6)$$

where the added term vanishes, since $\Sigma\kappa\varpi v=0$ on account of the constancy of the mean radius (ϖ_0).

163. Let us now consider, in particular, the case of an isolated vortex-ring the dimensions of whose cross-section are small compared with the radius (ϖ_0). It has been shown that

$$\psi=-\frac{1}{\pi}\iint\left\{F_1\left(\frac{r_2-r_1}{r_2+r_1}\right)-E_1\left(\frac{r_2-r_1}{r_2+r_1}\right)\right\}(r_1+r_2)\,\omega' dx' d\varpi',\quad\ldots\ldots(1)$$

where r_1, r_2 are defined by Art. 161 (6). For points (x, ϖ) in or near the substance of the vortex, the ratio r_1/r_2 is small, and the modulus (λ) of the elliptic integrals is accordingly nearly equal to unity. We then have

$$F_1(\lambda)=\tfrac{1}{2}\log\frac{16}{\lambda'^2},\qquad E_1(\lambda)=1,\quad\ldots\ldots\ldots\ldots\ldots\ldots\ldots(2)$$

approximately*, where λ' denotes the complementary modulus, viz.

$$\lambda'^2=1-\lambda^2=\frac{4r_1r_2}{(r_1+r_2)^2},\quad\ldots\ldots\ldots\ldots\ldots\ldots\ldots\ldots(3)$$

or $\lambda'^2=5r_1/r_2$, nearly.

Hence at points within the substance of the vortex the value of ψ is of the order $\kappa\varpi_0\log(\varpi_0/\epsilon)$, where ϵ is a small linear magnitude comparable with the dimensions of the section. The velocities at such points, depending (Art. 94) on the differential coefficients of ψ, will be of the order κ/ϵ.

We can now estimate the magnitude of the velocity dx_0/dt of translation of the vortex-ring. By Art. 162 (1), T is of the order $\rho\kappa^2\varpi_0\log(\varpi_0/\epsilon)$, and v is, as we have seen, of the order κ/ϵ; whilst $x-x_0$ is of course of the order ϵ. Hence the second term on the right-hand side of the formula (6) of the preceding Art. is, in the present case, small compared with the first, and the velocity of translation of the ring is of the order $\kappa/\varpi_0\,.\,\log(\varpi_0/\epsilon)$, and approximately constant.

An isolated vortex-ring moves then, without sensible change of size, parallel to its rectilinear axis with nearly constant velocity. This velocity is small compared with that of the fluid in the immediate neighbourhood of the circular axis, but may be greater or less than $\tfrac{1}{2}\kappa/\varpi_0$, the velocity of the fluid at the centre of the ring, with which it agrees in direction.

* See Cayley, *Elliptic Functions*, Arts. 72, 77; and Maxwell, *l.c.*

For the case of a *circular* section more definite results can be obtained as follows. If we neglect the variations of ϖ and ω over the section, the formulae (1) and (2) give

$$\psi = -\frac{\omega}{2\pi}\varpi_0 \iint \left(\log\frac{8\varpi_0}{r_1} - 2\right) dx'\,d\varpi',$$

or, if we introduce polar co-ordinates (s, χ) in the plane of the section,

$$\psi = -\frac{\omega}{2\pi}\varpi_0 \int_0^a \int_0^{2\pi} \left(\log\frac{8\varpi_0}{r_1} - 2\right) s'\,ds'\,d\chi', \quad \text{.......................(4)}$$

where a is the radius of the section. Now

$$\int_0^{2\pi} \log r_1\,d\chi' = \int_0^{2\pi} \log\{s^2 + s'^2 - 2ss'\cos(\chi - \chi')\}^{\frac{1}{2}}\,d\chi',$$

and this definite integral is known to be equal to $2\pi\log s'$, or $2\pi\log s$, according as $s' \gtrless s$. Hence, for points within the section,

$$\psi = -\omega\varpi_0 \int_0^s \left(\log\frac{8\varpi_0}{s} - 2\right) s'\,ds' - \omega\varpi_0 \int_s^a \left(\log\frac{8\varpi_0}{s'} - 2\right) s'\,ds'$$

$$= -\tfrac{1}{2}\omega\varpi_0 a^2 \left\{\log\frac{8\varpi_0}{a} - \tfrac{3}{2} - \tfrac{1}{2}\frac{s^2}{a^2}\right\}. \quad \text{...............................(5)}$$

The only variable part of this is the term $\frac{1}{4}\omega\varpi_0 s^2$; this shews that *to our order of approximation* the stream-lines within the section are concentric circles, the velocity at a distance s from the centre being $\frac{1}{2}\omega s$.

Substituting in Art. 162 (1) we find

$$\frac{T}{2\pi\rho} = -\tfrac{1}{2}\omega \int_0^a \int_0^{2\pi} \psi s\,ds\,d\chi = \frac{\kappa^2\varpi_0}{4\pi}\left\{\log\frac{8\varpi_0}{a} - \tfrac{7}{4}\right\}. \quad \text{.....................(6)}$$

The last term in Art. 162 (6) is equivalent to

$$\tfrac{3}{2}\varpi_0\omega\Sigma\kappa(x - x_0)^2.$$

In our present notation, where κ denotes the strength of the whole vortex, this is equal to $\frac{3}{8}\kappa^2\varpi_0/\pi$. Hence the formula for the velocity of translation of the vortex becomes*

$$\frac{dx_0}{dt} = \frac{\kappa}{4\pi\varpi_0}\left\{\log\frac{8\varpi_0}{a} - \tfrac{1}{4}\right\}. \quad \text{......................................(7)}$$

The vortex-ring carries with it a certain body of irrotationally moving fluid in its career; cf. Art. 155, 2°. According to the formula (7) the velocity of translation of the vortex will be equal to the velocity of the fluid at its centre when $\varpi_0/a = 86$, about. The accompanying mass will be ring-shaped or not, according as ϖ_0/a exceeds or falls short of this critical value.

The ratio of the fluid velocity at the periphery of the vortex to the velocity at the centre of the ring is $2\omega a\varpi_0/\kappa$, or $\varpi_0/\pi a$. For $a = \frac{1}{100}\varpi_0$, this is equal to 32, about.

The conditions under which a vortex-ring of *finite* section and uniform vorticity can travel unchanged have been investigated by Lichtenstein†. The shape of the section, when small, is found to be approximately elliptic, with the minor axis in the direction of translation. He has also discussed the analogous question relating to a vortex-pair (Art. 155).

* This result was given without proof by Sir W. Thomson in an appendix to a translation of Helmholtz' paper, *Phil. Mag.* (4), xxxiii. 511 (1867) [*Papers*, iv. 67]. It was verified by Hicks, *Phil. Trans.* A, clxxvi. 756 (1885); see also Gray, "Notes on Hydrodynamics," *Phil. Mag.* (6), xxviii. 13 (1914).

† *Math. Zeitsch.* xxiii. 89, 310 (1925). See also his *Grundlagen der Hydrodynamik*, Berlin, 1829.

164. If we have any number of circular vortex-rings, coaxal or not, the motion of any one of these may be conceived as made up of two parts, one due to the ring itself, the other due to the influence of the remaining rings. The preceding considerations shew that the second part is insignificant compared with the first, except when two or more rings approach within a very small distance of one another. Hence each ring will move, without sensible change of shape or size, with nearly uniform velocity in the direction of its rectilinear axis, until it passes within a short distance of a second ring.

A general notion of the result of the encounter of two rings may, in particular cases, be gathered from the result given in Art. 149 (3). Thus, let us suppose that we have two circular vortices having the same rectilinear axis. If the sense of the rotation be the same for both, the two rings will advance, on the whole, in the same direction. One effect of their mutual influence will be to increase the radius of the one in front, and to contract the radius of the one in the rear. If the radius of the one in front becomes larger than that of the one in the rear, the motion of the former ring will be retarded, and that of the latter accelerated. Hence if the conditions as to relative size and strength of the two rings be favourable, it may happen that the second ring will overtake and pass through the first. The parts played by the two rings will then be reversed; the one which is now in the rear will in turn overtake and pass through the other, and so on, the rings alternately passing one through the other*.

If the rotations be opposite, and such that the rings approach one another, the mutual influence will be to enlarge the radius of each. If the two rings be moreover equal in size and strength, the velocity of approach will continually diminish. In this case the motion at all points of the plane which is parallel to the two rings, and half-way between them, is tangential to this plane. We may therefore, if we please, regard the plane as a fixed boundary to the fluid on either side, and so obtain the case of a single vortex-ring moving directly towards a fixed rigid wall.

The foregoing remarks are taken from Helmholtz' paper. He adds, in conclusion, that the mutual influence of vortex-rings may easily be studied experimentally in the case of the (roughly) semicircular rings produced by drawing rapidly the point of a spoon for a short space through the surface of a liquid, the spots where the vortex-filaments meet the surface being marked by dimples. (Cf. Art. 27.) The method of experimental illustration by means of smoke-rings† is too well-known to need description here. A beautiful

* Cf. Hicks, "On the Mutual Threading of Vortex Rings," *Proc. Roy. Soc.* A, cii. 111 (1922). The corresponding case in two dimensions was worked out and illustrated graphically by Gröbli, *l.c. ante* p. 224; see also Love, "On the Motion of Paired Vortices with a Common Axis," *Proc. Lond. Math. Soc.* xxv. 185 (1894), and Hicks, *l.c.*

† Reusch, "Ueber Ringbildung der Flüssigkeiten," *Pogg. Ann.* cx. (1860); Tait, *Recen Advances in Physical Science*, London, 1876, c. xii.

variation of the experiment consists in forming the rings in water, the substance of the vortices being coloured*.

The motion of a vortex-ring in a fluid limited (whether internally or externally) by a fixed spherical surface, in the case where the rectilinear axis of the ring passes through the centre of the sphere, has been investigated by Lewis†, by the method of 'images.' The following simplified proof is due to Larmor‡. The vortex-ring is equivalent (Art. 150) to a spherical sheet of double-sources of uniform density, concentric with the fixed sphere. The 'image' of this sheet will, by Art. 96, be another uniform concentric double-sheet, which is, again, equivalent to a vortex-ring coaxal with the first. It easily follows from the Art. last cited that the strengths (κ, κ') and the radii (ϖ, ϖ') of the vortex-ring and its image are connected by the relation

$$\kappa\varpi^{\frac{1}{2}}+\kappa'\varpi'^{\frac{1}{2}}=0. \qquad \text{...(1)}$$

The argument obviously applies to the case of a re-entrant vortex of any form, provided it lie on a sphere concentric with the boundary.

The interest attaching to Kármán's stable configuration of a system of line-vortices of small section (Art. 156) has led to the discussion of analogous arrangements in three dimensions.

Considering, in the first instance, a procession of equal vortex-rings of infinitesimal section, spaced at equal intervals with a common axis, Levi and Forsdyke§ find that the arrangement is unstable for a type of disturbance in which the radii and the intervals vary simultaneously, the rings remaining accurately plane and circular. On the other hand, provided the ratio of the interval between successive rings to the common radius exceeds 1·20, periodic vibrations about the circular form are possible, of types discussed by J. J. Thomson and Dyson in the case of an isolated ring||.

They examine next the case of a helical vortex¶. If undisturbed this will have a certain angular velocity about its axis, and a certain velocity of advance. They find that there is stability if, and only if, the pitch of the helix exceeds 0·3.

The Conditions for Steady Motion.

165. In steady motion, *i.e.* when

$$\frac{\partial u}{\partial t}=0, \quad \frac{\partial v}{\partial t}=0, \quad \frac{\partial w}{\partial t}=0,$$

the equations (2) of Art. 6 may be written

$$u\frac{\partial u}{\partial x}+v\frac{\partial v}{\partial x}+w\frac{\partial w}{\partial x}-(v\zeta-w\eta)=-\frac{\partial\Omega}{\partial x}-\frac{1}{\rho}\frac{\partial p}{\partial x}, \qquad \text{...........(1)}$$

* Reynolds, "On the Resistance encountered by Vortex Rings &c.," *Brit. Ass. Rep.* 1876; *Nature*, xiv. 477.

† "On the Images of Vortices in a Spherical Vessel," *Quart. Journ. Math.* xvi. 338 (1879).

‡ "Electro-magnetic and other Images in Spheres and Planes," *Quart. Journ. Math.* xxiii. 94 (1889).

§ *Proc. Roy. Soc.* A, cxiv. 594; A, cxvi. 352 (1927).

|| For references see p. 246.

¶ *Proc. Roy. Soc.* A, cxx. 670 (1928).

Hence, if as in Art. 146 we put

$$\chi' = \int \frac{dp}{\rho} + \tfrac{1}{2}q^2 + \Omega, \quad \text{.........................(2)}$$

we have $$\frac{\partial \chi'}{\partial x} = v\zeta - w\eta, \quad \frac{\partial \chi'}{\partial y} = w\xi - u\zeta, \quad \frac{\partial \chi'}{\partial z} = u\eta - v\xi. \quad \text{........(3)}$$

It follows that $$u\frac{\partial \chi'}{\partial x} + v\frac{\partial \chi'}{\partial y} + w\frac{\partial \chi'}{\partial z} = 0,$$

$$\xi\frac{\partial \chi'}{\partial x} + \eta\frac{\partial \chi'}{\partial y} + \zeta\frac{\partial \chi'}{\partial z} = 0,$$

so that each of the surfaces $\chi' =$ const. contains both stream-lines and vortex-lines. If further δn denote an element of the normal at any point of such a surface, we have

$$\frac{\partial \chi'}{\partial n} = q\omega \sin \beta, \quad \text{............................(4)}$$

where q is the current velocity, ω the vorticity, and β the angle between the stream-line and the vortex-line at that point.

Hence the conditions that a given state of motion of a fluid may be a possible state of steady motion are as follows. It must be possible to draw in the fluid an infinite system of surfaces each of which is covered by a network of stream-lines and vortex-lines, and the product $q\omega \sin \beta \delta n$ must be constant over each such surface, δn denoting the length of the normal drawn to a consecutive surface of the system*.

These conditions may also be deduced from the considerations that the stream-lines are, in steady motion, the actual paths of the particles, that the product of the angular velocity into the cross-section is the same at all points of a vortex, and that this product is, for the same vortex, constant with regard to the time.

The theorem that the function χ', defined by (2), is constant over each surface of the above kind is an extension of that of Art. 21, where it was shewn that χ' is constant along a stream-line.

The above conditions are satisfied identically in all cases of irrotational motion, provided of course the boundary-conditions be such as are consistent with the steady motion.

In the motion of a liquid in two dimensions (xy) the product $q\delta n$ is constant along a stream-line; the conditions in question then reduce to this, that the vorticity ζ must be constant along each stream-line, or, by Art. 59 (5),

$$\frac{\partial^2 \psi}{\partial x^2} + \frac{\partial^2 \psi}{\partial y^2} = f(\psi), \quad \text{.........................(5)}$$

where $f(\psi)$ is an arbitrary function of ψ†.

* See a paper "On the Conditions for Steady Motion of a Fluid," *Proc. Lond. Math. Soc.* (1) ix. 91 (1878).

† Cf. Lagrange, *Nouv. Mém. de l'Acad. de Berlin*, 1781 [*Oeuvres*, iv. 720]; and Stokes, "On the Steady Motion of Incompressible Fluids," *Camb. Trans.* vii. (1842) [*Papers*, i. 15].

This condition is satisfied in all cases of motion in concentric circles about the origin. Another obvious solution of (5) is

$$\psi=\tfrac{1}{2}(Ax^2+2Bxy+Cy^2), \quad\text{.....................}(6)$$

in which case the stream-lines are similar and coaxal conics. The angular velocity at any point is $\tfrac{1}{2}(A+C)$, and is therefore uniform.

Again, if we put $f(\psi)=-k^2\psi$, where k is a constant, and transform to polar co-ordinates r, θ, we get

$$\frac{\partial^2\psi}{\partial r^2}+\frac{1}{r}\frac{\partial\psi}{\partial r}+\frac{1}{r^2}\frac{\partial^2\psi}{\partial\theta^2}+k^2\psi=0, \quad\text{.....................}(7)$$

which is satisfied (Art. 101) by $\psi=CJ_s(kr)\left.\begin{matrix}\cos\\ \sin\end{matrix}\right\} s\theta.$ ....(8)

This gives various solutions consistent with a fixed circular boundary of radius a, the admissible values of k being determined by

$$J_s(ka)=0. \quad\text{.....................}(9)$$

Suppose, for example, that in an unlimited mass of fluid the stream-function is

$$\psi=CJ_1(kr)\sin\theta, \quad\text{.....................}(10)$$

within the circle $r=a$, whilst outside this circle we have

$$\psi=U\left(r-\frac{a^2}{r}\right)\sin\theta. \quad\text{.....................}(11)$$

These two values of ψ agree for $r=a$, provided $J_1(ka)=0$. Moreover, the tangential velocity at this circle will be continuous, provided the two values of $\partial\psi/\partial r$ are equal, *i.e.* if

$$C=\frac{2U}{kJ_1'(ka)}=-\frac{2U}{kJ_0(ka)}. \quad\text{.....................}(12)$$

If we now impress on everything a velocity U parallel to Ox, we get a species of cylindrical vortex travelling with velocity U through a liquid which is at rest at infinity. The smallest of the possible values of k is given by $ka/\pi=1\cdot2197$; the relative stream-lines *inside* the vortex are then given by the lower diagram on p. 288, provided the *dotted* circle be taken as the boundary ($r=a$). It is easily proved, by Art. 157 (1), that the 'impulse' of the vortex is represented by $2\pi\rho a^2U$.

In the case of motion symmetrical about an axis (x), we have $q\,.\,2\pi\varpi\delta n$ constant along a stream-line, ϖ denoting as in Art. 94 the distance of any point from the axis of symmetry. The condition for steady motion then is that the ratio ω/ϖ must be constant along any stream-line. Hence, if ψ be the stream-function, we must have, by Art. 161 (2),

$$\frac{\partial^2\psi}{\partial x^2}+\frac{\partial^2\psi}{\partial\varpi^2}-\frac{1}{\varpi}\frac{\partial\psi}{\partial\varpi}=\varpi^2 f(\psi), \quad\text{.....................}(13)$$

where $f(\psi)$ denotes an arbitrary function of ψ*.

An interesting example is furnished by Hill's 'Spherical Vortex†.' If we assume

$$\psi=\tfrac{1}{2}A\varpi^2(a^2-r^2), \quad\text{.....................}(14)$$

where $r^2=x^2+\varpi^2$, for all points within the sphere $r=a$, the formula (2) of Art. 161 makes

$$\omega=-\tfrac{5}{2}A\varpi,$$

so that the condition of steady motion is satisfied. Again it is evident, on reference to Arts. 96, 97, that the irrotational flow of a stream with the general velocity $-U$ parallel to the axis, past a fixed spherical surface $r=a$, is given by

$$\psi=\tfrac{1}{2}U\varpi^2\left(1-\frac{a^3}{r^3}\right). \quad\text{.....................}(15)$$

* This result is due to Stokes, *l.c.*

† "On a Spherical Vortex," *Phil. Trans.* A, clxxxv. (1894).

The two values of ψ agree when $r=a$; this makes the normal velocity zero on both sides. In order that the tangential velocity may be continuous, the values of $\partial\psi/\partial r$ must also agree. Remembering that $\varpi=r\sin\theta$, this gives $A=-\frac{3}{2}U/a^2$, and therefore

$$\omega=\tfrac{15}{2}U\varpi/a^2. \qquad (16)$$

The sum of the strengths of the vortex-filaments composing the spherical vortex is $5\,Ua$.

The figure shews the stream-lines, both inside and outside the vortex; they are drawn, as usual, for equidistant values of ψ.

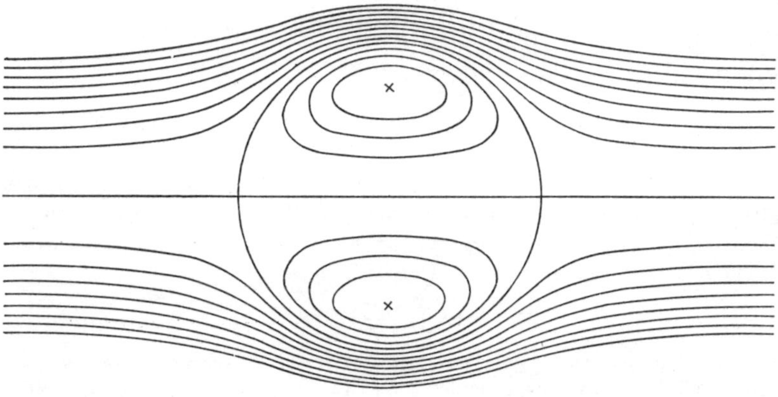

If we impress on everything a velocity U parallel to x, we get a spherical vortex advancing with constant velocity U through a liquid which is at rest at infinity.

By the formulae of Art. 162, we readily find that the square of the 'mean-radius' of the vortex is $\frac{2}{5}a^2$, the 'impulse' $2\pi\rho a^3U$, and the energy is $\frac{10}{7}\pi\rho a^3U^2$.

As explained in Art. 146, it is quite unnecessary to calculate formulae for the pressure, in order to assure ourselves that this is continuous at the surface of the vortex. The continuity of the pressure is already secured by the continuity of the velocity, and the constancy of the circulation in any moving circuit.

166. As already stated, the theory of vortex motion was originated by Helmholtz in 1858. It acquired additional interest when, in 1867, Kelvin suggested* the theory of vortex atoms. As a physical theory, this has long been abandoned, but it gave rise to a great number of interesting investigations, to which some reference should be made. We may mention the investigations as to the stability and the periods of vibration of rectilinear† and annular‡ vortices; the similar investigations relating to hollow vortices (where the rotationally moving core is replaced by a vacuum§); and the calculations of the forms of boundary of a hollow vortex which are consistent with steady motion‖. A summary of some of the leading results has been given by Love¶.

* *l.c. ante* p. 222.

† Sir W. Thomson, *l.c. ante* p. 230.

‡ J. J. Thomson, *l.c. ante* p. 216; Dyson, *Phil. Trans.* A, clxxxiv. 1041 (1893).

§ Sir W. Thomson, *l.c.*; Hicks, "On the Steady Motion and the Small Vibrations of a Hollow Vortex," *Phil. Trans.* 1884; Pocklington, "The Complete System of the Periods of a Hollow Vortex Ring," *Phil. Trans.* A, clxxxvi. 603 (1895); Carslaw, "The Fluted Vibrations of a Circular Vortex-Ring with a Hollow Core," *Proc. Lond. Math. Soc.* (1) xxviii. 97 (1896).

‖ Hicks, *l.c.*; Pocklington, "Hollow Straight Vortices," *Camb. Proc.* viii. 178 (1894).

¶ *l.c. ante* p. 192.

166 a. The dynamical theorems of the present chapter all depend on the constancy of the circulation in a moving circuit. It is postulated (Art. 146) that the extraneous forces if any are conservative, and also that the fluid is either homogeneous and incompressible, or subject to a definite relation between the pressure and the density.

There are of course many natural conditions, especially in Meteorology, in which this latter assumption does not hold. If we proceed as in Art. 33 without making this assumption we find, for the rate of change of the circulation in a moving circuit,

$$\frac{D}{Dt}\int(udx+vdy+wdz) = -\int s\left(\frac{\partial p}{\partial x}dx+\frac{\partial p}{\partial y}dy+\frac{\partial p}{\partial z}dz\right), \quad \text{......(1)}$$

where $s\,(=1/\rho)$ is the reciprocal of the density, or the 'bulkiness', of the fluid. The line-integral on the right hand may be converted into a surface-integral over any area bounded by the circuit, by Stokes' theorem; thus

$$\frac{D}{Dt}\int(udx+vdy+wdz) = \iint(lP+mQ+nR)\,dS, \quad \text{............(2)}$$

where
$$P=\frac{\partial(p,s)}{\partial(y,z)},\quad Q=\frac{\partial(p,s)}{\partial(z,x)},\quad R=\frac{\partial(p,s)}{\partial(x,y)}. \quad \text{..............(3)}$$

Now consider the vector whose components are P, Q, R. It is solenoidal, in virtue of the relation

$$\frac{\partial P}{\partial x}+\frac{\partial Q}{\partial y}+\frac{\partial R}{\partial z}=0, \quad \text{..........................(4)}$$

and its direction is given by the intersections of the surfaces $p=$ const., $s=$ const. If we imagine a series of surfaces of equal pressure to be drawn for equal infinitesimal intervals δp, and likewise a series of surfaces of equal bulkiness for equal infinitesimal intervals δs, these will divide the field into a system of tubes whose cross-sections are infinitesimal parallelograms. It is easy to shew that if $\delta\Sigma$ is the area of one of these parallelograms

$$\sqrt{(P^2+Q^2+R^2)}\,\delta\Sigma=\delta p\,\delta s. \quad \text{......................(5)}$$

Hence the product of the vector $(P, Q, R,)$ into the cross-section is not only uniform along any tube, but is the same for all the tubes. The equation (2) then shews that the rate of change of the circulation round a moving circuit is proportional to the number of the aforesaid tubes which it embraces*.

* V. Bjerknes, *Vid.-Selsk. Skrifter*, Kristiania, 1918. An independent proof is attributed to Silberstein (1896). Another theorem of a less simple character is given by Bjerknes, relating to the circulation of *momentum*

$$\int\rho\,(udx+vdy+wdz).$$

Some applications of the theorems to meteorological and other phenomena are explained in Stockholm, *Ak. Handl.* xxxi. (1898).

Clebsch's Transformation.

167. Another matter of some interest, which can however only be briefly touched upon, is Clebsch's transformation of the hydrodynamical equations*.

It is easily seen that the component velocities at any one instant can be expressed in the forms

$$u=-\frac{\partial\phi}{\partial x}+\lambda\frac{\partial\mu}{\partial x},\quad v=-\frac{\partial\phi}{\partial y}+\lambda\frac{\partial\mu}{\partial y},\quad w=-\frac{\partial\phi}{\partial z}+\lambda\frac{\partial\mu}{\partial z},\quad\text{.................(1)}$$

where ϕ, λ, μ are functions of x, y, z, provided the component rotations can be put in the forms

$$\xi=\frac{\partial(\lambda,\mu)}{\partial(y,z)},\quad\eta=\frac{\partial(\lambda,\mu)}{\partial(z,x)},\quad\zeta=\frac{\partial(\lambda,\mu)}{\partial(x,y)}.\quad\text{.........................(2)}$$

Now if the differential equations of the vortex-lines, viz.

$$\frac{dx}{\xi}=\frac{dy}{\eta}=\frac{dz}{\zeta},\quad\text{...(3)}$$

be supposed integrated in the form

$$\alpha=\text{const.},\quad\beta=\text{const.},\quad\text{.......................................(4)}$$

where α, β are functions of x, y, z, we must have

$$\xi=P\frac{\partial(\alpha,\beta)}{\partial(y,z)},\quad\eta=P\frac{\partial(\alpha,\beta)}{\partial(z,x)},\quad\zeta=P\frac{\partial(\alpha,\beta)}{\partial(x,y)},\quad\text{.......................(5)}$$

where P is some function of x, y, z†. Substituting these expressions in the identity

$$\frac{\partial\xi}{\partial x}+\frac{\partial\eta}{\partial y}+\frac{\partial\zeta}{\partial z}=0,$$

we find

$$\frac{\partial(P,\alpha,\beta)}{\partial(x,y,z)}=0,\quad\text{..(6)}$$

which shews that P is of the form $f(\alpha,\beta)$. If λ, μ be any two functions of α, β, we have

$$\frac{\partial(\lambda,\mu)}{\partial(y,z)}=\frac{\partial(\lambda,\mu)}{\partial(\alpha,\beta)}\times\frac{\partial(\alpha,\beta)}{\partial(y,z)},\ \text{\&c., \&c.};$$

and the equations (5) will therefore reduce to the form (2), provided λ, μ be chosen so that

$$\frac{\partial(\lambda,\mu)}{\partial(\alpha,\beta)}=f(\alpha,\beta),\quad\text{...(7)}$$

which can obviously be satisfied in an infinity of ways.

It is evident from (2) that the intersections of the surfaces $\lambda=$const., $\mu=$const. are the vortex-lines. This suggests that the functions λ, μ which occur in (1) may be supposed to vary continuously with t in such a way that the surfaces in question move with the fluid‡. Various analytical proofs of the possibility of this have been given; the simplest, perhaps, is by means of the equations (2) of Art. 15, which give (as in Art. 17)

$$u\,dx+v\,dy+w\,dz=u_0\,da+v_0\,db+w_0\,dc-d\chi.\quad\text{.......................(8)}$$

It has been proved that we may assume, initially,

$$u_0\,da+v_0\,db+w_0\,dc=-d\phi_0+\lambda\,d\mu.\quad\text{...............................(9)}$$

Hence, considering space-variations at time t, we shall have

$$u\,dx+v\,dy+w\,dz=-d\phi+\lambda\,d\mu,\quad\text{...............................(10)}$$

* "Ueber eine allgemeine Transformation d. hydrodynamischen Gleichungen," *Crelle*, liv. (1857) and lvi. (1859). See also Hill, *Quart. Journ. Math.* xvii. (1881), and *Camb. Trans.* xiv. (1883).

† Cf. Forsyth, *Differential Equations*, Art. 174.

‡ It must not be overlooked that on account of the insufficient determinacy of λ, μ these functions may vary continuously with t without relating always to the same particles of fluid.

where $\phi = \phi_0 + \chi$, and λ, μ have the same values as in (9), but are now expressed in terms of x, y, z, t. Since, in the 'Lagrangian' method, the independent space-variables relate to the individual particles, this proves the theorem.

On this understanding the equations of motion can be integrated, provided the extraneous forces have a potential, and that p is a function of ρ only. We have

$$\frac{\partial u}{\partial t} - 2v\zeta + 2w\eta = \frac{\partial u}{\partial t} + \left(u\frac{\partial \lambda}{\partial x} + v\frac{\partial \lambda}{\partial y} + w\frac{\partial \lambda}{\partial z}\right)\frac{\partial \mu}{\partial x} - \left(u\frac{\partial \mu}{\partial x} + v\frac{\partial \mu}{\partial y} + w\frac{\partial \mu}{\partial z}\right)\frac{\partial \lambda}{\partial x}$$

$$= \frac{\partial}{\partial x}\left(-\frac{\partial \phi}{\partial t} + \lambda\frac{\partial \mu}{\partial t}\right) + \frac{D\lambda}{Dt}\frac{\partial \mu}{\partial x} - \frac{D\mu}{Dt}\frac{\partial \lambda}{\partial x}; \qquad (11)$$

and therefore, on the present assumption that $D\lambda/Dt = 0$, $D\mu/Dt = 0$,

$$\int \frac{dp}{\rho} + \tfrac{1}{2}q^2 + \Omega = \frac{\partial \phi}{\partial t} - \lambda\frac{\partial \mu}{\partial t}, \qquad (12)$$

by Art. 146 (5), (6). An arbitrary function of t is here supposed incorporated in $\partial\phi/\partial t$.

If the above condition be not imposed on λ, μ, we have, writing

$$H = \int \frac{dp}{\rho} + \tfrac{1}{2}q^2 + \Omega - \frac{\partial \phi}{\partial t} + \lambda\frac{\partial \mu}{\partial t}, \qquad (13)$$

$$\frac{D\lambda}{Dt}\frac{\partial \mu}{\partial x} - \frac{D\mu}{Dt}\frac{\partial \lambda}{\partial x} = -\frac{\partial H}{\partial x}, \quad \frac{D\lambda}{Dt}\frac{\partial \mu}{\partial y} - \frac{D\mu}{Dt}\frac{\partial \lambda}{\partial y} = -\frac{\partial H}{\partial y}, \quad \frac{D\lambda}{Dt}\frac{\partial \mu}{\partial z} - \frac{D\mu}{Dt}\frac{\partial \lambda}{\partial z} = -\frac{\partial H}{\partial z}. \qquad (14)$$

Hence

$$\frac{\partial(H, \lambda, \mu)}{\partial(x, y, z)} = 0, \qquad (15)$$

shewing that H is of the form $f(\lambda, \mu, t)$; and

$$\frac{D\lambda}{Dt} = -\frac{\partial H}{\partial \mu}, \quad \frac{D\mu}{Dt} = \frac{\partial H}{\partial \lambda}. \qquad (16)^*$$

* The author is informed that these equations were given in a Fellowship dissertation (Dublin) by Mr T. Stuart (1900).

CHAPTER VIII

TIDAL WAVES

168. ONE of the most interesting and successful applications of hydrodynamical theory is to the small oscillations, under gravity, of a liquid having a free surface. In certain cases, which are somewhat special as regards the theory, but very important from a practical point of view, these oscillations may combine to form progressive waves travelling with (to a first approximation) no change of form over the surface.

The term 'tidal,' as applied to waves, has been used in various senses, but it seems most natural to confine it to gravitational oscillations possessing the characteristic feature of the oceanic tides produced by the action of the sun and moon. We have therefore ventured to place it at the head of this Chapter, as descriptive of waves in which the motion of the fluid is mainly horizontal, and therefore (as will appear) sensibly the same for all particles in a vertical line. This latter circumstance greatly simplifies the theory.

It will be convenient to recapitulate, in the first place, some points in the general theory of small oscillations which will receive constant exemplification in the investigations which follow*. The theory has reference in the first instance to a system of finite freedom, but the results, when properly interpreted, hold good without this restriction†.

Let $q_1, q_2, \ldots q_n$ be n generalized co-ordinates serving to specify the configuration of a dynamical system, and let them be so chosen as to vanish in the configuration of equilibrium. The kinetic energy T will be a homogeneous quadratic function of the generalized velocities $\dot{q}_1, \dot{q}_2, \ldots \dot{q}_n$, say

$$2T = a_{11}\dot{q}_1^2 + a_{22}\dot{q}_2^2 + \ldots + 2a_{12}\dot{q}_1\dot{q}_2 + \ldots, \quad \ldots\ldots\ldots\ldots\ldots(1)$$

where the coefficients are in general functions of the co-ordinates $q_1, q_2, \ldots q_n$, but may in the application to *small* motions be supposed constant, and to have the values corresponding to $q_1, q_2, \ldots q_n = 0$. Again, if (as we shall suppose) the system is 'conservative,' the potential energy V of a small displacement is a homogeneous quadratic function of the component displacement $q_1, q_2, \ldots q_n$, with (on the same understanding) constant coefficients, say

$$2V = c_{11}q_1^2 + c_{22}q_2^2 + \ldots + 2c_{12}q_1q_2 + \ldots. \quad \ldots\ldots\ldots\ldots\ldots(2)$$

* For a fuller account of the general theory see Thomson and Tait, Arts. 337, ...; Rayleigh, *Theory of Sound*, c. iv.; Routh, *Elementary Rigid Dynamics* (6th ed.), London, 1897, c. ix.; Whittaker, *Analytical Dynamics*, c. vii.; Lamb, *Higher Mechanics*, 2nd ed., Cambridge, 1929.

† The steps by which a rigorous transition can be made to the case of infinite freedom have been investigated by Hilbert, *Gött. Nachr.* 1904.

By a real* linear transformation of the co-ordinates $q_1, q_2, \dots q_n$ it is possible to reduce T and V simultaneously to sums of squares; the new variables thus introduced are called the 'normal co-ordinates' of the system. In terms of these we have

$$2T = a_1\dot{q}_1^2 + a_2\dot{q}_2^2 + \dots + a_n\dot{q}_n^2, \quad \text{.....................(3)}$$

$$2V = c_1 q_1^2 + c_2 q_2^2 + \dots + c_n q_n^2. \quad \text{.....................(4)}$$

The coefficients $a_1, a_2, \dots a_n$ are called the 'principal coefficients of inertia'; they are necessarily positive. The coefficients $c_1, c_2, \dots c_n$ may be called the 'principal coefficients of stability'; they are all positive when the undisturbed configuration is stable.

When given extraneous forces act on the system, the work done by these during an arbitrary infinitesimal displacement $\Delta q_1, \Delta q_2, \dots \Delta q_n$ may be expressed in the form

$$Q_1\Delta q_1 + Q_2\Delta q_2 + \dots + Q_n\Delta q_n. \quad \text{.....................(5)}$$

The coefficients $Q_1, Q_2, \dots Q_n$ are then called the 'normal components of disturbing force.'

In the application to infinitely small motions Lagrange's equations

$$\frac{d}{dt}\frac{\partial T}{\partial \dot{q}_r} - \frac{\partial T}{\partial q_r} = -\frac{\partial V}{\partial q_r} + Q_r \qquad [r = 1, 2, \dots n] \quad \text{...(6)}$$

take the form

$$a_{1r}\ddot{q}_1 + a_{2r}\ddot{q}_2 + \dots + c_{1r}q_1 + c_{2r}q_2 + \dots = Q_r \quad \text{..............(7)}$$

or, in the case of normal co-ordinates,

$$a_r\ddot{q}_r + c_r q_r = Q_r. \quad \text{..............................(8)}$$

It is easily seen from this that the dynamical characteristics of the normal co-ordinates are (1°) that an impulse of any normal type produces an initial motion of that type only, and (2°) that a steady disturbing force of any type maintains a displacement of that type only.

To obtain the *free* motions of the system we put $Q_r = 0$. Solving (8), we find

$$q_r = A_r \cos(\sigma_r t + \epsilon_r), \quad \text{..........................(9)}$$

where
$$\sigma_r = (c_r/a_r)^{\frac{1}{2}}, \quad \text{..............................(10)}$$

and A_r, ϵ_r are arbitrary constants†. Hence a mode of free motion is possible in which any normal co-ordinate q_r varies alone, and the motion of any particle of the system, since it depends linearly on q_r, will be simple-harmonic, of period $2\pi/\sigma_r$; moreover the particles will keep step with one another, passing simultaneously through their equilibrium positions. The several modes of this character are called the 'normal modes' of vibration of the system; their

* The algebraic proof of this involves the assumption that one at least of the functions T, V is essentially positive. In the present case T of course fulfils this condition.

† The ratio $\sigma/2\pi$ measures the 'frequency' of the oscillation. It is convenient to have a name for the quantity σ itself; the term 'speed' has been used in this sense by Kelvin and G. H. Darwin in their researches on the Tides.

number is equal to that of the degrees of freedom, and any free motion whatever of the system may be obtained from them by superposition, with a proper choice of the 'amplitudes' (A_r) and 'epochs' (ϵ_r). It is seen from (10) that in any normal mode the mean values (with respect to time) of the kinetic and potential energies are equal.

In certain cases, viz. when two or more of the free periods ($2\pi/\sigma$) of the system are equal, the normal co-ordinates are to a certain extent indeterminate, *i.e.* they can be chosen in an infinite number of ways. By compounding the corresponding modes, with arbitrary amplitudes and epochs, we obtain a small oscillation in which the motion of each particle is the resultant of simple-harmonic vibrations in different directions, and is therefore, in general, elliptic-harmonic, with the same period. This is exemplified in the spherical pendulum; an important instance in our own subject is that of progressive waves in deep water (Chapter IX.).

If any of the coefficients of stability (c_r) be negative, the value of σ_r is a pure imaginary. The circular function in (9) is then replaced by real exponentials, and an arbitrary displacement will in general increase until the assumptions on which the approximate equation (8) is based becomes untenable. The undisturbed configuration is then reckoned as unstable. The necessary and sufficient condition of stability (in the present sense) is that the potential energy V should be a minimum in the configuration of equilibrium.

To find the effect of disturbing forces, it is sufficient to consider the case where Q_r varies as a simple-harmonic function of the time, say

$$Q_r = C_r \cos(\sigma t + \epsilon), \quad \text{.........................(11)}$$

where the value of σ is now prescribed. Not only is this the most interesting case in itself, but we know from Fourier's Theorem that, whatever the law of variation of Q_r with the time, it can be expressed by a series of terms such as (11) A particular integral of (8) is then

$$q_r = \frac{C_r}{c_r - \sigma^2 a_r} \cos(\sigma t + \epsilon). \quad \text{.......................(12)}$$

This represents the 'forced oscillation' due to the periodic force Q_r. In it the motion of every particle is simple-harmonic, of the prescribed period $2\pi/\sigma$, and the extreme displacements coincide in time with the maxima and minima of the force.

A constant force equal to the instantaneous value of the actual force (11) would maintain a displacement

$$\bar{q}_r = \frac{C_r}{c_r} \cos(\sigma t + \epsilon), \quad \text{.........................(13)}$$

the same, of course, as if the inertia-coefficient a_r were null. Hence (12) may be written

$$q_r = \frac{1}{1 - \sigma^2/\sigma_r^2} \bar{q}_r, \quad \text{.........................(14)}$$

where σ_r has the value (10). This very useful formula enables us to write down the effect of a periodic force when we know that of a steady force of the same type. It is to be noticed that q_r and Q_r have the same or opposite phases according as $\sigma \lesseqgtr \sigma_r$, that is, according as the period of the disturbing force is greater or less than the free period. A simple example of this is furnished by a simple pendulum acted on by a periodic horizontal force. Other important illustrations will present themselves in the theory of the tides*.

When σ is very great in comparison with σ_r, the formula (12) becomes

$$q_r = -\frac{C_r}{\sigma^2 a_r}\cos(\sigma t + \epsilon)\,; \quad \text{.......................(15)}$$

the displacement is now always in the opposite phase to the force, and depends only on the *inertia* of the system.

If the period of the impressed force be nearly equal to that of the normal mode of order r, the amplitude of the forced oscillation, as given by (14), is very great compared with $\bar{q}_r$. In the case of exact equality, the solution (12) fails, and must be replaced by

$$q_r = \frac{C_r t}{2\sigma a_r}\sin(\sigma t + \epsilon). \quad \text{.......................(16)}$$

This gives an oscillation of continually increasing amplitude, and can therefore only be accepted as a representation of the initial stages of the disturbance.

Another very important property of the normal modes may be noticed. If by the introduction of frictionless constraints the system be compelled to oscillate in any other prescribed manner, the configuration at any instant can be specified by one variable, which we will denote by θ. In terms of this we shall have

$$q_r = B_r \theta,$$

where the quantities B_r are certain constants. This makes

$$2T = (B_1^2 a_1 + B_2^2 a_2 + \ldots + B_n^2 a_n)\,\dot{\theta}^2, \quad \text{..........................(17)}$$

$$2V = (B_1^2 c_1 + B_2^2 c_2 + \ldots + B_n^2 c_n)\,\theta^2. \quad \text{..........................(18)}$$

If $\theta \propto \cos(\sigma t + \epsilon)$, the constancy of the energy $(T+V)$ requires

$$\sigma^2 = \frac{B_1^2 c_1 + B_2^2 c_2 + \ldots + B_n^2 c_n}{B_1^2 a_1 + B_2^2 a_2 + \ldots + B_n^2 a_n}. \quad \text{.............................(19)}$$

Hence σ^2 is intermediate in value between the greatest and least of the quantities cr/a_r; in other words, the frequency of the constrained oscillation is intermediate between the greatest and least frequencies corresponding to the normal modes of the system. In particular, when a system is modified by the introduction of a constraint, the frequency of the slowest natural oscillation is *increased.* Moreover, if the constrained type differ but slightly from a normal type (r), σ^2 will differ from c_r/a_r by a small quantity *of the second order.* This gives a method of estimating approximately the frequency in cases where the normal types cannot be accurately determined†. Examples will be found in Arts. 191, 259.

* Cf. T. Young, "A Theory of Tides," *Nicholson's Journal,* xxxv. (1813) [*Miscellaneous Works*, London, 1854, ii. 262].

† Rayleigh, "Some General Theorems relating to Vibrations," *Proc. Lond. Math. Soc.* iv. 357 (1874) [*Papers*, i. 170], and *Theory of Sound*, c. iv. The method was elaborated by Ritz, *Journ. für Math.* cxxxv. 1 (1908), and *Ann. der Physik*, xxviii. (1909) [*Gesammelte Werke*, Paris, 1911, pp. 192, 265].

It may further be shewn that in the case of a *partial* constraint, which merely reduces the degree of freedom from n to $n-1$, the periods of the modified system separate those of the original one*.

It had been already remarked by Lagrange† that if in the equations of type (7), where the co-ordinates are not assumed to be normal, we put $Q_r=0$, and assume

$$q_r = A_r \cos(\sigma t+\epsilon), \qquad (20)$$

the resulting equations are identical with those which determine the stationary values of the expression

$$\sigma^2 = \frac{c_{11}A_1^2 + c_{22}A_2^2 + \ldots + 2c_{12}A_1A_2 + \ldots}{a_{11}A_1^2 + a_{22}A_2^2 + \ldots + 2a_{12}A_1A_2 + \ldots}, = \frac{V(A, A)}{T(A, A)}, \qquad (21)$$

say. Since $T(A, A)$ is essentially positive the denominator cannot vanish, and the expression has therefore a minimum value. It is moreover possible, starting from this property, to construct a proof that the n values of σ^2 are all real‡. They are obviously all positive if V be essentially positive.

Rayleigh's theorem is also closely related to the Hamiltonian formula (3) of Art. 135, as we may see by assuming

$$q_r = A_r \sin\sigma t, \qquad (22)$$

and taking $t_0=0$, $t_1=2\pi/\sigma$. Cf. Art. 205 *a*.

The modifications which are introduced into the theory of small oscillations by the consideration of viscous forces will be noticed in Chapter XI.

Long Waves in Canals.

169. Proceeding now to the special problem of this chapter, let us begin with the case of waves travelling along a straight canal, with horizontal bed, and parallel vertical sides. Let the axis of x be parallel to the length of the canal, that of y vertical and upwards, and let us suppose that the motion takes place in these two dimensions x, y. Let the ordinate of the free surface, corresponding to the abscissa x, at time t, be denoted by $y_0+\eta$, where y_0 is the ordinate in the undisturbed state.

As already indicated, we shall assume in all the investigations of this Chapter that the vertical acceleration of the fluid particles may be neglected, or, more precisely, that the pressure at any point (x, y) is sensibly equal to the statical pressure due to the depth below the free surface, viz.

$$p - p_0 = g\rho(y_0+\eta-y), \qquad (1)$$

where p_0 is the (uniform) external pressure.

Hence
$$\frac{\partial p}{\partial x} = g\rho\frac{\partial \eta}{\partial x}. \qquad (2)$$

This is independent of y, so that the horizontal acceleration is the same for all particles in a plane perpendicular to x. It follows that all particles which once lie in such a plane always do so; in other words, the horizontal velocity u is a function of x and t only.

* Routh, *Elementary Rigid Dynamics*, Art. 67; Rayleigh, *Theory of Sound* (2nd ed.), Art. 92 *a*; Whittaker, *Analytical Dynamics*, Art. 81.

† *Mécanique Analytique* (Bertrand's ed.), i. 331; *Oeuvres*, xi. 380.

‡ See Poincaré, *Journ. de Math.* (5), ii. 83 (1896); Lamb, *Higher Mechanics*, 2nd ed., Art. 92.

The equation of horizontal motion, viz.

$$\frac{\partial u}{\partial t} + u\frac{\partial u}{\partial x} = -\frac{1}{\rho}\frac{\partial p}{\partial x},$$

is further simplified in the case of infinitely small motions by the omission of the term $u\partial u/\partial x$, which is of the second order, so that

$$\frac{\partial u}{\partial t} = -g\frac{\partial \eta}{\partial x}. \qquad\qquad (3)$$

Now let

$$\xi = \int u\,dt;$$

i.e. ξ is the time-integral of the displacement past the plane x, up to the time t. In the case of *small* motions this will, to the first order of small quantities, be equal to the displacement of the particles which originally occupied that plane, or again to that of the particles which actually occupy it at time t. The equation (3) may now be written

$$\frac{\partial^2 \xi}{\partial t^2} = -g\frac{\partial \eta}{\partial x}. \qquad\qquad (4)$$

The equation of continuity may be found by calculating the volume of fluid which has, up to time t, entered the space bounded by the planes x and $x + \delta x$; thus, if h be the depth and b the breadth of the canal,

$$-\frac{\partial}{\partial x}(\xi h b)\,\delta x = \eta b\,\delta x,$$

or

$$\eta = -h\frac{\partial \xi}{\partial x}. \qquad\qquad (5)$$

The same result comes from the ordinary form of the equation of continuity, viz.

$$\frac{\partial u}{\partial x} + \frac{\partial v}{\partial y} = 0. \qquad\qquad (6)$$

Thus

$$v = -\int_0^y \frac{\partial u}{\partial x}dy = -y\frac{\partial u}{\partial x}, \qquad\qquad (7)$$

if the origin be (for the moment) taken in the bottom of the canal. This formula is of interest as shewing, as a consequence of our primary assumption, that the vertical velocity of any particle is simply proportional to its height above the bottom. At the free surface we have $y = h + \eta$, $v = \partial\eta/\partial t$, whence (neglecting a product of small quantities)

$$\frac{\partial \eta}{\partial t} = -h\frac{\partial^2 \xi}{\partial x\,\partial t}. \qquad\qquad (8)$$

From this (5) follows by integration with respect to t.

Eliminating η between (4) and (5), we obtain

$$\frac{\partial^2 \xi}{\partial t^2} = gh\frac{\partial^2 \xi}{\partial x^2}. \qquad\qquad (9)$$

The elimination of ξ gives an equation of the same form, viz.

$$\frac{\partial^2 \eta}{\partial t^2} = gh\frac{\partial^2 \eta}{\partial x^2}. \qquad\qquad (10)$$

The above investigation can readily be extended to the case of a uniform canal of any form of section*. If the sectional area of the undisturbed fluid be S, and the breadth at the free surface b, the equation of continuity is

$$-\frac{\partial}{\partial x}(\xi S)\,\delta x = \eta b\,\delta x, \quad \text{.........................(11)}$$

whence

$$\eta = -h\frac{\partial \xi}{\partial x}, \quad \text{.................................(12)}$$

as before, provided $h = S/b$, *i.e.* h now denotes the *mean* depth of the canal. The dynamical equation (4) is of course unaltered.

170. The equation (9) is of a well-known type which occurs in several physical problems, *e.g.* the transverse vibrations of strings, and the motion of sound-waves in one dimension.

To integrate it, let us write, for shortness,

$$c = \sqrt{(gh)}, \quad \text{.................................(13)}$$

and

$$x - ct = x_1, \quad x + ct = x_2.$$

In terms of x_1 and x_2 as independent variables, the equation takes the form

$$\frac{\partial^2 \xi}{\partial x_1 \partial x_2} = 0.$$

The complete solution is therefore

$$\xi = F(x - ct) + f(x + ct), \quad \text{...................(14)}$$

where F, f are arbitrary functions.

The corresponding values of the particle-velocity and of the surface-elevation are given by

$$\left.\begin{aligned} \frac{\dot{\xi}}{c} &= -F'(x - ct) + f'(x + ct), \\ \frac{\eta}{h} &= -F'(x - ct) - f'(x + ct). \end{aligned}\right\} \quad \text{.....................(15)}$$

The interpretation of these results is simple. Take first the motion represented by the first term in (14), alone. Since $F(x - ct)$ is unaltered when t and x are increased by τ and $c\tau$, respectively, it is plain that the disturbance which existed at the point x at time t has been transferred at time $t + \tau$ to the point $x + c\tau$. Hence the disturbance advances unchanged with a constant velocity c in space. In other words we have a 'progressive wave' travelling with velocity c in the direction of x-positive. In the same way the second term of (14) represents a progressive wave travelling with velocity c in the direction of x-negative. And it appears, since (14) is the *complete* solution of (9), that any motion whatever of the fluid, which is subject to the conditions laid down in the preceding Art., may be regarded as made up of waves of these two kinds.

* Kelland, *Trans. R. S. Edin.* xiv. (1839).

The velocity (c) of propagation is, by (13), that 'due to' half the depth of the undisturbed fluid*.

The following table giving, in round numbers and assuming $g=32$ f/s, the velocity of wave-propagation for various depths, will be of interest later in connection with the theory of the tides.

The last column gives the time a wave would take to travel over a distance equal to the earth's circumference ($2\pi a$). In order that a 'long' wave should traverse this distance in 24 hours, the depth would have to be about 14 miles. It must be borne in mind that these numerical results are only applicable to waves satisfying the conditions above postulated. The meaning of these conditions will be examined more particularly in Art. 172.

h (feet)	c (feet per sec.)	c (sea-miles per hour)	$2\pi a/c$ (hours)
$312\frac{1}{2}$	100	60	360
1250	200	120	180
5000	400	240	90
11250†	600	360	60
20000	800	480	45

171. To trace the effect of an arbitrary initial disturbance, let us suppose that when $t=0$ we have

$$\frac{\dot{\xi}}{c} = \phi(x), \quad \frac{\eta}{h} = \psi(x). \qquad \text{.........................(16)}$$

The functions F', f' which occur in (15) are then given by

$$\left.\begin{aligned} F'(x) &= -\tfrac{1}{2}\{\phi(x)+\psi(x)\}, \\ f'(x) &= \quad \tfrac{1}{2}\{\phi(x)-\psi(x)\}. \end{aligned}\right\} \qquad \text{.....................(17)}$$

Hence if we draw the curves $y=\eta_1$, $y=\eta_2$, where

$$\left.\begin{aligned} \eta_1 &= \tfrac{1}{2}h\{\psi(x)+\phi(x)\}, \\ \eta_2 &= \tfrac{1}{2}h\{\psi(x)-\phi(x)\}, \end{aligned}\right\} \qquad \text{.........................(18)}$$

the form of the wave-profile at any subsequent instant t is found by displacing these curves parallel to x, through spaces $\pm ct$, respectively, and adding (algebraically) the ordinates. If, for example, the original disturbance is confined to a length l of the axis of x, then after a time $l/2c$ it will have broken up into two progressive waves of length l, travelling in opposite directions.

In the particular case where in the initial state $\dot{\xi}=0$, and therefore $\phi(x)=0$, we have $\eta_1=\eta_2$; the elevation in each of the derived waves is then exactly half what it was, at corresponding points, in the original disturbance.

It appears from (16) and (17) that if the initial disturbance be such that $\dot{\xi} = \pm\,\eta/h\,.\,c$, the motion will consist of a wave system travelling in one direction only, since one or other of the functions F' and f' is then zero.

* Lagrange, *Nouv. mém. de l'Acad. de Berlin*, 1781 [*Oeuvres*, i. 747].

† This is probably comparable in order of magnitude with the mean depth of the ocean.

It is easy to trace the motion of a surface-particle as a progressive wave of either kind passes it. Suppose, for example, that

$$\xi = F(x - ct), \qquad \text{..............................(19)}$$

and therefore

$$\dot{\xi} = c\frac{\eta}{h}. \qquad \text{..............................(20)}$$

The particle is at rest until it is reached by the wave; it then moves forward with a velocity proportional at each instant to the elevation above the mean level, the velocity being in fact less than the wave-velocity c, in the ratio of the surface-elevation to the depth of the water. The total displacement at any time is given by

$$\xi = \frac{1}{h}\int \eta c\, dt.$$

This integral measures the volume, per unit breadth of the canal, of the portion of the wave which has up to the instant in question passed the particle. Finally, when the wave has passed away, the particle is left at rest in advance of its original position at a distance equal to the total volume of the elevated water divided by the sectional area of the canal.

172. We can now examine under what circumstances the solution expressed by (14) will be consistent with the assumptions made provisionally in Art. 169.

The exact equation of vertical motion, viz.

$$\rho\frac{Dv}{Dt} = -\frac{\partial p}{\partial y} - g\rho,$$

gives, on integration with respect to y,

$$p - p_0 = g\rho(y_0 + \eta - y) - \rho\int_y^{y_0+\eta} \frac{Dv}{Dt}\, dy. \qquad \text{..............(21)}$$

This may be replaced by the approximate equation (1), provided βh be small compared with $g\eta$, where β denotes the maximum vertical acceleration. Now in a progressive wave, if λ denote the distance between two consecutive nodes (*i.e.* points at which the wave-profile meets the undisturbed level), the time which the corresponding portion of the wave takes to pass a particle is λ/c, and therefore, provided the gradient $\partial\eta/\partial x$ is everywhere small, the vertical velocity will be of the order $\eta c/\lambda$*, and the vertical acceleration of the order $\eta c^2/\lambda^2$, where η is the maximum elevation (or depression). Hence βh will be small compared with $g\eta$, provided h^2/λ^2 is a small quantity.

Waves whose slope is gradual, and whose length λ is large compared with the depth h of the fluid, are called 'long' waves.

Again, the restriction to infinitely small motions, made in equation (3), consisted in neglecting $u\partial u/\partial x$ in comparison with $\partial u/\partial t$. In a progressive

* Hence, comparing with (20), we see that the ratio of the maximum vertical to the maximum horizontal velocity is of the order h/λ.

wave we have $\partial u/\partial t = \pm c\partial u/\partial x$; so that u must be small compared with c, and therefore, by (20), η must be small compared with h. It is to be observed that this condition is altogether distinct from the former one, which may be legitimate in cases where the motion cannot be regarded as infinitely small. See Art. 187.

The preceding conditions will of course be satisfied in the general case represented by equation (14), provided they are satisfied for each of the two progressive waves into which the disturbance can be analysed.

173. There is another, although on the whole a less convenient, method of investigating the motion of 'long' waves, in which the Lagrangian plan is adopted of making the co-ordinates refer to the individual particles of the fluid. For simplicity, we will consider only the case of a canal of rectangular section*. The fundamental assumption that the vertical acceleration may be neglected implies as before that the horizontal motion of all particles in a plane perpendicular to the length of the canal will be the same. We therefore denote by $x + \xi$ the abscissa at time t of the plane of particles whose undisturbed abscissa is x. If η denote the elevation of the free surface, in this plane, the equation of motion of unit breadth of a stratum whose thickness (in the undisturbed state) is δx will be

$$\rho h \delta x \frac{\partial^2 \xi}{\partial t^2} = -\frac{\partial p}{\partial x} \delta x (h + \eta),$$

where the factor $(\partial p/\partial x) . \delta x$ represents the pressure-difference for any two opposite particles x and $x + \delta x$ on the two faces of the stratum, while the factor $h + \eta$ represents the area of the stratum. Since we assume that the pressure about any particle depends only on its depth below the free surface we may write

$$\frac{\partial p}{\partial x} = g\rho \frac{\partial \eta}{\partial x},$$

so that our dynamical equation is

$$\frac{\partial^2 \xi}{\partial t^2} = -g\left(1 + \frac{\eta}{h}\right)\frac{\partial \eta}{\partial x}. \quad \text{.........................(1)}$$

The equation of continuity is obtained by equating the volumes of a stratum, consisting of the same particles, in the disturbed and undisturbed conditions respectively, viz.

$$\left(\delta x + \frac{\partial \xi}{\partial x}\delta x\right)(h + \eta) = h\delta x,$$

or

$$1 + \frac{\eta}{h} = \left(1 + \frac{\partial \xi}{\partial x}\right)^{-1}. \quad \text{.........................(2)}$$

* Airy, *Encyc. Metrop.* "Tides and Waves," Art. 192 (1845); see also Stokes, "On Waves," *Camb. and Dub. Math. Journ.* iv. 219 (1849) [*Papers*, ii. 222]. The case of a canal with sloping sides has been treated by McCowan, "On the Theory of Long Waves...," *Phil. Mag.* (5), xxxv. 250 (1892).

Between equations (1) and (2) we may eliminate either η or ξ; the result in terms of ξ is the simpler, being

$$\frac{\partial^2 \xi}{\partial t^2} = gh \frac{\dfrac{\partial^2 \xi}{\partial x^2}}{\left(1 + \dfrac{\partial \xi}{\partial x}\right)^3}. \quad \text{.............................(3)}$$

This is the general equation of 'long' waves in a uniform canal with vertical sides*.

So far the only assumption in the present investigation is that the vertical acceleration of the particles may be neglected in calculating the pressure. If we now assume, in addition, that η/h is a small quantity, the equations (2) and (3) reduce to

$$\eta = -h \frac{\partial \xi}{\partial x}, \quad \text{.................................(4)}$$

and

$$\frac{\partial^2 \xi}{\partial t^2} = gh \frac{\partial^2 \xi}{\partial x^2}. \quad \text{.................................(5)}$$

The elevation η now satisfies an equation of the same form, viz.

$$\frac{\partial^2 \eta}{\partial t^2} = gh \frac{\partial^2 \eta}{\partial x^2}. \quad \text{.................................(6)}$$

These are in conformity with our previous results; for the smallness of $\partial\xi/\partial x$ means that the relative displacement of any two particles is never more than a minute fraction of the distance between them, so that (to a first approximation) it is now immaterial whether the variable x be supposed to refer to a plane fixed in space, or to one moving with the fluid.

174. The potential energy of a wave, or system of waves, due to the elevation or depression of the fluid above or below the mean level is, per unit breadth, $g\rho \iint y\,dx\,dy$, where the integration with respect to y is to be taken between the limits 0 and η, and that with respect to x over the whole length of the waves. Effecting the former integration, we get

$$\tfrac{1}{2} g\rho \int \eta^2 dx. \quad \text{.................................(1)}$$

The kinetic energy is

$$\tfrac{1}{2} \rho h \int \dot{\xi}^2 dx. \quad \text{.................................(2)}$$

In a system of waves travelling in one direction only we have

$$\dot{\xi} = \pm \frac{c}{h} \eta,$$

so that the expressions (1) and (2) are equal; or the total energy is half potential, and half kinetic.

This result may be obtained in a more general manner, as follows†. Any progressive wave may be conceived as having been originated by the splitting

* Airy, *l.c.*

† Rayleigh, "On Waves," *Phil. Mag.* (5), i. 257 (1876) [*Papers*, i. 251].

up, into two waves travelling in opposite directions, of an initial disturbance in which the particle-velocity was everywhere zero, and the energy therefore wholly potential. It appears from Art. 171 that the two derived waves are symmetrical in every respect, so that each must contain half the original store of energy. Since, however, the elevation at corresponding points is for each derived wave exactly half that of the original disturbance, the potential energy of each will by (1) be one-fourth of the original store. The remaining (kinetic) part of the energy of each derived wave must therefore also be one-fourth of the original quantity.

175. If in any case of waves travelling in one direction only, without change of form, we impress on the whole mass a velocity equal and opposite to that of propagation, the motion becomes *steady*, whilst the forces acting on any particle remain the same as before. With the help of this artifice, the laws of wave-propagation can be investigated with great ease*. Thus, in the present case we shall have by Art. 22 (5), at the free surface,

$$\frac{p}{\rho} = \text{const.} - g(h+\eta) - \tfrac{1}{2}q^2, \qquad (1)$$

where q is the velocity. If the slope of the wave-profile be everywhere gradual, and the depth h small compared with the length of a wave, the horizontal velocity may be taken to be uniform throughout the depth, and approximately equal to q. Hence the equation of continuity is

$$q(h+\eta) = ch, \qquad (2)$$

c being the velocity, in the steady motion, at places where the depth of the stream is uniform and equal to h. Substituting for q in (1), we have

$$\frac{p}{\rho} = \text{const.} - gh\left(1+\frac{\eta}{h}\right) - \tfrac{1}{2}c^2\left(1+\frac{\eta}{h}\right)^{-2}. \qquad (3)$$

Hence if η/h be small, the condition for a free surface, viz. $p = \text{const.}$, is satisfied approximately, provided

$$c^2 = gh, \qquad (4)$$

which agrees with our former result.

The present method also accounts very simply for the relation between particle-velocity and surface-elevation already found in Art. 171. From (2) we have, approximately,

$$q = c\left(1-\frac{\eta}{h}\right). \qquad (5)$$

Hence in the wave-motion the particle-velocity relative to the undisturbed water is $c\eta/h$ in the direction of propagation.

When the elevation η, though small compared with the wave-length, is not

* Rayleigh, *l.c.*

regarded as infinitely small, a closer approximation to the wave-velocity is secured if in (4) we replace h by $\eta + h$. This gives a wave-velocity

$$c_0\left(1+\frac{1}{2}\frac{\eta}{h}\right),$$

approximately, where $c_0 = \sqrt{(gh)}$, relative to the fluid in the immediate neighbourhood. Since this fluid has itself a velocity $c_0\eta/h$, the velocity of propagation *in space* is approximately

$$c_0\left(1+\frac{3}{2}\frac{\eta}{h}\right), \qquad \text{.................................(6)}$$

a result due substantially to Airy*. It follows that a wave of the type now under consideration cannot be propagated entirely without change of profile, since the speed varies with the height. Another proof of (6) will be given presently when we come to consider specially the theory of waves of finite amplitude (Art. 187).

176. It appears from the linearity of the approximate equations that, in the case of sufficiently low waves, any number of independent solutions may be superposed. For example, having given a wave of any form travelling in one direction, if we superpose its *image* in the plane $x=0$, travelling in the opposite direction, it is obvious that in the resulting motion the horizontal velocity will vanish at the origin, and the circumstances are therefore the same as if there were a fixed barrier at this point. We can thus understand the reflection of a wave at a barrier; the elevations and depressions are reflected unchanged, whilst the horizontal velocity is reversed. The same results follow from the formula

$$\xi = F(ct-x) - F(ct+x), \qquad \text{.........................(1)}$$

which is evidently the most general value of ξ subject to the condition that $\xi = 0$ for $x = 0$.

We can further investigate without much difficulty the partial reflection of a wave at a point where there is an abrupt change in the section of the canal. Taking the origin at the point in question, we may write, for the negative side,

$$\eta_1 = F\left(t-\frac{x}{c_1}\right)+f\left(t+\frac{x}{c_1}\right), \qquad u_1=\frac{g}{c_1}F\left(t-\frac{x}{c_1}\right)-\frac{g}{c_1}f\left(t+\frac{x}{c_1}\right), \qquad \text{............(2)}$$

and for the positive side

$$\eta_2=\phi\left(t-\frac{x}{c_2}\right), \qquad u_2=\frac{g}{c_2}\phi\left(t-\frac{x}{c_2}\right), \qquad \text{...........................(3)}$$

where the function F represents the original wave, and f, ϕ the reflected and transmitted portions respectively. The constancy of mass requires that at the point $x=0$ we should have $b_1h_1u_1=b_2h_2u_2$, where b_1, b_2 are the breadths at the surface, and h_1, h_2 are the mean depths. We must also have at the same point $\eta_1=\eta_2$, on account of the continuity of pressure†. These conditions give

$$\frac{b_1h_1}{c_1}\{F(t)-f(t)\}=\frac{b_2h_2}{c_2}\phi(t), \qquad F(t)+f(t)=\phi(t).$$

* "Tides and Waves," Art. 208.

† It will be understood that the problem admits only of an approximate treatment, on account of the rapid change in the character of the motion near the point of discontinuity. The nature

We thence find that the ratios of the elevations in corresponding parts of the reflected and incident waves, and of the transmitted and incident waves, are

$$\frac{f}{F}=\frac{b_1c_1-b_2c_2}{b_1c_1+b_2c_2}, \qquad \frac{\phi}{F}=\frac{2b_1c_1}{b_1c_1+b_2c_2}, \qquad\qquad(4)$$

respectively. The reader may easily verify that the energy contained in the reflected and transmitted waves is equal to that of the original incident wave.

177. Our investigations, so far, relate to cases of *free* waves. When, in addition to gravity, small disturbing forces X, Y act on the fluid, the equation of motion is obtained as follows.

We assume that within distances comparable with the depth h these forces vary only by a small fraction of their total value. On this understanding we have, in place of Art. 169 (1),

$$\frac{p-p_0}{\rho}=(g-Y)(y_0+\eta-y), \qquad\qquad(1)$$

and therefore
$$\frac{1}{\rho}\frac{\partial p}{\partial x}=(g-Y)\frac{\partial\eta}{\partial x}-(y_0+\eta-y)\frac{\partial Y}{\partial x}.$$

We assume that Y is small compared with g, and (for the reason just stated) that $h\partial Y/\partial x$ is small compared with X. Hence, with sufficient approximation, the equation of horizontal motion, viz.

$$\frac{\partial^2\xi}{\partial t^2}=-\frac{1}{\rho}\frac{\partial p}{\partial x}+X, \qquad\qquad(2)$$

reduces to the form

$$\frac{\partial^2\xi}{\partial t^2}=-g\frac{\partial\eta}{\partial x}+X, \qquad\qquad(3)$$

where, moreover, X may be regarded as a function of x and t only. The equation of continuity is the same as in Art. 169, viz.

$$\eta=-h\frac{\partial\xi}{\partial x}. \qquad\qquad(4)$$

Hence, on elimination of η,

$$\frac{\partial^2\xi}{\partial t^2}=gh\frac{\partial^2\xi}{\partial x^2}+X. \qquad\qquad(5)$$

The *horizontal* component of the disturbing force is alone important.

If the disturbing influence consists of a variable surface-pressure (p_0), the equation (3) is replaced by

$$\frac{\partial^2\xi}{\partial t^2}=-g\frac{\partial\eta}{\partial x}-\frac{1}{\rho}\frac{\partial p_0}{\partial x}, \qquad\qquad(6)$$

of the approximation implied in the above assumptions will become more evident if we suppose the suffixes to refer to two sections S_1 and S_2, one on each side of the origin O, at distances from O which, though very small compared with the wave-length, are yet moderate multiples of the transverse dimensions of the canal. The motion of the fluid will be sensibly uniform over each of these sections, and parallel to the length. The condition in the text then expresses that there is no sensible change of level between S_1 and S_2.

whilst (4) is unaltered. In the case of a travelling pressure, say

$$\frac{p_0}{\rho} = f(Ut - x), \quad \text{.............................(7)}$$

we find

$$\frac{\eta}{h} = \frac{p_0}{\rho(U^2 - gh)}. \quad \text{.............................(8)}$$

The surface depression is in the same phase with the pressure, or the opposite, according as $U \lessgtr \sqrt{(gh)}$.

On the other hand, when it is the bottom which is disturbed, we have $X = 0$ in (2), whilst the equation of continuity becomes

$$\eta - \eta_0 = -h\frac{\partial \xi}{\partial x}, \quad \text{.............................(9)}$$

where η_0 is the elevation of the bottom above the mean level. Thus in the case of a seismic wave

$$\eta_0 = f(Ut - x), \quad \text{.............................(10)}$$

we find

$$\frac{\eta}{\eta_0} = \frac{U^2}{U^2 - gh}. \quad \text{.............................(11)}$$

178. The oscillations of water in a canal of uniform section, closed at both ends, may, as in the corresponding problem of Acoustics, be obtained by superposition of progressive waves travelling in opposite directions. It is more instructive, however, with a view to subsequent more difficult investigations, to treat the problem as an example of the general theory sketched in Art. 168.

We have to determine ξ so as to satisfy

$$\frac{\partial^2 \xi}{\partial t^2} = c^2 \frac{\partial^2 \xi}{\partial x^2} + X, \quad \text{.............................(1)}$$

together with the terminal conditions that $\xi = 0$ for $x = 0$ and $x = l$, say.

To find the free oscillations we put $X = 0$, and assume that

$$\xi \propto \cos(\sigma t + \epsilon),$$

where σ is to be found. On substitution we obtain

$$\frac{\partial^2 \xi}{\partial x^2} + \frac{\sigma^2}{c^2}\xi = 0, \quad \text{.............................(2)}$$

whence, omitting the time-factor,

$$\xi = A \sin\frac{\sigma x}{c} + B\cos\frac{\sigma x}{c}.$$

The terminal conditions give $B = 0$, and

$$\sigma l / c = r\pi, \quad \text{.............................(3)}$$

where r is integral. Hence the normal mode of order r is given by

$$\xi = A_r \sin\frac{r\pi x}{l}\cos\left(\frac{r\pi c t}{l} + \epsilon_r\right), \quad \text{..................... (4)}$$

where the amplitude A_r and epoch ϵ_r are arbitrary.

In the slowest oscillation ($r = 1$), the water sways to and fro, heaping itself up alternately at the two ends, and there is a node at the middle ($x = \frac{1}{2}l$). The period ($2l/c$) is equal to the time a progressive wave would take to traverse twice the length of the canal.

The periods of the higher modes are respectively $\frac{1}{2}$, $\frac{1}{3}$, $\frac{1}{4}$, ... of this, but it must be remembered, in this and in other similar problems, that our theory ceases to be applicable when the length l/r of a semi-undulation becomes comparable with the depth h.

On comparison with the general theory of Art. 168, it appears that the normal co-ordinates of the present system are quantities $q_1, q_2, \ldots q_n$ such that when the system is displaced according to any one of them, say q_r, we have

$$\xi = q_r \sin \frac{r\pi x}{l};$$

and we infer that the most general displacement of which the system is capable (subject to the conditions presupposed) is given by

$$\xi = \Sigma q_r \sin \frac{r\pi x}{l}, \quad \text{.........................(5)}$$

where $q_1, q_2, \ldots q_n$ are arbitrary. This is in accordance with Fourier's Theorem.

When expressed in terms of the normal velocities and the normal co-ordinates, the expressions for T and V must reduce to sums of squares. This is easily verified, in the present case, from the formula (5). Thus if S denote the sectional area of the canal, we find

$$2T = \rho S \int_0^l \dot{\xi}^2 dx = \Sigma a_r \dot{q}_r^2, \quad 2V = g\rho \frac{S}{h} \int_0^l \eta^2 dx = \Sigma c_r q_r^2, \quad \text{......(6)}$$

where
$$a_r = \tfrac{1}{2}\rho S l, \quad c_r = \tfrac{1}{2} r^2 \pi^2 g\rho h\, S/l. \quad \text{....................(7)}$$

It is to be noted that, on the present reckoning, the coefficients of stability (c_r) increase with the depth.

Conversely, if we assume from Fourier's Theorem that (5) is a sufficiently general expression for the value of ξ at any instant, the calculation just indicated shews that the coefficients q_r are the normal co-ordinates; and the frequencies can then be found from the general formula (10) of Art. 168; viz. we have

$$\sigma_r = (c_r/a_r)^{\frac{1}{2}} = r\pi\,(gh)^{\frac{1}{2}}/l, \quad \text{......................(8)}$$

in agreement with (3).

179. As an example of forced waves we take the case of a uniform horizontal force

$$X = f \cos(\sigma t + \epsilon). \quad \text{............................(9)}$$

This will illustrate, to a certain extent, the generation of tides in a land-locked sea of small dimensions.

Assuming that ξ varies as $\cos(\sigma t + \epsilon)$, and omitting the time-factor, the equation (1) becomes

$$\frac{\partial^2 \xi}{\partial x^2} + \frac{\sigma^2}{c^2}\xi = -\frac{f}{c^2},$$

the solution of which is

$$\xi = -\frac{f}{\sigma^2} + D\sin\frac{\sigma x}{c} + E\cos\frac{\sigma x}{c}. \quad \text{.................(10)}$$

The terminal conditions give

$$E = \frac{f}{\sigma^2}, \qquad D\sin\frac{\sigma l}{c} = \left(1 - \cos\frac{\sigma l}{c}\right)\frac{f}{\sigma^2}. \quad \text{..............(11)}$$

Hence, unless $\sin \sigma l/c = 0$, we have $D = f/\sigma^2 . \tan \sigma l/2c$, so that

$$\left.\begin{aligned} \xi &= \frac{2f}{\sigma^2 \cos(\frac{1}{2}\sigma l/c)} \sin\frac{\sigma x}{2c} \sin\frac{\sigma(l-x)}{2c} . \cos(\sigma t + \epsilon), \\ \text{and} \qquad \eta &= \frac{hf}{\sigma c \cos(\frac{1}{2}\sigma l/c)} \sin\frac{\sigma(x - \frac{1}{2}l)}{c} . \cos(\sigma t + \epsilon). \end{aligned}\right\} \quad \text{......(12)}$$

If the period of the disturbing force be large compared with that of the slowest free mode, $\sigma l/2c$ will be small, and the formula for the elevation becomes

$$\eta = \frac{f}{g}(x - \tfrac{1}{2}l)\cos(\sigma t + \epsilon), \quad \text{....................(13)}$$

approximately, exactly as if the water were devoid of inertia. The horizontal displacement of the water is always in the same phase with the force, so long as the period is greater than that of the slowest free mode, or $\sigma l/c < \pi$. If the period be diminished until it is less than the above value, the phase is reversed.

When the period is exactly equal to that of a free mode of *odd* order ($r = 1, 3, 5, \ldots$), the above expressions for ξ and η become infinite, and the solution fails. As pointed out in Art. 168, the interpretation of this is that, in the absence of dissipative forces, the amplitude of the motion becomes so great that our fundamental approximations are no longer justified.

If, on the other hand, the period coincide with that of a free mode of *even* order ($r = 2, 4, 6, \ldots$), we have $\sin \sigma l/c = 0$, $\cos \sigma l/c = 1$, and the terminal conditions are satisfied independently of the value of D. The forced motion may then be represented by*

$$\xi = -\frac{2f}{\sigma^2}\sin^2\frac{\sigma x}{2c}\cos(\sigma t + \epsilon). \quad \text{....................(14)}$$

This example illustrates the fact that the effect of a disturbing force may sometimes be conveniently calculated without resolving the force into its 'normal components.'

* In the language of the general theory, the impressed force has here no component of the particular type with which it synchronizes, so that a vibration of this type is not excited at all. In the same way a periodic pressure applied at any point of a stretched string will not excite any fundamental mode which has a *node* there, even though it synchronize with it.

Another very simple case of forced oscillations, of some interest in connection with tidal theory, is that of a canal closed at one end and communicating at the other with an open sea in which a periodic oscillation

$$\eta = a\cos(\sigma t + \epsilon) \quad \text{.........................(15)}$$

is maintained. If the origin be taken at the closed end, the solution is obviously

$$\eta = a\frac{\cos(\sigma x/c)}{\cos(\sigma l/c)}\cdot\cos(\sigma t+\epsilon), \quad \text{.....................(16)}$$

l denoting the length. If $\sigma l/c$ be small the tide has sensibly the same amplitude at all points of the canal. For particular values of l (determined by $\cos\sigma l/c = 0$) the solution fails through the amplitude becoming infinite.

Canal Theory of the Tides.

180. The theory of forced oscillations in canals, or on open sheets of water, owes most of its interest to its bearing on the phenomena of the tides. The 'canal theory,' in particular, has been treated very fully by Airy*. We will consider a few of the more interesting problems.

The calculation of the disturbing effect of a distant body on the waters of the ocean is placed for convenience in an Appendix at the end of this Chapter. It appears that the disturbing effect of the moon, for example, at a point P of the earth's surface, may be represented by a potential Ω whose approximate value is

$$\Omega = \tfrac{3}{2}\frac{\gamma M a^2}{D^3}(\tfrac{1}{3} - \cos^2\vartheta), \quad \text{.........................(1)}$$

where M denotes the mass of the moon, D its distance from the earth's centre, a the earth's radius, γ the 'constant of gravitation,' and ϑ the moon's zenith distance at the place P. This gives a horizontal acceleration $\partial\Omega/a\partial\vartheta$, or

$$f\sin 2\vartheta, \quad \text{...............................(2)}$$

towards the point of the earth's surface which is vertically beneath the moon, where

$$f = \tfrac{3}{2}\frac{\gamma M a}{D^3}. \quad \text{...............................(3)}$$

If E be the earth's mass, we may write $g = \gamma E/a^2$, whence

$$\frac{f}{g} = \frac{3}{2}\cdot\frac{M}{E}\cdot\left(\frac{a}{D}\right)^3.$$

Putting $M/E = \frac{1}{81}$, $a/D = \frac{1}{60}$, this gives $f/g = 8{\cdot}57\times 10^{-8}$. When the sun is the disturbing body, the corresponding result is $f/g = 3{\cdot}78\times 10^{-8}$.

It is convenient, for some purposes, to introduce a linear magnitude H, defined by

$$H = af/g. \quad \text{...............................(4)}$$

* *Encycl. Metrop.* "Tides and Waves," Section VI. (1845). Several of the leading features of the theory had been made out, by very simple methods, by Young, in 1813 and 1823 [*Works*, ii. 262, 291].

If we put $a = 21 \times 10^6$ feet, this gives, for the lunar tide, $H = 1{\cdot}80$ ft., and for the solar tide $H = {\cdot}79$ ft. It is shewn in the Appendix that H measures the maximum range of the tide, from high water to low water, on the 'equilibrium theory.'

181. Take now the case of a uniform canal coincident with the earth's equator, and let us suppose for simplicity that the moon describes a circular orbit in the same plane. Let ξ be the displacement, relative to the earth's surface, of a particle of water whose mean position is in longitude ϕ, measured eastwards from some fixed meridian. If ω be the angular velocity of the earth's rotation, the actual displacement of the particle at time t will be $\xi + a\omega t$, so that the tangential acceleration will be $\partial^2\xi/\partial t$. If we suppose the 'centrifugal force' to be as usual allowed for in the value of g, the processes of Arts. 169, 177 will apply without further alteration.

If n denote the angular velocity of the moon westward, relative to the fixed meridian*, we may write in Art. 180 (2)

$$\vartheta = nt + \phi + \epsilon,$$

so that the equation of motion is

$$\frac{\partial^2 \xi}{\partial t^2} = c^2 \frac{\partial^2 \xi}{a^2 \partial \phi^2} - f \sin 2(nt + \phi + \epsilon). \quad\text{.....................(1)}$$

The *free* oscillations are determined by the consideration that ξ is necessarily a periodic function of ϕ, its value recurring whenever ϕ increases by 2π. It may therefore be expressed, by Fourier's Theorem, in the form

$$\xi = \sum_0^\infty (P_r \cos r\phi + Q_r \sin r\phi). \quad\text{.....................(2)}$$

Substituting in (1), with the last term omitted, it is found that P_r and Q_r must satisfy the equation

$$\frac{d^2 P_r}{dt^2} + \frac{r^2 c^2}{a^2} P_r = 0. \quad\text{..........................(3)}$$

The motion, in any normal mode, is therefore simple-harmonic, of period $2\pi a/rc$.

For the *forced* waves, or tides, we find

$$\xi = -\tfrac{1}{4} \frac{fa^2}{c^2 - n^2 a^2} \sin 2(nt + \phi + \epsilon), \quad\text{.................(4)}$$

whence

$$\eta = \tfrac{1}{2} \frac{c^2 H}{c^2 - n^2 a^2} \cos 2(nt + \phi + \epsilon). \quad\text{.....................(5)}$$

The tide is therefore semi-diurnal (the *lunar* day being of course understood), and is 'direct' or 'inverted,' *i.e.* there is high or low water beneath the moon, according as $c \gtrless na$, in other words according as the velocity, relative to the

* That is, $n = \omega - n_1$, if n_1 be the angular velocity of the moon in her orbit.

earth's surface, of a point which moves so as to be always vertically beneath the moon, is less or greater than that of a free wave. In the actual case of the earth we have

$$\frac{c^2}{n^2 a^2} = \frac{g}{n^2 a} \cdot \frac{h}{a} = 311 \frac{h}{a},$$

so that unless the depth of the canal were to greatly exceed such depths as actually occur in the ocean, the tides would be inverted.

This result, which is sometimes felt as a paradox, comes under a general principle referred to in Art. 168. It is a consequence of the comparative slowness of the free oscillations in an equatorial canal of moderate depth. It appears from the rough numerical table on p. 257 that with a depth of 11250 feet a free wave would take about 30 hours to describe the earth's semi-circumference, whereas the period of the tidal disturbing force is only a little over 12 hours.

The formula (5) is, in fact, a particular case of Art. 168 (14), for it may be written

$$\eta = \frac{1}{1 - \sigma^2/\sigma_0{}^2} \bar{\eta}, \qquad \text{.............................(6)}$$

where $\bar{\eta}$ is the elevation given by the 'equilibrium theory,' viz.

$$\bar{\eta} = \tfrac{1}{2} H \cos 2 (nt + \phi + \epsilon), \qquad \text{.......................(7)}$$

and $\sigma = 2n$, $\sigma_0 = 2c/a$.

For such moderate depths as 10000 feet and under, $n^2 a^2$ is large compared with gh; the amplitude of the horizontal motion, as given by (4), is then $f/4n^2$, or $g/4n^2 a \, . \, H$, nearly, being approximately independent of the depth. In the case of the lunar tide this amplitude is about 140 feet. The maximum elevation is obtained by multiplying by $2h/a$; this gives, for a depth of 10000 feet, a height of only ·133 of a foot.

For greater depths the tides would be higher, but still inverted, until we reach the critical depth $n^2 a^2/g$, which is about 13 miles. For depths beyond this limit, the tides become direct, and approximate more and more to the value given by the equilibrium theory*.

182. The case of a circular canal parallel to the equator can be worked out in a similar manner. If the moon's orbit be still supposed to lie in the plane of the equator, we find by spherical trigonometry

$$\cos \vartheta = \sin \theta \cos (nt + \phi + \epsilon), \qquad \text{.....................(1)}$$

where θ is the co-latitude, and ϕ the longitude. The disturbing force in longitude is therefore

$$-\frac{\partial \Omega}{a \sin \theta \partial \phi} = -f \sin \theta \sin 2 (nt + \phi + \epsilon). \qquad \text{..............(2)}$$

This leads to

$$\eta = \tfrac{1}{2} \frac{c^2 H \sin^2 \theta}{c^2 - n^2 a^2 \sin^2 \theta} \cos 2 (nt + \phi + \epsilon). \qquad \text{..............(3)}$$

* Cf. Young, *l.c. ante* p. 253.

Hence if $na > c$ the tide will be direct or inverted according as $\sin\theta \lesseqgtr c/na$. If the depth be so great that $c > na$, the tides will be direct for all values of θ.

If the moon be not in the plane of the equator, but have a co-declination Δ, the formula (1) is replaced by

$$\cos\vartheta = \cos\theta\cos\Delta + \sin\theta\sin\Delta\cos\alpha, \qquad (4)$$

where α is the hour-angle of the moon from the meridian of P. For simplicity, we will neglect the moon's motion in declination in comparison with the earth's angular velocity of rotation; thus we put

$$\alpha = nt + \phi + \epsilon,$$

and treat Δ as constant. The resulting expression for the disturbing force along the parallel is found to be

$$-\frac{\partial\Omega}{a\sin\partial\theta\phi} = -f\cos\theta\sin 2\Delta\sin(nt+\phi+\epsilon) - f\sin\theta\sin^2\Delta\sin 2(nt+\phi+\epsilon). \qquad (5)$$

We thence obtain

$$\eta = \tfrac{1}{2}\frac{c^2H}{c^2 - n^2a^2\sin^2\theta}\sin 2\theta\sin 2\Delta\cos(nt+\phi+\epsilon) + \tfrac{1}{2}\frac{c^2H}{c^2 - n^2a^2\sin^2\theta}\sin^2\theta\sin^2\Delta\cos 2(nt+\phi+\epsilon). \qquad (6)$$

The first term gives a 'diurnal' tide of period $2\pi/n$; this vanishes and changes sign when the moon crosses the equator, *i.e.* twice a month. The second term represents a semi-diurnal tide of period π/n, whose amplitude is now less than before in the ratio of $\sin^2\Delta$ to 1.

183. In the case of a canal coincident with a meridian we should have to take account of the fact that the undisturbed figure of the free surface is one of relative equilibrium under gravity and centrifugal force, and is therefore not exactly circular. We shall have occasion later on to treat the question of displacements relative to a rotating globe somewhat carefully; for the present we will assume by anticipation that *in a narrow canal* the disturbances are sensibly the same as if the earth were at rest, and the disturbing body were to revolve round it with the proper relative motion.

If the moon be supposed to move in the plane of the equator, the hour-angle from the meridian of the canal may be denoted by $nt + \epsilon$, and if θ be the co-latitude of any point P on the canal, we find

$$\cos\vartheta = \sin\theta \,.\, \cos(nt+\epsilon). \qquad (1)$$

The equation of motion is therefore

$$\frac{\partial^2\xi}{\partial t^2} = c^2\frac{\partial^2\xi^2}{a^2\partial\theta^2} - \frac{\partial\Omega}{a\partial\theta} = c^2\frac{\partial^2\xi}{a^2\partial\theta^2} - \tfrac{1}{2}f\sin 2\theta \,.\, \{1 + \cos 2(nt+\epsilon)\}. \qquad (2)$$

Solving, we find

$$\eta = -\tfrac{1}{4}H\cos 2\theta - \tfrac{1}{4}\frac{c^2H}{c^2 - n^2a^2}\cos 2\theta \,.\, \cos 2(nt+\epsilon). \qquad (3)$$

The first term represents a permanent change of mean level to the extent

$$\eta = -\tfrac{1}{4}H\cos 2\theta. \quad \text{.............................(4)}$$

The fluctuations above and below the disturbed mean level are given by the second term in (3). This represents a semi-diurnal tide; and we notice that if, as in the actual case of the earth, c be less than na, there will be high water in latitudes above 45°, and low water in latitudes below 45°, when the moon is in the meridian of the canal, and *vice versâ* when the moon is 90° from that meridian. These circumstances would be all reversed if c were greater than na.

When the moon is not on the equator, but has a given declination, the mean level, as indicated by the term corresponding to (4), has a coefficient depending on the declination, and the consequent variations in it indicate a fortnightly (or, in the case of the sun, a semi-annual) tide. There is also introduced a diurnal tide whose sign depends on the declination. The reader will have no difficulty in examining these points, by means of the general value of Ω given in the Appendix.

184. In the case of a uniform canal encircling the globe (Arts. 181, 182) there is necessarily everywhere exact agreement (or exact opposition) of phase between the tidal elevation and the forces which generate it. This no longer holds, however, in the case of a canal or ocean of limited extent.

Let us take for instance the case of an equatorial canal of finite length*. Neglecting the moon's declination we have, if the origin of time be suitably chosen,

$$\frac{\partial^2 \xi}{\partial t^2} = c^2 \frac{\partial^2 \xi^2}{a^2 \partial \phi^2} - f\sin 2(nt+\phi), \quad \text{.....................(1)}$$

with the condition that $\xi = 0$ at the ends, where $\phi = \pm\alpha$, say.

If we neglect the inertia of the water the term $\partial^2\xi/\partial t^2$ is to be omitted and we find

$$\xi = \tfrac{1}{4}\frac{fa^2}{c^2}\left\{\sin 2nt\cos 2\alpha + \frac{\phi}{\alpha}\cos 2nt\sin 2\alpha - \sin 2(nt+\phi)\right\}. \quad \text{...(2)}$$

Hence
$$\eta = -\frac{h}{a}\frac{\partial\xi}{\partial\phi} = \tfrac{1}{2}H\left\{\cos 2(nt+\phi) - \frac{\sin 2\alpha}{2\alpha}\cos 2nt\right\}, \quad \text{.........(3)}$$

where $H = af/g$, as in Art. 180. This is the elevation on the (corrected) 'equilibrium' theory referred to in the Appendix to this Chapter. At the centre ($\phi = 0$) of the canal we have

$$\eta = \tfrac{1}{2}H\cos 2nt\left(1 - \frac{\sin 2\alpha}{2\alpha}\right). \quad \text{.....................(4)}$$

If α be small the range is here very small, but there is not a *node* in the absolute

* H. Lamb and Miss Swain, *Phil. Mag.* (6), xxix. 737 (1915). A similar effect of variable depth is discussed by Goldsbrough, *Proc. Lond. Math. Soc.* (2) xv. 64 (1915).

sense of the term. The times of high water coincide with the transits of moon and 'anti-moon*.' At the ends $\phi = \pm\alpha$ we have

$$\eta = \tfrac{1}{2}H\left\{\left(1-\frac{\sin 4\alpha}{4\alpha}\right)\cos 2(nt\pm\alpha) \mp \frac{1-\cos 4\alpha}{4\alpha}\sin 2(nt\pm\alpha)\right\}$$

$$= \tfrac{1}{2}HR_0\cos 2(nt\pm\alpha\mp\epsilon_0), \qquad (5)$$

if

$$R_0\cos 2\epsilon_0 = 1-\frac{\sin 4\alpha}{4\alpha}, \quad R_0\sin 2\epsilon_0 = -\frac{1-\cos 4\alpha}{4\alpha}. \qquad (6)$$

Here ϵ_0 denotes the hour-angle of the moon W. of the meridian when there is high water at the eastern end of the canal, or E. of the meridian when there is high water at the western end. When α is small we have

$$R_0 = 2\alpha, \quad \epsilon_0 = -\tfrac{1}{4}\pi + \tfrac{2}{3}\alpha, \qquad (7)$$

approximately.

When the inertia of the water is taken into account we have

$$\xi = \tfrac{1}{4}\frac{fa^2}{(m^2-1)c^2}\left[\sin 2(nt+\phi) - \frac{1}{\sin 4m\alpha}\{\sin 2(nt+\alpha)\sin 2m(\phi+\alpha) - \sin 2(nt-\alpha)\sin 2m(\phi-\alpha)\}\right], \qquad (8)$$

where $m = na/c$. Hence†

$$\eta = -\tfrac{1}{2}\frac{H}{m^2-1}\left[\cos 2(nt+\phi) - \frac{m}{\sin 4m\alpha}\{\sin 2(nt+\alpha)\cos 2m(\phi+\alpha) - \sin 2(nt-\alpha)\cos 2m(\phi-\alpha)\}\right]. \qquad (9)$$

If we imagine m to tend to the limit 0 we obtain the formula (3) of the equilibrium theory. It may be noticed that the expressions do not become infinite for $m \rightarrow 1$ as they would in the case of an endless canal. In all cases which are at all comparable with oceanic conditions m is, however, considerably greater than unity.

At the centre of the canal we have

$$\eta = -\tfrac{1}{2}\frac{H}{m^2-1}\cos 2nt\left(1-\frac{m\sin 2\alpha}{\sin 2m\alpha}\right). \qquad (10)$$

As in the equilibrium theory, the range is very small if α be small, but there is not a true node. At the ends we find

$$\eta = \tfrac{1}{2}\frac{H}{m^2-1}\left\{\left(\frac{m\sin 4\alpha}{\sin 4m\alpha}-1\right)\cos 2(nt\pm\alpha) \pm \frac{m(\cos 4m\alpha - \cos 4\alpha)}{\sin 4m\alpha}\sin 2(nt\pm\alpha)\right\}$$

$$= \tfrac{1}{2}HR_1\cos 2(nt\pm\alpha\mp\epsilon_1), \qquad (11)$$

if

$$R_1\cos 2\epsilon_1 = \frac{m\sin 4\alpha - \sin 4m\alpha}{(m^2-1)\sin 4m\alpha}, \quad R_1\sin 2\epsilon_1 = \frac{m(\cos 4m\alpha-\cos 4\alpha)}{(m^2-1)\sin 4m\alpha}. \qquad (12)$$

* This term is explained in the Appendix to this Chapter.

† Cf. Airy, "Tides and Waves," Art. 301.

When α is small we have

$$R_1 = 2\alpha, \quad \epsilon_1 = -\tfrac{1}{4}\pi + \tfrac{2}{3}\alpha, \quad \text{......................(13)}$$

approximately, as in the case of the equilibrium theory.

The value of R_1 becomes infinite when $\sin 4m\alpha = 0$. This determines the critical lengths of the canal for which there is a free period equal to π/n, or half a lunar day. The limiting value of ϵ_1 in such a case is given by

$$\tan 2\epsilon_1 = -\cot 2\alpha, \quad \text{or} \quad = \tan 2\alpha,$$

according as $4m\alpha$ is an odd or even multiple of π.

		Corrected Equilibrium Theory			Dynamical Theory		
2α (degrees)	$2a\alpha$ (miles)	Range at centre	Range at ends	ϵ_0 (degrees)	Range at centre	Range at ends	ϵ_1 (degrees)
0	0	0	0	*−45*	0	0	*−45*
9	540	·004	·157	*−42*	·004	·165	*−41·9*
18	1080	·016	·311	*−39*	·018	·396	*−38·5*
27	1620	·037	·460	*−36*	·044	·941	*−33·9*
31·5	1890	·050	·531	*−34·5*	·063	1·945	*−30·9*
36	2160	·065	·601	*−33*	·089	∞	{ *−27* *+63*
40·5	2430	·081	·668	*−31·6*	·125	1·956	*+68·2*
45	2700	·100	·733	*−30·1*	·174	·987	*+75·7*
54	3240	·142	·853	*−27·2*	·354	·660	*−83·5*
63	3780	·100	·050	*24·4*	·018	1·141	*−05·1*
72	4320	·243	1·051	*−21·6*	∞	∞	{ *−54* *+36*
81	4860	·301	1·127	*−18·9*	1·459	1·112	*+44·5*
90	5400	·363	1·185	*−16·2*	·864	·513	*+55·9*

The table illustrates the case of $m = 2·5$. If $\pi/n = 12$ lunar hours this implies a depth of 10820 ft., which is of the same order of magnitude as the mean depth of the ocean. The corresponding wave-velocity is about 360 sea-miles per hour. The first critical length is 2160 miles ($\alpha = \frac{1}{10}\pi$). The unit in terms of which the range is expressed is the quantity H, whose value for the lunar tide is about 1·80 ft. The hour-angles ϵ_0 and ϵ_1 are adjusted so as to lie always between $\pm 90°$, and the positive sign indicates position W. of the meridian in the case of the eastern end of the canal, and E. of the meridian for the western end.

Wave-Motion in a Canal of Variable Section.

185. When the section (S, say) of the canal is not uniform but varies gradually from point to point, the equation of continuity is by Art. 169 (11),

$$\eta = -\frac{1}{b}\frac{\partial}{\partial x}(S\xi), \quad \text{..............................(1)}$$

where b denotes the breadth at the surface. If h denote the mean depth over the width b, we have $S - bh$, and therefore

$$\eta = -\frac{1}{b}\frac{\partial}{\partial x}(hb\xi), \quad \text{..............................(2)}$$

where h, b are now functions of x.

The dynamical equation has the same form as before, viz.

$$\frac{\partial^2 \xi}{\partial t^2} = -g\frac{\partial \eta}{\partial x}. \qquad (3)$$

Between (2) and (3) we may eliminate either η or ξ; the equation in η is

$$\frac{\partial^2 \eta}{\partial t^2} = \frac{g}{b}\frac{\partial}{\partial x}\left(hb\frac{\partial \eta}{\partial x}\right). \qquad (4)$$

The laws of propagation of waves in a canal of gradually varying rectangular section were investigated by Green*. His result, freed from the restriction to the special form of section, may be obtained as follows.

If we introduce a variable τ defined by

$$\frac{dx}{d\tau} = (gh)^{\frac{1}{2}}, \qquad (5)$$

in place of x, the equation (4) transforms into

$$\frac{\partial^2 \eta}{\partial t^2} = \eta'' + \left(\frac{b'}{b} + \frac{1}{2}\frac{h'}{h}\right)\eta', \qquad (6)$$

where the accents denote differentiations with respect to τ. If b and h were constants, the equation would be satisfied by $\eta = F(\tau - t)$, as in Art. 170; in the present case we assume for trial,

$$\eta = \Theta \,.\, F(\tau - t), \qquad (7)$$

where Θ is a function of τ only. Substituting in (6), we find

$$2\frac{\Theta'}{\Theta}\cdot\frac{F'}{F} + \frac{\Theta''}{\Theta} + \left(\frac{b'}{b} + \frac{1}{2}\frac{h'}{h}\right)\left(\frac{F'}{F} + \frac{\Theta'}{\Theta}\right) = 0. \qquad (8)$$

The terms of this which involve F will cancel provided

$$2\frac{\Theta'}{\Theta} + \frac{b'}{b} + \frac{1}{2}\frac{h'}{h} = 0,$$

or

$$\Theta = Cb^{-\frac{1}{2}}h^{-\frac{1}{4}}, \qquad (9)$$

C being a constant. Hence, provided the remaining terms in (8) may be neglected, the equation (4) will be satisfied.

The above approximation is justified, provided we can neglect Θ''/Θ' and Θ'/Θ in comparison with F''/F. As regards Θ'/Θ, it appears from (9) and (7) that this is equivalent to neglecting $b^{-1} . db/dx$ and $h^{-1} . dh/dx$ in comparison with $\eta^{-1} . \partial\eta/\partial x$. If, now, λ denote a wave-length, in the general sense of Art. 172, $\partial\eta/\partial x$ is of the order η/λ, so that the assumption in question is that $\lambda db/dx$ and $\lambda dh/dx$ are small compared with b and h, respectively. In other words, it is assumed that the transverse dimensions of the canal vary only by small fractions of themselves within the limits of a wave-length. It is easily seen, in like manner, that the neglect of Θ''/Θ' in comparison with F''/F implies a similar limitation to the rates of change of db/dx and dh/dx.

Since the equation (4) is unaltered when we reverse the sign of t, the complete solution, subject to the above restrictions, is

$$\eta = b^{-\frac{1}{2}}h^{-\frac{1}{4}}\{F(\tau - t) + f(\tau + t)\}, \qquad (10)$$

where F and f are arbitrary functions.

The first term in this represents a wave travelling in the direction of x-positive; the velocity of propagation past any point is determined by the consideration that any particular phase is recovered when $\delta\tau$ and δt have equal values, and is therefore equal to $\sqrt{(gh)}$, by

* "On the Motion of Waves in a Variable Canal of small depth and width," *Camb. Trans.* vi. (1837) [*Papers*, p. 225]; see also Airy, "Tides and Waves," Art. 260.

(5), as we should expect from the case of a uniform section. In like manner the second term in (10) represents a wave travelling in the direction of x-negative. In each case the elevation of any particular part of the wave alters, as it proceeds, according to the law $b^{-\frac{1}{2}} h^{-\frac{1}{4}}$.

The reflexion of a progressive wave at a point where the section of a canal suddenly changes has been considered in Art. 176. The formulae there given shew, as we should expect, that the smaller the change in the dimensions of the section, the smaller will be the amplitude of the reflected wave. The case where the change from one section to the other is continuous, instead of abrupt, has been investigated by Rayleigh for a special law of transition*. It appears that if the space within which the transition is completed be a moderate multiple of a wave-length there is practically no reflexion; whilst in the opposite extreme the results agree with those of Art. 176.

If we assume, on the basis of these results, that when the change of section within a wave-length may be neglected a progressive wave suffers no appreciable disintegration by reflexion, the law of amplitude easily follows from the principle of energy†. It appears from Art. 174 that the energy of the wave varies as the length, the breadth, and the square of the height, and it is easily seen that the length of the wave, in different parts of the canal, varies as the corresponding velocity of propagation, and therefore as the square root of the mean depth. Hence in the above notation, $\eta^2 b h^{\frac{1}{2}}$ is constant, or

$$\eta \propto b^{-\frac{1}{2}} h^{-\frac{1}{4}},$$

which is Green's law above found.

186. In the case of simple harmonic motion, where $\eta \propto \cos(\sigma t + \epsilon)$, the equation (4) of the preceding Art. becomes

$$\frac{g}{b}\frac{\partial}{\partial x}\left(hb\frac{\partial \eta}{\partial x}\right) + \sigma^2 \eta = 0. \qquad \text{.........................(1)}$$

Some particular cases of considerable interest can be solved with ease.

1°. For example, let us take the case of a canal whose breadth varies as the distance from the end $x=0$, the depth being uniform; and let us suppose that at its mouth ($x=a$) the canal communicates with an open sea in which a tidal oscillation

$$\eta = C\cos(\sigma t + \epsilon) \qquad \text{...(2)}$$

is maintained. Putting $h=$const., $b \propto x$, in (1), we find

$$\frac{\partial^2 \eta}{\partial x^2} + \frac{1}{x}\frac{\partial \eta}{\partial x} + k^2 \eta = 0, \qquad \text{......................................(3)}$$

provided

$$k^2 = \sigma^2/gh. \qquad \text{..(4)}$$

Hence

$$\eta = C\frac{J_0(kx)}{J_0(ka)}\cos(\sigma t + \epsilon). \qquad \text{..................................(5)}$$

* "On Reflection of Vibrations at the Confines of two Media between which the Transition is gradual," *Proc. Lond. Math. Soc.* (1) xi. 51 (1880) [*Papers*, i. 460]; *Theory of Sound*, 2nd ed., London, 1894, Art. 148 *b*.

† Rayleigh, *l.c. ante* p. 260.

The curve $y=J_0(x)$ is figured on p. 286; it indicates how the amplitude of the forced oscillation increases, whilst the wave-length is practically constant, as we proceed up the canal from the mouth.

2°. Let us suppose that the variation is in the *depth* only, and that this increases uniformly from the end $x=0$ of the canal to the mouth, the remaining circumstances being as before. If, in (1), we put $h=h_0x/a$, $\kappa=\sigma^2a/gh_0$, we obtain

$$\frac{\partial}{\partial x}\left(x\frac{\partial\eta}{\partial x}\right)+\kappa\eta=0. \qquad (6)$$

The solution of this which is finite for $x=0$ is

$$\eta=A\left(1-\frac{\kappa x}{1^2}+\frac{\kappa^2x^2}{1^2.\,2^2}-\ldots\right), \qquad (7)$$

or

$$\eta=AJ_0(2\kappa^{\frac{1}{2}}x^{\frac{1}{2}}), \qquad (8)$$

whence finally, restoring the time-factor and determining the constant,

$$\eta=C\frac{J_0(2\kappa^{\frac{1}{2}}x^{\frac{1}{2}})}{J_0(2\kappa^{\frac{1}{2}}a^{\frac{1}{2}})}\cos(\sigma t+\epsilon). \qquad (9)$$

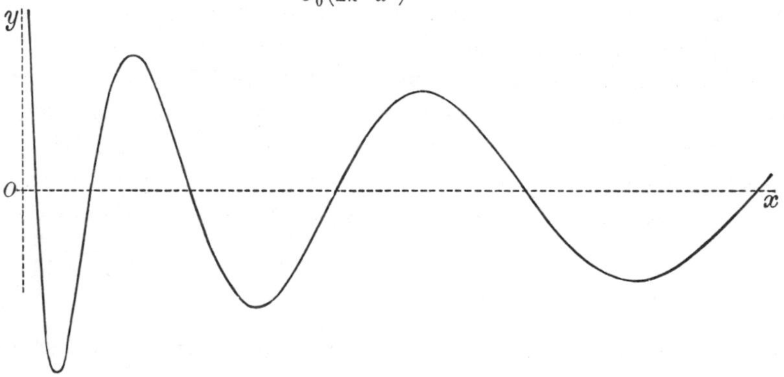

The annexed diagram of the curve $y=J_0(\sqrt{x})$, where, for clearness, the scale adopted for y is 200 times that of x, shews how the amplitude continually increases, and the wave-length diminishes, as we travel up the canal.

These examples may serve to illustrate the exaggeration of oceanic tides which takes place in shallow seas and in estuaries.

3°. If the breadth and depth both vary as the distance from the end $x=0$, we have, writing $b=b_0x/a$, $h=h_0x/a$,

$$x\frac{\partial^2\eta}{\partial x^2}+2\frac{\partial\eta}{\partial x}+\kappa\eta=0, \qquad (10)$$

where $\kappa=\sigma^2a/gh_0$ as before. Hence

$$\eta=A\left(1-\frac{\kappa x}{1\,.\,2}+\frac{\kappa^2x^2}{1\,.\,2\,.\,2\,.\,4}-\ldots\right)\cos(\sigma t+\epsilon). \qquad (11)$$

The series is equal to $J_1(2\kappa^{\frac{1}{2}}x^{\frac{1}{2}})/\kappa^{\frac{1}{2}}x^{\frac{1}{2}}$, and the constant A is determined by comparison with (2). The present assumption gives a fair representation of the case of the Bristol Channel, and the tides observed at various stations are found to be in good agreement with the formula*.

We add one or two simple problems of free oscillations.

* G. I. Taylor, *Camb. Proc.* xx. 320 (1921).

4°. Let us take the case of a canal of uniform breadth, of length $2a$, whose bed slopes uniformly from either end to the middle. If we take the origin at one end, the motion in the first half of the canal will be determined, as above, by

$$\eta = AJ_0(2\kappa^{\frac{1}{2}}x^{\frac{1}{2}}), \quad \text{.......................................(12)}$$

where $\kappa = \sigma^2 a/gh_0$, h_0 denoting the depth at the middle.

It is evident that the normal modes will fall into two classes. In the first of these η will have opposite values at corresponding points of the two halves of the canal, and will therefore vanish at the centre ($x=a$). The values of σ are then determined by

$$J_0(2\kappa^{\frac{1}{2}}a^{\frac{1}{2}}) = 0, \quad \text{.......................................(13)}$$

viz. κ being any root of this, we have

$$\sigma = \frac{(gh_0)^{\frac{1}{2}}}{a} \cdot (\kappa a)^{\frac{1}{2}}. \quad \text{.......................................(14)}$$

In the second class, the value of η is symmetrical with respect to the centre, so that $\partial\eta/\partial x = 0$ at the middle. This gives

$$J_0'(2\kappa^{\frac{1}{2}}a^{\frac{1}{2}}) = 0. \quad \text{.......................................(15)}$$

It appears that the slowest oscillation is of the asymmetrical class, and corresponds to the smallest root of (13), which is $2\kappa^{\frac{1}{2}}a^{\frac{1}{2}} = \cdot 7655\pi$, whence

$$\frac{2\pi}{\sigma} = 1\cdot 306 \times \frac{4a}{(gh_0)^{\frac{1}{2}}}.$$

5°. Again, let us suppose that the depth of the canal varies according to the law

$$h = h_0\left(1 - \frac{x^2}{a^2}\right), \quad \text{.......................................(16)}$$

where x now denotes the distance from the middle. Substituting in (1), with $b=$const., we find

$$\frac{\partial}{\partial x}\left\{\left(1 - \frac{x^2}{a^2}\right)\frac{\partial \eta}{\partial x}\right\} + \frac{\sigma^2}{gh_0}\eta = 0. \quad \text{..............................(17)}$$

If we put

$$\sigma^2 = n(n+1)\frac{gh_0}{a^2}, \quad \text{.......................................(18)}$$

this is of the same form as the general equation of zonal harmonics, Art. 84 (1).

In the present problem n is determined by the condition that η must be finite for $x/a = \pm 1$. This requires (Art. 85) that n should be integral; the normal modes are therefore of the type

$$\eta = CP_n\left(\frac{x}{a}\right) \cdot \cos(\sigma t + \epsilon), \quad \text{..................................(19)}$$

where P_n is a zonal harmonic, the value of σ being determined by (18).

In the slowest oscillation ($n=1$), the profile of the free surface is a straight line. For a canal of *uniform* depth h_0, and of the same length ($2a$), the corresponding value of σ would be $\pi c/2a$, where $c = (gh_0)^{\frac{1}{2}}$. Hence in the present case the frequency is less, in the ratio $2\sqrt{2}/\pi$, or $\cdot 9003$*.

The forced oscillations due to a uniform disturbing force

$$X = f\cos(\sigma t + \epsilon) \quad \text{.......................................(20)}$$

* For extensions, and applications to the theory of 'seiches' in lochs, see Chrystal, "Some Results in the Mathematical Theory of Seiches," *Proc. R. S. Edin.* xxv. 328 (1904), and *Trans. R. S. Edin.* xli. 599 (1905). For more recent investigations see Proudman, *Proc. Lond. Math. Soc.* (2) xiv. 240 (1914); Doodson, *Trans. R. S. Edin.* lii. 629 (1920); Jeffreys, *M. N. R. A. S., Geophys. Suppt.* i. 495 (1928).

can be obtained by the rule of Art. 168 (14). The equilibrium form of the free surface is evidently

$$\bar{\eta}=\frac{f}{g}x\cos(\sigma t+\epsilon), \quad\text{.....................................(21)}$$

and, since the given force is of the normal type $n=1$, we have

$$\eta=\frac{f}{g(1-\sigma^2/\sigma_0^2)}x\cos(\sigma t+\epsilon), \quad\text{...............................(22)}$$

where $$\sigma_0^2=2gh_0/a^2.$$

Waves of Finite Amplitude.

187. When the elevation η is not small compared with the mean depth h, waves, even in an uniform canal of rectangular section, are no longer propagated without change of type. The question was first investigated by Airy*, by methods of successive approximation. He found that in a progressive wave different parts will travel with different velocities, the wave-velocity corresponding to an elevation η being given approximately by Art. 175 (6).

A more complete view of the matter can be obtained by a method similar to that adopted by Riemann in treating the analogous problem in Acoustics. (See Art. 282.)

The sole assumption on which we are now proceeding is that the vertical acceleration may be neglected. It follows, as explained in Art. 168, that the horizontal velocity may be taken to be uniform over any section of the canal. The dynamical equation is

$$\frac{\partial u}{\partial t}+u\frac{\partial u}{\partial x}=-g\frac{\partial \eta}{\partial x}, \quad\text{..............................(1)}$$

as before, and the equation of continuity, in the case of a rectangular section, is easily seen to be

$$\frac{\partial}{\partial x}\{(h+\eta)u\}=-\frac{\partial \eta}{\partial t}, \quad\text{..........................(2)}$$

where h is the depth. This may be written

$$\frac{\partial \eta}{\partial t}+u\frac{\partial \eta}{\partial x}=-(h+\eta)\frac{\partial u}{\partial x}. \quad\text{........................(3)}$$

Multiplying this equation by $f'(\eta)$, where $f(\eta)$ is a function to be determined, and adding to (1), we have

$$\left(\frac{\partial}{\partial t}+u\frac{\partial}{\partial x}\right)\{f(\eta)+u\}=-(h+\eta)f'(\eta)\frac{\partial u}{\partial x}-g\frac{\partial \eta}{\partial x}$$

$$=-(h+\eta)f'(\eta)\frac{\partial}{\partial x}\{f(\eta)+u\}, \quad\text{........(4)}$$

provided $$(h+\eta)\{f'(\eta)\}^2=g.$$

* *l.c. ante* p. 267.

This is satisfied by

$$f(\eta)=2c_0\left\{\left(1+\frac{\eta}{h}\right)^{\frac{1}{2}}-1\right\}, \quad\text{.......................(5)}$$

where $c_0=\sqrt{(gh)}$. Hence, writing

$$P=f(\eta)+u, \quad Q=f(\eta)-u, \quad\text{....................(6)}$$

we have

$$\frac{\partial P}{\partial t}+(u+v)\frac{\partial P}{\partial x}=0, \quad\text{..........................(7)}$$

and, by similar steps,

$$\frac{\partial Q}{\partial t}+(u-v)\frac{\partial Q}{\partial x}=0, \quad\text{..........................(8)}$$

where
$$v=(h+\eta)f'(\eta)=c_0\left(1+\frac{\eta}{h}\right)^{\frac{1}{2}}. \quad\text{....................(9)}$$

It appears, therefore, that P is constant for a geometrical point moving in the positive direction of x with the velocity

$$c_0\left(1+\frac{\eta}{h}\right)^{\frac{1}{2}}+u, \quad\text{..........................(10)}$$

whilst Q is constant for a point moving in the negative direction with the velocity

$$c_0\left(1+\frac{\eta}{h}\right)^{\frac{1}{2}}-u. \quad\text{..........................(11)}$$

Hence any given value of P travels forwards, and any given value of Q travels backwards, with the velocities given by (10) and (11) respectively. The values of P and Q are determined by those of η and u, and conversely.

As an example, let us suppose that the initial disturbance is confined to the space for which $a<x<b$, so that P and Q are initially zero for $x<a$ and $x>b$. The region within which P differs from zero therefore advances, whilst that within which Q differs from zero recedes, so that after a time these regions separate, and leave between them a space within which $P=0$, $Q=0$, and the fluid is therefore at rest. The original disturbance has now been resolved into two progressive waves travelling in opposite directions.

In the advancing wave we have

$$Q=0, \quad \tfrac{1}{2}P=u=2c_0\left\{\left(1+\frac{\eta}{h}\right)^{\frac{1}{2}}-1\right\}, \quad\text{..............(12)}$$

so that the elevation and the particle-velocity are connected by a definite relation (cf. Art. 171). The wave-velocity is given by (10) and (12), viz. it is

$$c_0\left\{3\left(1+\frac{\eta}{h}\right)^{\frac{1}{2}}-2\right\}. \quad\text{..........................(13)}$$

To the first order of η/h, this is in agreement with Airy's result quoted on p. 262.

Similar conclusions can be drawn in regard to the receding wave*.

* The above results can also be deduced from the equation (3) of Art. 173, by a method due to Earnshaw; see Art. 283.

Since the wave-velocity increases with the elevation, it appears that in a progressive wave-system the slopes will become continually steeper in front, and more gradual behind, until at length a state of things is reached in which we are no longer justified in neglecting the vertical acceleration. As to what happens after this point we have at present no guide from theory; observation shews, however, that the crests tend ultimately to curl over and break.

The case of a 'bore,' where there is a transition from one uniform level to another, may be investigated by the artifice of steady motion (Art. 175). If Q denote the volume per unit breadth which crosses each section in unit time we have

$$u_1 h_1 = u_2 h_2 = Q, \qquad\qquad (14)$$

where the suffixes refer to the two uniform states, h_1 and h_2 denoting the depths. Considering the mass of fluid which is at a given instant contained between two cross-sections, one on each side of the transition, we see that in unit time it gains momentum to the amount $\rho Q (u_2 - u_1)$, the second section being supposed to lie to the right of the first. Since the mean pressures over the sections are $\frac{1}{2} g\rho h_1$ and $\frac{1}{2} g\rho h_2$, we have

$$Q(u_2 - u_1) = \tfrac{1}{2} g (h_1^2 - h_2^2). \qquad\qquad (15)$$

Hence, and from (14),

$$Q^2 = \tfrac{1}{2} g h_1 h_2 (h_1 + h_2). \qquad\qquad (16)$$

If we impress on everything a velocity $-u_1$ we get the case of a wave invading still water with a velocity of propagation

$$u_1 = \sqrt{\left\{\frac{g h_2 (h_1 + h_2)}{2h_1}\right\}} \qquad\qquad (17)$$

in the negative direction. The particle-velocity in the advancing wave is $u_1 - u_2$ in the direction of propagation. This is positive or negative according as $h_2 \gtrless h_1$, *i.e.* according as the wave is one of elevation or depression.

The equation of energy is however violated, unless the difference of level be regarded as infinitesimal. If, in the steady motion, we consider a particle moving along the surface stream-line, its loss of energy in passing the place of transition is

$$\tfrac{1}{2}\rho (u_1^2 - u_2^2) + g\rho (h_1 - h_2) \qquad\qquad (18)$$

per unit volume. In virtue of (14) and (16) this takes the form

$$\frac{g\rho (h_2 - h_1)^3}{4h_1 h_2}. \qquad\qquad (19)$$

Hence, so far as this investigation goes, a bore of elevation ($h_2 > h_1$) can be propagated unchanged on the assumption that dissipation of energy takes place to a suitable extent at the transition. If however $h_2 < h_1$, the expression (19) is negative, and a *supply* of energy would be necessary. It follows that a negative bore of finite height cannot in any case travel unchanged*.

188. In the detailed application of the equations (1) and (3) to tidal phenomena, it is usual to follow the method of successive approximation. As an example, we will take the case of a canal communicating at one end ($x = 0$) with an open sea, where the elevation is given by

$$\eta = a \cos \sigma t. \qquad\qquad (20)$$

* Rayleigh, "On the Theory of Long Waves and Bores," *Proc. Roy. Soc.* A, xc. 324 (1914) [*Papers*, vi. 250].

For a first approximation we have

$$\frac{\partial u}{\partial t} = -g\frac{\partial \eta}{\partial x}, \qquad \frac{\partial \eta}{\partial t} = -h\frac{\partial u}{\partial x}, \qquad \text{..............................(21)}$$

the solution of which, consistent with (20), is

$$\eta = a\cos\sigma\left(t - \frac{x}{c}\right), \qquad u = \frac{ga}{c}\cos\sigma\left(t - \frac{x}{c}\right). \qquad \text{....................(22)}$$

For a second approximation we substitute these values of η and u in (1) and (3), and obtain

$$\frac{\partial u}{\partial t} = -g\frac{\partial \eta}{\partial x} - \frac{g^2\sigma a^2}{2c^3}\sin 2\sigma\left(t - \frac{x}{c}\right), \qquad \frac{\partial \eta}{\partial t} = -h\frac{\partial u}{\partial x} - \frac{g\sigma a^2}{c^2}\sin 2\sigma\left(t - \frac{x}{c}\right). \quad \text{...(23)}$$

Integrating these by the usual methods, we find, as the solution consistent with (20),

$$\left.\begin{aligned} \eta &= a\cos\sigma\left(t - \frac{x}{c}\right) - \tfrac{3}{4}\frac{g\sigma a^2}{c^3}x\sin 2\sigma\left(t - \frac{x}{c}\right), \\ u &= \frac{ga}{c}\cos\sigma\left(t - \frac{x}{c}\right) - \tfrac{1}{8}\frac{g^2a^2}{c^3}\cos 2\sigma\left(t - \frac{x}{c}\right) - \tfrac{3}{4}\frac{g^2\sigma a^2}{c^4}x\sin 2\sigma\left(t - \frac{x}{c}\right). \end{aligned}\right\} \text{......(24)}$$

The annexed figure shews, with, of course, exaggerated amplitude, the profile of the waves in a particular case, as determined by the first of these equations. It is to be noted that if we fix our attention on a particular point of the canal, the rise and fall of the water do not take place symmetrically, the fall occupying a longer time than the rise.

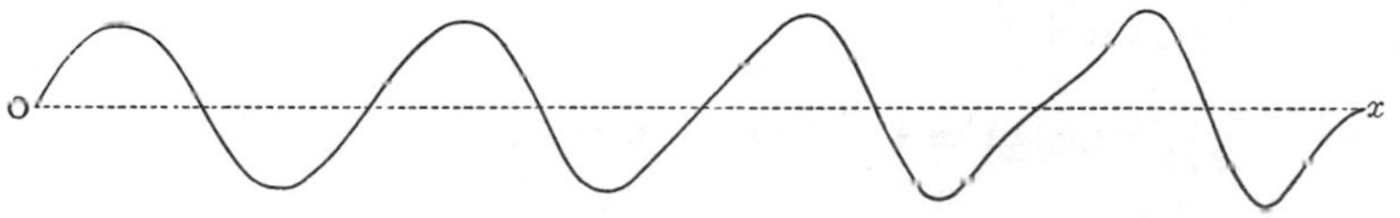

The occurrence of the factor x outside trigonometrical terms in (24) shews that there is a limit beyond which the approximation breaks down. The condition for the success of the approximation is evidently that $g\sigma ax/c^3$ should be small. Putting $c^2 = gh$, $\lambda = 2\pi c/\sigma$, this fraction becomes equal to $2\pi(a/h).(x/\lambda)$. Hence however small the ratio of the original elevation (a) to the depth, the fraction ceases to be small when x is a sufficient multiple of the wave-length (λ).

It is to be noticed that the limit here indicated is already being overstepped in the right-hand portions of the figure; and that the peculiar features which are beginning to shew themselves on the rear slope are an indication rather of the imperfections of the analysis than of any actual property of the waves. If we were to trace the curve further, we should find a secondary maximum and minimum of elevation developing themselves on the rear slope. In this way Airy attempted to explain the phenomenon of a double high-water which is observed in some rivers; but, for the reason given, the argument cannot be sustained*.

The same difficulty does not necessarily present itself in the case of a canal closed by a fixed barrier at a distance from the mouth, or, again, in the case of the forced waves due to a periodic horizontal force in a canal closed at both ends (Art. 179). Enough has, however, been given to shew the general character of the results to be expected in such cases. For further details we must refer to Airy's treatise †.

When analysed, as in (24), into a series of simple-harmonic functions of the time, the expression for the elevation of the water at any particular place (x) consists of two terms,

* McCowan, *l.c. ante* p. 259.

† "Tides and Waves," Arts. 198, ... and 308. See also G. H. Darwin, "Tides," *Encyc. Britann.* (9th ed.) xxiii. 362, 363 (1888).

of which the second represents an 'over-tide,' or 'tide of the second order,' being proportional to a^2; its frequency is double that of the primary disturbance (20). If we were to continue the approximation we should obtain tides of higher orders, whose frequencies are 3, 4, ... times that of the primary.

If, in place of (20), the disturbance at the mouth of the canal were given by

$$\zeta = a\cos\sigma t + a'\cos(\sigma' t + \epsilon),$$

it is easily seen that in the second approximation we should in like manner obtain tides of periods $2\pi/(\sigma+\sigma')$ and $2\pi/(\sigma-\sigma')$; these are called 'compound tides.' They are analogous to the 'combination-tones' in Acoustics which were first investigated by Helmholtz*.

Propagation in Two Dimensions.

189. Let us suppose, in the first instance, that we have a plane sheet of water of uniform depth h. If the vertical acceleration be neglected, the horizontal motion will as before be the same for all particles in the same vertical line. The axes of x, y being horizontal, let u, v be the component horizontal velocities at the point (x, y), and let ζ be the corresponding elevation of the free surface above the undisturbed level. The equation of continuity may be obtained by calculating the flux of matter into the columnar space which stands on the elementary rectangle $\delta x \delta y$; thus we have, neglecting terms of the second order,

$$\frac{\partial}{\partial x}(uh\delta y)\,\delta x + \frac{\partial}{\partial y}(vh\delta x)\,\delta y = -\frac{\partial}{\partial t}\{(\zeta + h)\,\delta x\delta y\},$$

whence

$$\frac{\partial \zeta}{\partial t} = -h\left(\frac{\partial u}{\partial x} + \frac{\partial v}{\partial y}\right). \quad \text{.........................(1)}$$

The dynamical equations are, in the absence of disturbing forces,

$$\rho\frac{\partial u}{\partial t} = -\frac{\partial p}{\partial x}, \quad \rho\frac{\partial v}{\partial t} = -\frac{\partial p}{\partial y},$$

where we may write

$$p - p_0 = g\rho(z_0 + \zeta - z),$$

if z_0 denote the ordinate of the free surface in the undisturbed state. We thus obtain

$$\frac{\partial u}{\partial t} = -g\frac{\partial \zeta}{\partial x}, \quad \frac{\partial v}{\partial t} = -g\frac{\partial \zeta}{\partial y}. \quad \text{.......................(2)}$$

If we eliminate u and v, we find

$$\frac{\partial^2 \zeta}{\partial t^2} = c^2\left(\frac{\partial^2 \zeta}{\partial x^2} + \frac{\partial^2 \zeta}{\partial y^2}\right), \quad \text{.........................(3)}$$

where $c^2 = gh$ as before.

In the application to simple-harmonic motion, the equations are shortened if we assume a complex time-factor $e^{i(\sigma t + \epsilon)}$, and reject in the end, the

* "Ueber Combinationstöne," *Berl. Monatsber.* May 22, 1856 [*Wiss. Abh.* i. 256]; and "Theorie der Luftschwingungen in Röhren mit offenen Enden," *Crelle*, lvii. 14 (1859) [*Wiss. Abh.* i. 318].

imaginary parts of our expressions. This is legitimate so long as we have to deal solely with linear equations. We have then, from (2),

$$u = \frac{ig}{\sigma}\frac{\partial\zeta}{\partial x}, \quad v = \frac{ig}{\sigma}\frac{\partial\zeta}{\partial y}, \quad \text{.........................(4)}$$

whilst (3) becomes

$$\frac{\partial^2\zeta}{\partial x^2} + \frac{\partial^2\zeta}{\partial y^2} + k^2\zeta = 0, \quad \text{.........................(5)}$$

where

$$k^2 = \sigma^2/c^2. \quad \text{..................................(6)}$$

The condition to be satisfied at a vertical bounding wall is obtained at once from (4), viz. it is

$$\frac{\partial\zeta}{\partial n} = 0, \quad \text{....................................(7)}$$

if δn denote an element of the normal to the boundary.

When the fluid is subject to small disturbing forces whose variation within the limits of the depth may be neglected, the equations (2) are replaced by

$$\frac{\partial u}{\partial t} = -g\frac{\partial\zeta}{\partial x} - \frac{\partial\Omega}{\partial x}, \quad \frac{\partial v}{\partial t} = -g\frac{\partial\zeta}{\partial y} - \frac{\partial\Omega}{\partial y}, \quad \text{..............(8)}$$

where Ω is the potential of these forces.

If we put

$$\bar{\zeta} = -\Omega/g, \quad \text{...............................(9)}$$

so that $\bar{\zeta}$ denotes the equilibrium-elevation corresponding to the potential Ω these may be written

$$\frac{\partial u}{\partial t} = -g\frac{\partial}{\partial x}(\zeta - \bar{\zeta}), \quad \frac{\partial v}{\partial t} = -g\frac{\partial}{\partial y}(\zeta - \bar{\zeta}). \quad \text{...........(10)}$$

In the case of simple-harmonic motion, these take the forms

$$u = \frac{ig}{\sigma}\frac{\partial}{\partial x}(\zeta - \bar{\zeta}), \quad v = \frac{ig}{\sigma}\frac{\partial}{\partial y}(\zeta - \bar{\zeta}), \quad \text{.................(11)}$$

whence, substituting in the equation of continuity (1) we obtain

$$(\nabla_1^2 + k^2)\zeta = \nabla_1^2\bar{\zeta}, \quad \text{.........................(12)}$$

if

$$\nabla_1^2 = \frac{\partial^2}{\partial x^2} + \frac{\partial^2}{\partial y^2}, \quad \text{.............................(13)}$$

and $k^2 = \sigma^2/gh$, as before. The condition to be satisfied at a vertical boundary is now

$$\frac{\partial}{\partial n}(\zeta - \bar{\zeta}) = 0. \quad \text{............................(14)}$$

190. The equation (3) of Art. 189 is identical in form with that which presents itself in the theory of the transverse vibrations of a uniformly stretched membrane. A still closer analogy, when regard is had to the boundary-conditions, is furnished by the theory of cylindrical waves of sound*. Indeed many of the results obtained in this latter theory can be at once transferred to our present subject.

* Rayleigh, *Theory of Sound*, Art. 338.

Thus, to find the free oscillations of a sheet of water bounded by vertical walls, we require a solution of

$$(\nabla_1^2 + k^2)\,\zeta = 0, \quad \text{.............................(1)}$$

subject to the boundary-condition

$$\frac{\partial \zeta}{\partial n} = 0. \quad \text{..................................(2)}$$

Just as in Art. 178 it will be found that such a solution is possible only for certain values of k, which accordingly determine the periods $(2\pi/kc)$ of the various normal modes.

Thus, in the case of a *rectangular* boundary, if we take the origin at one corner, and the axes of x, y along two of the sides, the boundary-conditions are that $\partial\zeta/\partial x = 0$ for $x = 0$ and $x = a$, and $\partial\zeta/\partial y = 0$ for $y = 0$ and $y = b$, where a, b are the lengths of the edges parallel to x, y respectively. The general value of ζ subject to these conditions is given by the double Fourier's series

$$\zeta = \Sigma\Sigma A_{m,n} \cos\frac{m\pi x}{a} \cos\frac{n\pi y}{b}, \quad \text{....................(3)}$$

where the summations include all integral values of m, n from 0 to ∞. Substituting in (1) we find

$$k^2 = \pi^2\left(\frac{m^2}{a^2} + \frac{n^2}{b^2}\right). \quad \text{.........................(4)}$$

If $a > b$, the component oscillation of longest period is got by making $m = 1$, $n = 0$, whence $ka = \pi$. The motion is then everywhere parallel to the longer side of the rectangle. Cf. Art. 178.

191. In the case of a *circular* sheet of water, it is convenient to take the origin at the centre, and to transform to polar co-ordinates, writing

$$x = r\cos\theta, \quad y = r\sin\theta.$$

The equation (1) of the preceding Art. becomes

$$\frac{\partial^2 \zeta}{\partial r^2} + \frac{1}{r}\frac{\partial \zeta}{\partial r} + \frac{1}{r^2}\frac{\partial^2 \zeta}{\partial \theta^2} + k^2\zeta = 0. \quad \text{...................(1)}$$

This might of course have been established independently.

As regards dependence on θ, the value of ζ may, by Fourier's Theorem, be supposed expanded in a series of cosines and sines of multiples of θ; we thus obtain a series of terms of the form

$$f(r)\left.\begin{matrix}\cos\\ \sin\end{matrix}\right\} s\theta. \quad \text{...............................(2)}$$

It is found on substitution in (1) that each of these terms must satisfy the equation independently, and that

$$f''(r) + \frac{1}{r}f'(r) + \left(k^2 - \frac{s^2}{r^2}\right)f(r) = 0. \quad \text{................(3)}$$

This is of the same form as Art. 101 (14). Since ζ must be finite for $r=0$, the various normal modes are given by

$$\zeta = A_s J_s(kr) \left.\begin{matrix}\cos\\ \sin\end{matrix}\right\} s\theta \,.\, \cos(\sigma t+\epsilon), \qquad \text{.................(4)}$$

where s may have any of the values 0, 1, 2, 3, ..., and A_s is an arbitrary constant. The admissible values of k are determined by the condition that $\partial\zeta/dr=0$ at the boundary $r=a$, say, or

$$J_s'(ka)=0. \qquad \text{..............................(5)}$$

The corresponding 'speeds' (σ) of the oscillations are then given by $\sigma=kc$, where $c=\surd(gh)$.

In the case $s=0$, the motion is symmetrical about the origin, so that the waves have annular ridges and furrows. The lowest roots of

$$J_0'(ka)=0, \text{ or } J_1(ka)=0, \qquad \text{......................(6)}$$

are given by

$$ka/\pi = 1{\cdot}2197, \quad 2{\cdot}2330, \quad 3{\cdot}2383, \ldots, \qquad \text{..............(7)}$$

these numbers tending ultimately to the form $ka/\pi=m+\frac{1}{4}$, where m is integral*. Hence

$$\sigma a/c = 3{\cdot}832, \quad 7{\cdot}016, \quad 10{\cdot}173, \ldots. \qquad \text{..................(7a)}$$

In the mth mode of the symmetrical class there are m nodal circles whose radii are given by $\zeta=0$ or

$$J_0(kr)=0. \qquad \text{...............................(8)}$$

The roots of this are†

$$kr/\pi = {\cdot}7655, \quad 1{\cdot}7571, \quad 2{\cdot}7546, \ldots. \qquad \text{..................(9)}$$

For example, in the first symmetrical mode there is one nodal circle $r={\cdot}628a$. The form of the section of the free surface by a plane through the axis of z, in any of these modes, will be understood from the drawing of the curve $y=J_0(x)$, which is given on the next page.

When $s>0$ there are s equidistant nodal diameters, in addition to the nodal circles

$$J_s(kr)=0. \qquad \text{...............................(10)}$$

It is to be noticed that, owing to the equality of the frequencies of the two modes represented by (4), the normal modes are now to a certain extent indeterminate; viz. in place of $\cos s\theta$ or $\sin s\theta$ we might substitute $\cos s(\theta-\alpha_s)$, where α_s is arbitrary. The nodal diameters are then given by

$$\theta-\alpha_s = \frac{2m+1}{2s}\pi, \qquad \text{.........................(11)}$$

where $m=0, 1, 2, \ldots, s-1$. The indeterminateness disappears, and the frequencies become unequal, if the boundary deviate, however slightly, from the circular form.

* Stokes, "On the Numerical Calculation of a class of Definite Integrals and Infinite Series." *Camb. Trans.* ix. (1850) [*Papers*, ii. 355].

It is to be noticed that ka/π is equal to τ_0/τ, where τ is the actual period, and τ_0 is the time a progressive wave would take to travel with the velocity $\surd(gh)$ over a space equal to the diameter $2a$.

† Stokes, *l.c.*

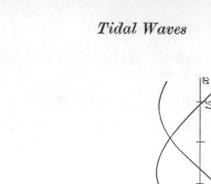

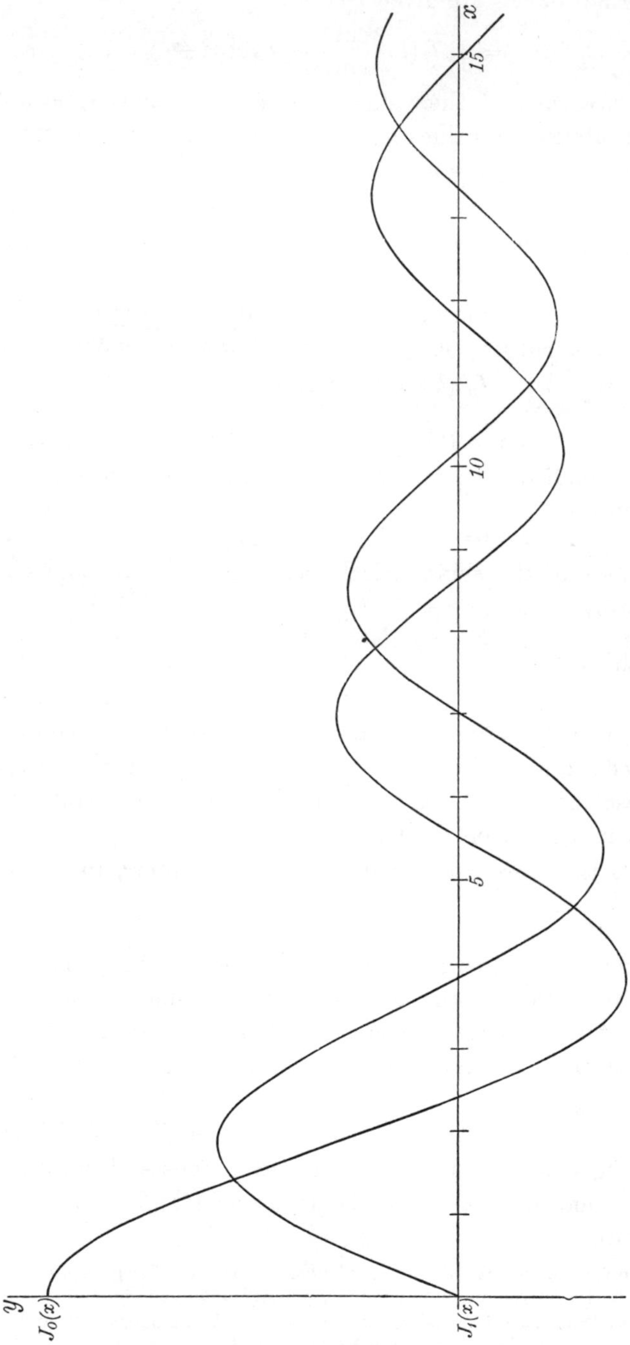

For the sake of clearness, the scale of the ordinates in this figure has been taken five times as great as that of the abscissae.

In the case of the circular boundary, we obtain by superposition of two fundamental modes of the same period, in different phases, a solution

$$\zeta = C_s J_s(kr) \, . \cos(\sigma t \mp s\theta + \epsilon). \qquad \text{.....................(12)}$$

This represents a system of waves travelling unchanged round the origin with an angular velocity σ/s in the positive or negative direction of θ. The motion of the individual particles is easily seen from Art. 189 (4) to be elliptic-harmonic, one principal axis of each elliptic orbit being along the radius vector. All this is in accordance with the general theory recapitulated in Art. 168.

The most interesting modes of the unsymmetrical class are those corresponding to $s=1$, *e.g.*

$$\zeta = AJ_1(kr)\cos\theta \, . \cos(\sigma t + e), \qquad \text{.....................(13)}$$

where k is determined by

$$J_1'(ka) = 0. \qquad \text{..............................(14)}$$

The roots of this are*

$$ka/\pi = \cdot586, \quad 1\cdot697, \quad 2\cdot717, \ldots, \qquad \text{..................(15)}$$

whence $\qquad \sigma a/c = 1\cdot841, \quad 5\cdot332, \quad 8\cdot536, \ldots. \qquad$ ....(15a)

We have now one nodal diameter ($\theta = \frac{1}{2}\pi$), whose position is, however, indeterminate, since the origin of θ is arbitrary. In the corresponding modes for an elliptic boundary, the nodal diameter would be fixed, viz. it would coincide with either the major or the minor axis, and the frequencies would be unequal.

The diagrams on the next page shew the contour-lines of the free surface in the first two modes of the present species. These lines meet the boundary at right angles, in conformity with the general boundary-condition (Art. 190 (2)). The simple-harmonic vibrations of the individual particles take place in straight lines perpendicular to the contour-lines, by Art. 189 (4). The form of the sections of the free surface by planes through the axis of z is given by the curve $y = J_1(x)$ on the opposite page.

The first of the two modes here figured has the longest period of all the normal types. In it, the water sways from side to side, much as in the slowest mode of a canal closed at both ends (Art. 178). In the second mode there is a nodal circle, whose radius is given by the lowest root of $J_1(kr) = 0$; this makes $r = \cdot719a$†.

* See Rayleigh's treatise, Art. 339. A general formula for calculating the roots of $J_s'(ka) = 0$, due to Prof. J. McMahon, is given in the special treatises.

† The oscillations of a liquid in a circular basin of any uniform depth were discussed by Poisson, "Sur les petites oscillations de l'eau contenue dans un cylindre," *Ann. de Gergonne*, xix. 225 (1828–9); the theory of Bessel's Functions had not at that date been worked out, and the results were consequently not interpreted. The full solution of the problem, with numerical details, was given independently by Rayleigh, *Phil. Mag.* (5), i. 257 (1876) [*Papers*, i. 25].

The investigation in the text is limited, of course, to the case of a depth small in comparison with the radius a. Poisson's and Rayleigh's solution for the case of finite depth will be noticed in Chapter IX.

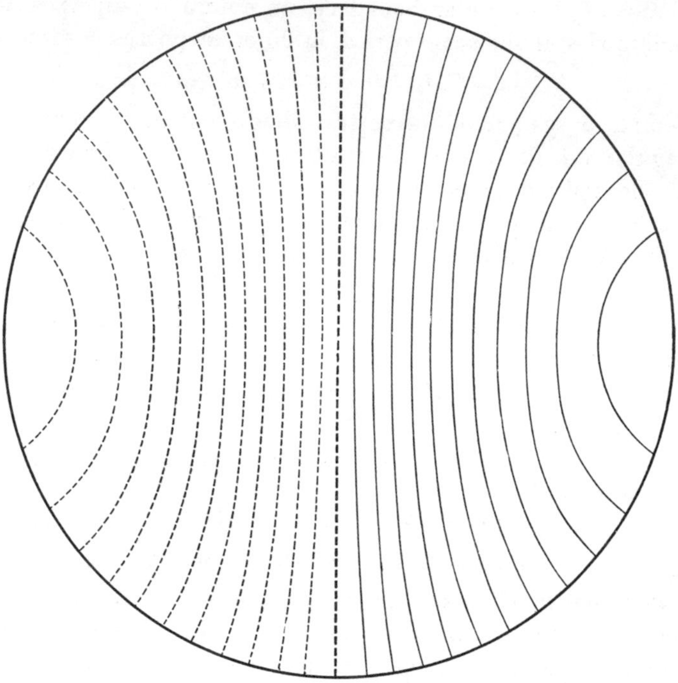

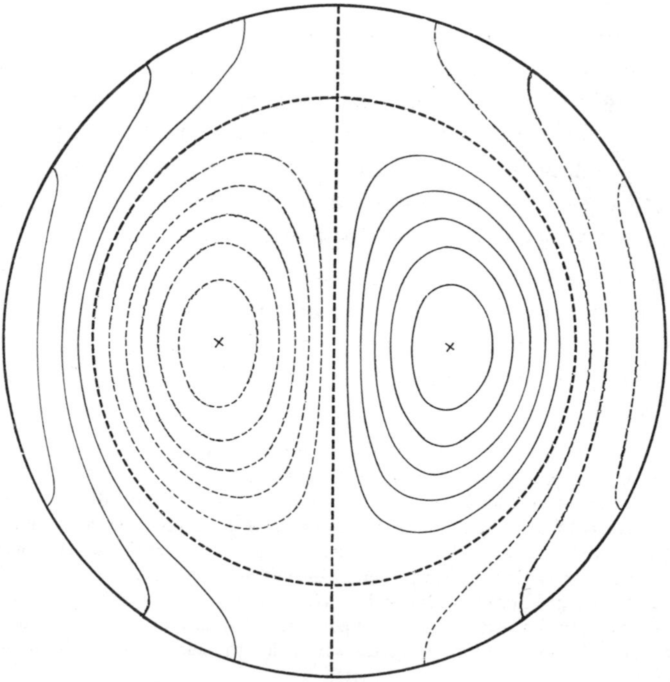

A comparison of the preceding investigation with the general theory of small oscillations referred to in Art. 168 leads to several important properties of Bessel's Functions.

In the first place, since the total mass of water is unaltered, we must have

$$\int_0^{2\pi}\int_0^a \zeta r\,d\theta\,dr=0, \quad\text{.....................................(16)}$$

where ζ has any one of the forms given by (4). For $s>0$ this is satisfied in virtue of the trigonometrical factor $\cos s\theta$ or $\sin s\theta$; in the symmetrical case it gives

$$\int_0^a J_0(kr)\,r\,dr=0. \quad\text{.....................................(17)}$$

Again, since the most general free motion of the system can be obtained by superposition of the normal modes, each with an arbitrary amplitude and epoch, it follows that any value whatever of ζ, which is subject to the condition (16), can be expanded in a series of the form

$$\zeta=\Sigma\Sigma\,(A_s\cos s\theta+B_s\sin s\theta)\,J_s(kr), \quad\text{..........................(18)}$$

where the summations embrace all integral values of s (including 0) and, for each value of s, all the roots k of (5). If the coefficients A_s, B_s be regarded as functions of t, the equation (18) may be regarded as giving the value of the surface-elevation at any instant. The quantities A_s, B_s are then the normal co-ordinates of the present system (Art. 168); and in terms of them the formulae for the kinetic and potential energies must reduce to sums of squares. Taking, for example, the potential energy

$$V=\tfrac{1}{2}g\rho\iint\zeta^2\,dx\,dy, \quad\text{.....................................(19)}$$

this requires that

$$\int_0^{2\pi}\int_0^a w_1 w_2 r\,d\theta\,dr=0, \quad\text{.....................................(20)}$$

where w_1, w_2 are any two terms of the expansion (18). If w_1, w_2 involve cosines or sines of different multiples of θ, this is verified at once by integration with respect to θ; but if we take

$$w_1\propto J_s(k_1r)\cos s\theta,\quad w_2\propto J_s(k_2r)\cos s\theta,$$

where k_1, k_2 are any two distinct roots of (5), we get

$$\int_0^a J_s(k_1r)\,J_s(k_2r)\,r\,dr=0. \quad\text{.....................................(21)}$$

The general results, of which (17) and (21) are particular cases, are

$$\int_0^a J_0(kr)\,r\,dr=-\frac{a}{k}J_0'(ka) \quad\text{.....................................(22)}$$

(cf. Art. 102 (10)), and

$$\int_0^a J_s(k_1r)\,J_s(k_2r)\,r\,dr=\frac{1}{k_1^2-k_2^2}\{k_2aJ_s'(k_2a)\,J_s(k_1a)-k_1aJ_s'(k_1a)\,J_s(k_2a)\}. \quad\text{...(23)}$$

In the case of $k_1=k_2$ the latter expression becomes indeterminate; the evaluation in the usual manner gives

$$\int_0^a \{J_s(ka)\}^2\,r\,dr=\frac{1}{2k^2}\left[k^2a^2\{J_s'(ka)\}^2+(k^2a^2-s^2)\{J_s(ka)\}^2\right]. \quad\text{..............(24)}$$

For the analytical proofs of these formulae we refer to the treatises cited on p. 136.

The small oscillations of an *annular* sheet of water bounded by concentric circles are easily treated, theoretically, with the help of Bessel's Functions of the second kind.' The only case of any special interest, however, is when the two radii are nearly equal; we then have practically a re-entrant canal, and the solution follows more simply by the method of Art. 178.

The analysis can also be applied to the case of a *circular sector* of any angle*, or to a sheet of water bounded by two concentric circular arcs and two radii.

An approximation to the frequency of the slowest mode in an *elliptic* basin of uniform depth can be obtained by Rayleigh's method, referred to in Art. 168.

The equation of the boundary being

$$\frac{x^2}{a^2}+\frac{y^2}{b^2}-1=0, \qquad (25)$$

let us assume, for the component displacements,

$$\left.\begin{aligned}\xi &= A\left(1-\frac{x^2}{a^2}-\frac{y^2}{b^2}\right)+B\frac{y^2}{b^2},\\ \eta &= -B\frac{xy}{a^2},\end{aligned}\right\} \qquad (26)$$

where the constants have been adjusted so as to make

$$\frac{x\xi}{a^2}+\frac{y\eta}{b^2}=0, \qquad (27)$$

at the boundary (25). The time-factor $\cos\sigma t$ is understood. The corresponding surface-elevation is

$$\zeta=-h\left(\frac{\partial\xi}{\partial x}+\frac{\partial\eta}{\partial y}\right)=\frac{h}{a^2}(2A+B)\,x. \qquad (28)$$

The assumption (26) is however too general for the present purpose, since it includes circulatory motions. The condition of zero vorticity requires

$$(2a^2+b^2)\,B=2a^2A. \qquad (29)$$

We find from (26)

$$2T=\rho h\iint(\dot\xi^2+\dot\eta^2)\,dx\,dy=2\pi\,\rho abh\sigma^2\left\{\tfrac{1}{6}A^2+\tfrac{1}{12}AB+\left(\tfrac{1}{16}+\tfrac{1}{48}\frac{b^2}{a^2}\right)B^2\right\}\sin^2\sigma t, \qquad (30)$$

$$2V=g\rho\iint\zeta^2\,dx\,dy=2\pi\,abgh^2\,.\,\frac{(2A+B)^2}{8a^2}\cos^2\sigma t. \qquad (31)$$

Expressing that the mean value of $T-V$ is zero, and introducing the relation (29), we find

$$\sigma^2=\frac{18a^2+6b^2}{5a^2+2b^2}\cdot\frac{c^2}{a^2}, \qquad (32)$$

where $c^2=gh$.

If we put $b=a$, this makes $\sigma a/c=1{\cdot}852$, the true value for the circular basin being 1·841. The approximate estimate is in excess, in accordance with a general principle (Art. 168). The various modes of longitudinal oscillations in an elliptic *canal* have been studied by Jeffreys† and Goldstein‡, and more recently by Hidaka§, by different methods. It appears that in the gravest mode $\sigma a/c=1{\cdot}8866$, whilst if we make $b/a\to 0$ in (32) we get $\sigma a/c=1{\cdot}8994$. It would appear that the formula gives a good approximation for values of b/a less than unity.

* See Rayleigh, *Theory of Sound*, Art. 339.

† *Proc. Lond. Math. Soc.* (2) xxiii. 455 (1924).

‡ *Ibid.* xxviii. 91 (1927).

§ *Mem. Imp. Mar. Obs.* (Japan), iv. 99 (1931). This paper includes the discussion of the free oscillations in basins with boundaries of various other shapes, and with various laws of depth.

192. As an example of *forced* oscillations in a circular basin, let us suppose that the disturbing forces are such that the equilibrium elevation would be

$$\bar{\zeta} = C\left(\frac{r}{a}\right)^s \cos s\theta \,.\, \cos(\sigma t + \epsilon). \qquad (33)$$

This makes $\nabla_1^2 \bar{\zeta} = 0$, so that the equation (12) of Art. 189 reduces to the form (1), above, and the solution is

$$\zeta = A J_s(kr) \cos s\theta \,.\, \cos(\sigma t + \epsilon), \qquad (34)$$

where A is an arbitrary constant. The boundary-condition (Art. 189 (14)) gives

$$AkaJ_s'(ka) = sC,$$

whence

$$\zeta = C \frac{sJ_s(kr)}{kaJ_s'(ka)} \cos s\theta \,.\, \cos(\sigma t + \epsilon). \qquad (35)$$

The case $s = 1$ is interesting as corresponding to a *uniform* horizontal force; and the result may be compared with that of Art. 179.

From the case $s = 2$ we could obtain a rough representation of the semi-diurnal tide in a polar basin bounded by a small circle of latitude, except that the rotation of the earth is not as yet taken into account.

We notice that the expression for the amplitude of oscillation becomes infinite when $J_s'(ka) = 0$. This is in accordance with a general principle, of which we have already had several examples; the period of the disturbing force being now equal to that of one of the free modes investigated in the preceding Art.

193*. When the sheet of water is of variable depth, the calculation at the beginning of Art. 189 gives, as the equation of continuity,

$$\frac{\partial \zeta}{\partial t} = -\frac{\partial (hu)}{\partial x} - \frac{\partial (hv)}{\partial y}. \qquad (1)$$

The dynamical equations (Art. 189 (2)) are of course unaltered. Hence, eliminating ζ, we find, for the free oscillations,

$$\frac{\partial^2 \zeta}{\partial t^2} = g\left\{\frac{\partial}{\partial x}\left(h\frac{\partial \zeta}{\partial x}\right) + \frac{\partial}{\partial y}\left(h\frac{\partial \zeta}{\partial y}\right)\right\}. \qquad (2)$$

If the time-factor be $e^{i(\sigma t + \epsilon)}$, we obtain

$$\frac{\partial}{\partial x}\left(h\frac{\partial \zeta}{\partial x}\right) + \frac{\partial}{\partial y}\left(h\frac{\partial \zeta}{\partial y}\right) + \frac{\sigma^2}{g}\zeta = 0. \qquad (3)$$

When h is a function of r, the distance from the origin, only, this may be written

$$h\nabla_1^2 \zeta + \frac{dh}{dr}\frac{\partial \zeta}{\partial r} + \frac{\sigma^2}{g}\zeta = 0. \qquad (4)$$

As a simple example we may take the case of a circular basin which shelves gradually from the centre to the edge, according to the law

$$h = h_0\left(1 - \frac{r^2}{a^2}\right). \qquad (5)$$

* This formed Art. 189 of the 2nd ed. of this work (1895). A similar investigation was given by Poincaré, *Leçons de mécanique céleste*, iii. 94 (Paris, 1910).

Introducing polar co-ordinates, and assuming that ζ varies as $\cos s\theta$ or $\sin s\theta$, the equation (4) takes the form

$$\left(1-\frac{r^2}{a^2}\right)\left(\frac{\partial^2\zeta}{\partial r^2}+\frac{1}{r}\frac{\partial\zeta}{\partial r}-\frac{s^2}{r^2}\zeta\right)-\frac{2}{a^2}r\frac{\partial\zeta}{\partial r}+\frac{\sigma^2}{gh_0}\zeta=0. \quad\text{.......................(6)}$$

That integral of this equation which is finite at the origin is easily found in the form of an ascending series. Thus, assuming

$$\zeta=\Sigma A_m\left(\frac{r}{a}\right)^m, \quad\text{.......................................(7)}$$

where the trigonometrical factors are omitted, for shortness, the relation between consecutive coefficients is found to be

$$(m^2-s^2)A_m=\left\{m(m-2)-s^2-\frac{\sigma^2a^2}{gh_0}\right\}A_{m-2},$$

or, if we write

$$\frac{\sigma^2a^2}{gh_0}=n(n-2)-s^2, \quad\text{.....................................(8)}$$

where n is not as yet assumed to be integral,

$$(m^2-s^2)A_m=(m-n)(m+n-2)A_{m-2}. \quad\text{..........................(9)}$$

The equation is therefore satisfied by a series of the form (7), beginning with the term $A_s(r/a)^s$, the succeeding coefficients being determined by putting $m=s+2, s+4, \ldots$ in (9). We thus find

$$\zeta=A_s\left(\frac{r}{a}\right)^s\left\{1-\frac{(n-s-2)(n+s)}{2(2s+2)}\frac{r^2}{a^2}+\frac{(n-s-4)(n-s-2)(n+s)(n+s+2)}{2.4(2s+2)(2s+4)}\frac{r^4}{a^4}-\ldots\right\}, \quad (10)$$

or in the usual notation of hypergeometric series

$$\zeta=A_s\frac{r^s}{a^s}.F\left(\alpha,\beta,\gamma,\frac{r^2}{a^2}\right), \quad\text{..............................(11)}$$

where $\alpha=\frac{1}{2}n+\frac{1}{2}s,\quad \beta=1+\frac{1}{2}s-\frac{1}{2}n,\quad \gamma=s+1.$

Since these make $\gamma-\alpha-\beta=0$, the series is not convergent for $r=a$, unless it terminate. This can only happen when n is integral, of the form $s+2j$. The corresponding values of σ are then given by (8).

In the symmetrical modes ($s=0$) we have

$$\zeta=A_0\left\{1-\frac{j(j-1)}{1^2}\frac{r^2}{a^2}+\frac{(j+1)j(j-1)(j-2)}{1^2.2^2}\frac{r^4}{a^4}-\ldots\right\}, \quad\text{..............(12)}$$

where j may be any integer greater than unity*. It may be shewn that this expression vanishes for $j-1$ values of r between 0 and a, indicating the existence of $j-1$ nodal circles. The value of σ is given by

$$\sigma^2=4j(j-1)\frac{gh_0}{a^2}, \quad\text{.......................................(13)}$$

whence $\sigma a/\surd(gh_0)=2{\cdot}828,\ 4{\cdot}899,\ 6{\cdot}928,\ \ldots$ ....(13 a)

The gravest symmetrical mode ($j=2$) has a nodal circle of radius $\cdot707\,a$.

Of the unsymmetrical modes, the slowest, for any given value of s, is that for which $n=s+2$, in which case we have

$$\zeta=A_s\frac{r^s}{a^s}\cos s\theta\cos(\sigma t+\epsilon),$$

the value of σ being given by

$$\sigma^2=2s.\frac{gh_0}{a^2}. \quad\text{..(14)}$$

In the case $s=1$ the various frequencies are given by

$$\sigma^2=(4j^2-2)\frac{gh_0}{a^2}, \quad\text{.................................(15)}$$

whence $\sigma a/\surd(gh_0)=1{\cdot}414,\ 3{\cdot}742,\ 5{\cdot}831,\ \ldots$ ....(16)

* If we put $r/a=\sin\frac{1}{2}\chi$, the series is identical with the expansion of $P_{j-1}(\cos\chi)$; see Art. 85 (4).

In the slowest of these modes, corresponding to $s=1$, $n=3$, the free surface is always *plane*. It appears from Art. 191 (15 *a*) that the frequency is ·768 of that of the corresponding mode in a circular basin of *uniform* depth h_0, and of the same radius*.

As in Art. 192 we could at once write down the formula for the tidal motion produced by a uniform horizontal periodic force; or, more generally, for the case where the disturbing potential is of the type

$$\Omega \propto r^s \cos s\theta \cos(\sigma t+\epsilon).$$

194. We may conclude this discussion of 'long' waves on plane sheets of water by an examination of the mode of propagation of disturbances from a centre in an unlimited sheet of uniform depth. For simplicity, we will consider only the case of symmetry, where the elevation ζ is a function of the distance r from the origin of disturbance. This will introduce us to some peculiar and rather important features which attend wave-propagation in two dimensions.

The investigation of a periodic disturbance involves the use of a Bessel's Function (of zero order) 'of the second kind,' as to which some preliminary notes may be useful.

To solve the equation

$$\frac{d^2\phi}{dz^2}+\frac{1}{z}\frac{d\phi}{dz}+\phi=0 \qquad (1)$$

by definite integrals, we assume†

$$\phi=\int e^{-zt}\,T\,dt, \qquad (2)$$

where T is a function of the complex variable t, and the limits of integration are constants as yet unspecified. This makes

$$z\frac{d^2\phi}{dz^2}+\frac{d\phi}{dz}+z\phi=-\left[(1+t^2)\,e^{-zt}\,T\right]+\int\left(\frac{d}{dt}\{(1+t^2)\,T\}-tT\right)e^{-zt}\,dt,$$

by a partial integration. The equation (1) is accordingly satisfied by

$$\phi=\int\frac{e^{-zt}\,dt}{\sqrt{(1+t^2)}}, \qquad (3)$$

provided the expression

$$\sqrt{(1+t^2)}\,e^{-zt}$$

vanishes at each limit of integration. Hence, on the supposition that z is real and positive, or at all events has its real part positive, the integral in (3) may be taken along a path joining any two of the points i, $-i$, $+\infty$ in the plane of the variable t; but two distinct paths joining the same points will not necessarily give the same result if they include between them one of the branch-points ($t=\pm i$) of the function under the integral sign.

Thus, for example, we have the solution

$$\phi_1=\int_{-i}^{i}\frac{e^{-zt}\,dt}{\sqrt{(1+t^2)}},$$

where the path is the portion of the imaginary axis which lies between the limits, and that value of the radical is taken which becomes $=1$ for $t=0$. If we write $t=\xi+i\eta$, we obtain

$$\phi_1=i\int_{-1}^{1}\frac{e^{-iz\eta}d\eta}{\sqrt{(1-\eta^2)}}=2i\int_0^{\frac{1}{2}\pi}\cos(z\cos\vartheta)\,d\vartheta=i\pi J_0(z), \qquad (4)$$

which is the solution already met with (Art. 100).

* For the oscillations in an *elliptic* basin with a similar law of depth see Goldsbrough, *Proc. Roy. Soc.* A, cxxx. 157 (1930).

† Forsyth, *Differential Equations*, c. vii. The systematic application of this method to the theory of Bessel's Functions is due to Hankel, "Die Cylinderfunktionen erster u. zweiter Art," *Math. Ann.* i. 467 (1869).

An independent solution is obtained if we take the integral (3) along the axis of η from the point $(0, i)$ to the origin, and thence along the axis of ξ to the point $(\infty, 0)$. This gives, with the same determination of the radical,

$$\phi_2 = \int_i^0 \frac{e^{-iz\eta}\, d(i\eta)}{\surd(1-\eta^2)} + \int_0^\infty \frac{e^{-z\xi}\, d\xi}{\surd(1+\xi^2)} = \int_0^\infty \frac{e^{-z\xi}\, d\xi}{\surd(1+\xi^2)} - i\int_0^1 \frac{e^{-iz\eta}\, d\eta}{\surd(1-\eta^2)}. \quad \text{.........(5)}$$

By adopting other pairs of limits, and other paths, we can obtain other forms of ϕ, but these must all be equivalent to ϕ_1 or ϕ_2, or to linear combinations of these. In particular, some other forms of ϕ_2 are important. It is known that the value of the integral (3) taken round any closed contour which excludes the branch-points $(t = \pm i)$ is zero. Let us first take as our contour a rectangle, two of whose sides coincide with the positive portions of the axes of ξ and η, except for a small semicircular indentation about the point $t = i$, whilst the remaining sides are at infinity. It is easily seen that the parts of the integral due to the infinitely distant sides will vanish, either through the vanishing of the factor $e^{-z\xi}$ when ξ is infinite, or through the infinitely rapid fluctuation of the function $e^{-iz\eta}/\eta$ when η is infinite. Hence for the path which gave us (5) we may substitute that which extends along the axis of η from the point $(0, i)$ to $(0, i\infty)$, provided the continuity of the radical be attended to. Now as the variable t travels counter-clockwise round the small semicircle, the radical changes continuously from $\surd(1-\eta^2)$ to $i\surd(\eta^2-1)$. We have therefore

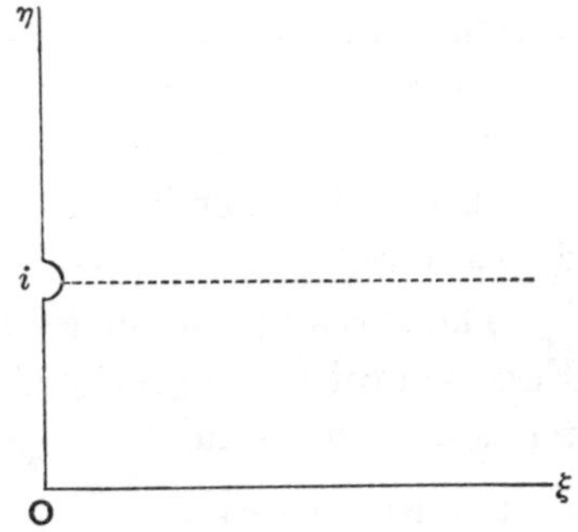

$$\phi_2 = \int_i^{i\infty} \frac{e^{-iz\eta}\, d(i\eta)}{i\surd(\eta^2-1)} = \int_1^\infty \frac{e^{-iz\eta}\, d\eta}{\surd(\eta^2-1)} = \int_0^\infty e^{iz\cosh u}\, du. \quad \text{....................(6)}$$

It will appear that this solution is the one which is specially appropriate to the case of diverging waves. Another method of obtaining it will be given in Chapter X.

If we equate the imaginary parts of (5) and (6) we obtain

$$J_0(z) = \frac{2}{\pi}\int_0^\infty \sin(z\cosh u)\, du, \quad \text{..............................(7)}$$

a form due to Mehler*.

On account of the physical importance of the solution (6) it is convenient to have a special notation for it. We write†

$$D_0(z) = \frac{2}{\pi}\int_0^\infty e^{-iz\cosh u}\, du. \quad \text{..............................(8)}$$

This is equivalent to

$$D_0(z) = -Y_0(z) - iJ_0(z), \quad \text{..............................(9)}$$

where ‡

$$Y_0(z) = -\frac{2}{\pi}\int_0^\infty \cos(z\cosh u)\, du. \quad \text{..............................(10)}$$

Equating the real parts of (5) and (6) we have, also,

$$Y_0(z) = -\frac{2}{\pi}\int_0^\infty e^{-z\sinh u}\, du + \frac{2}{\pi}\int_0^{\frac{1}{2}\pi} \sin(z\cos\vartheta)\, d\vartheta. \quad \text{..............(11)}$$

* *Math. Ann.* v. (1872).

† The use of a simple notation to meet the case of diverging waves seems justifiable. Our $D_0(z)$ is equivalent to $-iH_0^{(2)}(z)$ in Nielsen's notation, as slightly modified by Watson.

‡ This is the notation definitely recommended by Watson. The reader should be warned, however, that the same symbol has been employed by other writers in various senses. From a purely mathematical point of view the choice of a standard solution 'of the second kind' is largely a matter of convention, since the differential equation (1) is still satisfied if we add any constant multiple of $J_0(z)$. Tables of the function $Y_0(z)$ as defined by (10) are given in Watson's treatise.

For a like reason, the path adopted for ϕ_2 may be replaced by the line drawn from the point $(0, i)$ parallel to the axis of ξ (viz. the dotted line in the figure). To secure the continuity of $\sqrt{(1+t^2)}$, we note that as t describes the lower quadrant of the small semicircle, the value of the radical changes from $\sqrt{(1-\eta^2)}$ to $e^{\frac{1}{4}i\pi}\sqrt{(2\xi)}$, approximately. Hence along the dotted line we have, putting $t=i+\xi$,

$$\sqrt{(1+t^2)}=e^{\frac{1}{4}i\pi}\sqrt{(2\xi-i\xi^2)},$$

where that value of the radical is to be chosen which is real and positive when ξ is infinitesimal. Thus

$$\phi_2=\int_i^{\infty+i}\frac{e^{-z(\xi+i)}\,d(\xi+i)}{e^{\frac{1}{4}i\pi}\sqrt{(2\xi-i\xi^2)}}=\frac{1}{\sqrt{2}}e^{-i(z+\frac{1}{4}\pi)}\int_0^\infty\frac{e^{-z\xi}}{\xi^{\frac{1}{2}}}(1-\tfrac{1}{2}i\xi)^{-\frac{1}{2}}d\xi. \quad \text{.........(12)}$$

If we expand the binomial, and integrate term by term, we find

$$D_0(z)=\left(\frac{2}{\pi z}\right)^{\frac{1}{2}}e^{-i(z+\frac{1}{4}\pi)}\left\{1+\frac{1^2}{1\,!}\left(\frac{i}{8z}\right)+\frac{1^2.\,3^2}{2\,!}\left(\frac{i}{8z}\right)^2+\dots\right\}, \quad \text{..............(13)}$$

where use has been made of the formulae

$$\left.\begin{aligned}&\int_0^\infty e^{-z\xi}\,\xi^{-\frac{1}{2}}\,d\xi=\frac{\Pi(-\frac{1}{2})}{z^{\frac{1}{2}}}=\frac{\pi^{\frac{1}{2}}}{z^{\frac{1}{2}}},\\&\int_0^\infty e^{-z\xi}\,\xi^{m-\frac{1}{2}}\,d\xi=\frac{\Pi(m-\frac{1}{2})}{z^{m+\frac{1}{2}}}=\frac{1.3\dots(2m-1)}{2^m z^m}\frac{\pi^{\frac{1}{2}}}{z^{\frac{1}{2}}}.\end{aligned}\right\} \quad \text{..................(14)}$$

If we separate the real and imaginary parts of (13) we have, on comparison with (9),

$$J_0(z)=\left(\frac{2}{\pi z}\right)^{\frac{1}{2}}\{R\sin(z+\tfrac{1}{4}\pi)-S\cos(z+\tfrac{1}{4}\pi)\}, \quad \text{......................(15)}$$

$$Y_0(z)=-\left(\frac{2}{\pi z}\right)^{\frac{1}{2}}\{R\cos(z+\tfrac{1}{4}\pi)+S\sin(z+\tfrac{1}{4}\pi)\}, \quad \text{......................(16)}$$

where

$$\left.\begin{aligned}R&=1-\frac{1^2.\,3^2}{2\,!(8z)^2}+\frac{1^2.\,3^2.\,5^2.\,7^2}{4\,!(8z)^4}-\dots,\\S&=\frac{1^2}{1\,!(8z)}-\frac{1^2.\,3^2.\,5^2}{3\,!(8z)^3}+\dots.\end{aligned}\right\} \quad \text{...........................(17)}$$

The series in (13) and (17) are of the kind known as 'semi-convergent,' or 'asymptotic,' expansions; *i.e.* although for sufficiently large values of z the successive terms may for a while diminish, they ultimately increase again indefinitely, but if we stop at a small term we get an approximately correct result*. This may be established by an examination of the remainder after m terms in the process of evaluation of (12).

It follows from (15) that the large roots of the equation $J_0(z)=0$ approximate to those of

$$\sin(z+\tfrac{1}{4}\pi)=0. \quad \text{...(18)}$$

The series in (13) gives ample information as to the demeanour of the function $D_0(z)$ when z is large. When z is small, $D_0(z)$ is very great, as appears from (8). An approximate formula for this case can be obtained as follows. Referring to (11), we have

$$\int_0^\infty e^{-z\sinh u}du=\int_1^\infty e^{-\frac{1}{2}z\left(v-\frac{1}{v}\right)}\frac{dv}{v}=\int_1^\infty\frac{e^{-\frac{1}{2}zv}}{v}\left\{1+\frac{z}{2v}+\frac{1}{2\,!}\left(\frac{z}{2v}\right)^2+\dots\right\}dv$$

$$=\int_{\frac{1}{2}z}^\infty\frac{e^{-w}}{w}\left\{1+\frac{z^2}{4w}+\frac{1}{2\,!}\left(\frac{z^2}{4w}\right)^2+\dots\right\}dw. \quad \text{..............................(19)}$$

* Cf. Whittaker and Watson, *Modern Analysis*, c. viii.; Bromwich, *Theory of Infinite Series*, London, 1908, c. xi.; Watson, c. vii.; Gray and Mathews, c. iv. The semi-convergent expansion of $J_0(z)$ is due to Poisson, *Journ. de l'École Polyt.* cah. 19, p. 349 (1823); a rigorous investigation of this and other analogous expansions was given by Stokes, *l.c. ante* p. 285. The 'remainder' was examined by Lipschitz, *Crelle*, lvi. 189 (1859). Cf. Hankel, *l.c. ante* p. 293.

The first term gives*

$$\int_{\frac{1}{2}z}^{\infty} \frac{e^{-w}}{w}\,dw = -\gamma - \log \tfrac{1}{2}z + \ldots, \qquad (20)$$

and the remaining ones are small in comparison. Hence, by (9) and (11),

$$D_0(z) = -\frac{2}{\pi}(\log \tfrac{1}{2}z + \gamma + \tfrac{1}{2}i\pi + \ldots). \qquad (21)$$

It follows that

$$\lim_{z\to 0} z D_0'(z) = -\frac{2}{\pi}\dagger. \qquad (22)$$

The formula (21) is sufficient for our purposes, but the complete expression can now be obtained by comparison with the general solution of (1) in terms of ascending series, viz.‡

$$\phi = A J_0(z) + B\left\{J_0(z)\log z + \frac{z^2}{2^2} - s_2\frac{z^4}{2^2\,.\,4^2} + s_3\frac{z^6}{2^2\,.\,4^2\,.\,6^2} - \ldots\right\}, \qquad (23)$$

where

$$s_m = 1 + \frac{1}{2} + \frac{1}{3} + \ldots + \frac{1}{m}.$$

In order to identify this with (21), for small values of z, we must make

$$B = -\frac{2}{\pi}, \quad A = -\frac{2}{\pi}(\log \tfrac{1}{2}z + \gamma + \tfrac{1}{2}i\pi). \qquad (24)$$

Hence

$$D_0(z) = -\frac{2}{\pi}(\log \tfrac{1}{2}z + \gamma + \tfrac{1}{2}i\pi)J_0(z) - \frac{2}{\pi}\left\{\frac{z^2}{2^2} - s_2\frac{z^4}{2^2\,.\,4^2} + s_3\frac{z^6}{2^2\,.\,4^2\,.\,6^2} - \ldots\right\}. \qquad (25)$$

195. We can now proceed to the wave-problem stated at the beginning of Art. 194. For definiteness we will imagine the disturbance to be caused by a variable pressure p_0 applied to the surface. On this supposition the dynamical equations near the beginning of Art. 189 are replaced by

$$\frac{\partial u}{\partial t} = -g\frac{\partial \zeta}{\partial x} - \frac{1}{\rho}\frac{\partial p_0}{\partial x}, \quad \frac{\partial v}{\partial t} = -\frac{1}{\rho}\frac{\partial \zeta}{\partial y} - \frac{1}{\rho}\frac{\partial p_0}{\partial y}, \qquad (1)$$

whilst

$$\frac{\partial \zeta}{\partial t} = -h\left(\frac{\partial u}{\partial x} + \frac{\partial v}{\partial y}\right), \qquad (2)$$

as before.

If we introduce the velocity-potential in (1), we have, on integration,

$$\frac{\partial \phi}{\partial t} = g\zeta + \frac{p_0}{\rho}. \qquad (3)$$

We may suppose that p_0 refers to the *change* of pressure, and that the arbitrary function of t which has been incorporated in ϕ is chosen so that $\partial\phi/\partial t = 0$ in the regions not affected by the disturbance. Eliminating ζ by means of (2), we have

$$\frac{\partial^2 \phi}{\partial t^2} = gh\nabla_1^2\phi + \frac{1}{\rho}\frac{\partial p_0}{\partial t}. \qquad (4)$$

When ϕ has been determined, the value of ξ is given by (3).

* De Morgan, *Differential and Integral Calculus*, London, 1842, p. 653.

† The Bessel's Functions of the second kind were first thoroughly investigated and made available for the solution of physical problems in an arithmetically intelligible form by Stokes, in a series of papers published in the *Camb. Trans.* With the help of the modern Theory of Functions, some of the processes have been simplified by Lipschitz and others, and (especially from the physical point of view) by Rayleigh. These later methods have been used in the text.

‡ Forsyth, *Differential Equations*, c. vi. note 1; Watson, *Bessel Functions*, pp. 59, 60.

We will now assume that p_0 is sensible only over a small* area about the origin. If we multiply both sides of (4) by $\delta x \delta y$, and integrate over the area in question, the term on the left-hand side may be neglected (relatively), and we find

$$-\int \frac{\partial \phi}{\partial n} ds = \frac{1}{g\rho h}\frac{d}{dt}\iint p_0 \, dx\, dy, \qquad \text{.....................(5)}$$

where δs is an element of the boundary of the area, and δn refers to the horizontal normal to δs, drawn outwards. Hence the origin may be regarded as a two-dimensional source, of strength

$$f(t) = \frac{1}{g\rho h}\frac{dP_0}{dt}, \qquad \text{...........................(6)}$$

where P_0 is the total disturbing force.

Turning to polar co-ordinates, we have to satisfy

$$\frac{\partial^2 \phi}{\partial t^2} = c^2\left(\frac{\partial^2 \phi}{\partial r^2} + \frac{1}{r}\frac{\partial \phi}{\partial r}\right), \qquad \text{..........................(7)}$$

where $c^2 = gh$, subject to the condition

$$\lim_{r\to 0}\left(-2\pi r \frac{\partial \phi}{\partial r}\right) = f(t), \qquad \text{.......................(8)}$$

where $f(t)$ is the strength of the source, as above defined.

In the case of a simple-harmonic source $e^{i\sigma t}$ the equation (7) takes the form

$$\frac{\partial^2 \phi}{\partial r^2} + \frac{1}{r}\frac{\partial \phi}{\partial r} + k^2\phi = 0, \qquad \text{.........................(9)}$$

where $k = \sigma/c$, and a solution is

$$\phi = \tfrac{1}{4} D_0(kr)\, e^{i\sigma t}, \qquad \text{...........................(10)}$$

where the constant factor has been determined by Art. 194 (22). Taking the real part we have

$$\phi = \tfrac{1}{4}\{J_0(kr)\sin \sigma t - Y_0(kr)\cos \sigma t\}, \qquad \text{..............(11)}$$

corresponding to $f(t) = \cos \sigma t.$

For large values of kr the result (10) takes the form

$$\phi = \frac{1}{\sqrt{(8\pi kr)}} e^{i\sigma\left(t - \frac{r}{c}\right) - \frac{1}{4}i\pi}. \qquad \text{....................(12)}$$

The combination $t - r/c$ indicates that we have, in fact, obtained the solution appropriate to the representation of diverging waves.

It appears that the amplitude of the annular waves ultimately varies inversely as the square root of the distance from the origin.

* That is, the dimensions of the area are small compared with the 'length' of the waves generated, this term being understood in the general sense of Art. 172. On the other hand, the dimensions must be supposed large in comparison with h.

196. The solution we have obtained for the case of a simple-harmonic source $e^{i\sigma t}$ may be written

$$2\pi\phi = \int_0^\infty e^{i\sigma\left(t-\frac{r}{c}\cosh u\right)}\,du. \quad\text{.....................(13)}$$

This suggests generalization by Fourier's Theorem; thus the formula

$$2\pi\phi = \int_0^\infty f\left(t-\frac{r}{c}\cosh u\right)du \quad\text{.....................(14)}$$

should represent the disturbance due to a source $f(t)$ at the origin*. It is implied that the form of $f(t)$ must be such that the integral is convergent; this condition will as a matter of course be fulfilled whenever the source has been in action only for a finite time. A more complete formula, embracing both converging and diverging waves, is

$$2\pi\phi = \int_0^\infty f\left(t-\frac{r}{c}\cosh u\right)du + \int_0^\infty F\left(t+\frac{r}{c}\cosh u\right)du. \quad\text{......(15)}$$

The solution (15) may be verified, subject to certain conditions, by substitution in the differential equation (7). Taking the first term alone, we find

$$2\pi\left\{c^2\left(\frac{\partial^2\phi}{\partial r^2}+\frac{1}{r}\frac{\partial\phi}{\partial r}\right)-\frac{\partial^2\phi}{\partial t^2}\right\}$$

$$=\int_0^\infty\left\{\sinh^2 u\,.\,f''\left(t-\frac{r}{c}\cosh u\right)-\frac{c}{r}\cosh u\,.\,f'\left(t-\frac{r}{c}\cosh u\right)\right\}du$$

$$=\frac{c^2}{r^2}\int_0^\infty\frac{\partial^2}{\partial u^2}f\left(t-\frac{r}{c}\cosh u\right)du=-\frac{c}{r}\left[\sinh u\,.\,f'\left(t-\frac{r}{c}\cosh u\right)\right]_{u=0}^{u=\infty}.$$

This obviously vanishes whenever $f(t)=0$ for negative values of t exceeding a certain limit†.

Again
$$-2\pi r\frac{\partial\phi}{\partial r}=\frac{r}{c}\int_0^\infty\cosh u\,.\,f'\left(t-\frac{r}{c}\cosh u\right)du$$

$$=\frac{r}{c}\int_0^\infty(\sinh u+e^{-u})\,f'\left(t-\frac{r}{c}\cosh u\right)du$$

$$=-\left[f\left(t-\frac{r}{c}\cosh u\right)\right]_{u=0}^{u=\infty}+\frac{r}{c}\int_0^\infty e^{-u}f'\left(t-\frac{r}{c}\cosh u\right)du$$

$$=f\left(t-\frac{r}{c}\right)+\frac{r}{c}\int_0^\infty e^{-u}f'\left(t-\frac{r}{c}\cosh u\right)du,$$

under the same condition. The limiting value of this when $r\to 0$ is $f(t)$; and the statement made above as to the strength of the source in (14) is accordingly verified.

A similar process will apply to the second term of (15) provided $F(t)$ vanishes for positive values of t exceeding a certain limit.

197. We may apply (14) to trace the effect of a temporary source varying according to some simple prescribed law.

If we suppose that everything is quiescent until the instant $t=0$, so that

* The substance of Arts. 196, 197 is adapted from a paper "On Wave-Propagation in Two Dimensions," *Proc. Lond. Math. Soc.* (1), xxxv. 141 (1902). A result equivalent to (14) was obtained (in a different manner) by Levi-Civita, *Nuovo Cimento* (4), vi. (1897).

† The verification is very similar to that given by Levi-Civita.

$f(t)$ vanishes for negative values of t, we see from (14) or from the equivalent form

$$2\pi\phi = \int_{-\infty}^{t-\frac{r}{c}} \frac{f(\theta)\,d\theta}{\{(t-\theta)^2 - r^2/c^2\}^{\frac{1}{2}}} \qquad \text{.....................(16)}$$

that ϕ will be zero everywhere so long as $t < r/c$. If, moreover, the source acts only for a finite time τ, so that $f(t) = 0$ for $t > \tau$, we have, for $t > \tau + r/c$,

$$2\pi\phi = \int_0^{\tau} \frac{f(\theta)\,d\theta}{\{(t-\theta^2) - r^2/c^2\}^{\frac{1}{2}}}. \qquad \text{.....................(17)*}$$

This expression does not as a rule vanish; the wave accordingly is not sharply defined in the rear, as it is in front, but has, on the contrary, a sort of 'tail'† whose form, when $t - r/c$ is large compared with τ, is determined by

$$2\pi\phi = \frac{1}{(t^2 - r^2/c^2)^{\frac{1}{2}}} \int_0^{\tau} f(\theta)\,d\theta. \qquad \text{.....................(18)}$$

The elevation ζ at any point is given by (3), viz.

$$\zeta = \frac{1}{g}\frac{\partial\phi}{\partial t}. \qquad \text{..............................(19)}$$

It follows that

$$\int_{-\infty}^{\infty} \zeta dt = 0, \qquad \text{.............................(20)}$$

provided the initial and final values of ϕ vanish. It may be shewn that this will be the case when $f(t)$ is finite and the integral

$$\int_{-\infty}^{\infty} f(t)\,dt \qquad \text{...............................(21)}$$

is convergent. The meaning of these conditions appears from (6). It follows that even when dP_0/dt is always positive, so that the flux of liquid in the neighbourhood of the origin is altogether outwards, the wave which passes any point does not consist solely of an elevation (as it would in the corresponding one-dimensional problem) but, in the simplest case, of an elevation followed by a depression.

To trace in detail the progress of a solitary wave in a particular case we may assume

$$f(t) = \frac{\tau}{t^2 + \tau^2}, \qquad \text{...(22)}$$

which makes P_0 increase from one constant value to another according to the law

$$P_0 = A + B\tan^{-1}\frac{t}{\tau}. \qquad \text{...................................(23)}$$

* Analytically, it may be noticed that the equation (4), when $p_0 = 0$, may be written

$$\frac{\partial^2\phi}{\partial x^2} + \frac{\partial^2\phi}{\partial y^2} + \frac{\partial^2\phi}{\partial (ict)^2} = 0,$$

and that (17) consists of an aggregate of solutions of the known type

$$\{x^2 + y^2 + (ict)^2\}^{-\frac{1}{2}}.$$

† The existence of the 'tail' in the case of cylindrical electric waves was noted by Heaviside, *Phil. Mag.* (5), xxvi. (1888) [*Electrical Papers*, ii.].

The disturbing pressure has now no definite epoch of beginning or ending, but the range of time within which it is sensible can be made as small as we please by diminishing τ. For purposes of calculation it is convenient to assume

$$f(t)=\frac{1}{t-i\tau} \qquad \text{.......................................(24)}$$

in place of (22), and to retain in the end only the imaginary part. We have then

$$2\pi\phi=\int_0^\infty \frac{du}{t-\dfrac{r}{c}\cosh u-i\tau}=2\int_0^1 \frac{dz}{t-\dfrac{r}{c}-i\tau-\left(t+\dfrac{r}{c}-i\tau\right)z^2}, \qquad \text{...........(25)}$$

where $z=\tanh\frac{1}{2}u$. We now write

$$t-\frac{r}{c}-i\tau=a^2e^{-2i\alpha}, \quad t+\frac{r}{c}-i\tau=b^2e^{-2i\beta}, \qquad \text{.........................(26)}$$

where we may suppose that a, b are positive, and that the angles α, β lie between 0 and $\frac{1}{2}\pi$. Since

$$\left.\begin{aligned} a^4&=\left(t-\frac{r}{c}\right)^2+\tau^2, \quad b^4=\left(t+\frac{r}{c}\right)^2+\tau^2,\\ \tan 2\alpha&=\frac{c\tau}{ct-r}, \quad \tan 2\beta=\frac{ct}{ct+r}, \end{aligned}\right\} \qquad \text{...........................(27)}$$

it appears that $a \lesseqgtr b$ according as $t \gtrless 0$, and that $\alpha>\beta$ always. With this notation, we find

$$2\pi\phi=2\int_0^1\frac{dz}{a^2e^{-2i\alpha}-b^2e^{-2i\beta}z^2}=\frac{e^{i(\alpha+\beta)}}{ab}\log\frac{z+\dfrac{a}{b}e^{-i(\alpha-\beta)}}{z-\dfrac{a}{b}e^{-i(\alpha-\beta)}}. \qquad \text{...........(28)}$$

To interpret the logarithms, let us mark, in the plane of a complex variable z, the points

$$I=+1, \quad P=-\frac{a}{b}e^{-i(\alpha-\beta)}, \quad Q=\frac{a}{b}e^{-i(\alpha-\beta)}.$$

Since the integral in the second member of (28) is to be taken along the path OI, the proper value of the third member is

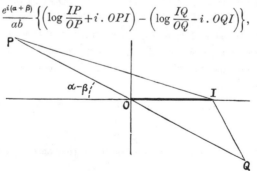

$$\frac{e^{i(\alpha+\beta)}}{ab}\left\{\left(\log\frac{IP}{OP}+i\,.\,OPI\right)-\left(\log\frac{IQ}{OQ}-i\,.\,OQI\right)\right\},$$

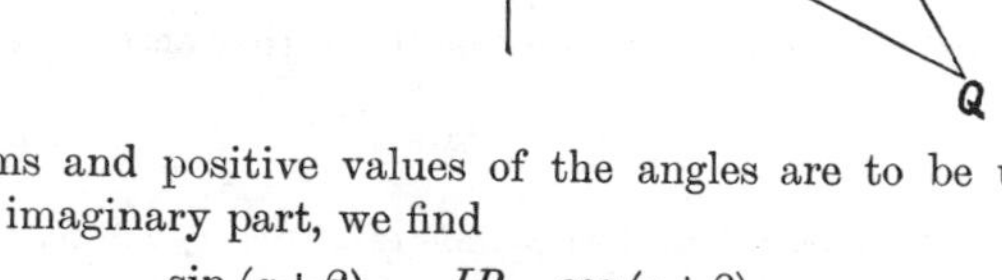

where real logarithms and positive values of the angles are to be understood. Hence, rejecting all but the imaginary part, we find

$$2\pi\phi=\frac{\sin(\alpha+\beta)}{ab}\log\frac{IP}{IQ}+\frac{\cos(\alpha+\beta)}{ab}(\pi-PIQ) \qquad \text{.....................(29)}$$

as the solution corresponding to a source of the type (22). Here

$$\frac{IP}{IQ}=\left(\frac{a^2+2ab\cos(\alpha-\beta)+b^2}{a^2-2ab\cos(\alpha-\beta)+b^2}\right)^{\frac{1}{2}}, \quad \tan PIQ=\frac{2ab\sin(\alpha-\beta)}{b^2-a^2}, \qquad \text{.........(30)}$$

and the values of a, b, α, β in terms of r and t are to be found from (27).

It will be sufficient to trace the effect of the most important part of the wave as it passes a point whose distance r from the origin is large compared with $c\tau$. If we confine ourselves to times at which $t - r/c$ is small compared with r/c, α will be small compared with b, PIQ will be a small angle, and IP/IQ will $=1$, nearly. If we put

$$t=\frac{r}{c}+\tau\tan\eta, \quad \text{...(31)}$$

we shall have

$$\alpha=\tfrac{1}{4}\pi-\tfrac{1}{2}\eta, \quad a=\surd(\tau\sec\eta), \quad \beta=\tfrac{1}{4}c\tau/r, \quad b=(2r/c)^{\frac{1}{2}}, \quad \text{..................(32)}$$

approximately; and the formula (29) will reduce to

$$2\pi\phi=\frac{\pi}{ab}\cos\alpha=\frac{\pi}{\surd 2\tau}\left(\frac{c\tau}{r}\right)^{\frac{1}{2}}\cos\left(\tfrac{1}{4}\pi-\tfrac{1}{2}\eta\right)\surd(\cos\eta). \quad \text{..................(33)}$$

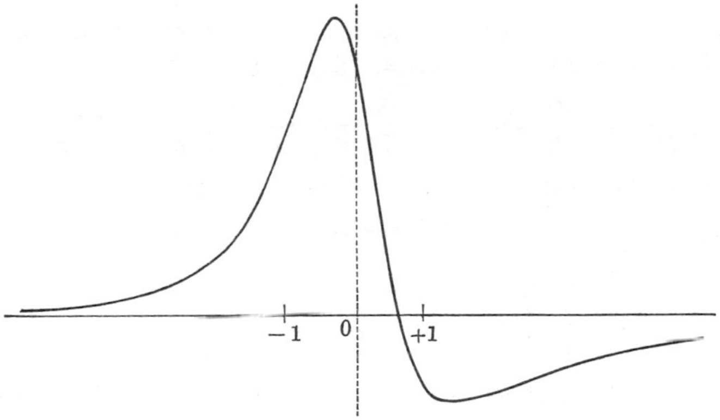

The elevation ζ is then given by

$$2\pi g\zeta=2\pi g\frac{d\phi}{d(\tau\tan\eta)}=\frac{1}{4\surd 2\tau^2}\left(\frac{c\tau}{r}\right)^{\frac{1}{2}}\sin\left(\tfrac{1}{4}\pi-\tfrac{3}{2}\eta\right)\cos^{\frac{3}{2}}\eta, \quad \text{............(34)}$$

approximately. The diagram shews the relation between ζ and t, as given by this formula*.

198. We proceed to consider the case of a spherical sheet, or ocean, of water covering a solid globe. We will suppose for the present that the globe does not rotate, and we will also in the first instance neglect the mutual attraction of the particles of the water. The mathematical conditions of the question are then exactly the same as in the acoustical problem of the vibrations of spherical layers of air†.

Let a be the radius of the globe, h the depth of the fluid; we assume that h is small compared with a, but not (as yet) that it is uniform. The position of any point on the sheet being specified by the angular co-ordinates θ, ϕ, let u be the component velocity of the fluid at this point along the meridian, in the direction of θ increasing, and v the component along the parallel of latitude, in the direction of ϕ increasing. Also let ζ denote the elevation of the free surface above the undisturbed level. The horizontal

* The points marked -1, 0, $+1$ correspond to the times $r/c-\tau$, r/c, $r/c+\tau$, respectively.

† Discussed in Rayleigh's *Theory of Sound*, c. xviii.

motion being assumed, for the reasons explained in Art. 172, to be the same at all points in a vertical line, the condition of continuity is

$$\frac{\partial}{\partial\theta}(uha\sin\theta\,\delta\phi)\,\delta\theta+\frac{\partial}{\partial\phi}(vha\,\delta\theta)\,\delta\phi=-a\sin\theta\,\delta\phi\,.\,a\delta\theta\,.\,\frac{\partial\zeta}{\partial t},$$

where the left-hand side measures the flux out of the columnar space standing on the element of area $a\sin\theta\,\delta\phi\,.\,a\delta\theta$, whilst the right-hand member expresses the rate of diminution of the volume of the contained fluid, owing to fall of the surface. Hence

$$\frac{\partial\zeta}{\partial t}=-\frac{1}{a\sin\theta}\left\{\frac{\partial\,(hu\sin\theta)}{\partial\theta}+\frac{\partial\,(hv)}{\partial\phi}\right\}.\quad\ldots\ldots\ldots\ldots(1)$$

If we neglect terms of the second order in u, v, the dynamical equations are, on the same principles as in Arts. 169, 189,

$$\frac{\partial u}{\partial t}=-g\frac{\partial\zeta}{a\partial\theta}-\frac{\partial\Omega}{a\partial\theta},\qquad\frac{\partial v}{\partial t}=-g\frac{\partial\zeta}{a\sin\theta\partial\phi}-\frac{\partial\Omega}{a\sin\theta\partial\phi},\quad\ldots\ldots(2)$$

where Ω denotes the potential of the extraneous forces.

If we put $$\bar{\zeta}=-\Omega/g,\quad\ldots\ldots\ldots\ldots(3)$$
these may be written

$$\frac{\partial u}{\partial t}=-\frac{g}{a}\frac{\partial}{\partial\theta}(\zeta-\bar{\zeta}),\qquad\frac{\partial v}{\partial t}=-\frac{g}{a\sin\theta}\frac{\partial}{\partial\phi}(\zeta-\bar{\zeta}).\quad\ldots\ldots\ldots(4)$$

Between (1) and (4) we can eliminate u, v, and so obtain an equation in ζ only.

In the case of simple-harmonic motion, the time-factor being $e^{i(\sigma t+\epsilon)}$, the equations take the forms

$$\zeta=\frac{i}{\sigma a\sin\theta}\left\{\frac{\partial\,(hu\sin\theta)}{\partial\theta}+\frac{\partial\,(hv)}{\partial\phi}\right\},\quad\ldots\ldots\ldots\ldots(5)$$

$$u=i\frac{g}{\sigma a}\frac{\partial}{\partial\theta}(\zeta-\bar{\zeta}),\qquad v=i\frac{g}{\sigma a\sin\theta}\frac{\partial}{\partial\phi}(\zeta-\bar{\zeta}).\quad\ldots\ldots\ldots(6)$$

199. We will now consider more particularly the case of *uniform* depth. To find the free oscillations we put $\bar{\zeta}=0$; the equations (5) and (6) of the preceding Art. then lead to

$$\frac{1}{\sin\theta}\frac{\partial}{\partial\theta}\left(\sin\theta\frac{\partial\zeta}{\partial\theta}\right)+\frac{1}{\sin^2\theta}\frac{\partial^2\zeta}{\partial\phi^2}+\frac{\sigma^2a^2}{gh}\zeta=0.\quad\ldots\ldots\ldots\ldots(1)$$

This is identical in form with the general equation of spherical surface-harmonics (Art. 83 (2)). Hence, if we put

$$\frac{\sigma^2a^2}{gh}=n\,(n+1),\quad\ldots\ldots\ldots\ldots(2)$$

a solution of (1) will be $$\zeta=S_n,\quad\ldots\ldots\ldots\ldots(3)$$
where S_n is the general surface-harmonic of order n.

It was pointed out in Art. 86 that S_n will not be finite over the whole sphere unless n be integral. Hence, for an ocean covering the whole globe, the form of the free surface at any instant is, in any fundamental mode, that of a 'harmonic spheroid'

$$r = a + h + S_n \cos(\sigma t + \epsilon), \quad \text{.......................(4)}$$

and the speed of the oscillation is given by

$$\sigma = \{n(n+1)\}^{\frac{1}{2}} \cdot \frac{(gh)^{\frac{1}{2}}}{a}, \quad \text{..........................(5)}$$

the value of n being integral.

The characters of the various normal modes are best gathered from a study of the nodal lines ($S_n = 0$) of the free surface. Thus, it is shewn in treatises on Spherical Harmonics* that the zonal harmonic $P_n(\mu)$ vanishes for n real and distinct values of μ lying between ± 1, so that in this case we have n nodal circles of latitude. When n is odd one of these coincides with the equator. In the case of the tesseral harmonic

$$(1-\mu)^{\frac{1}{2}s} \frac{d^s P_n(\mu)}{d\mu^s} \left.\begin{matrix}\cos\\ \sin\end{matrix}\right\} s\phi,$$

the second factor vanishes for $n - s$ values of μ, and the trigonometrical factor for $2s$ equidistant values of ϕ. The nodal lines therefore consist of $n - s$ parallels of latitude and $2s$ meridians. Similarly the sectorial harmonic

$$(1-\mu^2)^{\frac{1}{2}n} \left.\begin{matrix}\cos\\ \sin\end{matrix}\right\} n\phi$$

has as nodal lines $2n$ meridians.

These are, however, merely special cases, for since there are $2n + 1$ independent surface-harmonics of any integral order n, and since the frequency, determined by (5), is the same for each of these, there is a corresponding degree of indeterminateness in the normal modes, and in the configuration of the nodal lines.

We can also, by superposition, build up various types of progressive waves; *e.g.* taking a sectorial harmonic we get a solution in which

$$\zeta \propto (1-\mu^2)^{\frac{1}{2}n} \cos(n\phi - \sigma t + \epsilon); \quad \text{.....................(6)}$$

this gives a series of meridianal ridges and furrows travelling round the globe, the velocity of propagation, as measured at the equator, being

$$\frac{\sigma a}{n} = \left(\frac{n+1}{n}\right)^{\frac{1}{2}} \cdot (gh)^{\frac{1}{2}}. \quad \text{..........................(7)}$$

It is easily verified, on examination, that the orbits of the particles are now ellipses having their principal axes in the directions of the meridians and parallels, respectively. At the equator these ellipses reduce to straight lines.

In the case $n = 1$, the harmonic is always of the zonal type. The harmonic spheroid (4) is then, to our order of approximation, a sphere excentric to the globe. It is important to remark, however, that this case

* For references see p. 110.

is, strictly speaking, not included in our dynamical investigation, unless we imagine a constraint applied to the globe to keep it at rest; for the deformation in question of the free surface would involve a displacement of the centre of mass of the ocean, and a consequent reaction on the globe. A corrected theory for the case where the globe is free could easily be investigated, but the matter is hardly important, first because in such a case as that of the earth the inertia of the solid globe is so enormous compared with that of the ocean, and secondly because disturbing forces which can give rise to a deformation of the type in question do not as a rule present themselves in nature. It appears, for example, that the first term in the expression for the tide-generating potential of the sun or moon is a spherical harmonic of the *second* order (see the Appendix to this Chapter).

When $n=2$, the free surface at any instant is approximately ellipsoidal. The corresponding period, as found from (5), is then ·816 of that belonging to the analogous mode in an equatorial canal (Art. 181).

For large values of n the distance from one nodal line to another is small compared with the radius of the globe, and the oscillations then take place much as on a plane sheet of water. For example, the velocity of propagation, at the equator, of the sectorial waves represented by (6) tends with increasing n to the value $(gh)^{\frac{1}{2}}$, in agreement with Art. 170.

From a comparison of the foregoing investigation with the general theory of Art. 168 we are led to infer, on physical grounds alone, the possibility of the expansion of any arbitrary value of ζ in a series of surface-harmonics, thus

$$\zeta = \sum_0^\infty S_n,$$

the coefficients of the various independent harmonics being the normal co-ordinates of the system. Again, since the products of these coefficients must disappear from the expressions for the kinetic and potential energies, we are led to the 'conjugate' properties of spherical harmonics quoted in Art. 87. The actual calculation of the energies will be given in the next Chapter, in connection with an independent treatment of the same problem.

The effect of a simple-harmonic disturbing force can be written down at once from the formula (14) of Art. 168. If the surface value of Ω be expanded in the form

$$\Omega = \Sigma \Omega_n, \qquad \text{..........................(8)}$$

where Ω_n is a surface-harmonic of integral order n, the various terms are normal components of force, in the generalized sense of Art. 135; and the equilibrium value of ζ corresponding to any one term Ω_n is

$$\zeta_n = -\Omega_n/g. \qquad \text{..............................(9)}$$

Hence, for the forced oscillation due to this term, we have

$$\zeta_n = -\frac{1}{1-\sigma^2/\sigma_n^2}\frac{\Omega_n}{g}, \qquad \text{......................(10)}$$

where σ measures the 'speed' of the disturbing force, and σ_n that of the corresponding free oscillation, as given by (5). There is no difficulty, of course, in deducing (10) directly from the equations of the preceding Art.

200. We have up to this point neglected the mutual attraction of the parts of the liquid. In the case of an ocean covering the globe, and with such relations of density as we meet with in the actual earth and ocean, this is not insensible. To investigate its effect in the case of the free oscillations, we have only to substitute for Ω_n, in the last formula, the gravitation-potential of the displaced water. If the density of this be denoted by ρ, whilst ρ_0 represents the mean density of the globe and liquid combined, we have*

$$\Omega_n = -\frac{4\pi\gamma\rho a}{2n+1}\zeta_n, \quad \text{.........................(11)}$$

and

$$g = \tfrac{4}{3}\gamma\pi a\rho_0, \quad \text{...............................(12)}$$

γ denoting the gravitation-constant, whence

$$\Omega_n = -\frac{3}{2n+1}\cdot\frac{\rho}{\rho_0}\cdot g\zeta_n. \quad \text{......................(13)}$$

Substituting in (10) we find

$$\frac{\sigma_n^2}{\sigma_n'^2} = \left(1 - \frac{3}{2n+1}\frac{\rho}{\rho_0}\right), \quad \text{......................(14)}$$

where σ_n is now used to denote the actual speed of the oscillation, and σ_n' the speed calculated on the former hypothesis of no mutual attraction. Hence the corrected speed is given by

$$\sigma_n^2 = n(n+1)\left(1 - \frac{3}{2n+1}\frac{\rho}{\rho_0}\right)\frac{gh}{a^2}. \quad \text{.................(15)†}$$

For an ellipsoidal oscillation ($n = 2$), and for $\rho/\rho_0 = \cdot 18$ (as in the case of the Earth), we find from (14) that the effect of the mutual attraction is to *lower* the frequency in the ratio of ·94 to 1.

The slowest oscillation would correspond to $n = 1$, but, as already indicated, it would be necessary, in this mode, to imagine a constraint applied to the globe to keep it at rest. This being assumed, it appears from (15) that if $\rho > \rho_0$ the value of σ_1^2 is negative. The circular function of t is then replaced by real exponentials; this shews that the configuration in which the surface of the sea is a sphere concentric with the globe is one of unstable equilibrium. Since the effect of the constraint is merely to increase the inertia of the system, we infer that the equilibrium is still unstable when the globe is free. In the extreme case where the globe itself is supposed to have no gravitative

* See, for example, Routh, *Analytical Statics*, 2nd ed., Cambridge, 1902, ii. 146–7.

† This result was given by Laplace, *Mécanique Céleste*, Livre 1er, Art. 1 (1799). The free and the forced oscillations of the type $n=2$ had been previously investigated in his "Recherches sur quelques points du système du monde," *Mém. de l'Acad. roy. des Sciences*, 1775 [1778] [*Oeuvres Complètes*, ix. 109, ...].

power at all, it is obvious that the water, if disturbed, would tend ultimately, under the influence of dissipative forces, to collect itself into a spherical mass, the nucleus being expelled.

It is obvious from Art. 168, or it may easily be verified independently, that the forced vibrations due to a given periodic disturbing force, when the gravitation of the water is taken into account, will be given by the formula (10), provided Ω_n now denote the potential of the extraneous forces only, and σ_n have the value given by (15).

201. The oscillations of a sea bounded by meridians, or parallels of latitude, or both, can also be treated by the same method*. The spherical harmonics involved are however, as a rule, no longer of integral order, and it is accordingly difficult to deduce numerical results.

In the case of a zonal sea bounded by two parallels of latitude, we assume

$$\zeta=\{Ap(\mu)+Bq(\mu)\}\begin{matrix}\cos\\ \sin\end{matrix}\Big\} s\phi, \quad \text{.................................(1)}$$

where $\mu=\cos\theta$, and $p(\mu)$, $q(\mu)$ are the two functions of μ, containing $(1-\mu^2)^{\frac{1}{2}s}$ as a factor, which are given by the formula (2) of Art. 86. It will be noticed that $p(\mu)$ is an *even*, and $q(\mu)$ an *odd* function of μ.

If we distinguish the limiting parallels by suffixes, the boundary conditions are that $u=0$ for $\mu=\mu_1$ and $\mu=\mu_2$. For the free oscillations this gives, by Art. 198 (6),

$$Ap'(\mu_1)+Bq'(\mu_1)=0, \quad Ap'(\mu_2)+Bq'(\mu_2)=0, \quad \text{.....................(2)}$$

whence

$$\begin{vmatrix} p'(\mu_1), & q'(\mu_1) \\ p'(\mu_2), & q'(\mu_2) \end{vmatrix}=0, \quad \text{...................................(3)}$$

which is the equation to determine the admissible values of n, the order of the harmonics. The speeds (σ) corresponding to the various roots are given as before by Art. 199 (5).

If the two boundaries are equidistant from the equator, we have $\mu_2=-\mu_1$. The above solutions then break up into two groups; viz. for one of these we have

$$B=0, \quad p'(\mu_1)=0, \quad \text{.......................................(4)}$$

and for the other

$$A=0, \quad q'(\mu_1)=0. \quad \text{.......................................(5)}$$

In the former case ζ has the same value at two points symmetrically situated on opposite sides of the equator; in the latter the values at these points are numerically equal, but opposite in sign.

If we imagine one of the boundaries to be contracted to a point (say $\mu_2=1$), we pass to the case of a circular basin. The values of $p'(1)$ and $q'(1)$ are infinite, but their ratio can be evaluated by means of formulae given in Art. 84. This gives, by the second of equations (2), the ratio $A:B$, and substituting in the first we get the equation to determine n. A simpler method of treating this case consists, however, in starting with a solution which is known to be finite, whatever the value of n, at the pole $\mu=1$. This involves a change of variable, as to which there is some latitude of choice. We might take, for instance, the expression for $P_n^s(\cos\theta)$ in Art. 86 (6), and seek to determine n from the condition that

$$\frac{\partial}{\partial\theta}P_n^s(\cos\theta)=0 \quad \text{...(6)}$$

for $\theta=\theta_1$†. By making the radius of the sphere infinite, we can pass to the plane problem of Art. 191‡. The steps of the transition will be understood from Art. 100.

* Cf. Rayleigh, *l.c. ante* p. 301.

† This question has been discussed by Macdonald, *Proc. Lond. Math. Soc.* xxxi. 264 (1899).

‡ Cf. Rayleigh, *Theory of Sound*, Arts. 336, 338.

If the sheet of water considered have as boundaries two meridians (with or without parallels of latitude), say $\phi=0$ and $\phi=\alpha$, the condition that $v=0$ at these restricts us to the factor $\cos s\omega$, and gives $s\alpha=m\pi$, where m is integral. This determines the admissible values of s, which are not in general integral*. The diurnal and semi-diurnal tides in a non-rotating ocean of uniform depth bounded by two meridians have been studied by Proudman and Doodson, and worked out for special cases and for special depths†.

Dynamics of a Rotating System.

202. The theory of the tides on an open sheet of water is seriously complicated by the fact of the earth's rotation. If, indeed, we could assume that the periods of the free oscillations, and of the disturbing forces, were small compared with a day, the preceding investigations would apply as a first approximation, but these conditions are far from being fulfilled in the actual circumstances of the earth.

The difficulties which arise when we attempt to take the rotation into account have their origin in this, that a particle having a motion in latitude tends to keep its angular momentum about the earth's axis unchanged, and so to alter its motion in longitude. This point is of course familiar in connection with Hadley's theory of the trade-winds‡. Its bearing on tidal theory seems to have been first recognized by Maclaurin§.

Owing to the enormous inertia of the solid body of the earth compared with that of the ocean, the effect of tidal reactions in producing periodic changes of the angular velocity is quite insensible. This angular velocity will therefore for the present be treated as constant.

The theory of the small oscillations of a dynamical system about a state of equilibrium relative to a real or ideal rigid frame which rotates with constant angular velocity about a fixed axis differs in some important particulars from the theory of small oscillations about a state of absolute equilibrium, of which some account was given in Art. 168. It is therefore worth while to devote a little space to it before entering on the consideration of special problems. The system considered may be entirely free, or it may be connected with a rotating solid. In the latter case it is assumed that the connecting forces as well as the internal forces of the system are subject to the 'conservative' law.

203. The equations of motion of a particle m relative to rectangular axes Ox, Oy, Oz which rotate about Oz with angular velocity ω are

$$m(\ddot{x}-2\omega\dot{y}-\omega^2x)=X,\quad m(\ddot{y}+2\omega\dot{x}-\omega^2y)=Y,\quad m\ddot{z}=Z,\quad \ldots(1)$$

where X, Y, Z are the impressed forces.

* The reader who wishes to carry the study of the problem further in this direction is referred to Thomson and Tait, *Natural Philosophy* (2nd ed.), Appendix B, "Spherical Harmonic Analysis."

† *M. N. R. A. S., Geophy. Suppt.* i. 468 (1927), and ii. 209 (1929).

‡ "The Cause of the General Trade Winds," *Phil. Trans.* 1735.

§ *De Causâ Physicâ Fluxus et Refluxus Maris*, Prop. vii.: "Motus aquæ turbatur ex inæquali velocitate quâ corpora circa axem Terræ motu diurno deferuntur" (1740).

Let us now suppose that the relative co-ordinates (x, y, z) of each particle are expressed in terms of a certain number of independent quantities $q_1, q_2, \ldots q_r$. We write

$$\mathfrak{T} = \tfrac{1}{2}\Sigma m(\dot{x}^2 + \dot{y}^2 + \dot{z}^2), \quad T_0 = \tfrac{1}{2}\omega^2 \Sigma m(x^2 + y^2). \quad \ldots\ldots\ldots\ldots(2)$$

Hence $\mathfrak{T}$ denotes the kinetic energy of the relative motion, which we shall suppose expressed as a homogeneous quadratic function of the generalized velocities $\dot{q}_r$, with coefficients which are functions of the generalized co-ordinates q_r; whilst T_0 is the kinetic energy of the system when rotating, without relative motion, in the configuration $(q_1, q_2, \ldots q_n)$. Finally we put

$$\Sigma(X\delta x + Y\delta y + Z\delta z) = -\delta V + Q_1\delta q_1 + Q_2\delta q_2 + \ldots + Q_n\delta q_n, \quad \ldots(3)$$

where V is the potential energy and $Q_1, Q_2, \ldots Q_n$ are the generalized components of extraneous force.

If we multiply the three equations (1) by $\partial x/\partial q_r$, $\partial y/\partial q_r$, $\partial z/\partial q_r$, respectively, and add, and sum the result for all the particles of the system, and then proceed as in the 'direct' proof of Lagrange's equations, we obtain the following typical equation of motion in generalized co-ordinates*:

$$\frac{d}{dt}\frac{\partial\mathfrak{T}}{\partial\dot{q}_r} - \frac{\partial\mathfrak{T}}{\partial q_r} + \beta_{r1}\dot{q}_1 + \beta_{r2}\dot{q}_2 + \ldots + \beta_{rn}\dot{q}_n = -\frac{\partial}{\partial q_r}(V - T_0) + Q_r, \quad \ldots(4)$$

where
$$\beta_{rs} = 2\omega\Sigma m \frac{\partial(x, y)}{\partial(q_s, q_r)}. \quad \ldots\ldots\ldots\ldots\ldots\ldots\ldots\ldots(5)$$

It is to be noted that

$$\beta_{rs} = -\beta_{sr}, \quad \beta_{rr} = 0. \quad \ldots\ldots\ldots\ldots\ldots\ldots\ldots\ldots(6)$$

The equation (4) may also be derived from Art. 141 (23), with the help of Art. 142 (8), by supposing the rotating frame to be free, but to have an infinite moment of inertia.

The conditions for relative equilibrium, in the absence of disturbing forces, are found by putting $\dot{q}_1, \dot{q}_2, \ldots \dot{q}_r = 0$ in (4), whence

$$\frac{\partial}{\partial q_r}(V - T_0) = 0, \quad \ldots\ldots\ldots\ldots\ldots\ldots\ldots\ldots(7)$$

shewing that the equilibrium value of $V - T_0$ is 'stationary.'

Again, from (1) we have

$$\Sigma m(\dot{x}\ddot{x} + \dot{y}\ddot{y} + \dot{z}\ddot{z}) - \omega^2\Sigma m(x\dot{x} + y\dot{y} + z\dot{z}) = \Sigma(X\dot{x} + Y\dot{y} + Z\dot{z}), \quad \ldots(8)$$

or, by (2) and (3)

$$\frac{d}{dt}(\mathfrak{T} + V - T_0) = Q_1\dot{q}_1 + Q_2\dot{q}_2 + \ldots + Q_n\dot{q}_n. \quad \ldots\ldots\ldots\ldots(9)$$

This result may also be deduced from (4), taking account of the relations (6).

* Cf. Thomson and Tait, *Natural Philosophy* (2nd ed.), i. 310; Lamb, *Higher Mechanics*, 2nd ed., Art. 84.

When there are no disturbing forces we have

$$\mathfrak{T} + V - T_0 = \text{const.} \quad \text{.........................(10)}$$

The form assumed by the Hamiltonian theorem of Art. 135 is also to be noticed. The total kinetic energy of our system is

$$T = \tfrac{1}{2}\Sigma m\,\{(\dot{x} - \omega y)^2 + (\dot{y} + \omega x)^2 + \dot{z}^2\} = \mathfrak{T} + T_0 + \omega M, \quad \text{......(11)}$$

where

$$M = \Sigma m\,(x\dot{y} - y\dot{x}). \quad \text{.........................(12)}$$

If there are no extraneous forces we have

$$\Delta \int_{t_0}^{t_1} (T - V)\, dt = 0, \quad \text{.........................(13)}$$

subject to the usual terminal condition. Hence

$$\Delta \int_{t_0}^{t_1} (\mathfrak{T} + T_0 + \omega M - V)\, dt = 0, \quad \text{.................(14)}$$

with the condition

$$\Big[\Sigma m\,\{(\dot{x} - \omega y)\,\Delta x + (\dot{y} + \omega x)\,\Delta y + \dot{z}\Delta z\}\Big]_{t_0}^{t_1} = 0. \quad \text{.........(15)}$$

This theorem may also be deduced directly from (1) by the usual Hamiltonian procedure, and leads in turn to an independent proof of the equations (4), for the case of free motion. The inclusion of disturbing forces in the investigation presents no difficulty.

The condition (15) is fulfilled whenever the initial and final relative configurations are the same in the varied as in the actual motion.

204. We will now suppose the co-ordinates q_r to be chosen so as to vanish in the undisturbed state. In the case of a *small* disturbance, we may then write

$$2\mathfrak{T} = a_{11}\dot{q}_1^2 + a_{22}\dot{q}_2^2 + \ldots + 2a_{12}\dot{q}_1\dot{q}_2 + \ldots, \quad \text{..............(1)}$$

$$2(V - T_0) = c_{11}q_1^2 + c_{22}q_2^2 + \ldots + 2c_{12}q_1q_2 + \ldots, \quad \text{..............(2)}$$

where the coefficients may be treated as constants. The terms of the first degree in $V - T_0$ have been omitted, on account of the 'stationary' property.

In order to simplify the equations as much as possible, we will further suppose that, by a linear transformation, each of these expressions is reduced, as in Art. 168, to a sum of squares; viz.

$$2\mathfrak{T} = a_1\dot{q}_1^2 + a_2\dot{q}_2^2 + \ldots + a_n\dot{q}_n^2, \quad \text{....................(3)}$$

$$2(V - T_0) = c_1q_1^2 + c_2q_2^2 + \ldots + c_nq_n^2. \quad \text{....................(4)}$$

The quantities $q_1, q_2, \ldots q_n$ may be called the 'principal co-ordinates' of the system, but we must be on our guard against assuming that the same simplicity of properties attaches to them as in the case of no rotation. The coefficients $a_1, a_2, \ldots a_n$ and $c_1, c_2, \ldots c_n$ may be called the 'principal coefficients' of inertia and of stability, respectively. The latter coefficients are the same as if we were to ignore the rotation, and to introduce fictitious 'centrifugal' forces $(m\omega^2 x,\ m\omega^2 y,\ 0)$ acting on each particle in the direction outwards from the axis.

The equations (4) of the preceding Art. become, in the case of infinitely small motions,

$$\left.\begin{array}{l} a_1\ddot{q}_1 + c_1 q_1 \qquad\qquad + \beta_{12}\dot{q}_2 + \beta_{13}\dot{q}_3 + \ldots + \beta_{1n}\dot{q}_n = Q_1, \\ a_2\ddot{q}_2 + c_2 q_2 + \beta_{21}\dot{q}_1 \qquad\qquad + \beta_{23}\dot{q}_3 + \ldots + \beta_{2n}\dot{q}_n = Q_2, \\ \ldots\ldots\ldots\ldots\ldots\ldots\ldots\ldots\ldots\ldots\ldots\ldots \\ a_n\ddot{q}_n + c_n q_n + \beta_{n1}\dot{q}_1 + \beta_{n2}\dot{q}_2 + \beta_{n3}\dot{q}_3 + \ldots \qquad = Q_n, \end{array}\right\} \quad \ldots\ldots(5)$$

where the coefficients β_{rs} may be regarded as constants.

If we multiply these by $\dot{q}_1, \dot{q}_2, \ldots \dot{q}_n$ in order and add, we find, taking account of the relation $\beta_{rs} = -\beta_{sr}$,

$$\frac{d}{dt}(\mathfrak{T} + V - T_0) = Q_1\dot{q}_1 + Q_2\dot{q}_2 + \ldots + Q_n\dot{q}_n, \quad \ldots\ldots\ldots\ldots(6)$$

as has already been proved without approximation.

205. To investigate the *free* motions of the system, we put $Q_1, Q_2, \ldots Q_n = 0$, in (5), and assume, in accordance with the usual method of treating linear equations,

$$q_1 = A_1 e^{\lambda t}, \quad q_2 = A_2 e^{\lambda t}, \ldots q_n = A_n e^{\lambda t}. \quad \ldots\ldots\ldots\ldots\ldots\ldots(7)$$

Substituting, we find

$$\left.\begin{array}{l} (a_1\lambda^2 + c_1) A_1 \qquad + \beta_{12}\lambda A_2 + \ldots \qquad + \beta_{1n}\lambda A_n = 0, \\ \beta_{21}\lambda A_1 + (a_2\lambda^2 + c_2) A_2 + \ldots \qquad + \beta_{2n}\lambda A_n = 0, \\ \ldots\ldots\ldots\ldots\ldots\ldots\ldots\ldots\ldots\ldots\ldots\ldots \\ \beta_{n1}\lambda A_1 \qquad + \beta_{n2}\lambda A_2 + \ldots + (a_n\lambda^2 + c_n) A_n = 0. \end{array}\right\} \quad \ldots\ldots(8)$$

Eliminating the ratios $A_1 : A_2 : \ldots : A_n$, we get the equation

$$\begin{vmatrix} a_1\lambda^2 + c_1, & \beta_{12}\lambda, & \ldots & \beta_{1n}\lambda \\ \beta_{21}\lambda, & a_2\lambda^2 + c_2, & \ldots & \beta_{2n}\lambda \\ \ldots & \ldots & \ldots & \ldots \\ \beta_{n1}\lambda, & \beta_{n2}\lambda, & \ldots & a_n\lambda^2 + c_n \end{vmatrix} = 0, \quad \ldots\ldots\ldots\ldots(9)$$

or, as we shall occasionally write it, for shortness,

$$D(\lambda) = 0. \quad \ldots\ldots\ldots\ldots\ldots\ldots\ldots\ldots\ldots\ldots(10)$$

The determinant $D(\lambda)$ comes under the class called by Cayley 'skew-determinants,' in virtue of the relations (6) of Art. 203. If we reverse the sign of λ, the rows and columns are simply interchanged, and the value of the determinant is therefore unaltered. Hence the equation (10) will involve only *even* powers of λ, and the roots will be in pairs of the form

$$\lambda = \pm(\rho + i\sigma).$$

In order that the configuration of relative equilibrium should be stable it is essential that the values of ρ should all be zero, for otherwise terms of the forms $e^{\pm\rho t}\cos\sigma t$ and $e^{\pm\rho t}\sin\sigma t$ would present themselves in the realized expression for any co-ordinate q_r. This would indicate the possibility of an oscillation of continually increasing amplitude.

In the theory of absolute equilibrium, sketched in Art. 168, the necessary and sufficient condition of stability (in the above sense) was simply that the potential energy must be a minimum in the configuration of equilibrium. In the present case the conditions are more complicated*, but it is easily seen that if the expression for $V - T_0$ be essentially positive, in other words if the coefficients $c_1, c_2, \ldots c_n$ in (4) be all positive, the equilibrium must be stable. This follows at once from the equation

$$\mathfrak{T} + (V - T_0) = \text{const.}, \quad \text{.........................(11)}$$

proved in Art. 203, which shews that under the present supposition neither $\mathfrak{T}$ nor $V - T_0$ can increase beyond a certain limit depending on the initial circumstances†. It will be observed that this argument does not involve the use of approximate equations.

Hence stability is assured if $V - T_0$ is a minimum in the configuration of relative equilibrium. But this condition is not essential, and there may even be stability (from the present point of view) with $V - T_0$ a maximum, as will be shewn presently in the particular case of two degrees of freedom. It is to be remarked, however, that if the system be subject to dissipative forces, however slight, affecting the relative co-ordinates $q_1, q_2, \ldots q_n$, the equilibrium will be permanently or 'secularly' stable only if $V - T_0$ is a minimum. It is the characteristic of such forces that the work done by them on the system is always negative. Hence by (6) the expression $\mathfrak{T} + (V - T_0)$ will, so long as there is any relative motion of the system, continually diminish, in the algebraical sense. Hence if the system be started from relative rest in a configuration such that $V - T_0$ is negative, the above expression, and therefore *à fortiori* the part $V - T_0$, will assume continually increasing negative values, which can only take place by the system deviating more and more from its equilibrium-configuration.

This important distinction between 'ordinary' or kinetic, and secular' or practical stability was first pointed out by Thomson and Tait‡. It is to be observed that the above investigation presupposes a constant angular velocity (ω) maintained, if necessary, by a proper application of force to the rotating solid. When the solid is *free*, the condition of secular stability takes a somewhat different form, to be referred to later (Chapter XII.). In the

* They have been investigated by Routh, *l.c. ante* p. 195; see also his *Advanced Rigid Dynamics*, c. vi.

† The argument was originally applied to the theory of oscillations about a configuration of absolute equilibrium (Art. 168) by Dirichlet, "Ueber die Stabilität des Gleichgewichts," *Crelle*, xxxii. (1846) [*Werke*, Berlin, 1889–97, ii. 3]. An algebraic proof is indicated in *Higher Mechanics*, 2nd ed., Art. 99.

‡ *Natural Philosophy* (2nd ed.), Part I. p. 391. See also Poincaré, "Sur l'équilibre d'une masse fluide animée d'un mouvement de rotation," *Acta Mathematica*, vii. (1885), and *op. cit. ante* p. 146. Some simple mechanical illustrations are given in a paper "On Kinetic Stability," *Proc. Roy. Soc.* A, lxxx. 168 (1909), and in the author's *Higher Mechanics*, 2nd ed., p. 253.

practical applications we shall be concerned only with cases where $V - T_0$ is a minimum, and the coefficients $c_1, c_2, \dots c_n$ in Art. 204 (4) accordingly positive.

To examine the character of a free oscillation, in the case of stability, we remark that if λ be any root of (10), the equations (8) give

$$\frac{A_1}{\alpha_1} = \frac{A_2}{\alpha_2} = \dots = \frac{A_n}{\alpha_n} = C, \quad \text{.......................(12)}$$

where $\alpha_1, \alpha_2, \dots \alpha_n$ are the minors of any row in the determinant $D(\lambda)$, and C is arbitrary. These minors will as a rule involve odd as well as even powers of λ, and so assume unequal values for the two oppositely signed roots ($\pm \lambda$) of any pair. If we put $\lambda = \pm i\sigma$, the corresponding values of α_r will be of the forms $\mu_r \pm i\nu_r$, where μ_r, ν_r are real. Hence

$$q_r = C(\mu_r + i\nu_r)e^{i\sigma t} + C'(\mu_r - i\nu_r)e^{-i\sigma t}.$$

If we put $$C = \tfrac{1}{2}Ke^{i\epsilon}, \qquad C' = \tfrac{1}{2}Ke^{-i\epsilon},$$

we get a solution of our equations in real form, involving two arbitrary constants K, ϵ; thus

$$q_r = K\{\mu_r \cos(\sigma t + \epsilon) - \nu_r \sin(\sigma t + \epsilon)\}. \quad \text{..............(13)}$$

This formula expresses what may be called a 'natural mode' of oscillation of the system. The number of such possible modes is of course equal to the number of pairs of roots of (9), *i.e.* to the number of degrees of freedom of the system. It is to be noticed, as an effect of the rotation, that the various co-ordinates are no longer in the same phase.

If ξ, η, ζ denote the component displacements of any particle from its equilibrium position, we have

$$\left.\begin{aligned} \xi &= \frac{\partial x}{\partial q_1} q_1 + \frac{\partial x}{\partial q_2} q_2 + \dots + \frac{\partial x}{\partial q_n} q_n, \\ \eta &= \frac{\partial y}{\partial q_1} q_1 + \frac{\partial y}{\partial q_2} q_2 + \dots + \frac{\partial y}{\partial q_n} q_n, \\ \zeta &= \frac{\partial z}{\partial q_1} q_1 + \frac{\partial z}{\partial q_2} q_2 + \dots + \frac{\partial z}{\partial q_n} q_n \end{aligned}\right\} \quad \text{............................(14)}$$

Substituting from (13), we obtain a result of the form

$$\left.\begin{aligned} \xi &= P \,.\, K\cos(\sigma t + \epsilon) + P' \,.\, K\sin(\sigma t + \epsilon), \\ \eta &= Q \,.\, K\cos(\sigma t + \epsilon) + Q' \,.\, K\sin(\sigma t + \epsilon), \\ \zeta &= R \,.\, K\cos(\sigma t + \epsilon) + R' \,.\, K\sin(\sigma t + \epsilon), \end{aligned}\right\} \quad \text{.......................(15)}$$

where P, P', Q, Q', R, R' are determinate functions of the mean position of the particle, involving also the value of σ, and therefore different for the different normal modes, but independent of the arbitrary constants K, ϵ. These formulae represent an elliptic-harmonic motion of period $2\pi/\sigma$, the directions

$$\frac{\xi}{P} = \frac{\eta}{Q} = \frac{\zeta}{R}, \text{ and } \frac{\xi}{P'} = \frac{\eta}{Q'} = \frac{\zeta}{R'}, \quad \text{..........................(16)}$$

being those of two conjugate semi-diameters of the elliptic orbit, of lengths

$$(P^2 + Q^2 + R^2)^{\frac{1}{2}} \,.\, K, \quad \text{and} \quad (P'^2 + Q'^2 + R'^2)^{\frac{1}{2}} \,.\, K,$$

respectively. The positions and forms and relative dimensions of the elliptic orbits, as well as the relative phases of the particles in them, are accordingly in each natural mode determinate, the absolute dimensions and epochs being alone arbitrary.

205 a. When the angular velocity ω is small the normal modes will as a rule differ only slightly from the case of no rotation, and expressions for the altered types and frequencies can then be found as follows*. Since the determinantal equation (9) of Art. 205 is unaltered when we reverse the signs of all the β's, the frequencies will usually involve these quantities in the *second* order. Hence, considering for example the mode in which A_1 is finite, whilst $A_2, A_3, \dots A_n$ are relatively small, and writing $\lambda = i\sigma_1$, the rth equation of the system (8) gives, approximately,

$$\frac{A_r}{A_1} = \frac{i\beta_{r1}\sigma_1}{a_r(\sigma_1^2 - \sigma_r^2)}, \quad \text{.........................(17)}$$

where $\sigma_r^2 = c_r/a_r$. Hence, substituting in the first equation, we get a corrected value of σ_1^2; thus

$$\sigma_1^2 = \frac{c_1}{a_1}\left\{1 + \sum_r \frac{\beta_{1r}^2}{a_1 a_r(\sigma_1^2 - \sigma_r^2)}\right\}. \quad \text{..................(18)}$$

But these approximations fail if any denominator in the bracket vanishes or is even small. This case arises when two or more of the normal modes in the absence of rotation have the same or nearly the same period. Suppose, for instance, that σ_1^2 and σ_2^2 are nearly equal. We have then, from (8), with $\lambda = i\sigma$,

$$\left.\begin{aligned} (c_1^2 - \sigma^2 a_1) A_1 + i\beta_{12}\sigma A_2 = 0, \\ i\beta_{21} A_1 + (c_2 - \sigma^2 a_2) A_2 = 0, \end{aligned}\right\} \quad \text{.....................(19)}$$

so that A_1 and A_2 are comparable. Eliminating A_1/A_2, we have

$$(\sigma^2 - \sigma_1^2)(\sigma^2 - \sigma_2^2) = \frac{\beta_{12}^2}{a_1 a_2}\sigma^2. \quad \text{.....................(20)}$$

In the case of exact equality this gives

$$\sigma^2 - \sigma_1^2 = \pm \frac{\beta_{12}}{\sqrt{(a_1 a_2)}}\sigma, \quad \text{.......................(21)}$$

or

$$\sigma - \sigma_1 = \pm \frac{\beta_{12}}{2\sqrt{(a_1 a_2)}}, \quad \text{.........................(22)}$$

approximately. The change of frequency due to the rotation is now proportional to ω instead of ω^2.

The values of $A_3, A_4, \dots A_n$ in terms of A_1, A_2 are to be found from the remaining equations of the system (8), but would only affect the above conclusion by terms involving ω^2.

205 b. On account of the analytical difficulties which attend the determination of the free modes of oscillation, especially in the case of continuous systems, it is natural to look for an approximate method of calculating the more important frequencies, analogous to that employed by Rayleigh in the case of non-rotating systems (Art. 168).

* Rayleigh, *Phil. Mag.* (6) v. 293 (1903) [*Papers*, v. 89].

For this purpose we may have recourse to the variational formula (14) of Art. 203. In the application to small oscillations it is convenient to express this in terms of the *displacements* (ξ, η, ζ) of the particles from their positions of relative equilibrium. Writing $x_0+\xi$, $y_0+\eta$, $z_0+\zeta$ for x, y, z, where x_0, y_0, z_0 refer to the equilibrium position, we have

$$\Delta\int_{t_0}^{t_1} M\,dt = \Delta\int_{t_0}^{t_1} M'\,dt + \Big[\Sigma m\,(x_0\Delta\eta - y_0\Delta\xi)\Big]_{t_0}^{t_1}, \quad \text{........(1)}$$

where

$$M' = \Sigma m\,(\xi\dot\eta - \eta\dot\xi). \quad \text{........................(2)}$$

When the integrated terms in (1) are incorporated in the terminal condition (15) of Art. 203, the theorem becomes

$$\Delta\int_{t_0}^{t_1} (\mathfrak{T} + \omega M' + T_0 - V)\,dt = 0, \quad \text{....................(3)}$$

with the condition

$$\Big[\Sigma m\,\{(\dot\xi - \omega\eta)\,\Delta\xi + (\dot\eta + \omega\xi)\,\Delta\eta + \dot\zeta\Delta\zeta\}\Big]_{t_0}^{t_1} = 0. \quad \text{...........(4)}$$

Let us now suppose that the varied, as well as the natural motion, is simple-harmonic with the same period $2\pi/\sigma$, and that the limits of integration t_0, t_1 differ by an exact period. The terms in (4) which relate to the two limits will then cancel, so that the postulated condition is fulfilled. The result is that the mean value (with respect to time) of the expression

$$\mathfrak{T} + \omega M' - (V - T_0) \quad \text{.........................(5)}$$

is stationary for small arbitrary variations of the type of vibration, the period being kept constant.

In terms of generalized co-ordinates (assumed to vanish in relative equilibrium) M' will be a bilinear function of the two sets of variables

$$q_1, q_2, \ldots q_n \text{ and } \dot q_1, \dot q_2, \ldots \dot q_n,$$

whilst $\mathfrak{T}$ and $V - T_0$ are already by hypothesis homogeneous quadratic functions of the velocities and co-ordinates, respectively. Hence (5) is a homogeneous quadratic function of the variables q_r, $\dot q_r$.

If we now write

$$q_r = A_r \cos\sigma t + B_r \sin\sigma t, \quad \text{.......................(6)}$$

and denote the resulting mean value of the expression (5) by J, we have

$$J = \sigma^2 P + \sigma Q - R, \quad \text{............................(7)}$$

where P, Q, R are certain homogeneous quadratic functions of the variables A_r, B_r, whose precise forms are not required for the moment.

The stationary property asserts that

$$\sigma^2\Delta P + \sigma\Delta Q - \Delta R = 0 \quad \text{.........................(8)}$$

for all infinitesimal values of $\Delta A_r, \Delta B_r$. In particular, putting $\Delta A_r = \epsilon A_r$, $\Delta B_r = \epsilon B_r$, where ϵ is an infinitesimal constant independent of r, we have

$$J = 0, \quad \text{...................................(9)}$$

on account of the homogeneous character. The statement that in a free oscillation the mean value of the expression (5) is zero is a generalization of a result already pointed out in the case of $\omega = 0$, viz. that in oscillations about absolute equilibrium the mean values of the kinetic and potential energies are equal.

The present result can be expressed in another form. If for a moment we regard σ as a function of A_r, B_r, where these coefficients have general values, determined by the equation

$$\sigma^2 P + \sigma Q - R = 0, \quad \ldots\ldots\ldots\ldots(10)$$

we have

$$(2\sigma P + Q)\,\Delta\sigma + (\sigma^2 \Delta P + \sigma \Delta Q - \Delta R) = 0. \quad \ldots\ldots\ldots\ldots(11)$$

Hence when A_r, B_r have the special values appropriate to a free mode of oscillation, we have

$$\Delta\sigma = 0, \quad \ldots\ldots\ldots\ldots(12)$$

by (8). In other words, the values of σ determined by (10) are stationary.

It follows that if the values of P, Q, R in (10) are calculated on the basis of an assumed type of vibration which differs slightly from the truth, the error in the consequent values of σ will be of the second order.

These stationary values will include, as generally most important, the maxima and minima (in absolute value) of σ.

Applications of the above principle to particular cases will be found in Arts. 212 *a*, 216.

The general form of the functions P, Q, R in (7) may be noticed, although it is not essential to the argument. We have at once, on reference to Art. 204 (3) (4),

$$P = \tfrac{1}{4} S_r a_r (A_r^2 + B_r^2), \qquad R = \tfrac{1}{4} S_r c_r (A_r^2 + B_r^2), \quad \ldots\ldots\ldots(13)$$

where S_r denotes a summation of terms of the types indicated, with $r = 1, 2, \ldots n$. Again, from (2),

$$\omega M' = \omega \Sigma m \left\{ S_r \frac{\partial \xi}{\partial q_r} q_r \,.\, S_s \frac{\partial \eta}{\partial q_s} \dot{q}_s - S_r \frac{\partial \eta}{\partial q_r} q_r \,.\, S_s \frac{\partial \xi}{\partial q_s} \dot{q}_s \right\}$$

$$= \tfrac{1}{2} \{ \dot{q}_1 S_r \beta_{1r} q_r + \dot{q}_2 S_r \beta_{2r} q_r + \ldots + \dot{q}_n S_r \beta_{nr} q_r \}, \quad \ldots\ldots\ldots(14)$$

where

$$\beta_{rs} = 2\omega \Sigma m \frac{\partial(\xi, \eta)}{\partial(q_s, q_r)}. \quad \ldots\ldots\ldots\ldots(15)$$

Substituting from (6), and taking the mean value, we have

$$Q = \tfrac{1}{2} S_r S_s \beta_{rs} A_s B_r, \quad \ldots\ldots\ldots\ldots(16)$$

where, in the double summation, each permutation of suffixes is to be taken *once*.

As a verification we may note that if with these values of P, Q, R we form the equation (8) the coefficients of ΔA_r, ΔB_r will be found to be identical with the coefficients of $\cos \sigma t$ and $\sin \sigma t$, respectively, when we substitute from (6) in the typical equation of motion, Art. 204 (5).

206. The symbolical expressions for the *forced* oscillations due to a periodic disturbing force are easily written down. If we assume that $Q_1, Q_2, \ldots Q_n$ all vary as $e^{i\sigma t}$, where σ is prescribed, the equations (5) of Art. 204 give, if we omit the time-factors,

$$D(i\sigma)\, q_r = \alpha_{r1} Q_1 + \alpha_{r2} Q_2 + \ldots + \alpha_{rn} Q_n, \qquad (1)$$

where the coefficients on the right-hand side are the minors of the rth row in the determinant $D(i\sigma)$.

The most important point of contrast with the theory of the 'normal modes' in the case of no rotation is that the displacement of any one type is no longer affected solely by the disturbing force of that type. As a consequence, the motions of the individual particles are, as is easily seen from Art. 205 (14), now in general elliptic-harmonic. Again, there are in general differences of phase, variable with the frequency, between the displacements and the force.

As in Art. 168, the displacement becomes very great when $D(i\sigma)$ is very small, *i.e.* whenever the 'speed' σ of the disturbing force approximates to that of one of the natural modes of free oscillation.

When the period of the disturbing forces is infinitely long, the displacements tend to the 'equilibrium-values'

$$q_1 = Q_1/c_1,\ q_2 = Q_2/c_2,\ \ldots\ q_n = Q_n/c_n, \qquad (2)$$

as is seen directly from the equations (5) of Art. 204. This conclusion must be modified, however, when one or more of the coefficients of stability c_1, $c_2, \ldots c_n$ is zero. If, for example, $c_1 = 0$, the first row and column of the determinant $D(\lambda)$ are both divisible by λ, so that the determinantal equation has a pair of zero roots. In other words we have a possible free motion of infinitely long period. The coefficients of $Q_2, Q_3, \ldots Q_n$ on the right-hand side of (1) then become indeterminate for $\sigma = 0$, and the evaluated results do not as a rule coincide with (2). This point is of importance, because in some hydrodynamical applications, as we shall see, steady circulatory motions of the fluid, with a constant deformation of the free surface, are possible when no extraneous forces act; and as a consequence forced tidal oscillations of long period do not necessarily approximate to the values given by the equilibrium theory of the tides. Cf. Arts. 214, 217.

In order to elucidate the foregoing statements we may consider more in detail the case of two degrees of freedom. The equations of motion are then of the forms

$$a_1\ddot{q}_1 + c_1 q_1 + \beta\dot{q}_2 = Q_1, \qquad a_2\ddot{q}_2 + c_2 q_2 - \beta\dot{q}_1 = Q_2. \qquad (3)$$

The equation determining the periods of the free oscillations is

$$a_1 a_2 \lambda^4 + (a_1 c_2 + a_2 c_1 + \beta^2)\lambda^2 + c_1 c_2 = 0. \qquad (4)$$

For 'ordinary' stability it is sufficient that the roots of this quadratic in λ^2 should be real and negative. Since a_1, a_2 are essentially positive, it is easily seen that this condition is in any case fulfilled if c_1, c_2 are both positive, and that it will also be satisfied even when

c_1, c_2 are both negative, provided β^2 be sufficiently great. It will be shewn later, however, that in the latter case the equilibrium is rendered unstable by the introduction of dissipative forces. See Art. 322.

To find the forced oscillations when Q_1, Q_2 vary as $e^{i\sigma t}$, we have, omitting the time-factor,

$$(c_1 - \sigma^2 a_1)\, q_1 + i\sigma\beta q_2 = Q_1, \qquad -i\sigma\beta q_1 + (c_2 - \sigma^2 a_2)\, q_2 = Q_2, \quad \text{.............(5)}$$

whence

$$q_1 = \frac{(c_2 - \sigma^2 a_2)\, Q_1 - i\sigma\beta Q_2}{(c_1 - \sigma^2 a_1)(c_2 - \sigma^2 a_2) - \sigma^2\beta^2}, \qquad q_2 = \frac{i\sigma\beta Q_1 + (c_1 - \sigma^2 a_1)\, Q_2}{(c_1 - \sigma^2 a_1)(c_2 - \sigma^2 a_2) - \sigma^2\beta^2}. \quad \text{.........(6)}$$

Let us now suppose that $c_2 = 0$, or, in other words, that the displacement q_2 does not affect the value of $V - T_0$. We will also suppose that $Q_2 = 0$, *i.e.* that the extraneous forces do no work during a displacement of the type q_2. The above formulae then give

$$q_1 = \frac{a_2}{a_2(c_1 - \sigma^2 a_1) + \beta^2}\, Q_1, \qquad \dot{q}_2 = \frac{\beta}{a_2(c_1 - \sigma^2 a_1) + \beta^2}\, Q_1. \quad \text{.................(7)}$$

In the case of a disturbance of long period we have $\sigma = 0$, approximately, and therefore

$$q_1 = \frac{1}{c_1 + \beta^2/a_2}\, Q_1, \qquad \dot{q}_2 = \frac{\beta}{a_2 c_1 + \beta^2}\, Q_1. \quad \text{..........................(8)}$$

The displacement q_1 is therefore *less* than its equilibrium-value, in the ratio $1 : 1 + \beta^2/a_2 c_1$; and it is accompanied by a motion of the type q_2 although there is no extraneous force of the latter type (cf. Art. 217). We pass, of course, to the case of absolute equilibrium, considered in Art. 168, by putting $\beta = 0$*.

It should be added that the determination of the 'principal co-ordinates' of Art. 204 depends on the original forms of $\mathfrak{T}$ and $V - T_0$, and is therefore affected by the value of ω^2, which enters as a factor of T_0. The system of equations there given is accordingly not altogether suitable for a discussion of the question how the character and the frequencies of the respective principal modes of free vibration vary with ω. One remarkable point which is thus overlooked is that types of circulatory motion, which are of infinitely long period in the case of no rotation, may be converted by the slightest degree of rotation into oscillatory modes of periods comparable with that of the rotation. Cf. Arts. 212, 223.

To illustrate the matter in its simplest form, we may take the case of two degrees of freedom. If c_2 vanishes for $\omega = 0$, and so contains ω^2 as a factor in the general case, the two roots of equation (4) are

$$\lambda^2 = -c_1/a_1, \quad \lambda^2 = -c_2/a_2,$$

approximately, when ω^2 is small. The latter root makes $\lambda \propto \omega$, ultimately.

207. Proceeding to the hydrodynamical examples, we begin with the case of a plane horizontal sheet of water having in the undisturbed state a motion of uniform rotation about a vertical axis†. The results will apply without serious qualification to the case of a polar or other basin, of not too great dimensions, on a rotating globe.

* The preceding theory appeared in the 2nd ed. (1895) of this work. The effect of friction is considered in Art. 322.

† Sir W. Thomson, "On Gravitational Oscillations of Rotating Water," *Proc. R. S. Edin.* x. 92 (1879) [*Papers*, iv. 141].

Let the axis of rotation be taken as axis of z. The axes of x and y being now supposed to rotate in their own plane with the prescribed angular velocity ω, let us denote by u, v, w the velocities at time t, *relative to these axes*, of the particle which then occupies the position (x, y, z). The actual velocities of the same particle, parallel to the instantaneous positions of the axes, will be $u-\omega y$, $v+\omega x$, w, and the accelerations in the same directions will be

$$\frac{Du}{Dt}-2\omega v-\omega^2 x, \qquad \frac{Dv}{Dt}+2\omega u-\omega^2 y, \qquad \frac{Dw}{Dt}.$$

In the present application, the relative motion is assumed to be infinitely small, so that we may replace D/Dt by $\partial/\partial t$.

Now let z_0 be the ordinate of the free surface when there is relative equilibrium under gravity alone, so that

$$z_0=\tfrac{1}{2}\frac{\omega^2}{g}(x^2+y^2)+\text{const.}, \qquad \text{.......................(1)}$$

as in Art. 26. For simplicity we will suppose that the slope of this surface is everywhere very small; in other words, if r be the greatest distance of any part of the sheet from the axis of rotation, $\omega^2 r/g$ is assumed to be small.

If $z_0+\zeta$ denote the ordinate of the free surface when disturbed, then on the usual assumption that the vertical acceleration of the water is small compared with g, the pressure at any point (x, y, z) will be given by

$$p-p_0=g\rho(z_0+\zeta-z), \qquad \text{..........................(2)}$$

whence
$$-\frac{1}{\rho}\frac{\partial p}{\partial x}=-\omega^2 x-g\frac{\partial\zeta}{\partial x}, \qquad -\frac{1}{\rho}\frac{\partial p}{\partial y}=-\omega^2 y-g\frac{\partial\zeta}{\partial y}.$$

The equations of horizontal motion are therefore

$$\frac{\partial u}{\partial t}-2\omega v=-g\frac{\partial\zeta}{\partial x}-\frac{\partial\Omega}{\partial x}, \qquad \frac{\partial v}{\partial t}+2\omega u=-g\frac{\partial\zeta}{\partial y}-\frac{\partial\Omega}{\partial y}, \qquad \text{......(3)}$$

where Ω denotes the potential of the disturbing forces.

If we write
$$\bar{\zeta}=-\Omega/g, \qquad \text{.................................(4)}$$

i.e. $\bar{\zeta}$ is the 'equilibrium' value of the surface elevation, these become

$$\frac{\partial u}{\partial t}-2\omega v=-g\frac{\partial}{\partial x}(\zeta-\zeta), \qquad \frac{\partial v}{\partial t}+2\omega u=-g\frac{\partial}{\partial y}(\zeta-\bar{\zeta}). \qquad \text{......(5)}$$

The equation of continuity has the same form as in Art. 193, viz.

$$\frac{\partial\zeta}{\partial t}=-\frac{\partial(hu)}{\partial x}-\frac{\partial(hv)}{\partial y}, \qquad \text{..........................(6)}$$

where h denotes the depth, from the free surface to the bottom, in the undisturbed condition. This depth will not, of course, be uniform unless the bottom follows the curvature of the free surface as given by (1).

If we eliminate $\zeta - \bar{\zeta}$ from the equations (5), by cross-differentiation, we find

$$\frac{\partial}{\partial t}\left(\frac{\partial v}{\partial x} - \frac{\partial u}{\partial y}\right) + 2\omega\left(\frac{\partial u}{\partial x} + \frac{\partial v}{\partial y}\right) = 0, \qquad \text{.......(7)}$$

or, writing $u = \partial\xi/\partial t$, $v = \partial\eta/\partial t$, and integrating with respect to t,

$$\frac{\partial v}{\partial x} - \frac{\partial u}{\partial y} + 2\omega\left(\frac{\partial \xi}{\partial x} + \frac{\partial \eta}{\partial y}\right) = \text{const.} \qquad \text{.......(8)}$$

This is merely the expression of Helmholtz' theorem that the product of the vorticity

$$2\omega + \frac{\partial v}{\partial x} - \frac{\partial u}{\partial y} \text{ and the cross-section } \left(1 + \frac{\partial \xi}{\partial x} + \frac{\partial \eta}{\partial y}\right)\delta x\, \delta y,$$

of a vortex-filament, is constant.

In the case of a simple-harmonic disturbance, the time-factor being $e^{i\sigma t}$ the equations (5) and (6) become

$$i\sigma u - 2\omega v = -g\frac{\partial}{\partial x}(\zeta - \bar{\zeta}), \qquad i\sigma v + 2\omega u = -g\frac{\partial}{\partial y}(\zeta - \bar{\zeta}), \quad \text{......(9)}$$

and

$$i\sigma\zeta = -\frac{\partial(hu)}{\partial x} - \frac{\partial(hv)}{\partial y}. \qquad \text{.......(10)}$$

From (9) we find

$$u = \frac{g}{\sigma^2 - 4\omega^2}\left(i\sigma\frac{\partial}{\partial x} + 2\omega\frac{\partial}{\partial y}\right)(\zeta - \bar{\zeta}), \qquad v = \frac{g}{\sigma^2 - 4\omega^2}\left(i\sigma\frac{\partial}{\partial y} - 2\omega\frac{\partial}{\partial x}\right)(\zeta - \bar{\zeta}), \quad \text{.........(11)}$$

and if we substitute from these in (10), we obtain an equation in ζ only.

In the case of *uniform* depth the result takes the form

$$\nabla_1^2\zeta + \frac{\sigma^2 - 4\omega^2}{gh}\zeta = \nabla_1^2\bar{\zeta}, \qquad \text{.......(12)}$$

where $\nabla_1^2 = \partial^2/\partial x^2 + \partial^2/\partial y^2$, as before.

When $\bar{\zeta} = 0$, the equations (5) and (6) can be satisfied by *constant* values of u, v, ζ provided certain conditions are fulfilled. We must have

$$u = -\frac{g}{2\omega}\frac{\partial\zeta}{\partial y}, \qquad v = \frac{g}{2\omega}\frac{\partial\zeta}{\partial x}, \qquad \text{.......(13)}$$

and therefore

$$\frac{\partial(h, \zeta)}{\partial(x, y)} = 0. \qquad \text{.......(14)}$$

The latter condition shews that the contour-lines of the free surface must be everywhere parallel to the contour-lines of the bottom, but that the value of ζ is otherwise arbitrary. The flow of the fluid is everywhere parallel to the contour-lines, and it is therefore further necessary for the possibility of such steady motions that the depth should be uniform along the boundary (supposed to be a vertical wall). When the depth is everywhere the same, the condition (14) is satisfied identically, and the only limitation on the value of ζ is that it should be constant along the boundary.

208. A simple application of the preceding equations is to the case of free waves in an infinitely long uniform straight canal*.

If we assume

$$\zeta = ae^{ik(ct-x)+my}, \qquad v = 0, \qquad \text{.......(1)}$$

the axis of x being parallel to the length of the canal, the equations (5) of the preceding Art., with the terms in $\bar{\zeta}$ omitted, give

$$cu = g\zeta, \qquad 2\omega u = -gm\zeta, \qquad \text{.......(2)}$$

* Sir W. Thomson, *l.c. ante* p. 317.

whilst, from the equation of continuity (Art. 207 (6)),

$$c\zeta = hu. \qquad\qquad (3)$$

We thence derive $c^2 = gh, \qquad m = -2\omega/c.$ (4)

The former of these results shews that the wave-velocity is unaffected by the rotation.

When expressed in real form, the value of ζ is

$$\zeta = ae^{-2\omega y/c}\cos\{k(ct-x)+\epsilon\}. \qquad\qquad (5)$$

The exponential factor indicates that the wave-height increases as we pass from one side of the canal to the other, being least on the side which is *forward* in respect of the rotation. If we take account of the directions of motion of a water-particle, at a crest and at a trough, respectively, this result is seen to be in accordance with the tendency pointed out in Art. 202*.

It will be observed that there is, in the above solution, no limitation to the breadth of the canal, provided it be uniform.

The problem of determining the free oscillations in a rotating canal of *finite* length, or even the simpler one of reflection of a wave at a transverse barrier, does not however admit of a simple solution by superposition, as was the case in the investigations of Arts. 176, 178. For a wave travelling in the negative direction, we should find

$$\zeta = a'e^{2\omega y/c}\cos\{k(ct+x)+\epsilon'\}, \qquad\qquad (6)$$

but this cannot be combined with (5) so as to make $u=0$ at a barrier for all values of y†.

209. We take next the case of a circular sheet of water rotating about its centre‡.

If we introduce polar co-ordinates r, θ, and employ the symbols ξ, η to denote displacements along and perpendicular to the radius vector, then since $\dot{\xi} = i\sigma\xi$, $\dot{\eta} = i\sigma\eta$, the equations (9) of Art. 207 are equivalent to

$$\sigma^2\xi + 2i\omega\sigma\eta = g\frac{\partial}{\partial r}(\zeta-\bar{\zeta}), \qquad \sigma^2\eta - 2i\omega\sigma\xi = g\frac{\partial}{r\partial\theta}(\zeta-\bar{\zeta}), \qquad (1)$$

* For applications to tidal phenomena see Sir W. Thomson, *Nature*, xix. 154, 571 (1879), and G. I. Taylor, "Tidal Friction in the Irish Sea," *Phil. Trans.* A, ccxx. 1 (1918).

† Poincaré, *Leçons de Méc. Cél.* iii. 124. The problem here indicated has been solved by G. I. Taylor, *Proc. Lond. Math. Soc.* (2) xx. 148 (1920). He finds that, provided the wave-length $(2\pi/k)$ be sufficiently large compared with the breadth (b), there is regular reflection (with a change of phase), in the sense that at a distance from the barrier we have practically superposition of (5) and (6) above, with $a'=a$, the necessary condition being

$$k^2b^2 < \pi^2 + 4\omega^2 b^2/c^2.$$

The theory of the free oscillations in a rotating rectangular basin is also discussed in the paper cited. The case where the angular velocity of rotation is relatively small had been previously treated by Rayleigh, *Phil. Mag.* (6), v. 297 (1903) [*Papers*, v. 93], and *Proc. Roy. Soc.* A, lxxxii. 448 (1909) [*Papers*, v, 497].

‡ The investigation which follows is a development of some indications given by Kelvin in the paper cited on p. 317.

whilst the equation of continuity (10) becomes

$$\zeta = -\frac{\partial (h\xi r)}{r\partial r} - \frac{\partial (h\eta)}{r\partial \theta}. \quad \text{.........................(2)}$$

Hence

$$\xi = \frac{g}{\sigma^2 - 4\omega^2}\left(\frac{\partial}{\partial r} - \frac{2i\omega}{\sigma}\frac{\partial}{r\partial\theta}\right)(\zeta - \bar{\zeta}), \qquad \eta = \frac{ig}{\sigma^2 - 4\omega^2}\left(\frac{2\omega}{\sigma}\frac{\partial}{\partial r} - i\frac{\partial}{r\partial\theta}\right)(\zeta - \bar{\zeta}) \quad \text{........(3)}$$

and substituting in (2) we get the differential equation in ζ.

In the case of *uniform* depth we find

$$(\nabla_1^2 + \kappa^2)\,\zeta = \nabla_1^2 \bar{\zeta}, \quad \text{.........................(4)}$$

where

$$\nabla_1^2 = \frac{\partial^2}{\partial r^2} + \frac{1}{r}\frac{\partial}{\partial r} + \frac{1}{r^2}\frac{\partial^2}{\partial \theta^2}, \quad \text{.........................(5)}$$

and

$$\kappa^2 = \frac{\sigma^2 - 4\omega^2}{gh}. \quad \text{.........................(6)}$$

This might have been written down at once from Art. 207 (12).

The condition to be satisfied at the boundary ($r = a$, say) is $\xi = 0$, or

$$\left(r\frac{\partial}{\partial r} - \frac{2i\omega}{\sigma}\frac{\partial}{\partial\theta}\right)(\zeta - \bar{\zeta}) = 0. \quad \text{.........................(7)}$$

210. In the case of the *free* oscillations we have $\bar{\zeta} = 0$. The way in which the imaginary i enters into the above equations, taken in conjunction with Fourier's Theorem, suggests that θ occurs in the form of a factor $e^{is\theta}$, where s is integral. On this supposition, the differential equation (4) becomes

$$\frac{\partial^2 \zeta}{\partial r^2} + \frac{1}{r}\frac{\partial \zeta}{\partial r} + \left(\kappa^2 - \frac{s^2}{r^2}\right)\zeta = 0, \quad \text{.........................(8)}$$

and the boundary-condition (7) gives

$$r\frac{\partial \zeta}{\partial r} + \frac{2s\omega}{\sigma}\zeta = 0, \quad \text{.........................(9)}$$

for $r = a$.

The equation (8) is of Bessel's form, and the solution which is finite for $r = 0$ may therefore be written

$$\zeta = A J_s(\kappa r)\, e^{i(\sigma t + s\theta)}; \quad \text{.........................(10)}$$

but it is to be noticed that κ^2 is not, in the present problem, necessarily positive. When κ^2 is negative, we may replace $J_s(\kappa r)$ by $I_s(\kappa_1 r)$, where κ_1 is the positive square root of $(4\omega^2 - \sigma^2)/gh$, and

$$I_s(z) = \frac{z^s}{2^s . s!}\left\{1 + \frac{z^2}{2(2s+2)} + \frac{z^4}{2.4(2s+2)(2s+4)} + \dots\right\}^*. \quad \text{...(11)}$$

In the case of symmetry about the axis ($s = 0$), we have, in real form,

$$\zeta = A J_0(\kappa r) . \cos(\sigma t + \epsilon), \quad \text{.........................(12)}$$

* The functions $I_s(z)$ were tabulated by Prof. A. Lodge, *Brit. Ass. Rep.* 1889. The tables are reprinted by Dale, and by Jahnke and Emde. Extensive tables of the functions $e^{-z} I_0(z)$, $e^{-z} I_1(z)$ are given in Watson's treatise.

where κ is determined by

$$J_0'(\kappa a) = 0. \qquad (13)$$

The corresponding values of σ are then given by (6). The free surface has, in the various modes, the same forms as in Art. 191, but the frequencies are now greater. If we write

$$c^2 = gh, \qquad \beta = 4\omega^2 a^2/c^2, \qquad (14)$$

we have

$$\sigma^2 a^2/c^2 = \kappa^2 a^2 + \beta. \qquad (15)$$

It is easily seen, moreover, on reference to (3), that the relative motions of the fluid particles are no longer purely radial; the particles describe, in fact, ellipses whose major axes are in the direction of the radius vector.

For $s > 0$ we have

$$\zeta = A J_s(\kappa r) \,.\, \cos(\sigma t + s\theta + \epsilon), \qquad (16)$$

where the admissible values of κ, and thence of σ, are determined by (9), which gives

$$\kappa a J_s'(\kappa a) + \frac{2s\omega}{\sigma} J_s(\kappa a) = 0. \qquad (17)$$

The formula (16) represents a wave rotating relatively to the water with an angular velocity σ/s, the rotation of the wave being in the same direction with that of the water, or the opposite, according as σ/ω is negative or positive.

If κa is any real or pure imaginary root of (17), the corresponding value of σ is given by (15).

Some indications as to the values of σ may be gathered from a graphical construction. If we write $\kappa^2 a^2 = x$, we have, from (6),

$$\frac{\sigma}{2\omega} = \pm\left(1 + \frac{x}{\beta}\right)^{\frac{1}{2}}. \qquad (18)$$

If we further put

$$\frac{sJ_s(\kappa a)}{\kappa a J_s'(\kappa a)} = \phi(\kappa^2 a^2),$$

the equation (17) may be written

$$\phi(x) \pm \left(1 + \frac{x}{\beta}\right)^{\frac{1}{2}} = 0. \qquad (19)$$

The curve

$$y = -\phi(x) \qquad (20)$$

can be readily traced by means of the tables of the functions $J_s(z)$, $I_s(z)$; and its intersections with the parabola

$$y^2 = 1 + x/\beta \qquad (21)$$

will give, by their ordinates, the values of $\sigma/2\omega$. The constant β, on which the positions of the roots depend, is equal to the square of the ratio $2\omega a/(gh)^{\frac{1}{2}}$ which the period of a wave travelling round a circular canal of depth h and perimeter $2\pi a$ bears to the half-period (π/ω) of the rotation of the water.

The diagrams on the next page indicate the relative magnitudes of the lower roots, in the cases $s=1$ and $s=2$, when β has the values 2, 6, 40, respectively*.

* For clearness the scale of y has been taken to be 10 times that of x.

With the help of these figures we can trace, in a general way, the changes in the character of the free modes as β increases from zero. The results may be interpreted as due either to a continuous increase of ω, or to a continuous diminution of h. We will use

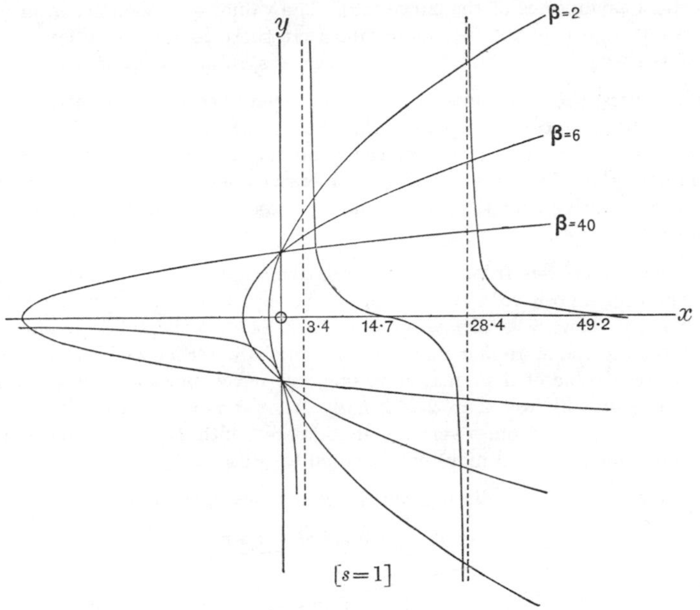

[$s=1$]

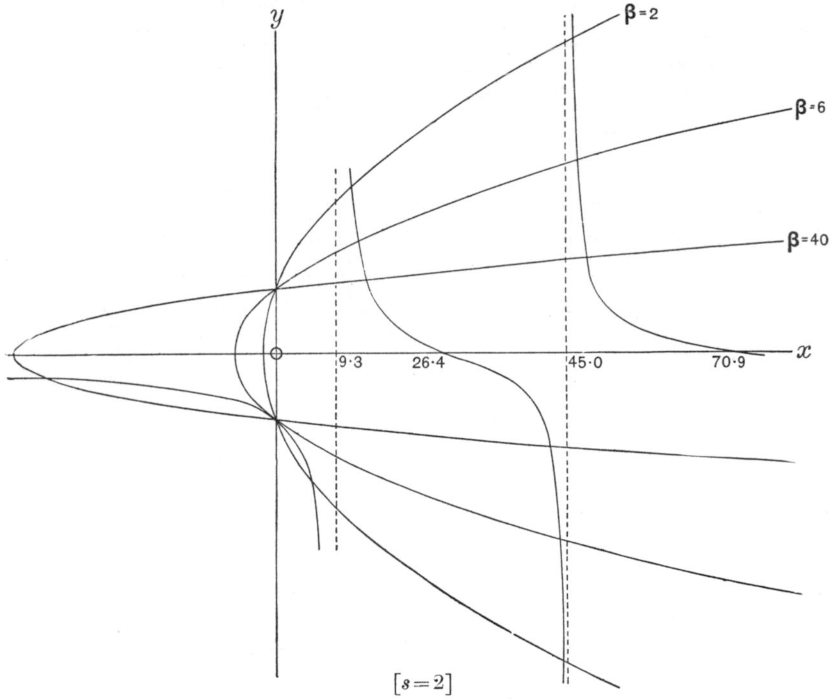

[$s=2$]

the terms 'positive' and 'negative' to distinguish waves which travel, relatively to the water, in the same direction as the rotation and the opposite.

When β is infinitely small, the values of x are given by $J_s'(x^{\frac{1}{2}})=0$; these correspond to the vertical asymptotes of the curve (20). The values of σ then occur in pairs of equal and oppositely-signed quantities, indicating that there is now no difference between the velocity of positive and negative waves. The case is, in fact, that of Art. 191 (12).

As β increases, the two values of σ forming a pair become unequal in magnitude, and the corresponding values of x separate, that being the greater for which $\sigma/2\omega$ is positive. When $\beta=s(s+1)$ the curve (20) and the parabola (21) *touch* at the point $(0, -1)$, the corresponding value of σ being -2ω. As β increases beyond this critical value, one value of x becomes negative, and the corresponding (negative) value of $\sigma/2\omega$ becomes smaller and smaller.

Hence, as β increases from zero, the relative angular velocity becomes greater for a negative than for a positive wave of (approximately) the same type; moreover the value of σ for a negative wave is always greater than 2ω. As the rotation increases, the two kinds of wave become more and more distinct in character as well as in 'speed.' With a sufficiently great value of β we may have one, but never more than one, positive wave for which σ is numerically less than 2ω. Finally, when β is very great, the value of σ corresponding to this wave becomes very small compared with 2ω, whilst the remaining values tend all to become more and more nearly equal to $\pm 2\omega$.

If we use a zero suffix to distinguish the case of $\omega=0$, we find

$$\frac{\sigma^2}{\sigma_0^{\,2}} = \frac{\kappa^2+4\omega^2/gh}{\kappa_0^{\,2}} = \frac{x+\beta}{x_0}, \qquad \text{.......................(22)}$$

where x_0 refers to the proper asymptote of the curve (20). This gives the 'speed' of any free mode in terms of that of the corresponding mode when there is no rotation.

The preceding statements are illustrated by the following table, which gives for the case of $s=1$ approximate values of κa within the range of the upper diagram on p. 323, together with the corresponding values of $\sigma/2\omega$ and $\sigma a/c$.

$\beta=0$	$\beta=2$			$\beta=6$			$\beta=40$			$\beta=\infty$		
$\kappa a=\sigma a/c$	κa	$\sigma/2\omega$	$\sigma a/c$	κa	$\sigma/2\omega$	$\sigma a/c$	κa	$\sigma/2\omega$	$\sigma a/c$	κa	$\sigma/2\omega$	$\sigma a/c$
$\pm 1·84$	2·19	+1·84	+2·61	2·29	+1·37	+3·35	2·38	+1·07	+6·76	2·40	+1·00	$+\beta^{\frac{1}{2}}$
	0	−1·00	−1·41	2·10i	−0·51	−1·26	6·23i	−0·17	−1·09	$i\beta^{\frac{1}{2}}$	$-\beta^{-\frac{1}{2}}$	−1·00
$\pm 5·33$	5·38	+3·93	+5·56	5·41	+2·42	+5·94	5·47	+1·32	+8·36	5·52	+1·00	$+\beta^{\frac{1}{2}}$
	5·28	−3·86	−5·47	5·25	−2·37	−5·79	5·18	−1·29	−8·17	5·14	−1·00	$-\beta^{\frac{1}{2}}$

211. As a sufficient example of *forced* oscillations we may assume

$$\bar{\zeta} = C\left(\frac{r}{a}\right)^s e^{i(\sigma t+s\theta+\epsilon)}, \qquad \text{.......................(23)}$$

where the value of σ is now prescribed.

This makes $\nabla_1^2\bar{\zeta}=0$, and the equation (4) then gives

$$\zeta = AJ_s(\kappa r)\, e^{i(\sigma t+s\theta+\epsilon)}, \qquad \text{.......................(24)}$$

where A is to be determined by the boundary-condition (7), viz.

$$A = \frac{s\left(1 + \frac{2\omega}{\sigma}\right)}{\kappa a J_s'(\kappa a) + \frac{2s\omega}{\sigma} J_s(\kappa a)} \cdot C. \quad \text{.................(25)}$$

This becomes very great when the frequency of the disturbance is nearly coincident with that of a free mode of corresponding type*.

From the point of view of tidal theory the most interesting cases are those of $s=1$ with $\sigma=\omega$, and $s=2$ with $\sigma=2\omega$, respectively. These would represent the diurnal and semidiurnal tides due to a distant disturbing body whose proper motion may be neglected in comparison with the rotation ω.

In the case of $s=1$ we have a *uniform* horizontal disturbing force. Putting, in addition, $\sigma=\omega$, we find without difficulty that the amplitude of the tide-elevation at the edge ($r=a$) of the basin has to its 'equilibrium-value' the ratio

$$\frac{3I_1(z)}{I_1(z)+zI_0(z)}, \quad \text{...(26)}$$

where $z=\frac{1}{2}\sqrt{(3\beta)}$. With the help of Lodge's tables we find that this ratio has the values

		1·000,	·638,	·396,	
for	$\beta=$	0,	12,	48,	respectively.

When $\sigma=2\omega$, we have $\kappa=0$, and thence, by (23), (24), (25),

$$\zeta=\bar{\zeta}, \quad \text{...(27)}$$

i.e. the tidal elevation has exactly the equilibrium-value.

This remarkable result can be obtained in a more general manner; it holds whenever the disturbing force is of the type

$$\bar{\zeta}=\chi(r)\, e^{i(2\omega t+s\theta+\epsilon)} \quad \text{.......................................(28)}$$

provided the depth h be a function of r only. If we revert to the equations (1), we notice that when $\sigma=2\omega$ they are satisfied by $\zeta=\bar{\zeta}$, $\eta=i\xi$. To determine ξ as a function of r, we substitute in the equation of continuity (2), which gives

$$\frac{\partial(h\xi)}{\partial r} - \frac{s-1}{r} h\xi = -\chi(r). \quad \text{..............................(29)}$$

The arbitrary constant which appears on integration of this equation is to be determined by the boundary-condition.

In the present case we have $\chi(r)=Cr^s/a^s$. Integrating, and making $\xi=0$ for $r=a$, we find

$$h\xi=\frac{Cr^{s-1}}{2a^s}(a^2-r^2)\, e^{i(2\omega t+s\theta+\epsilon)}. \quad \text{..............................(30)}$$

The relation $\eta=i\xi$ shews that the amplitudes of ξ and η are equal, while their phases differ by 90°; the relative orbits of the fluid particles are in fact circles of radii

$$\mathbf{r}=\frac{Cr^{s-1}}{2ha^s}(a^2-r^2), \quad \text{.......................................(31)}$$

described each about its centre with angular velocity 2ω in the negative direction. We may easily deduce that the path of any particle *in space* is an ellipse of semi-axes $r \pm \mathbf{r}$ described about the origin with harmonic motion in the positive direction, the period being $2\pi/\omega$. This accounts for the peculiar features of the case. For if ζ have always the

* The case of a *nearly* circular sheet is treated by Proudman, "On some Cases of Tidal Motion on Rotating Sheets of Water," *Proc. Lond. Math. Soc.* (2) xii. 453 (1913).

equilibrium-value, the horizontal forces due to the elevation exactly balance the disturbing force, and there remain only the forces due to the undisturbed form of the free surface (Art. 207 (1)). These give an acceleration $g\,dz_0/dr$, or $\omega^2 r$, to the centre, where r is the radius vector of the particle in its actual position. Hence all the conditions of the problem are satisfied by elliptic-harmonic motion of the individual particles provided the positions, the dimensions, and the 'epochs' of the orbits can be adjusted so as to satisfy the condition of continuity, with the assumed value of ζ. The investigation just given resolves this point.

When the sheet of water is bounded also by radial walls the problem is more difficult. The tidal oscillations (free and forced) in a *semicircular* basin of uniform depth are discussed by Proudman*, with an application to the tides of the Black Sea, the disturbing forces being of the idealized diurnal and semi-diurnal types.

The free and forced oscillations in a rotating *elliptic* basin of uniform depth are discussed by Goldstein †.

212‡. We may also notice the case of a circular basin of variable depth, the law of depth being the same as in Art. 193, viz.

$$h = h_0\left(1 - \frac{r^2}{a^2}\right). \quad\text{.............................(1)}$$

Assuming that ξ, η, ζ all vary as $e^{i(\sigma t + s\theta + \epsilon)}$, and that h is a function of r only, we find, from Art. 209 (2), (3),

$$(\sigma^2 - 4\omega^2)\,\zeta + g\frac{dh}{dr}\left(\frac{\partial}{\partial r} + \frac{2\omega s}{\sigma r}\right)(\zeta - \bar{\zeta}) + gh\left(\frac{\partial^2}{\partial r^2} + \frac{1}{r}\frac{\partial}{\partial r} - \frac{s^2}{r^2}\right)(\zeta - \bar{\zeta}) = 0. \quad\text{.........(2)}$$

Introducing the value of h from (1), we have, for the *free* oscillations,

$$\left(1 - \frac{r^2}{a^2}\right)\left(\frac{\partial^2 \zeta}{\partial r^2} + \frac{1}{r}\frac{\partial \zeta}{\partial r} - \frac{s^2}{r^2}\zeta\right) - \frac{2}{a^2}\left(r\frac{\partial \zeta}{\partial r} + \frac{2\omega s}{\sigma}\zeta\right) + \frac{\sigma^2 - 4\omega^2}{gh_0}\zeta = 0. \quad\text{............(3)}$$

This is identical with Art. 193 (6), except that we now have

$$\frac{\sigma^2 - 4\omega^2}{gh_0} - \frac{4\omega s}{\sigma a^2}$$

in place of σ^2/gh_0. The solution can therefore be written down from the results of that Art., viz. if we put

$$\frac{(\sigma^2 - 4\omega^2)\,a^2}{gh_0} - \frac{4\omega s}{\sigma} = n(n-2) - s^2, \quad\text{..............................(4)}$$

we have

$$\zeta = A_s\left(\frac{r}{a}\right)^s F\left(\alpha, \beta, \gamma, \frac{r^2}{a^2}\right) e^{i(\sigma t + s\theta + \epsilon)}, \quad\text{..........................(5)}$$

where

$$\alpha = \tfrac{1}{2}n + \tfrac{1}{2}s, \quad \beta = 1 + \tfrac{1}{2}s - \tfrac{1}{2}n, \quad \gamma = s + 1;$$

and the condition of convergence at the boundary $r = a$ requires that

$$n = s + 2j, \quad\text{...(6)}$$

where j is some positive integer. The values of σ are then given by (4).

The forms of the free surface are therefore the same as in the case of no rotation, but the motion of the water particles is different. The relative orbits are in fact now ellipses having their principal axes along and perpendicular to the radius vector; this follows easily from Art. 209 (3).

* *M. N. R. A. S., Geophys. Suppt.* ii. 32 (1928).

† *Ibid.* ii. 213 (1929).

‡ See the footnote to Art. 193.

In the symmetrical modes ($s=0$), the equation (4) gives

$$\sigma^2=\sigma_0{}^2+4\omega^2, \qquad\qquad (7)$$

where σ_0 denotes the 'speed' of the corresponding mode in the case of no rotation, as found in Art. 193.

For any value of s other than zero, the most important modes are those for which $n=s+2$. The equation (4) is then divisible by $\sigma+2\omega$, but this is an extraneous factor; discarding it, we have the quadratic

$$\sigma^2-2\omega\sigma=2s\frac{gh_0}{a^2}, \qquad\qquad (8)$$

whence

$$\sigma=\omega\pm\left(\omega^2+2s\frac{gh_0}{a^2}\right)^{\frac{1}{2}}. \qquad\qquad (9)$$

This gives two waves rotating round the origin, the relative wave-velocity being greater for the negative than for the positive wave, as in the case of uniform depth (Art. 210). With the help of (8) the formulae reduce to

$$\zeta=A_s\left(\frac{r}{a}\right)^s, \quad \xi=\tfrac{1}{2}\frac{a}{h_0}A_s\left(\frac{r}{a}\right)^{s-1}, \quad \eta=\tfrac{1}{2}i\frac{a}{h_0}A_s\left(\frac{r}{a}\right)^{s-1}, \qquad\qquad (10)$$

the factor $e^{i(\sigma t+s\theta+\epsilon)}$ being understood in each case. Since $\eta=i\xi$, the relative orbits are all circles.

The case $s=1$, $n=3$, is noteworthy; the free surface is then always *plane*, and the circular orbits have all the same radius. In the following table, which relates to this case, β stands for $4\omega^2a^2/c_0{}^2$, where $c_0=\surd(gh_0)$.

$\beta=0$	$\beta=2$		$\beta=6$		$\beta=40$	
$\sigma a/c_0$	$\sigma/2\omega$	$\sigma a/c_0$	$\sigma/2\omega$	$\sigma a/c_0$	$\sigma/2\omega$	$\sigma a/c_0$
$\pm1{\cdot}414$	$+1{\cdot}618$ $-0{\cdot}618$	$+2{\cdot}288$ $-0{\cdot}874$	$+1{\cdot}264$ $-0{\cdot}264$	$+3{\cdot}096$ $-0{\cdot}646$	$+1{\cdot}048$ $-0{\cdot}048$	$+6{\cdot}626$ $-0{\cdot}302$

When $n>s+2$, we have nodal circles. The equation (4) is then a *cubic* in $\sigma/2\omega$; it is easily seen that its roots are all real, lying between $-\infty$ and -1, -1 and 0, and $+1$ and $+\infty$, respectively. The following table is calculated for the case of $s=1$, $n=5$.

$\beta=0$	$\beta=2$		$\beta=6$		$\beta=40$	
$\sigma a/c_0$	$\sigma/2\omega$	$\sigma a/c_0$	$\sigma/2\omega$	$\sigma a/c_0$	$\sigma/2\omega$	$\sigma a/c_0$
$\pm3{\cdot}742$	$+2{\cdot}889$ $-0{\cdot}125$ $-2{\cdot}764$	$+4{\cdot}085$ $-0{\cdot}176$ $-3{\cdot}909$	$+1{\cdot}874$ $-0{\cdot}100$ $-1{\cdot}774$	$+4{\cdot}590$ $-0{\cdot}245$ $-4{\cdot}344$	$+1{\cdot}183$ $-0{\cdot}040$ $-1{\cdot}143$	$+7{\cdot}483$ $-0{\cdot}253$ $-7{\cdot}230$

The first and the last root of each triad give positive and negative waves of a somewhat similar character to those already obtained in the case of uniform depth. The smaller negative root gives a comparative slow oscillation which, when the angular velocity ω is infinitely small, becomes a *steady* rotational motion, without elevation or depression of the surface. The possibility of oscillations of this type was pointed out in Art. 206, *ad fin.* In the present case the transition is easily traced. It follows from (4) that the relevant limiting value of $\sigma/2\omega$, when ω is infinitesimal, is $-\frac{1}{7}$. We then find, from Art. 209 (2), (3),

$$\dot{\xi}=C\left(1-\frac{r^2}{a^2}\right)e^{i(\theta+\sigma t)},\quad \dot{\eta}=iC\left(1-5\frac{r^2}{a^2}\right)e^{i(\theta+\sigma t)}, \quad\text{.................(11)}$$

with

$$\zeta=-\frac{8i\sigma C}{g}r\left(1-\tfrac{3}{2}\frac{r^2}{a^2}\right)e^{i(\theta+\sigma t)}, \quad\text{.............................(12)}$$

ultimately, where

$$\sigma=-\tfrac{2}{7}\omega.$$

The most important type of *forced* oscillations is such that

$$\bar{\zeta}=C\left(\frac{r}{a}\right)^s e^{i(\sigma t+s\theta+\epsilon)}. \quad\text{...................................(13)}$$

We readily verify, on substitution in (3), that

$$\zeta=\frac{2sgh_0}{2sgh_0-(\sigma^2-2\omega\sigma)a^2}\bar{\zeta}. \quad\text{................................(14)}$$

We notice that when $\sigma=2\omega$ the tide-height has exactly the equilibrium-value, in agreement with Art. 211.

If σ_1, σ_2 denote the two roots of (8), the last formula may be written

$$\zeta=\frac{\bar{\zeta}}{(1-\sigma/\sigma_1)(1-\sigma/\sigma_2)}. \quad\text{...................................(15)}$$

The tidal oscillations in a semicircular basin with the above law of depth have been examined by Goldsbrough*. The difficulty of the problem consists in satisfying the conditions at the straight portion of the boundary.

212 a. Place may be found here for one or two illustrations of the approximate procedure outlined in Art. 205 a.

1°. To take first a known problem, that of the circular basin of uniform depth (Art. 210). Assuming as the polar co-ordinates of a displaced particle, relative to an initial line revolving with the angular velocity ω,

$$r'=r+\xi,\qquad \theta'=\theta+\eta/r, \quad\text{...................................(1)}$$

the equation of continuity is

$$\frac{\zeta}{h}=-\frac{\partial\xi}{\partial r}-\frac{\xi}{r}-\frac{\partial\eta}{r\partial\theta}; \quad\text{..(2)}$$

as in Art. 209 (2).

With our previous notation

$$\left.\begin{aligned}\mathfrak{T}=\tfrac{1}{2}\rho h\int_0^a\int_0^{2\pi}(\dot{\xi}^2+\dot{\eta}^2)\,r\,d\theta\,dr,\quad V-T_0=\tfrac{1}{2}g\rho\int_0^a\int_0^{2\pi}\zeta^2 r\,d\theta\,dr,\\ M'=\rho h\int_0^a\int_0^{2\pi}(\xi\dot{\eta}-\eta\dot{\xi})\,r\,d\theta\,dr.\end{aligned}\right\} \quad\text{...........(3)}$$

We take as our assumed type, for the gravest mode,

$$\xi=A\left(1-\frac{r^2}{a^2}\right)\cos(\sigma t+\theta),\quad \eta=\left(-A+B\frac{r^2}{a^2}\right)\sin(\sigma t+\theta), \quad\text{...........(4)}$$

which make

$$\frac{\zeta}{h}=(3A-B)\frac{r}{a^2}\cos(\sigma t+\theta). \quad\text{.................................(5)}$$

The constants in (4) have been adjusted so that ζ shall be finite for $r=0$.

* *Proc. Roy. Soc.* cxxii. 228 (1929).

Hence with the definitions of Art. 205 a, taking the mean values of the functions in (3), and performing the integrations,

$$P=\tfrac{1}{12}\pi\rho h a^2(4A^2-3AB+B^2),\quad Q=-\tfrac{1}{6}\pi\rho\omega h a^2(3A^2-AB),\quad R=\tfrac{1}{8}\pi g\rho h^2(3A-B)^2. \quad\text{......(6)}$$

If we write for shortness

$$c=\surd(gh),\qquad \sigma a/c=x,\qquad 4\omega^2a^2/c^2=\beta, \quad\text{..........................(7)}$$

the equation

$$\sigma^2P+\sigma Q-R=0 \quad\text{..(8)}$$

becomes

$$(4x^2-3\surd\beta x-\tfrac{27}{2})A^2-(3x^2-\surd\beta x-9)AB+(x^2-\tfrac{3}{2})B^2=0. \quad\text{...............(9)}$$

The stationary values of x are then given by

$$x^2(7x^2-6\surd\beta x-\beta-24)=0. \quad\text{..............................(10)}$$

The zero roots may be disregarded as corresponding to a merely circulatory motion, without change of surface-level. To compare with the numerical results of Art. 210 (p. 324) we put $\beta=2$, 6, 40 in succession. The finite roots of (10) are

$$\left.\begin{matrix}-1{\cdot}43\\+2{\cdot}65\end{matrix}\right\},\qquad \left.\begin{matrix}-1{\cdot}27\\+3{\cdot}27\end{matrix}\right\},\qquad \left.\begin{matrix}-1{\cdot}35\\+6{\cdot}77\end{matrix}\right\},$$

in the respective cases. It is only in the third case that there is any serious deviation from the correct value. It will be seen that the approximate method is fairly successful over a considerable range of the parameter β.

2°. In the case of a *rectangular* basin of uniform depth, we take axes Ox, Oy coincident with two of the sides, whose lengths are (say) a, b, respectively. Denoting by ξ, η the component displacements of a particle, we have

$$\left.\begin{aligned}\mathfrak{T}=\tfrac{1}{2}\rho h\int_0^a\int_0^b(\dot\xi^2+\dot\eta^2)\,dx\,dy,\qquad V-T_0=\tfrac{1}{2}g\rho\int_0^a\int_0^b\zeta^2\,dx\,dy,\\ M'=\rho h\int_0^a\int_0^b(\xi\dot\eta-\eta\dot\xi)\,dx\,dy.\end{aligned}\right\} \quad\text{.........(11)}$$

Let us assume as an approximate type

$$\xi=A\sin\frac{\pi x}{a}\cos\sigma t,\qquad \eta=B\sin\frac{\pi y}{b}\sin\sigma t. \quad\text{........................(12)}$$

This is suggested by the case of $\omega=0$, where either A or B is zero, and cannot be expected to give a good result for more than a limited range of ω. From (12) we derive

$$\frac{\zeta}{h}=-\frac{\partial\xi}{\partial x}-\frac{\partial\eta}{\partial y}=-\pi\left(\frac{A}{a}\cos\frac{\pi x}{a}\cos\sigma t+\frac{B}{b}\cos\frac{\pi y}{b}\sin\sigma t\right). \quad\text{...............(13)}$$

Hence

$$P=\tfrac{1}{8}\rho hab(A^2+B^2),\qquad Q=\frac{4\rho\omega hab}{\pi^2}AB,\qquad R=\tfrac{1}{8}\pi^2g\rho h^2\left(\frac{b}{a}A^2+\frac{a}{b}B^2\right). \quad\text{......(14)}$$

The equation (8) now takes the form

$$\left(\sigma^2-\frac{\pi^2c^2}{a^2}\right)A^2+\frac{32\omega\sigma}{\pi^2}AB+\left(\sigma^2-\frac{\pi^2c^2}{b^2}\right)B^2=0, \quad\text{..................(15)}$$

where $c^2=gh$ as before. The stationary values of σ are therefore given by

$$(\sigma^2-\sigma_1^2)(\sigma^2-\sigma_2^2)=\frac{256\omega^2\sigma^2}{\pi^4}, \quad\text{.................................(16)}$$

where σ_1, σ_2 are the values of σ corresponding to oscillations parallel to x and y, respectively, when there is no rotation.

If ω is small and a, b decidedly unequal, then in the type where $\sigma = \sigma_1$, nearly, we have

$$\sigma - \sigma_1 = \frac{128\omega^2\sigma_1}{\pi^4(\sigma_1^2 - \sigma_2^2)}, \quad \ldots\ldots(17)$$

approximately. The corresponding ratio B/A is then given by

$$\frac{16\omega\sigma_1}{\pi^2} A + (\sigma_1^2 - \sigma_2^2) B = 0, \quad \ldots\ldots(18)$$

and is accordingly small, as was to be expected.

For a *square* tank ($a = b$), on the other hand, (16) makes

$$\sigma^2 - \sigma_1^2 = \pm \frac{16\omega\sigma}{\pi^2}, \quad \ldots\ldots(19)$$

or

$$\sigma - \sigma_1 = \pm \frac{8\omega}{\pi^2}, \quad \ldots\ldots(20)$$

approximately. Then $B/A = \pm 1$.

Tides on a Rotating Globe.

213. We proceed to give some account of Laplace's problem of the tidal oscillations of an ocean of (comparatively) small depth covering a rotating globe*. In order to bring out more clearly the nature of the approximations which are made on various grounds, we adopt a method of establishing the fundamental equations somewhat different from that usually followed.

When in relative equilibrium, the free surface is of course a level-surface with respect to gravity and centrifugal force; we shall assume it to be a surface of revolution about the polar axis, but the ellipticity will not in the first instance be taken to be small.

We adopt this equilibrium-form of the free surface as a surface of reference, and denote by θ and ϕ the co-latitude (*i.e.* the angle which the normal makes with the polar axis) and the longitude, respectively, of any point upon it. We shall further denote by z the altitude, measured outwards along a normal, of any point above this surface.

The relative position of any particle of the fluid being specified by the three orthogonal co-ordinates θ, ϕ, z, the kinetic energy of unit mass is given by

$$2T = (R+z)^2 \dot{\theta}^2 + \varpi^2 (\omega + \dot{\phi})^2 + \dot{z}^2, \quad \ldots\ldots(1)$$

where R is the radius of curvature of the meridian-section of the surface of reference, and ϖ is the distance of the particle from the polar axis. It is to be noticed that R is a function of θ only, while ϖ is a function of both θ and z; and it easily follows from geometrical considerations that

$$\frac{\partial \varpi}{(R+z)\partial\theta} = \cos\theta, \qquad \frac{\partial \varpi}{\partial z} = \sin\theta. \quad \ldots\ldots(2)$$

* "Recherches sur quelques points du système du monde," *Mém. de l'Acad. roy. des Sciences*, 1775 [1778] and 1776 [1779]; *Oeuvres Complètes*, ix. 88, 187. The investigation is reproduced, with various modifications, in the *Mécanique Céleste*, Livre 4me, c. i. (1799).

The component accelerations are obtained at once from (1) by Lagrange's formula. Omitting terms of the second order, on account of the restriction to infinitely small motions, we have

$$\left.\begin{aligned}\frac{1}{R+z}\left(\frac{d}{dt}\frac{\partial T}{\partial \dot\theta}-\frac{\partial T}{\partial \theta}\right)&=(R+z)\,\ddot\theta-\frac{1}{R+z}(\omega^2+2\omega\dot\phi)\,\omega\frac{\partial \varpi}{\partial \theta},\\ \frac{1}{\varpi}\left(\frac{d}{dt}\frac{\partial T}{\partial \dot\phi}-\frac{\partial T}{\partial \phi}\right)&=\varpi\ddot\phi+2\omega\left(\frac{\partial \varpi}{\partial \theta}\dot\theta+\frac{\partial \varpi}{\partial z}\dot z\right),\\ \frac{d}{dt}\frac{\partial T}{\partial \dot z}-\frac{\partial T}{\partial z}&=\ddot z-(\omega^2+2\omega\dot\phi)\,\varpi\frac{\partial \varpi}{\partial z}.\end{aligned}\right\}\quad\text{......(3)}$$

Hence, if we write u, v, w for the component relative velocities of a particle, viz.

$$u=(R+z)\,\dot\theta,\qquad v=\varpi\dot\phi,\qquad w=\dot z,\qquad\text{..................(4)}$$

and make use of (2), the hydrodynamical equations may be put in the forms

$$\left.\begin{aligned}\frac{\partial u}{\partial t}-2\omega v\cos\theta\qquad\qquad&=-\frac{1}{R+z}\frac{\partial}{\partial \theta}\left(\frac{p}{\rho}+\Psi-\tfrac{1}{2}\omega^2\varpi^2+\Omega\right),\\ \frac{\partial v}{\partial t}+2\omega u\cos\theta+2\omega w\sin\theta&=\quad-\frac{1}{\varpi}\frac{\partial}{\partial \phi}\left(\frac{p}{\rho}+\Psi-\tfrac{1}{2}\omega^2\varpi^2+\Omega\right),\\ \frac{\partial w}{\partial t}-2\omega v\sin\theta\qquad\qquad&=\quad-\frac{\partial}{\partial z}\left(\frac{p}{\rho}+\Psi-\tfrac{1}{2}\omega^2\varpi^2+\Omega\right),\end{aligned}\right\}\quad\text{...(5)}$$

where Ψ is the gravitation-potential due to the earth's attraction, whilst Ω denotes the potential of the disturbing forces.

So far the only approximation has consisted in the omission of terms of the second order in u, v, w. In the present application, the depth of the sea being small compared with the dimensions of the globe, we may replace $R+z$ by R. We will further assume that the vertical velocity w is small compared with the horizontal components u, v and that $\partial w/\partial t$ may be neglected in comparison with ωv. As in the theory of 'long' waves, such assumptions are justified *à posteriori* if the results obtained are found to be consistent with them (cf. Art. 172)*.

Let us integrate the third of equations (5) between the limits z and ζ, where ζ denotes the elevation of the disturbed surface above the surface of reference. At the surface of reference ($z=0$) we have

$$\Psi-\tfrac{1}{2}\omega^2\varpi^2=\text{const.},$$

by hypothesis, and therefore at the free surface ($z=\zeta$)

$$\Psi-\tfrac{1}{2}\omega^2\varpi^2=\text{const.}+g\zeta,$$

approximately, provided $g=\left[\dfrac{\partial}{\partial z}\left(\Psi-\tfrac{1}{2}\omega^2\varpi^2\right)\right]_{z=0}$. ....(6)

* Thus in the simplified conditions of Arts. 219, 220 $\dot w/\omega v$ is of the order $m\,(=\omega^2 a/g)$.

Here g denotes the value of *apparent* gravity at the surface of reference; it is of course, in general, a function of θ, but its variation with z is neglected.

The integration in question then gives

$$\frac{p}{\rho} + \Psi - \tfrac{1}{2}\omega^2\varpi^2 = \text{const.} + g\zeta + 2\omega \sin\theta \int_z^\zeta v\,dz, \quad \text{.........(7)}$$

where the variation of the disturbing potential Ω with z has been neglected in comparison with g. The last term is of the order of $\omega h v \sin\theta$, where h is the depth of the fluid, and it may be shewn that in the subsequent applications this is of the order h/a as compared with $g\zeta$*. Hence, substituting in the first two of equations (5), we obtain, with the approximations indicated,

$$\frac{\partial u}{\partial t} - 2\omega v \cos\theta = -\frac{\partial}{R\partial\theta} g(\zeta - \bar{\zeta}), \quad \frac{\partial v}{\partial t} + 2\omega u \cos\theta = -\frac{\partial}{\varpi\partial\phi} g(\zeta - \bar{\zeta}), \quad \text{...(8)}$$

where

$$\bar{\zeta} = -\Omega/g. \quad \text{.................................(9)}$$

These equations are independent of z, so that the horizontal motion may be assumed to be sensibly the same for all particles in the same vertical line.

As in Art. 198, this last result greatly simplifies the equation of continuity. In the present case we find without difficulty

$$\frac{\partial\zeta}{\partial t} = -\frac{1}{\varpi}\left\{\frac{\partial(h\varpi u)}{R\partial\theta} + \frac{\partial(hv)}{\partial\phi}\right\}. \quad \text{.....................(10)}$$

It is important to notice that the preceding equations involve no assumptions beyond those expressly laid down; in particular, there is no restriction as to the ellipticity of the meridian, which may be of any degree of oblateness.

214. In order, however, to simplify the question as far as possible, without sacrificing any of its essential features, we now take advantage of the circumstance that in the actual case of the earth the ellipticity is a small quantity, being in fact comparable with the ratio $(\omega^2 a/g)$ of centrifugal force to gravity at the equator, which ratio is known to be about $\frac{1}{289}$. Subject to an error of this order of magnitude, we may put $R = a$, $\varpi = a \sin\theta$, $g = \text{const.}$, where a is the earth's mean radius. We thus obtain

$$\frac{\partial u}{\partial t} - 2\omega v \cos\theta = -\frac{g}{a}\frac{\partial}{\partial\theta}(\zeta - \bar{\zeta}), \quad \frac{\partial v}{\partial t} + 2\omega u \cos\theta = -\frac{g}{a \sin\theta}\frac{\partial}{\partial\phi}(\zeta - \bar{\zeta}), \quad \text{......(1)}$$

with

$$\frac{\partial\zeta}{\partial t} = -\frac{1}{a\sin\theta}\left\{\frac{\partial(hu\sin\theta)}{\partial\theta} + \frac{\partial(hv)}{\partial\phi}\right\}, \quad \text{................(2)}$$

this last equation being identical with Art. 198 (1)†.

* This, again, may be verified in the same cases. The upshot is that the vertical acceleration is neglected, as in the theory of 'long' waves.

† Except for the notation these are the equations arrived at by Laplace, *l.c. ante* p. 330.

Some conclusions of interest follow at once from the mere *form* of the equations (1). In the first place, if $\mathbf{u}$, $\mathbf{v}$ denote the velocities along and perpendicular to *any* horizontal direction s, we easily find, by transformation of co-ordinates,

$$\frac{\partial \mathbf{u}}{\partial t} - 2\omega \mathbf{v} \cos\theta = -g\frac{\partial}{\partial s}(\zeta - \bar{\zeta}). \quad \text{......................(3)}$$

In the case of a narrow canal, the transverse velocity $\mathbf{v}$ is zero, and the equation (3) takes the same form as in the case of no rotation; this has been assumed by anticipation in Art. 183. The only effect of the rotation in such cases is to produce a slight slope of the wave-crests and furrows in the direction *across* the canal, as investigated in Art. 208. In the general case, resolving at right angles to the direction of the relative velocity (q, say), we see that a fluid particle has an apparent acceleration $2\omega q \cos\theta$ towards the right of its path, in addition to that due to the forces.

Again, by comparison of (1) with Art. 207 (5), we see that the oscillations of a sheet of water of relatively small dimensions, in co-latitude θ, will take place according to the same laws as those of a *plane* sheet rotating about a normal to its plane with angular velocity $\omega \cos\theta$.

As in Art. 207, free steady motions are possible, subject to certain conditions. Putting $\bar{\zeta} = 0$, we find that the equations (1) and (2) are satisfied by constant values of u, v, ζ, provided

$$u = -\frac{g}{2\omega a \sin\theta\cos\theta}\frac{\partial\zeta}{\partial\phi}, \qquad v = \frac{g}{2\omega a\cos\theta}\frac{\partial\zeta}{\partial\theta}, \quad \text{...........(4)}$$

and

$$\frac{\partial(h\sec\theta, \zeta)}{\partial(\theta, \phi)} = 0. \quad \text{...........................(5)}$$

The latter condition is satisfied by any assumption of the form

$$\zeta = f(h\sec\theta), \quad \text{.............................(6)}$$

and the equations (4) then give the values of u, v. It appears from (4) that the velocity in these steady motions is everywhere parallel to the contour-lines of the disturbed surface.

If h is constant, or a function of the latitude only, the only condition imposed on ζ is that it should be independent of ϕ; in other words the elevation must be symmetrical about the polar axis.

215. We shall suppose henceforward that the depth h is a function of θ only, and that the barriers to the sea, if any, coincide with parallels of latitude.

We take first the cases where the disturbed form of the water-surface is one of revolution about polar axis. When the terms involving ϕ

are omitted, the equations (1) and (2) of the preceding Art. take the forms

$$\frac{\partial u}{\partial t} - 2\omega v \cos\theta = -\frac{g}{a}\frac{\partial}{\partial\theta}(\zeta - \bar{\zeta}), \qquad \frac{\partial v}{\partial t} + 2\omega u \cos\theta = 0, \quad \text{......(1)}$$

with

$$\frac{\partial\zeta}{\partial t} = -\frac{\partial(hu\sin\theta)}{a\sin\theta\,\partial\theta}. \quad \text{........................(2)}$$

Assuming a time-factor $e^{i\sigma t}$, and solving for u, v, we find

$$u = \frac{i\sigma g}{\sigma^2 - 4\omega^2\cos^2\theta}\frac{\partial}{a\partial\theta}(\zeta - \bar{\zeta}), \qquad v = -\frac{2\omega g\cos\theta}{\sigma^2 - 4\omega^2\cos^2\theta}\frac{\partial}{a\partial\theta}(\zeta - \bar{\zeta}), \text{...(3)}$$

with

$$i\sigma\zeta = -\frac{\partial(hu\sin\theta)}{a\sin\theta\,\partial\theta}. \quad \text{........................(4)}$$

The formulae for the component displacements (ξ, η, say) can be written down from the relations $u = \dot{\xi}$, $v = \dot{\eta}$, or $u = i\sigma\xi$, $v = i\eta\sigma$. It appears that the fluid particles describe ellipses having their principal axes along the meridians and the parallels of latitude, respectively, the ratio of the axes being $\sigma/2\omega \,.\, \sec\theta$. In the *forced* oscillations of the present type the ratio $\sigma/2\omega$ is very small; so that the ellipses are very elongated, with the greatest length from E. to W., except in the neighbourhood of the equator.

Eliminating u and v between (3) and (4), and writing, for shortness,

$$\zeta - \bar{\zeta} = \zeta', \qquad \frac{\sigma}{2\omega} = f, \qquad \frac{\omega^2 a}{g} = m, \quad \text{................(5)}$$

we find

$$\frac{\partial}{a\sin\theta\,\partial\theta}\left(\frac{h\sin\theta}{f^2 - \cos^2\theta}\frac{\partial\zeta'}{\partial\theta}\right) + 4m\zeta' = -4m\bar{\zeta}. \quad \text{.............(6)}$$

In the case of uniform depth, this become

$$\frac{\partial}{\partial\mu}\left(\frac{1-\mu^2}{f^2-\mu^2}\frac{\partial\zeta'}{\partial\mu}\right) + \beta\zeta' = -\beta\bar{\zeta}, \quad \text{...................(7)}$$

where $\mu = \cos\theta$, and

$$\beta = \frac{4ma}{h} = \frac{4\omega^2 a^2}{gh}. \quad \text{...........................(8)}$$

216. First, as regards the *free* oscillations. Putting $\bar{\zeta} = 0$, we have

$$\frac{\partial}{\partial\mu}\left(\frac{1-\mu^2}{f^2-\mu^2}\frac{\partial\zeta}{\partial\mu}\right) + \beta\zeta = 0, \quad \text{......................(9)}$$

and we notice that in the case of no rotation this is included in (1) of Art. 199, as may be seen by putting $\beta f^2 = \sigma^2 a^2/gh$, $f = \infty$. The general solution of (9) is necessarily of the form

$$\zeta = AF(\mu) + Bf(\mu), \quad \text{........................(10)}$$

where $F(\mu)$ is an even, and $f(\mu)$ an odd, function of μ, and the constants A, B are arbitrary. In the case of a zonal sea bounded by two parallels of latitude, the ratio $A : B$ and the admissible values of f (and thence of the frequency $\sigma/2\pi$) are determined by the conditions that $u = 0$ at each of these parallels. If the boundaries are symmetrically situated on opposite sides of the equator, the oscillations fall into two classes; viz. in one of these $B=0$, and in the other $A=0$. By supposing the boundaries to contract to points at the poles, we pass to the case of an unlimited ocean, and the admissible values of f are now determined by the condition that u must vanish for $\mu=\pm 1$. The argument is, in principle, exactly that of Art. 201, but the application of the last-mentioned condition is now more difficult, owing to the less familiar form in which the solution of the differential equation is obtained.

In the case of symmetry with respect to the equator, we assume, following the method of Kelvin* and Darwin†,

$$\frac{1}{\mu^2-f^2}\frac{\partial\zeta'}{\partial\mu}=B_1\mu+B_3\mu^3+\ldots+B_{2j+1}\mu^{2j+1}+\ldots. \qquad \ldots\ldots\ldots(11)$$

This leads to

$$\zeta'=A-\tfrac{1}{2}f^2B_1\mu^2+\tfrac{1}{4}(B_1-f^2B_3)\mu^4+\ldots+\frac{1}{2j}(B_{2j-3}-f^2B_{2j-1})\mu^{2j}+\ldots, \quad \ldots(12)$$

where A is arbitrary; and makes

$$\frac{\partial}{\partial\mu}\left(\frac{1-\mu^2}{\mu^2-f^2}\frac{\partial\zeta'}{\partial\mu}\right)=B_1+3(B_3-B_1)\mu^2+\ldots+(2j+1)(B_{2j+1}-B_{2j-1})\mu^{2j}+\ldots. \qquad \ldots\ldots(13)$$

Substituting in (9), and equating coefficients of the several powers of μ, we find

$$B_1-\beta A=0, \qquad \ldots\ldots\ldots\ldots\ldots\ldots\ldots\ldots(14)$$

$$B_3-\left(1-\frac{\beta f^2}{2\,.\,3}\right)B_1=0, \qquad \ldots\ldots\ldots\ldots\ldots\ldots(15)$$

and thenceforward

$$B_{2j+1}-\left(1-\frac{\beta f^2}{2j(2j+1)}\right)B_{2j-1}-\frac{\beta}{2j(2j+1)}B_{2j-3}=0. \qquad \ldots\ldots(16)$$

These equations determine $B_1, B_3, \ldots B_{2j+1}, \ldots$ in succession, in terms of A, and the solution thus obtained would be appropriate, as already explained, to the case of a zonal sea bounded by two parallels in equal N. and S. latitudes. In the case of an ocean covering the globe, it would, as we shall prove, give infinite velocities at the poles, except for certain definite values of f.

Let us write
$$B_{2j+1}/B_{2j-1}=N_{j+1}; \qquad \ldots\ldots\ldots\ldots\ldots\ldots\ldots(17)$$

we shall shew, in the first place, that as j increases N_j must tend either to the limit 0 or to the limit 1. The equation (16) may be written

$$N_{j+1}=1-\frac{\beta f^2}{2j(2j+1)}+\frac{\beta}{2j(2j+1)}\frac{1}{N_j}. \qquad \ldots\ldots\ldots\ldots(18)$$

* Sir W. Thomson, "Note on the 'Oscillations of the First Species' in Laplace's Theory of the Tides," *Phil. Mag.* (4), l. 279 (1875) [*Papers*, iv. 248].

† "On the Dynamical Theory of the Tides of Long Period," *Proc. Roy. Soc.* xli. 337 (1886) [*Papers*, i. 336].

Hence, when j is large, either

$$N_j = -\frac{\beta}{2j(2j+1)}, \qquad \text{.........................(19)}$$

approximately, or N_{j+1} is not small, in which case N_{j+2} will be nearly equal to 1, and the values of N_{j+3}, N_{j+4}, ... will tend more and more nearly to 1, the approximate formula being

$$N_{j+1} = 1 - \frac{\beta(f^2-1)}{2j(2j+1)}. \qquad \text{.........................(20)}$$

Hence, with increasing j, N_j tends to one or other of the forms (19) and (20).

In the former case (19), the series (11) will be convergent for $\mu = \pm 1$, and the solution will be valid over the whole globe.

In the other event (20), the product $N_3 N_4 \ldots N_{j+1}$, and therefore the coefficient B_{2j+1}, tends with increasing j to a finite limit other than zero. The series (11) will then, after some finite number of terms, become comparable with $1 + \mu^2 + \mu^4 + \ldots$, or $(1-\mu^2)^{-1}$, so that we may write

$$\frac{1}{\mu^2 - f^2}\frac{\partial \zeta'}{\partial \mu} = L + \frac{M}{1-\mu^2}, \qquad \text{.........................(21)}$$

where L and M are functions of μ which remain finite when $\mu = \pm 1$. Hence, from (3),

$$u = -\frac{i\sigma}{4m}\frac{(1-\mu^2)^{\frac{1}{2}}}{\mu^2 - f^2}\frac{\partial \zeta'}{\partial \mu} = -\frac{i\sigma}{4m}\{(1-\mu^2)^{\frac{1}{2}} L + (1-\mu^2)^{-\frac{1}{2}} M\}, \quad \text{...(22)}$$

which makes u infinite at the poles.

It follows that the conditions of our problem can be satisfied only if N_j tends to the limit zero; and this consideration, as we shall see, restricts us to a determinate series of values of f.

$$N_j = \frac{-\dfrac{\beta}{2j(2j+1)}}{1 - \dfrac{\beta f^2}{2j(2j+1)} - N_{j+1}}, \qquad \text{.........................(23)}$$

and by successive applications of this we obtain N_j in the form of a convergent continued fraction

$$N_j = \frac{-\dfrac{\beta}{2j(2j+1)}}{1 - \dfrac{\beta f^2}{2j(2j+1)} +} \; \frac{\dfrac{\beta}{(2j+2)(2j+3)}}{1 - \dfrac{\beta f^2}{(2j+2)(2j+3)} +} \; \frac{\dfrac{\beta}{(2j+4)(2j+5)}}{1 - \dfrac{\beta f^2}{(2j+4)(2j+5)} + \ldots}, \qquad \text{.........(24)}$$

on the present supposition that N_{j+k} tends with increasing k to the limit 0, in the manner indicated by (19). In particular, this formula (24) determines the value of N_2. Now from (15) we must have

$$N_2 = 1 - \frac{\beta f^2}{2 \cdot 3}, \qquad \text{.........................(25)}$$

whence

$$1 - \frac{\beta f^2}{2.3} + \frac{\dfrac{\beta}{4.5}}{1 - \dfrac{\beta f^2}{4.5} +} \, \frac{\dfrac{\beta}{6.7}}{1 - \dfrac{\beta f^2}{6.7} + \ldots} = 0, \quad \ldots\ldots\ldots\ldots(26)$$

which is equivalent to $N_1 = \infty$. This equation determines the admissible values of $f (= \sigma/2\omega)$. The constants in (11) are then given by

$$B_1 = \beta A, \qquad B_3 = N_2 \beta A, \qquad B_5 = N_2 N_3 \beta A, \ldots, \quad \ldots\ldots\ldots(27)$$

where A is arbitrary.

It is easily seen that when β is infinitesimal the roots of (26) are given by

$$\frac{\sigma^2 a^2}{gh} = \beta f^2 = n(n+1), \quad \ldots\ldots\ldots\ldots\ldots\ldots\ldots\ldots(28)$$

where n is an even integer; cf. Art. 199.

One arithmetically remarkable point remains to be noticed. It might appear at first sight that when a value of f has been found from (26) the coefficients $B_3, B_5, B_7, \ldots$ could be found in succession from (15) and (16), or by means of the equivalent formula (18). But this would require us to start with *exactly* the right value of f and to observe absolute accuracy in the subsequent stages of the work. The above argument shews, in fact, that any other value, differing by however little, if adopted as a starting point for the calculation will inevitably lead at length to values of N_j which approximate to the limit 1*.

An approximation to the longest free period may be attempted by the method of Art. 205 a.

Denoting by ξ, η the displacements southwards and eastwards, respectively, we have, in the notation of the Art. referred to,

$$\left.\begin{aligned} \mathfrak{T} = \pi\rho h a^2 \int_0^\pi (\dot\xi^2 + \dot\eta^2) \sin\theta \, d\theta, \qquad M' = 2\pi\rho h a^2 \int_0^\pi (\xi\cos\theta \, . \, \dot\eta - \eta \, . \, \dot\xi \cos\theta) \sin\theta \, d\theta, \\ V - T_0 = \pi g \rho a^2 \int_0^\pi \zeta^2 \sin\theta \, d\theta. \end{aligned}\right\} \ldots(29)$$

We will assume that as in the case of no rotation the surface elevation is represented by a zonal harmonic of the second order. The formulae (3) of Art. 215 then suggests for our assumed type

$$\xi = A \sin\theta \cos\theta \cos\sigma t, \qquad \eta = B \sin\theta \cos^2\theta \sin\sigma t, \quad \ldots\ldots\ldots\ldots\ldots(30)$$

which makes

$$\zeta = -\frac{a}{a \sin\theta} \frac{\partial}{\partial\theta} (\xi \sin\theta) = -\frac{h}{a} (3\cos^2\theta - 1) A \cos\sigma t. \quad \ldots\ldots\ldots\ldots(31)$$

We find

$$P = \pi\rho h a^2 (\tfrac{2}{15} A^2 + \tfrac{2}{35} B^2), \qquad Q = \tfrac{8}{35} \pi\rho\omega h a^2 AB, \qquad R = \tfrac{4}{5} \pi g \rho h^2 A^2. \quad \ldots\ldots\ldots(32)$$

The equation (10) of Art. 205 a becomes

$$(x^2 - 6) A^2 + \sqrt{\beta} \, . \, \tfrac{6}{7} x AB + \tfrac{3}{7} B^2 x^2 = 0, \quad \ldots\ldots\ldots\ldots\ldots\ldots(33)$$

where

$$x = \sigma a / \sqrt{(gh)}, \qquad \beta = 4\omega^2 a^2 / gh. \quad \ldots\ldots\ldots\ldots\ldots\ldots\ldots\ldots(34)$$

The stationary values of x are then given by

$$x^2 = 6 + \tfrac{3}{7}\beta. \quad \ldots\ldots\ldots\ldots\ldots\ldots\ldots\ldots\ldots\ldots(35)$$

* Sir W. Thomson, *l.c. ante* p. 335.

For example, taking $\beta=5$, which would correspond in the case of the earth to a depth of 58080 ft., we find

$$\sigma a/\sqrt{(gh)}=2{\cdot}854, \qquad \omega/\sigma=\cdot 3917.$$

The latter number gives the period in terms of the sidereal day. Hence in sidereal time $2\pi/\sigma=9$ h. 24 m. The true period, as calculated by Hough (see Art. 222) is 9 h. 52 m., but this allows for the mutual gravitation of the disturbed water, which we have neglected.

A correction is however easily made. Since we neglect effect of centrifugal force on gravity the influence of T_0 in (29) may be disregarded, whilst the value of V is altered in the ratio

$$1-\tfrac{3}{5}\frac{\rho_1}{\rho_0}=\cdot 892,$$

where $\rho_1/\rho_0\,(=\cdot 18)$ is the ratio of the density of the water to the mean density of the earth (see Art. 200). The result is to replace (35) by

$$x^2=5{\cdot}352+\tfrac{3}{7}\beta. \qquad \text{.......................................(36)}$$

For $\beta=5$ this gives a period of 9 h. 48 m., in close approximation to Hough's value.

For greater values of β, i.e. smaller depths of the ocean, or greater speeds of rotation, the approximation is less satisfactory, as we should expect from the nature of our assumed type.

217. It is shewn in the Appendix to this Chapter that the tide-generating potential, when expanded in simple-harmonic functions of the time, consists of terms of three distinct types.

The first type is such that the equilibrium tide-height would be given by

$$\bar{\zeta}=H'\left(\tfrac{1}{3}-\cos^2\theta\right).\cos(\sigma t+\epsilon). \qquad \text{..................(37)*}$$

The corresponding forced waves are called by Laplace the 'Oscillations of the First Species'; they include the lunar fortnightly and the solar semi-annual tides, and generally all the tides of long period. Their characteristic is symmetry about the polar axis, and they form accordingly the most important case of forced oscillations of the present type.

If we substitute from (37) in (7), and assume for

$$\frac{1-\mu^2}{\mu^2-f^2}\frac{\partial\zeta'}{\partial\mu} \text{ and } \zeta'$$

expressions of the forms (11) and (12), we have, in place of (14), (15),

$$B_1-\tfrac{1}{3}\beta H'-\beta A=0, \qquad \text{........................(38)}$$

$$B_3-\left(1-\frac{\beta f^2}{2\,.\,3}\right)B_1+\tfrac{1}{3}\beta H'=0, \qquad \text{.....................(39)}$$

whilst (16) and its consequences hold for all the higher coefficients. It may be noticed that (39) may be included under the general formula (16), provided we write $B_{-1}=-2H'$. It appears by the same argument as before that the only admissible solution for an ocean covering the globe is the one that makes $N_\infty=0$, and that accordingly N_j must have the value given by the continued fraction in (24), where f is now prescribed by the frequency of the disturbing forces.

* In strictness, θ here denotes the *geocentric* latitude, but the difference between this and the geographical latitude may be neglected consistently with the assumptions introduced in Art. 214.

In particular, this formula determines the value of N_1. Now

$$B_1 = N_1 B_{-1} = -2N_1 H',$$

and the equation (38) then gives

$$A = -\tfrac{1}{3}H' - \frac{2}{\beta}N_1 H'; \qquad \text{.......................(40)}$$

in other words, this is the only value of A which is consistent with a zero limit of N_j, and therefore with a finite velocity at the poles. Any other value of A, if adopted as a starting point for the calculation of B_1, B_3, B_5, ... in succession, by means of (38), (39), and (16), would lead ultimately to values of N_j approximating to the limit 1. Moreover, since *absolute* accuracy in the initial choice of A and in the subsequent computations would be essential to avoid this, the only practical method of calculating the coefficients is to use the formulae

$$B_1/H' = -2N_1, \qquad B_3 = N_2 B_1, \qquad B_5 = N_3 B_3, \ldots,$$

or

$$B_1/H' = -2N_1, \qquad B_3/H' = -2N_1 N_2, \qquad B_5/H' = -2N_1 N_2 N_3, \ldots \qquad \text{......(41)}$$

where the values of N_1, N_2, N_3, ... are to be computed from the continued fraction (24). It is evident *à posteriori* that the solution thus obtained will satisfy all the conditions of the problem, and that the series (12) will converge with great rapidity. The most convenient plan of conducting the calculation is to assume a roughly approximate value, suggested by (19), for one of the ratios N_j of sufficiently high order, and thence to compute

$$N_{j-1}, N_{j-2}, \ldots N_2, N_1$$

in succession by means of the formula (23). The values of the constants A, B_1, B_3, ..., in (12), are then given by (40) and (41). For the tidal elevation we find

$$\zeta/H' = -2N_1/\beta - (1 - f^2 N_1)\mu^2 - \tfrac{1}{2}N_1(1 - f^2 N_2)\mu^4 - \ldots$$
$$- \frac{1}{j}N_1 N_2 \ldots N_{j-1}(1 - f^2 N_j)\mu^{2j} - \ldots . \qquad \text{.........(42)}$$

In the case of the lunar fortnightly tide, f is the ratio of a sidereal day to a lunar month, and is therefore equal to about $\frac{1}{28}$, or more precisely ·0365. This makes $f^2 = {\cdot}00133$. It is evident that a fairly accurate representation of this tide, and *à fortiori* of the solar semi-annual tide, and of the remaining tides of long period, will be obtained by putting $f = 0$; this materially shortens the calculations.

The results will involve the value of β, $= 4\omega^2 a^2/gh$. For $\beta = 40$, which corresponds to a depth of 7260 feet, we find in this way

$$\zeta/H' = {\cdot}1515 - 1{\cdot}0000\mu^2 + 1{\cdot}5153\mu^4 - 1{\cdot}2120\mu^6 + {\cdot}6063\mu^8 - {\cdot}2076\mu^{10}$$
$$+ {\cdot}0516\mu^{12} - {\cdot}0097\mu^{14} + {\cdot}0018\mu^{16} - {\cdot}0002\mu^{18}, \qquad \text{.........(43)*}$$

* The coefficients in (43) and (44) differ only slightly from the numerical values obtained by Darwin for the case $f = {\cdot}0365$.

whence, at the poles ($\mu = \pm 1$),

$$\zeta = -\tfrac{2}{3}H' \times \cdot 154,$$

and, at the equator ($\mu = 0$),

$$\zeta = \tfrac{1}{3}H' \times \cdot 455.$$

Again, for $\beta = 10$, or a depth of 29040 feet, we get

$$\zeta/H' = \cdot 2359 - 1\cdot 0000\mu^2 + \cdot 5898\mu^4 - \cdot 1623\mu^6 + \cdot 0258\mu^8 - \cdot 0026\mu^{10} + \cdot 0002\mu^{12}. \quad \text{.........(44)}$$

This makes, at the poles,

$$\zeta = -\tfrac{2}{3}H' \times \cdot 470,$$

and, at the equator,

$$\zeta = \tfrac{1}{3}H' \times \cdot 708.$$

For $\beta = 5$, or a depth of 58080 feet, we find

$$\zeta/H' = \cdot 2723 - 1\cdot 0000\mu^2 + 3404\mu^4 - \cdot 0509\mu^6 + \cdot 0043\mu^8 - \cdot 0004\mu^{10}. \quad \text{...............(45)}$$

This gives, at the poles,

$$\zeta = -\tfrac{2}{3}H' \times \cdot 651,$$

and, at the equator,

$$\zeta = \tfrac{1}{3}H' \times \cdot 817.$$

Since the polar and equatorial values of the equilibrium tide are $-\tfrac{2}{3}H'$ and $\tfrac{1}{3}H'$, respectively, these results shew that for the depths in question the long-period tides are, on the whole, *direct*, though the nodal circles will, of course, be shifted more or less from the positions assigned by the equilibrium theory. It appears, moreover, that, for depths comparable with the actual depth of the sea, the tide has less than half the equilibrium value. It is easily seen from the form of equation (7) that with increasing depth, and consequent diminution of β, the tide-height will approximate more and more closely to the equilibrium value. This tendency is illustrated by the above numerical results.

It is to be remarked that the kinetic theory of the long-period tides was passed over by Laplace, under the impression that practically, owing to the operation of dissipative forces, they would have the values given by the equilibrium theory. He proved, indeed, that the tendency of frictional forces must be in this direction, but it has been maintained by Darwin* that in the case of the fortnightly tide, at all events, it is doubtful whether the effect would be nearly so great as Laplace supposed. We shall return to this point later.

218. When the disturbance is no longer restricted to be symmetrical about the polar axis, we must recur to the general equations (1) and (2) of Art. 214. We retain, however, the assumptions as to the law of depth and the nature of the boundaries introduced in Art. 215.

* *l.c. ante* p. 335.

If we assume that Ω, u, v, ζ all vary as $e^{i(\sigma t+s\phi+\epsilon)}$, where s is integral, th equations referred to give

$$i\sigma u - 2\omega v\cos\theta = -\frac{g}{a}\frac{\partial}{\partial\theta}(\zeta-\bar{\zeta}), \qquad i\sigma v + 2\omega u\cos\theta = -\frac{isg}{a\sin\theta}(\zeta-\bar{\zeta}), \quad \text{...(1)}$$

with

$$i\sigma\zeta = -\frac{1}{a\sin\theta}\left\{\frac{\partial(hu\sin\theta)}{\partial\theta} + ishv\right\}. \quad \text{.................(2)}$$

Solving for u, v, we find

$$\left.\begin{aligned} u &= \frac{i\sigma}{4m(f^2-\cos^2\theta)}\left(\frac{\partial\zeta'}{\partial\theta} + \frac{s}{f}\zeta'\cot\theta\right), \\ v &= -\frac{\sigma}{4m(f^2-\cos^2\theta)}\left(\frac{\cos\theta}{f}\frac{\partial\zeta'}{\partial\theta} + s\zeta'\operatorname{cosec}\theta\right), \end{aligned}\right\} \quad \text{...........(3)}$$

where we have written

$$\zeta - \bar{\zeta} = \zeta', \qquad \frac{\sigma}{2\omega} = f, \qquad \frac{\omega^2 a}{g} = m, \quad \text{.................(4)}$$

as before.

It appears that in all cases of simple-harmonic oscillation the fluid particles describe ellipses having their principal axes along the meridians and parallels of latitude, respectively.

Substituting from (3) in (2) we obtain the differential equation in ζ':

$$\frac{\partial}{\sin\theta\,\partial\theta}\left\{\frac{h\sin\theta}{f^2-\cos^2\theta}\left(\frac{\partial\zeta'}{\partial\theta} + \frac{s}{f}\zeta'\cot\theta\right)\right\}$$
$$- \frac{h}{f^2-\cos^2\theta}\left(\frac{s}{f}\cot\theta\frac{\partial\zeta'}{\partial\theta} + s^2\zeta'\operatorname{cosec}^2\theta\right) + 4ma\zeta' = -4ma\bar{\zeta}. \quad \text{...(5)}$$

219. The case $s = 1$ includes, as forced oscillations, Laplace's 'Oscillations of the Second Species,' where the disturbing potential is a tesseral harmonic of the second order; viz.

$$\bar{\zeta} = H''\sin\theta\cos\theta\,.\cos(\sigma t + \phi + \epsilon), \quad \text{.................(1)}$$

where σ differs not very greatly from ω. This includes the lunar and solar diurnal tides.

In the case of a disturbing body whose proper motion could be neglected, we should have $\sigma = \omega$, exactly, and therefore $f = \frac{1}{2}$. In the case of the moon, the orbital motion is so rapid that the actual period of the principal lunar diurnal tide is very appreciably longer than a sidereal day*; but the supposition that $f = \frac{1}{2}$ simplifies the formulae so materially that we adopt it in the following investigation†. We find that it enables us to calculate the forced oscillations when the depth follows the law

$$h = (1 - q\cos^2\theta)\,h_0, \quad \text{.........................(2)}$$

where q is any given constant.

* It is to be remarked, however, that there is an important term in the harmonic development of Ω for which $\sigma = \omega$ exactly, provided we neglect the changes in the plane of the disturbing body's orbit. This period is the same for the sun as for the moon, and the two partial tides thus produced combine into what is called the 'luni-solar' diurnal tide.

† Taken with very slight alteration from Airy, "Tides and Waves," Arts. 95 ..., and Darwin, *Encyc. Brit.* (9th ed.), xxiii. 359.

Taking an exponential factor $e^{i(\omega t+\phi+\epsilon)}$, and therefore putting $s=1$, $f=\frac{1}{2}$, in Art. 218 (3), and assuming

$$\zeta' = C\sin\theta\cos\theta, \qquad \text{.........(3)}$$

we find

$$u = -i\sigma\frac{C}{m}, \qquad v = \sigma\frac{C}{m}.\cos\theta. \qquad \text{.........(4)}$$

Substituting in the equation of continuity (Art. 318 (2)), we get

$$\zeta' + \bar{\zeta} = \frac{C}{ma}\frac{dh}{d\theta}, \qquad \text{.........(5)}$$

which is consistent with the law of depth (2), provided

$$C = -\frac{1}{1-2qh_0/ma}H''. \qquad \text{.........(6)}$$

This gives

$$\zeta = -\frac{2qh_0/ma}{1-2qh_0/ma}\bar{\zeta}. \qquad \text{.........(7)}$$

One remarkable consequence of this formula is that in the case of uniform depth ($q=0$) there is no diurnal tide, so far as the rise and fall of the surface is concerned. This result was first established (in a different manner) by Laplace, who attached great importance to it as shewing that his kinetic theory was able to account for the relatively small values of the diurnal tide as then (imperfectly) known, in striking contrast to what would be demanded by the equilibrium theory.

But, although with a uniform depth there is no rise and fall, there are tidal currents. It appears from (4) that every particle describes an ellipse whose major axis is in the direction of the meridian, and of the same length in all latitudes. The ratio of the minor to the major axis is $\cos\theta$, and so varies from 1 at the poles to 0 at the equator, where the motion is wholly N. and S.

220. In the case $s=2$, the forced oscillations of most importance are where the disturbing potential is a sectorial harmonic of the second order. These constitute Laplace's 'Oscillations of the Third Species,' for which

$$\bar{\zeta} = H'''\sin^2\theta.\cos(\sigma t + 2\phi + \epsilon), \qquad \text{.........(1)}$$

where σ is nearly equal to 2ω. This includes the most important of all the tidal oscillations, viz. the lunar and solar semi-diurnal tides.

If the orbital motion of the disturbing body were infinitely slow we should have $\sigma = 2\omega$, and therefore $f=1$; for simplicity we follow Laplace in making this approximation, although it is a somewhat rough one in the case of the principal lunar tide*.

A solution similar to that of the preceding Art can be obtained for the special law of depth†

$$h = h_0\sin^2\theta. \qquad \text{.........(2)}$$

* There is, however, a 'luni-solar' semi-diurnal tide whose speed is exactly 2ω if we neglect the changes in the planes of the orbits. Cf. p. 341, first footnote.

† Cf. Airy and Darwin, *ll. cc.*

Adopting an exponential factor $e^{i(2\omega t+2\phi+\epsilon)}$, and putting therefore $f=1$, $s=2$, we find that if we assume

$$\zeta' = C\sin^2\theta, \qquad \text{.................................(3)}$$

the equations (3) of Art. 218 give

$$u = \frac{i\sigma}{m} C\cot\theta, \qquad v = -\frac{\sigma}{2m} C\frac{1+\cos^2\theta}{\sin\theta}, \qquad \text{..............(4)}$$

whence, substituting in Art. 218 (2),

$$\zeta = \frac{2h_0}{ma} . C\sin^2\theta. \qquad \text{..............................(5)}$$

Putting $\zeta = \zeta' + \bar{\zeta}$, and substituting from (1) and (3), we find

$$C = -\frac{1}{1-2h_0/ma} H''', \qquad \text{......................(6)}$$

and therefore

$$\zeta = -\frac{2h_0/ma}{1-2h_0/ma}\bar{\zeta}. \qquad \text{..............................(7)}$$

For such depths as actually occur in the ocean $2h_0 < ma$, and the tide is therefore inverted. It may be noticed that the formulae (4) make the velocity infinite at the poles, as was to be expected, since the depth there is zero.

221. For any other law of depth a solution can only be obtained in the form of a series. In the case of *uniform* depth, we find, putting $s=2$, $f=1$, $4ma/h=\beta$ in Art. 218 (5),

$$(1-\mu^2)^2\frac{d^2\zeta'}{d\mu^2} + \{\beta(1-\mu^2)^2 - 2\mu^2 - 6\}\zeta' = -\beta(1-\mu^2)^2\bar{\zeta}, \qquad \text{......(8)}$$

where μ is written for $\cos\theta$. In this form the equation is somewhat intractable, since it contains terms of four different dimensions in μ. It simplifies a little, however, if we transform to

$$\nu, \; = (1-\mu^2)^{\frac{1}{2}}, \; = \sin\theta,$$

as independent variable; viz. we find

$$\nu^2(1-\nu^2)\frac{d^2\zeta'}{d\nu^2} - \nu\frac{d\zeta'}{d\nu} - (8 - 2\nu^2 - \beta\nu^4)\zeta' = -\beta\nu^4\bar{\zeta} = -\beta H'''\nu^6. \qquad \text{...(9)}$$

which is of *three* different dimensions in ν.

To obtain a solution for the case of an ocean covering the globe, we assume

$$\zeta' = B_0 + B_2\nu^2 + B_4\nu^4 + \ldots + B_{2j}\nu^{2j} + \ldots \qquad \text{...............(10)}$$

Substituting in (9), and equating coefficients, we find

$$B_0 = 0, \quad B_2 = 0, \quad 0 . B_4 = 0, \qquad \text{.....................(11)}$$

$$16B_6 - 10B_4 + \beta H''' = 0, \qquad \text{.....................(12)}$$

and thenceforward

$$2j(2j+6)B_{2j+4} - 2j(2j+3)B_{2j+2} + \beta B_{2j} = 0. \qquad \text{...........(13)}$$

These equations give B_6, B_8, ... B_{2j}, ... in succession, in terms of B_4, which is so far undetermined. It is obvious, however, from the nature of the problem, that, except for certain special values of h (and therefore of β), which are such that there is a free oscillation of corresponding type ($s=2$) having the speed 2ω, the solution must be unique. We shall see, in fact, that unless B_4 have a certain definite value the solution above indicated will make the meridian component (u) of the velocity discontinuous at the equator*.

The argument is in some respects similar to that of Art. 217. If we denote by N_j the ratio B_{2j+2}/B_{2j} of consecutive coefficients, we have, from (13),

$$N_{j+1} = \frac{2j+3}{2j+6} - \frac{\beta}{2j(2j+6)}\frac{1}{N_j}, \quad \text{.................(14)}$$

from which it appears that, with increasing j, N_j must tend to one or other of the limits 0 and 1. More precisely, unless the limit of N_j be zero, the limiting form of N_{j+1} will be

$$(2j+3)/(2j+6), \text{ or } 1 - \frac{3}{2j},$$

approximately. The latter is identical with the limiting form of the ratio of the coefficients of ν^{2j} and ν^{2j-2} in the expansion of $(1-\nu^2)^{\frac{1}{2}}$. We infer that, unless B_4 have such a value as to make $N_\infty = 0$, the terms of the series (10) will become ultimately comparable with those of $(1-\nu^2)^{\frac{1}{2}}$, so that we may write

$$\zeta' = L + (1-\nu^2)^{\frac{1}{2}} M, \quad \text{.........................(15)}$$

where L, M are functions of ν which do not vanish for $\nu = 1$. Near the equator ($\nu = 1$) this makes

$$\frac{d\zeta'}{d\theta} = \mp (1-\nu^2)^{\frac{1}{2}} \frac{d\zeta'}{d\nu} = \pm M. \quad \text{....................(16)}$$

Hence, by Art. 218 (3), u would change from a certain definite value to an equal but opposite value as we cross the equator.

It is therefore essential, for our present purpose, to choose the value of B_4 so that $N_\infty = 0$. This is effected by the same method as in Art. 217. Writing (13) in the form

$$N_j = \frac{\dfrac{\beta}{2j(2j+6)}}{\dfrac{2j+3}{2j+6} - N_{j+1}}, \quad \text{.........................(17)}$$

we see that N_j must be given by the converging continued fraction

$$N_j = \cfrac{\dfrac{\beta}{2j(2j+6)}}{\dfrac{2j+3}{2j+6} - \cfrac{\dfrac{\beta}{(2j+2)(2j+8)}}{\dfrac{2j+5}{2j+8} - \cfrac{\dfrac{\beta}{(2j+4)(2j+10)}}{\dfrac{2j+7}{2j+10} - \&c.}}}. \quad \text{......(18)}$$

* In the case of a polar sea bounded by a small circle of latitude whose angular radius is $<\frac{1}{2}\pi$, the value of B_4 is determined by the condition that $u=0$, or $\partial\zeta'/\partial\nu = 0$, at the boundary.

This holds from $j = 2$ upwards, but it appears from (12) that it will give also the value of N_1 (not hitherto defined), provided we use this symbol for B_4/H'''. We have then

$$B_4 = N_1 H''', \qquad B_6 = N_2 B_4, \qquad B_8 = N_3 B_6, \ldots .$$

Finally, writing $\zeta = \bar{\zeta} + \zeta'$, we obtain

$$\zeta/H''' = \nu^2 + N_1\nu^4 + N_1N_2\nu^6 + N_1N_2N_3\nu^8 + \ldots \quad \text{.........(19)}$$

As in Art. 217, the practical method of conducting the calculation is to assume an approximate value for N_{j+1}, where j is a moderately large number, and then to deduce N_j, N_{j-1}, ... N_2, N_1 in succession by means of the formula (17).

The above investigation is taken substantially from the very remarkable paper written by Kelvin* in vindication of Laplace's treatment of the problem, as given in the *Mécanique Céleste*. In the passage more especially in question, Laplace determines the constant B_4 by means of the continued fraction for N_1, without, it must be allowed, giving any adequate justification of the step; and the soundness of this procedure had been disputed by Airy†, and after him by Ferrel‡.

Laplace, unfortunately, was not in the habit of giving specific references, so that few of his readers appear to have become acquainted with the original presentment§ of the kinetic theory, where the solution for the case in question is put in a very convincing, though somewhat different, form. Aiming in the first instance at an approximate solution by means of a *finite* series, thus:

$$\zeta' = B_4\nu^4 + B_6\nu^6 + \ldots + B_{2k+2}\nu^{2k+2}, \quad \text{..............................(20)}$$

Laplace remarks|| that in order to satisfy the differential equations, the coefficients would have to fulfil the conditions

$$\left.\begin{aligned} &16B_6 - 10B_4 + \beta H''' = 0,\\ &40B_8 - 28B_6 + \beta B_4 = 0,\\ &\ldots\ldots\ldots\ldots\ldots\ldots\ldots\ldots\ldots\\ &(2k-2)(2k+4)B_{2k+2} - (2k-2)(2k+1)B_{2k} + \beta B_{2k-2} = 0,\\ &\qquad\qquad -2k(2k+3)B_{2k+2} + \beta B_{2k} = 0,\\ &\qquad\qquad\qquad\qquad \beta B_{2k+2} = 0, \end{aligned}\right\} \quad \text{...........(21)}$$

as is seen at once by putting $B_{2k+4} = 0$, $B_{2k+6} = 0$, ... in the general relation (13).

We have here $k+1$ equations between k constants. The method followed is to determine the constants by means of the first k relations; we thus obtain an exact solution, not of the proposed differential equation (9), but of the equation as modified by the addition of a term $\beta B_{2k+2}\nu^{2k+6}$ to the right-hand side. This is equivalent to an alteration of the disturbing force, and if we can obtain a solution such that the required alteration is very small, we may accept it as an approximate solution of the problem in its original form¶.

* Sir W. Thomson, "On an Alleged Error in Laplace's Theory of the Tides," *Phil. Mag.* (4), l. 227 (1875) [*Papers*, iv. 231].

† "Tides and Waves," Art. 111.

‡ "Tidal Researches," *U.S. Coast Survey Rep.* 1874, p. 154.

§ "Recherches sur quelques points du système du monde," *Mém. de l'Acad. roy. des Sciences*, 1776 [1779] [*Oeuvres*, ix. 187...].

|| *Oeuvres*, ix. 218. The notation has been altered.

¶ It is remarkable that this argument is of a kind constantly employed by Airy himself in his researches on waves.

Now, taking the first k relations of the system (21) in reverse order, we obtain B_{2k+2} in terms of B_{2k}, thence B_{2k} in terms of B_{2k-1}, and so on, until, finally, B_4 is expressed in terms of H'''; and it is obvious that if k be large enough the value of B_{2k+2}, and the consequent adjustment of the disturbing force which is required to make the solution exact, will be very small. This will be illustrated presently, after Laplace, by a numerical example.

The process just given is plainly equivalent to the use of the continued fraction (18) in the manner already explained, starting with $j+1=k$, and $N_k=\beta/2k(2k+3)$. The continued fraction, as such, does not, however, make its appearance in the memoir here referred to, but was introduced in the *Mécanique Céleste*, probably as an after-thought, as a condensed expression of the method of computation originally employed.

The table below gives the numerical values of the coefficients of the several powers of ν in the formula (19) for ζ/H''', in the cases $\beta=40$, 20, 10, 5, 1, which correspond to depths of 7260, 14520, 29040, 58080, 290400 feet, respectively*. The last line gives the value of ζ/H''' for $\nu=1$, i.e. the ratio of the amplitude at the equator to its equilibrium-value. At the poles ($\nu=0$), the tide has in all cases the equilibrium-value zero.

	$\beta=40$	$\beta=20$	$\beta=10$	$\beta=5$	$\beta=1$
ν^2	+ 1·0000	+ 1·0000	+ 1·0000	+ 1·0000	+ 1·0000
ν^4	+ 20·1862	− 0·2491	+ 6·1915	+ 0·7504	+ 0·1062
ν^6	+ 10·1164	− 1·4056	+ 3·2447	+ 0·1566	+ 0·0039
ν^8	− 13·1047	− 0·8594	+ 0·7234	+ 0·0157	+ 0·0001
ν^{10}	− 15·4488	− 0·2541	+ 0·0919	+ 0·0009	
ν^{12}	− 7·4581	− 0·0462	+ 0·0076		
ν^{14}	− 2·1975	− 0·0058	+ 0·0004		
ν^{16}	− 0·4501	− 0·0006			
η^{18}	− 0·0687				
ν^{20}	− 0·0082				
ν^{22}	− 0·0008				
ν^{24}	− 0·0001				
	− 7·434	− 1·821	+ 11·259	+ 1·924	+ 1·110

We may use the above numerical results to estimate the closeness of the approximation in each case. For example, when $\beta=40$, Laplace finds $B_{26}=-\cdot000004H'''$; the addition to the disturbing force which is necessary to make the solution exact would then be $-\cdot00002H'''\nu^{30}$, and would therefore bear to the actual force the ratio $-\cdot00002\nu^{28}$.

It appears from (19) that near the poles, where ν is small, the tides are in all cases direct. For sufficiently great depths, β will be very small, and the formulae (17) and (19) then shew that the tide has everywhere sensibly the equilibrium-value, all the coefficients being small except the first, which is unity. As h is diminished, β increases, and the formula (17) shews that each of the ratios N_j will continually increase, except when it changes sign

* The first three cases were calculated by Laplace, *l.c. ante* p. 330; the last by Kelvin. The numbers relating to the third case have been slightly corrected, in accordance with the computations of Hough; see p. 347.

from + to − by passing through the value ∞. No singularity in the solution attends this passage of N_j through ∞, except in the case of N_1, since, as is easily seen, the product $N_{j-1}N_j$ remains finite, and the coefficients in (19) are therefore all finite. But when $N_1 = \infty$, the expression for ζ becomes infinite, shewing that the depth has then one of the critical values already referred to.

The table on p. 346 indicates that for depths of 29040 feet, and upwards, the tides are everywhere direct, but that there is some critical depth between 29040 feet and 14520 feet, for which the tide at the equator changes from direct to inverted. The largeness of the second coefficient in the case $\beta = 40$ indicates that the depth could not be reduced much below 7260 feet before reaching a second critical value.

Whenever the equatorial tide is inverted, there must be one or more pairs of nodal circles ($\zeta = 0$), symmetrically situated on opposite sides of the equator. In the case of $\beta = 40$, the position of the nodal circles is given by $\nu = \cdot 95$, or $\theta = 90° \pm 18°$, approximately*.

222. The dynamical theory of the tides, in the case of an ocean covering the globe, with depth uniform along each parallel of latitude, has been greatly improved and developed by Hough†, who, taking up an abandoned attempt of Laplace, substituted expansions in spherical harmonics for the series of powers of μ (or ν). This has the advantage of more rapid convergence, especially, as might be expected, in cases where the influence of the rotation is relatively small; and it also enables us to take account of the mutual attraction of the particles of water, which, as we have seen in the simpler problem of Art. 200, is by no means insignificant.

If the surface-elevation ζ, and the conventional equilibrium tide-height $\bar{\zeta}$ (in which the effect of mutual attraction is not included), be expanded in series of spherical harmonics, thus

$$\zeta = \Sigma\zeta_n, \qquad \bar{\zeta} = \Sigma\bar{\zeta}_n, \qquad \text{.........................(1)}$$

the complete expression for the disturbing potential will be

$$\Omega = -g\Sigma\left(\bar{\zeta}_n + \frac{3}{2n+1}\frac{\rho}{\rho_0}\zeta_n\right);$$

cf. Art. 200. The series on the right hand is to be substituted for $\bar{\zeta}$ in the equations of Arts. 214...; this will be allowed for if we write

$$\zeta' = \Sigma(\alpha_n\zeta_n - \bar{\zeta}_n), \qquad \text{.........................(2)}$$

where

$$\alpha_n = 1 - \frac{3}{2n+1}\frac{\rho}{\rho_0}, \qquad \text{.........................(3)}$$

in modification of the notation of Art. 215 (5) or Art. 218 (4).

* For a fuller discussion of these points reference may be made to the original investigation of Laplace, and to Kelvin's papers.

† "On the Application of Harmonic Analysis to the Dynamical Theory of the Tides," *Phil. Trans.* A, clxxxix. 201, and cxci. 139 (1897). See also Darwin's *Papers*, i. 349.

In the oscillations of the 'First Species,' the differential equation may be written

$$\frac{\partial}{\partial\mu}\left(\frac{1-\mu^2}{f^2-\mu^2}\frac{\partial\zeta'}{\partial\mu}\right)+\beta\zeta=0. \qquad \text{.....................(4)}$$

If we assume

$$\zeta=\Sigma C_n P_n(\mu), \qquad \bar{\zeta}=\Sigma\gamma_n P_n(\mu), \qquad \text{.................(5)}$$

we have

$$\zeta'=\Sigma(\alpha_n C_n-\gamma_n)P_n(\mu). \qquad \text{.....................(6)}$$

Substituting in (4), and integrating between the limits -1 and μ, we find

$$\Sigma(\alpha_n C_n-\gamma_n)(1-\mu^2)\frac{dP_n}{d\mu}+\Sigma\beta C_n\{(f^2-1)+(1-\mu^2)\}\int_{-1}^{\mu}P_n d\mu=0. \ldots(7)$$

Now, by known formulae of zonal harmonics*,

$$\int_{-1}^{\mu}P_n d\mu=-\frac{1}{n(n+1)}(1-\mu^2)\frac{dP_n}{d\mu}, \qquad \text{.................(8)}$$

and

$$\int_{-1}^{\mu}P_n d\mu=\frac{1}{2n+1}(P_{n+1}-P_{n-1})$$

$$=\frac{1}{2n+1}\left\{\frac{1}{2n+3}\left(\frac{dP_{n+2}}{d\mu}-\frac{dP_n}{d\mu}\right)-\frac{1}{2n-1}\left(\frac{dP_n}{d\mu}-\frac{dP_{n-2}}{d\mu}\right)\right\}$$

$$=\frac{1}{(2n+1)(2n+3)}\frac{dP_{n+2}}{d\mu}-\frac{2}{(2n-1)(2n+3)}\frac{dP_n}{d\mu}+\frac{1}{(2n-1)(2n+1)}\frac{dP_{n-2}}{d\mu}. \ldots(9)$$

Substituting in (7), and equating to zero the coefficient of $(1-\mu^2)\dfrac{dP_n}{d\mu}$, we find

$$\frac{1}{(2n+3)(2n+5)}C_{n+2}-L_n C_n+\frac{1}{(2n-3)(2n-1)}C_{n-2}=\frac{\gamma_n}{\beta}, \qquad \ldots(10)$$

where

$$L_n=\frac{f^2-1}{n(n+1)}+\frac{2}{(2n-1)(2n+3)}-\frac{\alpha_n}{\beta}. \qquad \text{..............(11)}$$

The relation (10) will hold from $n=1$ onwards, provided we put

$$C_{-1}=3, \qquad C_0=0.$$

The further theory is based substantially on the argument of Laplace, given in Art. 221; and the work follows much the same lines as in Arts. 216, 217, 221.

In the *free* oscillations we have $\gamma_n=0$, and the admissible values of f are determined by the transcendental equation

$$L_2-\frac{\frac{1}{5.7^2.9}}{L_4-}\frac{\frac{1}{9.11^2.13}}{L_6-\&c.}=0, \qquad \text{.......................(12)}$$

or

$$L_1-\frac{\frac{1}{3.5^2.7}}{L_3-}\frac{\frac{1}{7.9^2.11}}{L_5-\&c.}=0, \qquad \text{.......................(13)}$$

* See Todhunter, *Functions of Laplace, &c.* c. v.; Whittaker and Watson, *Modern Analysis*, p. 306.

according as the mode is symmetrical or asymmetrical with respect to the equator. Alternative forms of the period equations are given by Hough, suitable for computation of the higher roots, and it is shewn that close approximations are given by the equations $L_n = 0$ or

$$\frac{\sigma^2}{4\omega^2} = 1 + n(n+1)\left\{\left(1 - \frac{3}{2n+1}\frac{\rho}{\rho_0}\right)\frac{gh}{4\omega^2 a^2} - \frac{2}{(2n-1)(2n+3)}\right\}, \quad \text{...(14)}$$

except for the first two or three values of n *.

The following table gives the periods (in sidereal time) of the slowest symmetrical oscillation (i.e. the one in which the surface-elevation would vary as $P_2(\mu)$ if there were no rotation), corresponding to various depths †.

β	Depth (feet)	$\frac{\sigma^2}{4\omega^2}$	Period h. m.	Period when $\omega=0$ h. m.
40	7260	·44155	18 3·5	32 49
20	14520	·62473	15 11·0	23 12
10	29040	·92506	12 28·6	16 25
5	58080	1·4785	9 52·1	11 35

The results obtained for the *forced* oscillations of the 'First Species' are very similar to those of Art. 217. The limiting form of the long-period tides when $\sigma=0$ shews the following results:

β	$\rho/\rho_0 = $·181		$\rho/\rho_0 = 0$	
	Pole	Equator	Pole	Equator
40	·140	·426	·154	·455
20	·266	·551		
10	·443	·681	·470	·708
5	·628	·796	·651	·817

The second and third columns give the ratio of the polar and equatorial tides to the respective equilibrium-values ‡. The numbers in the fourth and fifth columns are repeated from Art. 217. The comparison shews the effect of the mutual gravitation of the water in reducing the amplitude.

223. In the more general case, where symmetry about the axis is not imposed, the surface-elevation ζ is expanded by Hough in a series of tesseral harmonics of the type

$$P_n^s(\mu)\, e^{i(\sigma t + s\phi + \epsilon)}. \quad \text{.............................(1)}$$

* Reference may also be made to Poole, *Proc. Lond. Math. Soc.* (2) xix. 299.

† The slowest asymmetrical mode has a much longer period. It involves a displacement of the centre of mass of the water, so that a correction would be necessary if the nucleus were free; cf. Art. 199.

‡ The numbers are deduced from Hough's results. The paper referred to contains discussions of other interesting points, including an examination of cases of varying depth, with numerical illustrations.

In relation to tidal theory the most important cases are where the disturbing potential is of the form (1), with $n=2$ and $s=1$ or $s=2$.

The calculations are necessarily somewhat intricate*, and it may suffice here to mention a few of the more interesting results, which will indicate how the gaps in the previous investigations have been filled.

To understand the nature of the *free* oscillations, it is best to begin with the case of no rotation ($\omega=0$). As ω is increased, the pairs of numerically equal, but oppositely signed, values of σ which were obtained in Art. 199 begin to diverge in absolute value, that being the greater which has the same sign with ω. The character of the fundamental modes is also gradually altered. These oscillations are distinguished as 'of the First Class.'

At the same time certain steady motions which are possible, without change of level, when there is no rotation, are converted into long-period oscillations with change of level, the speeds being initially comparable with ω. The corresponding modes are designated as 'of the Second Class'†; cf. Art. 206.

The following table gives the speeds of those modes of the First Class which are of most importance in relation to the diurnal and semi-diurnal tides, respectively, and the corresponding periods, in sidereal time. The last column repeats the corresponding periods in the case of no rotation, as calculated from the formula (15) of Art. 200.

	Second Species [$s=1$]		Third Species [$s=2$]		
Depth (feet)	$\frac{\sigma}{\omega}$	Period h. m.	$\frac{\sigma}{\omega}$	Period h. m.	Period when $\omega=0$ h. m.
7260	1·6337 −0·9834	14 41 24 24	1·3347 −0·6221	17 59 38 34	32 49
14520	1·8677 −1·2450	12 51 19 16	1·6133 −0·8922	14 52 26 54	23 12
29040	2·1641 −1·6170	11 5 14 50	1·9968 −1·2855	12 1 18 40	16 25
58080	2·6288 −2·1611	9 8 11 6	2·5535 −1·8575	9 24 12 55	11 35

The quickest oscillation of the Second Class has in each case a period of over a day; and the periods of the remainder are very much longer.

* A simplification is made by Love, "Notes on the Dynamical Theory of the Tides," *Proc. Lond. Math. Soc.* (2) xii. 309 (1913). He writes

$$u=-\frac{\partial\chi}{a\,\partial\theta}-\frac{\partial\psi}{a\sin\theta\,\partial\phi},\quad v=-\frac{\partial\chi}{a\sin\theta\,\partial\phi}+\frac{\partial\psi}{a\,\partial\theta};$$

cf. Art. 154 (1). The values of χ, ψ are expanded in series of spherical harmonics.

† These two classes of oscillations have been already encountered in the plane problem of Art. 212.

As regards the *forced* oscillations of the 'Second Species,' Laplace's conclusion that when $\sigma = \omega$, exactly, the diurnal tide vanishes in the case of uniform depth, still holds. The computation for the most important lunar diurnal tide, for which $\sigma/\omega = \cdot 92700$, shews that with such depths as we have considered the tides are small compared with the equilibrium heights, and are in the main inverted.

Of the forced oscillations of the 'Third Species,' we may note first the case of the solar semi-diurnal tide, for which $\sigma = 2\omega$ with sufficient accuracy. For the four depths given in our tables, the ratio of the dynamical tide-height to the conventional equilibrium tide-height at the equator is found to be

$$+7\cdot 9548, \quad -1\cdot 5016, \quad -234\cdot 87, \quad +2\cdot 1389,$$

respectively.

"The very large coefficients which appear when $hg/4\omega^2a^2 = \frac{1}{10}$ indicate that for this depth there is a period of free oscillation of semi-diurnal type whose period differs but slightly from half-a-day. On reference to the tables ... it will be seen that we have, in fact, evaluated this period as 12 hours 1 minute, while for the case $hg/4\omega^2a^2 = \frac{1}{40}$ we have found a period of 12 hours 5 minutes*. We see then that though, when the period of forced oscillation differs from that of one of the types of free oscillation by as little as one minute, the forced tide may be nearly 250 times as great as the corresponding equilibrium tide, a difference of 5 minutes between these periods will be sufficient to reduce the tide to less than ten times the corresponding equilibrium tide. It seems then that the tides will not tend to become excessively large unless there is very close agreement with the period of one of the free oscillations.

"The critical depths for which the forced tides here treated of become infinite are those for which a period of free oscillation coincides exactly with 12 hours. They may be ascertained by putting [$\sigma = 2\omega$] in the period-equation for the free oscillations and treating this equation as an equation for the determination of h.......The two largest roots are..., and the corresponding critical depths are about 28,182 feet and 7375 feet.......

"It will be seen that in three cases out of the four here considered the effect of the mutual gravitation of the waters is to increase the ratio of the tide to the equilibrium tide [cf. Art. 221]. In two of the cases the sign is also reversed. This of course results from the fact that whereas when [$\rho/\rho_1 = 0\cdot 18093$] one of the periods of free oscillation is rather greater than 12 hours, when [$\rho/\rho_1 = 0$] the corresponding period will be less than 12 hours†."

Hough has also computed the lunar semi-diurnal tides for which

$$\frac{\sigma}{2\omega} = 0\cdot 96350.$$

* [Belonging to a mode which comes next in sequence to the one having a period of 17 h. 59 m.]

† Hough, *Phil. Trans.* A, cxci. 178, 179.

For the four depths aforesaid the ratios of the equatorial tide-heights to their equilibrium-values are found to be

$$-2{\cdot}4187, \quad -1{\cdot}8000, \quad +11{\cdot}0725, \quad +1{\cdot}9225,$$

respectively.

"On comparison of these numbers with those obtained for the solar tides..., we see that for a depth of 7260 feet the solar tides will be direct while the lunar tides will be inverted, the opposite being the case when the depth is 29,040 feet. This is of course due to the fact that in each of these cases there is a period of free oscillation intermediate between twelve solar (or, more strictly, sidereal) hours and twelve lunar hours. The critical depths for which the lunar tides become infinite are found to be 26,044 feet and 6448 feet.

"Consequently this phenomenon will occur if the depth of the ocean be between 29,182 feet and 26,044 feet, or between 7375 feet and 6448 feet. An important consequence would be that for depths lying between these limits the usual phenomena of spring and neap tides would be reversed, the higher tides occurring when the moon is in quadrature, and the lower at new and full moon*."

223 a. Some important contributions to the dynamical theory have been made by Goldsbrough. Considering, first, the tides in an ocean of uniform depth bounded by one or two parallels of latitude, he finds, in the case of a polar basin of angular radius 30°, for instance, that for such depths as have been considered in Arts. 217, 221 the long-period tides and the semi-diurnal tides do not deviate very widely from the values given by the equilibrium theory, when this is corrected as explained in the Appendix†. The case is different with the diurnal tides, which vary considerably with the size of the basin and the depth, and are as a rule considerable, whereas we have seen that in a uniform ocean covering the globe they are negligible.

In the case of an equatorial belt‡, the long-period tides again approximate to the equilibrium values, whilst the diurnal and semi-diurnal deviate widely, to an extent which varies considerably with the latitudes of the boundaries.

The variations here met with are doubtless conditioned by the relation between the imposed period and the natural periods of free oscillation. This question has been examined by Goldsbrough with reference to the semi-diurnal tides of the Atlantic ocean, which forms a more or less limited and isolated system. Taking the case of an ocean limited by two meridians 60° apart, and assuming the law of depth

$$h = h_0 \sin^2 \theta,$$

* Hough, *l.c.*, where reference is made to Kelvin's *Popular Lectures and Addresses*, London, 1894, ii. 22 (1868).

† *Proc. Lond. Math. Soc.* (2) xiv. 31 (1913). ‡ *Ibid.* xiv. 207 (1914).

he finds* that there will be a free oscillation with $\sigma = 2\omega$ exactly, provided $h_0 = 23{,}200$ ft., which means a mean depth of 15,500 ft. With $h_0 = 25{,}320$ ft., or a mean depth of 16,880 ft., he finds that the forced tides of the above period are still very large compared with the equilibrium values.

In a more recent paper† by Goldsbrough and Colborne the depth is taken to be uniform and equal to the estimated mean depth (12,700 ft.) of the Atlantic. For the imposed frequency they take that of the principal semi-diurnal constituent (usually denoted by M_2) of the lunar disturbing force ($\sigma/2\omega = \cdot 9625$). The amplitudes, though not so great as before, prove to be largely in excess of the equilibrium values. The diurnal tide in an ocean of this type has been investigated by Colborne‡.

224. It is not easy to estimate, in any but the most general way, the extent to which the foregoing conclusions of the dynamical theory would have to be modified if account could be taken of the actual configuration of the ocean, with its irregular boundaries and irregular variation of depth§. One or two points may however be noticed.

In the first place, the formulae (1) of Art. 206 would lead us to expect for any given tide a phase-difference, variable from place to place, between the tide-height and the disturbing force‖. Thus, in the case of the lunar semi-diurnal tides, for example, high-water or low-water need not synchronize with the transit of the moon or anti-moon across the meridian. More precisely, in the case of a disturbing force of given type for which the equilibrium tide-height at a particular place would be

$$\bar{\zeta} = a \cos \sigma t, \qquad \text{.......(1)}$$

the dynamical tide-height will be

$$\zeta = A \cos (\sigma t - \epsilon), \qquad \text{.......(2)}$$

where the ratio A/a, and the phase-difference ϵ, will be functions of the speed σ, as well as of the position of the station.

Again, consider the superposition of two oscillations of the same type but of slightly different speeds, *e.g.* the lunar and solar semi-diurnal tides. If the origin of t be taken at a syzygy, we have

$$\bar{\zeta} = a \cos \sigma t + a' \cos \sigma' t, \qquad \text{.......(3)}$$

and

$$\zeta = A \cos (\sigma t - \epsilon) + A' \cos (\sigma' t - \epsilon'). \qquad \text{.......(4)}$$

This may be written

$$\zeta = (A + A' \cos \phi) \cos (\sigma t - \epsilon) + A' \sin \phi \sin (\sigma t - \epsilon), \qquad \text{.......(5)}$$

where

$$\phi = (\sigma - \sigma') t - \epsilon + \epsilon'. \qquad \text{.......(6)}$$

* *Proc. Roy. Soc.* A, cxvii. 692 (1927).

† *Ibid.* cxxvi. 1 (1929).

‡ *Ibid.* cxxxi. 38 (1931).

§ As to the general mathematical problem reference may be made to Poincaré, "Sur l'équilibre et les mouvements des mers," *Liouville* (5), ii. 57, 217 (1896), and to his *Leçons de mécanique céleste*, iii.

‖ This is illustrated by the canal problem of Art. 184.

If the first term in the second member of (4) represents the lunar, and the second the solar tide, we shall have $\sigma < \sigma'$, and $A > A'$. If we write

$$A + A' \cos\phi = C\cos\alpha, \qquad A'\sin\phi = C\sin\alpha, \quad \text{...........(7)}$$

we get

$$\zeta = C\cos(\sigma t - \epsilon - \alpha), \quad \text{..........................(8)}$$

where

$$C = (A^2 + 2AA'\cos\phi + A'^2)^{\frac{1}{2}}, \qquad \alpha = \tan^{-1}\frac{A'\sin\phi}{A + A'\cos\phi}. \quad \text{...(9)}$$

This may be described as a simple-harmonic oscillation of slowly varying amplitude and phase. The amplitude ranges between the limits $A \pm A'$, whilst α may be supposed to lie always between $\pm\frac{1}{2}\pi$. The 'speed' must also be regarded as variable, viz. we find

$$\frac{d}{dt}(\sigma t - \alpha) = \frac{\sigma A^2 + (\sigma + \sigma')AA'\cos\phi + \sigma' A'^2}{A^2 + 2AA'\cos\phi + A'^2}. \quad \text{.........(10)}$$

This ranges between

$$\frac{A\sigma + A'\sigma'}{A + A'} \text{ and } \frac{A\sigma - A'\sigma'}{A - A'}. \quad \text{.....................(11)*}$$

The above is the well-known explanation of the phenomena of the spring- and neap-tides†; but we are now concerned further with the question of phase. On the equilibrium theory, the maxima of the amplitude C would occur whenever

$$(\sigma' - \sigma)t = 2n\pi,$$

where n is integral. On the dynamical theory the corresponding times of maximum are given by

$$(\sigma' - \sigma)t - (\epsilon' - \epsilon) = 2n\pi,$$

i.e. the dynamical maxima follow the statical by an interval‡

$$(\epsilon' - \epsilon)/(\sigma' - \sigma).$$

If the difference between σ' and σ were infinitesimal, this would be equal to $d\epsilon/d\sigma$.

The fact that the time of high-water, even at syzygy, may follow or precede the transit of the moon or anti-moon by an interval of several hours is well known§. The interval, when reckoned as a retardation, is, moreover, usually greater for the solar than for the lunar semi-diurnal tide, with the result that the spring-tides are in many places highest a day or two after the corresponding syzygy. The latter circumstance has been ascribed‖ to the operation of Tidal Friction (for which see Chapter XI.), but it is evident

* Helmholtz, *Lehre von den Tonempfindungen* (2e Aufl.), Braunschweig, 1870, p. 622.

† Cf. Thomson and Tait, Art. 60.

‡ This interval may of course be negative.

§ The values of the retardations (which we have denoted by ϵ) for the various tidal components, at a number of ports, are given by Baird and Darwin, "Results of the Harmonic Analysis of Tidal Observations," *Proc. R. S.* xxxix. 135 (1885), and Darwin, "Second Series of Results...," *Proc. R. S.* xlv. 556 (1889).

‖ Airy, "Tides and Waves," Art. 459.

that the phase-differences which are incidental to a complete dynamical theory, even in the absence of friction, cannot be ignored in this connection. There is reason to believe that they are, indeed, far more important than those due to the latter cause.

Lastly, it was shewn in Arts. 206, 217 that the long-period tides may deviate very considerably from the values given by the equilibrium theory, owing to the possibility of certain *steady* motions in the absence of disturbance. It has been pointed out by Rayleigh* that these steady motions may be impossible in certain cases where the ocean is limited by perpendicular barriers. Referring to Art. 214 (6), it appears that if the depth h be uniform, ζ must (in the steady motion) be a function of the co-latitude θ only, and therefore by (4) of the same Art., the eastward velocity v must be uniform along each parallel of latitude. This is inconsistent with the existence of a perpendicular barrier extending along a meridian. The objection would not necessarily apply to the case of a sea shelving gradually from the central parts to the edge†.

225. We may complete the investigation of Art. 200 by a brief notice of the question of the stability of the ocean, in the case of rotation.

It has been shewn in Art. 205 that the condition of secular stability is that $V - T_0$ should be a minimum in the equilibrium configuration. If we neglect the mutual attraction of the elevated water, the application to the present problem is very simple. The excess of the quantity $V - T_0$ over its undisturbed value is evidently

$$\iint\left\{\int_0^\zeta (\Psi - \tfrac{1}{2}\omega^2\varpi^2)\, dz\right\} dS, \qquad \text{.......................(1)}$$

where Ψ denotes the potential of the earth's attraction, δS is an element of the oceanic surface, and the rest of the notation is as before. Since $\Psi - \frac{1}{2}\omega^2\varpi^2$ is constant over the undisturbed level ($z = 0$), its value at a small altitude z may be taken to be $gz +$ const., where, as in Art. 213,

$$g = \left[\frac{\partial}{\partial z}(\Psi - \tfrac{1}{2}\omega^2\varpi^2)\right]_{z=0}. \qquad \text{.......................(2)}$$

Since $\iint \zeta\, dS = 0$, on account of the constancy of volume, we find from (1) that the increment of $V - T_0$ is

$$\tfrac{1}{2}\iint g\zeta^2\, dS. \qquad \text{................................(3)}$$

This is essentially positive, and the equilibrium is therefore 'secularly' stable‡.

* "Note on the Theory of the Fortnightly Tide," *Phil. Mag.* (6) v. 136 (1903) [*Papers*, iv. 84].

† The theory of the limiting forms of long-period tides in oceans of various types is discussed by Proudman, *Proc. Lond. Math. Soc.* (2) xiii. 273 (1913).

‡ Cf. Laplace, *Mécanique Céleste*, Livre 4me, Arts. 13, 14.

It is to be noticed that this proof does not involve any restriction as to the depth of the fluid, or as to smallness of the ellipticity, or even as to symmetry of the undisturbed surface with respect to the axis of rotation.

If we wish to take into account the mutual attraction of the water, the problem can only be solved without difficulty when the undisturbed surface is nearly spherical, and we neglect the variation of g. The question (as to secular stability) is then exactly the same as in the case of no rotation. The calculation for this case will find an appropriate place in the next chapter (Art. 264). The result, as we might anticipate from Art. 200, is that the necessary and sufficient condition of stability of the ocean is that its density should be less than the mean density of the earth*.

226. This is perhaps the most suitable occasion for a few additional remarks on the general question of stability of dynamical systems. We have in the main followed the ordinary usage which pronounces a state of equilibrium, or of steady motion, to be stable or unstable according to the character of the solution of the approximate equations of disturbed motion. If the solution consists of series of terms of the type $Ce^{\pm\lambda t}$, where all the values of λ are pure imaginary (*i.e.* of the form $i\sigma$), the undisturbed state is usually reckoned as stable; whilst if any of the λ's are real, it is accounted unstable. In the case of disturbed *equilibrium*, this leads algebraically to the usual criterion of a minimum value of V as a necessary and sufficient condition of stability.

It has in recent times been questioned whether this conclusion is, from a practical point of view, altogether warranted. It is pointed out that since the approximate dynamical equations become less and less accurate as the deviation from the equilibrium configuration increases, it is a matter for examination how far rigorous conclusions as to the ultimate extent of the deviation can be drawn from them†.

The argument of Dirichlet, which establishes that the occurrence of a minimum value of V is a *sufficient* condition of stability, in any practical sense, has already been referred to. No such simple proof is available to shew without qualification that this condition is *necessary*. If, however, we recognize the existence of dissipative forces, which are called into play by any motion whatever of the system, the conclusion can be drawn as in Art. 205.

A little consideration will shew that a good deal of the obscurity which attaches to the question arises from the want of a sufficiently precise mathematical definition of what is meant by 'stability.' The difficulty is encountered in an aggravated form when we pass to the question of

* Cf. Laplace, *l.c.*

† See papers by Liapounoff and Hadamard, *Liouville* (5), iii. (1897).

stability of motion. The various definitions which have been propounded by different writers are examined critically by Klein and Sommerfeld in their book on the theory of the top*. Rejecting previous definitions, they base their criterion on the character of the changes produced in the *path* of the system by small arbitrary disturbing impulses. If the undisturbed path be the *limiting form* of the disturbed path when the impulses are indefinitely diminished, it is said to be stable, but not otherwise. For instance, the vertical fall of a particle under gravity is reckoned as stable, although for a *given* impulsive disturbance, however small, the deviation of the particle's position at any time t from the position which it occupied in the original motion increases indefinitely with t. Even this criterion, as the writers referred to themselves recognize, is not free from ambiguity unless the phrase 'limiting form,' as applied to a path, be strictly defined. It appears moreover that a definition which is analytically precise may not in all cases be easy to reconcile with geometrical prepossessions†.

The foregoing considerations have reference, of course, to the question of 'ordinary' stability. The more important theory of 'secular' stability (Art. 205) is not affected. We shall meet with the criterion for this, under a somewhat modified form, at a later stage in our subject‡.

* *Ueber die Theorie des Kreisels*, Leipzig, 1897..., p. 342.

† Some good illustrations are furnished by Particle Dynamics. Thus a particle moving in a circle about a centre of force varying inversely as the cube of the distance will if slightly disturbed either fall into the centre, or recede to infinity, after describing in either case a spiral with an infinite number of convolutions. Each of these spirals has, analytically, the circle as its 'limiting form,' although the motion in the latter is most naturally described as unstable. Cf. Korteweg, *Wiener Ber.* May 20, 1886.

A narrower definition has been given by Love, and applied by Bromwich to several dynamical and hydrodynamical problems; see *Proc. Lond. Math. Soc.* (1) xxxiii. 325 (1901).

‡ This summary is taken substantially from the Art. "Dynamics, Analytical," in *Encyc. Brit.* 10th ed. xxvii. 566 (1902), and 11th ed. viii. 756 (1910).

APPENDIX

TO CHAPTER VIII

ON TIDE-GENERATING FORCES

a. If, in the annexed figure, O and C be the centres of the earth and of the disturbing body (say the moon), the potential of the moon's attraction at a point P near the earth's surface will be $-\gamma M/CP$, where M denotes the moon's mass, and γ the gravitation-constant. If we put $OC=D$, $OP=r$, and denote the moon's (geocentric) zenith-distance at P, viz. the angle POC, by ϑ, this potential is equal to

$$-\frac{\gamma M}{(D^2-2rD\cos\vartheta+r^2)^{\frac{1}{2}}}.$$

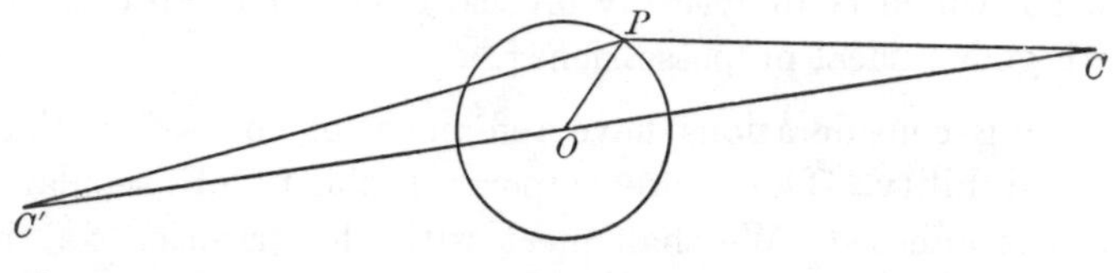

We require, however, not the absolute accelerative effect at P, but the acceleration relative to the earth. Now the moon produces in the whole mass of the earth an acceleration $\gamma M/D^2$* parallel to OC, and the potential of a uniform field of force of this intensity is evidently

$$-\frac{\gamma M}{D^2}\cdot r\cos\vartheta.$$

Subtracting this from the former result we get, for the potential of the relative attraction at P,

$$\Omega=-\frac{\gamma M}{(D^2-2rD\cos\vartheta+r^2)^{\frac{1}{2}}}+\frac{\gamma M}{D^2}\cdot r\cos\vartheta. \qquad (1)$$

This function Ω is identical with the 'disturbing-function' of planetary theory.

Expanding in powers of r/D, which is in our case a small quantity, and retaining only the most important term, we find

$$\Omega=\frac{3}{2}\frac{\gamma Mr^2}{D^3}\left(\tfrac{1}{3}-\cos^2\vartheta\right). \qquad (2)$$

Considered as a function of the position of P, this is a zonal harmonic of the second degree, with OC as axis.

The reader will easily verify that, to the order of approximation adopted, Ω is equal to the joint potential of two masses, each equal to $\frac{1}{2}M$, placed, one at C, and the other at a point C' in CO produced such that $OC'=OC$†.

b. In the 'equilibrium-theory' of the tides it is assumed that the free surface takes at each instant the equilibrium-form which might be maintained if the disturbing body were to retain unchanged its actual position relative to the rotating earth. In other

* The effect of this is to produce a monthly inequality in the motion of the earth's centre about the sun. The amplitude of the inequality in radius vector is about 3000 miles; that of the inequality in longitude is about 7″; see Laplace, *Mécanique Céleste*, Livre 6me, Art. 30, and Livre 13me, Art. 10.

† Thomson and Tait, Art. 804. These two fictitious bodies are designated as 'moon' and 'anti-moon,' respectively.

words, the free surface is assumed to be a level-surface under the combined action of gravity, of centrifugal force, and of the disturbing force. The equation to this level-surface is

$$\Psi - \tfrac{1}{2}\omega^2\varpi^2 + \Omega = \text{const.}, \quad \text{.........(3)}$$

where ω is the angular velocity of the rotation, ϖ denotes the distance of any point from the earth's axis, and Ψ is the potential of the earth's attraction. If we use square brackets [ ] to distinguish the values of the enclosed quantities at the undisturbed level, and denote by $\bar{\zeta}$ the elevation of the water above this level due to the disturbing potential Ω, the above equation is equivalent to

$$[\Psi - \tfrac{1}{2}\omega^2\varpi^2] + \left[\frac{\partial}{\partial z}(\Psi - \tfrac{1}{2}\omega^2\varpi^2)\right]\bar{\zeta} + \Omega = \text{const.}, \quad \text{.........(4)}$$

approximately, where $\partial/\partial z$ is used to indicate a space-differentiation along the normal outwards. The first term is of course constant, and we therefore have

$$\bar{\zeta} = -\frac{\Omega}{g} + C, \quad \text{.........(5)}$$

where, as in Art. 213,

$$g = \left[\frac{\partial}{\partial z}(\Psi - \tfrac{1}{2}\omega^2\varpi^2)\right]. \quad \text{.........(6)}$$

Evidently, g denotes the value of 'apparent gravity'; it will of course vary more or less with the position of P on the earth's surface.

It is usual, however, in the theory of the tides, to ignore the slight variations in the value of g, and the effect of the ellipticity of the undisturbed level on the surface-value of Ω. Putting, then, $r = a$, $g = \gamma E/a^2$, where E denotes the earth's mass, and a the mean radius of the surface, we have, from (2) and (5),

$$\bar{\zeta} = H(\cos^2 \vartheta - \tfrac{1}{3}) + C, \quad \text{.........(7)}$$

where

$$H = \tfrac{3}{2} \cdot \frac{M}{E} \cdot \left(\frac{a}{D}\right)^3 \cdot a, \quad \text{.........(8)}$$

as in Art. 180. Hence the equilibrium-form of the free surface is a harmonic spheroid of the second order, of the zonal type, whose axis passes through the disturbing body.

c. Owing to the diurnal rotation, and also to the orbital motion of the disturbing body, the position of the tidal spheroid relative to the earth is continually changing, so that the level of the water at any particular place will continually rise and fall. To analyse the character of these changes, let θ be the co-latitude, and ϕ the longitude, measured eastward from some fixed meridian, of any place P, and let Δ be the north-polar-distance, and α the hour-angle west of the same meridian, of the disturbing body. We have, then,

$$\cos \vartheta = \cos \Delta \cos \theta + \sin \Delta \sin \theta \cos(\alpha + \phi), \quad \text{.........(9)}$$

and thence, by (7),

$$\begin{aligned}\bar{\zeta} = {} & \tfrac{3}{2}H(\cos^2 \Delta - \tfrac{1}{3})(\cos^2 \theta - \tfrac{1}{3}) \\ & + \tfrac{1}{2}H \sin 2\Delta \sin 2\theta \cos(\alpha + \phi) \\ & + \tfrac{1}{2}H \sin^2 \Delta \sin^2 \theta \cos 2(\alpha + \phi) + C. \quad \text{.........(10)}\end{aligned}$$

Each of these terms may be regarded as representing a partial tide, and the results superposed.

Thus, the *first* term is a zonal harmonic of the second order, and gives a tidal spheroid symmetrical with respect to the earth's axis, having as nodal lines the parallels for which $\cos^2 \theta = \tfrac{1}{3}$, or $\theta = 90° \pm 35° 16'$. The amount of the tidal elevation in any particular latitude varies as $\cos^2 \Delta - \tfrac{1}{3}$. In the case of the moon the chief fluctuation in this quantity has a period of about a fortnight; we have here the origin of the 'lunar fortnightly' or 'declinational' tide. When the sun is the disturbing body, we have a 'solar semi-annual' tide. It is to be noticed that the mean value of $\cos^2 \Delta - \tfrac{1}{3}$ with respect to the time is not zero, so that the inclination of the orbit of the disturbing body to the equator involves as a consequence a permanent change of mean level. Cf. Art. 183.

The *second* term in (10) is a spherical harmonic of the type obtained by putting $n=2$, $s=1$ in Art. 86 (7). The corresponding tidal spheroid has as nodal lines the meridian which is distant 90° from that of the disturbing body, and the equator. The disturbance of level is greatest in the meridian of the disturbing body, at distances of 45° N. and S. of the equator. The oscillation at any one place goes through its period with the hour-angle, α, *i.e.* in a lunar or solar day. The amplitude is, however, not constant, but varies slowly with Δ, changing sign when the disturbing body crosses the equator. This term accounts for the lunar and solar 'diurnal' tides.

The *third* term is a sectorial harmonic ($n=2$, $s=2$), and gives a tidal spheroid having as nodal lines the meridians which are distant 45° E. and W. from that of the disturbing body. The oscillation at any one place goes through its period with 2α, *i.e.* in half a (lunar or solar) day, and the amplitude varies as $\sin^2 \Delta$, being greatest when the disturbing body is on the equator. We have here the origin of the lunar and solar 'semi-diurnal' tides.

The 'constant' C is to be determined by the consideration that, on account of the invariability of volume, we must have

$$\iint \bar{\zeta}\, dS = 0, \qquad \text{.......................................(11)}$$

where the integration extends over the surface of the ocean. If the ocean cover the whole earth we have $C=0$, by the general property of spherical surface-harmonics quoted in Art. 87. It appears from (7) that the greatest elevation above the undisturbed level is then at the points $\vartheta=0$, $\vartheta=180°$, *i.e.* at the points where the disturbing body is in the zenith or nadir, and the amount of this elevation is $\frac{2}{3}H$. The greatest depression is at places where $\vartheta=90°$, *i.e.* the disturbing body is on the horizon, and is $\frac{1}{3}H$. The greatest possible range is therefore equal to H.

In the case of a limited ocean, C does not vanish, but has at each instant a definite value depending on the position of the disturbing body relative to the earth. This value may be easily written down from equations (10) and (11); it is a sum of spherical harmonic functions of Δ, α, of the second order, with constant coefficients in the form of surface-integrals whose values depend on the distribution of land and water over the globe. The changes in the value of C, due to relative motion of the disturbing body, give a *general* rise and fall of the free surface, with (in the case of the moon) fortnightly, diurnal, and semi-diurnal periods. This 'correction to the equilibrium-theory' as usually presented, was first fully investigated by Thomson and Tait*. The necessity for a correction of the kind, in the case of a limited sea, had however been recognized by D. Bernoulli†.

The correction has an influence on the time of high water, which is no longer synchronous with the maximum of the disturbing potential. The interval, moreover, by which high water is accelerated or retarded differs from place to place‡.

d. We have up to this point neglected the mutual attraction of the particles of the water. To take this into account, we must add to the disturbing potential Ω the gravitation-potential of the elevated water. In the case of an ocean covering the earth, the correction can be easily applied, as in Art. 200. If we put $n=2$ in the formulae of

* *Natural Philosophy*, Art. 808; see also Darwin, "On the Correction to the Equilibrium Theory of the Tides for the Continents," *Proc. Roy. Soc.* April 1, 1886 [*Papers*, i. 328]. It appears as the result of a numerical calculation by Prof. H. H. Turner, appended to this paper, that with the actual distribution of land and water the correction is of little importance.

† *Traité sur le Flux et Reflux de la Mer*, c. xi. (1740). This essay, as well as the one by Maclaurin cited on p. 307, and another on the same subject by Euler, is reprinted in Le Seur and Jacquier's edition of Newton's *Principia*.

‡ Thomson and Tait, Art. 810. The point is illustrated by the formula (3) of Art. 184 *supra*.

that Art., the addition to the value of Ω is $-\frac{3}{5}\rho/\rho_0 \,.\, g\bar{\zeta}$; and we thence find without difficulty

$$\bar{\zeta}=\frac{H}{1-\frac{3}{5}\rho/\rho_0}(\cos^2\vartheta-\tfrac{1}{3}). \qquad\text{.......(12)}$$

It appears that all the tides are *increased*, in the ratio $(1-\frac{3}{5}\rho/\rho_0)^{-1}$. If we assume $\rho/\rho_0=\cdot 18$, this ratio is 1·12.

e. So much for the equilibrium theory. For the purposes of the kinetic theory of Arts. 213–224, it is necessary to suppose the value (10) of $\bar{\zeta}$ to be expanded in a series of simple-harmonic functions of the time. The actual expansion, taking account of the variations of Δ and α, and of the distance D of the disturbing body (which enters into the value of H), is a somewhat complicated problem of Physical Astronomy, into which we do not enter*.

Disregarding the constant C, which disappears in the dynamical equations (1) of Art. 215, the constancy of volume being now secured by the equation of continuity (2), it is easily seen that the terms in question will be of three distinct types.

First, we have the tides of long period, for which

$$\bar{\zeta}=H'(\cos^2\theta-\tfrac{1}{3})\,.\,\cos(\sigma t+\epsilon). \qquad\text{.......(13)}$$

The most important tides of this class are the 'lunar fortnightly' for which, in degrees per mean solar hour, $\sigma=1^\circ\cdot 098$, and the 'solar-annual' for which $\sigma=0^\circ\cdot 082$.

Secondly, we have the diurnal tides, for which

$$\bar{\zeta}=H''\sin\theta\cos\theta\,.\,\cos(\sigma t+\phi+\epsilon), \qquad\text{.......(14)}$$

where σ differs but little from the angular velocity ω of the earth's rotation. These include the 'lunar diurnal' ($\sigma=13^\circ\cdot 943$), the 'solar diurnal' ($\sigma=14^\circ\cdot 959$), and the 'luni-solar diurnal' ($\sigma=\omega=15^\circ\cdot 041$).

Lastly, we have the semi-diurnal tides, for which

$$\bar{\zeta}=H'''\sin^2\theta\,.\,\cos(\sigma t+2\phi+\epsilon), \qquad\text{.......(15)†}$$

where σ differs but little from 2ω. These include the 'lunar semi-diurnal' ($\sigma=28^\circ\cdot 984$), the 'solar semi-diurnal' ($\sigma=30^\circ$), and the 'luni-solar semi-diurnal' ($\sigma=2\omega=30^\circ\cdot 082$).

For a complete enumeration of the more important partial tides, and for the values of the coefficients H', H'', H''' in the several cases, we must refer to the investigations of Darwin, already cited. In the Harmonic Analysis of Tidal Observations, which is the special object of these investigations, the only result of dynamical theory which is made use of is the general principle that the tidal elevation at any place must be equal to the sum of a series of simple-harmonic functions of the time, whose periods are the same as those of the several terms in the development of the disturbing potential, and are therefore known *à priori*. The amplitudes and phases of the various partial tides, for any particular port, are then determined by comparison with tidal observations extending over a

* Reference may be made to Laplace, *Mécanique Céleste*, Livre 13me, Art. 2. The more complete development which has served as the basis of all recent accurate tidal work is due to Darwin, and is reprinted in his *Papers*, i. This development is only quasi-harmonic, certain elements which are only slowly variable being treated as constants, but adjustable from time to time. A strict harmonic development has recently been carried out by Doodson, *Proc. Roy. Soc.* A, c. 305 (1921).

† It is evident that over a small area, near the poles, which may be treated as sensibly plane, the formulae (14) and (15) make

$$\bar{\zeta}\propto r\cos(\sigma t+\phi+\epsilon), \text{ and } \bar{\zeta}\propto r^2\cos(\sigma t+2\phi+\epsilon),$$

respectively, where r, ω are plane polar co-ordinates. These forms have been used by anticipation in Arts. 211, 212.

sufficiently long period*. We thus obtain a practically complete expression which can be used for the systematic prediction of the tides at the port in question.

f. One point of special interest in the Harmonic Analysis is the determination of the long-period tides. It has been already stated that under the influence of dissipative forces these must tend to approximate more or less closely to their equilibrium values. In the case of an ocean covering the globe it is at least doubtful whether the dissipative forces would be sufficient to produce an appreciable effect in the direction indicated. The amplitudes might therefore be expected to fall below those given by the equilibrium theory, for the dynamical reason explained in Arts. 206, 214. In the actual ocean, on the other hand, this consideration does not apply†, whilst the influence of friction is much greater. We may assume, then, that if the earth were absolutely rigid the long-period tides would have their full equilibrium values. As a matter of fact the lunar fortnightly, which is the only one whose amplitude can be inferred with any certainty from the observations, appears to fall short by about one-third. The discrepancy is attributed to elastic yielding of the solid body of the earth to the tidal distorting forces exerted by the moon.

* It is of interest to note, in connection with Art. 187, that the tide-gauges, being situated in relatively shallow water, are sensibly affected by certain tides of the second order, which therefore have to be taken account of in the general scheme of Harmonic Analysis.

† See the paper by Rayleigh cited on p. 355 *ante*.

CHAPTER IX

SURFACE WAVES

227. We have now to investigate, as far as possible, the laws of wave-motion in liquids when the vertical acceleration is no longer neglected. The most important case not covered by the preceding theory is that of waves on relatively deep water, where, as will be seen, the agitation rapidly diminishes in amplitude as we pass downwards from the surface; but it will be understood that there is a continuous transition to the state of things investigated in the preceding chapter, where the horizontal motion of the fluid was sensibly the same from top to bottom.

We begin with the oscillations of a horizontal sheet of water, and we will confine ourselves in the first instance to cases where the motion is in two dimensions, of which one (x) is horizontal, and the other (y) vertical. The elevations and depressions of the free surface will then present the appearance of a series of parallel straight ridges and furrows, perpendicular to the plane xy.

The motion, being assumed to have been generated originally from rest by the action of ordinary forces, will necessarily be irrotational, and the velocity-potential ϕ will satisfy the equation

$$\frac{\partial^2 \phi}{\partial x^2} + \frac{\partial^2 \phi}{\partial y^2} = 0, \quad \text{.................................(1)}$$

with the condition

$$\frac{\partial \phi}{\partial n} = 0 \quad \text{....................................(2)}$$

at a fixed boundary.

To find the condition which must be satisfied at the free surface ($p =$ const.), let the origin O be taken at the undisturbed level, and let Oy be drawn vertically upwards. The motion being assumed to be infinitely small, we find, putting $\Omega = gy$ in the formula (4) of Art. 20, and neglecting the square of the velocity (q),

$$\frac{p}{\rho} = \frac{\partial \phi}{\partial t} - gy + F(t). \quad \text{...........................(3)}$$

Hence if η denote the elevation of the surface at time t above the point (x, 0), we shall have, since the pressure there is uniform,

$$\eta = \frac{1}{g}\left[\frac{\partial \phi}{\partial t}\right]_{y=\eta}, \quad \text{..................................(4)}$$

provided the function $F(t)$, and the additive constant, be supposed merged in the value of $\partial\phi/\partial t$. Subject to an error of the order already neglected, this may be written

$$\eta = \frac{1}{g}\left[\frac{\partial\phi}{\partial t}\right]_{y=0}. \quad \text{.............................(5)}$$

Since the normal to the free surface makes an infinitely small angle $(\partial\eta/\partial x)$ with the vertical, the condition that the normal component of the fluid velocity at the free surface must be equal to the normal velocity of the surface itself gives, with sufficient approximation,

$$\frac{\partial\eta}{\partial t} = -\left[\frac{\partial\phi}{\partial y}\right]_{y=0}. \quad \text{.............................(6)}$$

This is in fact what the general surface condition (Art. 9 (3)) becomes, if we put $F(x, y, z, t) \equiv y - \eta$, and neglect small quantities of the second order.

Eliminating η between (5) and (6), we obtain the condition

$$\frac{\partial^2\phi}{\partial t^2} + g\frac{\partial\phi}{\partial y} = 0, \quad \text{.............................(7)}$$

to be satisfied when $y = 0$. This is equivalent to $Dp/Dt = 0$.

In the case of simple-harmonic motion, the time-factor being $e^{i(\sigma t+\epsilon)}$, this condition becomes

$$\sigma^2\phi = g\frac{\partial\phi}{\partial y}. \quad \text{.............................(8)}$$

228. Let us apply this to the free oscillations of a sheet of water, or a straight canal, of uniform depth h, and let us suppose for the present that there are no limits to the fluid in the direction of x, the fixed boundaries, if any, being vertical planes parallel to xy.

Since the conditions are uniform in respect to x, the simplest supposition we can make is that ϕ is a simple-harmonic function of x; the most general case consistent with the above assumptions can be derived from this by superposition, in virtue of Fourier's Theorem.

We assume then

$$\phi = P\cos kx \,.\, e^{i(\sigma t+\epsilon)}, \quad \text{.........................(1)}$$

where P is a function of y only. The equation (1) of Art. 227 gives

$$\frac{d^2P}{dy^2} - k^2P = 0, \quad \text{.............................(2)}$$

whence

$$P = Ae^{ky} + Be^{-ky}. \quad \text{.............................(3)}$$

The condition of no vertical motion at the bottom is $\partial\phi/\partial y = 0$ for $y = -h$, whence

$$Ae^{-kh} = Be^{kh}, = \tfrac{1}{2}C, \text{ say.}$$

This leads to

$$\phi = C\cosh k(y+h)\cos kx \,.\, e^{i(\sigma t+\epsilon)}. \quad \text{.................(4)}$$

The value of σ is then determined by Art. 227 (8), which gives

$$\sigma^2 = gk\tanh kh. \quad \text{.............................(5)}$$

Substituting from (4) in Art. 227 (5), we find

$$\eta = \frac{i\sigma C}{g} \cosh kh \cos kx \,.\, e^{i(\sigma t+\epsilon)}, \quad \text{.....................(6)}$$

or, writing $$a = -\frac{\sigma C}{g} \,.\cosh kh,$$

and retaining only the real part of the expression,

$$\eta = a \cos kx \,.\, \sin(\sigma t + \epsilon). \quad \text{.......................(7)}$$

This represents a system of 'standing waves,' of wave-length $\lambda = 2\pi/k$, and vertical amplitude a. The relation between the period $(2\pi/\sigma)$ and the wave-length is given by (5). Some numerical examples of this dependence are given on p. 369.

In terms of a we have

$$\phi = -\frac{ga}{\sigma}\frac{\cosh k(y+h)}{\cosh kh} \cos kx \,.\cos(\sigma t + \epsilon), \quad \text{...............(8)}$$

and it is easily seen from Art. 62 that the corresponding value of the stream-function is

$$\psi = \frac{ga}{\sigma}\frac{\sinh k(y+h)}{\cosh kh} \sin kx \,.\cos(\sigma t + \epsilon). \quad \text{...............(9)}$$

If $\mathbf{x}$, $\mathbf{y}$ be the co-ordinates of a particle relative to its mean position (x, y), we have

$$\frac{d\mathbf{x}}{dt} = -\frac{\partial\phi}{\partial x}, \quad \frac{d\mathbf{y}}{dt} = -\frac{\partial\psi}{\partial y}, \quad \text{.......................(10)}$$

if we neglect the differences between the component velocities at the points (x, y) and $(x+\mathbf{x}, y+\mathbf{y})$, as being small quantities of the second order. Substituting from (8), and integrating with respect to t, we find

$$\left.\begin{aligned} \mathbf{x} &= -a\frac{\cosh k(y+h)}{\sinh kh} \sin kx \,.\sin(\sigma t + \epsilon), \\ \mathbf{y} &= a\frac{\sinh k(y+h)}{\sinh kh} \cos kx \,.\sin(\sigma t + \epsilon), \end{aligned}\right\} \quad \text{...........(11)}$$

where a slight reduction has been effected by means of (5). The motion of each particle is rectilinear, and simple-harmonic, the direction of motion varying from vertical, beneath the crests and hollows $(kx = m\pi)$, to horizontal, beneath the nodes $(kx = (m+\frac{1}{2})\pi)$. As we pass downwards from the surface to the bottom the amplitude of the vertical motion diminishes from $a \cos kx$ to 0, whilst that of the horizontal motion diminishes in the ratio $\cosh kh : 1$.

When the wave-length is very small compared with the depth, kh is large, and therefore $\tanh kh = 1$*. The formulae (11) then reduce to

$$\mathbf{x} = -ae^{ky} \sin kx \,.\sin(\sigma t + \epsilon), \qquad \mathbf{y} = ae^{ky} \cos kx \,.\sin(\sigma t + \epsilon), \quad \text{...(12)}$$

with $$\sigma^2 = gk. \quad \text{.................................(13)}$$

* This case may of course be more easily investigated independently.

The motion now diminishes rapidly from the surface downwards; thus at a depth of a wave-length the diminution of amplitude is in the ratio $e^{-2\pi}$ or 1/535. The forms of the lines of (oscillatory) motion ($\psi = \text{const.}$), for this case, are shewn in the annexed figure.

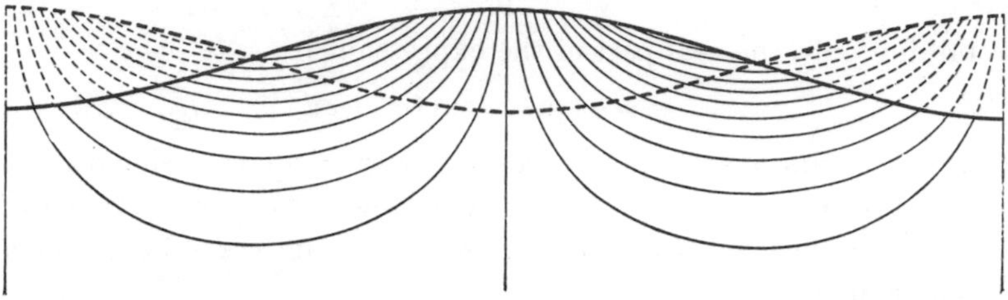

In the above investigation the fluid is supposed to extend to infinity in the direction of x, and there is consequently no restriction to the value of k. The formulae also give, however, the longitudinal oscillations in a canal of finite length, provided k have the proper values. If the fluid be bounded by the vertical planes $x = 0$, $x = l$ (say), the condition $\partial\phi/\partial x = 0$ is satisfied at both ends provided $\sin kl = 0$, or $kl = m\pi$, where $m = 1, 2, 3, \ldots\ldots$ The wave-lengths of the normal modes are therefore given by the formula $\lambda = 2l/m$. Cf. Art. 178.

229. The investigation of the preceding Art. relates to the case of 'standing' waves; it naturally claimed the first place, as a straightforward application of the usual method of ascertaining the normal modes of oscillation of a system about a state of equilibrium.

In the case, however, of a sheet of water, or a canal, of uniform depth, extending horizontally to infinity in both directions, we can, by superposition of two systems of standing waves of the same wave-length, obtain a system of progressive waves which advance unchanged with constant velocity. For this, it is necessary that the crests and troughs of one component system should coincide (horizontally) with the nodes of the other, that the amplitudes of the two systems should be equal, and that their phases should differ by a quarter-period.

Thus if we put
$$\eta = \eta_1 \pm \eta_2, \qquad \text{(1)}$$
where
$$\eta_1 = a \sin kx \cos \sigma t, \qquad \eta_2 = a \cos kx \sin \sigma t, \qquad \text{(2)}$$
we get
$$\eta = a \sin (kx \pm \sigma t), \qquad \text{(3)}$$
which represents an infinite train of waves travelling in the negative or positive direction of x, respectively, with the velocity c given by
$$c = \frac{\sigma}{k} = \left(\frac{g}{k} \tanh kh\right)^{\frac{1}{2}}, \qquad \text{(4)}$$
where the value of σ has been substituted from Art. 228 (5). In terms of

the wave-length (λ) we have

$$c = \left(\frac{g\lambda}{2\pi}\tanh\frac{2\pi h}{\lambda}\right)^{\frac{1}{2}}. \quad \text{.........................(5)}$$

When the wave-length is anything less than double the depth, we have $\tanh kh = 1$, sensibly, and therefore*

$$c = \left(\frac{g}{k}\right)^{\frac{1}{2}} = \left(\frac{g\lambda}{2\pi}\right)^{\frac{1}{2}}. \quad \text{.........................(6)}$$

On the other hand when λ is moderately large compared with h we have $\tanh kh = kh$, nearly, so that the velocity is independent of the wave-length, being given by

$$c = (gh)^{\frac{1}{2}}, \quad \text{.........................(7)}$$

as in Art. 170. This result is here obtained on the assumption that the wave-profile is a curve of sines, but Fourier's Theorem shews that the restriction is now to a great extent unnecessary.

It appears, on tracing the curve $y = (\tanh x)/x$, or from a numerical table to be given presently, that for a given depth h the wave-velocity increases constantly with the wave-length, from zero to the asymptotic value (7).

Let us now fix our attention, for definiteness, on a train of simple-harmonic waves travelling in the positive direction, *i.e.* we take the lower sign in (1) and (3). It appears, on comparison with Art. 228 (7), that the value of η_1 is deduced by putting $\epsilon = \frac{1}{2}\pi$, and subtracting $\frac{1}{2}\pi$ from the value of kx†, and that of η_2 by putting $\epsilon = 0$, simply. This proves a statement made above as to the relation between the component systems of standing waves, and also enables us to write down at once the proper modifications of the remaining formulae of the preceding Art.

Thus, we find, for the component displacements of a particle,

$$\left.\begin{aligned} \mathbf{x} = \mathbf{x}_1 - \mathbf{x}_2 = a\frac{\cosh k(y+h)}{\sinh kh}\cos(kx - \sigma t), \\ \mathbf{y} = \mathbf{y}_1 - \mathbf{y}_2 = a\frac{\sinh k(y+h)}{\sinh kh}\sin(kx - \sigma t). \end{aligned}\right\} \quad \text{...........(8)}$$

This shews that the motion of each particle is elliptic-harmonic, the period ($2\pi/\sigma, = \lambda/c$) being that in which the disturbance travels over a wave-length. The semi-axes, horizontal and vertical, of the elliptic orbits are

$$a\frac{\cosh k(y+h)}{\sinh kh} \quad \text{and} \quad a\frac{\sinh k(y+h)}{\sinh kh},$$

respectively. These both diminish from the surface to the bottom ($y = -h$), where the latter vanishes. The distance between the foci is the same for all

* Green, "Note on the Motion of Waves in Canals," *Camb. Trans.* vii. (1839) [*Papers*, p. 279].

† This is merely equivalent to a change of the origin from which x is measured.

the ellipses, being equal to a cosech kh. It easily appears, on comparison of (8) with (3), that a surface-particle is moving in the direction of wave-propagation when it is at a crest, and in the opposite direction when it is in a trough*.

When the depth exceeds half a wave-length, e^{-kh} is very small, and the formulae (8) reduce to

$$\mathbf{x} = ae^{ky}\cos(kx - \sigma t), \qquad \mathbf{y} = ae^{ky}\sin(kx - \sigma t), \quad \text{............(9)}$$

so that each particle describes a circle, with constant angular velocity $\sigma, = (2\pi g/\lambda)^{\frac{1}{2}}$†. The radii of these circles are given by the formula ae^{ky}, and therefore diminish rapidly downwards.

In the table given below, the second column gives the values of sech kh corresponding to various values of the ratio h/λ. This quantity measures the ratio of the horizontal motion at the bottom to that at the surface. The third column gives the ratio of the vertical to the horizontal diameter of the elliptic orbit of a surface-particle. The fourth and fifth columns give the ratios of the wave-velocity to that of waves of the same length on water of infinite depth, and to that of 'long' waves on water of the actual depth, respectively.

The tables of *absolute* values of periods and wave-velocities, on the opposite page, are abridged from Airy's treatise‡. The value of g adopted by him is 32·16 ft./sec.^{2}

The possibility of progressive waves advancing with unchanged form is limited, theoretically, to the case of uniform depth; but the numerical results shew that a variation in the depth will have no appreciable influence, provided the depth everywhere exceeds (say) half the wave-length.

h/λ	sech kh	tanh kh	$c/(gk^{-1})^{\frac{1}{2}}$	$c/(gh)^{\frac{1}{2}}$
0·00	1·000	0·000	0·000	1·000
·01	·998	·063	·250	·999
·02	·992	·125	·354	·997
·03	·983	·186	·432	·994
·04	·969	·246	·496	·990
·05	·953	·304	·552	·984
·06	·933	·360	·600	·977
·07	·911	·413	·643	·970
·08	·886	·464	·681	·961
·09	·859	·512	·715	·951
·10	·831	·557	·746	·941
·20	·527	·850	·922	·823
·30	·297	·955	·977	·712
·40	·161	·987	·993	·627
·50	·086	·996	·998	·563
·60	·046	·999	·999	·515
·70	·025	1·000	1·000	·477
·80	·013	1·000	1·000	·446
·90	·007	1·000	1·000	·421
1·00	·004	1·000	1·000	·399
∞	·000	1·000	1·000	·000

* The results of Arts. 228, 229, for the case of finite depth, were given, substantially, by Airy, "Tides and Waves," Arts. 160... (1845).

† Green, *l.c.*

‡ "Tides and Waves," Arts. 169, 170.

Depth of water, in feet	Length of wave, in feet: 1	10	100	1000	10,000	
	Period of wave, in seconds					
1	0·442	1·873	17·645	176·33	1763·3	
10	0·442	1·398	5·923	55·80	557·62	
100	0·442	1·398	4·420	18·73	176·45	
1000	0·442	1·398	4·420	13·98	59·23	
10,000	0·442	1·398	4·420	13·98	44·20	

Depth of water, in feet	Length of wave, in feet: 1	10	100	1000	10,000	
	Wave-velocity, in feet per second					
1	2·262	5·339	5·667	5·671	5·671	5·671
10	2·262	7·154	16·88	17·92	17·93	17·93
100	2·262	7·154	22·62	53·39	56·67	56·71
1000	2·262	7·154	22·62	71·54	168·8	179·3
10,000	2·262	7·154	22·62	71·54	226·2	567·1

We remark, finally, that the theory of progressive waves may be obtained, without the intermediary of standing waves, by assuming at once, in place of Art. 228 (1),

$$\phi = Pe^{i(\sigma t - kx)}. \qquad \text{.................................(10)}$$

The conditions to be satisfied by P are exactly the same as before, and we easily find, in real form,

$$\eta = a \sin(kx - \sigma t), \qquad \text{...........................(11)}$$

$$\phi = \frac{ga}{\sigma} \frac{\cosh k(y+h)}{\cosh kh} \cos(kx - \sigma t), \qquad \text{..................(12)}$$

with the same determination of σ as before. From (12) all the preceding results as to the motion of the individual particles can be inferred without difficulty.

230. The *energy* of a system of standing waves of the simple-harmonic type is easily found. If we imagine two vertical planes to be drawn at unit distance apart, parallel to xy, the potential energy per wave-length of the fluid between these planes is

$$\tfrac{1}{2} g\rho \int_0^\lambda \eta^2 dx.$$

Substituting the value of η from Art. 228 (7), we obtain

$$\tfrac{1}{4} g\rho a^2 \lambda \, . \sin^2(\sigma t + \epsilon). \qquad \text{...........................(1)}$$

The kinetic energy is, by the formula (1) of Art. 61,

$$\tfrac{1}{2}\rho\int_0^\lambda\left[\phi\frac{\partial\phi}{\partial y}\right]_{y=0}dx.$$

Substituting from Art. 228 (8), and remembering the relation between σ and k, we obtain

$$\tfrac{1}{4}g\rho a^2\lambda\,.\cos^2(\sigma t+\epsilon). \quad\text{..........................(2)}$$

The total energy, being the sum of (1) and (2), is constant, and equal to $\frac{1}{4}g\rho a^2\lambda$. We may express this by saying that the total energy per unit area of the water-surface is $\frac{1}{4}g\rho a^2$.

A similar calculation may be made for the case of progressive waves, or we may apply the more general argument explained in Art. 174. In either way we find that the energy at any instant is half potential and half kinetic, and that the total amount, per unit area, is $\frac{1}{2}g\rho a^2$. In other words, the energy of a progressive wave-system of amplitude a is equal to the work which would be required to raise a stratum of the fluid, of thickness a, through a height $\frac{1}{2}a$.

231. We next consider the oscillations of the common boundary of two superposed liquids which are otherwise unlimited.

Taking the origin at the mean level of the interface we may write

$$\phi=Ce^{ky}\cos kx\,e^{i\sigma t},\qquad \phi'=C'e^{-ky}\cos kx\,e^{i\sigma t}, \quad\text{..............(1)}$$

where the accents relate to the upper fluid. For these satisfy Art. 227 (1) and vanish for $y=-\infty$ and $y=+\infty$, respectively. Hence if the equation of the disturbed surface is

$$\eta=a\cos kx\,e^{i\sigma t} \quad\text{............................(2)}$$

we must have

$$-kC=kC'=i\sigma a \quad\text{............................(3)}$$

by Art. 227 (6). Again, the formulae

$$\frac{p}{\rho}=\frac{\partial\phi}{\partial t}-gy,\qquad \frac{p'}{\rho'}=\frac{\partial\phi'}{\partial t}-gy \quad\text{....................(4)}$$

give

$$\rho\,(i\sigma C-ga)=\rho'\,(i\sigma C'-ga) \quad\text{......................(5)}$$

as the condition for continuity of pressure at the interface. Substituting the values of C and C' from (3) we have

$$\sigma^2=gk\,.\frac{\rho-\rho'}{\rho+\rho'}. \quad\text{............................(6)}$$

The velocity of propagation of waves of length $2\pi/k$ is therefore given by

$$c^2=\frac{g}{k}\cdot\frac{\rho-\rho'}{\rho+\rho'}. \quad\text{..............................(7)}$$

The presence of the upper fluid has therefore the effect of diminishing the velocity of propagation of waves of any given length in the ratio $\{(1-s)/(1+s)\}^{\frac{1}{2}}$, where s is the ratio of the density of the upper to that of the lower fluid. This diminution has a two-fold cause; the potential energy of a given deformation of the common surface is diminished in the ratio $1-s$, whilst the inertia is increased in the ratio $1+s$*. As a numerical example, in the case of water over mercury ($s^{-1}=13{\cdot}6$) the wave-velocity is diminished in the ratio ·929.

It is to be noticed, in this and in other problems of the kind, that there is a discontinuity of motion at the common surface. The normal velocity $(-\partial\phi/\partial y)$ is of course continuous, but the tangential velocity $(-\partial\phi/\partial x)$ changes sign as we cross the surface; in other words we have (Art. 151) a *vortex-sheet*. This is an extreme illustration of the remark, made in Art. 17, that the free oscillations of a liquid of variable density are not necessarily irrotational. In reality the discontinuity, if it could ever be originated, would be immediately abolished by viscosity, and the vortex-sheet replaced by a film of vorticity†.

If $\rho<\rho'$, the value of σ is imaginary. The undisturbed equilibrium-arrangement is then unstable.

If the two fluids are confined between rigid horizontal planes $y=-h$, $y=h'$, we assume in place of (1)

$$\phi=C\cosh k(y+h)\cos kx\, e^{i\sigma t},\quad \phi'=C'\cosh k(y-h')\cos kx\, e^{i\sigma t}, \quad \text{.........(8)}$$

since these make $\partial\phi/\partial y=0$, $\partial\phi'/\partial y=0$ at the respective planes. Hence

$$-kC\sinh kh=kC'\sinh kh'=i\sigma a. \quad \text{..............................(9)}$$

The continuity of pressure requires

$$\rho\,(i\sigma C\cosh kh-ga)=\rho'\,(i\sigma C'\cosh kh'-ga). \quad \text{....................(10)}$$

Eliminating C, C',

$$\sigma^2=\frac{gk\,(\rho-\rho')}{\rho\coth kh+\rho'\coth kh'}. \quad \text{..............................(11)}$$

When kh and kh' are both very great this reduces to the form (6). When kh' is large and kh small we find

$$c^2=\sigma^2/k^2=\left(1-\frac{\rho'}{\rho}\right)gh, \quad \text{..................................(12)}$$

approximately, the main effect of the presence of the upper fluid being the change in the potential energy of a given deformation. Its *kinetic* energy is small compared with that of the lower fluid.

* This explains why the natural periods of oscillation of the common surface of two liquids of very nearly equal density are very long compared with those of a free surface of similar extent. The fact was noticed by Benjamin Franklin in the case of oil over water; see a letter dated 1762 (*Complete Works*, London, n. d., ii. 142).

Again, near the mouths of some of the Norwegian fiords there is a layer of fresh over salt water. Owing to the comparatively small potential energy involved in a given deformation of the common boundary, waves of considerable height in this boundary are easily produced. To this cause is ascribed the abnormal resistance occasionally experienced by ships in those waters. See Ekman, "On Dead-Water," *Scientific Results of the Norwegian North Polar Expedition*, pt. xv. Christiania, 1904. Reference may also be made to a paper by the author, "On Waves due to a Travelling Disturbance, with an application to Waves in Superposed Fluids," *Phil. Mag.* (6), xxxi. 386 (1916).

† The solution, taking account of viscosity, is given by Harrison, *Proc. Lond. Math. Soc.* (2), vi. 396 (1908).

When the upper surface of the upper fluid is *free* we may assume

$$\phi = C\cosh k(y+h)\cos kx\, e^{i\sigma t}, \quad \phi' = (A\cosh ky + B\sinh ky)\cos kx\, e^{i\sigma t}. \qquad\text{......(13)}$$

The kinematical condition is then

$$-kC\sinh kh = -B = i\sigma a. \qquad\text{......(14)}$$

The condition for continuity of pressure at the interface is

$$\rho\,(i\sigma C\cosh kh - ga) = \rho'\,(i\sigma A - ga). \qquad\text{......(15)}$$

The condition for constancy of pressure at the free surface is given by Art. 227 (8) provided we put $y = h'$ after the differentiations. Thus

$$\sigma^2 (A\cosh kh' + B\sinh kh') = gk\,(A\sinh kh' + B\cosh kh'). \qquad\text{......(16)}$$

The elimination of A, B, C between (14), (15), (16) leads to the equation

$$\sigma^4 (\rho\coth kh\coth kh' + \rho') - \sigma^2 \rho\,(\coth kh' + \coth kh)\,gk + (\rho - \rho')\,g^2k^2 = 0. \qquad\text{......(17)}$$

Since this is a quadratic in σ^2, there are *two* possible systems of waves of any given period $(2\pi/\sigma)$. This is as we should expect, for when the wave-length is prescribed the system has virtually two degrees of freedom, so that there are two independent modes of oscillation about the state of equilibrium. For example, in the extreme case where ρ'/ρ is small, one mode consists mainly in an oscillation of the upper fluid which is almost the same as if the lower fluid were solidified, whilst the other mode may be described as an oscillation of the lower fluid which is almost the same as if its upper surface were free.

The ratio of the amplitude at the upper to that at the lower surface is found to be

$$\frac{kc^2}{kc^2\cosh kh' - g\sinh kh'}. \qquad\text{......(18)}$$

Of the various special cases that may be considered, the most interesting is that in which kh is large; *i.e.* the depth of the lower fluid is great compared with the wave-length. Putting $\coth kh = 1$, we see that one root of (17) is now

$$\sigma^2 = gk, \qquad\text{......(19)}$$

exactly as in the case of a single fluid of infinite depth, and that the ratio of the amplitudes is $e^{kh'}$. This is merely a particular case of a general result stated near the end of Art. 233; it will in fact be found on examination that there is now no slipping at the common boundary of the two fluids.

The second root of (17) is, on the same supposition,

$$\sigma^2 = \frac{\rho - \rho'}{\rho\coth kh' + \rho'}\cdot gk, \qquad\text{......(20)}$$

and for this the ratio (18) assumes the value

$$-\left(\frac{\rho}{\rho'} - 1\right) e^{-kh'}. \qquad\text{......(21)}$$

If in (20) and (21) we put $kh' = \infty$, we fall back on a former case. If on the other hand we make kh' small, we find

$$\frac{\sigma^2}{k^2} = \left(1 - \frac{\rho'}{\rho}\right) gh', \qquad\text{......(22)}$$

and the ratio of the amplitudes is

$$-\left(\frac{\rho}{\rho'} - 1\right). \qquad\text{......(23)}$$

These problems were first investigated by Stokes*. The case of any number of super posed strata of different densities has been treated by Webb† and Greenhill‡.

* "On the Theory of Oscillatory Waves," *Camb. Trans.* viii. (1847) [*Papers*, i. 212].

† *Math. Tripos Papers*, 1884.

‡ "Wave Motion in Hydrodynamics," *Amer. Journ. of Math.* ix. (1887).

232. Let us next suppose that we have two fluids of densities ρ, ρ', one beneath the other, moving parallel to x with velocities U, U', respectively, the common surface (when undisturbed) being of course plane and horizontal. This is virtually a problem of small oscillations about a state of steady motion.

We write, then,

$$\phi = -Ux + \phi_1, \quad \phi' = -U'x + \phi_1', \qquad \text{.................(1)}$$

where ϕ_1, ϕ_1' are by hypothesis small.

The velocity of either fluid at the interface may be regarded as made up of the velocity of this surface itself, and the velocity of the fluid relative to it. Hence if η be the ordinate of the displaced surface we have, considering vertical components,

$$\frac{\partial \eta}{\partial t} + U\frac{\partial \eta}{\partial x} = -\frac{\partial \phi}{\partial y}, \quad \frac{\partial \eta}{\partial t} + U'\frac{\partial \eta}{\partial x} = -\frac{\partial \phi'}{\partial y}, \qquad \text{...........(2)*}$$

as the kinematical conditions to be satisfied for $y = 0$.

Again, the formula for the pressure in the lower fluid is

$$\frac{p}{\rho} = \frac{\partial \phi_1}{\partial t} - \tfrac{1}{2}\left\{\left(U - \frac{\partial \phi_1}{\partial x}\right)^2 + \left(\frac{\partial \phi_1}{\partial y}\right)^2\right\} - gy + \ldots$$

$$= \frac{\partial \phi_1}{\partial t} - U\frac{\partial \phi_1}{\partial x} - gy + \ldots, \qquad \text{...............................(3)}$$

the terms omitted being either of the second order, or irrelevant to the present purpose. Hence the condition of continuity of pressure is

$$\rho\left(\frac{\partial \phi_1}{\partial t} + U\frac{\partial \phi_1}{\partial x} - g\eta\right) = \rho'\left(\frac{\partial \phi_1'}{\partial t} + U'\frac{\partial \phi_1'}{\partial x} - g\eta\right). \qquad \text{.........(4)}$$

We have seen, in various connections, that in oscillations about steady motion there is not necessarily uniformity of phase throughout the system, and in the present case it would not be found possible to satisfy the conditions on such an assumption. Assuming both fluids to be of unlimited depth, the appropriate course is to write

$$\phi_1 = Ce^{ky + i(\sigma t - kx)}, \quad \phi_1' = C'e^{-ky + i(\sigma t - kx)}, \qquad \text{.................(5)}$$

and

$$\eta = ae^{i(\sigma t - kx)}. \qquad \text{...............................(6)}$$

The conditions (2) then give

$$i(\sigma - kU)a = -kC, \quad i(\sigma - kU')a = kC', \qquad \text{..............(7)}$$

whilst, from (4),

$$\rho\{i(\sigma - kU)C - ga\} = \rho'\{i(\sigma - kU')C' - ga\}. \qquad \text{...........(8)}$$

Hence

$$\rho(\sigma - kU)^2 + \rho'(\sigma - kU')^2 = gk(\rho - \rho') \qquad \text{..............(9)}$$

or

$$\frac{\sigma}{k} = \frac{\rho U + \rho' U'}{\rho + \rho'} \pm \left\{\frac{g}{k}\cdot\frac{\rho - \rho'}{\rho + \rho'} - \frac{\rho\rho'}{(\rho + \rho')^2}(U - U')^2\right\}^{\frac{1}{2}}. \qquad \text{......(10)}$$

The first term on the right-hand side may be called the mean velocity of the

* These are particular cases of the general boundary-condition (3) of Art. 9, as is seen by writing $F = y - \eta$, and neglecting small terms of the second order.

two currents. Relatively to this there are waves travelling with velocities $\pm c$, given by

$$c^2 = c_0{}^2 - \frac{\rho\rho'}{(\rho+\rho')^2}(U-U')^2, \quad \text{.....................(11)}$$

where c_0 denotes the wave-velocity in the absence of currents (Art. 231). It is to be noticed however that the values of σ given by (9) are imaginary if

$$(U-U')^2 > \frac{g}{k}\cdot\frac{\rho^2-\rho'^2}{\rho\rho'}. \quad \text{.....................(12)}$$

The common boundary is therefore unstable for sufficiently small wave-lengths. This result would indicate that, if there were no modifying circumstances, the slightest breath of wind would ruffle the surface of water. A more complete investigation will be given later, taking account of capillary forces, which act in the direction of stability. If $\rho=\rho'$, or if $g=0$, the plane form of the surface is (on the present reckoning) unstable for *all* wave-lengths. This result illustrates the statement, as to the instability of surfaces of discontinuity in a liquid, made in Art. 79*.

The case of $\rho=\rho'$, with $U=U'$, is of some interest, as illustrating the flapping of sails and flags†. We may conveniently simplify the question by putting $U=U'=0$; any common velocity may be superposed afterwards if desired. On these suppositions the equation (8) reduces to $\sigma^2=0$. On account of the double root the solution has to be completed by the method explained in books on Differential Equations. In this way we obtain the two independent solutions

$$\eta = ae^{ikx}, \qquad \phi_1 = 0, \qquad \phi_1' = 0, \quad \text{..................(13)}$$

and

$$\eta = ate^{ikx}, \qquad \phi_1 = -\frac{a}{k}e^{ky}\,.\,e^{ikx}, \qquad \phi_1' = \frac{a}{k}e^{-ky}\,.\,e^{ikx}. \quad \text{......(14)}$$

The former solution represents a state of equilibrium; the latter gives a system of stationary waves with amplitude increasing proportionally to the time. In this form of the problem there is no physical surface of separation to begin with; but if a slight discontinuity of motion be artificially produced, *e.g.* by impulses applied to a thin membrane which is afterwards dissolved, the discontinuity will persist, and, as we have seen, the height of the corrugations will continually increase.

An interesting application of the same method is to the case of a jet of thickness $2b$ moving through still fluid of the same density‡. Taking the origin in the medial plane we write, for the disturbed jet $\phi = -Ux+\phi_0$, and for the fluid on the two sides $\phi=\phi_1$ for $y>b$, and $\phi=\phi_2$ for $y<-b$. We also denote by η_1, η_2 the normal displacements of the two surfaces $y=b$ and $y=-b$, respectively. The proper assumptions are then

$$\left.\begin{array}{ll} \phi_1 = A_1 e^{-ky} e^{i(\sigma t-kx)}, & \phi_2 = A_2 e^{ky} e^{i(\sigma t-kx)}, \\ \eta_1 = C_1 e^{i(\sigma t-kx)}, & \eta_2 = C_2 e^{i(\sigma t-kx)}, \\ \phi_0 = (A_0 \cosh ky + B_0 \sinh ky)\, e^{i(\sigma t-kx)}. & \end{array}\right\} \quad \text{......................(15)}$$

* This instability was first remarked by Helmholtz, *l.c. ante* p. 22.

† Rayleigh, *Proc. Lond. Math. Soc.* (1) x. 4 (1879) [*Papers* i. 361].

‡ Rayleigh *l.c.*

There are obviously two types of disturbance, in which $\eta_1=\eta_2$, and $\eta_1=-\eta_2$, respectively. In the former case we have $C_1=C_2$, $A_0=0$, $A_2=-A_1$. The kinematical conditions (2) at the surface $y=b$ then give

$$i\sigma C_1=kA_1e^{-kh},\quad i(\sigma-kU)C=-kB_0\cosh kh, \qquad (16)$$

whilst the continuity of pressure requires, gravity being omitted,

$$(\sigma-kU)B_0\sinh kh=\sigma A_1e^{-kh}. \qquad (17)$$

Hence

$$(\sigma-kU)^2\tanh kh+\sigma^2=0. \qquad (18)$$

If the thickness $2b$ is small compared with the wave-length of the disturbance, we have

$$\sigma=\pm ikU\surd(kh), \qquad (19)$$

approximately, indicating a very gradual instability, as is often observed in the case of filaments of smoke.

In the case of symmetry $(\eta_1=-\eta_2)$, we should find

$$(\sigma-kU)^2\coth kh+\sigma^2=0 \qquad (20)$$

in place of (18).

233. The theory of progressive waves may also be investigated, in a very compact manner, by the method of Art. 175*.

Thus if ϕ, ψ be the velocity- and stream-functions when the problem has been reduced to one of steady motion, we assume

$$\frac{\phi+i\psi}{c}=-(x+iy)+i\alpha e^{ik(x+iy)}+i\beta e^{-ik(x+iy)},$$

whence

$$\left.\begin{aligned}\frac{\phi}{c}&=-x-(\alpha e^{-ky}-\beta e^{ky})\sin kx,\\ \frac{\psi}{c}&=-y+(\alpha e^{-ky}+\beta e^{ky})\cos ky.\end{aligned}\right\} \qquad (1)$$

This represents a motion which is periodic in respect to x, superposed on a uniform current of velocity c. We assume that $k\alpha$ and $k\beta$ are small quantities; in other words, that the amplitude of the disturbance is small compared with the wave-length.

The profile of the free surface must be a stream-line; we take it to be the line $\psi=0$. Its form is then given by (1), viz. to a first approximation we have

$$y=(\alpha+\beta)\cos kx, \qquad (2)$$

shewing that the origin is at the mean level of the surface. Again, at the bottom $(y=-h)$ we must also have $\psi=\text{const.}$; this requires

$$\alpha e^{kh}+\beta e^{-kh}=0.$$

The equations (1) may therefore be put in the forms

$$\left.\begin{aligned}\frac{\phi}{c}&=-x+C\cosh k(y+h)\sin kx,\\ \frac{\psi}{c}&=-y+C\sinh k(y+h)\cos kx.\end{aligned}\right\} \qquad (3)$$

* Rayleigh, *l.c. ante* p. 260.

The formula for the pressure is

$$\frac{p}{\rho} = \text{const.} - gy - \tfrac{1}{2}\left\{\left(\frac{\partial\phi}{\partial x}\right)^2 + \left(\frac{\partial\phi}{\partial y}\right)^2\right\}$$

$$= \text{const.} - gy - \frac{c^2}{2}\{1 - 2kC\cosh k(y+h)\cos kx\},$$

if we neglect k^2C^2. Since the equation to the stream-line $\psi = 0$ is

$$y = C\sinh kh \cos kx, \quad \text{.........................(4)}$$

approximately, we have, along this line,

$$\frac{p}{\rho} = \text{const.} + (kc^2\coth kh - g)\,y.$$

The condition for a free surface is therefore satisfied, provided

$$c^2 = gh\,.\,\frac{\tanh kh}{kh}. \quad \text{.........................(5)}$$

This determines the wave-length ($2\pi/k$) of possible stationary undulations on a stream of given uniform depth h, and velocity c. It is easily seen that the value of kh is real or imaginary according as c is less or greater than $(gh)^{\frac{1}{2}}$.

If we impress on everything the velocity $-c$ parallel to x, we get progressive waves on still water, and (5) is then the formula for the wave-velocity, as in Art. 229.

When the ratio of the depth to the wave-length is sufficiently great, the formulae (1) become

$$\frac{\phi}{c} = -x + \beta e^{ky}\sin kx, \quad \frac{\psi}{c} = -y + \beta e^{ky}\cos kx, \quad \text{...........(6)}$$

leading to

$$\frac{p}{\rho} = \text{const.} - gy - \frac{c^2}{2}\{1 - 2k\beta e^{ky}\cos kx + k^2\beta^2 e^{2ky}\}. \quad \text{.........(7)}$$

If we neglect $k^2\beta^2$, the latter equation may be written

$$\frac{p}{\rho} = \text{const.} + (kc^2 - g)\,y + kc\psi. \quad \text{....................(8)}$$

Hence if

$$c^2 = g/k, \quad \text{.................................(9)}$$

the pressure will be uniform not only at the upper surface, but along *every* stream-line $\psi = \text{const.}$* This point is of some importance; for it shews that the solution expressed by (6) and (9) can be extended to the case of any number of liquids of different densities, arranged one over the other in horizontal strata, provided the uppermost surface be free, and the total depth infinite. And, since there is no limitation to the thinness of the strata, we may even include the case of a heterogeneous liquid whose density varies continuously with the depth. Cf. Art. 235.

* This conclusion, it must be noted, is limited to the case of infinite depth. It was first remarked by Poisson, *l.c. post* p. 384.

Again, to find the velocity of propagation of waves over the common horizontal boundary of two masses of fluid which are otherwise unlimited, we may assume

$$\frac{\psi}{c} = -y + \beta e^{ky} \cos kx, \quad \frac{\psi'}{c} = -y + \beta e^{-ky} \cos kx, \quad \text{.....................(10)}$$

where the accent relates to the upper fluid. For these satisfy the condition of irrotational motion, $\nabla^2\psi = 0$; and they give a uniform velocity c at a great distance above and below the common surface, at which we have $\psi = \psi'$, $=0$, say, and therefore $y = \beta \cos kx$, approximately.

The pressure-equations are

$$\left.\begin{aligned} \frac{p}{\rho} &= \text{const.} - gy - \frac{c^2}{2}(1 - 2k\beta e^{ky} \cos kx), \\ \frac{p'}{\rho'} &= \text{const.} - gy - \frac{c^2}{2}(1 + 2k\beta e^{-ky} \cos kx), \end{aligned}\right\} \quad \text{.......................(11)}$$

which give, at the common surface,

$$\frac{p}{\rho} = \text{const.} - (g - kc^2)\, y, \quad \frac{p'}{\rho'} = \text{const.} - (g + kc^2)\, y, \quad \text{.....................(12)}$$

the usual approximations being made. The condition $p = p'$ thus leads to

$$c^2 = \frac{g}{k} \cdot \frac{\rho - \rho'}{\rho + \rho'}, \quad \text{...(13)}$$

as in Art. 231.

234. As a further example of the method we take the case of two superposed currents, already treated by the direct method in Art. 232.

The fluids being unlimited vertically, we assume

$$\psi = -U\{y - \beta e^{ky} \cos kx\}, \quad \psi' = -U'\{y - \beta e^{-ky} \cos kx\}, \quad \text{......(1)}$$

for the lower and upper fluids respectively. The origin is taken at the mean level of the common surface, which is assumed to be stationary, and to have the form

$$y = \beta \cos kx. \quad \text{.................................(2)}$$

The pressure-equations give

$$\left.\begin{aligned} \frac{p}{\rho} &= \text{const.} - gy - \tfrac{1}{2}U^2(1 - 2k\beta e^{ky} \cos kx), \\ \frac{p'}{\rho'} &= \text{const.} - gy - \tfrac{1}{2}U'^2(1 + 2k\beta e^{-ky} \cos kx), \end{aligned}\right\} \quad \text{............(3)}$$

whence, at the common surface,

$$\frac{p}{\rho} = \text{const.} + (kU^2 - g)\, y, \quad \frac{p'}{\rho'} = \text{const.} - (kU'^2 + g)\, y. \quad \text{.........(4)}$$

Since we must have $p = p'$ over this surface, we get

$$\rho U^2 + \rho' U'^2 = \frac{g}{k}(\rho - \rho'). \quad \text{.......................(5)}$$

This is the condition for stationary waves on the common surface of the two currents U, U'. It may be written

$$\left(\frac{\rho U + \rho' U'}{\rho + \rho'}\right)^2 = \frac{g}{k} \cdot \frac{\rho - \rho'}{\rho + \rho'} - \frac{\rho\rho'}{(\rho + \rho')^2}(U - U')^2, \quad \text{............(6)}$$

which is easily seen to be equivalent to Art. 232 (10).

When the currents are confined by fixed horizontal planes $y=-h$, $y=h'$, we assume

$$\psi=-U\left\{y-\beta\frac{\sinh k(y+h)}{\sinh kh}\cos kx\right\},\quad \psi'=-U'\left\{y+b\frac{\sinh k(y-h')}{\sinh kh'}\cos kx\right\}. \quad\text{.........(7)}$$

The condition for stationary waves on the common surface is then found to be

$$\rho U^2\coth kh+\rho' U'^2\coth kh'=\frac{g}{k}(\rho-\rho'). \quad\text{..........................(8)*}$$

235. The theory of waves in a *heterogeneous* liquid may be noticed, for the sake of comparison with the case of homogeneity.

The equilibrium value ρ_0 of the density will be a function of the vertical co-ordinate (y) only. Hence, writing

$$p=p_0+p',\qquad \rho=\rho_0+\rho', \quad\text{....................................(1)}$$

where p_0 is the equilibrium pressure, the equations of motion, viz.

$$\rho\frac{\partial u}{\partial t}=-\frac{\partial p}{\partial x},\quad \rho\frac{\partial v}{\partial t}=-\frac{\partial p}{\partial y}-g\rho, \quad\text{...............................(2)}$$

$$\frac{\partial \rho}{\partial t}+u\frac{\partial \rho}{\partial x}+v\frac{\partial \rho}{\partial y}=0, \quad\text{.................................(3)}$$

become

$$\rho_0\frac{\partial u}{\partial t}=-\frac{\partial p'}{\partial x},\quad \rho_0\frac{\partial v}{\partial t}=-\frac{\partial p'}{\partial y}-g\rho', \quad\text{.............................(4)}$$

$$\frac{\partial \rho'}{\partial t}+v\frac{\partial \rho_0}{\partial y}=0, \quad\text{...(5)}$$

small quantities of the second order being omitted. The fluid being incompressible, the equation of continuity retains the form

$$\frac{\partial u}{\partial x}+\frac{\partial v}{\partial y}=0, \quad\text{...(6)}$$

so that we may write

$$u=-\frac{\partial \psi}{\partial y},\quad v=\frac{\partial \psi}{\partial x}. \quad\text{.......................................(7)}$$

Eliminating p' and ρ' we find†

$$\nabla^2\ddot{\psi}+\frac{1}{\rho_0}\frac{d\rho_0}{dy}\left\{\frac{\partial\ddot{\psi}}{\partial y}-g\frac{\partial^2\psi}{\partial x^2}\right\}=0. \quad\text{..............................(8)}$$

At a free surface we must have $Dp/Dt=0$, or

$$\frac{\partial p'}{\partial t}=-v\frac{\partial p_0}{\partial y}=g\rho_0\frac{\partial \psi}{\partial x}. \quad\text{.......................................(9)}$$

Hence, and from (4), we must have

$$\frac{\partial\ddot{\psi}}{\partial y}=g\frac{\partial^2\psi}{\partial x^2} \quad\text{..(10)}$$

at such a surface.

To investigate cases of wave-motion we assume that

$$\psi\propto e^{i(\sigma t-kx)}. \quad\text{...(11)}$$

The equation (8) becomes

$$\frac{\partial^2\psi}{\partial y^2}-k^2\psi+\frac{1}{\rho_0}\frac{d\rho_0}{dy}\left(\frac{\partial\psi}{\partial y}-\frac{gk^2}{\sigma^2}\psi\right)=0; \quad\text{..........................(12)}$$

whilst the condition (10) takes the form

$$\frac{\partial\psi}{\partial y}-\frac{gk^2}{\sigma^2}\psi=0. \quad\text{......................................(13)}$$

* Greenhill, *l.c. ante* p. 372.

† Cf. Love, "Wave Motion in a Heterogeneous Heavy Liquid," *Proc. Lond. Math. Soc.* xxii. 307 (1891).

These are satisfied, whatever the vertical distribution of density, by the assumption that ψ varies as e^{ky}, provided

$$\sigma^2 = gk. \qquad (14)$$

For a fluid of infinite depth the relation between wave-length and period is then the same as in the case of homogeneity (cf. Art. 229), and the motion is irrotational.

For further investigations it is necessary to make some assumption as to the relation between ρ_0 and y. The simplest is that

$$\rho_0 \propto e^{-\beta y}, \qquad (15)$$

in which case (12) takes the form

$$\frac{\partial^2 \psi}{\partial y^2} - k^2 \psi - \beta \left(\frac{\partial \psi}{\partial y} - \frac{gk^2}{\sigma^2} \right) \psi = 0. \qquad (16)$$

The solution is

$$\psi = (Ae^{\lambda_1 y} + Be^{\lambda_2 y})\, e^{i(\sigma t - kx)}, \qquad (17)$$

where λ_1, λ_2 are the roots of

$$\lambda^2 - \beta\lambda + \left(\frac{g\beta}{\sigma^2} - 1 \right) k^2 = 0. \qquad (18)$$

We first apply this to the oscillations of liquid filling a closed rectangular vessel*. The quantity k may be any multiple of π/l, where l denotes the length. If the equations to the horizontal boundaries be $y=0$, $y=h$, the condition $\partial\psi/\partial x = 0$ gives

$$A + B = 0, \quad Ae^{\lambda_1 h} + Be^{\lambda_2 h} = 0, \qquad (19)$$

whence

$$e^{(\lambda_1 - \lambda_2)h} = 1, \quad \text{or} \quad \lambda_1 - \lambda_2 = 2is\pi/h, \qquad (20)$$

where s is integral. Hence, from (18),

$$\lambda_1 = \tfrac{1}{2}\beta + is\pi/h, \quad \lambda_2 = \tfrac{1}{2}\beta - is\pi/h, \qquad (21)$$

and therefore

$$\left(\frac{g\beta}{\sigma^2} - 1 \right) k^2 = \lambda_1 \lambda_2 = \tfrac{1}{4}\beta^2 + \frac{s^2\pi^2}{h^2}. \qquad (22)$$

We verify that σ is real or imaginary, *i.e.* the equilibrium arrangement is stable or unstable, according as β is positive or negative, *i.e.* according as the density diminishes or increases upwards†.

The case where the fluid (of depth h) has a free surface may serve as an illustration of the theory of 'temperature seiches' in lakes‡. Assuming the roots of (18) to be complex, say

$$\lambda = \tfrac{1}{2}\beta \pm im, \qquad (23)$$

with

$$m^2 = \left(\frac{g\beta}{\sigma^2} - 1 \right) k^2 - \frac{\beta^2}{4}, \qquad (24)$$

we have

$$\psi = Ce^{\frac{1}{2}\beta y} \sin my, \qquad (25)$$

the origin of y being taken at the bottom. The surface-condition (13) gives

$$\tfrac{1}{2}\beta \sin mh + m \cos mh = \frac{gk^2}{\sigma^2} \sin mh. \qquad (26)$$

With the help of (24) this may be written

$$\tan mh = \beta h \cdot \frac{mh}{m^2h^2 + k^2h^2 - \tfrac{1}{4}\beta^2 h^2}, \qquad (27)$$

* Rayleigh, "Investigation of the Character of the Equilibrium of an Incompressible Heavy Liquid of Variable Density," *Proc. Lond. Math. Soc.* (1) xiv. 170 [*Papers*, ii. 200]. Reference may also be made to a paper by the author "On Atmospheric Oscillations," *Proc. Roy. Soc.* lxxxiv. 566, 571 (1910), where another law of density is considered.

† The case of waves on a liquid of finite depth is discussed by Love (*l.c.*). See also Burnside, "On the Small Wave-Motions of a Heterogeneous Fluid under Gravity," *Proc. Lond. Math. Soc.* (1) xx. 392 (1889).

‡ Discussed by Wedderburn, *Trans. R. S. Edin.* xlvii. 619 (1910) and xlviii. 629 (1912).

from which the values of mh are to be found. They are given graphically by the intersections of the curves

$$y = \tan x, \quad y = \frac{\mu x}{x^2 + \alpha^2}, \qquad \text{.........(28)}$$

where $\mu = \beta h$, $\alpha^2 = k^2h^2 - \frac{1}{4}\beta^2h^2$.—The only case of interest is when βh is small. We have, then, $mh = s\pi$, approximately, and thence

$$\sigma^2 = g\beta \cdot \frac{k^2h^2}{s^2\pi^2 + k^2h^2}, \qquad \text{.........(29)}$$

which is seen to be identical with (22) when the square of βh is neglected. It appears in fact from (25) that the vertical motion at the free surface is very slight. The maximum vertical disturbance is at the levels $y = (s - \frac{1}{2})\pi$.

When the roots of (18) are *real* we should get only a slight correction to the formula $\sigma^2 = gk \tanh kh$ which holds for a homogeneous fluid.

236. The investigations of Arts. 227–234 relate to a special type of waves; the profile is simple-harmonic, and the train extends to infinity in both directions. But since all our equations are linear (so long as we confine ourselves to a first approximation), we can, with the help of Fourier's Theorem, build up by superposition a solution which shall represent the effect of arbitrary initial conditions. Since the subsequent motion is in general made up of systems of waves, of all possible lengths, travelling in either direction, each with the velocity proper to its own wave-length, the form of the free surface will continually alter. The only exception is when the wave-length of every system which is present in sensible amplitude is large compared with the depth of the fluid. The velocity of propagation, viz. $\sqrt{(gh)}$, is then independent of the wave-length, so that in the case of waves travelling in one direction only, the wave-profile remains unchanged in form as it advances (Art. 170).

The effect of a local disturbance of the surface, in the case of infinite depth, will be considered presently; but it is convenient to introduce in the first place the very important conception of 'group-velocity,' which has application, not only to water-waves, but to every case of wave-motion where the velocity of propagation of a simple-harmonic train varies with the wave-length.

It has often been noticed that when an isolated group of waves, of sensibly the same length, is advancing over relatively deep water, the velocity of the group as a whole is less than that of the individual waves composing it. If attention be fixed on a particular wave, it is seen to advance through the group, gradually dying out as it approaches the front, whilst its former place in the group is occupied in succession by other waves which have come forward from the rear*.

The simplest analytical representation of such a group is obtained by the superposition of two systems of waves of the same amplitude, and of nearly

* Scott Russell, "Report on Waves," *Brit. Ass. Rep.* 1844, p. 369. There is an interesting letter on this point from W. Froude, printed in Stokes' *Scientific Correspondence*, Cambridge, 1907, ii. 156.

but not quite the same wave-length. The corresponding equation of the free surface will be of the form

$$\eta = a \sin(kx - \sigma t) + a \sin(k'x - \sigma' t)$$
$$= 2a \cos\{\tfrac{1}{2}(k-k')x - \tfrac{1}{2}(\sigma - \sigma')t\} \sin\{\tfrac{1}{2}(k+k')x - \tfrac{1}{2}(\sigma+\sigma')t\}. \quad \ldots(1)$$

If k, k' be very nearly equal, the cosine in this expression varies very slowly with x; so that the wave-profile at any instant has the form of a curve of sines in which the amplitude alternates gradually between the values 0 and $2a$. The surface therefore presents the appearance of a series of groups of waves, separated at equal intervals by bands of nearly smooth water. The motion of each group is then sensibly independent of the presence of the others. Since the distance between the centres of two successive groups is $2\pi/(k-k')$, and the time occupied by the system in shifting through this space is $2\pi/(\sigma - \sigma')$, the group-velocity (U, say) is $=(\sigma-\sigma')/(k-k')$, or

$$U = \frac{d\sigma}{dk}, \quad \ldots\ldots(2)$$

ultimately. In terms of the wave-length $\lambda\ (=2\pi/k)$, we have

$$U = \frac{d(kc)}{dk} = c - \lambda \frac{dc}{d\lambda}, \quad \ldots\ldots(3)$$

where c is the wave-velocity.

This result holds for any case of waves travelling through a uniform medium. In the present application we have

$$c = \left(\frac{g}{k}\tanh kh\right)^{\frac{1}{2}}, \quad \ldots\ldots(4)$$

and therefore, for the group-velocity,

$$\frac{d(kc)}{dk} = \tfrac{1}{2}c\left(1 + \frac{2kh}{\sinh 2kh}\right) \quad \ldots\ldots(5)$$

The ratio which this bears to the wave-velocity c increases as kh diminishes, being $\frac{1}{2}$ when the depth is very great, and unity when it is very small, compared with the wave-length.

The above explanation seems to have been first given by Stokes*. The extension to a more general type of group was made by Rayleigh† and Gouy‡.

Another derivation of (3) can be given which is, perhaps, more intuitive. In a medium such as we are considering, where the wave-velocity varies with the frequency, a limited initial disturbance gives rise in general to a wave-system in which the different wave-lengths, travelling with different velocities,

* Smith's Prize Examination, 1876 [*Papers*, v. 362]. See also Rayleigh, *Theory of Sound*, Art. 191.

† *Nature*, xxv. 52 (1881) [*Papers*, i. 540].

‡ "Sur la vitesse de la lumière," *Ann. de Chim. et de Phys.* xvi. 262 (1889). It has recently been pointed out that the theory had been to some extent anticipated by Hamilton, working from the optical point of view, in 1839; see Havelock, *Cambridge Tracts*, No. 17 (1914), p. 6.

are gradually sorted out (Arts. 238, 239). If we regard the wave-length λ as a function of x and t, we have

$$\frac{\partial\lambda}{\partial t}+U\frac{\partial\lambda}{\partial x}=0, \qquad\qquad (6)$$

since λ does not vary in the neighbourhood of a geometrical point travelling with velocity U; this is, in fact, the definition of U. Again, if we imagine another geometrical point to travel with the *waves*, we have

$$\frac{\partial\lambda}{\partial t}+c\frac{\partial\lambda}{\partial x}=\lambda\frac{\partial c}{\partial x}=\lambda\frac{dc}{d\lambda}\frac{\partial\lambda}{dx}, \qquad\qquad (7)$$

the second member expressing the rate at which two consecutive wave-crests are separating from one another. Combining (6) and (7), we are led, again, to the formula (3)*.

This formula admits of a simple geometrical representation†. If a curve be constructed with λ as abscissa and c as ordinate, the group-velocity will be represented by

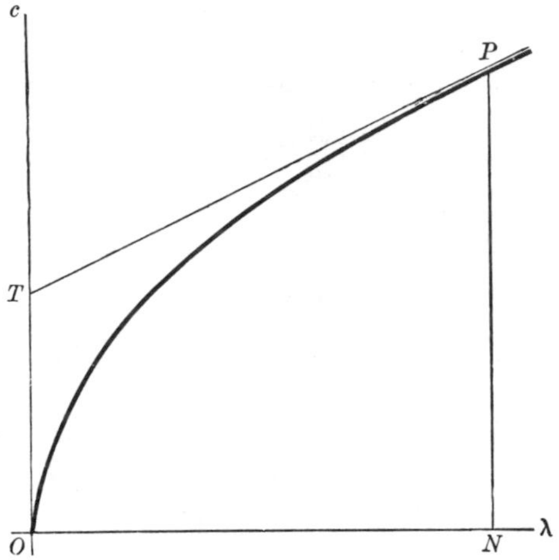

the intercept made by the tangent on the axis of c. Thus, in the figure, PN represents the wave-velocity for the wave-length ON, and OT represents the group-velocity. The frequency of vibration, it may be noticed, is represented by the tangent of the angle PON.

In the case of gravity-waves on deep water, $c \propto \lambda^{\frac{1}{2}}$; the curve has the form of the parabola $y^2=4ax$, and $OT=\frac{1}{2}PN$, *i.e.* the group-velocity is one-half the wave-velocity.

237. The group-velocity has moreover a dynamical, as well as a geometrical, significance. This was first shown by Osborne Reynolds‡, in the case of deep-water waves, by a calculation of the energy propagated across a

* See a paper "On Group-Velocity," *Proc. Lond. Math. Soc.* (2) i. 473 (1904). The subject is further discussed by G. Green, "On Group-Velocity, and on the Propagation of Waves in a Dispersive Medium," *Proc. R. S. Edin.* xxix. 445 (1909).

† *Manch. Mem.* xliv. No. 6 (1900).

‡ "On the Rate of Progression of Groups of Waves, and the Rate at which Energy is Transmitted by Waves," *Nature*, xvi. 343 (1877) [*Papers*, i. 198]. Reynolds also constructed a model which exhibits in a very striking manner the distinction between wave-velocity and group-velocity in the case of the transverse oscillations of a row of equal pendulums whose bobs are connected by a string.

vertical plane. In the case of infinite depth, the velocity-potential corresponding to a simple-harmonic train

$$\eta = a \sin k(x - ct) \quad \text{.........................(8)}$$

is

$$\phi = ac\, e^{ky} \cos k(x - ct), \quad \text{.........................(9)}$$

as may be verified by the consideration that for $y=0$ we must have $\partial\eta/\partial t = -\partial\phi/\partial y$. The variable part of the pressure is $\rho\,\partial\phi/\partial t$, if we neglect terms of the second order. The rate at which work is being done on the fluid to the right of the plane x is therefore

$$-\int_{-\infty}^{0} p\frac{\partial\phi}{\partial x}\,dy = \rho a^2k^2c^3 \sin^2 k(x-ct)\int_{-\infty}^{0} e^{2ky}\,dy$$

$$= \tfrac{1}{2}g\rho a^2 c \sin^2 k(x-ct), \quad \text{.........................(10)}$$

since $c^2 = g/k$. The mean value of this expression is $\frac{1}{4}g\rho a^2 c$. It appears on reference to Art. 230 that this is exactly one-half of the energy of the waves which cross the plane in question per unit time. Hence in the case of an isolated group the supply of energy is sufficient only if the group advance with *half* the velocity of the individual waves.

It is readily proved in the same manner that in the case of a finite depth h the average energy transmitted per unit time is*

$$\tfrac{1}{4}g\rho a^2 c\left(1 + \frac{2kh}{\sinh 2kh}\right), \quad \text{.........................(11)}$$

which is, by (5), the same as

$$\tfrac{1}{2}g\rho a^2 \times \frac{d(kc)}{dk}. \quad \text{.........................(12)}$$

Hence the rate of transmission of energy is equal to the group-velocity, $d(kc)/dk$, found independently by the former line of argument.

This identification of the kinematical group-velocity of the preceding Art. with the rate of transmission of energy may be extended to all kinds of waves. It follows indeed from the theory of interference groups (p. 381), which is of a general character. For let P be the centre of one of these groups, Q that of the quiescent region next in advance of P. In a time τ which extends over a number of periods, but is short compared with the time of transit of a group, the centre of the group will have moved to P', such that $PP' = U\tau$, and the space between P and Q will have gained energy to a corresponding amount. Another investigation, not involving the notion of 'interference,' was given by Rayleigh (*l.c.*).

From a physical point of view the group-velocity is perhaps even more important and significant than the wave-velocity. The latter may be greater or less than the former, and it is even possible to imagine mechanical media in which it would have the opposite direction; *i.e.* a disturbance might be

* Rayleigh, "On Progressive Waves," *Proc. Lond. Math. Soc.* (1) ix. 21 (1877) [*Papers*, i. 322]; *Theory of Sound*, i. Appendix.

propagated outwards from a centre in the form of a group, whilst the individual waves composing the group were themselves travelling backwards, coming into existence at the front, and dying out as they approach the rear*. Moreover, it may be urged that even in the more familiar phenomena of Acoustics and Optics the wave-velocity is of importance chiefly so far as it coincides with the group-velocity. When it is necessary to emphasize the distinction we may borrow the term 'phase-velocity' from modern Physics to denote what is more usually referred to in the present subject as 'wave-velocity.'

238. The theory of the waves produced in deep water by a local disturbance of the surface was investigated in two classical memoirs by Cauchy† and Poisson‡. The problem was long regarded as difficult, and even obscure, but in its two-dimensional form, at all events, it can be presented in a comparatively simple aspect.

It appears from Arts. 40, 41 that the initial state of the fluid is determinate when we know the form of the boundary, and the boundary-values of the normal velocity $\partial\phi/\partial n$, or of the velocity-potential ϕ. Hence two forms of the problem naturally present themselves; we may start with an initial elevation of the free surface, without initial velocity, or we may start with the surface undisturbed (and therefore horizontal) and an initial distribution of surface-impulse $(\rho\phi_0)$.

If the origin be in the undisturbed surface, and the axis of y be drawn vertically upwards, the typical solution for the case of initial rest is

$$\eta = \cos\sigma t\cos kx, \quad\text{.........................(1)}$$

$$\phi = g\frac{\sin\sigma t}{\sigma}e^{ky}\cos kx, \quad\text{.........................(2)}$$

provided

$$\sigma^2 = gk, \quad\text{.........................(3)}$$

in accordance with the ordinary theory of 'standing waves' of simple-harmonic profile (Art. 228).

If we generalize this by Fourier's double-integral theorem

$$f(x) = \frac{1}{\pi}\int_0^\infty dk\int_{-\infty}^\infty f(\alpha)\cos k(x-\alpha)\,d\alpha, \quad\text{...............(4)}$$

then, corresponding to the initial conditions

$$\eta = f(x), \qquad \phi_0 = 0, \quad\text{.........................(5)}$$

where the zero suffix indicates surface-value $(y = 0)$, we have

$$\eta = \frac{1}{\pi}\int_0^\infty \cos\sigma t\,dk\int_{-\infty}^\infty f(\alpha)\cos k(x-\alpha)\,d\alpha, \quad\text{...........(6)}$$

$$\phi = \frac{g}{\pi}\int_0^\infty \frac{\sin\sigma t}{\sigma}e^{ky}\,dk\int_{-\infty}^\infty f(\alpha)\cos k(x-\alpha)\,d\alpha. \quad\text{.........(7)}$$

* *Proc. Lond. Math. Soc.* (2) i. 473. † *l.c. ante* p. 17.

‡ "Mémoire sur la théorie des ondes," *Mém. de l'Acad. Roy. des Sciences*, i. (1816).

If the initial elevation be confined to the immediate neighbourhood of the origin, so that $f(\alpha)$ vanishes for all but infinitesimal values of α, we have, assuming

$$\int_{-\infty}^{\infty} f(\alpha)\,d\alpha = 1, \qquad \text{.............................(8)}$$

$$\phi = \frac{g}{\pi}\int_0^\infty \frac{\sin\sigma t}{\sigma} e^{ky}\cos kx\,dk. \qquad \text{.......................(9)}$$

This may be expanded in the form

$$\phi = \frac{gt}{\pi}\int_0^\infty \left\{1 - \frac{gt^2}{3!}k + \frac{(gt^2)^2}{5!}k^2 - \ldots\right\} e^{ky}\cos kx\,dk, \qquad \text{.........(10)}$$

where use is made of (3). If we write

$$-y = r\cos\theta, \qquad x = r\sin\theta, \qquad \text{....................(11)}$$

we have, y being negative,

$$\int_0^\infty e^{ky}\cos kx\,k^n\,dk = \frac{n!}{r^{n+1}}\cos(n+1)\theta, \qquad \text{..............(12)*}$$

so that (10) becomes

$$\phi = \frac{gt}{\pi}\left\{\frac{\cos\theta}{r} - \frac{1}{3}(\tfrac{1}{2}gt^2)\frac{\cos 2\theta}{r^2} + \frac{1}{3\,.\,5}(\tfrac{1}{2}gt^2)^2\frac{\cos 3\theta}{r^3} - \ldots\right\}, \qquad \text{...(13)}$$

a result which is easily verified. From this the value of η is obtained by Art. 227 (5), putting $\theta = \pm\frac{1}{2}\pi$. Thus, for $x > 0$,

$$\eta = \frac{1}{\pi x}\left\{\frac{gt^2}{2x} - \frac{1}{3\,.\,5}\left(\frac{gt^2}{2x}\right)^3 + \frac{1}{3\,.\,5\,.\,7\,.\,9}\left(\frac{gt^2}{2x}\right)^5 - \ldots\right\}. \qquad \text{......(14)†}$$

It is evident at once that any particular phase of the surface disturbance, *e.g.*, a zero or a maximum or a minimum of η, is associated with a definite value of $\frac{1}{2}gt^2/x$, and therefore that the phase in question travels over the surface with a constant *acceleration*. The meaning of this somewhat remarkable result will appear presently (Art. 240).

The series in (14) is virtually identical with one (usually designated by M‡) which occurs in the theory of Fresnel's diffraction-integrals. In its present form it is convenient only when we are dealing with the initial stages of the disturbance; it converges very slowly when $\frac{1}{2}gt^2/x$ is no longer small. An alternative form may, however, be obtained as follows.

* This formula may be dispensed with. It is sufficient to calculate the value of ϕ at points on the vertical axis of symmetry; its value at other points can then be written down at once by a property of harmonic functions (cf. Thomson and Tait, Art. 498).

† That the effect of a concentrated initial elevation of sectional area Q must be of the form

$$\eta = \frac{Q}{x} f(gt^2/x)$$

is evident from consideration of 'dimensions.'

‡ Cf. Rayleigh, *Papers*, iii. 129.

The surface-value of ϕ is, by (9),

$$\phi_0 = \frac{g}{\pi}\int_0^\infty \frac{\sin\sigma t}{\sigma}\cos kx\,dk$$

$$= \frac{1}{\pi}\left\{\int_0^\infty \sin\left(\frac{\sigma^2 x}{g} + \sigma t\right) d\sigma - \int_0^\infty \sin\left(\frac{\sigma^2 x}{g} - \sigma t\right) d\sigma\right\}. \quad \text{......(15)}$$

Putting

$$\zeta = \frac{x^{\frac{1}{2}}}{g^{\frac{1}{2}}}\left(\sigma \pm \frac{gt}{2x}\right), \quad \text{..........................(16)}$$

we find

$$\int_0^\omega \sin\left(\frac{\sigma^2 x}{g} + \sigma t\right) d\sigma = \frac{g^{\frac{1}{2}}}{x^{\frac{1}{2}}}\int_\omega^\infty \sin(\zeta^2 - \omega^2)\,d\zeta, \quad \text{...........(17)}$$

$$\int_0^\infty \sin\left(\frac{\sigma^2 x}{g} - \sigma t\right) d\sigma = \frac{g^{\frac{1}{2}}}{x^{\frac{1}{2}}}\int_{-\omega}^\infty \sin(\zeta^2 - \omega^2)\,d\zeta, \quad \text{........(18)}$$

where

$$\omega = \left(\frac{gt^2}{4x}\right)^{\frac{1}{2}}. \quad \text{...........................(19)}$$

Hence

$$\phi_0 = -\frac{2g^{\frac{1}{2}}}{\pi x^{\frac{1}{2}}}\int_0^\omega \sin(\zeta^2 - \omega^2)\,d\zeta. \quad \text{....................(20)}$$

From this the value of η is derived by Art. 227 (5); thus

$$\eta = \frac{g^{\frac{1}{2}} t}{\pi x^{\frac{3}{2}}}\int_0^\omega \cos(\zeta^2 - \omega^2)\,d\zeta$$

$$= \frac{g^{\frac{1}{2}} t}{\pi x^{\frac{3}{2}}}\left\{\cos\omega^2 \int_0^\omega \cos\zeta^2 d\zeta + \sin\omega^2 \int_0^\omega \sin\zeta^2 d\zeta\right\}. \quad \text{.........(21)}$$

This agrees with a result given by Poisson. The definite integrals are practically of Fresnel's forms*, and may be considered as known functions.

Lommel, in his researches on Diffraction†, has given a table of the function

$$1 - \frac{z^2}{3.5} + \frac{z^4}{3.5.7.9} - \ldots, \quad \text{.................(22)}$$

which is involved in (14), for values of z ranging from 0 to 60. We are thus enabled to delineate the first nine or ten waves with great ease. The figure on the next page shews the variation of η with the time, at a particular place; for different places the intervals between assigned phases vary as $\sqrt{x}$, whilst the corresponding elevations vary inversely as x. The diagrams on p. 388, on

* In terms of a usual notation we have

$$\int_0^\omega \cos\zeta^2 d\zeta = \sqrt{(\tfrac{1}{2}\pi)}\,C(u), \quad \int_0^\omega \sin\zeta^2 d\zeta = \sqrt{(\tfrac{1}{2}\pi)}\,S(u),$$

where

$$C(u) = \int_0^u \cos\tfrac{1}{2}\pi u^2 du, \quad S(u) = \int_0^u \sin\tfrac{1}{2}\pi u^2 du,$$

the upper limit of integration being $u = \sqrt{(2/\pi)}\,.\,\omega$. Tables of C (u) and S (u) computed by Gilbert and others are given in most books on Physical Optics. More extensive tables, due to Lommel, are reproduced by Watson, *Theory of Bessel Functions*, pp. 744, 745.

† "Die Beugungserscheinungen geradlinig begrenzter Schirme," *Abh. d. k. Bayer. Akad. d. Wiss.* 2e Cl. xv. (1886).

the other hand, shew the wave-profile at a particular instant; at different times, the horizontal distances between corresponding points vary as the square of the time that has elapsed since the beginning of the disturbance, whilst corresponding elevations vary inversely as the square of this time.

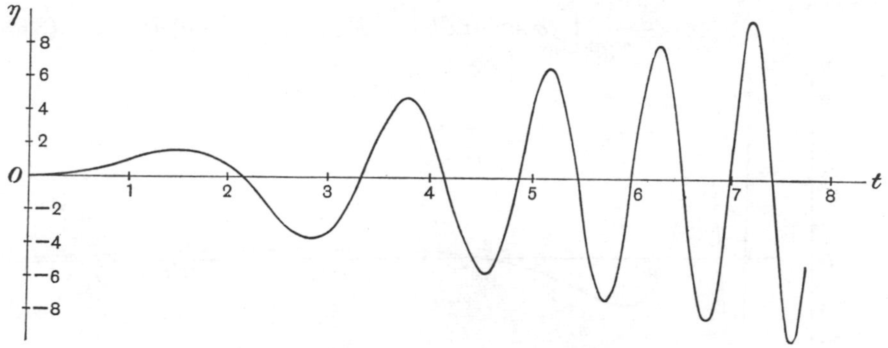

[The unit of the horizontal scale is $\surd(2x/g)$. That of the vertical scale is $Q/\pi x$, if Q be the sectional area of the initially elevated fluid.]

When $gt^2/4x$ is large, we have recourse to the formula (21), which makes

$$\eta = \frac{g^{\frac{1}{2}} t}{2^{\frac{3}{2}} \pi^{\frac{1}{2}} x^{\frac{3}{2}}} \left(\cos \frac{gt^2}{4x} + \sin \frac{gt^2}{4x} \right), \quad \text{.....................(23)}$$

approximately, as found by Poisson and Cauchy. This is in virtue of the known formulae

$$\int_0^\infty \cos \zeta^2 d\zeta = \int_0^\infty \sin \zeta^2 d\zeta = \frac{\surd\pi}{2\surd 2}. \quad \text{.................(24)}$$

Expressions for the remainder are also given by these writers. Thus Poisson obtains, substantially, the semi-convergent expansion

$$\eta = \frac{g^{\frac{1}{2}} t}{2^{\frac{3}{2}} \pi^{\frac{1}{2}} x^{\frac{3}{2}}} \left(\cos \frac{gt^2}{4x} + \sin \frac{gt^2}{4x} \right)$$
$$- \frac{1}{\pi x} \left\{ \frac{2x}{gt^2} - 1\,.\,3\,.\,5 \left(\frac{2x}{gt^2} \right)^3 + 1\,.\,3\,.\,5\,.\,7\,.\,9 \left(\frac{2x}{gt^2} \right)^5 - \ldots \right\}. \quad \text{......(25)}$$

This is derived as follows. We have

$$\int_0^\omega e^{i(\zeta^2 - \omega^2)}\, d\zeta = \int_0^\infty e^{i(\zeta^2 - \omega^2)}\, d\zeta - \int_\omega^\infty e^{i(\zeta^2 - \omega^2)}\, d\zeta$$
$$= \tfrac{1}{2}\surd\pi e^{-i(\omega^2 - \frac{1}{4}\pi)} + \frac{1}{2i\omega} + \frac{1}{(2i)^2 \omega^3} + \frac{1\,.\,3}{(2i)^3 \omega^5} + \ldots, \quad \text{.................(26)}$$

by a series of partial integrations. Taking the real part, and substituting in the first line of (21), we obtain the formula (25).

239. In the case of initial *impulses* applied to the surface, supposed undisturbed, the typical solution is

$$\rho\phi = \cos \sigma t\, e^{ky} \cos kx, \quad \text{..........................(27)}$$

$$\eta = -\frac{\sigma}{g\rho} \sin \sigma t \cos kx, \quad \text{........................(28)}$$

with $\sigma^2 = gk$ as before. Hence, if the initial conditions be

$$\rho\phi_0 = F(x), \qquad \eta = 0, \qquad \text{......................(29)}$$

we have

$$\phi = \frac{1}{\pi\rho}\int_0^\infty \cos\sigma t\, e^{ky}\, dk \int_{-\infty}^{\infty} F(\alpha)\cos k(x-\alpha)\, d\alpha, \qquad \text{.........(30)}$$

$$\eta = -\frac{1}{\pi g\rho}\int_0^\infty \sigma \sin\sigma t\, dk \int_{-\infty}^{\infty} F(\alpha)\cos k(x-\alpha)\, d\alpha. \qquad \text{......(31)}$$

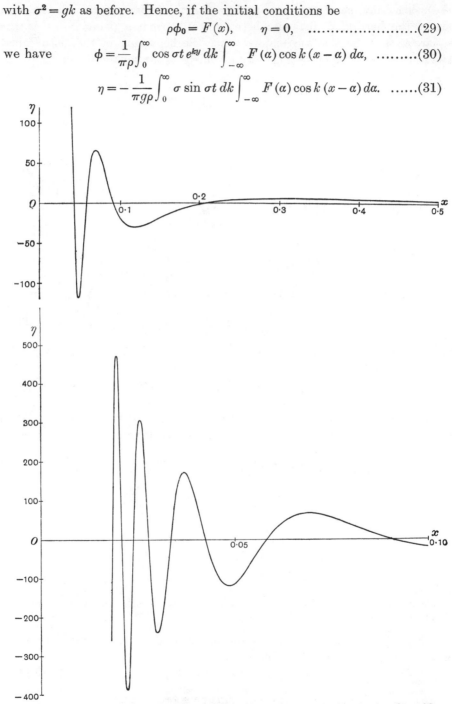

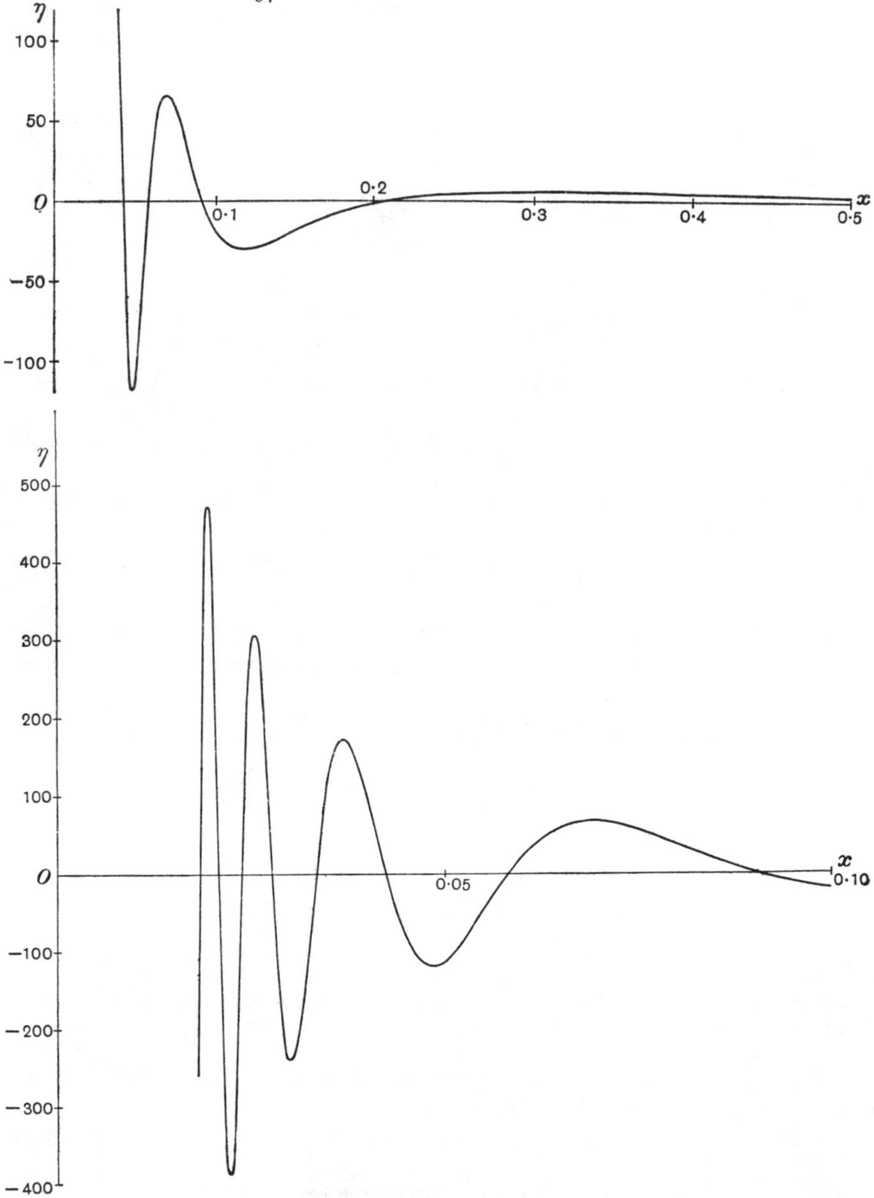

[The unit of the horizontal scales is $\frac{1}{2}gt^2$. That of the vertical scales is $2Q/\pi gt^2$.]

For a concentrated impulse acting at the point $x=0$ of the surface, we have, putting

$$\int_{-\infty}^{\infty} F(\alpha)\,d\alpha = 1, \qquad\qquad (32)$$

$$\phi = \frac{1}{\pi\rho}\int_0^\infty \cos\sigma t\, e^{ky} \cos kx\, dk. \qquad\qquad (33)$$

This integral may be treated in the same manner as (9); but it is evident that the results may be obtained immediately by performing the operation $1/g\rho\,.\,\partial/\partial t$ upon those of Art. 238. Thus from (13) and (14) we derive

$$\phi = \frac{1}{\pi\rho}\left\{\frac{\cos\theta}{r} - \tfrac{1}{2}gt^2\frac{\cos 2\theta}{r^2} + \frac{1}{1\,.\,3}(\tfrac{1}{2}gt^2)^2\frac{\cos 3\theta}{r^3} - \dots\right\}, \qquad (34)$$

$$\eta = \frac{t}{\pi\rho x^2}\left\{\frac{1}{1} - \frac{3}{1\,.\,3\,.\,5}\left(\frac{gt^2}{2x}\right)^2 + \frac{5}{1\,.\,3\,.\,5\,.\,7\,.\,9}\left(\frac{gt^2}{2x}\right)^4 - \dots\right\}. \qquad (35)^*$$

The series in (35) is related to the function

$$\frac{z}{1\,.\,3} - \frac{z^3}{1\,.\,3\,.\,5\,.\,7} + \frac{z^5}{1\,.\,3\,.\,5\,.\,7\,.\,9\,.\,11} - \dots, \qquad (36)$$

which has also been tabulated by Lommel. If we denote the series (22) and (36) by Σ_1 and Σ_2, respectively, we find

$$1 - \frac{3z^2}{1\,.\,3\,.\,5} + \frac{5z^4}{1\,.\,3\,.\,5\,.\,7\,.\,9} - \dots = \tfrac{1}{2}(1 + \Sigma_1 - z\Sigma_2), \qquad (37)$$

so that the forms of the first few waves can be traced without difficulty.

The annexed figure shews the rise and fall of the surface at a particular

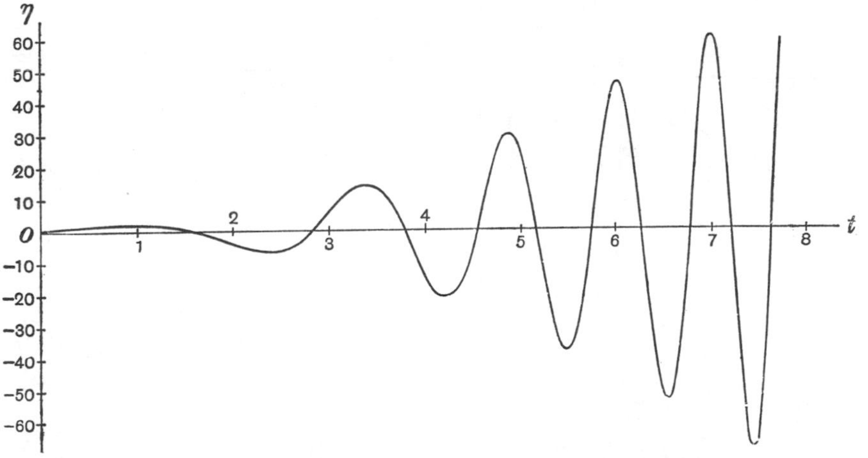

[The unit of the horizontal scale is $\sqrt{(2x/g)}$. That of the vertical scale is $\dfrac{P}{\pi\rho x}\sqrt{\dfrac{2}{gx}}$, where P represents the total initial impulse.]

* With the help of the theory of 'dimensions' it is easily seen *à priori* that the effect of a concentrated initial impulse P (per unit breadth) is necessarily of the form

$$\eta = \frac{Pt}{\rho x^2} f(gt^2/x).$$

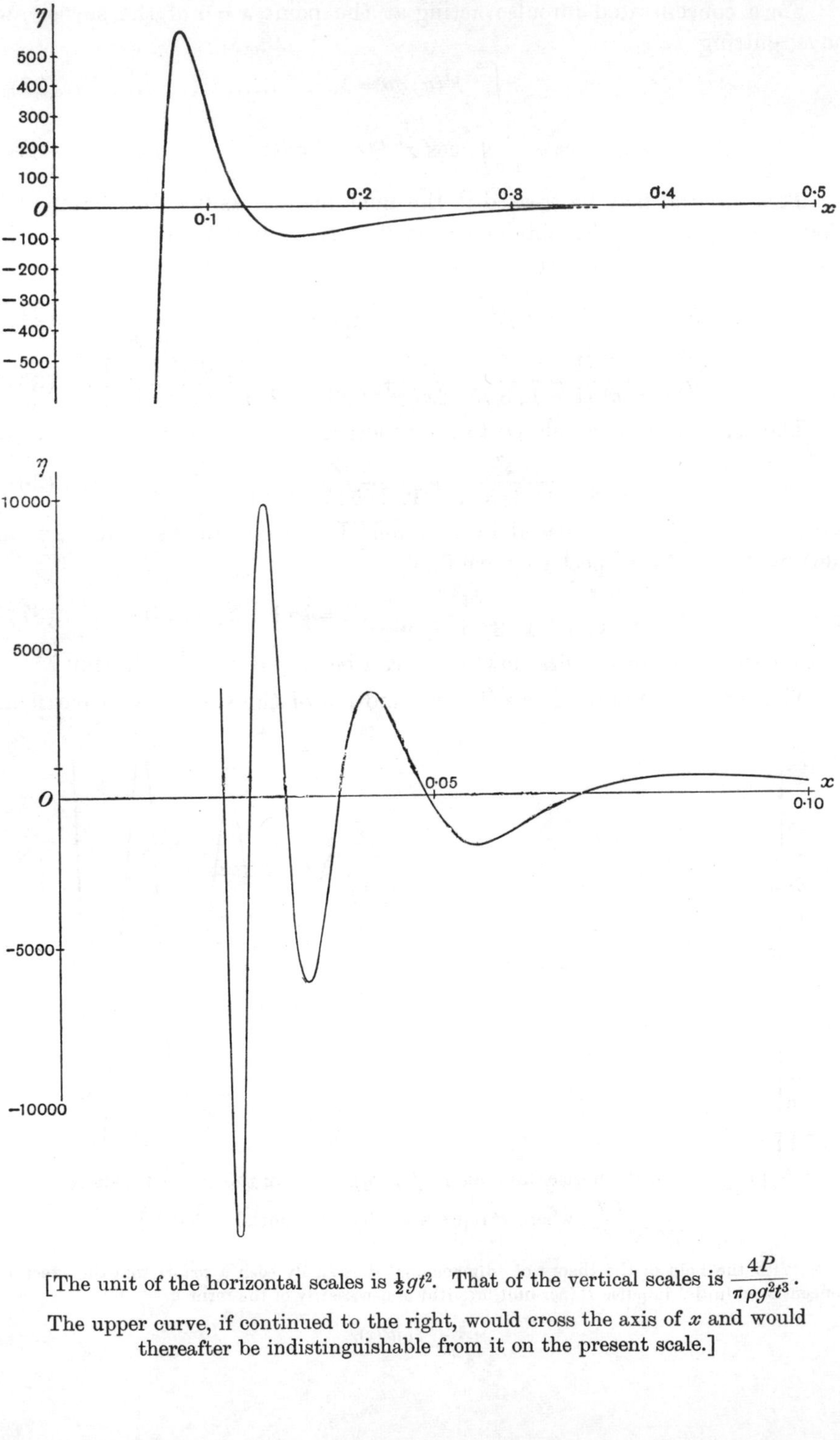

[The unit of the horizontal scales is $\frac{1}{2}gt^2$. That of the vertical scales is $\dfrac{4P}{\pi \rho g^2 t^3}$. The upper curve, if continued to the right, would cross the axis of x and would thereafter be indistinguishable from it on the present scale.]

place; for different places the time-intervals between assigned phases vary as $\sqrt{x}$, as in the former case, but the corresponding elevations now vary inversely as $x^{\frac{3}{2}}$. In the diagrams on the opposite page, which give an instantaneous view of the wave-profile, the horizontal distances between corresponding points vary as the square of the time, whilst corresponding ordinates vary inversely as the cube of the time.

For large values of $\frac{1}{2}gt^2/x$, we find, performing the operation $1/g\rho \,.\, \partial/\partial t$ upon (23),

$$\eta = \frac{g^{\frac{1}{2}} t^2}{2^{\frac{5}{2}} \pi^{\frac{1}{2}} \rho x^{\frac{5}{2}}} \left(\cos \frac{gt^2}{4x} - \sin \frac{gt^2}{4x} \right), \quad \text{.................(38)}$$

approximately.

240. It remains to examine the meaning and the consequences of the results above obtained. It will be sufficient to consider, chiefly, the case of Art. 238, where an initial *elevation* is supposed to be concentrated on a line of the surface.

At any subsequent time t the surface is occupied by a wave-system whose advanced portions are delineated on p. 388. For sufficiently small values of x the form of the waves is given by (23); hence as we approach the origin the waves are found to diminish continually in length, and to increase continually in height, in both respects without limit.

As t increases, the wave-system is stretched out horizontally, proportionally to the square of the time, whilst the vertical ordinates are correspondingly diminished, in such a way that the area

$$\int \eta dx$$

included between the wave-profile, the axis of x, and the ordinates corresponding to any two assigned phases (*i.e.* two assigned values of ω) is constant*. The latter statement may be verified immediately from the mere form of (14) or (21).

The oscillations of level, on the other hand, at any particular place, are represented on p. 387. These follow one another more and more rapidly, with ever increasing amplitude. For sufficiently great values of t, the course of these oscillations is given by (23).

In the region where this formula holds, at any assigned epoch, the changes in length and height from wave to wave are very gradual, so that a considerable number of consecutive waves may be represented approxi-

* This statement does not apply to the case of an initial *impulse*. The corresponding proposition then is that

$$\int \phi_0 dx,$$

taken between assigned values of ω, is constant. This appears from (34).

mately by a curve of sines. The circumstances are, in fact, all approximately reproduced when

$$\Delta \frac{gt^2}{4x} = 2\pi. \qquad \text{.............................(39)}$$

Hence, if we vary t alone, we have, putting $\Delta t = \tau$, the period of oscillation,

$$\tau = \frac{4\pi x}{gt}; \qquad \text{.............................(40)}$$

whilst, if we vary x alone, putting $\Delta x = -\lambda$, where λ is the wave-length, we find

$$\lambda = \frac{8\pi x^2}{gt^2}. \qquad \text{.............................(41)}$$

The wave-velocity is to be found from

$$\Delta \frac{gt^2}{4x} = 0; \qquad \text{.............................(42)}$$

this gives

$$\frac{\Delta x}{\Delta t} = \frac{2x}{t} = \sqrt{\frac{g\lambda}{2\pi}}, \qquad \text{.........................(43)}$$

by (41), as in the case of an infinitely long train of simple-harmonic waves of length λ.

We can now see something of a reason why each wave should by continually accelerated. The waves in front are longer than those behind, and are accordingly moving faster. The consequence is that all the waves are continually being drawn out in length, so that their velocities of propagation continually increase as they advance. But the higher the rank of a wave in the sequence, the smaller is its acceleration.

So far, we have been considering the progress of individual waves. But, if we fix our attention on a *group* of waves, characterized as having (approximately) a given wave-length λ, the position of this group is regulated according to (43) by the formula

$$\frac{x}{t} = \tfrac{1}{2} \sqrt{\frac{g\lambda}{2\pi}}; \qquad \text{.............................(44)}$$

i.e. the group advances with a constant velocity equal to *half* that of the component waves. The group does not, however, maintain a constant amplitude as it proceeds; it is easily seen from (23) that for a given value of λ the amplitude varies inversely as $\sqrt{x}$.

It appears that the region in the immediate neighbourhood of the origin may be regarded as a kind of source, emitting on each side an endless succession of waves of continually increasing amplitude and frequency, whose subsequent careers are governed by the laws above explained. This persistent activity of the source is not paradoxical; for our assumed initial accumulation of a finite volume of elevated fluid on an infinitely narrow base implies an unlimited store of energy.

In any practical case, however, the initial elevation is distributed over a band of finite breadth; we will denote this breadth by l. The disturbance at any point P is made up of parts due to the various elements, $\delta\alpha$, say, of the breadth l; these are to be calculated by the preceding formulae, and integrated over the breadth of the band. In the result, the mathematical infinity and other perplexing peculiarities, which we meet with in the case of a concentrated line-source, disappear. It would be easy to write down the requisite formulae, but, as they are not very tractable, and contain nothing not implied in the preceding statement, they may be passed over. It is more instructive to examine, in a general way, how the previous results will be modified.

The initial stages of the disturbance at a distance x, such that l/x is small, will evidently be much the same as on the former hypothesis; the parts due to the various elements $\delta\alpha$ will simply reinforce one another, and the result will be sufficiently expressed by (14) or (23) provided we multiply by

$$\int_{-8}^{\infty} f(\alpha)\, d\alpha,$$

i.e. by the sectional area of the initially elevated fluid. The formula (23), in particular, will hold when $\frac{1}{2}gt^2/x$ is large, so long as the wave-length λ at the point considered is large compared with l, *i.e.* by (41), so long as $\frac{1}{2}gt^2/x \,.\, l/x$ is small. But when, as t increases, the length of the waves at x becomes comparable with or smaller than l, the contributions from the different parts of l are no longer sensibly in the same phase, and we have something analogous to 'interference' in the optical sense. The result will, of course, depend on the special character of the initial distribution of the values of $f(\alpha)$ over the space l*, but it is plain that the increase of amplitude must at length be arrested, and that ultimately we shall have a gradual dying out of the disturbance.

There is one feature generally characteristic of the later stages which must be more particularly adverted to, as it has been the cause of some perplexity; viz. a fluctuation in the amplitude of the waves. This is readily accounted for on 'interference' principles. As a sufficient illustration, let us suppose that the initial elevation is uniform over the breadth l, and that we are considering a stage of the disturbance so late that the value of λ in the neighbourhood of the point x under consideration has become small compared with l. We shall evidently have a series of groups of waves separated by bands of comparatively smooth water, the centres of these bands occurring wherever l is an exact multiple of λ, say $l = n\lambda$. Substituting in (41), we find

$$\frac{x}{t} = \tfrac{1}{2}\sqrt{\frac{gl}{2n\pi}}, \qquad \text{..........................(45)}$$

i.e. the bands in question move forward with a constant velocity, which is, in

* Cf. Burnside, "On Deep-water Waves resulting from a Limited Original Disturbance," *Proc. Lond. Math. Soc.* (1) xx. 22 (1888).

fact, the group-velocity corresponding to the average wave-length in the neighbourhood*.

The ideal solution of Art. 238 necessarily fails to give any information as to what takes place at the origin itself. To illustrate this point in a special case, we may assume

$$f(\alpha)=\frac{Q}{\pi}\frac{b}{b^2+\alpha^2}; \quad \text{.......................(46)}$$

the formula (7) then gives

$$\phi=\frac{gQ}{\pi}\int_0^\infty \frac{\sin\sigma t}{\sigma} e^{k(y-b)}\cos kx\,dk. \quad \text{.......................(47)}$$

The surface-elevation at the origin is

$$\eta=\frac{Q}{\pi}\int_0^\infty \cos\sigma t\, e^{-kb}\,dk=\frac{2Q}{\pi g}\int_0^\infty \cos\sigma t\; e^{-\sigma^2 b/g}\,\sigma d\sigma=\frac{2Q}{\pi g}\frac{d}{dt}\int_0^\infty \sin\sigma t e^{-\sigma^2 b/g}\,d\sigma. \quad \text{...(48)}$$

By a known formula we have†

$$\int_0^\infty e^{-x^2}\sin 2\beta x\,dx=e^{-\beta^2}\int_0^\infty e^{x^2}\,dx. \quad \text{.......................(49)}$$

Hence, putting

$$\omega^2=gt^2/4b, \quad \text{.......................(50)}$$

we find

$$\eta=\frac{Q}{\pi b}\frac{d}{d\omega}\cdot e^{-\omega^2}\int_0^\omega e^{x^2}\,dx=\frac{Q}{\pi b}\left(1-2\omega e^{-\omega^2}\int_0^\omega e^{x^2}\,dx\right). \quad \text{...............(51)}$$

Hence

$$\frac{d}{d\omega}(\eta e^{\omega^2})=-\frac{2Q}{\pi b}\int_0^\omega e^{x^2}\,dx, \quad \text{.......................(52)}$$

shewing that ηe^{ω^2} steadily diminishes as t increases. Hence η can only change sign once. The form of the integrals in (48) shews that η tends finally to the limit zero; and it may be proved that the leading term in its asymptotic value is $-2Q/\pi g t^2$‡.

One noteworthy feature in the above problems is that the disturbance is propagated *instantaneously* to all distances from the origin, however great. Analytically, this might be accounted for by the fact that we have to deal with a synthesis of waves of all possible lengths, and that for infinite lengths the wave-velocity is infinite. It has been shewn, however, by Rayleigh § that the instantaneous character is preserved even when the water is of finite depth, in which case there is an upper limit to the wave-velocity. The physical reason of the peculiarity is that the fluid is treated as incompressible, so that changes of pressure are propagated with infinite velocity (cf. Art. 20). When compressibility is taken into account a finite, though it may be very short, interval elapses before the disturbance manifests itself at any point||.

* This fluctuation was first pointed out by Poisson, in the particular case where the initial elevation (or rather depression) has a parabolic outline.

The preceding investigations have an interest extending beyond the present subject, as shewing how widely the effects of a single initial impulse in a *dispersive* medium (*i.e.* one in which wave-velocity varies with wave-length) may differ from what takes place in the case of sound, or in the vibrations of an elastic solid. The above discussion is taken, with some modifications, from a paper "On Deep-Water Waves," *Proc. Lond. Math. Soc.* (2) ii. 371 (1904), where also the effect of a local periodic pressure is investigated.

† This formula presents itself as a subsidiary result in the process of evaluating

$$\int_0^\infty e^{-x^2}\cos 2\beta x\,dx$$

by a contour integration.

‡ The definite integral in (52) has been tabulated by Dawson, *Proc. Lond. Math. Soc.* (1) xxix. 519 (1898), and the function in (49) by Terazawa, *Science Reports of the Univ. of Tokio*, vi. 171 (1917).

§ "On the Instantaneous Propagation of Disturbance in a Dispersive Medium, ...," *Phil. Mag.* (6), xviii. 1 (1909) [*Papers*, v. 514]. See also Pidduck, "On the Propagation of a Disturbance in a Fluid under Gravity," *Proc. Roy. Soc.* A, lxxxiii. 347 (1910).

|| Pidduck, "The Wave-Problem of Cauchy and Poisson for Finite Depth and slightly Compressible Fluid," *Proc. Roy. Soc.* A, lxxxvi. 396 (1912).

241. The space which has been devoted to the above investigation may be justified by its historical interest, and by the consideration that it deals with one of the few problems of the kind which can be solved completely. It was shewn, however, by Kelvin that an approximate representation of the more interesting features can be obtained by a simpler process, which has moreover a very general application*.

The method depends on the approximate evaluation of integrals of the type

$$u = \int_a^b \phi(x)\, e^{if(x)}\, dx. \qquad \text{.........................(1)}$$

It is assumed that the circular function goes through a large number of periods within the range of integration, whilst $\phi(x)$ changes comparatively slowly; more precisely it is assumed that, when $f(x)$ changes by 2π, $\phi(x)$ changes by only a small fraction of itself. Under these conditions the various elements of the integral will for the most part cancel by annulling interference, except in the neighbourhood of those values of x, if any, for which $f(x)$ is stationary. If we write $x = \alpha + \xi$, where α is a value of x, within the range of integration, such that $f'(\alpha) = 0$, we have, for small values of ξ,

$$f(x) = f(\alpha) + \tfrac{1}{2}\xi^2 f''(\alpha), \qquad \text{.......................(2)}$$

approximately. The important part of the integral, corresponding to values of x in the neighbourhood of α, is therefore equal to

$$\phi(\alpha)\, e^{if(\alpha)} \int_{-\infty}^{\infty} e^{\frac{1}{2}if''(\alpha)\cdot\xi^2}\, d\xi, \qquad \text{......................(3)}$$

approximately, since, on account of the fluctuation of the integrand, the extension of the limits to $\pm\infty$ causes no appreciable error. Now by a known formula (Art. 238 (24)) we have

$$\int_{-\infty}^{\infty} e^{\pm im^2\xi^2}\, d\xi = \frac{\sqrt{\pi}}{m}\cdot\frac{1 \pm i}{\sqrt{2}} = \frac{\sqrt{\pi}}{m}\cdot e^{\pm\frac{1}{4}i\pi}. \qquad \text{.................(4)}$$

Hence (3) becomes

$$\frac{\sqrt{\pi}\,\phi(\alpha)}{\sqrt{|\frac{1}{2}f''(\alpha)|}}\cdot e^{i\{f(\alpha)\pm\frac{1}{4}\pi\}}, \qquad \text{.........................(5)}$$

where the upper or lower sign is to be taken in the exponential according as $f''(\alpha)$ is positive or negative.

If α coincides with one of the limits of integration in (1), the limits in (3) will be replaced by 0 and ∞, or $-\infty$ and 0, and the result (5) is to be halved.

If the approximation in (2) were continued, the next term would be $\frac{1}{6}\xi^3 f'''(\alpha)$; the foregoing method is therefore only valid under the condition

* Sir W. Thomson, "On the Waves produced by a Single Impulse in Water of any Depth, or in a Dispersive Medium," *Proc. R. S.* xlii. 80 (1887) [*Papers*, iv. 303]. The method of treating integrals of the type (1) had however been suggested by Stokes in his paper "On the Numerical Calculation of a Class of Definite Integrals and Infinite Series," *Camb. Trans.* ix. (1850) [*Papers*, ii. 341, footnote].

that $\xi f'''(\alpha)/f''(\alpha)$ must be small even when $\xi^2 f''(\alpha)$ is a moderate multiple of 2π. This requires that the quotient

$$f'''(\alpha)/\{|f''(\alpha)|\}^{\frac{3}{2}}$$

should be small.

Suppose now that, in a medium of any kind, an initial disturbance, whether of the nature of impulse or displacement, of amount $\cos kx$ per unit length, gives rise to an oscillation of the type

$$\eta = \phi(k) \cos kx\, e^{i\sigma t}, \qquad \text{.........................(6)}$$

where σ is a function of k determined by the theory of free waves. The effect of a concentrated unit initial disturbance is then given by the Fourier expression

$$\eta = \frac{1}{2\pi}\int_0^\infty \phi(k)\, e^{i(\sigma t - kx)}\, dk + \frac{1}{2\pi}\int_0^\infty \phi(k)\, e^{i(\sigma t + kx)}\, dk. \qquad \text{.........(7)}$$

It is understood that in the end only the real part of the expressions is to be retained.

The two terms in (7) represent the result of superposing trains of simple-harmonic waves of all possible lengths, travelling in the positive and negative directions of x, respectively. If, taking advantage of the symmetry, we confine our attention to the region lying to the right of the origin, the exponential in the first integral will alone, as a rule*, admit of a stationary value or values, viz. when

$$t\frac{d\sigma}{dk} = x. \qquad \text{.................................(8)}$$

This determines k, and therefore also σ, as a function of x and t, and we then find, in accordance with (5),

$$\eta = \frac{\phi(k)}{\sqrt{|2\pi t d^2\sigma/dk^2|}} \cdot \cos(\sigma t - kx \pm \tfrac{1}{4}\pi), \qquad \text{.................(9)}$$

where the ambiguous sign follows that of $d^2\sigma/dk^2$. The approximation postulates the smallness of the ratio

$$d^3\sigma/dk^3 \div \sqrt{\{t\,|\,d^2\sigma/dk^2\,|^3\}}. \qquad \text{........................(10)}$$

Since

$$\left.\begin{aligned} \frac{\partial}{\partial x}(\sigma t - kx) &= \left(t\frac{d\sigma}{dk} - x\right)\frac{\partial k}{\partial x} - k = -k, \\ \frac{\partial}{\partial t}(\sigma t - kx) &= \left(t\frac{d\sigma}{dk} - x\right)\frac{\partial k}{\partial t} + \sigma = \sigma, \end{aligned}\right\} \qquad \text{..............(11)}$$

by (8), it appears that the wave-length and the period in the neighbourhood of the point x at time t are $2\pi/k$ and $2\pi/\sigma$, respectively. The relation (8) shews that the wave-length is such that the corresponding *group*-velocity (Art. 236) is x/t.

* If the group-velocity were negative, as in some of the artificial cases referred to in Art. 237, the second integral would be the important one.

The above process, and the result, may be illustrated by various graphical constructions*. The simplest, in some respects, is based on a slight modification of the diagram of Art. 236. We construct a curve with λ as abscissa and ct as ordinate, where t denotes the time that has elapsed since the beginning of the disturbance. To ascertain the nature of the wave-system in the neighbourhood of any point x, we measure off a length OQ, equal to x, along the axis of ordinates. If PN be the ordinate corresponding to any given abscissa λ, the phase of the disturbance at x, due to the elementary wave-train whose wave-length is λ, will be given by the gradient of the line QP; for if we draw QR parallel to ON, we have

$$\frac{PR}{QR} = \frac{PN - OQ}{ON} = \frac{ct - x}{\lambda} = \frac{\sigma t - kx}{2\pi}. \qquad (12)$$

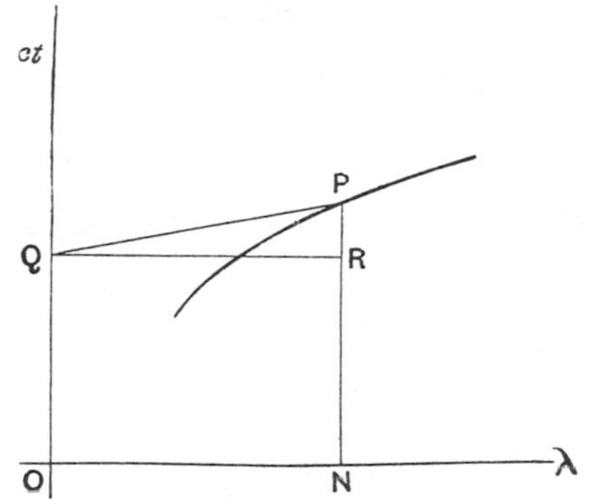

Hence the phase will be stationary if QP be a tangent to the curve; and the predominant wave-lengths at the point x are accordingly given by the abscissae of the points of contact of the several tangents which can be drawn from Q. These are characterized by the property that the group velocity has a given value x/t.

If we imagine the point Q to travel along the straight line on which it lies, we get an indication of the distribution of wave-lengths at the instant t for which the curve has been constructed. If we wish to follow the changes which take place in *time* at a given point x, we may either imagine the ordinates to be altered in the ratio of the respective times, or we may imagine the point Q to approach O in such a way that OQ varies inversely as t.

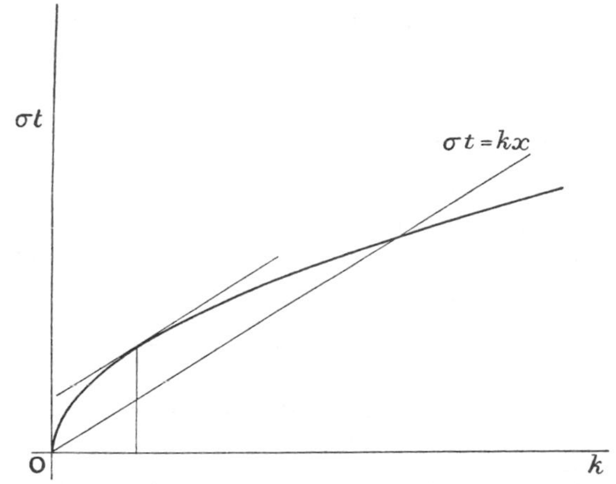

* *Proc. of the 5th Intern. Congress of Mathematicians*, Cambridge, 1912, p. 281.

The foregoing construction has the defect that it gives no indication of the relative amplitudes in different parts of the wave-system. For this purpose we may construct the curve which gives the relation between σt as ordinate and k as abscissa. If we draw a line through the origin whose gradient is x, the phase due to a particular elementary wave-train, viz. $\sigma t - kx$, will be represented by the difference of the ordinates of the curve and the straight line. This difference will be stationary when the tangent to the curve is parallel to the straight line, *i.e.* when $t\,d\sigma/dk = x$, as already found. It is further evident that the phase-difference, for elementary trains of slightly different wave-lengths, will vary ultimately as the square of the increment of k. Also that the range of values of k for which the phase is sensibly the same will be greater, and consequently the resulting disturbance will be more intense, the greater the vertical chord of curvature of the curve. This explains the occurrence of the quantity $t\,d^2\sigma/dk^2$ in the denominator of the formula (9).

In the hydrodynamical problem of Art. 238 we have*

$$\phi(k) = 1, \qquad \sigma^2 = gk, \qquad \text{(13)}$$

whence

$$d\sigma/dk = \tfrac{1}{2}g^{\frac{1}{2}}k^{-\frac{1}{2}}, \quad d^2\sigma/dk^2 = -\tfrac{1}{4}g^{\frac{1}{2}}k^{-\frac{3}{2}}, \quad d^3\sigma/dk^3 = \tfrac{3}{8}g^{\frac{1}{2}}k^{-\frac{5}{2}}. \qquad \text{(14)}$$

Hence, from (8),

$$k = gt^2/4x^2, \qquad \sigma = gt/2x, \qquad \text{(15)}$$

and therefore

$$\eta = \frac{g^{\frac{1}{2}}t}{\sqrt{(2\pi)}\,x^{\frac{3}{2}}}\, e^{i(gt^2/4x - \frac{1}{4}\pi)},$$

or, on rejecting the imaginary part,

$$\eta = \frac{g^{\frac{1}{2}}t}{2\pi^{\frac{1}{2}}x^{\frac{3}{2}}} \cos\left(\frac{gt^2}{4x} - \tfrac{1}{4}\pi\right). \qquad \text{(16)}$$

The quotient in (10) is found to be comparable with $(2x/gt^2)^{\frac{1}{2}}$, so that the approximation holds only for times and places such that $\frac{1}{2}gt^2$ is large compared with x.

These results are in agreement with the more complete investigation of Art. 238. The case of Art. 239 can be treated in a similar manner.

It appears from (16), or from the above geometrical construction (the curve being now a parabola as in Art. 236), that in the procession of waves at any instant the wave-length diminishes continually from front to rear; and that the waves which pass any assigned point will have their wave-lengths continually diminishing†.

242. We may next calculate the effect of an arbitrary, but steady, application of pressure to the surface of a stream. We shall consider only the state of steady motion which, under the influence of dissipative forces,

* The difficulty as to convergence in this case is met by the remark that the formula (9) of Art. 238 gives

$$\eta = \frac{1}{g}\frac{\partial\phi_0}{\partial t} = \lim_{y\to 0} \frac{1}{\pi}\int_0^\infty e^{ky}\cos\sigma t\cos kx\,dk,$$

where y is negative before the limit.

† For further applications reference may be made to Havelock, "The Propagation of Waves in Dispersive Media...," *Proc. Roy. Soc.* lxxxi. 398 (1908).

however small, will ultimately establish itself*. The question is in the first instance treated directly; a briefer method of obtaining the principal result is explained in Art. 248.

It is to be noted that in the absence of dissipative forces, the problem is to a certain extent indeterminate, for we can always superpose an endless train of free waves of arbitrary amplitude, and of wave-length such that their velocity relative to the water is equal and opposite to that of the stream, in which case they will maintain a fixed position in space.

To avoid this indeterminateness, we may avail ourselves of an artifice due to Rayleigh, and assume that the deviation of any particle of the fluid from the state of uniform flow is resisted by a force proportional to the *relative* velocity.

This law of friction does not profess to be altogether a natural one, but it serves to represent in a rough way the effect of small dissipative forces; and it has the great mathematical convenience that it does not interfere with the irrotational character of the motion. For if we write, in the equations of Art. 6,

$$X = -\mu(u - c), \qquad Y = -g - \mu v, \qquad Z = -\mu w, \quad \ldots\ldots\ldots(1)$$

where c denotes the velocity of the stream in the direction of x-positive, the method of Art. 33, when applied to a closed circuit, gives

$$\left(\frac{D}{Dt} + \mu\right)\int(u\,dx + v\,dy + w\,dz) = 0, \quad \ldots\ldots\ldots\ldots\ldots\ldots(2)$$

whence

$$\int(u\,dx + v\,dy + w\,dz) = Ce^{-\mu t}. \quad \ldots\ldots\ldots\ldots\ldots\ldots\ldots\ldots(3)$$

Hence the circulation in a circuit moving with the fluid, if once zero, is always zero. We now have

$$\frac{p}{\rho} = \text{const.} - gy + \mu(cx + \phi) - \tfrac{1}{2}q^2, \quad \ldots\ldots\ldots\ldots\ldots\ldots\ldots(4)$$

this being, in fact, the form assumed by Art. 21 (2) when we write

$$\Omega = gy - \mu(cx + \phi) \quad \ldots\ldots\ldots\ldots\ldots\ldots\ldots\ldots\ldots\ldots\ldots(5)$$

in accordance with (1) above.

To calculate, in the first place, the effect of a simple-harmonic distribution of pressure we assume

$$\frac{\phi}{c} = -x + \beta e^{ky}\sin kx, \qquad \frac{\psi}{c} = -y + \beta e^{ky}\cos kx. \quad \ldots\ldots\ldots\ldots(6)$$

* The first steps of the following investigation are adapted from a paper by Rayleigh, "The Form of Standing Waves on the Surface of Running Water," *Proc. Lond. Math. Soc.* xv. 69 (1883) [*Papers*, ii. 258], being simplified by the omission, for the present, of all reference to Capillarity. The definite integrals involved are treated, however, in a somewhat more general manner, and the discussion of the results necessarily follows a different course.

The problem had been treated by Popoff, "Solution d'un problème sur les ondes permanentes," *Liouville* (2), iii. 251 (1858); his analysis is correct, but regard is not had to the indeterminate character of the problem (in the absence of friction), and the results are consequently not pushed to a practical interpretation.

The equation (4) becomes, on neglecting the square of $k\beta$,

$$\frac{p}{\rho} = \ldots - gy + \beta e^{ky} (kc^2 \cos kx + \mu c \sin kx). \qquad (7)$$

This gives for the variable part of the pressure at the upper surface ($\psi = 0$)

$$p_0 = \rho\beta \{(kc^2 - g) \cos kx + \mu c \sin kx\}, \qquad (8)$$

which is equal to the real part of

$$\rho\beta (kc^2 - g - i\mu c) e^{ikx}.$$

If we equate the coefficient of e^{ikx} to C, we may say that to the pressure

$$p_0 = Ce^{ikx} \qquad (9)$$

corresponds the surface-form

$$g\rho y = \frac{\kappa}{k - \kappa - i\mu_1} Ce^{ikx}, \qquad (10)$$

where we have written κ for g/c^2, so that $2\pi/\kappa$ is the wave-length of the free waves which could maintain their position against the flow of the stream. We have also put $\mu/c = \mu_1$, for shortness.

Hence, taking the real parts, we find that the surface-pressure

$$p_0 = C \cos kx \qquad (11)$$

produces the wave-form

$$g\rho y = \kappa C \cdot \frac{(k - \kappa) \cos kx - \mu_1 \sin kx}{(k - \kappa)^2 + \mu_1^2} \qquad (12)$$

This shews that if μ be small the wave-crests will coincide in position with the maxima, and the troughs with the minima, of the applied pressure, when the wave-length is less than $2\pi/\kappa$; whilst the reverse holds in the opposite case. This is in accordance with a general principle. If we impress on everything a velocity $-c$ parallel to x, the result obtained by putting $\mu_1 = 0$ in (12) is seen to be a special case of Art. 168 (14).

In the critical case of $k = \kappa$, we have

$$g\rho y = -\frac{\kappa C}{\mu_1} . \sin kx, \qquad (13)$$

shewing that the excess of pressure is now on the slopes which face down the stream. This explains roughly how a system of progressive waves may be maintained against our assumed dissipative forces by a properly adjusted distribution of pressure over their slopes.

243. The solution expressed by (12) may be generalized, in the first place by the addition of an arbitrary constant to x, and secondly by a summation with respect to k. In this way we may construct the effect of any arbitrary distribution of pressure, say

$$p_0 = f(x), \qquad (14)$$

with the help of Fourier's Theorem (Art. 238 (4)).

We will suppose, in the first instance, that $f(x)$ vanishes for all but infinitely small values of x, for which it becomes infinite in such a way that

$$\int_{-\infty}^{\infty} f(x)\,dx = P; \quad \text{..........................(15)}$$

this will give us the effect of an integral pressure P concentrated on an infinitely narrow band of the surface at the origin. Replacing C in (12) by $P/\pi \,.\, \delta k$, and integrating with respect to k between the limits 0 and ∞, we obtain

$$g\rho y = \frac{\kappa P}{\pi} \cdot \int_0^{\infty} \frac{(k-\kappa)\cos kx - \mu_1 \sin kx}{(k-\kappa)^2 + \mu_1{}^2}\,dk. \quad \text{...............(16)}$$

If we put $\zeta = k + im$, where k, m are taken to be the rectangular co-ordinates of a variable point in a plane, the properties of the expression (16) are contained in those of the complex integral

$$\int \frac{e^{ix\zeta}}{\zeta - c}\,d\zeta. \quad \text{..(17)}$$

It is known that the value of this integral, taken round the boundary of any area which does not include the singular point ($\zeta = c$), is zero. In the present case we have $c = \kappa + i\mu_1$, where κ and μ_1 are both positive.

Let us first suppose that x is positive, and let us apply the above theorem to the region which is bounded externally by the line $m=0$ and by an infinite semicircle, described with the origin as centre on the side of this line for which m is positive, and internally by a small circle surrounding the point (κ, μ_1). The part of the integral due to the infinite semicircle obviously vanishes, and it is easily seen, putting $\zeta - c = re^{i\theta}$, that the part due to the small circle is

$$-2\pi i e^{i(\kappa + i\mu_1)x},$$

if the direction of integration be chosen in accordance with the rule of Art. 32. We thus obtain

$$\int_{-\infty}^{0} \frac{e^{ikx}}{k-(\kappa+i\mu_1)}\,dk + \int_0^{\infty} \frac{e^{ikx}}{k-(\kappa+i\mu_1)}\,dk - 2\pi i e^{i(\kappa+i\mu_1)x} = 0,$$

which is equivalent to

$$\int_0^{\infty} \frac{e^{ikx}}{k-(\kappa+i\mu_1)}\,dk = 2\pi i e^{i(\kappa+i\mu_1)x} + \int_0^{\infty} \frac{e^{-ikx}}{k+(\kappa+i\mu_1)}\,dk. \quad \text{..................(18)}$$

On the other hand, when x is negative we may take the integral (17) round the contour made up of the line $m=0$ and an infinite semicircle lying on the side for which m is negative. This gives the same result as before, with the omission of the term due to the singular point, which is now external to the contour. Thus, for x negative,

$$\int_0^{\infty} \frac{e^{ikx}}{k-(\kappa+i\mu_1)}\,dk = \int_0^{\infty} \frac{e^{-ikx}}{k+(\kappa+i\mu_1)}\,dk. \quad \text{.........................(19)}$$

An alternative form of the last term in (18) may be obtained by integrating round the contour made up of the negative portion of the axis of k, and the positive portion of the axis of m, together with an infinite quadrant. We thus find

$$\int_{-\infty}^{0} \frac{e^{ikx}}{k-(\kappa+i\mu_1)}\,dk + \int_0^{\infty} \frac{e^{-mx}}{im-(\kappa+i\mu_1)}\,i\,dm = 0,$$

which is equivalent to

$$\int_0^{\infty} \frac{e^{-ikx}}{k+(\kappa+i\mu_1)}\,dk = \int_0^{\infty} \frac{e^{-mx}}{m-\mu_1+i\kappa}\,dm. \quad \text{......................(20)}$$

This is for x positive. In the case of x negative, we must take as our contour the negative portions of the axes of k, m, and an infinite quadrant. This leads to

$$\int_0^\infty \frac{e^{-ikx}}{k+(\kappa+i\mu_1)}\,dk = \int_0^\infty \frac{e^{mx}}{m+\mu_1-i\kappa}\,dm, \quad\text{.........................(21)}$$

as the transformation of the second member of (19).

In the foregoing argument μ_1 is positive. The corresponding results for the integral

$$\int \frac{e^{ix\zeta}}{\zeta-(\kappa-i\mu_1)}\,d\zeta \quad\text{....................................(22)}$$

are not required for our immediate purpose, but it will be convenient to state them for future reference. For x positive, we find

$$\int_0^\infty \frac{e^{ikx}}{k-(\kappa-i\mu_1)}\,dk = \int_0^\infty \frac{e^{-ikx}}{k+(\kappa-i\mu_1)}\,dk = \int_0^\infty \frac{e^{-mx}}{m+\mu_1+i\kappa}\,dm\,; \quad\text{...........(23)}$$

whilst, for x negative,

$$\int_0^\infty \frac{e^{ikx}}{k-(\kappa-i\mu_1)}\,dk = -2\pi i e^{i(\kappa-i\mu_1)x} + \int_0^\infty \frac{e^{-ikx}}{k+(\kappa-i\mu_1)}\,dk$$

$$= -2\pi i e^{i(\kappa-i\mu_1)x} + \int_0^\infty \frac{e^{mx}}{m-\mu_1-i\kappa}\,dm. \quad\text{..............(24)}$$

The verification is left to the reader*.

If we take the real parts of the formulae (18), (20), and (19), (21), respectively, we obtain the results which follow.

The formula (16) is equivalent, for x positive, to

$$\frac{\pi g\rho}{\kappa P}\cdot y = -2\pi e^{-\mu_1 x}\sin\kappa x + \int_0^\infty \frac{(k+\kappa)\cos kx - \mu_1\sin kx}{(k+\kappa)^2+\mu_1^2}\,dk$$

$$= -2\pi e^{-\mu_1 x}\sin\kappa x + \int_0^\infty \frac{(m-\mu_1)\,e^{-mx}\,dm}{(m-\mu_1)^2+\kappa^2}, \quad\text{..............(25)}$$

and, for x negative, to

$$\frac{\pi g\rho}{\kappa P}\cdot y = \int_0^\infty \frac{(m+\mu_1)\,e^{mx}\,dm}{(m+\mu_1)^2+\kappa^2}. \quad\text{.....................(26)}$$

The interpretation of these results is simple. The first term of (25) represents a train of simple-harmonic waves, on the down-stream side of the origin, of wave-length $2\pi c^2/g$, with amplitudes gradually diminishing according to the law $e^{-\mu_1 x}$. The remaining part of the deformation of the free surface, expressed by the definite integrals in (25) and (26), though very great for small values of x, diminishes very rapidly as x increases in absolute value, however small the value of the frictional coefficient μ_1.

When μ_1 is infinitesimal, our results take the simpler forms

$$\frac{\pi g\rho}{\kappa P}\cdot y = -2\pi\sin\kappa x + \int_0^\infty \frac{\cos kx}{k+\kappa}\,dk$$

$$= -2\pi\sin\kappa x + \int_0^\infty \frac{me^{-mx}}{m^2+\kappa^2}\,dm, \quad\text{.............(27)}$$

* For another treatment of these integrals, see Dirichlet, *Vorlesungen ueber d. Lehre v. d. einfachen u. mehrfachen bestimmten Integralen* (ed. Arendt), Braunschweig, 1904, p. 170.

for x positive, and

$$\frac{\pi g\rho}{\kappa P}\cdot y=\int_0^\infty \frac{\cos kx}{k+\kappa}\,dk=\int_0^\infty \frac{me^{mx}}{m^2+\kappa^2}\,dm, \qquad (28)$$

for x negative. The part of the disturbance of level which is represented by the definite integrals in these expressions is now symmetrical with respect to the origin, and diminishes constantly as the distance from the origin increases. When κx is moderately large we find, by usual methods, the semi-convergent expansion

$$\int_0^\infty \frac{me^{-mx}}{m^2+\kappa^2}\,dm=\frac{1}{\kappa^2x^2}-\frac{3\,!}{\kappa^4x^4}+\frac{5\,!}{\kappa^6x^6}-\ldots \qquad (29)$$

It appears that at a distance of about half a wave-length from the origin, on the down-stream side, the simple-harmonic wave-profile is fully established.

The definite integrals in (27) and (28) can be reduced to known functions as follows. If we put $(k+\kappa)\,x=u$, we have, for x positive,

$$\int_0^\infty \frac{\cos kx}{k+\kappa}\,dk=\int_{\kappa x}^\infty \frac{\cos(\kappa x-u)}{u}\,du$$

$$=-\operatorname{Ci}\kappa x\cos\kappa x+(\tfrac{1}{2}\pi-\operatorname{Si}\kappa x)\sin\kappa x, \qquad (30)$$

where, in conformity with the usual notation,

$$\operatorname{Ci}u=-\int_u^\infty \frac{\cos u}{u}\,du,\quad \operatorname{Si}u=\int_0^u \frac{\sin u}{u}\,du. \qquad (31)$$

The functions $\operatorname{Ci}u$ and $\operatorname{Si}u$ have been tabulated by Glaisher*. It appears that as u increases from zero they tend very rapidly to their asymptotic values 0 and $\frac{1}{2}\pi$, respectively. For small values of u we have

$$\left.\begin{aligned}\operatorname{Ci}u&=\gamma+\log u-\frac{u^2}{2\,.\,2\,!}+\frac{u^4}{4\,.\,4\,!}-\ldots,\\ \operatorname{Si}u&=u-\frac{u^3}{3\,.\,3\,!}+\frac{u^5}{5\,.\,5\,!}-\ldots;\end{aligned}\right\} \qquad (32)$$

where γ is Euler's constant ·5772....

It is easily found from (25) and (26) that when μ_1 is infinitesimal, the integral depression of the surface is

$$-\int_{-\infty}^{\infty} y\,dx=\frac{P}{g\rho}, \qquad (33)$$

exactly as if the fluid were at rest.

244. The expressions (25), (26) and (27), (28) alike make the elevation infinite at the origin, but this difficulty disappears when the pressure, which we have supposed concentrated on a mathematical line of the surface, is diffused over a band of finite breadth.

* "Tables of the Numerical Values of the Sine-Integral, Cosine-Integral, and Exponential Integral," *Phil. Trans.* 1870; abridgments are given by Dale and by Jahnke and Emde. The expression of the last integral in (27) in terms of the sine- and cosine-integrals was obtained, in a different manner from the above, by Schlömilch, "Sur l'intégrale définie $\int_0^\infty \frac{d\theta}{\theta^2+a^2}e^{-x\theta}$," *Crelle*, xxxiii. (1846); see also De Morgan, *Differential and Integral Calculus*, London, 1842, p. 654, and Dirichlet, *Vorlesungen*, p. 208.

To calculate the effect of a distributed pressure

$$p_0 = f(x), \qquad \qquad (34)$$

it is only necessary to write $x - \alpha$ for x in (27) and (28), to replace P by $f(\alpha)\,\delta\alpha$, and to integrate the resulting value of y with respect to α between the proper limits. It follows from known principles of the Integral Calculus that if p_0 be finite the integrals will be finite for all values of x.

In the case of a uniform pressure p_0, applied to the part of the surface extending from $-\infty$ to the origin, we easily find by integration of (25), for $x > 0$,

$$g\rho y = -2p_0 \cos \kappa x + \frac{\kappa p_0}{\pi}\int_0^\infty \frac{e^{-mx}\,dm}{m^2+\kappa^2}, \qquad \qquad (35)$$

where μ_1 has been put $= 0$. Again, if the pressure p_0 be applied to the part of the surface extending from 0 to $+\infty$, we find, for $x < 0$,

$$g\rho y = \frac{\kappa p_0}{\pi}\int_0^\infty \frac{e^{mx}\,dm}{m^2+\kappa^2}. \qquad \qquad (36)$$

From these results we can easily deduce the requisite formulae for the case of a uniform pressure acting on a band of finite breadth. The definite integral in (35) and (36) can be evaluated in terms of the functions Ci u, Si u; thus in (35)

$$\kappa\int_0^\infty \frac{e^{-mx}\,dm}{m^2+\kappa^2} = \int_0^\infty \frac{\sin kx}{k+\kappa}\,dk = (\tfrac{1}{2}\pi - \mathrm{Si}\,\kappa x)\cos \kappa x + \mathrm{Ci}\,\kappa x \sin \kappa x. \quad (37)$$

In this way the diagram on p. 405 was constructed; it represents the case where the band (AB) has a breadth κ^{-1}, or ·159 of the length of a standing wave.

The circumstances in any such case might be realized approximately by dipping the edge of a slightly inclined board into the surface of a stream, except that the pressure on the wetted area of the board would not be uniform, but would diminish from the central parts towards the edges. To secure a uniform pressure, the board would have to be curved towards the edges, to the shape of the portion of the wave-profile included between the points A, B in the figure.

It will be noticed that if the breadth of the band be an exact multiple of the wave-length $(2\pi/\kappa)$, we have zero elevation of the surface at a distance, on the down-stream as well as on the up-stream side of the source of disturbance.

The diagram shews certain peculiarities at the points A, B due to the discontinuity in the applied pressure. A more natural representation of a local pressure is obtained if we assume

$$p_0 = \frac{P}{\pi}\frac{b}{b^2+x^2}. \qquad \qquad (38)$$

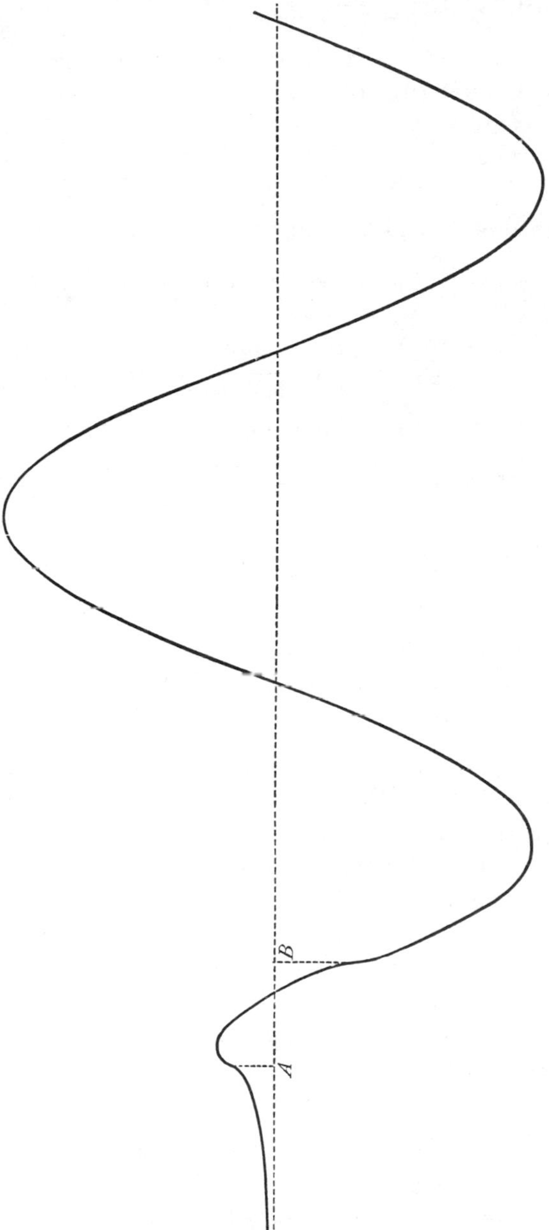
B
A

We may write this in the form

$$p_0 = \frac{P}{\pi} \cdot \frac{1}{b - ix} = \frac{P}{\pi} \int_0^\infty e^{-kb+ikx} \, dk, \quad \text{.....................(39)}$$

provided it is understood that, in the end, only the real part is to be retained. On reference to Art. 242 (9), (10), we see that the corresponding elevation of the free surface is given by

$$g\rho y = \frac{\kappa P}{\pi} \int_0^\infty \frac{e^{-kb+ikx}}{k - \kappa - i\mu_1} \, dk. \quad \text{.....................(40)}$$

By the method of Art. 243, we find that this is equivalent, for $x > 0$, to

$$g\rho y = \frac{\kappa P}{\pi} \left\{ 2\pi i e^{(\kappa+i\mu_1)(ix-b)} + \int_0^\infty \frac{e^{-imb-mx}}{m - \mu_1 + i\kappa} \, dm \right\}, \quad \text{.........(41)}$$

and, for $x < 0$, to

$$g\rho y = \frac{\kappa P}{\pi} \int_0^\infty \frac{e^{imb+mx}}{m + \mu_1 - i\kappa} \, dm. \quad \text{.....................(42)}$$

Hence, taking real parts, and putting $\mu_1 = 0$, we find

$$g\rho y = -2\kappa P e^{-\kappa b} \sin \kappa x + \frac{\kappa P}{\pi} \int_0^\infty \frac{m \cos mb - \kappa \sin mb}{m^2 + \kappa^2} e^{-mx} \, dm, \quad [x > 0], \quad \text{......(43)}$$

$$g\rho y = \frac{\kappa P}{\pi} \int_0^\infty \frac{m \cos mb - \kappa \sin mb}{m^2 + \kappa^2} e^{mx} \, dm, \quad [x < 0]. \quad \text{......(44)}$$

The factor $e^{-\kappa b}$ in the first term of (43) shews the effect of diffusing the pressure. It is easily proved that the values of y and dy/dx given by these formulae agree when $x = 0$*.

245. If in the problem of Art. 242 we suppose the depth to be finite and equal to h, there will be, in the absence of dissipation, indeterminateness or not, according as the velocity c of the stream is less or greater than $(gh)^{\frac{1}{2}}$, the maximum wave-velocity for the given depth; see Art. 229. The difficulty presented by the former case can be evaded by the introduction of small frictional forces; but it may be anticipated from the preceding investigation that the main effect of these will be to annul the elevation of the surface at a distance on the up-stream side of the region of disturbed pressure, and if we assume this at the outset we need not complicate our equations by retaining the frictional terms†.

For the case of a simple-harmonic distribution of pressure we assume

$$\left.\begin{aligned} \frac{\phi}{c} &= -x + \beta \cosh k(y+h) \sin kx, \\ \frac{\psi}{c} &= -y + \beta \sinh k(y+h) \cos kx, \end{aligned}\right\} \quad \text{.............................(1)}$$

* A different treatment of the problem of Arts. 243, 244 is given in a paper by Kelvin, "Deep Water Ship-Waves," *Proc. R. S. Edin.* xxv. 562 (1905) [*Papers*, iv. 368].

† There is no difficulty in so modifying the investigation as to take the frictional forces into account, when these are very small.

as in Art. 233 (3). Hence, at the surface

$$y=\beta \sinh kh \cos kx, \qquad\qquad (2)$$

we have

$$\frac{p_0}{\rho}=-gy-\tfrac{1}{2}(q^2-c^2)=\beta(kc^2\cosh kh-g\sinh kh)\cos kx, \qquad\qquad (3)$$

so that to the imposed pressure

$$p_0=C\cos kx \qquad\qquad (4)$$

will correspond the surface-form

$$y=\frac{C}{\rho}\cdot\frac{\sinh kh}{kc^2\cosh kh-g\sinh kh}\cos kx. \qquad\qquad (5)$$

As in Art. 242, the pressure is greatest over the troughs, and least over the crests, of the waves, or *vice versâ*, according as the wave-length is greater or less than that corresponding to the velocity c, in accordance with general theory.

The generalization of (5) by Fourier's method gives

$$y=\frac{P}{\pi\rho}\int_0^\infty \frac{\sinh kh\cos kx}{kc^2\cosh kh-g\sinh kh}dk \qquad\qquad (6)$$

as the representation of the effect of a pressure of integral amount P applied to a narrow band of the surface at the origin. This may be written

$$\frac{\pi\rho c^2}{P}\cdot y=\int_0^\infty \frac{\cos(xu/h)}{u\coth u-gh/c^2}du. \qquad\qquad (7)$$

Now consider the complex integral

$$\int\frac{e^{ix\zeta/h}}{\zeta\coth\zeta-gh/c^2}d\zeta, \qquad\qquad (8)$$

where $\zeta=u+iv$. The function under the integral sign has a singular point at $\zeta=\mp i\infty$, according as x is positive or negative, and the remaining singular points are given by the roots of

$$\frac{\tanh\zeta}{\zeta}=\frac{c^2}{gh}. \qquad\qquad (9)$$

Since (6) is an even function of x, it will be sufficient to take the case of x positive.

Let us first suppose that $c^2>gh$. The roots of (9) are then all pure imaginaries; viz. they are of the form $\pm i\beta$, where β is a root of

$$\frac{\tan\beta}{\beta}=\frac{c^2}{gh}. \qquad\qquad (10)$$

The smallest positive root of this lies between 0 and $\frac{1}{2}\pi$, and the higher roots approximate with increasing closeness to the values $(s+\frac{1}{2})\pi$, where s is integral. We will denote these roots in order by $\beta_0, \beta_1, \beta_2, \ldots$. Let us now take the integral (8) round the contour made up of the axis of u, an infinite semicircle on the positive side of this axis, and a series of small circles surrounding the singular points $\zeta=i\beta_0, i\beta_1, i\beta_2, \ldots$. The part due to the infinite semicircle obviously vanishes. Again, it is known that if α be a simple root of $f(\zeta)=0$ the value of the integral

$$\int\frac{F(\zeta)}{f(\zeta)}d\zeta$$

taken in the positive direction round a small circle enclosing the point $\zeta=\alpha$ is equal to

$$2\pi i\frac{F(\alpha)}{f'(\alpha)}. \qquad\qquad (11)$$

Now in the case of (8) we have

$$f'(\alpha)=\coth\alpha-\alpha(\coth^2\alpha-1)=\frac{1}{\alpha}\left\{\frac{gh}{c^2}\left(1-\frac{gh}{c^2}\right)+\alpha^2\right\}, \qquad\qquad (12)$$

whence, putting $\alpha=i\beta_s$, the expression (11) takes the form

$$2\pi B_s e^{-\beta_s x/h}, \qquad\qquad (13)$$

where
$$B_s = \frac{\beta_s}{\beta_s^2 - \dfrac{gh}{c^2}\left(1 - \dfrac{gh}{c^2}\right)}. \quad \text{.......................(14)}$$

The theorem in question then gives

$$\int_{-\infty}^{0} \frac{e^{ixu/h}}{u \coth u - gh/c^2}\, du + \int_{0}^{\infty} \frac{e^{ixu/h}}{u \coth u - gh/c^2}\, du - 2\pi \, \Sigma_0^\infty B_s e^{-\beta_s x/h} = 0. \quad \text{......(15)}$$

If in the former integral we write $-u$ for u, this becomes

$$\int_0^\infty \frac{\cos(xu/h)}{u \coth u - gh/c^2}\, du = \pi \, \Sigma_0^\infty B_s e^{-\beta_s x/h}. \quad \text{.......................(16)}$$

The surface-form is then given by

$$y = \frac{P}{\rho c^2} \cdot \Sigma_0^\infty B_s e^{-\beta_s x/h}. \quad \text{.......................(17)}$$

It appears that the surface-elevation (which is symmetrical with respect to the origin) is insensible beyond a certain distance from the seat of disturbance.

When, on the other hand, $c^2 < gh$, the equation (9) has a pair of real roots ($\pm\alpha$, say), the lowest roots ($\pm\beta_0$) of (10) having now disappeared. The integral (7) is then indeterminate, owing to the function under the integral sign becoming infinite within the range of integration. One of its values, viz. the 'principal value,' in Cauchy's sense, can however be found by the same method as before, provided we exclude the points $\zeta = \pm\alpha$ from the contour by drawing semicircles of small radius ϵ round them, on the side for which v is positive. The parts of the complex integral (8) due to these semicircles will be

$$-i\pi \frac{e^{\pm i\alpha x/h}}{f'(\pm\alpha)},$$

where $f'(\alpha)$ is given by (12); and their sum is therefore equal to

$$2\pi A \sin\frac{\alpha x}{h}, \quad \text{.......................(18)}$$

where
$$A = \frac{\alpha}{\alpha^2 - \dfrac{gh}{c^2}\left(\dfrac{gh}{c^2} - 1\right)}. \quad \text{.......................(19)}$$

The equation corresponding to (16) now takes the form

$$\left\{\int_0^{\alpha-\epsilon} + \int_{\alpha+\epsilon}^\infty\right\} \frac{\cos(xu/h)}{u \coth u - gh/c^2}\, du = -\pi A \sin\frac{\alpha x}{h} + \pi \, \Sigma_1^\infty B_s e^{-\beta_s x/h}, \quad \text{......(20)}$$

so that, if we take the principal value of the integral in (7), the surface-form on the side of x positive is

$$y = -\frac{P}{\rho c^2} A \sin\frac{\alpha x}{h} + \frac{P}{\rho c^2} \Sigma_1^\infty B_s e^{-\beta_s x/h}. \quad \text{.......................(21)}$$

Hence at a distance from the origin the deformation of the surface consists of the simple-harmonic train of waves indicated by the first term, the wave-length $2\pi h/\alpha$ being that corresponding to a velocity of propagation c relative to still water.

Since the function (7) is symmetrical with respect to the origin, the corresponding result for negative values of x is

$$y = \frac{P}{\rho c^2} A \sin\frac{\alpha x}{h} + \frac{P}{\rho c^2} \Sigma_1^\infty B_s e^{\beta_s x/h}. \quad \text{.......................(22)}$$

The general solution of our indeterminate problem is completed by adding to (21) and (22) terms of the form

$$C \cos\frac{\alpha x}{h} + D \sin\frac{\alpha x}{h}. \quad \text{.......................(23)}$$

The practical solution, including the effect of infinitely small dissipative forces, is obtained by so adjusting these terms as to make the deformation of the surface insensible at a distance on the up-stream side. We thus get, finally, for positive values of x,

$$y=-\frac{2P}{\rho c^2}A\sin\frac{\alpha x}{h}+\frac{P}{\rho c^2}\Sigma_1^\infty B_s e^{-\beta_s x/h}, \quad\text{.........................(24)}$$

and, for negative values of x,

$$y=\frac{P}{\rho c^2}\Sigma_1^\infty B_s e^{\beta_s x/h}. \quad\text{.........................(25)}$$

For a different method of reducing the definite integral in this problem we must refer to the paper by Kelvin cited below.

246. The same method can be employed to investigate the effect on a uniform stream of slight inequalities in the bed*.

Thus, in the case of a simple-harmonic corrugation given by

$$y=-h+\gamma\cos kx, \quad\text{.........................(1)}$$

the origin being as usual in the undisturbed surface, we assume

$$\left.\begin{aligned}\frac{\phi}{c}&=-x+(\alpha\cosh ky+\beta\sinh ky)\sin kx,\\ \frac{\psi}{c}&=-y+(\alpha\sinh ky+\beta\cosh ky)\cos kx.\end{aligned}\right\} \quad\text{.........................(2)}$$

The condition that (1) should be a stream-line is

$$\gamma=-\alpha\sinh kh+\beta\cosh kh. \quad\text{.........................(3)}$$

The pressure-formula is

$$\frac{p}{\rho}=\text{const.}-gy+kc^2(\alpha\cosh ky+\beta\sin ky)\cos kx, \quad\text{.....................(4)}$$

approximately, and therefore along the stream-line $\psi=0$

$$\frac{p}{\rho}=\text{const.}+(kc^2\alpha-g\beta)\cos kx,$$

so that the condition for a free surface gives

$$kc^2\alpha-g\beta=0. \quad\text{.........................(5)}$$

The equations (3) and (5) determine α and β. The profile of the free surface is given by

$$y=\beta\cos kx=\frac{\gamma}{\cosh kh-g/kc^2\,.\,\sinh kh}\cos kx. \quad\text{.........................(6)}$$

If the velocity of the stream be less than that of waves in still water of uniform depth h, of the same length as the corrugations, as determined by Art. 229 (4), the denominator is negative, so that the undulations of the free surface are inverted relatively to those of the bed. In the opposite case, the undulations of the surface follow those of the bed, but with a different vertical scale. When c has precisely the value given by Art. 229 (4), the solution fails, as we should expect, through the vanishing of the denominator. To obtain an intelligible result in this case we should be compelled to take special account of dissipative forces.

The above solution may be generalized, by Fourier's Theorem, so as to apply to the case where the inequalities of the bed follow any arbitrary law. Thus, if the profile of the bed be given by

$$y=-h+f(x)=-h+\frac{1}{\pi}\int_0^\infty dk\int_{-\infty}^\infty f(\xi)\cos k(x-\xi)\,d\xi, \quad\text{..............(7)}$$

* Sir W. Thomson, "On Stationary Waves in Flowing Water," *Phil. Mag.* (5) xxii. 353, 445, 517 (1886), and xxiii. 52 (1887) [*Papers*, iv. 270]. The effect of an abrupt change of level in the bed is discussed by Wien, *Hydrodynamik*, p. 201.

that of the free surface will be obtained by superposition of terms of the type (6) due to the various elements of the Fourier-integral; thus

$$y=\frac{1}{\pi}\int_0^\infty dk\int_{-\infty}^\infty \frac{f(\xi)\cos k(x-\xi)}{\cosh kh-g/kc^2.\sinh kh}\,d\xi. \quad \text{.......................(8)}$$

In the case of a single isolated inequality at the point of the bed vertically beneath the origin, this reduces to

$$y=\frac{Q}{\pi}\int_0^\infty \frac{\cos kx}{\cosh kh-g/kc^2.\sinh kh}\,dk$$

$$=\frac{Q}{\pi h}\int_0^\infty \frac{u\cos(xu/h)}{u\cosh u-gh/c^2.\sinh u}\,du, \quad \text{..........................(9)}$$

where Q represents the area included by the profile of the inequality above the general level of the bed. For a depression Q will of course be negative.

The discussion of the integral

$$\int \frac{\zeta e^{ix\zeta/h}\,d\zeta}{\zeta\cosh\zeta-gh/c^2.\sinh\zeta} \quad \text{...............................(10)}$$

can be conducted exactly as in Art. 245. The function to be integrated differs only by the factor $\zeta/(\sinh\zeta)$; the singular points therefore are the same as before, and we can at once write down the results.

Thus when $c^2>gh$ we find, for the surface-form,

$$y=\frac{Q}{h}\Sigma_0^\infty B_s\frac{B_s}{\sin\beta_s}\,e^{\mp\beta_s x/h}, \quad \text{...............................(11)}$$

the upper or the lower sign being taken according as x is positive or negative.

When $c^2<gh$, the 'practical' solution is, for x positive,

$$y=-\frac{2Q}{h}A\frac{a}{\sinh a}\sin\frac{ax}{h}+\frac{Q}{h}\Sigma_1^\infty B_s\frac{\beta_s}{\sin\beta_s}e^{-\beta_s x/h}, \quad \text{.................(12)}$$

and, for x negative,

$$y=\frac{Q}{h}\Sigma_1^\infty B_s\frac{\beta_s}{\sin\beta_s}\,e^{\beta_s x/h}. \quad \text{...............................(13)}$$

The symbols a, β_s, A, B_s have here exactly the same meanings as in Art. 245*.

247. We may calculate, in a somewhat similar manner, the disturbance produced in the flow of a uniform stream by a submerged cylindrical obstacle whose radius b is small compared with the depth f of its axis†. The cylinder is supposed placed horizontally athwart the stream.

We write

$$\phi=-cx\left(1+\frac{b^2}{r^2}\right)+\chi, \quad \text{........................(1)}$$

where c denotes as before the general velocity of the stream, and r denotes distance from the axis of the cylinder, viz.

$$r=\sqrt{\{x^2+(y+f)^2\}}, \quad \text{........................(2)}$$

* A very interesting drawing of the wave-profile produced by an isolated inequality in the bed is given in Kelvin's paper, *Phil. Mag.* (5) xxii. 517 (1886) [*Papers*, iv. 295].

† The investigation is taken from a paper "On some cases of Wave-Motion on Deep Water," *Ann. di matematica* (3), xxi. 237 (1913). I find that the problem had been suggested by Kelvin, *Phil. Mag.* (6) ix. 733 (1905) [*Papers*, iv. 369].

the origin being in the undisturbed level of the surface, vertically above the axis. This makes $\partial\phi/\partial r = 0$ for $r = b$, provided χ be negligible in the neighbourhood of the cylinder.

We assume

$$\chi = \int_0^\infty \alpha(k)\, e^{ky} \sin kx\, dk, \quad \text{.......................(3)}$$

where $\alpha(k)$ is a function of k, to be determined. For the equation of the free surface, assumed to be steady, we put

$$\eta = \int_0^\infty \beta(k) \cos kx\, dk. \quad \text{..........................(4)}$$

The geometrical condition to be satisfied at the free surface is

$$-\frac{\partial\phi}{\partial y} = c\frac{d\eta}{dx}, \quad \text{...............................(5)}$$

wherein we may put $y = 0$. Since (1) is equivalent to

$$\phi = -cx - b^2 c \int_0^\infty e^{-k(y+f)} \sin kx\, dk + \chi, \quad \text{.................(6)}$$

for positive values of $y + f$, this condition is satisfied if

$$b^2 c e^{-kf} + \alpha(k) = c\beta(k). \quad \text{..........................(7)}$$

Again, the variable part of the pressure at the free surface is given by

$$\begin{aligned}
\frac{p}{\rho} &= -gy - \tfrac{1}{2}\left(\frac{\partial\phi}{\partial x}\right)^2 \\
&= -g\eta - \tfrac{1}{2}c^2 - b^2c^2 \int_0^\infty e^{-kf} \cos kx\, k\, dk + c\frac{\partial\chi}{\partial x} \\
&= -g\eta - \tfrac{1}{2}c^2 - b^2c^2 \int_0^\infty e^{-kf} \cos kx\, k\, dk + c\int_0^\infty \alpha(k) \cos kx\, k\, dk, \quad \text{......(8)}
\end{aligned}$$

where terms of the second order in the disturbance have been omitted. This expression will be independent of x provided

$$g\beta(k) + kb^2c^2e^{-kf} - kc\,\alpha(k) = 0. \quad \text{....................(9)}$$

Combined with (7), this gives

$$\alpha(k) = \frac{k+\kappa}{k-\kappa} b^2 c e^{-kf}, \qquad \beta(k) = \frac{2b^2 k e^{-kf}}{k-\kappa}, \quad \text{..............(10)}$$

where

$$\kappa = g/c^2, \quad \text{...............................(11)}$$

as in Art. 242. Hence

$$\begin{aligned}
\eta &= 2b^2 \int_0^\infty \frac{ke^{-kf} \cos kx\, dk}{k-\kappa} \\
&= \frac{2b^2 f}{x^2+f^2} + 2\kappa b^2 \int_0^\infty \frac{e^{-kf} \cos kx\, dk}{k-\kappa}. \quad \text{..................(12)}
\end{aligned}$$

The integral is indeterminate, but if x be positive its principal value is equal to the real part of the expression

$$i\pi e^{-\kappa f+i\kappa x}+i\int_0^\infty \frac{e^{-imf-mx}}{im-\kappa}\,dm. \qquad (13)$$

Adopting this we have

$$\eta=\frac{2b^2f}{x^2+f^2}-2\pi\kappa b^2e^{-\kappa f}\sin\kappa x$$
$$-2\kappa b^2\int_0^\infty \frac{(\kappa\sin mf-m\cos mf)\,e^{-mx}}{m^2+\kappa^2}\,dm. \qquad (14)$$

For large values of x the second term is alone sensible.

Since the value of η in (12) is an even function of x we must have, for x negative,

$$\eta=\frac{2b^2f}{x^2+f^2}+2\pi\kappa b^2e^{-\kappa f}\sin\kappa x-2\kappa b^2\int_0^\infty \frac{(\kappa\sin mf-m\cos mf)\,e^{mx}}{m^2+\kappa^2}\,dm. \qquad (15)$$

On the disturbances represented by these formulae we can superpose any system of stationary waves of length $2\pi/\kappa$, since these could maintain their position in space, in spite of the motion of the stream; and if we choose as our additional system

$$\eta=-2\pi\kappa b^2e^{-\kappa f}\sin\kappa x \qquad (16)$$

we shall annul the disturbance at a distance on the up-stream side ($x<0$), as is required for a physical solution. The result is

$$\left.\begin{aligned}\eta&=\frac{2b^2f}{x^2+f^2}-4\pi\kappa b^2e^{-\kappa f}\sin\kappa x+\&c. &&[x>0],\\ \eta&=\frac{2b^2f}{x^2+f^2}+\&c. &&[x<0].\end{aligned}\right\} \qquad (17)$$

It appears that there is a local disturbance immediately above the obstacle, followed by a train of waves of length $2\pi c^2/g$ on the down-stream side*.

The investigation is easily adapted to the case where the section of the cylinder has any arbitrary form. The assumption really made above is that, to a first approximation, the effect of the cylinder at a distance is that of a suitably adjusted double source. In the more general case, referring to Art. 72 *a*, we may write

$$\phi=-cx+\phi_1+\chi, \qquad (18)$$

where

$$\phi_1=-\frac{(A+Q)\,x+H\,(y+f)}{2\pi\,\{x^2+(y+f)^2\}}\,c. \qquad (19)$$

It is convenient to work with complex quantities, and to write

$$\phi_1=C\int_0^\infty e^{-k(y+f)+ikx}\,dk \qquad (20)$$

with

$$C=\frac{i\,(A+Q)-H}{2\pi}\,c. \qquad (21)$$

* If we investigate the asymptotic expansion of the definite integral in (13), when κf is large, we find on substitution in (12) that the most important term gives $-2b^2f/(x^2+f^2)$, and so cancels the first term in the above values of η. The approximation has been carried further, for moderate values of κf, by Havelock, *Proc. Roy. Soc.* A, cxv. 274 (1927).

The real part of (20) is of course alone to be retained in the end. The steps of the calculation may be supplied by the reader. The final result is, for large values of $|x|$,

$$\left.\begin{aligned}\eta&=\frac{(A+Q)f-Hx}{\pi(f^2+x^2)}-\{2(A+Q)\kappa\sin\kappa x+2\kappa H\cos\kappa x\}e^{-\kappa f}+\&c. \quad [x>0],\\ \eta&=\frac{(A+Q)f-Hx}{\pi(f^2+x^2)}+\&c. \quad [x<0].\end{aligned}\right\}\quad\text{......(22)}$$

The local disturbance near the origin is not symmetrical unless $H=0$.

For an elliptic section whose major axis makes an angle a with the direction of the stream, we have

$$A=\pi(a^2\sin^2 a+b^2\cos^2 a),\quad Q=\pi ab,\quad H=\pi(a^2-b^2)\sin a\cos a. \quad\text{.........(23)}$$

The square of the amplitude of the waves is then

$$4\kappa^2(A+Q)^2+4\kappa^2H^2=4\pi^2\kappa^2(a+b)^2(a^2\sin^2 a+b^2\cos^2 a). \quad\text{...............(24)}$$

248. If in the problems of Arts. 243, 245 we impress on everything a velocity $-c$ parallel to x, we get the case of a pressure-disturbance advancing with constant velocity c over the surface of otherwise still water. In this form of the question it is not difficult to understand, in a general way, the origin of the train of waves following the disturbance.

If, for example, equal infinitesimal impulses be applied in succession to a series of infinitely close equidistant parallel lines of the surface, at equal intervals of time, each impulse will produce on its own account a system of waves of the character investigated in Art. 239. The systems due to the different impulses will be superposed, with the result that the only parts which reinforce one another will be those whose wave-velocity is equal to the velocity c with which the disturbing influence advances over the surface, and which are (moreover) travelling in the direction of this advance. And the investigations of Arts. 236, 237 shew that in the present problem the groups of waves of this particular length which are produced are continually being left behind. When capillary waves come to be considered, the latter statement will need to be modified.

The question can be investigated from a general standpoint, independent of the particular kind of waves considered, as follows*.

We take the origin at the instantaneous position of the disturbing influence, which is supposed to travel with velocity c in the direction of x-negative. The effect of an impulse δt delivered at an antecedent time t is given by Art. 241 (7) if we replace x by $ct-x$ and multiply by δt. Introducing the hypothesis of a small frictional force varying as the velocity, and integrating from $t=0$ to $t=\infty$, we get

$$\eta=\frac{1}{2\pi}\int_0^\infty\left\{\int_0^\infty\phi(k)\,e^{i\sigma t-ik(ct-x)}\,dk+\int_0^\infty\phi(k)\,e^{i\sigma t+ik(ct-x)}\,dk\right\}e^{-\frac{1}{2}\mu t}\,dt. \quad\text{...(1)}$$

The integration with respect to t gives

$$\eta=\frac{1}{2\pi}\int_0^\infty\frac{\phi(k)\,e^{ikx}\,dk}{\frac{1}{2}\mu-i(\sigma-kc)}+\frac{1}{2\pi}\int_0^\infty\frac{\phi(k)\,e^{-ikx}\,dk}{\frac{1}{2}\mu-i(\sigma+kc)}. \quad\text{.........(2)}$$

* *Phil. Mag.* (6) xxxi. 386 (1916).

The quantity μ is by hypothesis small, and will in the limit be made to vanish. The most important part of the result will therefore be due to values of k in the first integral which make

$$\sigma = kc \quad \text{.................................}(3)$$

approximately. Writing $k = \kappa + k'$, where κ is a root of this equation, we have

$$\sigma - kc = \left(\frac{d\sigma}{dk} - c\right) k' = (U - c)\, k', \quad \text{....................}(4)$$

nearly, where U denotes the group-velocity corresponding to the wave-length $2\pi/\kappa$. The important part of (2) for large values of x is therefore

$$\eta = \frac{1}{2\pi}\, \phi(\kappa)\, e^{i\kappa x} \int_{-\infty}^{\infty} \frac{e^{ik'x}\, dk'}{\frac{1}{2}\mu - i(U - c)\, k'}, \quad \text{.................}(5)$$

since the extension of the range of integration to $k' = \pm\infty$ makes no serious difference. Now if a be positive we have*

$$\int_{-\infty}^{\infty} \frac{e^{imx}\, dm}{a + im} = \begin{cases} 2\pi e^{-ax}, & [x > 0] \\ 0, & [x < 0] \end{cases} \quad \text{......................}(6)$$

whilst

$$\int_{-\infty}^{\infty} \frac{e^{imx}\, dm}{a - im} = \begin{cases} 0, & [x > 0] \\ 2\pi e^{ax}. & [x < 0] \end{cases} \quad \text{......................}(7)$$

Hence if $U < c$

$$\eta = \frac{\phi(\kappa)\, e^{i\kappa x}}{c - U}\, e^{-\frac{1}{2}\mu x/(c - U)}, \quad \text{or} \quad 0, \quad \text{....................}(8)$$

according as $x \gtrless 0$; whilst if $U > c$

$$\eta = 0, \quad \text{or} \quad \frac{\phi(\kappa)\, e^{i\kappa x}}{U - c}\, e^{-\frac{1}{2}\mu x/(U - c)} \quad \text{....................}(9)$$

in the respective cases. If we now make $\mu \to 0$ we have the simple expression

$$\eta = \frac{\phi(\kappa)\, e^{i\kappa x}}{|c - U|}, \quad \text{...........................}(10)$$

for the wave-train generated by the travelling disturbance. This train follows or precedes the disturbing agent acccording as $U \lesseqgtr c$. Examples of the two cases are furnished by gravity waves on water, and capillary waves, respectively (Arts. 236, 266).

The approximation in (4) is valid only if the quotient

$$d^2\sigma/dk^2 \,.\, k' \div (U - c) \quad \text{.........................}(11)$$

is small even when $k'x$ is a moderate multiple of 2π. This requires that

$$d^2\sigma/dk^2 \div (U - c)\, x \quad \text{.........................}(12)$$

should be small. Unless $U = c$, exactly, the condition is always fulfilled if x be sufficiently great. It may be added that the results (8), (9) are accurate, in the sense that they give the leading term in the evaluation of (2) by Cauchy's method of residues. Cf. Art. 242.

* The results quoted are equivalent to the familiar formulae

$$a \int_{-\infty}^{\infty} \frac{\cos mx\, dm}{a^2 + m^2} = \pm \int_{-\infty}^{\infty} \frac{m \sin mx\, dm}{a^2 + m^2} = \pi e^{\mp ax}$$

(where the upper or lower sign is to be chosen according as x is positive or negative), but can be obtained directly by contour integration.

In the case of waves on deep water, due to a concentrated pressure of integral amount P, we put

$$\phi(k) = i\sigma P/g\rho, \qquad \text{.........................(13)}$$

to conform to Art. 239 (28). Since $U = \frac{1}{2}c$, we obtain, on taking the real part,

$$\eta = -\frac{2P\kappa}{g\rho}\sin\kappa x, \qquad \text{.........................(14)}$$

in agreement with (27) of Art. 243*.

If there is more than one value of k satisfying (3), there will be a term of the type (10) for each such value. This happens in the case of water-waves due to gravity and capillarity combined (Art. 269), and in the case of superposed fluids, to be referred to presently.

249. The preceding results have a bearing on the theory of 'wave-resistance.' Taking the two-dimensional form of the question, let us imagine two fixed vertical planes to be drawn, one in front, and the other in the rear, of the disturbing body. If $U < c$ the region between the planes gains energy at the rate cE, where E is the mean energy per unit area of the free surface. This is due partly to the work done at the rear plane, at the rate UE (Art. 237), and partly to the reaction of the disturbing body. Hence if R be the resistance experienced by the latter, so far as it is due to the formation of waves, we have

$$Rc + UE = cE, \quad \text{or} \quad R = \frac{c-U}{c}E. \qquad \text{....................(1)}$$

On the other hand, if $U > c$, so that the wave-train precedes the body, the space between the planes loses energy at the rate cE. Since the loss at the first plane is UE, we have

$$Rc - UE = -cE, \quad \text{or} \quad R = \frac{U-c}{c}E. \qquad \text{................(2)}$$

Thus, in the case of a disturbance advancing with velocity $c\,[< \sqrt{(gh)}]$ over still water of depth h, we find, on reference to Art. 237,

$$R = \tfrac{1}{4}g\rho a^2\left(1 - \frac{2\kappa h}{\sinh 2\kappa h}\right), \qquad \text{......................(3)}$$

where a is the amplitude of the waves. As c increases from 0 to $\sqrt{(gh)}$, κh diminishes from ∞ to 0, so that R diminishes from $\frac{1}{4}g\rho a^2$ to 0. When $c > \sqrt{(gh)}$, the effect is merely local, and $R = 0$†. It must be remarked, however, that the amplitude a due to a disturbance of given type will also vary with c. For instance, in the case of the submerged cylinder, Art. 244 (43), a varies as $\kappa e^{-\kappa b}$, where $\kappa = g/c^2$, the depth being infinite. Hence R varies as

$$c^{-4}\,e^{-2gb/c^2}. \qquad \text{..............................(4)‡}$$

* It is not difficult to derive from (2) the complete formula referred to.

† Cf. Sir W. Thomson, "On Ship Waves," *Proc. Inst. Mech. Eng.* Aug. 3, 1887 [*Popular Lectures and Addresses*, London, 1889–94, iii. 450]. A formula equivalent to (3) was given in a paper by the same author, *Phil. Mag.* (5) xxii. 451 [*Papers*, iv. 279].

‡ The vertical force on the cylinder is calculated by Havelock, *Proc. Roy. Soc.* A, cxxii. 387 (1928).

An interesting variation of the general question is presented when we have a layer of one fluid on the top of another of somewhat greater density. If ρ, ρ' be the densities of the lower and upper fluids, respectively, and if the depth of the upper layer be h', whilst that of the lower fluid is practically infinite, the results of Stokes quoted in Art. 231 shew that two wave-systems may be generated, whose lengths $(2\pi/\kappa)$ are related to the velocity c of the disturbance by the formulae

$$c^2=\frac{g}{\kappa}, \qquad c^2=\frac{\rho-\rho'}{\rho\coth\kappa h'+\rho'}\cdot\frac{g}{\kappa}. \qquad \text{.........(5)}$$

It is easily proved that the value of κ determined by the second equation is real only if

$$c^2<\frac{\rho-\rho'}{\rho}gh'. \qquad \text{.........(6)}$$

If c exceeds the critical value thus indicated, only one type of waves will be generated, and if the difference of densities be slight the resistance will be practically the same as in the case of a single fluid. But if c fall below the critical value, a second type of waves may be produced, in which the amplitude at the common boundary greatly exceeds that at the upper surface; and it is to these waves that the 'dead-water resistance' referred to in Art. 231 is attributed*.

The problem of the submerged cylinder (Art. 247) furnishes an instance where the wave-resistance to the motion of a solid can be calculated. The mean energy, per unit area of the water surface, of the waves represented by the second term in equation (14) of that Art. is

$$E=\tfrac{1}{2}g\rho\,(4\pi\kappa b^2e^{-\kappa f})^2.$$

Since $U=\frac{1}{2}c$, we have from (1)

$$R=4\pi^2g\rho b^4\kappa^2e^{-2\kappa f}. \qquad \text{.........(7)}$$

For a given depth (f) of immersion, this is greatest when $\kappa f=1$, or

$$c=\surd(gf). \qquad \text{.........(8)}$$

In terms of the velocity c we have

$$R=4\pi^2g^3\rho b^4\,.\,c^{-4}e^{-2gf/c^2}. \qquad \text{.........(9)}$$

The graph of R as a function of c is appended†.

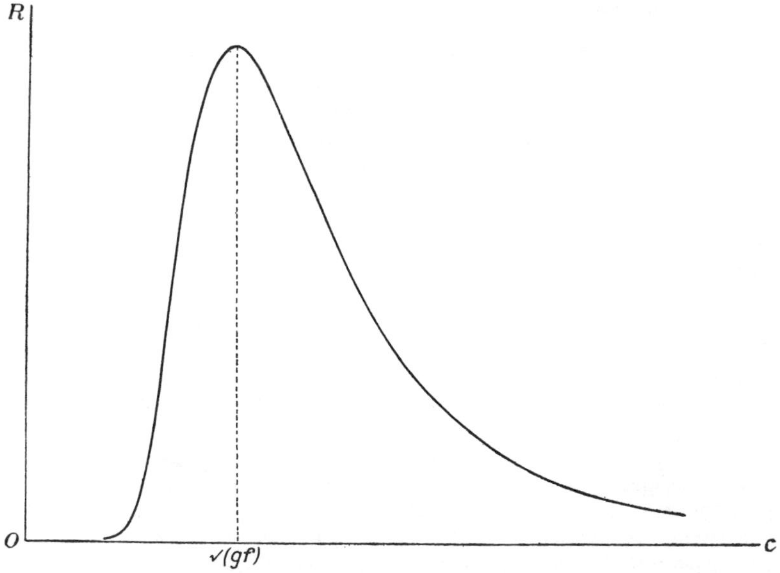

* Ekman, *l.c. ante* p. 371. See also the paper by the author, there quoted.

† *Ann. di mat., l.c.*

Waves of Finite Amplitude.

250. The restriction to 'infinitely small' motions, in the investigations of Arts. 227, ..., implies that the ratio (a/λ) of the maximum elevation to the wave-length must be small. The determination of the wave-forms which satisfy the conditions of uniform propagation without change of type, when this restriction is abandoned, forms the subject of a classical research by Stokes* and of many subsequent investigations.

The problem is most conveniently treated as one of steady motion. It was pointed out by Rayleigh† that if we neglect small quantities of the order a^3/λ^3, the solution in the case of infinite depth is contained in the formulae

$$\frac{\phi}{c} = -x + \beta e^{ky} \sin kx, \qquad \frac{\psi}{c} = -y + \beta e^{ky} \cos kx. \qquad \text{............(1)}$$

The equation of the wave-profile $(\psi = 0)$ is found by successive approximations to be

$$\begin{aligned} y &= \beta e^{ky} \cos kx = \beta\,(1 + ky + \tfrac{1}{2}k^2 y^2 + \ldots) \cos kx \\ &= \tfrac{1}{2}k\beta^2 + \beta\,(1 + \tfrac{9}{8}k^2\beta^2) \cos kx + \tfrac{1}{2}k\beta^2 \cos 2kx + \tfrac{3}{8}k^2\beta^3 \cos 3kx + \ldots; \quad \text{...(2)} \end{aligned}$$

or, if we put

$$\beta\,(1 + \tfrac{9}{8}k^2\beta^2) = a,$$

$$y - \tfrac{1}{2}ka^2 = a \cos kx + \tfrac{1}{2}ka^2 \cos 2kx + \tfrac{3}{8}k^2 a^3 \cos 3kx + \ldots. \qquad \text{......(3)}$$

So far as we have developed it, this coincides with the equation of a trochoid, in which the circumference of the rolling circle is $2\pi/k$, or λ, and the length of the arm of the tracing point is a.

We have still to shew that the condition of uniform pressure along this stream-line can be satisfied by a suitably chosen value of c. We have, from (1), without approximation,

$$\frac{p}{\rho} = \text{const.} - gy - \tfrac{1}{2}c^2 \{1 - 2k\beta e^{ky} \cos kx + k^2\beta^2 e^{2ky}\}, \qquad \text{.........(4)}$$

and therefore, at points of the line $y = \beta e^{ky} \cos kx$,

$$\begin{aligned} \frac{p}{\rho} &= \text{const.} + (kc^2 - g)\,y - \tfrac{1}{2}k^2c^2\beta^2 e^{2ky} \\ &= \text{const.} + (kc^2 - g - k^3c^2\beta^2)\,y + \ldots. \qquad \text{...............(5)} \end{aligned}$$

Hence the condition for a free surface is satisfied, to the present order of approximation, provided

$$c^2 = \frac{g}{k} + k^2c^2\beta^2 = \frac{g}{k}(1 + k^2a^2). \qquad \text{.......................(6)}$$

* "On the theory of Oscillatory Waves," *Camb. Trans.* viii. (1847) [*Papers*, i. 197]. The method was one of successive approximation based on the exact equations of Arts. 9 and 20 *ante*. In a supplement of date 1880 the space-co-ordinates x, y are regarded as functions of the independent variables ϕ, ψ [*Papers*, i. 314].

† *l.c. ante* p. 260. The method was subsequently extended so as to include all Stokes' results, *Phil. Mag.* (6) xxi. 183 [*Papers*, vi. 11].

This determines the velocity of progressive waves of permanent type, and shews that it increases somewhat with the amplitude a.

The figure shews the wave-profile, as given by (3), in the case of $ka = \frac{1}{2}$, or $a/\lambda = \cdot 0796$ *.

The approximately trochoidal form gives an outline which is sharper near the crests, and flatter in the troughs, than in the case of the simple-harmonic waves of infinitely small amplitude investigated in Art. 229, and these features become accentuated as the amplitude is increased. If the trochoidal form were exact, instead of merely approximate, the limiting form would have cusps at the crests, as in the case of Gerstner's waves to be considered presently.

In the actual problem, which is one of irrotational motion, the extreme form has been shewn by Stokes†, in a very simple manner, to have sharp angles of 120°. The question being still treated as one of steady motion, the motion near the angle will be given by the formulae of Art. 63; viz. if we introduce polar co-ordinates r, θ with the crest as origin, and the initial line of θ drawn vertically downwards, we have

$$\psi = Cr^m \cos m\theta, \qquad (7)$$

with the condition that $\psi = 0$ when $\theta = \pm\alpha$ (say), so that $m\alpha = \frac{1}{2}\pi$. This formula leads to

$$q = mCr^{m-1}, \qquad (8)$$

where q is the resultant fluid-velocity. But since the velocity vanishes at the crest, its value at a neighbouring point of the free surface will be given by

$$q^2 = 2gr \cos\alpha, \qquad (9)$$

as in Art. 24 (2). Comparing (8) and (9), we see that we must have $m = \frac{3}{2}$, and therefore $\alpha = \frac{1}{3}\pi$‡.

In the case of progressive waves advancing over still water, the particles at the crests, when these have their extreme forms, are moving forwards with exactly the velocity of the wave.

Another point of interest in connection with these waves of permanent type is that they possess, relatively to the undisturbed water, a certain

* The approximation in (3) is hardly adequate for so large a value of ka; see equation (17) below. The figure serves however to indicate the general form of the wave-profile.

† *Papers*, i. 227 (1880).

‡ The wave-profile has been investigated and traced by Michell, "The Highest Waves in Water," *Phil. Mag.* (5) xxxvi. 430 (1893). He finds that the extreme height is 142 λ, and that the wave-velocity is greater than in the case of infinitely small height in the ratio of 1·2 to 1. See also Wilton, *Phil. Mag.* (6) xxvi. 1053 (1913).

momentum in the direction of wave-propagation. The momentum, per wave-length, of the fluid contained between the free surface and a depth h (beneath the level of the origin), which we will suppose to be great compared with λ, is

$$-\rho \iint \frac{\partial \psi}{\partial y}\, dx\, dy = \rho c h \lambda, \quad \text{.......................(10)}$$

since $\psi = 0$, by hypothesis, at the surface, and $= ch$, by (1), at the great depth h. In the absence of waves, the equation to the upper surface would be $y = \frac{1}{2}ka^2$, by (3), and the corresponding value of the momentum would therefore be

$$\rho c\,(h + \tfrac{1}{2}ka^2)\,\lambda. \quad \text{..............................(11)}$$

The difference of these results is equal to

$$\pi \rho a^2 c, \quad \text{....................................(12)}$$

which gives therefore the momentum, per wave-length, of a system of progressive waves of permanent type, moving over water which is at rest at a great depth.

To find the vertical distribution of this momentum, we remark that the equation of a stream-line $\psi = ch'$ is found from (2) by writing $y + h'$ for y and $\beta e^{-kh'}$ for β. The mean-level of this stream-line is therefore given by

$$y = -h' + \tfrac{1}{2}k\beta^2 e^{-2kh'}. \quad \text{...........................(13)}$$

Hence the momentum, in the case of undisturbed flow, of the stratum of fluid included between the surface and the stream-line in question would be, per wave-length,

$$\rho c \lambda \,\{h' + \tfrac{1}{2}k\beta^2 (1 - e^{-2kh'})\}. \quad \text{.......................(14)}$$

The actual momentum being $\rho c h' \lambda$, we have, for the momentum of the same stratum in the case of waves advancing over still water,

$$\pi \rho a^2 c\,(1 - e^{-2kh'}). \quad \text{..............................(15)}$$

It appears therefore that the motion of the individual particles, in these progressive waves of permanent type, is not purely oscillatory, and that there is, on the whole, a slow but continued advance in the direction of wave-propagation*. The rate of this flow at a depth h' is found approximately by differentiating (15) with respect to h', and dividing by $\rho\lambda$, viz. it is

$$k^2 a^2 c e^{-2kh'}. \quad \text{..................................(16)}$$

This diminishes rapidly from the surface downwards.

The further approximation by Stokes, confirmed by the independent calculations of Rayleigh and others, gives as the equation of the wave-profile

$$y = \text{const.} + a \cos kx - (\tfrac{1}{2}ka^2 + \tfrac{17}{24}k^3 a^4) \cos 2kx + \tfrac{3}{8}k^2 a^3 \cos 3kx$$
$$- \tfrac{1}{3}k^3 a^4 \cos 4kx + \ldots, \quad \text{......(17)}$$

* Stokes, *l.c. ante* p. 417. Another very simple proof of this statement has been given by Rayleigh, *l.c. ante* p. 260.

with, for the wave-velocity,

$$c^2 = \frac{g}{k}(1 + k^2a^2 + \tfrac{5}{4}k^4a^4 + \ldots). \qquad (18)$$

A question as to the convergence, both of the series which form the coefficients of the successive cosines when the approximation is continued, and of the resulting series of cosines, was raised by Burnside*, who even expressed a doubt as to the possibility of waves of rigorously permanent type. This led Rayleigh to undertake an extended investigation†, which shewed that the condition of uniformity of pressure at the surface could be satisfied, for sufficiently small values of ka, to a very high degree of accuracy. He inferred that the existence of permanent types up to the highest wave of Michell was practically, if not demonstrably, certain. The existence has at length been definitely established by an investigation of Prof. Levi Cività‡, which puts an end to an historic controversy.

There are one or two simple properties of these permanent waves which come easily from first principles§. The problem being reduced to one of steady motion, let the origin be taken in the mean level, beneath (say) a crest, and let λ be the wave-length. Denoting by η surface-elevation above the mean level, we have, then,

$$\int_0^\lambda \eta\, dx = 0. \qquad (19)$$

Also, if q be the surface velocity, and q_0 its value at the mean level, we have

$$q^2 = q_0^2 - 2g\eta,$$

and therefore

$$\int_0^\lambda q^2\, dx = q_0^2 \lambda. \qquad (20)$$

Again, consider the mass of fluid contained between vertical planes through two successive crests, and bounded below by a plane $y = -h_1$ at which the velocity is sensibly horizontal and equal to c. It is easily seen that the total vertical mass-acceleration is zero, since there is no flux of vertical momentum across the boundaries. Hence if p be the surface-pressure, and p_1 that at the depth h_1,

$$\int_0^\lambda (p_1 - p)\, dx = g\rho \int_0^\lambda (h_1 + \eta)\, dx = g\rho h_1 \lambda. \qquad (21)$$

But, comparing pressures in the same vertical we have

$$p_1 - p = g\rho(h_1 + \eta) + \tfrac{1}{2}(q^2 - c^2),$$

and thence

$$\int_0^\lambda q^2 dx = c^2 \lambda. \qquad (22)$$

We may express this by saying that the mean square of the surface velocity, per equal increments of x, is equal to c^2. It follows also from (20) that $q_0 = c$, *i.e.* the velocity at the points where the wave-profile meets the mean level is equal to c.

* *Proc. Lond. Math. Soc.* (2) xv. 26 (1916).

† *Phil. Mag.* (6) xxxiii. 381 (1917) [*Papers*, vi. 478].

‡ "Détermination rigoureuse des ondes permanentes d'ampleur finie," *Math. Ann.* xciii. 264 (1925). The extension to waves in a canal of finite depth has been made by Struik, *Math. Ann.* xcv. 595 (1926).

§ Levi Cività, *l.c.*

251. A system of *exact* equations, expressing a possible form of wave-motion when the depth of the fluid is infinite, was given so long ago as 1802 by Gerstner*, and at a later period independently by Rankine†. The circumstance, however, that the motion in these waves is not irrotational detracts somewhat from the physical interest of the results.

If the axis of x be horizontal, and that of y be drawn vertically upwards, the formulae in question may be written

$$x = a + \frac{1}{k} e^{kb} \sin k(a + ct), \qquad y = b - \frac{1}{k} e^{kb} \cos k(a + ct), \quad \text{......(1)}$$

where the specification is on the Lagrangian plan (Art. 16), viz. a, b are two parameters serving to identify a particle, and x, y are the co-ordinates of this particle at time t. The constant k determines the wave-length, and c is the velocity of the waves which are travelling in the direction of x-negative.

To verify this solution, and to determine the value of c, we remark, in the first place, that

$$\frac{\partial (x, y)}{\partial (a, b)} = 1 - e^{2kb}, \quad \text{.............................(2)}$$

so that the Lagrangian equation of continuity (Art. 16 (2)) is satisfied. Again, substituting from (1) in the equations of motion (Art. 13), we find

$$\left.\begin{aligned} \frac{\partial}{\partial a}\left(\frac{p}{\rho} + gy\right) &= kc^2 e^{kb} \sin k(a + ct), \\ \frac{\partial}{\partial b}\left(\frac{p}{\rho} + gy\right) &= -kc^2 e^{kb} \cos k(a + ct) + kc^2 e^{2kb}; \end{aligned}\right\} \quad \text{...........(3)}$$

whence

$$\frac{p}{\rho} = \text{const.} - g\left\{b - \frac{1}{k} e^{kb} \cos k(a + ct)\right\} - c^2 e^{kb} \cos k(a + ct) + \tfrac{1}{2} c^2 e^{2kb}. \quad \text{...(4)}$$

For a particle on the free surface the pressure must be constant; this requires

$$c^2 = g/k, \quad \text{....................................(5)}$$

as in Art. 229. This makes

$$\frac{p}{\rho} = \text{const.} - gb + \tfrac{1}{2} c^2 e^{2kb}. \quad \text{........................(6)}$$

It is obvious from (1) that the path of any particle (a, b) is a circle of radius $k^{-1} e^{kb}$.

It has already been stated that the motion of the fluid in these waves is rotational. To prove this we remark that

$$\begin{aligned} u\,\delta x + v\,\delta y &= \left(\dot{x}\frac{\partial x}{\partial a} + \dot{y}\frac{\partial y}{\partial a}\right)\delta a + \left(\dot{x}\frac{\partial x}{\partial b} + \dot{y}\frac{\partial y}{\partial b}\right)\delta b \\ &= \frac{c}{k}\,\delta\left\{e^{kb} \sin k(a + ct)\right\} + ce^{2kb}\,\delta a, \quad \text{..............(7)} \end{aligned}$$

which is not an exact differential.

* Professor of Mathematics at Prague, 1789–1823. His paper, "Theorie der Wellen," was published in the *Abh. d. k. böhm. Ges. d. Wiss.* 1802 [Gilbert's *Annalen d. Physik*, xxxii. (1809)].

† "On the Exact Form of Waves near the Surface of Deep Water," *Phil. Trans.* 1863 [*Papers*, p. 481].

The circulation in the boundary of the parallelogram whose vertices coincide with the particles

$$(a, b),\ (a+\delta a, b),\ (a, b+\delta b),\ (a+\delta a, b+\delta b)$$

is, therefore,

$$-\frac{\partial}{\partial b}(ce^{2kb}\delta a)\,\delta b,$$

and the area of the circuit is

$$\frac{\partial(x, y)}{\partial(a, b)}\delta a\,\delta b = (1-e^{2kb})\,\delta a\,\delta b.$$

Hence the vorticity (ω) of the element (a, b) is

$$\omega = -\frac{2kce^{2kb}}{1-e^{2kb}}. \qquad \text{.............................(8)}$$

This is greatest at the surface, and diminishes rapidly with increasing depth. Its *sense* is opposite to that of the revolution of the particles in their circular orbits.

A system of waves of the present type cannot therefore be originated from rest, or destroyed, by the action of forces of the kind contemplated in the general theorem of Arts. 17, 33. We may however suppose that by properly adjusted pressures applied to the surface of the waves the liquid is gradually reduced to a state of flow in horizontal lines, in which the velocity (u') is a function of the ordinate (y') only*. In this state we shall have $\partial x'/\partial a = 1$, while y' is a function of b determined by the condition

$$\frac{\partial(x', y')}{\partial(a, b)} = \frac{\partial(x, y)}{\partial(a, b)}, \qquad \text{..........................(9)}$$

or

$$\frac{\partial y'}{\partial b} = 1 - e^{2kb}. \qquad \text{..............................(10)}$$

This makes

$$\frac{\partial u'}{\partial b} = \frac{\partial u'}{\partial y'}\frac{\partial y'}{\partial b} = -2\omega\frac{\partial y'}{\partial b} = 2kce^{2kb}, \qquad \text{................(11)}$$

and therefore

$$u' = ce^{2kb}. \qquad \text{................................(12)}$$

Hence, for the genesis of the waves by ordinary forces, we require as a foundation an initial horizontal motion, in the direction *opposite* to that of propagation of the waves ultimately set up, which diminishes rapidly from the surface downwards, according to the law (12), where b is a function of y' determined by

$$y' = b - \tfrac{1}{2}k^{-1}e^{2kb}. \qquad \text{..........................(13)}$$

It is to be noted that these rotational waves, when established, have zero momentum.

The figure shews the forms of the lines of equal pressure $b = \text{const.}$, for a series of equidistant values of b†. These curves are trochoids, obtained by

* For a fuller statement of the argument see Stokes' *Papers*, i. 222.

† The diagram is very similar to the one given originally by Gerstner, and copied more or less closely by subsequent writers. A version of Gerstner's investigation, including in one respect a correction, was given in the second edition of this work, Art. 233.

rolling circles of radii k^{-1} on the under sides of the lines $y = b + k^{-1}$, the distances of the tracing points from the respective centres being $k^{-1}e^{kb}$. Any one of these lines may be taken as representing the free surface, the extreme admissible form being that of the cycloid. The dotted lines represent the successive forms taken by a line of particles which is vertical when it passes through a crest or a trough.

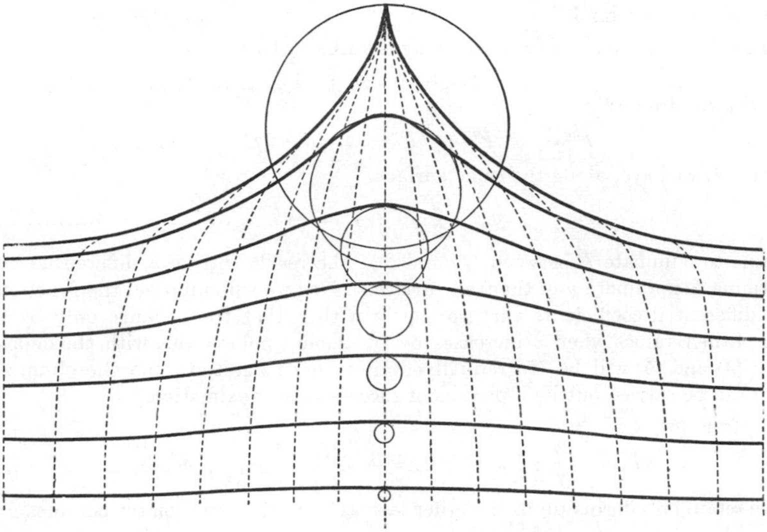

252. Scott Russell, in his interesting experimental investigations*, was led to pay great attention to a particular type which he called the 'solitary wave.' This is a wave consisting of a single elevation, of height not necessarily small compared with the depth of the fluid, which, if properly started, may travel for a considerable distance along a uniform canal, with little or no change of type. Waves of depression, of similar relative amplitude, were found not to possess the same character of permanence, but to break up into series of shorter waves.

Russell's 'solitary' type may be regarded as an extreme case of Stokes' oscillatory waves of permanent type, the wave-length being great compared with the depth of the canal, so that the widely separated elevations are practically independent of one another. The methods of approximation employed by Stokes become, however, unsuitable when the wave-length much exceeds the depth; and subsequent investigations of solitary waves of permanent type have proceeded on different lines.

The first of these was given independently by Boussinesq† and Rayleigh‡. The latter writer, treating the problem as one of steady motion, starts virtually from the formula

$$\phi + i\psi = F(x + iy) = e^{iy\frac{d}{dx}} F(x), \quad \text{.................................}(1)$$

* "Report on Waves," *Brit. Ass. Rep.* 1844.

† *Comptes Rendus*, June 19, 1871. ‡ *l.c. ante* p. 260.

where $F(x)$ is real. This is especially appropriate to cases, such as the present, where one of the family of stream-lines is straight. We derive from (1)

$$\phi = F - \frac{y^2}{2!}F'' + \frac{y^4}{4!}F^{\text{iv}} - \dots, \qquad \psi = yF' - \frac{y^3}{3!}F''' + \frac{y^5}{5!}F^{\text{v}} - \dots, \quad \text{.........(2)}$$

where the accents denote differentiations with respect to x. The stream-line $\psi = 0$ here forms the bed of the canal, whilst at the free surface we have $\psi = -ch$, where c is the uniform velocity, and h the depth, in the parts of the fluid at a distance from the wave, whether in front or behind.

The condition of uniform pressure along the free surface gives

$$u^2 + v^2 = c^2 - 2g(y - h), \quad \text{.........(3)}$$

or, substituting from (2),

$$F'^2 - y^2 F' F''' + y^2 F''^2 + \dots = c^2 - 2g(y - h). \quad \text{.........(4)}$$

But, from (2) we have, along the same surface,

$$yF' - \frac{y^3}{3!}F''' + \dots = -ch. \quad \text{.........(5)}$$

It remains to eliminate F between (4) and (5); the result will be a differential equation to determine the ordinate y of the free surface. If (as we will suppose) the function $F'(x)$ and its differential coefficients vary so slowly with x that they change only by a small fraction of their values when x increases by an amount comparable with the depth h, the terms in (4) and (5) will be of gradually diminishing magnitude, and the elimination in question can be carried out by a process of successive approximation.

Thus, from (5),

$$F' = -\frac{ch}{y} + \frac{1}{6}y^2 F''' + \dots = -ch\left\{\frac{1}{y} + \frac{1}{6}y^2\left(\frac{1}{y}\right)'' + \dots\right\}; \quad \text{.........(6)}$$

and if we retain only terms up to the order last written, the equation (4) becomes

$$\frac{1}{y^2} - \frac{2}{3}y\left(\frac{1}{y}\right)'' + y^2\left(\frac{1}{y}\right)'^2 = \frac{1}{h^2} - \frac{2g(y-h)}{c^2h^2},$$

or, on reduction,

$$\frac{1}{y^2} + \frac{2}{3}\frac{y''}{y} - \frac{1}{3}\frac{y'^2}{y^2} = \frac{1}{h^2} - \frac{2g(y-h)}{c^2h^2}. \quad \text{.........(7)}$$

If we multiply by y', and integrate, determining the arbitrary constant so as to make $y' = 0$ for $y = h$, we obtain

$$-\frac{1}{y} + \frac{1}{3}\frac{y'^2}{y} = -\frac{1}{h} + \frac{y-h}{h^2} - \frac{g(y-h)^2}{c^2h^2},$$

or

$$y'^2 = 3\,\frac{(y-h)^2}{h^2}\left(1 - \frac{gy}{c^2}\right). \quad \text{.........(8)}$$

Hence y' vanishes only for $y = h$ and $y = c^2/g$, and since the last factor must be positive, it appears that c^2/g is a *maximum* value of y. Hence the wave is necessarily one of elevation only, and, denoting by a the maximum height above the undisturbed level, we have

$$c^2 = g(h + a), \quad \text{.........(9)}$$

which is exactly the empirical formula for the wave-velocity adopted by Russell.

The extreme form of the wave must, as in Art. 250, have a sharp crest of 120°; and since the fluid is there at rest we shall have $c^2 = 2ga$. If the formula (9) were applicable to such an extreme case, it would follow that $a = h$.

If we put, for shortness,

$$y - h = \eta, \qquad \frac{h^2(h+a)}{3a} = b^2, \quad \text{.........(10)}$$

we find, from (8),

$$\eta' = \pm\frac{\eta}{b}\left(1 - \frac{\eta}{a}\right)^{\frac{1}{2}}, \quad \text{.........(11)}$$

the integral of which is

$$\eta = a \operatorname{sech}^2 \tfrac{1}{2}\frac{x}{b}, \quad \text{.....................................(12)}$$

if the origin of x be taken beneath the summit.

There is no definite 'length' of the wave, but we may note, as a rough indication of its extent, that the elevation has one-tenth of its maximum value when $x/b = 3{\cdot}636$.

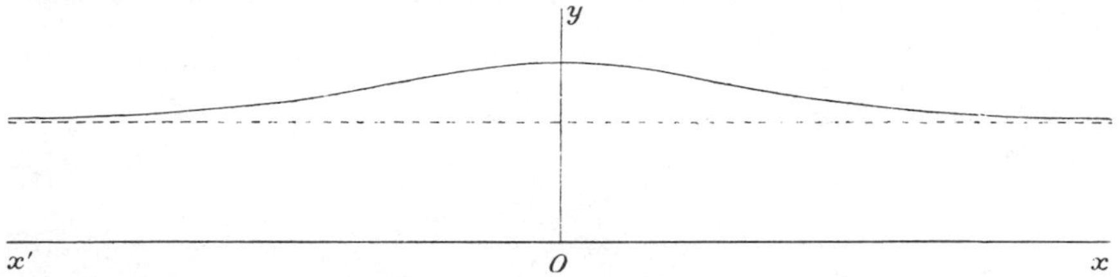

The annexed drawing of the curve

$$y = 1 + \tfrac{1}{2}\operatorname{sech}^2 \tfrac{1}{2}x$$

represents the wave-profile in the case $a = \tfrac{1}{2}h$. For lower waves the scale of y must be contracted, and that of x enlarged, as indicated by the annexed table giving the ratio b/h, which determines the horizontal scale, for various values of a/h.

It will be found, on reviewing the above investigation, that the approximations consist in neglecting the fourth power of the ratio $(h+a)/2b$*.

a/h	b/h
·1	1·915
·2	1·414
·3	1·202
·4	1·080
·5	1·000
·6	·943
·7	·900
·8	·866
·9	·839
1·0	·816

If we impress on the fluid a velocity $-c$ parallel to x we get the case of a progressive wave on still water. It is not difficult to shew that, if the ratio a/h be small, the path of each particle is then an arc of a parabola having its axis vertical and apex upwards†.

It might appear, at first sight, that the above theory is inconsistent with the results of Art. 187, where it was argued that a wave of finite height whose length is great compared with the depth must inevitably suffer a continual change of form as it advances, the changes being the more rapid the greater the elevation above the undisturbed level. The investigation referred to postulates, however, a length so great that the vertical acceleration may be neglected, with the result that the horizontal velocity is sensibly uniform from top to bottom (Art. 169). The numerical table above given shews, on the other hand, that the longer the 'solitary wave' is, the lower it is. In other words, the more nearly it approaches to the character of a 'long' wave, in the sense of Art. 169, the more easily is the change of type averted by a slight adjustment of the particle-velocities‡.

The motion at the outskirts of the solitary wave can be represented by a very simple formula. Considering a progressive wave travelling in the direction of x-positive, and taking the origin in the bottom of the canal, at a point in the front part of the wave, we assume

$$\phi = Ae^{-m(x-ct)}\cos my. \quad \text{...................................(13)}$$

This satisfies $\nabla^2\phi = 0$, and the surface-condition

$$\frac{\partial^2\phi}{\partial t^2} + g\frac{\partial\phi}{\partial y} = 0 \quad \text{...(14)}$$

* The theory of the solitary wave has been treated by Weinstein, *Lincei* (6) iii. 463 (1926), by the method of Levi Cività referred to in Art. 250. He finds that the formula (9) is a very close approximation.

† Boussinesq, *l.c.*

‡ Stokes, "On the Highest Wave of Uniform Propagation," *Proc. Camb. Phil. Soc.* iv. 361 (1883) [*Papers*, v. 140].

will also be satisfied for $y=h$, provided

$$c^2=gh\frac{\tan mh}{mh}. \qquad \text{.......(15)}$$

This will be found to agree approximately with Rayleigh's investigation if we put $m=b^{-1}$.

The above remark, which was communicated to the author by the late Sir George Stokes*, was suggested by an investigation by McCowan†, who shewed that the formula

$$\frac{\phi+i\psi}{c}=-(x+iy)+\alpha\tanh\tfrac{1}{2}m(x+iy) \qquad \text{.......(16)}$$

satisfies the conditions very approximately, provided

$$c^2=\frac{g}{m}\tan mh, \qquad \text{.......(17)}$$

and

$$m\alpha=\tfrac{2}{3}\sin^2 m(h+\tfrac{2}{3}a), \qquad a=\alpha\tan\tfrac{1}{2}m(h+a), \qquad \text{.......(18)}$$

where a denotes the maximum elevation above the mean level, and α is a subsidiary constant. In a subsequent paper‡ the extreme form of the wave when the crest has a sharp angle of 120° was examined. The limiting value of the ratio a/h was found to be ·78, in which case the wave-velocity is given by $c^2=1{\cdot}56gh$.

253. By a slight modification the investigation of Rayleigh and Boussinesq can be made to give the theory of a system of *oscillatory* waves of finite height in a canal of limited depth§.

In the steady-motion form of the problem the momentum per wave-length (λ) is represented by

$$\iint\rho u\,dx\,dy=-\rho\iint\frac{\partial\psi}{\partial x}\,dx\,dy=-\rho\psi_1\lambda, \qquad \text{.......(19)}$$

where ψ_1 corresponds to the free surface. If h be the mean depth, this momentum may be equated to $\rho ch\lambda$, where c denotes (in a sense) the mean velocity of the stream. On this understanding we have, at the surface, $\psi_1=-ch$, as before. The arbitrary constant in (3), on the other hand, must be left for the moment undetermined, so that we write

$$u^2+v^2=C-2gy. \qquad \text{.......(20)}$$

We then find, in place of (8),

$$y'^2=\frac{3g}{c^2h^2}(y-l)(h_1-y)(y-h_2), \qquad \text{.......(21)}$$

where h_1, h_2 are the upper and lower limits of y, and

$$l=\frac{c^2h^2}{gh_1h_2}. \qquad \text{.......(22)}$$

It is implied that l cannot be greater than h_2.

If we now write

$$y=h_1\cos^2\chi+h_2\sin^2\chi, \qquad \text{.......(23)}$$

we find

$$\beta\frac{d\chi}{dx}=\surd\{1-k^2\sin^2\chi\}, \qquad \text{.......(24)}$$

* Cf. *Papers*, v. 62.

† "On the Solitary Wave," *Phil. Mag.* (5) xxxii. 45 (1891).

‡ "On the Highest Wave of Permanent Type," *Phil. Mag.* (5) xxxviii. 351 (1894).

§ Korteweg and De Vries, "On the Change of Form of Long Waves advancing in a Rectangular Canal, and on a New Type of Long Stationary Waves," *Phil. Mag.* (5) xxxix. 422 (1895). The method adopted by these writers is somewhat different. Moreover, as the title indicates, the paper includes an examination of the manner in which the wave-profile is changing at any instant, if the conditions for permanency of type are not satisfied.

For other modifications of Rayleigh's method reference may be made to Gwyther, *Phil. Mag.* (5) l. 213, 308, 349 (1900).

where $$\beta=\sqrt{\left\{\frac{4h_1h_2l}{3(h_1-l)}\right\}},\qquad k^2=\frac{h_1-h_2}{h_1-l}. \qquad\text{.........................(25)}$$

Hence, if the origin of x be taken at a crest, we have

$$x=\beta\int_0^{\chi}\frac{d\chi}{\sqrt{(1-k^2\sin^2\chi}}=\beta F(\chi, k), \qquad\text{..........................(26)}$$

and $$y=h_2+(h_1-h_2)\,\mathrm{cn}^2\,\frac{x}{\beta}.\ \ [\text{mod. } k] \qquad\text{..........................(27)*}$$

The wave-length is given by

$$\lambda=2\beta\int_0^{\frac{1}{2}\pi}\frac{d\chi}{\sqrt{(1-k^2\sin^2\chi)}}=2\beta F_1(k). \qquad\text{..........................(28)}$$

Again, from (23) and (24),

$$\int_0^{\lambda}y\,dx=2\beta\int_0^{\frac{1}{2}\pi}\frac{h_1\cos^2\chi+h_2\sin^2\chi}{\sqrt{(1-k^2\sin^2\chi)}}\,d\chi=2\beta\,\{lF_1(k)+(h_1-l)\,E_1(k)\}. \qquad\text{......(29)}$$

Since this must be equal to $h\lambda$, we have

$$(h-l)\,F_1(k)=(h_1-l)\,E_1(k). \qquad\text{................................(30)}$$

In equations (25), (28), (30) we have four relations connecting the six quantities h_1, h_2, l, k, λ, β, so that if two of these be assigned the rest are analytically determinate. The wave-velocity c is then given by (22)†. For example, the form of the waves, and their velocity, are determined by the length λ, and the height h_1 of the crests above the bottom.

The solitary wave of Art. 252 is included as a particular case. If we put $l=h_2$, we have $k=1$, and the formulae (28) and (30) then shew that $\lambda=\infty$, $h_2=h$.

254. The theory of waves of permanent type has been brought into relation with general dynamical principles by Helmholtz‡.

If in the equations of motion of a 'gyrostatic' system, Art. 141 (23), we put

$$Q_1=-\frac{\partial V}{\partial q_1},\qquad Q_2=-\frac{\partial V}{\partial q_2},\qquad \ldots,\qquad Q_n=-\frac{\partial V}{\partial q_n}, \qquad\text{...........(1)}$$

where V is the potential energy, it appears that the conditions for steady motion, with $q_1, q_2, \ldots q_n$ constant, are

$$\frac{\partial}{\partial q_1}(V+K)=0,\qquad \frac{\partial}{\partial q_2}(V+K)=0,\qquad \ldots,\qquad \frac{\partial}{\partial q_n}(V+K)=0, \qquad\text{...(2)}$$

where K is the energy of the motion corresponding to any given values of the co-ordinates $q_1, q_2, \ldots q_n$ when these are prevented from varying by the application of suitable extraneous forces.

This energy is here supposed expressed in terms of the constant momenta corresponding to the ignored co-ordinates $\chi, \chi', \ldots$, and of the palpable co-ordinates $q_1, q_2, \ldots q_n$. It may however also be expressed in terms of the

* The waves represented by (27) are called 'cnoidal waves' by the authors cited. For the method of proceeding to a higher approximation we must refer to the original paper.

† When the depth is finite, a question arises as to what is meant exactly by the 'velocity of propagation.' The velocity adopted in the text is that of the wave-profile relative to the centre of inertia of the mass of fluid included between two vertical planes at a distance apart equal to the wave-length. Cf. Stokes, *Papers*, i. 202.

‡ "Die Energie der Wogen und des Windes," *Berl. Monatsber.* July 17, 1890 [*Wiss. Abh.* iii. 333].

velocities $\dot{\chi}, \dot{\chi}', \ldots$ and the co-ordinates $q_1, q_2, \ldots q_n$; in this form we denote it by T_0. It may be shewn, exactly as in Art. 142, that $\partial T_0/\partial q_r = -\partial K/\partial q_r$, so that the conditions (2) are equivalent to

$$\frac{\partial}{\partial q_1}(V - T_0) = 0, \quad \frac{\partial}{\partial q_2}(V - T_0) = 0, \quad \ldots, \quad \frac{\partial}{\partial q_n}(V - T_0) = 0. \quad \ldots(3)$$

Hence the condition for free steady motion with any assigned constant values of $q_1, q_2, \ldots q_n$ is that the corresponding value of $V + K$, or of $V - T_0$, should be stationary. Cf. Art. 203 (7).

Further, if in the equations of Art. 141 we write $-\partial V/\partial q_r + Q_r$ for Q_r, so that Q_r now denotes a component of extraneous force, we find, on multiplying by $\dot{q}_1, \dot{q}_2, \ldots \dot{q}_n$ in order, and adding,

$$\frac{d}{dt}(\mathfrak{T} + V + K) = Q_1\dot{q}_1 + Q_2\dot{q}_2 + \ldots + Q_n\dot{q}_n, \quad \ldots\ldots\ldots\ldots(4)$$

where $\mathfrak{T}$ is the part of the energy which involves the velocities $\dot{q}_1, \dot{q}_2, \ldots \dot{q}_n$. It follows, by the same argument as in Art. 205, that the condition for 'secular' stability, when there are dissipative forces affecting the co-ordinates $q_1, q_2, \ldots q_n$, but not the ignored co-ordinates $\chi, \chi', \ldots$, is that $V + K$ should be a minimum.

In the application to the problem of stationary waves, it will tend to clearness if we eliminate all infinities from the question by imagining that the fluid circulates in a ring-shaped canal of uniform rectangular section (the sides being horizontal and vertical), of very large radius. The generalized velocity $\dot{\chi}$ corresponding to the ignored co-ordinate may be taken to be the flux per unit breadth of the channel, and the constant momentum of the circulation may be replaced by the cyclic constant κ. The co-ordinates $q_1, q_2, \ldots q_n$ of the general theory are now represented by the value of the surface-elevation (η) considered as a function of the longitudinal space-co-ordinate x. The corresponding components of extraneous force are represented by arbitrary pressures applied to the surface.

If l denote the whole length of the circuit, then considering unit breadth of the canal we have

$$V = \tfrac{1}{2} g\rho \int_0^l \eta^2 dx, \quad \ldots\ldots\ldots\ldots(5)$$

where η is subject to the condition

$$\int_0^l \eta\, dx = 0. \quad \ldots\ldots\ldots\ldots(6)$$

If we could with the same ease obtain a general expression for the kinetic energy of the steady motion corresponding to any prescribed form of the surface, the condition in either of the forms above given would, by the usual processes of the Calculus of Variations, lead to a determination of the possible forms, if any, of stationary waves*.

* For some general considerations bearing on the problem of stationary waves on the common surface of two currents reference may be made to Helmholtz' paper. This also contains, at the end, some speculations, based on calculations of energy and momentum, as to the length of the waves which would be excited in the first instance by a wind of given velocity. These appear to involve the assumption that the waves will necessarily be of permanent type, since it is only on some such hypothesis that we get a determinate value for the momentum of a train of waves of small amplitude.

Practically, this is not feasible, except by methods of successive approximation, but we may illustrate the question by reproducing, on the basis of the present theory, the results already obtained for 'long' waves of infinitely small amplitude.

If h be the depth of the canal, the velocity in any section when the surface is maintained at rest, with arbitrary elevation η, is $\dot{\chi}/(h+\eta)$, where $\dot{\chi}$ is the flux. Hence, for the cyclic constant,

$$\kappa = \dot{\chi}\int_0^l (h+\eta)^{-1}\,dx = \frac{l\dot{\chi}}{h}\left(1+\frac{1}{h^2 l}\int_0^l \eta^2\,dx\right), \quad \text{.......................(7)}$$

approximately, where the term of the first order in η has been omitted, in virtue of (6).

The kinetic energy, $\frac{1}{2}\rho\kappa\dot{\chi}$, may be expressed in terms of either $\dot{\chi}$ or κ. We thus obtain the forms

$$T_0 = \tfrac{1}{2}\frac{\rho l\dot{\chi}^2}{h}\left(1+\frac{1}{h^2 l}\int_0^l \eta^2\,dx\right), \quad \text{............................(8)}$$

$$K = \tfrac{1}{2}\frac{\rho h\kappa^2}{l}\left(1-\frac{1}{h^2 l}\int_0^l \eta^2\,dx\right). \quad \text{............................(9)}$$

The variable part of $V-T_0$ is

$$\tfrac{1}{2}\rho\left(g-\frac{\dot{\chi}^2}{h^3}\right)\int_0^l \eta^2\,dx, \quad \text{..................................(10)}$$

and that of $V+K$ is

$$\tfrac{1}{2}\rho\left(g-\frac{\kappa^2}{hl^2}\right)\int_0^l \eta^2\,dx. \quad \text{..................................(11)}$$

It is obvious that these are both stationary for $\eta=0$; and that they will be stationary for *any* infinitely small values of η, provided $\dot{\chi}^2 = gh^3$, or $\kappa^2 = ghl^2$. If we put $\dot{\chi} = ch$, or $\kappa = cl$, this condition gives

$$c^2 = gh, \quad \text{..(12)}$$

in agreement with Art. 175.

It appears, moreover, that $\eta=0$ makes $V+K$ a maximum or a minimum according as c^2 is greater or less than gh. In other words, the plane form of the surface is secularly stable if, and only if, $c<\sqrt{(gh)}$. It is to be remarked, however, that the dissipative forces here contemplated are of a special character, viz. they affect the *vertical* motion of the surface, but not (directly) the flow of the liquid. It is otherwise evident from Art. 175 that if pressures be applied to maintain any given constant form of the surface, then if $c^2>gh$ these pressures must be greatest over the elevations and least over the depressions. Hence if the pressures be removed, the inequalities of the surface will tend to increase.

Wave-Propagation in Two Dimensions.

255. We may next consider some cases of wave-propagation in two horizontal dimensions x, y. The axis of z being drawn vertically upwards, we have, on the hypothesis of infinitely small motions,

$$\frac{p}{\rho} = \frac{\partial\phi}{\partial t} - gz + F(t), \quad \text{.........................(1)}$$

where ϕ satisfies

$$\nabla^2\phi = 0. \quad \text{...............................(2)}$$

The arbitrary function $F(t)$ may be supposed merged in the value of $\partial\phi/\partial t$.

If the origin be taken in the undisturbed surface, and if ζ denote the elevation at time t above this level, the pressure-condition to be satisfied at the surface is

$$\zeta = \frac{1}{g}\left[\frac{\partial\phi}{\partial t}\right]_{z=0}, \quad \text{.............................(3)}$$

and the kinematical surface-condition is

$$\frac{\partial \zeta}{\partial t} = -\left[\frac{\partial \phi}{\partial z}\right]_{z=0}; \qquad \text{.........................(4)}$$

cf. Art. 227. Hence, for $z = 0$, we must have

$$\frac{\partial^2 \phi}{\partial t^2} + g\frac{\partial \phi}{\partial z} = 0, \qquad \text{.............................(5)}$$

or, in the case of simple-harmonic motion,

$$\sigma^2 \phi = g\frac{\partial \phi}{\partial z}, \qquad \text{................................(6)}$$

if the time-factor be $e^{i(\sigma t + \epsilon)}$.

The fluid being supposed to extend to infinity, horizontally and downwards, we may briefly examine, in the first place, the effect of a local initial disturbance of the surface, in the case of symmetry about the origin.

The typical solution for the case of initial *rest* is easily seen, on reference to Art. 100, to be

$$\left.\begin{aligned} \phi &= g\frac{\sin \sigma t}{\sigma} e^{kz} J_0(k\varpi), \\ \zeta &= \cos \sigma t\, J_0(k\varpi), \end{aligned}\right\} \qquad \text{......................(7)}$$

provided $\qquad \sigma^2 = gk, \qquad$ ....(8)

as in Art. 228.

To generalize this, subject to the condition of symmetry, we have recourse to the theorem

$$f(\varpi) = \int_0^\infty J_0(k\varpi)\, k\, dk \int_0^\infty f(\alpha)\, J_0(k\alpha)\, \alpha\, d\alpha \qquad \text{.............(9)}$$

of Art. 100 (12). Thus, corresponding to the initial conditions,

$$\zeta = f(\varpi), \qquad \phi_0 = 0, \qquad \text{.......................(10)}$$

we have

$$\left.\begin{aligned} \phi &= g\int_0^\infty \frac{\sin \sigma t}{\sigma} e^{kz} J_0(k\varpi)\, k\, dk \int_0^\infty f(\alpha)\, J_0(k\alpha)\, \alpha\, d\alpha, \\ \zeta &= \int_0^\infty \cos \sigma t\, J_0(k\varpi)\, k\, dk \int_0^\infty f(\alpha)\, J_0(k\alpha)\, \alpha\, d\alpha. \end{aligned}\right\} \qquad \text{......(11)}$$

If the initial elevation be concentrated in the immediate neighbourhood of the origin, then, assuming

$$\int_0^\infty f(\alpha)\, 2\pi\alpha\, d\alpha = 1, \qquad \text{........................(12)}$$

we have

$$\phi = \frac{g}{2\pi}\int_0^\infty \frac{\sin \sigma t}{\sigma} e^{kz} J_0(k\varpi)\, k\, dk. \qquad \text{................(13)}$$

Expanding, and making use of (8), we get

$$\phi = \frac{gt}{2\pi}\int_0^\infty \left\{k - \frac{gt^2}{3!}k^2 + \frac{(gt^2)^2}{5!}k^3 - \dots\right\} e^{kz} J_0(k\varpi)\, dk. \qquad \text{......(14)}$$

If we put $z = -r\cos\theta, \quad \varpi = r\sin\theta,$ ....(15)

we have
$$\int_0^\infty e^{kz} J_0(k\varpi)\,dk = \frac{1}{r}, \qquad\qquad (16)$$

by Art. 102 (9), and thence*

$$\int_0^\infty e^{kz} J_0(k\varpi)\,k^n\,dk = \left(\frac{\partial}{\partial z}\right)^n \frac{1}{r} = n!\,\frac{P_n(\mu)}{r^{n+1}}, \qquad\qquad (17)$$

where $\mu = \cos\theta$ (cf. Art. 85). Hence

$$\phi = \frac{gt}{2\pi}\left\{\frac{P_1(\mu)}{r^2} - \frac{gt^2}{3!}\frac{2!\,P_2(\mu)}{r^3} + \frac{(gt^2)^2}{5!}\frac{3!\,P_3(\mu)}{r^4} - \ldots\right\}. \qquad (18)$$

From this the value of ζ is to be obtained by (3). It appears from Arts. 84, 85 that

$$P_{2n+1}(0) = 0, \qquad P_{2n}(0) = (-)^n \frac{1.3\ldots(2n-1)}{2.4\ldots 2n}, \qquad\qquad (19)$$

whence

$$\zeta = \frac{1}{2\pi\varpi^2}\left\{\frac{1^2}{2!}\frac{gt^2}{\varpi} - \frac{1^2.3^2}{6!}\left(\frac{gt^2}{\varpi}\right)^3 + \frac{1^2.3^2.5^2}{10!}\left(\frac{gt^2}{\varpi}\right)^5 - \ldots\right\}. \qquad (20)†$$

It follows that any particular phase of the motion is associated with a particular value of gt^2/ϖ, and thence that the various phases travel radially outwards from the origin, each with a constant acceleration.

No exact equivalent for (20), analogous to the formula (21) of Art. 238 which was obtained in the two-dimensional form of the problem, and accordingly suitable for discussion in the case where gt^2/ϖ is large, has been discovered. An approximate value may however be obtained by Kelvin's method (Art. 241). Since $J_0(z)$ is a fluctuating function which tends as z increases to have the same period 2π as $\sin z$, the elements of the integral in (13) will for the most part cancel one another with the exception of those for which

$$t\,d\sigma/dk = \varpi, \quad \text{or} \quad k\varpi = gt^2/4\varpi, \qquad\qquad (21)$$

nearly. Now when $k\varpi$ is large we have

$$J_0(k\varpi) = \left(\frac{2}{\pi k\varpi}\right)^{\frac{1}{2}} \sin\left(k\varpi + \tfrac{1}{4}\pi\right), \qquad\qquad (22)$$

approximately, by Art. 194 (15), and we may therefore replace (13) by

$$\phi = \frac{g^{\frac{1}{2}}}{2^{\frac{3}{2}}\pi^{\frac{3}{2}}\varpi^{\frac{1}{2}}}\int_0^\infty e^{kz}\cos\left(\sigma t - k\varpi - \tfrac{1}{4}\pi\right)dk. \qquad\qquad (23)$$

Comparing with (7) and (9) of Art. 241, and putting now $z = 0$, we find as the surface value of ϕ

$$\phi_0 = \frac{g^{\frac{1}{2}}}{2\pi\varpi^{\frac{1}{2}}\sqrt{|t\,d^2\sigma/dk^2|}}\sin(\sigma t - k\varpi), \qquad\qquad (24)$$

* Hobson, *Proc. Lond. Math. Soc.* xxv. 72, 73 (1893). This formula may, however, be dispensed with; see the first footnote on p. 385 *ante*.

† This result was given by Cauchy and Poisson.

where k and σ are to be expressed in terms of ϖ and t by means of (8) and (21). Note has here been taken of the fact that $d^2\sigma/dk^2$ is negative. Since

$$\sigma t = (gkt^2)^{\frac{1}{2}} = 2k\varpi, \qquad td^2\sigma/dk^2 = -\tfrac{1}{4}g^{\frac{1}{2}}tk^{-\frac{3}{2}} = -2\varpi^3/gt^2, \quad \text{...(25)}$$

we have

$$\phi_0 = \frac{gt}{2^{\frac{3}{2}}\pi\varpi^2}\sin\frac{gt^2}{4\varpi}. \quad \text{.........................(26)}$$

The surface elevation is then given by (3). Keeping, for consistency, only the most important term, we find

$$\zeta = \frac{gt^2}{2^{\frac{5}{2}}\pi\varpi^3}\cos\frac{gt^2}{4\varpi}, \quad \text{.........................(27)}$$

which agrees with the result obtained, in other ways, by Cauchy and Poisson.

It is not necessary to dwell on the interpretation, which will be readily understood from what has been said in Art. 240 with respect to the two-dimensional case. The consequences were worked out in some detail by Poisson on the hypothesis of an initial paraboloidal depression.

When the initial data are of *impulse*, the typical solution is

$$\left.\begin{aligned} \rho\phi &= \cos\sigma t\, e^{kz} J_0(k\varpi), \\ \zeta &= -\frac{\sigma}{g\rho}\sin\sigma t\, J_0(k\varpi), \end{aligned}\right\} \quad \text{.......................(28)}$$

which, being generalized, gives, for the initial conditions

$$\rho\phi_0 = F(\varpi), \qquad \zeta = 0, \quad \text{.......................(29)}$$

the solution

$$\left.\begin{aligned} \phi &= \frac{1}{\rho}\int_0^\infty \cos\sigma t\, e^{kz} J_0(k\varpi)\, k\, dk \int_0^\infty F(\alpha)\, J_0(k\alpha)\, \alpha\, d\alpha, \\ \zeta &= -\frac{1}{g\rho}\int_0^\infty \sigma\sin\sigma t\, J_0(k\varpi)\, k\, dk \int_0^\infty F(\alpha)\, J_0(k\alpha)\, \alpha\, d\alpha. \end{aligned}\right\} \quad \text{...(30)}$$

In particular, for a concentrated impulse at the origin, such that

$$\int_0^\infty F(\alpha)\, 2\pi\alpha\, d\alpha = 1, \quad \text{.........................(31)}$$

we find

$$\phi = \frac{1}{2\pi\rho}\int_0^\infty \cos\sigma t\, e^{kz} J_0(k\varpi)\, k\, dk. \quad \text{.................(32)}$$

Since this may be written

$$\phi = \frac{1}{2\pi\rho}\frac{\partial}{\partial t}\int_0^\infty \frac{\sin\sigma t}{\sigma} e^{kz} J_0(k\varpi)\, k\, dk, \quad \text{..............(33)}$$

we find, performing $1/g\rho\,.\,\partial/\partial t$ on the results contained in (18) and (20),

$$\left.\begin{aligned} \phi &= \frac{1}{2\pi\rho}\left\{\frac{P_1(\mu)}{r^2} - \frac{gt^2}{2!}\frac{2!\,P_2(\mu)}{r^3} + \frac{(gt^2)^2}{4!}\frac{3!\,P_3(\mu)}{r^4} - \dots\right\}, \\ \zeta &= \frac{t}{2\pi\rho\varpi^3}\left\{1 - \frac{1^2.3^2}{5!}\left(\frac{gt^2}{\varpi}\right)^2 + \frac{1^2.3^2.5^2}{9!}\left(\frac{gt^2}{\varpi}\right)^4 - \dots\right\}. \end{aligned}\right\} \quad \text{...(34)}$$

Again, when $\frac{1}{2}gt^2/\varpi$ is large, we have, in place of (27),

$$\zeta = -\frac{gt^3}{2^{\frac{7}{2}}\pi\rho\varpi^4}\sin\frac{gt^2}{4\varpi}. \qquad\qquad (35)^*$$

256. We proceed to consider the effect of a local disturbance of pressure advancing with constant velocity over the surface†. This will give us, at all events as to the main features, an explanation of the peculiar system of waves which is seen to accompany a ship moving through sufficiently deep water.

A *complete* investigation, after the manner of Arts. 242, 243, would be somewhat difficult; but the general characteristics can readily be made out with the help of preceding results, the procedure being similar to that of Art 249.

Let us suppose that we have a pressure-point moving with velocity c along the axis of x, in the negative direction, and that at the instant under consideration it has reached the point O. The elevation ζ at any point P may

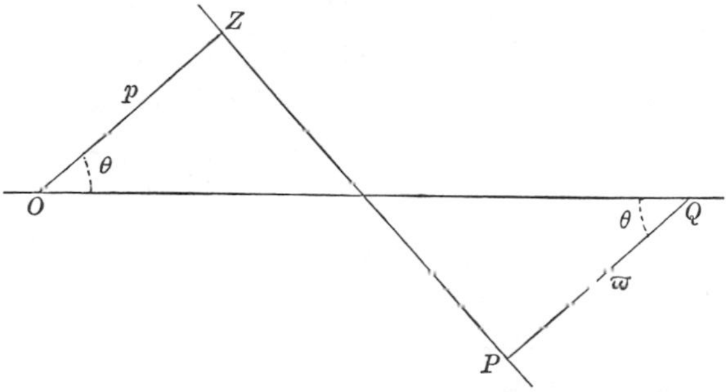

be regarded as due to a series of infinitely small impulses applied at equal infinitely short intervals at points of the axis of x to the right of O. Of the annular wave-systems thus successively generated, those only will combine to produce a sensible effect at P which had their origin in the neighbourhood of certain points Q, which are determined by the consideration that the phase at P is 'stationary' for variations in the position of Q. Now if t is the time which the source of disturbance has taken to travel from Q to O, the phase of the waves at P, originated at Q, is

$$\frac{gt^2}{4\varpi} + \frac{1}{2}\pi, \qquad\qquad (1)$$

where $\varpi = QP$ (Art. 255 (35)). Hence the condition for stationary phase is

$$\dot{\varpi} = \frac{2\varpi}{t}. \qquad\qquad (2)$$

* The waves due to various types of explosive action beneath the surface have been studied by Terazawa, *Proc. Roy. Soc.* A, xcii. 57 (1915), and by the author of this work, *l.c. ante* p. 410, and *Proc. Lond. Math. Soc.* (2) xxi. 359 (1922).

† For a more general treatment of such questions reference may be made to a paper by the author, "On Wave-Patterns due to a Travelling Disturbance," *Phil. Mag.* (6) xxxi. 539 (1916).

Since, in this differentiation, O and P are regarded as fixed, we have

$$\dot{\varpi} = c \cos \theta,$$

where $\theta = OQP$; hence

$$OQ = ct = 2\varpi \sec \theta. \qquad (3)$$

It is further evident that the points in the immediate neighbourhood of P, for which the resultant phase is the same as at P, will lie in a line perpendicular to QP. A glance at the figure on p. 433 then shews that a curve of uniform phase is characterized by the property that the tangent bisects the interval between the origin and the foot of the normal. If p denote the perpendicular from the origin to the tangent, and θ the angle which p makes with the axis of x, we have, by a known formula,

$$PZ = -\frac{dp}{d\theta};$$

whence

$$2p = -\frac{dp}{d\theta} \cot \theta, \qquad (4)$$

$$p = a \cos^2 \theta. \qquad (5)$$

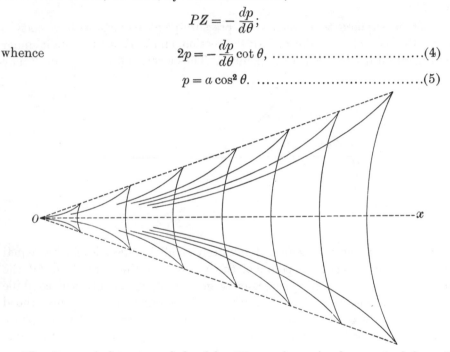

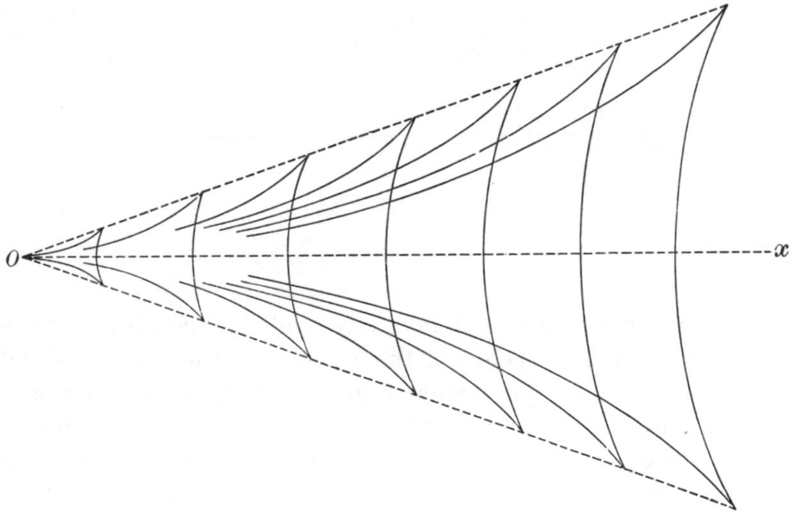

The forms of the curves defined by (5) are shewn in the annexed figure*, which is traced from the equations

$$\left.\begin{aligned} x &= p \cos \theta - \frac{dp}{d\theta} \sin \theta = \tfrac{1}{4} a (5 \cos \theta - \cos 3\theta), \\ y &= p \sin \theta + \frac{dp}{d\theta} \cos \theta = -\tfrac{1}{4} a (\sin \theta + \sin 3\theta). \end{aligned}\right\} \qquad (6)$$

* Cf. Sir W. Thomson, "On Ship Waves," *Proc. Inst. Mech. Eng.* Aug. 3, 1887 [*Popular Lectures*, iii. 482], where a similar drawing is given. The investigation there referred to, based apparently on the theory of 'group-velocity,' was not published. See also R. E. Froude, "On Ship Resistance," *Papers of the Greenock Phil. Soc.* Jan. 19, 1894. It is shewn immediately that there is a difference of phase between the two branches meeting at a cusp, so that the drawing does not represent quite accurately the configuration of the wave-ridges.

The phase-difference from one curve to the corresponding portion of the next is 2π. This implies a difference $2\pi c^2/g$ in the parameter a.

Since two curves of the above kind pass through any assigned point P within the boundaries of the wave-system, it is evident that there are *two* corresponding effective positions of Q in the foregoing discussion. These are determined by a very simple construction. If the line OP be bisected in C, and a circle be drawn on CP as diameter, meeting the axis of x in R_1, R_2, the perpendiculars PQ_1, PQ_2 to PR_1, PR_2, respectively, will meet the axis in the required points, Q_1, Q_2. For CR_1 is parallel to PQ_1 and equal to $\frac{1}{2}PQ_1$; the perpendicular from O on PR_1 produced is therefore equal to PQ_1. Similarly, the perpendicular from O on PR_2 produced is equal to PQ_2.

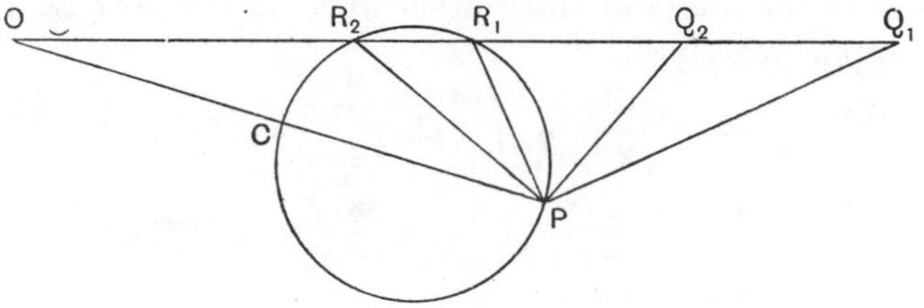

The points Q_1, Q_2 coincide when OP makes an angle $\sin^{-1}\frac{1}{3}$, or 19° 28′, with the axis of symmetry. For greater inclinations of OP they are imaginary. It appears also from (6) that the values of x, y are stationary when $\sin^2\theta = \frac{1}{3}$; this gives a series of cusps lying on the straight lines

$$\frac{y}{x} = \pm\frac{1}{2\sqrt{2}} = \pm\tan 19^\circ\, 28'. \qquad\qquad (7)$$

To obtain an approximate estimate of the actual height of the waves, in the different parts of the system, we have recourse to the formula (35) of Art. 255. If P_0 denote the total disturbing pressure, the elevation at P due to the annular wave-system started at a point Q to the right of O may be written

$$\delta\zeta = -\frac{gt^3}{8\sqrt{2\pi}\rho\varpi^4}\cdot\sin\frac{gt^2}{4\varpi}\cdot P_0\,\delta t, \qquad\qquad (8)$$

where $\qquad \varpi = PQ, \qquad t = OQ/c.$

This is to be integrated with respect to t, but (as already explained) the only parts of the integral which contribute appreciably to the final result will be those for which t has very nearly the values (τ_1, τ_2) corresponding to the special points Q_1, Q_2 above mentioned.

As regards the phase, we have, writing $t = \tau + t'$,

$$\frac{gt^2}{4\varpi} = \left[\frac{gt^2}{4\varpi}\right] + t'\left[\frac{d}{dt}\left(\frac{gt^2}{4\varpi}\right)\right] + \frac{t'^2}{1\,.\,2}\left[\frac{d^2}{dt^2}\left(\frac{gt^2}{4\varpi}\right)\right] + \ldots, \qquad (9)$$

where, in the terms in [ ], t is to be put equal to τ_1 or τ_2 as the case may be.

The second term vanishes by hypothesis, since the phase at P for waves started near Q_1 or Q_2 is 'stationary.' Again, we find

$$\frac{d^2}{dt^2}\left(\frac{gt^2}{4\varpi}\right)=\frac{g}{2\varpi}-\frac{gt}{\varpi^2}\dot{\varpi}+\frac{gt^2}{4}\left(\frac{2\dot{\varpi}^2}{\varpi^3}-\frac{\ddot{\varpi}}{\varpi^2}\right).$$

Since $$\dot{\varpi}=c\cos\theta,\qquad \ddot{\varpi}=\frac{c^2\sin^2\theta}{\varpi},\quad\text{.....................(10)}$$

this gives, with the help of (2),

$$\left[\frac{d^2}{2t^2}\left(\frac{gt^2}{4\varpi}\right)\right]=\frac{g}{\varpi}(\tfrac{1}{2}-\tan^2\theta).\quad\text{.....................(11)}$$

Owing to the fluctuations of the trigonometrical term no great error will be committed if we neglect the variation of the first factor in (8), or if, further, we take the limits of integration with respect to t' to be $\pm\infty$. We have then, approximately,

$$\zeta=-\frac{g\tau_1{}^3P_0}{8\sqrt{2\pi}\rho\varpi_1{}^4}\int_{-\infty}^{\infty}\sin\left(\frac{g\tau_1{}^2}{4\varpi_1}+m_1{}^2t'^2\right)dt'$$
$$-\frac{g\tau_2{}^3P_0}{8\sqrt{2\pi}\rho\varpi_2{}^4}\int_{-\infty}^{\infty}\sin\left(\frac{g\tau_2{}^2}{4\varpi_2}+m_2{}^2t'^2\right)dt',\quad\text{............(12)}$$

where $$m_1{}^2=\frac{g}{2\varpi_1}(\tfrac{1}{2}-\tan^2\theta_1),\qquad m_2{}^2=\frac{g}{2\varpi_2}(\tan^2\theta_2-\tfrac{1}{2}),\quad\text{.........(13)}$$

and the suffixes refer to the points Q_1, Q_2 of the last figure.

Since $$\int_{-\infty}^{\infty}\cos m^2t'^2dt'=\int_{-\infty}^{\infty}\sin m^2t'^2dt'=\sqrt{(\tfrac{1}{2}\pi)}/m,\quad\text{............(14)}$$

where the positive value of m is understood, we find

$$\zeta=-\frac{g\tau_1{}^3P_0}{8\sqrt{2}\pi^{\frac{1}{2}}\rho\varpi_1{}^4m_1}\cdot\sin\left(\frac{g\tau_1{}^2}{4\varpi_1}+\tfrac{1}{4}\pi\right)-\frac{g\tau_2{}^3P_0}{8\sqrt{2}\pi^{\frac{1}{2}}\rho\varpi_2{}^4m_2}\cdot\sin\left(\frac{g\tau_2{}^2}{4\varpi_2}-\tfrac{1}{4}\pi\right).$$
......(15)

The two terms give the parts due to the transverse and lateral waves respectively. Since $\varpi_1=PQ_1=\frac{1}{2}c\tau_1\cos\theta_1$, $\varpi_2=PQ_2=\frac{1}{2}c\tau_2\cos\theta_2$, it appears that if we consider either term by itself, the phase is constant along the corresponding part of the curve

$$p=\varpi=a\cos^2\theta,$$

whilst the elevation varies as

$$\frac{\sqrt{2}g^{\frac{1}{2}}P_0}{\pi^{\frac{1}{2}}\rho c^3a^{\frac{1}{2}}}\cdot\frac{\sec^3\theta}{\sqrt{|1-3\sin^2\theta|}}.\quad\text{.....................(16)}$$

At the cusps, where the two systems combine, there is a phase-difference of a quarter-period between them.

The formulae make ζ infinite at a cusp, where $\sin^2\theta=\frac{1}{3}$, but this is merely an indication of the failure of our approximation. That the elevation at a point P in the neighbourhood of a cusp would be relatively great might have been foreseen, since, as appears from (9) and (11), the range of points on the axis of x which have sent waves to P in sensibly the same phase is then abnormally extended. The infinity which occurs when $\theta = \frac{1}{2}\pi$ is of a somewhat different character, being due to the artificial nature of the assumption we have made, of a pressure concentrated at a *point*. With a diffused pressure this difficulty would disappear*.

It is to be noticed, moreover, that the whole of this investigation applies only to points for which $gt^2/4\varpi$ is large; cf. Arts. 240, 255. It will be found on examination that this restriction is equivalent to an assumption that the parameter a is large compared with $2\pi c^2/g$. The argument therefore does not apply without reserve to the parts of the wave-pattern near the origin.

256 a. As already indicated, wave-systems of the above type are generated by other forms of travelling disturbance. Some of these cases are amenable to calculation. The translation of a submerged sphere, for instance, has been dealt with by Havelock†, and the wave-resistance determined. The writer‡ has discussed by another method the translation of a submerged solid, without restriction as to its precise form or orientation. The results are naturally simplest when the direction of motion coincides with one of the three directions of 'permanent translation' considered in Art. 124. The resistance is then given by the formula

$$R = \frac{g^4 (\mathbf{A} + \rho Q)^2 I}{\pi \rho c^6}. \qquad (17)$$

Here $\mathbf{A}$ denotes the appropriate inertia-coefficient from Art. 121, Q is the volume of the solid, c its velocity, and

$$I = \int_0^{\frac{1}{2}\pi} \sec^5 \theta e^{-2gf/c^2 . \sec^2 \theta} \, d\theta, \qquad (18)$$

where f is the depth of immersion. Another form of this integral (due to Havelock) is

$$I = \tfrac{1}{4} e^{-\alpha} \left\{ K_0(\alpha) + \left(1 + \frac{1}{2\alpha}\right) K_1(\alpha) \right\}, \qquad (19)$$

in the accepted notation of Bessel Functions§, with $\alpha = gf/c^2$. For a sphere we have $\mathbf{A} = \frac{2}{3}\pi\rho a^3$, $Q = \frac{4}{3}\pi a^3$, where a is the radius. Hence if M' be the mass of fluid displaced,

$$R = 3M'g \cdot \left(\frac{a}{f}\right)^3 \left(\frac{gf}{c^2}\right)^3 . I, \qquad (20)||$$

in agreement with Havelock's result. As an example, if $c = \sqrt{(gf)}$,

$$R = \cdot 365 \, M'g \, (a/f)^3.$$

* More elaborate investigations have been carried out by Hopf in a dissertation of date Munich, 1909, and Hogner, *Arkiv för Matem.* xvii. (1923). The latter writer examines in particular the shape of the waves near the 'cusps,' where the two systems cross.

† *Proc. Roy. Soc.* A, xciii. 520 (1917); xcv. 354 (1918). See also Green, *Phil. Mag.* (6) xxxvi. 48 (1918).

‡ *Proc. Roy. Soc.* A, cxi. 14 (1926).

§ Watson, p. 172.

|| This formula was given incorrectly in the author's paper.

A graph of R as a function of c is given by Havelock; it has a general resemblance to the curve on p. 416.

In a subsequent paper* the same method is applied by Havelock to a travelling disturbance consisting of various arrangements of (double) sources, with important applications to the wave-resistance of ships.

Some further reference to the theoretical literature of wave-resistance may be in place here. Although the mode of disturbance is different, the action of the bows of a ship may be compared to that of a pressure-point. The diagram on p. 434 accounts for the two systems of transverse and lateral waves which are observed, and for the especially conspicuous 'echelon' waves near the cusps, where the two systems cross. If in addition we imagine a negative pressure-point at the stern we get a rough representation of the action of the ship as a whole. With varying speeds the stern waves may tend partially to annul, or to reinforce, the effect of the bow waves, with the result that the resistance may be expected to fluctuate up and down as the length of the ship is increased, or the speed varied†. It is found in fact that the curve of resistance as a function of the speed exhibits several maxima (or 'humps') with the corresponding minima, as well as a general increase.

To obtain an improved representation of what happens in the immediate neighbourhood of a ship and to calculate the consequent resistance is of course a difficult matter, but attempts have been made with considerable success. A beginning was made by J. H. Michell‡ with an idealized ship form, which differs mainly from that of a real ship in that the inclination of the surface to the medial plane is everywhere small. This plan has recently been followed up by Wigley§, who has discussed a variety of forms (subject to the same limitation), calculated their resistance, and compared it with the results of model experiments, with a considerable measure of qualitative agreement. Havelock, in a long series of papers‖ has discussed the effect of various features in the design of a ship, such as length of 'parallel middle body,' mean draught, and so on. His method consists (in part) in the choice of a suitable arrangement of travelling sources, and is accordingly free from the special restriction above mentioned¶.

A general formula for the wave-resistance of geometrically similar bodies, similarly immersed (wholly or partially), was given long ago by Froude. Since the resistance can only depend on the speed, the density of the fluid, the intensity of gravity, and on some linear magnitude which fixes the scale, considerations of dimensions shew that it must satisfy a relation of the form

$$R = \rho l^2 c^2 f\left(\frac{gl}{c^2}\right), \qquad \text{.........................(21)}$$

where c is the speed, and l the characteristic linear magnitude. It will be

* *Proc. Roy. Soc.* A, cxviii. 24 (1927).

† W. Froude, "On the Effect on the Wave-Making Resistance of Ships of Length of Parallel Middle Body," *Trans. Inst. Nav. Arch.* xvii. (1877). Also R. E. Froude, "On the Leading Phenomena of the Wave-Making Resistance of Ships," *Trans. Inst. Nav. Arch.* xxii. (1881), where drawings of actual wave-patterns under varied conditions of speed are given, which are, as to their main features, in striking agreement with the results of the above theory. Some of these drawings are reproduced in Kelvin's paper in the *Proc. Inst. Mech. Eng.* above cited.

‡ *Phil. Mag.* (5) xlv. 106 (1898).

§ *Trans. Inst. Nav. Arch.* lxviii. 124 (1926); lxix. 27 (1927); lxxii. (1930).

‖ In the *Proc. Roy. Soc.* from 1909 onwards.

¶ Excellent accounts of the development of the subject are given by Hogner, *Proc. Congress. App. Math.* Delft, 1924, p. 146, and Wigley, *Congress for techn. Mechanics*, Stockolm, 1930.

noticed that (17) is a particular case of this. It follows from (21) that the wave-resistance of a ship can be inferred from a model experiment provided the value of l/c^2 is the same on the model as on the full scale.

256 b. To examine the modification produced in the wave-pattern when the depth of the water has to be taken into account, the argument on p. 433 must be put in a more general form. If, as before, t is the time the pressure-point has taken to travel from Q to O, it may be shewn that the phase of the disturbance at P, due to the impulse delivered at Q, will differ only by a constant from

$$k\,(Vt - \varpi), \qquad \text{.................................(22)}$$

where $2\pi/k$ is the predominant wave-length in the neighbourhood of P, and V the corresponding wave-velocity*. This predominant wave-length is determined by the condition that the phase is stationary for variations of the wave-length only, *i.e.*

$$\frac{\partial}{\partial k}\cdot k\,(Vt - \varpi) = 0, \quad \text{or} \quad \varpi = Ut, \qquad \text{.................(23)}$$

where $U, = d\,(kV)/dk$, is the group-velocity (Art. 236).

For the effective part of the disturbance at P, the phase (22) must further be stationary as regards variations in the position of Q; hence, differentiating partially with respect to t, we have

$$\dot{\varpi} = V, \quad \text{or} \quad V = c\cos\theta, \qquad \text{.......................(24)}$$

since $\dot{\varpi} = c\cos\theta$. Now, referring to the figure on p. 433, we have

$$p = ct\cos\theta - \varpi = Vt - \varpi. \qquad \text{.......................(25)}$$

Hence for a given wave-ridge p will bear a constant ratio to the wave-length λ, and in passing from one wave-ridge to the next this ratio will increase (or decrease) by unity. Since λ is determined as a function of θ by (24), this gives the relation between p and θ.

Thus in the case of infinite depth, the formula (24) gives

$$c^2\cos^2\theta = V^2 = \frac{g\lambda}{2\pi}, \qquad \text{..........................(26)}$$

and the required relation is of the form

$$p = a\cos^2\theta, \qquad \text{............................(27)}$$

as above.

When the depth (h) is finite, we have

$$c^2\cos^2\theta = V^2 = \frac{g\lambda}{2\pi}\tanh\frac{2\pi h}{\lambda}, \qquad \text{....................(28)}$$

and the relation is

$$\frac{p}{a}\tanh\frac{a}{p} = \frac{c^2}{gh}\cos^2\theta, \qquad \text{..........................(29)}$$

* The symbol c, which was previously employed in this sense, now denotes the velocity of the pressure-point over the water.

where the values of a for successive wave-ridges are in arithmetic progression. Since the expression on the left-hand side cannot exceed unity, it appears that if $c^2 > gh$ there will be an inferior limit to the value of θ, determined by

$$\cos^2\theta = gh/c^2, \qquad (30)$$

the curve then extending to infinity.

It follows that when the speed of the disturbing influence exceeds $\sqrt{(gh)}$ the transverse waves disappear, and we have only the lateral waves. This tends to diminish the wave-making resistance (cf. Art. 249)*.

The changes in the configuration of the wave-pattern as the ratio c^2/gh increases from zero to infinity are traced by Havelock†.

Standing Waves in Limited Masses of Water.

257. The problem of free oscillations in two horizontal dimensions (x, y), in the case where the depth is uniform and the fluid is bounded laterally by vertical walls, can be reduced to the same analytical form as in Art. 190.

If the origin be taken in the undisturbed surface, and if ζ denote the elevation at time t above this level, the conditions to be satisfied at the free surface are as in Art. 255 (3), (4).

The equation of continuity, $\nabla^2\phi = 0$, and the condition of zero vertical motion at the depth $z = -h$, are both satisfied by

$$\phi = \phi_1 \cosh k(z+h), \qquad (1)$$

where ϕ_1 is a function of x, y, such that

$$\frac{\partial^2\phi_1}{\partial x^2} + \frac{\partial^2\phi_1}{\partial y^2} + k^2\phi_1 = 0. \qquad (2)$$

The form of ϕ_1 and the admissible values of k are determined by this equation, and by the condition that

$$\frac{\partial\phi_1}{\partial n} = 0, \qquad (3)$$

at the vertical walls. The corresponding values of the 'speed' (σ) of the oscillations are then given by the surface-condition (6), of Art. 255; viz. we have

$$\sigma^2 = gk \tanh kh. \qquad (4)$$

This makes

$$\zeta = \frac{ik}{\sigma} \sinh kh \,.\, \phi_1. \qquad (5)$$

* It is found that the power required to propel a torpedo-boat in relatively shallow water increases with the speed up to a certain critical velocity, dependent on the depth, then decreases, and finally increases again. See papers by Rasmussen, *Trans. Inst. Nav. Arch.* xli. 12 (1899); Rota, *ibid.* xlii. 239 (1900); Yarrow and Marriner, *ibid.* xlvii. 339, 344 (1905).

† *Proc. Roy. Soc.* lxxxi. 426 (1908). See also Ekman, *l.c. ante* p. 371.

The conditions (2) and (3) are of the same form as in the case of small depth, and we could therefore at once write down the results for a rectangular or a circular* tank. The values of k, and the forms of the free surface, in the various fundamental modes, are the same as in Arts. 190, 191†, but the amplitude of the oscillation now diminishes with increasing depth below the surface, according to the law (1); whilst the speed of any particular mode is given by (4).

When kh is small, we have $\sigma^2 = k^2gh$, as in the Arts. referred to.

We may also notice in this connection the case of a long and narrow rectangular tank having near its centre one or more cylindrical obstacles, whose generating lines are vertical.

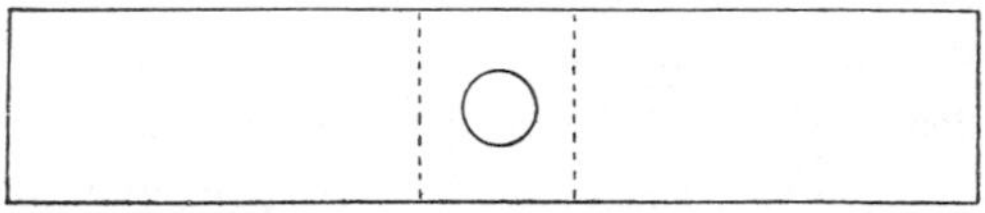

The origin being taken at the centre of the free surface, and the axis of x parallel to the length l, we imagine two planes $x = \pm x'$ to be drawn, such that x' is moderately large compared with the horizontal dimensions of the obstacles, whilst still small in comparison with the length (l). Beyond these planes we shall have

$$\frac{\partial^2\phi_1}{\partial x^2} + k^2\phi_1 = 0, \quad \text{...(6)}$$

approximately, and therefore, for $x > x'$,

$$\phi_1 = A\sin kx + B\cos kx, \quad \text{...................................(7)}$$

whilst, for $x < -x'$,

$$\phi_1 = A\sin kx - B\cos kx, \quad \text{...................................(8)}$$

since in the gravest mode, which is alone here considered, ϕ must be an odd function of x.

In the region *between* the planes $x = \pm x'$ the configuration of the lines $\phi_1 = \text{const.}$ is, for a reason to be explained in Art. 290 in connection with other questions, sensibly the same as if in (2) we were to put $k = 0$. So far as this region is concerned, the problem is in fact the same as that of conduction of electricity along a bar of metal which has the same form as the actual mass of water, and has accordingly one or more cylindrical perforations occupying the place of the obstacles. The electrical resistance between the two planes is then equivalent to that of a certain length $2x' + \alpha$ of an unperforated bar of the same section. The difference of potential between the planes may be taken to be $2(kAx' + B)$, by (7), since kx' is small; and the current per unit sectional area is kA, approximately. Thus

$$2(kAx' + B) = (2x' + \alpha)kA, \quad \text{................................(9)}$$

whence

$$B/A = \tfrac{1}{2}k\alpha \quad \text{..(10)}$$

and

$$\phi_1 = A(\sin kx + \tfrac{1}{2}k\alpha\cos kx), \quad \text{................................(11)}$$

for $x > x'$.

* For references to the original investigations by Poisson and Rayleigh of waves in a circular tank see p. 287. The problem was also treated by Merian, *Ueber die Bewegung tropfbarer Flüssigkeiten in Gefässen*, Basel, 1828 [see VonderMühll, *Math. Ann.* xxvii. 575], and by Ostrogradsky, "Mémoire sur la propagation des ondes dans un bassin cylindrique," *Mém. des Sav. Étrang.* iii. (1862).

† It may be remarked that either of the two modes figured on p. 288 may easily be excited by properly timed horizontal agitation of a tumbler containing water.

The condition $\partial\phi/\partial x=0$, to be satisfied for $x=\frac{1}{2}l$, gives

$$\cos\tfrac{1}{2}kl-\tfrac{1}{2}k\alpha\sin kl, \qquad\qquad (12)$$

or, since $k\alpha$ is a small quantity,

$$\cos\tfrac{1}{2}k(l+\alpha)=0. \qquad\qquad (13)$$

The introduction of the obstacles has therefore the effect of virtually increasing the length of the tank by α. The period of the gravest mode is accordingly

$$\frac{2\pi}{\sigma}=2\sqrt{\left(\frac{\pi l'}{g}\,.\coth\frac{\pi h}{l'}\right)}, \qquad\qquad (14)$$

where $l'=l+\alpha$.

The value of α is known for one or two cases. In the case of a circular column of radius b, in the centre of the tank, the formulae (11) and (13) of Art. 64 shew that ϕ_1 varies as $x+C$, or $x+\pi b^2/a$, practically, when x is large compared with the breadth a of the tank. Comparing with (11) above we see that

$$\alpha=2\pi b^2/a, \qquad\qquad (15)$$

subject to the condition that the ratio b/a must not exceed about $\frac{1}{4}$*.

When the plane $x=0$ is occupied by a thin rigid diaphragm of breadth a, having a central vertical slit of breadth c, the formula is

$$\alpha=\frac{2a}{\pi}\log\sec\frac{\pi(a-c)}{2a}. \qquad\qquad (16)$$

258. The number of cases of motion with a *variable* depth, of which the solution has been obtained, is very small.

1°. We may notice, first, the two-dimensional oscillations of water across a channel whose section consists of two straight lines inclined at 45° to the vertical†.

The axes of y, z being respectively horizontal and vertical, in the plane of a cross-section, we assume

$$\phi+i\psi=A\{\cosh k(y+iz)+\cos k(y+iz)\}, \qquad\qquad (1)$$

the time-factor $\cos(\sigma t+\epsilon)$ being understood. This gives

$$\phi=A(\cosh ky\cos kz+\cos ky\cosh kz),\qquad \psi=A(\sinh ky\sin kz-\sin ky\sinh kz). \quad (2)$$

The latter formula shews at once that the lines $y=\pm z$ constitute the stream-line $\psi=0$, and may therefore be taken as fixed boundaries.

The condition to be satisfied at the free surface is, as in Art. 227,

$$\sigma^2\phi=g\frac{\partial\phi}{\partial z}. \qquad\qquad (3)$$

Substituting from (2) we find, if h denote the height of the surface above the origin,

$$\sigma^2(\cosh ky\cos kh+\cos ky\cosh kh)=gk(-\cosh ky\sin kh+\cos ky\sinh kh).$$

This will be satisfied for all values of y, provided

$$\sigma^2\cos kh=-gk\sin kh,\qquad \sigma^2\cosh kh=gk\sin kh, \qquad\qquad (4)$$

whence

$$\tanh kh=-\tan kh. \qquad\qquad (5)$$

This determines the admissible values of k; the corresponding values of σ are then given by either of the equations (4).

* The formula (14) was in this case found to be in good agreement with experiment (Lamb and Cooke, *Phil. Mag.* (6) xx. 303 (1910)). The experiments were made chiefly with a view to test the above method of approximation, which has other more important applications; see Arts. 306, 307.

† Kirchhoff, "Ueber stehende Schwingungen einer schweren Flüssigkeit," *Berl. Monatsber.* May 15, 1879 [*Ges. Abh.* p. 428]; Greenhill, *l.c. ante* p. 372.

Since (2) makes ϕ an *even* function of y, the oscillations which it represents are symmetrical with respect to the medial plane $y=0$.

The asymmetrical oscillations are given by

$$\phi+i\psi=iA\{\cosh k(y+iz)-\cos k(y+iz)\}, \quad \text{.........................(6)}$$

or $\phi=-A(\sinh ky \sin kz+\sin ky \sinh kz)$, $\quad \psi=A(\cosh ky \cos kz-\cos ky \cosh kz)$. ...(7)

The stream-line $\psi=0$ consists, as before, of the lines $y=\pm z$; and the surface-condition (3) gives

$$\sigma^2(\sinh ky \sin kh+\sin ky \sinh kh)=gk(\sinh ky \cos kh+\sin ky \cosh kh).$$

This requires

$$\sigma^2 \sin kh=gk \cos kh, \qquad \sigma^2 \sinh kh=gk \cosh kh, \quad \text{.....................(8)}$$

whence

$$\tanh kh=\tan kh. \quad \text{.......................................(9)}$$

The equations (5) and (9) present themselves in the theory of the lateral vibrations of a bar free at both ends; viz. they are both included in the equation

$$\cos m \cosh m=1, \quad \text{.......................................(10)*}$$

where $m=2kh$.

The root $kh=0$, of (9), which is extraneous in the theory referred to, is now important; it corresponds in fact to the slowest mode of oscillation in the present problem. Putting $Ak^2=B$, and making k infinitesimal, the formulae (7) become, on restoring the time-factor, and taking the real parts,

$$\phi=-2Byz\,.\cos(\sigma t+\epsilon), \qquad \psi=B(y^2-z^2)\,.\cos(\sigma t+\epsilon), \quad \text{..............(11)}$$

whilst from (8)

$$\sigma^2=\frac{g}{h}. \quad \text{..(12)}$$

The corresponding form of the free surface is

$$\zeta=\frac{1}{g}\left[\frac{\partial\phi}{\partial t}\right]_{z=h}=2\sigma Bhy\,.\sin(\sigma t+\epsilon). \quad \text{..........................(13)}$$

The surface in this mode is therefore always plane. The annexed figure shews the lines of motion ($\psi=$const.) for a series of equidistant values of ψ.

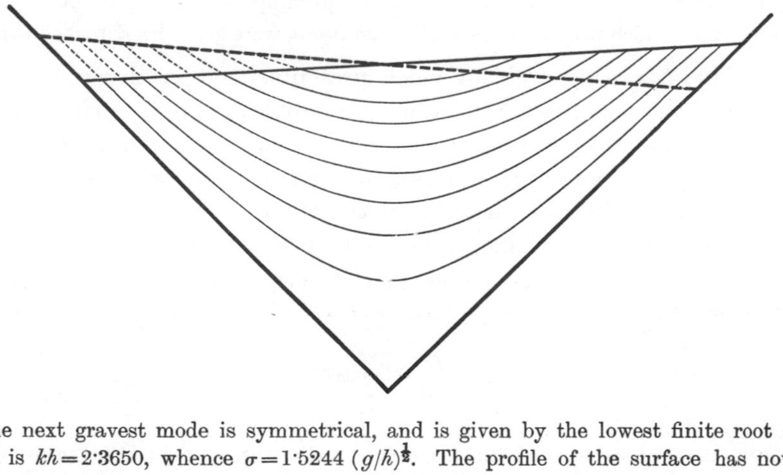

The next gravest mode is symmetrical, and is given by the lowest finite root of (5), which is $kh=2{\cdot}3650$, whence $\sigma=1{\cdot}5244\,(g/h)^{\frac{1}{2}}$. The profile of the surface has now two

* Cf. Rayleigh, *Theory of Sound*, i. 277, where the numerical solution of the equation is fully discussed.

nodes, whose positions are determined by putting $\phi=0$, $z=h$, in (2); whence it is found that

$$\frac{y}{h}=\pm\cdot 5516^*.$$

The next mode corresponds to the lowest finite root of (9), and so on†.

2°. Greenhill, in the paper already cited, has investigated the *symmetrical* oscillations of the water across a channel whose section consists of two straight lines inclined at 60° to the vertical. In the (analytically) simplest mode of this kind we have, omitting the time-factor.

$$\phi+i\psi=iA(y+iz)^3+B, \qquad \text{.......(14)}$$

or

$$\phi=Az(z^2-3y^2)+B, \qquad \psi=Ay(y^2-3z^2), \qquad \text{.......(15)}$$

the latter formula making $\psi=0$ along the boundary $y=\pm\sqrt{3}\,.\,z$. The surface-condition (3) is satisfied for $z=h$, provided

$$\sigma^2=g/h, \qquad B=2Ah^3. \qquad \text{.......(16)}$$

The corresponding form of the free surface, viz.

$$\zeta=\frac{1}{g}\left[\frac{\partial\phi}{\partial t}\right]_{z=h}=-\frac{3A}{\sigma}(h^2-y^2)\sin(\sigma t+\epsilon), \qquad \text{.......(17)}$$

is a parabolic cylinder, with two nodes at distances of ·5774 of the half-breadth from the centre. The slowest mode, which must evidently be of asymmetrical type, has not yet been determined.

3°. If in any of the above cases we transfer the origin to either edge of the canal, and then make the breadth infinite, we get a system of standing waves on a sea bounded by a sloping bank. This may be regarded as made up of an incident and a reflected system. The reflection is complete, but there is in general a change of phase.

When the inclination of the bank is 45° the solution is

$$\phi=H\{e^{kz}(\cos ky-\sin ky)+e^{-ky}(\cos kz+\sin kz)\}\cos(\sigma t+\epsilon). \qquad \text{.......(18)}$$

For an inclination of 30° to the horizontal we have

$$\phi=H\{e^{kz}\sin ky+e^{-\frac{1}{2}k(\sqrt{3}y+z)}\sin\tfrac{1}{2}k(y-\sqrt{3}z)$$
$$-\sqrt{3}e^{-\frac{1}{2}k(\sqrt{3}y-z)}\cos\tfrac{1}{2}k(y+\sqrt{3}z)\}\cos(\sigma t+\epsilon). \qquad \text{.......(19)}$$

In each case $\sigma^2=gk$, as in the case of waves on an unlimited sheet of deep water.

These results, which may easily be verified *ab initio*, were given by Kirchhoff (*l.c.*).

259. An interesting problem which presents itself in this connection is that of the transversal oscillations of water contained in a canal of *circular* section. This has not yet been solved, but it may be worth while to point out that an approximate determination of the frequency of the slowest mode, in the case where the free surface is at the level of the axis, can be effected by Rayleigh's method, explained near the end of Art. 168.

If we assume as an 'approximate type' that in which the free surface remains always plane, making a small angle θ (say) with the horizontal, it appears, from Art. 72 (17), that the kinetic energy T is given by

$$2T=\left(\frac{4}{\pi}-\frac{\pi}{4}\right)\rho a^4\dot{\theta}^2, \qquad \text{.......(1)}$$

* Rayleigh, *Theory of Sound*, Art. 178.

† An experimental verification of the frequencies, and of the positions of the loops (places of maximum vertical amplitude), in various fundamental modes, was made by Kirchhoff and Hansemann, "Ueber stehende Schwingungen des Wassers," *Wied. Ann.* x. (1880) [Kirchhoff, *Ges. Abh.* p. 442].

where a is the radius, whilst for the potential energy V we have

$$2V = \tfrac{2}{3} g\rho a^3 \theta^2. \quad \text{.........(2)}$$

If we assume that θ varies as $\cos(\sigma t + \epsilon)$, this gives

$$\sigma^2 = \frac{8\pi}{48 - 3\pi^2} \frac{g}{a}, \quad \text{.........(3)}$$

whence $\sigma = 1{\cdot}169\,(g/a)^{\frac{1}{2}}$*.

In the case of a rectangular section of breadth $2a$, and depth a, the speed is given by Art. 257 (4), where we must put $k = \pi/2a$ from Art. 178, and $h = a$. This gives

$$\sigma^2 = \tfrac{1}{2}\pi \tanh \tfrac{1}{2}\pi \,.\, \frac{g}{a}, \quad \text{.........(4)}$$

or $\sigma = 1{\cdot}200\,(g/a)^{\frac{1}{2}}$. The frequency in the actual problem is less, since the kinetic energy due to a given motion of the surface is greater, whilst the potential energy for a given deformation is the same. Cf. Art. 45.

260. We may next consider the free oscillations of the water included between two transverse partitions in a uniform horizontal canal. Before proceeding to particular cases, we may examine for a moment the nature of the analytical problem.

If the axis of x be parallel to the length, and the origin be taken in one of the ends, the velocity-potential in any one of the fundamental modes referred to may, by Fourier's Theorem, be supposed expressed in the form

$$\phi = (P_0 + P_1 \cos kx + P_2 \cos 2kx + \ldots + P_s \cos skx + \ldots) \cos(\sigma t + \epsilon), \quad \text{...(1)}$$

where $k = \pi/l$, if l denote the length of the compartment. The coefficients P_s are here functions of y, z. If the axis of z be drawn vertically upwards, and that of y be therefore horizontal and transverse to the canal, the forms of these functions, and the admissible values of σ, are to be determined from the equation of continuity

$$\nabla^2 \phi = 0, \quad \text{.........(2)}$$

with the conditions that

$$\frac{\partial \phi}{\partial n} = 0 \quad \text{.........(3)}$$

at the sides, and

$$\sigma^2 \phi = g \frac{\partial \phi}{\partial z} \quad \text{.........(4)}$$

at the free surface. Since $\partial\phi/\partial x$ must vanish for $x = 0$ and $x = l$, it follows from known principles† that each term in (1) must satisfy the conditions (2), (3), (4) independently; viz. we must have

$$\frac{\partial^2 P_s}{\partial y^2} + \frac{\partial^2 P_s}{\partial z^2} - s^2 k^2 P_s = 0, \quad \text{.........(5)}$$

with

$$\frac{\partial P_s}{\partial n} = 0 \quad \text{.........(6)}$$

* *Hydrodynamics*, 2nd ed. (1895). Rayleigh finds, as a closer approximation, $\sigma = 1{\cdot}1644\,(g/a)^{\frac{1}{2}}$; see *Phil. Mag.* (5) xlvii. 566 (1899) [*Papers*, iv. 407].

† See Stokes, "On the Critical Values of the Sums of Periodic Series," *Camb. Trans.* viii. (1847) [*Papers*, i. 236].

at the lateral boundary, and

$$\sigma^2 P_s = g\,\frac{\partial P_s}{\partial z} \quad\text{..............................(7)}$$

at the free surface.

The term P_0 gives purely transverse oscillations such as have been discussed in Art. 258. Any other term $P_s \cos skx$ gives a series of fundamental modes with s nodal lines transverse to the canal, and 0, 1, 2, 3, ... nodal lines parallel to the length.

It will be sufficient for our purpose to consider the term $P_1 \cos kx$. It is evident that the assumption

$$\phi = P_1 \cos kx \,.\, \cos(\sigma t + \epsilon), \quad\text{......................(8)}$$

with a proper form of P_1 and the corresponding value of σ determined as above, gives the velocity-potential of a possible system of standing waves, of arbitrary wave-length $2\pi/k$, in an unlimited canal of the given form of section. Now, as explained in Art. 229, by superposition of two properly adjusted systems of standing waves of this type we can build up a system of progressive waves

$$\phi = P_1 \cos(kx \mp \sigma t). \quad\text{........................(9)}$$

We infer that progressive waves of simple-harmonic profile, of any assigned wave-length, are possible in an infinitely long canal of any uniform section.

We might go further, and assert the possibility of an infinite number of types, of any given wave-length, with wave-velocities ranging from a certain lowest value to infinity. The types, however, in which there are longitudinal nodes at a distance from the sides are from the present point of view of subordinate interest.

Two extreme cases call for special notice, viz. where the wave-length is very great or very small compared with the dimensions of the transverse section.

The most interesting types of the former class have no longitudinal nodes, and are covered by the general theory of 'long' waves given in Arts. 169, 170. The only additional information we can look for is as to the shapes of the wave-ridges in the direction transverse to the canal.

In the case of relatively short waves, the most important type is one in which the ridges extend across the canal with gradually varying height, and the wave-velocity is that of free waves on deep water as given by Art. 229 (6).

There is another type of short waves which may present itself when the banks are inclined, and which we may distinguish by the name of 'edge-waves,' since the amplitude diminishes exponentially as the distance from the bank increases. In fact, if the amplitude at the edges be within the limits imposed by our approximations, it will become altogether insensible at a distance whose projection on the slope exceeds a wave-length. The wave-velocity is *less* than that of waves of the same length on deep water. It does not appear that the type of motion here referred to is very important.

A general formula for these edge-waves has been given by Stokes*. Taking the origin in one edge, the axis of z vertically upwards, and that of y transverse to the canal, and treating the breadth as relatively infinite, the formula in question is

$$\phi = He^{-k(y\cos\beta - z\sin\beta)}\cos k(x-ct), \quad \text{.................(10)}$$

where β is the slope of the bank to the horizontal, and

$$c = \left(\frac{g}{k}\sin\beta\right)^{\frac{1}{2}}. \quad \text{..............................(11)}$$

The reader will have no difficulty in verifying this result.

261. We proceed to the consideration of some special cases. We shall treat the question as one of standing waves in an infinitely long canal, or in a compartment bounded by two transverse partitions whose distance apart is a multiple of half the arbitrary wave-length ($2\pi/k$), but the investigations can easily be modified as above so as to apply to progressive waves, and we shall occasionally state results in terms of the wave-velocity.

1°. The solution for the case of a rectangular section, with horizontal bed and vertical sides, could be written down at once from the results of Arts. 190, 257. The nodal lines are transverse and longitudinal, except in the case of a coincidence in period between two distinct modes, when more complex forms are possible. This will happen, for instance, in the case of a square tank.

2°. In the case of a canal whose section consists of two straight lines inclined at 45° to the vertical we have, first, a type discovered by Kelland, viz. if the axis of x coincide with the bottom line of the canal,

$$\phi = A\cosh\frac{ky}{\sqrt{2}}\cosh\frac{kz}{\sqrt{2}}\cos kx\,.\,\cos(\sigma t+\epsilon). \quad \text{.......................(1)}$$

This evidently satisfies $\nabla^2\phi = 0$, and makes

$$\frac{\partial\phi}{\partial y} = \pm\frac{\partial\phi}{\partial z}, \quad \text{...(2)}$$

for $y = \pm z$, respectively. The surface-condition (Art. 260 (4)) then gives

$$\sigma^2 = \frac{gk}{\sqrt{2}}\tanh\frac{kh}{\sqrt{2}}, \quad \text{..(3)}$$

where h is the height of the free surface above the bottom line. If we put $\sigma = kc$, the wave-velocity c is given by

$$c^2 = \frac{g}{\sqrt{2}k}\tanh\frac{kh}{\sqrt{2}}, \quad \text{..(4)}$$

where $k = 2\pi/\lambda$, if λ be the wave-length.

When h/λ is small, this reduces to

$$c = (\tfrac{1}{2}gh)^{\frac{1}{2}}, \quad \text{...(5)}$$

in agreement with Art. 170 (13), since the mean depth is now denoted by $\frac{1}{2}h$†.

* "Report on Recent Researches in Hydrodynamics," *Brit. Ass. Rep.* 1846 [*Papers*, i. 167].
† Kelland, "On Waves," *Trans. R. S. Edin.* xiv (1839).

When, on the other hand, h/λ is moderately large, we have

$$c^2=\frac{g}{k\sqrt{2}}. \qquad \text{.........(6)}$$

The formula (1) indicates now a rapid increase of amplitude towards the sides. We have here, in fact, an instance of 'edge-waves,' and the wave-velocity agrees with that obtained by putting $\beta=45°$ in Stokes' formula.

The remaining types of oscillation which are symmetrical with respect to the medial plane $y=0$ are given by the formula

$$\phi=C(\cosh\alpha y\cos\beta z+\cos\beta y\cosh\alpha z)\cos kx\,.\cos(\sigma t+\epsilon), \qquad \text{.........(7)}$$

provided α, β, σ are properly determined. This evidently satisfies (2), and the equation of continuity gives

$$\alpha^2-\beta^2=k^2. \qquad \text{.........(8)}$$

The surface-condition, Art. 260 (4), to be satisfied for $z=h$, requires

$$\sigma^2\cosh\alpha h=g\alpha\sinh\alpha h, \qquad \sigma^2\cos\beta h=-g\beta\sin\beta h. \qquad \text{.........(9)}$$

Hence

$$\alpha h\tanh\alpha h+\beta h\tan\beta h=0. \qquad \text{.........(10)}$$

The values of α, β are determined by (8) and (10), and the corresponding values of σ are then given by either of the equations (9). If, for a moment, we write

$$x=\alpha h, \qquad y=\beta h, \qquad \text{.........(11)}$$

the roots are given by the intersections of the curve

$$x\tanh x+y\tan y=0, \qquad \text{.........(12)}$$

whose general form can be easily traced, with the hyperbola

$$x^2-y^2=k^2h^2. \qquad \text{.........(13)}$$

There are an infinite number of real solutions, with values of βh lying in the second, fourth, sixth, ... quadrants. These give respectively 2, 4, 6, ... longitudinal nodes of the free surface. When h/λ is moderately large, we have $\tanh\alpha h=1$, nearly, and βh is (in the simplest mode of this class) a little greater than $\frac{1}{2}\pi$. The two longitudinal nodes in this case approach very closely to the edges as λ is diminished, whilst the wave-velocity becomes practically equal to that of waves of length λ on deep water. As a numerical example, assuming $\beta h=1{\cdot}1\times\frac{1}{2}\pi$, we find

$$\alpha h=10{\cdot}910, \qquad kh=10{\cdot}772, \qquad c=1{\cdot}0064\left(\frac{g}{k}\right)^{\frac{1}{2}}.$$

The distance of either nodal line from the nearest edge is then $\cdot 12h$.

We may next consider the asymmetrical modes. The solution of this type which is analogous to Kelland's was noticed by Greenhill (*l.c.*). It is

$$\phi=A\sinh\frac{ky}{\sqrt{2}}\sinh\frac{kz}{\sqrt{2}}\cos kx\,.\cos(\sigma t+\epsilon), \qquad \text{.........(14)}$$

with

$$\sigma^2=\frac{gk}{\sqrt{2}}\coth\frac{kh}{\sqrt{2}}. \qquad \text{.........(15)}$$

When kh is small, this makes $\sigma^2=g/h$, so that the 'speed' is very great compared with that given by the theory of 'long' waves. The oscillation is in fact mainly transversal, with a very gradual variation of phase as we pass along the canal. The middle line of the surface is of course nodal. When on the other hand kh is great, we get 'edge-waves,' as in the case of Kelland's solution.

The remaining asymmetrical oscillations are given by

$$\phi=A(\sinh\alpha y\sin\beta z+\sin\beta y\sinh\alpha z)\cos kx\,.\cos(\sigma t+\epsilon). \qquad \text{.........(16)}$$

This leads in the same manner as before to

$$\alpha^2 - \beta^2 = k^2, \quad \text{...(17)}$$

and $$\sigma^2 \sinh \alpha h = g\alpha \cosh \alpha h, \qquad \sigma^2 \sin \beta h = g\beta \cos \beta h, \quad \text{.................(18)}$$

whence $$\alpha h \coth \alpha h = \beta h \cot \beta h. \quad \text{...(19)}$$

There are an infinite number of solutions, with values of βh in the third, fifth, seventh, ... quadrants, giving 3, 5, 7, ... longitudinal nodes, one of which is central.

3°. The case of a canal with plane sides inclined at 60° to the vertical has been treated by Macdonald*. He has discovered a very comprehensive type, which may be verified as follows.

The assumption $$\phi = P \cos kx \,.\, \cos(\sigma t + \epsilon), \quad \text{................................(20)}$$

where $$P = A \cosh kz + \beta \sinh kz + \cosh \frac{ky\sqrt{3}}{2}\left(C \cosh \frac{kz}{2} + D \sinh \frac{kz}{2}\right), \quad \text{.........(21)}$$

evidently satisfies the equation of continuity; and it is easily shewn that it makes

$$\frac{\partial \phi}{\partial y} = \pm\sqrt{3}\frac{\partial \phi}{\partial z}$$

for $y = \pm\sqrt{3}z$, provided $$C = 2A, \qquad D = -2B. \quad \text{...(22)}$$

The surface-condition, Art. 260 (4), is then satisfied, provided

$$\left.\begin{aligned} &\frac{\sigma^2}{gk}(A \cosh kh + B \sinh kh) = A \sinh kh + B \cosh kh, \\ &\frac{2\sigma^2}{gk}\left(A \cosh \frac{kh}{2} - B \sinh \frac{kh}{2}\right) = A \sinh \frac{kh}{2} - B \cosh \frac{kh}{2}. \end{aligned}\right\} \quad \text{...............(23)}$$

The former of these is equivalent to

$$A = H\left(\cosh kh - \frac{\sigma^2}{gk} \sinh kh\right), \qquad B = H\left(\frac{\sigma^2}{gk} \cosh kh - \sinh kh\right), \quad \text{.........(24)}$$

and the latter then leads to

$$2\left(\frac{\sigma^2}{gk}\right)^2 - 3\frac{\sigma^2}{gk} \coth 3\frac{kh}{2} + 1 = 0. \quad \text{...............................(25)}$$

Also, substituting from (22) and (24) in (21), we find

$$P = H\left\{\cosh k(z-h) + \frac{\sigma^2}{gk} \sinh k(z-h)\right\}$$
$$+ 2H \cosh \frac{ky\sqrt{3}}{2}\left\{\cosh k\left(\frac{z}{2}+h\right) - \frac{\sigma^2}{gk} \sinh k\left(\frac{z}{2}+h\right)\right\}. \quad \text{...(26)}$$

The equations (25) and (26) were arrived at by Macdonald, by a different process. The surface-value of P is

$$P = H\left\{1 + 2 \cosh \frac{ky\sqrt{3}}{2}\left(\cosh \frac{3kh}{2} - \frac{\sigma^2}{gk} \sinh \frac{3kh}{2}\right)\right\}. \quad \text{..................(27)}$$

The equation (25) is a quadratic in σ^2/gk. In the case of a wave whose length $(2\pi/k)$ is great compared with h, we have

$$\coth \frac{3kh}{2} = \frac{2}{3kh},$$

nearly, and the roots of (25) are then

$$\frac{\sigma^2}{gk} = \tfrac{1}{2}kh \quad \text{and} \quad \frac{\sigma^2}{gk} = 1/kh, \quad \text{.................................(28)}$$

* "Waves in Canals," *Proc. Lond. Math. Soc.* xxv. 101 (1894).

approximately. If we put $\sigma = kc$, the former result gives $c^2 = \frac{1}{2}gh$, in accordance with the usual theory of 'long' waves (Arts. 169, 170). The formula (27) now makes $P = 3H$, approximately; this is independent of y, so that the wave-ridges are nearly straight. The second of the roots (28) makes $c^2 = g/k^2h$, giving a much greater 'phase-velocity'; but there is nothing paradoxical in this. The *group*-velocity is in fact relatively small. It will be found on examination that the cross-sections of the waves are parabolic in form, and that there are two nodal lines parallel to the length of the canal. The period is almost exactly that of the symmetrical transverse oscillation discussed in Art. 258.

When, on the other hand, the wave-length is short compared with the transverse dimensions of the canal, kh is large, and $\coth \frac{3}{2}kh = 1$, nearly. The roots of (25) are then

$$\frac{\sigma^2}{gk} = 1 \text{ and } \frac{\sigma^2}{gk} = \frac{1}{2}, \quad \text{.................................(29)}$$

approximately. The former result makes $P = H$, nearly, so that the wave-ridges are straight, experiencing only a slight change of altitude towards the sides. The speed, $\sigma = (gk)^{\frac{1}{2}}$, is exactly what we should expect from the general theory of waves on relatively deep water.

If in this case we transfer the origin to one edge of the water-surface, writing $z + h$ for z, and $y - \sqrt{3}h$ for y, and then make h infinite, we get the case of a system of waves travelling parallel to a shore which slopes downwards at an angle of 30° to the horizon. The result is

$$\phi = H\{e^{kz} + e^{-\frac{1}{2}k(\sqrt{3}y+z)} - 3e^{-\frac{1}{2}k(\sqrt{3}y-z)}\} \cos kx \,.\, \cos(\sigma t + \epsilon), \quad \text{...........(30)}$$

where $c = (g/k)^{\frac{1}{2}}$. This admits of immediate verification. At a distance of a wave-length or so from the shore, the value of ϕ, near the surface, reduces to

$$\phi = He^{kz} \cos kx \,.\, \cos(\sigma t + \epsilon), \quad \text{...............................(31)}$$

practically, in conformity with Art. 228. Near the edge the elevation changes sign, there being a longitudinal node for which

$$\frac{\sqrt{3}}{2} ky = \log_e 2, \quad \text{...(32)}$$

or $y/\lambda = \cdot 127$.

The second of the two roots (29) gives a system of edge-waves, the results being equivalent to those obtained by making $\beta = 30°$ in Stokes' formula*.

Oscillations of a Spherical Mass of Liquid.

262. The theory of the gravitational oscillations of a mass of liquid about the spherical form is due to Kelvin†.

Taking the origin at the centre, and denoting the radius vector at any point of the surface by $a + \zeta$, where a is the radius in the undisturbed state we assume

$$\zeta = \sum_1^\infty \zeta_n, \quad \text{..................................(1)}$$

where ζ_n is a surface-harmonic of integral order n. The equation of continuity $\nabla^2\phi = 0$ is satisfied by

$$\phi = \sum_1^\infty \frac{r^n}{a^n} S_n, \quad \text{..............................(2)}$$

* For extensions to other angles of inclination of the shore see Hanson, *Proc. Roy. Soc.* A, cxi. 491 (1926).

† Sir W. Thomson, "Dynamical Problems regarding Elastic Spheroidal Shells and Spheroids of Incompressible Liquid," *Phil. Trans.* 1863 [*Papers*, iii. 384].

where S_n is a surface-harmonic, and the kinematical condition

$$\frac{\partial \zeta}{\partial t} = -\frac{\partial \phi}{\partial r}, \quad \text{..............................(3)}$$

to be satisfied when $r = a$, gives

$$\frac{\partial \zeta_n}{\partial t} = -\frac{n}{a} S_n. \quad \text{..............................(4)}$$

The gravitation-potential at the free surface is (see Art. 200)

$$\Omega = -\frac{4\pi\gamma\rho a^3}{3r} - \sum_1^\infty \frac{4\pi\gamma\rho a}{2n+1} \zeta_n, \quad \text{......................(5)}$$

where γ is the gravitation-constant. Putting

$$g = \tfrac{4}{3}\pi\gamma\rho a, \qquad r = a + \Sigma\zeta_n,$$

we find

$$\Omega = \text{const.} + g \sum_1^\infty \frac{2(n-1)}{2n+1} \zeta_n. \quad \text{....................(6)}$$

Substituting from (2) and (6) in the pressure-equation

$$\frac{p}{\rho} = \frac{\partial \phi}{\partial t} - \Omega + \text{const.}, \quad \text{........................(7)}$$

we find, since p must be constant over the surface,

$$\frac{\partial S_n}{\partial t} = \frac{2(n-1)}{2n+1} g\zeta_n. \quad \text{............................(8)}$$

Eliminating S_n between (4) and (8), we obtain

$$\frac{\partial^2 \zeta_n}{\partial t^2} + \frac{2n(n-1)}{2n+1} \frac{g}{a} \zeta_n = 0. \quad \text{......................(9)}$$

This shews that $\zeta_n \propto \cos(\sigma_n t + \epsilon)$, where

$$\sigma_n{}^2 = \frac{2n(n-1)}{2n+1} \frac{g}{a}. \quad \text{........................(10)}$$

For the same density of liquid, $g \propto a$, and the frequency is therefore independent of the dimensions of the globe.

The formula makes $\sigma_1 = 0$, as we should expect, since in a small deformation expressed by a surface-harmonic of the first order the surface remains spherical, and the period is therefore infinitely long.

"For the case $n = 2$, or an ellipsoidal deformation, the length of the isochronous simple pendulum becomes $\frac{5}{4}a$, or one and a quarter times the earth's radius, for a homogeneous liquid globe of the same mass and diameter as the earth; and therefore for this case, or for any homogeneous liquid globe of about $5\frac{1}{2}$ times the density of water, the half-period is 47 m. 12 s."

"A steel globe of the same dimensions, without mutual gravitation of its parts, could scarcely oscillate so rapidly, since the velocity of plane waves of distortion in steel is only about 10,140 feet per second, at which rate a space equal to the earth's diameter would not be travelled in less than 1 h. 8 m. 40 s.*"

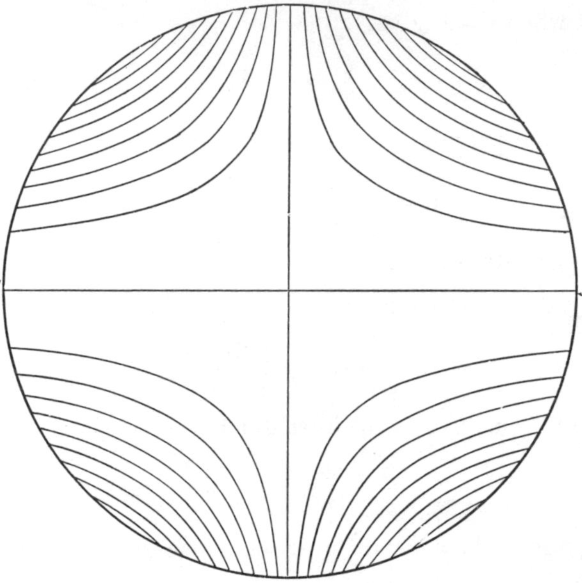

When the surface oscillates in the form of a *zonal* harmonic spheroid of the second order, the equation of the lines of motion is $x\varpi^2 = \text{const.}$, where ϖ denotes the distance of any point from the axis of symmetry, which is taken as axis of x (see Art. 95 (11)). The forms of these lines, for a series of equidistant values of the constant, are shewn in the figure.

263. This problem may also be treated very compactly by the method of 'normal co-ordinates' (Art. 168).

The kinetic energy is given by the formula

$$T = \tfrac{1}{2}\rho \iint \phi \frac{\partial \phi}{\partial r}\, dS, \quad \text{..........................(11)}$$

where δS is an element of the surface $r = a$. Hence, when the surface oscillates in the form $r = a + \zeta_n$, we find, on substitution from (2) and (4),

$$T = \tfrac{1}{2}\frac{\rho a}{n} \iint \dot{\zeta}_n{}^2 dS. \quad \text{..........................(12)}$$

To find the potential energy, we may suppose that the external surface is constrained to assume in succession the forms $r = a + \theta\zeta_n$, where θ varies

* Sir W. Thomson, *l.c.* The exact theory of the vibrations of an elastic sphere gives, for the slowest oscillation of a steel globe of the dimensions of the earth, a period of 1 h. 18 min. See a paper "On the Vibrations of an Elastic Sphere," *Proc. Lond. Math. Soc.* xiii. 212 (1882). The vibrations of a sphere of incompressible substance, under the joint influence of gravity and elasticity, have been discussed by Bromwich, *Proc. Lond. Math. Soc.* xxx. 98 (1898). The influence of compressibility is examined by Love, *Some Problems of Geodynamics* (Adams Prize Essay), Cambridge, 1911, p. 126.

from 0 to 1. At any stage of this process, the gravitation potential at the surface is, by (6),

$$\Omega = \text{const.} + \frac{2(n-1)}{2n+1} g\theta\zeta_n. \qquad \text{.....................(13)}$$

Hence the work required to add a film of thickness $\zeta_n \delta\theta$ is

$$\theta\delta\theta . \frac{2(n-1)}{2n+1} g\rho \iint \zeta_n^2 dS. \qquad \text{.........................(14)}$$

Integrating this from $\theta = 0$ to $\theta = 1$, we find

$$V = \frac{n-1}{2n+1} g\rho \iint \zeta_n^2 dS. \qquad \text{.........................(15)}$$

The results corresponding to the general deformation (1) are obtained by prefixing the sign Σ of summation with respect to n, in (12) and (15); since the terms involving products of surface-harmonics of different orders vanish, by Art. 87.

The fact that the general expressions for T and V thus reduce to sums of squares shews that any spherical-harmonic deformation is of a 'normal type.' Also, assuming that $\zeta_n \propto \cos(\sigma_n t + \epsilon)$, the consideration that the total energy $T + V$ must be constant leads us again to the result (10).

In the case of the forced oscillations due to a disturbing potential $\Omega' \cos(\sigma t + \epsilon)$ which satisfies the equation $\nabla^2 \Omega' = 0$ at all points of the fluid, we must suppose Ω' to be expanded in a series of solid harmonics. If $\bar{\zeta}_n$ be the equilibrium-elevation corresponding to the term of order n, we have, by Art. 168 (14), for the forced oscillation,

$$\zeta_n = \frac{1}{1 - \sigma^2/\sigma_n^2} \bar{\zeta}_n, \qquad \text{............................(16)}$$

where σ is the imposed speed, and σ_n that of the free oscillations of the same type, as given by (10).

The numerical result given above for the case $n = 2$ shews that, in a non-rotating liquid globe of the same dimensions and mean density as the earth, forced oscillations having the characters and periods of the actual lunar and solar tides would practically have the amplitudes assigned by the equilibrium-theory.

264. The investigation is easily extended to the case of an ocean of any uniform depth, covering a symmetrical spherical nucleus.

Let b be the radius of the nucleus, a that of the external surface. The surface-form being

$$r = a + \Sigma_1^\infty \zeta_n, \qquad \text{...(1)}$$

we assume, for the velocity-potential,

$$\phi = \left\{(n+1)\frac{r^n}{b^n} + n\frac{b^{n+1}}{r^{n+1}}\right\} S_n, \qquad \text{..................................(2)}$$

where the coefficients have been adjusted so as to make $\partial\phi/\partial r = 0$ for $r = b$.

The condition that
$$\frac{\partial \zeta}{\partial t} = -\frac{\partial \phi}{\partial r}, \qquad \text{.........(3)}$$

for $r=a$, gives
$$\frac{\partial \zeta_n}{\partial t} = -n(n+1)\left\{\left(\frac{a}{b}\right)^n - \left(\frac{b}{a}\right)^{n+1}\right\}\frac{S_n}{a}. \qquad \text{.........(4)}$$

For the gravitation-potential at the free surface (1) we have
$$\Omega = -\frac{4\pi\gamma\rho_0 a^3}{3r} - \Sigma_1^\infty \frac{4\pi\gamma\rho a}{2n+1}\zeta_n, \qquad \text{.........(5)}$$

where ρ_0 is the mean density of the whole mass. Hence, putting $g=\frac{4}{3}\pi\gamma\rho_0 a$, we find
$$\Omega = \text{const.} + g\Sigma_1^\infty\left(1 - \frac{3}{2n+1}\frac{\rho}{\rho_0}\right)\zeta_n. \qquad \text{.........(6)}$$

The pressure-condition at the free surface then gives
$$\left\{(n+1)\left(\frac{a}{b}\right)^n + n\left(\frac{b}{a}\right)^{n+1}\right\}\frac{\partial S_n}{\partial t} = \left(1 - \frac{3}{2n+1}\frac{\rho}{\rho_0}\right)g\zeta_n. \qquad \text{.........(7)}$$

The elimination of S_n between (4) and (7) leads to
$$\frac{\partial^2 \zeta_n}{\partial t^2} + \sigma_n^2\zeta_n = 0, \qquad \text{.........(8)}$$

where
$$\sigma_n^2 = \frac{n(n+1)\left\{\left(\frac{a}{b}\right)^n - \left(\frac{b}{a}\right)^{n+1}\right\}}{(n+1)\left(\frac{a}{b}\right)^n + n\left(\frac{b}{a}\right)^{n+1}}\left(1 - \frac{3}{2n+1}\frac{\rho}{\rho_0}\right)\frac{g}{a}. \qquad \text{.........(9)}$$

The case $n=1$ is exceptional, since the calculation assumes that the nucleus is fixed. It suggests, however, that σ_1 vanishes when $\rho=\rho_0$, and is imaginary when $\rho > \rho_0$, as we should expect. The correction is hardly important, but it may be shewn that when the nucleus is free the result given by (9), viz.
$$\sigma_1^2 = \frac{1 - b^3/a^3}{1 + b^3/2a^3}\left(1 - \frac{\rho}{\rho_0}\right)\frac{g}{a}, \qquad \text{.........(10)}$$

must be increased in the ratio
$$1 + \frac{b^3}{2a^3} : 1 + \frac{b^3}{2a^3} - \frac{m}{M},$$

where M is the total mass, and m that of the ocean alone. The conclusions as to stability are unaffected.

If in (9) we put $b=0$, we reproduce the result of the preceding Art. If, on the other hand, the depth of the ocean be small compared with the radius, we find, putting $b=a-h$, and neglecting the square of h/a,
$$\sigma_n^2 = n(n+1)\left(1 - \frac{3}{2n+1}\frac{\rho}{\rho_0}\right)\frac{gh}{a^2}, \qquad \text{.........(11)}$$

provided n be small compared with a/h. This agrees with Laplace's result, obtained in a more direct manner in Art. 200.

But if n be comparable with a/h, we have, putting $n=ka$,
$$\left(\frac{a}{b}\right)^n = \left(1 - \frac{h}{a}\right)^{-ka} = e^{kh},$$

so that (9) reduces to
$$\sigma^2 = gk\tanh kh, \qquad \text{.........(12)}$$

as in Art. 228. Moreover, the expression (2) for the velocity-potential becomes, if we write $r=a+z$,
$$\phi = \phi_1 \cosh k(z+h), \qquad \text{.........(13)}$$

where ϕ_1 is a function of the co-ordinates in the surface, which may now be treated as plane. Cf. Art. 257.

The formulae for the kinetic and potential energies, in the general case, are easily found by the same method as in the preceding Art. to be

$$T=\tfrac{1}{2}\rho a\Sigma_1^\infty \frac{(n+1)\left(\frac{a}{b}\right)^n+n\left(\frac{b}{a}\right)^{n+1}}{n(n+1)\left\{\left(\frac{a}{b}\right)^n-\left(\frac{b}{a}\right)^{n+1}\right\}}\iint\dot{\zeta}_n^2dS, \quad\text{.................(14)}$$

and

$$V=\tfrac{1}{2}g\rho\Sigma_1^\infty\left(1-\frac{3}{2n+1}\frac{\rho}{\rho_0}\right)\iint\zeta_n^2dS. \quad\text{..........................(15)}$$

The latter result shews, again, that the equilibrium configuration is one of minimum potential energy, and therefore thoroughly stable, provided $\rho<\rho_0$.

In the case where the depth is relatively small, whilst n is finite, we obtain, putting $b=a-h$,

$$T=\tfrac{1}{2}\frac{\rho a^2}{h}\Sigma_1^\infty\frac{1}{n(n+1)}\iint\dot{\zeta}_n^2dS, \quad\text{..............................(16)}$$

whilst the expression for V is of course unaltered.

If the amplitudes of the harmonics ζ_n be regarded as generalized co-ordinates, the formula (16) shews that for relatively small depths the 'inertia-coefficients' vary inversely as the depth. We have had other illustrations of the effect of constraint in our discussions of tidal waves.

Capillarity.

265. The part played by Cohesion in certain cases of fluid motion has long been recognized in a general way, but it is only within comparatively recent years that the question has been subjected to exact mathematical treatment. We proceed to give some account of the remarkable investigations of Kelvin and Rayleigh in this field.

It is beyond our province to discuss the physical theory of the matter*. It is sufficient, for our purpose, to know that the free surface of a liquid, or, more generally, the common surface of two fluids which do not mix, behaves as if it were in a state of uniform *tension*, the stress between two adjacent portions of the surface, estimated at per unit length of the common boundary-line, depending only on the nature of the two fluids and on the temperature. We shall denote this 'surface-tension,' as it is called, by the symbol T_1. The 'dimensions' of T_1 are $\mathbf{MT^{-2}}$ on the absolute system of measurement. Its value in C.G.S. units (dynes per linear centimetre) appears to be about 74 for a water-air surface at 20° C.†; it diminishes somewhat with rise of temperature. The corresponding value for a mercury-air surface is about 540.

An equivalent statement is that the 'free' energy of any system, of which the surface in question forms part, contains a term proportional to the area of the surface, the amount of this 'superficial energy' (as it is usually termed)

* For this, see Maxwell, *Encyc. Britann.* Art. "Capillary Action" [*Papers*, Cambridge, 1890, ii. 541], where references to the older writers are given. Also, Rayleigh, "On the Theory of Surface Forces," *Phil. Mag.* (5) xxx. 285, 456 (1890) [*Papers*, iii. 397].

† Rayleigh, "On the Tension of Water-Surfaces, Clean and Contaminated, investigated by the method of Ripples," *Phil. Mag.* (5) xxx. 386 (1890) [*Papers*, iii. 394]; Pedersen, *Phil. Trans.* A, ccvii. 341 (1907); Bohr, *Phil. Trans.* A, ccix. 281 (1909).

per unit area being equal to T_1*. Since the condition of stable equilibrium is that the free energy should be a minimum, the surface tends to contract as much as is consistent with the other conditions of the problem.

The chief modification which the consideration of surface-tension will introduce into our previous methods is contained in the theorem that the fluid pressure is now discontinuous at a surface of separation, viz. we have

$$p - p' = T_1\left(\frac{1}{R_1} + \frac{1}{R_2}\right),$$

where p, p' are the pressures close to the surface on the two sides, and R_1, R_2 are the principal radii of curvature of the surface, to be reckoned negative when the corresponding centres of curvature lie on the side to which the accent refers. This formula is readily obtained by resolving along the normal the forces acting on a rectangular element of a superficial film, bounded by lines of curvature; but it seems unnecessary to give here the proof, which may be found in most modern treatises on Hydrostatics.

266. The simplest problem we can take, to begin with, is that of waves on a plane surface forming the common boundary of two fluids at rest.

If the origin be taken in this plane, and the axis of y normal to it, the velocity-potentials corresponding to a simple-harmonic deformation of the common surface may be assumed to be

$$\left.\begin{aligned}\phi &= Ce^{ky}\cos kx\,.\cos(\sigma t+\epsilon),\\ \phi' &= C'e^{-ky}\cos kx\,.\cos(\sigma t+\epsilon),\end{aligned}\right\} \quad\ldots\ldots\ldots\ldots\ldots\ldots(1)$$

where the former equation relates to the side on which y is negative, and the latter to that on which y is positive. For these values satisfy $\nabla^2\phi = 0$, $\nabla^2\phi' = 0$, and make the velocity zero for $y = \mp\infty$, respectively.

The corresponding displacement of the surface in the direction of y will be of the type

$$\eta = a\cos kx\,.\sin(\sigma t+\epsilon)\,; \quad\ldots\ldots\ldots\ldots\ldots\ldots(2)$$

and the conditions that

$$\frac{\partial\eta}{\partial t} = -\frac{\partial\phi}{\partial y} = -\frac{\partial\phi'}{\partial y},$$

for $y = 0$, give

$$\sigma a = -kC = kC'. \quad\ldots\ldots\ldots\ldots\ldots\ldots(3)$$

If, for the moment, we ignore gravity, the variable part of the pressure is therefore given by

$$\left.\begin{aligned}\frac{p}{\rho} &= \frac{\partial\phi}{\partial t} = \frac{\sigma^2 a}{k}e^{ky}\cos kx\,.\sin(\sigma t+\epsilon),\\ \frac{p'}{\rho'} &= \frac{\partial\phi'}{\partial t} = -\frac{\sigma^2 a}{k}e^{-ky}\cos kx\,.\sin(\sigma t+\epsilon).\end{aligned}\right\} \quad\ldots\ldots\ldots\ldots(4)$$

* The distinction between 'free' and 'intrinsic' energy depends on thermo-dynamical principles. In the case of changes made at constant temperature with free communication of heat, it is with the 'free' energy that we are concerned.

To find the pressure-condition at the common surface, we may calculate the forces which act in the direction of y on a strip of breadth δx. The fluid pressures on the two sides have a resultant $(p'-p)\,\delta x$, and the difference of the tensions parallel to y on the two edges gives $\delta\,(T_1 \partial\eta/\partial x)$. We thus get the equation

$$p - p' + T_1 \frac{\partial^2 \eta}{\partial x^2} = 0, \qquad \text{.........(5)}$$

to be satisfied when $y=0$ approximately. This might have been written down at once as a particular case of the general surface-condition (Art. 265). Substituting in (5) from (2) and (4), we find

$$\sigma^2 = \frac{T_1 k^3}{\rho + \rho'}, \qquad \text{.........(6)}$$

which determines the speed of the oscillations of wave-length $2\pi/k$.

The energy of motion, per wave-length, of the fluid included between two planes parallel to xy, at unit distance apart, is

$$T = \tfrac{1}{2}\rho \int_0^\lambda \left[\phi \frac{\partial \phi}{\partial y}\right]_{y=0} dx - \tfrac{1}{2}\rho' \int_0^\lambda \left[\phi' \frac{\partial \phi'}{\partial y}\right]_{y=0} dx. \qquad \text{.........(7)}$$

If we assume $\qquad \eta = a \cos kx, \qquad$ ....(8)

where a depends on t only, and therefore, having regard to the kinematical conditions,

$$\phi = -k^{-1} \dot{a} e^{ky} \cos kx, \qquad \phi' = k^{-1} \dot{a} e^{-ky} \cos kx, \qquad \text{.........(9)}$$

we find $\qquad T = \tfrac{1}{4}(\rho+\rho')\, k^{-1} \dot{a}^2 . \lambda. \qquad$ ....(10)

Again, the energy of extension of the surface of separation is

$$V = T_1 \int_0^\lambda \left\{1 + \left(\frac{\partial \eta}{\partial x}\right)^2\right\}^{\frac{1}{2}} dx - T_1 \lambda = \tfrac{1}{2} T_1 \int_0^\lambda \left(\frac{\partial \eta}{\partial x}\right)^2 dx. \qquad \text{.........(11)}$$

Substituting from (8), this gives

$$V = \tfrac{1}{4} T_1 k^2 a^2 . \lambda. \qquad \text{.........(12)}$$

To find the mean energy, of either kind, per unit area of the common surface, we must omit the factor λ.

If we assume that $a \propto \cos(\sigma t + \epsilon)$, where σ is determined by (6), we verify that the total energy $T+V$ is constant. Conversely, if we assume that

$$\eta = \Sigma\,(a \cos kx + \beta \sin kx), \qquad \text{.........(13)}$$

it is easily seen that the expressions for T and V will reduce to sums of squares of $\dot{a}$, $\dot{\beta}$ and a, β, respectively, with constant coefficients, so that the quantities a, β are 'normal co-ordinates.' The general theory of Art. 168 then leads independently to the formula (6) for the speed.

By compounding two systems of standing waves, as in Art. 229, we obtain a progressive wave-system

$$\eta = a \cos(kx \mp \sigma t), \qquad \text{.........(14)}$$

travelling with the velocity

$$c = \frac{\sigma}{k} = \left(\frac{T_1 k}{\rho + \rho'}\right)^{\frac{1}{2}}, \qquad \text{.........(15)}$$

or, in terms of the wave-length,

$$c = \left(\frac{2\pi T_1}{\rho + \rho'}\right)^{\frac{1}{2}} . \lambda^{-\frac{1}{2}}. \qquad \text{.........(16)}$$

The contrast with Art. 229 is noteworthy; as the wave-length is diminished, the period diminishes in a more rapid ratio, so that the wave-velocity *increases.*

Since c varies as $\lambda^{-\frac{1}{2}}$, the group-velocity is, by Art. 236 (3),

$$U = c - \lambda \frac{dc}{d\lambda} = \tfrac{3}{2} c. \qquad \text{.........................(17)}$$

The verification of the relation between group-velocity and transmission of energy is of some interest. Taking

$$\eta = \alpha \cos k(ct - x), \qquad \text{.........................(18)}$$

we find that the total energy per unit area of the surface is

$$\tfrac{1}{4}(\rho + \rho')kc^2\alpha^2 + \tfrac{1}{4}T_1 k^2 \alpha^2 = \tfrac{1}{2}(\rho + \rho')kc^2\alpha^2, \qquad \text{.....................(19)}$$

by (10), (12). The mean rate at which work is done by fluid pressure at a plane perpendicular to x is found by a calculation similar to that of Art. 237 to be

$$\tfrac{1}{4}(\rho + \rho')kc^3\alpha^2. \qquad \text{.........................(20)}$$

The rate at which surface-tension does work at such a plane is

$$T_1 \frac{\partial \eta}{\partial x} \dot{\eta} = T_1 k^2 c \alpha^2 \sin^2 k(ct - x),$$

the mean value of which is

$$\tfrac{1}{2} T_1 k^2 c \alpha^2 = \tfrac{1}{2}(\rho + \rho')kc^3\alpha^2. \qquad \text{.........................(21)}$$

If we add this to (20), and divide by the second member of (19), the quotient is $\frac{3}{2}c$, in agreement with (17).

The fact that the group-velocity for capillary waves exceeds the wave-velocity helps to explain some interesting phenomena to be referred to later (Arts. 271, 272).

For numerical illustration we may take the case of a free water-surface; thus, putting $\rho = 1$, $\rho' = 0$, $T_1 = 74$, we have the following results, the units being the centimetre and second*.

Wave-length	Wave-velocity	Frequency
·50	30	61
·10	68	680
·05	96	1930

267. When gravity is to be taken into account, the common surface, in equilibrium, will of course be horizontal. Taking the positive direction of y upwards, the pressure at the disturbed surface will be given by

$$\left.\begin{aligned} \frac{p}{\rho} &= \frac{\partial \phi}{\partial t} - gy = \left(\frac{\sigma^2}{k} - g\right) a \cos kx \,.\, \sin(\sigma t + \epsilon), \\ \frac{p'}{\rho'} &= \frac{\partial \phi'}{\partial t} - gy = -\left(\frac{\sigma^2}{k} + g\right) a \cos kx \,.\, \sin(\sigma t + \epsilon), \end{aligned}\right\} \qquad \text{............(1)}$$

* Cf. Sir W. Thomson, *Papers*, iii. 520.

The above theory gives the explanation of the crispations observed on the surface of water contained in a finger-bowl set into vibration by stroking the rim with a wetted finger. It is to be observed, however, that the frequency of the capillary waves in this experiment is *double* that of the vibrations of the bowl; see Rayleigh, "On Maintained Vibrations," *Phil. Mag.* (5) xv. 229 (1883) [*Papers*, ii. 188; *Theory of Sound*, 2nd ed., c. xx.].

approximately. Substituting in Art. 266 (5), we find

$$\sigma^2 = \frac{\rho - \rho'}{\rho + \rho'} gk + \frac{T_1 k^3}{\rho + \rho'}. \quad \text{.........................(2)}$$

Putting $\sigma = kc$, we find, for the velocity of a train of progressive waves,

$$c^2 = \frac{\rho - \rho'}{\rho + \rho'} \frac{g}{k} + \frac{T_1}{\rho + \rho'} k = \frac{1-s}{1+s}\left(\frac{g}{k} + T'k\right), \quad \text{..............(3)}$$

where we have written

$$\frac{\rho'}{\rho} = s, \quad \frac{T_1}{\rho - \rho'} = T'. \quad \text{..........................(4)}$$

In the particular cases of $T_1 = 0$ and $g = 0$, respectively, we fall back on the results of Arts. 231 and 266.

There are several points to be noticed with respect to the formula (3). In the first place, although, as the wave-length $(2\pi/k)$ diminishes from ∞ to 0, the speed (σ) continually increases, the wave-velocity, after falling to a certain minimum, begins to increase again. This minimum value (c_m, say) is given by

$${c_m}^2 = \frac{1-s}{1+s} \cdot 2\,(gT')^{\frac{1}{2}}, \quad \text{............................(5)}$$

and corresponds to a wave-length

$$\lambda_m = \frac{2\pi}{k_m} = 2\pi \sqrt{\left(\frac{T'}{g}\right)}. \quad \text{.......................(6)*}$$

In terms of λ_m and c_m the formula (3) may be written

$$\frac{c^2}{{c_m}^2} = \frac{1}{2}\left(\frac{\lambda}{\lambda_m} + \frac{\lambda_m}{\lambda}\right), \quad \text{............................(7)}$$

shewing that for any prescribed value of c, greater than c_m, there are two admissible values (reciprocals) of λ/λ_m. For example, corresponding to

$$\frac{c}{c_m} = \quad 1{\cdot}2 \qquad 1{\cdot}4 \qquad 1{\cdot}6 \qquad 1{\cdot}8 \qquad 2{\cdot}0$$

we have

$$\frac{\lambda}{\lambda_m} = \begin{cases} 2{\cdot}476 \qquad 3{\cdot}646 \qquad 4{\cdot}917 \qquad 6{\cdot}322 \qquad 7{\cdot}873 \\ \;\;{\cdot}404 \qquad\; {\cdot}274 \qquad\; {\cdot}203 \qquad\; {\cdot}158 \qquad\; {\cdot}127, \end{cases}$$

to which we add, for future reference,

$$\sin^{-1}\frac{c_m}{c} = 56^\circ\,26' \qquad 45^\circ\,35' \qquad 38^\circ\,41' \qquad 33^\circ\,45' \qquad 30^\circ.$$

For sufficiently large values of λ the first term in the formula (3) for c^2 is large compared with the second; the force governing the motion of the waves being mainly that of gravity. On the other hand, when λ is very small, the second term preponderates, and the motion is mainly governed

* The theory of the minimum wave-velocity, together with most of the substance of Arts. 266, 267, was given by Sir W. Thomson, "Hydrokinetic Solutions and Observations," *Phil. Mag.* (4) xlii. 374 (1871) [*Papers*, iv. 76]; see also *Nature*, v. 1 (1871).

by cohesion, as in Art. 266. As an indication of the actual magnitudes here in question, we may note that if $\lambda/\lambda_m > 3$, the influence of cohesion on the wave-velocity amounts only to about 5 per cent., whilst gravity becomes relatively ineffective to a like degree if $\lambda/\lambda_m < \frac{1}{3}$.

It has been proposed by Kelvin to distinguish by the name of 'ripples' waves whose length is less than λ_m.

If we substitute from (7) in the general formula (Art. 236 (3)) for the group-velocity, we find

$$U = c - \lambda \frac{dc}{d\lambda} = c\left(1 - \tfrac{1}{2}\frac{\lambda^2 - {\lambda_m}^2}{\lambda^2 + {\lambda_m}^2}\right). \qquad \text{(8)}$$

Hence the group-velocity is greater or less than the wave-velocity, according as $\lambda \lessgtr \lambda_m$. For sufficiently long waves the group-velocity is practically equal to $\frac{1}{2}c$, whilst for very short waves it tends to the value $\frac{3}{2}c$*.

The relative importance of gravity and cohesion, as depending on the value of λ, may be traced to the form of the expression for the potential energy of a deformation of the type

$$\eta = \alpha \cos kx. \qquad \text{(9)}$$

The part of this energy due to the extension of the bounding surface is, per unit area,

$$\frac{\pi^2 T_1 \alpha^2}{\lambda^2}, \qquad \text{(10)}$$

whilst the part due to gravity is

$$\tfrac{1}{2} g (\rho - \rho') \alpha^2. \qquad \text{(11)}$$

As λ diminishes, the former becomes more and more important compared with the latter.

For a water-surface, using the same data as before, with $g = 981$, we find from (5) and (6)

$$\lambda_m = 1{\cdot}73, \qquad c_m = 23{\cdot}2,$$

the units being the centimetre and the second. That is to say, roughly, the minimum wave-velocity is about nine inches per second, or ·45 sea-mile per hour, with a wave-length of two-thirds of an inch. Combined with the numerical results already obtained, this gives,

for	$c=$	27·8	32·5	37·1	41·8	46·4
the values	$\lambda=$	4·3	6·3	8·5	10·9	13·6
		·70	·47	·35	·27	·22

in centimetres and centimetres per second, respectively.

The relations between wave-length and wave-velocity are shewn graphically on the next page, where the dotted curves refer to the cases where gravity and capillarity act separately, whilst the full curve exhibits the joint effect. As explained in Art. 236, the group-velocity is represented by the intercept made by the tangent on the axis of ordinates. Since two tangents can be drawn to the curve from any point on this axis (beyond a certain distance from O), there are two values of the wave-length corresponding to any prescribed value of the *group*-velocity U. These two values of λ coincide when U has a certain (minimum) value, indicated by the point where the tangent to the curve at the point of inflexion cuts Oc; and it may be easily shewn that we then have

$$\frac{\lambda}{\lambda_m} = \sqrt{(3 + 2\sqrt{3})} = 2{\cdot}542, \qquad U = {\cdot}767 c_m,$$

where c_m is the minimum *wave*-velocity as above.

* Cf. Rayleigh, *ll.cc. ante* p. 383.

A further consequence of (2) is to be noted. We have hitherto tacitly supposed that the lower fluid is the denser (*i.e.* $\rho > \rho'$), as is indeed necessary for stability when T_1 is neglected. The formula referred to shews, however, that there is stability even when $\rho < \rho'$, provided

$$\lambda < 2\pi \left(\frac{T_1}{g(\rho' - \rho)}\right)^{\frac{1}{2}}, \qquad \qquad (12)$$

i.e. provided λ be less than the wave-length λ_m of minimum velocity when the denser fluid is below. Hence in the case of water above and air below the maximum wave-length consistent with stability is 1·73 cm. If the fluids be included between two parallel vertical walls, this imposes a superior limit to the admissible wave-length, and we learn that there is stability (in the two-dimensional problem) provided the interval between the walls does not exceed ·86 cm. We have here an explanation, in principle, of a familiar experiment in

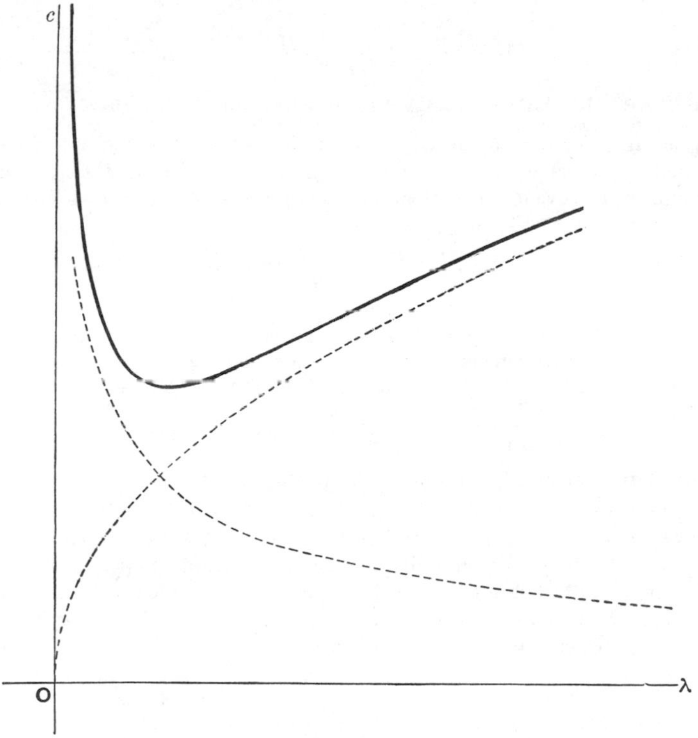

which water is retained by atmospheric pressure in an inverted tumbler, or other vessel, whose mouth is covered by a gauze with sufficiently fine meshes*.

268. We next consider the case of waves on a horizontal surface forming the common boundary of two parallel currents U, U'†.

* The case where the fluids are contained in a cylindrical tube was solved by Maxwell, *Encyc. Britann.* Art. "Capillary Action" [*Papers*, ii. 585], and compared with some experiments of Duprez. The agreement is better than might have been expected when we consider that the special condition to be satisfied at the line of contact of the surface with the wall of the tube has been left out of account.

† Cf. Sir W. Thomson, *l.c. ante* p. 459.

If we apply the method of Art. 234, we find without difficulty that the condition for a stationary wave-profile is now

$$\rho U^2 + \rho' U'^2 = \frac{g}{k}(\rho - \rho') + kT_1, \quad \text{.....................(1)}$$

the last term being due to the altered form of the pressure-condition which has to be satisfied at the surface. This may be written

$$\left(\frac{\rho U + \rho' U'}{\rho + \rho'}\right)^2 = \frac{g}{k} \cdot \frac{\rho - \rho'}{\rho + \rho'} + \frac{kT_1}{\rho + \rho'} - \frac{\rho\rho'}{(\rho + \rho')^2}(U - U')^2. \quad \text{.........(2)}$$

The relative velocity of the waves, which is superposed on the mean velocity of the currents (Art. 232), is $\pm c$, provided

$$c^2 = c_0^2 - \frac{\rho\rho'}{(\rho + \rho')^2}(U - U')^2, \quad \text{.......................(3)}$$

where c_0 denotes the wave-velocity in the absence of currents.

The various inferences to be drawn from (3) are much as in the Art. cited, with the important qualification that, since c_0 has now a minimum value, viz. the c_m of Art. 267 (5), the equilibrium of the surface when plane is stable for disturbances of all wave-lengths so long as

$$|U - U'| < \frac{1+s}{s^{\frac{1}{2}}} \cdot c_m, \quad \text{.................................(4)}$$

where $s = \rho'/\rho$.

When the relative velocity of the two currents exceeds this value, c becomes imaginary for wave-lengths lying between certain limits. It is evident that in the alternative method of Art. 232 the time-factor $e^{i\sigma t}$ will now take the form $e^{\pm \alpha t + i\beta t}$, where

$$\alpha = \left\{\frac{s}{(1+s)^2}(U - U')^2 - c_0^2\right\}^{\frac{1}{2}} k, \qquad \beta = \frac{s}{1+s} k\,|U - U'|. \quad \text{...............(5)}$$

The real part of the exponential indicates the possibility of a disturbance of continually increasing amplitude.

For the case of air over water we have $s = \cdot 00129$, $c_m = 23\cdot 2$ (C. S.), whence the maximum value of $|U - U'|$ consistent with stability is about 646 centimetres per second, or (roughly) 12·5 sea-miles per hour*. For slightly greater values the instability will manifest itself by the formation, in the first instance, of wavelets of about two-thirds of an inch in length, which will continually increase in amplitude until they transcend the limits implied in our approximation.

269. The waves due to a local impulse on the surface of still water may be investigated to a certain extent by Kelvin's method (Art. 241).

Since $\dot{\eta} = -\partial\phi/\partial y$ at the surface, the effect of a unit impulse at the origin is

$$\phi = \frac{1}{\pi\rho}\int_0^\infty \cos \sigma t\, e^{ky} \cos kx\, dk, \quad \eta = -\frac{1}{\pi\rho}\int_0^\infty \frac{\sin \sigma t}{\sigma} \cos kx\, k\, dk. \quad \text{...(1)}$$

Hence to conform to (6) of Art. 241 we must put

$$\phi(k) = ik/\rho\sigma. \quad \text{................................(2)}$$

* The wind-velocity at which the surface of water actually begins to be ruffled so as to lose the power of distinct reflection is much less than this, and is determined by other causes. This question is considered later (Chapter XI.).

If in Art. 267 (2) we put $\rho' = 0$ and write, for shortness,

$$T_1/\rho = T', \qquad \text{.............................(3)}$$

we have

$$\sigma^2 = gk + T'k^3. \qquad \text{.............................(4)}$$

Let us first suppose that capillarity alone is operative, so that

$$\sigma^2 = T'k^3. \qquad \text{.............................(5)}$$

Since

$$\frac{d\sigma}{dk} = \tfrac{3}{2}T'^{\frac{1}{2}}k^{\frac{1}{2}}, \quad \frac{d^2\sigma}{dk^2} = \tfrac{3}{4}T'^{\frac{1}{2}}k^{-\frac{1}{2}}, \qquad \text{.....................(6)}$$

we find

$$k = \tfrac{4}{9}\frac{x^2}{T't^2}, \quad \sigma = \tfrac{8}{27}\frac{x^3}{T't^3}, \quad t\frac{d^2\sigma}{dk^2} = \tfrac{9}{8}\frac{T't^2}{x}. \qquad \text{...............(7)}$$

The procedure of Art. 241 then gives

$$\eta = \frac{1}{\pi^{\frac{1}{2}}\rho T'^{\frac{1}{2}}x^{\frac{1}{2}}}\sin\left(\frac{4x^3}{27\,T't^2} - \tfrac{1}{4}\pi\right) \qquad \text{.....................(8)}$$

The test-fraction (10) of Art. 241 is now comparable with $T'^{\frac{1}{2}}t/x^{\frac{3}{2}}$, and the approximation therefore cannot claim great accuracy except as regards the *earlier* stages of the disturbance at any point. It appears also from (8) that the wave-length and period at any point begin by being infinitesimal, and continually increase. These several circumstances are in contrast with what holds in the case of gravity waves (Art. 240).

We have seen (Art. 267) that when gravity is taken into account there are two wave-lengths corresponding to any assigned value of the group-velocity U which exceeds the minimum U_0. The particular wave-lengths corresponding to given values of x and t may be found by the geometrical methods of Art. 241. Analytically, putting $d\sigma/dk = U = x/t$, they are determined by the real values of k satisfying the equation

$$(g + 3T'k^2)^2 = 4\sigma^2\left(\frac{d\sigma}{dk}\right)^2 = \frac{4x^2}{t^2}(gk + T'k^3). \qquad \text{..............(9)}$$

The approximate expression for η will accordingly consist of two terms of the type (9) of Art. 241, so that we have two systems of waves superposed. For $x < U_0 t$, Kelvin's method indicates that the disturbance is unimportant*.

When $x/U_0 t$ is sufficiently large the real solutions of (9) are

$$k = \tfrac{1}{4}\frac{gt^2}{x^2}, \quad k = \tfrac{4}{9}\frac{x^2}{T't^2}, \qquad \text{.......................(10)}$$

approximately, as if gravity and capillarity were respectively alone operative. The conditions for the validity of Kelvin's approximation in this case, viz. that gt^2/x and $x^3/T't^2$ should both be large, are to some extent opposed, but admit of being reconciled if x and t are both sufficiently great. The wave-length must in each case be small compared with x.

The effect of a travelling disturbance can be written down from the general formulae of Art. 248. If $2\pi/\kappa_1$, $2\pi/\kappa_2$ be the two wave-lengths corresponding

* Rayleigh, *Phil. Mag.* (6) xxi. 180 (1911) [*Papers*, vi. 9].

to the wave-velocity c, it appears from the figure on p. 461 that if $\kappa_1 < \kappa_2$, we shall have $U_1 < c$, $U_2 > c$. The result will be

$$\left.\begin{aligned} \eta &= \frac{\phi(\kappa_1)}{c - U_1} e^{i\kappa_1 x}, \quad [x > 0] \\ \eta &= \frac{\phi(\kappa_2)}{U_2 - c} e^{i\kappa_2 x}. \quad [x < 0] \end{aligned}\right\} \qquad (11)$$

If we put

$$\phi(k) = iP/\rho\sigma, \qquad (12)$$

this will be found to agree, as an approximation, with the result of the more complete investigation which follows.

270. We resume the more formal investigation of the effect of a steady pressure-disturbance on the surface of a running stream, by the methods of Arts. 242, 243, including now the effect of capillary forces. This will give, in addition to the former results, the explanation (in principle) of the fringe of ripples which is seen in advance of a solid moving at a moderate speed through still water, or on the up-stream side of any disturbance in a uniform current.

Beginning with a simple-harmonic distribution of pressure, we assume

$$\frac{\phi}{c} = -x + \beta e^{ky} \sin kx, \quad \frac{\psi}{c} = -y + \beta e^{ky} \cos kx, \qquad (1)$$

the upper surface coinciding with the stream-line $\psi = 0$, whose equation is

$$y = \beta \cos kx, \qquad (2)$$

approximately. At a point just beneath this surface we find, as in Art. 242 (8), for the variable part of the pressure,

$$p_0 = \beta\rho\{(kc^2 - g)\cos kx + \mu c \sin kx\}, \qquad (3)$$

where μ is the frictional coefficient. At an adjacent point just above the surface we must have

$$p_0' = p_0 + T_1 \frac{d^2 y}{dx^2} = \beta\rho\{(kc^2 - g - k^2 T')\cos kx + \mu c \sin kx\}, \qquad (4)$$

where T' is written for T_1/ρ. This is equal to the real part of

$$\beta\rho(kc^2 - g - k^2 T' - i\mu c) e^{ikx}.$$

We infer that to the imposed pressure

$$p_0 = C \cos kx \qquad (5)$$

will correspond the surface-form

$$\rho y = C \frac{(kc^2 - g - k^2 T')\cos kx - \mu c \sin kx}{(kc^2 - g - k^2 T')^2 + \mu^2 c^2}. \qquad (6)$$

Let us first suppose that the velocity c of the stream exceeds the minimum wave-velocity (c_m) investigated in Art. 267. We may then write

$$kc^2 - g - k^2 T' = T'(k - \kappa_1)(\kappa_2 - k), \qquad (7)$$

where κ_1, κ_2 are the two values of k corresponding to the wave-velocity c on still water; in other words, $2\pi/\kappa_1$, $2\pi/\kappa_2$ are the lengths of the two systems of free waves which could maintain a stationary position in space, on the surface of the flowing stream. We will suppose that $\kappa_2 > \kappa_1$.

In terms of these quantities, the formula (6) may be written

$$\rho y = \frac{C}{T'} \cdot \frac{(k-\kappa_1)(\kappa_2 - k)\cos kx - \mu' \sin kx}{(k-\kappa_1)^2(\kappa_2-k)^2 + \mu'^2}, \qquad (8)$$

where $\mu' = \mu c/T'$. This shews that if μ' be small the pressure is least over the crests, and greatest over the troughs of the waves when k is greater than κ_2 or less than κ_1, whilst the reverse is the case when k is intermediate to κ_1, κ_2. In the case of a progressive disturbance advancing over still water, these results are seen to be in accordance with Art. 168 (14).

271. From (8) we can infer as in Art. 243 the effect of a pressure of integral amount P concentrated on a line of the surface at the origin, viz. we find

$$y = \frac{P}{\pi T_1}\int_0^\infty \frac{(k-\kappa_1)(\kappa_2-k)\cos kx - \mu' \sin kx}{(k-\kappa_1)^2(\kappa_2-k)^2+\mu'^2}\,dk. \qquad (9)$$

The definite integral is the real part of

$$\int_0^\infty \frac{e^{ikx}\,dk}{(k-\kappa_1)(\kappa_2-k)-i\mu'}. \qquad (10)$$

The dissipation-coefficient μ' has been introduced solely for the purpose of making the problem determinate; we may therefore avail ourselves of the slight gain in simplicity obtained by supposing μ' to be infinitesimal. In this case the two roots of the denominator in (10) are

$$k = \kappa_1 + i\nu, \qquad k = \kappa_2 - i\nu,$$

where

$$\nu = \frac{\mu'}{\kappa_2 - \kappa_1}.$$

The integral (10) is therefore equivalent to

$$\frac{1}{\kappa_2-\kappa_1-2i\nu}\left\{\int_0^\infty \frac{e^{ikx}\,dk}{k-(\kappa_1+i\nu)} - \int_0^\infty \frac{e^{ikx}\,dk}{k-(\kappa_2-i\nu)}\right\}. \qquad (11)$$

These integrals are of the forms discussed in Art. 243. Since $\kappa_2 > \kappa_1$, ν is positive, and it appears that when x is positive the former integral is equal to

$$2\pi i e^{i\kappa_1 x} + \int_0^\infty \frac{e^{-ikx}}{k+\kappa_1}\,dk, \qquad (12)$$

and the latter to

$$\int_0^\infty \frac{e^{-ikx}}{k+\kappa_2}\,dk. \qquad (13)$$

On the other hand, when x is negative, the former reduces to

$$\int_0^\infty \frac{e^{-ikx}}{k+\kappa_1}\,dk, \qquad (14)$$

and the latter to

$$-2\pi i e^{i\kappa_2 x} + \int_0^\infty \frac{e^{-ikx}}{k+\kappa_2}\,dk. \qquad (15)$$

We have here simplified the formulae by putting $\nu = 0$ *after* the transformations.

If we now discard the imaginary parts of our expressions, we obtain the results which immediately follow.

When μ' is infinitesimal, the equation (9) gives, for x positive,

$$\frac{\pi T_1}{P}\cdot y = -\frac{2\pi}{\kappa_2-\kappa_1}\sin \kappa_1 x + F(x), \qquad (16)$$

and, for x negative,

$$\frac{\pi T_1}{P} \cdot y = -\frac{2\pi}{\kappa_2 - \kappa_1} \sin \kappa_2 x + F(x), \quad \ldots(17)$$

where $$F(x) = \frac{1}{\kappa_2 - \kappa_1} \left\{ \int_0^\infty \frac{\cos kx}{k + \kappa_1} dk - \int_0^\infty \frac{\cos kx}{k + \kappa_2} dk \right\}. \quad \ldots\ldots(18)$$

This function $F(x)$ can be expressed in terms of the known functions Ci $\kappa_1 x$, Si $\kappa_1 x$, Ci $\kappa_2 x$, Si $\kappa_2 x$, by Art. 243 (30). The disturbance of level represented by it is very small for values of x, whether positive or negative, which exceed, say, half the greater wave-length $(2\pi/\kappa_1)$.

Hence, beyond some such distance, the surface is covered on the down-stream side by a regular train of simple-harmonic waves of length $2\pi/\kappa_1$, and on the up-stream side by a train of the shorter wave-length $2\pi/\kappa_2$. It appears from the numerical results of Art. 267 that when the velocity c of the stream much exceeds the minimum wave-velocity (c_m) the former system of waves is governed mainly by gravity, and the latter by cohesion.

It is worth notice that, in contrast with the case of Art. 234, the elevation is now finite when $x = 0$, viz. we have

$$\frac{\pi T_1}{P} \cdot y = \frac{1}{\kappa_2 - \kappa_1} \log \frac{\kappa_2}{\kappa_1}. \quad \ldots\ldots\ldots(19)$$

This follows easily from (16) and (18).

The figure shews the transition between the two sets of waves, in the case of $\kappa_2 - 5\kappa_1$.

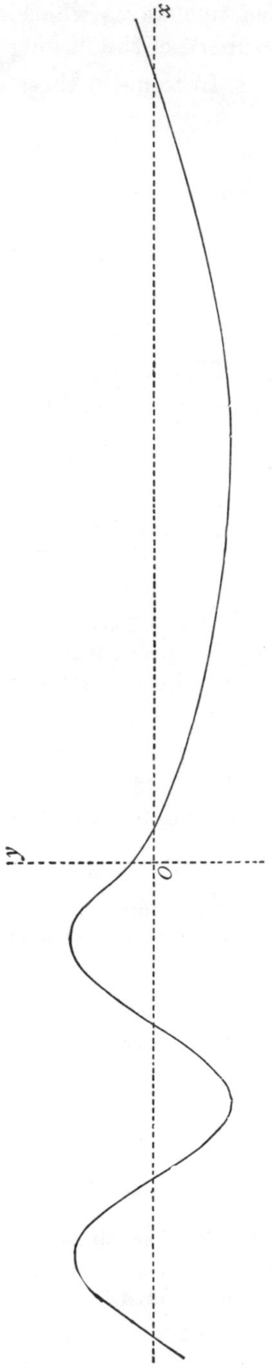

The general explanation of the effects of an isolated pressure-disturbance advancing over still water is now modified by the fact that there are *two* wave-lengths corresponding to the given velocity c. For one of these (the shorter) the group-velocity is greater, whilst for the other it is less, than c, We can thus understand why the waves of shorter wave-length should be found ahead, and those of longer wave-length in the rear, of the disturbing pressure.

It will be noticed that the formulae (16), (17) make the height of the up-steam capillary waves the same as that of the down-stream gravity waves; but this result will be greatly modified when the pressure is diffused over a band of sensible breadth, instead of being concentrated on a mathematical line. If, for example, the breadth of the band do not exceed one-fourth of the wave-length on the down-stream side, whilst it considerably exceeds the wave-length of the up-stream ripples, as may happen with a very moderate velocity, the different parts of the breadth will on the whole reinforce one another as regards their action on the down-stream side, whilst on the up-stream side we shall have 'interference,' with a comparatively small residual amplitude.

This point may be illustrated by assuming that the integral surface-pressure P has the distribution

$$p' = \frac{P}{\pi}\frac{b}{b^2+x^2}, \qquad \text{.......................................(20)}$$

which is more diffused, the greater the value of b.

The method of calculation will be understood from Art. 244. The result is that on the down-stream side

$$y = -\frac{2P}{\rho T'(\kappa_2 - \kappa_1)} e^{-\kappa_1 b}\sin\kappa_1 x + \ldots, \qquad \text{...........................(21)}$$

and on the up-stream side

$$y = -\frac{2P}{\rho T'(\kappa_2 - \kappa_1)} e^{-\kappa_2 b}\sin\kappa_2 x + \ldots, \qquad \text{...........................(22)}$$

where the terms which are insensible at a distance of half a wave-length or so from the origin are omitted. The exponential factors shew the attenuation due to diffusion; this is greater on the side of the capillary waves, since $\kappa_2 > \kappa_1$.

When the velocity c of the stream is less than the minimum wave-velocity, the factors of

$$kc^2 - g - k^2T'$$

are imaginary. There is now no indeterminateness caused by putting $\mu = 0$ *ab initio*. The surface-form is given by

$$y = -\frac{P}{\pi\rho}\int_0^\infty \frac{\cos kx}{k^2T' - kc^2 + g}\,dk. \qquad \text{.....................(23)}$$

The integral might be transformed by the previous method, but it is evident *à priori* that its value tends rapidly, with increasing x, to zero, on account of the more and more rapid fluctuations in sign of $\cos kx$. The disturbance of level is now confined to the neighbourhood of the origin. For $x = 0$ we find

$$y = -\frac{P}{(c_m{}^4 - c^4)^{\frac{1}{2}}\rho}\left(1 + \frac{2}{\pi}\sin^{-1}\frac{c^2}{c_m{}^2}\right). \qquad \text{..............(24)}$$

Finally we have the critical case where c is exactly equal to the minimum wave-velocity, and therefore $\kappa_2 = \kappa_1$. The first term in (16) or (17) is now infinite, whilst the remainder of the expression, when evaluated, is finite. To get an intelligible result in this case it is necessary to retain the frictional coefficient μ'.

If we put $\mu' = 2\varpi^2$, we have

$$(k-\kappa)^2 + i\mu' = \{k - (\kappa + \varpi - i\varpi)\}\{k - (\kappa - \varpi + i\varpi)\}, \qquad \text{..................(25)}$$

so that the integral (10) may now be equated to

$$\frac{1+i}{4\varpi}\left\{\int_0^\infty \frac{e^{ikx}}{k-(\kappa-\varpi+i\varpi)}\,dk - \int_0^\infty \frac{e^{ikx}}{k-(\kappa+\varpi-i\varpi)}\,dk\right\}. \qquad (26)$$

The formulae of Art. 243 shew that when ϖ is small the most important part of this expression, for points at a distance from the origin on either side, is

$$\frac{1+i}{4\varpi}\cdot 2\pi i e^{i\kappa x}. \qquad (27)$$

It appears that the surface-elevation is now given by

$$\frac{\pi T_1}{P}\cdot y = -\frac{\pi}{\mu'^{\frac{1}{2}}}\cos\left(\kappa x - \tfrac{1}{4}\pi\right). \qquad (28)$$

272. The investigation by Rayleigh*, from which the foregoing differs principally in the manner of treating the definite integrals, was undertaken with a view to explaining more fully some phenomena described by Scott Russell† and Kelvin‡.

"When a small obstacle, such as a fishing line, is moved forward slowly through still water, or (which of course comes to the same thing) is held stationary in moving water, the surface is covered with a beautiful wave-pattern, fixed relatively to the obstacle. On the up-stream side the wave-length is short, and, as Thomson has shewn, the force governing the vibrations is principally cohesion. On the down-stream side the waves are longer, and are governed principally by gravity. Both sets of waves move with the same velocity relatively to the water; namely, that required in order that they may maintain a fixed position relatively to the obstacle. The same condition governs the velocity, and therefore the wave-length, of those parts of the pattern where the fronts are oblique to the direction of motion. If the angle between this direction and the normal to the wave-front be called θ, the velocity of propagation of the waves must be equal to $v_0 \cos\theta$, where v_0 represents the velocity of the water relatively to the fixed obstacle.

"Thomson has shewn that, whatever the wave-length may be, the velocity of propagation of waves on the surface of water cannot be less than about 23 centimetres per second. The water must run somewhat faster than this in order that the wave-pattern may be formed. Even then the angle θ is subject to a limit defined by $v_0 \cos\theta = 23$, and the curved wave-front has a corresponding asymptote.

"The immersed portion of the obstacle disturbs the flow of the liquid independently of the deformation of the surface, and renders the problem in its original form one of great difficulty. We may however, without altering the essence of the matter, suppose that the disturbance is produced by the application to one point of the surface of a slightly abnormal pressure, such

* *l.c. ante* p. 399.

† "On Waves," *Brit. Ass. Rep.* 1844.

‡ *l.c. ante* p. 459.

as might be produced by electrical attraction, or by the impact of a small jet of air. Indeed, either of these methods—the latter especially—gives very beautiful wave-patterns*."

The character of the wave-pattern can be made out by the method explained near the end of Art. 256. If we take account of capillarity alone, the formula (19) of that Art. gives

$$c^2 \cos^2\theta = V^2 = \frac{2\pi T'}{\lambda}, \quad \text{.........................(1)}$$

by Art. 266, and the form of the wave-ridges is accordingly determined by the equation

$$p = a \sec^2\theta. \quad \text{.................................(2)†}$$

This leads to

$$x = a \sec\theta\,(1 - 2\tan^2\theta), \qquad y = 3a \sec\theta \tan\theta. \quad \text{...........(3)}$$

When gravity and capillarity are both regarded, we have, by Art. 267,

$$c^2 \cos^2\theta = V^2 = \frac{g\lambda}{2\pi} + \frac{2\pi T'}{\lambda}. \quad \text{.......................(4)}$$

Hence, if we put

$$c_m = (4gT')^{\frac{1}{4}}, \qquad b = 2\pi\left(\frac{T'}{g}\right)^{\frac{1}{2}}, \quad \text{....................(5)}$$

we have

$$\frac{\cos^2\theta}{\cos^2\alpha} = \tfrac{1}{2}\left(\frac{\lambda}{b} + \frac{b}{\lambda}\right), \quad \text{..........................(6)}$$

where

$$\cos\alpha = c_m/c. \quad \text{.................................(7)}$$

The relation between p and θ is therefore of the form

$$\frac{\cos^2\theta}{\cos^2\alpha} = \tfrac{1}{2}\left(\frac{p}{a\cos^2\alpha} + \frac{a\cos^2\alpha}{p}\right), \quad \text{....................(8)}$$

or

$$\frac{p}{a} = \cos^2\theta \pm \sqrt{(\cos^4\theta - \cos^4\alpha)}. \quad \text{....................(9)}$$

The four straight lines for which $\theta = \pm\alpha$ are asymptotes of the curve thus determined. The values of $\frac{1}{2}\pi - \alpha$ for several values of the ratio c/c_m have been given in Art. 267.

When the ratio c/c_m is at all considerable, α is nearly equal to $\frac{1}{2}\pi$, and the asymptotes make very acute angles with the axis of x. The upper figure on the following page gives the part of the curve which is relevant to the physical problem in the case of $c = 10\,c_m$‡. The ratio between the wave-lengths of the 'waves' and the 'ripples' in the line of symmetry is then, of course, very great. The curve should be compared with that which forms the basis of the figure on p. 434.

As the ratio c/c_m is diminished, the asymptotes open out, whilst the two cusps on either side of the axis approach one another, coincide, and finally

* Rayleigh, *l.c.*

† Since U is now $> V$, it appears from Art. 256 (20) that the constant a must be negative.

‡ The necessary calculations were made by Mr H. J. Woodall. The scale of the figure does not admit of the asymptotes being shewn distinct from the curve.

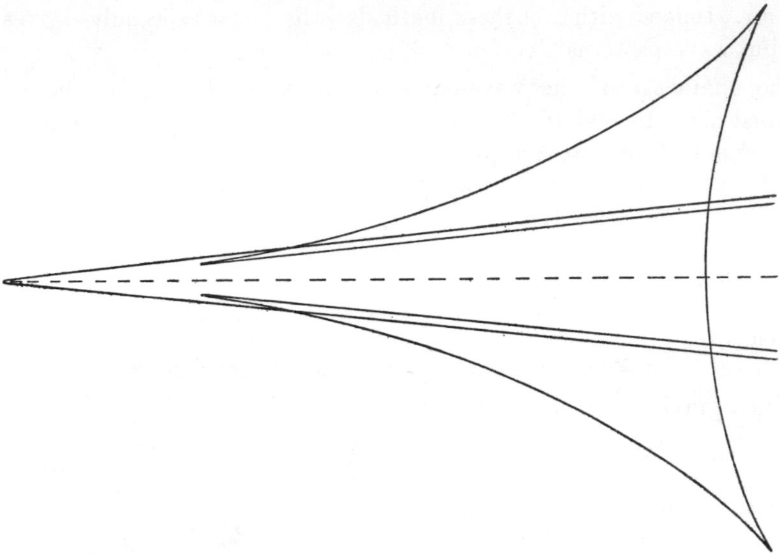

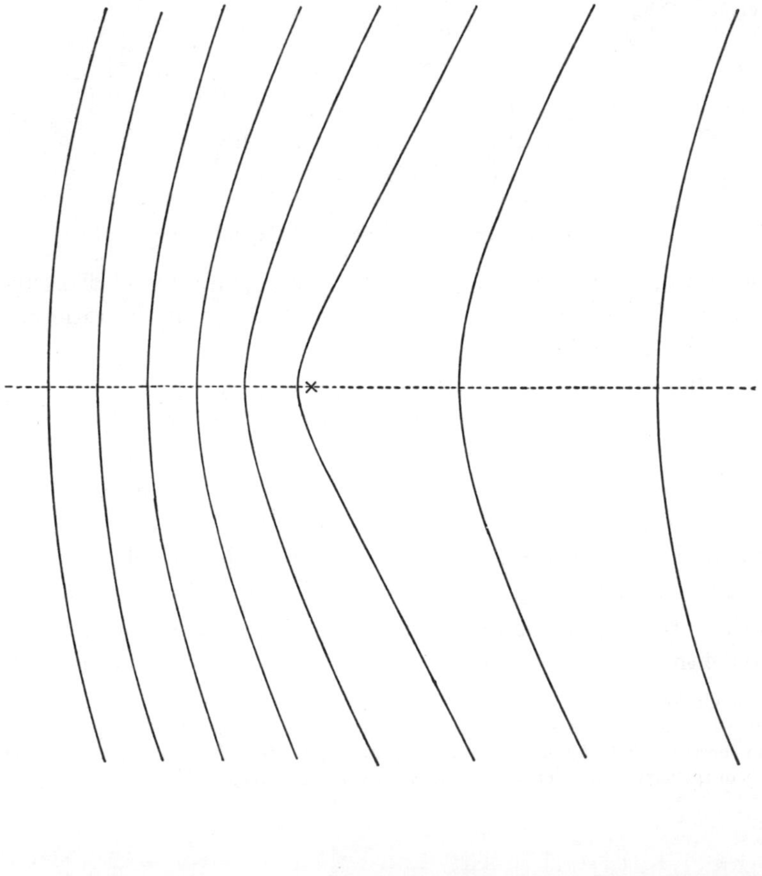

disappear*. The wave-system has then a configuration of the kind shewn in the lower diagram, which is drawn for the case where the ratio of the wave-lengths in the line of symmetry is 4 : 1. This corresponds to $\alpha = 26° 34'$, or $c = 1{\cdot}12 c_m$†. When $c < c_m$, the wave-pattern disappears.

273. Another problem of great interest is the determination of the nature of the equilibrium of a cylindrical column of liquid, of circular section. This contains the theory of the well-known experiments of Bidone, Savart, and others, on the behaviour of a jet issuing under pressure from a small orifice in the wall of a containing vessel. It is obvious that the uniform velocity in the direction of the axis of the jet does not affect the dynamics of the question, and may be disregarded in the analytical treatment.

We will take first the two-dimensional vibrations of the column, the motion being supposed to be the same in each section. Using polar co-ordinates r, θ in the plane of a section, with the origin in the axis, we may write, in accordance with Art. 63,

$$\phi = A\frac{r^s}{a^s}\cos s\theta \,.\, \cos(\sigma t + \epsilon), \qquad (1)$$

where a is the mean radius. The equation of the boundary at any instant will then be

$$r = a + \zeta, \qquad (2)$$

where

$$\zeta = -\frac{sA}{\sigma a}\cos s\theta \,.\, \sin(\sigma t + \epsilon), \qquad (3)$$

the relation between the coefficients being determined by

$$\frac{\partial \zeta}{\partial t} = -\frac{\partial \phi}{\partial r}, \qquad (4)$$

or $r = a$. For the variable part of the pressure inside the column, close to the surface, we have

$$\frac{p}{\rho} = \frac{\partial \phi}{\partial t} = -\sigma A \cos s\theta \,.\, \sin(\sigma t + \epsilon). \qquad (5)$$

The curvature of a curve which differs infinitely little from a circle having its centre at the origin is found by elementary methods to be

$$\frac{1}{R} = \frac{1}{r} - \frac{1}{r^2}\frac{d^2 r}{d\theta^2},$$

or, in the notation of (2),

$$\frac{1}{R} = \frac{1}{a} - \frac{1}{a^2}\left(\zeta + \frac{\partial^2 \zeta}{\partial \theta^2}\right). \qquad (6)$$

Hence the surface-condition

$$p = \frac{T_1}{R} + \text{const.} \qquad (7)$$

* A tentative diagram shewed that they were nearly coincident for $c = 2c_m$ ($\alpha = 60°$).

† The figure may be compared with the drawing, from observation, given by Scott Russell, *l.c.*

gives, on substitution from (5),

$$\sigma^2 = s(s^2-1)\frac{T_1}{\rho a^3}. \qquad\text{(8)*}$$

For $s=1$, we have $\sigma=0$; to our order of approximation the section remains circular, being merely displaced, so that the equilibrium is neutral. For all other integral values of s, σ^2 is positive, so that the equilibrium is thoroughly stable for two-dimensional deformations. This is evident *à priori*, since the circle is the form of least perimeter, and therefore least energy, for given sectional area.

In the case of a jet issuing from an orifice in the shape of an ellipse, an equilateral triangle, or a square, prominence is given to the disturbance of the type $s=2$, 3, or 4, respectively. The motion being steady, the jet exhibits a system of stationary waves, whose length is equal to the velocity of the jet multiplied by the period $(2\pi/\sigma)$†.

274. Abandoning now the restriction to two dimensions, we assume that

$$\phi = \phi_1 \cos kz \,.\, \cos(\sigma t+\epsilon), \qquad\text{(9)}$$

where the axis of z coincides with that of the cylinder, and ϕ_1 is a function of the remaining co-ordinates x, y. Substituting in the equation of continuity, $\nabla^2\phi=0$, we get

$$(\nabla_1^2 - k^2)\,\phi_1 = 0, \qquad\text{(10)}$$

where $\nabla_1^2 = \partial^2/\partial x^2 + \partial^2/\partial y^2$. If we put $x=r\cos\theta$, $y=r\sin\theta$, this may be written

$$\frac{\partial^2\phi_1}{\partial r^2} + \frac{1}{r}\frac{\partial\phi_1}{\partial r} + \frac{1}{r^2}\frac{\partial^2\phi_1}{\partial\theta^2} - k^2\phi_1 = 0. \qquad\text{(11)}$$

This equation is of the form considered in Arts. 101, 191, except for the sign of k^2; the solutions which are finite for $r=0$ are therefore of the type

$$\phi_1 = BI_s(kr)\left.\begin{matrix}\cos\\ \cdot\\ \sin\end{matrix}\right\} s\theta, \qquad\text{(12)}$$

where, as in Art. 210 (11),

$$I_s(z) = \frac{z^s}{2^s\,.\,s!}\left\{1 + \frac{z^2}{2(2s+2)} + \frac{z^4}{2\,.\,4(2s+2)(2s+4)} + \dots\right\}. \qquad\text{(13)}$$

Hence, writing

$$\phi = BI_s(kr)\cos s\theta \cos kz \,.\, \cos(\sigma t+\epsilon), \qquad\text{(14)}$$

we have, by (4),

$$\zeta = -B\frac{kaI_s'(ka)}{\sigma a}\cos s\theta\cos kz \,.\, \sin(\sigma t+\epsilon). \qquad\text{(15)}$$

To find the sum of the principal curvatures, we remark that, as an obvious consequence of Euler's and Meunier's theorems on curvature of surfaces, the

* For the original investigation, by the method of energy, see Rayleigh, "On the Instability of Jets," *Proc. Lond. Math. Soc.* x. 4 (1878), and "On the Capillary Phenomena of Jets," *Proc. Roy. Soc.* xxix. 71 (1879) [*Papers*, i. 361, 377; *Theory of Sound*, 2nd ed. c. xx.]. The latter paper contains a comparison of the theory with experiment.

† It is assumed that this wave-length is large compared with the circumference of the jet. Otherwise, the formula (18) must be employed, with $\sigma=kc$, where c is the velocity of the jet.

curvature of any section differing infinitely little from a principal normal section is, to the first order of small quantities, the same as that of the principal section itself. It is sufficient therefore in the present problem to calculate the curvatures of a transverse section of the cylinder, and of a section through the axis. These are the principal sections in the undisturbed state, and the principal sections of the deformed surface will make infinitely small angles with them. For the transverse section the formula (6) applies, whilst for the axial section the curvature is $-\partial^2\zeta/\partial z^2$; so that the required sum of the principal curvature is

$$\frac{1}{R_1}+\frac{1}{R_2}=\frac{1}{a}-\frac{1}{a^2}\left(\zeta+\frac{\partial^2\zeta}{\partial\theta^2}\right)-\frac{\partial^2\zeta}{\partial z^2}$$

$$=\frac{1}{a}-B\frac{kaI_s'(ka)}{\sigma a^3}(k^2a^2+s^2-1)\cos s\theta\cos kz\,.\sin(\sigma t+\epsilon). \quad\ldots\ldots(16)$$

Also, at the surface,

$$\frac{p}{\rho}=\frac{\partial\phi}{\partial t}=-\sigma BI_s(ka)\cos s\theta\cos kz\,.\sin(\sigma t+\epsilon). \quad\ldots\ldots\ldots(17)$$

The surface-condition of Art. 265 then gives

$$\sigma^2-\frac{kaI_s'(ka)}{I_s(ka)}(k^2a^2+s^2-1)\,.\,\frac{T_1}{\rho a^3}. \quad\ldots\ldots\ldots\ldots\ldots\ldots(18)$$

For $s>0$, σ^2 is positive; but in the case $(s=0)$ of symmetry about the axis σ^2 will be negative if $ka<1$; that is, the equilibrium is unstable for disturbances whose wave-length $(2\pi/k)$ exceeds the circumference of the jet. To ascertain the type of disturbance for which the instability is greatest, we require to know the value of ka which makes

$$\frac{kaI_0'(ka)}{I_0(ka)}(1-k^2a^2)$$

a maximum. For this Rayleigh finds $k^2a^2=\cdot 4858$, whence, for the wave-length of maximum instability,

$$2\pi/k=4{\cdot}508\times 2a.$$

There is a tendency therefore to the production of bead-like swellings and contractions, of this wave-length, with continually increasing amplitude, until finally the jet breaks up into detached drops*.

275. This leads naturally to the discussion of the small oscillations of a drop of liquid about the spherical form†. We will slightly generalize the

* The argument here is that if we have a series of possible types of disturbance, with time-factors $e^{\alpha_1 t}$, $e^{\alpha_2 t}$, $e^{\alpha_3 t}$, ..., where $\alpha_1>\alpha_2>\alpha_3>\ldots$, and if these be excited simultaneously, the amplitude of the first will increase relatively to those of the other components in the ratios $e^{(\alpha_1-\alpha_2)t}$, $e^{(\alpha_1-\alpha_3)t}$, ... The component with the greatest α will therefore ultimately predominate.

The instability of a cylindrical jet surrounded by other fluid has been discussed by Rayleigh, "On the Instability of Cylindrical Fluid Surfaces," *Phil. Mag.* (5) xxxiv. 177 (1892) [*Papers*, iii. 594]. For a jet of air in water the wave-length of maximum instability is found to be $6{\cdot}48\times 2a$.

† Rayleigh, *l.c.*; Webb, *Mess. of Math.* ix. 177 (1880).

question by supposing that we have a sphere of liquid, of density ρ, surrounded by an infinite mass of other liquid of density ρ'.

Taking the origin at the centre, let the shape of the common surface at any instant be given by

$$r = a + \zeta = a + S_n \,.\, \sin(\sigma t + \epsilon), \qquad \text{.....................(1)}$$

where a is the mean radius, and S_n is a surface-harmonic of order n. The corresponding values of the velocity-potential will be, at internal points,

$$\phi = -\frac{\sigma a}{n} \frac{r^n}{a^n} S_n \,.\, \cos(\sigma t + \epsilon), \qquad \text{.....................(2)}$$

and, at external points,

$$\phi' = \frac{\sigma a}{n+1} \frac{a^{n+1}}{r^{n+1}} S_n \,.\, \cos(\sigma t + \epsilon), \qquad \text{.....................(3)}$$

since these make
$$\frac{\partial \zeta}{\partial t} = -\frac{\partial \phi}{\partial r} = -\frac{\partial \phi'}{\partial r},$$

for $r = a$. The variable parts of the internal and external pressures at the surface are then given by

$$p = \ldots + \frac{\rho \sigma^2 a}{n} S_n \,.\, \sin(\sigma t + \epsilon), \quad p' = \ldots - \frac{\rho' \sigma^2 a}{n+1} S_n \,.\, \sin(\sigma t + \epsilon). \quad \text{...(4)}$$

To find the sum of the curvatures we make use of the theorem of Solid Geometry that if λ, μ, ν be the direction-cosines of the normal at (x, y, z) to that surface of the family

$$F(x, y, z) = \text{const.}$$

which passes through the point, viz.

$$\lambda, \mu, \nu = \frac{F_x, F_y, F_z}{\sqrt{(F_x^2 + F_y^2 + F_z^2)}},$$

then
$$\frac{1}{R_1} + \frac{1}{R_2} = \frac{\partial \lambda}{\partial x} + \frac{\partial \mu}{\partial y} + \frac{\partial \nu}{\partial z}. \qquad \text{.....................(5)}$$

Since the square of ζ is to be neglected, the equation (1) of the harmonic spheroid may also be written

$$r = a + \zeta_n, \qquad \text{.....................(6)}$$

where
$$\zeta_n = \frac{r^n}{a^n} S_n \,.\, \sin(\sigma t + \epsilon), \qquad \text{.....................(7)}$$

i.e. ζ_n is a *solid* harmonic of degree n. We thus find

$$\left.\begin{aligned} \lambda &= \frac{x}{r} - \frac{\partial \zeta_n}{\partial x} + n \frac{x}{r^2} \zeta_n, \\ \mu &= \frac{y}{r} - \frac{\partial \zeta_n}{\partial y} + n \frac{y}{r^2} \zeta_n, \\ \nu &= \frac{z}{r} - \frac{\partial \zeta_n}{\partial z} + n \frac{z}{r^2} \zeta_n, \end{aligned}\right\} \qquad \text{.....................(8)}$$

whence

$$\frac{1}{R_1}+\frac{1}{R_2}=\frac{2}{r}+\frac{n(n+1)}{r^2}\zeta_n=\frac{2}{a}+\frac{(n-1)(n+2)}{a^2}S_n\,.\,\sin(\sigma t+\epsilon).\ \ldots(9)$$

Substituting from (4) and (9) in the general surface-condition of Art. 265, we find

$$\sigma^2=n(n+1)(n-1)(n+2)\frac{T_1}{\{(n+1)\rho+n\rho'\}\,a^3}.\ \ \ldots\ldots\ldots(10)$$

If we put $\rho'=0$, this gives

$$\sigma^2=n(n-1)(n+2)\frac{T_1}{\rho a^3}.\ \ \ldots\ldots\ldots\ldots\ldots\ldots\ldots\ldots(11)$$

The most important mode of vibration is that for which $n=2$; we then have

$$\sigma^2=\frac{8T_1}{\rho a^3}.$$

Hence for a drop of water, putting $T_1=74$, $\rho=1$, we find, for the frequency,

$$\sigma/2\pi=3{\cdot}87a^{-\frac{3}{2}}\text{ vibrations per second,}$$

if a be the radius in centimetres. The radius of the sphere which would vibrate seconds is $a=2{\cdot}47$ cm. or a little less than an inch.

The case of a spherical bubble of air, surrounded by liquid, is obtained by putting $\rho=0$ in (10), viz. we have

$$\sigma^2=(n+1)(n-1)(n+2)\frac{T_1}{\rho' a^3}.\ \ \ldots\ldots\ldots\ldots\ldots\ldots(12)$$

For the same density of the liquid, the frequency of any given mode is greater than in the case represented by (11), on account of the diminished inertia: cf. Art. 91 (7), (8).

CHAPTER X

WAVES OF EXPANSION

276. A TREATISE on Hydrodynamics would hardly be complete without some reference to this subject, if merely for the reason that all actual fluids are more or less compressible, and that it is only when we recognize this compressibility that we escape such apparently paradoxical results as that of Art. 20, where a change of pressure was found to be propagated *instantaneously* through a liquid mass.

We shall accordingly investigate in this Chapter the general laws of propagation of small disturbances, passing over, however, for the most part, such details as belong more properly to the Theory of Sound.

In most cases which we shall consider, the changes of pressure are small, and may be taken to be proportional to the changes in density, thus

$$\Delta p = \kappa \,.\, \frac{\Delta \rho}{\rho},$$

where $\kappa \,(= \rho\, dp/d\rho)$ is a certain coefficient, called the 'elasticity of volume.' For a given liquid the value of κ varies with the temperature, and (very slightly) with the pressure. For water at 15° C., $\kappa = 2{\cdot}045 \times 10^{10}$ dynes per square centimetre. The case of gases will be considered presently.

Plane Waves.

277. We take first the case of plane waves in a uniform medium.

The motion being in one dimension (x), the dynamical equation is, in the absence of extraneous forces,

$$\frac{\partial u}{\partial t} + u\frac{\partial u}{\partial x} = -\frac{1}{\rho}\frac{\partial p}{\partial x} = -\frac{1}{\rho}\frac{dp}{d\rho}\frac{\partial \rho}{\partial x}, \quad \text{.....................(1)}$$

whilst the equation of continuity, Art. 7 (5), reduces to

$$\frac{\partial \rho}{\partial t} + \frac{\partial}{\partial x}(\rho u) = 0. \quad \text{..........................(2)}$$

If we put
$$\rho = \rho_0(1 + s), \quad \text{...............................(3)}$$

where ρ_0 is the density in the undisturbed state, s may be called the 'condensation' in the plane x. Substituting in (1) and (2), we find, on the supposition that the motion is infinitely small,

$$\frac{\partial u}{\partial t} = -\frac{\kappa}{\rho_0}\frac{\partial s}{\partial x}, \quad \text{...............................(4)}$$

and
$$\frac{\partial s}{\partial t} = -\frac{\partial u}{\partial x}, \quad \text{.................................(5)}$$

if
$$\kappa = \left[\rho \frac{dp}{d\rho}\right]_{\rho=\rho_0}, \quad \text{.........(6)}$$
as above. Eliminating s we have
$$\frac{\partial^2 u}{\partial t^2} = c^2 \frac{\partial^2 u}{\partial x^2}, \quad \text{.........(7)}$$
where
$$c^2 = \frac{\kappa}{\rho_0} = \left[\frac{dp}{d\rho}\right]_{\rho=\rho_0}. \quad \text{.........(8)}$$

The equation (7) is of the form treated in Art. 170, and the complete solution is
$$u = f(ct - x) + F(ct + x), \quad \text{.........(9)}$$
representing two systems of waves travelling with the constant velocity c, one in the positive and the other in the negative direction of x. It appears from (5) that the corresponding value of s is given by
$$cs = f(ct - x) - F(ct + x). \quad \text{.........(10)}$$

For a single wave we have $u = \pm cs$, ...(11)

since one or other of the functions f, F is zero. The upper or the lower sign is to be taken according as the wave is travelling in the positive or the negative direction. It is easily shewn in this case that the approximations involved in (4) and (5) are valid provided u is everywhere small compared with c.

There is an exact correspondence between the above approximate theory and that of 'long' gravity-waves on water. If we write η/h for s, and gh for κ/ρ_0, the equations (4) and (5), above, become identical with Art. 169 (3), (5).

278. With the value of κ given in Art. 276, we find for water at 15° C.
$$c = 1430 \text{ metres per second.}$$
The number obtained directly by Colladon and Sturm* in their experiments on the lake of Geneva was 1437, at a temperature of 8° C.†

In the case of a gas, if we assume that the temperature is constant, the value of κ is determined by Boyle's Law
$$p/p_0 = \rho/\rho_0, \quad \text{.........(1)}$$
viz.
$$\kappa = p_0, \quad \text{.........(2)}$$
so that
$$c = \sqrt{(p_0/\rho_0)}. \quad \text{.........(3)}$$

This is known as the 'Newtonian' velocity of sound‡. If we denote by H the height of a 'homogeneous atmosphere' of the gas, we have $p_0 = g\rho_0 H$, and therefore
$$c = (gH)^{\frac{1}{2}}, \quad \text{.........(4)}$$

* *Ann. de Chim. et de Phys.* xxxvi. (1827). It may be mentioned that the velocity of sound in water contained in a *tube* is liable to be appreciably diminished by the yielding of the wall. See Helmholtz, *Fortschritte d. Physik*, iv. 119 (1848) [*Wiss. Abh.* i. 242]; Korteweg, *Wied. Ann.* v. 526 (1878); Lamb, *Manch. Mem.* xlii. No. 1 (1898).

† Recent experiments in sea-water give a velocity of 4956 ft. per sec. at a temperature of 17° C., with an increment of about 11 ft. per sec. for each degree centigrade (Wood and others, *Proc. Roy. Soc.* A, ciii. 284 (1923)).

‡ *Principia*, Lib. ii. Sect. viii. Prop. 48.

which may be compared with the formula (13) of Art. 170 for the velocity of 'long' gravity-waves in liquids. For air at 0° C. we have as corresponding values

$$p_0 = 76 \times 13{\cdot}60 \times 981, \quad \rho_0 = {\cdot}00129,$$

in absolute C.G.S. units; whence

$$c = 280 \text{ metres per second.}$$

This is considerably below the value found by direct observation.

The reconciliation of theory and fact is due to Laplace*. When a gas is suddenly compressed, its temperature rises, so that the pressure is increased more than in proportion to the diminution of volume; and a similar statement applies of course to the case of a sudden expansion. The formula (1) is appropriate only to the case where the expansions and rarefactions are so gradual that there is ample time for equalization of temperature by thermal conduction and radiation. In most cases of interest, the alternations of density are exceedingly rapid; the flow of heat from one element to another has hardly set in before its direction is reversed, so that practically each element behaves as if it neither gained nor lost heat.

On this view we have, in place of (1), the 'adiabatic' law

$$p/p_0 = (\rho/\rho_0)^\gamma, \qquad \text{.............................(5)}$$

where γ is the ratio of the two specific heats of the gas. This makes

$$\kappa = \gamma p_0, \qquad \text{.................................(6)}$$

and therefore

$$c = \sqrt{(\gamma p_0/\rho_0)} = \sqrt{(\gamma g H)}. \qquad \text{......................(7)}$$

If we put $\gamma = 1{\cdot}402$†, the former result is to be multiplied by 1·184, whence

$$c = 332 \text{ metres per second,}$$

which agrees very closely with the best direct determinations.

The confidence felt by physicists in the soundness of Laplace's view is so complete that it is now usual to apply the formula (7) in the inverse manner, and to infer the values of γ for various gases and vapours from observation of wave-velocities in them.

In strictness, a similar distinction should be made between the 'adiabatic' and 'isothermal' coefficients of elasticity of a liquid or a solid, but practically the difference is unimportant. Thus in the case of water the ratio of the two volume-elasticities is calculated to be 1·0012‡.

The effects of thermal radiation and conduction on air-waves have been studied theoretically by Stokes§ and Rayleigh‖. When the oscillations are too rapid for complete

* The usual reference is to a paper "Sur la vitesse du son dans l'air et dans l'eau," *Ann. de Chim. et de Phys.* iii. 238 (1816) [*Mécanique Céleste*, Livre 12me, c. iii. (1823)]. But Poisson in a memoir of date 1807 (quoted below on p. 484) refers to this explanation as having been already given by Laplace.

† The value found by the most recent direct experiments.

‡ Everett, *Units and Physical Constants*.

§ "An Examination of the possible effect of the Radiation of Heat on the Propagation of Sound," *Phil. Mag.* (4) i. 305 (1851) [*Papers*, iii. 142].

‖ *Theory of Sound*, Art. 247. See *infra* Art. 360. In a paper "On the Cooling of Air by Radiation and Conduction, and on the Propagation of Sound," *Phil. Mag.* (5) xlvii. 308 (1899) [*Papers*, iv. 376], Rayleigh concludes on experimental grounds that conduction is much more effective in this respect than radiation.

equalization of temperature, but not so rapid as to exclude communication of heat between adjacent elements, the waves diminish in amplitude as they advance, owing to the dissipation of energy which takes place in the thermal processes. The effect of conduction will be noticed, along with that of viscosity, in the next Chapter.

According to the law of Charles and Dalton

$$p = R\rho\theta, \quad \text{.................................(8)}$$

where θ is the absolute temperature, and R is a constant depending on the nature of the gas. The velocity of sound will therefore vary as the square root of θ. For several of the more permanent gases, which have sensibly the same value of γ, the formula (7) shews that the velocity varies inversely as the square root of the density, provided the relative densities be determined under the same conditions of pressure and temperature.

279. The theory of plane waves can also be treated very simply by the Lagrangian method (Arts. 13, 14).

If ξ denote the displacement at time t of the particles whose undisturbed abscissa is x, the stratum of matter originally included between the planes x and $x + \delta x$ is at the time $t + \delta t$ bounded by the planes

$$x + \xi \text{ and } x + \xi + \left(1 + \frac{\partial \xi}{\partial x}\right) \delta x,$$

so that the equation of continuity is

$$\rho \left(1 + \frac{\partial \xi}{\partial x}\right) = \rho_0, \quad \text{............................(1)}$$

where ρ_0 is the density in the undisturbed state. Hence if s denote the 'condensation' $(\rho - \rho_0)/\rho_0$, we have

$$s = -\frac{\partial \xi}{\partial x} \Big/ \left(1 + \frac{\partial \xi}{\partial x}\right). \quad \text{.........................(2)}$$

The dynamical equation, obtained by considering the forces acting on unit area of the above stratum, is

$$\rho_0 \frac{\partial^2 \xi}{\partial t^2} = -\frac{\partial p}{\partial x}. \quad \text{...............................(3)}$$

These equations are exact, but in the case of *small* motions we may write

$$p = p_0 + \kappa s, \quad \text{.................................(4)}$$

and

$$s = -\frac{\partial \xi}{\partial x}. \quad \text{.................................(5)}$$

Substituting in (3) we find

$$\frac{\partial^2 \xi}{\partial t^2} = c^2 \frac{\partial^2 \xi}{\partial x^2}, \quad \text{...............................(6)}$$

where $c^2 = \kappa/\rho_0$. The solution of (6) is the same as in Arts. 170, 277.

280. The *kinetic energy* of a system of plane waves is given by

$$T = \tfrac{1}{2}\rho_0 \iiint u^2 \, dx\, dy\, dz, \quad \text{.........................(1)}$$

where u is the velocity at the point (x, y, z) at time t.

The calculation of the *intrinsic energy* requires a little care. The work done by unit mass in expanding through a small range, from the actual volume v to the standard volume v_0, is given to the second order of small quantities by the expression

$$\tfrac{1}{2}(p+p_0)(v_0-v),$$

as is obvious on inspection from Watt's diagram. Putting

$$p=p_0+\kappa s, \quad v_0-v=sv_0, \quad \text{.........(2)}$$

we have

$$\tfrac{1}{2}(p+p_0)(v_0-v)=p_0(v_0-v)+\tfrac{1}{2}(p-p_0)(v_0-v)$$
$$=p_0(v_0-v)+\tfrac{1}{2}\kappa s^2 v_0. \quad \text{......(3)}$$

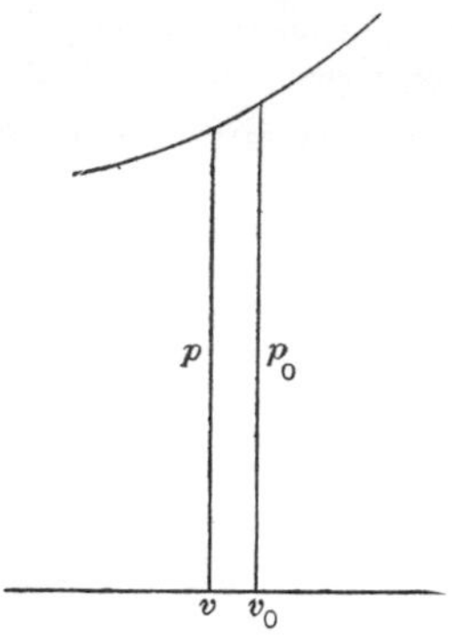

If we take the sum of the corresponding expressions for all the mass-elements of the system, the term $p_0(v_0-v)$ will disappear whenever the conditions are such that the total change of volume is zero. This being assumed, we have, for the work done by the gas contained in any given region, in passing from its actual state to the normal state, the expression

$$W=\tfrac{1}{2}\kappa\iiint s^2\,dx\,dy\,dz. \quad \text{.........................(4)}$$

So far, no assumption has been made as to the precise manner in which the transition takes place; this will affect the value of κ. It is only in the case of *adiabatic* expansion that the expression (4) can be identified with the 'intrinsic energy' in the strict sense of the term. When the expansion is *isothermal*, the expression gives what is known in Thermodynamics as the 'free energy.'

In a progressive plane wave we have $cs=\pm u$, and therefore $T=W$. The equality of the two kinds of energy, in this case, may also be inferred from the more general line of argument given in Art. 174.

In the Theory of Sound special interest attaches, of course, to the case of simple-harmonic vibrations. If a be the amplitude of a progressive wave of period $2\pi/\sigma$, we may assume, in conformity with Art. 279 (6),

$$\xi=a\cos(kx-\sigma t+\epsilon), \quad \text{.........................(5)}$$

where $k=\sigma/c$, and the wave-length is accordingly $\lambda=2\pi/k$. The formulae (1) and (4) then give, for the energy contained in a prismatic space of sectional area unity and length λ (in the direction x),

$$T+W=\tfrac{1}{2}\rho_0\sigma^2a^2\lambda, \quad \text{.........................(6)}$$

the same as the kinetic energy of the whole mass when animated with the maximum velocity σa.

The rate of transmission of energy across unit area of a plane moving with the particles situate in it is

$$p\frac{\partial\xi}{\partial t}=p\sigma a\sin(kx-\sigma t+\epsilon). \quad \text{..................................(7)}$$

The work done by the constant part of the pressure in a complete period is zero. For the variable part we have

$$\Delta p = \kappa s = -\kappa \frac{\partial \xi}{\partial x} = \kappa ka \sin(kx - \sigma t + \epsilon). \qquad (8)$$

Substituting in (7), we find, for the mean rate of transmission of energy,

$$\tfrac{1}{2}\kappa\sigma k a^2 = \tfrac{1}{2}\rho_0 \sigma^2 a^2 \times c. \qquad (9)$$

Hence the energy transmitted in any number of complete periods is exactly that corresponding to the waves which pass the plane in the same time, as we should expect, since, c being independent of λ, the group-velocity is identical with the wave-velocity (cf. Art. 237).

Waves of Finite Amplitude.

281. If p be a function of ρ only, the equations (1) and (3) of Art. 279 give, without approximation,

$$\frac{\partial^2 \xi}{\partial t^2} = \frac{\rho^2}{\rho_0^2}\frac{dp}{d\rho} \cdot \frac{\partial^2 \xi}{\partial x^2}. \qquad (1)$$

On the 'isothermal' hypothesis that

$$p/p_0 = \rho/\rho_0, \qquad (2)$$

this becomes

$$\frac{\partial^2 \xi}{\partial t^2} = \frac{p_0}{\rho_0} \frac{\dfrac{\partial^2 \xi}{\partial x^2}}{\left(1 + \dfrac{\partial \xi}{\partial x}\right)^2}. \qquad (3)$$

In the same way, the 'adiabatic' relation

$$p/p_0 = (\rho/\rho_0)^\gamma \qquad (4)$$

leads to

$$\frac{\partial^2 \xi}{\partial t^2} = \frac{\gamma p_0}{\rho_0} \frac{\dfrac{\partial^2 \xi}{\partial x^2}}{\left(1 + \dfrac{\partial \xi}{\partial x}\right)^{\gamma+1}}. \qquad (5)$$

These exact equations (3) and (5) may be compared with the similar equation for 'long' waves in a uniform canal, Art. 173 (3).

It appears from (1) that the equation (6) of Art. 279 could be regarded as exact if the relation between p and ρ were such that

$$\rho^2 \frac{dp}{d\rho} = \rho_0^2 c^2. \qquad (6)$$

Hence plane waves of finite amplitude can be propagated without change of type if, and only if,

$$p - p_0 = \rho_0 c^2 \left(1 - \frac{\rho_0}{\rho}\right). \qquad (7)$$

A relation of this form does not hold for any known substance, whether at constant temperature or when free from gain or loss of heat by conduction and radiation*. Hence sound-waves of finite amplitude must inevitably undergo a change of type as they proceed.

282. The laws of propagation of waves of finite amplitude, on the above assumption that p is a definite function of ρ, have been investigated independently by Earnshaw and Riemann. It is proposed to give here a brief

* The relation would make p negative when ρ falls below a certain value.

account of the results, referring for further details to the original papers, and to the very full discussion of the matter by Rayleigh *.

The Eulerian equations (1) and (2) of Art. 277 may be written

$$\frac{\partial u}{\partial t} + u\frac{\partial u}{\partial x} = -\frac{\partial \varpi}{\partial x}, \quad \frac{\partial \varpi}{\partial t} + u\frac{\partial \varpi}{\partial x} = -c^2\frac{\partial u}{\partial x}, \quad \text{..............(1)}$$

where

$$\varpi = \int_{\rho_0}^{\rho}\frac{dp}{\rho}, \quad c = \sqrt{\left(\frac{dp}{d\rho}\right)}. \quad \text{....................(2)}$$

The quantity c is the wave-velocity for small amplitudes; it is in general a function of ρ, and therefore variable. If we now write

$$d\varpi = c\,d\omega, \quad \text{or} \quad \omega = \int_{\rho_0}^{\rho}\left(\frac{dp}{d\rho}\right)^{\frac{1}{2}}\frac{d\rho}{\rho}, \quad \text{....................(3)}$$

the equations (1) become

$$\frac{\partial u}{\partial t} + u\frac{\partial u}{\partial x} = -c\frac{\partial \omega}{\partial x}, \quad \frac{\partial \omega}{\partial t} + u\frac{\partial \omega}{\partial x} = -c\frac{\partial u}{\partial x}. \quad \text{..............(4)}$$

Hence by addition and subtraction

$$\left\{\frac{\partial}{\partial t} + (u+c)\frac{\partial}{\partial x}\right\}(\omega + u) = 0, \quad \text{......................(5)}$$

and

$$\left\{\frac{\partial}{\partial t} + (u-c)\frac{\partial}{\partial x}\right\}(\omega - u) = 0. \quad \text{......................(6)}$$

Hence $\omega + u$ is constant for a geometrical point moving with the velocity

$$\left(\frac{dp}{d\rho}\right)^{\frac{1}{2}} + u, \quad \text{..............................(7)}$$

whilst $\omega - u$ is constant for a point whose velocity is

$$-\left(\frac{dp}{d\rho}\right)^{\frac{1}{2}} + u. \quad \text{..............................(8)}$$

Hence, any given value of $\omega + u$ moves forward, and any value of $\omega - u$ moves backward, with the velocity given by (7) or (8), as the case may be.

These are Riemann's results †. They enable us to understand, in a general way, the nature of the motion in any given case. Thus if the initial disturbance be confined to the space between the two planes $x = a$, $x = b$, we may suppose that ω and u both vanish for $x < a$ and for $x > b$. The region within which $\omega + u$ is variable will advance, and that within which $\omega - u$ is variable will recede, until after a time these regions separate and leave between them a space for which $\omega = 0$, $u = 0$, and in which the fluid is therefore at rest and of the normal density ρ_0. The original disturbance has thus been split up into two progressive waves travelling in opposite directions. In the advancing wave we have $\omega = u$, so that both the density and the particle-velocity are

* "Aerial Plane Waves of Finite Amplitude," *Proc. Roy. Soc.* A, lxxxiv. 247 (1910) [*Papers*, v. 573]. See also *Theory of Sound*, c. xi.

† "Ueber die Fortpflanzung ebener Luftwellen von endlicher Schwingungsweite," *Gött. Abh.* viii. 43 (1858–9) [*Werke*, 2te Aufl., Leipzig, 1892, p. 157].

propagated forwards at the rate given by (7). Whether we adopt the isothermal or the adiabatic law of expansion, this velocity of propagation will be greater, the greater the value of ρ. The law of progress of the wave may be illustrated by drawing a curve with x as abscissa and ρ as ordinate, and making each point of this curve move forward with the appropriate velocity, as given by (7). Since those parts move faster which have the greater ordinates, the curve will eventually become at some point perpendicular to x. The quantities $\partial u/\partial x$, $\partial \rho/\partial x$ are then infinite; and the preceding method fails to yield any information as to the subsequent course of the motion. Cf. Art. 187.

283. Similar results can be deduced from Earnshaw's investigation*, which is, however, somewhat less general in that it applies only to a progressive wave supposed already established.

Assuming for definiteness the adiabatic relation between p and ρ, and writing $y=x+\xi$, so that y denotes the absolute co-ordinate at time t of the particle whose undisturbed position was x, we have from Art. 281 (5)

$$\frac{\partial^2 y}{\partial t^2}=c_0^2\frac{\partial^2 y}{\partial x^2}\Big/\left(\frac{\partial y}{\partial x}\right)^{\gamma+1}, \qquad \text{.......(1)}$$

where $c_0^2=\gamma p_0/\rho_0$. This is satisfied by

$$\frac{\partial y}{\partial t}=f\left(\frac{\partial y}{\partial x}\right), \qquad \text{.......(2)}$$

provided

$$f'\left(\frac{\partial y}{\partial x}\right)=\pm c_0\Big/\left(\frac{\partial y}{\partial x}\right)^{\frac{\gamma+1}{2}}. \qquad \text{.......(3)}$$

Hence a first integral of (1) is

$$\frac{\partial y}{\partial t}=C\mp\frac{2c_0}{\gamma-1}\Big/\left(\frac{\partial y}{\partial x}\right)^{\frac{\gamma-1}{2}}. \qquad \text{.......(4)}$$

Determining C so that $\partial y/\partial t=0$ at the confines of the wave, where $\partial y/\partial x=1$, we have, since $\partial y/\partial x=\rho_0/\rho$,

$$u=\frac{\partial y}{\partial t}=\mp\frac{2c_0}{\gamma-1}\left\{\left(\frac{\rho}{\rho_0}\right)^{\frac{\gamma-1}{2}}-1\right\}. \qquad \text{.......(5)}$$

To find the rate at which any particular value of u is propagated, we note that the value of u which holds for the particle x at time t will be transmitted to the particle $x+\delta x$ at time $t+\delta t$, provided

$$\frac{\partial^2 y}{\partial t^2}\delta t+\frac{\partial^2 y}{\partial x\,\partial t}\delta x=0, \qquad \text{.......(6)}$$

whence, from (2) and (3),

$$\delta x\pm c_0\left(\frac{\rho}{\rho_0}\right)^{\frac{\gamma+1}{2}}\delta t=0. \qquad \text{.......(7)}$$

The values of u and ρ are therefore propagated *from particle to particle* at the rate

$$\mp c_0\left(\frac{\rho}{\rho_0}\right)^{\frac{\gamma+1}{2}}.$$

To deduce the rate of propagation *in space* we have

$$\delta y=\frac{\partial y}{\partial x}\delta x+\frac{\partial y}{\partial t}\delta t=\left\{\mp c_0\left(\frac{\rho}{\rho_0}\right)^{\frac{\gamma-1}{2}}+u\right\}\delta t. \qquad \text{.......(8)}$$

* "On the Mathematical Theory of Sound," *Phil. Trans.* cl. 133 (1858).

The lower sign relates to waves travelling in the direction of x positive; the upper to that of x negative. The speed of propagation is greater the greater the value of ρ, as was found also from Riemann's investigation. It follows from (8) that in a positive wave the relation between u and y is of the form

$$u = F\left\{y - \left(c_0 + \frac{\gamma+1}{2} u\right) t\right\}, \qquad (9)$$

which is Rayleigh's generalization of a formula obtained by Poisson* in 1807 on the isothermal hypothesis ($\gamma=1$). That Poisson's formula involves a change in the type of a wave as it proceeds was pointed out by Stokes†. It is to be noted that if we make $\gamma \to 1$ in (5) we get

$$u = \pm c_0 \log\left(\frac{\rho_0}{\rho}\right), \quad \text{or} \quad \rho = \rho_0 e^{\mp u/c}. \qquad (10)‡$$

284. The conditions for a wave of permanent type have been investigated in a very simple manner by Rankine§.

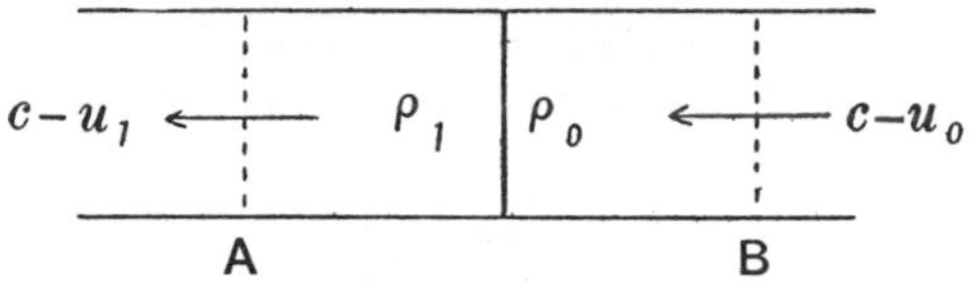

Let A, B be two points of an ideal tube of unit section drawn in the direction of propagation, which is (say) that of x positive, and let the values of the pressure, density, and particle-velocity at A and B be denoted by p_1, ρ_1, u_1 and p_0, ρ_0, u_0, respectively.

If, as in Art. 175, we impress on everything a velocity c equal and opposite to that of the wave, we reduce the problem to one of steady motion. Since the same amount of matter now crosses in unit time each section of the tube, we have

$$\rho_1(c - u_1) = \rho_0(c - u_0) = m, \qquad (1)$$

say, where m denotes the mass swept past in unit time by a plane moving with the wave, in the original form of the problem. This quantity m is called by Rankine the 'mass-velocity' of the wave.

Again, the total force acting on the mass included between A and B at any instant is $p_0 - p_1$, in the direction BA, and the rate at which this mass is gaining momentum in the same direction is

$$m(c - u_1) - m(c - u_0).$$

Hence

$$p_0 - p_1 = m(u_0 - u_1). \qquad (2)$$

Combined with (1) this gives

$$p_1 + \frac{m^2}{\rho_1} = p_0 + \frac{m^2}{\rho_0}. \qquad (3)$$

* Mémoire sur la théorie du son," *Journ. de l'École Polytechn.* vii. 367.

† "On a Difficulty in the Theory of Sound," *Phil. Mag.* (3) xxiii. 349 (1848) [*Papers*, ii. 51].

‡ This result, together with an analogous one for 'long' waves in water, seems to have been first noticed by De Morgan. See Airy, *Phil. Mag.* (3) xxxiv. 401 (1849).

§ "On the Thermodynamic Theory of Waves of Finite Longitudinal Disturbance," *Phil. Trans.* clx. 277 (1870) [*Papers*, p. 530].

Hence a wave of finite amplitude could not be propagated unchanged except in a medium such that

$$p+\frac{m^2}{\rho}=\text{const.} \quad \text{or} \quad p+m^2v=\text{const.}, \qquad \text{..............(4)}$$

if v be the volume of unit mass. This conclusion has already been arrived at, in a different manner, in Art. 281. It may be noticed that the relation (4) is represented on Watt's diagram by a *straight line.*

If the variation of density be slight, the relation (4) may, however, be regarded as holding approximately for actual fluids, provided m have the proper value. Putting

$$\rho=\rho_0(1+s), \qquad p=p_0+\kappa s, \qquad m=\rho_0 c, \qquad \text{...........(5)}$$

we find

$$c^2=\kappa/\rho_0, \qquad \text{..................................(6)}$$

as in Art. 277.

The fact that in actual fluids a progressive wave of finite amplitude continually alters its type, so that the variations of density towards the front become more and more abrupt, has led various writers to speculate on the possibility of a wave of discontinuity, analogous to a 'bore' in water-waves (cf. Art. 187).

It was shewn, first by Stokes*, and afterwards by several other writers, that the conditions of constancy of mass and of constancy of momentum can both be satisfied for such a wave. The simplest case is when there is no variation in the values of ρ and u except at the plane of discontinuity. If, in the preceding argument, the sections A, B be taken, one behind, and the other in front of this plane, we have, by (3),

$$m=\left(\frac{p_1-p_0}{\rho_1-\rho_0}\cdot\rho_1\rho_0\right)^{\frac{1}{2}}, \qquad \text{..........................(7)}$$

$$c-u_0=\frac{m}{\rho_0}=\left(\frac{p_1-p_0}{\rho_1-\rho_0}\cdot\frac{\rho_1}{\rho_0}\right)^{\frac{1}{2}}, \qquad \text{........................(8)}$$

and

$$u_1-u_0=\frac{m}{\rho_0}-\frac{m}{\rho_1}=\pm\left(\frac{(p_1-p_0)(\rho_1-\rho_0)}{\rho_1\rho_0}\right)^{\frac{1}{2}}. \qquad \text{..............(9)}$$

The upper or the lower sign is to be taken according as ρ_1 is greater or less than ρ_0, *i.e.* according as the wave is one of condensation or of rarefaction. The results involve *differences* of velocity, as we should expect, since any uniform velocity of the whole medium may be superposed.

We may assume, for instance, that the quantities p_0, ρ_0, u_0, which define the condition of the medium ahead of the wave, are given arbitrarily; also that the density ρ_1 of the air in the advancing wave is prescribed. Further, some definite relation between p_1, ρ_1 and p_0, ρ_0, based on physical considerations, is presupposed. The remaining quantities m, c, u_1 are then determined

* *l.c. ante* p. 484.

by (7), (8), (9). The formula (8) gives the velocity with which the wave invades the region in front of it.

These results are, however, open to the criticism* that in actual fluids the equation of energy cannot be satisfied consistently with (1) and (2). Calculating the excess of the work done per unit time on the fluid entering the space AB at B over that done by the fluid leaving at A, and subtracting the gain of kinetic energy, we obtain

$$p_0(c-u_0)-p_1(c-u_1)-\tfrac{1}{2}m\{(c-u_1)^2-(c-u_0)^2\},$$

or

$$p_1u_1-p_0u_0-\tfrac{1}{2}m(u_1^2-u_0^2),$$

or

$$\tfrac{1}{2}(p_1+p_0)(u_1-u_0), \quad \text{.........................(10)}$$

these forms being equivalent in virtue of the dynamical equation (2). The corresponding result *per unit mass* is obtained by dividing by m. If we substitute for u_1-u_0 from (1), we obtain

$$\tfrac{1}{2}(p_1+p_0)(v_0-v_1), \quad \text{.........................(11)}$$

where v is written as before for $1/\rho$.

If the two states of the medium be represented by two points A, B on Watt's diagram, the expression (11) is equal to the area included between the straight line AB, the axis of v, and the ordinates of A, B. If the transition from B to A could be affected without gain or loss of heat at any stage of the process, the points in question would lie on the same 'adiabatic curve,' and the gain of intrinsic energy would be represented by the area included between this curve, the axis of v, and the extreme ordinates. For an actual gas, the adiabatic is concave upwards; and the latter area is accordingly less (in absolute value) than the former. If we have regard to the signs to be attributed to the areas, we find that for a wave of condensation ($v_1 < v_0$) the work done on the medium would do more than is accounted for by the increase of the kinetic and intrinsic energies; whilst in a wave of rarefaction ($v_1 > v_0$) the work given out is more than the equivalent of the apparent loss of energy†.

It follows that the equation of energy cannot be satisfied for discontinuous waves, except in the case of a hypothetical medium whose adiabatic

* Rayleigh, *Theory of Sound*, Art. 253. The comparison with p. 280 *ante* is interesting.

† In some investigations by Hugoniot, which are expounded by Hadamard in his *Leçons sur la propagation des ondes et les équations de l'hydrodynamique*, Paris, 1903, the argument given in the text is inverted. The possibility of a wave of discontinuity being *assumed*, it is pointed out that the equation of energy will be satisfied if we equate the expression (10) to the increment of the intrinsic energy (for which see Art. 11 (8)). On this ground the formula

$$\tfrac{1}{2}(p_1+p_0)(v_0-v_1)=\frac{1}{\gamma-1}(p_1v_1-p_0v_0)$$

is propounded, as governing the transition from one state to the other: "Telle est la relation qu'Hugoniot a substituée à [$pv^\gamma=$const.] pour exprimer que la condensation ou dilatation brusque se fait sans absorption ni dégagement de chaleur. On lui donne actuellement le nom de *loi adiabatique dynamique*, la relation [$pv^\gamma=$const.], qui convient aux changements lents, étant désignée sous le nom de *loi adiabatique statique*" (Hadamard, p. 192). But no physical evidence is adduced in support of the proposed law.

lines are straight. This is identical with the condition already obtained for permanency of type in continuous waves.

In the above investigation no account has been taken of dissipative forces, such as viscosity and thermal conduction and radiation. Practically, a wave of discontinuity would imply a finite difference of temperature between the portions of the fluid on the two sides of the plane of discontinuity, so that, to say nothing of viscosity, there would necessarily be a dissipation of energy due to thermal action at the junction. The fact that a permanent wave of rarefaction would involve a supply of energy indicates that a wave of this type is impossible. It is easily seen moreover that such a wave, even if once established, would be unstable.

The question whether when dissipation is allowed for the relation between the two states can be reconciled with the equation of energy, in a wave of condensation, has been discussed by Rankine and (more fully) by Rayleigh*. In these investigations the transition from one uniform state to another is supposed to be continuous, though possibly very rapid. Since the temperature-gradient ($d\theta/dx$) is zero in front of and behind the wave, the total gain of heat by unit mass in its passage from state B to state A must be zero. The heat required to effect infinitesimal changes δp, δv is given by the thermodynamical formula

$$\delta Q = \frac{v\,\delta p + \gamma p\,\delta v}{\gamma - 1}. \qquad \text{.........................(12)}$$

By hypothesis $\delta p = -m^2 \delta v$, by (4), and therefore

$$\delta Q = \frac{\delta p}{(\gamma - 1)\,m^2}\{p + m^2 v - (\gamma + 1)\,p\}. \qquad \text{..............(13)}$$

Hence, expressing that $\int dQ = 0$,

$$p + m^2 v = \tfrac{1}{2}(\gamma + 1)(p_0 + p_1). \qquad \text{....................(14)}$$

In particular

$$\left.\begin{aligned} m^2 v_1 &= \tfrac{1}{2}(\gamma - 1)\,p_1 + \tfrac{1}{2}(\gamma + 1)\,p_0, \\ m^2 v_0 &= \tfrac{1}{2}(\gamma + 1)\,p_1 + \tfrac{1}{2}(\gamma - 1)\,p_0. \end{aligned}\right\} \qquad \text{..............(15)}$$

From (13) and (14) we derive

$$\delta Q = \frac{(\gamma + 1)\,\delta p}{2(\gamma - 1)\,m^2}(p_0 + p_1 - 2p), \qquad \text{................(16)}$$

and thence

$$Q = \frac{\gamma + 1}{2(\gamma - 1)\,m^2}(p_1 - p)(p - p_0). \qquad \text{.............(17)}$$

This gives the total heat absorbed by unit mass up to the stage to which p refers.

If conduction alone be considered the flux of heat into the region lying to the left of a plane situate between A and B is $k\,d\theta/dx$, where k is the

* *l.c. ante* p. 482.

conductivity, whilst in unit time an amount mQ is carried across the plane by convection. Since the region to the left neither gains nor loses heat we have

$$k\frac{d\theta}{dx} = -mQ. \qquad \text{.............................(18)}$$

Eliminating θ by means of the formula $pv = R\theta$, Rankine proceeds to find the relation between x and p. Combined with (14), this formula gives

$$\theta = \frac{\{\tfrac{1}{2}(\gamma+1)(p_0+p_1)-p\}p}{m^2R}. \qquad \text{.....................(19)}$$

Hence

$$\frac{dx}{dp} = \frac{d\theta}{dp}\Big/\frac{d\theta}{dx} = -\frac{(\gamma-1)k}{(\gamma+1)mR}\cdot\frac{(\gamma+1)(p_0+p_1)-4p}{(p_1-p)(p-p_0)}, \qquad \text{.........(20)}$$

by (18). At some point in the wave we must have $p = \frac{1}{2}(p_0+p_1)$ in virtue of the assumed continuity, and at this point the above value of dx/dp is negative. Moreover dx/dp cannot change sign, since otherwise we should have two different values of p for the same value of x. Hence $p_0 < p_1$, and therefore $v_0 > v_1$, *i.e.* the wave must be one of condensation. As x increases, p falls steadily from p_1 to p_0, and the denominator of the second fraction in (20) is accordingly positive. In order that the numerator should be positive we must have

$$\frac{p_1}{p_0} < \frac{\gamma+1}{3-\gamma}. \qquad \text{...............................(21)}$$

For air this limiting ratio is about $\frac{3}{2}$.

The integral of (20) is

$$x = \frac{k}{(\gamma+1)mC_v}\left\{\frac{(\gamma-1)(p_0+p_1)}{p_1-p_0}\log\frac{p_1-p}{p-p_0} - 2\log\frac{4(p_1-p)(p-p_0)}{(p_1-p_0)^2}\right\}, \qquad \text{.........(22)}$$

if the origin of x be taken at the point where $p = \frac{1}{2}(p_0+p_1)$. We have here utilized the thermodynamic relation

$$R = (\gamma-1)C_v, \qquad \text{............................(23)}$$

where C_v is the specific heat at constant volume. As p changes from p_1 to p_0, x increases from $-\infty$ to $+\infty$, but if the ratio p_1/p_0 differs appreciably from unity the space within which the transition is practically effected is very minute, so that the circumstances closely approach those of a discontinuity*.

In the case of air we may take

$$k = 5{\cdot}22\times10^{-5}, \quad \gamma = 1{\cdot}40, \quad C_v = {\cdot}1715, \quad \rho_0 = {\cdot}00129, \quad p_0 = 1{\cdot}013\times10^6$$

in C.G.S. units. Hence, assuming for example $p_1/p_0 = 1{\cdot}4$, we find from (15) $m = 49{\cdot}6$, and thence

$$k/(\gamma+1)mC_v = 2{\cdot}559\times10^{-6}.$$

* Rayleigh, *l.c.* Similar conclusions were arrived at independently by G. I. Taylor, "The Conditions Necessary for Discontinuous Motion in Gases," *Proc. Roy. Soc.* A, lxxxiv. 371 (1910).

From these data we deduce that the pressure changes from

$$\frac{9}{10}p_1+\frac{1}{10}p_0 \text{ to } \frac{9}{10}p_0+\frac{1}{10}p_1$$

in a space of $2{\cdot}7\times10^{-5}$ cm.

The velocity of advance of the disturbance, relatively to quiescent air, is

$$m/\rho_0=3{\cdot}84\times10^4 \text{ cm./sec.}$$

When viscosity, as well as conductivity, is taken into account the investigation becomes very complicated. It was found by Rayleigh that the general character of the results is unaltered, except that the range of admissible values of p_1/p_0 is greatly extended. His solution for the case where viscosity alone is considered will be given later (Art. 360 *a*).

Spherical Waves.

285. The general equations of small motion are

$$\rho_0\frac{\partial u}{\partial t}=-\frac{\partial p}{\partial x},\quad \rho_0\frac{\partial v}{\partial t}=-\frac{\partial p}{\partial y},\quad \rho_0\frac{\partial w}{\partial t}=-\frac{\partial p}{\partial z}. \qquad\text{.........(1)}$$

Writing

$$p=p_0+\kappa s,\quad c^2=\kappa/\rho_0, \qquad\text{.........(2)}$$

and integrating with respect to t, we have

$$u=-c^2\frac{\partial}{\partial x}\int_0^t s\,dt+u_0,\quad v=-c^2\frac{\partial}{\partial y}\int_0^t s\,dt+v_0,\quad w=-c^2\frac{\partial}{\partial z}\int_0^t s\,dt+w_0, \ \text{...(3)}$$

where u_0, v_0, w_0 are the values u, v, w at the point (x, y, z) at the instant $t=0$. If this initial motion is irrotational, with a velocity-potential ϕ_0, we have

$$u=-\frac{\partial\phi}{\partial x},\quad v=-\frac{\partial\phi}{\partial y},\quad w=-\frac{\partial\phi}{\partial z}, \qquad\text{.........(4)}$$

where

$$\phi=c^2\int_0^t s\,dt+\phi_0. \qquad\text{.........(5)}$$

This continued existence of a velocity-potential has been proved more generally in Arts. 17 and 33.

From (5) we have

$$c^2 s=\frac{\partial\phi}{\partial t}. \qquad\text{.........(6)}$$

We will now suppose that the disturbance is symmetrical about a fixed point, which we take as origin. The motion is then necessarily irrotational, so that a velocity-potential ϕ exists, which is here a function of r, the distance from the origin, and t, only.

To form the equation of continuity we remark that, owing to the difference of flux across the inner and outer surfaces, the space included between the spheres r and $r+\delta r$ is gaining mass at the rate

$$\frac{\partial}{\partial r}\left(4\pi r^2\rho\frac{\partial\phi}{\partial r}\right)\delta r.$$

Since the same rate is also expressed by $\partial\rho/\partial t \,.\, 4\pi r^2 \delta r$ we have

$$r^2 \frac{\partial \rho}{\partial t} = \frac{\partial}{\partial r}\left(\rho r^2 \frac{\partial \phi}{\partial r}\right). \quad \text{.............................(7)}$$

This might also have been arrived at by direct transformation of the general equation of continuity, Art. 7 (5). In the case of infinitely small motions, it becomes

$$\frac{\partial s}{\partial t} = \frac{1}{r^2}\frac{\partial}{\partial r}\left(r^2 \frac{\partial \phi}{\partial r}\right), \quad \text{.............................(8)}$$

whence, substituting from (6),

$$\frac{\partial^2 \phi}{\partial t^2} = c^2 \left(\frac{\partial^2 \phi}{\partial r^2} + \frac{2}{r}\frac{\partial \phi}{\partial r}\right). \quad \text{.........................(9)*}$$

This may be put into the more convenient form

$$\frac{\partial^2 \,.\, r\phi}{\partial t^2} = c^2 \frac{\partial^2 \,.\, r\phi}{\partial r^2}, \quad \text{..........................(10)}$$

the solution of which is

$$r\phi = f(r - ct) + F(r + ct). \quad \text{....................(11)}$$

Hence the motion is made up of two systems of spherical waves, travelling, one outwards, the other inwards, with velocity c. Considering for a moment the first system alone, we have by (6)

$$cs = -\frac{1}{r} f'(r - ct),$$

which shews that a condensation is propagated outwards with velocity c, but diminishes as it proceeds, its amount varying inversely as the distance from the origin. The velocity in the same train of waves is

$$-\frac{\partial \phi}{\partial r} = -\frac{1}{r} f'(r - ct) + \frac{1}{r^2} f(r - ct).$$

As r increases the second term becomes less and less important compared with the first, so that ultimately the velocity is propagated according to the same law as the condensation.

We notice that whenever diverging or converging waves are alone present we have from (11)

$$\frac{1}{r}\frac{\partial}{\partial r}(r\phi) = \mp cs\,; \quad \text{............................(12)}$$

this corresponds to Art. 277 (11).

For some purposes the formula for a system of divergent waves is more conveniently written

$$4\pi r\phi = f\left(t - \frac{r}{c}\right). \quad \text{.........................(13)}$$

* If we assume Boyle's Law the exact equation of symmetrical spherical waves is

$$\frac{\partial^2 \phi}{\partial t^2} - 2\frac{\partial \phi}{\partial r}\frac{\partial^2 \phi}{\partial r \partial t} + \left(\frac{\partial \phi}{\partial r}\right)^2 \frac{\partial^2 \phi}{\partial r^2} = c^2 \left(\frac{\partial^2 \phi}{\partial r^2} + \frac{2}{r}\frac{\partial \phi}{\partial r}\right).$$

Since this makes

$$\lim_{r\to 0}\left[-4\pi r^2\frac{\partial\phi}{\partial r}\right]=f(t), \quad\text{......................(14)}$$

the waves in question may be regarded as due to a source of strength $f(t)$ at the origin; cf. Art. 196.

If the source is in action only for a finite time the value of ϕ as given by (13) will vanish outside the limits of the wave. Hence from (6)

$$\int s\,dt=0, \quad\text{..................................(15)}$$

where the integral extends over the whole time of transit of the disturbance past the point considered. The fact that a diverging spherical wave must necessarily contain both condensed and rarefied portions was first remarked by Stokes*. Cf. Art. 197.

As in the case of plane progressive waves (Art. 280), the energy of a finite system of divergent spherical waves is half kinetic and half potential.

This follows from the general argument of Art. 174, and may be verified independently as follows. We have, identically,

$$r^2\left(\frac{\partial\phi}{\partial r}\right)^2=\left\{\frac{\partial(r\phi)}{\partial r}\right\}^2-\frac{\partial}{\partial r}(r\phi^2).$$

If we write

$$q=-\frac{\partial\phi}{\partial r},\qquad c^2s=\frac{\partial\phi}{\partial t}, \quad\text{......................................(16)}$$

this gives, by (12), in the case of a divergent wave-system,

$$r^2q^2=c^2r^2s^2-\frac{\partial}{\partial r}(r\phi^2).$$

Hence

$$\int_0^\infty \tfrac{1}{2}\rho q^2\,.\,4\pi r^2\,dr=\int_0^\infty \tfrac{1}{2}\rho c^2s^2\,.\,4\pi r^2\,dr, \quad\text{......................(17)}$$

if $r\phi^2$ vanishes at the inner and outer boundaries of the system†.

286. The determination of the functions f and F in (11), in terms of the initial conditions, for an unlimited space, can be effected as follows.

Let us suppose that the distributions of velocity and condensation at time $t=0$ are determined by the formulae

$$\phi=\psi(r),\qquad \frac{\partial\phi}{\partial t}=\chi(r), \quad\text{.........................(18)}$$

where ψ, χ are arbitrary functions. Comparing with (11), we have

$$f(z)+F(z)=z\psi(z), \quad\text{......................(19)}$$

$$-f'(z)+F'(z)=\frac{z}{c}\int\chi(z),$$

the latter of which gives on integration

$$-f(z)+F(z)=\frac{1}{c}\int_0^z z\chi(z)+C. \quad\text{......................(20)}$$

* "On Some Points in the Received Theory of Sound," *Phil. Mag.* (3) xxxiv. 52 (1849) [*Papers*, ii. 82]. See also Rayleigh, *Theory of Sound*, Art. 279.

† *Proc. Lond. Math. Soc.* (1) xxxv. 160 (1902).

Again, the condition that there is no creation or annihilation of fluid at the origin, viz. $r^2\partial\phi/\partial r \to 0$ for $r \to 0$ gives

$$f(-z)+F(z)=0. \qquad \text{.........................(21)}$$

The formulae (19) and (20) determine the functions f and F for positive values of z; and (21) then determines f for negative values of z*.

The final result may be written

$$r\phi = \tfrac{1}{2}(r-ct)\,\psi(r-ct)+\tfrac{1}{2}(r+ct)\,\psi(r+ct)+\frac{1}{2c}\int_{r-ct}^{r+ct} z\chi(z)\,dz, \text{...(22)}$$

or
$$r\phi = -\tfrac{1}{2}(ct-r)\,\psi(ct-r)+\tfrac{1}{2}(ct+r)\,\psi(ct+r)+\frac{1}{2c}\int_{ct-r}^{ct+r} z\chi(z)\,dz, \text{...(23)}$$

according as r is greater or less than ct. This may be immediately verified.

As a very simple example we may suppose that the air is initially at rest, and that the initial disturbance consists of a uniform condensation s_0 extending through a sphere of radius a. We have then $\psi(z)=0$, whilst $\chi(z)=c^2 s_0$ or 0 according as $z \lessgtr a$. At a distance $r\,(>a)$ from the origin, the motion will not begin until $t=(r-a)/c$, and will cease when $t=(r+a)/c$. For intermediate instants we shall have

$$r\phi = \tfrac{1}{4}cs_0\,\{a^2-(r-ct)\}^2, \qquad \text{...................................(24)}$$

and thence
$$\frac{s}{s_0}=\frac{r-ct}{2r}. \qquad \text{...(25)}$$

The disturbance is now confined to a spherical shell of thickness $2a$; and the condensation s is positive through the outer half, and negative through the inner half, of the thickness.

We shall require, shortly, an expression for the value of ϕ at the origin, for all values of t, in terms of the initial circumstances. We have, by (11) and (21),

$$[\phi]_{r=0}=\lim_{r\to 0}\frac{f(r-ct)+F(r+ct)}{r}$$

$$=\lim_{r\to 0}\frac{F(ct+r)-F(ct-r)}{r}=2F'(ct),$$

or, by (19) and the consecutive equation,

$$[\phi]_{r=0}=\frac{d}{dt}\cdot t\psi(ct)+t\chi(ct). \qquad \text{....................(26)}$$

For instance, in the special problem just considered, we have $\psi=0$ for all values of the variable, whilst $\chi(r)=c^2 s_0$ or 0 according as $r \lessgtr a$. Hence at the origin we have $\phi=c^2 s_0 t$ or 0 according as $ct \lessgtr a$. When $ct=a$, ϕ changes abruptly from acs_0 to 0, so that the value of s at the centre becomes for an instant negative infinite. The infinity is avoided if we imagine the initial value of s to change gradually but rapidly from s_0 to 0 in the neighbourhood of $r=a$.

General Equation of Sound-Waves.

287. We proceed to the general case of propagation of expansion-waves. We neglect, as before, small quantities of the second order, so that the dynamical equation is, as in Art. 285,

$$c^2 s=\frac{\partial\phi}{\partial t}. \qquad \text{..............................(1)}$$

* Rayleigh, *Theory of Sound*, Art. 279.

Also, writing $\rho = \rho_0(1+s)$ in the general equation of continuity, Art. 7 (5), we have, with the same approximation,

$$\frac{\partial s}{\partial t} = \frac{\partial^2 \phi}{\partial x^2} + \frac{\partial^2 \phi}{\partial y^2} + \frac{\partial^2 \phi}{\partial z^2}. \quad \text{..........(2)}$$

The elimination of s between (1) and (2) gives

$$\frac{\partial^2 \phi}{\partial t^2} = c^2 \left(\frac{\partial^2 \phi}{\partial x^2} + \frac{\partial^2 \phi}{\partial y^2} + \frac{\partial^2 \phi}{\partial z^2} \right), \quad \text{..........(3)}$$

or, in our former notation,

$$\frac{\partial^2 \phi}{\partial t^2} = c^2 \nabla^2 \phi. \quad \text{..........(4)}$$

Since this equation is linear, it will be satisfied by the arithmetic mean of any number of separate solutions $\phi_1, \phi_2, \phi_3, \ldots$. As in Art. 38, let us imagine an infinite number of systems of rectangular axes to be arranged uniformly about any point P as origin, and let $\phi_1, \phi_2, \phi_3, \ldots$ be the velocity-potentials of motions which are the same with respect to these systems as the original motion ϕ is with respect to the system x, y, z. In this case the arithmetic mean ($\bar{\phi}$, say) of the functions $\phi_1, \phi_2, \phi_3, \ldots$ will be the velocity-potential of a motion symmetrical with respect to the point P, and will therefore come under the investigation of Art. 286, provided r denote the distance of any point from P. In other words, if $\bar{\phi}$ be a function of r and t, defined by the equation

$$\bar{\phi} = \frac{1}{4\pi} \iint \phi \, d\varpi, \quad \text{..........(5)}$$

where ϕ is any solution of (4), and $\delta\varpi$ is the solid angle subtended at P by an element of the surface of a sphere of radius r having this point as centre, then

$$\frac{\partial^2 . r\bar{\phi}}{\partial t^2} = c^2 \frac{\partial^2 . r\bar{\phi}}{\partial r^2}. \quad \text{..........(6)*}$$

Hence

$$r\bar{\phi} = f(r - ct) + F(r + ct). \quad \text{..........(7)}$$

The mean value of ϕ over a sphere having any point P of the medium as centre is therefore subject to the same laws as the velocity-potential of a symmetrical spherical disturbance. We see at once that the value of ϕ at P at the time t depends on the means of the values which ϕ and $\partial\phi/dt$ originally had at points of a sphere of radius ct described about P as centre, so that the disturbance is propagated in all directions with uniform velocity c. Thus if the original disturbance extend only through a finite portion Σ of space, the disturbance at any point P external to Σ will begin after a time r_1/c, will last for a time $(r_2 - r_1)/c$, and will then cease altogether; r_1, r_2 denoting the radii of two spheres described with P as centre, the one just excluding, the other just including Σ.

* This result was obtained, in a different manner, by Poisson, "Mémoire sur la théorie du son," *Journ. de l'École Polytechn.* vii. 334–338 (1807). The remark that it leads at once to the complete solution of (4) is due to Liouville, *Journ. de Math.* i. 1 (1856).

To express the solution of (4), already virtually obtained, in an analytical form, let the values of ϕ and $\partial\phi/\partial t$, when $t=0$, be

$$\phi = \psi(x, y, z), \qquad \frac{\partial\phi}{\partial t} = \chi(x, y, z). \quad \text{..................(8)}$$

The mean values of these functions over a sphere of radius r described about (x, y, z) as centre are

$$\bar{\phi} = \frac{1}{4\pi}\iint \psi(x + lr, y + mr, z + nr)\, d\varpi,$$

$$\frac{\partial\bar{\phi}}{\partial t} = \frac{1}{4\pi}\iint \chi(x + lr, y + mr, z + nr)\, d\varpi,$$

where l, m, n denote the direction-cosines of any radius of this sphere, and $\delta\varpi$ the corresponding elementary solid angle. If we put

$$l = \sin\theta\cos\omega, \qquad m = \sin\theta\sin\omega, \qquad n = \cos\theta,$$

we shall have $\qquad \delta\varpi = \sin\theta\,\delta\theta\,\delta\omega.$

Hence, comparing with Art. 286 (26), we see that the value of ϕ at the point (x, y, z), at any subsequent time t, is

$$\phi = \frac{1}{4\pi}\frac{\partial}{\partial t}\cdot t\iint \psi(x + ct\sin\theta\cos\omega,\ y + ct\sin\theta\sin\omega,\ z + ct\cos\theta)\sin\theta d\, d\omega$$
$$+ \frac{t}{4\pi}\iint \chi(x + ct\sin\theta\cos\omega,\ y + ct\sin\theta\sin\omega,\ z + ct\cos\theta)\sin\theta\, d\theta\, d\omega, \quad \text{......(9)}$$

which is the form given by Poisson*.

A simple application is to the special problem considered in Art. 286, where the initial condition was one of uniform condensation s_0 through a sphere of radius a having the origin as centre. If a spherical surface of radius $PQ = ct$ described about an external

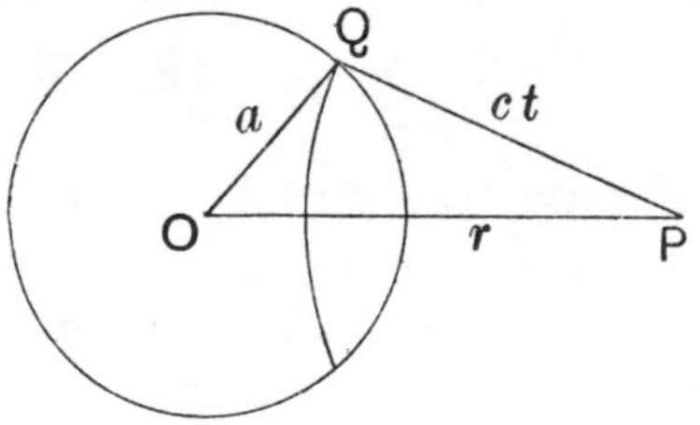

point P as centre intersects the sphere $r=a$, the area of the portion included within the latter is $2\pi \,.\, PQ^2(1-\cos OPQ)$ and the average of the initial values of s over the whole spherical surface $4\pi \,.\, PQ^2$ is therefore

$$\tfrac{1}{2}(1-\cos OPQ)\, s_0 = \frac{a^2-(ct-r)^2}{4ctr} s_0, \quad \text{............................(10)}$$

where $r = OP$. Hence

$$\phi_P = \frac{cs_0}{4r}\{a^2-(ct-r)^2\}, \quad \text{..................................(11)}$$

as in Art. 286 (24).

* "Mémoire sur l'intégration de quelques équations linéaires aux différences partielles, et particulièrement de l'équation générale du mouvement des fluides élastiques," *Mém. de l'Acad. des Sciences*, iii. 121 (1819).

For other proofs see Kirchhoff, *Mechanik*, c. xxiii, and Rayleigh, *Theory of Sound*, Art. 273.

In the case of a transient disturbance in infinite space the mean value (with respect to time) of the condensation s at any point is zero. For we have from (1)

$$c^2 \frac{\partial s}{\partial x} = \frac{\partial^2 \phi}{\partial x \partial t} = -\frac{\partial u}{\partial t},$$

with two similar equations. Hence

$$\frac{\partial}{\partial x} \int s\, dt = -\left[\frac{u}{c^2}\right] = 0, \text{ \&c., \&c.,}$$

if $u, v, w = 0$ at both limits of the integration with respect to t. The integral

$$\int s\, dt$$

has therefore the same value at all points of space, and considering infinitely distant points where the waves are attenuated by divergence we see that this value is zero. Cf. Art. 285 (15).

288. The expression for the kinetic energy of the fluid contained in any given region is

$$T = \tfrac{1}{2}\rho_0 \iiint \left\{\left(\frac{\partial \phi}{\partial x}\right)^2 + \left(\frac{\partial \phi}{\partial y}\right)^2 + \left(\frac{\partial \phi}{\partial z}\right)^2\right\} dx\, dy\, dz. \quad \text{............(1)}$$

Hence
$$\frac{dT}{dt} = \rho_0 \iiint \left(\frac{\partial \phi}{\partial x}\frac{\partial \dot{\phi}}{\partial x} + \frac{\partial \phi}{\partial y}\frac{\partial \dot{\phi}}{\partial y} + \frac{\partial \phi}{\partial z}\frac{\partial \dot{\phi}}{\partial z}\right) dx\, dy\, dz,$$

where $\dot{\phi}$ stands for $\partial\phi/\partial t$. By Green's Theorem (Art. 43), this may be put in the form

$$\frac{dT}{dt} = -\rho_0 \iint \dot{\phi} \frac{\partial \phi}{\partial n} dS - \rho_0 \iiint \dot{\phi} \nabla^2 \phi\, dx\, dy\, dz$$

$$= -\rho_0 \iint \dot{\phi} \frac{\partial \phi}{\partial n} dS - \frac{\rho_0}{c^2} \iiint \dot{\phi} \ddot{\phi}\, dx\, dy\, dz.$$

Hence if
$$W = \tfrac{1}{2}\kappa \iiint s^2\, dx\, dy\, dz = \tfrac{1}{2}\frac{\rho_0}{c^2} \iiint \dot{\phi}^2\, dx\, dy\, dz, \quad \text{............(2)}$$

we have
$$\frac{d}{dt}(T + W) = -\rho_0 \iint \dot{\phi} \frac{\partial \phi}{\partial n} dS. \quad \text{.....................(3)}$$

We have seen (Art. 280) that, subject to a certain condition, W represents the intrinsic energy.

The complete interpretation of (3) may be left to the reader. In various important cases, *e.g.* when the boundary is fixed ($\partial\phi/\partial n = 0$), or free ($\dot{\phi} = 0$), the surface-integral vanishes, and we have

$$T + W = \text{const.} \quad \text{.............................(4)}$$

This leads to a proof of the determinateness of the motion consequent on a given initial distribution of velocity and condensation. For if ϕ_1, ϕ_2 were two distinct forms of the velocity-potential satisfying the prescribed initial conditions, then, in the motion for which $\phi = \phi_1 - \phi_2$, $T + W$ would be constantly null, since it vanishes initially. Since every element of T and W

is essentially positive, this requires that the derivatives of ϕ with respect to x, y, z, t should all vanish; *i.e.* ϕ_1 and ϕ_2 can only differ by an absolute constant*. The argument applies, of course, to all cases where we can predicate the vanishing of the surface-integral in (3).

Simple-Harmonic Vibrations.

289. In the case of simple-harmonic motion, the time-factor being $e^{i\sigma t}$, the equation (4) of Art. 287 takes the form

$$(\nabla^2 + k^2)\,\phi = 0, \qquad\text{.......................(1)}$$

provided

$$k = \sigma/c. \qquad\text{.......................(2)}$$

It appears on comparison with Art. 280 that $2\pi/k$ is the wave-length of plane waves of the assumed period $(2\pi/\sigma)$.

In the case of symmetry with respect to the origin, we have by Art. 285 (10), or by transformation of (1),

$$\frac{\partial^2 . r\phi}{\partial r^2} + k^2 . r\phi = 0. \qquad\text{.......................(3)}$$

The solution of this may be written†

$$\phi = A\,\frac{\sin kr}{kr} + B\,\frac{\cos kr}{kr}. \qquad\text{.......................(4)}$$

If there is no source at the origin we must have $B = 0$, and (4) reduces to

$$\phi = A\,\frac{\sin kr}{kr}. \qquad\text{.......................(5)}$$

It may be noticed that this solution may be obtained by superposition of systems of plane waves, the directions of propagation being distributed uniformly. Thus, for a system of plane waves whose direction of propagation makes an angle θ with a given radius vector r, we have

$$\phi = e^{-ikr\cos\theta}, \qquad\text{.......................(6)}$$

and the mean-value of this for all directions through the origin is

$$\phi = \frac{1}{4\pi}\int_0^\pi e^{-ikr\cos\theta} . 2\pi\sin\theta\,d\theta = \frac{\sin kr}{kr}. \qquad\text{..............(7)}$$

We can draw from (5) a conclusion applicable to the general case, to which the equation (1) refers. It follows from Art. 287 (6) that the mean value of ϕ over a sphere of radius r, described with any point O as centre, satisfies an equation of the form (3). Hence in the notation of the Art. quoted, we have

$$\bar{\phi} = \frac{\sin kr}{kr} . \phi_0, \qquad\text{.......................(8)}$$

where ϕ_0 denotes the value of ϕ at O. This assumes that ϕ has no singularities within the sphere to which r refers ‡. Cf. Art. 38.

* Kirchhoff, *Mechanik*, c. xxiii.

† The time-factor is omitted here and elsewhere for shortness.

‡ The theorem was given by H. Weber, *Crelle*, lxix. (1868).

Returning to the case of symmetry, we note that the solution (4) may also be written

$$\phi = C\frac{e^{-ikr}}{r} + D\frac{e^{ikr}}{r}. \quad \text{.........................(9)}$$

It is evident on reference to Art. 285 (13) that the formula

$$\phi = \frac{e^{-ikr}}{4\pi k} \quad \text{.................................(10)}$$

represents the system of diverging waves due to a unit source at the origin.

To calculate the emission of energy by an isolated source in free space we must use an expression in real form, say

$$4\pi\phi = \frac{\cos k(ct - r)}{r}. \quad \text{...........................(11)}$$

The work done at a spherical surface of radius r, on the matter outside this surface, is

$$\left(p_0 + \rho_0\frac{\partial\phi}{\partial t}\right)\left(-\frac{\partial\phi}{\partial r}\right)4\pi r^2 \quad \text{.......................(12)}$$

per unit time. If we substitute from (11), and take the mean values of the trigonometrical terms, the result is

$$\frac{\rho_0 k^2 c}{8\pi} \quad \text{....................................(13)}$$

This may also be deduced from (9) of Art. 280, since the waves tend to become ultimately plane*.

The second term in (9) represents in a similar way the case of a *sink*, where energy is *absorbed* at the rate (13). The conception of a sink of energy is however hardly a natural one in Acoustics, and is in fact not employed.

The velocity-potential of a *double* source may be derived as in Art. 56. Thus if the axis of symmetry be coincident with that of x, we may write

$$4\pi\phi = -\frac{\partial}{\partial x}\frac{e^{-ikr}}{r}, \quad \text{..........................(14)}$$

or, in real form,

$$4\pi\phi = -\frac{\partial}{\partial x}\frac{\cos k(ct - r)}{r} = -\frac{\partial}{\partial r}\frac{\cos k(ct - r)}{r}.\cos\theta, \quad \text{.........(15)}$$

where θ is the inclination of the radius vector r to the axis of x. For large values of kr we have, approximately,

$$4\pi\phi = -\frac{k\sin k(ct - r)}{r}.\cos\theta. \quad \text{..................(16)}$$

The mean rate of emission of energy is now

$$\frac{\rho_0 k^4 c}{24\pi}. \quad \text{................................(17)}$$

This may be calculated as before, or inferred from the theory of plane waves.

* The a of Art. 280 is now equal to $1/4\pi cr$. Substituting this in the formula referred to, and multiplying by $4\pi r^2$, we obtain the result (13).

These calculations, it may be repeated, refer only to the case of an isolated source in free space. The presence of an obstacle may greatly modify the results. For instance, in the case of a simple source near an infinite plane wall the amplitude of vibration at any point is doubled by reflection as from an image of the source, and the power required to maintain the source is quadrupled. On the other hand, a source completely surrounded by rigid walls does no work on the whole, since the energy of the gas is constant.

290. The general theory of functions satisfying the equation

$$(\nabla^2 + k^2)\,\phi = 0 \qquad \text{..............................(1)}$$

has been developed by Helmholtz*, Rayleigh†, and others‡. It has many points of analogy with that of Laplace's equation $\nabla^2\phi = 0$, which is, indeed, a particular case, obtained by making either $c = \infty$, or $\sigma = 0$.

The typical solution of (1), from which all others can be derived, is that which corresponds to a unit source, viz.

$$\phi = \frac{e^{-ikr}}{4\pi r}, \qquad \text{...................................(2)}$$

where r denotes distance from the source.

It appears from Green's Theorem (Art. 43) that if ϕ, ϕ' be any two functions which, together with their first and second derivatives, are finite and single-valued throughout any finite region, we have

$$\iint \left(\phi \frac{\partial \phi'}{\partial n} - \phi' \frac{\partial \phi}{\partial n}\right) dS = \iiint (\phi' \nabla^2 \phi - \phi \nabla^2 \phi')\, dx\,dy\,dz. \qquad \text{........(3)}$$

If, in addition, ϕ and ϕ' both satisfy (1), the right-hand member vanishes, and we have

$$\iint \phi \frac{\partial \phi'}{\partial n}\, dS = \iint \phi' \frac{\partial \phi}{\partial n}\, dS. \qquad \text{..........................(4)}$$

From this we deduce, by the same method§ as in Art. 57, the formula

$$\phi_P = -\frac{1}{4\pi} \iint \frac{e^{-ikr}}{r} \frac{\partial \phi}{\partial n}\, dS + \frac{1}{4\pi} \iint \phi \frac{\partial}{\partial n}\left(\frac{e^{-ikr}}{r}\right) dS, \qquad \text{.........(5)}$$

giving the value of ϕ at any point P of a region in terms of the values of ϕ and $\partial\phi/\partial n$ at the boundary. The symbol r here denotes the distances of the respective surface-elements from P, and we see that the value of ϕ may be regarded as due to certain distributions of simple and double sources over the boundary||.

* "Theorie der Luftschwingungen in Röhren mit offenen Enden," *Crelle*, lvii. 1 (1859) [*Wiss. Abh.* ii. 303].

† *Theory of Sound*, ii.

‡ For an account of the more recent mathematical theory, see Pockels, "Ueber die partielle Differentialgleichung $\Delta u + k^2 u = 0$," Leipzig, 1891, and Sommerfeld, *l.c. ante* p. 61.

§ Viz. we put $\phi' = e^{-ikr}/r$, where r denotes distance from a fixed point, and isolate this point (when it falls within the region considered), by drawing a small spherical surface about it.

|| Helmholtz, *l.c.*

Again, if r' denote distance from a point P' *external* to the region, we have

$$0 = -\frac{1}{4\pi}\iint \frac{e^{-ikr'}}{r'}\frac{\partial\phi}{\partial n}\,dS + \frac{1}{4\pi}\iint \phi\frac{\partial}{\partial n}\left(\frac{e^{-ikr'}}{r'}\right) dS. \quad \text{...........(6)}$$

It is to be noticed that, as in Art. 58, the particular distribution of sources over the boundary which is exhibited in (5) is only one out of an infinite number which would give exactly the same value of ϕ at points *within* the region. For instance, by addition of (5) and (6) we get another such distribution, which may, moreover, be varied indefinitely by varying the position of P'*.

The theorems (5) and (6) will apply also to the case of an infinite region bounded internally by one or more closed surfaces, provided that at an infinite distance R from the origin ϕ tends to the form

$$\phi = C\frac{e^{-ikR}}{R}. \quad \text{................................(7)}$$

We may express this by saying that there are no sources of sound at infinity.

We can, *under certain conditions*, carry the analogy with the theory of the ordinary potential a step further, and express the value of ϕ at any point of a given region in terms of simple sources only, or double sources only, distributed over the boundary; thus

$$\phi_P = \frac{1}{4\pi}\iint \frac{e^{-ikr}}{r}\left(\frac{\partial\phi}{\partial n} + \frac{\partial\phi'}{\partial n'}\right) dS, \quad \text{................(8)}$$

$$\phi_P = \frac{1}{4\pi}\iint (\phi - \phi')\frac{\partial}{\partial n}\left(\frac{e^{-ikr}}{r}\right) dS, \quad \text{................(9)}$$

where the auxiliary function ϕ' together with its first and second derivatives is assumed to be finite and to satisfy (1) throughout the space external to the given region, whilst at the boundary

$$\phi' = \phi, \text{ or } \frac{\partial\phi'}{\partial n} = \frac{\partial\phi}{\partial n}, \quad \text{.......................(10)}$$

as the case may be. It is also assumed that ϕ' tends ultimately to the form (7) when the region to which it relates extends to infinity. It is unnecessary to give the proof, which follows closely the lines of Art. 58.

It would be wrong, however, to assume that, as in the case of the ordinary potential, a function ϕ necessarily exists which satisfies (1) throughout a given finite region, and also fulfils the condition that ϕ or $\partial\phi/\partial n$ shall assume *arbitrarily prescribed* values over the boundary. The supposed existence-theorem holds, it is true, *as a rule*, but it fails for a series of definite values of k, which correspond to the normal modes of vibration of the mass of air occupying the region, when the boundary-condition is $\phi = 0$, or $\partial\phi/\partial n = 0$, respectively.

* Larmor, *l.c.* *ante* p. 60.

For the same reason, the formulae (8) and (9) cannot be applied without reservation to the case of an infinite region, since the determination of the auxiliary function ϕ' may be impossible.

To illustrate the theory, let us suppose that throughout a sphere of radius a, having its centre at the origin O, we have

$$\phi = \frac{\sin kR}{R}, \qquad (11)$$

where R now denotes distance from O. If we put

$$\phi' = \sin ka \,.\, \frac{e^{-ik(R-a)}}{R}, \qquad (12)$$

in the external space, the conditions of validity of the formula (8) are satisfied, and we find

$$\phi = \frac{ke^{ika}}{4\pi a} \iint \frac{e^{-ikr}}{r} \, dS. \qquad (13)$$

It is not difficult to verify, *à posteriori*, that this is equivalent to (11) for $R < a$ and to (12) for $R > a$.

Again, let us seek a surface-distribution of simple sources which will make

$$\phi = \frac{e^{-ikR}}{R}, \qquad (14)$$

in the space external to the sphere. The value of ϕ' for the internal space, which coincides with this at the boundary, is

$$\phi' = \frac{e^{-ika}}{\sin ka} \cdot \frac{\sin kR}{R}, \qquad (15)$$

and we get

$$\phi' = \frac{k}{4\pi a \sin ka} \iint \frac{e^{-ikr}}{r} \, dS. \qquad (16)$$

But the determination of ϕ' fails whenever k is a root of $\sin ka = 0$. It appears in fact that when this condition is satisfied, a uniform distribution of simple sources over a sphere of radius a produces *no* effect at external points.

A special case is where the region considered is 'semi-infinite,' being bounded only by a plane. Suppose for instance that it is the space on the positive side of the plane $x = 0$. If we assume $\phi'(-x, y, z) = \phi(x, y, z)$, then at the boundary $\phi' = \phi$, and $\partial\phi'/\partial n' = \partial\phi/\partial n$, so that (8) reduces to the form

$$\phi_P = -\frac{1}{2\pi} \iint \frac{e^{-ikr}}{r} \frac{\partial\phi}{\partial n} \, dS. \qquad (17)$$

On the other hand, if we make $\phi'(-x, y, z) = -\phi(x, y, z)$, we have $\phi' = -\phi$, $\partial\phi'/\partial n = \partial\phi/\partial n$ at the boundary, and therefore

$$\phi_P = \frac{1}{2\pi} \iint \phi \frac{\partial}{\partial n} \left(\frac{e^{-ikr}}{r} \right) dS. \qquad (18)$$

If all the dimensions of the region referred to are small compared with the wave-length, we may put $e^{-ikr} = 1$, approximately, in (5), and the formula assumes the shape

$$\phi_P = -\frac{1}{4\pi} \iint \frac{1}{r} \frac{\partial\phi}{\partial n} \, dS + \frac{1}{4\pi} \iint \phi \frac{\partial}{\partial n} \left(\frac{1}{r} \right) dS, \qquad (19)$$

as in Art. 57. Hence, within distances small compared with the wave-length, the variations of ϕ may be calculated as if it satisfied the equation $\nabla^2\phi = 0$.

This principle is of great service in the approximate treatment of various acoustical questions (cf. Arts. 299, 300).

Finally, we may remark that, if we restore the time-factor, the formula (5) may be written

$$\phi_P = -\frac{1}{4\pi}\iint \frac{e^{i\sigma\left(t-\frac{r}{c}\right)}}{r}\frac{\partial\phi}{\partial n}\,dS + \frac{1}{4\pi}\iint \phi\,\frac{\partial}{\partial n}\left(\frac{e^{i\sigma\left(t-\frac{r}{c}\right)}}{r}\right)dS. \quad \text{...(20)}$$

This may be generalized by Fourier's double-integral theorem, which may be written in the form

$$F(t) = \frac{1}{2\pi}\int_{-\infty}^{\infty} d\sigma \int_{-\infty}^{\infty} F(\tau)\, e^{i\sigma(t-\tau)}\, d\tau. \quad \text{..............(21)}$$

If we denote by $\phi(t)$ the value which ϕ has at a point (x, y, z) of the boundary at the instant t, and by $f(t)$ the corresponding value of $\partial\phi/\partial n$, we obtain as the value of ϕ at an internal point P

$$\phi_P(t) = -\frac{1}{4\pi}\iint \frac{f\left(t-\frac{r}{c}\right)}{r}\,dS + \frac{1}{4\pi}\iint \frac{\partial}{\partial n}\frac{\phi\left(t-\frac{r}{c}\right)}{r}\,dS, \quad \text{......(22)}$$

provided that in the last term the space-differentiation applies only to r as it appears explicitly. This remarkable formula, which gives the value of ϕ at any instant, at a point P, in terms of antecedent values of ϕ and $\partial\phi/\partial n$ at points of a closed surface surrounding P, was first obtained by Kirchhoff*, in a different manner, from the general equation (4) of Art. 287. It has been supposed by various writers to contain *the* precise mathematical formulation of 'Huygens' principle' in Acoustics; but as has been already pointed out, in connection with the special case (5), the representation of ϕ in this manner is largely arbitrary and indeterminate.

291. The solution of the equation

$$(\nabla^2 + k^2)\,\phi = \Phi, \quad \text{............................(1)}$$

where Φ is a given function of x, y, z which vanishes outside a finite region Σ, is also treated by the writers referred to on p. 498.

The solution is indicated by the analogy with the theory of the ordinary gravitation potential. The equation is satisfied by

$$\phi_P = -\frac{1}{4\pi}\iiint \Phi' \frac{e^{-ikr}}{r}\,dx'\,dy'\,dz', \quad \text{.....................(2)}$$

where Φ' is the value of Φ at (x', y', z'), r denotes the distance of this point from the point P where the value of ϕ is required, and the integration extends over the region Σ. When P lies outside Σ this is obvious, since the second member of (2) is the potential of a distribution of simple sources with volume

* "Zur Theorie der Lichtstrahlen," *Berl. Ber.* 1882, p. 641 [*Ges. Abh.* ii. 22]. Various other proofs have been given; cf Larmor, *l.c. ante* p. 60, and Love, *Proc. Lond. Math. Soc.* (2) i. 37 (1903).

density $-\Phi$ throughout Σ. To verify the solution when P lies within Σ we divide this region into two portions Σ_1, Σ_2 of which Σ_2 encloses P and is ultimately taken to be infinitely small (in comparison with k^{-1}) in its linear dimensions. Since P is external to Σ_1, we need only consider those elements of the integral in (2) which correspond to the space within Σ_2. As regards these we have developing the exponential,

$$\phi_P = -\frac{1}{4\pi}\iiint \frac{\Phi'}{r}\,dx'\,dy'\,dz' + \frac{ik}{4\pi}\iiint \Phi'\,dx'\,dy'\,dz' + \ldots \quad \ldots\ldots(3)$$

As in the case of the ordinary potential, the first term satisfies $\nabla^2\phi = \Phi$, but contributes an ultimately vanishing amount to the value of ϕ itself. The second and further terms contribute nothing, ultimately, to the value of ϕ or $\nabla^2\phi$.

It may be shewn that (2) is the only solution of (1) which holds at all points of space and vanishes at infinity. In the case of a limited region we may add an arbitrary solution of $(\nabla^2 + k^2)\phi = 0$; this enables us to satisfy the boundary-conditions.

We may apply the above theory to the determination of the effect of periodic extraneous forces (X, Y, Z) acting on the medium. The equations of motion are, by an obvious extension of Art. 277 (4), (5),

$$\frac{\partial u}{\partial t} = -c^2\frac{\partial s}{\partial x} + X, \quad \frac{\partial v}{\partial t} = -c^2\frac{\partial s}{\partial y} + Y, \quad \frac{\partial w}{\partial t} = -c^2\frac{\partial s}{\partial z} + Z, \quad \ldots\ldots\ldots\ldots\ldots(4)$$

with

$$\frac{\partial s}{\partial t} = -\left(\frac{\partial u}{\partial x} + \frac{\partial v}{\partial y} + \frac{\partial w}{\partial z}\right). \quad \ldots\ldots\ldots\ldots\ldots(5)$$

Hence

$$\frac{\partial^2 s}{\partial t^2} = c^2\nabla^2 s - \left(\frac{\partial X}{\partial x} + \frac{\partial Y}{\partial y} + \frac{\partial Z}{\partial z}\right), \quad \ldots\ldots\ldots\ldots\ldots(6)$$

or, assuming a time-factor e^{ikct},

$$(\nabla^2 + k^2)\,s = \frac{1}{c^2}\left(\frac{\partial X}{\partial x} + \frac{\partial Y}{\partial y} + \frac{\partial Z}{\partial z}\right). \quad \ldots\ldots\ldots\ldots\ldots(7)$$

In the case of an unlimited region the solution is

$$s = -\frac{1}{4\pi c^2}\iiint\left(\frac{\partial X'}{\partial x'} + \frac{\partial Y'}{\partial y'} + \frac{\partial Z'}{\partial z'}\right)\frac{e^{-ikr}}{r}\,dx'dy'dz', \quad \ldots\ldots\ldots\ldots\ldots(8)$$

it being assumed that X, Y, Z vanish beyond a certain finite distance from the origin. Since $\partial/\partial x' . r^{-1} = -\partial/\partial x . r^{-1}$, &c., this takes the form

$$s = \frac{1}{4\pi c^2}\iiint\left(X'\frac{\partial}{\partial x} + Y'\frac{\partial}{\partial y} + Z'\frac{\partial}{\partial z}\right)\frac{e^{-ikr}}{r}\,dx'dy'dz'. \quad \ldots\ldots\ldots\ldots\ldots(9)$$

By comparison with (4) we see that at points *outside* the region where the forces act the motion is irrotational, with the velocity-potential

$$\phi = -ics/k \quad \ldots\ldots\ldots\ldots\ldots(10)$$

due to a certain distribution of double sources.

For instance, if we imagine the forces to be concentrated on an infinitely small space surrounding the origin, and to be in the direction of x, we have, writing

$$F = \rho\iiint X'dx'dy'dz', \quad \ldots\ldots\ldots\ldots\ldots(11)$$

$$\phi = -\frac{iF}{4\pi kc\rho}\frac{\partial}{\partial x}\frac{e^{-ikr}}{r}, \quad \ldots\ldots\ldots\ldots\ldots(12)$$

where r now denotes distance from the origin. A concentrated force Fe^{ikct} is therefore equivalent to a double source of strength $iF/kc\rho$.

From (9) and (11) we have, on restoring the time-factor,

$$s = \frac{F}{4\pi\rho c^2}\frac{\partial}{\partial x}\frac{e^{i\sigma\left(t-\frac{r}{c}\right)}}{r}, \qquad (13)$$

corresponding to a force $Fe^{i\sigma t}$. This can obviously be generalized so as to apply to any law of force as a function of the time. Denoting this by $F(t)$, we have

$$s = \frac{1}{4\pi\rho c^2}\frac{\partial}{\partial x}\frac{F\left(t-\frac{r}{c}\right)}{r}. \qquad (14)$$

Applications of Spherical Harmonics.

292. The solution of the equation

$$(\nabla^2 + k^2)\phi = 0, \qquad (1)$$

when the boundary-conditions have reference to *spherical* surfaces, may be conducted as follows.

We may suppose the value of ϕ over any sphere of radius r, having its centre at the origin, to be expanded in a series of spherical surface-harmonics whose coefficients are functions of r. We therefore write

$$\phi = \Sigma R_n \phi_n, \qquad (2)$$

where ϕ_n is a solid harmonic of degree n, and R_n is a function of r only.

Now

$$\nabla^2(R_n\phi_n) = \nabla^2 R_n \,.\, \phi_n + 2\left(\frac{\partial R_n}{\partial x}\frac{\partial \phi_n}{\partial x} + \frac{\partial R_n}{\partial y}\frac{\partial \phi_n}{\partial y} + \frac{\partial R_n}{\partial z}\frac{\partial \phi_n}{\partial z}\right) + R_n\nabla^2\phi_n$$

$$= \nabla^2 R_n \,.\, \phi_n + \frac{2}{r}\frac{dR_n}{dr}\left(x\frac{\partial \phi_n}{\partial x} + y\frac{\partial \phi_n}{\partial y} + z\frac{\partial \phi_n}{\partial z}\right) + R_n\nabla^2\phi_n. \qquad (3)$$

And, by the definition of a solid harmonic, we have

$$\nabla^2\phi_n = 0,$$

and

$$x\frac{\partial \phi_n}{\partial x} + y\frac{\partial \phi_n}{\partial y} + z\frac{\partial \phi_n}{\partial z} = n\phi_n.$$

Hence

$$\nabla^2(R_n\phi_n) = \left(\nabla^2 R_n + \frac{2n}{r}\frac{dR_n}{dr}\right)\phi_n = \left(\frac{d^2R_n}{dr^2} + \frac{2(n+1)}{r}\frac{dR_n}{dr}\right)\phi_n. \qquad (4)$$

If we substitute from (2) in (1), the terms in ϕ_n must satisfy the equation independently, whence

$$\frac{d^2R_n}{dr^2} + \frac{2(n+1)}{r}\frac{dR_n}{dr} + k^2R_n = 0. \qquad (5)$$

This can be integrated by series. Thus, assuming that

$$R_n = \Sigma A_m (kr)^m,$$

the relation between consecutive coefficients is found to be

$$m(2n+1+m)A_m + A_{m-2} = 0.$$

This gives two ascending series, one beginning with $m=0$, and the other with $m=-2n-1$; thus

$$R_n = A\left\{1-\frac{k^2r^2}{2(2n+3)}+\frac{k^4r^4}{2.4(2n+3)(2n+5)}-\dots\right\}$$
$$+Br^{-2n-1}\left\{1-\frac{k^2r^2}{2(1-2n)}+\frac{k^4r^4}{2.4(1-2n)(3-2n)}-\dots\right\},$$

where A, B are arbitrary constants. Hence if we put $\phi_n = r^n S_n$, so that S_n is a surface-harmonic of order n, the general solution of (1) may be written

$$\phi = \Sigma\{A\psi_n(kr)+B\Psi_n(kr)\}\,r^n S_n, \quad\text{.................(6)}$$

provided

$$\left.\begin{aligned}\psi_n(\zeta) &= \frac{1}{1.3\dots(2n+1)}\left\{1-\frac{\zeta^2}{2(2n+3)}+\frac{\zeta^4}{2.4(2n+3)(2n+5)}-\dots\right\},\\ \Psi_n(\zeta) &= \frac{1.3\dots(2n-1)}{\zeta^{2n+1}}\left\{1-\frac{\zeta^2}{2(1-2n)}+\frac{\zeta^4}{2.4(1-2n)(3-2n)}-\dots\right\}.\end{aligned}\right\}$$

......(7)*

The first term of (6) is alone to be retained when the motion is finite at the origin.

The functions $\psi_n(\zeta)$, $\Psi_n(\zeta)$ can also be expressed in finite terms, as follows:

$$\psi_n(\zeta) = \left(-\frac{d}{\zeta d\zeta}\right)^n\frac{\sin\zeta}{\zeta}, \quad \Psi_n(\zeta) = \left(-\frac{d}{\zeta d\zeta}\right)^n\frac{\cos\zeta}{\zeta}. \quad\text{.........(8)}$$

These are readily identified by (7) by expanding $\sin\zeta$, $\cos\zeta$, and performing the differentiations. As particular cases we have

$$\psi_0(\zeta) = \frac{\sin\zeta}{\zeta}, \quad \psi_1(\zeta) = \frac{\sin\zeta}{\zeta^3}-\frac{\cos\zeta}{\zeta^2}, \quad \psi_2(\zeta) = \left(\frac{3}{\zeta^5}-\frac{1}{\zeta^3}\right)\sin\zeta-\frac{3\cos\zeta}{\zeta^4}.$$

...............(9)

The formulae (6) and (8) shew that the general solution of the equation

$$\frac{d^2R_n}{d\zeta^2}+\frac{2(n+1)}{\zeta}\frac{dR_n}{d\zeta}+R_n=0, \quad\text{.................(10)}$$

which is obtained by writing ζ for kr in (5), is

$$R_n = \left(\frac{d}{\zeta d\zeta}\right)^n\frac{Ae^{i\zeta}+Be^{-i\zeta}}{\zeta}. \quad\text{....................(11)}$$

* There is a slight deviation here from the notation adopted by Heine, *Kugelfunktionen*, i. 82. It may be noted that the formula (6) gives an immediate proof of the theorem (8) of Art. 289.

The functions in (7) are related to Bessel's Functions of fractional order as follows:

$$\zeta^n\psi_n(\zeta) = \sqrt{\left(\frac{\pi}{2\zeta}\right)}J_{n+\frac{1}{2}}(\zeta),$$
$$\zeta^n\Psi_n(\zeta) = (-)^n\sqrt{\left(\frac{\pi}{2\zeta}\right)}J_{-n-\frac{1}{2}}(\zeta).$$

Tables of Bessel's Functions of order $\pm\frac{1}{2}(2m+1)$, where m is integral, were computed by Lommel for unit intervals of ζ; they are reprinted by Jahnke and Emde, and in Watson's treatise. Closer tables (at intervals of ·2) are given by Dinnick, *Archiv d. Math. u. Phys.* (3) xx. (1912).

This is easily verified; for if R_n be any solution of (10) we find that the corresponding equation for R_{n+1} is satisfied by

$$R_{n+1}=\frac{dR_n}{\zeta d\zeta},$$

and by repeated applications of this result it appears that (10) is satisfied by

$$R_n=\left(\frac{d}{\zeta d\zeta}\right)^n R_0, \quad \text{.........(12)}$$

where R_0 is the solution of

$$\frac{d^2(\zeta R_0)}{d\zeta^2}+\zeta R_0=0,$$

that is

$$R_0=\frac{Ae^{i\zeta}+Be^{-i\zeta}}{\zeta}. \quad \text{.........(13)*}$$

It will be convenient to have a special notation for that combination of the functions $\psi_n(\zeta)$, $\Psi_n(\zeta)$ which is appropriate to the expression of *diverging* waves. We write

$$f_n(\zeta)=\left(-\frac{d}{\zeta d\zeta}\right)^n\frac{e^{-i\zeta}}{\zeta}=\Psi_n(\zeta)-i\psi_n(\zeta). \quad \text{.........(14)}$$

As particular cases:

$$f_0(\zeta)=\frac{e^{-i\zeta}}{\zeta},\quad f_1(\zeta)=\left(\frac{i}{\zeta^2}+\frac{1}{\zeta^3}\right)e^{-i\zeta},\quad f_2(\zeta)=\left(-\frac{1}{\zeta^3}+\frac{3i}{\zeta^4}+\frac{3}{\zeta^5}\right)e^{-i\zeta}. \quad \text{...(15)}$$

The general formula is

$$f_n(\zeta)=\frac{i^n e^{-i\zeta}}{\zeta^{n+1}}\left\{1+\frac{n(n+1)}{2i\zeta}+\frac{(n-1)n(n+1)(n+2)}{2.4.(i\zeta)^2}+\dots+\frac{1.2.3\dots 2n}{2.4.6\dots 2n(i\zeta)^n}\right\}. \quad \text{.........(16)}$$

This may be proved by 'mathematical induction,' or by means of the differential equation satisfied by $f_n(\zeta)$†. If we equate, separately, real and imaginary parts, expressions for $\psi_n(\zeta)$, $\Psi_n(\zeta)$, in terms of $\cos\zeta$, $\sin\zeta$, and finite algebraical series, can be deduced by (14).

The functions $\psi_n(\zeta)$, $\Psi_n(\zeta)$, $f_n(\zeta)$ all satisfy recurrence-formulae of the types

$$\psi_n'(\zeta)=-\zeta\psi_{n+1}(\zeta), \quad \text{.........(17)}$$

$$\zeta\psi_n'(\zeta)+(2n+1)\psi_n(\zeta)=\psi_{n-1}(\zeta); \quad \text{.........(18)}$$

these are frequently useful in reductions.

We have also the relation

$$\{\psi_n'(\zeta)\Psi_n(\zeta)-\psi_n(\zeta)\Psi_n'(\zeta)\}\zeta^{2n+2}=1, \quad \text{.........(19)}$$

or the equivalent formula

$$\{\psi_{n-1}(\zeta)\Psi_n(\zeta)-\psi_n(\zeta)\Psi_{n-1}(\zeta)\}\zeta^{2n+1}=1. \quad \text{.........(20)}$$

* The above analysis, which has a wide application in mathematical physics, has been given, in one form or another, by various writers, from Laplace, "Sur la diminution de la durée du jour par le refroidissement de la Terre," *Conn. des Tems* pour l'An 1823, p. 245 (1820) [*Méc. Céleste*, Livre 11me, c. iv.], downwards. For references to the history of the matter, considered as a problem in Differential Equations, see Glaisher, "On Riccati's Equation and its Transformations," *Phil. Trans.* 1881.

† Cf. Stokes, *l.c. post* p. 508. The notation is different.

It follows easily from (17) and (18) that the left-hand member of (19) is unaltered in value when n is replaced by $n-1$, and the proof is completed by examining the case of $n=0$. The formula may also be derived from Art. 290 (4), the region considered being that included between two concentric spherical surfaces*. If in the formula quoted we put

$$\phi = \Psi_n(kr)\, r^n S_n, \quad \phi' = \psi_n(kr)\, r^n S_n, \quad \ldots\ldots\ldots\ldots\ldots(21)$$

it appears that the expression

$$\{\psi_n'(kr)\,\Psi_n(kr) - \psi_n(kr)\,\Psi_n'(kr)\}\, r^{2n+2} . \iint S_n^2 d\varpi, \quad \ldots\ldots(22)$$

where the integration includes all the elementary solid angles $\delta\varpi$ having their vertices at the origin, is independent of r. Evaluating for r infinitesimal, we are led again to the relation (19).

293. A simple application of the foregoing analysis is to the vibrations of air contained in a spherical envelope.

1°. Let us first consider the free vibrations when the envelope is rigid. Since the motion is finite at the origin, we have

$$\phi = A\psi_n(kr)\, r^n S_n . e^{i\sigma t}, \quad \ldots\ldots\ldots\ldots\ldots\ldots\ldots\ldots(1)$$

with the boundary-condition $\quad ka\psi_n'(ka) + n\psi_n(ka) = 0, \quad \ldots\ldots\ldots\ldots\ldots\ldots\ldots\ldots(2)$

a being the radius. This determines the admissible values of k and thence of $\sigma\ (=kc)$.

It is evident from Art. 292 (8) that this equation reduces always to the form

$$\tan ka = F(ka), \quad \ldots\ldots\ldots\ldots\ldots\ldots\ldots\ldots\ldots\ldots(3)$$

where $F(ka)$ is a rational algebraic function. The roots can then be calculated without difficulty, either by means of a series, or by a method devised by Fourier†.

In the case of the purely radial vibrations ($n=0$), we have

$$\phi = A\,\frac{\sin kr}{kr}\, e^{i\sigma t}, \quad \ldots\ldots\ldots\ldots\ldots\ldots\ldots\ldots(4)$$

with the boundary-condition $\quad \tan ka = ka, \quad \ldots\ldots\ldots\ldots\ldots\ldots\ldots\ldots\ldots\ldots(5)$

which determines the frequencies of the normal modes. The roots of this equation, which presents itself in various physical problems, can be calculated most readily by means of a series‡. The values obtained by Schwerd§ for the first few roots are

$$\frac{ka}{\pi} = 1{\cdot}4303,\quad 2{\cdot}4590,\quad 3{\cdot}4709,\quad 4{\cdot}4774,\quad 5{\cdot}4818,\quad 6{\cdot}4844, \quad \ldots\ldots\ldots(6)$$

approximating to the form $m+\frac{1}{2}$, where m is integral. These numbers give the ratio $(2a/\lambda)$ of the diameter of the sphere to the wave-length. Taking the reciprocals we find

$$\frac{\lambda}{2a} = {\cdot}6992,\quad {\cdot}4067,\quad {\cdot}2881,\quad {\cdot}2233,\quad {\cdot}1824,\quad {\cdot}1542. \quad \ldots\ldots\ldots\ldots(7)$$

In the case of the second and higher roots of (5) the roots of lower order give the positions of the spherical nodes ($\partial\phi/\partial r = 0$). Thus in the second mode there is a spherical node whose radius is given by

$$\frac{r}{a} = \frac{1{\cdot}4303}{2{\cdot}4590} = {\cdot}5817.$$

* Cf. Rayleigh, *Theory of Sound*, Art. 327.

† *Théorie analytique de la Chaleur*, Paris, 1822, Art. 286.

‡ Euler, *Introductio in Analysin Infinitorum*, Lausannae, 1748, ii. 319; Rayleigh, *Theory of Sound*, Art. 207.

§ Quoted by Verdet, *Leçons d'Optique Physique*, Paris, 1869–70, i. 266.

In the case $n=1$, if we take the axis of x coincident with that of the harmonic S_1, and write $x=r\cos\theta$, we have

$$\phi = A\left(\frac{\sin kr}{k^2r^2} - \frac{\cos kr}{kr}\right)\cos\theta \,.\, e^{i\sigma t}; \quad \text{.........................(8)}$$

and the equation (2) becomes

$$\tan ka = \frac{2ka}{2-k^2a^2}. \quad \text{.......................................(9)}$$

The zero root is irrelevant. The next root gives, for the ratio of the diameter to the wave-length,

$$ka/\pi = \cdot 6625,$$

and the higher values of this ratio approximate to the successive integers 2, 3, 4, ... In the case of the lowest root, we have, inverting,

$$\lambda/2a = 1\cdot 509.$$

In this, the gravest of all the normal modes, the air sways to and fro much in the same manner as in a doubly-closed pipe. In the case of any one of the higher roots, the roots of lower order give the positions of the spherical nodes ($\partial\phi/\partial r=0$). For the further discussion of the problem we must refer to the original investigation by Rayleigh*.

2°. To find the motion of the enclosed air due to a prescribed normal vibration of the boundary, say

$$\frac{\partial\phi}{\partial r} = S_n \,.\, e^{i\sigma t}, \quad \text{..(10)}$$

we have

$$\phi = A\psi_n(kr)\, r^n S_n \,.\, e^{i\sigma t}, \quad \text{..................................(11)}$$

with the condition

$$A\{ka\psi_n'(ka) + n\psi_n(ka)\}\, a^{n-1} = 1,$$

and therefore

$$\phi = \frac{\psi_n(kr)}{ka\psi_n'(ka) + n\psi_n(ka)} \,.\, a\left(\frac{r}{a}\right)^n S_n \,.\, e^{i\sigma t}. \quad \text{......................(12)}$$

This expression becomes infinite, as we should expect, whenever ka is a root of (2); *i.e.* whenever the period of the imposed vibration coincides with that of one of the natural periods, of the same spherical-harmonic type.

By putting $ka=0$ we pass to the case of an incompressible fluid. The formula (12) then reduces to

$$\phi = \frac{a}{n}\left(\frac{r}{a}\right)^n S_n \,.\, e^{i\sigma t}, \quad \text{....................................(13)}$$

as in Art. 91. It is important to notice that the same result holds approximately, even in the case of a gas, whenever ka is small, *i.e.* whenever the wave-length ($2\pi/k$) corresponding to the actual period is large compared with the circumference of the sphere. We have here an illustration of a general principle stated in Art. 290, of which considerable use will be made presently (Arts. 299, 300).

3°. To determine the motion of a gas within a space bounded by two concentric spheres, we require the complete formula (6) of Art. 292. The only interesting case, however, is where the two radii are nearly equal; and this can be solved more easily by an independent process†.

In terms of polar co-ordinates r, θ, ω, the equation $(\nabla^2+k^2)\,\phi=0$ becomes

$$\frac{\partial^2\phi}{\partial r^2} + \frac{2}{r}\frac{\partial\phi}{\partial r} + \frac{1}{r^2}\left[\frac{\partial}{\partial\mu}\left\{(1-\mu^2)\frac{\partial\phi}{\partial\mu}\right\} + \frac{1}{1-\mu^2}\frac{\partial^2\phi}{\partial\omega^2}\right] + k^2\phi = 0, \quad \text{...........(14)}$$

* "On the Vibrations of a Gas contained within a Rigid Spherical Envelope," *Proc. Lond. Math. Soc.* (1) iv. 93 (1872); *Theory of Sound*, Art. 331.

† Rayleigh, *Theory of Sound*, Art. 333. The direct solution is given by Chree, *Mess. of Math.* xv. 20 (1886); it depends on the formula (19) of Art. 292.

where $\mu = \cos\theta$. If, now, $\partial\phi/\partial r = 0$ for $r = a$ and $r = b$, where a and b are nearly equal, we may neglect the radial motion altogether, so that the equation reduces to

$$\frac{\partial}{\partial\mu}\left\{(1-\mu^2)\frac{\partial\phi}{\partial\mu}\right\} + \frac{1}{1-\mu^2}\frac{\partial^2\phi}{\partial\omega^2} + k^2a^2\phi = 0. \quad\text{......(15)}$$

It appears, exactly as in Art. 199, that the only solutions which are finite over the whole spherical surface are of the type

$$\phi \propto S_n, \quad\text{......(16)}$$

where S_n is a surface-harmonic of integral order n, and that the corresponding values of k are given by

$$k^2a^2 = n(n+1). \quad\text{......(17)}$$

In the gravest mode ($n=1$) the gas sways to and fro across the equator of the harmonic S_1, being, in the extreme phases of the oscillation, condensed at one pole and rarefied at the other. Since $ka = \sqrt{2}$ in this case, we have for the equivalent wave-length $\lambda/2a = 2{\cdot}221$.

In the next mode ($n=2$) the type of the vibration depends on that of the harmonic S_2. If this be zonal, the equator is a node. The frequency is determined by $ka = \sqrt{6}$, or $\lambda/2a = 1{\cdot}283$.

294. We may next consider the propagation of waves *outwards* from a spherical surface in an unlimited medium*.

If at the surface ($r=a$) we have a prescribed normal velocity

$$\dot{r} = S_n \,.\, e^{i\sigma t}, \quad\text{......(1)}$$

the appropriate solution of $(\nabla^2 + k^2)\phi = 0$ is, in the notation of Art. 292,

$$\phi = C_n f_n(kr) \,.\, r^n S_n \,.\, e^{i\sigma t}. \quad\text{......(2)}$$

The condition

$$-\frac{\partial\phi}{\partial r} = S_n \,.\, e^{i\sigma t}, \quad\text{......(3)}$$

which is to be satisfied at the surface of the sphere ($r=a$), gives

$$C_n = -\frac{1}{\{kaf_n'(ka) + nf_n(ka)\}\,a^{n-1}}. \quad\text{......(4)}$$

At distances r which are large compared with the wave-length ($2\pi/k$), we have

$$f_n(kr) = \frac{i^n e^{-ikr}}{(kr)^{n+1}}, \quad\text{......(5)}$$

approximately, so that (2) becomes

$$\phi = \frac{i^n C_n}{k^{n+1}}\frac{e^{ik(ct-r)}}{r} S_n, \quad\text{......(6)}$$

or, in real form,

$$\phi = \frac{|C_n|}{k^{n+1}}\frac{\cos k(ct - r + \epsilon)}{r} S_n. \quad\text{......(7)}$$

The rate of propagation of energy outwards is

$$-\iint p\frac{\partial\phi}{\partial r} r^2 d\varpi, \quad\text{......(8)}$$

where $\delta\varpi$ is an elementary solid angle, and r may conveniently be taken to be very great. Since

$$p = p_0 + \rho_0\frac{\partial\phi}{\partial t}, \quad\text{......(9)}$$

we find, for the mean value of (8),

$$\tfrac{1}{2}\frac{\rho_0 c}{k^{2n}}|C_n{}^2| \,.\, \iint S_n{}^2 d\varpi. \quad\text{......(10)}$$

* This problem was solved, in a somewhat different manner, by Stokes, "On the Communication of Vibrations from a Vibrating Body to a surrounding Gas," *Phil. Trans.* 1868 [*Papers*, iv. 299].

This might have been obtained at once from the result (9) of Art. 280, since the waves propagated in any assigned direction tend to become ultimately plane.

When $n > 0$, the normal velocity is in opposite phases over any two regions of the spherical surface $r = a$ which are separated by a nodal line $S_n = 0$. The lateral motion of the air near the sphere, from places which are moving outwards to others which are moving inwards will consequently, if the wave-length be not too small, have the effect of diminishing the intensity of the disturbance propagated to a distance, as compared with what it would have been if the normal velocity had been everywhere in the same phase; and this effect will be more marked the higher the order n of the harmonic involved, owing to the greater number of compartments into which the surface of the sphere is divided by the nodal lines. Moreover, for the same harmonic S_n, and for an assigned frequency $(\sigma/2\pi)$, the influence of the lateral motion will increase with great rapidity as the wave-velocity c, and (consequently) the wave-length $2\pi/k$, is increased. This accounts for the feeble character of the sound emitted by a bell in an atmosphere of hydrogen, as compared with what is observed in the case of air*.

To illustrate these statements, we note that if the lateral motion of the air had been prevented by a multitude of conical partitions extending indefinitely outwards in the directions of the radii of the sphere, the expression (10) would have been replaced by

$$\tfrac{1}{2}\rho_0 c\,|\,C_0\,|^2 . \iint S_n^2\, d\varpi. \quad \text{..................................(11)}$$

The ratio (I_n, say) which this bears to (10) is equal to the 'absolute value' of the expression

$$\frac{(ka)^{2n}\{kaf_n'(ka) + nf_n(ka)\}^2}{\{kaf_0'(ka)\}^2}. \quad \text{..............................(12)}$$

From the values of f_0, f_1, f_2 given in Art. 292 (15), we easily obtain

$$I_0 = 1, \qquad I_1 = \frac{4 + k^4a^4}{k^2a^2(1 + k^2a^2)}, \qquad I_2 = \frac{81 + 9k^2a^2 - 2k^4a^4 + k^6a^6}{k^4a^4(1 + k^2a^2)}. \quad \text{.........(13)}$$

The following numerical examples are given (with others) by Stokes:

ka	I_0	I_1	I_2
4	1	0·95588	0·87523
2	1	1	1·8625
1	1	2·5	44·5
0·5	1	13	1064·2
0·25	1	60·294	19650

Again, to compare the rates of communication of energy under similar circumstances to two different gases, we have, for the ratio of these rates, the absolute value of the expression

$$\frac{(k'a)^{2n-1}\{k'af_n'(ka) + nf_n(k'a)\}^2}{(ka)^{2n-1}\{kaf_n'(ka) + nf_n(ka)\}^2}, \quad \text{..............................(14)}$$

where the accent attached to k refers to the second gas. This is easily deduced from (10) and (4), with the help of the relations

$$\frac{\rho_0 c}{\rho_0' c'} = \frac{c'}{c} = \frac{k}{k'},$$

* Stokes, *l.c.*

the frequency being taken to be the same in the two cases*. For $n=2$, the ratio comes out equal to

$$\frac{(ka)^7\,(81+9k'^2a^2-2k'^4a^4+k'^6a^6)}{(k'a)^7\,(81+9k^2a^2-2k^4a^4+k^6a^6)}. \qquad (15)$$

Thus, supposing the two gases to be oxygen and hydrogen, and taking $ka=\cdot 5$, $k'a=\cdot 125$, we find that the rates of propagation of energy outwards are as 16000 : 1, nearly†.

295. The case $n=1$ of the preceding Art. is specially interesting from the point of view of the theory of the pendulum, since it corresponds to an oscillation of the sphere, as rigid, to and fro in a straight line. It should be noticed, however, that the neglect of the terms of the second order in the dynamical equations involves the assumption that the amplitude of vibration of the sphere is small compared with the radius.

For this problem we hardly need to have recourse to the general theory, the motion of the fluid being that due to a double source (Art. 289) at the centre of the sphere.

Assuming that the centre oscillates in the axis of x, with a velocity $U=\alpha e^{i\sigma t}$, say, we write

$$\phi=C\frac{\partial}{\partial x}\frac{e^{-ikr}}{r}=C\frac{d}{dr}\frac{e^{-ikr}}{r}.\cos\theta, \qquad (16)$$

if $x=r\cos\theta$. The condition that $-\partial\phi/\partial r=U\cos\theta$, for $r=a$, gives

$$C\frac{d^2}{da^2}\frac{e^{-ika}}{a}=-\alpha, \qquad (17)$$

whence

$$C=\frac{(2-k^2a^2-2ika)\,\alpha a^3 e^{ika}}{4+k^4a^4}. \qquad (18)$$

The resultant force on the sphere is

$$X=-\int_0^\pi p\cos\theta\,.\,2\pi a^2\sin\theta\,d\theta, \qquad (19)$$

where p denotes the pressure at the surface, viz.

$$p=p_0+\rho_0\dot{\phi}=p_0+i\sigma\rho_0 C\frac{d}{da}\frac{e^{-ika}}{a}.\cos\theta. \qquad (20)$$

Performing the integration, and substituting the value of C from (18), we find

$$X=-\tfrac{4}{3}\pi\rho_0 a^3.\frac{2+k^2a^2-ik^3a^3}{4+k^4a^4}\,i\sigma\alpha e^{i\sigma t}. \qquad (21)$$

This may be written in the form

$$X=-\tfrac{4}{3}\pi\rho_0 a^3.\frac{2+k^2a^2}{4+k^4a^4}.\frac{dU}{dt}-\tfrac{4}{3}\pi\rho_0 a^3.\frac{k^3a^3}{4+k^4a^4}.\sigma U. \qquad (22)‡$$

If we reverse the sign of X, we get the extraneous force which must be applied to the sphere in order to maintain the assumed simple-harmonic vibration.

* It is also assumed that the ratio γ of the specific heats is the same for the two gases.

† The distribution of energy in the space surrounding a vibrating sphere has been studied by Lennard-Jones, *Proc. Lond. Math. Soc.* (2) xx. 347 (1921). The energy in a region immediately surrounding the sphere is mainly kinetic, the fluid moving almost as if it were incompressible; cf. Art. 290. This region is more extensive the lower the frequency, and the higher the order of the harmonic involved. Considering the whole wave-system it is found that there is a finite excess of kinetic over potential energy.

‡ This formula is given by Rayleigh, *Theory of Sound*, Art. 325. For another treatment of the problem of the vibrating sphere, see Poisson, "Sur les mouvements simultanés d'un pendule et de l'air environnant," *Mém. de l'Acad. des Sciences*, xi. 521 (1832), and Kirchhoff, *Mechanik*, c. xxiii.

The first term of the expression in (22) is the same as if the inertia of the sphere were increased by the amount

$$\frac{2+k^2a^2}{4+k^4a^4}\times\tfrac{4}{3}\pi\rho_0 a^3; \quad\text{.......................(23)}$$

whilst the second is the same as if the sphere were subject to a frictional force varying as the velocity, the coefficient being

$$\frac{k^3a^3}{4+k^4a^4}\times\tfrac{4}{3}\pi\rho_0 a^3\sigma. \quad\text{.......................(24)}$$

In the case of an incompressible fluid, and, more generally, whenever the wave-length $2\pi/k$ is large compared with the circumference of the sphere, we may put $ka=0$. The addition to the inertia is then *half* that of the fluid displaced; whilst the frictional coefficient vanishes*. Cf. Art. 92.

The frictional coefficient is in any case of high order in ka, so that the vibrations of a sphere whose circumference is moderately small compared with the wave-length are only slightly affected in this way. To find the energy expended per unit time in generating waves in the surrounding medium, we must multiply the frictional term in (22), now regarded as an equation in real quantities, by U, and take the mean value; this is found to be

$$\tfrac{2}{3}\pi\rho_0 a^3 . \frac{k^3a^3}{4+k^4a^4} . \sigma\alpha^2. \quad\text{.......................(25)}$$

In other words, if ρ_1 be the mean density of the sphere, the fraction of its energy which is expended in one period is

$$2\pi\frac{\rho_0}{\rho_1} . \frac{k^3a^3}{4+k^4a^4} . \quad\text{.......................(26)}$$

296. The analysis of Art. 292 can be applied to calculate the scattering of waves by a spherical obstacle. In particular we shall consider the case of an incident system of plane waves, travelling in the direction of x-negative, and represented, apart from the time-factor, by

$$\phi = e^{ikx}. \quad\text{.......................(1)}$$

Since this satisfies $(\nabla^2+k^2)\phi=0$, and has no singularities in the neighbourhood of the origin, and is (further) symmetrical about the axis of x, it must admit of being expanded in a series of terms of the type

$$\psi_n(kr)\, r^n . P_n(\cos\theta), \quad\text{.......................(2)}$$

provided $x=r\cos\theta=r\mu$, say. We assume, then,

$$e^{ikr\mu} = A_0\psi_0(kr) + A_1\psi_1(kr)\,krP_1(\mu) + \ldots + A_n\psi_n(kr)(kr)^nP_n(\mu) + \ldots. \quad\text{......(3)}$$

If we differentiate this n times with respect to μ, the first n terms will disappear, since $P_s(\mu)$ is algebraic of degree s. Dividing the result by $(kr)^n$, and noting that

$$\frac{d^n}{d\mu^n}P_n(\mu) = 1.2.5\ldots(2n-1), \quad\text{.......................(4)}$$

by Art. 85 (1), we have

$$i^n e^{ikr\mu} = 1.3\ldots(2n-1)\,A_n\psi_n(kr) + \ldots. \quad\text{.......................(5)}$$

* Poisson, *l.c.*

Hence, putting $r=0$,

$$A_n=(2n+1)\,i^n, \qquad\text{.............................(6)}$$

by Art. 292 (7). Hence

$$e^{ikx}=\sum_0^\infty (2n+1)\,\psi_n(kr)\,(ikr)^n P_n(\mu). \qquad\text{..............(7)*}$$

This gives in spherical harmonics the velocity-potential of a source at an infinite distance. A similar expansion for the case of a source at a finite distance from the origin O is obtained as follows. Let P' be the position of the source, and P the point at which the velocity-potential is required. We write

$$OP=r, \qquad OP'=r', \qquad \rho^2=r^2-2rr'\mu+r'^2, \qquad\text{.......................(8)}$$

where $\mu=\cos POP'$. If $r<r'$ we may assume

$$f_0(k\rho)=\sum_0^\infty A_n\psi_n(kr)\,(kr)^n P_n(\mu). \qquad\text{..............................(9)}$$

If we vary ρ and μ only, we have $\rho\,d\rho=-rr'd\mu$, and therefore

$$-\frac{1}{k\rho}\frac{d}{d(k\rho)}=\frac{1}{kr\,.\,kr'}\frac{d}{d\mu}. \qquad\text{................................(10)}$$

Performing this operation n times on (9), we have by Art. 292 (14)

$$f_n(k\rho)=\frac{1\,.\,3\ldots(2n-1)}{(kr')^n}\,\psi_n(kr)\,A_n+\ldots. \qquad\text{........................(11)}$$

Now putting $r=0$ we have

$$A_n=(2n+1)\,(kr')^n f_n(kr'), \qquad\text{...............................(12)}$$

and therefore

$$f_0(k\rho)=\sum_0^\infty (2n+1)\,(kr)^n\,(kr')^n f_n(kr')\,\psi_n(kr)\,P_n(\mu). \qquad\text{.................(13)}$$

If $r>r'$ we have merely to interchange r and r' in the formula, since ρ is symmetrical with respect to these variables. Thus

$$f_0(k\rho)=\sum_0^\infty (2n+1)\,(kr)^n\,(kr')^n\,\psi_n(kr')\,f_n(kr)\,P_n(\mu). \qquad\text{.................(14)†}$$

We may utilize the formula (7) to shew how the typical solution of

$$(\nabla^2+k^2)\,\phi=0, \qquad\text{.......................................(15)}$$

which is finite at the origin, viz.

$$\phi=\psi_n(kr)\,r^n S_n, \qquad\text{.......................................(16)}$$

may be obtained by superposition of plane waves. The case of $n=0$ has already been considered (Art. 289).

We have, by the conjugate property of surface-harmonics (Art. 87),

$$\iint e^{ikr\mu} S_n\,d\varpi=(2n+1)\,(ikr)^n\,\psi_n(kr)\iint P_n(\mu)\,S_n\,d\varpi, \qquad\text{..............(17)}$$

if $\delta\varpi$ denote an element of area of a spherical surface of unit radius described about the origin. The symbol μ is here taken to denote the cosine of the angular distance of $\delta\varpi$ from the point Q where the surface is met by an arbitrary radius-vector r. Now, by a known formula of Laplace ‡,

$$\iint P_n(\mu)\,S_n\,d\varpi=\frac{4\pi}{2n+1}\,S_n', \qquad\text{.................................(18)}$$

where S_n' denotes the value of S_n at Q. Hence

$$(ikr)^n\,\psi_n(kr)\,S_n'=\frac{1}{4\pi}\iint e^{ikr\mu} S_n\,d\varpi. \qquad\text{..........................(19)}$$

* Rayleigh, *Proc. Lond. Math. Soc.* (1) iv. 253 (1873). See also Heine, *Kugelfunktionen*, i. 82 (1878). The above proof is adapted from Heine's proof of (13) below.

† The formula (13) (except for the notation) was proved in the above manner by Heine, i. 346. Equivalent results had been obtained by Clebsch (1863) in the paper quoted *ante* p. 110.

‡ Ferrers, *Spherical Harmonics*, p. 89.

The typical solution is thus expressed as the mean of a series of plane waves of unit amplitude whose normals are distributed about the origin with a variable density expressed by the harmonic S_n.

It follows that the motion in any region free from sources can be resolved into superposed trains of plane waves.

297. We proceed to the special problem of the incidence of air-waves on a spherical obstacle.

Consider a constituent

$$\phi = B_n \psi_n(kr)\, r^n S_n, \quad \text{.........(1)}$$

of an incident wave-system, and let the corresponding constituent of the scattered waves be

$$\phi' = B_n' f_n(kr)\, r^n S_n. \quad \text{.........(2)}$$

If the spherical obstacle be *fixed*, the condition

$$\frac{\partial}{\partial r}(\phi + \phi') = 0, \quad \text{.........(3)}$$

to be satisfied for $r = a$, gives

$$\frac{B_n'}{B_n} = -\frac{ka\psi_n'(ka) + n\psi_n(ka)}{kaf_n'(ka) + nf_n(ka)}. \quad \text{.........(4)}$$

This result can only be interpreted with facility when the wave-length is large compared with the perimeter of the sphere, *i.e.* when ka is small. Now for small values of ζ we have, by Art. 292 (7), (16),

$$\psi_n(\zeta) = \frac{1}{1.3\ldots(2n+1)}, \quad f_n(\zeta) = \frac{1.3\ldots(2n-1)}{\zeta^{2n+1}}, \quad \text{.........(5)}$$

approximately, whence, for $n > 0$,

$$\frac{B_n'}{B_n} = \frac{n}{n+1} \cdot \frac{(ka)^{2n+1}}{\{1.3\ldots(2n-1)\}^2(2n+1)}. \quad \text{.........(6)}$$

The case $n = 0$ is exceptional; we find

$$\frac{B_0'}{B_0} = -\tfrac{1}{3}(ka)^3, \quad \text{.........(7)}$$

approximately.

If the incident waves be plane, and represented by e^{ikx}, we have $S_n = P_n$, and by Art. 296 (7),

$$B_n = (2n+1)\, i^n k^n, \quad \text{.........(8)}$$

whence

$$B_0' = -\tfrac{1}{3}(ka)^2, \quad B_1' = \tfrac{1}{2} ik(ka)^3. \quad \text{.........(9)}$$

The most important part of the scattered waves, at a distance r which is large compared with the wave-length, is accordingly represented by

$$\phi' = B_0' f_0(kr) + B_1' f_1(kr)\, r\cos\theta = -(ka)^3\left(\tfrac{1}{3} + \tfrac{1}{2}\cos\theta\right)\frac{e^{-ikr}}{kr}, \quad \text{.........(10)}$$

by Art. 292 (15). The physical origin of the two terms is explained near the end of Art. 300.

As in Art. 294, the rate at which energy is propagated outwards in the scattered waves is

$$\Sigma \tfrac{1}{2}\frac{\rho_0 c}{k^{2n}} |B_n'|^2 \cdot \iint P_n^2\, d\varpi. \quad \text{.........(11)}$$

The proper standard of comparison here is the energy-flux across unit area of a wave-front in the incident system. On the present scale, this is $\tfrac{1}{2}\rho_0 k^2 c$, by Art. 280, and the ratio of (11) to this is

$$\Sigma \frac{4\pi}{(2n+1)\, k^{2n+2}} |B_n'^2|, \quad \text{.........(12)}$$

by Art. 87 (5). The terms of lowest order, when ka is small, are those for which $n=0$, $n=1$. Substituting from (9), and taking the sum, we obtain

$$\tfrac{7}{9}(ka)^4 \, . \, \pi a^2. \qquad \text{.......................................(13)}$$

The rate at which energy is scattered varies therefore inversely as the fourth power of the wave-length*.

As a numerical example, a spherule $\frac{1}{1000}$ of an inch in diameter scatters only $1{\cdot}43 \times 10^{-17}$ of the incident energy, if the wave-length be four feet. There is therefore no difficulty in understanding how a fog which is quite opaque optically may transmit ordinary sounds with great freedom.

298. We take next the case of plane waves incident on a *moveable* sphere. Its equation of motion will be of the form

$$M\ddot{\xi} = -\iint p \cos\theta \, a^2 d\varpi + X, \qquad \text{.......................................(1)}$$

where X denotes the extraneous force, if any.

If the time-factor be e^{ikct} we have

$$p = p_0 + \rho_0 \frac{\partial}{\partial t}(\phi + \phi') = p_0 + ikc\rho_0(\phi + \phi'). \qquad \text{.......................(2)}$$

Again, the kinematic surface-condition is

$$-\frac{\partial}{\partial r}(\phi + \phi') = \dot{\xi}\cos\theta = ikc\xi\cos\theta. \qquad \text{.............................(3)}$$

1°. Let us first suppose the sphere to be perfectly free to move under the impact of the air-waves, so that $X=0$. Writing $M = \frac{4}{3}\pi\rho_1 a^3$, and substituting from (2) in (1), we find

$$kc\rho_1\xi = i\{B_1\psi_1(ka) + B_1'f_1(ka)\}\rho_0, \qquad \text{.............................(4)}$$

since the products of harmonics of different orders disappear when integrated over the sphere. Again, from (3),

$$-ikc\xi = B_1\{ka\psi_1'(ka) + \psi_1(ka)\} + B_1'\{kaf_1'(ka) + f_1(ka)\}, \qquad \text{...........(5)}$$

whilst Art. 297 (4) holds for $n>1$. Eliminating ξ between (4) and (5) we have

$$\frac{B_1'}{B_1} = -\frac{\{ka\psi_1'(ka) + \psi_1(ka)\}\rho_1 - \psi_1(ka)\rho_0}{\{kaf_1'(ka) + f_1(ka)\}\rho_1 - f_1(ka)\rho_0}. \qquad \text{.....................(6)}$$

If ka be small the approximate values of $\psi_1(ka)$ and $f_1(ka)$ make

$$\frac{B_1'}{B_1} = \frac{\rho_1 - \rho_0}{6\rho_1 + 3\rho_0} k^3 a^3. \qquad \text{.......................................(7)}$$

The scattered waves of type $n=1$ disappear as we should expect, to this degree of approximation, when $\rho_1 = \rho_0$. The sphere merely drifts to and fro with the air.

If ξ_0 be the displacement of the air at the origin when the sphere is absent we have, on our present scale,

$$ikc\xi_0 = -ike^{ikct}. \qquad \text{.......................................(8)}$$

Hence, substituting from (7) in (5), and recalling that $B_1 = 3ik$, we find

$$\frac{\xi}{\xi_0} = \frac{3\rho_0}{2\rho_1 + \rho_0}. \qquad \text{.......................................(9)}$$

As we should expect, this ratio is less or greater than unity according as $\rho_1 \gtrless \rho_0$.

* The above problem was investigated, by a somewhat different analysis, by Rayleigh, *Proc. Lond. Math. Soc.* (1) iv. 253 (1872); see also *Theory of Sound*, Arts. 296, 334, 335. The result (13) is given by him in a paper "On the Transmission of Light through an Atmosphere containing Small Particles in Suspension," *Phil. Mag.* (5) xlvii. 375 (1899) [*Papers*, iv. 397].

2°. As an illustration of the theory of *resonance* we may also consider the case where the sphere is urged towards a fixed position by a force varying as the displacement. If $2\pi/\sigma_0$ be the natural period of vibration of the sphere when the influence of the air is neglected, we write in (1)

$$X = -M\sigma_0^2\xi. \qquad (10)$$

We have then in place of (4)

$$(\sigma_0^2 - k^2c^2)\,\rho_1\xi = -ikc\,\rho_0\,\{B_1\psi_1(ka) + B_1'f_1(ka)\}. \qquad (11)$$

Hence, and from (5),

$$\frac{\sigma_0^2 - k^2c^2}{k^2c^2}\,\rho_1 = -\frac{B_1\psi_1(ka) + B_1'f_1(ka)}{B_1\{ka\psi_1'(ka) + \psi_1(ka)\} + B_1'\{kaf_1'(ka) + f_1(ka)\}}\cdot\rho_0. \qquad (12)$$

When there are no extraneous sources, $B_1 = 0$, and

$$\frac{\sigma_0^2 - k^2c^2}{k^2c^2}\,\rho_1 = -\frac{f_1(ka)}{kaf_1'(ka) + f_1(ka)}\cdot\rho_0. \qquad (13)$$

This is an equation to determine k, and thence the character of the 'free' motion of the sphere, as influenced by the surrounding medium. When reduced to an algebraical form by Art. 292 (15), it is a biquadratic* in k, viz.

$$(k^2c^2 - \sigma_0^2)(k^2a^2 - 2ika - 2) + 2\beta k^2c^2(ika + 1) = 0, \qquad (14)$$

where $\beta = \frac{1}{2}\rho_0/\rho_1$. The two smaller roots are alone important from the present point of view. These are given approximately by

$$k^2c^2 = \sigma_0^2/(1+\beta). \qquad (15)$$

We recognize that the main effect of the presence of the fluid is to increase the inertia of the sphere by half that of the fluid displaced; cf. Arts. 92, 295. To find the rate of decay of the oscillation it would be necessary to carry the approximations further. It will be found, in agreement with the investigation of Art. 295, that the 'free' oscillations are of the type

$$\xi = Ce^{-\nu t}\cos(\sigma' t + \epsilon), \qquad (16)$$

where, if we retain only the most important parts,

$$\sigma' = \frac{\sigma_0}{\sqrt{(1+\beta)}}, \qquad \nu = \frac{\beta}{4(1+\beta)}\,\frac{\sigma'^4 a^3}{c^3}. \qquad (17)$$

In the *forced* oscillations, where the value of k is prescribed, we have from (12)

$$\frac{B_1'}{B_1} = -\frac{\{ka\psi_1'(ka) + \psi_1(ka)\}(\sigma_0^2 - k^2c^2) + 2\beta k^2c^2\psi_1(ka)}{\{kaf_1'(ka) + f_1(ka)\}(\sigma_0^2 - k^2c^2) + 2\beta k^2c^2 f_1(ka)}. \qquad (18)$$

If ka be small, the approximate values of $\psi_1(ka)$ and $f_1(ka)$ make

$$\frac{B_1'}{B_1} = \frac{\sigma_0^2 - (1-2\beta)k^2c^2}{\sigma_0^2 - (1+\beta)k^2c^2}\cdot\tfrac{1}{6}k^3a^3, \qquad (19)$$

but the approximation is plainly illusory when kc is nearly equal to $\sigma_0/(1+\beta)^{\frac{1}{2}}$, *i.e.* when the frequency of the incident waves is nearly coincident with that of the free vibration.

To examine more closely the case of approximate synchronism, we write, in the exact formula (18),

$$f_1(ka) = \Psi_1(ka) - i\psi_1(ka), \qquad (20)$$

and obtain

$$\frac{B_1'}{B_1} = -\frac{g_1(ka)}{G_1(ka) - ig_1(ka)}, \qquad (21)$$

where

$$\left.\begin{aligned} g_1(ka) &= \{ka\psi_1'(ka) + \psi_1(ka)\}\left(\frac{\sigma_0^2a^2}{c^2} - k^2a^2\right) + 2\beta k^2a^2\psi_1(ka), \\ G_1(ka) &= \{ka\Psi_1'(ka) + \Psi_1(ka)\}\left(\frac{\sigma_0^2a^2}{c^2} - k^2a^2\right) + 2\beta k^2a^2\Psi_1(ka). \end{aligned}\right\} \qquad (22)$$

* An equivalent result is obtained by putting

$$ae^{i\sigma t} = i\sigma\xi, \qquad X = M(\sigma_0^2 - \sigma^2)\,\xi,$$

in Art. 295 (21).

The modulus of the right-hand member of (21) is never greater than unity, but it attains the value unity, and the amplitude of the scattered waves is therefore a maximum, when

$$G_1(ka)=0, \qquad \text{.......(23)}$$

in which case

$$B_1' = -iB_1. \qquad \text{.......(24)}$$

In the case of the plane system of waves represented by Art. 296 (1) we have then

$$B_1' = 3k, \qquad \text{.......(25)}$$

and the velocity-potential of the scattered waves at a distance is, in real form,

$$\phi' = -3\frac{\sin k(ct-r)}{kr}\cos\theta, \qquad \text{.......(26)}$$

corresponding to the incident waves

$$\phi = \cos k(ct+x). \qquad \text{.......(27)}$$

This result, it may be noted, is independent of the magnitude of ka.

When ka is small we substitute for $\Psi_1(ka)$ from Art. 292 (7). The equation (23) takes the form

$$-(2+\tfrac{1}{4}k^4a^4+\ldots)\left(\frac{\sigma_0{}^2a^2}{c^2}-k^2a^2\right)+2\beta k^2a^2(1+\tfrac{1}{2}k^2a^2+\ldots)=0, \qquad \text{.......(28)}$$

and it is easily ascertained that when $\sigma_0 a/c$ is small this is satisfied by a real value of ka which is a very little less than that corresponding to the free vibrations, viz.

$$ka = \frac{\sigma_0 a}{(1+\beta)^{\frac{1}{2}}c}. \qquad \text{.......(29)}$$

Again, on reference to (3), we find

$$\xi = \frac{6}{k^3a^3c}\sin kct, \qquad \text{.......(30)}$$

approximately. The amplitude of vibration of the air-particles in the original wave is $1/c$ on the present scale. The amplitude of the sphere exceeds this in the ratio $6/k^3a^3$. Moreover it appears from (10) that the dissipation of energy in the scattered waves, when a maximum, is $6\pi\rho_0 c$, or, in terms of the energy-flux in the primary waves,

$$3\lambda^2/\pi, \qquad \text{.......(31)}$$

where λ is the wave-length. The ratio of this to the dissipation produced by a fixed sphere is $\frac{108}{7}(ka)^{-6}$.

On the other hand, it is to be noticed that the wave-length of maximum dissipation is very sharply defined. It may be shewn without much difficulty that the dissipation sinks to one-half the maximum when the wave-length of the incident vibration deviates from the critical value by the fraction

$$\frac{\beta k^3a^3}{4(1+\beta)}$$

of itself. In any acoustical application this will be an exceedingly minute fraction. In practice, massive bodies are not usually set into vigorous sympathetic vibration by the *direct* impact of air-waves, but rather through the intermediary of resonance-boxes and sounding-boards.

The occurrence of the factor 3 in (31) calls for some remark. The result is independent of the direction of the incident waves, owing to the three degrees of freedom which the sphere possesses. If the sphere were restricted to vibration in a definite straight line, the amount of scattering would depend on the direction of incidence, and the mean for all such directions would be λ^2/π*.

* The investigations of this Art. are taken from a paper entitled "A Problem in Resonance, illustrative of the Theory of Selective Absorption of Light," *Proc. Lond. Math. Soc.* xxxii. 11 (1900). The concluding remark is due to Rayleigh, "Some General Theorems concerning Forced Vibrations and Resonance," *Phil. Mag.* (6) iii. 97 (1902) [*Papers*, v. 8].

299. The diffraction of plane waves of sound by a lamina, or by an aperture in a plane screen, can be treated by approximate methods provided the dimensions of the obstacle or of the aperture be small compared with the wave-length*. This relation is of course the exact opposite of that which usually prevails in Optics, and the results are accordingly quite different in character. In particular we meet with nothing of the nature of sound-shadows or sound-rays under the present condition.

1°. Let us first take the case where a train of waves, travelling in the direction of x-negative, impinges on a flat disk in the plane $x=0$. If the disk were absent the motion would be represented by

$$\phi = e^{ikx}, \qquad \text{.........(1)}$$

everywhere. This gives a normal velocity $-ik$ at the surface of the disk; and the complete solution is therefore

$$\phi = e^{ikx} + \chi, \qquad \text{.........(2)}$$

where χ represents the motion which would be produced in the surrounding air by the oscillation of the disk normal to its plane with the velocity ik. The formula (18) of Art. 290 gives

$$\chi_P = \frac{1}{2\pi}\iint \chi \frac{\partial}{\partial n}\left(\frac{e^{-ikr}}{r}\right) dS, \qquad \text{.........(3)}$$

where the integration is taken over the positive side only of the disk. If x, y, z be the co-ordinates of P relative to an origin in the disk, we may write $\partial/\partial n = -\partial/\partial x$; and if the distance of P from any point of the disk be large compared with the linear dimensions of the latter, we have further

$$\chi_P = -\frac{1}{2\pi}\iint \chi dS \cdot \frac{\partial}{\partial x}\left(\frac{e^{-ikr}}{r}\right), \qquad \text{.........(4)}$$

where r may now be taken to denote distance from the origin. The scattered waves are therefore such as would be produced by a double-source of suitable strength.

Under the fundamental condition above stated, the variation of χ in the immediate neighbourhood of the disk is very approximately the same as if the fluid were incompressible (Art. 290). In the latter case, if the density of the fluid, and the velocity of the disk normal to its plane, were each taken equal to unity, the expression $2\iint \chi dS$ would be equal to the 'inertia-coefficient' of the disk (Art. 121 (3)). Denoting this coefficient, which is determined solely by the size and shape of the disk, by M, we have, in the present case,

$$\iint \chi dS = \tfrac{1}{2} ikM, \qquad \text{.........(5)}$$

and therefore

$$\chi_P = -\frac{ikM}{4\pi}\frac{\partial}{\partial x}\left(\frac{e^{-ikr}}{r}\right) = -\frac{k^2 M}{4\pi}\cdot\frac{e^{-ikr}}{r}\cos\theta, \qquad \text{.........(6)}$$

approximately, where θ is the angle which OP makes with Ox.

For a *circular* disk of radius a, we have, by Arts. 102, 108,

$$M = \tfrac{8}{3}a^3, \qquad \text{.........(7)}$$

and therefore

$$\chi_P = -\tfrac{8}{3}\frac{\pi a^3}{\lambda^2}\cdot\frac{e^{-ikr}}{r}\cos\theta. \qquad \text{.........(8)}$$

2°. When plane waves are incident directly upon a screen in the plane $x=0$, we should have, if the screen were complete,

$$\phi = e^{ikx} + e^{-ikx}, \text{ or } = 0, \qquad \text{.........(9)}$$

* Rayleigh, "On the Passage of Waves through Apertures in Plane Screens, and Allied Problems," *Phil. Mag.* (5) xliii. 259 (1897) [*Papers*, iv. 283].

according as $x \gtrless 0$, the term e^{-ikx} representing the reflected waves. When there is an aperture, we assume

$$\phi = e^{ikx} + e^{-ikx} + \chi, \text{ and } \phi = \chi', \qquad \text{.......(10)}$$

for the two sides, respectively. The continuity of pressure and velocity requires

$$2 + \chi = \chi', \qquad \frac{\partial \chi}{\partial x} = \frac{\partial \chi'}{\partial x}, \qquad \text{.......(11)}$$

over the aperture, whilst

$$\frac{\partial \chi}{\partial x} = 0, \qquad \frac{\partial \chi'}{\partial x} = 0, \qquad \text{.......(12)}$$

over the rest of the plane $x = 0$.

These conditions are all fulfilled if we take χ and χ' to be the potentials of the distributions of simple sources over the area of the aperture which will make

$$\chi = -1, \qquad \chi' = +1, \qquad \text{.......(13)}$$

respectively, over this area.

Now from (17) of Art. 290 we have

$$\chi_P = -\frac{1}{2\pi} \iint \frac{e^{-ikr}}{r} \frac{\partial \chi}{\partial n} dS. \qquad \text{.......(14)}$$

In the present case the integration may be confined to the area of the aperture, in virtue of (12). On this understanding we have, at distances r which are great compared with the dimensions of the area,

$$\chi_P = -\frac{1}{2\pi} \iint \frac{\partial \chi}{\partial n} dS \,.\, \frac{e^{-ikr}}{r}. \qquad \text{.......(15)}$$

If k were $= 0$, the determination of χ in accordance with (13) would be identical with the problem of finding the flow of an incompressible fluid through the aperture; and for points in the immediate neighbourhood the flow will in the actual problem have sensibly the same configuration. Hence we may write

$$\iint \frac{\partial \chi}{\partial n} dS = 2C, \qquad \text{.......(16)}$$

where C is the conductivity of the aperture*. Thus (15) becomes

$$\chi_P = -C \frac{e^{-ikr}}{\pi r}, \qquad \text{.......(17)}$$

approximately. From this the value of χ' follows by the obvious relation

$$\chi'(-x, y, z) = -\chi(x, y, z). \qquad \text{.......(18)}$$

It appears that the transmitted waves are such as would be produced by a *simple* source of suitable strength.

The value of C for an elliptic aperture has been given in Art. 113 (8). For the circular form

$$C = 2a \qquad \text{.......(19)}$$

and

$$\chi_P = -\frac{2a}{\pi} \frac{e^{-ikr}}{r}. \qquad \text{.......(20)}$$

Comparison with (8) shews that, under the assumed condition, the amplitude of the waves scattered by a disk is, at like distances, much less than that of the waves transmitted by an aperture of the same size and shape. It is readily seen that the total energy transmitted per second through a circular aperture bears to the energy-flux in the primary waves the ratio

$$8a^2/\pi^2 \text{ or } \cdot 816\pi a^2. \qquad \text{.......(21)}$$

The ratio of the amplitude of the scattered waves, at any distant point, to that of the primary waves, is independent of the wave-length, so long as this is large compared with the greatest breadth of the aperture.

* Cf. Arts. 102, 3°; 108, 1°; and 113.

300. A similar calculation can be applied to the scattering of sound-waves by an obstacle of any form, under the same fundamental condition that the dimensions of the obstacle are all small compared with the wave-length*.

The origin being taken in or near the obstacle, we assume

$$\phi = e^{ikx} + \chi, \qquad \text{.........(1)}$$

where the first term represents the incident, and the second the scattered waves. At the surface of the obstacle, supposed rigid and fixed, we must have

$$\frac{\partial \chi}{\partial n} = -\frac{\partial}{\partial n} e^{ikx} = -ikle^{ikx}, \qquad \text{.........(2)}$$

provided l, m, n be the direction-cosines of the normal, drawn outwards.

The formula (5) of Art. 290 gives

$$\chi_P = -\frac{1}{4\pi}\iint \frac{e^{-ikr}}{r}\frac{\partial \chi}{\partial n}\,dS + \frac{1}{4\pi}\iint \chi \frac{\partial}{\partial n}\left(\frac{e^{-ikr}}{r}\right) dS, \qquad \text{.........(3)}$$

where the integrations extend over the surface of the obstacle. We proceed to obtain an approximate value of the expression on the right-hand side when the distances r are large compared with the dimensions of the obstacle. The co-ordinates of any point on the surface will be denoted by x, y, z, whilst those of the point P are distinguished as x_1, y_1, z_1.

Taking the first term, we write

$$\frac{e^{-ikr}}{r} = \left(\frac{e^{-ikr}}{r}\right)_0 + x\left(\frac{\partial}{\partial x}\frac{e^{-ikr}}{r}\right)_0 + y\left(\frac{\partial}{\partial y}\frac{e^{-ikr}}{r}\right)_0 + z\left(\frac{\partial}{\partial z}\frac{e^{-ikr}}{r}\right)_0 + \ldots,$$

where the zero-suffix implies that x, y, z are to be put $=0$ in the expressions to which it is attached. This may also be written

$$\frac{e^{-ikr}}{r} = \left(1 - x\frac{\partial}{\partial x_1} - y\frac{\partial}{\partial y_1} - z\frac{\partial}{\partial z_1} + \ldots\right)\frac{e^{-ikr_0}}{r_0}, \qquad \text{.........(4)}$$

where r_0 denotes the distance P from the origin. Again, from (2),

$$\frac{\partial \chi}{\partial n} = -ikl + k^2 xl + \ldots. \qquad \text{.........(5)}$$

Taking the product of (4) and (5), and integrating over the surface, we obtain

$$\iint \frac{e^{-ikr}}{r}\frac{\partial \chi}{\partial n}\,dS = k^2 Q\,\frac{e^{-ikr_0}}{r_0} + ikQ\,\frac{\partial}{\partial x_1}\frac{e^{-ikr_0}}{r_0}, \qquad \text{.........(6)}$$

approximately, where Q is the volume of the obstacle. We have here made use of the obvious relations

$$\iint l\,dS = 0, \quad \iint xl\,dS = Q, \quad \iint yl\,dS = 0, \quad \iint zl\,dS = 0. \qquad \text{.........(7)}$$

The terms retained on the right-hand side of (6) are of the same order of magnitude, whilst those which are omitted are small in comparison.

As regards the second term in (3), we have

$$\frac{\partial}{\partial n}\frac{e^{-ikr}}{r} - \left(l\frac{\partial}{\partial x} + m\frac{\partial}{\partial y} + n\frac{\partial}{\partial z}\right)\frac{e^{-ikr}}{r} = -\left(l\frac{\partial}{\partial x_1} + m\frac{\partial}{\partial y_1} + n\frac{\partial}{\partial z_1}\right)\frac{e^{-ikr}}{r}. \qquad \text{......(8)}$$

We may, consistently with our former approximation, write r_0 for r, and remove the space-derivatives of e^{-ikr_0}/r_0 outside the signs of integration. The result then involves the surface-integrals

$$\iint l\chi\,dS, \quad \iint m\chi\,dS, \quad \iint n\chi\,dS. \qquad \text{.........(9)}$$

It appears from (2) or (5), and from a general principle stated in Art. 290, that the function χ is, in the immediate neighbourhood of the obstacle, sensibly identical with the velocity-potential of the motion of a liquid produced by a translation of the obstacle through it with

* Rayleigh, "On the Incidence of Aerial and Electric Waves upon Small Obstacles in the Form of Ellipsoids or Elliptic Cylinders...," *Phil. Mag.* (5) xliv. 28 (1897) [*Papers*, iv. 305].

the velocity ik parallel to x. Hence the integrals (9) are recognized as components of 'impulse' under the imagined circumstances; and we may write, in conformity with Art. 121,

$$\iint l\chi\, dS = ik\mathbf{A}, \quad \iint m\chi\, dS = ik\mathbf{C}', \quad \iint n\chi\, dS = ik\mathbf{B}', \quad \text{..................(10)}$$

provided the density of the hypothetical liquid be taken to be unity. Hence

$$\iint \chi \frac{\partial}{\partial n}\left(\frac{e^{-ikr}}{r}\right) dS = -ik\left(\mathbf{A}\frac{\partial}{\partial x_1} + \mathbf{C}'\frac{\partial}{\partial y_1} + \mathbf{B}'\frac{\partial}{\partial z_1}\right)\frac{e^{-ikr_0}}{r_0}. \quad \text{............(11)}$$

The final approximate formula is therefore

$$\chi_P = -\frac{k^2 Q}{4\pi}\frac{e^{-ikr}}{r} - \frac{ik}{4\pi}\left\{(\mathbf{A}+Q)\frac{\partial}{\partial x_1} + \mathbf{C}'\frac{\partial}{\partial y_1} + \mathbf{B}'\frac{\partial}{\partial z_1}\right\}\frac{e^{-ikr}}{r}, \quad \text{........(12)*}$$

where the zero-suffix attached to r has been omitted, as no longer necessary. When kr is large, this may be written

$$\chi_P = -\frac{k^2 Q}{4\pi}\frac{e^{-ikr}}{r} - \frac{k^2}{4\pi}\{(\mathbf{A}+Q)\lambda_1 + \mathbf{C}'\mu_1 + \mathbf{B}'\nu_1\}\frac{e^{-ikr}}{r}, \quad \text{..............(13)}$$

where λ_1, μ_1, ν_1 are the direction-cosines of r.

For a sphere of radius a, $\mathbf{A} = \frac{2}{3}\pi a^3$, $Q = \frac{4}{3}\pi a^3$, $\mathbf{B}' = 0$, $\mathbf{C}' = 0$, leading again to the result (10) of Art. 297.

The scattered waves may be regarded as due to the combination of a simple and a double source. The axis of the latter is not in general coincident with the direction of the incident waves.

A more symmetrical formula is obtained if we suppose the primary waves to come from any arbitrary direction (λ, μ, ν), so that (1) is replaced by

$$\phi = e^{ik(\lambda x + \mu y + \nu z)} + \chi. \quad \text{..................................(14)}$$

On reviewing the steps of the preceding investigation, we find without difficulty

$$\chi_P = -\frac{k^2 Q}{4\pi}\frac{e^{-ikr}}{r} - \frac{k^2 Q}{4\pi}(\lambda\lambda_1 + \mu\mu_1 + \nu\nu_1)\frac{e^{-ikr}}{r} - \frac{k^2}{4\pi}\{\mathbf{A}\lambda\lambda_1 + \mathbf{B}\mu\mu_1 + \mathbf{C}\nu\nu_1$$
$$+ \mathbf{A}'(\mu\nu_1 + \mu_1\nu) + \mathbf{B}'(\nu\lambda_1 + \nu_1\lambda) + \mathbf{C}'(\lambda\mu_1 + \lambda_1\mu)\}\frac{e^{-ikr}}{r}, \quad \text{..............(15)}$$

in place of (13). As in Art. 124, the directions of the co-ordinate axes can be chosen so that $\mathbf{A}'$, $\mathbf{B}'$, $\mathbf{C}' = 0$, and the formula then reduces to

$$\chi_P = -\frac{k^2 Q}{4\pi}\frac{e^{-ikr}}{r} - \frac{k^2}{4\pi}\{(\mathbf{A}+Q)\lambda\lambda_1 + (\mathbf{B}+Q)\mu\mu_1 + (\mathbf{C}+Q)\nu\nu_1\}\frac{e^{-ikr}}{r}. \quad \text{......(16)}$$

In the case of an ellipsoid of semi-axes a, b, c, we have, by Art. 121 (4),

$$\mathbf{A} + Q = \frac{2}{2-\alpha_0}Q, \qquad \mathbf{B} + Q = \frac{2}{2-\beta_0}Q, \qquad \mathbf{C} + Q = \frac{2}{2-\gamma_0}Q, \quad \text{............(17)}$$

where α_0, β_0, γ_0 are defined by Art. 114 (6). In the case of the circular disk ($a = b$, $c = 0$), we have $Q = 0$, $\mathbf{A} = \frac{8}{3}a^3$, $\mathbf{B} = 0$, $\mathbf{C} = 0$; and (16) reduces to

$$\chi_P = -\frac{2}{3}\frac{k^3 a^3}{\pi}\lambda\lambda_1\frac{e^{-ikr}}{r}. \quad \text{..................................(18)}$$

The effect of obliquity of the disk to the incident waves is to diminish the amplitude of the scattered waves in the ratio of the cosine (λ) of the obliquity.

The explanation of the two types of disturbance in (13) or (16) may easily be given in general terms. In the first place, if the obstacle were absent, the space which it occupies would be the seat of alternate condensations and rarefactions. By its resistance to these, the obstacle exerts a certain reaction on the medium; the waves at a great distance, thus

* If we divide by k and then make $k \to 0$ we reproduce the result obtained in Art. **121** *a* for the case of incompressibility.

produced, are in fact such as would be caused in an otherwise quiescent medium by a periodic variation in the volume of the obstacle, just sufficient to compensate the variations of density referred to. The result is equivalent to a 'simple source' of sound. Superposed on this disturbance, we have a second wave-system due to the *immobility* of the obstacle. If the latter were freely moveable, and had (moreover) the same inertia as the air which it displaces, it would sway backwards and forwards in the sound-vibrations, and this second wave-system would be absent. This system is, in fact, that which would be produced if the obstacle were to vibrate to and fro in a straight line, with a motion equal and opposite to that of the air-particles in the undisturbed waves. This is equivalent to a 'double source.'

The problem of Diffraction, when the wave-length is *small* (instead of large) compared with the dimensions of the obstacle, presents as a rule great analytical difficulties. The only case which can be regarded as completely solved is that of the semi-infinite plane screen, where nothing depends on the magnitude of the wave-length. This is considered in Art. 308 below. In the case of plane waves incident on a fixed sphere, which at first sight appears promising, the complete expression for the disturbance due to the incident and scattered waves is given by the formulae of Art. 297; thus

$$\phi+\phi' = \Sigma\,(2n+1)\,(ikr)^n \left\{\psi_n(kr) - \frac{ka\psi_n'(ka)+n\psi_n(ka)}{kaf_n'(ka)+nf_n(ka)}\,f_n(kr)\right\} P_n(\mu). \quad \text{......(19)}$$

At points on the surface of the sphere this reduces to

$$\phi+\phi' = -\,\Sigma\,\frac{(2n+1)\,i^n P_n(\mu)}{(ka)^{n+1}\{kaf_n'(ka)+nf_n(ka)\}}. \quad \text{......................(20)}$$

Unfortunately, when the wave-length is small compared with the circumference $2\pi a$ of the sphere, ka is large, and the series in (20) is found to converge very slowly, so that a large number of terms have to be taken in order to secure a satisfactory approximation. The process has been carried out by Rayleigh* in the case of $ka=10$, which suffices to shew the incipient formation of a sound-shadow in the rear of the sphere (*i.e.* in the neighbourhood of $\mu=-1$)†.

301. If, no longer restricting ourselves to simple-harmonic vibrations, we seek to integrate the equation

$$\frac{\partial^2\phi}{\partial t^2} = c^2\nabla^2\phi \quad \text{..............................(1)}$$

in a series of spherical harmonics, say

$$\phi = \Sigma R_n \phi_n, \quad \text{..............................(2)}$$

where ϕ_n is a solid harmonic of order n, we have by Art. 292 (4)

$$\frac{\partial^2 R_n}{\partial t^2} = c^2\left\{\frac{\partial^2 R_n}{\partial r^2} + \frac{2(n+1)}{r}\frac{\partial R_n}{\partial r}\right\}. \quad \text{....................(3)}$$

If R_n be a solution of this, it is easily verified that the corresponding equation for R_{n+1} is satisfied by

$$R_{n+1} = \frac{1}{r}\frac{\partial R_n}{\partial r}, \quad \text{..........................(4)}$$

* "On the Acoustic Shadow of a Sphere," *Phil. Trans.* A, cciii. 87 (1904) [*Papers*, v. 149].

† The cognate optical and electrical problems, which are of importance in relation to such widely diverse questions as the theory of the rainbow and the influence of the curvature of the earth in wireless telegraphy, have been discussed by Debye, L. Lorenz, Macdonald, Nicholson, Poincaré, and others, not without controversy as to the legitimacy of the mathematical processes employed. Full references are given by Love, "On the Transmission of Electric Waves over the Surface of the Earth," *Phil. Trans.* A, ccxv. 105 (1914). See also Watson, *Proc. Roy. Soc.* xcv. 83 (1918).

and hence that (3) is satisfied by

$$R_n = \left(\frac{\partial}{r\partial r}\right)^n R_0 = \left(\frac{\partial}{r\partial r}\right)^n . \frac{f(r-ct)+F(r+ct)}{r} \quad \text{.............(5)*}$$

In the case $n = 1$, we have the solution

$$\phi = \frac{\partial}{\partial r} . \frac{f(r-ct)+F(r+ct)}{r} \cos\theta. \quad \text{......................(6)}$$

This has been employed by Kirchhoff, and more fully by Love, to examine the rather interesting question how the front of a system of waves, started by the motion of a sphere, is propagated through the surrounding medium.

In Kirchhoff's investigation† the motion of the sphere is prescribed, its velocity being a given function of the time, and the solution is comparatively simple.

Love discusses‡ the waves started by an instantaneous impulse given to a ball-pendulum. The equation of motion of the pendulum being

$$M\left(\frac{d^2\xi}{dt^2} + \sigma_0^2\xi\right) = -\iint p\cos\theta\, a^2 d\varpi, \quad \text{..........................(7)}$$

as in Art. 298, we assume

$$\phi = \frac{\partial}{\partial r} \frac{f(ct-r)}{r} . \cos\theta, \quad \text{....................................(8)}$$

the term in (6) which corresponds to waves travelling inwards being omitted. This leads to

$$\frac{d^2\xi}{dt^2} + \sigma_0^2\xi = \frac{2\beta c}{a^2}\left\{f''(ct-a) + \frac{1}{a}f'(ct-a)\right\}, \quad \text{........................(9)}$$

where

$$\beta = \tfrac{2}{3}\pi\rho_0 a^3/M. \quad \text{..(10)}$$

The kinematical condition to be satisfied at the surface of the sphere ($r=a$) gives

$$\frac{d\xi}{dt} = -\frac{1}{a}f''(ct-a) - \frac{2}{a^2}f'(ct-a) - \frac{2}{a^3}f(ct-a). \quad \text{.....................(11)}$$

To solve the simultaneous equations (9), (11), we assume

$$f(ct-r) = Ae^{\lambda(ct-r+a)}, \quad \xi = Be^{\lambda ct}, \quad \text{.............................(12)}$$

whence

$$(\lambda^2c^2+\sigma_0^2)B = \frac{2\beta c}{a^3}(\lambda a+1)\lambda A, \quad \lambda cB = -\frac{1}{a^3}(\lambda^2a^2+2\lambda a+2)A. \quad \text{...........(13)}$$

Eliminating the ratio A/B, we obtain the biquadratic§ in λ:

$$(\lambda^2c^2+\sigma_0^2)(\lambda^2a^2+2\lambda a+2) + 2\beta c^2\lambda^2(\lambda a+1) = 0. \quad \text{....................(14)}$$

Distinguishing the several roots by suffixes, we have

$$\xi = -\Sigma_1^4 \frac{\lambda_s^2a^2+2\lambda_s a+2}{ca^3\lambda_s} A_s e^{\lambda_s ct}, \quad \text{.............................(15)}$$

$$\phi = -\Sigma_1^4 (\lambda_s r+1)\frac{A_s}{r^2} e^{\lambda_s(ct-r+a)}\cos\theta. \quad \text{.......................(16)}$$

* Cf. Clebsch, *l.c. ante* p. 110; C. Niven, *Solutions of the Senate House Problems...for* 1878, p. 158.

† *l.c. ante* p. 510.

‡ "Some Illustrations of Modes of Decay of Vibratory Motions," *Proc. Lond. Math. Soc.* (2) ii. 88 (1904).

§ If we put $\lambda = ik$ this becomes identical with Art. 298 (14).

If we start with arbitrary values of ξ and $d\xi/dt$, the medium being previously at rest, this solution presupposes that $t>0$ and $r<ct+a$. The initial circumstances supply two conditions to be satisfied by the four constants A_s. Thus, assuming that for $t=0$

$$\xi\equiv 0,\quad \frac{d\xi}{dt}=U_0, \qquad \text{.........(17)}$$

we have

$$\Sigma_1^4\left(\lambda_s a+2+\frac{2}{\lambda_s a}\right)A_s=0,\quad \Sigma_1^4(\lambda_s^2a^2+2\lambda_s a+2)\,A_s=-U_0a^3. \qquad \text{.........(18)}$$

The remaining conditions result from a consideration of the discontinuity at the spherical boundary of the advancing wave. Let δS be an element of this boundary, and through the contour of δS draw normals outwards to meet a parallel surface at a distance $c\delta t$; we thus mark out an element of volume $\delta S\,.\,c\delta t$. In time δt the fluid contained in this element has its normal velocity changed from 0 to $-\partial\phi/\partial r$, the normal velocity just within the boundary, by the action of the excess of pressure $c^2\rho_0 s$, where s is the condensation, on the inner face. Hence

$$-\frac{\partial\phi}{\partial r}\,.\,\rho_0\,\delta S\,.\,c\delta t=c^2\rho_0 s\,.\,\delta S\,.\,\delta t,$$

or, since $c^2s=\partial\phi/\partial t$,

$$\frac{\partial\phi}{\partial t}=-c\,\frac{\partial\phi}{\partial r}, \qquad \text{.........(19)}$$

which is to be satisfied for $r=ct+a$*. Substituting from (16) we find

$$\Sigma\,(\lambda_s r+2)\,A_s=0.$$

This equation cannot hold generally unless

$$\Sigma\lambda_s A_s=0,\quad \Sigma A_s=0, \qquad \text{.........(20)}$$

which (it will be noticed) at the same time secure the continuity of ϕ, and thence of the velocity-components tangential to the wave-front.

The four conditions (18), (20) may now be written

$$\Sigma\lambda_s^2A_s=-U_0a,\quad \Sigma\lambda_sA_s=0,\quad \Sigma A_s=0,\quad \Sigma\frac{A_s}{\lambda_s}=0, \qquad \text{.........(21)}$$

whence

$$A_1=-\frac{\lambda_1}{(\lambda_1-\lambda_2)(\lambda_1-\lambda_3)(\lambda_1-\lambda_4)}\,.\,U_0a,\ \ldots,\ \ldots,\ \ldots. \qquad \text{.........(22)}$$

The motion of the air is accordingly given by

$$\left.\begin{aligned}\phi&=\frac{\partial}{\partial r}\Sigma_1^4\frac{A_s}{r}e^{\lambda_s(ct-r+a)}\cos\theta \qquad &[r<ct+a],\\ \phi&=0 &[r>ct+a].\end{aligned}\right\} \qquad \text{.........(23)}$$

In practice β is a very minute fraction, and the roots of (14) are, to a first approximation,

$$\lambda_1=\frac{i\sigma_0}{c},\quad \lambda_2=-\frac{i\sigma_0}{c},\quad \lambda_3=\frac{-1+i}{a},\quad \lambda_4=\frac{-1-i}{a}. \qquad \text{.........(24)}$$

If the distance travelled by a sound-wave in the period of vibration be a considerable multiple of the circumference of the sphere, λ_3, λ_4 will be large compared with λ_1, λ_2. Hence, substituting in (22) and (23), we find, for $r<ct+a$,

$$\phi=-\frac{\partial}{\partial r}\left\{\frac{U_0a^3}{2r}\cos\sigma_0\left(t-\frac{r-a}{c}\right)-\frac{\sqrt{2}\,U_0a^3}{2r}e^{-(ct+a-r)/a}\cos\left(\frac{ct+a-r}{a}-\tfrac{1}{4}\pi\right)\right\}\cos\theta \qquad \text{......(25)}$$

* The theory of discontinuities at wave-fronts has been treated systematically by Christoffel, "Untersuchungen über die mit dem Fortbestehen linearer partieller Differential-Gleichungen verträglichen Unstetigkeiten," *Ann. di Matemat.* viii. 81 (1876); and by Love, "Wave-Motions with Discontinuities at Wave-Fronts," *Proc. Lond. Math. Soc.* (2) i. 37 (1903).

The first part of this expression is the same as if the sphere had been executing simple-harmonic vibrations of period $2\pi/\sigma_0$ and amplitude U_0/σ_0 for an indefinite time. The second part is insensible at a distance of several diameters of the sphere from the inner side of the boundary of the advancing wave; but near this boundary it becomes comparable with the first part. To trace the decay of the oscillations, it would be necessary to proceed to a second approximation; but this part of the question has been already dealt with in Arts. 295, 298. It will be sufficient to remark that the most important part of the disturbance, well within the advancing wave, will be given by an expression of the form

$$\phi = \frac{C}{r} e^{-m(ct-r)} \cos \sigma_0 \left(t - \frac{r}{c} + \epsilon\right) . \cos \theta. \qquad \text{......................(26)}$$

The factor e^{-mct} exhibits the decay of the vibration at any place as the original energy of the pendulum is gradually spent in the generation of waves. To account for the factor e^{mr}, we note that within the region occupied by the waves the amplitude of any point Q will (except for spherical divergence) be greater than that at a point P, nearer to the centre on the same radius vector, in the ratio $e^{m.PQ}$, for the reason that it represents a disturbance which started earlier by an interval PQ/c, during which the vibration of the pendulum has been decaying according to the law e^{-mct} *.

Sound-Waves in Two Dimensions.

302. When ϕ is independent of z, we have

$$\frac{\partial^2 \phi}{\partial t^2} = c^2 \nabla_1^2 \phi, \qquad \text{..............................(1)}$$

where

$$\nabla_1^2 = \frac{\partial^2}{\partial x^2} + \frac{\partial^2}{\partial y^2}. \qquad \text{..............................(2)}$$

In the case of symmetry about the origin this becomes

$$\frac{\partial^2 \phi}{\partial t^2} = c^2 \left(\frac{\partial^2 \phi}{\partial r^2} + \frac{1}{r}\frac{\partial \phi}{\partial r}\right), \qquad \text{......................(3)}$$

where $r = \sqrt{(x^2 + y^2)}$. The general solution has been obtained in Art. 196, in the form

$$2\pi\phi = \int_0^\infty f\left(t - \frac{r}{c}\cosh u\right) du + \int_0^\infty F\left(t + \frac{r}{c}\cosh u\right) du; \qquad \text{......(4)}$$

and it was further shewn that the solution

$$2\pi\phi = \int_0^\infty f\left(t - \frac{r}{c}\cosh u\right) du \qquad \text{...................(5)}$$

represents the system of diverging waves produced by a source $f(t)$ at the origin.

We are now able to give another derivation of these results. It appears from Art. 285 (13) that if a point-source $f(t)\,\delta z$ be situate at the point $(0, 0, z)$ its effect at a point in the plane xy at a distance r from the origin is represented by

$$\frac{1}{4\pi\sqrt{(r^2 + z^2)}} f\left(t - \frac{\sqrt{(r^2 + z^2)}}{c}\right) \delta z.$$

* Cf. a paper "On a Peculiarity of the Wave-System due to the Free Vibrations of a Nucleus in an Extended Medium," *Proc. Lond. Math. Soc.* (1) xxxii. 208 (1900).

If we integrate this with respect to z between the limits $\mp\infty$, we get the effect of a system of point-sources distributed over the axis of z with uniform line-density $f(t)$; thus

$$\phi = \frac{1}{4\pi}\int_{-\infty}^{\infty} f\left(t - \frac{\sqrt{(r^2+z^2)}}{c}\right)\frac{dz}{\sqrt{(r^2+z^2)}} = \frac{1}{2\pi}\int_0^{\infty} f\left(t - \frac{r}{c}\cosh u\right)du. \quad \text{...(6)}$$

The same method can of course be applied to obtain the second term in (4).

The equations of sound-motion in one, two, or three dimensions, subject to the restriction of symmetry, are all included in the form

$$\frac{\partial^2\phi}{\partial t^2} = c^2\left(\frac{\partial^2\phi}{\partial r^2} + \frac{m-1}{r}\frac{\partial\phi}{\partial r}\right). \quad \text{.....(7)}$$

The complicated and somewhat intractable form in which the solution for the case $m=2$ has been obtained is in striking contrast to the analytical simplicity, and outward formal resemblance, of the solutions for the cases $m=1$, $m=3$; but this circumstance must not mislead us as to the true physical relations. For the sake of a definite comparison between the three cases, we may examine the effect (A) of a plane-source, (B) of a line-source, and (C) of a point-source, whose 'strength' is in each case

$$f(t) = \frac{\tau}{t^2+\tau^2}. \quad \text{.....(8)}$$

This gives a convenient representation of a source of a more or less transient character, since the time during which it is sensible can be made as short as we please by diminishing τ, whilst the time-integral is unaffected.

The results may be conveniently expressed in terms of the condensation s.

(A) In the case $m=1$, we find, for $x>0$,

$$s = \frac{\tau}{2c}\frac{1}{\left(t-\dfrac{x}{c}\right)^2+\tau^2}. \quad \text{.....(9)}$$

(B) When $m=2$, the analytical work is similar to that of Art. 197. The result is, for the most important part of the wave,

$$s = \frac{1}{4\sqrt{2c^2\tau^2}}\sqrt{\left(\frac{cr}{r}\right)}\sin\left(\tfrac{1}{4}\pi - \tfrac{3}{2}\eta\right)\cos^{\frac{3}{2}}\eta. \quad \text{.....(10)}$$

where η is determined by

$$t = \frac{r}{c} + \tau\tan\eta.$$

(C) In three dimensions we have

$$s = \frac{\tau}{2\pi c^2}\frac{\dfrac{r}{c}-t}{r\left\{\left(t-\dfrac{r}{c}\right)^2+\tau^2\right\}^2}. \quad \text{.....(11)}$$

The three cases are represented, with s as ordinate and t as abscissa, on the next page. The scale of t is the same in each case, but there is, of course, no relation between the vertical scales. In (A) we have a wave of pure condensation; in (B) the primary condensation is followed by a rarefaction of less amount, but lasting for a longer time; whilst in (C) the condensation and rarefaction are anti-symmetrical. In (B) and (C) alike we necessarily have, at any point,

$$\int_{-\infty}^{\infty} s\,dt = 0; \quad \text{.....(12)}$$

cf. Art. 288. If the source had been strictly limited in duration, the medium, in the case of three dimensions, would have remained absolutely at rest after the passage of the wave,

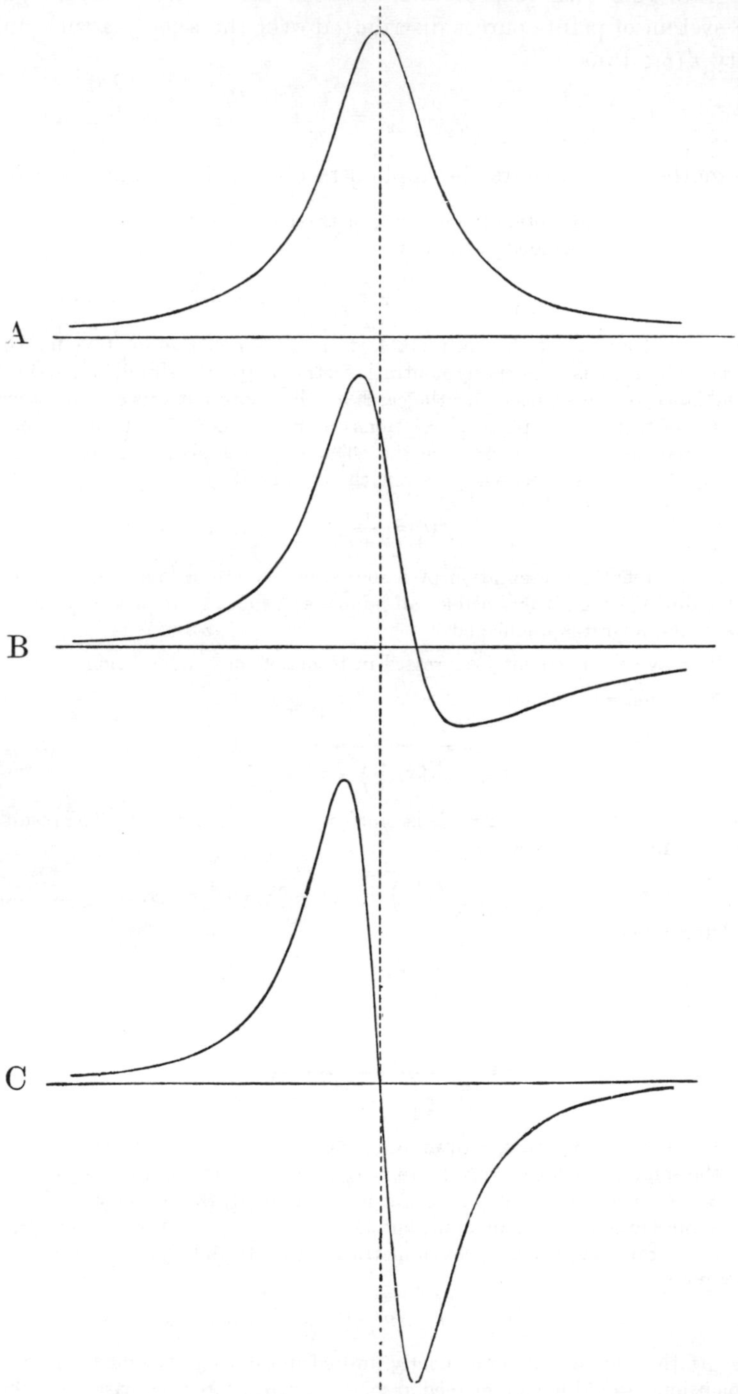
A
B
C

as in the case of one dimension, although for a different reason. In the intermediate case of two dimensions, the wave has an indefinitely extended 'tail,' and there is only an asymptotic approach to rest.

It appears that from a physical standpoint the cases $m=1$, $m=2$, $m=3$ form a sequence with a regular gradation of properties due to the increasing mobility of the medium.

When we abandon the restriction to symmetry, the general solution of (1) is, in polar co-ordinates,

$$\phi = \Sigma\,(Q_s r^s \cos s\theta + R_s r^s \sin s\theta), \qquad \text{.................(13)}$$

where Q_s, R_s are functions of r and t satisfying

$$\frac{\partial^2 Q_s}{\partial t^2} = c^2\left(\frac{\partial^2 Q_s}{\partial r^2} + \frac{2s+1}{r}\frac{\partial Q_s}{\partial r}\right), \qquad \text{.....................(14)}$$

and the corresponding equation in R_s. The solution of (14) is

$$Q_s = \left(\frac{\partial}{r\partial r}\right)^s Q_0, \qquad \text{..........................(15)}$$

where $$Q_0 = \int_0^\infty f\left(t - \frac{r}{c}\cosh u\right)du + \int_0^\infty F\left(t + \frac{r}{c}\cosh u\right)du. \qquad \text{......(16)}$$

The proof is similar to that of Art. 301 (5)*.

303. In the case of simple-harmonic motion ($e^{i\sigma t}$), we have, in polar co-ordinates,

$$\frac{\partial^2\phi}{\partial r^2} + \frac{1}{r}\frac{\partial\phi}{\partial r} + \frac{1}{r^2}\frac{\partial^2\phi}{\partial\theta^2} + k^2\phi = 0, \qquad \text{.......................(1)}$$

where $k=\sigma/c$. The solution of this equation subject to the condition of finiteness at the origin is, as in Art. 191,

$$\phi = \Sigma\,(A_s \cos s\theta + B_s \sin s\theta)\,J_s(kr), \qquad \text{.....................(2)}$$

where s may have all integral values from 0 to ∞.

From this we derive at once the theorem that the mean value ($\bar{\phi}$) of ϕ, over a circle of radius r described with the origin as centre, is

$$\bar{\phi} = J_0(kr)\,.\,\phi_0, \qquad \text{...............................(3)}$$

where ϕ_0 is the value at the origin†. This theorem (which is subject, of course, to the condition above stated) is analogous to that of Art. 289 (8), and might have been proved in a similar manner.

In the transverse oscillations of the air contained in a circular cylinder, the normal modes are given by the several terms of (2), where the admissible values of k, and thence of σ, are determined by

$$J_s'(ka) = 0, \qquad \text{...(4)}$$

a being the radius. The interpretation of the results will be understood from Art. 191, where the mathematical problem is identical. The figures on p. 288 shew the forms of the lines of equal pressure, to which the motions of the particles are orthogonal, in two of the more important modes‡.

* This Art. is taken, with slight alterations, from a paper cited on p. 298 *ante*.

† H. Weber, *Math. Ann.* i. (1868).

‡ The problem is fully discussed by Rayleigh, *Theory of Sound*, Art. 339.

The Bessel's Functions $J_s(\zeta)$ are subject to the recurrence-formula

$$\frac{d}{d\zeta}\frac{J_s(\zeta)}{\zeta^s} = -\frac{J_{s+1}(\zeta)}{\zeta^s}, \qquad \text{.........................(5)}$$

which corresponds to Art. 292 (17). This easily follows from the series-expression for $J_s(\zeta)$, given in Art. 101. From (5), and from the differential equation of $J_s(\zeta)$, viz.

$$f''(\zeta) + \frac{1}{\zeta}f'(\zeta) + \left(1 - \frac{s^2}{\zeta^2}\right)f(\zeta) = 0, \qquad \text{................(6)}$$

various other recurrence-formulae may be derived, *e.g.*

$$\zeta J_s'(\zeta) + sJ_s(\zeta) = \zeta J_{s-1}(\zeta), \qquad \text{.......................(7)}$$

corresponding to Art. 298 (18).

By successive applications of (5) we obtain

$$J_s(\zeta) = \zeta^s\left(-\frac{d}{\zeta d\zeta}\right)^s J_0(\zeta). \qquad \text{.......................(8)}$$

It is easily verified, by the method of 'mathematical induction,' that the expression on the right-hand side of (8) is in fact a solution of the differential equation (6) provided $J_0(\zeta)$ is a solution of the same equation with s put $= 0$. This suggests a convenient choice, for our present purpose, of the Bessel's Functions 'of the second kind.' We write

$$D_s(\zeta) = \zeta^s\left(-\frac{d}{\zeta d\zeta}\right)^s D_0(\zeta), \qquad \text{.......................(9)}$$

where $D_0(\zeta)$ is the function introduced in Art. 194*, viz.

$$D_0(\zeta) = \frac{2}{\pi}\int_0^\infty e^{-i\zeta\cosh u}\,du. \qquad \text{.......................(10)}$$

It is evident without further proof that $D_s(\zeta)$ will satisfy the differential equation (6), and will have the same system of recurrence-formulae as $J_s(\zeta)$. As an important special case of (9), we have

$$D_1(\zeta) = -D_0'(\zeta). \qquad \text{.........................(11)}$$

The following approximations are useful. When ζ is small, we have, by Arts. 100, 194,

$$J_0(\zeta) = 1 - \frac{\zeta^2}{4} + \ldots, \quad D_0(\zeta) = -\frac{2}{\pi}(\log\tfrac{1}{2}\zeta + \gamma + \tfrac{1}{2}i\pi + \ldots)\ \ldots, \qquad \text{...(12)}$$

and thence, by (8) and (9), for $s > 0$,

$$J_s(\zeta) = \frac{\zeta^s}{2^s \cdot s!} + \ldots, \quad D_s(\zeta) = \frac{2^s(s-1)!}{\pi\zeta^s} + \ldots. \qquad \text{..........(13)}$$

* It may be shewn, by the method of Art. 302, that $D_0(kr)$ is the potential of a uniform distribution of simple-harmonic point-sources along the axis of z. See Rayleigh, "On Point-, Line-, and Plane-Sources of Sound," *Proc. Lond. Math. Soc.* xix. 504 (1888) [*Papers*. iii. 44; *Theory of Sound*, Art. 342].

Again, for large values of ζ, we have

$$J_s(\zeta) = \left(\frac{2}{\pi\zeta}\right)^{\frac{1}{2}} \sin(\zeta + \tfrac{1}{4}\pi - \tfrac{1}{2}s\pi) + \ldots, \quad D_s(\zeta) = \left(\frac{2}{\pi\zeta}\right)^{\frac{1}{2}} i^s e^{-i(\zeta+\frac{1}{4}\pi)} + \ldots, \quad \text{.........(14)}$$

unless the order s of the functions be itself comparable with or greater than the variable ζ.

The formulae may be used to investigate the communication of vibrations from an oscillating cylinder (*e.g.* a piano string) to the surrounding air. The velocity of translation of the cylinder being

$$U = \alpha e^{i\sigma t}, \quad \text{.........(15)}$$

the radial velocity at the surface $r = a$ will be

$$-\frac{\partial\phi}{\partial r} = \alpha e^{i\sigma t} \cdot \cos\theta. \quad \text{.........(16)}$$

The corresponding value of ϕ is

$$\phi = A D_1(kr) \cos\theta \cdot e^{i\sigma t}, \quad \text{.........(17)}$$

with the condition

$$-kD_1'(ka) \cdot A = \alpha. \quad \text{.........(18)}$$

If, as we will suppose, the circumference of the cylinder is very small compared with the wave-length of the sound, ka will be a small fraction, and we find from (13)

$$A = \tfrac{1}{2}\pi k a^2 \alpha.$$

Hence at distances r which are large compared with k^{-1}, we have, by (14),

$$\phi = \sqrt{(\tfrac{1}{2}\pi)} \cdot \frac{ik^{\frac{1}{2}} a^2}{r^{\frac{1}{2}}} \alpha \cos\theta e^{i(\sigma t - kr + \frac{1}{4}\pi)}. \quad \text{.........(19)}$$

If the velocity at the boundary $r = a$ had been everywhere radial, with the amplitude α, the value of ϕ at a distance would have been

$$\phi = \sqrt{(\tfrac{1}{2}\pi)} \cdot \frac{a}{k^{\frac{1}{2}} r^{\frac{1}{2}}} \alpha e^{i(\sigma t - kr - \frac{1}{4}\pi)}. \quad \text{.........(20)}$$

In the actual case the intensity, as measured by the square of the amplitude, is less in the ratio k^2a^2, which is by hypothesis very small. This illustrates the effect of lateral motion near the surface of the cylinder, in reducing the amplitude of the waves propagated to a distance; cf. Art. 294. For example, by far the greater part of the sound due to a piano string comes, not directly from the wire, but from the sounding-board, which is set into forced vibration by the alternating pressures at the supports.

The reaction of the air on the vibrating cylinder is

$$-\int_0^{2\pi} p\cos\theta \cdot a\,d\theta = -\rho_0 a \int_0^{2\pi} \frac{\partial\phi}{\partial t}\cos\theta = -\pi\rho_0 a \cdot i\sigma A D_1(ka)\, e^{i\sigma t}$$

$$= \pi\rho_0 a^2 \cdot \frac{D_1(ka)}{ka\, D_1'(ka)} \cdot \frac{dU}{dt}, \quad \text{.........(21)}$$

by (18). When ka is small, this reduces to

$$-\pi\rho_0 a^2 \cdot \frac{dU}{dt}, \quad \text{.........(22)}$$

approximately. The most important part of the effect is that the inertia of the cylinder is increased by an amount equal to that of the air displaced; cf. Art. 68*.

304. We may also investigate the scattering of a system of plane waves by a fixed cylindrical obstacle whose axis is parallel to the wave-fronts.

Assuming, for the potential of the incident waves,

$$\phi = e^{ikx}, \quad \text{.........(1)}$$

* A fuller investigation is given by Stokes, *l.c. ante* p. 508.

as in Art. 296, we require in the first place to expand this in a series of the type (2) of Art. 303. The requisite formula is

$$e^{ikx} = J_0(kr) + 2iJ_1(kr)\cos\theta + \ldots + 2i^s J_s(kr)\cos s\theta + \ldots. \quad \ldots\ldots(2)$$

This may be proved directly*, by expanding $e^{ikr\cos\theta}$, making use of the formula

$$\cos^n\theta = \frac{1}{2^{n-1}}\left\{\cos n\theta + \frac{n}{1}\cos(n-2)\theta + \frac{n(n-1)}{1.2}\cos(n-4)\theta + \ldots\right\}, \quad \ldots\ldots\ldots(3)$$

and picking out the coefficient of $\cos s\theta$ in the result.

The expansion (2) involves the equality

$$\frac{1}{\pi}\int_0^\pi e^{ikr\cos\theta}\cos s\theta\, d\theta = i^s J_s(kr), \quad \ldots\ldots\ldots(4)$$

which is a known formula in Bessel's Functions†. Conversely if we assume this, as otherwise established, we may derive another proof of (2).

The scattered waves being represented by

$$\phi' = \Sigma B_s D_s(kr)\cos s\theta, \quad \ldots\ldots\ldots(5)$$

the surface-condition $$\frac{\partial}{\partial r}(\phi + \phi') = 0 \quad [r=a] \quad \ldots\ldots\ldots(6)$$

gives $$B_s = -\frac{2i^s J_s'(ka)}{D_s'(ka)}, \quad \ldots\ldots\ldots(7)$$

except in the case $s=0$, when the factor 2 is to be omitted.

If ka be small, we have, approximately,

$$J_0'(ka) = -\tfrac{1}{2}ka, \qquad D_0'(ka) = -\frac{2}{\pi ka}, \quad \ldots\ldots\ldots(8)$$

and, for $s>0$, $$J_s'(ka) = \frac{(ka)^{s-1}}{2^s(s-1)!}, \qquad D_s'(ka) = -\frac{2^s s!}{\pi(ka)^{s+1}}. \quad \ldots\ldots\ldots(9)$$

Hence $$B_0 = -\tfrac{1}{4}\pi k^2a^2, \qquad B_s = \frac{\pi i^s (ka)^{2s}}{2^{2s-1}s!(s-1)!} \quad [s>0]. \quad \ldots\ldots(10)$$

The most important terms correspond to $s=0$, $s=1$. Neglecting the rest, we have, for the scattered waves,

$$\phi' = -\tfrac{1}{4}\pi k^2a^2\{D_0(kr) - 2iD_1(kr)\cos\theta\}. \quad \ldots\ldots\ldots(11)$$

For large values of kr this becomes, on restoring the time-factor,

$$\phi' = -\tfrac{1}{2}\sqrt{(\tfrac{1}{2}\pi)}\frac{k^{\frac{3}{2}}a^2}{r^{\frac{1}{2}}}(1 + 2\cos\theta)\, e^{i(\sigma t - kr - \frac{1}{4}\pi)}. \quad \ldots\ldots(12)‡$$

The rate (per unit length of the cylinder) at which energy is carried outwards by the scattered waves is

$$-\int_0^{2\pi} p\frac{\partial\phi'}{\partial r}.r\,d\theta = -\rho_0 r\int_0^{2\pi}\frac{\partial\phi'}{\partial t}\frac{\partial\phi'}{\partial r}d\theta,$$

* Heine, *Kugelfunktionen*, i. 82. The method employed in the proof of Art. 296 (7) is also available.

† Watson, p. 21. The case $s=0$ has already been met with in Art. 100; it may be interpreted as shewing how the potential $J_0(kr)$ can be obtained by superposition of systems of plane waves travelling in directions uniformly distributed about the origin in the plane xy; cf. Art. 289 (7).

‡ Cf. Rayleigh, *Theory of Sound*, Art. 343.

where r may conveniently be taken very great. If we substitute the real part of ϕ' from (12), the mean value is found to be

$$\tfrac{3}{8}\pi^2\rho_0\sigma\,(ka)^4. \qquad\qquad(13)$$

The energy-flux in the primary waves is, as in Art. 297, $\frac{1}{2}\rho_0 k^2 c$. The ratio of (13) to this is (since $\sigma = kc$)

$$\tfrac{3}{8}\pi^2\,(ka)^3\,.\,2a. \qquad\qquad(14)$$

Thus a wire $\frac{1}{50}$ of an inch in diameter scatters only $6{\cdot}63\times10^{-8}$ of the incident energy, when the wave-length is four feet.

305. The approximate methods of Arts. 299, 300 can be applied to the corresponding problems in two dimensions*. The formula (5) of Art. 290 is now replaced by

$$\phi_P = -\tfrac{1}{4}\int D_0(kr)\frac{\partial\phi}{\partial n}\,ds + \tfrac{1}{4}\int\phi\frac{\partial}{\partial n}D_0(kr)\,ds, \qquad\qquad(1)$$

which is established in an analogous manner. In the case of a region extending to infinity the line-integrals can be restricted to the internal boundary, provided that at a great distance R from the origin ϕ tends to the form $D_0(kR)$ or $e^{-ikR}/R^{\frac{1}{2}}$.

In the same way we have

$$0 = -\tfrac{1}{4}\int D_0(kr')\frac{\partial\phi}{\partial n}\,ds + \tfrac{1}{4}\int\phi\frac{\partial}{\partial n}D_0(kr')\,ds, \qquad\qquad(2)$$

where r' denotes distance from a point P' external to the region considered.

Within a region whose dimensions in the plane xy are small compared with the wave length kr will be small, and the formula (1) reduces to

$$\phi_P = \frac{1}{2\pi}\int\log r\,\frac{\partial\phi}{\partial n}\,ds - \frac{1}{2\pi}\int\phi\frac{\partial}{\partial n}\log r\,ds, \qquad\qquad(3)$$

where a constant term has been omitted. This satisfies the equation

$$\nabla_1^2\phi = 0 \qquad\qquad(4)$$

appropriate to the case of an incompressible fluid.

1°. Taking first the direct impact of waves on a plane lamina, we write

$$\phi = e^{ikx} + \chi, \qquad\qquad(5)$$

where χ is the potential of the scattered waves. If the lamina be supposed to occupy that portion of the plane $x=0$ which lies between the lines $y=\pm b$, the condition to be satisfied by χ is

$$\frac{\partial\chi}{\partial x} = -ik, \quad [x=0,\ b>y>-b]. \qquad\qquad(6)$$

If we apply the formulae (1), (2) to the region lying to the right of the axis of y, and take P' at the *image* of P with respect to this boundary, we find by subtraction

$$\chi_P = \tfrac{1}{2}\int_{-b}^{b}\chi\frac{\partial}{\partial n}D_0(kr)\,dy, \qquad\qquad(7)$$

where the values of χ and $\partial D_0/\partial n$ on the positive face of the lamina are to be understood. If x, y refer to the position of P we may write $\partial/\partial n = -\partial/\partial x$; and at a distance r from the origin, large compared with $2b$, we have

$$\chi_P = -\tfrac{1}{2}\int_{-b}^{b}\chi\,dy\,.\,\frac{\partial}{\partial x}D_0(kr)\,; \qquad\qquad(8)$$

* Rayleigh, *ll.cc. ante* pp. 517, 519.

cf. Art. 299 (4). The definite integral is one-half the 'impulse' of the lamina (per unit length) when moving broadside-on with velocity ik in an incompressible fluid of unit density; hence by comparison with Art. 71 (11)

$$\int_{-b}^{b} \chi dy = \tfrac{1}{2} ik \,.\, \pi b^2, \quad \text{.........(9)}$$

and therefore

$$\chi_P = -\tfrac{1}{4} i\pi k b^2 \frac{\partial}{\partial x} D_0(kr) = \tfrac{1}{4} i\pi k^2 b^2 D_1(kr) \,.\, \cos\theta. \quad \text{.........(10)}$$

When kr is large this reduces to

$$\chi_P = -\frac{1}{2\sqrt{2}} \frac{\pi^{\frac{1}{2}} k^{\frac{3}{2}} b^2}{r^{\frac{1}{2}}} e^{-i(kr+\frac{1}{4}\pi)} \cos\theta, \quad \text{.........(11)}$$

by Art. 303 (14).

The ratio of the energy scattered per second to the energy-flux in the primary waves is easily found to be

$$\tfrac{1}{16} \pi^2 (kb)^3 \,.\, 2b, \quad \text{.........(12)}$$

which is exactly one-sixth of the corresponding ratio in the case of a circular cylinder of radius b.

2°. In the case of an aperture bounded by parallel straight edges ($y = \pm b$) in a plane screen ($x = 0$), we assume as in Art. 299, 2°

$$\phi = e^{ikx} + e^{-ikx} + \chi, \quad \text{and} \quad \phi = \chi', \quad \text{.........(13)}$$

for the two sides respectively, and seek to determine χ, χ' so that

$$\chi = -1, \quad \chi' = +1 \quad \text{.........(14)}$$

over the aperture, whilst

$$\frac{\partial \chi}{\partial x} = 0, \quad \frac{\partial \chi'}{\partial x} = 0 \quad \text{.........(15)}$$

over the screen. Now if we apply (1) and (2) to the part of the plane lying to the right of the axis of y, and if we further take P' at the image of P, we have by addition

$$\chi_P = -\tfrac{1}{2} \int_{-b}^{b} D_0(kr) \frac{\partial \chi}{\partial n} dy, \quad \text{.........(16)}$$

where δn is drawn from the positive face. At distances r from the origin which are large compared with $2b$, this becomes

$$\chi_P = -\tfrac{1}{2} \int_{-b}^{b} \frac{\partial \chi}{\partial n} dy \,.\, D_0(kr). \quad \text{.........(17)}$$

In the immediate neighbourhood of the aperture the motion represented by the functions χ, χ' must resemble the flow of a liquid through the same aperture, and an approximate value of the definite integral in (17) is accordingly obtained by comparison with the results of Art. 66, 1°. It appears that corresponding to a flux unity through the aperture the increment of χ in passing from the aperture itself to a distance r which is large compared with $2b$ is

$$\frac{1}{\pi} \log \frac{2r}{b}.$$

We may still suppose r to be small compared with the wave-length, and the formulae (14) and (17) then shew that the corresponding increment of χ in the actual problem is

$$1 + \frac{1}{\pi} \int_{-b}^{b} \frac{\partial \chi}{\partial n} dy \,.\, (\log \tfrac{1}{2} kr + \gamma + \tfrac{1}{2} i\pi), \quad \text{.........(18)}$$

by Art. 303 (12). Equating this to

$$\frac{1}{\pi} \int_{-b}^{b} \frac{\partial \chi}{\partial n} dy \,.\, \log \frac{2r}{b}, \quad \text{.........(19)}$$

we find

$$\int_{-b}^{b} \frac{\partial \chi}{\partial n} dy = -\frac{\pi}{\log \frac{1}{4} kb + \gamma + \frac{1}{2} i\pi}. \quad \text{.........(20)}$$

Hence when kr is large, we have from (17)

$$\chi_P=\frac{\frac{1}{2}\pi}{\log\frac{1}{4}kb+\gamma+\frac{1}{2}i\pi}D_0(kr)=\frac{1}{\log\frac{1}{4}kb+\gamma+\frac{1}{2}i\pi}\left(\frac{\pi}{2kr}\right)^{\frac{1}{2}}e^{-i(kr+\frac{1}{4}\pi)}. \quad\text{......(21)}$$

The value of χ' at any point P on the negative side of the plane $x=0$ is equal and opposite to the value of χ at the image of P with respect to the plane.

The ratio of the energy transmitted through the aperture to the energy-flux in the primary waves is found to be

$$\frac{\frac{1}{4}\pi^2}{kb\{(\log\frac{1}{4}kb+\gamma)^2+\frac{1}{4}\pi^2\}}\cdot 2b. \quad\text{......(22)}$$

If the wave-length be 10, or 100, or 1000 times the breadth of the aperture, the factor of $2b$ comes out equal to 1·240, or 3·795, or 17·20, respectively.

3°. The two-dimensional problem of the diffraction of plane waves by a cylindrical obstacle of any form of cross-section can be treated by the method of Art. 300*, the formula (1) above taking the place of Art. 290 (5) in the investigation. As no new point arises, it will be sufficient to state the chief result. The waves are supposed to be incident from the direction $(\lambda, \mu, 0)$, and we write, accordingly,

$$\phi=e^{ik(\lambda x+\mu y)}+\chi, \quad\text{......(23)}$$

where χ is to represent the scattered waves. We assume also that the axes of x, y have special directions in the plane of the cross-section, such that the kinetic energy (per unit length parallel to z) of an incompressible fluid of unit density, when the cylinder moves through it with velocity $(u, v, 0)$, would be given by an expression of the form

$$\tfrac{1}{2}(\mathbf{A}u^2+\mathbf{B}v^2), \quad\text{......(24)}$$

the term in uv being absent. The dimensions of the section being supposed small compared with the wave-length, the waves scattered in the direction $(\lambda_1, \mu_1, 0)$ are given by

$$\chi_P=-\frac{k^2S}{(8\pi kr)^{\frac{1}{2}}}e^{-i(kr+\frac{1}{4}\pi)}-\frac{k^2}{(8\pi kr)^{\frac{1}{2}}}\{(\mathbf{A}+S)\lambda\lambda_1+(\mathbf{B}+S)\mu\mu_1\}e^{-i(kr+\frac{1}{4}\pi)}, \quad\text{...(25)}$$

where S is the area of the cross-section.

For an elliptic section whose semi-axes in the directions of x, y are a, b, we have (see Art. 71 (11))

$$S=\pi ab, \quad \mathbf{A}=\pi b^2, \quad \mathbf{B}=\pi a^2. \quad\text{......(26)}$$

In the cases of a circular cylinder $(a=b)$, and of a flat lamina $(a=0)$, we reproduce results already obtained.

306. We may also investigate the disturbance produced in a train of plane waves by a thin screen which is interrupted by a series of parallel, equal, and equidistant slits. As before, the treatment is approximate, and involves the assumption that the wave-length is large compared with the distance between the centres of successive apertures.

As a preliminary question, we require to determine the flow of an incompressible fluid through a fixed rigid grating of the above kind. This can be solved by Schwarz' method (Art. 73); but for the present purpose it will be sufficient to state, and verify, the result. The axis of x being taken normal to the plane of the grating, and that of y in this plane, at right angles to the lengths of the apertures, we write

$$\cosh w=\mu\cosh z, \quad\text{......(1)}$$

where, for the moment,

$$w=\phi+i\psi, \qquad z=x+iy, \quad\text{......(2)}$$

* Cf. Rayleigh, *l.c. ante* p. 519.

and the constant μ is supposed greater than unity. This makes w a cyclic function; but we avoid all indeterminateness if in the first instance we confine ourselves to that half of the plane xy for which $x>0$, and if further we fix the value of w at some one point. We will assume that at the origin $\psi=0$, whilst ϕ is equal to the real positive value of $\cosh^{-1}\mu$.

The formula (1) gives

$$\cosh\phi\cos\psi=\mu\cosh x\cos y, \qquad \sinh\phi\sin\psi=\mu\sinh x\sin y. \quad \text{............(3)}$$

The locus $\phi=0$ consists of those portions of the axis of y for which

$$1>\mu\cos y>-1;$$

these represent the apertures, so that on the scale of our formulae the half-breadth of an aperture is $\sin^{-1}(1/\mu)$. For other portions of the region $x>0$, ϕ will be positive. Again, the lines $\psi=0$, $\psi=\pm\pi$, $\psi=\pm 2\pi$, ... will consist partly of the lines $y=0$, $y=\pm\pi$, $y=\pm 2\pi$, ..., respectively, and partly of those portions of the axis of y for which

$$|\mu\cos y|>1;$$

these latter correspond to the parts of the screen between the apertures.

The curves $\phi=$ const., $\psi=$ const. are traced, for a particular case, on the opposite page, the value of μ adopted for convenience of calculation being

$$\mu=\cosh\tfrac{1}{5}\pi=1\cdot 2040,$$

whence
$$\sin^{-1}\frac{1}{\mu}=\cdot 312\pi, \qquad \cos^{-1}\frac{1}{\mu}=\cdot 188\pi.$$

The latter numbers give the relative breadths of the apertures and of the intervening portions of the screen.

The formulae (3), and the diagram*, admit of a variety of interpretations in Electrostatics and other mathematically cognate subjects. In the present application we must suppose that at points symmetrically situated on opposite sides of the axis of y the values of ψ are identical, whilst those of ϕ are equal in magnitude but of opposite signs.

It appears from (3), or by inspection of the figure, that the function ϕ in (3) is a periodic even function of y, the period being π. It can therefore be expanded by Fourier's Theorem in a series of cosines of multiples of $2y$, the coefficients being functions of x whose general form is to be determined by substitution in the equation

$$\nabla_1^2\phi=0. \quad \text{..(4)}$$

Thus, for x-positive we find, having regard to the condition to be satisfied for large values of x,

$$\phi=\log\mu+x+\sum_1^\infty C_s e^{-2sx}\cos 2sy. \quad \text{..............................(5)†}$$

If we introduce a more general linear unit, and denote the breadth of each aperture by a, and that of each intervening strip by b, we may write

$$\cosh\phi\cos\psi=\mu\cosh\frac{\pi x}{a+b}\cos\frac{\pi y}{a+b}, \quad \sinh\phi\sin\psi=\mu\sinh\frac{\pi x}{a+b}\sin\frac{\pi y}{a+b}, \quad \text{...(6)}$$

where
$$\mu=\sec\frac{\pi b}{2(a+b)}=\operatorname{cosec}\frac{\pi a}{2(a+b)}. \quad \text{............................(7)}$$

* Taken from a paper cited on p. 538 below. A formula equivalent to (1) was given by Larmor in the Mathematical Tripos, Part II, 1895.

† The precise values of the coefficients C_s are not required for our purpose. It may be shown that

$$C_s=\frac{(-1)^{s-1}}{s}F\left(s,\ -s,\ 1,\ \frac{1}{\mu^2}\right)=\frac{(-1)^{s-1}}{s}\left(1-\frac{1}{\mu^2}\right)F\left(1+s,\ 1-s,\ 1,\ \frac{1}{\mu^2}\right),$$

in the hypergeometric notation. See the paper cited.

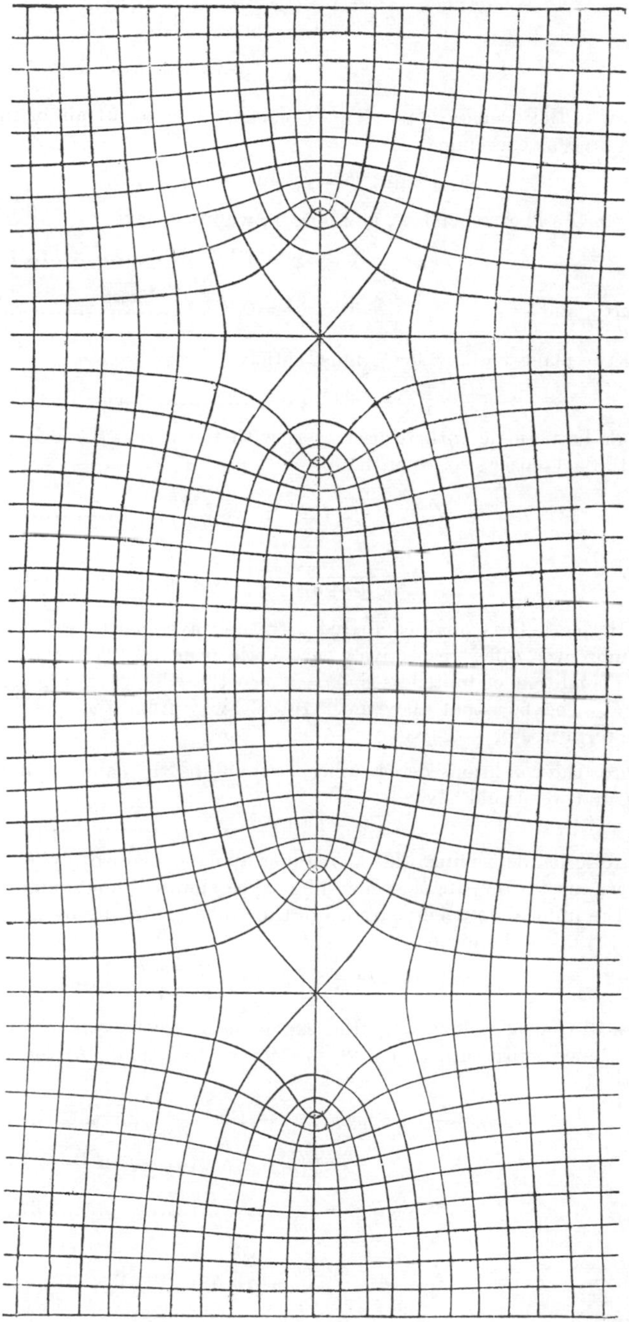

The expansion (5) is now replaced by

$$\phi = \log \mu + \frac{\pi x}{a+b} + \sum_1^\infty C_s e^{-\kappa_s x} \cos \frac{2s\pi y}{a+b}, \quad \text{.........(8)}$$

where

$$\kappa_s = \frac{2s\pi}{a+b}. \quad \text{.........(9)}$$

We turn now to the acoustical problem. Corresponding to a train of incident waves whose potential is e^{ikx}, we assume*

$$\Phi = e^{ikx} + e^{-ikx} + \chi, \quad \text{or} \quad \Phi = \chi', \quad \text{.........(10)}$$

according as $x \gtrless 0$. As in Arts. 299, 2°, and 305, 2°, we must have

$$\chi = -1, \qquad \chi' = +1, \quad \text{.........(11)}$$

over the apertures, and

$$\frac{\partial \chi}{\partial x} = 0, \quad \frac{\partial \chi'}{\partial x} = 0, \quad \text{.........(12)}$$

over the rest of the plane $x=0$. Since χ must satisfy

$$(\nabla_1^2 + k^2)\chi = 0, \quad \text{.........(13)}$$

and must further be periodic with respect to y, with the period $a+b$, it must admit of being expanded in a Fourier's series of the form

$$\chi = B_0 e^{-ikx} + \sum_1^\infty B_s e^{-\lambda_s x} \cos \frac{2s\pi y}{a+b}, \quad \text{.........(14)}$$

provided

$$\lambda_s = \left\{ \frac{4s^2\pi^2}{(a+b)^2} - k^2 \right\}^{\frac{1}{2}}. \quad \text{.........(15)}$$

Since, by hypothesis, $a+b$ is small compared with the wave-length $2\pi/k$, the quantities λ_s are real, and, moreover, differ respectively very little from κ_s. Terms involving $e^{\lambda_s x}$ are excluded by the condition of finiteness for $x=\infty$, so that the waves represented by χ are ultimately plane. The fact that they must travel away from the grating justifies the omission of the term in e^{ikx}.

If k were zero, the conditions determining χ would be the same as if the fluid were incompressible, and we should have

$$\chi = -1 + C\phi, \quad \text{.........(16)}$$

where ϕ is the function determined (in the manner above explained) by (6), and C is some constant; and we may anticipate that the same expression will hold approximately in the actual case for the immediate neighbourhood of the grating. Again, for small values of kx, the expansion (14) takes the form

$$\chi = B_0(1 - ikx) + \sum_1^\infty B_s e^{-\kappa_s x} \cos \frac{2s\pi y}{a+b}, \quad \text{.........(17)}$$

where the substitution of κ_s for λ_s in the exponential involves an error of the order $k^2(a+b)^2/4\pi^2$. Hence, substituting from (8) in (16), we find that (16) and (17) are in fact identical, provided

$$B_0 = -1 + C\log\mu, \qquad -ikB_0 = \frac{\pi C}{a+b}, \quad \text{.........(18)}$$

and, for $s > 0$,

$$B_s = CC_s. \quad \text{.........(19)}$$

Hence

$$B_0 = -\frac{1}{1+ikl}, \quad \text{.........(20)}$$

where

$$l = \frac{a+b}{\pi} \log \sec \frac{\pi b}{2(a+b)}. \quad \text{.........(21)}$$

* The symbol Φ is here used for the acoustic velocity-potential, as ϕ is at present used in a different sense.

As regards χ', all the conditions are satisfied if we suppose that its value at any point P' on the negative side of the grating is equal in absolute magnitude, but opposite in sign, to that of χ at the image P of P' on the positive side. Hence

$$\chi' = -B_0 e^{ikx} - \sum_1^\infty B_s e^{\lambda_s x} \cos \frac{2s\pi y}{a+b}. \quad \text{.........................(22)}$$

At a distance of several wave-lengths from the grating the last terms in (14) and (22) may be neglected, and the waves are sensibly plane. On reference to (10) we see that the coefficients of the reflected and transmitted waves are $1+B_0$ and $-B_0$, or

$$\frac{ikl}{1+ikl} \text{ and } \frac{1}{1+ikl}, \quad \text{...................................(23)}$$

respectively, that of the primary waves being taken as unit. Hence the intensities I, I' of the waves are given by

$$I = \frac{k^2l^2}{1+k^2l^2}, \qquad I' = \frac{1}{1+k^2l^2}. \quad \text{.............................(24)}$$

For sufficiently great wave-lengths there is very little reflection, even when the apertures occupy only a small fraction of the area of the screen. As corresponding numerical values we have

$$\frac{a}{a+b} = 0, \quad \cdot1, \quad \cdot2, \quad \cdot3, \quad \cdot4, \quad \cdot5, \quad \cdot6, \quad \cdot7, \quad \cdot8, \quad \cdot9, \quad 1\cdot0.$$

$$\frac{l}{a+b} = \infty, \ \cdot590, \ \cdot374, \ \cdot251, \ \cdot169, \ \cdot110, \ \cdot067, \ \cdot037, \ \cdot016, \ \cdot004, \ 0.$$

Let us suppose, for example, that the wave-length is ten times the interval $a+b$, and that the apertures occupy one-tenth of the area of the grating. It will be found that the reflected and transmitted intensities are

$$I = \cdot121, \qquad I' = \cdot879,$$

respectively. Notwithstanding the comparative narrowness of the openings, 88 per cent. of the sound gets through.

307. A similar method applies to the case of a grating composed of parallel equidistant wires.

It was shewn in Art. 64 that the potential- and stream-functions for a liquid flowing through a grating of parallel cylindrical bars of radius b are given approximately by

$$w = z + \frac{\pi b^2}{a} \coth \frac{\pi z}{a}, \quad \text{......................................(1)}$$

where a is the interval between the axes of consecutive bars, provided $b < \frac{1}{4}a$.

If the real part of z be positive, we have

$$w = z + \frac{\pi b^2}{a}\left(1 + 2\sum_1^\infty e^{-\frac{2s\pi z}{a}}\right), \quad \text{...............................(2)}$$

whence

$$\phi = x + \frac{\pi b^2}{a}\left(1 + 2\sum_1^\infty e^{-\frac{2s\pi x}{a}} \cos \frac{2s\pi y}{a}\right). \quad \text{..........................(3)}$$

Similarly, if x be negative, we find

$$\phi = x - \frac{\pi b^2}{a}\left(1 + 2\sum_1^\infty e^{\frac{2s\pi x}{a}} \cos \frac{2s\pi y}{a}\right). \quad \text{..........................(4)}$$

In the acoustical problem the velocity-potential will be of the form

$$\Phi = e^{ikx} + Ae^{-ikx} + \sum_1^\infty C_s e^{-\lambda_s x} \cos \frac{2s\pi y}{a}, \quad \text{..........................(5)}$$

or

$$\Phi = \quad Be^{ikx} - \sum_1^\infty C_s e^{-\lambda_s x} \cos \frac{2s\pi y}{a}, \quad \text{..........................(6)}$$

according as $x \gtrless 0$, where λ_s is the positive quantity defined by

$$\lambda_s^2 = \frac{4s^2\pi^2}{a^2} - k^2. \quad \text{.....(7)}$$

For values of x which are small compared with the wave-length we may ignore the difference between λ_s and $2s\pi/a$, provided the wave-length be large compared with a Under these circumstances the formulae (5), (6) reduce to

$$\Phi = 1 + A + ik(1-A)x + \sum_1^\infty C_s e^{-\frac{2s\pi x}{a}} \cos\frac{2s\pi y}{a}, \quad \text{.....(8)}$$

and

$$\Phi = B + ikBx - \sum_1^\infty C_s e^{\frac{2s\pi x}{a}} \cos\frac{2s\pi y}{a}, \quad \text{.....(9)}$$

respectively. The function Φ accordingly assumes the form

$$\Phi = \alpha\phi + \beta, \quad \text{.....(10)}$$

where ϕ is determined by (3) and (4), and α, β are constants, provided

$$1 + A = \alpha\frac{\pi b^2}{a} + \beta, \quad B = -\alpha\frac{\pi b^2}{a} + \beta, \quad ik(1-A) = \alpha, \quad ikB = \alpha, \quad C_s = 2\alpha\frac{\pi b^2}{a}. \quad \text{...(11)}$$

These make

$$A = \frac{ikl}{1+ikl}, \quad B = \frac{1}{1+ikl}, \quad \text{.....(12)}$$

where

$$l = \pi b^2/a. \quad \text{.....(13)}$$

The intensities of the reflected and transmitted waves are therefore

$$I = \frac{k^2l^2}{1+k^2l^2}, \quad I' = \frac{1}{1+k^2l^2}. \quad \text{.....(14)}$$

If the half-wave-length be large compared with b^2/a, we have free transmission, with hardly any reflection. This further illustrates the "extreme smallness of the obstruction offered by fine wires or fibres to the passage of sound*."

308. The diffraction of plane waves of sound by the edge of a semi-infinite screen, and the formation of a sound-shadow, have been investigated by Sommerfeld† and, with some extensions, by Carslaw‡. It is to be remarked that the data, in this problem, involve no special linear magnitude except the wave-length, and that the general character of the results is accordingly independent of the latter. The case of normal incidence can be treated very simply as follows§.

We will suppose that the screen occupies that half of the plane xz for which x is positive. It is convenient to introduce the 'parabolic' co-ordinates of Hankel and others. We write

$$k(x+iy) = (\xi + i\eta)^2, \quad \text{.....(1)}$$

or

$$kx = \xi^2 - \eta^2, \quad ky = 2\xi\eta, \quad \text{.....(2)}$$

and therefore

$$kr = \xi^2 + \eta^2, \quad \text{.....(3)}$$

* Rayleigh, *Theory of Sound*, Art. 343.

The investigations of Arts. 306, 307 are adapted from a paper "On the Reflection and Transmission of Electric Waves by a Metallic Grating," *Proc. Lond. Math. Soc.* (1) xxix. 523 (1898).

† "Mathematische Theorie der Diffraktion," *Math. Ann.* xlvii. 317 (1895).

‡ "Some Multiform Solutions of the Partial Differential Equations of Physics...," *Proc. Lond. Math. Soc.* xxx. 121 (1899).

§ The method is taken from a paper "On Sommerfeld's Diffraction Problem, and on Reflection by a Parabolic Mirror," *Proc. Lond. Math. Soc.* (2) iv. 190 (1906).

if r denote distance from the origin. The curves

$$\xi = \text{const.}, \qquad \eta = \text{const.}$$

form a system of confocal parabolas, the common focus being at the origin.

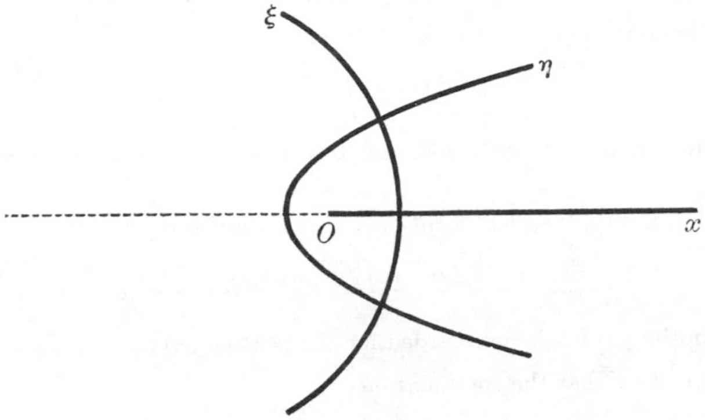

The co-ordinate η may be taken to be everywhere positive, except at the two surfaces of the screen, where it vanishes. The co-ordinate ξ will then have opposite signs on the two sides of the axis of x, and will vanish over the free portion of this axis. We easily find

$$\left.\begin{aligned} \frac{\partial \xi}{\partial x} &= \frac{1}{2}\frac{\xi}{r}, & \frac{\partial \eta}{\partial x} &= -\frac{1}{2}\frac{\eta}{r}, \\ \frac{\partial \xi}{\partial y} &= \frac{1}{2}\frac{\eta}{r}, & \frac{\partial \eta}{\partial y} &= \frac{1}{2}\frac{\xi}{r}. \end{aligned}\right\} \qquad (4)$$

The velocity-potential ϕ must satisfy the equation

$$\frac{\partial^2 \phi}{\partial x^2} + \frac{\partial^2 \phi}{\partial y^2} + k^2\phi = 0, \qquad (5)$$

the time-factor being e^{ikct} as usual.

The primary waves being represented by

$$\phi = e^{iky}, \qquad (6)$$

we seek for solutions of the types

$$e^{iky} u, \qquad e^{-iky} v. \qquad (7)$$

Taking the first of these, we have

$$\frac{\partial^2 u}{\partial x^2} + \frac{\partial^2 u}{\partial y^2} + 2ik\frac{\partial u}{\partial y} = 0. \qquad (8)$$

In virtue of the relations (4) this takes the form

$$\frac{\partial^2 u}{\partial \xi^2} + \frac{\partial^2 u}{\partial \eta^2} + 4i\left(\eta\frac{\partial u}{\partial \xi} + \xi\frac{\partial u}{\partial \eta}\right) = 0, \qquad (9)$$

which is satisfied by

$$u = f(\xi + \eta) = f(\zeta), \qquad (10)$$

say, provided

$$\frac{d^2 f}{d\zeta^2} + 2i\zeta\frac{df}{d\zeta} = 0, \qquad (11)$$

i.e. provided

$$u = A + B\int_0^{\xi+\eta} e^{-i\zeta^2}\, d\zeta. \qquad (12)$$

Now when ω is large and positive we have the asymptotic formula

$$\int_0^{\omega} e^{-i\zeta^2}\, d\zeta = \tfrac{1}{2}\sqrt{\pi}\, e^{-\frac{1}{4}i\pi} + \frac{i}{2\omega}e^{-i\omega^2} + \dots \qquad (13)$$

Hence at a great distance from the origin, on the side of y-positive, we have

$$ue^{iky}=(A+\tfrac{1}{2}\sqrt{\pi}Be^{-\frac{1}{4}i\pi})\,e^{iky}+\frac{i}{2\,(\xi+\eta)}\,e^{-ikr}, \qquad\text{.....................(14)}$$

approximately. The last term represents waves diverging from the origin.

In a similar manner we could obtain a solution

$$u'=A'+B'\int_0^{\xi-\eta}e^{i\zeta^2}d\zeta, \qquad\text{.................................(15)}$$

but as this would include waves converging towards the origin, it is unsuitable for the present purpose.

Again, starting from the second form in (7), we obtain the solution

$$v=C+D\int_0^{\xi-\eta}e^{-i\zeta^2}d\zeta, \qquad\text{.................................(16)}$$

together with another which is disregarded for the reason given.

We proceed to shew that the combination

$$\phi=Ae^{iky}+Ce^{-iky}+Be^{iky}\int_0^{\xi+\eta}e^{-i\zeta^2}d\zeta+De^{-iky}\int_0^{\xi-\eta}e^{-i\zeta^2}d\zeta \qquad\text{...........(17)}$$

can be made to satisfy all the conditions of the question.

In the first place, when x is large and negative, whilst $y=0$, ϕ must reduce to the form (6). Hence, putting $\xi=0$, $\eta=\infty$, we find, having regard to (13),

$$A+\tfrac{1}{2}\sqrt{\pi}e^{-\frac{1}{4}i\pi}B=1, \qquad C-\tfrac{1}{2}\sqrt{\pi}e^{-\frac{1}{4}i\pi}D=0. \qquad\text{.....................(18)}$$

Again, when $y=0$ and $x>0$, we must have $\partial\phi/\partial y=0$. Making use of (4), and putting $\eta=0$, we find that this condition is satisfied if

$$A=C, \qquad B=D. \qquad\text{..(19)}$$

Hence

$$\phi=\tfrac{1}{2}(e^{iky}+e^{-iky})+\frac{e^{\frac{1}{4}i\pi}}{\sqrt{\pi}}\left\{e^{iky}\int_0^{\xi+\eta}e^{-i\zeta^2}d\zeta+e^{-iky}\int_0^{\xi-\eta}e^{-i\zeta^2}d\zeta\right\}. \qquad\text{.........(20)}$$

This is equivalent to the form which Sommerfeld's result assumes in the case of normal incidence*.

When $\xi+\eta$, $\xi-\eta$ are both large and positive, the formula reduces to

$$\phi=e^{iky}+e^{-iky}. \qquad\text{..(21)}$$

This refers to the region in front of the screen, at a distance to the right, the second term indicating complete reflection.

When $\xi+\eta$ is large and positive, whilst $\xi-\eta$ is large and negative, we have

$$\phi=e^{iky}, \qquad\text{..(22)}$$

approximately. This refers to the region lying well to the left of the axis of y, where the primary waves predominate.

When $\xi+\eta$, $\xi-\eta$ are both large and negative, we have

$$\phi=0; \qquad\text{..(23)}$$

this refers to the sound-shadow behind the screen.

On each side of the plane $y=0$ there is an intermediate region in which the transition takes place from the state of things represented approximately by (22) to that represented by (21) or (23) respectively. To frame a criterion for the validity of our approximations,

* For the case of oblique incidence reference may be made to papers cited on p. 538.

we choose a quantity ω such that $\omega\sqrt{\pi}$ may be regarded as large*. The results (21) and (22) were obtained on the supposition that $|\xi-\eta|$, as well as $|\xi+\eta|$, is large. The boundary of the region in question, on the side of y-positive, is therefore found by putting $|\xi-\eta|=\omega$. This leads, by (2), to the parabola

$$y=\frac{k}{2\omega}x^2-\frac{\omega^2}{2k}, \qquad \text{.........(24)}$$

whose latus-rectum is proportional to the wave-length. The corresponding region on the side of y-negative is bounded by the parabola

$$y=-\frac{k}{2\omega}x^2+\frac{\omega^2}{2k}. \qquad \text{.........(25)}$$

These intermediate regions are the seat of the diffraction phenomena which are important in the optical analogy, but must here be passed over. It is not difficult to shew that at points near the boundary of the geometrical shadow, whose distances from the edge of the screen are large compared with the wave-length, the results will be in practical agreement with those obtained by Fresnel's methods†.

Atmospheric Waves.

309. The theory of waves travelling vertically in the atmosphere is of some interest as an example of wave-propagation in a variable medium‡.

Let the axis of x be drawn vertically upwards, and let ξ denote the vertical displacement at time t of the plane of particles whose undisturbed position is x. Let p and ρ be the corresponding values of the pressure and density, the equilibrium values being denoted by p_0 and ρ_0. These latter quantities are subject to the statical relation

$$\frac{\partial p_0}{\partial x}=-g\rho_0. \qquad \text{.........(1)}$$

The dynamical equation is

$$\rho_0\frac{\partial^2\xi}{\partial t^2}=-\frac{\partial p}{\partial x}-g\rho_0=-\frac{\partial}{\partial x}(p-p_0); \qquad \text{.........(2)}$$

and the equation of continuity is

$$\rho\left(1+\frac{\partial\xi}{\partial x}\right)=\rho_0. \qquad \text{.........(3)}$$

If we ignore conduction and radiation of heat, the values of p and ρ at any point are connected by the 'adiabatic' relation

$$p/p_0=(\rho/\rho_0)^\gamma, \qquad \text{.........(4)}$$

where γ denotes the ratio of the two specific heats. Hence, to the first order,

$$p-p_0=-\gamma p_0\frac{\partial\xi}{\partial x}. \qquad \text{.........(5)}$$

* The value of ω need not be more than moderately large. If, for instance, we put $\omega=6$, the error in the approximation does not amount to more than 10 per cent.

† See the paper quoted. The diffraction of a 'solitary wave' is discussed by the author in *Proc. Lond. Math. Soc.* (2) viii. 422 (1910).

‡ The question has been treated by Poisson, *l.c. ante* p. 493 and Rayleigh, "On Vibrations of an Atmosphere," *Phil. Mag.* (4) xxix. 173 (1890) [*Papers*, iii. 335]. The investigation in the text appeared in the *Proc. Lond. Math. Soc.* (2) vii. 122 (1908).

Substituting in (2), we find

$$\frac{\partial^2\xi}{\partial t^2} = c^2\frac{\partial^2\xi}{\partial x^2} - \gamma g\frac{\partial\xi}{\partial x}, \qquad \text{.........................(6)}$$

provided

$$c^2 = \gamma p_0/\rho_0, \qquad \text{.................................(7)}$$

i.e. c denotes the (usually variable) velocity of sound corresponding to the properties of the medium at the plane x. If H denote the height of the 'homogeneous atmosphere' corresponding to the temperature at this plane we have

$$c^2 = \gamma g H. \qquad \text{.................................(8)}$$

Let us first suppose that the equilibrium temperature is *uniform*, so that H and c are constants, and

$$\rho_0 = Ce^{-x/H}. \qquad \text{.................................(9)}$$

It is convenient then to take $2H$ as unit of length, and to adjust the unit of time so that $c = 1$. On this convention we have

$$\rho_0 = Ce^{-2x} \qquad \text{.................................(10)}$$

and

$$\frac{\partial^2\xi}{\partial t^2} = \frac{\partial^2\xi}{\partial x^2} - 2\frac{\partial\xi}{\partial x}. \qquad \text{..........................(11)}$$

If we write

$$\xi = ue^x, \qquad \text{.................................(12)}$$

this becomes

$$\frac{\partial^2 u}{\partial t^2} = \frac{\partial^2 u}{\partial x^2} - u. \qquad \text{............................(13)}$$

In the case of simple-harmonic vibrations, the time-factor being $e^{i\sigma t}$, we have, if $\sigma^2 > 1$,

$$u = Ae^{i\sigma t + i\sqrt{(\sigma^2-1)}x} + Be^{i\sigma t - i\sqrt{(\sigma^2-1)}x}, \qquad \text{.................(14)}$$

the first and second terms representing wave-systems which are propagated downwards and upwards respectively.

When $\sigma^2 < 1$, we have

$$u = \{Ae^{\sqrt{(1-\sigma^2)}x} + Be^{-\sqrt{(1-\sigma^2)}x}\} e^{i\sigma t}. \qquad \text{.................(15)}$$

Each term represents a standing vibration such as would ultimately be established in consequence of the continued action of a simple-harmonic plane source, the two terms applying respectively to the regions below and above the source.

Hence the disturbance due to a prescribed vibration

$$\xi = e^{i\sigma t} \qquad \text{.................................(16)}$$

maintained at the plane $x = 0$ will be given by

$$\xi = e^x \,.\, e^{i[\sigma t \mp \sqrt{(\sigma^2-1)}x]}, \quad [\sigma^2 > 1] \qquad \text{....................(17)}$$

or

$$\xi = e^{[1 \mp \sqrt{(1-\sigma^2)}]x} \,.\, e^{i\sigma t}, \quad [\sigma^2 < 1], \qquad \text{.................(18)}$$

where the upper or lower sign is to be taken according as $x \gtrless 0$.

Again, the disturbance due to a periodic *force* $e^{i\sigma t}$ concentrated on an infinitely thin stratum at $x=0$ is found to be

$$\xi = -\frac{i}{2\sqrt{(\sigma^2-1)}} e^x \,.\, e^{i[\sigma t \mp \sqrt{(\sigma^2-1)}x]}, \quad [\sigma^2>1] \quad \text{.............(19)}$$

or

$$\xi = \frac{1}{2\sqrt{(1-\sigma^2)}} e^{[1\mp\sqrt{(1-\sigma^2)}]x} \,.\, e^{i\sigma t}, \quad [\sigma^2<1], \quad \text{.............(20)}$$

provided the density at $x=0$ be taken as the unit of density. To verify this we remark that, in terms of our present units, the formula (5) becomes

$$p - p_0 = -e^{-2x}\frac{\partial \xi}{\partial x}. \quad \text{.............(21)}$$

The formulae (19) and (20) accordingly give, for the difference of pressure on the two sides of the plane $x=0$, the value $e^{i\sigma t}$. It will be noticed that the amplitudes given by (19) and (20) increase indefinitely as σ approaches the critical value unity. In general units the critical value of σ is $c/2H$ and the corresponding period is $4\pi H/c$. For air at ordinary temperatures this is (very roughly) about 5 minutes.

It appears from (14), on restoring the general units and taking the real part, that in a progressive train of waves of length $2\pi/k$ we have

$$\xi = ae^{\frac{1}{2}x/H}\cos(\sigma t \pm kx), \quad \text{.............(22)}$$

with

$$\sigma^2 = k^2c^2 + \tfrac{1}{4}c^2/H^2. \quad \text{.............(23)}$$

The wave-velocity is therefore

$$V = \frac{\sigma}{k} = c\sqrt{\left(1+\frac{1}{4k^2H^2}\right)}. \quad \text{.............(24)}$$

This varies with the frequency, but so long as the wave-length is small compared with $4\pi H$, it is approximately constant, differing from c by a small quantity of the second order. The main effect of the variation of density is on the amplitude, which increases as the waves travel upwards into the rarer regions, according to the law indicated by the exponential factor in (22). This increase might have been foreseen without calculation; for when the variation of density within the space of a wave-length is small, there is no sensible reflection, and the mean energy per unit volume, which varies as $a^2\rho_0$ (a being the amplitude), must therefore remain unaltered as the waves proceed. Since $\rho_0 \propto e^{-x/H}$, this shews that $a \propto e^{\frac{1}{2}x/H}$.

It is easily verified that the mean energy per unit volume in either of the trains represented by (22) is

$$\tfrac{1}{2}\rho\sigma^2a^2, \quad \text{.............(25)}$$

and that the rate of propagation of energy is equal to the kinematical group-velocity

$$U = \frac{d\sigma}{dk} = \frac{c}{\sqrt{(1+1/4k^2H^2)}}. \quad \text{.............(26)}$$

The atmosphere has been supposed unlimited, upwards and downwards, but the effect of rigid horizontal boundaries is easily ascertained. For instance, in the case of an atmosphere resting on a plane $x=-h$, if a prescribed vibration

$$\xi = a\cos\sigma t \qquad [\sigma > c/2H]$$

is maintained at the plane $x=0$, we have

$$\xi = ae^{\frac{1}{2}x/H}\cos(\sigma t - kx), \qquad [x > c]$$

and

$$\xi = ae^{\frac{1}{2}x/H}\frac{\sin k(x+h)}{\sin kh}\cos\sigma t, \qquad [x < c],$$

where the relation between k and σ is as in (23).

To find the free motion consequent on arbitrary initial conditions we start with the typical solution*

$$u=\left\{A(k)\cos\sqrt{(k^2+1)}\,t+B(k)\frac{\sin\sqrt{(k^2+1)}\,t}{\sqrt{(k^2+1)}}\right\}e^{ikx} \quad \text{..................(27)}$$

of the equation (13). This makes

$$u=A(k), \quad \frac{\partial u}{\partial t}=B(k), \quad [t\to 0]. \quad \text{..............................(28)}$$

Generalizing by Fourier's Theorem, we have

$$u=\int_{-\infty}^{\infty}A(k)\cos\sqrt{(k^2+1)}\,t\,.\,e^{ikx}\,dk+\int_{-\infty}^{\infty}B(k)\frac{\sin\sqrt{(k^2+1)}\,t}{\sqrt{(k^2+1)}}e^{ikx}\,dk, \quad \text{......(29)}$$

where

$$A(k)=\frac{1}{2\pi}\int_{-\infty}^{\infty}f(\alpha)\,e^{-ik\alpha}\,d\alpha, \quad B(k)=\frac{1}{2\pi}\int_{-\infty}^{\infty}F(\alpha)\,e^{-ik\alpha}\,d\alpha. \quad \text{............(30)}$$

This satisfies (13) and makes

$$u=f(x), \quad \frac{\partial u}{\partial t}=F(x), \quad [t\to 0]. \quad \text{..............................(31)}$$

As an example, let us suppose that there is no initial displacement, but that an initial momentum is concentrated in the neighbourhood of the plane $x=0$. We have then $f(x)=0$, whilst $F(x)$ is sensible only for infinitesimal values of x, for which it becomes infinite in such a way that

$$\int_{-\infty}^{\infty}F(\alpha)\,d\alpha=1, \quad \text{....................................(32)}$$

say. If the density at the plane $x=0$ be taken as the unit of density, this makes the impressed momentum to be unity (per unit area of the plane). Hence $A(k)=0$, $B(k)=1/2\pi$; and

$$u=\frac{1}{2\pi}\int_{-\infty}^{\infty}e^{ikx}\frac{\sin\sqrt{(k^2+1)}\,t}{\sqrt{(k^2+1)}}\,dk. \quad \text{..........................(33)}$$

The integral can be evaluated. We have, by a change of variable,

$$u=\frac{1}{2\pi}\int_{-\infty}^{\infty}\sin(t\cosh\omega+x\sinh\omega)\,d\omega.$$

If $t^2>x^2$, we write

$$t=\sqrt{(t^2-x^2)}\cosh\beta, \quad x=\sqrt{(t^2-x^2)}\sinh\beta, \quad \omega+\beta=\omega',$$

and obtain

$$u=\frac{1}{2\pi}\int_{-\infty}^{\infty}\sin\{\sqrt{(t^2-x^2)}\cosh\omega'\}\,d\omega'=\tfrac{1}{2}J_0\{\sqrt{(t^2-x^2)}\}, \quad \text{..............(34)}$$

by Mehler's formula (Art. 194 (7)).

* The method is similar to that adopted by Poincaré in the case of the 'équation des télégraphistes'; see his *Théorie analytique de la propagation de la chaleur*, Paris, 1895, c. viii.

On the other hand, if $t^2 < x^2$, we write

$$t = \sqrt{(x^2 - t^2)} \sinh \beta, \quad x = \pm\sqrt{(x^2 - t^2)} \cosh \beta, \quad \omega \pm \beta = \omega',$$

and obtain

$$u = \frac{1}{2\pi} \int_{-\infty}^{\infty} \sin \{\sqrt{(x^2 - t^2)} \sinh \omega'\} \, d\omega' = 0. \quad \text{.......................(35)}$$

The disturbance does not reach the position x until after the lapse of a time $t = \pm x$, and the subsequent displacement is given by

$$\xi = \tfrac{1}{2} e^x J_0 \{\sqrt{(t^2 - x^2)}\} \quad \text{..................................(36)}$$

or, in terms of general units,

$$\xi = \frac{1}{2\sqrt{(\rho_0 \rho_0')}} J_0 \left\{ \frac{\sqrt{(c^2 t^2 - x^2)}}{2H} \right\}, \quad \text{.............................(37)}$$

where ρ_0 is the density at the position x, and ρ_0' the density at the place $(x = 0)$ where the unit impulse was applied. The structure of this formula is in accordance with a well-known law of reciprocity*. It will be observed that the displacement ξ at any point does not remain constant after the wave has passed it, as it would in the case of a uniform medium, but fluctuates in sign with a continually decreasing amplitude. There is, moreover, a tendency to a definite periodicity in these fluctuations; viz. the period tends to the limit 2π, or $4\pi H/c$ in general units.

The verification of the conservation of momentum in this problem may be noticed. Consider the linear moment (Σmx), with respect to the plane $x = 0$, of the mass of air included between the upper and lower boundaries of the wave-system. As compared with the equilibrium state, this has been increased by

$$\int_{-t}^{t} \rho_0 \xi \, dx = \tfrac{1}{2} \int_{-t}^{t} e^{-x} J_0 \{\sqrt{(t^2 - x^2)}\} \, dx = \int_0^t \cosh x J_0 \{\sqrt{(t^2 - x^2)}\} \, dx. \quad \text{......(38)}$$

It may be shewn that this definite integral is equal to t†. Differentiating, we verify that the total momentum is unity.

310. Let us next suppose that the equilibrium temperature, instead of being uniform, diminishes upwards with a uniform gradient. This implies an upper boundary to the atmosphere, and it is therefore convenient to take the origin in this boundary and to measure x *downwards*. Accordingly, if θ_0 be the equilibrium temperature (absolute), we write

$$\theta_0 = \beta x, \quad \text{....................................(1)}$$

where β is the uniform temperature gradient. Since p_0, ρ_0, θ_0 are connected by the relation

$$p_0 = R \rho_0 \theta_0, \quad \text{...............................(2)}$$

we have

$$\frac{1}{\rho_0} \frac{d\rho_0}{dx} = \frac{1}{p_0} \frac{dp_0}{dx} - \frac{1}{\theta_0} \frac{d\theta_0}{dx} = \frac{g\rho_0}{p_0} - \frac{\beta}{\theta_0} = \frac{m}{x},$$

provided

$$m = \frac{g}{R\beta} - 1. \quad \text{...............................(3)}$$

Hence

$$\rho_0 \propto x^m, \qquad p_0 \propto x^{m+1}. \quad \text{..........................(4)}$$

* See *Proc. Lond. Math. Soc.* (1) xix. 144.

† By direct multiplication of series we find

$$\cosh(t \cos\theta) J_0(t \sin\theta) = 1 + \frac{t^2}{2!} P_2(\cos\theta) + \frac{t^4}{4!} P_4(\cos\theta) + \dots,$$

whence

$$\int_0^{\frac{1}{2}\pi} \cosh(t\cos\theta) J_0(t\sin\theta) \sin\theta \, d\theta = 1,$$

by Art. 87 (3). The expansion is due to Hobson, *Proc. Lond. Math. Soc.* (1) xxv. 66 (1893).

In the case of an atmosphere in 'convective equilibrium*,' where $p_0 \propto \rho_0{}^\gamma$, we have $m\gamma = m+1$, or

$$m = \frac{1}{\gamma - 1}, \qquad \beta = \frac{(\gamma-1)g}{\gamma R} = \beta_1, \text{ say.} \qquad \text{..............(5)}$$

If $\beta = g/R$ we have $m=0$, and the density is uniform. This condition, however, like all others in which m has a smaller value than that given by (5), would become unstable if the restriction to vertical motion were abandoned.

For the equation of motion we have, reversing the sign of x in Art. 309 (6),

$$\frac{\partial^2 \xi}{\partial t^2} = c^2 \frac{\partial^2 \xi}{\partial x^2} + \gamma g \frac{\partial \xi}{\partial x}, \qquad \text{..........................(6)}$$

where

$$c^2 = \gamma p_0/\rho_0 = \gamma R \beta x. \qquad \text{..........................(7)}$$

Now let us write

$$\tau = \int_0^x \frac{dx}{c} = \sqrt{\left(\frac{4x}{\gamma R \beta}\right)}, \quad \text{or} \quad x = \tfrac{1}{4}\gamma R \beta \tau^2, \text{...............(8)}$$

so that τ denotes the time which a point moving always with the local velocity of sound would take to travel from the top of the atmosphere to the position x. If in (6) we replace x as independent variable by τ, we get

$$\frac{\partial^2 \xi}{\partial t^2} = \frac{\partial^2 \xi}{\partial \tau^2} + \frac{2m+1}{\tau}\frac{\partial \xi}{\partial \tau}, \qquad \text{..........................(9)}$$

where m is given by (3).

In the case of simple-harmonic vibrations ($e^{i\sigma t}$), we have

$$\frac{\partial^2 \xi}{\partial \tau^2} + \frac{2m+1}{\tau}\frac{\partial \xi}{\partial \tau} + \sigma^2 \xi = 0, \qquad \text{........................(10)}$$

the solution of which is

$$\xi = \tau^{-m}\{A J_m(\sigma\tau) + B J_{-m}(\sigma\tau)\}. \qquad \text{..................(11)}$$

It follows from Art. 309 (5) that

$$p - p_0 \propto p_0 \frac{\partial \xi}{\partial x} \propto \tau^{2m+1}\frac{\partial \xi}{\partial \tau}; \qquad \text{........................(12)}$$

and in order that this may tend to the limit 0 for $x \to 0$ we must have $B=0$.

The solution corresponding to a vibration

$$\xi = e^{i\sigma t}, \qquad \text{.................................(13)}$$

maintained at the plane for which $\tau = \tau_1$, is accordingly

$$\xi = \left(\frac{\tau_1}{\tau}\right)^m \frac{J_m(\sigma\tau)}{J_m(\sigma\tau_1)} e^{i\sigma t}. \qquad \text{........................(14)}$$

For large values of $\sigma\tau$ we have, from Art. 303 (14),

$$\xi \propto \frac{1}{\tau^{m+\frac{1}{2}}} \sin(\sigma\tau + \tfrac{1}{4}\pi - \tfrac{1}{2}m\pi)\, e^{i\sigma t}. \qquad \text{..................(15)}$$

* Sir W. Thomson, *Manch. Memoirs* (3) ii. 125 (1862) [*Papers*, iii. 255].

Hence (14) represents a standing oscillation due to superposition of two wave-trains of equal amplitude, travelling upwards and downwards respectively. If Δx, $\Delta\tau$ be the changes in x and τ corresponding to a wave-length (λ), we have $\Delta(\sigma\tau) = 2\pi$, and therefore ultimately, when x is large,

$$\lambda = \Delta x = \tfrac{1}{2}\gamma R\beta\tau \,.\, \Delta\tau = \pi\gamma R\beta\tau/\sigma = 2\pi c/\sigma, \quad \text{...........(16)}$$

as we should expect.

The expression (14) becomes infinite when

$$J_m(\sigma\tau_1) = 0. \quad \text{...........................(17)}$$

This determines the periods $2\pi/\sigma$ of free oscillation of the air lying above a fixed rigid horizontal plane for which $\tau = \tau_1$*.

311. Proceeding to the consideration of disturbances propagated horizontally, we take the axes of x and y horizontal, and that of z vertical, with the positive direction downwards. The equations of small motion are then, in the 'Eulerian' notation,

$$\rho_0 \frac{\partial u}{\partial t} = -\frac{\partial p}{\partial x}, \quad \rho_0 \frac{\partial v}{\partial t} = -\frac{\partial p}{\partial y}, \quad \rho_0 \frac{\partial w}{\partial t} = -\frac{\partial p}{\partial z} + g\rho, \quad \text{........(1)}$$

$$\frac{D\rho}{Dt} + \rho_0 \chi = 0, \quad \text{..............................(2)}$$

where

$$\chi = \frac{\partial u}{\partial x} + \frac{\partial v}{\partial y} + \frac{\partial w}{\partial z}. \quad \text{........................(3)}$$

We assume again, for the most part, that the *variations* of pressure and density from the equilibrium values are connected by the adiabatic relation

$$\frac{Dp}{Dt} = c^2 \frac{D\rho}{Dt}, \quad \text{..............................(4)}$$

where

$$c^2 = \gamma p_0/\rho_0 = \gamma R\theta_0, \quad \text{..........................(5)}$$

i.e. c denotes the velocity of sound corresponding to the equilibrium temperature at the level z.

Writing

$$p = p_0 + p', \quad \rho = \rho_0 + \rho', \quad \text{.........................(6)}$$

and continuing to neglect small quantities of the second order, we have

$$\rho_0 \frac{\partial u}{\partial t} = -\frac{\partial p'}{\partial x}, \quad \rho_0 \frac{\partial v}{\partial t} = -\frac{\partial p'}{\partial y}, \quad \rho_0 \frac{\partial w}{\partial t} = -\frac{\partial p'}{\partial z} + g\rho', \quad \text{........(7)}$$

and

$$\frac{\partial \rho'}{\partial t} + w\frac{\partial \rho_0}{\partial z} = -\rho_0 \chi. \quad \text{.........................(8)}$$

Also, from (4) and (2),

$$\frac{\partial p'}{\partial t} + g\rho_0 w = -\gamma p_0 \chi. \quad \text{.........................(9)}$$

* For an investigation of the effect of arbitrary initial conditions see the paper by the author cited in the last footnote on p. 541.

Eliminating p' and ρ', we find

$$\left.\begin{aligned}\frac{\partial^2 u}{\partial t^2}=\frac{\partial}{\partial x}(c^2\chi+gw),\qquad \frac{\partial^2 v}{\partial t^2}=\frac{\partial}{\partial y}(c^2\chi+gw),\\ \frac{\partial^2 w}{\partial t^2}=\frac{\partial}{\partial z}(c^2\chi+gw)-\left\{\frac{d\,.\,c^2}{dz}-(\gamma-1)\,g\right\}\chi.\end{aligned}\right\}\quad\ldots\ldots\ldots\ldots(10)$$

Hence, if ξ, η, ζ are the components of vorticity,

$$\frac{\partial^2\xi}{\partial t^2}=-\left\{\frac{d\,.\,c^2}{dz}-(\gamma-1)\,g\right\}\frac{\partial\chi}{\partial y},\quad \frac{\partial^2\eta}{\partial t^2}=\left\{\frac{d\,.\,c^2}{dz}-(\gamma-1)\,g\right\}\frac{\partial\chi}{\partial x},\quad\ldots\ldots(11)$$

and $\partial^2\zeta/\partial t^2=0$. So far, the equations are general; they shew that irrotational motion is impossible, except under one of two conditions. We must either have

$$c=\text{const.},\quad \gamma=1,\quad\ldots\ldots\ldots\ldots\ldots\ldots\ldots\ldots\ldots(12)$$

which is the case of uniform equilibrium temperature with (moreover) isothermal expansions, or else

$$\frac{d\,.\,c^2}{dz}=(\gamma-1)\,g,\quad\ldots\ldots\ldots\ldots\ldots\ldots\ldots\ldots\ldots\ldots(13)$$

or

$$\frac{d\theta_0}{dz}=\frac{(\gamma-1)\,g}{\gamma R},\quad\ldots\ldots\ldots\ldots\ldots\ldots\ldots\ldots\ldots\ldots(14)$$

which is the case of convective equilibrium. These inferences are in accordance with Art. 17. In either of these special cases the equations (10) are satisfied by

$$u=-\frac{\partial\phi}{\partial x},\quad v=-\frac{\partial\phi}{\partial y},\quad w=-\frac{\partial\phi}{\partial z},\quad\ldots\ldots\ldots\ldots\ldots\ldots(15)$$

provided

$$\frac{\partial^2\phi}{\partial t^2}=-(c^2\chi+gw)=c^2\nabla^2\phi+g\frac{\partial\phi}{\partial z}\,.\ldots\ldots\ldots\ldots\ldots\ldots(16)$$

There is also the possibility of steady rotational motions, as we might expect, since either of the two physical states contemplated is, under the relative condition, one of neutral equilibrium*.

311 a. We proceed to consider various assumptions as to the vertical distribution of temperature. In the case of an *isothermal* atmosphere, where c is constant, the appropriate solution is

$$u=e^{-(\gamma-1)gz/c^2}f(ct-x),\quad v=0,\quad w=0,\quad\ldots\ldots\ldots\ldots(1)$$

or more generally

$$u=e^{-(\gamma-1)gz/c^2}\frac{\partial P}{\partial x},\quad v=e^{-(\gamma-1)gz/c^2}\frac{\partial P}{\partial y},\quad w=0,\quad\ldots\ldots\ldots\ldots(2)$$

if P is a function of the horizontal co-ordinates x, y, and the time, satisfying

$$\frac{\partial^2 P}{\partial t^2}=c^2\nabla_1^2 P,\quad\ldots\ldots\ldots\ldots\ldots\ldots\ldots\ldots\ldots\ldots(3)$$

where

$$\nabla_1^2=\partial^2/\partial x^2+\partial^2/\partial y^2.$$

* For further details as to this, reference may be made to a paper "On Atmospheric Oscillations," *Proc. Roy. Soc.* A, lxxxiv. 551 (1890), from which most of Arts. 311, 311 *a*, 312 is derived.

These equations represent systems of waves spreading horizontally with the constant velocity c, or $\sqrt{(\gamma g H)}$, where H is the height of the 'homogeneous atmosphere.' Since on the present hypothesis ρ_0 varies as $e^{z/H}$, or $e^{\gamma g z/c^2}$, it follows from (4) and (2) of Art. 311 that Dp/Dt will vary as e^{gz/c^2}. The condition of zero variation of pressure in the upper regions, where $z \to -\infty$, is therefore fulfilled. The velocity increases with altitude, but the *momentum* per unit volume diminishes. The expansions have been assumed to be adiabatic. If they are isothermal, we must put $\gamma = 1$; the particle-velocity is then independent of altitude.

In the case of *convective* equilibrium we take the origin in the upper boundary of the atmosphere, and write

$$c^2 = (\gamma - 1)\, gz = gz/m, \quad \text{.........................(4)}$$

in accordance with Art 310 (5).

To examine the propagation of waves horizontally we assume that ϕ, in Art. 311 (16), varies as $e^{i(\sigma t - kx)}$, or more generally that, as regards its dependence on the horizontal co-ordinates, it satisfies

$$(\nabla_1^2 + k^2)\,\phi = 0, \quad \text{.............................(5)}$$

the time-factor being $e^{i\sigma t}$. In either case the equation becomes

$$z\frac{\partial^2 \phi}{\partial z^2} + m\frac{\partial \phi}{\partial z} + \left(\frac{m\sigma^2}{gk} - kz\right) k\phi = 0, \quad \text{.................(6)}$$

This is simplified if we put

$$\phi = e^{-kz}\psi; \quad \text{...............................(7)}$$

thus

$$z\frac{\partial^2 \psi}{\partial z^2} + (m - 2kz)\frac{\partial \psi}{\partial z} - m\left(1 - \frac{\sigma^2}{gk}\right) k\psi = 0. \text{..............(8)}$$

If we put

$$m\left(1 - \frac{\sigma^2}{gk}\right) = 2\alpha, \quad \text{..........................(9)}$$

the solution which is finite for $z \to 0$ is

$$\psi_1 = A\left\{1 + \frac{\alpha}{1m}(2kz) + \frac{\alpha(\alpha+1)}{1\,.\,2m\,(m+1)}(2kz)^2 + \ldots\right\} \quad \text{......(10)}$$

or

$$\psi_1 = A_1F_1(\alpha;\, m;\, 2kz), \text{......................(11)}$$

in an accepted notation*. The remaining solution is of the form

$$\psi_2 = B\psi_1 \int_0^z \frac{e^{2kz}\,dz}{z^m \psi_1^2}. \text{.........................(12)}$$

This makes $\partial\psi_2/\partial z$ vary as z^{-m} for $z \to 0$, whilst ρ_0 varies as z^m. Now from Art. 311 (2), (4),

$$\frac{Dp}{Dt} = -\rho_0 c^2 \chi = \rho_0(\ddot{\phi} + gw), \quad \text{.....................(13)}$$

and it follows that Dp/Dt will not vanish for $z \to 0$ unless $B = 0$.

* See for example Barnes, *Camb. Trans.* xx. 253 (1906), where references to other papers are given.

The condition that $\partial\phi/\partial z = 0$ for $z = h$, the depth of the atmosphere, now gives

$$\frac{2\alpha}{m}\left\{1 + \frac{\alpha+1}{1\,(m+1)}(2kh) + \frac{(\alpha+1)(\alpha+2)}{1\,.\,2\,(m+1)\,(m+2)}(2kh)^2 + \ldots\right\}$$
$$= 1 + \frac{\alpha}{1m}(2kh) + \frac{\alpha\,(\alpha+1)}{1\,.\,2m\,(m+1)}(2kh)^2 + \ldots\Big\} \quad \text{......(14)}$$

or
$$\frac{2\alpha}{m}\,{}_1F_1(\alpha+1;\ m+1;\ 2kh) = {}_1F_1(\alpha;\ m;\ 2kh). \quad \text{.........(15)}$$

This determines α in terms of the wave-length $2\pi/k$. The corresponding value of σ, and thence the wave-velocity, follows from (9).

The chief interest is in waves which are long in comparison with h. When kh is small, a first approximation to a root of (14) is $2\alpha/m = 1$, and a second is

$$\frac{2\alpha}{m} = (1+kh)\Big/\left(1 + \frac{m+2}{m+1}kh\right) = 1 - \frac{kh}{m+1}, \quad \text{............(16)}$$

whence
$$\frac{\sigma^2}{gk} = \frac{kh}{m+1}. \quad \text{..............................(17)}$$

Now if we denote by H_1 the 'reduced height' of the atmosphere, *i.e.* the height to which it would extend if it had a uniform density equal to that of the lowest stratum, we have

$$H_1 = h^{-m}\int_0^h z^m\,dz = \frac{h}{m+1}. \quad \text{.....................(18)}$$

The velocity of propagation of long waves therefore tends to the value

$$V = \sigma/k = \sqrt{(gH_1)}. \quad \text{..........................(19)}$$

This may be compared with the isothermal case to which the formula (7) of Art. 278 relates. At 15° C. the value of H_1 is about 27640 ft., whence $V = 943$ ft./sec.

From (7) and (10) we find, as regards dependence on z,

$$\phi = A\left\{1 + \frac{k^2}{m+1}\left(\tfrac{1}{2}z^2 - hz\right)\right\}, \quad \text{.....................(20)}$$

approximately. For simplicity suppose the remaining factor to be $e^{i(\sigma t - kx)}$. Then

$$u = ikA, \quad v = 0, \quad w = \frac{k^2A}{m+1}(h-z), \quad \text{..............(21)}$$

the same factor being implied. Since the ratio w/u is of the order kh, the oscillations are mainly horizontal. The horizontal amplitude, again, is sensibly uniform from top to bottom. It will appear, presently, that this feature is peculiar to the present hypothesis of oscillations about convective equilibrium, and to that of an isothermal atmosphere with isothermal expansions.

We notice also that the formula (20), with the implied factor, makes

$$c^2\left(\frac{\partial u}{\partial x}+\frac{\partial v}{\partial y}+\frac{\partial w}{\partial z}\right)+gw=\frac{gk^2h}{m+1}, \qquad (22)$$

approximately, and therefore independent of altitude.

The remaining relevant solutions of (14), when kh is small, involve finite as distinguished from small values of akh. The corresponding modes of oscillation approximate to the type of waves propagated vertically, as in Art. 310, but with a gradual variation of phase in the horizontal sense.

312. In the more general case where the vertical distribution of temperature is arbitrary, we have recourse to the equations (10) of Art. 311. We derive by differentiation

$$\frac{\partial^2\chi}{\partial t^2}=c^2\nabla^2\chi+\left(\frac{d\,.\,c^2}{dz}+\gamma g\right)\frac{\partial\chi}{\partial z}+g\left(\frac{\partial\xi}{\partial y}-\frac{\partial\eta}{\partial x}\right), \qquad (1)$$

since

$$\nabla^2 w=\frac{\partial\chi}{\partial z}+\frac{\partial\xi}{\partial y}-\frac{\partial\eta}{\partial x}, \qquad (2)$$

identically. Hence from Art. 311 (11),

$$\frac{\partial^4\chi}{\partial t^4}=c^2\nabla^2\frac{\partial^2\chi}{\partial t^2}+\left(\frac{d\,.\,c^2}{dz}+\gamma g\right)\frac{\partial^3\chi}{\partial t^2\partial z}-g\left\{\frac{d\,.\,c^2}{dz}-(\gamma-1)g\right\}\nabla_1^2\chi. \qquad (3)$$

If we assume that χ varies as $e^{i(\sigma t-kx)}$, or more generally that χ satisfies

$$(\nabla_1^2+k^2)\chi=0, \qquad (4)$$

with a time-factor $e^{i\sigma t}$, where k is a constant (to be determined if necessary by lateral boundary-conditions), we find

$$c^2\frac{\partial^2\chi}{\partial z^2}+\left(\frac{d\,.\,c^2}{dz}+\gamma g\right)\frac{\partial\chi}{\partial z}+\left[\sigma^2-k^2c^2-\frac{gk^2}{\sigma^2}\left\{\frac{d\,.\,c^2}{dz}-(\gamma-1)g\right\}\right]\chi=0, \qquad (5)$$

which is the differential equation to be satisfied by χ.

Again, from the first two of equations (10) of Art. 311, we find

$$\sigma^2\frac{\partial w}{\partial z}+gk^2w=(\sigma^2-k^2c^2)\chi, \qquad (6)$$

and from the third equation

$$g\frac{\partial w}{\partial z}+\sigma^2 w=-c^2\frac{\partial\chi}{\partial z}-(\gamma-1)g\chi, \qquad (7)$$

Hence, by elimination of $\partial w/\partial z$,

$$(\sigma^4-g^2k^2)w=-\sigma^2c^2\frac{\partial\chi}{\partial z}-g(\gamma\sigma^2-k^2c^2)\chi, \qquad (8)^*$$

which will be needed presently.

We will only attempt to carry this further in the case where the equilibrium temperature gradient is uniform, say

$$\theta_0=\beta z, \qquad (9)$$

* The complete elimination of w between (6) and (7) leads again to (5).

if the origin is at the level of zero temperature. As in Art. 310 the values of ρ_0 and p_0 will vary as z^m and z^{m+1}, respectively, where

$$m = \frac{g}{R\beta} - 1, \quad \text{.............................(10)}$$

and consequently

$$c^2 = \gamma R \beta z = \frac{\gamma g z}{m+1}. \quad \text{.........................(11)}$$

Hence

$$\frac{d \,.\, c^2}{dz} - (\gamma - 1) g = -\frac{\gamma g}{m+1}\left(\frac{\beta_1}{\beta} - 1\right), \quad \text{...............(12)}$$

where β_1, as in Art. 310 (5), denotes the temperature gradient in convective equilibrium.

The equation (5) now reduces to the form

$$z\frac{\partial^2 \chi}{\partial z^2} + (m+2)\frac{\partial \chi}{\partial z} + \left\{\frac{m+1}{\gamma}\frac{\sigma^2}{gk} + \left(\frac{\beta_1}{\beta} - 1\right)\frac{gk}{\sigma^2} - kz\right\} k\chi = 0, \quad \text{......(13)}$$

or, if we write

$$\chi = e^{-kz}\psi, \quad \text{...............................(14)}$$

$$z\frac{\partial^2 \psi}{\partial z^2} + (m + 2 - 2kz)\frac{\partial \psi}{\partial z} + 2\alpha k \psi = 0, \quad \text{.............(15)}$$

where

$$2\alpha = \frac{m+1}{\gamma}\frac{\sigma^2}{gk} + \left(\frac{\beta_1}{\beta} - 1\right)\frac{gk}{\sigma^2} - (m+2). \quad \text{.............(16)}$$

The solution of (15) which is finite for $z \to 0$ is

$$\psi = 1 - \frac{\alpha}{1\,(m+2)}(2kz) + \frac{\alpha(\alpha - 1)}{1\,.\,2\,(m+2)(m+3)}(2kz)^2 - \ldots$$
$$= {}_1F_1(-\alpha;\; m+2;\; 2kz). \quad \text{...(17)}$$

Then, substituting from (14) and (17) in (8),

$$(\sigma^4 - g^2k^2)\,w = -\frac{\gamma g^2 k}{m+1}\left[\frac{\sigma^2}{gk}\left\{z\frac{\partial \psi}{\partial z} + (m+1)\psi\right\} - \left(1 + \frac{\sigma^2}{gk}\right)kz\psi\right]e^{kz}. \quad \text{......(18)}$$

We may anticipate that for long waves ($kh \to 0$) the wave-velocity will be comparable with $\sqrt{(gh)}$. Hence

$$\frac{\sigma^2}{gk} = \frac{\sigma^2}{k^2} \cdot \frac{kh}{gh}$$

may be assumed provisionally to be a small quantity. The 'convective' case of $\beta = \beta_1$ has already been discussed, and we may expect that if the ratio β/β_1 falls only slightly below unity the results will not be very different. But when this ratio differs appreciably from unity, the middle term in the expression (16) for 2α will preponderate, and α may be large, whilst αkh is finite. We have then, approximately,

$$\psi = 1 - \frac{2\alpha kz}{1\,(m+2)} + \frac{(2\alpha kz)^2}{1\,.\,2\,(m+2)(m+3)} - \ldots = {}_0F_1(m+2;\; -2\alpha kz), \quad \text{......(19)}$$

and thence

$$z\frac{\partial\psi}{\partial z}+(m+1)\psi=(m+1)\left\{1-\frac{2\alpha kz}{1\,(m+1)}+\frac{(2\alpha kz)^2}{1\,.\,2\,(m+1)\,(m+2)}-\ldots\right\}$$
$$=(m+1)\,{}_0F_1(m+1;\,-2\alpha kz). \quad\text{.........(20)}$$

These series can be expressed in terms of the Bessel Functions. If we put

$$\eta^2=8\alpha kz, \qquad \omega^2=8\alpha kh, \quad\text{..................(21)}$$

we have

$$\psi=2^{m+1}\Pi\,(m+1)\,\eta^{-m-1}J_{m+1}(\eta), \quad\text{................(22)}$$

$$z\frac{\partial\psi}{\partial z}+(m+1)\,\psi=2^m\,\Pi\,(m+1)\,\eta^{-m}\,J_m(\eta). \quad\text{.........(23)}$$

Since

$$\frac{gk}{\sigma^2}=2\alpha\Big/\left(\frac{\beta_1}{\beta}-1\right), \quad\text{..........................(24)}$$

approximately, from (16), the condition $w=0$ for $\eta=\omega$ reduces to

$$\left(\frac{\beta_1}{\beta}-1\right)J_m(\omega)=\tfrac{1}{2}\omega\,J_{m+1}(\omega). \quad\text{.................(25)}$$

This determines ω, and thence α. For the wave-velocity we have

$$V^2=\frac{\sigma^2}{k^2}=\left(\frac{\beta_1}{\beta}-1\right)\frac{4gh}{\omega^2}=\left(\frac{\beta_1}{\beta}-1\right)\frac{4\,(m+1)\,gH_1}{\omega^2}, \quad\text{.........(26)}$$

where, as in Art. 311 a, H_1 is the height of a homogeneous atmosphere having the temperature of the lowest stratum. The formula makes V imaginary when $\beta<\beta_1$, the atmosphere being then unstable.

As a numerical example, suppose the temperature-gradient to have half the convective value. Putting $\gamma=1{\cdot}40$, we have $m=6$, and the equation (25) becomes

$$J_6(\omega)=\tfrac{1}{2}\omega\,J_7(\omega). \quad\text{..........................(27)}$$

The lowest root of this is $\omega=4{\cdot}96$, approximately, whence

$$V=1{\cdot}07\,\sqrt{(gH_1)}. \quad\text{..........................(28)}$$

The result must in any case lie between $\sqrt{(gH_1)}$ and $\sqrt{(\gamma gH_1)}$, or $1{\cdot}18\,\sqrt{(gH_1)}$. If we assume the value of H_1 at 15° C. to be 27640 ft., (28) makes

$$V=1010 \text{ ft./sec.}^*$$

To compare the horizontal and vertical velocities in a simple case, assume that u varies as $e^{i(\sigma t-kx)}$, and $v=0$. Going back to the equations (10) of Art. 311 we have

$$\left.\begin{aligned}\sigma^2u-igkw&=ikc^2\chi,\\ igku+\sigma^2w&=-c^2\frac{\partial\chi}{\partial z}-\gamma g\chi.\end{aligned}\right\} \quad\text{..................(29)}$$

* Prof. G. I. Taylor has calculated the velocity of long waves on assumptions more nearly representative of the actual atmosphere. Assuming that the temperature falls uniformly from 283° (absolute) at the ground to 220° at a height of 3 km., and is uniform above this height, he finds $V=1024$ ft./sec., which differs little from the speed of the air-waves caused by the great Krakatoa explosion of 1883 (*Proc. Roy. Soc.* cxxvi. 169, 728 (1929)).

Hence

$$(\sigma^4 - g^2k^2)\,u = -\,ik\left\{gc^2\frac{\partial\chi}{\partial z} + (\gamma g^2 - \sigma^2c^2)\,\chi\right\}$$

$$= -\,\frac{i\gamma g^2k}{m+1}\left\{z\frac{\partial\chi}{\partial z} + (m+1)\,\chi - \frac{\sigma^2}{gk}kz\chi\right\}$$

$$= \frac{i\gamma g^2k}{m+1}\left\{z\frac{\partial\psi}{\partial z} + (m+1)\,\psi - \left(1 + \frac{\sigma^2}{gk}\right)kz\psi\right\}e^{-kz}. \qquad \text{......(30)}$$

Elimination of u in (29) would reproduce (13). If we compare with (18) we see that for $z=0$ the ratio of the vertical to the horizontal motion is σ^2/gk, and since $w=0$ for $z=h$, we infer that the vertical velocity is everywhere relatively small.

If we express our results in terms of Bessel Functions, omitting all common factors, and keeping only the most important terms, we find

$$u = J_m(\eta)/\eta^m, \qquad \text{......(31)}$$

$$w = \frac{1}{2\alpha}\left\{\left(\frac{\beta_1}{\beta} - 1\right)J_m(\eta) - \tfrac{1}{2}\eta\,J_{m+1}(\eta)\right\}\Big/\eta^m. \qquad \text{......(32)}$$

The horizontal velocity now varies with the altitude; the ratio of the velocity at the top to that at the bottom being

$$\frac{\omega^m}{2^m\,\Pi(m)\,J_m(\omega)}. \qquad \text{......(33)}$$

In the case above considered where $\beta = \tfrac{1}{2}\beta_1$, $m = 6$, $\omega = 4{\cdot}96$ this works out as 2·55.

As an example of *forced* oscillations we might introduce a disturbing potential of tidal type, say

$$\Omega = gHe^{-kz + i(\sigma t - kx)}; \qquad \text{......(34)}$$

cf. Art. 181. The leading features of the result may however be inferred at once from the Theory of Vibrations outlined in Chap. VIII. If the prescribed period $2\pi/\sigma$ differs but little from the free period proper to the wave-length $2\pi/k$, the motion will have the general character of the corresponding free oscillation, with the vertical distribution of velocity just referred to. But with a wider divergence from the free period the horizontal velocity may be practically uniform from top to bottom.

313. The general equations of small motion of a gas about a state of equilibrium in any constant field of force (X, Y, Z) are obtained by a slight generalization of the procedure of Art. 311.

In the undisturbed condition

$$\frac{\partial p_0}{\partial x} = \rho_0 X, \quad \frac{\partial p_0}{\partial y} = \rho_0 Y, \quad \frac{\partial p_0}{\partial z} = \rho_0 Z. \qquad \text{......(1)}$$

Hence, with previous notation,

$$\rho_0\frac{\partial u}{\partial t} = -\frac{\partial p'}{\partial x} + \rho' X, \quad \rho_0\frac{\partial v}{\partial t} = -\frac{\partial p'}{\partial y} + \rho' Y, \quad \rho_0\frac{\partial w}{\partial t} = -\frac{\partial p'}{\partial z} + \rho' Z, \qquad \text{......(2)}$$

with
$$\frac{D\rho}{Dt} = -\rho_0 \chi, \qquad \text{.............................(3)}$$

where
$$\chi = \frac{\partial u}{\partial x} + \frac{\partial v}{\partial y} + \frac{\partial w}{\partial z}.$$

We assume as before that the variations of pressure and density are connected by the relation
$$\frac{Dp}{Dt} = c^2 \frac{D\rho}{Dt}, \qquad \text{.............................(4)}$$

where $c^2 = \gamma p_0/\rho_0 = \gamma R\theta_0$, *i.e.* c is the velocity of sound corresponding to the equilibrium temperature at the point (x, y, z).

Hence
$$\frac{\partial p'}{\partial t} + \rho_0 (Xu + Yv + Zw) = -\rho_0 c^2 \chi, \qquad \text{..............(5)}$$

Eliminating p' and ρ' between (2), (3), and (5), we obtain
$$\rho_0 \frac{\partial^2 u}{\partial t^2} = \frac{\partial}{\partial x}\{\rho_0 c^2 \chi + \rho_0 (Xu + Yv + Zw)\}$$
$$- X\left\{\frac{\partial(\rho_0 u)}{\partial x} + \frac{\partial(\rho_0 v)}{\partial y} + \frac{\partial(\rho_0 w)}{\partial z}\right\}, \qquad \text{..............(6)}$$

with two similar equations.

We will now suppose that the forces X, Y, Z have a potential, in which case the equilibrium pressure p_0 will be a function of ρ_0, say
$$p_0 = f(\rho_0). \qquad \text{.................................(7)}$$

Hence, from (1),
$$X = \frac{1}{\rho_0}\frac{\partial \rho_0}{\partial x} f'(\rho_0), \quad Y = \frac{1}{\rho_0}\frac{\partial \rho_0}{\partial y} f'(\rho_0), \quad Z = \frac{1}{\rho_0}\frac{\partial \rho_0}{\partial z} f'(\rho_0), \qquad \text{......(8)}$$

and therefore
$$\frac{1}{\rho_0}\frac{\partial \rho_0}{\partial x}(Xu + Yv + Zw) = X\left(\frac{u}{\rho_0}\frac{\partial \rho_0}{\partial x} + \frac{v}{\rho_0}\frac{\partial \rho_0}{\partial y} + \frac{w}{\rho_0}\frac{\partial \rho_0}{\partial z}\right). \qquad \text{......(9)}$$

The equation (6) may therefore be written
$$\frac{\partial^2 u}{\partial t^2} = \frac{\partial}{\partial x}(c^2\chi + Xu + Yv + Zw) + \frac{1}{\rho_0}\frac{\partial \rho_0}{\partial x}\{c^2 - f'(\rho_0)\}\chi. \qquad \text{......(10)}$$

An equivalent form is
$$\frac{\partial^2 u}{\partial t^2} = \frac{\partial}{\partial x}(c^2\chi + Xu + Yv + Zw) - \left\{\frac{\partial \cdot c^2}{\partial x} - (\gamma - 1)X\right\}\chi. \qquad \text{......(11)}$$

The disturbed motion is therefore not in general irrotational. If however the distribution of temperature in the undisturbed state is such that the gas is in *convective* equilibrium, so that p_0 varies as ρ_0^{γ}, we have
$$f'(\rho_0) = \gamma p_0/\rho_0 = c^2,$$

and the second part of (10) disappears. The three equations of the type (10) are then satisfied by
$$u = -\frac{\partial \phi}{\partial x}, \quad v = -\frac{\partial \phi}{\partial y}, \quad w = -\frac{\partial \phi}{\partial z}, \qquad \text{..............(12)}$$

provided
$$\frac{\partial^2 \phi}{\partial t^2} = c^2 \nabla^2 \phi + \left(X \frac{\partial \phi}{\partial x} + Y \frac{\partial \phi}{\partial y} + Z \frac{\partial \phi}{\partial z} \right). \quad \text{......(13)}$$

The same conclusion holds if the equilibrium state be one of uniform temperature, provided the *expansions* are also isothermal. The wave-velocity c is then a constant.

These results are more readily obtained if we introduce the special hypothesis from the outset. If we assume that the pressure and density remain connected by the same law (7) as in the equilibrium state, we have in place of (5)
$$p' = \rho' f'(\rho_0) = c^2 \rho'. \quad \text{......(14)}$$

The equations (2) may therefore be written
$$\rho_0 \frac{\partial u}{\partial t} = -\frac{\partial p'}{\partial x} + \frac{p'}{\rho_0} \frac{\partial \rho_0}{\partial x}. \quad \text{......(15)}$$

Hence
$$\frac{\partial u}{\partial t} = -\frac{\partial}{\partial x}\left(\frac{p'}{\rho_0}\right), \quad \frac{\partial v}{\partial t} = -\frac{\partial}{\partial y}\left(\frac{p'}{\rho_0}\right), \quad \frac{\partial w}{\partial t} = -\frac{\partial}{\partial z}\left(\frac{p'}{\rho_0}\right) \quad \text{......(16)}$$

These have the irrotational solution (11) with
$$p' = \rho_0 \frac{\partial \phi}{\partial t}. \quad \text{......(17)}$$

Eliminating p' and ρ' between (5), (11), and (16), we obtain the equation (12).

So far, the motions contemplated have been 'free,' in the sense that no forces are operative except those of the constant field (X, Y, Z). In the case of a small disturbing force whose potential is Ω a term $-\rho_0 \partial^2 \Omega / \partial x \partial t$ is to be added to the right-hand side of (10). The equation (12) is then replaced by
$$p' = \rho_0 \left(\frac{\partial \phi}{\partial t} - \Omega \right), \quad \text{......(18)}$$

and we have
$$\frac{\partial^2 \phi}{\partial t^2} = c^2 \nabla^2 \phi + \left(X \frac{\partial \phi}{\partial x} + Y \frac{\partial \phi}{\partial y} + Z \frac{\partial \phi}{\partial z} \right) + \frac{\partial \Omega}{\partial t}. \quad \text{......(19)}$$

314. The theory of such questions as the large-scale oscillations of the earth's atmosphere is still imperfect. One difficulty is that of taking account of the physical conditions which prevail in the upper regions.

The results of Arts. 311 *a*, 312 indicate that in the slower modes of oscillation the motion of the air will be mainly horizontal. Taking first the case of an isothermal atmosphere surrounding a non-rotating globe, and subject to an isothermal law of expansion, the equation (13) of Art. 313 becomes, in terms of polar co-ordinates r, θ, ϕ,
$$\frac{\partial^2 \Phi}{\partial t^2} = c^2 \left\{ \frac{\partial^2 \Phi}{\partial r^2} + \frac{2}{r} \frac{\partial \Phi}{\partial r} + \frac{1}{r^2 \sin \theta} \frac{\partial}{\partial \theta} \left(\sin \theta \frac{\partial \Phi}{\partial \theta} \right) + \frac{1}{r^2 \sin^2 \theta} \frac{\partial^2 \Phi}{\partial \phi^2} \right\} - g \frac{\partial \Phi}{\partial r}, \quad \text{......(1)}$$

the velocity potential being now denoted by Φ. If, guided by the result of Art. 311 a (with $\gamma = 1$), we neglect the radial motion, and put $r = a$ (the radius of the globe), we have, in the case of simple-harmonic vibrations,

$$\frac{c^2}{a^2}\left\{\frac{1}{\sin\theta}\frac{\partial}{\partial\theta}\left(\sin\theta\frac{\partial\Phi}{\partial\theta}\right) + \frac{1}{\sin^2\theta}\frac{\partial^2\Phi}{\partial\phi^2}\right\} + \sigma^2\Phi = 0. \qquad\text{.........(2)}$$

As in the problem of Art. 199, Φ will, in any normal mode, vary as a surface-harmonic of integral order n, whence

$$\sigma^2 a^2/c^2 = n(n+1). \qquad\text{.........(3)}$$

The interpretation follows the same lines as in the Art. referred to. The condensation ($s = c^{-2}\partial\phi/\partial t$) corresponds to the ζ/h of that Art., whilst c^2 takes the place of gh. Since we now have $c^2 = gH$, where H is the height of the homogeneous atmosphere, it appears that the free oscillations follow the same laws as those of a liquid ocean of uniform depth H covering the globe*.

For numerical illustration we may put

$$c = 2{\cdot}80 \times 10^4\ \text{cm./sec.}, \quad 2\pi a = 4 \times 10^9\ \text{cm.}$$

In the cases $n = 1$, $n = 2$ this gives, by (3), free periods of 28·1 and 16·2 hours, respectively, for a temperature of 0° C. For a temperature of 15° C. the periods would be 27·4 hours and 15·8 hours.

315. The hypothesis of convective equilibrium, with (for consistency) adiabatic expansion, lends itself with equal ease to calculation. Mathematically, it has the advantage of the definite condition at the upper boundary.

The equation (1) of the preceding Art. will still apply, provided it be remembered that c^2 now varies with the depth below the upper boundary of the atmosphere. Assuming that the velocity potential varies as a spherical harmonic of order n, we have, in a free oscillation,

$$c^2\left\{\frac{\partial^2\Phi}{\partial r^2} + \frac{2}{r}\frac{\partial\Phi}{\partial r} - \frac{n(n+1)}{r^2}\Phi\right\} - g\frac{\partial\Phi}{\partial r} + \sigma^2\Phi = 0. \qquad\text{.........(1)}$$

The depth h of the atmosphere is assumed to be small in comparison with the earth's radius. Hence, putting $r = a - z$, where a refers to the outer boundary, and writing

$$c^2 = gz/m, \qquad\text{.........(2)}$$

in conformity with Art. 310 (5), we find

$$z\frac{\partial^2\Phi}{\partial z^2} + m\frac{\partial\Phi}{\partial z} + \left\{\frac{m\sigma^2}{g} - \frac{n(n+1)z}{a^2}\right\}\Phi = 0, \qquad\text{.........(3)}$$

since c^2/a may be neglected in comparison with g. If we write

$$k^2 = n(n+1)/a^2, \qquad\text{.........(4)}$$

* Rayleigh, *l.c. ante* p. 541.

this becomes identical with the equation (6) of Art. 311 *a*, and it follows that

$$\sigma^2 = n(n+1)\frac{gH_1}{a^2}. \qquad\qquad (5)$$

The free oscillations about convective equilibrium are therefore analogous to those of a liquid ocean whose depth is equal to the reduced depth H_1 of the atmosphere.

316. This analogy still holds when we proceed to the case of a *rotating* globe. If, for a moment, we suppose the axis of z to coincide with the axis of rotation, whilst the axes of x, y revolve with the angular velocity ω of the globe, the equations (2) of Art. 313 are replaced by

$$\left.\begin{aligned} \rho_0\left(\frac{\partial u}{\partial t} - 2\omega v\right) &= -\frac{\partial p'}{\partial x} + \rho' X - \rho_0\frac{\partial \Omega}{\partial x}, \\ \rho_0\left(\frac{\partial v}{\partial t} + 2\omega u\right) &= -\frac{\partial p'}{\partial y} + \rho' Y - \rho_0\frac{\partial \Omega}{\partial y}, \\ \rho_0\frac{\partial w}{\partial t} \qquad &= -\frac{\partial p'}{\partial z} + \rho' Z - \rho_0\frac{\partial \Omega}{\partial z}, \end{aligned}\right\} \qquad (6)$$

provided the centrifugal force be supposed included in (X, Y, Z). The symbols u, v, w here denote apparent velocities, *i.e.* velocities relative to the rotating globe. For the sake of generality, terms have been introduced to represent the effect of disturbing forces, whose potential is Ω. The equation of continuity is unaltered in form.

Proceeding as in the Art. referred to, we have

$$\frac{\partial^2 u}{\partial t^2} - 2\omega\frac{\partial v}{\partial t} = \frac{\partial P}{\partial x}, \qquad \frac{\partial^2 v}{\partial t^2} + 2\omega\frac{\partial u}{\partial t} = \frac{\partial P}{\partial y}, \qquad \frac{\partial^2 w}{\partial t^2} = \frac{\partial P}{\partial z}, \qquad (7)$$

where

$$P = c^2\chi + Xu + Yv + Zw - \frac{\partial \Omega}{\partial t}. \qquad (8)$$

If we now change the meaning of our symbols, taking u to be the velocity along the meridian, v that along a parallel, and w that along the vertical, we have, in analogy with Art. 213 (5),

$$\left.\begin{aligned} \frac{\partial^2 u}{\partial t^2} - 2\omega\frac{\partial v}{\partial t}\cos\theta = \frac{\partial P}{r\partial\theta}, \quad \frac{\partial^2 v}{\partial t^2} + 2\omega\frac{\partial u}{\partial t}\cos\theta + 2\omega\frac{\partial w}{\partial t}\sin\theta = \frac{\partial P}{r\sin\theta\partial\phi}, \\ \frac{\partial^2 w}{\partial t^2} - 2\omega\frac{\partial v}{\partial t}\sin\theta = \frac{\partial P}{\partial r}, \end{aligned}\right\} \qquad (9)$$

where θ, ϕ denote co-latitude and longitude, respectively.

In the application to tidal motions various simplifications can be introduced, as in the discussion of the oceanic problem (Art. 213). In particular, neglecting the vertical acceleration, we infer from the last equation that P may be regarded as approximately independent of r, and consequently that the horizontal velocities u, v are sensibly the same for all particles in the same vertical*. Now putting $r = a - z$, we have, in polar co-ordinates,

$$P = \frac{c^2}{a \sin\theta} \left\{ \frac{\partial}{\partial\theta}(u \sin\theta) + \frac{\partial v}{\partial\phi} \right\} - c^2 \frac{\partial w}{\partial z} - gw - \frac{\partial\Omega}{\partial t}. \quad \text{.........(10)}$$

If we put $c^2 = gz/m$, multiply by z^{m-1}, and integrate with respect to z between the limits 0 and h, we find

$$P = \frac{gH_1}{a \sin\theta} \left\{ \frac{\partial}{\partial\theta}(u \sin\theta) + \frac{\partial v}{\partial\phi} \right\} - \frac{\partial\Omega}{\partial t}, \quad \text{..............(11)}$$

on the supposition that $z^m w$ vanishes at both limits.

The equations, as simplified, now stand as follows:

$$\frac{\partial^2 u}{\partial t^2} - 2\omega \cos\theta \frac{\partial v}{\partial t} = \frac{\partial P}{a \partial\theta}, \qquad \frac{\partial^2 v}{\partial t^2} + 2\omega \cos\theta \frac{\partial u}{\partial t} = \frac{\partial P}{a \sin\theta \partial\phi}, \quad \text{......(12)}$$

where P is given by (11). If we put

$$P = -g \frac{\partial\zeta}{\partial t}, \qquad \Omega = -g \frac{\partial\bar{\zeta}}{\partial t}, \quad \text{...................(13)}$$

the equations (11) and (12) resemble those found in Art. 214 for the case of an aqueous ocean of uniform depth H_1. The theory of the oceanic tides on a rotating globe, discussed in Chap. VIII, can therefore be at once applied to the gravitational tides of an atmosphere of the type here considered.

The results apply also to the case of an isothermal atmosphere (with isothermal expansion) if in (10) we put $c^2 = gH$.

Some calculations of free periods of an isothermal atmosphere have been made by Margules†. He assumes a temperature of 0° C., and (virtually) $c = 2{\cdot}84 \times 10^4$ cm./sec. for the velocity of sound. His results may also be interpreted as the periods of an aqueous ocean whose depth is 7980 metres, or 26,240 ft., provided we neglect the mutual gravitation of the water.

For the first three oscillations of zonal type ($s = 0$ in the notation of Art. 223) which are symmetrical with respect to the equator he finds periods of

12·28, 7·88, 6·37

sidereal hours; and for the first three asymmetrical modes

20·44, 9·59, 6·67.

* On a more general view as to the constitution of the atmosphere the approximate assumption $\partial P/\partial r = 0$ would be replaced by

$$\frac{\partial P}{\partial r} = \left\{ \frac{\partial \,.\, c^2}{\partial r} + (\gamma - 1) g \right\} \chi,$$

and the resemblance with the oceanic tides becomes less exact.

† *Wiener Sitzber., Math. nat. wiss. Classe*, ci. 597 (1892) and cii. 11 (1895). The author is indebted for these references to Prof. S. Chapman. In the second paper quoted free periods are calculated for a series of values of $2\pi/\omega$ other othan 24 hours; these include examples of the type referred to at the end of Art. 206. Both papers include a discussion of the modification introduced by frictional forces varying as the velocity.

The results for symmetrical oscillations of sectorial type ($s=1$) are given in pairs, corresponding to waves travelling E. and W., respectively, relatively to the rotating globe; thus

$$\left.\begin{matrix}13{\cdot}87\\36{\cdot}57\end{matrix}\right\},\quad \left.\begin{matrix}9{\cdot}22\\10{\cdot}22\end{matrix}\right\},\quad \left.\begin{matrix}6{\cdot}63\\6{\cdot}77\end{matrix}\right\}.$$

For the tesseral type ($s=2$) he gives $\left.\begin{matrix}11{\cdot}94\\18{\cdot}42\end{matrix}\right\}.$

These results may be compared with those obtained by Hough (see pp. 349, 350), for an ocean of depth 29,040 feet, except that they do not allow for mutual attraction. Cf. also Arts. 210, 212.

Having regard to possible exaggeration by 'resonance,' it is a matter of some interest to inquire whether the atmosphere may have a free period nearly equal to 12 lunar or solar hours. We notice that Margules finds, for the most important free oscillation having the same general character as a semi-diurnal tide-wave, a period of 11·94 sidereal hours, on the assumption of a uniform temperature of 0° C. Again, Hough*, in his researches on tidal theory, finds that the depth h of an ocean for which the period is exactly 12 sidereal hours is given by

$$gh/4\omega^2a^2 = 0{\cdot}10049.$$

This is evaluated at 29,182 feet. It is to be remarked however that in the calculation the mutual attraction of the disturbed fluid was taken into account, whereas in the aerial ocean this influence must be quite insensible. Allowing for this, and calculating for a period of 12 mean *solar* hours, it appears that

$$gh/4\omega^2a^2 = 0{\cdot}08911,$$

or $h=25{,}710$ feet†. The mean temperature of the air near the earth's surface is usually estimated at 15° C., which gives $H_1=27{,}640$ feet. Without pressing too far conclusions based on the hypothesis of an atmosphere uniform over the earth, and approximately in convective equilibrium, we may assert with some probability the existence of a free oscillation of the atmosphere, of semi-diurnal type, with a period not very different from, but somewhat less than, 12 mean solar hours.

As a matter of observation the most regular oscillations of the barometer have *solar* diurnal and semi-diurnal periods, whilst the corresponding lunar tides are almost insensible‡. The amplitude of the solar semi-diurnal oscillation at places on the equator is about ·937 mm. or ·0375 in., whilst the amplitude given by the 'equilibrium' theory of the tides is only ·00043 in. Some numerical results given by Hough in illustration of the kinetic theory of oceanic tides would indicate that in order that this amplitude should be increased by dynamical action some eighty- or ninety-fold, the free period must differ from the imposed period by not more than 2 or 3 minutes. Since the difference between the lunar and solar semi-diurnal periods amounts to 26 minutes, it is quite conceivable that the solar influence might in this way be rendered much more effective than the lunar. There remains however the difficulty that the *phase* of the observed semi-diurnal inequality is accelerated instead of retarded (as it would be by tidal friction) relatively to the sun's transit.

The observed oscillations have been ascribed by Kelvin to a different cause, viz. to the daily variation in temperature, which, when analysed into simple-harmonic constituents, will have components whose periods are respectively 1, $\frac{1}{2}$, $\frac{1}{3}$, $\frac{1}{4}$, ... of a solar day. It is very remarkable that the second (viz. the semi-diurnal) barometric oscillation has a considerably

* *l.c. ante* p. 347.

† See the paper cited on p. 548.

‡ Thus Chapman finds an amplitude of ·00036 in. of mercury for the lunar semi-diurnal atmospheric tide at Greenwich (*Q.J.R. Met. Soc.* xliv. 271 (1918)).

greater amplitude than the first. It was suggested by Kelvin that the explanation of this peculiarity is to be sought for in a much closer agreement of the period of the semi-diurnal component with a free period of the earth's atmosphere than is the case with the diurnal component*.

On either hypothesis it is necessary to postulate a free period of almost exactly 12 solar hours in order to account for the requisite degree of selective resonance. The most recent estimate†, based on a comparison of the speed of atmospheric waves with tidal theory, points to a free period definitely too short.

* Kelvin, "On the Thermodynamic Acceleration of the Earth's Rotation," *Proc. R. S. Edin.* xi. (1882) [*Papers*, iii. 341]. For a full discussion see Chapman, *Q.J.R. Met. Soc.* l. 165 (1923). The forced tides due to variation of temperature were discussed by Margules, *Wiener Ber.* xcix. 204 (1890).

† G. I. Taylor, *Proc. Roy. Soc.* cxxvi. 169, 728 (1929–30).

CHAPTER XI

VISCOSITY

317. THE main theme of this Chapter is the resistance to distortion, known as 'viscosity' or 'internal friction,' which is exhibited more or less by all real fluids, but which we have hitherto neglected.

It will be convenient, following a plan already adopted on several occasions, to recall briefly the outlines of the general theory of a dynamical system subject to dissipative forces which are linear functions of the generalized velocities*. This will not only be useful as tending to bring under one point of view most of the special investigations which follow; it will sometimes indicate the general character of the results to be expected in cases which are beyond our powers of calculation.

We begin with the case of one degree of freedom. The equation of motion is of the type

$$a\ddot{q} + b\dot{q} + cq = Q. \quad \text{.............................(1)}$$

Here q is a generalized co-ordinate specifying the deviation from a position of equilibrium; a is the coefficient of inertia, and is necessarily positive; c is the coefficient of stability, and is positive in the applications which we shall consider; b is a coefficient of friction, and is positive. Since the terms on the left-hand side of (1) are differently affected by changing the sign of t, the motion of a system subject to an equation of this type is not reversible.

If we put $$T = \tfrac{1}{2}a\dot{q}^2, \quad V = \tfrac{1}{2}cq^2, \quad F = \tfrac{1}{2}b\dot{q}^2, \quad \text{....................(2)}$$
the equation may be written

$$\frac{d}{dt}(T + V) = -2F + Q\dot{q}. \quad \text{..........................(3)}$$

This shews that the energy $T + V$ is increasing at a rate less than that at which the extraneous force is doing work on the system. The difference $2F$ represents the rate at which energy is being dissipated; this is always positive.

In *free* motion we have

$$a\ddot{q} + b\dot{q} + cq = 0. \quad \text{.............................(4)}$$

If we assume that $q \propto e^{\lambda t}$, the solution takes different forms according to the relative importance of the frictional term. If $b^2 < 4ac$, we have

$$\lambda = -\tfrac{1}{2}\frac{b}{a} \pm i\left(\frac{c}{a} - \tfrac{1}{4}\frac{b^2}{a^2}\right)^{\frac{1}{2}}, \quad \text{.......................(5)}$$

or, say, $$\lambda = -\tau^{-1} \pm i\sigma. \quad \text{............................(6)}$$

* For a fuller account of the theory reference may be made to Rayleigh, *Theory of Sound*, cc. iv., v.; Thomson and Tait, *Natural Philosophy* (2nd ed.), Arts. 340–345; Routh, *Advanced Rigid Dynamics*, cc. vi., vii.

Hence the full solution, expressed in real form, is

$$q = Ae^{-t/\tau}\cos(\sigma t + \epsilon), \quad \text{.........................(7)}$$

where A, ϵ are arbitrary. The type of motion which this represents may be described as a simple-harmonic vibration with amplitude diminishing asymptotically to zero, according to the law $e^{-t/\tau}$. The time τ in which the amplitude sinks to $1/e$ of its original value is called the 'modulus of decay' of the oscillations.

If $b/2a$ be small compared with $(c/a)^{\frac{1}{2}}$, $b^2/4ac$ is a small quantity of the second order, and the 'speed' σ is then practically unaffected by the friction. This is the case whenever the time $(2\pi\tau)$ in which the amplitude sinks to $e^{-2\pi}$ $(=\frac{1}{535})$ of its initial value is large compared with the period $(2\pi/\sigma)$.

When, on the other hand, $b^2 > 4ac$, the values of λ are real and negative. Denoting them by $-\alpha_1$, $-\alpha_2$, we have

$$q = A_1e^{-\alpha_1 t} + A_2e^{-\alpha_2 t}. \quad \text{.........................(8)}$$

This represents 'aperiodic motion'; viz. the system never passes more than once through its equilibrium position, towards which it finally creeps asymptotically.

In the critical case $b^2 = 4ac$, the two values of λ are equal; we then find by usual methods

$$q = (A + Bt)\,e^{-\alpha t}, \quad \text{.........................(9)}$$

which may be similarly interpreted.

As the frictional coefficient b is increased, the two quantities α_1, α_2 become more and more unequal; viz. one of them (α_2, say) tends to the value b/a, and the other to the value c/b. The effect of the second term in (8) then rapidly disappears, and the residual motion is the same as if the inertia-coefficient (a) were zero.

318. We consider next the effect of a periodic extraneous force. Assuming that

$$Q = Ce^{i(\sigma t + \epsilon)}, \quad \text{..............................(10)}$$

the equation (1) gives

$$q = \frac{Q}{c - \sigma^2 a + i\sigma b}. \quad \text{............................(11)}$$

If we put

$$1 - \frac{\sigma^2 a}{c} = R\cos\epsilon_1, \qquad \frac{\sigma b}{c} = R\sin\epsilon_1, \quad \text{..............(12)}$$

where ϵ_1 lies between 0 and 180°, we have

$$q = \frac{Q}{Rc}e^{-i\epsilon_1}. \quad \text{............................(13)}$$

Taking real parts, we may say that the force

$$Q = C\cos(\sigma t + \epsilon) \quad \text{.........................(14)}$$

will maintain the oscillation

$$q = \frac{C}{Rc} \cos(\sigma t + \epsilon - \epsilon_1). \quad \text{.......................(15)}$$

Since

$$R^2 = \left(1 - \frac{\sigma^2 a}{c}\right)^2 + \frac{\sigma^2 b^2}{c^2}, \quad \text{.......................(16)}$$

it is easily found that if $b^2 < 2ac$ the amplitude is greatest when

$$\sigma = \left(\frac{c}{a}\right)^{\frac{1}{2}} \cdot \left(1 - \tfrac{1}{2}\frac{b^2}{ac}\right)^{\frac{1}{2}}, \quad \text{.......................(17)}$$

its value then being

$$\frac{C}{b}\left(\frac{a}{c}\right)^{\frac{1}{2}}\left(1 - \tfrac{1}{4}\frac{b^2}{ac}\right)^{-\frac{1}{2}}. \quad \text{..........................(18)}$$

In the case of relatively small friction, where $b^2/4ac$ may be neglected as of the second order, the amplitude is greatest when the period of the imposed force coincides with that of the free oscillation (cf. Art. 168). The formula (18) then shews that the amplitude when a maximum bears to its 'equilibrium-value' (C/c) the ratio $(ac)^{\frac{1}{2}}/b$, which is by hypothesis large.

On the other hand, when $b^2 > 2ac$ the amplitude continually increases as the speed σ diminishes, tending ultimately to the 'equilibrium-value' C/c.

It also appears from (15) and (12) that the maximum displacement follows the maximum of the force at an interval of phase equal to ϵ_1, where

$$\tan \epsilon_1 = \frac{\sigma b}{c - \sigma^2 a}. \quad \text{..........................(19)}$$

If the period be longer than the free-period in the absence of friction this difference of phase lies between 0 and 90°; in the opposite case it lies between 90° and 180°. If the frictional coefficient b be relatively small, the interval differs very little from 0 or 180°, as the case may be, unless σ be very nearly equal to the critical speed $(c/a)^{\frac{1}{2}}$. For the critical speed the phase-difference is 90°.

The rate of dissipation is $b\dot{q}^2$, the mean value of which is easily found to be

$$\tfrac{1}{2}\frac{bC^2}{(\sigma a - c/\sigma)^2 + b^2}. \quad \text{..........................(20)}$$

This is greatest when $\sigma = (c/a)^{\frac{1}{2}}$ exactly.

As in Art. 168, when the oscillations are very rapid the formula (11) gives

$$q = -Q/\sigma^2 a, \quad \text{............................(21)}$$

approximately; the *inertia* only of the system being operative.

On the other hand, when σ is small, the displacement has very nearly the equilibrium-value

$$q = Q/c. \quad \text{................................(22)}$$

319. An interesting example is furnished by the tides in an equatorial canal*.

The equation of motion, as modified by the introduction of a frictional term, is

$$\frac{\partial^2 \xi}{\partial t^2} = -\mu \frac{\partial \xi}{\partial t} + c^2 \frac{\partial^2 \xi}{a^2 \partial \phi^2} + X, \quad \text{.........................(1)}$$

where the notation is as in Art. 181†, a denoting the earth's radius.

In the case of *free* waves, putting $X = 0$, and assuming that

$$\xi \propto e^{\lambda t + ik a \phi}, \quad \text{.................................(2)}$$

we find

$$\lambda^2 + \mu\lambda + k^2 c^2 = 0,$$

whence

$$\lambda = -\tfrac{1}{2}\mu \pm i\,(k^2c^2 - \tfrac{1}{4}\mu^2)^{\frac{1}{2}}. \quad \text{........................(3)}$$

If we neglect the square of μ/kc, this gives, in real form,

$$\xi = A e^{-\frac{1}{2}\mu t} \cos\{k\,(ct \pm a\phi) + \epsilon\}. \quad \text{......................(4)}$$

The modulus of decay is $2\mu^{-1}$, and the wave-velocity is (to the first order) unaffected by the friction.

To find the forced waves due to the attraction of the moon we write, in conformity with Art. 181,

$$X = ife^{2i(nt+\phi+\epsilon)}, \quad \text{.............................(5)}$$

where n is the angular velocity of the moon relative to a fixed point on the canal. We find, assuming the same time-factor,

$$\xi = \tfrac{1}{4} \frac{ifa^2}{c^2 - n^2a^2 + \frac{1}{2} i\mu n a^2} e^{2i(nt+\phi+\epsilon)}. \quad \text{.....................(6)}$$

Hence, for the surface-elevation, we have

$$\eta = -h \frac{\partial \xi}{a \partial \phi} = \tfrac{1}{2} \frac{Hc^2}{c^2 - n^2a^2 + \frac{1}{2} i\mu n a^2} e^{2i(nt+\phi+\epsilon)}, \quad \text{..............(7)}$$

where $H = af/g$, as in Art. 180.

To put these expressions in real form, we write

$$\tan 2\chi = \tfrac{1}{2} \frac{\mu n a^2}{c^2 - n^2 a^2}, \quad \text{..........................(8)}$$

where $0 < \chi < 90°$. We thus find that to the tidal disturbing force

$$X = -f \sin 2\,(nt + \phi + \epsilon) \quad \text{.......................(9)}$$

correspond the horizontal displacement

$$\xi = -\tfrac{1}{4} \frac{fa^2}{\{(c^2 - n^2a^2)^2 + \frac{1}{4}\mu^2 n^2 a^4\}^{\frac{1}{2}}} \sin 2\,(nt + \phi + \epsilon - \chi), \quad \text{......(10)}$$

and the surface-elevation

$$\eta = \tfrac{1}{2} \frac{Hc^2}{\{(c^2 - n^2a^2)^2 + \frac{1}{4}\mu^2 n^2 a^4\}^{\frac{1}{2}}} \cos 2\,(nt + \phi + \epsilon - \chi). \quad \text{......(11)}$$

* Airy, "Tides and Waves," Arts. 315....

† In particular, c^2 now stands for gh, where h is the depth.

Since in these expressions $nt+\phi+\epsilon$ measures the hour-angle of the moon past the meridian of any point (ϕ) on the canal, it appears that high-water will follow the moon's transit at an interval t_1 given by $nt_1=\chi$.

If $c^2<n^2a^2$, or $h/a<n^2a/g$, we should in the case of infinitesimal friction have $\chi=90°$, *i.e.* the tides would be *inverted* (cf. Art. 181). With sensible friction, χ will lie between 90° and 45°, and the time of high-water is *accelerated* by the time-equivalent of the angle $90°-\chi$.

On the other hand, when $h/a>n^2a/g$, so that in the absence of friction the tides would be *direct*, the value of χ lies between 0° and 45°, and the time of high-water is *retarded* by the time-equivalent of this angle.

The figures on the next page shew the two cases. The letters M, M' indicate the positions of the moon and 'anti-moon' (see p. 358), supposed situate in the plane of the equator, and the curved arrows shew the direction of the earth's rotation.

It is evident that in each case the attraction of the disturbing system on the elevated water is equivalent to a couple tending to *diminish* the angular momentum of the system composed of the earth and sea.

In the present problem the amount of the couple can be easily calculated. We find, from (9) and (11), for the mean tangential force on the elevated water, per unit area of the surface,

$$\frac{1}{2\pi}\int_0^{2\pi}\rho X\eta\,d\phi=-\tfrac{1}{2}\rho\mathbf{h}f\sin 2\chi, \quad\text{..................(12)}$$

where $\mathbf{h}$ is the vertical amplitude. Since the positive direction of X is eastwards, this shews that there is on the whole a balance of westward force. If we multiply by the area of the water-surface and by the radius a we get the amount of the retarding couple.

The effect of phase-differences in the composition of two tides of slightly different speeds has been already mentioned in Art. 224. To apply the formulae there given to the present case we must write $\sigma=2n$, $\epsilon=2\chi$. We find, from (8), above,

$$\frac{d\epsilon}{d\sigma}=\frac{d\chi}{dn}=\frac{\mu a^2(c^2+n^2a^2)}{4(c^2-n^2a^2)^2+\mu^2n^2a^4}. \quad\text{..................(13)*}$$

If we have two tide-generating bodies with very nearly equal periods, this expression gives the interval of time at which the spring-tides would follow the instant of conjunction (or opposition). The ratio of this value of $d\epsilon/d\sigma$ to a day ($2\pi/n$) cannot exceed

$$\frac{n^2a^2+c^2}{8\pi\,|\,n^2a^2-c^2\,|}.$$

The above investigation is reproduced on account of its theoretic interest, but it has only a restricted application to the actual circumstances of the earth. Even in the case of a broad equatorial oceanic belt of (say) 11,250 ft. depth, the phase-differences which it

* Cf. Airy, "Tides and Waves," Arts. 328....

is capable of explaining appear to be quite insignificant. We have from (8), and from Art. 181,

$$\tan 2\chi = -\frac{1}{1-311\,(h/a)} \cdot \frac{1}{n\tau} = -\cdot 191 \times \frac{2\pi}{n\tau},$$

where $\tau = 2/\mu$, the modulus of decay of free oscillations. It seems rational to suppose that the modulus of decay in such a case would be a considerable multiple of the lunar day $(2\pi/n)$, in which event the change produced by friction in the time of high-water would be comparable with

$$2\pi/n\tau \times 22 \text{ minutes}.$$

Hence we cannot account in this way for a phase-acceleration of more than a few minutes.

There is a similar limit to the amount of lagging of the spring-tides as calculated from the formula (13).

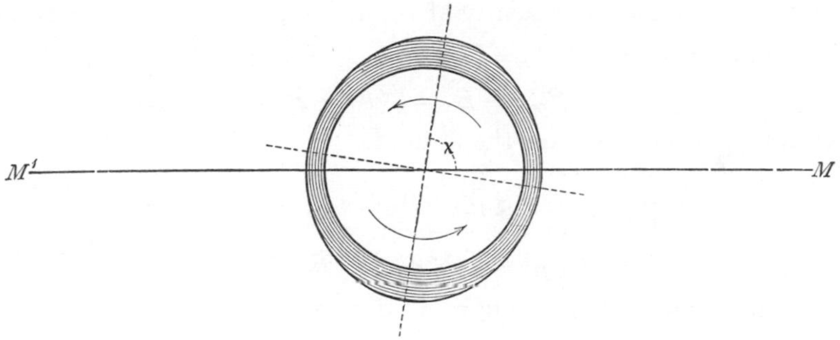

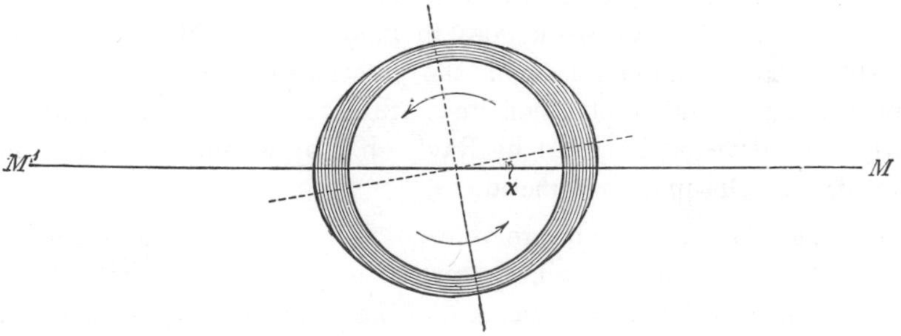

The *tidal* currents in mid-ocean are in fact so slight that their frictional effects are unimportant, even from an astronomical standpoint. In shallow water and in narrow seas and estuaries, on the other hand, they become enormously exaggerated as a result of the inertia of the water and the configuration of the ocean bed and the coasts. It appears now to be established* that the total dissipation of energy in such regions, at the expense ultimately of the earth's rotation, is comparable with that which is inferred on astronomical evidence. See Art. 371.

320. Returning to the general theory, let $q_1, q_2, \ldots q_n$ be the co-ordinates of a dynamical system, which we will suppose subject to conservative forces depending on its configuration, to 'motional' forces varying as the velocities,

* G. I. Taylor, *l.c. ante* p. 320; H. Jeffreys, *Phil. Trans.* A, ccxxi. 239 (1921).

and to given extraneous forces. The equations of small motion of such a system, on the most general assumptions we can make, will be of the type

$$\frac{d}{dt}\frac{\partial T}{\partial \dot{q}_r} + B_{r1}\dot{q}_1 + B_{r2}\dot{q}_2 + \ldots = -\frac{\partial V}{\partial q_r} + Q_r, \quad \ldots\ldots\ldots\ldots\ldots(1)$$

where the kinetic and potential energies T V are given by expressions of the forms

$$2T = a_{11}\dot{q}_1^2 + a_{22}\dot{q}_2^2 + \ldots + 2a_{12}\dot{q}_1\dot{q}_2 + \ldots, \quad \ldots\ldots\ldots\ldots\ldots(2)$$

$$2V = c_{11}q_1^2 + c_{22}q_2^2 + \ldots + 2c_{12}q_1q_2 + \ldots. \quad \ldots\ldots\ldots\ldots\ldots(3)$$

It is to be remembered that

$$a_{rs} = a_{sr}, \quad c_{rs} = c_{sr}, \quad \ldots\ldots\ldots\ldots\ldots\ldots\ldots\ldots\ldots\ldots(4)$$

but we do not assume the equality of B_{rs} and B_{sr}.

If we now write $\qquad b_{rs} = b_{sr} = \frac{1}{2}(B_{rs} + B_{sr}), \quad \ldots\ldots\ldots\ldots\ldots\ldots\ldots\ldots(5)$

and $\qquad \beta_{rs} = -\beta_{sr} = \frac{1}{2}(B_{rs} - B_{sr}), \quad \ldots\ldots\ldots\ldots\ldots\ldots\ldots(6)$

the typical equation (1) takes the form

$$\frac{d}{dt}\frac{\partial T}{\partial \dot{q}_r} + \frac{\partial F}{\partial \dot{q}_r} + \beta_{r1}\dot{q}_1 + \beta_{r2}\dot{q}_2 + \ldots = -\frac{\partial V}{\partial q_r} + Q_r, \quad \ldots\ldots\ldots\ldots(7)$$

provided $\qquad 2F = b_{11}\dot{q}_1^2 + b_{22}\dot{q}_2^2 + \ldots + 2b_{12}\dot{q}_1\dot{q}_2 + \ldots. \quad \ldots\ldots\ldots\ldots\ldots(8)$

From the equations in this form we derive

$$\frac{d}{dt}(T+V) + 2F = \Sigma Q_r\dot{q}_r. \quad \ldots\ldots\ldots\ldots\ldots\ldots\ldots\ldots(9)$$

The right-hand side expresses the rate at which the extraneous forces are doing work. Part of this work goes to increase the total energy $T+V$ of the system; the remainder is, from the present point of view, dissipated, at the rate $2F$. In the application to natural problems the function F is essentially positive; it is called by Rayleigh*, by whom it was first formally employed, the 'Dissipation-Function.'

The terms in (7) which are due to F may be distinguished as the 'frictional terms.' The remaining terms in $\dot{q}_1, \dot{q}_2, \ldots \dot{q}_n$, with coefficients subject to the relation $\beta_{rs} = -\beta_{sr}$, are of the type we have already met with in the general equations of a 'gyrostatic' system (Art. 141); they may therefore be referred to as the 'gyrostatic terms.'

321. When the gyrostatic terms are absent, the equation (7) reduces to

$$\frac{d}{dt}\frac{\partial T}{\partial \dot{q}_r} + \frac{\partial F}{\partial \dot{q}_r} + \frac{\partial V}{\partial q_r} = Q_r. \quad \ldots\ldots\ldots\ldots\ldots\ldots\ldots\ldots(10)$$

As in Art. 168, we may suppose that by transformation of co-ordinates the expressions for T and V are reduced to sums of squares, thus:

$$2T = a_1\dot{q}_1^2 + a_2\dot{q}_2^2 + \ldots + a_n\dot{q}_n^2, \quad \ldots\ldots\ldots\ldots\ldots\ldots\ldots(11)$$

$$2V = c_1q_1^2 + c_2q_2^2 + \ldots + c_nq_n^2. \quad \ldots\ldots\ldots\ldots\ldots\ldots\ldots(12)$$

* "Some General Theorems relating to Vibrations," *Proc. Lond. Math. Soc.* (1) iv. 357 (1873) [*Papers*, i. 170]; *Theory of Sound*, Art. 81.

It occasionally, but by no means necessarily, happens that the same transformation also reduces F to this form, say

$$2F = b_1\dot{q}_1^2 + b_2\dot{q}_2^2 + \ldots + b_n\dot{q}_n^2. \qquad (13)$$

The typical equation (10) then assumes the simple form

$$a_r\ddot{q}_r + b_r\dot{q}_r + c_r q_r = Q_r, \qquad (14)$$

which has been discussed in Art. 317. Each co-ordinate q_r now varies independently of the rest.

When F is not reduced by the same transformation as T and V, the equations (10) take the form

$$\left.\begin{array}{l} a_1\ddot{q}_1 + b_{11}\dot{q}_1 + b_{12}\dot{q}_2 + \ldots + b_{1n}\dot{q}_n + c_1 q_1 = Q_1, \\ a_2\ddot{q}_2 + b_{21}\dot{q}_1 + b_{22}\dot{q}_2 + \ldots + b_{2n}\dot{q}_n + c_2 q_2 = Q_2, \\ \ldots\ldots\ldots\ldots\ldots\ldots\ldots\ldots\ldots\ldots \\ a_n\ddot{q}_n + b_{n1}\dot{q}_1 + b_{n2}\dot{q}_2 + \ldots + b_{nn}\dot{q}_n + c_n q_n = Q_n, \end{array}\right\} \qquad (15)$$

where $b_{rs} = b_{sr}$.

The motion is now more complicated; for example, in the case of free oscillations about stable equilibrium, each particle executes (in any fundamental type) an elliptic-harmonic vibration, with the axes of the orbit contracting according to the law $e^{-\alpha t}$.

The question becomes somewhat simpler when the frictional coefficients b_{rs} are *small*, since the modes of motion will then be almost the same as in the case of no friction. Thus it appears from (15) that a mode of free motion is possible in which the main variation is in *one* co-ordinate, say q_r. The rth equation then reduces to

$$a_r\ddot{q}_r + b_{rr}\dot{q}_r + c_r q_r = 0, \qquad (16)$$

where we have omitted terms in which the relatively small quantities $\dot{q}_1, \dot{q}_2, \ldots \dot{q}_n$ (other than $\dot{q}_r$) are multiplied by the small coefficients $b_{r1}, b_{r2}, \ldots b_{rn}$. We have seen in Art. 317 that if b_{rr} be small the solution of (16) is of the type

$$q_r = Ae^{-t/\tau}\cos(\sigma t + \epsilon), \qquad (17)$$

where

$$\tau^{-1} = \tfrac{1}{2}b_{rr}/a_r, \quad \sigma = (c_r/a_r)^{\frac{1}{2}}. \qquad (18)$$

The relatively small variations of the remaining co-ordinates are then given by the remaining equations of the system (15). For example, with the same approximations,

$$a_s\ddot{q}_s + b_{rs}\dot{q}_r + c_s q_s = 0, \qquad (19)$$

whence

$$q_s = \frac{\sigma b_{rs}}{c_s - \sigma^2 a_s} Ae^{-t/\tau}\sin(\sigma t + \epsilon). \qquad (20)$$

Except in the case of approximate quality of period between two fundamental modes, the elliptic orbits of the particles will on the present suppositions be very flat.

If we were to assume that

$$q_r = \alpha\cos(\sigma t + \epsilon), \qquad (21)$$

where σ has the same value as in the case of no friction, whilst α varies slowly with the time, and that the variations of the other co-ordinates are relatively small, we should find

$$T + V = \tfrac{1}{2}a_r\dot{q}_r^2 + \tfrac{1}{2}c_r q_r^2 = \tfrac{1}{2}\sigma^2 a_r \alpha^2, \qquad (22)$$

nearly. Again, the dissipation is $\quad 2F = b_{rr}\dot{q}_r^2,$

the mean value of which is

$$\tfrac{1}{2}\sigma^2 b_{rr}\alpha^2, \qquad (23)$$

approximately. Hence equating the rate of decay of the energy to the mean value of the dissipation, we get

$$\frac{d\alpha}{dt} = -\frac{1}{2}\frac{b_{rr}}{a_r}\alpha, \qquad (24)$$

whence

$$\alpha = \alpha_0 e^{-t/\tau}, \qquad (25)$$

if

$$\tau^{-1} = \tfrac{1}{2}b_{rr}/a_r \qquad (26)$$

as in (18). This method of ascertaining the rate of decay of the oscillations is sometimes useful when the complete determination of the character of the motion, as affected by friction, would be more difficult (cf. Arts. 348, 355).

When the frictional coefficients are relatively great, the inertia of the system becomes ineffective; and the most appropriate system of co-ordinates is that which reduces F and V simultaneously to sums of squares, say

$$\left.\begin{aligned}2F&=b_1\dot{q}_1^2+b_2\dot{q}_2^2+\ldots+b_n\dot{q}_n^2,\\ 2V&=c_1q_1^2+c_2q_2^2+\ldots+c_nq_n^2.\end{aligned}\right\} \qquad (27)$$

The equations of free-motion are then of the type

$$b_r\dot{q}_r+c_rq_r=0, \qquad (28)$$

whence

$$q_r=Ce^{-t/\tau}, \qquad (29)$$

if

$$\tau=b_r/c_r. \qquad (30)$$

322. When gyrostatic as well as frictional terms are present in the fundamental equations, the theory is naturally more complicated. It will be sufficient here to consider the case of *two* degrees of freedom, by way of further elucidation of a point discussed in Art. 206*.

The equations of motion are now of the types

$$\left.\begin{aligned}a_1\ddot{q}_1+b_{11}\dot{q}_1+(b_{12}+\beta)\dot{q}_2+c_1q_1&=Q_1,\\ a_2\ddot{q}_2+(b_{21}-\beta)\dot{q}_1+b_{22}\dot{q}_2+c_2q_2&=Q_2.\end{aligned}\right\} \qquad (1)$$

To determine the modes of free motion we put $Q_1=0$, $Q_2=0$, and assume that q_1 and q_2 vary as $e^{\lambda t}$. This leads to the biquadratic in λ:

$$a_1a_2\lambda^4+(a_2b_{11}+a_1b_{22})\lambda^3+(a_2c_1+a_1c_2+\beta^2+b_{11}b_{22}-b_{12}^2)\lambda^2 \\ +(b_{11}c_2+b_{22}c_1)\lambda+c_1c_2=0. \qquad (2)$$

There is no difficulty in shewing, with the help of criteria given by Routh†, that if, as in our case, the quantities

$$a_1,\ a_2,\ b_{11},\ b_{22},\ b_{11}b_{22}-b_{12}^2$$

are all positive, the necessary and sufficient conditions that this biquadratic should have the real parts of its roots all negative are that c_1, c_2 should both be positive.

If we neglect terms of the second order in the frictional coefficients, the same conclusion may be attained more directly as follows. On this hypothesis the roots of (2) are, approximately,

$$\lambda=-\alpha_1\pm i\sigma_1,\qquad -\alpha_2\pm i\sigma_2, \qquad (3)$$

where σ_1, σ_2 are, to the first order, the same as in the case of no friction, viz. they are the roots of

$$a_1a_2\sigma^4-(a_2c_1+a_1c_2+\beta^2)\sigma^2+c_1c_2=0, \qquad (4)$$

whilst α_1, α_2 are determined by

$$\alpha_1+\alpha_2=\tfrac{1}{2}\left(\frac{b_{11}}{a_1}+\frac{b_{22}}{a_2}\right),\qquad \frac{\alpha_1}{\sigma_1^2}+\frac{\alpha_2}{\sigma_2^2}=\tfrac{1}{2}\left(\frac{b_{11}}{c_1}+\frac{b_{22}}{c_2}\right). \qquad (5)$$

It is evident that, if σ_1 and σ_2 are to be real, c_1, c_2 must have the same sign, and that if α_1, α_2 are to be positive, this sign must be $+$. Conversely, if c_1, c_2 are both positive, the values of σ_1^2, σ_2^2 are real and positive, and the quantities c_1/a_1, c_2/a_2 both lie in the interval between them. It then easily follows from (5) that α_1, α_2 are both positive‡.

* For a fuller treatment reference may be made to the works cited *ante* p. 250.

† *Advanced Rigid Dynamics*, Art. 287.

‡ A simple example of the above theory is supplied by the case of a particle in an ellipsoidal bowl rotating about a principal axis, which is vertical. If the bowl be frictionless, the equilibrium of the particle when at the lowest point will be stable unless the period of the rotation lie

If one of the coefficients c_1, c_2 (say c_2) be zero, one of the values of σ (say σ_2) is zero, indicating a free mode of infinitely long period. We then have

$$\sigma_1{}^2=\frac{c_1}{a_1}+\frac{\beta^2}{a_1a_2},\qquad \alpha_2=\frac{b_{22}c_1}{a_2c_1+\beta^2}. \quad\text{.....(6)}$$

As in Art. 206 we could easily write down the expressions for the forced oscillations in the general case where Q_1, Q_2 vary as $e^{i\sigma t}$, but we shall here consider more particularly the case where $c_2=0$ and $Q_2=0$. The equations (1) then give

$$\left.\begin{array}{l}(c_1-\sigma^2a_1+i\sigma b_{11})\,q_1+(b_{12}+\beta)\,\dot{q}_2=Q_1,\\ i\sigma\,(b_{12}-\beta)\,q_1+(i\sigma a_2+b_{22})\,\dot{q}_2=0.\end{array}\right\} \quad\text{.....(7)}$$

Hence

$$q_1=\frac{i\sigma a_2+b_{22}}{-a_1a_2i\sigma^3-(a_2b_{11}+a_1b_{22})\,\sigma^2+(a_2c_1+\beta^2+b_{11}b_{22}-b_{12}{}^2)\,i\sigma+b_{22}c_1}\,Q_1. \quad\text{......(8)}$$

This may also be written $$q_1=\frac{i\sigma a_2+b_{22}}{a_1a_2\{(i\sigma+\alpha_1)^2+\sigma_1{}^2\}\,(i\sigma+\alpha_2)}\,Q_1. \quad\text{.....(9)}$$

Our main object is to examine the case of a disturbing force of long period, for the sake of its bearing on Laplace's argument as to the fortnightly tide (Art. 217). We will therefore suppose that the ratio σ_1/σ, as well as σ_1/α_1, is large. The formula then reduces to

$$q_1=\frac{i\sigma a_2+b_{22}}{a_1a_2\sigma_1{}^2\,(i\sigma+\alpha_2)}\,Q_1=\frac{i\sigma a_2+b_{22}}{b_{22}c_1\,(i\sigma/\alpha_2+1)}\,Q_1. \quad\text{.....(10)}$$

Everything now turns on the values of the ratios σ/α_2 and $\sigma a_2/b_{22}$. If σ be so small that these may be both neglected, we have

$$q_1=Q_1/c_1, \quad\text{.....(11)}$$

in agreement with the equilibrium theory. The assumption here made is that the period of the imposed force is long compared with the modulus of decay. If, on the other hand, we assume σ/α_2 and $\sigma a_2/b_{22}$ to be large, we obtain

$$q_1=\frac{a_2\alpha_2}{b_{22}c_1}\,Q_1=\frac{a_2}{a_2c_1+\beta^2}\,Q_1, \quad\text{.....(12)}$$

as in Art. 206 (8).

Viscosity.

323. We proceed to consider the special kind of resistance which is met with in fluids. The methods we shall employ are of necessity the same as are applicable to the resistance to distortion, known as 'elasticity,' which is characteristic of solid bodies. The two classes of phenomena are physically distinct, the latter depending on the actual changes of shape produced, the former on the *rate* of change of shape, but the mathematical methods appropriate to them are to a great extent identical.

If we imagine three planes to be drawn through any point P perpendicular to the axes of x, y, z respectively, the three components of the stress,

between the periods of the two fundamental modes of oscillation (one in each principal plane) of the particle when the bowl is at rest. But if there be friction of motion between the particle and the bowl, there will be 'secular' stability only so long as the speed of the rotation is less than that of the slower of the two modes referred to. If the rotation be more rapid, the particle will gradually work its way outwards into a position of relative equilibrium in which it rotates with the bowl like the bob of a conical pendulum. In this state the system made up of the particle and the bowl has *less* energy for the same angular momentum than when the particle was at the bottom. Cf. Art. 254. Some further illustrations are given in a paper "On Kinetic Stability," *Proc. Roy. Soc.* A, lxxx. 168 (1907).

per unit area, exerted across the first of these planes may be denoted by p_{xx}, p_{xy}, p_{xz}, respectively; those of the stress across the second plane by p_{yx}, p_{yy}, p_{yz}; and those of the stress across the third plane by p_{zx}, p_{zy}, p_{zz}*. If we fix our attention on an element $\delta x \delta y \delta z$ having its centre at P, we find, on taking moments, and dividing by $\delta x \delta y \delta z$,

$$p_{yz} = p_{zy}, \quad p_{zx} = p_{xz}, \quad p_{xy} = p_{yx},$$

the extraneous forces and the kinetic reactions being omitted, since they are of a higher order of small quantities than the surface tractions. These equalities reduce the nine components of stress to six; in the case of a viscous fluid they will also follow independently from the expressions for p_{yz}, p_{zx}, p_{xy} in terms of the rates of distortion, to be given presently (Art. 325).

324. It appears from Arts. 1, 2 that in a fluid the deviation of the state of stress denoted by $p_{xx}, p_{xy}, \ldots$ from one of pressure uniform in all directions depends entirely on the motion of distortion in the neighbourhood of P, *i.e.* on the six quantities a, b, c, f, g, h by which this distortion was in Art. 30 shewn to be specified. Before endeavouring to express $p_{xx}, p_{xy}, \ldots$ as functions of these quantities, it is convenient to establish certain formulae of transformation.

Let us draw Px', Py', Pz' in the directions of the principal axes of distortion at P, and let a', b', c' be the rates of extension along these lines. Further let the mutual configuration of the two sets of axes x, y, z and x', y', z', be specified in the usual manner by the annexed scheme of direction-cosines. We have, then,

	x	y	z
x'	l_1,	m_1,	n_1,
y'	l_2,	m_2,	n_2,
z'	l_3,	m_3,	n_3.

$$\frac{\partial u}{\partial x} = \left(l_1 \frac{\partial}{\partial x'} + l_2 \frac{\partial}{\partial y'} + l_3 \frac{\partial}{\partial z'}\right)(l_1 u' + l_2 v' + l_3 w') = l_1^2 \frac{\partial u'}{\partial x'} + l_2^2 \frac{\partial v'}{\partial y'} + l_3^2 \frac{\partial w'}{\partial z'}.$$

Hence

$$\left.\begin{aligned} a &= l_1^2 a' + l_2^2 b' + l_3^2 c', \\ b &= m_1^2 a' + m_2^2 b' + m_3^2 c', \\ c &= n_1^2 a' + n_2^2 b' + n_3^2 c', \end{aligned}\right\} \quad \ldots\ldots\ldots\ldots\ldots\ldots(1)$$

the last two relations being written down from symmetry. We notice that

$$a + b + c = a' + b' + c', \quad \ldots\ldots\ldots\ldots\ldots\ldots(2)$$

as we should expect, since either side measures the 'expansion' (Art. 7).

Again

$$\frac{\partial w}{\partial y} + \frac{\partial v}{\partial z} = \left(m_1 \frac{\partial}{\partial x'} + m_2 \frac{\partial}{\partial y'} + m_3 \frac{\partial}{\partial z'}\right)(n_1 u' + n_2 v' + n_3 w')$$
$$+ \left(n_1 \frac{\partial}{\partial x'} + n_2 \frac{\partial}{\partial y'} + n_3 \frac{\partial}{\partial z'}\right)(m_1 u' + m_2 v' + m_3 w');$$

* In conformity with the usual practice in the theory of Elasticity, we reckon a *tension* as positive, a *pressure* as negative. Thus in the case of a frictionless fluid we have

$$p_{xx} = p_{yy} = p_{zz} = -p.$$

and this, with the two corresponding formulae, gives

$$\left.\begin{aligned} f &= 2\,(m_1 n_1 a' + m_2 n_2 b' + m_3 n_3 c'), \\ g &= 2\,(n_1 l_1 a' \;\; + n_2 l_2 b' \;\; + n_3 l_3 c'), \\ h &= 2\,(l_1 m_1 a' \;\; + l_2 m_2 b' \;\; + l_3 m_3 c'). \end{aligned}\right\} \qquad \text{.................(3)}$$

325. From the symmetry of the circumstances it is plain that the stresses exerted at P across the planes $y'z'$, $z'x'$, $x'y'$ must be wholly perpendicular to these planes. Let us denote them by p_1, p_2, p_3, respectively. In the figure of Art. 2 let ABC now represent a plane drawn perpendicular to x, infinitely close to P, meeting the axes of x', y', z' in A, B, C, respectively; and let Δ denote the area ABC. The areas of the remaining faces of the tetrahedron $PABC$ will then be $l_1\Delta$, $l_2\Delta$, $l_3\Delta$. Resolving parallel to x the forces acting on the tetrahedron, we find

$$p_{xx}\Delta = p_1 l_1 \Delta \,.\, l_1 + p_2 l_2 \Delta \,.\, l_2 + p_3 l_3 \Delta \,.\, l_3,$$

the external impressed forces and the resistances to acceleration being omitted for the same reason as before. Hence, and by similar reasoning,

$$\left.\begin{aligned} p_{xx} &= p_1 l_1^2 \;\; + p_2 l_2^2 \;\; + p_3 l_3^2, \\ p_{yy} &= p_1 m_1^2 + p_2 m_2^2 + p_3 m_3^2, \\ p_{zz} &= p_1 n_1^2 \;\; + p_2 n_2^2 \;\; + p_3 n_3^2. \end{aligned}\right\} \qquad \text{....................(1)}$$

We notice that

$$p_{xx} + p_{yy} + p_{zz} = p_1 + p_2 + p_3. \qquad \text{....................(2)}$$

Hence the arithmetic mean of the normal pressures on any three mutually perpendicular planes through the point P is the same. We shall denote this mean pressure by p*.

Again, resolving parallel to y, we obtain the third of the following symmetrical system of equations:

$$\left.\begin{aligned} p_{yz} &= p_1 m_1 n_1 + p_2 m_2 n_2 + p_3 m_3 n_3, \\ p_{zx} &= p_1 n_1 l_1 \;\; + p_2 n_2 l_2 \;\; + p_3 n_3 l_3, \\ p_{xy} &= p_1 l_1 m_1 \;\; + p_2 l_2 m_2 \;\; + p_3 l_3 m_3. \end{aligned}\right\} \qquad \text{.................(3)}$$

These shew that

$$p_{yz} = p_{zy}, \quad p_{zx} = p_{xz}, \quad p_{xy} = p_{yx},$$

as was proved independently in Art. 323.

If in the same figure we suppose PA, PB, PC to be drawn parallel to x, y, z, respectively, whilst ABC is any plane drawn near P, whose direction-cosines are l, m, n, we find in the same way that the components (p_{hx}, p_{hy}, p_{hz}) of the stress exerted across this plane are

$$\left.\begin{aligned} p_{hx} &= l p_{xx} + m p_{xy} + n p_{xz}, \\ p_{hy} &= l p_{yx} + m p_{yy} + n p_{yz}, \\ p_{hz} &= l p_{zx} + m p_{zy} + n p_{zz}. \end{aligned}\right\} \qquad \text{......................(4)}$$

* The question remains open as to whether, in the case of a gas, the mean pressure is a function of the density and temperature only (as in the statical condition to which Boyle's and Dalton's laws in the first instance relate), or whether it depends also on the rate of expansion at the point (x, y, z). See *infra* Art. 358.

326. Now p_1, p_2, p_3 differ from $-p$ by quantities depending on the motion of distortion, which must therefore be functions of a', b', c', only. The simplest hypothesis we can frame on this point is that these functions are linear. We write therefore

$$\left.\begin{aligned} p_1 &= -p + \lambda(a'+b'+c') + 2\mu a', \\ p_2 &= -p + \lambda(a'+b'+c') + 2\mu b', \\ p_3 &= -p + \lambda(a'+b'+c') + 2\mu c', \end{aligned}\right\} \quad \text{.....................(1)}$$

where λ, μ are constants depending on the nature of the fluid, and on its physical state, this being the most general assumption consistent with the above suppositions, and with symmetry. Substituting these values of p_1, p_2, p_3 in (1) and (3) of Art. 325, and making use of the results of Art. 324, we find

$$\left.\begin{aligned} p_{xx} &= -p + \lambda(a+b+c) + 2\mu a, \\ p_{yy} &= -p + \lambda(a+b+c) + 2\mu b, \\ p_{zz} &= -p + \lambda(a+b+c) + 2\mu c, \end{aligned}\right\} \quad \text{.....................(2)}$$

$$p_{yz} = \mu f, \quad p_{zx} = \mu g, \quad p_{xy} = \mu h. \quad \text{.....................(3)}$$

The definition of p adopted in Art. 325 implies the relation

$$3\lambda + 2\mu = 0, \quad \text{................................(4)}$$

whence, finally, introducing the values of a, b, c, f, g, h from Art. 30,

$$\left.\begin{aligned} p_{xx} &= -p - \tfrac{2}{3}\mu\left(\frac{\partial u}{\partial x} + \frac{\partial v}{\partial y} + \frac{\partial w}{\partial z}\right) + 2\mu\frac{\partial u}{\partial x}, \\ p_{yy} &= -p - \tfrac{2}{3}\mu\left(\frac{\partial u}{\partial x} + \frac{\partial v}{\partial y} + \frac{\partial w}{\partial z}\right) + 2\mu\frac{\partial u}{\partial y}, \\ p_{zz} &= -p - \tfrac{2}{3}\mu\left(\frac{\partial u}{\partial x} + \frac{\partial v}{\partial y} + \frac{\partial w}{\partial z}\right) + 2\mu\frac{\partial u}{\partial z}, \end{aligned}\right\} \quad \text{..............(5)}$$

$$\left.\begin{aligned} p_{yz} &= \mu\left(\frac{\partial w}{\partial y} + \frac{\partial v}{\partial z}\right) = p_{zy}, \\ p_{zx} &= \mu\left(\frac{\partial u}{\partial z} + \frac{\partial w}{\partial x}\right) = p_{xz}, \\ p_{xy} &= \mu\left(\frac{\partial v}{\partial x} + \frac{\partial u}{\partial y}\right) = p_{yx}. \end{aligned}\right\} \quad \text{.......................(6)}$$

The constant μ is called the 'coefficient of viscosity.' Its physical meaning may be illustrated by reference to the case of a fluid in what is called 'laminar' motion (Art. 30); *i.e.* the fluid moves in a system of parallel planes, the velocity being in direction everywhere the same, and in magnitude proportional to the distance from some fixed plane of the system. Each stratum of fluid will then exert on the one next to it a tangential traction, opposing the relative motion, whose amount per unit area is μ times the velocity-gradient in the direction perpendicular to the planes. In symbols, if $u = \alpha y$, $v = 0$, $w = 0$, we have

$$p_{xx} = p_{yy} = p_{zz} = -p, \quad p_{yz} = 0, \quad p_{zx} = 0, \quad p_{xy} = \mu\alpha.$$

If M, L, T denote the units of mass, length, and time, the unit of stress varies as $\mathsf{M}\mathsf{L}^{-1}\mathsf{T}^{-2}$, and that of the rates of distortion $(a, b, c, \ldots)$ as T^{-1}, so that the dimensions of μ are $\mathsf{M}\mathsf{L}^{-1}\mathsf{T}^{-1}$.

The stresses in different fluids, under similar circumstances of motion, will be proportional to the corresponding values of μ; but if we wish to compare their effects in modifying the existing motion we have to take account of the ratio of these stresses to the inertia of the fluid. From this point of view, the determining quantity is the ratio μ/ρ; it is therefore usual to denote this by a special symbol ν, called by Maxwell the 'kinematic' coefficient of viscosity. The dimensions of ν are $\mathsf{L}^2\mathsf{T}^{-1}$*.

It will be noticed that the hypothesis made above that the stresses $p_{xx}, p_{xy}, \ldots$ are *linear* functions of the rates of strain $a, b, c, \ldots$ is of a purely tentative character, and that although there is considerable *à priori* probability that it will represent the facts accurately in the case of infinitely small motions, we have so far no assurance that it will hold generally. It was however pointed out by Reynolds† that this hypothesis has been put to a very severe test in the experiments of Poiseuille and others, to be referred to presently (Art. 331). Considering the very wide range of values of the rates of distortion over which these experiments extend, we can hardly hesitate to accept the equations in question as a complete statement of the laws of viscosity. In the case of gases we have additional grounds for this assumption in the investigations of the kinetic theory by Maxwell‡.

The practical determination of μ (or ν) is a matter of some difficulty. Without entering into the details of experimental methods, we quote a few of the best-established results. Poiseuille's observations, as reduced by Helmholtz§, give for water

$$\mu = \frac{\cdot 01779}{1 + \cdot 033680\theta + \cdot 000220990\theta^2},$$

in C.G.S. units, where θ is the temperature on the Centigrade scale. The viscosity, as in the case of all liquids as yet investigated, diminishes rapidly as the temperature rises; thus at 10° C. the value is $\mu_{10} = \cdot 0131$. The results of more recent experiments are in good agreement with the above formula‖. For mercury Koch¶ found $\mu_0 = \cdot 01697$, and $\mu_{10} = \cdot 01633$, respectively. It should be added that in the case of some liquids, the mineral oils especially, the value of μ is considerably increased under pressures of the order of hundreds of atmospheres**.

In gases, the value of μ is found to be sensibly independent of the pressure, within very

* In compressible fluids there may, on a certain view, be a second coefficient of viscosity, involved in the expression for the mean pressure p as depending on the physical state and the rate of expansion. See Arts. 325, 358.

† "On the Theory of Lubrication, &c.," *Phil. Trans.* clxxvii. 157 (1886) [*Papers*, ii. 228].

‡ "On the Dynamical Theory of Gases," *Phil. Trans.* clvii. 49 (1866) [*Papers*, ii. 26].

§ "Ueber Reibung tropfbarer Flüssigkeiten," *Wien. Sitzungsber.* xl. 607 (1860) [*Wiss. Abh.* i. 218].

‖ Hosking, *Phil. Mag.* (6) xvii. 502 (1909).

¶ *Wied. Ann.* xiv. (1881).

** Hyde, *Proc. Roy. Soc.* A, xcvii. 240 (1919).

wide limits, but to *increase* somewhat with rise of temperature. An empirical formula* for the case of air is

$$\mu = \cdot 0001702\,(1 + \cdot 003290\theta + \cdot 0000070\theta^2).$$

At atmospheric pressure, assuming $\rho = \cdot 00129$, this gives

$$\nu_0 = \cdot 132.$$

The value of ν varies inversely as the pressure.

327. We have still to inquire into the dynamical conditions to be satisfied at the boundaries.

At a free surface, or at the surface of contact of two dissimilar fluids, the three components of stress across the surface must be continuous†. The resulting conditions can easily be written down with the help of Art. 325 (4).

A more difficult question arises as to the state of things at the surface of contact of a fluid with a solid. It appears probable that in all ordinary cases there is no motion, relative to the solid, of the fluid immediately in contact with it. The contrary supposition would imply an infinitely greater resistance to the sliding of one portion of the fluid past another than to the sliding of the fluid over a solid‡.

If however we wish, temporarily, to leave this point open, the most natural supposition to make is that the slipping is resisted by a tangential force proportional to the relative velocity. If we consider the motion of a small film of fluid, of thickness infinitely small compared with its lateral dimensions, in contact with the solid, it is evident that the tangential traction on its inner surface must ultimately balance the force exerted on its outer surface by the solid. The former force may be calculated from Art. 325 (4); the latter is in a direction opposite to the relative velocity, and proportional to it. The constant (β, say) which expresses the ratio of the tangential force to the relative velocity may be called the 'coefficient of sliding friction.'

328. The equations of motion of a viscous fluid are obtained by considering, as in Art. 6, a rectangular element $\delta x\,\delta y\,\delta z$ having its centre at (x, y, z). Taking, for instance, the resolution parallel to x, the difference of the normal tractions on the two yz-faces gives $(\partial p_{xx}/\partial x)\,\delta x\,.\,\delta y\,\delta z$. The tangential tractions on the two zx-faces contribute $(\partial p_{yx}/\partial y)\,\delta y\,.\,\delta z\,\delta x$, and the two xy-faces give in like manner $(\partial p_{zx}/\partial z)\,\delta z\,.\,\delta x\,\delta y$. Hence, with our usual notation,

$$\left.\begin{aligned}\rho\frac{Du}{Dt} &= \rho X + \frac{\partial p_{xx}}{\partial x} + \frac{\partial p_{yx}}{\partial y} + \frac{\partial p_{zx}}{\partial z},\\ \rho\frac{Dv}{Dt} &= \rho Y + \frac{\partial p_{xy}}{\partial x} + \frac{\partial p_{yy}}{\partial y} + \frac{\partial p_{zy}}{\partial z},\\ \rho\frac{Dw}{Dt} &= \rho Z + \frac{\partial p_{xz}}{\partial x} + \frac{\partial p_{yz}}{\partial y} + \frac{\partial p_{zz}}{\partial z}.\end{aligned}\right\} \qquad (1)$$

* Grindley and Gibson, *Proc. Roy. Soc.* A, lxxx. 114 (1907).

† This statement requires an obvious modification when capillarity is taken into account. Cf. Art. 265.

‡ Stokes, "On the Theories of the Internal Friction of Fluids in Motion, &c.," *Camb. Trans.* viii. 287 (1845) [*Papers*, i. 75].

Substituting the values of $p_{xx}, p_{xy}, \ldots$ from Art. 326 (5), (6), we find

$$\left.\begin{aligned} \rho\frac{Du}{Dt} &= \rho X - \frac{\partial p}{\partial x} + \tfrac{1}{3}\mu\frac{\partial \theta}{\partial x} + \mu\nabla^2 u, \\ \rho\frac{Dv}{Dt} &= \rho Y - \frac{\partial p}{\partial y} + \tfrac{1}{3}\mu\frac{\partial \theta}{\partial y} + \mu\nabla^2 v; \\ \rho\frac{Dw}{Dt} &= \rho Z - \frac{\partial p}{\partial z} + \tfrac{1}{3}\mu\frac{\partial \theta}{\partial z} + \mu\nabla^2 w, \end{aligned}\right\} \quad\ldots\ldots\ldots\ldots(2)$$

where $$\theta = \frac{\partial u}{\partial x} + \frac{\partial v}{\partial y} + \frac{\partial w}{\partial z}, \quad\ldots\ldots\ldots\ldots(3)$$

and ∇^2 has its usual meaning.

When the fluid is incompressible, these reduce to

$$\left.\begin{aligned} \rho\frac{Du}{Dt} &= \rho X - \frac{\partial p}{\partial x} + \mu\nabla^2 u, \\ \rho\frac{Dv}{Dt} &= \rho Y - \frac{\partial p}{\partial y} + \mu\nabla^2 v, \\ \rho\frac{Dw}{Dt} &= \rho Z - \frac{\partial p}{\partial z} + \mu\nabla^2 w. \end{aligned}\right\} \quad\ldots\ldots\ldots\ldots(4)$$

These dynamical equations were first obtained by Navier* and Poisson† on various considerations as to the mutual action of the ultimate molecules of fluids. The method above adopted, which does not involve any hypothesis of this kind, appears to be due in principle to de Saint-Venant‡ and Stokes§.

The equations (4) admit of an interesting interpretation. The first of them, for example, may be written

$$\frac{Du}{Dt} = X - \frac{1}{\rho}\frac{\partial p}{\partial x} + \nu\nabla^2 u. \quad\ldots\ldots\ldots\ldots(5)$$

The first two terms on the right-hand side express the rate of variation of u in consequence of the external forces and of the instantaneous distribution of pressure, and have the same forms as in the case of a frictionless liquid. The remaining term $\nu\nabla^2 u$, due to viscosity, gives an additional variation following the same law as that of temperature in Thermal Conduction, or of density in the theory of Diffusion. This variation is in fact proportional to the (positive or negative) excess of the mean value of u through a small sphere of given radius surrounding the point (x, y, z) over its value at that point‖. In connection with the thermal analogy it is interesting to note that the value of ν for water is of the same order of magnitude as that (·01249) found by Everett for the thermometric conductivity of the Greenwich gravel.

* "Mémoire sur les Lois du Mouvement des Fluides," *Mém. de l'Acad. des Sciences*, vi. 389 (1822).

† "Mémoire sur les Équations générales de l'Équilibre et du Mouvement des Corps solides élastiques et des Fluides," *Journ. de l'École Polytechn.* xiii. 1 (1829).

‡ *Comptes Rendus*, xvii. 1240 (1843).

§ "On the Theories of the Internal Friction of Fluids in Motion, &c." *Camb. Trans.* viii. 287 (1845) [*Papers*, i. 75].

‖ Maxwell, "On the Mathematical Classification of Physical Quantities," *Proc. Lond. Math. Soc.* (1) iii. 224 (1871) [*Papers*, ii. 257]; *Electricity and Magnetism*, Art. 26.

When the forces X, Y, Z have a potential Ω, the equations (4) may be written

$$\left.\begin{aligned}\frac{\partial u}{\partial t}-v\zeta+w\eta&=-\frac{\partial\chi'}{\partial x}+\nu\nabla^2u,\\ \frac{\partial v}{\partial t}-w\xi+u\zeta&=-\frac{\partial\chi'}{\partial y}+\nu\nabla^2v,\\ \frac{\partial w}{\partial t}-u\eta+v\xi&=-\frac{\partial\chi'}{\partial z}+\nu\nabla^2w,\end{aligned}\right\}\qquad(6)$$

where
$$\chi'=\frac{p}{\rho}+\tfrac{1}{2}q^2+\Omega,\qquad(7)$$

q denoting the resultant velocity, and ξ, η, ζ the components of vorticity. If we eliminate χ' by cross-differentiation, we find

$$\left.\begin{aligned}\frac{D\xi}{Dt}&=\xi\frac{\partial u}{\partial x}+\eta\frac{\partial u}{\partial y}+\zeta\frac{\partial u}{\partial z}+\nu\nabla^2\xi,\\ \frac{D\eta}{Dt}&=\xi\frac{\partial v}{\partial x}+\eta\frac{\partial v}{\partial y}+\zeta\frac{\partial v}{\partial z}+\nu\nabla^2\eta,\\ \frac{D\zeta}{Dt}&=\xi\frac{\partial w}{\partial x}+\eta\frac{\partial w}{\partial y}+\zeta\frac{\partial w}{\partial z}+\nu\nabla^2\zeta.\end{aligned}\right\}\qquad(8)$$

The first three terms on the right-hand side of each of these equations express, as in the case of Art. 146 (4), the rates at which ξ, η, ζ vary for a given particle, when the vortex-lines move with the fluid, and the strengths of the vortices remain constant. The additional variation of these quantities, due to viscosity, is given by the last terms, and follows the law of conduction of heat. It is evident from this analogy that vortex-motion cannot originate in the interior of a viscous liquid, but must be diffused inwards from the boundary.

328 a. In the two-dimensional case the equations (6) of the preceding Art. reduce to

$$\left.\begin{aligned}\frac{\partial u}{\partial t}-v\zeta&=-\frac{\partial\chi'}{\partial x}-\nu\frac{\partial\zeta}{\partial y},\\ \frac{\partial v}{\partial t}+u\zeta&=-\frac{\partial\chi'}{\partial y}+\nu\frac{\partial\zeta}{\partial x}.\end{aligned}\right\}\qquad(1)$$

Hence, or from (8) of the preceding Art.,

$$\frac{D\zeta}{Dt}=\nu\nabla_1^2\zeta,\qquad(2)$$

where the thermal analogy is obvious.

From (1) we derive a simple expression for the rate of change of the circulation in a *fixed* circuit. Thus

$$\frac{d}{dt}\int(u\,dx+v\,dy)=\int(lu+mv)\,\zeta\,ds+\nu\int\frac{\partial\zeta}{\partial n}\,ds,\qquad(3)$$

where (l, m) is the direction of the inward normal. The first term on the right-hand side gives the effect of the transport of vortices into the region enclosed by the circuit; the second shews the effect of viscosity.

For instance, in the case of motion in circles about an axis, we have

$$r\frac{\partial q}{\partial t}=\nu r\frac{\partial}{\partial r}\left(\frac{\partial q}{\partial r}+\frac{q}{r}\right),\qquad(4)$$

or
$$\frac{\partial q}{\partial t}=\nu\left(\frac{\partial^2 q}{\partial r^2}+\frac{1}{r}\frac{\partial q}{\partial r}-\frac{q}{r^2}\right).\qquad(5)$$

It is sometimes convenient to have at hand the principal formulae in plane polar co-ordinates. When we denote by u, v the component velocities along and at right angles to the radius vector the kinematical formulae are, as in the Appendix to Chapter VIII,

$$\frac{\partial u}{\partial t}+\frac{u}{r}+\frac{\partial v}{r\partial\theta}=0, \quad \text{.........(6)}$$

$$\zeta=\frac{\partial v}{\partial r}+\frac{v}{r}-\frac{\partial u}{r\partial\theta}. \quad \text{.........(7)}$$

The expressions for the component accelerations are given on p. 158. The equations (1) accordingly transform into

$$\left.\begin{aligned}\frac{\partial u}{\partial t}+u\frac{\partial u}{\partial r}+v\frac{\partial u}{r\partial\theta}-\frac{v^2}{r}&=R-\frac{1}{\rho}\frac{\partial p}{\partial r}-\nu\frac{\partial\zeta}{r\partial\theta},\\ \frac{\partial v}{\partial t}+u\frac{\partial v}{\partial r}+v\frac{\partial v}{r\partial\theta}+\frac{uv}{r}&=\Theta-\frac{1}{\rho}\frac{\partial p}{r\partial\theta}+\nu\frac{\partial\zeta}{\partial r}.\end{aligned}\right\} \quad \text{.........(8)}$$

Where R, Θ denote the radial and transverse components of extraneous force.

To find the component stresses, we denote by (u_1, v_1) the velocity referred to fixed Cartesian axes Ox_1, Oy_1, so that

$$u_1=u\cos\theta-v\sin\theta, \qquad v_1=u\sin\theta+v\cos\theta, \quad \text{.........(9)}$$

$$\frac{\partial}{\partial x_1}=\cos\theta\frac{\partial}{\partial r}-\sin\theta\frac{\partial}{r\partial\theta}, \qquad \frac{\partial}{\partial y_1}=\sin\theta\frac{\partial}{\partial r}+\cos\theta\frac{\partial}{r\partial\theta}. \quad \text{.........(10)}$$

Hence, if after the differentiations we take the axis of x coincident with the instantaneous position of the radius vector, putting $\theta=0$, we find

$$\frac{\partial u_1}{\partial x_1}=\frac{\partial u}{\partial r}, \qquad \frac{\partial v_1}{\partial y_1}=\frac{\partial v}{r\partial\theta}+\frac{u}{r},$$

$$\frac{\partial v_1}{\partial x_1}+\frac{\partial u_1}{\partial y_1}=\frac{\partial v}{\partial r}+\frac{\partial u}{r\partial\theta}-\frac{v}{r}.$$

The viscous stresses are accordingly, from Art. 326 (5),

$$\left.\begin{aligned}p_{rr}=-p+2\mu\frac{\partial u}{\partial r}, \qquad p_{\theta\theta}=-p+2\mu\left(\frac{\partial v}{r\partial\theta}+\frac{u}{r}\right),\\ p_{r\theta}=\mu\left(\frac{\partial v}{\partial r}+\frac{\partial u}{r\partial\theta}-\frac{v}{r}\right).\end{aligned}\right\} \quad \text{.........(11)}$$

If we resolve in the directions of r and θ the stresses on the sides of an element $r\partial\theta\partial r$, we reproduce the equations (8).

329. To compute the rate of dissipation of energy, due to viscosity, we consider first the portion of fluid which at time t occupies a rectangular element $\delta x\,\delta y\,\delta z$ having its centre at (x, y, z). Calculating the rates at which work is being done by the tractions on the pairs of opposite faces, we obtain

$$\left\{\frac{\partial}{\partial x}(p_{xx}u+p_{xy}v+p_{xz}w)+\frac{\partial}{\partial y}(p_{yx}u+p_{yy}v+p_{yz}w)\right.$$
$$\left.+\frac{\partial}{\partial z}(p_{zx}u+p_{zy}v+p_{zz}w)\right\}\delta x\,\delta y\,\delta z. \quad \text{......(1)}$$

The terms

$$\left\{\left(\frac{\partial p_{xx}}{\partial x}+\frac{\partial p_{yx}}{\partial y}+\frac{\partial p_{zx}}{\partial z}\right)u+\left(\frac{\partial p_{xy}}{\partial x}+\frac{\partial p_{yy}}{\partial y}+\frac{\partial p_{zy}}{\partial z}\right)v\right.$$
$$\left.+\left(\frac{\partial p_{xz}}{\partial x}+\frac{\partial p_{yz}}{\partial y}+\frac{\partial p_{zz}}{\partial z}\right)w\right\}\delta x\,\delta y\,\delta z \quad \text{.........(2)}$$

express, by Art. 328 (1), the rate at which the tractions on the faces are doing work on the element as a whole, in increasing its kinetic energy and in compensating the work done against the extraneous forces X, Y, Z. The remaining terms express the rate at which work is being done in changing the volume and shape of the element. They may be written

$$(p_{xx}a + p_{yy}b + p_{zz}c + p_{yz}f + p_{zx}g + p_{xy}h)\,\delta x\,\delta y\,\delta z, \quad \text{.........(3)}$$

where a, b, c, f, g, h have the same meanings as in Arts. 30, 324. Substituting from Art. 326 (2), (3), we get

$$-p(a+b+c)\,\delta x\,\delta y\,\delta z$$
$$+\{-\tfrac{2}{3}\mu(a+b+c)^2 + \mu(2a^2+2b^2+2c^2+f^2+g^2+h^2)\}\,\delta x\,\delta y\,\delta z. \quad \text{...(4)}$$

It will be sufficient for the present to consider the case where there is no variation of density, so that

$$a + b + c = 0. \quad \text{..................................(5)}$$

The expression (4) then reduces to

$$\mu(2a^2+2b^2+2c^2+f^2+g^2+h^2)\,\delta x\,\delta y\,\delta z, \quad \text{...............(6)}$$

which accordingly represents the rate at which mechanical energy is disappearing. On the principles established by Joule, the energy thus apparently lost takes the form of heat, developed in the element.

If we integrate over the whole volume of the liquid, we find, for the total rate of dissipation,

$$2F = \iiint \Phi\, dx\,dy\,dz, \quad \text{.............................(7)}$$

where

$$\Phi = \mu\left\{2\left(\frac{\partial u}{\partial x}\right)^2 + 2\left(\frac{\partial v}{\partial y}\right)^2 + 2\left(\frac{\partial w}{\partial z}\right)^2 + \left(\frac{\partial w}{\partial y} + \frac{\partial v}{\partial z}\right)^2 + \left(\frac{\partial u}{\partial z} + \frac{\partial w}{\partial x}\right)^2 + \left(\frac{\partial v}{\partial x} + \frac{\partial u}{\partial y}\right)^2\right\}. \quad \text{.....(8)*}$$

If we subtract from this the expression

$$2\mu\left(\frac{\partial u}{\partial x} + \frac{\partial v}{\partial y} + \frac{\partial w}{\partial z}\right)^2,$$

which is zero on the present hypothesis, we obtain

$$\Phi = \mu\left\{\left(\frac{\partial w}{\partial y} - \frac{\partial v}{\partial z}\right)^2 + \left(\frac{\partial u}{\partial z} - \frac{\partial w}{\partial x}\right)^2 + \left(\frac{\partial v}{\partial x} - \frac{\partial u}{\partial y}\right)^2\right\}$$
$$-4\mu\left(\frac{\partial v}{\partial y}\frac{\partial w}{\partial z} - \frac{\partial v}{\partial z}\frac{\partial w}{\partial y} + \frac{\partial w}{\partial z}\frac{\partial u}{\partial x} - \frac{\partial w}{\partial x}\frac{\partial u}{\partial z} + \frac{\partial u}{\partial x}\frac{\partial v}{\partial y} - \frac{\partial u}{\partial y}\frac{\partial v}{\partial x}\right). \quad \text{......(9)}$$

If we integrate this over a region such that u, v, w vanish at every point of the boundary, as in the case of a liquid filling a closed vessel, on the hypothesis of no slipping, the terms due to the second line cancel (after partial integration), and we obtain

$$2F = \iiint \Phi\, dx\,dy\,dz = \mu \iiint (\xi^2 + \eta^2 + \zeta^2)\, dx\,dy\,dz. \quad \text{.....................(10)†}$$

* Stokes, "On the Effect of the Internal Friction of Fluids on the Motion of Pendulums," *Camb. Trans.* ix. [8] (1851) [*Papers*, iii. 1].

† Bobyleff, "Einige Betrachtungen über die Gleichungen der Hydrodynamik," *Math. Ann.* vi. 72 (1873); Forsyth, "On the Motion of a Viscous Incompressible Fluid," *Mess. of Math.* ix. (1880).

A more immediate proof of this formula is obtained if we note that on the present assumptions the equation of energy, Art. 10 (5), is replaced by

$$\frac{D}{Dt}(T+V)=\mu\iiint(u\nabla^2 u+v\nabla^2 v+w\nabla^2 w)\,dx\,dy\,dz$$

$$=\mu\iiint\left\{u\left(\frac{\partial\eta}{\partial z}-\frac{\partial\zeta}{\partial y}\right)+v\left(\frac{\partial\zeta}{\partial x}-\frac{\partial\xi}{\partial z}\right)+w\left(\frac{\partial\xi}{\partial y}-\frac{\partial\eta}{\partial x}\right)\right\}dx\,dy\,dz$$

$$=-\mu\iiint(\xi^2+\eta^2+\zeta^2)\,dx\,dy\,dz. \quad\text{.........(11)}$$

In the general case, where no limitation is made as to the boundary-conditions, the formula (9) leads to

$$2F=\mu\iiint(\xi^2+\eta^2+\zeta^2)\,dx\,dy\,dz-\mu\iint\frac{\partial\,.\,q^2}{\partial n}\,dS+2\mu\iint\begin{vmatrix} l, & m, & n \\ u, & v, & w \\ \xi, & \eta, & \zeta\end{vmatrix}dS, \quad\text{...(12)}$$

where, in the former of the two surface-integrals, δn denotes an element of the normal, and in the latter, l, m, n are the direction-cosines of the normal, drawn inwards in each case from the surface-element δS.

When the motion considered is irrotational, this formula reduces to

$$2F=-\mu\iint\frac{\partial\,.\,q^2}{\partial n}\,dS, \quad\text{.........(13)}$$

simply. In the particular case of a spherical boundary this expression follows independently from Art. 44 (5).

It appears from (6) that F cannot vanish unless

$$a=b=c=0, \quad\text{and}\quad f=g=h=0,$$

at every point of the fluid. It follows, on reference to Art. 30, that the only condition under which a liquid can be in motion without dissipation of energy by viscosity is that there must be nowhere any extension or contraction of linear elements; in other words, the motion must consist of a translation and a rotation of the mass as a whole, as in the case of a rigid body.

Problems of Steady Motion.

330. Proceeding now to the consideration of special problems, it may be well to state at the outset that although the equations of motion of viscous fluids are well established, the calculations based on them are often subject to serious limitations. The reason is partly to be sought in the omission, for the sake of mathematical simplicity, of small terms of the second order in the Eulerian expressions for the accelerations, which terms are often at least as important as those due to viscosity. Another reason is that even when the investigations are rigorous the types of motion obtained are often unstable. Attention is occasionally called in the sequel to these points, which will be discussed more fully in Arts. 365, *et seq.*

The first application which we shall consider is to the steady motion of liquid, under pressure, between two fixed parallel planes. Let the origin be taken in one of these planes, and the axis of z perpendicular to them. We assume, in the first instance, that u is a function of z only, and that v, $w=0$. Since the traction parallel to x on any plane perpendicular to z is equal to $\mu\partial u/\partial z$, the difference of the tractions on the two faces of a stratum of unit area and thickness δz gives a resultant $\mu\partial^2 u/\partial z^2 . \delta z$. This must be balanced by the pressures, which give a resultant $-\partial p/\partial x$ per unit volume of the stratum. Hence

$$\mu \frac{\partial^2 u}{\partial z^2} = \frac{\partial p}{\partial x}. \quad \text{.................................(1)}$$

Also, since there is no motion parallel to z, $\partial p/\partial z$ must vanish. These results also follow immediately from the general equations of Art. 328.

It follows that the pressure-gradient $\partial p/\partial x$ is an absolute constant. Hence (1) gives

$$u = A + Bz + \frac{1}{2\mu} z^2 \frac{\partial p}{\partial x}, \quad \text{..........................(2)}$$

and determining the constants so as to make $u = 0$ for $z = 0$ and $z = h$, we find

$$u = -\frac{1}{2\mu} z (h - z) \frac{\partial p}{\partial x}. \quad \text{..........................(3)}$$

Hence

$$\int_0^h u\,dz = -\frac{h^3}{12\mu}\frac{\partial p}{\partial x}. \quad \text{..........................(4)}$$

When, as in Prof. Hele Shaw's experiments*, a liquid flows in two dimensions between close parallel plates, we may write

$$\mu \frac{\partial^2 u}{\partial z^2} = \frac{\partial p}{\partial x}, \quad \mu \frac{\partial^2 v}{\partial z^2} = \frac{\partial p}{\partial y}, \quad \text{.................................(5)}$$

provided we neglect the rates of variation of u, v with respect to x, y in comparison with their rates of variation with respect to z. Also, assuming that $w=0$ everywhere, we have $\partial p/\partial z = 0$, *i.e.* p is a function of x and y only. The conditions of no slipping at the planes $z=0$, $z=h$ are satisfied if we write

$$u = \frac{6z(h-z)}{h^2} u', \quad v = \frac{6z(h-z)}{h^2} v'. \quad \text{.................................(6)}$$

The quantities u', v' here denote the mean velocities in the stratum, and are assumed to be functions of x, y only. Substituting in (5) we find

$$\frac{\partial p}{\partial x} = -\frac{12\mu}{h^2} u', \quad \frac{\partial p}{\partial y} = -\frac{12\mu}{h^2} v'. \quad \text{.................................(7)}$$

Hence u', v' may be regarded as the components of an irrotational motion of a liquid in two dimensions, in which the velocity-potential is

$$\phi = ph^2/12\mu. \quad \text{..(8)}$$

The kinematical conditions, when the liquid is forced by pressure past an obstacle having the form of a lamina of thickness h placed between the plates, are accordingly identical, for the most part, with those relating to the two-dimensional flow of a *frictionless* fluid past a cylinder whose section has the shape of the lamina. The statement is made with a slight qualification, since the equations (5) must cease to hold at distances from the obstacle comparable with h, owing to the fact that the viscous fluid cannot glide past the edge of the obstacle, as a perfect fluid would do. But the configurations of the stream-lines in the two problems can be made as nearly the same as we choose by taking the plates sufficiently close together†.

* Referred to in the footnote on p. 86.

† Stokes, "Mathematical Proof of the Identity of the Stream-Lines obtained by means of a Viscous Film with those of a Perfect Fluid moving in Two Dimensions," *Brit. Ass. Rep.* 1898, p. 143 [*Papers*, v. 278].

330 a. If the boundary $z=0$ has a velocity U parallel to x, we have in place of (3)

$$u=\frac{h-z}{h}U-\frac{z(h-z)}{2\mu}\frac{dp}{dx}, \qquad (9)$$

and the total flux per unit breadth across a plane perpendicular to x is

$$\int_0^h u\,dz=\tfrac{1}{2}hU-\frac{h^3}{12\mu}\frac{dp}{dx}. \qquad (10)$$

These formulae may be taken as approximately valid even if the interval h between the two surfaces is variable, provided the gradient dh/dx is small, and even if both surfaces are curved, provided h be everywhere small compared with the radii of curvature. In the case of cylindrical surfaces x may be taken to be the arc measured along the circumference, perpendicular to the generating lines.

The above results, as thus generalized, have an important application in the theory of Lubrication, which was initiated by Osborne Reynolds in a classical paper*. That two parallel or nearly parallel surfaces can slide one over the other with but slight frictional resistance, even under great normal pressure, provided a film of viscous fluid is maintained between them, is of course familiar. The problem was to explain how in practical cases this is effected automatically in spite of the pressure. The arrangement must be such that the interval between the two surfaces is of variable thickness, and that the tendency of the relative motion must be continually to drag a supply of the lubricant from the thicker to the thinner portions.

A simple typical case is that of a block sliding over a plane surface. Since the relative motion alone is important we will suppose that it is the latter surface ($z=0$) which is in motion, whilst the block itself is at rest. For simplicity it is assumed, further, that both surfaces are unlimited in the direction of y, so that the motion of the fluid is strictly two-dimensional. The lower surface of the block will be supposed to extend from $x=0$ to $x=a$. Assuming it to be plane, but slightly tilted, we write

$$h=h_1+mx, \quad h_2=h_1+ma, \qquad (11)$$

where m is small.

Since the flux across all planes perpendicular to x must be the same we have from (10)

$$h^3\frac{dp}{dx}=6\mu U(h-h_0), \qquad (12)$$

where h_0 corresponds to the maximum of p. Hence,

$$\frac{dp}{dh}=\frac{6\mu U}{m}\left(\frac{1}{h^2}-\frac{h_0}{h^3}\right), \qquad (13)$$

$$p=\frac{6\mu U}{m}\left(-\frac{1}{h}+\frac{h_0}{2h^2}+C\right). \qquad (14)$$

Determining h_0 and C so that $p=0$ for $h=h_1$ and $h=h_2$, we find

$$h_0=\frac{2h_1h_2}{h_1+h_2}, \qquad (15)$$

$$p=\frac{6\mu Ua}{h_1^2-h_2^2}\cdot\frac{(h_1-h)(h-h_2)}{h^2}. \qquad (16)$$

* Quoted on p. 575.

A general addition of a constant to p will of course make no difference to the essential results.

We see at once that if U is positive, as we will suppose, a positive pressure in the film is impossible unless $h_1 > h_2$; *i.e.* the interval must contract in the direction of the velocity U, as above stated.

For the total pressure we find

$$P = \int_0^a p\,dx = \frac{1}{m}\int_{h_1}^{h_2} p\,dh = \frac{6\mu U a^2}{(k-1)^2 h_2^2}\left(\log k - \frac{2(k-1)}{k+1}\right), \quad \text{.............(17)}$$

where $k = h_1/h_2$. The frictional resistance on the moving plane is

$$F = -\int_0^a \mu\frac{du}{dz}\,dx = \frac{\mu U}{m}\int_{h_1}^{h_2}\left(\frac{4}{h} - \frac{3h_0}{h^2}\right)dh = \frac{\mu U a}{(k-1)\,h_2}\left(4\log k - \frac{6(k-1)}{k+1}\right). \quad \text{...(18)}$$

It was found by Reynolds, and confirmed by Rayleigh*, that P, considered as a function of k, is a maximum for $k = 2{\cdot}2$, about. This makes

$$P = {\cdot}16\frac{\mu U a^2}{h_2^2}, \quad F = {\cdot}75\frac{\mu U a}{h_2}. \quad \text{..............................(19)}$$

The coefficient of friction (F/P) is of the order h_2/a and can therefore be made very small.

The co-ordinate $(\bar{x})$ of the centre of pressure is given by

$$P\bar{x} = \int_0^a xp\,dx = \frac{1}{m^2}\int_{h_1}^{h_2}(h - h_1)\,p\,dh = \frac{kPa}{k-1} - \frac{1}{2m^2}\int_{h_1}^{h_2} h^2\frac{dp}{dh}\,dh$$

$$= \frac{kPa}{k-1} - \frac{3\mu U a^3}{(k-1)^2 h_2^2}\left(1 - \frac{2k}{k^2-1}\log k\right), \quad \text{.................................(20)}$$

or

$$\frac{\bar{x}}{\frac{1}{2}a} = \frac{2k}{k-1} - \frac{k^2 - 1 - 2k\log k}{(k^2-1)\log k - 2(k-1)^2}. \quad \text{..........................(21)†}$$

For the application of (13) to the case of a shaft revolving (slightly eccentrically) in a fixed bearing reference may be made to the papers cited below‡.

When there is flow in the direction of y as well as x, we have in addition to (10)

$$\int_0^h v\,dz = \tfrac{1}{2}hV - \frac{h^3}{12\mu}\frac{\partial p}{\partial y}, \quad \text{.......................(22)}$$

and the equation of continuity is

$$\frac{\partial}{\partial x}\int_0^h u\,dz + \frac{\partial}{\partial y}\int_0^h v\,dz = 0, \quad \text{....................(23)}$$

or

$$\frac{\partial}{\partial x}\left(h^3\frac{\partial p}{\partial x}\right) + \frac{\partial}{\partial y}\left(h^3\frac{\partial p}{\partial y}\right) = 6\mu\left\{\frac{\partial}{\partial x}(hU) + \frac{\partial}{\partial y}(hV)\right\}. \quad \text{.........(24)}$$

This has been applied by Michell to the case of a rectangular block of finite dimensions sliding over a plane surface§.

* "Notes on the Theory of Lubrication," *Phil. Mag.* (6) xxxv. 1 (1918) [*Papers*, vi. 523].

† Rayleigh, *l.c.* For $k = 2{\cdot}2$ this makes $\bar{x} = {\cdot}580a$.

‡ Reynolds, *l.c.*; Sommerfeld, *Zeitschrift f. Math.* l. 97 (1904); Harrison, *Camb. Trans.* xxii. 39 (1913), and xxii. 373 (1920). Also A. G. M. Michell in *Mechanical Properties of Fluids*, London, 1923, p. 134; and Stanton, *Friction*, London, 1923, p. 93.

§ *Zeitschr. f. Math.* liii. 123 (1905). Some account of this very elegant investigation is given in the two books just cited.

331. We consider next the steady flow of a liquid through a straight pipe of uniform circular section.

If we take the axis of z coincident with the axis of the tube, and assume that the velocity is everywhere parallel to z, and a function of the distance (r) from this axis, the tangential stress across a plane perpendicular to r will be $\mu \partial w/\partial r$. Hence, considering a cylindrical shell of fluid whose bounding radii are r and $r + \delta r$, and whose length is l, the difference of the tangential tractions on the two curved surfaces gives a retarding force

$$-\frac{\partial}{\partial r}\left(\mu \frac{\partial w}{\partial r} . 2\pi r l\right) \delta r.$$

On account of the steady character of the motion, this must be balanced by the normal pressures on the plane ends of the shell. Since $\partial w/\partial z = 0$, the difference of these two normal pressures is equal to

$$(p_1 - p_2)\, 2\pi r\, \delta r,$$

where p_1, p_2 are the values of p (the mean pressure) at the two ends. Hence

$$\frac{\partial}{\partial r}\left(r \frac{\partial w}{\partial r}\right) = -\frac{p_1 - p_2}{\mu l} . r. \qquad \text{.......................(1)}$$

Again, if we resolve along the radius the forces acting on a rectangular element, we find $\partial p/\partial r = 0$, so that the mean pressure is uniform over each section of the pipe.

The integral of (1) is

$$w = -\frac{p_1 - p_2}{4\mu l} r^2 + A \log r + B. \qquad \text{....................(2)}$$

Since the velocity must be finite at the axis, we must have $A = 0$; and if we determine B on the hypothesis that there is no slipping at the wall of the pipe ($r = a$, say), we obtain

$$w = \frac{p_1 - p_2}{4\mu l}(a^2 - r^2). \qquad \text{.............................(3)}$$

This gives, for the flux across any section,

$$\int_0^a w . 2\pi r dr = \frac{\pi a^4}{8\mu} . \frac{p_1 - p_2}{l}. \qquad \text{...................... (4)}$$

It has been assumed, for shortness, that the flow takes place under pressure only. If we have an extraneous force X acting parallel to the length of the pipe, the flux will be

$$\frac{\pi a^4}{8\mu}\left(\frac{p_1 - p_2}{l} + \rho X\right). \qquad \text{.........................(5)}$$

In practice, X is the component of gravity in the direction of the length.

The formula (4) contains exactly the laws found experimentally by Poiseuille* in his researches on the flow of water through capillary tubes;

* "Recherches expérimentales sur le mouvement des liquides dans les tubes de très petits diamètres," *Comptes Rendus*, xi. xii. (1840–1), *Mém. des Sav. Étrangers*, ix. (1846).

viz. that the time of efflux of a given volume of water is directly as the length of the tube, inversely as the difference of pressure at the two ends, and inversely as the fourth power of the diameter.

This last result is of importance as furnishing a conclusive proof that there is in these experiments no appreciable slipping of the fluid in contact with the wall. If we were to assume a slipping-coefficient β, as explained in Art. 327, the surface-condition would be

$$-\mu \frac{\partial w}{\partial r} = \beta w,$$

or

$$w = -\lambda \frac{\partial w}{\partial r}, \quad \text{.......(6)}$$

if $\lambda = \mu/\beta$. This determines B, in (2), so that

$$w = \frac{p_1 - p_2}{4\mu l}(a^2 - r^2 + 2\lambda a). \quad \text{.......(7)}$$

If λ/a be small, this gives sensibly the same law of velocity as in a tube of radius $a+\lambda$, on the hypothesis of no slipping. The corresponding value of the flux is

$$\frac{\pi a^4}{8\mu} \cdot \frac{p_1 - p_2}{l} \cdot \left(1 + 4\frac{\lambda}{a}\right). \quad \text{.......(8)}$$

If λ were more than a very minute fraction of a in the narrowest tubes employed by Poiseuille [$a = {\cdot}0015$ cm.] a deviation from the law of the fourth power of the diameter, which was found to hold very exactly, would become apparent. This is sufficient to exclude the possibility of values of λ such as ·235 cm., which were inferred by Helmholtz and Piotrowski from their experiments on the torsional oscillations of a metal globe filled with water, described in the paper already cited*.

The assumption of no slipping being thus justified, the comparison of the formula (4) with experiment gives a very direct means of determining the value of the coefficient μ for various fluids†.

It follows from (3) and (4) that the rate of shear close to the wall of the tube is equal to $4w_0/a$, where w_0 is the mean velocity over the cross-section. As a numerical example, we may take a case given by Poiseuille, where a mean velocity of 126·6 cm./sec. was obtained in a tube of ·01134 cm. diameter. This makes $4w_0/a = 89300$, if the unit of time be the second.

For values of w_0 exceeding certain limits, depending on the relation between the diameter of the pipe and the viscosity, the linear type of flow here investigated becomes unstable, at all events for disturbances exceeding a certain amplitude; see Art. 365. There are analogous limitations to the results of Arts. 330, 331, and indeed to many of the calculations which follow.

332. Some theoretical results for sections other than circular may be noticed.

1°. The solution for a channel of *annular* section is readily deduced from equation (2) of the preceding Art., with A retained. Thus if the boundary-conditions be that $w=0$ for $r=a$ and $r=b$, we find

$$w = \frac{p_1 - p_2}{4\mu l}\left\{a^2 - r^2 + \frac{b^2 - a^2}{\log(b/a)} \log\frac{r}{a}\right\}, \quad \text{.......(1)}$$

* For a fuller discussion of this point see Whetham, "On the alleged Slipping at the Boundary of a Liquid in Motion," *Phil. Trans.* A, clxxxi. 559 (1890).

† Corrections are required in practice owing to the deviation from the theoretical flow near the ends of the tube; see Stanton, *Friction*, p. 15.

giving a flux $$\int_a^b w \,.\, 2\pi r\,dr = \frac{\pi}{8\mu}\,.\,\frac{p_1 - p_2}{l}\,.\,\left\{b^4 - a^4 - \frac{(b^2 - a^2)^2}{\log(b/a)}\right\}. \quad\text{.....................(2)}$$

2°. It has been pointed out by Greenhill* that the analytical conditions of the present problem are similar to those which determine the motion of a frictionless liquid in a rotating prismatic vessel of the same form of section (Art. 72). If the axis of z be parallel to the length of the pipe, and if we assume that w is a function of x, y only, then in the case of steady motion the equations reduce to

$$\left.\begin{aligned} \frac{\partial p}{\partial x} = 0, \quad \frac{\partial p}{\partial y} = 0, \\ \mu \nabla_1^2 w = \frac{\partial p}{\partial z}, \end{aligned}\right\} \quad\text{.....................................(3)}$$

where $\nabla_1^2 = \partial^2/\partial x^2 + \partial^2/\partial y^2$. Hence, denoting by P the constant pressure-gradient $(-\partial p/\partial z)$, we have

$$\nabla_1^2 w = -P/\mu, \quad\text{..(4)}$$

with the condition that $w = 0$ at the boundary. If we write $\psi - \frac{1}{2}\omega(x^2 + y^2)$ for w, and 2ω for P/μ, we reproduce the conditions of the Art. referred to. This proves the analogy in question.

In the case of an elliptic section of semi-axes a, b, we assume

$$w = C\left(1 - \frac{x^2}{a^2} - \frac{y^2}{b^2}\right), \quad\text{....................................(5)}$$

which will satisfy (4) provided $$C = \frac{P}{2\mu}\,.\,\frac{a^2 b^2}{a^2 + b^2}. \quad\text{.......................................(6)}$$

The discharge per second is therefore

$$\iint w\,dx\,dy = \frac{P}{4\mu}\,.\,\frac{\pi a^3 b^3}{a^2 + b^2}. \quad\text{...................................(7)†}$$

This bears to the discharge through a circular pipe of the same sectional area the ratio $2ab/(a^2 + b^2)$. For small values of the eccentricity (e) this fraction differs from unity by a quantity of the order e^4. Hence considerable variations may exist in the shape of the section without seriously affecting the discharge, provided the sectional area be unaltered. Even when $a : b = 8 : 7$, the discharge is diminished by less than one per cent.

333. We consider next some simple cases of steady rotatory motion.

The first is that of two-dimensional rotation about the axis of z, the angular velocity being a function of the distance (r) from this axis. Writing

$$u = -\omega y, \quad v = \omega x, \quad\text{..........................(1)}$$

we find that the rates of extension along and perpendicular to the radius vector are zero, whilst the rate of shear in the plane xy is $r\,d\omega/dr$. Hence the moment, about the axis, of the tangential forces on a cylindrical surface of radius r is, per unit length of the axis, $= \mu r\,d\omega/dr \,.\, 2\pi r \,.\, r$. On account of the steady motion, the fluid included between two coaxal cylinders is neither gaining nor losing angular momentum, so that the above expression must be independent of r. This gives

$$\omega = A/r^2 + B. \quad\text{............................(2)}$$

* "On the Flow of a Viscous Liquid in a Pipe or Channel," *Proc. Lond. Math. Soc.* (1) xiii. 43 (1881).

† This, with corresponding results for some other forms of section, appears to have been obtained by Boussinesq in 1868: see Hicks, *Brit. Ass. Rep.* 1882, p. 63.

If the fluid extend to infinity, while the internal boundary is that of a solid cylinder of radius a, whose angular velocity is ω_0, we have

$$\omega/\omega_0 = a^2/r^2. \qquad (3)$$

The frictional couple on the cylinder is therefore

$$-4\pi\mu a^2\omega_0. \qquad (4)$$

If the fluid were bounded externally by a fixed coaxal cylindrical surface of radius b we should find

$$\omega = \frac{a^2}{r^2} \cdot \frac{b^2 - r^2}{b^2 - a^2} \cdot \omega_0, \qquad (5)$$

which gives a frictional couple

$$-4\pi\mu \,.\, \frac{a^2 b^2}{b^2 - a^2} \cdot \omega_0. \qquad (6)^*$$

The formulae will apply to the case where the outer cylinder is maintained in rotation whilst the inner one is at rest, if we interchange the meanings of a and b. Experiments on this plan have been made by Mallock†, Couette‡, and others, the couple on the inner cylinder being measured by the torsion of a suspending wire, or some similar contrivance. The results will be referred to later (Art. 366 a)§.

334. A similar solution to that of the preceding Art., restricted however to the case of infinitely small motions, can be obtained for the steady motion of a fluid surrounding a solid sphere which is made to rotate uniformly about a diameter. Taking the centre as origin, and the axis of rotation as axis of z, we assume

$$u = -\omega y, \quad v = \omega x, \quad w = 0, \qquad (1)$$

where ω is a function of the radius vector r, only. If we put

$$P = \int \omega r \, dr, \qquad (2)$$

these equations may be written

$$u = -\frac{\partial P}{\partial y}, \quad v = \frac{\partial P}{\partial x}, \quad w = 0; \qquad (3)$$

and it appears on substitution in Art. 328 (4) that, provided we neglect the terms of the second order in the velocities, the equations are satisfied by

$$p = \text{const.}, \quad \nabla^2 P = \text{const.} \qquad (4)$$

The latter equation may be written

$$\frac{d^2P}{dr^2} + \frac{2}{r}\frac{dP}{dr} = \text{const.}, \quad \text{or} \quad r\frac{d\omega}{dr} + 3\omega = \text{const.}, \qquad (5)$$

whence

$$\omega = A/r^3 + B. \qquad (6)$$

* This problem was first treated, not quite accurately, by Newton, *Principia*, Lib. II. Prop. 51. The above results were given substantially by Stokes, *ll.cc. ante* pp. 577, 580.

† "Determination of the Viscosity of Water," *Proc. Roy. Soc.* xlv. 126 (1888); "Experiments on Fluid Viscosity," *Phil. Trans.* A, clxxxvii. 41.

‡ "Études sur le frottement des liquides," *Ann. de chimie et phys.* xxi. 433 (1890).

§ A number of modified problems connected with the rotation of circular cylinders are discussed by Jeffery, *Proc. Roy. Soc.* A, ci. 169 (1922), and Frazer, *Phil. Trans.* ccxxv. 93 (1925).

If the fluid extend to infinity and is at rest there, whilst ω_0 is the angular velocity of the rotating sphere ($r = a$), we have

$$\omega/\omega_0 = a^3/r^3. \quad \text{.................................(7)}$$

If the external boundary be a fixed concentric sphere of radius b the solution is

$$\omega = \frac{a^3}{r^3} \cdot \frac{b^3 - r^3}{b^3 - a^3} \cdot \omega_0. \quad \text{..............................(8)}$$

The retarding couple on the sphere may be calculated directly by means of the formulae of Art. 326, or, perhaps more simply, by means of the Dissipation Function of Art. 329. We find without difficulty that the rate of dissipation of energy is

$$\mu \iiint (x^2 + y^2)\left(\frac{d\omega}{dr}\right)^2 dx\,dy\,dz = \tfrac{8}{3}\pi\mu \int_a^b r^4 \left(\frac{d\omega}{dr}\right)^2 dr = 8\pi\mu \frac{a^3 b^3}{b^3 - a^3}\omega_0^2. \quad \text{...(9)}$$

If N denote the couple which must be applied to the sphere to maintain the rotation, this expression must be equivalent to $N\omega_0$, whence

$$N = 8\pi\mu \frac{a^3 b^3}{b^3 - a^3}\omega_0, \quad \text{..........................(10)}$$

or, in the case corresponding to (7), where $b = \infty$,

$$N = 8\pi\mu a^3 \omega_0. \quad \text{..............................(11)*}$$

The neglect of the terms of the second order in this problem involves a more serious limitation of its practical value than might be expected. It is not difficult to ascertain that the assumption virtually made is that the ratio $\omega_0 a^2/\nu$ is small. If we put $\nu = {\cdot}018$ (water), and $a = 10$, we find that the equatorial velocity $\omega_0 a$ must be small compared with ·0018 (C.S.)†.

When the terms of the second order are sensible, steady motion of the above kind is impossible. The sphere acts like a centrifugal fan, the motion at a distance from the sphere consisting of a flow outwards from the equator and inwards towards the poles, superposed on a motion of rotation‡.

In the case to which the formulae (8) and (10) relate the condition for the validity of the approximation is that the expression

$$\frac{\omega_0 a^2}{\nu}\left(1 - \frac{b^3}{a^3}\right) \quad \text{......................................(12)}$$

should be small, it being assumed that a and b are not very different§.

* Kirchhoff, *Mechanik*, c. xxvi.

† Cf. Rayleigh, "On the Flow of Viscous Liquids, especially in two Dimensions," *Phil. Mag* (4) xxxvi. 354 (1893) [*Papers*, iv. 78].

‡ Stokes, *l.c. ante* p. 577.

§ Experiments on the viscosity of air have been made by Zemplèn (*Ann. der Phys.* (4) xxix. 869 (1909) and xxxviii. 71 (1912)) on this plan, except that the *outer* sphere was made to rotate, the couple N being measured by the torsion of a wire from which the inner sphere was suspended. He found that the formula analogous to (10) gives consistent results for a wide range of $\omega_0 a^2/\nu$, and remarked that criteria of this kind are to be taken as indicating an order of magnitude, rather than an absolute standard. This must be admitted; but it should be noted that the relevant criterion in the present case has rather the form (12).

334 a. Some simple cases of variable motion can be solved by means of the analogy with the Conduction of Heat, noticed in Art. 328*.

1°. Take for instance the case of 'laminar' motion where the flow is in parallel planes and uniform over each plane, the direction being everywhere the same. With a suitable choice of axes we have $v=0$, $w=0$, whilst u is a function of z only. The equations (4) of Art. 328 are then satisfied by $p=$ const., and

$$\frac{\partial u}{\partial t}=\nu\frac{\partial^2 u}{\partial z^2}. \qquad (1)$$

This is identical in form with the equation of linear motion of heat, so that known solutions of the latter problem can be at once transferred to our present subject.

For example, suppose that the fluid extends to infinity in both directions of z, and that we have initially $u=\pm U$, the upper or lower sign being taken according as z is positive or negative. This corresponds to the case of two media in contact, initially at different temperatures. Appropriating the known solution of this problem we have

$$u=\frac{2U}{\sqrt{\pi}}\int_0^\theta e^{-\theta^2}d\theta, \qquad (2)$$

where, in the upper limit,

$$\theta=z/\sqrt{(4\nu t)}. \qquad (3)$$

It is easily verified that (2) does in fact satisfy (1), and that it makes $u\to\pm U$ for $t\to 0$.

The function multiplied by U in (2) was tabulated by Encke†. It appears that $u=\frac{1}{2}U$ when $\theta=\cdot4769$. For water this gives, in seconds and centimetres, $t=61\cdot8z^2$. The corresponding result for air is $t=8\cdot3z^2$. These results indicate how rapidly a surface of discontinuity in a viscous fluid would be obliterated, if indeed it could ever be formed.

The vorticity is

$$\eta=\frac{\partial u}{\partial z}=\frac{U}{\sqrt{(\pi\nu t)}}e^{-z^2/4\nu t}. \qquad (4)$$

This formula represents the diffusion of vorticity, which was initially confined to a vortex sheet at $z=0$, into the fluid on either side.

2°. Again, suppose that the fluid on both sides of an infinite plane lamina ($z=0$) is initially at rest, and that the lamina is suddenly set in motion parallel to Ox with a velocity U which is then maintained constant. The result is, for $z>0$,

$$u=U\left\{1-\frac{2}{\sqrt{\pi}}\int_0^\theta e^{-\theta^2}d\theta\right\}, \qquad (5)$$

* This analogy has been utilized by Rayleigh, *Proc. Lond. Math. Soc.* (1) xi. 57 (1880) [*Papers*, i. 474], and by several subsequent writers, *e.g.* G. I. Taylor, *Aeronautical Research Committee, R. & M.* 598 (1918), and K. Terazawa, *Japanese Journ. of Phys.* i. 7 (1922).

† *Berl. Astr. Jahrbuch*, 1834. The table is reprinted in Kelvin's *Papers*, iii. 434, and (abbreviated) in the collections of Dale, and Jahnke and Emde.

where the upper limit is given by (3). The retarding force on the lamina, per unit area, is

$$-2\mu\left(\frac{\partial u}{\partial z}\right)_{z\to 0}=\frac{2\mu U}{\sqrt{(\pi\nu t)}}. \quad \text{.......................(6)}$$

Next suppose that the lamina is moved in any manner, its velocity at time t being $U(t)$. The contribution to the retarding force due to an increment δU of the velocity at an antecedent time τ is

$$\frac{2\mu\delta U}{\sqrt{\{\pi\nu(t-\tau)\}}}.$$

The force at time t is therefore

$$\frac{2\mu}{\sqrt{(\pi\nu)}}\int_{-\infty}^{t}\frac{U'(\tau)\,d\tau}{\sqrt{(t-\tau)}}=\frac{2\mu}{\sqrt{(\pi\nu)}}\int_0^\infty U'(t-t_1)\frac{dt_1}{\sqrt{t_1}}. \quad \text{...........(7)*}$$

The determination of the motion of the lamina under given *forces* is more difficult. The question is hardly a practical one, but it has been solved† for the case of a constant force, such as gravity, the plane of the lamina being vertical. It appears that there is no 'terminal velocity,' the asymptotic value being

$$\frac{g\sigma}{\rho}\sqrt{\left(\frac{t}{\pi\nu}\right)}, \quad \text{...........................(8)}$$

where σ is the mass per unit area of the lamina. The case of a lamina of finite breadth in the direction of motion would be quite different.

3°. Suppose the motion to be in circles about an axis, the velocity being a function of the distance r from this axis.

Taking the axis in question as axis of z, we have obviously $D\zeta/Dt=\partial\zeta/\partial t$ and therefore, by Art. 328 (8),

$$\frac{\partial\zeta}{\partial t}=\nu\nabla_1^2\zeta, \quad \text{...............................(9)}$$

where $\nabla_1^2=\partial^2/\partial x^2+\partial^2/\partial y^2$. Integrating this over the area of a circle of radius r we have

$$\frac{d}{dt}\int_0^r\zeta\,.\,2\pi r\,dr=\nu\iint\nabla_1^2\zeta\,dx\,dy=\nu\frac{\partial\zeta}{\partial r}\,.\,2\pi r. \quad \text{...........(10)}$$

Hence, differentiating with respect to r,

$$\frac{\partial\zeta}{\partial t}=\nu\left(\frac{\partial^2\zeta}{\partial r^2}+\frac{1}{r}\frac{\partial\zeta}{\partial r}\right), \quad \text{.........................(11)}$$

which is identical with the equation of radial flow of heat in two dimensions‡.

* Stokes, *Camb. Trans.* ix. (1850) [*Papers*, iii. 132].

† Boggio, *Rend. dell. Accad. d. Lincei*, xvi. (1907); Rayleigh, *Phil. Mag.* (6) xxi. 697 (1911) [*Papers*, vi. 29].

‡ Carslaw, *Conduction of Heat*, Cambridge, 1921, p. 113.

For instance, suppose we have initially an isolated vortex of strength κ concentrated in the axis of z. The thermal analogy is the diffusion of heat from an instantaneous line-source in an infinite medium*; and the solution is

$$\zeta = \frac{\kappa}{4\pi\nu t} e^{-r^2/4\nu t}. \qquad (12)$$

That this satisfies (11) is easily verified by differentiation. Moreover it gives, for the circulation in a circle of radius r, the value

$$\int_0^r \zeta . 2\pi r dr = \kappa(1 - e^{-r^2/4\nu t}), \qquad (13)$$

the limiting value of which for $t \to 0$ is κ. The velocity is

$$q = \frac{\kappa}{2\pi r}(1 - e^{-r^2/4\nu t}). \qquad (14)$$

As t increases from 0 to ∞ this diminishes from $\kappa/2\pi r$ to 0. The vorticity on the other hand increases (for $r > 0$) from zero to a maximum and then falls asymptotically to zero.

4°. Again, suppose that at the instant $t = 0$ a uniform tangential stress f begins to act on the surface of a liquid of depth h which is at rest.

If the origin be at the bottom the conditions to be satisfied, besides the equation (1), are

$$u = 0 \text{ for } t \to 0, \quad \mu \partial u/\partial z = f \text{ for } z = h.$$

We write

$$u = fz/\mu + u', \qquad (15)$$

where the first term represents the asymptotic condition ($t \to \infty$). The equation (1), and the conditions $u' = 0$ for $z = 0$ and $\partial u'/\partial z = 0$ for $z = h$, are satisfied by a series

$$u' = \Sigma A_m \sin mz\, e^{-\nu m^2 t}, \qquad (16)$$

provided

$$mh = \tfrac{1}{2}(2s + 1)\pi, \qquad (17)$$

where $s = 0, 1, 2, 3, \ldots$. We have to determine the coefficients A_m so that

$$fz/\mu + \Sigma A_m \sin mz = 0 \qquad (18)$$

indentically. We may proceed by the ordinary Fourier method, or we may quote at once the known expansion

$$\theta = \frac{4}{\pi}\left\{\sin\theta - \frac{1}{3^2}\sin 3\theta + \frac{1}{5^2}\sin 5\theta - \ldots\right\}, \qquad (19)$$

which holds from $\theta = -\tfrac{1}{2}\pi$ to $\theta = \tfrac{1}{2}\pi$, inclusive. The final result is

$$u = \frac{hf}{\mu}\left\{\frac{z}{h} - \sin kz\, e^{-\nu k^2 t} + \frac{1}{3^2}\sin 3kz\, e^{-9\nu k^2 t} - \ldots\right\}, \qquad (20)$$

where $k = \tfrac{1}{2}\pi/h$.

* Carslaw, p. 152.

Calculations of this kind have sometimes been designed to illustrate the action of wind in producing ocean currents; but if we insert numerical values of ν and h, the final state would according to the formula be approached with extraordinary slowness. For instance, if $\nu = \cdot 018$, $h = 10^5$, the coefficient of $\sin kz$ in (20) would be diminished in the ratio $1/e$ in a time

$$t = 1/\nu k^2 = 4h^2/\pi^2\nu = 7140 \text{ years!}$$

In reality the conditions are enormously modified by turbulence. A more practical interpretation is obtained if we replace μ by a 'coefficient of turbulence,' as to which see Art. 366 b*.

5°. As a variation of this question, we may examine the steady currents which would be produced when account is taken of the earth's rotation†.

We take the origin in the free surface, with the axis of z drawn upwards. If ω be the component of the earth's angular velocity about the vertical we have, on the assumption that the conditions are uniform in respect of x and y, and that the motion has become steady,

$$-2\omega v = \nu \frac{\partial^2 u}{\partial z^2}, \quad 2\omega u = \nu \frac{\partial^2 v}{\partial z^2}. \quad\text{.....................(21)}$$

These may be combined into the single equation

$$\frac{\partial^2}{\partial z^2}(u + iv) = \frac{2i\omega}{\nu}(u + iv). \quad\text{.........................(22)}$$

Writing

$$\omega/\nu = \beta^2, \quad\text{..............................(23)}$$

and taking the depth to be practically infinite, we have

$$u + iv = Ae^{(1+i)\beta z}. \quad\text{..........................(24)}$$

The condition $\mu \partial u/\partial z = f$, to be satisfied for $z = 0$, gives $(1+i)\beta A = f/\mu$, whence

$$u + iv = \frac{(1-i)f}{2\mu\beta} e^{(1+i)\beta z}, \quad\text{.......................(25)}$$

or

$$u = \frac{f}{\sqrt{2}\mu\beta} e^{\beta z} \cos(\beta z - \tfrac{1}{4}\pi), \quad v = \frac{f}{\sqrt{2}\mu\beta} e^{\beta z} \sin(\beta z - \tfrac{1}{4}\pi). \quad\text{......(26)}$$

The motion is practically confined to a surface stratum whose depth is of the order β^{-1} The direction of the flow at the surface deviates 45° to the right (in the northern hemisphere) from that of the force. The total momentum per unit area of the surface, on the other hand, is

$$\int_{-\infty}^{0} \rho(u + iv)\, dz = -\frac{if}{2\mu\beta^2} = -\frac{if}{2\omega\rho}, \quad\text{.................(27)}$$

the direction being at right angles to that of the force.

* The extreme slowness of diffusion in strictly laminar motion was remarked by Helmholtz, "Ueber atmospherische Bewegungen," *Sitzb. d. Berl. Akad.* 1888, p. 649 [*Wiss. Abh.* iii. 292].

See also Hough, "On the influence of Viscosity on Waves and Currents," *Proc. Lond. Math. Soc.* (1) xxviii. 264 (1896).

† Ekman, "On the influence of the earth's rotation on ocean currents," *Arkiv f. matematik...*, ii. (1905); see also xvii. (1923). These papers contain other important developments.

Here again it is to be remarked that the results have a practical value only if we replace μ by a coefficient of turbulence. With the ordinary value of μ for water β^{-1} would be only of the order of 20 cm.

335. The motion of a viscous incompressible fluid, when the effects of inertia are insensible, can be treated in a very general manner, in terms of spherical harmonic functions.

It will be convenient, in the first place, to investigate the general solution of the following system of equations:

$$\nabla^2 u' = 0, \quad \nabla^2 v' = 0, \quad \nabla^2 w' = 0, \qquad \text{.....................(1)}$$

$$\frac{\partial u'}{\partial x} + \frac{\partial v'}{\partial y} + \frac{\partial w'}{\partial z} = 0. \qquad \text{...........................(2)}$$

The functions u', v', w' may be expanded in series of solid harmonics, and it is plain that the terms of algebraical degree n in these expansions, say u_n', v_n', w_n', must separately satisfy (2). The equations (1) may therefore be put in the forms

$$\left.\begin{aligned} \frac{\partial}{\partial y}\left(\frac{\partial v_n'}{\partial x} - \frac{\partial u_n'}{\partial y}\right) &= \frac{\partial}{\partial z}\left(\frac{\partial u_n'}{\partial z} - \frac{\partial w_n'}{\partial x}\right), \\ \frac{\partial}{\partial z}\left(\frac{\partial w_n'}{\partial y} - \frac{\partial v_n'}{\partial z}\right) &= \frac{\partial}{\partial x}\left(\frac{\partial v_n'}{\partial x} - \frac{\partial u_n'}{\partial y}\right), \\ \frac{\partial}{\partial x}\left(\frac{\partial u_n'}{\partial z} - \frac{\partial w_n'}{\partial x}\right) &= \frac{\partial}{\partial y}\left(\frac{\partial w_n'}{\partial y} - \frac{\partial v_n'}{\partial z}\right). \end{aligned}\right\} \qquad \text{..............(3)}$$

Hence $$\frac{\partial w_n'}{\partial y} - \frac{\partial v_n'}{\partial z} = \frac{\partial \chi_n}{\partial x}, \quad \frac{\partial u_n'}{\partial z} - \frac{\partial w_n'}{\partial x} = \frac{\partial \chi_n}{\partial y}, \quad \frac{\partial v_n'}{\partial x} - \frac{\partial u_n'}{\partial y} = \frac{\partial \chi_n}{\partial z}, \qquad \text{......(4)}$$

where χ_n is some function of x, y, z; and it further appears from these relations that $\nabla^2 \chi_n = 0$, so that χ_n is a solid harmonic of degree n.

From (4) we also obtain

$$z\frac{\partial \chi_n}{\partial y} - y\frac{\partial \chi_n}{\partial z} = x\frac{\partial u_n'}{\partial x} + y\frac{\partial u_n'}{\partial y} + z\frac{\partial u_n'}{\partial z} + u_n' - \frac{\partial}{\partial x}(xu_n' + yv_n' + zw_n'), \quad \text{...(5)}$$

with two similar equations. Now it follows from (1) and (2) that

$$\nabla^2 (xu_n' + yv_n' + zw_n') = 0, \qquad \text{........................(6)}$$

so that we may write $$xu_n' + yv_n' + zw_n' = \phi_{n+1}, \qquad \text{........................(7)}$$

where ϕ_{n+1} is a solid harmonic of degree $n+1$. Hence (5) may be written

$$(n+1)\, u_n' = \frac{\partial \phi_{n+1}}{\partial x} + z\frac{\partial \chi_n}{\partial y} - y\frac{\partial \chi_n}{\partial z}. \qquad \text{..................(8)}$$

The factor $n+1$ may be dropped without loss of generality; and we obtain as the solution of the proposed system of equations:

$$\left.\begin{aligned} u' &= \Sigma\left(\frac{\partial\phi_n}{\partial x} + z\frac{\partial\chi_n}{\partial y} - y\frac{\partial\chi_n}{\partial z}\right), \\ v' &= \Sigma\left(\frac{\partial\phi_n}{\partial y} + x\frac{\partial\chi_n}{\partial z} - z\frac{\partial\chi_n}{\partial x}\right), \\ w' &= \Sigma\left(\frac{\partial\phi_n}{\partial z} + y\frac{\partial\chi_n}{\partial x} - x\frac{\partial\chi_n}{\partial y}\right). \end{aligned}\right\} \qquad \text{.....(9)}$$

where the harmonics ϕ_n, χ_n are arbitrary*.

336. If we neglect the inertia-terms, the equations of motion of a viscous liquid reduce, in the absence of extraneous forces, to the forms

$$\mu\nabla^2 u = \frac{\partial p}{\partial x}, \qquad \mu\nabla^2 v = \frac{\partial p}{\partial y}, \qquad \mu\nabla^2 w = \frac{\partial p}{\partial z}, \qquad \text{.....(1)}$$

with

$$\frac{\partial u}{\partial x} + \frac{\partial v}{\partial y} + \frac{\partial w}{\partial z} = 0. \qquad \text{.....(2)}$$

By differentiation we obtain

$$\nabla^2 p = 0, \qquad \text{.....(3)}$$

so that p can be expanded in a series of solid harmonics, thus

$$p = \Sigma p_n. \qquad \text{.....(4)}$$

The terms of the solution which involve harmonics of different algebraical degrees will be independent. To obtain the terms in p_n we assume

$$\left.\begin{aligned} u &= Ar^2\frac{\partial p_n}{\partial x} + Br^{2n+3}\frac{\partial}{\partial x}\frac{p_n}{r^{2n+1}}, \\ v &= Ar^2\frac{\partial p_n}{\partial y} + Br^{2n+3}\frac{\partial}{\partial y}\frac{p_n}{r^{2n+1}}, \\ w &= Ar^2\frac{\partial p_n}{\partial z} + Br^{2n+3}\frac{\partial}{\partial z}\frac{p_n}{r^{2n+1}}, \end{aligned}\right\} \qquad \text{.....(5)}$$

where $r^2 = x^2 + y^2 + z^2$. The terms multiplied by B are solid harmonics of degree $n+1$, by Arts. 81, 83. Now

$$\nabla^2\left(r^2\frac{\partial p_n}{\partial x}\right) = r^2\nabla^2\frac{\partial p_n}{\partial x} + 4\left(x\frac{\partial}{\partial x} + y\frac{\partial}{\partial y} + z\frac{\partial}{\partial z}\right)\frac{\partial p_n}{\partial x} + \frac{\partial p_n}{\partial x}\nabla^2 r^2 = 2(2n+1)\frac{\partial p_n}{\partial x}.$$

Hence the equations (1) are satisfied, provided

$$A = \frac{1}{2(2n+1)\mu}. \qquad \text{.....(6)}$$

Also, substituting in (2), we find

$$2nA - (n+1)(2n+3)B = 0,$$

whence

$$B = \frac{n}{(n+1)(2n+1)(2n+3)\mu}. \qquad \text{.....(7)}$$

* Cf. Borchardt, "Untersuchungen über die Elasticität fester Körper unter Berücksichtigung der Wärme," *Berl. Monatsber.* Jan. 9, 1873 [*Gesammelte Werke*, Berlin, 1888, p. 245]. The investigation in the text is from a paper "On the Oscillations of a Viscous Spheroid," *Proc. Lond. Math. Soc.* (1) xiii. 51 (1881).

Hence the general solution of the system (1) and (2) is

$$\left.\begin{aligned} u &= \frac{1}{\mu}\Sigma\left\{\frac{r^2}{2(2n+1)}\frac{\partial p_n}{\partial x} + \frac{nr^{2n+3}}{(n+1)(2n+1)(2n+3)}\frac{\partial}{\partial x}\frac{p_n}{r^{2n+1}}\right\} + u', \\ v &= \frac{1}{\mu}\Sigma\left\{\frac{r^2}{2(2n+1)}\frac{\partial p_n}{\partial y} + \frac{nr^{2n+3}}{(n+1)(2n+1)(2n+3)}\frac{\partial}{\partial y}\frac{p_n}{r^{2n+1}}\right\} + v', \\ w &= \frac{1}{\mu}\Sigma\left\{\frac{r^2}{2(2n+1)}\frac{\partial p_n}{\partial z} + \frac{nr^{2n+3}}{(n+1)(2n+1)(2n+3)}\frac{\partial}{\partial z}\frac{p_n}{r^{2n+1}}\right\} + w', \end{aligned}\right\} \quad \ldots(8)$$

where u', v', w' have the forms given in (9) of the preceding Art*.

The formulae (8) make

$$xu + yv + zw = \frac{1}{\mu}\Sigma\frac{nr^2}{2(2n+3)}p_n + \Sigma n\phi_n. \quad \ldots\ldots\ldots\ldots\ldots(9)$$

Also, if we denote by ξ, η, ζ the components of vorticity, we find

$$\left.\begin{aligned} \xi &= \frac{1}{\mu}\Sigma\frac{1}{(n+1)}\left(y\frac{\partial p_n}{\partial z} - z\frac{\partial p_n}{\partial y}\right) + \Sigma(n+1)\frac{\partial \chi_n}{\partial x}, \\ \eta &= \frac{1}{\mu}\Sigma\frac{1}{(n+1)}\left(z\frac{\partial p_n}{\partial x} - x\frac{\partial p_n}{\partial z}\right) + \Sigma(n+1)\frac{\partial \chi_n}{\partial y}, \\ \zeta &= \frac{1}{\mu}\Sigma\frac{1}{(n+1)}\left(x\frac{\partial p_n}{\partial y} - y\frac{\partial p_n}{\partial x}\right) + \Sigma(n+1)\frac{\partial \chi_n}{\partial z}. \end{aligned}\right\} \quad \ldots\ldots(10)$$

These make

$$x\xi + y\eta + z\zeta = \Sigma n(n+1)\chi_n. \quad \ldots\ldots\ldots\ldots\ldots\ldots\ldots(11)$$

The components of stress across the surface of a sphere of radius r are, by Art. 325 (4),

$$p_{rx} = \frac{x}{r}p_{xx} + \frac{y}{r}p_{xy} + \frac{z}{r}p_{xz}, \ \ldots, \ \ldots \quad \ldots\ldots\ldots\ldots\ldots\ldots(12)$$

If we substitute the values of p_{xx}, p_{xy}, p_{xz}, ... from Art. 326 (5), (6), we find

$$\left.\begin{aligned} rp_{rx} &= -xp + \mu\left(r\frac{\partial}{\partial r} - 1\right)u + \mu\frac{\partial}{\partial x}(xu + yv + zw), \\ rp_{ry} &= -yp + \mu\left(r\frac{\partial}{\partial r} - 1\right)v + \mu\frac{\partial}{\partial y}(xu + yv + zw), \\ rp_{rz} &= -zp + \mu\left(r\frac{\partial}{\partial r} - 1\right)w + \mu\frac{\partial}{\partial z}(xu + yv + zw). \end{aligned}\right\} \quad \ldots\ldots(13)$$

These formulae are of course general. In the present case, substituting from (8), and making use of the relation

$$xp_n = \frac{r^2}{2n+1}\left(\frac{\partial p_n}{\partial x} - r^{2n+1}\frac{\partial}{\partial x}\frac{p_n}{r^{2n+1}}\right), \quad \ldots\ldots\ldots\ldots\ldots\ldots(14)$$

* This solution is derived, with some modifications, from various sources. Cf. Thomson and Tait, Art. 736; Borchardt, *l.c.*; Oberbeck, "Ueber stationäre Flüssigkeitsbewegungen mit Berücksichtigung der inneren Reibung," *Crelle*, lxxxi. 62 (1876).

we obtain, after a little reduction,

$$rp_{rx} = \Sigma \left\{ \frac{n-1}{2n+1} r^2 \frac{\partial p_n}{\partial x} + \frac{2n^2+4n+3}{(n+1)(2n+1)(2n+3)} r^{2n+3} \frac{\partial}{\partial x} \frac{p_n}{r^{2n+1}} \right\}$$

$$+ 2\mu\Sigma(n-1)\frac{\partial \phi_n}{\partial x} - \mu\Sigma(n-1)\left(y\frac{\partial \chi_n}{\partial z} - z\frac{\partial \chi_n}{\partial y}\right). \quad \text{.........(15)}$$

The corresponding expressions for rp_{ry} and rp_{rz} can be derived by cyclical changes of letters.

337. The results of Arts. 335, 336 can be applied to the solution of a number of problems where the boundary-conditions have relation to spherical surfaces. The most interesting cases fall under one or other of two classes; viz. we have either

$$xu + yv + zw = 0, \quad \text{.............................(1)}$$

everywhere, and therefore $p_n = 0$, $\phi_n = 0$, or

$$x\xi + y\eta + z\zeta = 0, \quad \text{.............................(2)}$$

and therefore $\chi_n = 0$.

1°. Let us investigate the steady flow of a liquid past a fixed spherical obstacle. If we take the origin at the centre, and the axis of x parallel to the flow, the boundary-conditions are that $u=0$, $v=0$, $w=0$ for $r=a$ (the radius), and $u=U$ (say), $v=0$, $w=0$ for $r=\infty$. It is obvious that the vortex-lines will be circles about the axis of x, so that the relation (2) will be fulfilled. Again, the equation (9) of Art. 336, taken in conjunction with the condition to be satisfied at infinity, shews that as regards the functions p_n and ϕ_n we are limited to surface-harmonics of orders 0 and 1, and therefore to the cases $n=0$, $n=1$, $n=-2$. Also, we must evidently have $p_1=0$. Assuming, then,

$$p_{-2} = A\frac{x}{r^3}; \quad \phi_1 = Ux, \quad \phi_{-2} = B\frac{x}{r^3}, \quad \text{.......................(3)}$$

we find

$$\left.\begin{aligned} u &= U + \left(B - \frac{Ar^2}{6\mu}\right)\frac{\partial}{\partial x}\frac{x}{r^3} + \frac{2A}{3\mu r}, \\ v &= \left(B - \frac{Ar^2}{6\mu}\right)\frac{\partial}{\partial y}\frac{x}{r^3}, \\ w &= \left(B - \frac{Ar^2}{6\mu}\right)\frac{\partial}{\partial z}\frac{x}{r^3}. \end{aligned}\right\} \quad \text{.............................(4)}$$

These make

$$xu + yv + zw = \left(U - \frac{2B}{r^3} + \frac{A}{\mu r}\right)x. \quad \text{.............................(5)}$$

Also, from Art. 336 (15), or directly from (13),

$$\left.\begin{aligned} p_{rx} &= -\frac{x}{r}p_0 + \left(Ar - \frac{6\mu B}{r}\right)\frac{\partial}{\partial x}\frac{x}{r^3} - \frac{A}{r^2}, \\ p_{ry} &= -\frac{y}{r}p_0 + \left(Ar - \frac{6\mu B}{r}\right)\frac{\partial}{\partial y}\frac{x}{r^3}, \\ p_{rz} &= -\frac{z}{r}p_0 + \left(Ar - \frac{6\mu B}{r}\right)\frac{\partial}{\partial z}\frac{x}{r^3}. \end{aligned}\right\} \quad \text{.........................(6)}$$

The condition of no slipping at the surface $r=a$ gives

$$U + \frac{2A}{3\mu a} = 0, \qquad B - \frac{Aa^2}{6\mu} = 0, \quad \text{................................(7)}$$

whence

$$A = -\tfrac{3}{2}\mu U a, \qquad B = -\tfrac{1}{4}Ua^3. \quad \text{................................(8)}$$

The component tractions on this surface are therefore

$$p_{rx} = -\frac{x}{a}p_0 + \tfrac{3}{2}\mu\frac{U}{a}, \quad p_{ry} = -\frac{y}{a}p_0, \quad p_{rz} = -\frac{z}{a}p_0. \quad \text{..............(9)}$$

If δS be an element of the surface, we have

$$\iint p_{rx}dS = 6\pi\mu Ua, \quad \iint p_{ry}dS = 0, \quad \iint p_{rz}dS = 0. \quad \text{..................(10)}$$

The resultant force on the sphere is therefore $6\pi\mu a U$, in the direction of x-positive.

The formulae (4) now take the shape

$$\left.\begin{aligned} u &= U\left(1-\frac{a}{r}\right) + \tfrac{1}{4}Ua\,(r^2-a^2)\frac{\partial}{\partial x}\frac{x}{r^3}, \\ v &= \tfrac{1}{4}Ua\,(r^2-a^2)\frac{\partial}{\partial y}\frac{x}{r^3}, \\ w &= \tfrac{1}{4}Ua\,(r^2-a^2)\frac{\partial}{\partial z}\frac{x}{r^3}. \end{aligned}\right\} \quad \text{..........................(11)}$$

The character of the motion is most concisely expressed in terms of Stokes' stream-function (Art. 94). The radial velocity being

$$U\left(1-\tfrac{3}{2}\frac{a}{r}+\tfrac{1}{2}\frac{a^3}{r^3}\right)\cos\theta, \quad \text{....................................(12)}$$

the flux $(2\pi\psi)$ through a circle with Ox as axis, whose radius subtends an angle θ at O, is given by

$$\psi = -\tfrac{1}{2}U\left(1-\tfrac{3}{2}\frac{a}{r}+\tfrac{1}{2}\frac{a^3}{r^3}\right)r^2\sin^2\theta. \quad \text{..........................(13)}$$

If we impress on everything a velocity $-U$ in the direction of x, we get the case of a sphere moving steadily through a viscous fluid which is at rest at infinity. The stream-function is then

$$\psi = \tfrac{3}{4}Uar\left(1-\tfrac{1}{3}\frac{a^2}{r^2}\right)\sin^2\theta. \quad \text{.............................(14)*}$$

The diagram opposite shews the stream-lines $\psi=$const., in this case, for a series of equidistant values of ψ. The contrast with the case of a frictionless liquid, delineated on p. 128, is remarkable, but it must be remembered that the fundamental assumptions are very different. In the former case inertia was predominant, and viscosity neglected; in the present problem these circumstances are reversed.

The configuration of the stream-lines indicates that the existence of an outer rigid boundary, even at a distance of many diameters of the sphere, would greatly modify the results. The resistance would of course be increased†.

If P be the extraneous force acting on the sphere in the direction of x-negative, this must balance the resistance, whence

$$P = 6\pi\mu a U. \quad \text{..(15)}$$

It is to be noticed that the formula (14) makes the momentum and the energy of the fluid both infinite‡. The steady motion here investigated could therefore only be fully established by a constant force acting on the sphere through an infinite distance.

* This problem was first solved by Stokes, in terms of the stream-function; see Art. 338.

† The slow motion of a sphere in the neighbourhood of a plane rigid wall has been investigated by Lorentz, *Abhandlungen über theoretische Physik*, Leipzig, 1907, ..., i. 23. The case of a concentric spherical boundary is treated by Williams, by the method of Art. 338, below, and compared with experiment, in an interesting paper (*Phil. Mag.* (6) xxix. 526 (1915)). The fall of a sphere along the axis of a vertical tube filled with liquid is discussed by Ladenburg, *Ann. der Phys.* xxiii. 447 (1907).

‡ Rayleigh, *Phil. Mag.* (5) xxi. 374 (footnote) (1886) [*Papers*, ii. 480].

The whole of this investigation is based on the assumption that the inertia-terms $u\partial u/\partial x, \ldots$ in the fundamental equations (4) of Art. 328 may be neglected in comparison with $\nu\nabla^2 u, \ldots$. It easily follows from (11) above that Ua must be small compared with ν*. This condition can always be realized by making U or a sufficiently small, but in the case of mobile fluids like water it restricts us to velocities or dimensions which are, from a practical point of view, exceedingly minute. Thus even for a sphere of a millimetre radius moving through water ($\nu = {\cdot}018$), the velocity must be considerably less than ·18 cm. per sec.†

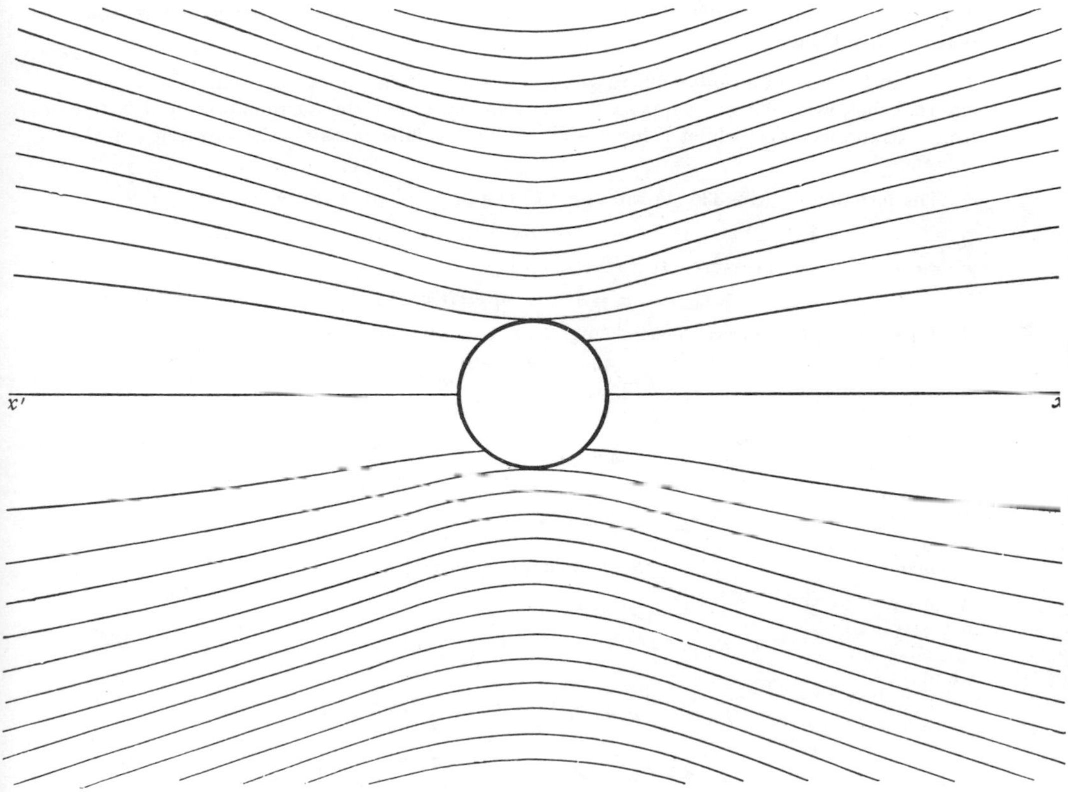

We may employ the formula (15) to find the 'terminal velocity' of a sphere falling vertically in a fluid‡. The force P is then the excess of the gravity of the sphere over its buoyancy, viz.

$$P = \tfrac{4}{3}\pi(\rho' - \rho)a^3 g, \qquad (16)$$

where ρ denotes the density of the fluid, and ρ' the mean density of the sphere. This gives

$$U = \tfrac{2}{9}\frac{\rho' - \rho}{\mu} g a^2. \qquad (17)$$

* The quantity Ua/ν, which is here assumed to be small, is of zero dimensions, and so independent of the fundamental units adopted. It may be called the 'Reynolds number' appropriate to the present problem; see *post* Art. 366.

† Rayleigh, *l.c. ante* p. 589. For experimental researches bearing on this point, see Allen, "The Motion of a Sphere in a Viscous Fluid," *Phil. Mag.* (5) l. 323, 519 (1900); Arnold, *Phil. Mag.* (6) xxii. 755 (1911); and Williams, *l.c.*

‡ Stokes, *l.c. ante* p. 580.

This will only apply, as already stated, provided Ua/ν is small. For a particle of sand descending in water, we may put (roughly)

$$\rho'=2\rho, \qquad \nu=\cdot018, \qquad g=981,$$

whence it appears that a must be small compared with ·0114 cm. Subject to this condition, the terminal velocity is $U=12000a^2$.

For a globule of water falling through the air, we have

$$\rho'=1, \qquad \rho=\cdot00129, \qquad \mu=\cdot00017.$$

This gives a terminal velocity $U=1280000a^2$, subject to the condition that a is small compared with ·006 cm.

2°. In the case of a *liquid* sphere we have to take account of the internal as well as the external motion*. We will suppose, in the first place, that the surrounding fluid is free from extraneous force, whilst a force $-K$, per unit volume, acts on the substance of the sphere.

The formulae of Art. 336 (1) will apply to the internal space, provided

$$p=p'+Kx, \qquad (18)$$

where p' is the true pressure. If we further write

$$p_1=A'x, \qquad \phi_1=B'x, \qquad (19)$$

the formulae (8) of the Art. referred to give for the internal motion

$$\left.\begin{aligned} u&=\frac{A'r^5}{30\mu'}\frac{\partial}{\partial x}\frac{x}{r^3}+\frac{A'r^2}{6\mu'}+B', \\ v&=\frac{A'r^5}{30\mu'}\frac{\partial}{\partial y}\frac{x}{r^3}, \\ w&=\frac{A'r^5}{30\mu'}\frac{\partial}{\partial z}\frac{x}{r^3}, \end{aligned}\right\} \qquad (20)$$

whence

$$xu+yv+zw=\left(\frac{A'r^2}{10\mu'}+B'\right)x. \qquad (21)$$

Also, since

$$x^2=-\tfrac{1}{3}r^5\frac{\partial}{\partial x}\frac{x}{r^3}+\tfrac{1}{3}r^2, \qquad xy=-\tfrac{1}{3}r^5\frac{\partial}{\partial y}\frac{x}{r^3}, \qquad xz=-\tfrac{1}{3}r^5\frac{\partial}{\partial z}\frac{x}{r^3}, \qquad (22)$$

we find

$$\left.\begin{aligned} p_{rx}&=-\frac{x}{r}p_0+(\tfrac{3}{10}A'-\tfrac{1}{3}K)r^4\frac{\partial}{\partial x}\frac{x}{r^3}+\tfrac{1}{3}Kr, \\ p_{ry}&=-\frac{y}{r}p_0+(\tfrac{3}{10}A'-\tfrac{1}{3}K)r^4\frac{\partial}{\partial y}\frac{x}{r^3}, \\ p_{rz}&=-\frac{z}{r}p_0+(\tfrac{3}{10}A'-\tfrac{1}{3}K)r^4\frac{\partial}{\partial z}\frac{x}{r^3}. \end{aligned}\right\} \qquad (23)$$

The corresponding formulae for the external space will be as in (4) and (6), above.

Expressing that the radial velocity vanishes for $r=a$, we have

$$\frac{A}{\mu a}-\frac{2B}{a^3}+U=0, \qquad \frac{A'a^2}{10\mu'}+B'=0. \qquad (24)$$

The continuity of the velocity requires further

$$-\frac{Aa^2}{6\mu}+B=\frac{A'a^5}{30\mu'}. \qquad (25)$$

* This problem was investigated by Rybczynski, *Bull. Acad. d. Sciences de Cracovie*, 1911, p. 40, and independently by Hadamard, *Comptes Rendus*, clii. 1735 (1911). These references are taken from a paper by Smoluchowski, "On the Practical Applicability of Stokes' Law of Resistance...," *Proc. of Math. Congress*, Cambridge, 1912, ii. 192.

Again, on comparison of (6) and (23) it appears that the continuity of stress requires

$$Aa - \frac{6\mu B}{a} = \tfrac{3}{10} A' a^4 - \tfrac{1}{3} K a^4, \qquad A = -\tfrac{1}{3} K a^3. \quad \text{.....................(26)}$$

We have thus five equations to determine A, B, A', B', U when K is given. Solving, we find

$$U = \tfrac{2}{3} \frac{Ka^2}{\mu} \cdot \frac{\mu + \mu'}{2\mu + 3\mu'}. \quad \text{....................................(27)}$$

The total force which must act on the sphere in the direction of x-negative, in order to maintain it at rest in the stream, is

$$\tfrac{4}{3} \pi a^3 K = 6\pi a \mu U \cdot \frac{2\mu + 3\mu'}{3\mu + 3\mu'}. \quad \text{................................(28)}$$

The internal motion is given by the formula

$$\psi = -\tfrac{1}{2} B' \left(1 - \frac{r^2}{a^2}\right) r^2 \sin^2\theta, \quad \text{................................(29)}$$

where

$$B' = -\frac{Ka^2}{6\mu + 9\mu'}. \quad \text{...(30)}$$

If we put $\mu' = \infty$ we reproduce the results relating to a solid sphere.

To adapt the results to the case of motion under gravity (supposed to act in the direction of x-negative), we must put

$$K = g(\rho' - \rho), \quad \text{..(31)}$$

where ρ' is the internal density. The terminal velocity is then given by (28). If $\rho' < \rho$, U is negative, indicating that the globule *ascends* relatively to the surrounding fluid. In the case of a bubble of gas ascending through water we may put, with sufficient accuracy, $\rho' = 0$, $\mu' = 0$, whence

$$U = -\tfrac{1}{3} g\rho a^2/\mu. \quad \text{...(32)}$$

3°. A variation of the problem of the solid sphere is afforded if we allow for the possibility of slipping of the fluid over the surface, assuming the empirical law referred to in Art. 327.

The formulae (6) give, for the *normal* stress on a sphere of radius r, the expression

$$-p_0 - 3A\frac{x}{r^3} + 12\mu B \frac{x}{r^5}, \quad \text{...................................(33)}$$

the three components of which may be written, in virtue of Art. 336 (14),

$$\left.\begin{aligned} &-p_0 \frac{x}{r} + \left(Ar - \frac{4\mu B}{r}\right) \frac{\partial}{\partial x} \frac{x}{r^3} - \frac{A}{r^2} + \frac{4\mu B}{r^4}, \\ &-p_0 \frac{y}{r} + \left(Ar - \frac{4\mu B}{r}\right) \frac{\partial}{\partial y} \frac{x}{r^3}, \\ &-p_0 \frac{z}{r} + \left(Ar - \frac{4\mu B}{r}\right) \frac{\partial}{\partial z} \frac{x}{r^3}. \end{aligned}\right\} \quad \text{......................(34)}$$

Subtracting these from the expressions in (6) we find, for the components of *tangential* stress,

$$-\frac{2\mu B}{r} \frac{\partial}{\partial x} \frac{x}{r^3} - \frac{4\mu B}{r^4}, \qquad -\frac{2\mu B}{r} \frac{\partial}{\partial y} \frac{x}{r^3}, \qquad -\frac{2\mu B}{r} \frac{\partial}{\partial z} \frac{x}{r^3}. \quad \text{..............(35)}$$

At the surface $r = a$, the radial velocity must vanish, and the expressions in (4) will become components of tangential velocity. We must have, therefore,

$$-\frac{2\mu B}{a} = \beta\left(B - \frac{Aa^2}{6\mu}\right), \qquad -\frac{4\mu B}{a^4} = \beta\left(U + \frac{2A}{3\mu a}\right), \quad \text{.................(36)}$$

where β is the coefficient of sliding friction. Hence

$$A = -\tfrac{3}{2}\mu Ua \,.\, \frac{2\mu+\beta a}{3\mu+\beta a}, \qquad B = -\frac{\beta Ua^4}{12\mu+4\beta a}; \quad \text{.....................(37)}$$

and it appears on reference to (5) that these satisfy the condition of zero radial velocity.

The resultant force on the sphere is, by (6),

$$X = \iint p_{rx}\, dS = -4\pi A = 6\pi\mu Ua \,.\, \frac{2\mu+\beta a}{3\mu+\beta a}. \quad \text{.....................(38)*}$$

For $\beta=\infty$, this agrees with (15). If β were $=0$, the resultant would be $4\pi\mu Ua$.

4°. The problem of a rotating sphere in an infinite mass of liquid is solved by assuming

$$u = z\frac{\partial\chi_{-2}}{\partial y} - y\frac{\partial\chi_{-2}}{\partial z}, \qquad v = x\frac{\partial\chi_{-2}}{\partial z} - z\frac{\partial\chi_{-2}}{\partial x}, \qquad w = y\frac{\partial\chi_{-2}}{\partial x} - x\frac{\partial\chi_{-2}}{\partial y}, \quad \text{......(39)}$$

where

$$\chi_{-2} = Az/r^3, \quad \text{...(40)}$$

the axis of z being that of rotation. At the surface $r=a$ we must have

$$u = -\omega y, \qquad v = \omega x, \qquad w = 0,$$

if ω be the angular velocity of the sphere. This gives $A = \omega a^3$; cf. Art. 334.

338. Problems relating to flow about a sphere, in planes through an axis of symmetry, have been usually treated, as by Stokes originally, by means of the current-function ψ. It may be useful, therefore, to give a few indications of the method.

Putting $y = \varpi\cos\vartheta$, $z = \varpi\sin\vartheta$, and accordingly

$$\left.\begin{aligned} &v = \upsilon\cos\vartheta, && w = \upsilon\sin\vartheta, \\ \xi = 0, \quad &\eta = -\omega\sin\vartheta, && \zeta = \omega\cos\vartheta, \end{aligned}\right\} \quad \text{........................(1)}$$

we have

$$\begin{aligned} \nabla^2\eta &= -\left[\frac{\partial^2}{\partial x^2} + \frac{\partial^2}{\partial\varpi^2} + \frac{1}{\varpi}\frac{\partial}{\partial\varpi} + \frac{\partial^2}{\varpi^2\partial\vartheta^2}\right]\omega\sin\vartheta \\ &= -\sin\vartheta\left[\frac{\partial^2}{\partial x^2} + \frac{\partial^2}{\partial\varpi^2} + \frac{1}{\varpi}\frac{\partial}{\partial\varpi} - \frac{1}{\varpi^2}\right]\omega \\ &= -\frac{\sin\vartheta}{\varpi}\left[\frac{\partial^2}{\partial x^2} + \frac{\partial^2}{\partial\varpi^2} - \frac{1}{\varpi}\frac{\partial}{\partial\varpi}\right]\varpi\omega, \quad \text{............................(2)} \end{aligned}$$

and similarly

$$\nabla^2\zeta = \frac{\cos\vartheta}{\varpi}\left[\frac{\partial^2}{\partial x^2} + \frac{\partial^2}{\partial\varpi^2} - \frac{1}{\varpi}\frac{\partial}{\partial\varpi}\right]\varpi\omega. \quad \text{..........................(3)}$$

In the case of steady motion, we have, from Art. 336 (1), $\nabla^2\eta=0$, $\nabla^2\zeta=0$, and therefore

$$\left[\frac{\partial^2}{\partial x^2} + \frac{\partial^2}{\partial\varpi^2} - \frac{1}{\varpi}\frac{\partial}{\partial\varpi}\right]\varpi\omega = 0, \quad \text{...............................(4)}$$

or, substituting the value of ω from Art. 161 (2),

$$\left[\frac{\partial^2}{\partial x^2} + \frac{\partial^2}{\partial\varpi^2} - \frac{1}{\varpi}\frac{\partial}{\partial\varpi}\right]^2\psi = 0. \quad \text{...............................(5)}$$

In questions relating to a spherical boundary we put

$$x = r\cos\theta, \qquad \varpi = r\sin\theta. \quad \text{...................................(6)}$$

Since

$$\frac{\partial}{\partial\varpi} = \sin\theta\frac{\partial}{\partial r} + \cos\theta\frac{\partial}{r\partial\theta},$$

we obtain

$$\left[\frac{\partial^2}{\partial r^2} + \frac{\sin\theta}{r^2}\frac{\partial}{\partial\theta}\left(\frac{1}{\sin\theta}\right)\frac{\partial}{\partial\theta}\right]^2\psi = 0, \quad \text{...........................(7)}$$

the equation given by Stokes.

* Basset, *Hydrodynamics*, ii. 271.

This is satisfied by $\psi = \sin^2\theta f(r)$ ....(8)

provided $\left[\dfrac{d^2}{dr^2} - \dfrac{2}{r^2}\right]^2 f(r) = 0,$ ....(9)

the solution of which is

$$f(r) = \frac{A}{r} + Br + Cr^2 + Dr^4. \quad \text{..........(10)}$$

In the case of uniform flow at infinity we must have $\psi = -\frac{1}{2}Ur^2\sin^2\theta$ for $r=\infty$, whence

$$D=0, \qquad C = -\tfrac{1}{2}U. \quad \text{..........(11)}$$

If we now denote by u the velocity along the radius vector, and by v the velocity at right angles to the radius, in a meridian plane, we have

$$\left.\begin{aligned} u &= -\frac{1}{r\sin\theta}\frac{\partial\psi}{r\partial\theta} = U\cos\theta - 2\left(\frac{A}{r^3} + \frac{B}{r}\right)\cos\theta, \\ v &= \frac{1}{r\sin\theta}\frac{\partial\psi}{\partial r} \quad = -U\sin\theta - \left(\frac{A}{r^3} - \frac{B}{r}\right)\sin\theta. \end{aligned}\right\} \quad \text{..........(12)}$$

The rates of elongation in the directions of r and θ, and at right angles to these two, are found by superposition of the amounts due to u and v separately to be

$$\left.\begin{aligned} \frac{\partial u}{\partial r} = 2\left(\frac{3A}{r^4} + \frac{B}{r^2}\right)\cos\theta, \qquad \frac{\partial v}{r\partial\theta} + \frac{u}{r} = -\left(\frac{3A}{r^4} + \frac{B}{r^2}\right)\cos\theta, \\ \frac{u}{r} + \frac{v}{r\tan\theta} = -\left(\frac{3A}{r^4} + \frac{B}{r^2}\right)\cos\theta, \end{aligned}\right\} \quad \text{..........(13)}$$

and the rate of shear in the plane of r and θ is

$$\frac{\partial u}{r\partial\theta} + \frac{\partial v}{\partial r} - \frac{v}{r} = \frac{6A}{r^4}\sin\theta. \quad \text{..........(14)}$$

The vorticity is $\omega = \dfrac{\partial v}{\partial r} + \dfrac{v}{r} - \dfrac{\partial u}{r\partial\theta} = -\dfrac{2B\sin\theta}{r^2}.$ ....(15)

The force on the sphere may be calculated directly from the stress-formulae, or may be inferred more simply from the rate of dissipation of energy. It follows from (13) and (14) that the function Φ of Art. 329 (8) takes the form

$$\Phi = 12\mu\left(\frac{3A}{r^4} + \frac{B}{r^2}\right)^2\cos^2\theta + 36\mu\frac{A^2}{r^8}\sin^2\theta. \quad \text{..........(16)}$$

To find the total rate of dissipation of energy in the fluid we must multiply this by $2\pi r\sin\theta\, r\delta\theta\,\delta r$, and integrate from $\theta=0$ to $\theta=\pi$, and from $r=a$ to $r=\infty$. The result is

$$16\pi\mu\left(\frac{3A^2}{a^5} + \frac{2AB}{a^3} + \frac{B^2}{a}\right) \quad \text{..........(17)}$$

On the hypothesis of no slipping at the surface $r=a$ we find from (12)

$$A = -\tfrac{1}{4}Ua^3, \qquad B = \tfrac{3}{4}Ua, \quad \text{..........(18)}$$

and therefore $\psi = -\frac{1}{2}U\left(1 - \frac{3}{2}\dfrac{a}{r} + \frac{1}{2}\dfrac{a^3}{r^3}\right)r^2\sin^2\theta,$ ....(19)

or, if the sphere be regarded as moving with velocity $-U$ through a liquid which is at rest at infinity,

$$\psi = \tfrac{3}{4}Uar\left(1 - \tfrac{1}{3}\frac{a^2}{r^2}\right)\sin^2\theta, \quad \text{..........(20)}$$

as in Art. 337 (14).

The force ($-P$, say) which must be applied to the sphere to maintain the motion is found by equating the rate of dissipation of energy to PU. Substituting in (17) from (18) we find

$$P = 6\pi\mu a U, \quad \text{..........(21)}$$

as before.

If there is slipping, with a coefficient β of sliding friction, the conditions to be satisfied for $r=a$ are

$$u=0, \quad \beta v=\mu\left(\frac{\partial v}{\partial r}+\frac{\partial u}{r\partial\theta}-\frac{v}{r}\right), \quad \text{.........................(22)}$$

in the original form of the problem where the sphere is regarded as at rest. Substituting from (12) and (14) we find

$$A=-\tfrac{1}{4}Ua^3\div\left(1+\frac{3\mu}{\beta a}\right), \qquad B=\tfrac{3}{4}Ua\left(1+\frac{2\mu}{\beta a}\right)\div\left(1+\frac{3\mu}{\beta a}\right). \quad \text{............(23)}$$

There is in this case an additional dissipation of energy by sliding friction at the surface of the sphere, of amount βv^2 per unit area. If we integrate this over the surface, the result is, by (22) and (14),

$$\frac{96\pi\mu^2A^2}{\beta a^6}. \quad \text{..................................(24)}$$

If we add this to (17), and insert the values of A and B from (23), we find, on equating the total dissipation to PU,

$$P=6\pi\mu a U.\frac{\beta a+2\mu}{\beta a+3\mu}, \quad \text{..................................(25)}$$

in agreement wlth Basset's result (Art. 337 (38)).

339. The problem of the steady translation of an ellipsoid in a viscous liquid can be solved in terms of the gravitation-potential of the solid, regarded as homogeneous and of unit density.

The equation of the surface being

$$\frac{x^2}{a^2}+\frac{y^2}{b^2}+\frac{z^2}{c^2}=1, \quad \text{..(1)}$$

the gravitation-potential is given, at external points, by Dirichlet's formula*

$$\Omega=\pi abc\int_\lambda^\infty\left(\frac{x^2}{a^2+\lambda}+\frac{y^2}{b^2+\lambda}+\frac{z^2}{c^2+\lambda}-1\right)\frac{d\lambda}{\Delta}, \quad \text{.....................(2)}$$

where

$$\Delta=\{(a^2+\lambda)(b^2+\lambda)(c^2+\lambda)\}^{\frac{1}{2}}, \quad \text{................................(3)}$$

and the lower limit is the positive root of

$$\frac{x^2}{a^2+\lambda}+\frac{y^2}{b^2+\lambda}+\frac{z^2}{c^2+\lambda}=1. \quad \text{....................................(4)}$$

This makes

$$\frac{\partial\Omega}{\partial x}=2\pi\alpha x, \quad \frac{\partial\Omega}{\partial y}=2\pi\beta y, \quad \frac{\partial\Omega}{\partial z}=2\pi\gamma z, \quad \text{..........................(5)}$$

where

$$\alpha=abc\int_\lambda^\infty\frac{d\lambda}{(a^2+\lambda)\Delta}, \quad \beta=abc\int_\lambda^\infty\frac{d\lambda}{(b^2+\lambda)\Delta}, \quad \gamma=abc\int_\lambda^\infty\frac{d\lambda}{(c^2+\lambda)\Delta}. \quad \text{......(6)}$$

We will also write

$$\chi=abc\int_\lambda^\infty\frac{d\lambda}{\Delta}; \quad \text{..(7)}$$

it has been shewn in Art. 114 that this satisfies $\nabla^2\chi=0$.

If the fluid be streaming past the ellipsoid, regarded as fixed, with the general velocity U in the direction of x, we assume†

$$u=A\frac{\partial^2\Omega}{\partial x^2}+B\left(x\frac{\partial\chi}{\partial x}-\chi\right)+U, \quad v=A\frac{\partial^2\Omega}{\partial x\partial y}+Bx\frac{\partial\chi}{\partial y}, \quad w=A\frac{\partial^2\Omega}{\partial x\partial z}+Bx\frac{\partial\chi}{\partial z}. \quad \text{......(8)}$$

* *Crelle*, xxxii. 80 (1846) [*Werke*, ii. 11]; see also Kirchhoff, *Mechanik*, c. xviii., and Thomson and Tait (2nd ed.), Art. 494 *m*.

† Oberbeck, *l.c. ante* p. 596.

These satisfy the equation of continuity, in virtue of the relations

$$\nabla^2\Omega=0, \qquad \nabla^2\chi=0;$$

and they evidently make $u=U$, $v=0$, $w=0$ at infinity. Again, they make

$$\nabla^2 u=2B\frac{\partial^2\chi}{\partial x^2}, \quad \nabla^2 v=2B\frac{\partial^2\chi}{\partial x\partial y}, \quad \nabla^2 w=2B\frac{\partial^2\chi}{\partial x\partial z}, \quad \text{.................(9)}$$

so that the equations (1) of Art. 336 are satisfied by

$$p=2B\mu\frac{\partial\chi}{\partial x}+\text{const.} \quad \text{.......................................(10)}$$

It remains to shew that by a proper choice of A, B we can make $u, v, w=0$ at the surface (1). The conditions $v=0$, $w=0$ require

$$\left[2\pi A\frac{d\alpha}{d\lambda}+B\frac{d\chi}{d\lambda}\right]_{\lambda=0}=0, \quad \text{or} \quad 2\pi\frac{A}{a^2}+B=0. \quad \text{....................(11)}$$

With the help of this relation, the condition $u=0$ reduces to

$$2\pi A\alpha_0-B\chi_0+U=0, \quad \text{.......................................(12)}$$

where the suffix denotes that the lower limit in the integrals (6) and (7) is to be replaced by zero. Hence

$$\pi A=-\tfrac{1}{2}Ba^2, \quad B=\frac{U}{\chi_0+\alpha_0 a^2}. \quad \text{...............................(13)}$$

At a great distance r from the origin we have

$$\Omega=-\tfrac{4}{3}\pi abc/r, \qquad \chi=2abc/r,$$

whence it appears, on comparison with the equations (4) of Art. 337, that the disturbance is the same as would be produced by a sphere of radius R, determined by

$$\tfrac{3}{4}UR=2abcB, \quad \text{or} \quad R=\tfrac{8}{3}\frac{abc}{\chi_0+\alpha_0 a^2}. \quad \text{...........................(14)}$$

The resistance experienced by the ellipsoid will therefore be

$$6\pi\mu RU. \quad \text{..(15)}$$

In the case of a circular disk moving broadside-on, we have $a=0$, $b=c$; whence $\alpha_0=2$, $\chi_0=\pi ac$, so that

$$R=8c/3\pi=\cdot 85c.$$

If the disk move edgeways we have

$$R=16c/9\pi=\cdot 566c^*.$$

340. As a variation on the preceding problems we may investigate the steady motion of a liquid in a given constant field of force.

Omitting terms of the second order, we have

$$-\frac{1}{\rho}\frac{\partial p}{\partial x}+X+\nu\nabla^2 u=0, \quad -\frac{1}{\rho}\frac{\partial p}{\partial y}+Y+\nu\nabla^2 v=0, \quad -\frac{1}{\rho}\frac{\partial p}{\partial z}+Z+\nu\nabla^2 w=0, \quad \text{......(1)}$$

and

$$\frac{\partial u}{\partial x}+\frac{\partial v}{\partial y}+\frac{\partial w}{\partial z}=0. \quad \text{...(2)}$$

Hence

$$\nabla^2\frac{p}{\rho}=\frac{\partial X}{\partial x}+\frac{\partial Y}{\partial y}+\frac{\partial Z}{\partial z}, \quad \text{.................................(3)}$$

* Other limiting cases are those of the circular cylinder, and of an infinitely long flat blade, either end-on or broadside-on to the stream. These have been examined by A. Berry and Miss L. M. Swain, *Proc. Roy. Soc.* A, cii. 766 (1923). The velocity at infinity does not vanish, but is logarithmically infinite. Cf. Art. 343 below.

which is satisfied by

$$\frac{p}{\rho}=-\frac{1}{4\pi}\iiint\left(\frac{\partial X'}{\partial x'}+\frac{\partial Y'}{\partial y'}+\frac{\partial Z'}{\partial z'}\right)\frac{dx'dy'dz'}{r}, \quad\text{.......................(4)}$$

if

$$r=\sqrt{\{(x-x')^2+(y-y')^2+(z-z')^2\}}. \quad\text{...............................(5)}$$

If the forces X, Y, Z are confined to a certain region we have, by partial integration,

$$\frac{p}{\rho}=\frac{1}{4\pi}\iiint\left(X'\frac{\partial}{\partial x'}+Y'\frac{\partial}{\partial y'}+Z'\frac{\partial}{\partial z'}\right)\frac{1}{r}\,dx'dy'dz'$$

$$=-\frac{1}{4\pi}\iiint\left(X'\frac{\partial}{\partial x}+Y'\frac{\partial}{\partial y}+Z'\frac{\partial}{\partial z}\right)\frac{1}{r}\,dx'dy'dz'. \quad\text{.................(6)}$$

Hence, in the case of a force concentrated at the origin, writing

$$P=\rho\iiint X'dx'dy'dz', \qquad Q=\rho\iiint Y'dx'dy'dz', \qquad R=\rho\iiint Z'dx'dy'dz', \quad\text{.........(7)}$$

we have

$$p=-\frac{1}{4\pi}\left(P\frac{\partial}{\partial x}+Q\frac{\partial}{\partial y}+R\frac{\partial}{\partial z}\right)\frac{1}{r}. \quad\text{.............................(8)}$$

To this we may add any solution of $\nabla^2 p=0$.

We now have, except at the origin,

$$\nabla^2 u=\frac{1}{\mu}\frac{\partial p}{\partial x}, \qquad \nabla^2 v=\frac{1}{\mu}\frac{\partial p}{\partial y}, \qquad \nabla^2 w=\frac{1}{\mu}\frac{\partial p}{\partial z}. \quad\text{.......................(9)}$$

If we substitute from (8), the integration comes under Art. 336, with $n=-2$. Thus

$$\mu u=\frac{r^2}{24\pi}\frac{\partial}{\partial x}\left(P\frac{\partial}{\partial x}+Q\frac{\partial}{\partial y}+R\frac{\partial}{\partial z}\right)\frac{1}{r}+\frac{P}{6\pi r}, \quad\text{.....................(10)}$$

with similar formulae for v, w. If we add to (8) terms

$$Ax+By+Cz, \quad\text{..(11)}$$

the corresponding terms in μu will be, by Art. 336, with $n=1$,

$$A'+\tfrac{1}{6}Ar^2-\tfrac{1}{30}r^5\frac{\partial}{\partial x}\left(A\frac{\partial}{\partial x}+B\frac{\partial}{\partial y}+C\frac{\partial}{\partial z}\right)\frac{1}{r}. \quad\text{....................(12)}$$

The complete solution is obtained by addition from (10) and (12)

Thus if we have a fixed spherical boundary $r=b$, we find

$$A=\frac{5P}{4\pi b^3}, \qquad A'=-\frac{3P}{8\pi b}, \quad\text{................................(13)}$$

whence

$$6\pi\mu u=\frac{P}{r}\left(1-\frac{9r}{4b}+\frac{5r^3}{4b^3}\right)+\tfrac{1}{4}r^2\left(1-\frac{r^3}{b^3}\right)\frac{\partial}{\partial x}\left(P\frac{\partial}{\partial x}+Q\frac{\partial}{\partial y}+R\frac{\partial}{\partial z}\right)\frac{1}{r}. \quad\text{.........(14)}$$

If we make $b=\infty$, we get

$$6\pi\mu u=\frac{P}{r}+\tfrac{1}{4}r^2\frac{\partial}{\partial x}\left(P\frac{\partial}{\partial x}+Q\frac{\partial}{\partial y}+R\frac{\partial}{\partial z}\right)\frac{1}{r}. \quad\text{....................(15)}$$

If we put $P=-6\pi\mu Ua$, $Q=0$, $R=0$, this is seen to be consistent with the results of Art. 337 for large values of r/a.

341. The analogous problem in two dimensions may conveniently be treated by means of the stream-function.

Putting

$$u=-\frac{\partial\psi}{\partial y}, \qquad v=\frac{\partial\psi}{\partial x} \quad\text{.................................(1)}$$

in the equations

$$X-\frac{1}{\rho}\frac{\partial p}{\partial x}+\nu\nabla_1^2 u=0, \qquad Y-\frac{1}{\rho}\frac{\partial p}{\partial y}+\nu\nabla_1^2 v=0, \quad\text{...................... (2)}$$

and eliminating p, we have

$$\nu\nabla_1^4\psi=\frac{\partial X}{\partial y}-\frac{\partial Y}{\partial x}, \quad \text{.......(3)}$$

where

$$\nabla_1^2=\partial^2/\partial x^2+\partial^2/\partial y^2. \quad \text{.......(4)}$$

Hence

$$\nabla_1^2\psi=\frac{1}{2\pi\nu}\iint\left(\frac{\partial X'}{\partial y'}-\frac{\partial Y'}{\partial x'}\right)\log r\,dx'dy'+\chi, \quad \text{.......(5)}$$

where

$$r=\surd\{(x-x')^2+(y-y')^2\}, \quad \text{.......(6)}$$

and χ is a solution of $\nabla_1^2\chi=0$.

If we suppose the forces X, Y to vanish outside a certain region, we have, by partial integration,

$$\nabla_1^2\psi=-\frac{1}{2\pi\nu}\iint\left(X'\frac{\partial}{\partial y'}-Y'\frac{\partial}{\partial x'}\right)\log r\,dx'dy'+\chi$$

$$=\frac{1}{2\pi\nu}\iint\left(X'\frac{\partial}{\partial y}-Y'\frac{\partial}{\partial x}\right)\log r\,dx'dy'+\chi. \quad \text{.......(7)}$$

In particular, in the case of a force concentrated on a small area at the origin, writing

$$P=\rho\iint X'dx'dy', \qquad Q=\rho\iint Y'dx'dy', \quad \text{.......(8)}$$

we have

$$\nabla_1^2\psi=\frac{1}{2\pi\mu r^2}(Py-Qx)+\chi. \quad \text{.......(9)}$$

As an example, suppose that the fluid is enclosed by a fixed boundary $r=a$ and is subject to a force P at the origin. The appropriate form of (9) is, in polar co-ordinates,

$$\frac{\partial^2\psi}{\partial r^2}+\frac{1}{r}\frac{\partial\psi}{\partial r}+\frac{1}{r^2}\frac{\partial^2\psi}{\partial\theta^2}=\frac{P}{2\pi\mu}\frac{\sin\theta}{r}+Ar\sin\theta. \quad \text{.......(10)}$$

Hence, integrating,

$$\psi=\frac{P}{4\pi\mu}(r\log r+A'r^3+Br)\sin\theta. \quad \text{.......(11)}$$

At the boundary we must have $\psi=$ const., $\partial\psi/\partial r=0$, whence

$$\log a+A'a^2+B=0, \qquad 1+\log a+3A'a^2+B=0, \quad \text{.......(12)}$$

or

$$A'=-1/2a^2, \qquad B=-\log a+\tfrac{1}{2}. \quad \text{.......(13)}$$

Thus, finally,

$$\psi=\frac{Pr}{4\pi\mu}\left\{\log\frac{r}{a}+\tfrac{1}{2}\left(1-\frac{r^2}{a^2}\right)\right\}\sin\theta. \quad \text{.......(14)}$$

It will be noticed that this fails to give a definite result for $a=\infty$.

Returning to the general formula (7), and putting

$$F=\rho\iint X'\log r\,dx'dy', \qquad G=\rho\iint Y'\log r\,dx'dy', \quad \text{.......(15)}$$

we have

$$\nabla_1^2\psi=\frac{1}{2\pi\mu}\left(\frac{\partial F}{\partial y}-\frac{\partial G}{\partial x}\right)+\chi. \quad \text{.......(16)}$$

Since

$$\nabla_1^2F=2\pi\rho X, \qquad \nabla_1^2G=2\pi\rho Y, \quad \text{.......(17)}$$

we find, from (2),

$$\left.\begin{aligned}\frac{\partial p}{\partial x}&=\frac{1}{2\pi}\frac{\partial}{\partial x}\left(\frac{\partial F}{\partial x}+\frac{\partial G}{\partial y}\right)+\mu\frac{\partial\chi'}{\partial x},\\ \frac{\partial p}{\partial y}&=\frac{1}{2\pi}\frac{\partial}{\partial y}\left(\frac{\partial F}{\partial x}+\frac{\partial G}{\partial y}\right)+\mu\frac{\partial\chi'}{\partial y},\end{aligned}\right\} \quad \text{.......(18)}$$

where χ' is the function 'conjugate' to χ, the relations being

$$\frac{\partial\chi'}{\partial x}=-\frac{\partial\chi}{\partial y}, \qquad \frac{\partial\chi'}{\partial y}=\frac{\partial\chi}{\partial x}. \quad \text{.......(19)}$$

Hence

$$p=\frac{1}{2\pi}\left(\frac{\partial F}{\partial x}+\frac{\partial G}{\partial y}\right)+\chi'+\text{const.} \quad \text{.......(20)}$$

There is a remarkable analogy between the theory of the steady motion of a viscous liquid in two dimensions and that of the flexure of an elastic plate*. If w be the normal displacement in the latter problem, we have†

$$A\nabla_1^4 w = Z + \frac{\partial M}{\partial x} - \frac{\partial L}{\partial y},$$

where Z stands for normal force per unit area, and L, M are components of impressed couple per unit length about lines in the plate parallel to x, y respectively, whilst A is a constant depending on the elastic properties and the thickness. If we put $Z=0$ the analogy with (3) is complete; the couples L, M correspond to the forces X, Y, and the displacement w to the stream-function ψ, so that the contour-lines of the deformed plate are identical with the stream-lines. Since in the hydrodynamical problem we have $\psi=$const., $\partial\psi/\partial n=0$ at a fixed boundary, the plate, in the elastic analogue, must be supposed fixed and clamped at the edge. Thus the formula (14) corresponds to the case of a circular plate clamped at the edge and subject to a couple, concentrated at the centre, in a perpendicular plane

The analogy enables us to form a good idea of the general distribution of velocity in some cases where the actual calculation would be difficult‡.

We must not delay longer over problems of this type which, for reasons already given, have only a limited application except to fluids of great viscosity. We can therefore only advert to the mathematically very elegant investigations which have been given of the steady rotation of an ellipsoid§, and of the flow through a channel bounded by a hyperboloid of revolution of one sheet||.

342. The formula of Stokes for the resistance experienced by a slowly moving sphere has been employed in physical researches of fundamental importance, as a means of estimating the size of minute globules of water, and thence the number of globules contained in a cloud of given mass¶. Consequently the conditions of its validity have been much discussed both from the experimental** and from the theoretical side.

We have seen (Art. 328) that the accurate equations of motion may be written

$$\frac{\partial u}{\partial t} - v\zeta + w\eta = X - \frac{\partial \chi'}{\partial x} + \nu\nabla^2 u, \ldots, \ldots, \quad \ldots\ldots(1)$$

where

$$\chi' = \frac{p}{\rho} + \tfrac{1}{2}q^2. \quad \ldots\ldots(2)$$

* Rayleigh, "On the Flow of Viscous Fluids, especially in Two Dimensions," *Phil. Mag.* (5) xxxvi. 354 (1893) [*Papers*, iv. 78]. The analogy is slightly extended in the text.

† *Proc. Lond. Math. Soc.* xxi. 77.

‡ Some interesting applications are made by Rayleigh in the paper cited.

§ Edwardes, *Quart. Journ. Math.* xxvi. 70, 157 (1892); Jeffery, "On the Steady Rotation of a Solid of Revolution in a Viscous Fluid," *Proc. Lond. Math. Soc.* (2) xiv. 327 (1915).

|| Sampson, *l.c. ante* p. 126.

¶ Townsend, *Camb. Proc.* ix. 244 (1897); J. J. Thomson, *Phil. Mag.* (5) xlvi. 528 (1898); see also the latter author's *Conduction of Electricity through Gases*, Cambridge, 1903, p. 120. The mutual influence of the globules forming a cloud is considered by Cunningham, *Proc. Roy. Soc.* A, lxxxiii. 357 (1910), and Smoluchowski, *l.c. ante* p. 600.

** See the papers by Allen and Arnold cited on p. 599.

Hence a distribution of velocity which satisfies the conditions of any given problem, when the terms of the second order are neglected, will hold when these are retained, provided we introduce constraining forces

$$X_1 = w\eta - v\zeta, \quad Y_1 = u\zeta - w\xi, \quad Z_1 = v\xi - u\eta, \qquad \text{.........(3)}$$

and at the same time suppose the pressure to be diminished by $\frac{1}{2}\rho q^2$. These forces are everywhere perpendicular to the stream-lines and to the vortex-lines, and their intensity is

$$R_1 = q\omega \sin \vartheta, \qquad \text{.........(4)}$$

where ϑ is the angle between the direction of the velocity q and the axis of the vorticity ω*.

The magnitude of the hypothetical constraining forces, as compared with the viscous forces

$$\nu\nabla^2 u, \quad \nu\nabla^2 v, \quad \nu\nabla^2 w, \qquad \text{.........(5)}$$

gives some indication as to the validity of an approximation in which the inertia terms are neglected.

In the case of Stokes' formulae, Art. 337 (11), for the steady motion past a sphere held at rest, we have

$$\xi = 0, \quad \eta = \frac{3}{2}\frac{Uaz}{r^3}, \quad \zeta = -\frac{3}{2}\frac{Uay}{r^3}, \qquad \text{.........(6)}$$

and therefore in the distant parts of the fluid where $u = U$, $v = 0$, $w = 0$, ultimately,

$$X_1 = 0, \quad Y_1 = -\frac{3}{2}\frac{U^2ay}{r^3}, \quad Z_1 = -\frac{3}{2}\frac{U^2az}{r^3}. \qquad \text{.........(7)}$$

On the other hand, for the viscous forces (5) we find

$$\frac{3}{2}\nu Ua\frac{\partial^2}{\partial x^2}\frac{1}{r}, \quad \frac{3}{2}\nu Ua\frac{\partial^2}{\partial x\partial y}\frac{1}{r}, \quad \frac{3}{2}\nu Ua\frac{\partial^2}{\partial x\partial z}\frac{1}{r}. \qquad \text{.........(8)}$$

The ratio of the former to the latter is of the order Ur/ν, which increases indefinitely with r, however small U may be. For this reason the formulae in question cannot be regarded as valid at points distant from the sphere. Since, however, both the constraining forces and the viscous forces are in these regions relatively small, it does not necessarily follow that the character of the motion in the immediate neighbourhood of the sphere will be seriously affected. At points near the sphere the constraining forces tend to vanish, whilst the viscous forces are of the order $\nu Ua/r^3$.

The above criticism is due to Prof. Oseen† of Upsala, who has made an interesting innovation in the treatment of the question by writing $U + u$

* Rayleigh, "On the Flow of Viscous Fluids, especially in Two Dimensions," *Phil. Mag.* (5) xxxvi. 354 (1893) [*Papers*, iv. 78].

† "Ueber die Stokes'sche Formel, und über eine verwandte Aufgabe in der Hydrodynamik," *Arkiv för matematik*, ...," Bd. vi. no. 29 (1910). The same remark was made independently by F. Noether, "Ueber den Gültigkeitsbereich der Stokes'schen Widerstandsformel," *Zeitschr. f. Math. u. Phys.* lxii. (1911).

for u, and neglecting terms of the second order in u, v, w only. These latter symbols now denote the components of the velocity which would remain if a translation $-U$ were superposed on the whole system. The hydrodynamical equations accordingly take the forms

$$\left.\begin{aligned} U\frac{\partial u}{\partial x} &= -\frac{1}{\rho}\frac{\partial p}{\partial x} + \nu\nabla^2 u, \\ U\frac{\partial v}{\partial x} &= -\frac{1}{\rho}\frac{\partial p}{\partial y} + \nu\nabla^2 v, \\ U\frac{\partial w}{\partial x} &= -\frac{1}{\rho}\frac{\partial p}{\partial z} + \nu\nabla^2 w, \end{aligned}\right\} \qquad (9)$$

with

$$\frac{\partial u}{\partial x} + \frac{\partial v}{\partial y} + \frac{\partial w}{\partial z} = 0. \qquad (10)$$

The inertia terms are thus to some extent taken into account, but it is to be remarked that although the approximation is undoubtedly improved at infinity, where $u, v, w = 0$, it is in some degree impaired at the surface of the sphere where we now have $u = -U$. This is a matter for subsequent examination.

The solution of the equations (9) and (10) for the purpose in hand can be effected very simply*. In the first place we have

$$\nabla^2 p = 0, \qquad (11)$$

and a particular solution is therefore obtained if we write

$$p = \rho U\frac{\partial \phi}{\partial x}, \qquad (12)$$

$$u = -\frac{\partial \phi}{\partial x}, \quad v = -\frac{\partial \phi}{\partial y}, \quad w = -\frac{\partial \phi}{\partial z}, \qquad (13)$$

where ϕ satisfies

$$\nabla^2 \phi = 0. \qquad (14)$$

The solution is completed if we write

$$u = -\frac{\partial \phi}{\partial x} + u', \quad v = -\frac{\partial \phi}{\partial y} + v', \quad w = -\frac{\partial \phi}{\partial z} + w', \qquad (15)$$

where u', v', w' are solutions of the equations

$$\left(\nabla^2 - 2k\frac{\partial}{\partial x}\right)u' = 0, \quad \left(\nabla^2 - 2k\frac{\partial}{\partial y}\right)v' = 0, \quad \left(\nabla^2 - 2k\frac{\partial}{\partial z}\right)w' = 0, \qquad (16)$$

and

$$\frac{\partial u'}{\partial x} + \frac{\partial v'}{\partial y} + \frac{\partial w'}{\partial z} = 0. \qquad (17)$$

We have here written, for shortness,

$$k = U/2\nu. \qquad (18)$$

* The method, which differs from that of Prof. Oseen, and the subsequent interpretation, are reproduced from a paper "On the Uniform Motion of a Sphere through a Viscous Fluid," *Phil. Mag.* (6) xxi. 112 (1911).

Since the vortex-lines must be circles having the axis of x as a common axis, we may assume

$$\xi=0, \quad \eta=-\frac{\partial\chi}{\partial z}, \quad \zeta=\frac{\partial\chi}{\partial y}, \qquad \text{.....................(19)}$$

where χ is a function of x and ϖ (the distance from the axis of x) only. It follows from (16) that we must have

$$\left(\nabla^2-2k\frac{\partial}{\partial x}\right)\chi=0, \qquad \text{..........................(20)}$$

an additive function of x only being obviously irrelevant. Hence

$$\left.\begin{aligned} 2k\frac{\partial u'}{\partial x}&=\nabla^2 u'=\frac{\partial\eta}{\partial z}-\frac{\partial\zeta}{\partial y}=-\left(\frac{\partial^2\chi}{\partial y^2}+\frac{\partial^2\chi}{\partial z^2}\right)=\frac{\partial^2\chi}{\partial x^2}-2k\frac{\partial\chi}{\partial x},\\ 2k\frac{\partial v'}{\partial x}&=\nabla^2 v'=\frac{\partial\zeta}{\partial x}-\frac{\partial\xi}{\partial z}=\frac{\partial^2\chi}{\partial x\partial y},\\ 2k\frac{\partial w'}{\partial x}&=\nabla^2 w'=\frac{\partial\xi}{\partial y}-\frac{\partial\eta}{\partial x}=\frac{\partial^2\chi}{\partial x\partial z}.\end{aligned}\right\} \text{......(21)}$$

We thus obtain the solution

$$u'=\frac{1}{2k}\frac{\partial\chi}{\partial x}-\chi, \quad v'=\frac{1}{2k}\frac{\partial\chi}{\partial y}, \quad w'=\frac{1}{2k}\frac{\partial\chi}{\partial z}, \qquad \text{...........(22)}$$

which is easily verified.

The equation (20) may be written

$$(\nabla^2-k^2)\,e^{-kx}\chi=0, \qquad \text{..........................(23)}$$

the solution of which is well known, the simplest type being $e^{-kx}\chi=Ce^{-kr}/r$; cf. Art. 289. Adopting this, we have finally

$$\left.\begin{aligned} u&=-\frac{\partial\phi}{\partial x}+\frac{1}{2k}\frac{\partial\chi}{\partial x}-\chi,\\ v&=-\frac{\partial\phi}{\partial y}+\frac{1}{2k}\frac{\partial\chi}{\partial y},\\ w&=-\frac{\partial\phi}{\partial z}+\frac{1}{2k}\frac{\partial\chi}{\partial z},\end{aligned}\right\} \qquad \text{......................(24)}$$

where

$$\chi=\frac{Ce^{-k(r-x)}}{r}. \qquad \text{.............................(25)}$$

Since ϕ must obviously involve only zonal harmonics of negative degree, we write

$$\phi=\frac{A_0}{r}+A_1\frac{\partial}{\partial x}\frac{1}{r}+A_2\frac{\partial^2}{\partial x^2}\frac{1}{r}+\ldots. \qquad \text{.................(26)}$$

For small values of kr we have

$$\chi=C\left(\frac{1}{r}-k+\frac{kx}{r}+\ldots\right), \qquad \text{......................(27)}$$

which leads to

$$\left.\begin{aligned}\frac{1}{2k}\frac{\partial\chi}{\partial x}-\chi&=-\frac{C}{2k}\left(\tfrac{4}{3}\frac{k}{r}-\frac{\partial}{\partial x}\frac{1}{r}+\tfrac{1}{3}kr^2\frac{\partial^2}{\partial x^2}\frac{1}{r}+\ldots\right),\\ \frac{1}{2k}\frac{\partial\chi}{\partial y}&=-\frac{C}{2k}\left(-\frac{\partial}{\partial y}\frac{1}{r}+\tfrac{1}{3}kr^2\frac{\partial^2}{\partial x\partial y}\frac{1}{r}+\ldots\right),\\ \frac{1}{2k}\frac{\partial\chi}{\partial z}&=-\frac{C}{2k}\left(-\frac{\partial}{\partial z}\frac{1}{r}+\tfrac{1}{3}kr^2\frac{\partial^2}{\partial x\partial z}\frac{1}{r}+\ldots\right).\end{aligned}\right\}\quad\ldots\ldots(28)$$

Hence the relations

$$u=-U,\quad v=0,\quad w=0,$$

which are to hold for $r=a$, will be satisfied provided

$$C=\tfrac{3}{2}Ua;\quad A_0=\tfrac{3}{2}\nu a,\quad A_1=-\tfrac{1}{4}Ua^3,\quad\ldots\ldots\ldots\ldots\ldots(29)$$

approximately; and it will be noticed that the condition for the success of the approximation is again that ka, or the 'Reynolds number' Ua/ν, should be small.

To find the distribution of velocity in the neighbourhood of the sphere we may use the formulae (24), (26) and (28), with the values of the constants given in (29). The result is identical with Art. 337 (11), if regard be had to the altered meaning of u. The resistance experienced by the sphere has therefore the same value ($6\pi\mu aU$) as on Stokes' theory. The same results follow also from a consideration of the stream-function (ψ), which takes a comparatively simple form. When the sphere is regarded as in motion, and the fluid at rest at infinity, the radial velocity is

$$-\frac{\partial\phi}{\partial r}+\frac{1}{2k}\frac{\partial\chi}{\partial r}-\chi\cos\theta,\quad\ldots\ldots\ldots\ldots\ldots\ldots\ldots\ldots(30)$$

where θ denotes the inclination of the radius vector to the axis of x. Hence

$$\psi=r^2\int_0^\theta\left(\frac{\partial\phi}{\partial r}-\frac{1}{2k}\frac{\partial\chi}{\partial r}+\chi\cos\theta\right)\sin\theta\,d\theta.\quad\ldots\ldots\ldots\ldots(31)$$

Substituting from (25), (26), and (29), and performing the integration, we find

$$\psi=\tfrac{3}{2}\nu a(1+\cos\theta)\{1-e^{-kr(1-\cos\theta)}\}-\tfrac{1}{4}\frac{Ua^3}{r}\sin^2\theta.\quad\ldots\ldots\ldots(32)$$

For small values of kr this becomes

$$\psi=\tfrac{3}{4}Ua\left(r-\tfrac{1}{3}\frac{a^2}{r}\right)\sin^2\theta,\quad\ldots\ldots\ldots\ldots\ldots\ldots\ldots\ldots(33)$$

in agreement with Art. 337 (14).

In other respects the motion differs widely from that represented by Stokes' formulae. In the first place, as pointed out by Oseen, the stream-lines are no longer symmetrical with respect to the plane $x=0$, the motion being in fact no longer reversible. Again, the vorticity is

$$\omega=-\frac{\partial\chi}{\partial\varpi}=\tfrac{3}{2}Ua(1+kr)\frac{\varpi}{r^3}e^{-k(r-x)},\quad\ldots\ldots\ldots\ldots\ldots(34)$$

and is therefore insensible, on account of the exponential factor alone, except within a region bounded more or less vaguely by a paraboloidal surface having its focus at O, for which $k(r-x)$ has a moderate constant value. This region may here be referred to as the 'wake,' although it includes a certain space ahead of the sphere. The velocity (u, v, w) relative to the distant parts of the fluid tends, for large values of r, and for points outside the wake, to become purely radial, as if due to a source of strength $4\pi A_0$, or $6\pi\nu a$, at the origin. This is compensated by an inward flow in the wake; thus for points along the axis of the wake, to the right, where $x = r$, we find

$$u = -\tfrac{3}{2}\frac{Ua}{r} + \tfrac{1}{2}\frac{Ua^3}{r^3}. \qquad \text{.........................(35)}$$

This indicates a velocity following the sphere (when this is regarded as in motion), which ultimately varies inversely as the distance instead of as the square of the distance.

It remains to estimate the degree of approximation which the preceding results afford in various parts of the field. For this we have recourse again to a comparison of the 'constraining forces,' which would be necessary to make the solution exact, with the viscous forces. The former are given by the formulae (3), with the new meanings of u, v, w, and the alteration of the pressure is

$$\tfrac{1}{2}\rho\,(u^2 + v^2 + w^2). \qquad \text{.........................(36)}$$

This is constant $(= \tfrac{1}{2}\rho U^2)$ over the surface of the sphere, and so does not affect the resultant force on the latter.

At distant points well outside the wake, the terms in (24) which depend on χ may be neglected, and we have, ultimately,

$$u = \tfrac{3}{2}\nu a\frac{x}{r^3}, \quad v = \tfrac{3}{2}\nu a\frac{y}{r^3}, \quad w = \tfrac{3}{2}\nu a\frac{z}{r^3}. \qquad \text{..............(37)}$$

Also, from (34),

$$\xi = 0, \quad \eta = \tfrac{3}{2}Uka\frac{z}{r^2}e^{-k(r-x)}, \quad \zeta = -\tfrac{3}{2}Uka\frac{y}{r^2}e^{-k(r-x)}. \qquad \text{......(38)}$$

Hence

$$X_1 = \tfrac{9}{8}U^2a^2\frac{\varpi^2}{r^5}e^{-k(r-x)}, \quad Y_1 = -\tfrac{9}{8}U^2a^2\frac{xy}{r^5}e^{-k(r-x)}, \quad Z_1 = -\tfrac{9}{8}U^2a^2\frac{xz}{r^5}e^{-k(r-x)}, \qquad \text{......(39)}$$

the resultant of which is

$$R_1 = \tfrac{9}{8}U^2a^2\frac{\varpi}{r^4}e^{-k(r-x)}, \qquad \text{.......................(40)}$$

at right angles to the radius vector in a plane through the axis of x. The viscous forces may be found from (24) and (25). If we retain only the terms which are most important when r is large, we find

$$\nu\nabla^2 u = -\nu k^2 C\frac{\varpi^2}{r^3}e^{-k(r-x)}, \quad \nu\nabla^2 v = -\nu k^2 C\frac{y(r-x)}{r^3}e^{-k(r-x)},$$

$$\nu\nabla^2 w = -\nu k^2 C\frac{z(r-x)}{r^3}e^{-k(r-x)}. \qquad \text{.....................(41)}$$

It follows from (29) that the ratio of (39) to (41) is of the order $(1/kr) . (a/r)$. The approximation in this part of the field is therefore amply sufficient.

At points well within the wake, where $k(r-x)$ is small, we find from (32), for large values of r/a,

$$u = -\tfrac{3}{2}\frac{Ua}{r}, \quad v = 0, \quad w = 0, \qquad \text{.....................(42)}$$

and from (34) $$\xi = 0, \quad \eta = \tfrac{3}{2}Uka\frac{z}{r^2}, \quad \zeta = -\tfrac{3}{2}Uka\frac{y}{r^2}, \qquad \text{..............(43)}$$

approximately. These make

$$X_1 = 0, \quad Y_1 = \tfrac{9}{4}U^2ka^2\frac{y}{r^3}, \quad Z_1 = \tfrac{9}{4}U^2ka^2\frac{z}{r^3}. \qquad \text{...........(44)}$$

The viscous forces are found to be

$$\nu\nabla^2 u = 2\nu kC\frac{x}{r^3}, \quad \nu\nabla^2 v = 2\nu kC\frac{y}{r^3}, \quad \nu\nabla^2 w = 2\nu kC\frac{z}{r^3}, \qquad \text{......(45)}$$

approximately. The ratio of the magnitudes is of the order ka.

Near the surface of the sphere we have $u = -U$, $v = 0$, $w = 0$, nearly, and therefore, from (3) and (19),

$$X_1 = 0, \quad Y_1 = -U\frac{\partial\chi}{\partial y}, \quad Z_1 = -U\frac{\partial\chi}{\partial z}, \qquad \text{..............(46)}$$

or, by (27) and (29),

$$X_1 = 0, \quad Y_1 = \tfrac{3}{2}U^2a\frac{y}{r^3}, \quad Z_1 = \tfrac{3}{2}U^2a\frac{z}{r^3}, \qquad \text{...........(47)}$$

These are of the order U^2/a. The viscous forces are obtained from (24) and (27); thus

$$\nu\nabla^2 u = \tfrac{3}{2}\nu Ua\frac{3x^2 - r^2}{r^5}, \quad \nu\nabla^2 v = \tfrac{3}{2}\nu Ua\frac{3xy}{r^5}, \quad \nu\nabla^2 w = \tfrac{3}{2}\nu Ua\frac{3xz}{r^5}, \ \text{...(48)}$$

giving a resultant of the order $\nu U/a^2$. The ratio of the magnitudes is therefore Ua/ν, which has already been assumed to be small. The approximation, though less perfect here than on Stokes' theory, is seen to be adequate.

343. If we attempt to find the steady motion produced by the translation of a *cylinder* with constant velocity through an infinite mass of liquid, on the basis of the equations (1) of Art. 336, it proves to be impossible to satisfy all the conditions*. This was pointed out by Stokes, who gave the following explanation: "The pressure of the cylinder on the fluid continually tends to increase the quantity of fluid which it carries with it, while the friction of the fluid at a distance from the cylinder continually tends to diminish it. In the case of a sphere, these two causes eventually counteract each other, and the motion becomes uniform. But in the case of a cylinder, the increase in the quantity of fluid carried continually gains on the decrease due to the friction of the surrounding fluid, and the quantity carried increases indefinitely as the cylinder moves on†."

* Cf. the remark following equation (14) of Art. 341.

† *Camb. Trans.* ix. (1850) [*Papers*, iii. 65].

It appears, however, that if the inertia-terms are partially taken into account, after the manner of Oseen (Art. 342), the above conclusion is modified, and a definite value for the resistance is obtained*.

The hydrodynamical equations are now satisfied by

$$u = -\frac{\partial\phi}{\partial x} + \frac{1}{2k}\frac{\partial\chi}{\partial x} - \chi, \quad v = -\frac{\partial\phi}{\partial y} + \frac{1}{2k}\frac{\partial\chi}{\partial y}, \quad\text{..............(1)}$$

and

$$p = \rho U\frac{\partial\phi}{\partial x}, \quad\text{..............................(2)}$$

provided

$$\nabla_1^2\phi = 0, \quad\text{..............................(3)}$$

and

$$\left(\nabla_1^2 - 2k\frac{\partial}{\partial x}\right)\chi = 0. \quad\text{..........................(4)}$$

The appropriate solution of (4) is

$$\chi = Ce^{kx}\int_0^\infty e^{-kr\cosh\omega}\,d\omega. \quad\text{........................(5)}$$

For the definite integral we have the expansions†

$$\int_0^\infty e^{-kr\cosh\omega}\,d\omega = -(\gamma + \log\tfrac{1}{2}kr)\,I_0(kr) + \frac{k^2r^2}{2^2} + s_2\frac{k^4r^4}{2^2.4^2} + s_3\frac{k^6r^6}{2^2.4^2.6^2} + \dots$$

$$= \sqrt{\left(\frac{\pi}{2kr}\right)}\,e^{-kr}\left\{1 - \frac{1^2}{8kr} + \frac{1^2.3^2}{1.2(8kr)^2} - \dots\right\}, \quad\text{...........(6)}$$

the latter form, which is semiconvergent, being suitable for large values of kr.

For small values of kr we have

$$\chi = -C(1 + kx)(\gamma + \log\tfrac{1}{2}kr), \quad\text{.....................(7)}$$

whence

$$\left.\begin{aligned}\frac{1}{2k}\frac{\partial\chi}{\partial x} - \chi &= -\frac{C}{2k}\left\{k(\tfrac{1}{2} - \gamma - \log\tfrac{1}{2}kr) + \frac{\partial}{\partial x}\log r - \tfrac{1}{2}kr^2\frac{\partial^2}{\partial x^2}\log r + \dots\right\},\\ \frac{1}{2k}\frac{\partial\chi}{\partial y} &= -\frac{C}{2k}\left\{\frac{\partial}{\partial y}\log r - \tfrac{1}{2}kr^2\frac{\partial^2}{\partial x\partial y}\log r + \dots\right\}.\end{aligned}\right\} \quad\text{......(8)}$$

Hence if we put

$$\phi = A_0\log r + A_1\frac{\partial}{\partial x}\log r + \dots, \quad\text{.....................(9)}$$

we find that the conditions $u = -U$, $v = 0$, $w = 0$ will be satisfied for $r = a$, provided

$$C = \frac{2U}{\frac{1}{2} - \gamma - \log(\frac{1}{2}ka)}, \quad A_0 = -\frac{C}{2k}, \quad A_1 = \tfrac{1}{4}Ca^2, \quad\text{........(10)}$$

* Prof. Bairstow, applying the more usual approximate theory, finds a definite resistance for the case of a circular cylinder in a channel between parallel walls, *Proc. Roy. Soc.* A, c. 394 (1922).

† The proofs are analogous to those of Art. 194; cf. Watson, pp. 80, 202. The definite integral is a Bessel's function of imaginary argument, of the 'second kind.' It is tabulated in Watson's treatise, where it is denoted by $K_0(kr)$.

approximately. Hence near the cylinder we have

$$\left.\begin{aligned} u &= \tfrac{1}{2}C\left\{\gamma - \tfrac{1}{2} + \log \tfrac{1}{2}kr + \tfrac{1}{4}(r^2 - a^2)\frac{\partial^2}{\partial x^2}\log r\right\}, \\ v &= \tfrac{1}{4}C(r^2 - a^2)\frac{\partial^2}{\partial x \partial y}\log r. \end{aligned}\right\} \quad \text{.........(11)}$$

The vorticity is given by

$$\zeta = \frac{\partial v}{\partial x} - \frac{\partial u}{\partial y} = Ce^{kz}\frac{\partial}{\partial y}\int_0^\infty e^{-kr\cosh\omega}\,d\omega,$$

which for large values of kr takes the form

$$\zeta = -kC\frac{y}{r}\sqrt{\left(\frac{\pi}{2kr}\right)}\,e^{-k(r-x)}. \quad \text{....................(12)}$$

The general interpretation would follow the same lines as in the case of the sphere (Art. 342).

To calculate the force exerted by the fluid on the cylinder we have to integrate the expression

$$\begin{aligned} rp_{rx} &= \left(-p + 2\mu\frac{\partial u}{\partial x}\right)x + \mu\left(\frac{\partial v}{\partial x} + \frac{\partial u}{\partial y}\right)y \\ &= -px + \mu r\frac{\partial u}{\partial r} + \mu\left(x\frac{\partial u}{\partial x} + y\frac{\partial v}{\partial x}\right) \end{aligned} \quad \text{..................(13)}$$

with respect to the angular co-ordinate (θ) from 0 to 2π. The products of plane harmonics of different orders will disappear in this process. The first term of (13) gives, when r is put equal to a,

$$-\rho UA_0\int_0^{2\pi}\cos^2\theta\,d\theta = -\pi\rho\,UA_0 = \pi\mu C. \quad \text{..............(14)}$$

The second term contributes, on substitution from (11), $\pi\mu C$. The third term gives a zero result, to our order of approximation. The final value for the resistance per unit length is therefore

$$2\pi\mu C = \frac{4\pi\mu U}{\tfrac{1}{2} - \gamma - \log(\tfrac{1}{2}ka)}. \quad \text{....................(15)}$$

The investigation is subject, as in the case of Art. 342, to the condition that ka, or $Ua/2\nu$, is to be small*. It may be noted that the value of the expression in (15) does not vary rapidly with a. Thus for $ka = \frac{1}{10}$ we find $4{\cdot}31\mu U$, and, for $ka = \frac{1}{20}$, $3{\cdot}48\mu U$.

343 a. The 'linearized' equations of Oseen, Art. 342 (9), have been the starting point of a number of further investigations. It is to be noticed that even if we accept the equations as adequate the boundary-conditions have

* The above investigation is taken from the paper by the author cited on p. 610.
The formula (15) is stated to be in good agreement with experiment for sufficiently small values of Ua/ν; see Wieselsberger, *Phys. Zeitschr.* 1921, p. 321.

only been approximately satisfied. Oseen, continuing the approximation, found for the resistance of a sphere the formula*

$$6\pi\mu Ua\left(1+\tfrac{3}{8}R\right), \qquad\qquad (16)$$

where $R = Ua/\nu$. He has also investigated the case of the ellipsoid†, and ascertained a correction to Oberbeck's result (Art. 339). This problem includes the case of the elliptic cylinder, treated independently by Bairstow‡ and others, and by Harrison and Filon§. The formula for the resistance, given by these writers, is

$$\frac{4\pi\mu U}{\dfrac{a}{a+b}-\gamma-\log\{\tfrac{1}{4}k(a+b)\}}, \qquad\qquad (17)$$

per unit length. The cylinder is supposed placed symmetrically, and a denotes that semi-axis which is parallel to the stream. If we put $b = a$, we reproduce the result for the circular section. If $b = 0$ we have the case of a flat blade edgeways on to the stream. The exact solution for the circular cylinder has been discussed analytically by Faxén‖. The same writer has investigated¶, on the basis of Oseen's equations, the motion of a sphere along the axis of a tube, or parallel to a plane boundary, or between parallel walls, and compared his results as far as possible with experiment.

Some further, more problematical, calculations which have been based on the same equations will be referred to later (Art. 371 b).

344. Some interesting general theorems, relating to the dissipation of energy in the steady motion of a liquid under constant extraneous forces, have been given by Helmholtz and Korteweg. They involve the assumption that the inertia-terms in the dynamical equations may be neglected.

1°. Considering a given motion in a region bounded by any closed surface Σ, let u, v, w be the component velocities, and $u+u', v+v', w+w'$ the values of the corresponding components in any other motion which is subject only to the condition that u', v', w' vanish at all points of the boundary Σ. By Art. 329 (3), the dissipation in the altered motion is

$$\iiint\{(p_{xx}+p'_{xx})(a+a')+\ldots+\ldots+(p_{yz}+p'_{yz})(f+f')+\ldots+\ldots\}\,dx\,dy\,dz, \qquad (1)$$

where the accent attached to any symbol indicates the value which the function in question assumes when u, v, w are replaced by u', v', w'. Now the formulae (2), (3) of Art. 326 shew that, in the case of an incompressible fluid,

$$p_{xx}a'+p_{yy}b'+p_{zz}c'+p_{yz}f'+p_{zx}g'+p_{xy}h'$$
$$=p'_{xx}a+p'_{yy}b+p'_{zz}c+p'_{yz}f+p'_{zx}g+p'_{xy}h, \qquad (2)$$

* *Arkiv f. matemat.*... ix. (1913); also Burgess, *Amer. J. Math.* xxxviii. 81 (1916). The approximation is continued in a series of powers of the 'Reynolds number' R by Goldstein, *Proc. Roy. Soc.* A, cxxiii. 225 (1929).

† *Arkiv f. math. u. phys.* xxiv. (1915). An account of this and of many other researches in this connection is given in his *Hydrodynamik*, Leipzig (1927).

‡ L. Bairstow, B. M. Cave, E. D. Lenz, *Phil. Trans.* A, ccxxiii. 383 (1923). The more general case of a cylinder of arbitrary section is here discussed. A comparison with experimental results so far as these were available is added for the case of the circular section.

§ Harrison, *Camb. Trans.* xxiii. 71 (1924); Filon, *Proc. Roy. Soc.* A, cxiii. 7 (1926) and *Phil. Trans.* A, ccxxvii. 93 (1927).

‖ *K. Soc. d. Wiss.* Upsala (1926).

¶ *Dissertation*, Upsala, 1921; *Ann. d. Physik.* (4) lxviii. 89 (1922); *Arkiv f. matemat.* xvii. (1933), xviii (1924), xix (1925).

each side being a symmetric function of a, b, c, f, g, h and a', b', c', f', g', h'. Hence, and by Art. 329, the expression (1) reduces to the form

$$\iiint \Phi\, dx\, dy\, dz + 2\iiint (p_{xx}a' + p_{yy}b' + p_{zz}c' + p_{yz}f' + p_{zx}g' + p_{xy}h')\, dx\, dy\, dz + \iiint \Phi' dx\, dy\, dz.$$

The second integral may be written

$$\iiint \left(p_{xx}\frac{\partial u'}{\partial x} + p_{xy}\frac{\partial u'}{\partial y} + p_{xz}\frac{\partial u'}{\partial z} + \ldots + \ldots\right) dx\, dy\, dz;$$

and by a partial integration, remembering that u', v', w' vanish at the boundary, this becomes

$$-\iiint \left\{u'\left(\frac{\partial p_{xx}}{\partial x} + \frac{\partial p_{xy}}{\partial y} + \frac{\partial p_{xz}}{\partial z}\right) + \ldots + \ldots\right\} dx\, dy\, dz$$

or

$$\iiint \left\{u'\left(\frac{\partial p}{\partial x} - \mu\nabla^2 u\right) + \ldots + \ldots\right\} dx\, dy\, dz. \quad \text{.........(3)}$$

So far there is no restriction on the values of u, v, w except that they satify the equation of continuity. But if they satisfy the equations of motion when the inertia-terms are omitted, viz.

$$-\frac{\partial p}{\partial x} + \mu\nabla^2 u - \rho\frac{\partial \Omega}{\partial x} = 0, \quad \text{.........(4)}$$

where Ω is a single-valued potential, the integral (3) vanishes in virtue of the equation of continuity, by Art. 42 (4).

Under these conditions the dissipation in the altered motion is equal to

$$\iiint \Phi\, dx\, dy\, dz + \iiint \Phi' dx\, dy\, dz, \quad \text{.........(5)}$$

or $2(F+F')$, say. That is, it exceeds the dissipation in the steady motion by the essentially positive quantity $2F'$ which represents the dissipation in the motion (u', v', w').

In other words, provided the inertia-terms may be neglected, the motion of a liquid under constant forces having a single-valued potential is characterized by the property that the dissipation in any region is less than in any other motion consistent with the same values of u, v, w at the boundary.

It follows that, with prescribed velocities over the boundary, there is under the same condition only one type of steady motion in the region*.

It has been pointed out by Rayleigh† that the integral (3) vanishes, and the dissipation is accordingly a minimum, under somewhat wider conditions. The integral (3) may be replaced by

$$-\mu\iiint (u'\nabla^2 u + v'\nabla^2 v + w'\nabla^2 w)\, dx\, dy\, dz, \quad \text{.........(6)}$$

which will vanish whenever

$$\nabla^2 u = \frac{\partial H}{\partial x}, \quad \nabla^2 v = \frac{\partial H}{\partial y}, \quad \nabla^2 w = \frac{\partial H}{\partial z}, \quad \text{.........(7)}$$

where H is a single-valued function of x, y, z. This condition, which is purely kinematical, implies that

$$\nabla^2\xi = 0, \quad \nabla^2\eta = 0, \quad \nabla^2\zeta = 0, \quad \text{.........(8)}$$

and conversely. Under this head are included the case of steady motion between parallel planes, where

$$u = A + Bz + Cz^2, \quad v = 0, \quad w = 0 \quad \text{.........(9)}$$

(Art. 330), and that of motion in circles between coaxal cylinders (Art. 333). It is to be noticed that there is now no necessity, so far as the truth of the theorem is concerned, that the motion represented by u, v, w should be small, or even that it should be dynamically

* Helmholtz, "Zur Theorie der stationären Ströme in reibenden Flüssigkeiten," *Verh. d. naturhist.-med. Vereins*, Oct. 30, 1868 [*Wiss. Abh.* i. 223].

† "On the Motion of a Viscous Fluid," *Phil. Mag.* (6) xxvi. 776 (1913) [*Papers*, vi. 187].

possible as a steady motion, provided only that the relations (7) and the equation of continuity are satisfied. For instance, in any case of motion between concentric spheres the dissipation is necessarily greater than was found in Art. 334, and the couple N required to maintain the motion must therefore exceed the value there given.

2°. If u, v, w refer to any motion whatever in the given region, we have

$$2\dot{F}=\iiint\dot{\Phi}\,dx\,dy\,dz=2\iiint(p_{xx}\dot{a}+p_{yy}\dot{b}+p_{zz}\dot{c}+p_{yz}\dot{f}+p_{zx}\dot{g}+p_{xy}\dot{h})\,dx\,dy\,dz, \quad\text{......(10)}$$

since the formula (2) holds when dots take the place of accents.

The treatment of this integral is the same as before. If we suppose that $\dot{u}$, $\dot{v}$, $\dot{w}$ vanish over the bounding surface Σ, we find

$$\dot{F}=-\iiint\left\{\dot{u}\left(\frac{\partial p_{xx}}{\partial x}+\frac{\partial p_{xy}}{\partial y}+\frac{\partial p_{xz}}{\partial z}\right)+\ldots+\ldots\right\}dx\,dy\,dz$$

$$=-\rho\iiint(\dot{u}^2+\dot{v}^2+\dot{w}^2)\,dx\,dy\,dz+\rho\iiint(X\dot{u}+Y\dot{v}+Z\dot{w})\,dx\,dy\,dz, \quad\text{..................(11)}$$

in the case of slow motion.

When the extraneous forces have a single-valued potential the latter integral vanishes, so that

$$\dot{F}=-\rho\iiint(\dot{u}^2+\dot{v}^2+\dot{w}^2)\,dx\,dy\,dz. \quad\text{..............................(12)}$$

This is essentially negative, so that F continually diminishes, the process ceasing only when $\dot{u}=0$, $\dot{v}=0$, $\dot{w}=0$, that is, when the motion has become steady.

Hence when the velocities over the boundary Σ are maintained constant, the motion in the interior will tend to become steady. The type of steady motion ultimately attained is therefore stable, as well as unique*.

It has been shewn by Rayleigh† that the above theorem can be extended so as to apply to any dynamical system devoid of potential energy, in which the kinetic energy (T) and the dissipation-function (F) can be expressed as quadratic functions of the generalized velocities, with constant coefficients.

If the extraneous forces have not a single-valued potential, or if instead of given velocities we have given tractions over the boundary, the theorems require a slight modification. The excess of the dissipation over *double* the rate at which work is being done by the extraneous forces (including the tractions on the boundary) tends to a unique minimum, which is only attained when the motion is steady‡.

Periodic Motion.

345. We next examine the influence of viscosity in various problems of small oscillations.

We begin with the case of 'laminar' motion, as this will enable us to illustrate some points of great importance, without elaborate mathematics. If we assume that $v=0$, $w=0$, whilst u is a function of y only, the equations (4) of Art. 328 require that $p=$ const., and

$$\frac{\partial u}{\partial t}=\nu\frac{\partial^2 u}{\partial y^2}. \quad\text{.................................(1)}$$

* Korteweg, "On a General Theorem of the Stability of the Motion of a Viscous Fluid," *Phil. Mag.* (5) xvi. 112 (1883).

† *l.c.* p. 618.

‡ Cf. Helmholtz, *l.c.* p. 618.

This has the same form as the equation of linear motion of heat. In the case of simple-harmonic motion, assuming a time-factor $e^{i(\sigma t+\epsilon)}$, we have

$$\frac{\partial^2 u}{\partial y^2} = \frac{i\sigma}{\nu} u, \qquad \text{.................................(2)}$$

the solution of which is

$$u = Ae^{(1+i)\beta y} + Be^{-(1+i)\beta y}, \qquad \text{.......................(3)}$$

provided

$$\beta = \left(\frac{\sigma}{2\nu}\right)^{\frac{1}{2}}. \qquad \text{.................................(4)}$$

Let us first suppose that the fluid lies on the positive side of the plane xz, and that the motion is due to a prescribed oscillation

$$u = ae^{i(\sigma t+\epsilon)} \qquad \text{.................................(5)}$$

of a rigid surface coincident with this plane. If the fluid extend to infinity in the direction of y-positive, the first term in (3) is excluded, and, determining B by the boundary-condition (5), we have

$$u = ae^{-(1+i)\beta y + i(\sigma t+\epsilon)}, \qquad \text{..........................(6)}$$

or, taking the real part,

$$u = ae^{-\beta y}\cos(\sigma t - \beta y + \epsilon), \qquad \text{.......................(7)}$$

corresponding to a prescribed motion

$$u = a\cos(\sigma t + e) \qquad \text{.............................(8)}$$

at the boundary*.

The formula (7) represents a wave of transversal vibrations propagated inwards from the boundary with the velocity σ/β, but with rapidly diminishing amplitude, the falling off within a wave-length being in the ratio $e^{-2\pi}$, or $\frac{1}{535}$.

The linear magnitude β^{-1} is of great importance in all problems of oscillatory motion which do not involve changes of density, as indicating the extent to which the effects of viscosity penetrate into the fluid. In the case of air ($\nu = \cdot 13$) its value is $\cdot 21P^{\frac{1}{2}}$ centimetres, if P be the period of oscillation in seconds. For water the corresponding value is $\cdot 072P^{\frac{1}{2}}$. We shall have further illustrations, presently, of the fact that the influence of viscosity extends only to a short distance from the surface of a body performing small oscillations with sufficient frequency. A similar statement can be made with respect to the free surface of a liquid in wave-motion.

The retarding force on the rigid plane is, per unit area,

$$-\mu\left[\frac{\partial u}{\partial y}\right]_{y=0} = \mu\beta a\{\cos(\sigma t+\epsilon) - \sin(\sigma t+\epsilon)\}$$

$$= \rho\nu^{\frac{1}{2}}\sigma^{\frac{1}{2}} a\cos(\sigma t + \epsilon + \tfrac{1}{4}\pi). \qquad \text{.................(9)}$$

The force has its maxima at intervals of one-eighth of a period before the oscillating plane passes through its mean position†.

* Stokes, *l.c. ante* p. 580.

† For investigations relating to the case where the motion of the lamina is not restricted to be simple harmonic see Stokes, *l.c.*; Basset, *Quart. Journ. Math.* (1910); and Rayleigh, "On the Motion of Solid Bodies through a Viscous Fluid," *Phil. Mag.* (6) xxi. 697 (1911) [*Papers*, vi. 29]. See also Havelock, *Phil. Mag.* (6) xlii. 620 (1921).

On the forced oscillation above investigated we may superpose any of the normal modes of free motion of which the system is capable. If we assume that

$$u \propto A \cos my + B \sin my, \quad \text{.................................(10)}$$

and substitute in (1), we find

$$\frac{\partial u}{\partial t} = -\nu m^2 u, \quad \text{...(11)}$$

whence we obtain the solution

$$u = \Sigma (A \cos my + B \sin my)\, e^{-\nu m^2 t}. \quad \text{..........................(12)}$$

The admissible values of m, and the ratios $A : B$, are as a rule determined by the boundary-conditions. The arbitrary constants which remain are then to be found in terms of the initial conditions, by Fourier's methods.

In the case of a fluid extending from $y = -\infty$ to $y = +\infty$, all the real values of m are admissible. The solution, in terms of the initial conditions, can in this case be immediately written down by Fourier's Theorem (Art. 238 (4)). Thus

$$u = \frac{1}{\pi} \int_0^\infty dm \int_{-\infty}^\infty f(\lambda) \cos m (y - \lambda)\, e^{-\nu m^2 t}\, d\lambda, \quad \text{....................(13)}$$

if

$$u = f(y) \quad \text{...(14)}$$

be the arbitrary initial distribution of velocity.

The integration with respect to m can be effected by the known formula

$$\int_0^\infty e^{-\alpha x^2} \cos \beta x\, dx = \tfrac{1}{2} \left(\frac{\pi}{\alpha}\right)^{\frac{1}{2}} e^{-\beta^2/4\alpha}. \quad \text{..........................(15)}$$

We thus find

$$u = \frac{1}{2 (\pi \nu t)^{\frac{1}{2}}} \int_{-\infty}^\infty e^{-(y-\lambda)^2/4\nu t} f(\lambda)\, d\lambda. \quad \text{..........................(16)}$$

From this we may derive the solution (2) of Art. 334 a.

346. When the fluid does not extend to infinity, but is bounded by a fixed rigid plane $y = h$, then, in determining the motion due to a forced oscillation of the plane $y = 0$, both terms of (3) are required, and the boundary-conditions give

$$A + B = a, \quad A e^{(1+i)\beta h} + B e^{-(1+i)\beta h} = 0, \quad \text{..............(17)}$$

whence

$$u = a \frac{\sinh (1+i) \beta (h - y)}{\sinh (1+i) \beta h} \cdot e^{i(\sigma t + \epsilon)}, \quad \text{..................(18)}$$

as is easily verified. This gives for the retarding force per unit area on the oscillating plane

$$-\mu \left[\frac{\partial u}{\partial y}\right]_{y=0} = \mu (1+i) \beta a \coth (1+i) \beta h \,.\, e^{i(\sigma t + \epsilon)}. \quad \text{.........(19)}$$

The real part of this expression may be reduced to the form

$$\sqrt{2}\mu\beta a \frac{\sinh 2\beta h \cos (\sigma t + \epsilon + \frac{1}{4}\pi) + \sin 2\beta h \sin (\sigma t + \epsilon + \frac{1}{4}\pi)}{\cosh 2\beta h - \cos 2\beta h} \quad \text{....(20)}$$

When βh is moderately large this is equivalent to (9) above; whilst for small values of βh it reduces to

$$\frac{\mu a}{h} \,.\, \cos (\sigma t + \epsilon), \quad \text{.............................(21)}$$

as might have been foreseen.

This example contains the theory of the modification introduced by Maxwell* into Coulomb's method† of investigating the viscosity of fluids by the rotational oscillation of a circular disk in its own (horizontal) plane. The addition of fixed parallel disks at a short distance above and below greatly increases the influence of viscosity.

The free modes of motion are expressed by (12), with the conditions that $u=0$ for $y=0$ and $y=h$. This gives $A=0$ and $mh=s\pi$, where s is integral. The corresponding moduli of decay are then given by $\tau=1/\nu m^2$.

347. As a further example, let us take the case of a horizontal force

$$X=f\cos(\sigma t+\epsilon), \qquad (1)$$

acting uniformly on an infinite mass of water of uniform depth h.

The equation (1) of Art. 345 is now replaced by

$$\frac{\partial u}{\partial t}=\nu\frac{\partial^2 u}{\partial y^2}+X. \qquad (2)$$

If the origin be taken in the bottom, the boundary-conditions are $u=0$ for $y=0$, and $\partial u/\partial y=0$ for $y=h$, this latter condition expressing the absence of tangential force on the free surface. Replacing (1) by

$$X=fe^{i(\sigma t+\epsilon)}, \qquad (3)$$

we find

$$u=-\frac{if}{\sigma}\left\{1-\frac{\cosh(1+i)\beta(h-y)}{\cosh(1+i)\beta h}\right\}e^{i(\sigma t+\epsilon)}, \qquad (4)$$

if $\beta=(\sigma/2\nu)^{\frac{1}{2}}$, as before.

When βh is large, the expression in { } reduces practically to its first term for all points of the fluid whose height above the bottom exceeds a moderate multiple of β^{-1}. Hence, taking the real part,

$$u=\frac{f}{\sigma}\sin(\sigma t+\epsilon). \qquad (5)$$

This shews that the bulk of the fluid, with the exception of a stratum at the bottom, oscillates exactly like a free particle, the effect of viscosity being insensible. For points near the bottom the formula (4) becomes

$$u=-\frac{if}{\sigma}(1-e^{-(1+i)\beta y})\,e^{i(\sigma t+\epsilon)}, \qquad (5)$$

or, on rejecting the imaginary part,

$$u=\frac{f}{\sigma}\sin(\sigma t+\epsilon)-\frac{f}{\sigma}e^{-\beta y}\sin(\sigma t-\beta y+\epsilon). \qquad (7)$$

This might have been obtained directly, as the solution of (2) satisfying the conditions that $u=0$ for $y=0$, and

$$u=\frac{f}{\sigma}.\sin(\sigma t+\epsilon)$$

for large values of βy.

* *l.c. ante* p. 575. † *Mém. de l'Inst.* iii. (1800).

The curves A, B, C, D, E, F in the accompanying figure represent successive forms assumed by the same line of particles at intervals of one-tenth of a period. To complete the series it would be necessary to add the *images* of E, D, C, B with respect to the vertical through O. The whole system of curves may be regarded as successive aspects of a properly shaped spiral revolving uniformly about a vertical axis through O. The vertical range of the diagram is one wave-length $(2\pi/\beta)$ of the laminar disturbance.

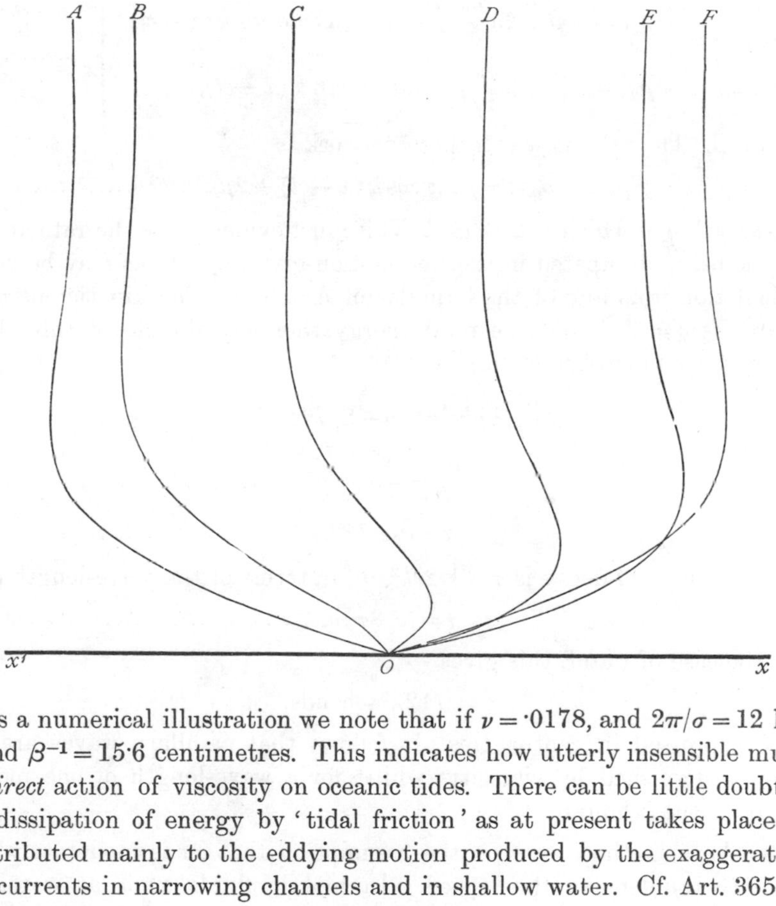

As a numerical illustration we note that if $\nu = \cdot 0178$, and $2\pi/\sigma = 12$ hours, we find $\beta^{-1} = 15{\cdot}6$ centimetres. This indicates how utterly insensible must be the *direct* action of viscosity on oceanic tides. There can be little doubt that such dissipation of energy by 'tidal friction' as at present takes place is to be attributed mainly to the eddying motion produced by the exaggeration of tidal currents in narrowing channels and in shallow water. Cf. Art. 365.

When βh is small the real part of (4) gives

$$u = \frac{f}{2\nu} y\,(2h - y) \,.\, \cos\,(\sigma t + \epsilon), \qquad \ldots\ldots\ldots\ldots\ldots\ldots\ldots(8)$$

the velocity being in the same phase with the force, and varying inversely as ν.

348. The effect of viscosity on free oscillatory waves on deep water may be estimated as follows.

When viscosity is neglected we have the solution

$$\phi = ace^{ky}\cos k(x-ct), \qquad\qquad (1)$$

$$u = kace^{ky}\sin k(x-ct), \quad v = -kace^{ky}\cos k(x-ct), \qquad\qquad (2)$$

$$\eta = a\sin k(x-ct), \qquad\qquad (3)$$

where η is the surface elevation, and c is the wave-velocity. This type of motion can persist even with viscosity, provided the requisite surface forces

$$\left.\begin{aligned} p_{yy} &= -p + 2\mu\frac{\partial v}{\partial y} = -p - 2\mu k^2 ac\cos k(x-ct), \\ p_{yx} &= \mu\left(\frac{\partial v}{\partial x}+\frac{\partial u}{\partial y}\right) = 2\mu k^2 ac\sin k(x-ct), \end{aligned}\right\} \qquad\qquad (4)$$

are applied. The rate at which these do work is

$$p_{yy}v + p_{yx}u = pkac\cos k(x-ct) + 2\mu k^3 a^2c^2, \qquad\qquad (5)$$

the mean value of which is $2\mu k^3 a^2 c^2$. This must evidently be the rate at which energy is being dissipated in the free motion given by (2), as may be verified by calculation from any of the formulae of Art. 329. The kinetic energy per unit area is $\frac{1}{4}\rho k a^2 c^2$, and the total energy therefore double of this. Hence in the absence of surface forces,

$$\frac{d}{dt}\left(\tfrac{1}{2}\rho k c^2 a^2\right) = -2\mu k^3 c^2 a^2, \qquad\qquad (6)$$

or

$$\frac{da}{dt} = -2\nu k^2 a, \qquad\qquad (7)$$

whence

$$a = a_0 e^{-2\nu k^2 t}. \qquad\qquad (8)$$

The modulus of decay is $\tau = 1/2\nu k^2$, or, in terms of the wave-length (λ),

$$\tau = \lambda^2/8\pi^2\nu. \qquad\qquad (9)*$$

In the case of water, this gives

$$\tau = \cdot712\lambda^2 \text{ seconds},$$

if λ be expressed in centimetres. It follows that capillary waves are very rapidly extinguished by viscosity; whilst for a wave-length of one metre τ would be about 2 hours.

The above method rests on the assumption that $\sigma\tau$ is moderately large, where $\sigma\,(=kc)$ denotes the 'speed.' In mobile fluids such as water this condition is fulfilled for all but excessively minute wave-lengths. The method fails for another reason when the depth is less than (say) half the wave-length. Owing to the practically infinite resistance to slipping at the bottom, the dissipation can no longer be calculated as if the motion were irrotational†.

* Stokes, *l.c. ante* p. 577, and *Papers*. iii. 74. (Through an oversight in the original calculation the value obtained for τ was too small by one-half.)

† The same point arises in the oscillations of superposed liquids (Art. 231) as investigated by Harrison, *Proc. Lond. Math. Soc.* (2) vi. 396; vii. 107 (1908). The modulus of decay was found to vary as $\nu^{-\frac{1}{2}}$ instead of as ν^{-1}. Cf. Art. 364.

The calculation can be modified so as to illustrate the generation and maintenance of waves by wind. It is not likely that the action of the wind even on a simple-harmonic wave-profile could be represented by any simple formula*. But, neglecting the tangential action, which seems to be of secondary importance, we may imagine the normal pressure to be expressed by a Fourier series of sines and cosines of multiples of $k(x-ct)$, and it is evident that the only constituent which does a net amount of work in a complete period has the form

$$\Delta p = C\cos k(x-ct). \qquad \text{.........................(10)}$$

The equation (6) is replaced by

$$\frac{d}{dt}\left(\tfrac{1}{2}\rho k c^2 \alpha^2\right) = \tfrac{1}{2}kc\alpha C - 2\mu k^3 c^2 \alpha^2, \qquad \text{.................(11)}$$

or

$$\frac{d\alpha}{dt} = \frac{C}{2\rho c} - 2\nu k^2 \alpha. \qquad \text{.........................(12)}$$

The amplitude will therefore increase or diminish according as

$$C \gtrless 4\mu k^2 c\alpha. \qquad \text{..............................(13)}$$

In recent papers† Dr Jeffreys assumes that the pressure of the relative wind on the advancing wave-profile may be represented roughly by an expression of the form

$$\beta\rho'(U-c)^2 \frac{\partial\eta}{\partial x}, \qquad \text{......................................(14)}$$

where U is the velocity of the wind, ρ' the density of the air, and β a numerical coefficient lying between 0 and 1, and probably less than $\frac{1}{2}$. This is equivalent to putting

$$C = \beta\rho'(U-c)^2 k\alpha$$

in (13). If we neglect capillarity $c^2 = g/k$, and the criterion takes the form

$$(U-c)^2 c \gtrless \frac{4\nu g\rho}{\beta\rho'}. \qquad \text{......................................(15)}$$

For a given wind this is least when $c = \frac{1}{3}U$; the least wind which could maintain waves is therefore, on the present assumption,

$$U = 3\left(\frac{\nu g\rho}{\beta\rho'}\right)^{\frac{1}{3}}. \qquad \text{......................................(16)}$$

If we assume $\nu = \cdot0178$, $g = 981$, $\rho'/\rho = \cdot00129$, $\beta = \cdot3$, we get $U = 107$, $\lambda = 8\cdot1$, in centimetres and seconds. Some observations by Jeffreys agree as to the order of magnitude of the least wind-velocity.

349. The direct calculation of the effect of viscosity on water-waves can be conducted as follows.

If the axis of y be drawn vertically upwards, and if we assume that the motion is confined to the dimensions x, y, we have

$$\frac{\partial u}{\partial t} = -\frac{1}{\rho}\frac{\partial p}{\partial x} + \nu\nabla^2 u, \qquad \frac{\partial v}{\partial t} = -\frac{1}{\rho}\frac{\partial p}{\partial y} - g + \nu\nabla^2 v, \qquad \text{...........(1)}$$

with

$$\frac{\partial u}{\partial x} + \frac{\partial v}{\partial y} = 0. \qquad \text{..............................(2)}$$

* It is hoped that an account will be published shortly of the experiments on which the late Sir T. Stanton had been engaged, relevant to this question.

† *Proc. Roy. Soc.* A, cvii. 189 (1924); cx. 241 (1925).

These are satisfied by

$$u = -\frac{\partial\phi}{\partial x} - \frac{\partial\psi}{\partial y}, \quad v = -\frac{\partial\phi}{\partial y} + \frac{\partial\psi}{\partial x}, \quad \text{.................(3)}$$

and

$$\frac{p}{\rho} = \frac{\partial\phi}{\partial t} - gy, \quad \text{................................(4)}$$

provided

$$\nabla_1^2\phi = 0, \quad \frac{\partial\psi}{\partial t} = \nu\nabla_1^2\psi, \quad \text{......................(5)}$$

where

$$\nabla_1^2 = \frac{\partial^2}{\partial x^2} + \frac{\partial^2}{\partial y^2}.$$

To determine the 'normal modes' which are periodic in respect of x with a prescribed wave-length $2\pi/k$, we assume a time-factor e^{nt} and a space-factor e^{ikx}. The solutions of (5) are then

$$\phi = (Ae^{ky} + Be^{-ky})\, e^{ikx+nt}, \quad \psi = (Ce^{my} + De^{-my})\, e^{ikx+nt}, \quad \text{......(6)}$$

with

$$m^2 = k^2 + n/\nu. \quad \text{......................(7)}$$

The boundary-conditions will supply equations which are sufficient to determine the nature of the various modes, and the corresponding values of n.

In the case of infinite depth one of these conditions takes the form that the motion must be finite for $y = -\infty$. Excluding for the present the cases where m is pure-imaginary, this requires that $B = 0$, $D = 0$, provided m denote that root of (7) which has its real part positive. Hence

$$u = -(ikAe^{ky} + mCe^{my})\, e^{ikx+nt}, \quad v = -(kAe^{ky} - ikCe^{my})\, e^{ikx+nt}. \quad \text{...(8)}$$

If η denote the elevation at the free surface, we must have $\partial\eta/\partial t = v$. If the origin of y be taken in the undisturbed level, this gives

$$\eta = -\frac{k}{n}(A - iC)\, e^{ikx+nt}. \quad \text{..........................(9)}$$

If T_1 denote the surface-tension, the stress-conditions at the surface are evidently

$$p_{yy} = T_1\frac{\partial^2\eta}{\partial x^2}, \quad p_{xy} = 0, \quad \text{......................(10)}$$

to the first order, since the inclination of the surface to the horizontal is assumed to be infinitely small.

Now

$$p_{yy} = -p + 2\mu\frac{\partial v}{\partial y}, \quad p_{xy} = \mu\left(\frac{\partial v}{\partial x} + \frac{\partial u}{\partial y}\right), \quad \text{..........(11)}$$

whence, by (4) and (9) we find, at the surface,

$$\frac{p_{yy}}{\rho} - T'\frac{\partial^2\eta}{\partial x^2} = -\frac{\partial\phi}{\partial t} + (g + T'k^2)\,\eta + 2\nu\frac{\partial v}{\partial y}$$

$$= -\frac{1}{n}\{(n^2 + 2\nu k^2 n + gk + T'k^3)\,A - i\,(gk + T'k^3 + 2\nu kmn)\,C\}, \quad \text{......(12)}$$

$$\frac{p_{xy}}{\rho} = -\{2i\nu k^2 A + (n + 2\nu k^2)\,C\}, \quad \text{.................(13)}$$

where $T' = T_1/\rho$, the common factor e^{ikx+nt} being understood.

Substituting in (10), and eliminating the ratio $A : C$, we obtain

$$(n + 2\nu k^2)^2 + gk + T'k^3 = 4\nu^2 k^3 m. \quad \text{.................(14)}$$

If we eliminate m by means of (7), we get a biquadratic in n, but only those roots are admissible which give a positive value to the real part of the left-hand member of (14), and so make the real part of m positive.

If we write, for shortness,

$$gk + T'k^3 = \sigma^2, \quad \nu k^2/\sigma = \theta, \quad n + 2\nu k^2 = x\sigma, \quad \text{...........(15)}$$

the biquadratic in question takes the form

$$(x^2+1)^2 = 16\theta^3 (x - \theta). \quad \text{......................(16)}$$

It is not difficult to shew that this has always two roots (both complex) which violate the restriction just stated, and two admissible roots which may be real or complex according to the magnitude of the ratio θ. If λ be the wave-length, and $c \,(= \sigma/k)$ the wave-velocity in the absence of friction, we have

$$\theta = \nu k/c = 2\pi\nu/c\lambda. \quad \text{.........................(17)}$$

Now, for water, if c_m denote the minimum wave-velocity of Art. 267, we find $2\pi\nu/c_m = \cdot 0048$ cm., so that except for very minute wave-lengths θ is a small number. Neglecting the square of θ, we have $x = \pm i$, and

$$n = -2\nu k^2 \pm i\sigma. \quad \text{............................(18)}$$

The condition $p_{xy} = 0$ shews that

$$\frac{C}{A} = -\frac{2i\nu k^2}{n + 2\nu k^2} = \mp \frac{2\nu k^2}{\sigma}, \quad \text{......................(19)}$$

which is, under the same circumstances, very small. Hence the motion is approximately irrotational, with a velocity-potential

$$\phi = A e^{-2\nu k^2 t + ky + i(kx \pm \sigma t)}. \quad \text{......................(20)}$$

If we put $a = \mp kA/\sigma$, the equation (9) of the free surface becomes, approximately, on taking the real part,

$$\eta = a e^{-2\nu k^2 t} \cos(kx \pm \sigma t). \quad \text{.........................(21)}$$

The wave-velocity is σ/k, or $(g/k + T'k)^{\frac{1}{2}}$, as in Art. 267, and the law of decay is that investigated independently in the preceding Art.*

To examine more closely the character of the motion, as affected by viscosity, we may calculate the vorticity (ω) at any point of the fluid. This is given by

$$\omega = \frac{\partial v}{\partial x} - \frac{\partial u}{\partial y} = \nabla_1^2 \psi = \frac{\alpha}{\nu}\psi = \frac{n}{\nu} C e^{my + ikx + nt}. \quad \text{......................(22)}$$

Now, from (7) and (18), we have, approximately,

$$m = (1 \pm i)\beta, \quad \text{where } \beta = (\sigma/2\nu)^{\frac{1}{2}}.$$

With the same notation as before, we find

$$\omega = \mp 2\sigma k a e^{-2\nu k^2 t + \beta y} \cos\{kx \pm (\sigma t + \beta y)\}. \quad \text{......................(23)}$$

* Similar results were obtained by Basset, *Hydrodynamics*, ii. Arts. 520–522 (1888), where the case of finite depth is also treated. Reference may also be made to Hough, *l.c. ante* p. 593, where the case of a spherical sheet of water is considered.

This diminishes rapidly from the surface downwards, in accordance with the thermal analogy pointed out in Art. 328. Owing to the oscillatory character of the motion, the sign of the vorticity which is being diffused inwards from the surface is continually being reversed, so that beyond a stratum of thickness comparable with $2\pi/\beta$ the effect is insensible, just as the fluctuations of temperature at the earth's surface cease to have any sensible influence at a depth of a few yards.

In the case of a very viscous fluid, such as treacle or pitch, θ may be large even when the wave-length is considerable. The admissible roots of (16) are then both real. One of them is evidently nearly equal to 2θ, and continuing the approximation we find

$$x = 2\theta - \frac{1}{2\theta} + \ldots,$$

whence, neglecting capillarity, we have, by (15),

$$n = -g/2k\nu. \qquad \text{(24)}$$

The remaining real root is $1{\cdot}09\theta$, nearly, which gives

$$n = -{\cdot}91\nu k^2. \qquad \text{(25)}$$

The former root is the more important. It represents a slow creeping of the fluid towards a state of equilibrium with a horizontal surface. The rate of recovery depends on the relation between the gravity of the fluid (which is proportional to $g\rho$) and the viscosity (μ), the influence of inertia being insensible. It appears from (7) and (15) that $m = k$, nearly, so that the motion is approximately irrotational*.

The type of motion corresponding to (25), on the other hand, depends, as to its persistence, on the relation between the inertia (ρ) and the viscosity (μ), the effect of gravity being unimportant. It dies out very rapidly.

The above investigation gives the most important of the normal modes, of the prescribed wave-length, of which the system is capable. We know *à priori* that there must be an infinity of others. These correspond to pure-imaginary values of m, and are of a less persistent character. If in place of (6) we assume

$$\phi = Ae^{ky} \,.\, e^{ikx+nt}, \quad \psi = (C\cos m'y + D\sin m'y)\, e^{ikx+nt}, \qquad \text{(26)}$$

with

$$m'^2 = -k^2 - \frac{n}{\nu}, \qquad \text{(27)}$$

and carry out the investigation as before, we find

$$\left.\begin{aligned} (n^2 + 2\nu k^2 n + gh + T'k^3)\, A - i\,(gk + T'k^3)\, C - 2i\nu k m' n D &= 0, \\ 2ik^2 A + (k^2 - m'^2)\, C &= 0. \end{aligned}\right\} \qquad \text{(28)}$$

Any real value of m' is admissible, these equations determining the ratios $A : C : D$; and the corresponding value of n is

$$n = -\nu\,(k^2 + m'^2). \qquad \text{(29)}$$

In any one of these modes the plane xy is divided horizontally and vertically into a series of quasi-rectangular compartments, within each of which the fluid circulates, gradually coming to rest as the original momentum is spent against viscosity.

By a proper synthesis of the various normal modes it must be possible to represent the decay of any arbitrary initial disturbance.

* Cf. Tait, "Note on Ripples in a Viscous Liquid," *Proc. R. S. Edin.* xvii. 110 (1890) [*Scientific Papers*, Cambridge, 1898–1900, ii. 313].

350. The equations (12) and (13) of the preceding Art. may be used to illustrate more in detail the generation and maintenance of water-waves against viscosity, by suitable forces applied to the surface.

If the external forces p'_{yy}, p'_{xy} be given multiples of e^{ikx+nt}, where k and n are prescribed, the equations in question determine A and C, and thence the values of η. Thus we find

$$\frac{p'_{yy}}{g\rho\eta}=\frac{(n^2+2\nu k^2 n+\sigma^2)A-i(\sigma^2+2\nu kmn)C}{gk(A-iC)}, \qquad (1)$$

$$\frac{p'_{xy}}{g\rho\eta}=\frac{n}{gk}\cdot\frac{2i\nu k^2 A+(n+2\nu k^2)C}{A-iC}, \qquad (2)$$

where σ^2 has been written for $gk+T'k^3$ as before.

Let us first examine the effect of a purely tangential force. Assuming $p'_{yy}=0$, we find

$$\frac{p'_{xy}}{g\rho\eta}=\frac{in}{gk}\cdot\frac{(n+2\nu k^2)^2+\sigma^2-4\nu^2k^3m}{n+2\nu k^2-2\nu km}. \qquad (3)$$

If, as we shall suppose for reasons already indicated, $\nu k^2/\sigma$ and $\nu km/\sigma$ are small, the elevation will be greatest when $n=\pm i\sigma$, nearly. To find the force necessary to maintain a train of waves of given amplitude, travelling in the direction of x-positive, we put $n=-i\sigma$. This makes

$$\frac{p'_{xy}}{g\rho\eta}=\frac{4\nu k\sigma}{g}, \quad \text{or} \quad p'_{xy}=4\mu k\sigma\eta, \qquad (4)$$

approximately. Hence the force acts forwards on the crests of the waves, and backwards at the troughs, changing sign at the nodes. A force having the same distribution, but less intensity in proportion to the height of the waves than that given by (4), would only retard, without preventing, the decay of the waves by viscosity. A force having the opposite sign would accelerate this decay.

The case of purely normal force can be investigated in a similar manner. If $p'_{xy}=0$, we have

$$\frac{p'_{yy}}{g\rho\eta}=\frac{(n+2\nu k^2)^2+\sigma^2-4\nu^2k^3m}{gk}. \qquad (5)$$

The reader may easily satisfy himself that when there is no viscosity this coincides with the result of Art. 242. If we put $n=-i\sigma$, we obtain, with the same approximations as before,

$$p'_{yy}=-4i\mu k\sigma\eta. \qquad (6)$$

Hence the wave-system

$$\eta=a\sin(kx-\sigma t) \qquad (7)$$

would be maintained without increase or decrease by the pressure-distribution

$$p=\text{const.}+4\mu ka\sigma\cos(kx-\sigma t), \qquad (8)$$

applied to the surface. The requisite pressure is greatest on the rear and least on the front slopes of the waves*.

If we call to mind the phases of the particles, revolving in their circular orbits, at different parts of a wave-profile, it is evident that the forces above investigated, whether normal or tangential, are on the whole urging the surface-particles in the directions in which they are already moving.

Owing to the irregular, eddying character of a wind blowing over a roughened surface, it is not easy to give more than a general explanation of the manner in which it generates and maintains waves. It is not difficult to see, however, that the action of the wind will tend to produce surface forces of the kinds above investigated. When the air is moving in the direction in which the wave-form is travelling, but with a greater velocity, there will evidently be an excess of pressure on the rear slopes, as well as a tangential drag on the exposed crests. The aggregate effect of these forces will be a surface drift, and the residual tractions, whether normal or tangential, will have on the whole the distribution above postulated. Hence the tendency will be to increase the amplitude of the waves to such a point that the dissipation balances the work done by the surface forces. In like manner waves travelling faster than the wind, or against the wind, will have their amplitude continually reduced†.

It has been shewn (Art. 267) that, under the joint influence of gravity and capillarity, there is a minimum wave-velocity of 23·2 cm. per sec., or 0·45 mile per hour. The velocity of a wind must at all events exceed this if it is to maintain waves against viscosity‡. Some observations of Scott Russell§ may be quoted here.

"Let [a spectator] begin his observations in a perfect calm, when the surface of the water is smooth and reflects like a mirror the images of surrounding objects. This appearance will not be affected by even a slight motion of the air, and a velocity of less than half a mile an hour ($8\frac{1}{2}$ in. per sec.) does not sensibly disturb the smoothness of the reflecting surface. A gentle zephyr flitting along the surface from point to point, may be observed to destroy the perfection of the mirror for a moment, and on departing, the surface remains polished as before; if the air have a velocity of about a mile an hour, the surface of the water becomes less capable of distinct reflexion, and on observing it in such a condition, it is to be noticed that the diminution of this reflecting power is owing to the presence of those minute corrugations of the superficial film which form waves of the *third order* [capillary waves]. ... This first stage of disturbance has this distinguishing circumstance, that the phenomena on the surface cease almost simultaneously with the intermission of the disturbing cause so that a spot which is sheltered from the direct action of the wind remains smooth, the waves of the third order being incapable of travelling spontaneously to any considerable distance, except when under the continued action of the original disturbing force. This condition is the indication of present force, not of that which is past.

* This agrees with the result given at the end of Art. 242 where, however, the dissipative forces were of a different kind.

† Cf. Airy, "Tides and Waves," Arts. 265–272; Stokes, *Camb. Trans.* ix. [62] [*Papers*, iii. 74]; Rayleigh, *l.c. ante* p. 174.

‡ Sir W. Thomson, *l.c. ante* p. 459.

§ *l.c. ante* p. 423.

While it remains it gives that deep blackness to the water which the sailor is accustomed to regard as the index of the presence of wind, and often as the forerunner of more.

"The second condition of wave motion is to be observed when the velocity of the wind acting on the smooth water has increased to two miles an hour. Small waves then begin to rise uniformly over the whole surface of the water; these are waves of the second order, and cover the water with considerable regularity. Capillary waves disappear from the ridges of these waves, but are to be found sheltered in the hollows between them, and on the anterior slopes of these waves. The regularity of the distribution of these secondary waves over the surface is remarkable; they begin with about an inch of amplitude, and a couple of inches long; they enlarge as the velocity or duration of the wave increases; by and by the coterminal waves unite; the ridges increase, and if the wind increase the waves become cusped, and are regular waves of the *second order* [gravity waves]*. They continue enlarging their dimensions, and the depth to which they produce the agitation increasing simultaneously with their magnitude, the surface becomes extensively covered with waves of nearly uniform magnitude."

This quotation is retained from previous editions, for the sake of its vivid descriptions, but the numerical estimates relating to the first appearance of waves may require qualification. In particular, the initial wave-length is apparently too great to justify the term 'capillary.'

351. The calming effect of oil on water-waves appears to be due to the *variations* of tension caused by the extensions and contractions of the contaminated surface†. The surface-tension of pure water is greater than the sum of the tensions of the surfaces of separation of oil and air, and oil and water, respectively, so that a drop of oil thrown on water is gradually drawn out into a thin film. When the film is sufficiently thin, say not more than two millionths of a millimetre in thickness, it is found that the tension is no longer constant but is increased when the thickness is reduced by stretching, and conversely. It is evident at once from the figure on p. 366 that in oscillatory waves the tendency is for any portion of the surface to be alternately contracted and extended, according as it is above or below the mean level. The consequent variations in tension produce an alternating tangential drag on the water, with a consequent increase in the rate of dissipation of energy.

The preceding formulae enable us to submit this explanation, to a certain extent, to the test of calculation.

It is evident beforehand that the effect of the quasi-elasticity of the oil-film will be greater the shorter the wave-length; and that if the wave-length be sufficiently small the surface will be practically inextensible, and the horizontal velocity at the surface will be annulled. We will assume this condition to be fulfilled.

The internal motion of the water will be given by the formulae (8) of Art. 349, but the determination of the constants is different. The condition to be satisfied by the normal stress is the same as in the Art. cited, and gives

$$(\alpha^2+2\nu k^2\alpha+\sigma^2)A-i(\sigma^2+2\nu km\alpha)C=0, \qquad (1)$$

* Scott Russell's wave of the *first order* is the 'solitary wave' discussed *ante* Art. 252.

† Reynolds, "On the Effect of Oil in destroying Waves on the Surface of Water," *Brit. Ass. Rep.* 1880 [*Papers*, i. 409]; Aitken, "On the Effect of Oil on a Stormy Sea," *Proc. Roy. Soc. Edin.* xii. 56 (1883).

where $$\sigma^2 = gk + T'k^3, \qquad\qquad (2)$$

T' referring now to the total tension of the oil-film. In place of the condition of vanishing tangential stress, we have the condition

$$u = 0 \quad \text{for} \quad y = 0, \qquad\qquad (3)$$

which gives $$ikA + mC = 0. \qquad\qquad (4)$$

Eliminating the ratio $A : C$, we find

$$m(\alpha^2 + \sigma^2) - k\sigma^2 = 0, \qquad\qquad (5)$$

or, on eliminating m by means of the equation

$$m^2 = k^2 + \alpha/\nu, \qquad\qquad (6)$$

$$\left(\frac{\alpha}{\nu} + k^2\right)(\alpha^2 + \sigma^2)^2 - k^2\sigma^4 = 0. \qquad\qquad (7)$$

This equation has an extraneous root $\alpha = 0$, and other roots are inadmissible as giving, when substituted in (5), negative values to the real part of m. If $\nu k^2/\sigma$ is small, the relevant roots are, to a first approximation, $\alpha = \pm i\sigma$, and to a second

$$\alpha = \pm i\sigma - \frac{\nu^{\frac{1}{2}} k\sigma^{\frac{1}{2}}}{2\sqrt{2}}, \qquad\qquad (8)$$

where the correction to the 'speed' σ of the oscillations is neglected. The modulus of decay is therefore

$$\tau = \frac{2\sqrt{2}}{\nu^{\frac{1}{2}} k\sigma^{\frac{1}{2}}}. \qquad\qquad (9)$$

The ratio of this to the modulus obtained on the hypothesis of *constant* surface-tension (viz. $1/2\nu k^2$) is $4\sqrt{2}\,(\nu k^2/\sigma)^{\frac{1}{2}}$, which is by hypothesis small*.

352. Problems of periodic motion in three dimensions, having special relation to *spherical* surfaces, may be treated in a general manner, as follows.

We investigate, first, the general solution of the system of equations:

$$(\nabla^2 + h^2)\,u' = 0, \qquad (\nabla^2 + h^2)\,v' = 0, \qquad (\nabla^2 + h^2)\,w' = 0, \qquad (1)$$

$$\frac{\partial u'}{\partial x} + \frac{\partial v'}{\partial y} + \frac{\partial w'}{\partial z} = 0, \qquad\qquad (2)$$

in terms of spherical harmonics. This is an extension of the problem considered in Art. 335. We will consider only, in the first instance, cases where u', v', w' are finite at the origin.

The solutions fall naturally into two distinct classes. If r denote the radius vector, the typical solution of the First Class is

$$\left.\begin{aligned} u' &= \psi_n(hr)\left(y\frac{\partial}{\partial z} - z\frac{\partial}{\partial y}\right)\chi_n, \\ v' &= \psi_n(hr)\left(z\frac{\partial}{\partial x} - x\frac{\partial}{\partial z}\right)\chi_n, \\ w' &= \psi_n(hr)\left(x\frac{\partial}{\partial y} - y\frac{\partial}{\partial x}\right)\chi_n, \end{aligned}\right\} \qquad\qquad (3)$$

where χ_n is a solid harmonic of positive degree n, and ψ_n is defined as in Art. 292 (7). It is easily verified that the above expressions do in fact satisfy (1) and (2).

* The investigation is abbreviated from that given in the second edition of this work.

It is to be noticed that this solution makes

$$xu' + yv' + zw' = 0. \qquad (4)$$

The typical solution of the Second Class is

$$\left.\begin{aligned} u' &= (n+1)\,\psi_{n-1}(hr)\frac{\partial\phi_n}{\partial x} - n\psi_{n+1}(hr)\,h^2 r^{2n+3}\frac{\partial}{\partial x}\frac{\phi_n}{r^{2n+1}}, \\ v' &= (n+1)\,\psi_{n-1}(hr)\frac{\partial\phi_n}{\partial y} - n\psi_{n+1}(hr)\,h^2 r^{2n+3}\frac{\partial}{\partial y}\frac{\phi_n}{r^{2n+1}}, \\ w' &= (n+1)\,\psi_{n-1}(hr)\frac{\partial\phi_n}{\partial z} - n\psi_{n+1}(hr)\,h^2 r^{2n+3}\frac{\partial}{\partial z}\frac{\phi_n}{r^{2n+1}}, \end{aligned}\right\} \qquad (5)$$

where ϕ_n is a solid harmonic of positive degree n. The coefficients of $\psi_{n-1}(hr)$ and $\psi_{n+1}(hr)$ in these expressions are solid harmonics of degrees $n-1$ and $n+1$ respectively, so that the equations (1) are satisfied. To verify that (2) is also satisfied we need the formulae of reduction

$$\psi_n'(\zeta) = -\zeta\psi_{n+1}(\zeta), \qquad (6)$$

$$\zeta\psi_n'(\zeta) + (2n+1)\,\psi_n(\zeta) = \psi_{n-1}(\zeta), \qquad (7)$$

which are repeated from Art. 292.

The formulae (5) make

$$xu' + yv' + zw' = n(n+1)(2n+1)\,\psi_n(hr)\,\phi_n, \qquad (8)$$

the reduction being effected by means of (6) and (7).

If we write

$$\xi' = \frac{\partial w'}{\partial y} - \frac{\partial v'}{\partial z}, \qquad \eta' = \frac{\partial u'}{\partial z} - \frac{\partial w'}{\partial x}, \qquad \zeta' = \frac{\partial v'}{\partial x} - \frac{\partial u'}{\partial y}, \qquad (9)$$

we find, in the solutions of the First Class,

$$\left.\begin{aligned} \xi' &= -\frac{1}{2n+1}\left\{(n+1)\,\psi_{n-1}(hr)\frac{\partial\chi_n}{\partial x} - n\psi_{n+1}(hr)\,h^2 r^{2n+3}\frac{\partial}{\partial x}\frac{\chi_n}{r^{2n+1}}\right\}, \\ \eta' &= -\frac{1}{2n+1}\left\{(n+1)\,\psi_{n-1}(hr)\frac{\partial\chi_n}{\partial y} - n\psi_{n+1}(hr)\,h^2 r^{2n+3}\frac{\partial}{\partial y}\frac{\chi_n}{r^{2n+1}}\right\}, \\ \zeta' &= -\frac{1}{2n+1}\left\{(n+1)\,\psi_{n-1}(hr)\frac{\partial\chi_n}{\partial z} - n\psi_{n+1}(hr)\,h^2 r^{2n+3}\frac{\partial}{\partial z}\frac{\chi_n}{r^{2n+1}}\right\}. \end{aligned}\right\} \qquad (10)$$

These make $\qquad x\xi' + y\eta' + z\zeta' = -n(n+1)\,\psi_n(hr)\,\chi_n. \qquad (11)$

In the solutions of the Second Class we have

$$\left.\begin{aligned} \xi' &= -(2n+1)\,h^2\psi_n(hr)\left(y\frac{\partial}{\partial z} - z\frac{\partial}{\partial y}\right)\phi_n, \\ \eta' &= -(2n+1)\,h^2\psi_n(hr)\left(z\frac{\partial}{\partial x} - x\frac{\partial}{\partial z}\right)\phi_n, \\ \zeta' &= -(2n+1)\,h^2\psi_n(hr)\left(x\frac{\partial}{\partial y} - y\frac{\partial}{\partial x}\right)\phi_n, \end{aligned}\right\} \qquad (12)$$

and therefore $\qquad x\xi' + y\eta' + z\zeta' = 0. \qquad (13)$

In the derivation of these results use has been made of (6), and of the formulae

$$\left.\begin{aligned} x\chi_n &= \frac{r^2}{2n+1}\left(\frac{\partial \chi_n}{\partial x} - r^{2n+1}\frac{\partial}{\partial x}\frac{\chi_n}{r^{2n+1}}\right), \\ y\chi_n &= \frac{r^2}{2n+1}\left(\frac{\partial \chi_n}{\partial y} - r^{2n+1}\frac{\partial}{\partial y}\frac{\chi_n}{r^{2n+1}}\right), \\ z\chi_n &= \frac{r^2}{2n+1}\left(\frac{\partial \chi_n}{\partial z} - r^{2n+1}\frac{\partial}{\partial z}\frac{\chi_n}{r^{2n+1}}\right), \end{aligned}\right\} \quad \text{..............(14)}$$

which hold whatever the form of χ_n.

To shew that the aggregate of the solutions of the types (3) and (5), with all integral values of n, and all possible forms of the harmonics ϕ_n, χ_n, constitutes the *complete* solution of the proposed system of equations (1) and (2), we remark in the first place that the equations in question imply

$$(\nabla^2 + h^2)(xu' + yv' + zw') = 0, \quad \text{.................(15)}$$

and

$$(\nabla^2 + h^2)(x\xi' + y\eta' + z\zeta') = 0. \quad \text{.................(16)}$$

It is evident from Art. 292 that the complete solution of these, subject to the condition of finiteness at the origin, is contained in the equations (8) and (11) above, if these be generalized by prefixing the sign Σ of summation with respect to n. Now when $xu' + yv' + zw'$ and $x\xi' + y\eta' + z\zeta'$ are given throughout any space, the values of u', v', w' are rendered by (2) completely determinate. For if there were two sets of values, say u', v', w' and u'', v'', w'', both satisfying the prescribed conditions, then, writing

$$u_1 = u' - u'', \quad v_1 = v' - v'', \quad w_1 = w' - w'',$$

we should have

$$\left.\begin{aligned} xu_1 + yv_1 + zw_1 &= 0, \\ x\xi_1 + y\eta_1 + z\zeta_1 &= 0, \\ \frac{\partial u_1}{\partial x} + \frac{\partial v_1}{\partial y} + \frac{\partial w_1}{\partial z} &= 0. \end{aligned}\right\} \quad \text{........................(17)}$$

If u_1, v_1, w_1 be regarded as the component velocities of a liquid, the first of these shews that the lines of flow are closed curves lying on a system of concentric spherical surfaces. Hence the 'circulation' (Art. 31) in any such line has a finite value. On the other hand, the second equation shews, by Art. 32, that the circulation in any circuit drawn on one of the above spherical surfaces is zero. These conclusions are irreconcileable unless u_1, v_1, w_1 are all zero.

Hence, in the present problem, whenever the functions ϕ_n and χ_n have been determined by (8) and (11), the values of u', v', w' follow uniquely as in (3) and (5).

When the region contemplated is bounded *internally* by a spherical surface, the condition of finiteness when $r = 0$ is no longer imposed, and we have an additional system of solutions in which the functions $\psi_n(\zeta)$ are replaced by $\Psi_n(\zeta)$, in accordance with Art. 292*.

353. The equations of small motion of an incompressible fluid are, in the absence of extraneous forces,

$$\frac{\partial u}{\partial t} = \frac{1}{\rho}\frac{\partial p}{\partial x} + \nu\nabla^2 u, \quad \frac{\partial v}{\partial t} = -\frac{1}{\rho}\frac{\partial p}{\partial y} + \nu\nabla^2 v, \quad \frac{\partial w}{\partial t} = -\frac{1}{\rho}\frac{\partial p}{\partial z} + \nu\nabla^2 w, \quad \text{...(1)}$$

with

$$\frac{\partial u}{\partial x} + \frac{\partial v}{\partial y} + \frac{\partial w}{\partial z} = 0. \quad \text{........................(2)}$$

If we assume that u, v, w all vary as $e^{\alpha t}$, the equations (1) may be written

$$(\nabla^2 + h^2)\, u = \frac{1}{\mu}\frac{\partial p}{\partial x}, \quad (\nabla^2 + h^2)\, v = \frac{1}{\mu}\frac{\partial p}{\partial y}, \quad (\nabla^2 + h^2)\, w = \frac{1}{\mu}\frac{\partial p}{\partial z}, \quad \text{...(3)}$$

where

$$h^2 = -\frac{\alpha}{\nu}. \quad \text{..................................(4)}$$

From (2) and (3) we deduce

$$\nabla^2 p = 0. \quad \text{..................................(5)}$$

Hence a particular solution of (3) and (2) is

$$u = \frac{1}{h^2\mu}\frac{\partial p}{\partial x}, \quad v = \frac{1}{h^2\mu}\frac{\partial p}{\partial y}, \quad w = \frac{1}{h^2\mu}\frac{\partial p}{\partial z}; \quad \text{...........(6)}$$

and the general solution is

$$u = \frac{1}{h^2\mu}\frac{\partial p}{\partial x} + u', \quad v = \frac{1}{h^2\mu}\frac{\partial p}{\partial y} + v', \quad w = \frac{1}{h^2\mu}\frac{\partial p}{\partial z} + w', \quad \text{......(7)}$$

where u', v', w' are determined by the conditions of the preceding Art.

Hence the solutions in spherical harmonics, subject to the condition of finiteness at the origin, fall into two classes.

In the First Class we have

$$\left.\begin{aligned} p &= \text{const.},\\ u &= \psi_n(hr)\left(y\frac{\partial}{\partial z} - z\frac{\partial}{\partial y}\right)\chi_n,\\ v &= \psi_n(hr)\left(z\frac{\partial}{\partial x} - x\frac{\partial}{\partial z}\right)\chi_n,\\ w &= \psi_n(hr)\left(x\frac{\partial}{\partial y} - y\frac{\partial}{\partial x}\right)\chi_n, \end{aligned}\right\} \quad \text{....................(8)}$$

and therefore

$$xu + yv + zw = 0. \quad \text{............................(9)}$$

* Advantage is here taken of an improvement introduced by Love, "The Free and Forced Vibrations of an Elastic Spherical Shell containing a given Mass of Liquid," *Proc. Lond. Math. Soc.* xix. 170 (1888).

The above investigation is taken, with slight changes of notation, from the following papers: "On the Oscillations of a Viscous Spheroid," *Proc. Lond. Math. Soc.* xiii. 51 (1881); "On the Vibrations of an Elastic Sphere," *Proc. Lond. Math. Soc.* xiii. 189 (1882); "On the Motion of a Viscous Fluid contained in a Spherical Vessel," *Proc. Lond. Math. Soc.* xvi. 27 (1884). The method has since been applied by the author, and by other writers, to a great variety of

In the Second Class we have

$$\left.\begin{aligned} p &= p_n, \\ u &= \frac{1}{h^2\mu}\frac{\partial p_n}{\partial x} + (n+1)\,\psi_{n-1}(hr)\frac{\partial \phi_n}{\partial x} - n\psi_{n+1}(hr)\,h^2 r^{2n+3}\frac{\partial}{\partial x}\frac{\phi_n}{r^{2n+1}}, \\ v &= \frac{1}{h^2\mu}\frac{\partial p_n}{\partial y} + (n+1)\,\psi_{n-1}(hr)\frac{\partial \phi_n}{\partial y} - n\psi_{n+1}(hr)\,h^2 r^{2n+3}\frac{\partial}{\partial y}\frac{\phi_n}{r^{2n+1}}, \\ w &= \frac{1}{h^2\mu}\frac{\partial p_n}{\partial z} + (n+1)\,\psi_{n-1}(hr)\frac{\partial \phi_n}{\partial z} - n\psi_{n+1}(hr)\,h^2 r^{2n+3}\frac{\partial}{\partial z}\frac{\phi_n}{r^{2n+1}}, \end{aligned}\right\} \quad \text{...(10)}$$

and

$$x\xi + y\eta + z\zeta = 0, \quad \text{.............................(11)}$$

where ξ, η, ζ denote the component rotations of the fluid at the point (x, y, z). The symbols χ_n, ϕ_n, p_n stand for solid harmonics of the degrees indicated.

The component tractions on the surface of a sphere of radius r are given as in Art. 336 by

$$\left.\begin{aligned} rp_{rx} &= -xp + \mu\left(r\frac{\partial}{\partial r} - 1\right)u + \mu\frac{\partial}{\partial x}(xu + yv + zw), \\ rp_{ry} &= -yp + \mu\left(r\frac{\partial}{\partial r} - 1\right)v + \mu\frac{\partial}{\partial y}(xu + yv + zw), \\ rp_{rz} &= -zp + \mu\left(r\frac{\partial}{\partial r} - 1\right)w + \mu\frac{\partial}{\partial z}(xu + yv + zw). \end{aligned}\right\} \quad \text{......(12)}$$

In the solutions of the First Class we find without difficulty

$$\left.\begin{aligned} rp_{rx} &= -xp + P_n\left(y\frac{\partial \chi_n}{\partial z} - z\frac{\partial \chi_n}{\partial y}\right), \\ rp_{ry} &= -yp + P_n\left(z\frac{\partial \chi_n}{\partial x} - x\frac{\partial \chi_n}{\partial z}\right), \\ rp_{rz} &= -zp + P_n\left(x\frac{\partial \chi_n}{\partial y} - y\frac{\partial \chi_n}{\partial x}\right), \end{aligned}\right\} \quad \text{.................(13)}$$

where

$$P_n = \mu\{hr\psi_n'(hr) + (n-1)\,\psi_n(hr)\}. \quad \text{..............(14)}$$

To obtain the corresponding formulae for the solutions of the Second Class, we remark first that the terms in p_n give

$$-xp_n + \frac{1}{h^2}\left(r\frac{\partial}{\partial r} - 1\right)\frac{\partial p_n}{\partial x} + \frac{n}{h^2}\frac{\partial p_n}{\partial x}$$
$$= \left\{\frac{2(n-1)}{h^2} - \frac{r^2}{2n+1}\right\}\frac{\partial p_n}{\partial x} + \frac{r^{2n+3}}{2n+1}\frac{\partial}{\partial x}\frac{p_n}{r^{2n+1}}. \quad \text{......(15)}$$

physical problems. It was long overlooked that substantially the same analysis had been given by Clebsch in the paper "Ueber die Reflexion an einer Kugelfläche," to which reference has already been made on pp. 110, 512. The fact that Clebsch failed (confessedly) in the primary object of his investigation, which was to treat a problem of Physical Optics independently of the assumptions of the 'geometrical' theory, perhaps contributed to the unjust neglect into which his paper had fallen. The analytical difficulties which he found insuperable, when the wave-length is small compared with the circumference of the sphere, are identical with those alluded to on p. 521 *ante*.

The remaining terms involve

$$\left(r\frac{\partial}{\partial r}-1\right)u' = (n+1)\{hr\psi'_{n-1}(hr)+(n-2)\psi_{n-1}(hr)\}\frac{\partial\phi_n}{\partial x}$$

$$-n\{hr\psi'_{n+1}(hr)+n\psi_{n+1}(hr)\}h^2r^{2n+3}\frac{\partial}{\partial x}\frac{\phi_n}{r^{2n+1}}, \quad \ldots(16)$$

and

$$\frac{\partial}{\partial x}(xu'+yv'+zw') = n(n+1)(2n+1)\frac{\partial}{\partial x}\psi_n(hr)\phi_n$$

$$= n(n+1)\left\{\psi_{n-1}(hr)\frac{\partial\phi_n}{\partial x}+\psi_{n+1}(hr)h^2r^{2n+3}\frac{\partial}{\partial x}\frac{\phi_n}{r^{2n+1}}\right\}. \quad \ldots\ldots\ldots(17)$$

Various reductions have here been effected by means of Art. 352 (6), (7), (14). Hence, and by symmetry, we obtain

$$\left.\begin{aligned}
rp_{rx} &= A_n\frac{\partial p_n}{\partial x}+B_nr^{2n+1}\frac{\partial}{\partial x}\frac{p_n}{r^{2n+1}}+C_n\frac{\partial\phi_n}{\partial x}+D_nr^{2n+1}\frac{\partial}{\partial x}\frac{\phi_n}{r^{2n+1}},\\
rp_{ry} &= A_n\frac{\partial p_n}{\partial y}+B_nr^{2n+1}\frac{\partial}{\partial y}\frac{p_n}{r^{2n+1}}+C_n\frac{\partial\phi_n}{\partial y}+D_nr^{2n+1}\frac{\partial}{\partial y}\frac{\phi_n}{r^{2n+1}},\\
rp_{rz} &= A_n\frac{\partial p_n}{\partial z}+B_nr^{2n+1}\frac{\partial}{\partial z}\frac{p_n}{r^{2n+1}}+C_n\frac{\partial\phi_n}{\partial z}+D_nr^{2n+1}\frac{\partial}{\partial z}\frac{\phi_n}{r^{2n+1}},
\end{aligned}\right\}\ldots(18)$$

where

$$\left.\begin{aligned}
A_n &= \frac{2(n-1)}{h^2}-\frac{r^2}{2n+1},\\
B_n &= \frac{r^2}{2n+1},\\
C_n &= \mu(n+1)\{hr\psi'_{n-1}(hr)+2(n-1)\psi_{n-1}(hr)\},\\
D_n &= -\mu nh^2r^2\{hr\psi'_{n+1}(hr)-\psi_{n+1}(hr)\}.
\end{aligned}\right\}\ldots\ldots\ldots(19)$$

354. The general formulae being once established, the application to special problems is easy.

1°. We may first investigate the decay of the motion of a viscous fluid contained in a spherical vessel which is at rest.

The boundary-conditions are

$$u=0, \qquad v=0, \qquad w=0, \quad \ldots\ldots(1)$$

for $r=a$, the radius of the vessel. In the modes of the First Class, represented by Art. 353 (8), these conditions are satisfied by

$$\psi_n(ha)=0. \quad \ldots\ldots(2)$$

The roots of this are all real, and the corresponding values of the modulus of decay (τ) are given by

$$\tau = -\frac{1}{\alpha} = \frac{a^2}{\nu}(ha)^{-2}. \quad \ldots\ldots(3)$$

The modes $n=1$ are of a rotatory character. The equation (2) then takes the form

$$\tan ha = ha, \quad \ldots\ldots(4)$$

the lowest root of which is $ha=4{\cdot}493$. Hence

$$\tau = {\cdot}0495a^2/\nu.$$

In the case of water, we have $\nu = \cdot 018$ C.S., and

$$\tau = 2\cdot 75 a^2 \text{ seconds},$$

if a be expressed in centimetres.

The modes of the Second Class are given by Art. 353 (10). The surface-conditions may be expressed by saying that the following three functions of x, y, z, viz.

$$\left.\begin{aligned} \mathbf{u} &= \frac{1}{h^2\mu}\frac{\partial p_n}{\partial x} + (n+1)\,\psi_{n-1}(ha)\frac{\partial \phi_n}{\partial x} - n\psi_{n+1}(ha)\,h^2 r^{2n+3}\frac{\partial}{\partial x}\frac{\phi_n}{r^{2n+1}}, \\ \mathbf{v} &= \frac{1}{h^2\mu}\frac{\partial p_n}{\partial y} + (n+1)\,\psi_{n-1}(ha)\frac{\partial \phi_n}{\partial y} - n\psi_{n+1}(ha)\,h^2 r^{2n+3}\frac{\partial}{\partial y}\frac{\phi_n}{r^{2n+1}}, \\ \mathbf{w} &= \frac{1}{h^2\mu}\frac{\partial p_n}{\partial z} + (n+1)\,\psi_{n-1}(ha)\frac{\partial \phi_n}{\partial z} - n\psi_{n+1}(ha)\,h^2 r^{2n+3}\frac{\partial}{\partial z}\frac{\phi_n}{r^{2n+1}}, \end{aligned}\right\} \quad \text{.........(5)}$$

must severally vanish when $r = a$. Now these functions, as they stand, are sums of solid harmonics, and so satisfy the equations

$$\nabla^2\mathbf{u} = 0, \qquad \nabla^2\mathbf{v} = 0, \qquad \nabla^2\mathbf{w} = 0; \quad \text{.........(6)}$$

and since they are finite throughout the sphere, and vanish at the boundary, they must everywhere vanish, by Art. 40. Hence, forming the equation

$$\frac{\partial \mathbf{u}}{\partial x} + \frac{\partial \mathbf{v}}{\partial y} + \frac{\partial \mathbf{w}}{\partial z} = 0, \quad \text{.........(7)}$$

we find

$$\psi_{n+1}(ha) = 0. \quad \text{.........(8)}$$

Again, since

$$x\mathbf{u} + y\mathbf{v} + z\mathbf{w} = 0, \quad \text{.........(9)}$$

for $r = a$, we find

$$\frac{1}{h^2\mu}p_n + (n+1)(2n+1)\,\psi_n(ha)\,\phi_n = 0, \quad \text{.........(10)}$$

where use has been made of Art. 352 (6), (7). This determines the ratio $p_n : \phi_n$*.

In the case $n = 1$, the equation (8) becomes

$$\tan ha = \frac{3ha}{3 - h^2a^2}, \quad \text{.........(11)}$$

the lowest root of which is $ha = 5\cdot 764$, leading to

$$\tau = \cdot 0301\, a^2/\nu.$$

For the method of combining the various solutions so as to represent the decay of any arbitrary initial motion we must refer to a paper by the author cited on p. 635.

2°. We take next the case of a hollow spherical shell containing liquid, and oscillating about the vertical diameter†.

The forced oscillations of the liquid will evidently be of the First Class, with $n = 1$. If the axis of z coincide with the vertical diameter of the shell, we find, putting $\chi_1 = Cz$ in Art. 353 (8),

$$u = C\psi_1(hr)\,y, \qquad v = -C\psi_1(hr)\,x, \qquad w = 0. \quad \text{.........(12)}$$

If ω denote the angular velocity of the shell, the surface-condition gives

$$C\psi_1(ha) = -\omega. \quad \text{.........(13)}$$

It appears that at any instant the particles situate on a spherical surface of radius r concentric with the boundary are rotating together with an angular velocity

$$\frac{\psi_1(hr)}{\psi_1(ha)}\,\omega. \quad \text{.........(14)}$$

* Another method of applying the surface-conditions is indicated in Art. 361.

† This was first treated, in a different manner, by Helmholtz, *l.c. ante* p. 575.

If we assume that
$$\omega=\alpha e^{i(\sigma t+\epsilon)}, \qquad\text{.......(15)}$$
and put
$$h^2=-i\sigma/\nu=(1-i)^2\beta^2, \qquad\text{.......(16)}$$
where, as in Art. 345,
$$\beta^2=\sigma/2\nu, \qquad\text{.......(17)}$$
the expression (14) for the angular velocity may be separated into its real and imaginary arts with the help of the formula
$$\psi_1(\zeta)=\frac{\sin\zeta}{\zeta^3}-\frac{\cos\zeta}{\zeta^2}. \qquad\text{.......(18)}$$

If the viscosity be so small that βa is considerable, then, keeping only the most important term, we have, for points near the surface,
$$\psi_1(hr)=-\frac{1}{2h^2r^2}e^{(1+i)\beta r}, \qquad\text{.......(19)}$$
and therefore, for the angular velocity (14),
$$\propto\frac{a^2}{r^2}e^{-\beta(a-r)}\,.\,e^{i\{\sigma t+\beta(r-a)+\epsilon\}}, \qquad\text{.......(20)}$$
the real part of which is
$$\propto\frac{a^2}{r^2}e^{-\beta(a-r)}\,.\,\cos\{\sigma t+\beta(r-a)+\epsilon\}. \qquad\text{.......(21)}$$
As in the case of laminar motion (Art. 345), this represents a system of waves travelling inwards from the surface with rapidly diminishing amplitude.

When, on the other hand, the viscosity is very great, βa is small, and the formula (14) reduces to
$$\omega\cos(\sigma t+\epsilon), \qquad\text{.......(22)}$$
nearly, when the imaginary part is rejected. This shews that the fluid now moves almost bodily with the sphere.

The stress-components at the surface of the sphere are given by Art. 353 (13). In the present case the formulae reduce to
$$p_{rx}=-\frac{x}{a}p+\mu Ch\psi_1'(ha)\,y,\qquad p_{ry}=-\frac{y}{a}p-\mu Ch\psi_1'(ha)\,x,\qquad p_{rz}=-\frac{z}{a}p. \quad\text{...(23)}$$
If δS denote an element of the surface, these give a couple
$$N=-\iint(xp_{ry}-yp_{rx})\,dS=C\mu h\psi_1'(ha)\iint(x^2+y^2)\,dS=\tfrac{8}{3}\pi\mu a^3\frac{h^2a^2\psi_2(ha)}{\psi_1(ha)}\,\omega, \quad\text{...(24)}$$
by (13) and Art. 352 (6).

In the case of small viscosity, where βa is large, we find, on reference to Art. 292 (8), putting $ha=(1-i)\beta a$, that
$$2i\psi_n(ha)=\left(-\frac{d}{\zeta d\zeta}\right)^n\frac{e^{i\zeta}}{\zeta}, \qquad\text{.......(25)}$$
approximately, where $\zeta=(1-i)\beta a$. This leads to
$$N=-\tfrac{8}{3}\pi\mu a^3(1+i)\beta a\omega. \qquad\text{.......(26)}$$
If we take account of the time-factor in (15), this is equivalent to
$$N=-\tfrac{4}{3}\pi\rho a^5(\beta a)^{-1}\frac{d\omega}{dt}-\tfrac{8}{3}\pi\mu a^3(\beta a)\,\omega. \qquad\text{.......(27)}$$
The first term has the effect of a slight addition to the inertia of the sphere; the second gives a frictional force varying as the velocity.

355. The general formulae of Arts. 352, 353 may be further applied to discuss the effect of viscosity on the small oscillations of a mass of liquid about the spherical form. The principal result of the investigation can, however, be obtained more simply by the method of Art. 348.

It appears from Art. 262 that when viscosity is neglected, the velocity-potential in any fundamental mode is of the form

$$\phi = A\,\frac{r^n}{a^n}\,S_n\,.\,\cos(\sigma t+\epsilon), \qquad \text{.......(1)}$$

where S_n is a surface-harmonic. This gives for twice the kinetic energy included within a sphere of radius r the expression

$$\rho\iint\phi\,\frac{\partial\phi}{\partial r}\,r^2\,d\varpi = \rho n a\left(\frac{r}{a}\right)^{2n+1}\iint S_n^2\,d\varpi\,.\,A^2\cos^2(\sigma t+\epsilon), \qquad \text{.......(2)}$$

if $\delta\varpi$ denote an elementary solid angle, and therefore for the total kinetic energy

$$T = \tfrac{1}{2}\rho n a\iint S_n^2\,d\varpi\,.\,A^2\cos^2(\sigma t+\epsilon). \qquad \text{.......(3)}$$

The potential energy must therefore be

$$V = \tfrac{1}{2}\rho n a\iint S_n^2\,d\varpi\;\; A^2\sin^2(\sigma t+\epsilon), \qquad \text{.......(4)}$$

and the total energy is

$$T+V = \tfrac{1}{2}\rho n a\iint S_n^2\,d\varpi\,.\,A^2. \qquad \text{.......(5)}$$

Again, the dissipation in a sphere of radius r, calculated on the assumption that the motion is irrotational, is, by Art. 329 (12),

$$\mu\iint\frac{\partial\,.\,q^2}{\partial r}\,r^2\,d\varpi = \mu r^2\,\frac{\partial}{\partial r}\iint q^2\,d\varpi. \qquad \text{.......(6)}$$

Now

$$r^2\iint q^2\,d\varpi = \frac{\partial}{\partial r}\iint\phi\,\frac{\partial\phi}{\partial r}\,r^2\,d\varpi, \qquad \text{.......(7)}$$

each side, when multiplied by $\rho\delta r$, being double the kinetic energy of the fluid contained between two spherical surfaces of radii r and $r+\delta r$. Hence, from (2),

$$\iint q^2\,d\varpi = \frac{n(2n+1)}{a^2}\left(\frac{r}{a}\right)^{2n-2}\iint S_n^2\,d\varpi\,.\,A^2\cos^2(\sigma t+\epsilon).$$

Substituting in (6), and putting $r=a$, we have, for the total dissipation,

$$2F = 2n(n-1)(2n+1)\,\frac{\mu}{a}\iint S_n^2\,d\varpi\,.\,A^2\cos^2(\sigma t+\epsilon), \qquad \text{.......(8)}$$

the mean value of which, per unit time, is

$$2\overline{F} = n(n-1)(2n+1)\,\frac{\mu}{a}\iint S_n^2\,d\varpi\,.\,A^2. \qquad \text{.......(9)}$$

If the effect of viscosity be represented by a gradual variation of the coefficient A, we must have

$$\frac{d}{dt}(T+V) = -2\overline{F}, \qquad \text{.......(10)}$$

whence, substituting from (5) and (9),

$$\frac{dA}{dt} = -(n-1)(2n+1)\,\frac{\nu}{a^2}\,A. \qquad \text{.......(11)}$$

This shews that $A \propto e^{-t/\tau}$, where

$$\tau = \frac{1}{(n-1)(2n+1)}\,\frac{a^2}{\nu}\,. \qquad \text{.......(12)*}$$

The most remarkable feature of this result is the excessively minute extent to which the oscillations of a globe of moderate dimensions are affected by such a degree of viscosity as is ordinarily met with in nature. For a globe of the size of the earth, and of the same kinematic viscosity as water, we have, on the C. G. S. system, $a = 6{\cdot}37\times 10^8$, $\nu = {\cdot}0178$, and the value of τ for the gravitational oscillation of longest period ($n=2$) is therefore

$$\tau = 1{\cdot}44\times 10^{11} \text{ years}.$$

* *Proc. Lond. Math. Soc.* (1) xiii. 61, 65 (1881).

Even with the value found by Darwin* for the viscosity of pitch near the freezing temperature, viz. $\mu = 1{\cdot}3 \times 10^8 \times g$, we find, taking $g = 980$, the value

$$\tau = 180 \text{ hours},$$

for the modulus of decay of the slowest oscillation of a globe of the size of the earth, having the density of water and the viscosity of pitch. Since this is still large compared with the period of 1 h. 34 m. found in Art. 262, it appears that such a globe would oscillate almost like a perfect fluid.

The above investigation does not involve any special assumption as to the nature of the forces which produce the tendency to the spherical form. The result applies, therefore, equally well to the vibrations of a liquid globule under the surface-tension of the bounding film. The modulus of decay of the slowest oscillation of a globule of water is, in seconds, $\tau = 11{\cdot}2a^2$, where the unit of a is the centimetre.

The same method, applied to the case of a spherical bubble, gives

$$\tau = \frac{1}{(n+2)(2n+1)} \frac{a^2}{\nu}, \quad \text{.................................}(13)$$

where ν is the viscosity of the surrounding liquid. If this be water we have, for $n = 2$, $\tau = 2{\cdot}8a^2$.

The formula (12) includes of course the case of waves on a plane surface. When n is very great we find, putting $\lambda = 2\pi a/n$,

$$\tau = \lambda^2/8\pi^2\nu, \quad \text{..}(14)$$

in agreement with Art. 348.

The above results all postulate that $2\pi\tau$ is a considerable multiple of the period. The opposite extreme, where the viscosity is so great that the motion is aperiodic, can be investigated by the method of Arts. 335, 336, the effects of inertia being disregarded. In the case of a highly viscous globe returning asymptotically to the spherical form under the influence of gravitation, it appears that

$$\tau = \frac{2(n+1)^2 + 1}{n} \frac{\nu}{ga}, \quad \text{....................................}(15)$$

a result first given by Darwin (*l.c.*). For a system of equal parallel corrugations on a plane surface we deduce

$$\tau = 4\pi\nu/g\lambda. \quad \text{...}(16)$$

Cf. Art. 349 (24).

356. Problems of periodic motion of a liquid in the space between two concentric spheres require for their treatment additional solutions of the equations of Art. 353, in which p is of the form p_{-n-1}, and the functions $\psi_n(hr)$ which occur in the complementary functions u', v', w' are to be replaced by $\Psi_n(hr)$.

The question is simplified, when the radius of the second sphere is infinite, by the condition that the fluid is at rest at infinity. It was shewn in Art. 292 that the functions $\psi_n(\zeta)$, $\Psi_n(\zeta)$ are both included in the form

$$\left(\frac{d}{\zeta d\zeta}\right)^n \frac{Ae^{i\zeta} + Be^{-i\zeta}}{\zeta} \quad \text{..........................}(1)$$

In the present applications, we have $\zeta = hr$, where h is defined by Art. 353 (4), and we will suppose, for definiteness, that that value of h is

* "On the Bodily Tides of Viscous and Semi-Elastic Spheroids, ...," *Phil. Trans.* clxx. 1 (1878).

adopted which makes the real part of ih positive. The condition of zero motion at infinity then requires that $A=0$, and we have to deal only with the function

$$f_n(\zeta)=\left(-\frac{d}{\zeta d\zeta}\right)^n\frac{e^{-i\zeta}}{\zeta}, \qquad\text{.........................(2)}$$

introduced in Art. 292. It was there pointed out that the formulae of reduction for $f_n(\zeta)$ are exactly the same as for $\psi_n(\zeta)$ and $\Psi_n(\zeta)$; and the general solution of the equations of small periodic motion of a viscous liquid, for the space external to a sphere, is therefore given at once by Art. 353 (8), (10), with p_{-n-1} written for p_n, and $f_n(hr)$ for $\psi_n(hr)$.

1°. The rotatory oscillations of a sphere surrounded by an infinite mass of liquid are included in the solutions of the First Class, with $n=1$. As in Art. 354, 2°, we put $\chi_1=Cz$, and find

$$u=Cf_1(hr)\,y, \qquad v=-Cf_1(hr)\,x, \qquad w=0, \qquad\text{.......................(3)}$$

with the condition

$$Cf_1(ha)=-\omega, \qquad\text{..(4)}$$

a being the radius, and ω the angular velocity of the sphere, which we suppose given by the formula

$$\omega=\alpha e^{i(\sigma t+\epsilon)}. \qquad\text{..(5)}$$

Putting $h=(1-i)\,\beta$, where $\beta=(\sigma/2\nu)^{\frac{1}{2}}$, we find that the particles on a concentric sphere of radius r are rotating together with the angular velocity

$$\frac{f_1(hr)}{f_1(ha)}\,\omega=\frac{\alpha a^3}{r^3}\,\frac{1+ihr}{1+iha}\,e^{-\beta(r-a)}\,.\,e^{i\{\sigma t-\beta(r-a)+\epsilon\}}, \qquad\text{.......................(6)}$$

where the values of $f_1(hr)$, $f_1(ha)$ have been substituted from Art. 292 (15). The real part of (6) is

$$\frac{\alpha}{1+2\beta a+2\beta^2a^2}\,\frac{a^3}{r^3}\,e^{-\beta(r-a)}\left[\{1+\beta(a+r)+2\beta^2ar\}\cos\{\sigma t-\beta(r-a)+\epsilon\}\right.$$
$$\left.-\beta(r-a)\sin\{\sigma t-\beta(r-a)+\epsilon\}\right], \qquad\text{..............(7)}$$

corresponding to an angular velocity

$$\omega=\alpha\cos(\sigma t+\epsilon) \qquad\text{..(8)}$$

of the sphere.

The couple on the sphere is found in the same way as in Art. 354 to be

$$N=-\tfrac{8}{3}\pi\mu a^3\omega\,\frac{h^2a^2f_2(ha)}{f_1(ha)}=-\tfrac{8}{3}\pi\mu a^3\omega\,\frac{3+3iha-h^2a^2}{1+iha}. \qquad\text{..................(9)}$$

Putting $ha=(1-i)\,\beta a$, and separating the real and imaginary parts, we find

$$N=-\tfrac{8}{3}\pi\mu a^3\omega\,\frac{(3+6\beta a+6\beta^2a^2+2\beta^3a^3)+2i\beta^2a^2(1+\beta a)}{1+2\beta a+2\beta^2a^2}. \qquad\text{...............(10)}$$

This is equivalent to

$$N=-\tfrac{8}{3}\pi\rho a^5\,\frac{1+\beta a}{1+2\beta a+2\beta^2a^2}\,\frac{d\omega}{dt}-\tfrac{8}{3}\pi\mu a^3\,\frac{3+6\beta a+6\beta^2a^2+2\beta^3a^3}{1+2\beta a+2\beta^2a^2}\,\omega. \qquad\text{.........(11)}$$

The interpretation is similar to that of Art. 354 (27)*.

When the period $(2\pi/\sigma)$ is infinitely long, this reduces to

$$N=-8\pi\mu a^3\omega, \qquad\text{..(12)}$$

in agreement with Art. 334 (11).

2°. In the case of a ball pendulum oscillating in an infinite mass of fluid, which we treat as incompressible, we take the origin at the mean position of the centre, and the axis of x in the direction of the oscillation.

* Another treatment of this problem is given by Kirchhoff, *Mechanik*, c. xxvi.

The conditions to be satisfied at the surface are then

$$u=U, \qquad v=0, \qquad w=0, \qquad \text{.................................(13)}$$

for $r=a$ (the radius), where U denotes the velocity of the sphere. It is evident that we are concerned only with a solution of the Second Class; and the formulae (10) of Art. 353, with the substitution of the functions f_n for ψ_n, make

$$xu+yv+zw=-\frac{n+1}{h^2\mu}p_{-n-1}+n(n+1)(2n+1)f_n(hr)\phi_n. \qquad \text{............(14)}$$

By comparison with (13), it appears that this must involve surface-harmonics of the *first order* only. We therefore put $n=1$, and assume

$$p_{-2}=Ax/r^3, \qquad \phi_1=Bx. \qquad \text{....................................(15)}$$

Hence

$$\left.\begin{aligned} u&=\frac{A}{h^2\mu}\frac{\partial}{\partial x}\frac{x}{r^3}+2Bf_0(hr)-Bf_2(hr)\,h^2r^5\frac{\partial}{\partial x}\frac{x}{r^3},\\ v&=\frac{A}{h^2\mu}\frac{\partial}{\partial y}\frac{x}{r^3}\qquad\qquad\qquad -Bf_2(hr)\,h^2r^5\frac{\partial}{\partial y}\frac{x}{r^3},\\ w&=\frac{A}{h^2\mu}\frac{\partial}{\partial z}\frac{x}{r^3}\qquad\qquad\qquad -Bf_2(hr)\,h^2r^5\frac{\partial}{\partial z}\frac{x}{r^3}.\end{aligned}\right\} \qquad \text{......................(16)}$$

The conditions (13) are therefore satisfied if

$$A=\mu h^4a^5f_2(ha)\,B, \qquad 2f_0(ha)\,B=U. \qquad \text{...........................(17)}$$

The character of the motion, which is evidently symmetrical about the axis of x, can be most concisely expressed by means of the stream-function. From (14) or (16) we find

$$xu+yv+zw=-\frac{2A}{h^2\mu}\frac{x}{r^3}+6Bf_1(hr)\,x=-\frac{Ux}{f_0(ha)}\left\{\frac{h^2a^5}{r^3}f_2(ha)-3f_1(hr)\right\}, \text{......(18)}$$

or, substituting from Art. 292 (15),

$$xu+yv+zw=\left\{\left(1-\frac{3i}{ha}-\frac{3}{h^2a^2}\right)\frac{a^3}{r^3}+3\left(\frac{i}{hr}+\frac{1}{h^2r^2}\right)\frac{a}{r}e^{-ih(r-a)}\right\}Ux. \qquad \text{......(19)}$$

If we put $x=r\cos\theta$, this leads, in terms of the stream-function ψ of Art. 94, to

$$\psi=-\tfrac{1}{2}Ua^2\sin^2\theta\left\{\left(1-\frac{3i}{ha}-\frac{3}{h^2a^2}\right)\frac{a}{r}+\frac{3}{ha}\left(i+\frac{1}{hr}\right)e^{-ih(r-a)}\right\}. \qquad \text{.........(20)}$$

Writing

$$U=\alpha e^{i(\sigma t+\epsilon)}, \qquad \text{..(21)}$$

and therefore $h=(1-i)\beta$, where $\beta=(\sigma/2\nu)^{\frac{1}{2}}$, we find, on rejecting the imaginary part of (20),

$$\begin{aligned}\psi=-\tfrac{1}{2}\alpha a^2\sin^2\theta\Big[&\left\{\left(1+\frac{3}{2\beta a}\right)\cos(\sigma t+\epsilon)+\frac{3}{2\beta a}\left(1+\frac{1}{\beta a}\right)\sin(\sigma t+\epsilon)\right\}\frac{a}{r}\\ &-\frac{3}{2\beta a}\left\{\cos\{\sigma t-\beta(r-a)+\epsilon\}+\left(1+\frac{1}{\beta r}\right)\sin\{\sigma t-\beta(r-a)+\epsilon\}\right\}\epsilon^{-\beta(r-a)}\Big]\end{aligned} \text{....(22)}$$

At a sufficient distance from the sphere, the part of the disturbance which is expressed by the terms in the first line of this expression is predominant. This part is irrotational, and differs only in amplitude and phase from the motion produced by a sphere oscillating in a frictionless liquid (Arts. 92, 96). The terms in the second line are of the type we have already met with in the case of laminar motion (Art. 345).

To calculate the resultant force (X) on the sphere, we have recourse to Art. 353 (18). Substituting from (15), and rejecting all but the constant terms in p_{rx}, since the surface-harmonics of other than zero order will disappear when integrated over the sphere, we find

$$X=\iint p_{rx}\,dS=4\pi\left(B_{-2}\frac{A}{a}+C_1Ba^2\right), \qquad \text{..........................(23)}$$

where

$$B_{-2}=-\tfrac{1}{3}a^2, \qquad C_1=2\mu haf_0'(ha), \qquad \text{..........................(24)}$$

by Art. 353 (19). Hence, by (17),

$$X=\frac{2\pi\mu Uha^2}{f_0(ha)}\{2f_0'(ha)-\tfrac{1}{3}h^3a^3f_2(ha)\}=2\pi\mu Uh^2a^3\left(\tfrac{1}{3}-\frac{3i}{ha}-\frac{3}{h^2a^2}\right)$$

$$=-2\pi\rho a^3\sigma U\left\{\left(\tfrac{1}{3}+\frac{3}{2\beta a}\right)i+\frac{3}{2\beta a}\left(1+\frac{1}{\beta a}\right)\right\}. \qquad (25)$$

This is equivalent to

$$X=-\tfrac{4}{3}\pi\rho a^3\left(\tfrac{1}{2}+\frac{9}{4\beta a}\right)\frac{dU}{dt}-3\pi\rho a^3\sigma\left(\frac{1}{\beta a}+\frac{1}{\beta^2a^2}\right)U. \qquad (26)$$

The first term gives the correction to the inertia of the sphere. This amounts to the fraction

$$\frac{1}{2}+\frac{9}{4\beta a}$$

of the mass of fluid displaced, instead of $\frac{1}{2}$ as in the case of a frictionless liquid (Art. 92). The second term gives a frictional force varying as the velocity*.

When the period $2\pi/\sigma$ is made infinitely long, the formula (26) reduces to

$$X=-6\pi\rho\nu aU, \qquad (27)$$

in agreement with Art. 337 (15), since $\beta^2=\sigma/2\nu$.

357. A few notes may be appended on the two-dimensional problems which are analogous to those of Arts. 354, 356.

Terms of the second order being neglected, the equations are

$$\frac{\partial u}{\partial t}=-\frac{1}{\rho}\frac{\partial p}{\partial x}+\nu\nabla_1^2u, \qquad \frac{\partial v}{\partial t}=-\frac{1}{\rho}\frac{\partial p}{\partial y}+\nu\nabla_1^2v, \qquad (1)$$

with

$$\frac{\partial u}{\partial x}+\frac{\partial v}{\partial y}=0.$$

As in Art. 349, these are satisfied by

$$u=-\frac{\partial\phi}{\partial x}-\frac{\partial\psi}{\partial y}, \qquad v=-\frac{\partial\phi}{\partial y}+\frac{\partial\psi}{\partial x}, \qquad (2)$$

and

$$p=\rho\frac{\partial\phi}{\partial t}, \qquad (3)$$

provided

$$\nabla_1^2\phi=0, \qquad \frac{\partial\psi}{\partial t}=\nu\nabla_1^2\psi. \qquad (4)$$

1°. It will be found that the modes of decay of an arbitrary initial motion of a liquid enclosed in a fixed circular cylinder are given in polar co-ordinates by

$$\Psi=\left\{\frac{J_s(kr)}{J_s(ka)}-\frac{r^s}{a^s}\right\}(A\cos s\theta+B\sin s\theta)\,e^{-\nu k^2t}, \qquad (5)$$

* This problem was first solved, in a different manner, by Stokes, *l.c. ante* p. 580. For other methods of treatment see O. E. Meyer, "Ueber die pendelnde Bewegung einer Kugel unter dem Einflusse der inneren Reibung des umgebenden Mediums," *Crelle*, lxxiii. (1871); Kirchhoff, *Mechanik*, xxvi.

The more general case where the velocity of the sphere is an arbitrary function of the time has been discussed by Basset, "On the Motion of a Sphere in a Viscous Liquid," *Phil. Trans.* clxxix. 43 (1887); *Hydrodynamics*, c. xxii. The question has been simplified in recent papers by Picciati and Boggio; see Basset, *Quart. J. of Math.* xli. 369 (1910), and Rayleigh, *l.c. ante* p. 591. See also Havelock, *Phil. Mag.* (6) xlii. 628 (1921).

when Ψ now stands for the stream-function of Art. 59. The condition of zero normal motion at the boundary ($r=a$) is already satisfied; and the tangential velocity $\partial\Psi/\partial r$ will also vanish there, provided

$$kaJ_s'(ka) - sJ_s(ka) = 0,$$

which is equivalent, by Art. 303 (5), to

$$J_{s+1}(ka) = 0. \qquad \text{.........(6)}$$

This determines the admissible values of k, and thence the values of the modulus of decay ($\tau = 1/\nu k^2$)*.

In the case of symmetry we have $s=0$. The lowest root of $J_1(ka)=0$ is $ka=3{\cdot}832$, which gives

$$\tau = {\cdot}0681 a^2/\nu.$$

If we put, for water, $\nu = {\cdot}014$ C.G.S., we find $\tau = 4{\cdot}9a^2$ seconds, provided a be expressed in centimetres.

For $s=1$, the lowest root is $ka=5{\cdot}135$, whence

$$\tau = {\cdot}0379 a^2/\nu,$$

or, for water, $\tau = 2{\cdot}7a^2$.

2°. In the case of periodic motion, with a time-factor $e^{i\sigma t}$, we have, from (4),

$$(\nabla_1^2 + h^2)\psi = 0, \qquad \text{.........(7)}$$

provided $h^2 = -i\sigma/\nu$, or (say)

$$h = (1-i)\beta, \qquad \beta = (\sigma/2\nu)^{\frac{1}{2}}. \qquad \text{.........(8)}$$

The solution of (7) in polar co-ordinates involves Bessel's Functions with the complex argument $(1-i)\beta r$. The selection of suitable functions for the various cases, and the working out of results in a practical form, involve some points of delicacy†. In view of the length of the necessary investigations, and of the fact that the problems in question are inferior in interest to those which relate to a spherical boundary, we content ourselves with a reference to the original papers by Stokes‡.

Viscosity in Gases.

358. When variations of density have to be taken into account, the most general supposition we can make with regard to the 'mean pressure' p, consistently with our previous assumptions, is, in the case of a 'perfect' gas,

$$p = R\rho\theta - \mu'(a+b+c), \qquad \text{.........(1)}$$

where θ is the absolute temperature, R is a constant depending on the nature of the gas, and μ' is a second coefficient of viscosity§. There does not appear to be any experimental evidence as to the precise value to be attributed to μ'; but according to the kinetic theory of gases $\mu' = 0$||, and we shall for simplicity adopt this hypothesis. If it is desired to retain μ' in the formulae, the necessary corrections can easily be made.

* This result is from the paper "On the Motion of a Viscous Fluid contained in a Spherical Vessel," cited on p. 635. The case of $s=0$ was discussed by Stearn, "On some Cases of the Varying Motion of a Viscous Fluid," *Quart. Journ. Math.* xvii. 90 (1880).

† The investigations of Art. 194 require revision when the argument is *complex*. The formulae (4), (5), (6) are valid, provided the real part of the argument be positive (as is secured by the choice of h in (8) above); but the derivation of the descending and ascending series (13) and (20) presents new points. Incidentally, the results obtained by equating separately real and imaginary parts would call for examination.

‡ *l.c. ante* p. 580. See also Watson, *Theory of Bessel Functions*, p. 201.

§ Cf. Kirchhoff, *Vorlesungen über die Theorie der Wärme.* Leipzig, 1894, c. xi.; Stokes, *Papers*, iii. 136.

|| Maxwell, *l.c. ante* p. 575.

It was shewn in Art. 329 that the work done in time δt by the tractions on the faces of an element $\delta x \delta y \delta z$, in changing the volume and shape of the element, is

$$-p(a+b+c)\,\delta x \delta y \delta z\,.\,\delta t + \Phi\,\delta x \delta y \delta z\,.\,\delta t, \quad \text{..............(2)}$$

where $\Phi = -\tfrac{2}{3}\mu(a+b+c)^2 + \mu(2a^2+2b^2+2c^2+f^2+g^2+h^2)$. ......(3)

Now, by Art. 7 (3),

$$a+b+c = -\frac{1}{\rho}\frac{D\rho}{Dt} = \rho\frac{D\mathbf{v}}{Dt}, \quad \text{........................(4)}$$

where $\mathbf{v}$ denotes the volume of unit mass. Hence if E is the intrinsic energy per unit mass, and DQ/Dt the rate per unit volume at which a fluid element is receiving heat by conduction from adjacent elements, or by radiation, we have the equation of energy of unit volume:

$$\frac{DE}{Dt}\rho = -p\frac{D\mathbf{v}}{Dt}\rho + \Phi + \frac{DQ}{Dt}. \quad \text{....................(5)}$$

The rate at which heat must actually be absorbed in order to effect the changes of density and temperature is, on thermodynamic principles,

$$\frac{DQ'}{Dt} = p\frac{D\mathbf{v}}{Dt}\rho + \frac{DE}{Dt}\rho. \quad \text{..........................(6)}$$

Comparing, we have

$$\frac{DQ'}{Dt} = \frac{DQ}{Dt} + \Phi. \quad \text{..............................(7)}$$

Hence in addition to the heat gained by conduction, &c., an amount measured by Φ per unit volume and unit time is generated in the element, at the expense (of course) of other forms of energy.

If we write (3) in the form

$$\Phi = \tfrac{2}{3}\mu\{(b-c)^2+(c-a)^2+(a-b)^2\} + \mu(f^2+g^2+h^2), \quad \text{......(8)}$$

it is seen that Φ is essentially positive, and (moreover) that it cannot vanish unless

$$a = b = c \quad \text{and} \quad f = g = h = 0,$$

i.e., unless the distortion of the fluid element consists of an expansion or contraction which is the same in all directions. The conclusion that there is no dissipation of energy in this case rests of course on the assumption that the value of μ' in (1) is zero.

359. We may notice the effect of viscosity on sound-waves. For consistency it is necessary to take account at the same time of heat-conduction, whose influence is of the same order of importance*; but in the first instance we follow Stokes† in examining the effect of viscosity alone.

* This was first remarked by Kirchhoff, "Ueber den Einfluss der Wärmeleitung in einem Gase auf die Schallbewegung," *Pogg. Ann.* cxxxiv. 177 (1868) [*Ges. Abh.* i. 540].

† *l.c. ante* p. 17 [*Papers*, i. 100].

In the case of plane waves in a laterally unlimited medium, we have, if we take the axis of x in the direction of propagation, and neglect terms of the second order in the velocity,

$$\frac{\partial u}{\partial t} = -\frac{1}{\rho_0}\frac{\partial p}{\partial x} + \tfrac{4}{3}\nu\frac{\partial^2 u}{\partial x^2}, \qquad \text{.........................(1)}$$

by Art. 328 (2), (3). If s denote the condensation, the equation of continuity is, as in Art. 277,

$$\frac{\partial s}{\partial t} = -\frac{\partial u}{\partial x}; \qquad \text{.................................(2)}$$

and the physical equation is, if the transfer of heat be neglected,

$$p = p_0 + c^2\rho_0 s, \qquad \text{..............................(3)}$$

where c is the velocity of sound in the absence of viscosity. Eliminating p and s, we have

$$\frac{\partial^2 u}{\partial t^2} = c^2\frac{\partial^2 u}{\partial x^2} + \tfrac{4}{3}\nu\frac{\partial^3 u}{\partial x^2 \partial t}. \qquad \text{.........................(4)}$$

To apply this to the case of forced waves, we may suppose that at the plane $x = 0$ a given vibration

$$u = ae^{i\sigma t} \qquad \text{..................................(5)}$$

is kept up. Assuming as the solution of (4)

$$u = ae^{i\sigma t + mx}, \qquad \text{................................(6)}$$

we find $$m^2(c^2 + \tfrac{4}{3}i\nu\sigma) = -\sigma^2, \qquad \text{.........................(7)}$$

whence $$m = \pm\frac{i\sigma}{c}\left(1 - \tfrac{4}{3}i\cdot\frac{\nu\sigma}{c^2}\right)^{-\frac{1}{2}}. \qquad \text{.......................(8)}$$

If we neglect the square of $\nu\sigma/c^2$, and take the lower sign, this gives

$$m = -\frac{i\sigma}{c} - \tfrac{3}{2}\frac{\nu\sigma^2}{c^3}. \qquad \text{............................(9)}$$

Substituting in (6), and taking the real part, we get, for the waves propagated in the direction of x-positive,

$$u = ae^{-x/l}\cos\sigma\left(t - \frac{x}{c}\right), \qquad \text{.......................(10)}$$

where $$l = \tfrac{3}{2}c^3/\nu\sigma^2. \qquad \text{................................(11)}$$

The amplitude of the waves diminishes exponentially as they proceed, the diminution being more rapid the greater the value of σ. The wave-velocity is, to the first order of $\nu\sigma/c^2$, unaffected by the friction.

The linear magnitude l measures the distance in which the amplitude falls to $1/e$ of its original value. If λ denote the wave-length $(2\pi c/\sigma)$, we have

$$\tfrac{2}{3}\nu\sigma/c^2 = \lambda/2\pi l; \qquad \text{..............................(12)}$$

it is assumed in the above calculation that this is a small ratio.

In the case of air-waves we have $c=3{\cdot}32\times10^4$, $\nu={\cdot}132$, C.G.S., whence

$$\nu\sigma/c^2=2\pi\nu/\lambda c=2{\cdot}50\lambda^{-1}\times10^{-5},\quad l=9{\cdot}56\lambda^2\times10^3,$$

if λ be expressed in centimetres. The effect on the amplitude is very slight except for sounds of very short wave-length.

To find the decay of *free* waves of any prescribed wave-length $(2\pi/k)$, we assume

$$u=ae^{ikx+nt}; \qquad (13)$$

and, substituting in (4), we obtain

$$n^2+\tfrac{4}{3}\nu k^2 a=-k^2c^2. \qquad (14)$$

If we neglect the square of $\nu k/c$, this gives

$$n=-\tfrac{2}{3}\nu k^2\pm ikc. \qquad (15)$$

Hence, in real form,

$$u=ae^{-t/\tau}\cos k\,(x\pm ct), \qquad (16)$$

where

$$\tau=3/2\nu k^2. \qquad (17)$$

360. When conductivity is to be allowed for, the dynamical equation (1), and the equation of continuity (2) are unaffected, but the physical relations must be modified.

The amount of heat required to produce small changes in the volume v and (absolute) temperature θ of unit mass of a gas is

$$\delta Q=p\,\delta v+C_v\,\delta\theta=\left\{(\gamma-1)\frac{\theta_0}{v_0}\,\delta v+\delta\theta\right\}C_v, \qquad (18)$$

where C_v is the specific heat at constant volume. If we multiply by $\rho_0\delta x$, the mass per unit area of a thin stratum, and divide by ∂t we get the rate at which heat must be supplied to the stratum. Equating this to $k\partial^2\theta/\partial x^2\,.\,\delta x$, where k is the thermal conductivity, we find*

$$\frac{\partial\theta}{\partial t}+(\gamma-1)\frac{\theta_0}{v_0}\frac{\partial v}{\partial t}=\nu'\frac{\partial^2\theta}{\partial x^2}, \qquad (19)$$

where

$$\nu'=k/\rho_0 C_v, \qquad (20)$$

i.e. ν' is the 'thermometric' conductivity†.

The relation between p, ρ, θ is

$$\frac{p}{p_0}=\frac{\rho\theta}{\rho_0\theta_0}. \qquad (21)$$

If we put

$$\rho=\rho_0(1+s),\qquad \theta=\theta_0(1+\eta), \qquad (22)$$

* The heat generated by internal friction (as explained in Art. 358) is here neglected, as being of the second order of small quantities.

† Maxwell, *Theory of Heat*, c. xviii. If *radiation* were important, a term proportional to $\theta-\theta_0$ would be introduced in (19). Cf. Stokes, *Phil. Mag.* (5) i. 305 (1851) [*Papers*, iii. 142]; also Rayleigh, *Theory of Sound*, Art. 247.

and neglect terms of the second order in s and η, the equations (19) and (21) may be written

$$\frac{\partial \eta}{\partial t} - (\gamma - 1)\frac{\partial s}{\partial t} = \nu' \frac{\partial^2 \eta}{\partial x^2}, \quad \text{.......................(23)}$$

and

$$p = p_0(1 + s + \eta). \quad \text{..........................(24)}$$

Substituting this value of p in (1), we have

$$\frac{\partial u}{\partial t} = -b^2 \frac{\partial s}{\partial x} - b^2 \frac{\partial \eta}{\partial x} + \tfrac{4}{3}\nu \frac{\partial^2 u}{\partial x^2}, \quad \text{.....................(25)}$$

where $b, = (p_0/\rho_0)^{\frac{1}{2}}$, is the Newtonian velocity of sound (Art. 278). Eliminating s by (2), we find

$$\frac{\partial^2 u}{\partial t^2} = b^2 \frac{\partial^2 u}{\partial x^2} - b^2 \frac{\partial^2 \eta}{\partial x \partial t} + \tfrac{4}{3}\nu \frac{\partial^3 u}{\partial x^2 \partial t}, \quad \text{..................(26)}$$

and

$$\frac{\partial \eta}{\partial t} + (\gamma - 1)\frac{\partial u}{\partial x} = \nu' \frac{\partial^2 \eta}{\partial x^2}, \quad \text{.......................(27)}$$

which are two simultaneous equations to determine u and η.

If we now assume that u and η both vary as

$$e^{i\sigma t + mx},$$

we find

$$\left.\begin{aligned}(\sigma^2 + m^2 b^2 + \tfrac{4}{3} i\nu\sigma m^2)\,u - i\sigma m b^2 \eta = 0,\\ (\gamma - 1)\,mu + (i\sigma - \nu' m^2)\,\eta = 0,\end{aligned}\right\} \quad \text{..............(28)}$$

whence

$$\sigma^3 + \{c^2\sigma + (\tfrac{4}{3}\nu + \nu')\,i\sigma^2\}\,m^2 + \nu'(ib^2 - \tfrac{4}{3}\nu\sigma)\,m^4 = 0, \quad \text{......(29)}$$

writing c^2 for γb^2.

We verify that if $\nu = 0$, $\nu' = 0$, we have $m = \pm\, i\sigma/c$. Also that if $\nu = 0$, $\nu' = \infty$ we have $m = \pm\, i\sigma/b$, since the conditions are now practically isothermal. Further, that if σ is very great, whilst $\nu = 0$, we have again $m = \pm\, i\sigma/c$, independently of the value of ν'. Cf. Art. 278.

According to Maxwell's kinetic theory of gases

$$\nu' = \tfrac{5}{2}\nu; \quad \text{.................................(30)}$$

but we shall only assume that ν' and ν are of the same order of magnitude.

We have seen that for ordinary sound-waves the ratio $\nu\sigma/c^2$ is small. The roots of the above quadratic in m^2 are therefore

$$m_1^2 = -\sigma^2/c^2, \qquad m_2^2 = i\sigma c^2/\nu' b^2 = i\gamma\sigma/\nu', \quad \text{..............(31)}$$

approximately. A more accurate value of the former root is

$$m_1^2 = -\frac{\sigma^2}{c^2}\left[1 - \left\{\tfrac{4}{3}\nu + \left(1 - \frac{b^2}{c^2}\right)\nu'\right\}\frac{i\sigma}{c^2}\right], \quad \text{..............(32)}$$

whence

$$m_1 = \pm\left(\frac{i\sigma}{c} + \frac{1}{l}\right), \quad \text{.........................(33)}$$

if

$$l = \frac{c^3}{\left\{\tfrac{2}{3}\nu + \tfrac{1}{2}\left(1 - \frac{b^2}{c^2}\right)\nu'\right\}\sigma^2}. \quad \text{....................(34)}$$

The complete solution for $x > 0$ is found to be, approximately,

$$\left.\begin{aligned} u &= A_1 e^{i\sigma t + m_1 x} + A_2 e^{i\sigma t + m_2 x}, \\ \eta &= \frac{\gamma - 1}{c} A_1 e^{i\sigma t + m_1 x} + \frac{m_2}{i\sigma} A_2 e^{i\sigma t + m_2 x}, \end{aligned}\right\} \qquad \text{.............(35)}$$

provided m_1, m_2 are chosen so as to have their real parts negative. The arbitrary constants A_1, A_2 enable us to represent the effect of prescribed periodic variations of u and η at the plane $x = 0$. For ordinary frequencies the ratio $m_2 c/\sigma$ is large, and the ratio A_2/A_1, as determined by the thermal conditions at $x = 0$, is accordingly usually small. The second term in the value of u is then unimportant, even near the origin, and in any case it becomes insignificant in comparison with the first term for sufficiently great values of x. Its use is to represent the purely local effect of a periodic source of heat at the origin.

If we adopt the value (30) of ν', and take $c^2/b^2 = \gamma = 1{\cdot}40$, we find from (34) that the value of l is diminished by the conductivity in the ratio ·65.

The investigation of this Art. is due in principle to Kirchhoff*, who further examined the effect on diverging spherical waves and on the propagation of sound-waves in a narrow tube. This problem has a bearing on the well-known experiments of Kundt.

360 a. Reference has already been made to the influence of viscosity in the theory of sound-waves of permanent type, Art. 284. When viscosity alone is allowed for, and thermal conduction ignored, the theory is specially simple, and is worth notice for the sake of the application (the only one in this book) of the principle of Art. 358.

The question being treated as one of steady motion, the dynamical equation is

$$\rho u \frac{\partial u}{\partial x} = -\frac{\partial p}{\partial x} + \tfrac{4}{3}\mu \frac{\partial^2 u}{\partial x^2}. \qquad \text{.........................(1)}$$

Putting, as in Art. 284, $\rho u = m$, or $u = mv$, where v is the volume of unit mass, this may be written

$$m^2 \frac{\partial v}{\partial x} = -\frac{\partial p}{\partial x} + \tfrac{4}{3}\mu m \frac{\partial^2 v}{\partial x^2}, \qquad \text{.......................(2)}$$

whence

$$\begin{aligned} p + m^2 v &= p_0 + m^2 v_0 + \tfrac{4}{3}\mu m \frac{\partial v}{\partial x} \\ &= p_1 + m^2 v_1 + \tfrac{4}{3}\mu m \frac{\partial v}{\partial x}, \end{aligned} \qquad \text{....................(3)}$$

since $\partial v/\partial x$ vanishes at the limits of the wave. This replaces (4) of Art. 284.

* *l.c. ante* p. 646. His investigations are reproduced in Rayleigh's *Theory of Sound*, 2nd ed. Arts. 348–350.

Hence if Q be the heat absorbed by unit mass up to any stage, we have

$$(\gamma-1)\frac{\partial Q}{\partial x}=v\frac{\partial p}{\partial x}+\gamma p\frac{\partial v}{\partial x}$$

$$=\gamma\left\{p_0+m^2v_0-m^2v+\tfrac{4}{3}\mu m\frac{\partial v}{\partial x}\right\}\frac{\partial v}{\partial x}$$

$$+v\left\{-m^2\frac{\partial v}{\partial x}+\tfrac{4}{3}\mu m\frac{\partial^2 v}{\partial x^2}\right\}, \quad\text{............(4)}$$

by Art. 284 (12). The rate at which heat is generated by viscosity is, by Art. 358, $\frac{4}{3}\mu v\,(\partial u/\partial x)^2$ per unit mass. This must be equal to $u\partial Q/\partial x$, whence

$$\frac{\partial Q}{\partial x}=\tfrac{4}{3}\mu m\left(\frac{\partial v}{\partial x}\right)^2, \quad\text{..............................(5)}$$

since $u=mv$. Substituting in (4) we find

$$\gamma\,(p_0+m^2v_0)\frac{\partial v}{\partial x}-(\gamma+1)\,m^2v\,\frac{\partial v}{\partial x}+\tfrac{4}{3}\mu m\frac{\partial}{\partial x}\left(v\,\frac{\partial v}{\partial x}\right)=0. \quad\text{......(6)}$$

Integrating between the limits at which $\partial v/\partial x=0$, and dividing by v_1-v_0, we have

$$\gamma\,(p_0+m^2v_0)=\tfrac{1}{2}(\gamma+1)\,m^2\,(v_0+v_1). \quad\text{.................(7)}$$

Hence (6) may be written

$$\tfrac{1}{2}(\gamma+1)\,m\,(v_0+v_1-v)\,\frac{\partial v}{\partial x}+\tfrac{4}{3}\mu\frac{\partial}{\partial x}\left(v\,\frac{\partial v}{\partial x}\right)=0. \quad\text{...........(8)}$$

The integral, adjusted so as to make $\partial v/\partial x=0$ for $v=v_0$, is

$$\tfrac{4}{3}\mu v\frac{\partial v}{\partial x}+\tfrac{1}{2}(\gamma+1)\,(v-v_1)\,(v_0-v)=0. \quad\text{.................(9)}$$

Hence, save for an additive constant,

$$x=\frac{8\mu}{3\,(\gamma+1)\,m\,(v_0-v_1)}\{v_1\log(v-v_1)-v_0\log(v_0-v)\}, \quad\text{......(10)}$$

where m is given by (7). There is now no restriction as to the magnitude of the ratio v_0/v_1*.

If in (10) we put $v=\alpha v_0+\beta v_1$, where $\alpha+\beta=1$, the value of x differs only by a constant from

$$\frac{8\mu}{3\,(\gamma+1)\,m}\cdot\frac{v_1\log\alpha-v_0\log\beta}{v_0-v_1}. \quad\text{..............................(11)}$$

For example, if we put $\alpha=\cdot 9$, $\beta=\cdot 1$, and again $\alpha=\cdot 1$, $\beta=\cdot 9$, the difference between the two values of x is

$$\frac{8\mu}{3\,(\gamma+1)\,m}\cdot\frac{v_0+v_1}{v_0-v_1}\log 9. \quad\text{................................(12)}$$

Thus if $v_0=2v_1$, we find from (7), and from the other numerical data at the end of Art. 284, $m=68\cdot 3$. Putting $\mu=\cdot 00018$, the expression (12) works out as $1\cdot 94\times 10^{-5}$ cm.

* The investigation is from Rayleigh's paper cited on p. 482.

360 b. The principles of Art. 360 have been applied by Rayleigh* to explain the action of porous bodies in absorption of sound. For the purpose of a general explanation we may simplify the matter by taking account of viscosity alone.

Referring to Art. 347 (5), we find that in the case of a fluid oscillating over a plane wall under a periodic force X the tangential drag on the fluid is, per unit area,

$$-\mu\left(\frac{\partial u}{\partial y}\right)_{y=0} = -(1-i)\frac{\mu\beta f}{\sigma}e^{i(\sigma t+\epsilon)} = -(1-i)\frac{\mu\beta}{\sigma}X. \quad \text{.........(1)}$$

This was obtained on the supposition of incompressibility, but it will hold as an approximation provided the wave-length be great in comparison with the other linear dimensions with which we shall be concerned. Among these is the linear magnitude $\beta^{-1}, = (2\nu/\sigma)^{\frac{1}{2}}$, which is a measure of the extent to which the retarding influence of viscosity penetrates into the fluid†.

In applying (1) to waves travelling along a tube, or between parallel walls, the force X (per unit mass) may be replaced by $-\partial p/\rho_0\partial x$. Taking the case of the tube, and assuming for the present that β^{-1} is small compared with the radius a, we have, calculating the forces on the fluid contained in a length δx,

$$\pi\rho_0 a^2\frac{\partial\bar{u}}{\partial t} = -\pi a^2\frac{\partial\bar{p}}{\partial x} + (1-i)\frac{\nu\beta}{\sigma}\frac{\partial\bar{p}}{\partial x}\times 2\pi a,$$

where $\bar{u}$, $\bar{p}$ denote the average velocity and pressure over the cross-section. Since $\sigma = 2\nu\beta^2$, this may be written

$$\frac{\partial\bar{u}}{\partial t} = -\left(1 - \frac{1-i}{\beta a}\right)\frac{\partial\bar{p}}{\rho_0\partial x}. \quad \text{.......................(2)}$$

We have also

$$\bar{p} = p_0 + c^2\rho_0\bar{s}, \quad \frac{\partial\bar{s}}{\partial t} = -\frac{\partial\bar{u}}{\partial x}, \quad \text{.......................(3)}$$

where s is the condensation. Hence, eliminating $\bar{s}$,

$$\frac{\partial^2\bar{u}}{\partial t^2} = \left(1 - \frac{1-i}{\beta a}\right)c^2\frac{\partial^2\bar{u}}{\partial x^2}. \quad \text{.......................(4)}$$

It is already assumed that $\bar{u}$ varies as $e^{i\sigma t}$. Hence, putting

$$\bar{u} = Ce^{i\sigma t + mx}, \quad \text{................................(5)}$$

we have

$$m^2 = -\frac{\sigma^2}{c^2}\left(1 - \frac{1-i}{\beta a}\right)^{-1},$$

or

$$m = \pm\frac{i\sigma}{c}\left(1 + \frac{1-i}{2\beta a}\right), \quad \text{.........................(6)}$$

* "On Porous Bodies in relation to Sound," *Phil. Mag.* (5) xvi. 181 (1883) [*Papers*, ii. 220]; *Theory of Sound*, Art. 351. See also the author's *Dynamical Theory of Sound*, London, 1910, p. 192.

† Taking $\nu = \cdot 132$, and denoting by $N\,(=\sigma/2\pi)$ the frequency, we find $\beta^{-1} = \cdot 207 N^{-\frac{1}{2}}$ cm.

on account of the assumed smallness of $1/\beta a$. This may be written

$$m = \pm\left(\frac{i\sigma}{c'} + \frac{1}{l'}\right), \quad \text{.........................(7)}$$

where $$c' = c\left(1 - \frac{1}{2\beta a}\right), \quad l' = ac/\nu\beta. \quad \text{....................(8)*}$$

Hence, taking the lower sign, and writing (5) in real form,

$$\bar{u} = Ce^{-x/l'}\cos\sigma\left(t - \frac{x}{c'}\right). \quad \text{.......................(9)}$$

In Art. 360 (34) we found, putting $\nu' = 0$, that $l = \frac{3}{2}c^3/\nu\sigma^2$. Hence

$$l'/l = \tfrac{2}{3}\sigma^2 a/\beta c^2 = \tfrac{2}{3}\cdot\left(\frac{2\pi a}{\lambda}\right)^2\cdot\frac{1}{\beta a}, \quad \text{..................(10)}$$

approximately, where λ is the wave-length. The rate of decay of the waves as they advance is therefore much greater than in the open, if the wave-length is comparable with, or greater than, the circumference of the cross-section.

When the tube is so narrow that the radius a is of the same order of magnitude as β^{-1} the character of the motion is altered. The friction has now a much greater hold on the vibrating mass, and the inertia of the latter becomes negligible. The mean velocity $\bar{u}$ is then related practically to the mean pressure-gradient by the formula (4) of Art. 331; thus

$$\bar{u} = -\frac{a^2}{8\mu}\frac{\partial\bar{p}}{\partial x}. \quad \text{..............................(11)}$$

Hence, referring to (3), $$\frac{\partial\bar{u}}{\partial t} = \frac{c^2 a^2}{8\nu}\frac{\partial^2 u}{\partial x^2}. \quad \text{.............................(12)}$$

This is identical in form with the equation of linear conduction of heat.

Substituting from (5) we have

$$m = \pm(1 + i)\,q, \quad \text{.............................(13)}$$

provided $$q^2 = 4\nu\sigma/c^2a^2 = 2\sigma^2/\beta^2a^2c^2. \quad \text{.....................(14)}$$

Hence, in real form, taking the lower sign,

$$\bar{u} = Ce^{-qx}\cos(\sigma t - qx). \quad \text{........................(15)}$$

The phase is repeated whenever x increases by $2\pi/q$, but in this interval the amplitude is diminished in the ratio $e^{-2\pi}$, or $\frac{1}{535}$. The ratio of this interval to the wave-length λ in the open is

$$2\pi/q\lambda = \beta a/\sqrt{2}\lambda, \quad \text{..........................(16)}$$

which is on the present suppositions a small fraction.

When a sound-wave impinges on the surface of a solid which is permeated by a large number of narrow channels part of the energy is lost, so far as

* Formulae equivalent to these were given (without proof) by Helmholtz in 1863; see his *Wiss. Abh.* i. 384. There is an error in the quotation by Kirchhoff.

sound is concerned, by dissipation in these channels, in the way above explained. The interstices in hangings and carpets act in a similar manner, and it is to this cause that the effect of such appliances in deadening echoes in a room is to be ascribed, a certain proportion of the energy being lost at each reflection. It is to be observed that it is only through the action of true dissipative forces, such as viscosity and thermal conduction, that sound can die out in an enclosed space.

361. In the investigations which follow the thermal processes are neglected for simplicity. We may infer from the preceding results that this will not affect the *order of magnitude* of the terms which represent the effect of dissipative action.

The general equations of sound-waves as affected by viscosity are, by Art. 328 (2),

$$\left.\begin{aligned}\frac{\partial u}{\partial t} &= -\frac{1}{\rho_0}\frac{\partial p}{\partial x} + \nu\nabla^2 u + \tfrac{1}{3}\nu\frac{\partial \vartheta}{\partial x},\\ \frac{\partial v}{\partial t} &= -\frac{1}{\rho_0}\frac{\partial p}{\partial y} + \nu\nabla^2 v + \tfrac{1}{3}\nu\frac{\partial \vartheta}{\partial y},\\ \frac{\partial w}{\partial t} &= -\frac{1}{\rho_0}\frac{\partial p}{\partial z} + \nu\nabla^2 w + \tfrac{1}{3}\nu\frac{\partial \vartheta}{\partial z},\end{aligned}\right\} \quad \text{.....................(1)}$$

where

$$\vartheta = \frac{\partial u}{\partial x} + \frac{\partial v}{\partial y} + \frac{\partial w}{\partial z}. \quad \text{..............................(2)}$$

If s denote the condensation we have in addition the equation of continuity

$$\frac{\partial s}{\partial t} = -\left(\frac{\partial u}{\partial x} + \frac{\partial v}{\partial y} + \frac{\partial w}{\partial z}\right), \quad \text{..........................(3)}$$

and the physical equation

$$p = p_0 + \rho_0 c^2 s, \quad \text{..................................(4)}$$

where c is the velocity of sound in the absence of viscosity.

Eliminating p and ϑ, we have

$$\left.\begin{aligned}\frac{\partial u}{\partial t} &= \nu\nabla^2 u - \left(c^2 + \tfrac{1}{3}\nu\frac{\partial}{\partial t}\right)\frac{\partial s}{\partial x},\\ \frac{\partial v}{\partial t} &= \nu\nabla^2 v - \left(c^2 + \tfrac{1}{3}\nu\frac{\partial}{\partial t}\right)\frac{\partial s}{\partial y},\\ \frac{\partial w}{\partial t} &= \nu\nabla^2 w - \left(c^2 + \tfrac{1}{3}\nu\frac{\partial}{\partial t}\right)\frac{\partial s}{\partial z}.\end{aligned}\right\} \quad \text{........................(5)}$$

From (5) and (3) we deduce by differentiation

$$\frac{\partial^2 s}{\partial t^2} = \left(c^2 + \tfrac{4}{3}\nu\frac{\partial}{\partial t}\right)\nabla^2 s. \quad \text{..........................(6)}$$

If we assume a time-factor $e^{i\sigma t}$, (6) takes the form

$$(\nabla^2 + k^2)\,s = 0, \quad \text{..................................(7)}$$

where $$k^2 = \frac{\sigma^2}{c^2 + \frac{4}{3} i\nu\sigma}, \qquad (8)$$

whilst (5) may be written

$$\left.\begin{aligned} (\nabla^2 + h^2)\, u &= (k^2 - h^2) \frac{\partial \phi}{\partial x}, \\ (\nabla^2 + h^2)\, v &= (k^2 - h^2) \frac{\partial \phi}{\partial y}, \\ (\nabla^2 + h^2)\, w &= (k^2 - h^2) \frac{\partial \phi}{\partial z}, \end{aligned}\right\} \qquad (9)$$

where $$h^2 = -i\sigma/\nu, \qquad (10)$$

and $$\phi = -i\sigma s/k^2. \qquad (11)$$

These equations are satisfied by

$$u = -\frac{\partial \phi}{\partial x}, \qquad v = -\frac{\partial \phi}{\partial y}, \qquad w = -\frac{\partial \phi}{\partial z}, \qquad (12)$$

where ϕ is any solution of (7).

In particular, in the case of waves diverging from a spherical surface $r = a$, where a prescribed radial velocity $e^{i\sigma t}$ is maintained, we have

$$\phi = A f_0(kr)\, e^{i\sigma t}, \qquad (13)$$

with the condition $$-kA f_0'(ka) = 1. \qquad (14)$$

Hence $$\phi = -\frac{f_0(kr)}{k f_0'(ka)} e^{i\sigma t}, \qquad (15)$$

or, in full, $$\phi = \frac{a^2}{1 + ika} \cdot \frac{e^{i(\sigma t - kr + ka)}}{r}. \qquad (16)$$

We have seen (Art. 359) that even in the case of acoustical frequencies the ratio $\nu\sigma/c^2$ is exceedingly small, so that

$$k = \frac{\sigma}{c}\left(1 - \frac{2}{3} \frac{i\nu\sigma}{c^2}\right), \qquad (17)$$

very approximately. The interpretation of (16) as regards the slight effect of viscosity on wave-velocity, and its influence in attenuation of the waves as they proceed, is the same as in the one-dimensional case of Art. 359. It appears that for distances of very many wave-lengths the attenuation due to viscosity is altogether negligible in comparison with that due to spherical divergence.

When the motion is not symmetrical about the origin, the solution of the equations (7) and (9) is to be completed by the analysis of Art. 352. Thus, in the case of diverging waves, we have solutions of the type

$$\left.\begin{aligned} u &= -\frac{\partial \phi}{\partial x} + (n+1) f_{n-1}(hr) \frac{\partial \chi_n}{\partial x} - n f_{n+1}(hr)\, h^2 r^{2n+3} \frac{\partial}{\partial x} \frac{\chi_n}{r^{2n+1}}, \\ v &= -\frac{\partial \phi}{\partial y} + (n+1) f_{n-1}(hr) \frac{\partial \chi_n}{\partial y} - n f_{n+1}(hr)\, h^2 r^{2n+3} \frac{\partial}{\partial y} \frac{\chi_n}{r^{2n+1}}, \\ w &= -\frac{\partial \phi}{\partial z} + (n+1) f_{n-1}(hr) \frac{\partial \chi_n}{\partial z} - n f_{n+1}(hr)\, h^2 r^{2n+3} \frac{\partial}{\partial z} \frac{\chi_n}{r^{2n+1}}, \end{aligned}\right\} \qquad (18)$$

where
$$\phi = f_n(kr)\,\phi_n, \qquad\qquad (19)$$
the functions ϕ_n, χ_n being solid harmonics of positive degree n*.

These formulae make
$$xu + yv + zw = -\{krf_n'(kr) + nf_n(kr)\}\,\phi_n + n(n+1)(2n+1)f_n(hr)\,\chi_n, \qquad (20)$$
and
$$\left.\begin{aligned}
yw - zv &= -f_n(kr)\left(y\frac{\partial}{\partial z} - z\frac{\partial}{\partial y}\right)\phi_n \\
&\quad + (2n+1)\{hrf_n'(hr) + (n+1)f_n(hr)\}\left(y\frac{\partial}{\partial z} - z\frac{\partial}{\partial y}\right)\chi_n, \\
zu - xw &= -f_n(kr)\left(z\frac{\partial}{\partial x} - x\frac{\partial}{\partial z}\right)\phi_n \\
&\quad + (2n+1)\{hrf_n'(hr) + (n+1)f_n(hr)\}\left(z\frac{\partial}{\partial x} - x\frac{\partial}{\partial z}\right)\chi_n, \\
xv - yu &= -f_n(kr)\left(x\frac{\partial}{\partial y} - y\frac{\partial}{\partial x}\right)\phi_n \\
&\quad + (2n+1)\{hrf_n'(hr) + (n+1)f_n(hr)\}\left(x\frac{\partial}{\partial y} - y\frac{\partial}{\partial x}\right)\chi_n,
\end{aligned}\right\} \qquad (21)$$
where use has been made of the reduction-formulae of Art. 292.

For a reason already given we may with ample accuracy treat k as real and equal to σ/c. As regards h we write
$$h = (1-i)\beta, \quad \text{where} \quad \beta = \sqrt{(\sigma/2\nu)}, \qquad (22)$$
as in Art. 345. The terms in (18) which involve χ_n will therefore contain a factor $e^{-\beta r}$, and will accordingly tend to become negligible at distances r which are large compared with the linear magnitude β^{-1}, whose value for air is about $\cdot 21/\sqrt{N}$ cm., if N be the number of vibrations per sec. (Art. 345). The motion at a distance which is a moderate multiple of β^{-1} will therefore be practically irrotational, with a velocity-potential given by (19). It is to be noticed also that the ratio $ka/\beta a$, being equal to $\sqrt{(2\nu\sigma/c^2)}$ approximately, is to be regarded as a small quantity.

To apply the formulae to the case of a sphere oscillating parallel to x with a velocity
$$U = e^{i\sigma t}, \qquad (23)$$
we put $n=1$ in (18), and assume
$$\phi_1 = A_1 x, \qquad \chi_1 = B_1 x. \qquad (24)$$
The conditions
$$u = U, \qquad v = 0, \qquad w = 0 \qquad (25)$$
to be satisfied at the surface $r=a$ give, by (20) and (21),
$$-\{kaf_1'(ka) + f_1(ka)\}\,A_1 + 6f_1(ha)\,B_1 = 1, \qquad (26)$$
$$-f_1(ka)\,A_1 + 3\{haf_1'(ha) + 2f_1(ha)\}\,B_1 = 1, \qquad (27)$$
whence
$$\left.\begin{aligned}
A_1 &= -\frac{haf_1'(ha)}{haf_1'(ha)\{kaf_1'(ka) + f_1(ka)\} + 2f_1(ha)\,kaf_1'(ka)}, \\
B_1 &= \frac{\tfrac{1}{3}kaf_1'(ka)}{haf_1'(ha)\{kaf_1'(ka) + f_1(ka)\} + 2f_1(ha)\,kaf_1'(ka)}.
\end{aligned}\right\} \qquad (28)$$

The solutions of the 'First Class' are of less interest from the present standpoint.

Substituting from Art. 292 (15), we find

$$\left.\begin{aligned} A_1 &= -\frac{(3+3iha-h^2a^2)\,k^3a^3e^{ika}}{k^2a^2(1+iha)+(2+2ika-k^2a^2)\,h^2a^2}, \\ B_1 &= \frac{\frac{1}{3}(3+3ika-k^2a^2)\,h^3a^3e^{iha}}{k^2a^2(1+iha)+(2+2ika-k^2a^2)\,h^2a^2}. \end{aligned}\right\} \qquad\qquad (29)$$

At distant points the motion is practically irrotational, with the velocity-potential

$$\phi = A_1 f_1(kr)\,xe^{i\sigma t}. \qquad\qquad (30)$$

From the acoustical point of view the most interesting case is where the radius a of the sphere is large compared with β^{-1}. If we retain only the highest power of ha in (29) we find

$$A_1 = \frac{k^3a^3e^{ika}}{2+2ika-k^2a^2}, \qquad\qquad (31)$$

exactly as if viscosity had been ignored from the outset (Art. 295). This shews that the conclusions of Stokes as to the influence of lateral motion in the communication of vibrations to a gas are in the main unaffected by viscosity. It is true that the lateral motion of the air close to a vibrating surface is modified, and may even be reversed in direction, but the effect extends only to a stratum whose thickness is of the order β^{-1}, and if this is small compared with the dimensions of the compartments into which the surface is divided by the nodal lines the general argument of Art. 294 still applies.

In the case of very slow oscillations, on the other hand, or obstacles of very small radius, where βa is not large, ka is necessarily small, and we have from (29) and (22)

$$A_1 = \left\{\tfrac{1}{2}\left(1+\frac{3}{2\beta a}\right) - \frac{3i}{4\beta a}\left(1+\frac{1}{\beta a}\right)\right\} k^3a^3, \qquad\qquad (32)$$

approximately. This is in consonance with Art. 350 (22). At distances r which are small compared with the wave-length, but moderately large in comparison with β^{-1}, the motion is in fact practically the same as if the fluid were incompressible.

362. We may further investigate the scattering of plane waves by a spherical obstacle. The question is the same as in Art. 297, except that viscosity is now taken into account. It is assumed that the circumference of the obstacle is small compared with the wave-length, so that ka is small*.

By Art. 296 we may write, for the velocity-potential of the incident waves,

$$\phi = e^{ikx} = \psi_0(kr) + 3ikr\,\psi_1(kr)\cos\theta + \ldots, \qquad\qquad (1)$$

where θ is the usual angular co-ordinate, and the factor $e^{i\sigma t}$ or e^{ikct} is understood. It is clear from Art. 297 that the terms involving harmonics of higher order than the first may be neglected. For small values of kr (1) takes the form

$$\phi = 1 - \tfrac{1}{6}k^2r^2 + \ldots + ikx + \ldots. \qquad\qquad (2)$$

We will first suppose the sphere to be fixed. The velocity at its surface due to (2) alone is made up chiefly of a uniform radial velocity $\frac{1}{3}k^2a$ and a uniform velocity $-ik$ parallel to x. Reversing these, the velocity-potential ϕ' of the scattered waves at a distance r which is large compared with β^{-1} is obtained by superposition from Art. 361 (16), (30), with the proper coefficients. Thus

$$\phi' = -\tfrac{1}{3}\frac{k^3a^3e^{ika}}{1+ika} f_0(kr) + (H+iK)\,f_1(kr)\,kr\cos\theta + \ldots, \qquad\qquad (3)$$

where
$$H+iK = iA_1 = -\frac{(3+3iha-h^2a^2)\,ik^3a^3e^{ika}}{k^2a^2(1+iha)+(2+2ika-k^2a^2)\,h^2a^2}. \qquad\qquad (4)$$

* This, with the corresponding problem in two dimensions, was treated by Sewell, "On the Extinction of Sound in a Viscous Atmosphere by Small Obstacles...," *Phil. Trans.* A, ccx. 239 (1910). I have somewhat modified and condensed the procedure.

The main interest of the investigation is to ascertain the rate at which energy is partly dissipated by friction, and partly diverted from the train of primary waves, owing to the presence of the obstacle. For this purpose the values of ϕ and ϕ' must be expressed in real form. This being presupposed, let us write

$$q=\frac{\partial\phi}{\partial r},\qquad q'=\frac{\partial\phi'}{\partial r},\qquad\text{.......................................(5)}$$

so that q, q' are the inward radial velocities due to the primary and secondary waves, respectively, at distances r which are large compared with β^{-1}; and let p, p' be the corresponding pressures, viz.

$$p=\rho_0\frac{\partial\phi}{\partial t},\qquad p'=\rho_0\frac{\partial\phi'}{\partial t}.\qquad\text{.......................................(6)}$$

The rate at which work is being done at a spherical surface of large radius r, on the included air, is given by the integral

$$\iint(p+p')(q+q')\,dS,\qquad\text{.......................................(7)}$$

taken over the surface. Since the mechanical energy in the enclosed space is constant, the mean value of this integral represents energy dissipated by fluid friction. To this we add the work spent in generating scattered waves, viz.

$$-\iint p'q'\,dS.\qquad\text{.......................................(8)}$$

Again, the term

$$\iint pq\,dS\qquad\text{.......................................(9)}$$

represents work dissipated in the primary waves alone when the obstacle is absent. Hence the total rate at which energy is withdrawn from the primary waves, in consequence of the presence of the obstacle, is equal to the time-average of the integral

$$\iint(pq'+p'q)\,dS,\qquad\text{.......................................(10)}$$

taken over the surface of a sphere of very large radius.

In forming the sum $pq'+p'q$ we need only include products of terms which involve spherical harmonics of equal order. Moreover, since k is taken to be real, the final result, so far as the harmonics of *zero* order are concerned, must be the same as when viscosity was neglected. In terms of the energy-flux in the primary waves the result in question is

$$\tfrac{4}{9}k^4a^4\,.\,\pi a^2,\qquad\text{.......................................(11)}$$

by Art. 297 (7), (11).

We may therefore confine our attention to the harmonics of order 1. Taking the real parts of the expressions in (1) and (3), when multiplied by $e^{i\sigma t}$, we have, then,

$$\phi=-3kr\psi_1(kr)\cos\theta\,.\sin\sigma t,\qquad\text{.......................................(12)}$$

$$\phi'=(H\cos\sigma t-K\sin\sigma t)\,kr\Psi_1(kr)\cos\theta+(H\sin\sigma t+K\cos\sigma t)\,kr\psi_1(kr)\cos\theta,\ \text{...(13)}$$

by Art. 292 (14). These make

$$p=-3\rho_0\sigma kr\psi_1(kr)\cos\theta\,.\cos\sigma t,\qquad\text{.......................................(14)}$$

$$p'=-\rho_0\sigma(H\sin\sigma t+K\cos\sigma t)\,kr\Psi_1(kr)\cos\theta$$
$$+\rho_0\sigma(H\cos\sigma t-K\sin\sigma t)\,kr\psi_1(kr)\cos\theta,\ \text{......(15)}$$

$$q=-3k\{kr\psi_1'(kr)+\psi_1(kr)\}\cos\theta\,.\sin\sigma t,\qquad\text{.......................................(16)}$$

$$q'=k(H\cos\sigma t-K\sin\sigma t)\{kr\Psi_1'(kr)+\Psi_1(kr)\}\cos\theta$$
$$+k(H\sin\sigma t+K\cos\sigma t)\{kr\psi_1'(kr)+\psi_1(kr)\}\cos\theta.\ \text{......(17)}$$

Hence

$$pq'+p'q=\tfrac{3}{2}\rho_0\sigma k^3r^2H\{\psi_1'(kr)\Psi_1(kr)-\psi_1(kr)\Psi_1'(kr)\}\cos^2\theta$$
$$+\text{terms in }\cos 2\sigma t,\ \sin 2\sigma t.\ \text{......(18)}$$

Since $\iint\cos^2\theta\,dS=\tfrac{4}{3}\pi r^2$, the mean value of that portion of the integral in (10) which is due to the harmonics of the first order is

$$2\pi\rho_0\sigma k^3r^4H\{\psi_1'(kr)\Psi_1(kr)-\psi_1(kr)\Psi_1'(kr)\}=2\pi\rho_0cH,\ \text{..............(19)}$$

the reduction depending on Art. 292 (19).

When βa is large we may expect the influence of viscosity to be negligible. We have in fact from (4)

$$H+iK=\frac{ik^3a^3e^{ika}}{2+2ika-k^2a^2}. \quad \text{......(20)}$$

Evaluating for small values of ka we find

$$H=\tfrac{1}{12}k^6a^6; \quad \text{......(21)}$$

and the result in (19) becomes, in terms of the primary energy-flux $(\frac{1}{2}\rho_0 k^2 c)$,

$$\tfrac{1}{3}k^4a^4\,.\,\pi a^2. \quad \text{......(22)}$$

Adding to (11), we reproduce the result (12) of Art. 297, there obtained by a much simpler process.

If on the other hand a is comparable with or less than β^{-1}, we have, on account of the smallness of ka,

$$H+iK=\tfrac{1}{2}i\left(1-\frac{3i}{ha}-\frac{3}{h^2a^2}\right)k^3a^3=\frac{3k^3a^3}{4\beta a}\left(1+\frac{1}{\beta a}\right)+\frac{ik^3a^3}{2}\left(1+\frac{3}{2\beta a}\right). \quad \text{......(23)}$$

The loss of energy is then, in terms of the primary flux,

$$\frac{4\pi c^2 H}{k^2}=\frac{3ka}{\beta a}\left(1+\frac{1}{\beta a}\right).\,\pi a^2. \quad \text{......(24)}$$

The term (11) is now altogether negligible in comparison.

When a is small compared with β^{-1} the result is, approximately,

$$\frac{6\nu}{ca}\,.\,\pi a^2. \quad \text{......(25)}$$

The fraction of the incident energy which is lost now varies inversely as the radius of the sphere. The total amount lost varies directly as the radius*. For air at 0° C. we have

$$6\nu/ca=2{\cdot}39\times 10^{-5}\times a^{-1},$$

the unit of a being the centimetre.

363. The foregoing calculations have some interest in relation to the transmission of sound by fog. Owing to its great inertia in comparison with that of an equal volume of air, a globule of water in suspension, if not too small, may remain practically at rest as the air-waves beat upon it. If, however, the radius be diminished, the inertia diminishes as a^3, whilst the surface on which viscosity acts diminishes as a^2, and it is to be expected that a stage will at length be reached when the globule will simply drift to and fro with the vibrating air, and so cause little or no loss of energy.

To examine this point a little more closely, we now regard the sphere as perfectly free to move. The velocity at its surface, in the scattered waves, will be made up of a radial velocity $-\frac{1}{3}k^2a$, as before, and a velocity $ik+d\xi/dt$, or $i(k+\sigma\xi)$, parallel to x, where ξ denotes the displacement of the centre from its mean position. Hence in place of Art. 361 (24) we must write

$$\phi_1=i(k+\sigma\xi)A_1x, \qquad \chi_1=i(k+\sigma\xi)B_1x, \quad \text{......(26)}$$

where A_1, B_1 have the values given by Art. 361 (29).

The direct calculation of the stresses on the surface of the sphere is somewhat troublesome, but may be evaded by considerations of momentum. It will be seen that in this

* Numerical results for a range of values of βa, based on a closer approximation, are given in Sewell's paper.

process we need only take account of the spherical harmonics of unit order. We write, therefore,

$$\phi = \ldots + 3ikr\,\psi_1(kr)\cos\theta + \ldots, \qquad (27)$$

$$\phi' = \ldots + i(k+\sigma\xi)\,A_1 r f_1(kr)\cos\theta + \ldots, \qquad (28)$$

for the incident, and the scattered waves at a distance, respectively.

We calculate the rate of change of momentum of the fluid contained between the sphere and a concentric spherical surface whose radius r is large compared with β^{-1}. The primary waves contribute the term

$$-i\sigma\rho_0 \iiint \frac{\partial\phi}{\partial x}\,dx\,dy\,dz = 4\pi\rho_0\sigma k\,\{r^3\psi_1(kr) - a^3\psi_1(ka)\}. \qquad (29)$$

As regards the secondary waves, the first part of the value of u in Art. 361 (18) gives

$$-i\sigma\rho_0 \iiint \frac{\partial\phi'}{\partial x}\,dx\,dy\,dz = \tfrac{4}{3}\pi\rho_0\sigma(k+\sigma\xi)\,A_1\{r^3 f_1(kr) - a^3 f_1(ka)\}. \qquad (30)$$

The remaining part, involving χ_1, gives

$$-2\sigma\rho_0(k+\sigma\xi)\,B_1\int_0^r f_0(hr)\,4\pi r^2 dr = -8\pi\rho_0\sigma(k+\sigma\xi)\,B_1\{r^3 f_1(hr) - a^3 f_1(ha)\}, \qquad (31)$$

where the first term in { } may be omitted, as tending ultimately to zero on account of the factor $e^{-\beta r}$ which is involved. The rate of change of momentum of the sphere itself is

$$-\tfrac{4}{3}\pi\rho_1 a^3\sigma^2\xi, \qquad (32)$$

where ρ_1 is its density.

The motion in the neighbourhood of the spherical surface of large radius r may ultimately be taken to be irrotational, and the resultant pressure on this surface is therefore

$$-\iint(p+p')\cos\theta\,dS = -i\sigma\rho_0\iint(\phi+\phi')\cos\theta\,dS$$
$$= 4\pi\rho_0\sigma k r^3\psi_1(kr) + \tfrac{4}{3}\pi\rho_0\sigma(k+\sigma\xi)\,A_1 r^3 f_1(kr). \qquad (33)$$

Equating the total rate of change of momentum to the resultant pressure, we have

$$\rho_0 k\psi_1(ka) + \tfrac{1}{3}\rho_0(k+\sigma\xi)\,A_1 f_1(ka) - 2\rho_0(k+\sigma\xi)\,B_1 f_1(ha) + \tfrac{1}{3}\rho_1\sigma\xi = 0. \qquad (34)$$

This reduces, in virtue of Art. 361 (26), to

$$\rho_0 k\psi_1(ka) - \tfrac{1}{3}\rho_0(k+\sigma\xi)\{1 + kaf_1'(ka)\,A_1\} + \tfrac{1}{3}\rho_1\sigma\xi = 0, \qquad (35)$$

whence

$$-\frac{\sigma\xi}{k} = 1 - \frac{\rho_1 - 3\rho_0\psi_1(ka)}{\rho_1 - \rho_0 - \rho_0 kaf_1'(ka)\,A_1}. \qquad (36)$$

This formula gives the ratio of the displacement of the sphere to that of the air at the position of its centre when the sphere is absent.

When viscosity is negligible we have $ha = \infty$. Hence if ka be small, we find

$$A_1 = \tfrac{1}{2}(ka)^3, \qquad -kaf_1'(ka) = 3\,(ka)^{-3}, \qquad (37)$$

approximately. The ratio in question is accordingly

$$1 - \frac{\rho_1 - \rho_0}{\rho_1 + \tfrac{1}{2}\rho_0}, \qquad (38)$$

in agreement with Art. 298 (21).

On the other hand, when a is comparable with or less than β^{-1}, we have, from Art. 362 (4),

$$A_1 = \left\{\tfrac{1}{2}\left(1 + \frac{3}{2\beta a}\right) - \frac{3i}{4\beta a}\left(1 + \frac{1}{\beta a}\right)\right\} k^3 a^3, \qquad (39)$$

higher powers of ka being neglected. For small values of βa this reduces to

$$A_1 = -\frac{3ik^3a^3}{4\beta^2 a^2}, \qquad (40)$$

approximately; and the formula (36) becomes

$$-\frac{\sigma\xi}{k}=1-\left\{1-\tfrac{9}{4}\frac{i\rho_0}{(\rho_1-\rho_0)\beta^2a^2}\right\}^{-1}. \quad\text{.........................(41)}$$

If β^2a^2, though itself small, be large compared with $\rho_0/(\rho_1-\rho_0)$, this expression is small, and the sphere remains nearly at rest, inertia being still predominant. But when β^2a^2 is small compared with $\rho_0/(\rho_1-\rho_0)$, the ratio approximates to unity, the globule now moving with the air.

Putting $\rho_0/\rho_1=\cdot00129$, $\nu=\cdot132$, the condition for this is that the radius a should be small compared with

$$1\cdot10\times10^{-2}\times N^{-\frac{1}{2}}\,\text{cm.},$$

where N is the frequency of the air-waves. Thus for a frequency of 256, a must be at most of the order ·001 mm.

To calculate the loss of energy we must suppose (28) to be put in the form

$$\phi'=\ldots+(H'+iK')\,f_1(kr)\,kr\cos\theta+\ldots, \quad\text{.......................(42)}$$

in analogy with (3). The result, so far as the harmonics of unit order are concerned, is then

$$2\pi\rho_0cH', \quad\text{..(43)}$$

in place of (19). To find H' we have the equation

$$H'+iK'=i\left(1+\frac{\sigma\xi}{k}\right)A_1=\frac{iA_1\{\rho_1-3\rho_0\psi_1(ka)\}}{\rho_1-\rho_0-\rho_0kaf_1'(ka)A_1}=\frac{iA_1(\rho_1-\rho_0)}{\rho_1-\rho_0+3\rho_0A_1/k^3a^3}, \quad\text{......(44)}$$

approximately.

When viscosity is negligible, or βa large, we have

$$A_1=\tfrac{1}{2}k^3a^3-\tfrac{1}{12}ik^6a^6, \quad\text{....................................(45)}$$

approximately, and therefore

$$H'=\tfrac{1}{12}\left(\frac{\rho_1-\rho_0}{\rho_1+\frac{1}{2}\rho_0}\right)^2.k^6a^6. \quad\text{...............................(46)}$$

In terms of the energy-flux in the primary waves, the energy diverted from these waves is

$$\left\{\tfrac{4}{9}+\tfrac{1}{3}\left(\frac{\rho_1-\rho_0}{\rho_1+\frac{1}{2}\rho_0}\right)^2\right\}k^4a^4.\pi a^2, \quad\text{...............................(47)}$$

where the part given by (11), due to the resistance of the sphere to compression, has been added. If we put $\rho_1/\rho_0=\infty$ we reproduce the result of Art. 297 (12).

When on the other hand βa is small, the approximation becomes troublesome; but it is evident that when the radius is so small that the globule simply drifts to and fro with the air the dissipation due to the terms of the first order will be negligible, and the total dissipation is then given practically by the formula (11) of Art. 362.

364. To examine the effect of viscosity on the free vibrations of air contained in a spherical vessel, the functions f_n which occur in the formulae (18) of Art. 361 must be replaced by ψ_n on account of the finiteness of the velocity at the centre.

The formulae (20) and (21) of Art. 361 then shew that at the boundary $r=a$ we must have

$$-\{ka\psi_n'(ka)+n\psi_n(ka)\}\phi_n+n(n+1)(2n+1)\psi_n(ha)\chi_n=0, \quad\text{...........(1)}$$

and

$$-\psi_n(ka)\phi_n+(2n+1)\{ha\psi_n'(ha)+(n+1)\psi_n(ha)\}\chi_n=0. \quad\text{...........(2)}$$

From these we deduce

$$ka\psi_n'(ka)\phi_n+n(2n+1)ha\psi_n'(ha)\chi_n=0. \quad\text{.......................(3)}$$

Hence

$$\frac{ka\psi_n'(ka)+n\psi_n(ka)}{ka\psi_n'(ka)}=-\frac{(n+1)\psi_n(ha)}{ha\psi_n'(ha)}. \quad\text{...........................(4)}$$

Assuming that a is large compared with β^{-1} we have to ascertain the correction to be applied to the results of Art. 293, 1°, where in the absence of viscosity ka was shewn to satisfy the equation

$$\zeta\psi_n'(\zeta)+n\psi_n(\zeta)=0. \qquad (5)$$

We write accordingly

$$ka=\zeta+\epsilon, \qquad (6)$$

where ζ satisfies (5), and ϵ is assumed to be small. The left-hand member of (4) becomes

$$=\frac{\zeta\psi_n''(\zeta)+(n+1)\psi_n'(\zeta)}{-n\psi_n(\zeta)}\cdot\epsilon=\frac{(n+1)\psi_n'(\zeta)+\zeta\psi_n(\zeta)}{n\psi_n(\zeta)}\cdot\epsilon=\frac{\zeta^2-n(n+1)}{n\zeta}\cdot\epsilon,$$

by Art. 292 (10). The right-hand member reduces to

$$-\frac{n+1}{ha}\tan(ha+\tfrac{1}{2}n\pi),$$

by Art. 292 (8), since the modulus of ha is assumed to be large. Moreover, writing

$$ha=(1-i)\beta a,$$

we have

$$\tan(ha+\tfrac{1}{2}n\pi)=-i,$$

approximately. Hence

$$\epsilon=\frac{n(n+1)\zeta}{\zeta^2-n(n+1)}\cdot\frac{-1+i}{2\beta a}. \qquad (7)$$

Since the time-factor implied in our formulae is e^{ikct}, it appears that the real part of (7) indicates a slight diminution of the frequency. The imaginary part shews that the modulus of decay of the oscillations is

$$\tau \doteq \frac{\zeta^2-n(n+1)}{n(n+1)\zeta}\cdot\frac{2\beta a^2}{c}=\frac{\zeta^2-n(n+1)}{n(n+1)\sqrt{\zeta}}\cdot\sqrt{\left(\frac{2a^3}{\nu c}\right)}, \qquad (8)$$

since $\beta=\sqrt{(c\zeta/2\nu a)}$, approximately.

In the case of $n=1$ we have, in the gravest mode of vibration, $\zeta=2{\cdot}081$, and accordingly

$$\tau=1{\cdot}143\sqrt{\left(\frac{a^3}{\nu c}\right)}.$$

Assuming $c=3{\cdot}32\times10^4$, $\nu={\cdot}132$, this gives $\tau={\cdot}0173a^{\frac{3}{2}}$. It is to be remembered, however, that these numerical estimates must be considerably under the mark, owing to the neglect of the thermal processes.

The foregoing investigation does not apply to the *radial* vibrations. When $n=0$ the formulae (12) of Art. 361 apply, with

$$\phi=C\psi_0(kr), \qquad (9)$$

and the boundary-condition gives

$$\psi_0'(ka)=0. \qquad (10)$$

If ka be a root of this, we have, from Art. 361 (17),

$$\sigma=kc\left(1+\tfrac{2}{3}\frac{i\nu k}{c}\right), \qquad (11)$$

approximately. The modulus of decay is accordingly

$$\tau=\frac{3a^2}{2\nu}\cdot(ka)^{-2}. \qquad (12)$$

It is to be noticed that the ratio of (8) to (12) is of the order $\sqrt{(ac/\nu)}$, numerical factors being omitted. In all cases to which our approximations apply this ratio is large, so that the radial vibrations are much more slowly extinguished, so far as viscosity alone is concerned, than those which correspond to values of n greater than 0. This is readily accounted for. In the latter modes the condition that there is to be no slipping of the fluid in contact with the vessel implies a relatively greater amount of distortion of the fluid elements, and consequent dissipation of energy, in the superficial layers of the gas.

The method of the dissipation function, which was applied in Art. 348 to the case of water waves, might be used to obtain the result (12) for the *radial* vibrations, but would lead to an erroneous result for $n > 0$, since the underlying assumption that the motion is only slightly modified by the friction is violated at the boundary.

In the gravest radial vibration we have $ka = 4{\cdot}493$, whence

$$\tau = {\cdot}0743a^2/\nu.$$

In the case of air at 0° C. this makes $\tau = {\cdot}56a^2$*.

Turbulent Motion.

365. It remains to call attention to the chief outstanding difficulty of our subject.

It has already been remarked that the neglect of the terms of the second order ($u\partial u/\partial x$, &c.) seriously limits the application of many of the preceding results to fluids possessed of ordinary degrees of mobility. Unless the velocities, or the linear dimensions involved, be very small the actual motion in such cases, so far as it admits of being observed, is found to be very different from that represented by our formulae. For example, when a solid of 'easy' shape moves through a liquid, an irregular eddying motion may be produced in a layer of the fluid next to the solid, and a trail of eddies left behind, whilst the motion at a distance laterally may be comparatively smooth and uniform.

The mathematical disability above pointed out does not apply to cases of *rectilinear* flow, such as have been discussed in Arts. 330, 331; but even here observation shews that the types of motion investigated, though theoretically possible, become under certain conditions practically unstable.

The case of flow through a pipe of circular section was made the subject of a careful experimental study by Reynolds†, by means of filaments of coloured fluid introduced into the stream. So long as the mean velocity (w_0) over the cross-section falls below a certain limit depending on the radius of the pipe and the nature of the fluid, the flow is smooth and in accordance with Poiseuille's laws; accidental disturbances are rapidly obliterated, and the *régime* appears to be thoroughly stable. As w_0 is gradually increased beyond this limit the flow becomes increasingly sensitive to small disturbances, but if care be taken to avoid these the smooth rectilinear character may for a while be preserved, until at length a stage is reached beyond which this is no longer possible. When the rectilinear *régime* definitely breaks down the motion becomes wildly irregular, and the tube appears to be filled with

* This Art. is derived with slight alteration from a paper cited on p. 635.

† "An Experimental Investigation of the Circumstances which determine whether the Motion of Water shall be Direct or Sinuous, and of the Law of Resistance in Parallel Channels." *Phil. Trans.* clxxiv. 935 (1883) [*Papers*, ii. 51]. For a historical account of the researches and partial anticipations of other writers, see Knibbs, *Proc. Roy. Soc. N.S.W.* xxxi. 314 (1897). Reference is there made in particular to Hagen, *Berl. Abh.* 1854, p. 17.

interlacing and constantly varying streams, crossing and recrossing the pipe. It was inferred by Reynolds, from considerations of dimensions, that the 'upper critical velocity,' *i.e.* the upper limit of smooth rectilinear flow, must be proportional to ν/D, where D is the diameter of the pipe, and ν the kinematic coefficient of viscosity. Since the dimensions of ν are $\mathsf{L}^2\mathsf{T}^{-1}$, this is in fact the only combination which is of the dimensions of velocity. As the result of his experiments, Reynolds gave for the upper critical velocity the formula

$$U = P/BD, \qquad \text{.................................(1)}$$

where P is the factor which expresses the variation of the viscosity of water with temperature (centigrade), as found by Poiseuille, viz.

$$P = (1 + \cdot 03368\theta + \cdot 00022099\theta^2)^{-1},$$

and $B = 43\cdot 79$, the unit of length being the metre.

Reducing to centimetres, putting $P = \nu/\nu_0$, and taking the value of ν_0 from p. 575, the critical ratio is found accordingly to be

$$w_0 D/\nu = 12830. \qquad \text{..............................(2)}$$

The dependence of the critical velocity on ν was tested by varying the temperature*. Subsequent observers have obtained considerably higher values for the numerical constant in (2); and much seems to depend on the success with which disturbing causes have been avoided†.

366. Simultaneously with the change in the character of the motion there is a change in the relation between the pressure-gradient $(-dp/dz)$ and the mean velocity w_0. So long as the rectilinear character is maintained the gradient varies as w_0, as found by Poiseuille, but when the irregular, turbulent‡, mode of flow has set in the gradient increases more rapidly, in many cases apparently as w_0^2, more or less approximately. This more rapid increase of resistance is no doubt due to the action of the eddies in continually bringing fresh fluid, moving with a considerable relative velocity, close up to the boundary, and so increasing the distortion-rate $(\partial w/\partial n)$ greatly beyond that which would obtain in strictly 'laminar' motion§.

It was found by Reynolds that the transition from the linear law of resistance to that of turbulent flow took place for a definite value of $w_0 D/\nu$. Since disturbing influences are in such experiments hardly to be excluded, the corresponding value of w_0 must be regarded as a 'lower' critical

* The proportionality to ν was confirmed by Barnes and Coker, for an extended range of temperature, *Proc. R. S.* lxxiv. 341 (1904).

† Cf. Barnes and Coker, *l.c.* and Ekman, *Arkiv för Matem.* vi. (1910). Ekman's experiments were made with Reynolds' original apparatus.

‡ This very descriptive term is due to Lord Kelvin.

§ Cf. Stokes, *Papers*, i. 99.

velocity, to be distinguished from that referred to in Art. 365. Reynolds' result is equivalent to

$$w_0 D/\nu = 2030. \qquad\qquad (3)$$

The dependence on ν was tested as before by varying the temperature*.

Some indications as to the possible forms of the tangential resistance per unit area of the walls of the tube are obtained by consideration of dimensions. If we assume that

$$p_{rz} \propto \rho^m \nu^n w_0{}^r a^s,$$

we must have

$$\mathsf{ML^{-1}T^{-2}} = (\mathsf{ML^{-3}})^m (\mathsf{L^2T^{-1}})^n (\mathsf{LT^{-1}})^r \mathsf{L}^s,$$

whence $\qquad m = 1, \qquad s = -n, \qquad r = 2 - n,$

so that

$$p_{rz} \propto \rho w_0{}^2 \left(\frac{\nu}{w_0 a}\right)^n. \qquad\qquad (4)$$

Generalizing this, we have the formula

$$p_{rz} = \rho w_0{}^2 f\left(\frac{w_0 a}{\nu}\right) \qquad\qquad (5)\dagger$$

If in (4) we put $n = 1$, we have Poiseuille's law for linear flow. If we put $n = 0$ we get the formula frequently adopted by writers on Hydraulics for the case of turbulent flow through pipes whose diameter exceeds a certain limit, viz.

$$p_{rz} = k\rho w_0{}^2, \qquad\qquad (6)$$

where k is a numerical constant depending on the nature of the surface. As a rough average value for the case of water moving over a clean iron surface we may take $k = \cdot 0025$‡. A more elaborate empirical formula for p_{rz}, taking account of the influence of the diameter, was given by Darcy as the result of very extensive observations on the flow of water through conduits§.

It is to be noticed that if the resistance were accurately proportional to the square of the velocity it would be independent of the viscosity and of the diameter of the pipe. This follows at once from (5)‖.

Reynolds and various other observers have found that a closer representation of the facts is obtained if in (4) we give to n a value different from zero. The value $n = \frac{1}{4}$ has been suggested, whilst Reynolds proposed $n = \cdot 277$. The

* Values of this lower 'Reynolds number' in the neighbourhood of 2000 have been obtained by various experimenters, *e.g.* Coker and Clement, *Phil. Trans.* A, cci. 45 (1902).

† Rayleigh, "On the Question of the Stability of the Flow of Fluids," *Phil. Mag.* (5) xxxiv. 59 (1892) [*Papers*, iii. 575].

The formula (5) has been tested experimentally over a wide range of conditions, and with fluids so different as water and air. It was verified that the resistance varied as $\rho w_0{}^2$ whenever the value of $\nu/w_0 a$ is the same. See Stanton and Pannell, "Similarity of Motion in Relation to the Surface Friction of Fluids," *Phil. Trans.* A, ccxiv. 199 (1913); Blasius, "Das Aehnlichkeitsgesetz bei Reibungsvorgängen," *Zeitschr. d. Ver. deutsch. Ingenieure*, 1912, p. 639.

‡ Rankine, *Applied Mechanics*, Art. 638; Unwin, *Encyc. Britann.* 11th ed. Art. "Hydraulics."

§ *Recherches expérimentales rélatives au mouvement de l'eau dans les tuyaux*, Paris, 1855. The formula is quoted by Rankine and Unwin.

‖ Rayleigh, *l.c.*

most suitable value of the index appears indeed to depend on the degree of smoothness of the surface; and probably no formula of the type (4) has more than a limited application. Blasius, from a collation of the best available experiments on turbulent flow in *smooth* tubes, gives a formula for the pressure-gradient equivalent to $\frac{1}{2}\lambda\rho w_0^2/D$, where $\lambda = {\cdot}316\,(\nu/w_0 D)^{\frac{1}{4}}$. Since

$$\pi D p_{rz} = -\frac{\partial p}{\partial z}\,.\,\tfrac{1}{4}\pi D^2,$$

this makes

$$p_{rz} = {\cdot}027\,\rho w_0^2\left(\frac{\nu}{w_0 D}\right)^{\frac{1}{4}}. \qquad \text{.......................(7)}$$

Rayleigh has pointed out that the form of the function f in (5) might be determined by experiments in which ν alone is varied. Experiment appears to indicate that with increasing values of $w_0 D/\nu$ the function f tends to a definite limit, so that (6) is a sort of asymptotic law of resistance*.

If we accept the formula (6) as the expression of observed facts, a conclusion of some interest may at once be drawn. Taking the axis of z in the general direction of the flow, if $\overline{w}$ denote the *mean* velocity (with respect to *time*) at any point of space, we have, at the surface,

$$\mu\frac{\partial\overline{w}}{\partial n} = k\rho w_0^2,$$

if w_0 denote the general velocity of the stream, and δn an element of the normal. If we take a linear magnitude l such that

$$\frac{w_0}{l} = \frac{\partial\overline{w}}{\partial n},$$

then l measures the distance between two planes moving with a relative velocity w_0 in the regular 'laminar' flow which would give the same tangential stress. We find

$$w_0 l = \nu/k. \qquad \text{................................(8)}$$

For example, putting $\nu = {\cdot}018$, $w_0 = 300$ [C.S.], $k = {\cdot}0025$, we obtain $l = {\cdot}024$ cm.† The smallness of this result suggests that in the turbulent flow of a fluid the value of $\overline{w}$ falls rapidly to zero within a very minute distance of the walls‡.

The distribution of the mean velocity ($\overline{w}$) over the cross-section has been examined by Stanton§ in some experiments on the flow of air through slightly roughened pipes, where the law of resistance proportional to the square of the velocity was found to hold. Up to a short distance from the walls the velocity followed approximately a parabolic law

$$\overline{w} = \overline{w}_c\left(1 - \beta\frac{r^2}{a^2}\right), \qquad \text{..........................(9)}$$

* Stanton, *Friction*, London, 1923, p. 55.

† Cf. Sir W. Thomson, *Phil. Mag.* (5) xxiv. 277 (1887).

‡ This was in fact found experimentally by Darcy, *l.c.*

§ *Proc. Roy. Soc.* A, lxxxv. 366 (1911).

where $\overline{w}_c$ denotes the mean velocity at the axis, and β is a constant. In some later experiments* he brings evidence for the view that close to the walls there is a region of strictly laminar flow. The depth of this region, in the case examined, was a fraction of a millimetre.

366a. In the experimental arrangement of Mallock and Couette, referred to on p. 588, we have another simple type of steady motion, which is more amenable to experimental investigation. When the inner cylinder was at rest, Mallock concluded that the steady motion represented by Art. 333 (5) was stable so long as the angular velocity of the outer cylinder did not exceed a certain limit, and definitely unstable when it exceeded a certain higher limit. In the intermediate stage there was a susceptibility to disturbing influences, much as in the case of the pipe†. When on the other hand the outer cylinder was fixed, the steady motion was found to be unstable for all speeds of revolution of the inner cylinder. These conclusions require qualification in the light of subsequent work, but the experiments referred to are of interest as the first attempt to study practically a case of turbulent motion other than in a pipe.

The effect of an unsymmetrical, but two-dimensional, disturbance has been discussed mathematically by Harrison‡, by the methods of Reynolds and Orr (Art. 369). He investigates the maximum relative angular velocity of the cylinders which is consistent with stability as regards disturbances of the above type.

The problem has recently been studied, both mathematically and experimentally, by Taylor§, with definite results. Starting with a stable condition and gradually increasing the ratio of the angular velocities, he finds that instability first manifests itself in the form of a three-dimensional and initially steady disturbance which is symmetrical about the axis of rotation, but periodic as regards distance parallel to this axis. The lines of flow when projected on a meridian plane present the appearance of a system of vortices contained in rectangular compartments, and rotating alternately in opposite directions. When the cylinders revolve in the same direction, each compartment extends over the whole radial space between them; in the opposite case there is an outer, but much feebler system of vortices. It was ascertained, both theoretically and experimentally, that when the inner cylinder was fixed, the steady motion was stable for *all* observed speeds of rotation of the outer one. When the outer cylinder was fixed there was stability for sufficiently low speeds of the inner one. In all cases the speed at which instability sets in was sharply defined.

* *Proc. Roy. Soc.* A, xcvii. 413 (1920), and *Friction*, p. 30.

† See a letter by Kelvin, quoted by Rayleigh, *Phil. Mag.* (6) xxviii. (1914) [*Papers*, vi. 266].

‡ *Camb. Trans.* xxii. 425 (1920), and *Proc. Camb. Phil. Soc.* xx. 455 (1921).

§ "Stability of a Viscous Liquid contained between Two Rotating Cylinders," *Phil. Trans.* A, ccxxiii. 289 (1922).

366 b. It has been emphasised repeatedly that unless the velocities involved are exceedingly small, or the spatial relations very constricted, calculations based on the hypothesis of rectilinear flow, such as those of Art. 330, lead to results which are in striking disaccord with experience. For instance, in the case considered in Art. 344 *a*, 3°, if we attribute to μ the usual value for water, it would take an enormous time for the effect of the surface forces to penetrate beyond a very small depth. What really happens is that eddies are formed, with the result that there is an interchange of momentum between adjacent layers of fluid. The conception is the same as in Maxwell's theory of gases, except that we are now concerned with molar, as distinguished from molecular momentum, *i.e.* with the momentum of elementary portions of fluid regarded as continuous (cf. Art. 369).

It has been proposed by various writers, from Reynolds* onwards, to allow for this process by the introduction of a coefficient $\bar{\mu}$ of 'molar' or 'mechanical' or 'eddy' viscosity in place of μ. That is, we assume for instance that the tangential stress on a plane perpendicular to Oz is made up of components

$$\bar{\mu}\frac{\partial \bar{u}}{\partial z}, \qquad \bar{\mu}\frac{\partial \bar{v}}{\partial z},$$

where $\bar{u}$, $\bar{v}$ are the *mean* values of u, v at the point considered, taken over a short interval of time. We thus abandon any attempt to follow in detail the rapid changes which take place, and concern ourselves only with mean effects, in the above sense.

Naturally, this coefficient $\bar{\mu}$ is not to be regarded as a physical constant characteristic of the fluid; its value will depend on the type and scale of motion considered, and will often vary considerably from one part of the fluid to another. It is accordingly not known *à priori*, though sometimes an estimate can be made from analogy, but is to be found by comparison of calculation with experiment. Its meaning is rather that it gives a measure of the degree of turbulence under the circumstances considered.

For instance, in Stanton's experiments, quoted on p. 666, considering the forces on unit length of a cylinder of air of radius r,

$$\bar{\mu}\frac{\partial \bar{w}}{\partial r}\,.\,2\pi r = \frac{\partial p}{\partial z}\,.\,\pi r^2, \qquad \text{.......................................(1)}$$

where $\partial p/\partial z$ is the pressure-gradient along the pipe. Also, assuming (6) of Art. 366,

$$k\rho w_0^2\,.\,2\pi a = -\frac{\partial p}{\partial z}\,.\,\pi a^2, \qquad \text{.......................................(2)}$$

whence

$$\frac{\bar{\mu}}{\rho}\frac{\partial \bar{w}}{\partial r} = -\frac{kr}{a}w_0^2. \qquad \text{.......................................(3)}$$

Hence

$$\frac{\bar{\mu}}{\rho} = \frac{kw_0^2 a}{2\beta \bar{w}_c}. \qquad \text{.......................................(4)}$$

* *l.c. ante* p. 627. (1886) [*Papers*, ii. 236].

This makes $\bar{\mu}$ uniform over the section of the pipe, but proportional to the average velocity over the cross-section, and the radius, conjointly. If we put*

$$\bar{w}_c = 1500, \quad w_0 = 1125, \quad a = 2{\cdot}5, \quad k = {\cdot}0025, \quad \beta = {\cdot}5,$$

we get

$$\bar{\mu}/\rho = 5{\cdot}3, \quad \text{or} \quad \bar{\mu} = {\cdot}0068.$$

Enormously greater values, as may be expected, are found in cases where the motion is on a larger scale†.

366 c. The same idea has been applied to the law of variation of wind with altitude over the earth's surface‡, the earth's rotation being taken into account as in the analogous problem of Art. 334*a*, 5°. We assume the axis of z to be drawn upwards in the direction opposite to that of apparent gravity, and the axes of x and y to revolve with the component (ω) about Oz of the earth's angular velocity. Assuming the motion relative to these axes to be steady, putting $w = 0$, and neglecting horizontal gradients of u and v, we have§ by Art. 203 (1)

$$\left.\begin{aligned} -2\omega v = -\frac{\partial p}{\rho \partial x} + \bar{\nu}\frac{\partial^2 u}{\partial z^2}, \qquad 2\omega u = -\frac{\partial p}{\rho \partial y} + \bar{\nu}\frac{\partial^2 v}{\partial z^2}, \\ 0 = -\frac{\partial p}{\partial z} + g\rho, \end{aligned}\right\} \quad \text{.........(1)}$$

where $\bar{\nu} = \bar{\mu}/\rho$. Now assume that the pressure-gradient in the neighbourhood of the origin is uniform, say

$$-\frac{\partial p}{\rho \partial x} = 0, \qquad -\frac{\partial p}{\rho \partial y} = f. \quad \text{.......................(2)}$$

Then

$$\bar{\nu}\frac{\partial^2}{\partial z^2}(u + iv) - 2i\omega(u + iv) = -if. \quad \text{.................(3)}$$

If we write

$$\beta^2 = \omega/\bar{\nu}, \qquad f/2\omega = V, \quad \text{.......................(4)}$$

the solution which is finite for $z = \infty$ is

$$u + iv = V + Ce^{-(1+i)\beta z}. \quad \text{.........................(5)}$$

At a great height we have $u = V$, $v = 0$; this is the 'gradient wind,' parallel to the isobars, which would prevail if there were no friction.

Suppose that at the ground ($z = 0$) the wind makes an angle α in the positive direction from the axis of x, so that

$$u_0 + iv_0 = V_0 e^{i\alpha}. \quad \text{.............................(6)}$$

Hence

$$\left.\begin{aligned} u &= V + e^{-\beta z}\{V_0\cos(\alpha - \beta z) - V\cos\beta z\}, \\ v &= \qquad e^{-\beta z}\{V_0\sin(\alpha - \beta z) + V\sin\beta z\}. \end{aligned}\right\} \quad \text{..............(7)}$$

* The data are of the same order as in one of Stanton's experiments.

† Cf. Jeffreys, "On Turbulence in the Ocean," *Phil. Mag.* (6) xxxix. 578 (1920).

‡ G. I. Taylor, "Eddy Motion in the Atmosphere," *Phil. Trans.* A, ccxv. 1 (1915).

§ The marks denoting mean velocities of u, v, and p (with respect to time) are omitted, as unnecessary for the moment.

We may assume that the tangential stress at the ground has the same direction as the velocity there; or

$$\frac{\partial u}{\partial z} : \frac{\partial v}{\partial z} = u : v, \qquad \text{for } z = 0, \quad \text{.......................(8)}$$

whence, after reduction,

$$V_0 = V(\cos\alpha - \sin\alpha). \quad \text{..........................(9)}$$

Substituting from (9) in (7), we have

$$\left.\begin{aligned} u/V &= 1 - \sin\alpha\,\{\cos(\alpha - \beta z) + \sin(\alpha - \beta z)\}\, e^{-\beta z}, \\ v/V &= \quad \sin\alpha\,\{\cos(\alpha - \beta z) - \sin(\alpha - \beta z)\}\, e^{-\beta z}. \end{aligned}\right\} \quad \text{.........(10)}$$

The heights at which the wind coincides in direction with the gradient wind are given by $v = 0$, or

$$\tan(\alpha - \beta z) = 1. \quad \text{..............................(11)}$$

The equation (9) shews that α must be $< \frac{1}{4}\pi$, so that the first value of z which satisfies (11) is

$$z = (\alpha + \tfrac{3}{4}\pi)/\beta. \quad \text{..............................(12)}$$

Comparisons of the theoretical results with observation have led to determinations of the kinematic eddy viscosity $\bar{\nu}$ of the order of 10^5 C.G.S.

367. Although much has been written on the subject, the explanation of the practical instability of linear flow under the conditions stated in Arts. 365, 366, and of the manner in which the irregular eddies are maintained against viscosity, has yet to be found. We can only attempt here a brief account of various attempts which have been made to elucidate the question.

Rayleigh, in a series of papers*, has examined the stability, for infinitely small disturbances, of various types of steady motion, such as might be produced by viscosity. Although viscosity is, in the *disturbed* motion, ignored, the results may be expected to throw some light on the question, except in cases where the influence of a boundary predominates. The exception is, however, important.

As the method is simple, and as the results have an independent interest, we may briefly notice the two-dimensional form of the problem.

Let us suppose that in a slight disturbance of the steady laminar motion

$$u = U, \qquad v = 0, \qquad w = 0,$$

where U is a function of y only, we have

$$u = U + u', \qquad v = v', \qquad w = 0. \quad \text{................................(1)}$$

The equation of continuity is

$$\frac{\partial u'}{\partial x} + \frac{\partial v'}{\partial y} = 0. \quad \text{..(2)}$$

* *Proc. Lond. Math. Soc.* x. 4 (1879), xi. 57 (1880), xix. 67 (1887), xxvii. 5 (1895); *Phil. Mag.* (5) xxxiv. 59, 177 (1892); (6) xxvi. 1001 (1913) [*Papers*, i. 361, 374, iii. 575, 594, iv. 203].

The dynamical equations reduce, by Art. 146 (4), to the condition of persistent vorticity $D\zeta/Dt=0$, or

$$\frac{\partial\zeta}{\partial t}+(U+u')\frac{\partial\zeta}{\partial x}+v'\frac{\partial\zeta}{\partial y}=0, \quad\text{.........(3)}$$

where

$$\zeta=\frac{\partial v'}{\partial x}-\frac{\partial u'}{\partial y}-\frac{dU}{dy}. \quad\text{.........(4)}$$

Hence, neglecting terms of the second order in u', v',

$$\left(\frac{\partial}{\partial t}+U\frac{\partial}{\partial x}\right)\left(\frac{\partial v'}{\partial x}-\frac{\partial u'}{\partial y}\right)-\frac{d^2U}{dy^2}v'=0. \quad\text{.........(5)}$$

Contemplating now a disturbance which is periodic in respect to x, we assume that u', v' vary as $e^{ikx+i\sigma t}$. Hence, from (2) and (5),

$$iku'+\frac{\partial v'}{\partial y}=0, \quad\text{.........(6)}$$

and

$$i(\sigma+kU)\left(ikv'-\frac{\partial u'}{\partial y}\right)-\frac{d^2U}{dy^2}v'=0. \quad\text{.........(7)}$$

Eliminating u', we find

$$(\sigma+kU)\left(\frac{\partial^2 v'}{\partial y^2}-k^2v'\right)-\frac{d^2U}{dy^2}kv'=0, \quad\text{.........(8)}$$

which is the fundamental equation.

If, for any value of y, dU/dy is discontinuous, the equation (8) must be replaced by

$$(\sigma+kU)\,\Delta\left(\frac{\partial v'}{\partial y}\right)-\Delta\left(\frac{dU}{dy}\right)kv'=0, \quad\text{.........(9)}$$

where Δ denotes the difference of the values of the respective quantities on the two sides of the plane of discontinuity. This is obtained from (8) by integration with respect to y, the discontinuity being regarded as the limit of an infinitely rapid variation. The equation may also be obtained as the condition of continuity of pressure, or as the condition that there should be no tangential slipping at the (displaced) boundary.

At a fixed boundary, we must have $v'=0$.

1°. Suppose that a layer of fluid of uniform vorticity bounded in the undisturbed state by the planes $y=\pm h$ is interposed between two masses of fluid moving irrotationally, the velocity being everywhere continuous. This forms an interesting variation of a problem discussed in Art. 234.

Assuming, then, $U=\mathbf{u}$ for $y>h$, $U=\mathbf{u}y/h$ for $h>y>-h$, and $U=-\mathbf{u}$ for $y<-h$, we notice that $d^2U/dy^2=0$ except at the surfaces of transition, so that (8) reduces to

$$\frac{\partial^2 v'}{\partial y^2}-k^2v'=0. \quad\text{.........(10)}$$

The appropriate solutions of this are:

$$\left.\begin{aligned} v'&=Ae^{-ky}, &&\text{for } y>h;\\ v'&=Be^{-ky}+Ce^{ky}, &&\text{for } h>y>-h;\\ v'&=De^{ky}, &&\text{for } y<-h. \end{aligned}\right\} \quad\text{.........(11)}$$

The continuity of v' requires

$$\left.\begin{aligned} Ae^{-kh}&=Be^{-kh}+Ce^{kh},\\ De^{-kh}&=Be^{kh}+Ce^{-kh}. \end{aligned}\right\} \quad\text{.........(12)}$$

With the help of these relations, the condition (9) gives

$$\left.\begin{aligned} 2(\sigma+k\mathbf{u})\,Ce^{kh}-\frac{\mathbf{u}}{h}(Be^{-kh}+Ce^{kh})&=0,\\ 2(\sigma-k\mathbf{u})\,Be^{kh}+\frac{\mathbf{u}}{h}(Be^{kh}+Ce^{-kh})&=0. \end{aligned}\right\} \quad\text{.........(13)}$$

Eliminating the ratio $B : C$, we obtain

$$\sigma^2=\frac{\mathbf{u}^2}{4h^2}\{(2kh-1)^2-e^{-4kh}\}. \qquad (14)$$

For small values of kh this makes $\sigma^2=-k^2\mathbf{u}^2$, as in the case of absolute discontinuity (Art. 234). For large values of kh, on the other hand, $\sigma=\pm k\mathbf{u}$, indicating stability. Hence the question as to the stability for disturbances of wave-length λ depends on the ratio $\lambda/2h$. The function in $\{\ \}$ on the right-hand side of (14) has been tabulated by Rayleigh. It appears that there is instability if $\lambda/2h>5$, about; and that the instability is a maximum for $\lambda/2h=8$.

2°. In the papers referred to, Rayleigh has further investigated various cases of flow between parallel walls, with the view of throwing light on the conditions of stability of linear motion in a pipe. The main result is that if d^2U/dy^2 does not change sign, in other words, if the curve with y as abscissa and U as ordinate is of one curvature throughout, the motion is stable. Since, however, the disturbed motion involves slipping at the walls, it remains doubtful how far the conclusions apply to the question at present under consideration, in which the condition of no slipping appears to be fundamental.

3°. The substitution of (10) for (8), when $d^2U/dy^2=0$, is equivalent to assuming that the vorticity ζ is the same as in the undisturbed motion; since on this hypothesis we have

$$\frac{\partial u'}{\partial y}=\frac{\partial v'}{\partial x}=ikv', \qquad (15)$$

which, with (6), leads to the equation in question.

It is to be observed, however, that when $d^2U/dy^2=0$ the equation (8) may be satisfied, for a particular value of y, by $\sigma+kU=0$. For example, we may suppose that at the plane $y=0$ a thin layer of (infinitely small) additional vorticity is introduced. We then have, on the hypothesis that the fluid is unlimited,

$$v'=Ae^{\mp ky+i(\sigma t+kx)}, \qquad (16)$$

the upper or the lower sign being taken according as y is positive or negative. The condition (9) is then satisfied by

$$\sigma+kU_0=0, \qquad \Delta\left(\frac{dU}{dy}\right)=0, \qquad (17)$$

where U_0 denotes the value of U for $y=0$. Since the superposition of a uniform velocity in the direction of x does not alter the problem, we may suppose $U_0=0$, and therefore $\sigma=0$. The disturbed motion is steady; in other words, the original state of flow is (to the first order of small quantities) *neutral* for a disturbance of this kind*.

368. Kelvin attacked directly the very difficult problem of determining the stability of laminar motion when viscosity is taken into account†. The cases specially considered are (i) the flow under pressure between fixed parallel planes (see Art. 330), (ii) the uniform shearing motion between parallel planes one of which has a constant velocity relative to the other, which is fixed, and (iii) the flow of a stream over an inclined plane bed. His general conclusion was that the laminar flow is in all cases stable for infinitely small disturbances,

* Cf. Sir W. Thomson, "On a Disturbing Infinity in Lord Rayleigh's solution for Waves in a plane Vortex Stratum," *Brit. Ass. Rep.* 1880, p. 492 [*Papers*, iv. 186], and Rayleigh's reply, *Proc. Lond. Math. Soc.* xxvii. 5 [*Papers*, iv. 203].

† "Rectilinear Motion of Viscous Fluid between two Parallel Planes," *Phil. Mag.* (5) xxiv. 188 (1887); "Broad River flowing down an Inclined Plane Bed," *Phil. Mag.* (5) xxiv. 272 (1887) [*Papers*, iv. 321].

but that for disturbances exceeding a certain limit the motion becomes unstable, these limits of stability being narrower the smaller the viscosity. The investigation is difficult, and portions of it have been questioned by Rayleigh* and Orr, to the latter of whom we are indebted for a detailed examination of the whole matter†. Most writers who have attacked the subject are disposed however to regard the conclusion as probable, though as yet hardly demonstrated. It will be noticed that it is in accordance with the observations of Reynolds and others, referred to in Arts. 365, 366.

In the case of uniform shearing motion between parallel planes $y=0$, $y=h$, the former of which is at rest, the first steps of the procedure are as follows. We assume, for the undisturbed motion,

$$u=\beta y, \qquad v=0, \qquad w=0, \quad \text{.................}(1)$$

and for the disturbed motion

$$u=\beta y-\frac{\partial\psi}{\partial y}, \qquad v=\frac{\partial\psi}{\partial x}, \qquad w=0. \quad \text{.................}(2)$$

The vorticity is therefore

$$\zeta=-\beta+\nabla_1^2\psi. \quad \text{.................}(3)$$

The third of equations (8) of Art. 328 gives

$$\frac{\partial\zeta}{\partial t}+u\frac{\partial\zeta}{\partial x}+v\frac{\partial\zeta}{\partial y}=\nu\nabla_1^2\zeta. \quad \text{.................}(4)$$

Substituting from (2), and neglecting terms of the second order in ψ, we have

$$\left(\frac{\partial}{\partial t}+\beta y\frac{\partial}{\partial x}\right)\nabla_1^2\psi=\nu\nabla_1^4\psi. \quad \text{.................}(5)$$

Assuming a disturbance of the type $e^{i(\sigma t+kx)}$, we have

$$\frac{\partial^2 S}{\partial y^2}=\left\{k^2+\frac{i(\sigma+k\beta y)}{\nu}\right\}S, \quad \text{.................}(6)$$

where

$$S=\nabla_1^2\psi=\frac{\partial^2\psi}{\partial y^2}-k^2\psi, \quad \text{.................}(7)$$

the exponential factor being omitted.

Since the conditions (1) must hold at the boundaries, we must have $\partial\psi/\partial x=0$, $\partial\psi/\partial y=0$, or

$$\psi=0, \qquad \frac{\partial\psi}{\partial y}=0, \quad \text{.................}(8)$$

for $y=0$ and for $y=h$.

If S be the complete solution of (6), the integration of (7) by the method of 'variation of parameters' gives

$$\psi=\frac{1}{2k}\left\{e^{ky}\int e^{-ky}S\,dy-e^{-ky}\int e^{ky}S\,dy\right\}, \quad \text{.................}(9)$$

whence

$$\frac{\partial\psi}{\partial y}=\frac{1}{2}\left\{e^{ky}\int e^{-ky}S\,dy+e^{-ky}\int e^{ky}S\,dy\right\}. \quad \text{.................}(10)$$

The indefinite integrals introduce of course two arbitrary additive constants, in addition to the two which are involved in S.

The conditions (8) are fulfilled for $y=0$ if we take the lower limit of the integrals to be zero. The conditions at $y=h$ lead to

$$\int_0^h e^{-kh}S\,dy=0, \qquad \int_0^h e^{kh}S\,dy=0. \quad \text{.................}(11)$$

* *l.c. ante* p. 670.

† "The Stability or Instability of the Steady Motions of a Perfect Liquid and of a Viscous Liquid," *Proc. Roy. Irish Acad.* xxvii. 9, 69 (1906–7).

Hence, if we write
$$S=C_1S_1+C_2S_2, \qquad \text{.........(12)}$$
where S_1, S_2 are any two independent solutions of (6), we have, on elimination of the arbitrary constants C_1, C_2,
$$\int_0^h e^{ky}S_1\,dy\,.\int_0^h e^{-ky}S_2\,dy-\int_0^h e^{-ky}S_1\,dy\,.\int_0^h e^{ky}S_2\,dy=0, \qquad \text{.........(13)}$$
an equation given by Orr*, and afterwards independently by Sommerfeld†. This determines the values of σ when k is given. For stability it is essential that if $\sigma=p+iq$, q should be positive.

If we put
$$k^2+\frac{i(\sigma+k\beta y)}{\nu}=\left(\frac{k\beta}{\nu}\right)^{\frac{2}{3}}\eta, \qquad \text{.........(14)}$$
the equation (6) takes the form
$$\frac{\partial^2 S}{\partial\eta^2}+\eta S=0, \qquad \text{.........(15)}$$
which is integrable by series‡. Thus
$$S=A_1\left\{1-\frac{\eta^3}{2\,.\,3}+\frac{\eta^6}{2\,.\,5\,.\,3\,.\,6}-\frac{\eta^9}{2\,.\,5\,.\,8\,.\,3\,.\,6\,.\,9}+\dots\right\}$$
$$+A_2\eta\left\{1-\frac{\eta^3}{3\,.\,4}+\frac{\eta^6}{3\,.\,6\,.\,4\,.\,7}-\frac{\eta^9}{3\,.\,6\,.\,9\,.\,4\,.\,7\,.\,10}+\dots\right\}, \qquad \text{.........(16)}$$
or, as it may also be written,
$$S=B_1\eta^{\frac{1}{2}}J_{-\frac{1}{3}}(\tfrac{2}{3}\eta^{\frac{3}{2}})+B_2\eta^{\frac{1}{2}}J_{\frac{1}{3}}(\tfrac{2}{3}\eta^{\frac{3}{2}}), \qquad \text{.........(17)}$$
in the notation of Bessel's Functions§.

The further investigation of the problem is difficult. It has been carried forward to some extent by Orr, and more recently by Rayleigh||, in whose papers other references will be found.

In a recent discussion¶ of the question Prof. Southwell, starting from the equation (5) assumes, in effect, $\sigma=p+iq$, so that the time-factor is $e^{-qt+i(pt+kx)}$, and proceeds to shew that if $p=0$, *i.e.* if the disturbance is non-oscillating, the admissible values of q are necessarily positive, and the shearing motion accordingly, so far, stable. He examines, further, the nature of the corresponding modes of decay, and illustrates them by an interesting series of diagrams of the relative stream-lines.

369. Reynolds, in a remarkable paper**, attacked the general question from a different point of view. Taking the turbulent motion as already existing, he sought to establish a criterion which shall decide whether the turbulent character will increase or diminish or be stationary.

For this purpose the velocity (u, v, w) is resolved in two constituents. We may, for instance, write
$$\bar{u}=\frac{1}{\tau}\int_{t-\frac{1}{2}\tau}^{t+\frac{1}{2}\tau}u\,dt,\quad \bar{v}=\frac{1}{\tau}\int_{t-\frac{1}{2}\tau}^{t+\frac{1}{2}\tau}v\,dt,\quad \bar{w}=\frac{1}{\tau}\int_{t-\frac{1}{2}\tau}^{t+\frac{1}{2}\tau}w\,dt; \qquad \text{.........(1)}$$

* *l.c. ante* p. 673. † *Atti del IV. Congr. intern. dei matematici*, Roma, 1909, ii. 116.

‡ Cf. Stokes, *Camb. Trans.* x. 106 (1857) [*Papers*, iv. 77].

§ For the relation between (15) and Riccati's and Bessel's equations see Forsyth, *Differential Equations*, Art. 111.

|| "Stability of Viscous Fluid Motion," *Phil. Mag.* (6) xxviii. (1914); "On the Stability of the Simple Shearing Motion of a Viscous Incompressible Fluid," *Phil. Mag.* (6) xxx. 329 (1915) [*Papers*, vi. 266, 341].

¶ *Phil. Trans.* A, ccix. 205 (1930).

** "On the Dynamical Theory of Incompressible Viscous Fluids and the Determination of the Criterion," *Phil. Trans.* A, clxxxvi. 123 (1894) [*Papers*, ii. 535].

so that $\bar{u}$, $\bar{v}$, $\bar{w}$ are the mean values of u, v, w at the point (x, y, z), taken over an interval of time extending from $t-\frac{1}{2}\tau$ to $t+\frac{1}{2}\tau$. Again, we might consider the mean values at the instant t over a space (*e.g.* a sphere) surrounding the point (x, y, z); thus

$$\bar{u}=\frac{1}{S}\iiint u\,dx\,dy\,dz, \quad \bar{v}=\frac{1}{S}\iiint v\,dx\,dy\,dz, \quad \bar{w}=\frac{1}{S}\iiint w\,dx\,dy\,dz. \qquad (2)$$

Or, again, we might take a double mean, for times ranging over an interval τ, and points ranging over a space S. The actual velocities are in each case denoted by

$$u=\bar{u}+u', \quad v=\bar{v}+v', \quad w=\bar{w}+w', \qquad (3)$$

when u', v', w' may be called the components of the turbulent motion. This implies that

$$\bar{u}'=0, \quad \bar{v}'=0, \quad \bar{w}'=0, \qquad (4)$$

where the bar placed over a symbol denotes the mean value, taken according to the particular convention adopted.

For simplicity we will adopt the definition of mean value which is embodied in the formulae (1).

Reynolds starts from the dynamical equations in the forms

$$\rho\frac{\partial u}{\partial t}=\rho X+\frac{\partial}{\partial x}(p_{xx}-\rho uu)+\frac{\partial}{\partial y}(p_{yx}-\rho uv)+\frac{\partial}{\partial z}(p_{zx}-\rho uw), \;\ldots,\; \ldots, \qquad (5)$$

which are seen to be equivalent to Art. 328 (1) in virtue of the equation of continuity

$$\frac{\partial u}{\partial x}+\frac{\partial v}{\partial y}+\frac{\partial w}{\partial z}=0. \qquad (6)$$

These forms are not essential to the argument, but are interesting as an application of the method employed by Maxwell* in the kinetic theory of gases. They express the rate of variation of the momentum contained in a fixed rectangular space $\delta x\,\delta y\,\delta z$, as a consequence partly of the forces acting on the substance which at the moment occupies this space, and partly of the flux of matter across the boundary, carrying its momentum with it. Thus the fluxes of x-momentum across unit areas perpendicular to Ox, Oy, Oz are $\rho u\,.\,u$, $\rho v\,.\,u$, and $\rho w\,.\,u$, respectively; and taking the difference of the fluxes across opposite faces of the elementary space $\delta x\,\delta y\,\delta z$, we obtain a gain of x-momentum equal to

$$-\frac{\partial}{\partial x}(\rho u\,\delta y\,\delta z\,.\,u)\,\delta x-\frac{\partial}{\partial y}(\rho v\,\delta z\,\delta x\,.\,u)\,\delta y-\frac{\partial}{\partial z}(\rho w\,\delta x\,\delta y\,.\,u)\,\delta z$$

per unit time.

We now take the mean value of each member of the equations (5), using the substitutions (3). It is assumed that we may, without sensible error, take the mean values of $\bar{u}$, $\bar{u}u'$, $\bar{u}v'$, $\bar{u}w'$, ... to be $\bar{u}$, 0, 0, 0, ..., respectively. This is not exact, but it is permissible provided the fluctuations of u, v, w about their mean values are sufficiently numerous within the time-interval τ. It follows that

$$\overline{uu}=\bar{u}\bar{u}+\overline{u'u'}, \quad \overline{uv}=\bar{u}\bar{v}+\overline{u'v'}, \quad \overline{uw}=\bar{u}\bar{w}+\overline{u'w'}, \;\ldots \qquad (7)$$

In this way we obtain

$$\rho\frac{\partial\bar{u}}{\partial t}=\rho X+\frac{\partial}{\partial x}(\bar{p}_{xx}-\rho\bar{u}\bar{u}-\rho\overline{u'u'})+\frac{\partial}{\partial y}(\bar{p}_{yx}-\rho\bar{u}\bar{v}-\rho\overline{u'v'})$$
$$+\frac{\partial}{\partial z}(\bar{p}_{zx}-\rho\bar{u}\bar{w}-\rho\overline{u'w'}), \;\ldots,\; \ldots, \qquad (8)$$

whilst the equation of continuity gives

$$\frac{\partial\bar{u}}{\partial x}+\frac{\partial\bar{v}}{\partial y}+\frac{\partial\bar{w}}{\partial z}=0. \qquad (9)$$

* *l.c. ante* p. 575.

These are the equations of mean motion*. It is to be noticed that the dynamical equations have the same form as the exact equations (5), provided we introduce additional stress-components

$$P_{xx} = -\rho\overline{u'u'}, \quad P_{yx} = -\rho\overline{u'v'}, \quad P_{zx} = -\rho\overline{u'w'}, \ \ldots \qquad \text{.................(10)}$$

This recalls the explanation of gaseous viscosity by Maxwell (*l.c.*).

The equations (8) may be written, in virtue of (9),

$$\left(\frac{\partial}{\partial t} + \bar{u}\frac{\partial}{\partial x} + \bar{v}\frac{\partial}{\partial y} + \bar{w}\frac{\partial}{\partial z}\right)\rho\bar{u} = \rho X + \frac{\partial}{\partial x}(\bar{p}_{xx} - \rho\overline{u'u'}) + \frac{\partial}{\partial y}(\bar{p}_{yx} - \rho\overline{u'v'})$$

$$+ \frac{\partial}{\partial z}(\bar{p}_{zx} - \rho\overline{u'w'}), \ \ldots, \ \ldots \qquad \text{........................(11)}$$

If we multiply these by $\bar{u}$, $\bar{v}$, $\bar{w}$ in order, and add, we obtain

$$\left(\frac{\partial}{\partial t} + \bar{u}\frac{\partial}{\partial x} + \bar{v}\frac{\partial}{\partial y} + \bar{w}\frac{\partial}{\partial z}\right)\tfrac{1}{2}\rho(\bar{u}^2 + \bar{v}^2 + \bar{w}^2)$$

$$= \rho(X\bar{u} + Y\bar{v} + Z\bar{w}) + \bar{u}\left\{\frac{\partial}{\partial x}(\bar{p}_{xx} - \rho\overline{u'u'}) + \frac{\partial}{\partial y}(\bar{p}_{yx} - \rho\overline{u'v'}) + \frac{\partial}{\partial z}(\bar{p}_{zx} - \rho\overline{u'w'})\right\}$$

$$+ \bar{v}\left\{\frac{\partial}{\partial x}(\bar{p}_{xy} - \rho\overline{v'u'}) + \frac{\partial}{\partial y}(\bar{p}_{yy} - \rho\overline{v'v'}) + \frac{\partial}{\partial z}(\bar{p}_{zy} - \rho\overline{v'w'})\right\}$$

$$+ \bar{w}\left\{\frac{\partial}{\partial x}(\bar{p}_{xz} - \rho\overline{w'u'}) + \frac{\partial}{\partial y}(\bar{p}_{yz} - \rho\overline{w'v'}) + \frac{\partial}{\partial z}(\bar{p}_{zz} - \rho\overline{w'w'})\right\}. \qquad \text{.........(12)}$$

Let us first suppose that there are no extraneous forces X, Y, Z; and let us apply (12) to the case of a region bounded by fixed walls at which u, v, w, and therefore also $\bar{u}$, $\bar{v}$, $\bar{w}$, all vanish. If we write

$$T_0 = \tfrac{1}{2}\rho\iiint(\bar{u}^2 + \bar{v}^2 + \bar{w}^2)\,dx\,dy\,dz, \qquad \text{.............................(13)}$$

we obtain, after some partial integrations,

$$\frac{dT_0}{dt} = -\iiint\Phi_0\,dx\,dy\,dz + \iiint\Psi\,dx\,dy\,dz, \qquad \text{........................(14)}$$

where

$$\Phi_0 = \bar{p}_{xx}\frac{\partial\bar{u}}{\partial x} + \bar{p}_{yy}\frac{\partial\bar{v}}{\partial y} + \bar{p}_{zz}\frac{\partial\bar{w}}{\partial z} + \bar{p}_{yz}\left(\frac{\partial\bar{w}}{\partial y} + \frac{\partial\bar{v}}{\partial z}\right) + \bar{p}_{zx}\left(\frac{\partial\bar{u}}{\partial z} + \frac{\partial\bar{w}}{\partial x}\right) + \bar{p}_{xy}\left(\frac{\partial\bar{v}}{\partial x} + \frac{\partial\bar{u}}{\partial y}\right)$$

$$= \mu\left\{2\left(\frac{\partial\bar{u}}{\partial x}\right)^2 + 2\left(\frac{\partial\bar{v}}{\partial y}\right)^2 + 2\left(\frac{\partial\bar{w}}{\partial z}\right)^2 + \left(\frac{\partial\bar{w}}{\partial y} + \frac{\partial\bar{v}}{\partial z}\right)^2 + \left(\frac{\partial\bar{u}}{\partial z} + \frac{\partial\bar{w}}{\partial x}\right)^2 + \left(\frac{\partial\bar{v}}{\partial x} + \frac{\partial\bar{u}}{\partial y}\right)^2\right\}, \qquad \text{...........(15)}$$

and

$$\Psi = \rho\left\{\overline{u'u'}\frac{\partial\bar{u}}{\partial x} + \overline{v'v'}\frac{\partial\bar{v}}{\partial y} + \overline{w'w'}\frac{\partial\bar{w}}{\partial z} + \overline{v'w'}\left(\frac{\partial\bar{w}}{\partial y} + \frac{\partial\bar{v}}{\partial z}\right)\right.$$

$$\left. + \overline{w'u'}\left(\frac{\partial\bar{u}}{\partial z} + \frac{\partial\bar{w}}{\partial x}\right) + \overline{u'v'}\left(\frac{\partial\bar{v}}{\partial x} + \frac{\partial\bar{u}}{\partial y}\right)\right\}. \qquad \text{...........(16)}$$

The formula (14) gives the rate of variation of the energy of the mean motion ($\bar{u}$, $\bar{v}$, $\bar{w}$). The first term on the right-hand side represents the dissipation due to the mean motion alone, and is essentially negative. The second term represents the rate at which work is being done by the fictitious stresses (10).

* Or rather 'mean-mean-motion,' in the phraseology of Reynolds. He applies the term 'mean-motion' to the system of velocities (u, v, w) to distinguish it from 'molecular motion.' The turbulent motion (u', v', w') is called by him 'relative-mean-motion.'

Now if T be the true kinetic energy, we may write, in virtue of assumptions already made,

$$\bar{T} = T_0 + \bar{T}', \qquad \text{.........(17)}$$

where

$$T' = \tfrac{1}{2}\rho \iiint (u'^2 + v'^2 + w'^2)\, dx\, dy\, dz, \qquad \text{.........(18)}$$

i.e. T' is the kinetic energy of the eddying motion. By the method of Art. 344 it may be shewn that on the present supposition of fixed boundaries at which there is no slipping, the total dissipation is, on the average, equal to the sum of the dissipations due to the mean motion and the eddying motion respectively. Thus

$$\frac{d\bar{T}}{dt} = -\iiint \Phi_0\, dx\, dy\, dz - \iiint \overline{\Phi'}\, dx\, dy\, dz, \qquad \text{.........(19)*}$$

where

$$\Phi' = \mu\left\{2\left(\frac{\partial u'}{\partial x}\right)^2 + 2\left(\frac{\partial v'}{\partial y}\right)^2 + 2\left(\frac{\partial w'}{\partial z}\right)^2 + \left(\frac{\partial w'}{\partial y} + \frac{\partial v'}{\partial z}\right)^2 + \left(\frac{\partial u'}{\partial z} + \frac{\partial w'}{\partial x}\right)^2 + \left(\frac{\partial v'}{\partial x} + \frac{\partial u'}{\partial y}\right)^2\right\}. \qquad \text{.........(20)}$$

Comparing with (14), we have

$$\frac{d\overline{T'}}{dt} = -\iiint \overline{\Phi'}\, dx\, dy\, dz - \iiint \Psi\, dx\, dy\, dz. \qquad \text{.........(21)}$$

The sign of the expression on the right-hand side determines whether the mean energy $\overline{T'}$ of the eddying motion (u', v', w') will increase or diminish. The first part, which alone involves the viscosity μ, is essentially negative; the second part depends on the inertia of the fluid, and may be positive or negative according to circumstances.

When there are extraneous forces X, Y, Z to be taken into account, and when the velocities u, v, w do not necessarily vanish at the boundary of the region considered, the equation (14) requires to be amended by the addition of terms which represent partly the convection of kinetic energy of mean motion into the region, partly the work done by the forces X, Y, Z, and partly the work done at the boundary by the mean stresses $\bar{p}_{xx}$, $\bar{p}_{yx}$, $\bar{p}_{zx}$, ..., and by the fictitious stresses P_{xx}, P_{yx}, P_{zx}, ...

The equation (21), on the other hand, requires only the addition of a term representing the convection of the energy of turbulent motion across the boundary.

The derivation of the remarkable formulae (14) and (21), and of the modifications just referred to, appears to be free from objection, on the convections adopted. But, in applying these formulae to actual conditions, the restrictions and assumptions which have been introduced as to the character of the turbulent motions must be borne in mind.

One or two consequences of the formula (21) may be noted†. In the first place, the relative magnitude of the two terms on the right-hand side is unaffected if we reverse the signs of u', v', w', or if we multiply them by any constant factor. The stability of a given state of mean motion should not therefore depend on the *scale* of the disturbance. On the other hand, certain combinations of u', v', w' appear to be more favourable to stability than others. Thus, in the case of disturbed laminar motion parallel to Ox, between two rigid planes $y = \pm b$, the formula (16) reduces to

$$\Psi = \rho\overline{u'v'}\frac{\partial \bar{u}}{\partial y}, \qquad \text{.........(22)}$$

* It should be noticed that we are here virtually taking the differential time-element δt to be of the order of magnitude of the interval τ employed in the definitions (1). The procedure in the text avoids the use of some very lengthy equations which appear in the original.

† Cf. Lorentz, "Ueber die Entstehung turbulenter Flüssigkeitsbewegungen und über den Einfluss dieser Bewegungen bei der Strömung durch Röhren," *Abhandlungen über theoretische Physik*, Leipzig, 1907, i. 43. The paper is a revised form of one published in 1897.

so that the types of disturbance which tend to increase are those in which (for $y>0$) combinations of u', v' with the same sign preponderate. This indicates a tendency to equalization of the velocity in the different strata. Again, the relative importance of the second term in (21), which alone can contribute to the increase of $\overline{T'}$, is greater the greater the rates of strain $\partial\bar{u}/\partial x$, ... in the mean motion. This suggests a reason why a given type of mean motion does not begin to break down until a certain critical velocity is reached.

If we apply the (modified) formulae to the case of flow in a uniform cylindrical pipe, on the supposition that the pressure gradient $(-d\bar{p}/dx)$ is zero, we find

$$\frac{dT_0}{dt}=\rho X\bar{u}\pi a^2-2\pi\int_0^a \Phi_0 r\,dr+2\pi\int_0^a \Psi r\,dr, \qquad (23)$$

and

$$\frac{d\overline{T'}}{dt}=-\iint\overline{\Phi'}\,dy\,dz-2\pi\int_0^a \Psi r\,dr, \qquad (24)$$

where

$$\Phi_0=\mu\left(\frac{\partial\bar{u}}{\partial r}\right)^2, \qquad \Psi=\overline{\rho u'q'}\,\frac{\partial\bar{u}}{\partial r}. \qquad (25)$$

The region here considered is that contained between two cross-sections (of area πa^2) at unit distance apart; the axis of x coincides with that of the pipe; and q denotes the velocity at right angles to this axis. It is assumed of course that $\bar{q}=0$ and $\partial\bar{u}/\partial x=0$; also that the mean state of things is in all respects the same at each section. The conditions of steady motion are obtained by equating the right-hand members of (23) and (24) to zero.

Reynolds discusses in detail the two-dimensional form of the problem, where there is a flow parallel to x between two fixed plane walls $y=\pm b$. Assuming that $\bar{u}$ varies as b^2-y^2, in conformity with Art. 330, he seeks to determine a minimum value of the flux consistent with the condition $d\overline{T'}/dt=0$; but for this we must refer to the original paper. The result obtained is that the critical ratio $u_0 b/\nu$, where u_0 is the mean value of $\bar{u}$ between the limits $y=\pm b$, must exceed 258*.

Resistance of Fluids.

370. This subject is important in relation to many practical questions, *e.g.* the propulsion of ships, the flight of projectiles, and the effect of wind on structures. Although it has recently been studied with renewed energy, owing to its bearing on the problems of artificial flight, our knowledge of it is still largely empirical.

It has been seen that in the case of an isolated body moving in frictionless liquid, at a distance from the boundaries (if any), there is no abstraction of energy; in particular, if the motion of the fluid has been started from rest, and is therefore irrotational and acyclic, its influence can be completely allowed for by a modification of the *inertia* of the solid† (Arts. 92, 117).

The first attempt to obtain, on exact theoretical lines, a result less opposed to ordinary experience is contained in the investigations of Kirchhoff and

* A different result is obtained by Sharpe, "On the Stability of the Motion of a Viscous Liquid," *Trans. Amer. Math. Soc.* vi. 496 (1905), where also the case of flow through a cylindrical pipe is investigated. These problems, together with that of uniform shearing motion between parallel planes, have been treated more fully by Orr, *l.c. ante* p. 673. The differences in the numerical results obtained appear to arise from differences in the types of disturbance considered. The last-mentioned problem has also been treated by Lorentz (*l.c.*).

† The absence of resistance, properly so called, in such cases is often referred to by continental writers as the 'paradox of d'Alembert.'

Rayleigh relating to the two-dimensional form of the problem of the motion of a plane lamina (Arts. 76, 77). It is to be noticed that the motion of the fluid in such problems is no longer strictly irrotational, a surface of discontinuity being equivalent to a vortex-sheet (Art. 151).

Apart from the fact that viscosity is ignored, this theory is open to the objection that the unlimited mass of 'dead-water' following the lamina implies an infinite kinetic energy, and on this as on other grounds it must be recognized that the proper application of the methods of Helmholtz and Kirchhoff is to the case of *free* surfaces, as of a jet*. The calculations of Kirchhoff and Rayleigh give, it is true, a resistance varying as the square of velocity, as is required, on their assumptions, by the principle of momentum†, and is found to hold within some limits in practice, but the distribution of pressure over the surfaces of the lamina is found to be widely different. There is not merely an excess of pressure on the anterior face, but a defect of pressure, or suction, at the rear, both circumstances contributing to the total resistance. This is exemplified in the annexed diagrams‡, where the

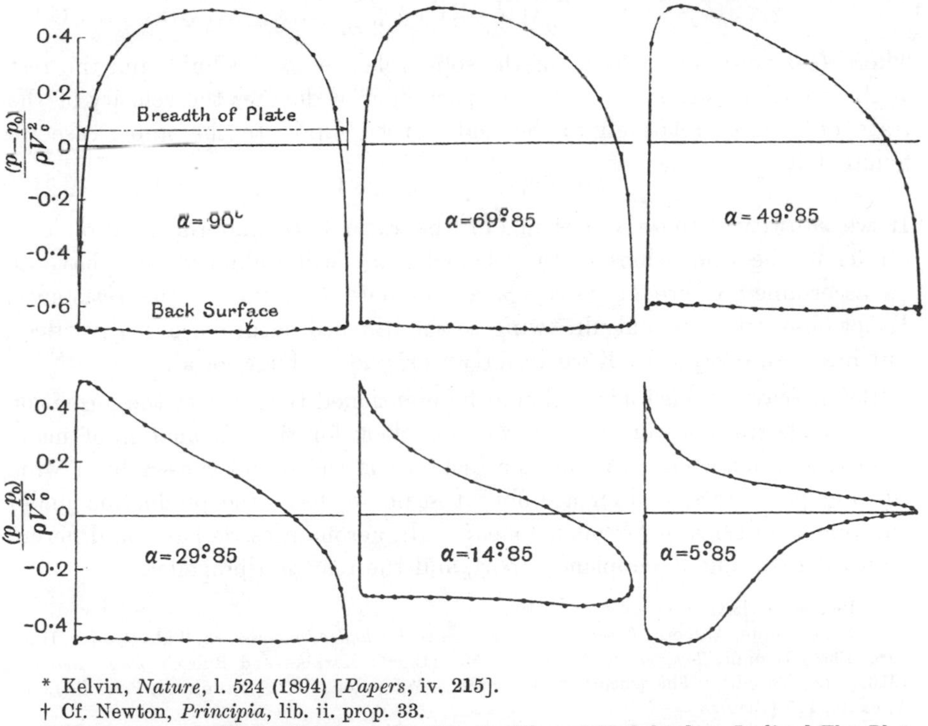

* Kelvin, *Nature*, l. 524 (1894) [*Papers*, iv. 215].

† Cf. Newton, *Principia*, lib. ii. prop. 33.

‡ From a paper by Fage and Johannsen, "On the Flow of Air behind an Inclined Flat Plate of Infinite Span," *Aeronautical Research Committee, R. and M.* No. 1104 [*Proc. Roy. Soc.* cxvi. 170 (1927)]. The diagrams are reproduced by permission of the Controller of H.M. Stationery Office. For some of the earlier measurements, reference may be made to Stanton, "On the Resistance of Plane Surfaces in a Uniform Current of Air," *Proc. Inst. Civ. Eng.* clvi. 78 (1904); Eiffel, *La Résistance de l'Air*, Paris, 1910.

ordinates indicate the distribution of pressure and suction over the breadth of the lamina at various inclinations (α) to the direction of the stream.

Attempts have been made by various writers to adapt the methods of Kirchhoff to the case of a curved lamina*, but for the reasons given they have little bearing on the practical problem.

370 a. The double trail of vortices with opposite rotations following in the wake of an elongated lamina or, generally any cylindrical solid, has been depicted, sometimes very effectively with the help of photography, by various observers†. Beyond a certain very moderate speed vortices appear to detach themselves from the solid on the two sides alternately, the general arrangement being similar to the unsymmetrical type discussed by Kármán (Art. 156), except that the vortices are not concentrated, as was there assumed for simplicity. This has been made the basis of a two-dimensional theory of resistance. Assuming that the motion is irrotational except for the concentrated vortices, Kármán deduces from considerations of momentum the formula‡

$$\frac{\rho\kappa b}{a}(U-2V)+\frac{\rho\kappa^2}{2\pi a}, \qquad\text{.........................(1)}$$

where U denotes the velocity of the solid relative to the fluid, and the rest of the notation is as in Art. 156. In particular V denotes the velocity of the train of vortices relatively to the undisturbed fluid. In the stable case we found that

$$b/a = \cdot 281, \qquad \kappa = \sqrt{8}\, Va. \qquad\text{.......................(2)}$$

If we substitute these values in (1), the ratio V/U, and the relation of b (or a) to the dimensions of the obstacle, are still unknown, and have to be ascertained before (1) becomes a complete formula for the resistance. Exact observation is difficult, owing to the diffused nature of actual vortices, but has been essayed by Kármán for water, and by Fage for air.

Before leaving this topic it should be mentioned that it is to the formation of a double train of vortices that we must look for the explanation of many acoustical phenomena. A familiar instance is the sound caused by a wind rushing past trees, and on a different scale we have the production under suitable conditions of 'Aeolian tones§.' In aeronautics we have on different scales the singing of aeroplane wires‖, and the roar of a propeller.

* References have been given on p. 106.

† For example, Ahlborn, *Ueber den Mechanismus des hydrodynamischen Widerstandes*, Hamburg, 1902; Bénard, *Comptes Rendus*, cxlvii. 839 (1908); Kármán and Rubach, *Phys. Zeitschr.* 1913, p. 49; Prandtl, "The generation of vortices...," London, 1927; Rosenhead, *Proc. Roy. Soc.* A, cxxix. 115 (1930).

‡ An independent calculation by Prof. Synge, *Proc. Roy. Irish Acad.* xxxvii. A, 95 (1929), assuming only a 'semi-infinite' trail of vortices, leads to an equivalent result. The corresponding formula when the fluid is confined by parallel walls is obtained by Rosenhead, and (approximately) by Glauert, *ll.c. ante* p. 229.

§ Rayleigh, *Phil. Mag.* (6) vi. 29 (1915) [*Papers*, vi. 315].

‖ Relf, *Phil. Mag.* (6) xlii. 173 (1921).

Rayleigh remarks that, from dimensional considerations, the frequency N of the note produced by air rushing past a cylindrical wire of diameter D must satisfy a formula of the type

$$N=\frac{U}{D}f\left(\frac{\nu}{UD}\right). \quad \text{.............................(3)}$$

The empirical formula which he constructed as a good representation of some observations of Strouhal is

$$\frac{ND}{U}=\cdot 195\left(1-\frac{20\cdot 1\,\nu}{UD}\right). \quad \text{......................(4)}$$

We may compare with this some observations of Kármán on the frequency with which vortices are detached from a circular cylinder in a stream of water. His results for two different speeds are equivalent to

$$N=\cdot 207\,U/D, \quad \text{and} \quad \cdot 198\,U/D.$$

Fage, experimenting in air, found that for a considerable range of speeds the frequency of the vortices detached from one edge of a *flat blade* broadside-on was given with great consistency by the formula

$$N=\cdot 146\,U/D,$$

where D is the breadth of the blade. Similar experiments by Kármán in water indicate a numerical factor ranging from ·139 to ·145.

370 b. The only case in which the action of a *uniform* * frictionless stream on an immersed solid has a resultant is that of the two-dimensional cylinder with circulation round it. There is then a 'lift' at right angles to the stream, of amount

$$L=\kappa\rho U, \quad \text{..............................(1)†}$$

per unit length, where U is the velocity of the stream and κ the circulation. This theorem, which is independent of the shape and dimensions of the cross-section, forms the basis of the modern theory of the lift of an aerofoil ‡. A proof has already been given incidentally in Art. 72 *b*, but the importance of the matter may warrant the insertion of the following demonstration, which is of a less artificial character.

If (u, v) be the fluid velocity, vanishing at infinity, the formula for the pressure is

$$\frac{p}{\rho}=\text{const.}-\tfrac{1}{2}\{(u-U)^2+v^2\}, \quad \text{...............................(2)}$$

since the motion relative to the body is steady. Hence if l, m be the direction-cosines of the outward normal to an element δs of the contour of the cross-section, the resultant pressure on the solid parallel to x is

$$X=\tfrac{1}{2}\rho\int(u^2+v^2)\,l\,ds-\rho U\int ul\,ds=-\rho\iint\left(u\frac{\partial u}{\partial x}+v\frac{\partial v}{\partial x}\right)dx\,dy-\rho U\int ul\,ds$$
$$=-\rho\iint\left(u\frac{\partial u}{\partial x}+v\frac{\partial u}{\partial y}\right)dx\,dy-\rho U\int ul\,ds=\rho\int(lu+mv)\,u\,ds-\rho U\int ul\,ds. \quad \text{...(3)}$$

* The necessity for this qualification is illustrated in Arts. 72 *b*, 143.

† Kutta, *l.c. ante* p. 79; the theorem was given in an unpublished dissertation of date 1902. A prior publication is attributed to Joukowski (1906).

‡ Lanchester, *Aerodynamics*, London, 1907; Prandtl, *Gött. Nachr. math. phys. Classe*, 1918, 1919.

We have here omitted two line-integrals taken round an infinite enclosing contour; these vanish since the velocity at infinity is of the order $1/r$, where r denotes distance from the origin. In the same way we find

$$Y = \rho \int (lu + mv) v\, ds - \rho U \int um\, ds. \qquad \text{.....(4)}$$

At the surface of the cylinder we have

$$lu + mv = lU, \qquad \text{.....(5)}$$

whence $X = 0$, and

$$Y = \rho U \int (lv - mu)\, ds = \rho\kappa U. \qquad \text{.....(6)}$$

The case of an elliptic cylinder, which includes as an extreme form that of a plane lamina, may be examined on the basis of the formula given at the end of Art. 72. With the notation there adopted, the fluid pressures on an elliptic cylinder of semi-axes a, b reduce (when $\omega = 0$) to a force

$$X = -\pi\rho\kappa V, \quad Y = \pi\rho\kappa U, \qquad \text{.....(7)}$$

and a couple

$$N = -\pi\rho U V (a^2 - b^2). \qquad \text{.....(8)}$$

371. It is remarkable that the empirical formula

$$k\rho U^2 \qquad \text{.....(1)}$$

for the tangential resistance per unit area (Art. 366 (6)), with about the same value of the coefficient k, is found to hold over a wide range of cases of turbulent flow over extended surfaces. It applies for instance to the friction of the wind blowing over level ground*, and to the resistance of the ocean bed to the flow of tidal streams. Taylor, in a remarkable paper†, has calculated on this basis, from the known velocities of the streams, the rate of dissipation of energy in the Irish Sea. This rate can also be calculated in an independent way, viz. from the velocities and heights at the N. and S. entrances, of the ingoing and outgoing tidal streams, and the work done by the lunar attraction. The result obtained in either way was of the order of 3×10^{17} ergs per second. An estimate, on the same principle, of the tidal dissipation in the whole ocean has been made by H. Jeffreys‡, the figure arrived at being $2 \cdot 2 \times 10^{19}$. The dissipation required to account for the acceleration of the moon's mean motion is reckoned at $1 \cdot 41 \times 10^{19}$.

As regards the total resistance to the translation through a liquid (or a gas whenever the compression is unimportant) of similar bodies, of any shape, in corresponding directions, we are led by consideration of dimensions to a formula of the type

$$F = \rho U^2 l^2 f\left(\frac{Ul}{\nu}\right), \qquad \text{.....(2)}$$

where l is any length defining the scale of the body (*e.g.* the radius, in the case of a sphere). The approximate proportionality to U^2 which is found in many cases indicates that the function f is then nearly constant, and that the resistance is accordingly almost independent of the viscosity. As in former cases this does not mean that viscosity is without influence; it plays its part,

* G. I. Taylor, *Proc. Roy. Soc.* A, xcii. 196 (1915).

† *l.c. ante* p. 567, and *Monthly Notices R.A.S.* lxxx. 308 (1920).

‡ *Phil. Trans.* A, ccxxi. 239 (1920).

along with the resistance to slipping over the surface of the solid, in bringing about the *régime* which is finally established.

The formula (2) is the basis of the method by which the forces on an air-ship or an aerofoil are estimated from model experiments in a wind-channel. The factor $f(Ul/\nu)$ is in fact what is determined as the 'drag-coefficient.' If the value of the 'Reynolds number' Ul/ν could be made the same in the case of a model as on the full scale, the forces should be proportional to the corresponding values of $\rho U^2 l^2$. The relative smallness of the linear dimension l in the model may be compensated to some extent by an increase in the velocity U or, as in a 'high-speed' wind-channel, by using highly compressed air, since for a gas at a given temperature the value of ν varies inversely as the density.

As the variable increases from zero the drag-coefficient is found to diminish at first, and afterwards to increase, tending apparently to a constant value.

The form of least resistance can only be found empirically. In the usual design of an air-ship, having a profile blunt at the nose and tapering towards the tail, the central stream-line follows the profile closely throughout, turbulence being sensible only in a thin stratum near the surface, and in the wake. A similar 'stream-line form,' as it is called, is adopted for the sections of aeroplane struts and wires.

The 'dimensional' argument which has been used above and in Arts. 365, 366 may be put in another form*. Taking any one of the dynamical equations of motion of an incompressible fluid, say

$$\frac{\partial u}{\partial t}+u\frac{\partial u}{\partial x}+v\frac{\partial u}{\partial y}+w\frac{\partial u}{\partial z}=-\frac{\partial p}{\rho\partial x}+\nu\nabla^2 u, \quad \text{..............(3)}$$

let us imagine another state of motion, of the same or another fluid, which differs only in the scales of length and time. That is, distinguishing the symbols relating to it by accents, we assume that x', y', z' are in a constant ratio to x, y, z, respectively, and t' in a constant ratio to t. The terms in the equation corresponding to (3) will differ only by the same factor throughout, provided the following equalities of ratios subsist, viz.

$$\frac{u'}{t'}:\frac{u}{t}=\frac{u'^2}{x'}:\frac{u}{x}=\frac{p'}{\rho' x'}:\frac{p}{\rho x}=\frac{\nu' u'}{x'^2}:\frac{\nu u}{x^2}. \quad \text{..................(4)}$$

These are equivalent to

$$u':u=\frac{x'}{t'}:\frac{x}{t};\quad p':p=\rho' u'^2:\rho u^2;\quad \frac{u'x'}{\nu'}=\frac{ux}{\nu}. \quad \text{...........(5)}$$

The equation of continuity will also obviously be satisfied by the new variables. We infer that the modified state of motion will also be dynamically possible provided the value of the Reynolds number Ul/ν is the same, where U and l

* Cf. Helmholtz, *Berl. Ber.* June 26, 1873 [*Wiss. Abh.* i. 158], where a number of interesting applications of the principles of dynamical similarity are given.

are a typical velocity and a typical length respectively. Also that the stresses at corresponding points will be proportional to ρU^2, and the forces on corresponding surfaces to $\rho U^2 l^2$.

The Boundary Layer.

371 a. It is plain that any rational theory of resistance must take account of the absolute resistance which a solid opposes to the sliding of a fluid over its surface. On the other hand, the slightest observation is enough to shew that the transition from the velocity of the surface to that of the fluid abreast of it is often effected within a very short space. In fact, when a solid of fair easy shape, such as a sphere, or a cylinder, or an aerofoil, moves through a mobile fluid such as water with a velocity definitely greater than that contemplated in such investigations as those of Arts. 337–343, the vorticity appears to be confined almost to a narrow band along the anterior portion of the surface, and to the wake. It is to the study, both dynamical and experimental, of this transition region that the efforts of many investigators have for some time been directed. It is not to be supposed, of course, that there is a definite surface of separation between the layer and the adjacent fluid, for the transition is necessarily continuous, but it is usually possible to assign a limit, and often a very narrow one, within which it is practically complete.

In what follows it is convenient to think of the solid as at rest, and the fluid as streaming past it with a velocity (U), uniform except so far as it is disturbed by the presence of the solid.

The conditions are simplest in the two-dimensional case of a plane lamina, or plate, placed edgeways to the stream. The boundary layer here begins at or near the leading edge, and gradually widens with increase of distance (x) from this. So long as the local Reynolds number Ux/ν is below a certain limit (apparently of the order 10^5) the motion within the layer is steady, and is often described as 'laminar,' in the sense that the stream-lines run nearly parallel to the surface. When the limit is exceeded, the layer becomes turbulent and its thickness increases more rapidly.

The laminar flow has been studied mathematically by various writers. The exact equations of steady motion, viz.

$$\left.\begin{aligned} u\frac{\partial u}{\partial x}+v\frac{\partial u}{\partial y} &= -\frac{\partial p}{\rho\partial x}+\nu\nabla_1^2 u,\\ u\frac{\partial v}{\partial x}+v\frac{\partial v}{\partial y} &= -\frac{\partial p}{\rho\partial y}+\nu\nabla_1^2 v,\end{aligned}\right\} \quad \ldots\ldots\ldots\ldots(1)$$

with

$$\frac{\partial u}{\partial x}+\frac{\partial v}{\partial y}=0, \quad \ldots\ldots\ldots\ldots(2)$$

are hardly tractable, but various simplifications are possible.

We take the origin in the leading edge, and the axis of x along the lamina in the direction of the stream. Since v is relatively small, the second equation shews that p is practically independent of y. Consequently we neglect $\partial p/\partial x$, since this vanishes for large values of y where the stream is unaffected by the presence of the lamina. Also, since u and $\partial u/\partial x$ vanish at the lamina, the value of $\partial^2 u/\partial x^2$ within the boundary layer may be neglected in comparison with that of $\partial^2 u/\partial y^2$. The equations thus reduce to

$$u\frac{\partial u}{\partial x}+v\frac{\partial u}{\partial y}=\nu\frac{\partial^2 u}{\partial y^2}, \qquad \text{.........................(3)}$$

with of course (2). This forms the starting-point of the work of Prandtl and Blasius*. The approximations are explained in greater detail by Blasius; they may be justified in the last resort by comparison with the results deduced. The boundary conditions to be satisfied are $u=0$, $v=0$, for $y=0$, and $u=U$ for $y\to\infty$.

After a rather intricate calculation, Blasius arrives at a result for the tangential drag on the lamina which may be written

$$(p_{xy})_{y=0}=\cdot 332\rho U^2\sqrt{\left(\frac{\nu}{Ux}\right)}. \qquad \text{.....................(4)}$$

It is assumed of course that the condition for laminar motion is fulfilled, viz. that the value of Ux/ν does not exceed the limit already referred to. If this proviso holds over the whole breadth (l) of the lamina, the total drag on one side is

$$\int_0^l p_{xy}\,dx=\cdot 664U^2l\sqrt{\left(\frac{\nu}{Ul}\right)}, \qquad \text{.....................(5)}$$

and varies therefore as $U^{\frac{3}{2}}$.

The question is approached in a different way by Kármán†. He calculates the flux of momentum parallel to x out of the space included between the lamina ($y=0$), two adjacent ordinates of a curve

$$y=\eta\,(x), \qquad \text{...................................(6)}$$

which is taken to represent the limit of the boundary layer, and the connecting arc of the curve.

In the annexed figure

$$QQ'=\delta x, \quad PQ=\eta, \quad P'Q'=\eta+\delta\eta.$$

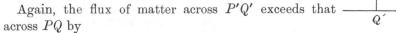

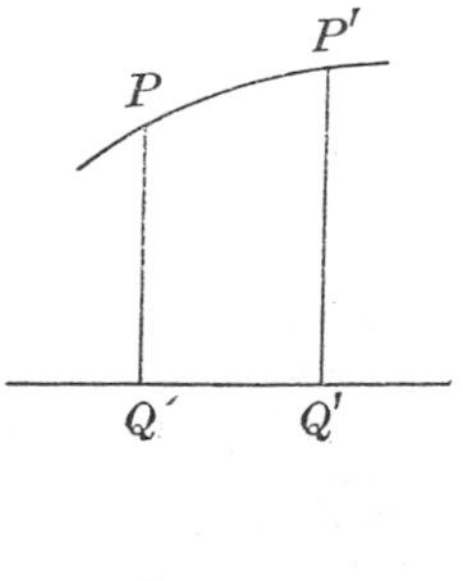

The flux of momentum across $P'Q'$ exceeds that across PQ by

$$\frac{d}{dx}\int_0^\eta \rho u^2 dy\,.\,\delta x.$$

Again, the flux of matter across $P'Q'$ exceeds that across PQ by

$$\frac{d}{dx}\int_0^\eta \rho u\,dy\,.\,\delta x,$$

* Prandtl, "Ueber Flüssigkeitsbewegung mit kleiner Reibung" (1904), reprinted in *Vier Abhandlungen zur Hydrodynamik...*, Göttingen, 1927; Blasius "Grenzschichten in Flüssigkeiten mit kleiner Reibung" (Dissertation), Leipzig, 1907. For an interesting independent treatment, see R. v. Mises, *Zeitschr. f. angew. Math. u. Mech.* vii. 425 (1927).

† *Abh. des aerodynamischen Instituts*, Aachen, 1921.

and this must therefore be the amount which in unit time crosses PP' inwards, where the velocity is sensibly U, parallel to x. The total flux of momentum out of the aforesaid region is therefore

$$\left\{\frac{d}{dx}\int_0^\eta \rho u^2 dy - U\frac{d}{dx}\int_0^\eta \rho u\, dy\right\}\delta x.$$

This is to be equated to the forces which act parallel to x on the matter which for the moment occupies the region. These consist of the pressure-component

$$-\frac{dp}{dx}\eta\delta x$$

and the drag

$$-\mu\left(\frac{\partial u}{\partial y}\right)_{y=0}.\delta x$$

exerted by the lamina. Hence

$$\frac{d}{dx}\int_0^\eta u^2 dy - U\frac{d}{dx}\int_0^\eta u\, dy = -\frac{dp}{\rho\, dx} - \nu\left(\frac{\partial u}{\partial y}\right)_{y=0}, \quad \text{.........(7)}$$

which is Kármán's 'integral equation*.' It is to be observed that, so far, the curve (6) may be drawn anywhere within the region where the transition from the zero velocity at the lamina to the velocity U of the stream is practically complete. The equation therefore can tell us nothing as to the 'thickness' of the boundary layer, or its mode of variation with x. For this, we must make some more or less plausible assumption as to the distribution of the velocity u within the range from 0 to η, and the result will necessarily depend to some extent on the particular assumption made. The conditions to be satisfied are $u = U$ and $\partial u/\partial y = 0$ for $y = \eta$, and $u = 0$, $\partial^2 u/\partial y^2 = 0$ for $y = 0$, the latter condition being required by (3). These are satisfied, for example, by

$$u = U\sin\frac{\pi y}{2\eta}. \quad \text{.........(8)}$$

Substituting in (7), with the omission of $\partial p/\partial x$, we find

$$\frac{d\eta}{dx} = \frac{\pi^2}{4-\pi}\frac{\nu}{U\eta}, \quad \text{.........(9)}$$

and thence

$$\eta = 4{\cdot}80\sqrt{\left(\frac{\nu x}{U}\right)}, \quad \text{.........(10)}$$

* It may be derived from (1) by integration with respect to y between the limits 0 and η, taking account of the equation of continuity. Thus

$$\int_0^\eta\left(u\frac{\partial u}{\partial x} + v\frac{\partial u}{\partial y}\right)dy = \int_0^\eta u\frac{\partial u}{\partial x}dy + [uv]_0^\eta - \int_0^\eta u\frac{\partial v}{\partial y}dy = \int_0^\eta \frac{\partial(u^2)}{\partial x}dy + Uv_\eta$$

$$= \int_0^\eta \frac{\partial(u^2)}{\partial x}dy + U\int_0^\eta \frac{\partial v}{\partial y}dy = \int_0^\eta \frac{\partial}{\partial x}(u^2 - Uu)\,dy.$$

This is equivalent to the left-hand side of (7).

on the assumption that the boundary layer begins at the leading edge. Hence

$$(p_{xy})_{y=0}=\frac{\pi}{2}\frac{\mu U}{\eta}=\cdot 328\rho U^2\sqrt{\left(\frac{\nu}{Ux}\right)}, \quad \text{..............(11)}$$

a near approximation to Blasius' result (4).

The particular assumption (8) makes

$$\frac{\partial v}{\partial y}=-\frac{\partial u}{\partial x}=\frac{\pi U y}{2\eta^2}\cos\frac{\pi y}{2\eta}\frac{d\eta}{dx},$$

whence

$$v=U\frac{d\eta}{dx}\left\{\frac{y}{\eta}\sin\frac{\pi y}{2\eta}-\frac{2}{\pi}\left(1-\cos\frac{\pi y}{2\eta}\right)\right\}. \quad \text{..............(12)}$$

371 b. When a non-turbulent stream encounters a solid of continuous curvature, the motion in the regions up-stream, as well as abreast at all events of the anterior portion of the surface, remains apparently irrotational, and has the same general configuration as in the diagrams of Arts. 68, 96 *. In particular there is a central stream-line abutting on the solid at the front 'stagnation point,' characterized by zero velocity. The boundary layer, starting here and at first laminar, follows the surface for some distance on either side; in the case of a cylinder for instance as far even as 70° or 80° from the stagnation point. In the case of an aerofoil it may extend almost to the trailing edge. The circumstances vary of course with the shape of the body, and also with the velocity U of the stream. Usually a point is reached at which the layer becomes turbulent and breaks away from the surface, leaving between it and the solid a region of turbulence on a larger scale, with a back-flow along the surface.

The two-dimensional case has been treated theoretically by Blasius and others, using, as curvilinear co-ordinates, the arc s of the profile and the normal n drawn from the surface towards the fluid. The equations adopted for the boundary layer are then

$$u\frac{\partial u}{\partial s}+v\frac{\partial u}{\partial n}=-\frac{\partial p}{\rho\partial s}+\nu\frac{\partial^2 u}{\partial n^2}, \quad \text{....................(13)}$$

$$\frac{\partial u}{\partial s}+\frac{\partial v}{\partial n}=0, \quad \text{............................(14)}$$

the effect of the curvature being neglected †. We neglect $\partial p/\partial n$, but $\partial p/\partial s$ no longer vanishes, as was assumed in the case of the plane plate. In the irrotational region near the surface we have Bernoulli's equation

$$\frac{p}{\rho}+\tfrac{1}{2}U^2=\text{const.}, \quad \text{..........................(15)}$$

* An elaborate experimental comparison of the normal pressures at various points of a prolate ellipsoid of revolution with the theoretical values calculated from Arts. 104, 105 has been carried out by R. Jones, *Phil. Trans.* A, ccxxvi. 231 (1927). When the ellipsoid was end-on to the stream the agreement was very close over almost the whole length.

† Cf. the effect of making $r\to\infty$ in the polar equations of Art. 328 *a*.

and thence
$$-\frac{1}{\rho}\frac{\partial p}{\partial s} = U\frac{\partial U}{\partial s}. \qquad \text{.........................(16)}$$

The point on the surface at which the 'break-away' occurs is determined by the condition $\partial u/\partial n = 0$.

Blasius, in the paper cited, has applied these equations to the case of a cylinder (of any form of section) placed symmetrically to the stream, and then proceeds to deal specially with the case of the circular section. In the steady *régime* the break-away is found to occur somewhere about 90° from the forward stagnation point. On the other hand if the cylinder starts from rest, either suddenly, or with a constant acceleration, the break begins at 180° and then travels forward. In the latter case he deduces formulae for the resistance, as depending partly on the normal pressure, and partly on the tangential stress*.

The calculations are remarkable for their analytical skill, but the results need some qualification, owing to the assumption that the velocity U outside the boundary layer is the same as if the fluid had been free to slip over the surface. Minute accuracy is indeed not claimed for them. Subsequent writers have assumed for U algebraical expressions in terms of the arc s of the profile, which can be adapted to fit experimental values of p†.

In the former investigations of Arts. 337, 342, etc. it was assumed that the Reynolds number ($R = Ul/\nu$) did not exceed a very small numerical value. In such cases as are now under consideration, on the other hand, the value of R may be very great, owing to the linear dimensions involved and the small kinematic viscosity of ordinary fluids. A very interesting question has been raised by Oseen: what would be the limiting character of the fluid motion, in any given case, for $\nu \to 0$, or $R \to \infty$? That it should be identical with the result obtained by assuming $\nu = 0$ *ab initio* is of course not to be expected.

The question is a difficult one, and almost hopeless, apparently, from the standpoint of the exact equations of hydrodynamics. Oseen adopts as his basis the linearized equations of Art. 342 (6), but here we are met by the *à priori* difficulty that these equations ignore certain quadratic terms which are only known to be negligible in cases where the Reynolds number involved is very small. They can hardly be accepted without misgiving as a foundation for ascertaining what happens in the case of a real fluid when R is increased indefinitely.

Apart from this point, Oseen's investigations have considerable mathematical interest. They cannot be reproduced here‡, but the general conclusions

* Similar calculations for the three-dimensional case of a solid of revolution, with its axis in the direction of the stream, have been carried out by Boltze, Göttingen, 1908 (Dissertation). A detailed application is made to the sphere.

† Pohlhausen, *Abh. d. aerodynam. Inst. Aachen*, 1921; Goldstein, *Camb. Proc.* xxvi. 1 (1930).

‡ A full account is given, with references, in the treatise referred to on p. 617, *ante*.

may be briefly stated. Taking for example the steady translation of a body through an infinite mass of liquid, he finds that the solution will have different analytical characters in the cylindrical region which has been traversed by the solid (which we may for the moment call the 'wake') and in the rest of infinite space. Throughout this latter region the motion is irrotational, and the fluid accordingly glides smoothly over the anterior face of the solid. In the wake, on the other hand, there is no slipping at the rear surface, and the motion is rotational (but not really 'turbulent'). Over the cylindrical surface where the two regions meet there is continuity of normal velocity, but a discontinuity in the tangential component, with an inadmissible discontinuity of pressure. The analytical solution, on these lines, has been worked out by Zeilon* for the special cases of a circular cylinder, a circular disk, and a hemisphere advancing with either the curved surface, or the flat base, in front. Some adjustments are proposed with a view to avoiding the awkward discontinuity referred to. It is claimed that the results give a fairly adequate picture of what takes place in actual cases. In particular, stress is laid on the fact that the theoretical distribution of pressure over the front portion of the cylinder is in general agreement with experimental tests. This would follow indeed from almost any reasonable configuration of irrotational motion in the adjacent region (cf. Art. 371 *a*). But the point at which the wake breaks away from the body, its definite boundary, and its internal constitution are quite different from what is observed. Still higher divergences would doubtless be revealed if the method could be applied to an air-ship or an aerofoil†.

371 c. When we pass to the consideration of *turbulent* flow over a solid the symbols u, v have to be understood in some statistical sense, as (for instance) time-averages over a very short interval. If as in Art. 369 we mark the changed meaning by bars placed over the respective letters, the equation (7) of Art. 371 *a* gives

$$\frac{\partial}{\partial x}\int_0^\eta \overline{u^2}\,dy - U\frac{\partial}{\partial x}\int_0^\eta \bar{u}\,dy = -\frac{1}{\rho}\frac{\partial p}{\partial x}\eta - \nu\left(\frac{\partial \bar{u}}{\partial y}\right)_{y=0}, \quad \text{.........(1)}$$

where it may be noticed that the mean ($\overline{u^2}$) of the square of the velocity is not identical with the square of the mean velocity ($\bar{u}$). This distinction, whatever its practical importance, has not been observed by writers on the present question. It may be noted, however, that when the velocity of a turbulent stream is investigated by means of a combination of Pitot and static-pressure tubes (Art. 24), it is rather the velocity of 'mean square' which is indicated.

* "On potential problems in the theory of fluid resistance," Stockholm, 1924. See also the appendix (by Zeilon) to Oseen's treatise quoted on p. 617 *ante*.

† I find that similar criticisms, but in greater detail, are made by F. Noether, *Handb. d. phys. u. techn. Mechanik*, Leipzig, 1928..., v. 792.

Some attempts have been made to investigate cases of turbulent flow on the basis of equation (1), but necessarily require a supplementary hypothesis as to the distribution of mean velocity within the boundary layer, or some equivalent assumption. Formulae of the type

$$\frac{\bar{u}}{U} = \left(\frac{y}{\eta}\right)^n \qquad \text{.................................}(2)$$

have been proposed, but need some qualification since they make $\partial \bar{u}/\partial y$ either zero or infinite when $y \to 0$, except in the inadmissible case of $n = 1$ *.

We are here at the limits of theory. Further knowledge on this part of the subject has to be derived from experiment, and recourse must be had to the publications of the various aeronautical laboratories. The literature is extensive, and constantly increasing in volume; it cannot be condensed or summarized here.

A note may be added on the question of the lift of an aerofoil, already referred to. The nature of the flow of a *frictionless* fluid past an aerofoil is indicated in the annexed figure *A*. The central stream-line, only, is shewn,

A

but an idea of the complete configuration may be gained from the diagram on p. 86. A real fluid cannot flow like this, owing to the resistance to slipping, and the infinite velocity and consequent infinite negative pressure involved at the sharp trailing edge. The hypothesis of a thin boundary layer to smooth off the infinities does not greatly improve the picture; the influence of viscosity, and consequent generation of vorticity, near the edge are too great to be ignored.

But if on the irrotational motion of figure *A* we superpose a clockwise circulation, it is possible to adjust this so that the velocity at the trailing edge shall be finite †. The elementary streams on the two sides then meet, and move smoothly from the edge, without discontinuity. The result is indicated

* A formula free from this difficulty, viz.

$$y = \frac{\bar{u}}{a} + \left(\eta - \frac{U}{a}\right)\left(\frac{\bar{u}}{U}\right)^n,$$

where

$$a = \left(\frac{\partial \bar{u}}{\partial y}\right)_{y \to 0},$$

has been employed by Hegge van der Zijnen (with $n = 7$), *Publications of the Delft aeronautical laboratory*, No. 6, 1924.

† This is exemplified in Art. 70, in the case of the circular arc.

in figure B; and it is now possible to understand how, by the introduction of a thin boundary layer and a narrow wake, the behaviour of an actual fluid can be mentally pictured.

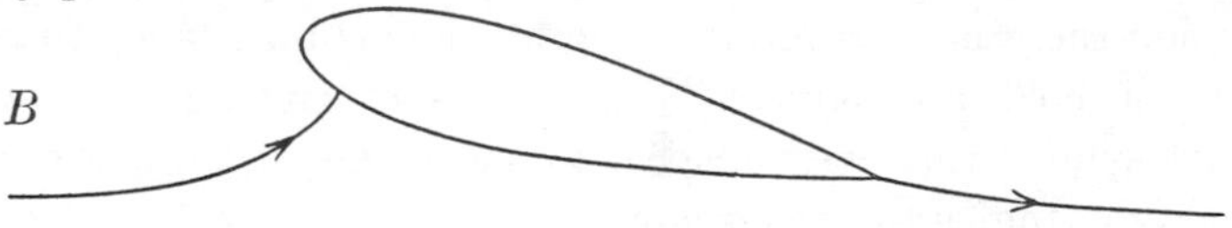

It is still not altogether easy, in spite of the attempts that have been made, to trace out deductively the stages by which the result is established when the relative flow is started*. Fortunately, some beautiful experiments† with small-scale models in a tank come to our help. A vortex with counter-clockwise sense is first formed, and detached from the edge, and then passes down the stream, leaving a complementary circulation round the aerofoil in the opposite sense. The boundary layers on the two sides creep along the surface, and contribute opposite vortices to the wake, which gradually diffuse and annul one another.

Influence of Compressibility.

371 d. The flow of a compressible fluid past an obstacle seems to have been first treated mathematically by Rayleigh‡. Assuming the adiabatic law we have

$$\frac{c^2}{c_0^2} = \left(\frac{\rho}{\rho_0}\right)^{\gamma-1}, \quad \text{.................................(1)}$$

where c denotes the velocity of sound corresponding to the local value of ρ, and the zero suffix relates to the undisturbed part of the stream. If, further, the motion is irrotational, we have from Art. 24 *a*

$$q^2 - U^2 = \frac{2}{\gamma-1}(c_0^2 - c^2). \quad \text{........................(2)}$$

Hence
$$\frac{d\rho}{\rho} = \frac{1}{\gamma-1}\frac{d(c^2)}{c^2} = -\frac{1}{2c^2}d(q^2). \quad \text{.....................(3)}$$

The equation of continuity, Art. 7, thus becomes, in steady motion,

$$\nabla^2\phi = \frac{1}{2c^2}\left\{\frac{\partial\phi}{\partial x}\frac{\partial(q^2)}{\partial x} + \frac{\partial\phi}{\partial y}\frac{\partial(q^2)}{\partial y} + \frac{\partial\phi}{\partial z}\frac{\partial(q^2)}{\partial z}\right\}, \quad \text{...........(4)}$$

where c is given in terms of q by (2).

In polar co-ordinates the two-dimensional form of this equation is

$$\frac{\partial^2\phi}{\partial r^2} + \frac{1}{r}\frac{\partial\phi}{\partial r} + \frac{1}{r^2}\frac{\partial^2\phi}{\partial\theta^2} = \frac{1}{2c^2}\left\{\frac{\partial\phi}{\partial r}\frac{\partial(q^2)}{\partial r} + \frac{\partial\phi}{r\partial\theta}\frac{\partial(q^2)}{r\partial\theta}\right\}, \quad \text{.........(5)}$$

with
$$q^2 = \left(\frac{\partial\phi}{\partial r}\right)^2 + \left(\frac{\partial\phi}{r\partial\theta}\right)^2. \quad \text{..........................(6)}$$

* Reference must however be made to the discussion by Jeffreys, *Proc. Roy. Soc.* A, cxxviii. 376 (1930).

† Prandtl, *l.c. ante* p. 680; Walker, *Aeronautical Research Comm., R. and M.* 1402 (1932). (A report on experiments under the direction of Prof. B. M. Jones and W. S. Farren.)

‡ *Phil. Mag.* (6) xxxii. 1 (1916) [*Papers*, vi. 402].

This was applied by Rayleigh to the flow past a circular cylinder. He first substitutes on the right-hand side the values of ϕ and q appropriate to the case of incompressibility, and then integrates. It appears that to a first approximation there is no resultant drag on the cylinder, and it is easily seen that this would hold however far the process were continued, the values of q^2 being always symmetrical with respect to the plane through the axis of the cylinder at right angles to the stream.

The conclusion is however conditional on the convergence of the successive results obtained, and there is evidence of more than one kind that this is not the case for values of U/c_0 exceeding a certain limit. Leaving this question for a moment, it is easy to adapt Rayleigh's process to the case of a cylinder of any form of cross-section, and also to include the effect of circulation*. If c were infinite the value of ϕ at a distance would tend to the form

$$\phi_1 = -Ur\cos\theta + \frac{\kappa\theta}{2\pi}, \qquad (7)$$

where the origin of r is taken in the immediate neighbourhood of the obstacle, and the initial line of θ parallel to the general direction of the stream. We adopt this as a first approximation, and substitute on the right-hand side of (5). We may also, for consistency, in the next approximation, replace c by its constant value (c_0) at infinity. From (7) we have

$$q_1^2 = U^2 + \frac{\kappa U}{\pi r}\sin\theta, \qquad (8)$$

$$\frac{\partial(q_1^2)}{\partial r}\frac{\partial\phi_1}{\partial r} + \frac{\partial(q_1^2)}{r\partial\theta}\frac{\partial\phi_1}{r\partial\theta} = \frac{\kappa U^2}{\pi r^2}\sin 2\theta, \qquad (9)$$

retaining only those terms which it is necessary to consider when r is increased indefinitely. Substituting in (5), integrating, and having regard to the conditions at infinity, we have, in the distant regions,

$$\phi = -Ur\cos\theta + \frac{\kappa\theta}{2\pi} - \frac{\kappa U^2}{8\pi c_0^2}\sin 2\theta. \qquad (10)$$

The 'complementary' terms, here omitted, would involve only negative powers of r, and would not affect the subsequent calculation of the forces. Hence, with sufficient approximation, the radial and transverse velocities are

$$-\frac{\partial\phi}{\partial r} = U\cos\theta, \quad -\frac{\partial\phi}{r\partial\theta} = -U\sin\theta - \frac{\kappa}{2\pi r} + \frac{\kappa U^2}{4\pi c_0^2 r}\cos 2\theta, \qquad (11)$$

whence

$$q^2 = U^2 + \frac{\kappa U}{2\pi r}\left(1 - \frac{U^2}{2c_0^2}\cos 2\theta\right)\sin\theta. \qquad (12)$$

The velocities parallel and at right angles to the stream are

$$\left.\begin{aligned} u &= U + \frac{\kappa}{2\pi r}\left(1 - \frac{U^2}{2c_0^2}\cos 2\theta\right)\sin\theta, \\ v &= \quad - \frac{\kappa}{2\pi r}\left(1 - \frac{U^2}{2c_0^2}\cos 2\theta\right)\sin\theta. \end{aligned}\right\} \qquad (13)$$

* *Aeronautical Research Comm.*, *R. and M.* 1156 (1928).

The forces on the obstacle can now be inferred from the modified flow at infinity, as in the original proofs of the Kutta-Joukowski theorem. The mass of fluid enclosed at any instant within a circle of large radius r is gaining momentum in the direction at right angles to the stream at the rate

$$\int_0^{2\pi}\left(-\frac{\partial\phi}{\partial r}\right)\rho v r\,d\theta = -\tfrac{1}{2}\kappa\rho_0 U\left(1-\frac{U^2}{4c_0^2}\right), \quad \text{.........(14)}$$

by (10) and (13). Again, from (3),

$$\log\frac{\rho}{\rho_0} = \int_q^U \frac{d\,(q^2)}{2c_0^2} = \frac{U^2-q^2}{2c_0^2}, \quad \text{.........(15)}$$

and therefore

$$\frac{\rho}{\rho_0} = 1 + \frac{U^2-q^2}{2c_0^2}, \quad \text{.........(16)}$$

with consistent approximation. Hence, since for large values of r the density tends to ρ_0, we may put

$$p = p_0 + c_0^2(\rho-\rho_0) = p_0 - \frac{\kappa\rho_0 U}{2\pi r}\left(1+\frac{U^2}{2c_0^2}\cos 2\theta\right)\sin\theta. \quad \text{......(17)}$$

The resultant pressure at right angles to the stream on the aforesaid mass of fluid is therefore

$$-\int_0^{2\pi} p\sin\theta\,r\,d\theta = \tfrac{1}{2}\kappa\rho_0 U\left(1+\frac{U^2}{4c_0^2}\right). \quad \text{.........(18)}$$

Comparing with (14), the 'lift' at right angles to the stream is given by the familiar Joukowski formula

$$L = \kappa\rho_0 U, \quad \text{.........(19)}$$

with (at most) a proportional error of the order $(U/c_0)^4$. Still more easily it may be shewn that subject to the same approximation the drag is zero.

The formula (19) was first extended to the case of a compressible fluid by Glauert*. His investigation involves no explicit limitation to the magnitude of the ratio U/c_0, so long as it does not exceed unity. His formulae for the motion at infinity are equivalent to

$$\phi = -Ur\cos\theta + \frac{\kappa}{2\pi}\tan^{-1}\left\{\sqrt{\left(1-\frac{U^2}{c_0^2}\right)}\tan\theta\right\}, \quad \text{.........(20)}$$

which may be compared with (10).

371 e. To examine under what limitation steady flow of a compressible fluid is possible past an obstacle of given shape, Prof. G. I. Taylor has had recourse to an electrical method which differs from that of Art. 60 *a* in that the thickness of the conducting sheet is variable.

The kinematical conditions of irrotational steady flow in two dimensions are comprised in the equation

$$u = -\frac{\partial\phi}{\partial x}, \quad v = -\frac{\partial\phi}{\partial y}, \quad \rho u = -\frac{\partial\psi}{\partial y}, \quad \rho v = \frac{\partial\psi}{\partial x}. \quad \text{.........(1)}$$

* *Proc. Roy. Soc.* A, cxviii. 113 (1927).

The equations of electric flow in a current-sheet of variable thickness h are

$$\sigma f = -\frac{\partial V}{\partial x}, \quad \sigma g = -\frac{\partial V}{\partial y}, \quad hf = -\frac{\partial W}{\partial y}, \quad hg = \frac{\partial W}{\partial x}, \quad \text{......(2)}$$

where (f, g) is the current density, σ is the specific resistance, V is the electric potential, and W the current function. The two sets of formulae become identical if we put

$$\phi = V, \quad \psi = W, \quad u = \sigma f, \quad v = \sigma g, \quad \rho u = hf, \quad \rho v = hg, \quad \text{......(3)}$$

which involve $h = \rho\sigma$; or, again, if we make

$$\phi = W, \quad \psi = -V, \quad u = -hg, \quad v = hf, \quad \rho u = -\sigma g, \quad \rho v = \sigma f, \text{......(4)}$$

which require $h = \sigma/\rho$. So far the correspondence is merely kinematical, and the condition (2) of Art. 371 *d* has to be satisfied. Hence in the first form of the analogy we must have

$$\frac{h}{h_0} = \frac{\rho}{\rho_0} = \left\{1 - \frac{\gamma - 1}{2}\frac{U^2}{c_0^2}\left(\frac{q^2}{U^2} - 1\right)\right\}^{\frac{1}{\gamma-1}}, \quad \text{..................(5)}$$

the zero suffix relating to the regions where the flow is sensibly undisturbed. The value of U/c_0 being fixed, the distribution of the values of q/U is first inferred electrically in an experiment with uniform depth. These being substituted in (5) give an amended value of h, and when the tank has been modified so as to have this variable depth, the process is repeated; and so on. For fuller details we must refer to the original papers*.

In the case of flow past a circular cylinder, Prof. Taylor found that the successive configurations converged rather rapidly for values of U/c_0 less than ·45, but ceased to converge after this limit.

The second analogy was employed in the case of an aerofoil section, with circulation adjusted so as to avoid infinite velocity at the trailing edge. Here the limit of convergence was found to be $U/c_0 = {\cdot}58$.

371 f. Another form of the equation of motion of compressible fluids is to be noticed. If we assume only that the motion is steady, and not necessarily irrotational, then in two dimensions we have (Art. 165)

$$u\frac{\partial \chi}{\partial x} + v\frac{\partial \chi}{\partial y} = 0, \quad \text{..............................(1)}$$

where

$$\chi = \int \frac{dp}{\rho} + \tfrac{1}{2}q^2. \quad \text{.............................(2)}$$

Hence

$$\frac{1}{\rho}\frac{dp}{d\rho}\left(u\frac{\partial \rho}{\partial x} + v\frac{\partial \rho}{\partial y}\right) - \tfrac{1}{2}\left(u\frac{\partial (q^2)}{\partial x} + v\frac{\partial (q^2)}{\partial y}\right) = 0. \quad \text{...........(3)}$$

* Taylor and Shearman, *Proc. Roy. Soc.* A, cxxi. 194 (1928); Taylor, *Journal of the Lond. Math. Soc.* v. 224 (1930).

Writing $dp/d\rho = c^2$, and referring to the equation of continuity, we have

$$c^2\left(\frac{\partial u}{\partial x}+\frac{\partial v}{\partial y}\right)-\tfrac{1}{2}\left(u\frac{\partial (q^2)}{\partial x}+v\frac{\partial (q^2)}{\partial y}\right)=0, \qquad (4)$$

or, in full,

$$\left(1-\frac{u^2}{c^2}\right)\frac{\partial u}{\partial x}-\frac{uv}{c^2}\left(\frac{\partial v}{\partial x}+\frac{\partial u}{\partial y}\right)+\left(1-\frac{v^2}{c^2}\right)\frac{\partial v}{\partial y}=0. \qquad (5)$$

In the case of irrotational motion this becomes

$$\left(1-\frac{u^2}{c^2}\right)\frac{\partial^2\phi}{\partial x^2}-\frac{2uv}{c^2}\frac{\partial^2\phi}{\partial x\partial y}+\left(1-\frac{v^2}{c^2}\right)\frac{\partial^2\phi}{\partial y^2}=0, \qquad (6)$$

which is equivalent to Rayleigh's equation (5).

The equation (6) is converted into a linear equation if we have recourse to the 'principle of duality*,' and adopt u, v as the independent variables. Assuming

$$\Phi = ux + vy - \phi, \qquad (7)$$

we find

$$\left(1-\frac{v^2}{c^2}\right)\frac{\partial^2\Phi}{\partial u^2}+\frac{2uv}{c^2}\frac{\partial^2\Phi}{\partial u\partial v}+\left(1-\frac{u^2}{c^2}\right)\frac{\partial^2\Phi}{\partial v^2}=0. \qquad (8)$$

Some interesting comments on the nature of the problems presented by this equation are made by Bateman†.

371 g. When compressibility has to be allowed for the formula (2) of Art. 371 requires modification. If κ denote the elasticity, the method of dimensions leads easily to the assumption

$$F=\rho U^2 l^2 f\left(\frac{Ul}{\nu},\ \frac{\rho U^2}{\kappa}\right). \qquad (1)$$

If U is small compared with the velocity of sound in the gas, viz. $\sqrt{(\kappa/\rho)}$, this approximates to the form

$$F=\rho U^2 l^2 f\left(\frac{Ul}{\nu},\ 0\right), \qquad (2)$$

already considered.

The law of resistance varying as the square of the velocity is found to hold fairly well in the case of a projectile moving through air up to velocities of about 800 ft. per sec. When the velocity approaches or exceeds that of sound the law changes, as we should expect. We have then a wave-making resistance, analogous to that discussed in Art. 249, in addition to the frictional type.

When $U > c_0$, the ordinary velocity of sound, a wave of (approximate) discontinuity is formed, as appears from the photographs of Mach, Boys‡, and others. The formulae of Rankine (Art. 284) appropriate to such a case have

* Forsyth, *Differential Equations*, Art. 242.

† *Proc. Roy. Soc.* A, cxxv. 598 (1929).

‡ *Nature*, xlvii. 440 (1893).

been applied by Rayleigh* to calculate the pressure at the nose of the projectile.

The problem being reduced to one of steady motion, we consider the motion in the line of symmetry. There are two stages to be considered. Denoting the relative velocity of the air by q, we have in front of the wave $q = U$, and $p = p_0$, $\rho = \rho_0$ (say). We denote the corresponding quantities just behind the wave by q_1, p_1, ρ_1. Hence, writing $m = q\rho$ in the equations (14) and (15) of Art. 284,

$$\rho_1 q_1^2 = \tfrac{1}{2}(\gamma - 1) p_1 + \tfrac{1}{2}(\gamma + 1) p_0, \qquad \text{.....................(3)}$$

$$\rho_0 U^2 = \tfrac{1}{2}(\gamma + 1) p_1 + \tfrac{1}{2}(\gamma - 1) p_0. \qquad \text{.....................(4)}$$

Since $c_0^2 = \gamma p_0/\rho_0$, the latter equation gives

$$\frac{p_1}{p_0} = \frac{2\gamma}{\gamma + 1}\frac{U^2}{c_0^2} - \frac{\gamma - 1}{\gamma + 1}, \qquad \text{.....................(5)}$$

thus determining p_1/p_0.

Again, the velocity of the air, in its passage from the rear of the wave to the nose falls continuously from q_1 to 0. Hence by Art. 25 (1)

$$q_1^2 = \frac{2\gamma}{\gamma - 1}\left(\frac{p_2}{\rho_2} - \frac{p_1}{\rho_1}\right) = \frac{2\gamma}{\gamma - 1}\frac{p_1}{\rho_1}\left\{\left(\frac{p_2}{p_1}\right)^{\frac{\gamma-1}{\gamma}} - 1\right\}, \qquad \text{...........(6)}$$

where p_2, ρ_2 refer to the nose. Substituting for $\rho_1 q_1^2/p_1$ from (3) we find

$$\left(\frac{p_2}{p_1}\right)^{\frac{\gamma-1}{\gamma}} = \frac{(\gamma + 1)^2}{4\gamma} + \frac{\gamma^2 - 1}{4\gamma}\frac{p_0}{p_1}. \qquad \text{.....................(7)}$$

Combined with (5) this gives the required value of p_2.

Taking $\gamma = 1{\cdot}41$ we have

$$p_2/p_0 = 1{\cdot}90, \quad 5{\cdot}67, \quad 11{\cdot}7, \quad 20{\cdot}7$$

in the cases $\qquad U/c_0 = 1, \quad 2, \quad 3, \quad 4$, respectively.

Conversely, the theory can be applied to the measurement of air velocities exceeding that of sound. The ratio p_2/p_0 is obtained from the readings of a Pitot tube whose nozzle points up the stream, and of a 'static pressure' tube. The equation (7) then determines the value of p_1/p_0, whence U is found by (5). In this way Stanton has measured velocities of two or three times that of sound, and found them to agree closely with independent, but more elaborate, experimental determinations†.

* *l.c. ante* p. 482.

† *Rep. of the Nat. Phys. Lab.* for 1921, p. 146.

CHAPTER XII

ROTATING MASSES OF LIQUID

372. THIS subject had its origin in the investigations on the theory of the Earth's Figure which began with Newton and Maclaurin, and were continued by the great French school of mathematicians which flourished near the end of the eighteenth and the beginning of the nineteenth century. It has in recent times undergone great development, at the hands, notably, of Thomson and Tait, Poincaré, Darwin, and Jeans.

The problem is to ascertain the possible forms of relative equilibrium of a homogeneous gravitating mass of liquid, when rotating about a fixed axis with constant angular velocity, and to determine the stability or instability of such forms.

We take the axis of rotation as axis of z, and the mass-centre, which must evidently lie on the axis, as origin. If ω be the angular velocity of rotation the component accelerations at (x, y, z) are $-\omega^2 x$, $-\omega^2 y$, $-\omega^2 z$, and the dynamical equations therefore reduce to

$$-\omega^2 x = -\frac{1}{\rho}\frac{\partial p}{\partial x} - \frac{\partial \Omega}{\partial x}, \quad -\omega^2 y = -\frac{1}{\rho}\frac{\partial p}{\partial y} - \frac{\partial \Omega}{\partial y}, \quad 0 = -\frac{1}{\rho}\frac{\partial p}{\partial z} - \frac{\partial \Omega}{\partial z}, \quad \ldots(1)$$

where Ω is the potential energy per unit mass. Hence

$$\frac{p}{\rho} = \tfrac{1}{2}\omega^2(x^2+y^2) - \Omega + \text{const.} \quad \ldots\ldots\ldots\ldots\ldots\ldots(2)$$

At the free surface we have $p = \text{const.}$

Some general properties of forms of equilibrium have been proved by Poincaré and Lichtenstein.

In the first place there is for a given fluid an upper limit to the angular velocity, if the external pressure is zero*. Considering any internal region we have by Art. 42 (3)

$$\iint \frac{\partial p}{\partial n}\, dS = -\iiint \nabla^2 p\, dx\, dy\, dz = 2\rho(2\pi\rho - \omega^2)\iiint dx\, dy\, dz, \quad \ldots\ldots(3)$$

where $\partial p/\partial n$ denotes the *inward* gradient of p, and ρ is expressed in 'astronomical' measure†. Applying this to any small spherical region we learn that the pressure cannot be a minimum at an internal point if $\omega^2 < 2\pi\rho$, and cannot be a maximum if $\omega^2 > 2\pi\rho$. If it vanishes over the boundary it cannot therefore be a negative anywhere in the interior in the former case, or

* It will be shewn presently that this proviso is unnecessary.

† This involves of course a special unit of pressure.

positive in the latter. In the intervening case of $\omega^2 = 2\pi\rho$ we have $\nabla^2 p = 0$ throughout the interior, and $p = 0$ at the boundary, and therefore $p = 0$ everywhere (Art. 40).

Hence in a fluid unable to sustain tensile stress there is an upper limit, viz. $\sqrt{(2\pi\rho)}$, to the angular velocity*. If the density be the earth's mean density, viz. $\rho = \frac{3}{4}\pi g a$, the limiting value of ω is given in terms of the earth's angular velocity (ω_0) by

$$\frac{\omega^2}{\omega_0^{\,2}} = \frac{3}{2}\frac{g}{\omega_0^{\,2} a} = 433.$$

The shortest possible period is accordingly 1 h. 7 m.

Again, an equilibrium form is necessarily symmetrical with respect to the plane through the mass-centre at right angles to the axis†. We conceive the fluid mass as made up of columnar portions of infinitesimal section having their lengths parallel to Oz. The centres of these columns will lie on a certain surface (which may consist however of several detached portions). Unless this surface is plane there will be some point M on it at which z is a maximum; let PQ be that line drawn in the fluid, parallel to Oz and terminated both ways at the boundary, which is bisected at M, and suppose that $z_P > z_Q$. It easily follows from the theory of the attraction of a straight line of matter that the potential (energy) per unit mass, due to any one of the elementary columns, cannot be less at P than it is at Q, and will as a rule be greater. Hence, on the whole, $\Omega_P > \Omega_Q$ and therefore by (2) $p_P < p_Q$, contrary to the hypothesis.

The points P and Q have so far been assumed to be distinct. If they coincide we find in a similar manner that in the absence of a plane of symmetry we should have $\partial\Omega/\partial z > 0$, and therefore $\partial p/\partial z < 0$, at M, which is now a point of the free surface. But if the tangent plane at M is parallel to Oz we must have $\partial p/\partial z = 0$ there, whilst if M were a singular point on the surface all the space-derivatives of p would vanish.

Incidentally we may note, as a result of the preceding argument, that if there is no rotation *every* plane through the mass-centre must be a plane of symmetry. We have thus a simple proof of the proposition that the *only* form of equilibrium of a mass of homogeneous liquid under its own attraction is a sphere‡.

We conclude that the middle points of all chords of the free surface drawn parallel to the axis lie in a plane normal to this axis, which we may call the equatorial plane. Hence no straight line parallel to the axis can meet the surface in more than two points. It follows that the z-component of the attraction at any internal or external point not on the plane of symmetry will be directed towards this plane. For the theory of the attraction of a

* Poincaré, *Bull. Astr.* 1885; *Figures d'Équilibre*, Paris, 1902, p. 11. The proof is modified.

† Lichtenstein, *Berl. Ber.* 1918, p. 1120. The argument is slightly simplified.

‡ Carleman, *Math. Zeitschrift*, iii. 1 (1918). A proof that the sphere is the only *stable* form, due to Liapounoff, is given by Poincaré, *Figures d'Équilibre*, c. ii.

uniform straight line, already appealed to, shews that this is true as regards each of the elementary columns into which we have supposed the mass to be divided. Hence $\partial\Omega/\partial z > 0$, and therefore $\partial p/\partial z < 0$, at all points on the positive side of the plane of symmetry. It follows that $\partial p/\partial n > 0$ at all points of the free surface, and that $p > 0$ at all internal points. The former of these statements is inconsistent with (3) if $\omega^2 > 2\pi\rho$. The limitation $\omega < \sqrt{(2\pi\rho)}$ is therefore imposed quite apart from any question of internal tension.

A narrower limit has been investigated by Crudeli*. His argument, slightly modified, is as follows. The theory of Attractions shews that a function whose value is $p - p_0$ (where p_0 is the surface pressure) throughout the fluid, and zero outside, can be regarded as the gravitation-potential of a suitable distribution of (positive or negative) matter, viz. with a surface density

$$-\frac{1}{4\pi}\frac{\partial p}{\partial n}$$

over the boundary, and a volume density

$$-\frac{1}{4\pi}\nabla^2 p = \frac{\rho}{2\pi}(2\pi\rho - \omega^2)$$

throughout the interior. Hence

$$4\pi(p - p_0) = -\iint\frac{\partial p}{\partial n}\frac{dS}{r} + 2\rho(2\pi\rho - \omega^2)\iiint\frac{dx\,dy\,dz}{r}. \quad \text{......(4)}$$

But at internal points

$$\iiint\frac{\rho\,dx\,dy\,dz}{r} = -\Omega = \frac{p}{\rho} - \tfrac{1}{2}\omega^2(x^2 + y^2) + \text{const.},$$

and thence

$$\frac{2\omega^2}{\rho}p = 4\pi p_0 - \iint\frac{\partial p}{\partial n}\frac{dS}{r} - \omega^2(2\pi\rho - \omega^2)(x^2 + y^2) + \text{const.} \quad \text{......(5)}$$

Now consider a tangent plane normal to the axis of rotation, and such that the region occupied by the fluid lies wholly on one side of it, and let P be a point of contact†. Let us form the derivatives in the direction of the inward normal at P, of the two equal sides of (5). Since we have seen that $\partial p/\partial n$ must be positive at all points of the boundary, it follows from the theory of Attractions and from the supposition just made that the normal derivative of

$$-\iint\frac{\partial p}{\partial n}\frac{dS}{r}$$

at P must be less than $2\pi\partial p/\partial n$. Hence

$$\frac{2\omega^2}{\rho}\frac{\partial p}{\partial n} < 2\pi\frac{\partial p}{\partial n}, \quad \text{..............................(6)}$$

or $\omega^2 < \pi\rho$. This alters the least possible period of rotation of a fluid mass having the Earth's mean density to 1 h. 35 m.

* *Accad. d. Lincei* (5) xix. 666 (1910).

† Crudeli appears to assume that the boundary is everywhere convex. From the argument as given above this seems to be unnecessary. For instance a ring-shaped figure is not excluded.

373. Proceeding now to the consideration of special forms, we begin with the case where the external boundary is ellipsoidal. We write down, in the first place, some formulae relating to the attraction of ellipsoids.

The gravitation-potential, at internal points, of a uniform mass enclosed by the surface

$$\frac{x^2}{a^2}+\frac{y^2}{b^2}+\frac{z^2}{c^2}=1 \quad \text{.........................(1)}$$

is

$$\Omega=\pi\rho abc\int_0^\infty\left(\frac{x^2}{a^2+\lambda}+\frac{y^2}{b^2+\lambda}+\frac{z^2}{c^2+\lambda}-1\right)\frac{d\lambda}{\Delta}, \quad \text{.........(2)*}$$

where

$$\Delta=\{(a^2+\lambda)(b^2+\lambda)(c^2+\lambda)\}^{\frac{1}{2}}. \quad \text{..................(3)}$$

This may be written

$$\Omega=\pi\rho(\alpha_0x^2+\beta_0y^2+\gamma_0z^2-\chi_0), \quad \text{..................(4)}$$

where, as in Art. 114,

$$\alpha_0=abc\int_0^\infty\frac{d\lambda}{(a^2+\lambda)\Delta},\quad \beta_0=abc\int_0^\infty\frac{d\lambda}{(b^2+\lambda)\Delta},\quad \gamma_0=abc\int_0^\infty\frac{d\lambda}{(c^2+\lambda)\Delta}, \quad \text{......(5)}$$

and

$$\chi_0=abc\int_0^\infty\frac{d\lambda}{\Delta}. \quad \text{.........................(6)}$$

The potential energy of the mass is given by

$$V=\tfrac{1}{2}\iiint\Omega\rho\,dx\,dy\,dz, \quad \text{.........................(7)}$$

where the integrations extend over the volume. Substituting from (4) we find

$$\begin{aligned}V&=\tfrac{2}{3}\pi^2\rho^2abc\,\{\tfrac{1}{5}(\alpha_0a^2+\beta_0b^2+\gamma_0c^2)-\chi_0\}\\&=\tfrac{2}{3}\pi^2\rho^2a^2b^2c^2\int_0^\infty\left\{\tfrac{1}{5}\left(\frac{a^2}{a^2+\lambda}+\frac{b^2}{b^2+\lambda}+\frac{c^2}{c^2+\lambda}\right)+1\right\}\frac{d\lambda}{\Delta}\\&=\tfrac{2}{3}\pi^2\rho^2a^2b^2c^2\int_0^\infty\left\{\tfrac{2}{5}\lambda d\left(\frac{1}{\Delta}\right)-\tfrac{2}{5}\frac{d\lambda}{\Delta}\right\}=-\tfrac{8}{15}\pi^2\rho^2a^2b^2c^2\int_0^\infty\frac{d\lambda}{\Delta}. \quad \text{...(8)}\end{aligned}$$

This expression is negative because the zero of reckoning corresponds to a state of infinite diffusion of the mass. If we adopt as zero of potential energy that of the mass when collected into a sphere of radius R, $=(abc)^{\frac{1}{3}}$, we must add the term

$$\tfrac{16}{15}\pi^2\rho^2R^5. \quad \text{..............................(9)}$$

If the ellipsoid be of revolution, the integrals reduce. If it be of the *planetary* form we may put, in the notation of Art. 107,

$$a=b=\frac{(\zeta^2+1)^{\frac{1}{2}}}{\zeta}c, \quad \text{.........................(10)}$$

and obtain†

$$\left.\begin{aligned}\alpha_0=\beta_0&=(\zeta^2+1)\zeta\cot^{-1}\zeta-\zeta^2,\\ \gamma_0&=2(\zeta^2+1)(1-\zeta\cot^{-1}\zeta),\end{aligned}\right\} \quad \text{..............(11)}$$

$$V=\tfrac{16}{15}\pi^2\rho^2R^5\left\{1-\left(\frac{\zeta^2+1}{\zeta^2}\right)^{\frac{1}{3}}\zeta\cot^{-1}\zeta\right\}, \quad \text{..............(12)}$$

* For references see p. 604. The sign of Ω has been changed from the more usual convention.

† Most simply by writing $c^2+\lambda=(a^2-c^2)u^2$. The results are expressed by Thomson and Tait (Art. 771) and other writers in terms of a quantity f, the reciprocal of ζ.

provided the zero of V correspond to the spherical form. If e be the eccentricity of the meridian, we have

$$e^2 = 1 - \frac{c^2}{a^2} = \frac{1}{\zeta^2 + 1}, \quad \text{.........................(13)}$$

and the formulae may be written

$$\left.\begin{aligned} \alpha_0 = \beta_0 &= \frac{\sqrt{(1-e^2)}}{e^3} \sin^{-1} e - \frac{1-e^2}{e^2}, \\ \gamma_0 &= \frac{2}{e^2}\left\{1 - \sqrt{(1-e^2)}\,\frac{\sin^{-1} e}{e}\right\}, \end{aligned}\right\} \quad \text{..............(14)}$$

$$V = \tfrac{16}{15}\pi^2\rho^2 R^5 \left\{1 - (1-e^2)^{\frac{1}{6}}\,\frac{\sin^{-1} e}{e}\right\}. \quad \text{..............(15)}$$

For an *ovary* ellipsoid we put (Art. 103)

$$a = b = \frac{(\zeta^2 - 1)^{\frac{1}{2}}}{\zeta}\,c, \quad \text{..........................(16)}$$

and obtain

$$\left.\begin{aligned} \alpha_0 = \beta_0 &= \zeta^2 - (\zeta^2 - 1)\,\zeta \coth^{-1}\zeta, \\ \gamma_0 &= 2\,(\zeta^2 - 1)\,(\zeta \coth^{-1}\zeta - 1), \end{aligned}\right\} \quad \text{.................(17)}$$

$$V = \tfrac{16}{15}\pi^2\rho^2 R^5 \left\{1 - \left(\frac{\zeta^2 - 1}{\zeta^2}\right)^{\frac{1}{3}} \zeta \coth^{-1}\zeta\right\}. \quad \text{...........(18)}$$

The case of an infinitely long elliptic cylinder may also be noticed. Putting $c = \infty$ in (5), we find

$$\alpha_0 = \frac{2b}{a+b}, \quad \beta_0 = \frac{2a}{a+b}, \quad \gamma_0 = 0. \quad \text{.................(19)}$$

The potential energy per unit length of the cylinder is

$$V_1 = \tfrac{4}{15}\pi^2\rho^2 a^2 b^2 \log \frac{(a+b)^2}{4ab}. \quad \text{.......................(20)}$$

Maclaurin's Ellipsoids.

374. Now suppose the ellipsoid to rotate in relative equilibrium about the axis of z, with angular velocity ω. Since

$$\frac{p}{\rho} = \tfrac{1}{2}\omega^2(x^2 + y^2) - \Omega + \text{const.}, \quad \text{....................(1)}$$

the surfaces of equal pressure are given by

$$\left(\alpha_0 - \frac{\omega^2}{2\pi\rho}\right)x^2 + \left(\beta_0 - \frac{\omega^2}{2\pi\rho}\right)y^2 + \gamma_0 z^2 = \text{const.} \quad \text{...........(2)}$$

In order that one of these may coincide with the external surface

$$\frac{x^2}{a^2} + \frac{y^2}{b^2} + \frac{z^2}{c^2} = 1, \quad \text{.............................(3)}$$

we must have

$$\left(\alpha_0 - \frac{\omega^2}{2\pi\rho}\right)a^2 = \left(\beta_0 - \frac{\omega^2}{2\pi\rho}\right)b^2 = \gamma_0 c^2. \quad \text{.................(4)}$$

In the case of an ellipsoid of revolution ($a = b$), these conditions reduce to one, viz.

$$\left(\alpha_0 - \frac{\omega^2}{2\pi\rho}\right) a^2 = \gamma_0 c^2. \quad \text{.......................(5)}$$

Since $a^2/(a^2+\lambda)$ is greater or less than $c^2/(c^2+\lambda)$, according as a is greater or less than c, it follows from the forms of α_0, γ_0 given in Art. 373 (5) that the above condition can be fulfilled by a suitable value of ω for any assigned planetry ellipsoid, but not for the ovary form. This important result is due to Maclaurin*.

If we substitute from Art. 373 (11), the condition (5) takes the form

$$\frac{\omega^2}{2\pi\rho} = (3\zeta^2 + 1)\,\zeta \cot^{-1}\zeta - 3\zeta^2, \quad \text{.....................(6)}$$

or, in the notation of Art. 107,

$$\omega^2/2\pi\rho = \zeta q_2(\zeta). \quad \text{.............................(7)}$$

It will be noticed that the value of ω corresponding to any prescribed ellipticity depends on the density ρ, and not on the actual size of the ellipsoid. It is easily seen that this is in accordance with the theory of 'dimensions.'

If M be the total mass, H its angular momentum about the axis of rotation, we have

$$M = \tfrac{4}{3}\pi\rho a^2 c, \qquad H = \tfrac{2}{5} M a^2 \omega, \quad \text{....................(8)}$$

whence

$$\frac{H^2}{M^3 R} = \tfrac{6}{25}\left(\frac{\zeta^2+1}{\zeta^2}\right)^{\frac{2}{3}} \{(3\zeta^2+1)\,\zeta\cot^{-1}\zeta - 3\zeta^2\}. \quad \text{...........(9)}$$

The formula (6) has been discussed, under different forms, by Simpson, d'Alembert, and (more fully) by Laplace†. It is easily proved that the right-hand side of (6) vanishes for $\zeta = 0$ and $\zeta = \infty$, but it is otherwise finite and positive; consequently that it has a greatest value for some intermediate value of ζ. There is thus, for given density ρ, an upper limit to the angular velocities for which an ellipsoid of revolution is a possible form of relative equilibrium. A more detailed investigation is required to shew that there is only one maximum, and consequently no minimum, value of the function on the right-hand side of (6) or (7).

Laplace also examined, from the same point of view, the formula for the angular momentum. It appears that the right-hand side of (9) increases continually from 0 to ∞ as ζ decreases from ∞ to 0. Hence for a given volume of given fluid there is one, and only one, form of Maclaurin's ellipsoid having a prescribed angular momentum.

* *l.c. ante* p. 307.

† *Mécanique Céleste*, Livre 3me, c. iii. For other references see Todhunter, *History of the Theories of Attraction...*, London, 1873, cc. x, xvi.

These questions may also be investigated by actual computation of the functions on the right-hand sides of (6) and (9). The table below giving numerical details of a series of Maclaurin's ellipsoids, is adapted from Thomson and Tait *. The unit of angular momentum in the last column is $M^{\frac{3}{2}} R^{\frac{1}{2}}$, where 'astronomical' units are of course implied.

The maximum value of $\omega^2/2\pi\rho$ is ·2247, corresponding to e = ·9299, a/c = 2·7198. For any smaller value of $\omega^2/2\pi\rho$ there are *two* possible ellipsoids of revolution, the eccentricity being in one case less and in the other greater than ·9299.

In the case of a homogeneous liquid mass of density equal to the mean density of the earth, we have

$$\tfrac{4}{3}\pi\rho R = 980, \qquad R = 6{\cdot}37 \times 10^8,$$

if the units of length and time be the centimetre and the second, whence it is found that the fastest rotation consistent with an ellipsoidal form of revolution has a period of 2 h. 25 m.

e	a/R	c/R	$\omega^2/2\pi\rho$	Angular momentum
0	1·0000	1·0000	0	0
·1	1·0016	·9967	·0027	·0255
·2	1·0068	·9865	·0107	·0514
·3	1·0159	·9691	·0243	·0787
·4	1·0295	·9435	·0436	·1085
·5	1·0491	·9086	·0690	·1417
·6	1·0772	·8618	·1007	·1804
·7	1·1188	·7990	·1387	·2283
·8	1·1856	·7114	·1816	·2934
·8127	1·1973	·6976	·1868	·3035
·9	1·3189	·5749	·2203	·4000
·91	1·341	·5560	·2225	·4156
·92	1·367	·5355	·2241	·4330
·93	1·396	·5131	·2247	·4525
·94	1·431	·4883	·2239	·4748
·95	1·474	·4603	·2213	·5008
·96	1·529	·4280	·2160	·5319
·97	1·602	·3895	·2063	·5692
·98	1·713	·3409	·1890	·6249
·99	1·921	·2710	·1551	·7121
1·00	∞	0	0	∞

When ζ is great, the right-hand side of (7) reduces to $\frac{4}{15}\zeta^{-2}$ approximately. Hence in the case of a planetary ellipsoid differing infinitely little from a sphere we have, for the *ellipticity*,

$$\epsilon = \frac{a-c}{a} = \tfrac{1}{2}\zeta^{-2} = \tfrac{15}{16}\frac{\omega^2}{\pi\rho}.$$

If g denote the value of gravity at the surface of a sphere of radius a, of the same density, we have $g = \frac{4}{3}\pi\rho a$, whence

$$\epsilon = \tfrac{5}{4}\frac{\omega^2 a}{g}.$$

Putting $\omega^2 a/g = \frac{1}{289}$, we find that a homogeneous liquid globe of the same size and mass as the earth, rotating in the same period, would have an ellipticity of $\frac{1}{231}$.

* *Natural Philosophy*, Art. 772.

Jacobi's Ellipsoids.

375. To ascertain whether an ellipsoid with three *unequal* axes is a possible form of relative equilibrium, we return to the conditions (4) of Art. 374. These are equivalent to

$$(\alpha_0 - \beta_0)\, a^2 b^2 + \gamma_0 c^2 (a^2 - b^2) = 0, \quad \ldots\ldots\ldots\ldots(1)$$

and

$$\frac{\omega^2}{2\pi\rho} = \frac{\alpha_0 a^2 - \beta_0 b^2}{a^2 - b^2}. \quad \ldots\ldots\ldots\ldots(2)$$

If we substitute from Art. 373, the condition (1) may be written

$$(a^2 - b^2) \int_0^\infty \left\{ \frac{a^2 b^2}{(a^2+\lambda)(b^2+\lambda)} - \frac{c^2}{c^2+\lambda} \right\} \frac{d\lambda}{\Delta} = 0. \quad \ldots\ldots\ldots\ldots(3)$$

The first factor, equated to zero, gives Maclaurin's ellipsoids, discussed in the preceding Art. The second factor gives

$$\int_0^\infty \{a^2 b^2 - (a^2 + b^2 + \lambda)\, c^2\} \frac{\lambda d\lambda}{\Delta^3} = 0, \quad \ldots\ldots\ldots\ldots(4)$$

which may be regarded as an equation determining c in terms of a, b. When $c^2 = 0$, every element of the integral is positive, and when

$$c^2 = a^2 b^2/(a^2 + b^2)$$

every element is negative. Hence there is some value of c, less than the smaller of the two semi-axes a, b, for which the integral vanishes.

The corresponding value of ω is given by (2), which takes the form

$$\frac{\omega^2}{2\pi\rho} = abc \int_0^\infty \frac{\lambda d\lambda}{(a^2+\lambda)(b^2+\lambda)\,\Delta}, \quad \ldots\ldots\ldots\ldots(5)$$

so that ω is real. It will be observed that the ratio $\omega^2/2\pi\rho$ depends as before only on the *shape* of the ellipsoid, and not on its absolute size*.

The equations (4) and (5) were carefully discussed by C. O. Meyer†, who shewed that when a, b are given there is only one value of c satisfying (4), and that, further, a maximum value (viz. ·1871)‡ of $\omega^2/2\pi\rho$ occurs for $a = b = 1{\cdot}7161\, c$. The Jacobian ellipsoid then coincides with one of Maclaurin's forms. This limiting form, which is shewn on the opposite page, may be determined by putting

$$a = b, \quad c^2 + \lambda = (a^2 - c^2)\, u^2, \quad c^2 = (a^2 + c^2)\, \zeta^2,$$

in the second factor of (3). We find

$$\int_\zeta^\infty \left\{ \left(\frac{1+\zeta^2}{1+u^2}\right)^2 - \frac{\zeta^2}{u^2} \right\} \frac{du}{1+u^2} = 0, \quad \ldots\ldots\ldots\ldots(6)$$

* The possibility of an ellipsoidal form with three unequal axes was first asserted by Jacobi, "Ueber die Figur des Gleichgewichts," *Pogg. Ann.* xxxiii. 229 (1834) [*Werke*, ii. 17]; see also Liouville, "Sur la figure d'une masse fluide homogène, en équilibre, et douée d'un mouvement de rotation," *Journ. de l'École Polytechn.* xiv. 290 (1834).

† "De aequilibrii formis ellipsoidicis," *Crelle*, xxiv. (1842).

‡ According to Thomson and Tait this should be ·1868. See the table on the preceding page.

whence
$$\cot^{-1}\zeta = \frac{13\zeta + 3\zeta^3}{3 + 14\zeta^2 + 3\zeta^4}. \quad \text{..........................(7)*}$$

There is only one finite root, viz. $\zeta = \cdot 7171$; this gives, for the eccentricity of the meridian, $e = \cdot 8127$.

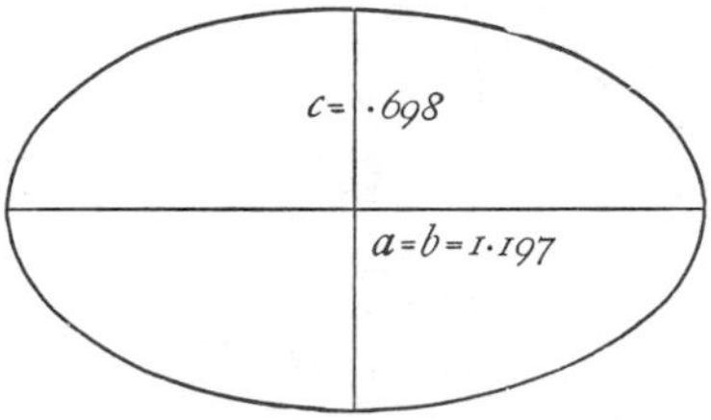

Since, in the general case, the two ratios $a:b:c$ are subject to the condition (4), there is virtually only one variable parameter, and the Jacobian ellipsoids form what may be called a 'linear' series. The sequence of figures in the series is illustrated by the following table, computed by Darwin†. As

Axes			$\frac{\omega^2}{2\pi\rho}$	Angular momentum
a/R	b/R	c/R		
1·197	1·197	·698	·1871	·304
1·216	1·179	·698	·187	·304
1·279	1·123	·696	·186	·306
1·383	1·045	·692	·181	·313
1·601	·924	·677	·166	·341
1·899	·811	·649	·141	392
2·346	·702	·607	·107	·481
3·136	·586	·545	·067	·644
5·04	·45	·44	·026	1·016
∞	0	0	0	∞

$\omega^2/2\pi\rho$ diminishes from its upper limit ·1871, the ratio of one equatorial axis of the ellipsoid to the polar axis increases, whilst that of the other diminishes, the asymptotic form being that of an infinitely long circular cylinder rotating about an axis perpendicular to its length ($a = \infty$, $b = c$). The figures on page 706 shew two intermediate forms, the unit of length being the radius (R) of the sphere of equal volume.

It may be noticed that an infinitely long elliptic cylinder may rotate in relative equilibrium about its longitudinal axis. It is easily proved, with the help of the formulae (19) of Art. 373, that the angular velocity is given by

$$\frac{\omega^2}{2\pi\rho} = \frac{2ab}{(a+b)^2}. \quad \text{..............................(8)‡}$$

* Thomson and Tait, Art. 778′.

† "On Jacobi's Figure of Equilibrium for a Rotating Mass of Fluid," *Proc. Roy. Soc.* xli. 319 (1886) [*Papers*, iii. 119].

‡ Matthiessen, "Neue Untersuchungen über frei rotirende Flüssigkeiten," *Schriften der Univ. zu Kiel*, vi. (1859). This paper contains a very complete list of previous writings on the subject.

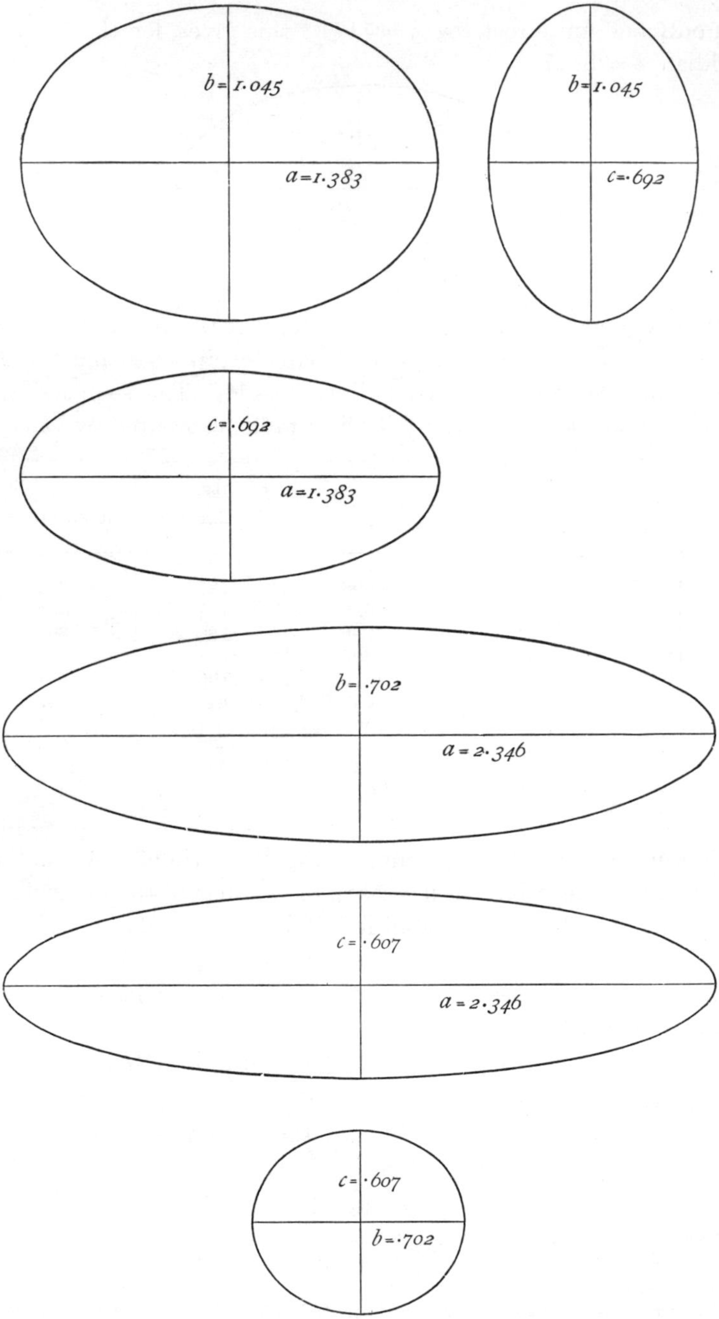
b = 1·045
a = 1·383
b = 1·045
c = ·692
c = ·692
a = 1·383
b = ·702
a = 2·346
c = ·607
a = 2·346
c = ·607
b = ·702

Other Special Forms.

376. The problem of relative equilibrium, of which Maclaurin's and Jacobi's ellipsoids are particular cases, has been the subject of many remarkable investigations, to which only slight reference can here be made.

The case of the annulus was first treated by Laplace*, with special reference to the theory of Saturn's rings.

The annulus is supposed to be a figure of revolution about the axis of z, and the origin is taken at the intersection of this axis with the equatorial plane of symmetry which we know must exist (Art. 372). Further, the cross-section is taken to be an ellipse whose semi-axes parallel to Ox and Oz are a and c respectively. If C be the centre of this section, we write $OC = D$; and it is assumed that the ratios a/D, c/D are both small.

Under these conditions, the component attractions at any point in the substance of the ring are, to a first approximation, the same as if the radius D were infinite, so that we may write, in accordance with Art. 373 (19),

$$\Omega = \pi\rho\,(\alpha_0 x^2 + \gamma_0 z^2) + \text{const.}, \quad \text{.......................(1)}$$

where

$$\alpha_0 = \frac{2c}{a+c}, \qquad \gamma_0 = \frac{2a}{a+c}, \quad \text{.......................(2)}$$

provided the origin of x be now transferred to C. The pressure-equation is, accordingly, for points of the cross-section,

$$\frac{p}{\rho} = \tfrac{1}{2}\omega^2 (D+x)^2 - \Omega + \frac{S}{\sqrt{\{(D+x)^2 + z^2\}}} + \text{const.}, \quad \text{...........(3)}$$

where S denotes the mass of the central attracting body at O. This may be expanded in the form

$$\frac{p}{\rho} = \tfrac{1}{2}\omega^2 (D^2 + 2Dx + x^2) - \pi\rho\,(\alpha_0 x^2 + \gamma_0 z^2) + \frac{S}{D}\left(1 - \frac{x}{D} + \frac{2x^2 - z^2}{2D^2} - \dots\right). \quad \text{......(4)}$$

If p is to be approximately constant over the circumference of the section

$$\frac{x^2}{a^2} + \frac{z^2}{c^2} = 1, \quad \text{..............................(5)}$$

the terms in x must cancel, and the coefficients of x^2 and z^2 must be in the ratio of c^2 to a^2. Hence

$$\omega^2 D^3 = S, \quad \text{..................................(6)}$$

and

$$a^2\left(\alpha_0 - \frac{3\omega^2}{2\pi\rho}\right) = c^2\left(\gamma_0 + \frac{\omega^2}{2\pi\rho}\right). \quad \text{.....................(7)}$$

* "Mémoire sur la théorie de l'anneau de Saturne," *Mém. de l'Acad. des Sciences*, 1789 [1787] [*Mécanique Céleste*, Livre 3me, c. vi].

The former of these equations shews that the period of revolution of the ring must be that proper to a satellite at the same distance; and the latter may be written

$$\frac{\omega^2}{2\pi\rho} = \frac{2ac(a-c)}{(3a^2+c^2)(a+c)}, \quad\text{.............................(8)}$$

whence it appears that the equatorial diameter of the section must be the greater.

The expression on the right-hand side of (8) has a maximum value ·1086, corresponding to $a/c = 2{\cdot}594$. Hence for a fluid ring at a given distance D from the central body there is an inferior limit to the destiny.

Laplace points out that a ring such as we have imagined would be unstable even if rigid, and must *à fortiori* be unstable when fluid. It is now generally held that the constitution of the Saturnian rings is meteoric.

When the central body is absent, or its mass relatively small, the attraction of the ring at points of its substance must be calculated to a higher degree of approximation. It easily appears that the cross-section must be nearly circular, and that the angular velocity must be much less than in the previous case. It is found that when $S=0$

$$\frac{\omega^2}{2\pi\rho} = \frac{a^2}{2D^2}\left(\log\frac{8D}{a} - \frac{5}{4}\right), \quad\text{.................................(9)*}$$

nearly, provided a/D be small.

This formula may be verified as follows. In cylindrical co-ordinates, with the origin at the centre of the ring, the potential at external points satisfies an equation of the type (1) of Art. 100, viz.

$$\frac{\partial^2\Omega}{\partial z^2} + \frac{\partial^2\Omega}{\partial\varpi^2} + \frac{1}{\varpi}\frac{\partial\Omega}{\partial\varpi} = 0. \quad\text{....................................(10)}$$

If we introduce polar co-ordinates in the plane of a cross-section, writing

$$z = r\sin\theta, \qquad \varpi = D + r\cos\theta, \quad\text{..............................(11)}$$

this becomes

$$\frac{\partial^2\Omega}{\partial r^2} + \frac{1}{r}\frac{\partial\Omega}{\partial r} + \frac{1}{r^2}\frac{\partial^2\Omega}{\partial\theta^2} + \frac{1}{D+r\cos\theta}\left(\frac{\partial\Omega}{\partial r}\cos\theta - \frac{1}{r}\frac{\partial\Omega}{\partial r}\sin\theta\right) = 0. \quad\text{.........(12)}$$

To obtain a solution which shall be valid for values of r which are small compared with D, we take, as a first approximation, $\Omega = \Omega_0$, where Ω_0 satisfies

$$\frac{\partial^2\Omega_0}{\partial r^2} + \frac{1}{r}\frac{\partial\Omega_0}{\partial r} = 0. \quad\text{...(13)}$$

Thus

$$\Omega_0 = A + B\log r. \quad\text{...(14)}$$

For a second approximation we put

$$\Omega = \Omega_0 + \Omega_1\cos\theta. \quad\text{...(15)}$$

Substituting we find

$$\frac{\partial^2\Omega_1}{\partial r^2} + \frac{1}{r}\frac{\partial\Omega_1}{\partial r} - \frac{\Omega_1}{r^2} = -\frac{1}{D}\frac{\partial\Omega_0}{\partial r} = -\frac{B}{Dr}, \quad\text{...........................(16)}$$

whence

$$\Omega_1 = Cr + \frac{C'}{r} - \frac{B}{2D}r\log r. \quad\text{.................................(17)}$$

* A slightly different result was given by Matthiessen, *l.c.* The formula (9) was obtained by Mme Sophie Kowalewsky, *Astr. Nachr.* cxi. 37 (1885); Poincaré, *l.c. infra*; Dyson, *l.c. ante* p. 156. See also Basset, *Amer. Journ. Math.* xi. (1888).

At distances r which, though small compared with D, are large compared with the radius a of the section, the result thus obtained must approximate to the potential of a circular line of matter, of radius D, and line-density $\pi\rho a^2$. This is given by

$$\Omega = -\pi\rho a^2 D \int_0^{2\pi} \frac{d\chi}{\sqrt{(r_1^2 \cos^2 \frac{1}{2}\chi + r_2^2 \sin^2 \frac{1}{2}\chi)}} = -\frac{4\pi\rho a^2 D}{r_2} F_1(k), \quad \text{.........(18)}$$

where, as in Art. 161, r_1, r_2 denote the least and greatest distances of the point considered from the circumference, and the modulus k of the elliptic integral is given by

$$k^2 = 1 - \frac{r_1^2}{r_2^2}. \quad \text{...(19)}$$

Since this is nearly equal to unity, we have*

$$F_1(k) = \log\frac{4r_2}{r_1} + \frac{1}{4}\frac{r_1^2}{r_2^2}\left(\log\frac{4r_2}{r_1} - 1\right) + \ldots, \quad \text{.......................(20)}$$

of which the first term will suffice for our purpose.

To accord with the present notation, we put

$$r_1 = r, \qquad r_2 = \sqrt{(4D^2 + 4rD\cos\theta + r^2)} = 2D\left(1 + \frac{r}{2D}\cos\theta\right), \quad \text{...........(21)}$$

approximately. Hence

$$\Omega = -2\pi\rho a^2\left(\log\frac{8D}{r} - \frac{r\cos\theta}{2D}\log\frac{8D}{r} + \frac{r\cos\theta}{2D}\right). \quad \text{.................(22)}$$

The result contained in (15), (14), and (17) will tend to coincide with (22) as r increases, while still remaining small compared with D, provided

$$B = 2\pi\rho a^2, \qquad C = \frac{\pi\rho a^2}{D}(\log 8D - 1). \quad \text{..........................(23)}$$

We therefore adopt, as the value of the external potential of the fluid annulus, at points near its surface, the expression

$$\Omega = -2\pi\rho a^2\left\{\log\frac{8D}{r} - \left(\log\frac{8D}{r} - 1\right)\frac{r\cos\theta}{2D}\right\} + \frac{C'\cos\theta}{r}. \quad \text{...............(24)}$$

To find the potential at internal points, we must replace the right-hand member of (12) by $4\pi\rho$. By the same process of approximation as before we find, having regard to the condition of finiteness for $r=0$,

$$\Omega = \text{const.} + \pi\rho r^2 + C'' r\cos\theta - \frac{\pi\rho r^3}{4D}\cos\theta. \quad \text{........................(25)}$$

The values of Ω and of $\partial\Omega/\partial r$ derived from (24) and (25) must be continuous for $r=a$. This gives

$$C'' = \frac{\pi\rho a^2}{D}\left(\log\frac{8D}{a} - 1\right), \qquad C' = -\frac{\pi\rho a^4}{4D}. \quad \text{........................(26)}$$

The condition for a free surface requires that the expression

$$\tfrac{1}{2}\omega^2(D + r\cos\theta)^2 - \Omega \quad \text{...................................(27)}$$

should be constant for $r=a$. Neglecting the square of r/D we find

$$\omega^2 D = C'' - \frac{\pi\rho a^2}{4D}. \quad \text{.......................................(28)}$$

Substituting the value of C'' from (26), we obtain the result (9).

It has been shewn by Dyson that a ring of the above kind would be unstable for types of disturbance in which the sectional area varies with the longitude, and for such types only. Its tendency would therefore be to break up into detached masses.

* Cayley, *Elliptic Functions*, p. 54.

Darwin has investigated* in great detail the case of two detached masses of liquid rotating in relative equilibrium about their common centre of gravity like the components of a double star. When the distance between the masses is large compared with the dimensions of either, the series of spherical harmonics in which the solution is expressed are rapidly convergent; but in other cases the approximations become very laborious†. The specially interesting case where one mass is much smaller than the other appears to have been first discussed by Roche in 1847‡.

General Problem of Relative Equilibrium.

377. The question as to the possible configurations of relative equilibrium of a rotating homogeneous liquid was taken up from a more general point of view by Poincaré, in a celebrated paper§.

Consider in the first place an ordinary dynamical system of n degrees of freedom, whose constitution depends on a variable parameter λ, the potential energy V being accordingly a function of the n generalized co-ordinates $q_1, q_2, \ldots q_n$ and of λ. The possible configurations of equilibrium corresponding to a prescribed value of λ are determined by n equations of the type

$$\frac{\partial V}{\partial q_r} = 0; \qquad\qquad (1)$$

and by varying λ we get one or more 'linear series' of equilibrium configurations. Such a series may be represented by a curve in an n-dimensional space, of which $q_1, q_2, \ldots q_n$ are the Cartesian co-ordinates.

Again considering small deviations from any equilibrium configuration, we have

$$V = c_{11}\delta q_1^2 + c_{22}\delta q_2^2 + \ldots + 2c_{12}\delta q_1 \delta q_2 + \ldots, \qquad\qquad (2)$$

where $c_{11}, c_{22}, c_{12}, \ldots$ are 'coefficients of stability' (Art. 168) defined by

$$c_{rr} = \frac{\partial^2 V}{\partial q_r^2}, \qquad c_{rs} = \frac{\partial^2 V}{\partial q_r \partial q_s}. \qquad\qquad (3)$$

By a linear transformation of the variations $\delta q_1, \delta q_2, \ldots \delta q_n$, the expression (2) can be reduced, in an infinite number of ways, to a sum of squares; but whatever mode of reduction be adopted, the number of positive as well as of negative coefficients is, by a theorem due to Sylvester, invariable.

* "On Figures of Equilibrium of Rotating Masses of Fluid," *Phil. Trans.* A, clxxviii. 379 (1887) [*Papers*, iii. 135].

† For a fuller investigation of the problems of Arts. 374–376 reference may be made to Tisserand, *Traité de Mécanique Céleste*, Paris, 1889–1896, ii.

‡ See Darwin, "On the Figure and Stability of a Liquid Satellite," *Phil. Trans.* A, ccvi. 161 (1906) [*Papers*, iii. 436]. For the application of Poincaré's methods to this problem reference may be made to Schwarzschild, "Die Poincaré'sche Theorie des Gleichgewichts...," *Ann. d. Münch. Sternwarte*, iii. 233 (1897), and Jeans, *Problems of Cosmogony...*, Cambridge, 1919.

§ "Sur l'équilibre d'une masse fluide animée d'un mouvement de rotation," *Acta Math.* vii. 259 (1885). See also his *Figures d'équilibre*. For an account of the earlier researches and partial anticipations by Liapounoff, see Lichtenstein, *Math. Zeitschrift*, i. 228 (1918).

The coefficients in the transformed expression may be called principal coefficients of stability. In order that the configuration in question may be stable it is necessary and sufficient that these should all be positive.

As we vary λ, the several linear series will remain distinct so long as the discriminant Δ of the quadratic form (2) does not vanish, *i.e.* so long as no principal coefficient of stability vanishes. But if, as we follow a linear series, Δ vanishes and changes sign for a particular value of λ, it appears that the configuration in question is a 'form of bifurcation,' *i.e.* it is (as it were) the meeting point with another linear series. The case may also arise where, as λ passes through a particular value, two linear series coalesce and then become imaginary. If the configuration in question does not belong to any other linear series, we have what is called a 'limiting form' of equilibrium, and it may be shewn that Δ has different signs in the two series, in the neighbourhood of the junction. A specially important case is where two series coalesce and afterwards become imaginary, whilst a third series passes continuously through the common point.

The foregoing statements may be illustrated by the case of a system of one degree of freedom. The positions of equilibrium are given by

$$\partial V/\partial q=0, \quad \ldots\ldots(4)$$

which determines one or more values of q in terms of λ. If we differentiate with respect to λ, we obtain

$$\frac{\partial^2 V}{\partial q^2}\frac{dq}{d\lambda}+\frac{\partial^2 V}{\partial q\partial\lambda}=0. \quad \ldots\ldots(5)$$

This gives, for each linear series, a unique value of $dq/d\lambda$, and so determines the succession of equilibrium configurations, unless $\partial^2 V/\partial q^2=0$. The several series therefore remain distinct so long as the coefficient of stability does not vanish; but if $\partial^2 V/\partial q^2=0$, $dq/d\lambda$ is infinite or indeterminate according as $\partial^2 V/\partial q\partial\lambda$ is or is not different from zero. In the former case, two series in general coalesce.

Writing $$\partial V/\partial q=\phi(\lambda, q), \quad \ldots\ldots(6)$$

let us consider the surface $$z=\phi(x, y), \quad \ldots\ldots(7)$$

where x, y, z are ordinary Cartesian co-ordinates. The curve $\phi(x, y)=0$ which separates

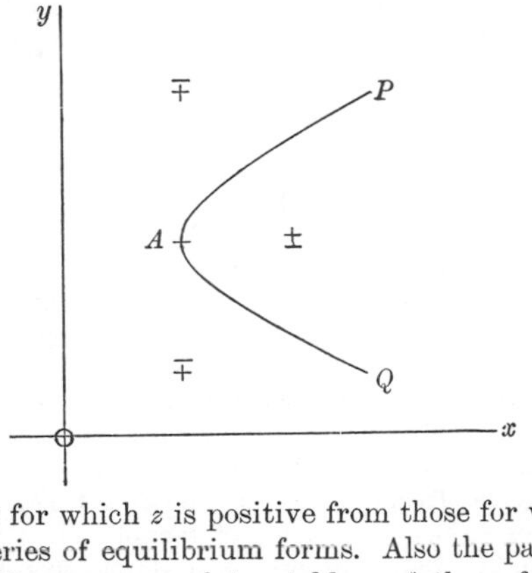

the parts of the plane xy for which z is positive from those for which z is negative, represents the various linear series of equilibrium forms. Also the parts of the curve for which the gradient $\partial z/\partial y$ is positive correspond to stable, and those for which $\partial z/\partial y$ is negative to unstable configurations.

The critical points ($\partial^2 V/\partial q^2 = 0$) correspond to $\partial z/\partial y = 0$; the tangent-line to the curve is then parallel to y, or else the point in question is a singular point on the curve. In the former case, if no other branch of the curve goes through the point of contact, we have a 'limiting form'; and it is evident that there is a change from stability to instability at this point. This case is represented in the preceding figure (p. 711), where the two series PA and QA coalesce in the limiting form A. If the upper signs in the figure refer to the values of z in the corresponding regions, the series PA is unstable and QA stable. If the lower signs obtain, these statements must be reversed.

If however we have also $\partial^2 V/\partial q \partial \lambda = 0$, or $\partial z/\partial x = 0$, we have a singular point. The case where two series (PA and QA) coalesce and become imaginary, whilst a third series (HAK) passes through the common point and remains real, is shewn below. In the latter series we have a transition from stability to instability, or *vice versâ*, whilst the other series are both stable or both unstable in the neighbourhood of A*.

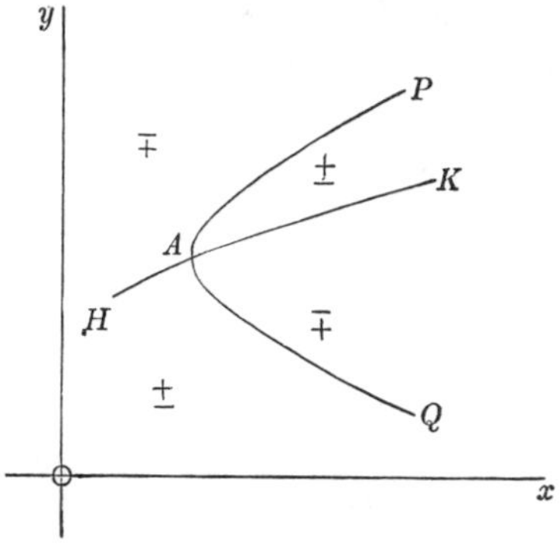

When there are n degrees of freedom, the equations of equilibrium are

$$\frac{\partial V}{\partial q_1} = 0, \quad \frac{\partial V}{\partial q_2} = 0, \quad \ldots, \quad \frac{\partial V}{\partial q_n} = 0. \qquad \text{..............................(8)}$$

We may utilize the $n-1$ equations following the first to determine $q_2, \ldots q_n$ in terms of q_1 and λ. Let us denote the result of substituting these values in the general expression for V by $\psi(q_1, \lambda)$. We have then,

$$\frac{\partial \psi}{\partial q_1} = \frac{\partial V}{\partial q_1} + \frac{\partial V}{\partial q_2}\frac{\partial q_2}{\partial q_1} + \ldots + \frac{\partial V}{\partial q_n}\frac{\partial q_n}{\partial q_1} = \frac{\partial V}{\partial q_1}, \qquad \text{..........................(9)}$$

by (8), so that the remaining condition of equilibrium may be written

$$\frac{\partial \psi}{\partial q_1} = 0. \qquad \text{..(10)}$$

From this we derive

$$\frac{\partial^2 \psi}{\partial q_1^2}\frac{dq_1}{d\lambda} + \frac{\partial^2 \psi}{\partial q_1 \partial \lambda} = 0, \qquad \text{.......................................(11)}$$

which shews that the sequence of equilibrium configurations is unique unless $\partial^2\psi/\partial q_1^2 = 0$. The rest of the argument is then as before, with ψ substituted for V. It is easily proved that the condition $\partial^2\psi/\partial q_1^2 = 0$ is analytically equivalent to $\Delta = 0$†.

* As a simple example, take the case of a particle free to move in a smooth curvilinear tube (having points of inflexion) in a vertical plane, the tube being capable of being set in different positions by rotation about an axis perpendicular to this plane. Other examples are furnished by investigations as to the positions of equilibrium of a floating log, as depending on the density, and their respective stabilities. The case of a log of square section is discussed in the author's *Statics*, Cambridge, 1912, pp. 221, 234. The case of a simple crossing between two series, both of which are real on either side of the intersection, may be illustrated in a similar manner.

† The argument is taken, with little alteration, from Poincaré's treatise.

378. The bearing of these considerations on the theory of relative equilibrium of a rotating system will be apparent.

In the case of equilibrium relative to a rigid frame which is constrained to rotate with constant angular velocity ω about a fixed axis, the conditions are most conveniently considered under the type

$$\frac{\partial}{\partial q_r}(V - T_0) = 0, \qquad \text{.............................(1)}$$

where V is the potential energy, and T_0 is the kinetic energy of the system when rotating as rigid in any assigned configuration $(q_1, q_2, \ldots q_n)$; cf. Art. 205. By varying ω we get the various linear series of equilibrium configurations. Moreover, if the system be subject to dissipative forces affecting all *relative* motions, the condition of secular stability is that $V - T_0$ should be a minimum.

When, on the other hand, the system is free, the case comes under the general theory of gyrostatic systems, and the more appropriate form of the conditions is

$$\frac{\partial}{\partial q_r}(V + K) = 0, \qquad \text{.............................(2)}$$

where K is the kinetic energy of the system when rotating, as rigid, in the configuration $(q_1, q_2, \ldots q_n)$ with the component momenta corresponding to the ignored co-ordinates unaltered (Art. 254); and the condition of secular stability is that $V + K$ should be a minimum. From the present point of view the only ignored co-ordinate which we need consider is an angular co-ordinate specifying the position in space of a plane of reference in the system, passing through the axis of rotation and therefore also through the centre of inertia. The corresponding component of momentum is the angular momentum about the axis; we shall denote this by κ. By varying κ we get the various linear series of equilibrium configurations.

In the case of a rotating liquid, the generalized co-ordinates $q_1, q_2, \ldots$ are infinite in number, but the theory is otherwise unaltered. Let us suppose, for a moment, that we have a liquid covering a rigid rotating nucleus. If the nucleus be constrained to rotate with constant angular velocity, or (what comes to the same thing) if it be of preponderant inertia, we have the first form of the problem; whereas if the nucleus be free, the second form applies. The distinction between the two forms disappears when we confine ourselves to disturbances which do not affect the moment of inertia of the system with respect to the axis of rotation.

The second form of the problem is from the present point of view the more important. We pass to the case of a homogeneous rotating liquid by imagining the nucleus to become infinitely small. In this case the solution of the problem of relative equilibrium is partially known. We have, first, the linear series of

Maclaurin's ellipsoids in which, as κ ranges from 0 to ∞, a/R ranges from 1 to ∞ (Art. 374). Again, we have the two* series of Jacobian ellipsoids in which, as κ ranges from $\cdot 304 M^{\frac{3}{2}} R^{\frac{1}{2}}$ to ∞, a/b ranges in one case from 1 to ∞, and in the other from 1 to 0, where a, b denote the two equatorial semi-axes (Art. 375). When $\kappa = \cdot 304 M^{\frac{3}{2}} R^{\frac{1}{2}}$, we have a form of bifurcation, and accordingly a change in the character of the stability.

379. As a simple application of the preceding theory we may examine the secular stability of Maclaurin's ellipsoid for those types of ellipsoidal disturbance in which the axis of rotation remains a principal axis†.

Let ω be the angular velocity in the state of equilibrium, and κ the angular momentum. If I denote the moment of inertia of the disturbed system, the angular velocity, if this were to rotate, as rigid, would be κ/I. Hence

$$V+K=V+\tfrac{1}{2}I\left(\frac{\kappa}{I}\right)^2=V+\tfrac{1}{2}\frac{\kappa^2}{I}, \quad\text{.................................(1)}$$

and the condition of secular stability is that this expression should be a minimum. We will suppose for definiteness that the zero of reckoning of V corresponds to the state of infinite diffusion. Then in any other configuration V will be negative.

In our previous notation we have

$$I=\tfrac{1}{5}M(a^2+b^2), \quad\text{...(2)}$$

c being the axis of rotation. Since $abc=R^3$, we may write

$$V+\tfrac{5}{2}\frac{\kappa^2}{M(a^2+b^2)}=f(a, b), \quad\text{...................................(3)}$$

where $f(a, b)$ is a symmetric function of the two independent variables a, b. If we consider the surface whose ordinate is $f(a, b)$, where a, b are regarded as rectangular co-ordinates of a point in a horizontal plane, the configurations of relative equilibrium will correspond to points whose altitude is stationary, whilst for secular stability the altitude must further be a minimum.

For $a=\infty$, or $b=\infty$, we have $f(a, b)=0$. For $a=0$, we have $V=0$, and $f(a, b) \propto 1/b^2$, and similarly for $b=0$. For $a=0$, $b=0$, simultaneously, we have $f(a, b)=\infty$. It is known that, whatever the value of κ, there is always one and only one possible form of Maclaurin's ellipsoid. Hence as we follow the section of the above-mentioned surface by the plane of symmetry ($a=b$), the ordinate varies from ∞ to 0, having one and only one stationary value in the interval. It is evident that this value is negative, and a minimum‡. Hence the altitude at this point of the surface cannot be a maximum. Moreover, since there is a limit to the negative value of V, viz. when the ellipsoid becomes a sphere, there is always at least one finite point of minimum (and negative) altitude on the surface.

Now it appears, on reference to the table on p. 705, that when $\kappa < \cdot 304 M^{\frac{3}{2}} R^{\frac{1}{2}}$, there is one and only one ellipsoidal form of equilibrium, viz. one of revolution. The preceding considerations shew that this corresponds to a point of minimum altitude, and is therefore secularly stable (for symmetrical ellipsoidal disturbances).

* The two series include the same succession of geometrical forms, but are from the present point of view to be regarded as analytically distinct.

† Poincaré, *l.c.* For a more analytical investigation see Basset, "On the Stability of Maclaurin's Liquid Spheroid," *Proc. Camb. Phil. Soc.* viii. 23 (1892).

‡ It follows that Maclaurin's ellipsoid is always stable for a deformation such that the surface remains an ellipsoid of revolution.

When $\kappa > \cdot 304 M^{\frac{2}{3}} R^{\frac{1}{2}}$, there are three points of stationary altitude, viz. one in the plane of symmetry, corresponding to a Maclaurin's ellipsoid, and two others symmetrically situated on opposite sides of this plane, corresponding to Jacobian forms. It is evident from topographical considerations that the altitude must be a minimum at the two last-named points, and neither maximum nor minimum at the former. Any other arrangement would involve the existence of additional points of stationary altitude.

The result of the investigation is that Maclaurin's ellipsoid is secularly stable or unstable, for ellipsoidal disturbances, according as the eccentricity e is less or greater than ·8127, the eccentricity of the ellipsoid of revolution which is the starting point of Jacobi's series; whilst the Jacobian ellipsoids are all stable for such disturbances*.

The further discussion of the stability of Maclaurin's ellipsoid would carry us too far. It was shewn by Poincaré that the equilibrium is secularly stable for deformations of *all* types so long as e falls below the above-mentioned limit. This is established by shewing that there is no form of bifurcation for any ellipsoid of revolution of smaller eccentricity. It follows, from the consideration of 'exchange of stabilities,' that Jacobi's series begin by being thoroughly stable.

380. Poincaré has further examined the coefficients of stability of the series of Maclaurin's and Jacobi's ellipsoids, by the method of Lamé's functions, with the view of ascertaining what members are forms of bifurcation. He finds that there are an infinite number of such forms, and consequently an infinite number of other linear series of equilibrium configurations. In each case it is possible to assign the form of the members of the new series in the neighbourhood of the bifurcation. The question has been further discussed by Darwin†, and by Poincaré himself in a subsequent paper‡.

The case which has attracted most interest is the first bifurcation which occurs in the series of Jacobi's ellipsoids. According to Darwin†, the critical ellipsoid is that for which $a/R = 1{\cdot}8858$, $b/R = {\cdot}8150$, $c/R = {\cdot}6507$. After this point Jacobi's ellipsoids are unstable.

In the figure§ on p. 716, in which the ratios a/c and b/c are taken as co-ordinates, the straight line HAK represents the series of Maclaurin's ellipsoids corresponding to different values of κ; whilst the branches AR, AS represent those of the Jacobian figures. The point H corresponds to the case of the sphere, when $\kappa = 0$; and the Maclaurin series is stable from

* This result, like the preceding, was stated, without proof, by Thomson and Tait, *Natural Philosophy* (2nd ed.), Art. 778″.

† "On the Pear-shaped Figure of Equilibrium of a Rotating Mass of Liquid," *Phil. Trans.* A, cxcviii. 301 (1901) [*Papers*, iii. 288].

‡ "Sur la Stabilité des Figures Pyriformes affectées par une Masse Fluide en Rotation," *Phil. Trans.* A, cxcviii. 333 (1901).

§ The diagram is constructed from the tables on pp. 703, 705. A sketch is given in Poincaré's treatise.

H to A, and afterwards unstable. The points P, Q indicate the stage at which the Jacobian ellipsoids become unstable. At these points new series branch off. The difficult question as to the stability of these has been discussed by

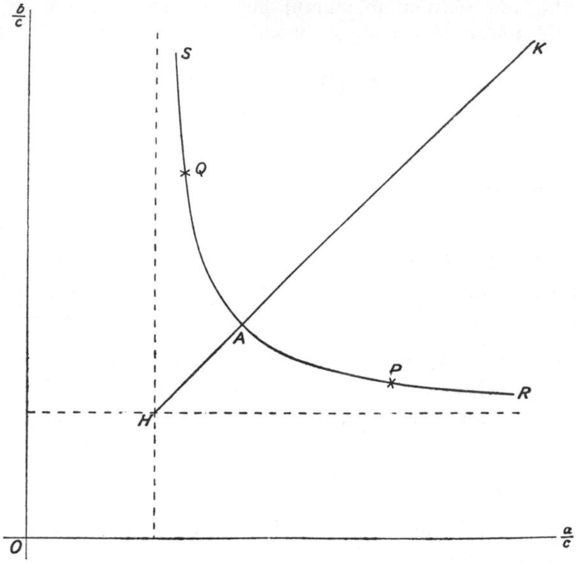

Darwin, Poincaré, and Jeans*. The latter writer concludes definitely that they are in the first instance unstable. The first members of these new series have the 'pear-shaped' form shewn in the annexed diagrams, which are taken from the paper by Darwin just referred to.

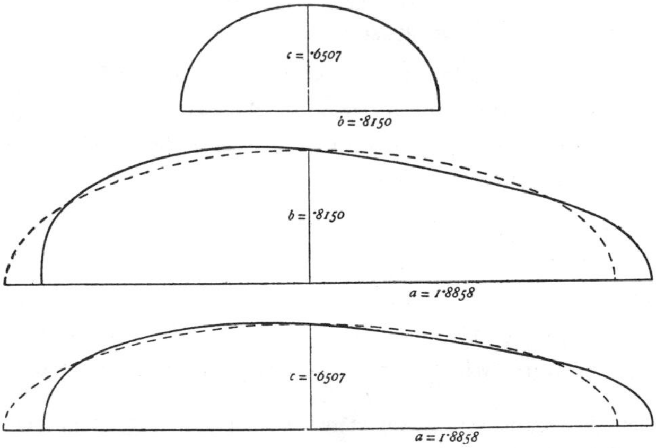

The corresponding two-dimensional problem has been discussed by Jeans†, by a special method.

* Poincaré, *l.c.*; Darwin, "The Stability of the Pear-shaped Figure of Equilibrium," *Phil. Trans.* A, cc. 251 (1902) [*Papers*, iii. 317]; Jeans, *l.c. ante* p. 710.

† "On the Equilibrium of Rotating Liquid Cylinders," *Phil. Trans.* A, cc. 67 (1902).

Small Oscillations.

381. The small oscillations of a rotating ellipsoid mass have been discussed by various writers.

The simplest types of disturbance which we can consider are those in which the surface remains ellipsoidal, with the axis of revolution as a principal axis. In the case of Maclaurin's ellipsoid, there are two distinct types of this character; in one of these the surface remains an ellipsoid of revolution, whilst in the other the equatorial axes become unequal, one increasing and the other decreasing, whilst the polar axis is unchanged. It was shewn by Riemann* that the latter type is unstable when the eccentricity (e) of the meridian section exceeds ·9529. In this investigation frictional forces are not contemplated, and the criterion is one of 'ordinary' stability. We have seen (Art. 379) that practically the equilibrium is unstable when e exceeds ·8127. The periods of Riemann's two types of oscillation (when $e <$ ·9529) have been calculated by Love†, who has also discussed the two-dimensional oscillations (of elliptic type) of a rotating elliptic cylinder‡.

The problem of small oscillations was treated in a more general manner by Poincaré§. It appears from Art. 207 that the equations of small motion relative to rotating axes may be written

$$\frac{\partial u}{\partial t} - 2\omega v = -\frac{\partial \psi}{\partial x}, \quad \frac{\partial v}{\partial t} + 2\omega u = -\frac{\partial \psi}{\partial y}, \quad \frac{\partial w}{\partial t} = -\frac{\partial \psi}{\partial z}, \quad \text{......(1)}$$

where

$$\psi = \frac{p}{\rho} + \Omega - \tfrac{1}{2}\omega^2 (x^2 + y^2), \quad \text{......................(2)}$$

if Ω denote the gravitation potential of the liquid mass. From these, and from the equation of continuity

$$\frac{\partial u}{\partial x} + \frac{\partial v}{\partial y} + \frac{\partial w}{\partial z} = 0, \quad \text{............................(3)}$$

we deduce

$$\frac{\partial^2}{\partial t^2} \nabla^2 \psi + 4\omega^2 \frac{\partial^2 \psi}{\partial z^2} = 0. \quad \text{..........................(4)}$$

If we assume that u, v, w vary as $e^{i\sigma t}$, we find

$$u = \frac{i\sigma \dfrac{\partial \psi}{\partial x} + 2\omega \dfrac{\partial \psi}{\partial y}}{\sigma^2 - 4\omega^2}, \quad v = \frac{-2\omega \dfrac{\partial \psi}{\partial x} + i\sigma \dfrac{\partial \psi}{\partial y}}{\sigma^2 - 4\omega^2}, \quad w = \frac{i}{\sigma}\frac{\partial \psi}{\partial z}, \quad \text{...(5)}$$

* "Beitrag zu den Untersuchungen über die Bewegung eines flüssigen gleichartigen Ellipsoides," *Gött. Abh.* ix. 3 (1860) [*Werke*, p. 192]. See also Basset, *Hydrodynamics*, Art. 367, Riemann also shews that Jacobi's ellipsoids are stable (in the above restricted sense) for ellipsoidal disturbances.

† "On the Oscillations of a Rotating Liquid Spheroid, and the Genesis of the Moon," *Phil. Mag.* (5) xxvii. 254 (1889). The symmetrical type is easily treated by means of the equation (23) of Art. 382, below.

‡ "On the Motion of a Liquid Elliptic Cylinder under its own Attraction," *Quart. Journ. Math.* xxiii. 153 (1888).

§ *l.c. ante* p. 710.

and therefore from (3), or immediately from (4),

$$\frac{\partial^2\psi}{\partial x^2}+\frac{\partial^2\psi}{\partial y^2}+\left(1-\frac{4\omega^2}{\sigma^2}\right)\frac{\partial^2\psi}{\partial z^2}=0. \qquad \text{.................(6)}$$

If we write
$$1-\frac{4\omega^2}{\sigma^2}=\tau^2, \qquad z=\tau z', \qquad \text{.........................(7)}$$

this takes the form
$$\frac{\partial^2\psi}{\partial x^2}+\frac{\partial^2\psi}{\partial y^2}+\frac{\partial^2\psi}{\partial z'^2}=0. \qquad \text{.........................(8)}$$

If the equation of the undisturbed ellipsoid be

$$\frac{x^2}{a^2}+\frac{y^2}{b^2}+\frac{z^2}{c^2}=1, \qquad \text{.............................(9)}$$

the appropriate solutions of (8) are those which involve the ellipsoidal harmonics corresponding to the surface

$$\frac{x^2}{a^2}+\frac{y^2}{b^2}+\frac{z'^2}{c^2/\tau^2}=1, \qquad \text{.........................(10)}$$

which is obtained from (9) by homogeneous strain*.

At the surface (9) we must have $p=$ const., and therefore

$$\psi=\Omega-\tfrac{1}{2}\omega^2(x^2+y^2). \qquad \text{.......................(11)}$$

The potential Ω of the disturbed form depends on the normal displacement (ζ) at the surface; this is connected with ψ by a relation of the form

$$lu+mv+nw=\frac{\partial\zeta}{\partial t}=i\sigma\zeta, \qquad \text{.......................(12)}$$

where the surface values of u, v, w are to be taken from (5).

The procedure is then as follows. Assuming that ζ is an ellipsoidal surface-harmonic relative to (9), the surface-value of Ω is calculated, and substituted in (11). The resulting surface-value of ψ is then expressed in terms of harmonics relative to the auxiliary surface (10); the corresponding expression of ψ in the interior can then be written down in ellipsoidal solid harmonics. The condition (12) then gives an equation to determine σ; it appears that this equation is always algebraic.

In the case of Maclaurin's ellipsoid the process is somewhat simplified, the harmonics involved being of the types studied in Arts. 104, 107. This problem has been fully worked out by Bryan†, who has in particular completed Riemann's investigation by shewing that the equilibrium is 'ordinarily' stable for *all* types of disturbance so long as the eccentricity of the meridian is less than ·9529.

* It appears that for some types of free oscillation τ is imaginary, and the surface (9) consequently a hyperboloid.

† "The Waves on a Rotating Liquid Spheroid of Finite Ellipticity," *Phil. Trans.* A, clxxx. 187 (1888).

Dirichlet's Ellipsoids.

382. The motion of a liquid mass under its own gravitation, with a *varying* ellipsoidal surface, was first studied by Dirichlet*. Adopting the Lagrangian method of Art. 13, he proposed as the subject of investigation the whole class of motions in which the displacements are linear functions of the co-ordinates. This was carried further, on the same lines, by Dedekind† and Riemann‡. More recently, it has been shewn by Greenhill§ and others that some branches of the problem can be treated very successfully by the Eulerian method.

We will take first the case where the ellipsoid does not change the directions of its axes, and the internal motion is irrotational. This is interesting as an example of *finite* oscillation of a liquid mass about the spherical form.

The expression for the velocity-potential has been given in Art. 110; viz. we have

$$\phi = -\tfrac{1}{2}\left(\frac{\dot{a}}{a}x^2 + \frac{\dot{b}}{b}y^2 + \frac{\dot{c}}{c}z^2\right), \qquad \text{.......................(1)}$$

with the condition of constant volume

$$\frac{\dot{a}}{a} + \frac{\dot{b}}{b} + \frac{\dot{c}}{c} = 0. \qquad \text{.................................(2)}$$

The pressure is then given by

$$\frac{p}{\rho} = \frac{\partial\phi}{\partial t} - \Omega - \tfrac{1}{2}q^2 + F(t), \qquad \text{.......................(3)}$$

by Art. 20 (4); and substituting the value of Ω from Art. 373 we find

$$\frac{p}{\rho} = -\tfrac{1}{2}\left(\frac{\ddot{a}}{a}x^2 + \frac{\ddot{b}}{b}y^2 + \frac{\ddot{c}}{c}z^2\right) - \pi\rho\,(\alpha_0 x^2 + \beta_0 y^2 + \gamma_0 z^2) + F(t). \quad \text{...(4)}$$

The conditions that the pressure may be uniform over the external surface

$$\frac{x^2}{a^2} + \frac{y^2}{b^2} + \frac{z^2}{c^2} = 1 \qquad \text{.............................(5)}$$

are therefore

$$\left(\frac{\ddot{a}}{a} + 2\pi\rho\alpha_0\right)a^2 = \left(\frac{\ddot{b}}{b} + 2\pi\rho\beta_0\right)b^2 = \left(\frac{\ddot{c}}{c} + 2\pi\rho\gamma_0\right)c^2. \qquad \text{.........(6)}$$

* "Untersuchungen über ein Problem der Hydrodynamik," *Gött. Abh.* viii. 3 (1860); *Crelle*, lviii. 181 [*Werke*, ii. 263]. The paper was posthumous, and was edited and amplified by Dedekind.

† *Crelle*, lviii. 217 (1861).

‡ *l.c. ante* p. 717.

§ "On the Rotation of a Liquid Ellipsoid about its Mean Axis," *Proc. Camb. Phil. Soc.* iii. 233 (1879); "On the general Motion of a liquid Ellipsoid under the Gravitation of its own parts," *Proc. Camb. Phil. Soc.* iv. 4 (1880).

These equations, with (2), determine the variations of a, b, c. If we multiply the three terms of (2) by the three equal magnitudes in (6), we obtain

$$\dot{a}\ddot{a} + \dot{b}\ddot{b} + \dot{c}\ddot{c} + 2\pi\rho\,(\alpha_0 a\dot{a} + \beta_0 b\dot{b} + \gamma_0 c\dot{c}) = 0. \quad\text{..............(7)}$$

If we substitute the values of α_0, β_0, γ_0 from Art. 373, this has the integral

$$\dot{a}^2 + \dot{b}^2 + \dot{c}^2 - 4\pi\rho\,abc\int_0^\infty \frac{d\lambda}{\Delta} = \text{const.} \quad\text{..................(8)}$$

It has already been proved (Art. 373) that the potential energy is

$$V = \text{const.} - \tfrac{8}{15}\pi^2\rho^2 a^2 b^2 c^2 \int_0^\infty \frac{d\lambda}{\Delta}, \quad\text{..................(9)}$$

and it easily follows from (1) that the kinetic energy is

$$T = \tfrac{2}{15}\,\pi\rho\,abc\,(\dot{a}^2 + \dot{b}^2 + \dot{c}^2). \quad\text{........................(10)}$$

Hence (8) is recognized as the equation of energy

$$T + V = \text{const.} \quad\text{.............................(11)}$$

When the ellipsoid is of revolution ($a = b$), the equation (8), with $a^2 c = R^3$, is sufficient to determine the motion. We find

$$\tfrac{2}{15}\pi\rho R^3\left(1 + \frac{R^3}{2c^3}\right)\dot{c}^2 + V = \text{const.} \quad\text{..................(12)}$$

The character of the motion depends on the total energy. If this be less than the potential energy in the state of infinite diffusion, the ellipsoid will oscillate regularly between the prolate and oblate forms, with a period depending on the amplitude; whilst if the energy exceed this limit it will not oscillate, but will tend to one or other of two extreme forms, viz. an infinite line of matter coinciding with the axis of z, or an infinite film coincident with the plane of xy*.

If, in the case of an ellipsoid of revolution, we superpose on the irrotational motion given by (1) a uniform rotation ω about the axis of z, the component velocities (parallel to fixed axes) are

$$u = \frac{\dot{a}}{a}x - \omega y, \qquad v = \frac{\dot{a}}{a}y + \omega x, \qquad w = \frac{\dot{c}}{c}z. \quad\text{........................(13)}$$

The Eulerian equations (Art. 6 (2)) then reduce to

$$\left.\begin{aligned} \frac{\ddot{a}}{a}x - \dot{\omega}y - 2\frac{\dot{a}}{a}\omega y - \omega^2 x &= -\frac{1}{\rho}\frac{\partial p}{\partial x} - \frac{\partial \Omega}{\partial x}, \\ \frac{\ddot{a}}{a}y + \dot{\omega}x + 2\frac{\dot{a}}{a}\omega x - \omega^2 y &= -\frac{1}{\rho}\frac{\partial p}{\partial y} - \frac{\partial \Omega}{\partial y}, \\ \frac{\ddot{c}}{c}z \qquad\qquad\qquad\qquad &= -\frac{1}{\rho}\frac{\partial p}{\partial z} - \frac{\partial \Omega}{\partial z}. \end{aligned}\right\} \quad\text{......................(14)}$$

* Dirichlet, *l.c.* When the amplitude of oscillation is small, the period must coincide with that obtained by putting $n = 2$ in the formula (10) of Art. 262. This has been verified by Hicks, *Proc. Camb. Phil. Soc.* iv. 309 (1883).

The first two equations give, by cross-differentiation,

$$\frac{\dot{\omega}}{\omega}+2\frac{\dot{a}}{a}=0, \qquad (15)$$

or

$$\omega a^2=\omega_0 a_0^2, \qquad (16)$$

which is simply the expression of von Helmholtz' theorem that the 'strength' of a vortex is constant (Art. 146). In virtue of (15), the equations (14) have the integral

$$\frac{p}{\rho}=-\tfrac{1}{2}\left(\frac{\ddot{a}}{a}-\omega^2\right)(x^2+y^2)-\tfrac{1}{2}\frac{\ddot{c}}{c}z^2-\Omega+\text{const.} \qquad (17)$$

Introducing the value of Ω from Art. 373 (4), we find that the pressure will be constant over the surface

$$\frac{x^2+y^2}{a^2}+\frac{z^2}{c^2}=1, \qquad (18)$$

provided

$$\left(\frac{\ddot{a}}{a}+2\pi\rho\alpha_0-\omega^2\right)a^2=\left(\frac{\ddot{c}}{c}+2\pi\rho\gamma_0\right)c^2. \qquad (19)$$

In virtue of the relation (15), and of the condition of constancy of volume

$$2\frac{\dot{a}}{a}+\frac{\dot{c}}{c}=0, \qquad (20)$$

this may be put in the form

$$2\dot{a}\ddot{a}+\dot{c}\ddot{c}+2(\omega^2 a\dot{a}+\omega\dot{\omega}a^2)+4\pi\rho\alpha_0 a\dot{a}+2\pi\rho\gamma_0 c\dot{c}=0, \qquad (21)$$

whence

$$2\dot{a}^2+\dot{c}^2+2\omega^2a^2-4\pi\rho a^2c\int_0^\infty\frac{d\lambda}{(a^2+\lambda)(c^2+\lambda)^{\frac{1}{2}}}=\text{const.} \qquad (22)$$

This, again, may be identified as the equation of energy.

In terms of c as dependent variable, (22) may be written

$$\tfrac{2}{15}\pi\rho R^3\left\{\left(1+\frac{R^3}{2c^3}\right)\dot{c}^2+\frac{2\omega_0^2a_0^4}{R^3}c\right\}+V=\text{const.} \qquad (23)$$

If the initial circumstances be favourable, the surface will oscillate regularly between two extreme forms. Since, for a prolate ellipsoid, V increases with c, it is evident that, whatever the initial conditions, there is a limit to the elongation in the direction of the axis which the rotating ellipsoid can attain. On the other hand, we may have an indefinite spreading out in the equatorial plane*.

If we write

$$K=\tfrac{4}{15}\pi\rho\omega_0^2a_0^4c, \qquad (24)$$

the condition of relative equilibrium, as obtained from (23), is

$$\frac{d}{dc}(V+K)=0, \qquad (25)$$

in accordance with Art. 378 (2). The small oscillations (of symmetrical type) about equilibrium may be investigated by writing $c=c_0+c'$, where c_0 is the solution of (25), and treating c' as small.

383. The study of the motion of a fluid mass bounded by a varying ellipsoidal surface was carried further by Riemann in the paper already quoted. The problem has since become the subject of an extensive literature, some references to which are given below†. The case where the ellipsoidal

* Dirichlet, *l.c.*

† Brioschi, "Développements rélatifs au § 3 des Recherches de Dirichlet sur un problème d'Hydrodynamique," *Crelle*, lix. 63 (1861); Lipschitz, "Reduction der Bewegung eines flüssigen

boundary is invariable in form, but rotates about a principal axis (Oz), can be treated very simply*.

If u, v, w denote the *apparent* velocities relative to axes x, y rotating in their own plane with constant angular velocity ω, the equations of motion are, by Art. 207†,

$$\left.\begin{aligned}\frac{Du}{Dt}-2\omega v-\omega^2 x&=-\frac{1}{\rho}\frac{\partial p}{\partial x}-\frac{\partial\Omega}{\partial x},\\ \frac{Dv}{Dt}+2\omega u-\omega^2 y&=-\frac{1}{\rho}\frac{\partial p}{\partial y}-\frac{\partial\Omega}{\partial y},\\ \frac{Dw}{Dt}\qquad\qquad&=-\frac{1}{\rho}\frac{\partial p}{\partial z}-\frac{\partial\Omega}{\partial z}.\end{aligned}\right\}\quad\text{.......(1)}$$

If the fluid have a uniform vorticity ζ whose axis is parallel to z, the actual velocities parallel to the instantaneous positions of the axes will be

$$\left.\begin{aligned}u-\omega y&=\frac{a^2-b^2}{a^2+b^2}(\omega-\tfrac{1}{2}\zeta)\,y-\tfrac{1}{2}\zeta y,\\ v+\omega x&=\frac{a^2-b^2}{a^2+b^2}(\omega-\tfrac{1}{2}\zeta)\,x+\tfrac{1}{2}\zeta x,\\ w\quad&=0,\end{aligned}\right\}\quad\text{.......(2)}$$

since the conditions are evidently satisfied by the superposition of the irrotational motion which would be produced by the revolution of a rigid ellipsoidal envelope with angular velocity $\omega-\frac{1}{2}\zeta$ on the uniform rotation $\frac{1}{2}\zeta$ (cf. Art. 110). Hence

$$u=\frac{2a^2}{a^2+b^2}(\omega-\tfrac{1}{2}\zeta)\,y,\qquad v=-\frac{2b^2}{a^2+b^2}(\omega-\tfrac{1}{2}\zeta)\,x,\qquad w=0.\quad\text{.......(3)}$$

Substituting in (1), and integrating, we find

$$\frac{p}{\rho}=\frac{2a^2b^2}{(a^2+b^2)^2}(\omega-\tfrac{1}{2}\zeta)^2(x^2+y^2)+\tfrac{1}{2}\omega^2(x^2+y^2)-\frac{2(b^2x^2+a^2y^2)}{a^2+b^2}\,\omega(\omega-\tfrac{1}{2}\zeta)-\Omega+\text{const.}\quad\text{...(4)}$$

Hence the conditions for a free surface are

$$\begin{aligned}&\left\{\frac{2a^2b^2}{(a^2+b^2)^2}(\omega-\tfrac{1}{2}\zeta)^2+\tfrac{1}{2}\omega^2-\frac{2b^2}{a^2+b^2}\,\omega(\omega-\tfrac{1}{2}\zeta)-\pi\rho\alpha_0\right\}a^2\\ &=\left\{\frac{2a^2b^2}{(a^2+b^2)^2}(\omega-\tfrac{1}{2}\zeta)^2+\tfrac{1}{2}\omega^2-\frac{2a^2}{a^2+b^2}\,\omega(\omega-\tfrac{1}{2}\zeta)-\pi\rho\beta_0\right\}b^2\\ &=-\pi\rho\gamma_0c^2.\quad\text{.......(5)}\end{aligned}$$

This includes a number of interesting cases.

1°. If we put $\omega=\frac{1}{2}\zeta$, we get the conditions of Jacobi's ellipsoid (Art. 374 (5)).

2°. If we put $\omega=0$, so that the external boundary is stationary in space, we get

$$\left\{\pi\rho\alpha_0-\frac{a^2b^2}{2(a^2+b^2)^2}\zeta^2\right\}a^2=\left\{\pi\rho\beta_0-\frac{a^2b^2}{2(a^2+b^2)^2}\zeta^2\right\}b^2=\pi\rho\gamma_0c^2.\quad\text{.......(6)}$$

homogenen Ellipsoids auf das Variations-problem eines einfachen Integrals, ...," *Crelle*, lxxviii. 245 (1874); Greenhill, *l.c. ante* p. 719; Basset, "On the Motion of a Liquid Ellipsoid under the Influence of its own Attraction," *Proc. Lond. Math. Soc.* xvii. 255 (1886) [*Hydrodynamics*, c. xv.]; Tedone, *Il moto di un ellissoide fluido secondo l'ipotesi di Dirichlet*, Pisa, 1894; Stekloff, "Problème du mouvement d'une masse fluide incompressible de la forme ellipsoïdale...," *Ann. de l'école normale* (3), xxvi. (1909); Hargreaves, *Camb. Trans.* xxii. 61 (1914).

* Greenhill, *l. c.*

† We might also employ the equations of Art. 12, regard being had to the different meaning of the symbols u, v, w.

These conditions are equivalent to

$$(\alpha_0-\beta_0)a^2b^2+\gamma_0c^2(a^2-b^2)=0, \qquad \text{.................................(7)}$$

and

$$\frac{\zeta^2}{2\pi\rho}=\frac{(a^2+b^2)^2}{a^2b^2}\cdot\frac{a^2\alpha_0-b^2\beta_0}{a^2-b^2}. \qquad \text{.................................(8)}$$

It is evident, on comparison with Art. 375, that c must be the least axis of the ellipsoid and that the value (8) of $\zeta^2/2\pi\rho$ is positive.

The paths of the particles are determined by

$$\dot{x}=-\frac{a^2}{a^2+b^2}\zeta y, \qquad \dot{y}=\frac{b^2}{a^2+b^2}\zeta x, \qquad \dot{z}=0, \qquad \text{..........................(9)}$$

whence

$$x=ka\cos(\sigma t+\epsilon), \qquad y=kb\sin(\sigma t+\epsilon), \qquad z=0, \qquad \text{.................(10)}$$

if

$$\sigma=\frac{ab}{a^2+b^2}\zeta, \qquad \text{...(11)}$$

and k, ϵ are arbitrary constants.

These results are due to Dedekind*. It is remarked by Love that as regards the external forms the series of Dedekind's and of Jacobi's ellipsoids are identical.

3°. Let $\zeta=0$, so that the motion is irrotational. The conditions (5) reduce to

$$\left\{\alpha_0-\frac{(a^2-b^2)(a^2+3b^2)}{(a^2+b^2)^2}\frac{\omega^2}{2\pi\rho}\right\}a^2=\left\{\beta_0-\frac{(b^2-a^2)(3a^2+b^2)}{(a^2+b^2)^2}\frac{\omega^2}{2\pi\rho}\right\}b^2=\gamma_0c^2. \qquad \text{...(12)}$$

These may be replaced by

$$\{\alpha_0(3a^2+b^2)+\beta_0(3b^2+a^2)\}a^2b^2-\gamma_0(a^4+6a^2b^2+b^4)c^2=0, \qquad \text{..............(13)}$$

and

$$\frac{\omega^2}{2\pi\rho}=\frac{(a^2+b^2)^2}{a^4+6a^2b^2+b^4}\cdot\frac{\alpha_0a^2-\beta_0b^2}{a^2-b^2}. \qquad \text{..............................(14)}$$

The equation (13) determines c in terms of a, b. Let us suppose that $a>b$. Then the left-hand side is easily seen to be negative for $c=a$, and positive for $c=b$. Hence there is some real value of c, *between* a and b, for which the condition is satisfied; and the value of ω given by (14) is then real, for the same reason as in Art. 375.

4°. In the case of an elliptic cylinder rotating about its axis the conditions (5) reduce, by Art. 373 (19), to

$$\omega^2+\frac{4a^2b^2}{(a^2+b^2)^2}(\omega-\tfrac{1}{2}\zeta)^2=\frac{4\pi\rho ab}{(a+b)^2}. \qquad \text{..........................(15)†}$$

If we put $\omega=\frac{1}{2}\zeta$, we get the case of Art. 375 (8).

If $\omega=0$, so that the external boundary is stationary, we have

$$\zeta^2=4\pi\rho\frac{(a^2+b^2)^2}{ab(a+b)^2}. \qquad \text{..................................(16)}$$

If $\zeta=0$, *i.e.* the motion is irrotational, we have

$$\omega^2=4\pi\rho\frac{ab(a^2+b^2)^2}{(a+b)^2(a^4+6a^2b^2+b^4)}. \qquad \text{.........................(17)}$$

* *l.c. ante* p. 719. See also Love, "On Dedekind's Theorem, ..." *Phil. Mag.* (5) xxv. 40 (1888).

† Greenhill, *Proc. Camb. Phil. Soc.* iii. 233 (1879).

384. The oscillations of a rotating ellipsoidal mass of liquid contained in a rigid envelope have been discussed by several writers*. We follow at first (with some amplifications) the very elegant treatment adopted by Poincaré.

It is assumed that the mass-centre and the principal axes of inertia of the envelope coincide with those of the fluid, and that the vorticity of the fluid is uniform.

Superposing a uniform rotation (p, q, r) on the formulae (13) of Art. 146, where the envelope was supposed fixed, we have, with a slight change in the notation,

$$\left.\begin{aligned} u &= \frac{a}{c} q_1 z - \frac{a}{b} r_1 y + qz - ry, \\ v &= \frac{b}{a} r_1 x - \frac{b}{c} p_1 z + rx - pz, \\ w &= \frac{c}{b} p_1 y - \frac{c}{a} q_1 x + py - qx. \end{aligned}\right\} \quad \text{.......(1)}$$

The components of vorticity are accordingly

$$\xi = 2p + \left(\frac{c}{b} + \frac{b}{c}\right) p_1, \qquad \eta = 2q + \left(\frac{a}{c} + \frac{c}{a}\right) q_1, \qquad \zeta = 2r + \left(\frac{b}{a} + \frac{a}{b}\right) r_1. \quad \text{.........(2)}$$

The kinetic energy of the whole system is given by

$$2T = Ap^2 + Bq^2 + Cr^2 + A_1 p_1^2 + B_1 q_1^2 + C_1 r_1^2 + 2Fpp_1 + 2Gqq_1 + 2Hrr_1, \quad \text{......(3)}$$

where A, B, C denote the principal moments of inertia of the whole system, whilst A_1, B_1, C_1, F, G, H refer to the fluid alone; thus

$$A_1 = \frac{b^2}{c^2} \Sigma (mz^2) + \frac{c^2}{b^2} \Sigma (my^2) = \tfrac{1}{5} \Sigma (m) (b^2 + c^2), \quad \text{etc., etc.,} \quad \text{..............(4)}$$

$$F = \frac{b}{c} \Sigma (mz^2) + \frac{c}{b} \Sigma (my^2) = \tfrac{2}{5} \Sigma (m)\, bc, \quad \text{etc., etc.,} \quad \text{....................(5)}$$

the summations extending over the mass of the fluid. The principal moments of the envelope will be

$$A_0 = A - A_1, \qquad B_0 = B - B_1, \qquad C_0 = C - C_1. \quad \text{........................(6)}$$

The angular momentum of the system about Ox is

$$\begin{aligned} A_0 p + \Sigma m (yw - zv) &= A_0 p + \left(p + \frac{c}{b} p_1\right) \Sigma (my^2) + \left(p + \frac{b}{c} p_1\right) \Sigma (mz^2) \\ &= Ap + Fp_1 = \frac{\partial T}{\partial p}. \quad \text{..(7)} \end{aligned}$$

The dynamical equations relating to the moving axes are therefore

$$\left.\begin{aligned} \frac{d}{dt}\frac{\partial T}{\partial p} - r\frac{\partial T}{\partial q} + q\frac{\partial T}{\partial r} &= L, \\ \frac{d}{dt}\frac{\partial T}{\partial q} - p\frac{\partial T}{\partial r} + r\frac{\partial T}{\partial p} &= M, \\ \frac{d}{dt}\frac{\partial T}{\partial r} - q\frac{\partial T}{\partial p} + p\frac{\partial T}{\partial q} &= N, \end{aligned}\right\} \quad \text{...............................(8)}$$

where L, M, N are the moments of the external forces.

* Greenhill, *l.c. ante* p. 11; Hough, "The Oscillations of a Rotating Ellipsoidal Shell containing Fluid," *Phil. Trans.* A, clxxxvi. 469 (1895); Poincaré, "Sur la précession des corps déformables," *Bull. Astr.* 1910; Basset, *Quart. J. of Math.* xlv. 223 (1914). The application to precessional problems seems to have been first made by Kelvin [*Papers*, iii. 322, and iv. 129]; the explicit solution is due to Hough and Poincare.

The equations of Helmholtz (Art. 146 (4)) when adapted to moving axes become

$$\frac{D\xi}{Dt} - r\eta + q\zeta = \xi\frac{\partial u}{\partial x} + \eta\frac{\partial u}{\partial y} + \zeta\frac{\partial u}{\partial z}, \quad \text{etc.,} \quad \text{etc.,} \qquad \text{(9)}$$

whence, on substitution from (1) and (2),

$$\frac{d\xi}{dt} = \frac{a}{c}q_1\zeta - \frac{a}{b}r_1\eta, \quad \text{etc.,} \quad \text{etc.,} \qquad \text{(10)}$$

the symbol of total differentiation (d/dt) being used, since by hypothesis ξ, η, ζ are functions of t only.

Now from (2) we have

$$\tfrac{1}{5}\Sigma(m)\, bc\xi = Fp + A_1p_1 = \frac{\partial T}{\partial p_1}, \qquad \text{(11)}$$

and the Helmholtz equations accordingly take the forms

$$\left.\begin{aligned} &\frac{d}{dt}\frac{\partial T}{\partial p_1} - q_1\frac{\partial T}{\partial r_1} + r_1\frac{\partial T}{\partial q_1} = 0, \\ &\frac{d}{dt}\frac{\partial T}{\partial q_1} - r_1\frac{\partial T}{\partial p_1} + p_1\frac{\partial T}{\partial r_1} = 0, \\ &\frac{d}{dt}\frac{\partial T}{\partial r_1} - p_1\frac{\partial T}{\partial q_1} + q_1\frac{\partial T}{\partial p_1} = 0. \end{aligned}\right\} \qquad \text{(12)}$$

If we substitute from (3) in (8) and (12) we obtain the following two systems of equations:

$$\left.\begin{aligned} &\frac{d}{dt}(Ap + Fp_1) - r(Bq + Gq_1) + q(Cr + Hr_1) = L, \\ &\frac{d}{dt}(Bq + Gq_1) - p(Cr + Hr_1) + r(Ap + Fp_1) = M, \\ &\frac{d}{dt}(Cr + Hr_1) - q(Ap + Fp_1) + p(Bq + Gq_1) = N, \end{aligned}\right\} \qquad \text{(13)}$$

$$\left.\begin{aligned} &\frac{d}{dt}(Fp + A_1p_1) + r_1(Gq + B_1q_1) - q_1(Hr + C_1r_1) = 0, \\ &\frac{d}{dt}(Gq + B_1q_1) + p_1(Hr + C_1r_1) - r_1(Fp + A_1p_1) = 0, \\ &\frac{d}{dt}(Hr + C_1r_1) + q_1(Fp + A_1p_1) - p_1(Gq + B_1q_1) = 0. \end{aligned}\right\} \qquad \text{(14)}$$

In the case of symmetry about the axis of z, to which we now confine ourselves*, we have

$$a = b, \qquad A = B, \qquad A_1 = B_1, \qquad C_1 = H, \qquad F = G. \qquad \text{(15)}$$

Hence if (as we shall suppose) the external forces have zero moment about the axis of symmetry, we have

$$C\frac{dr}{dt} + C_1\frac{dr_1}{dt} + F(pq_1 - p_1q) = 0, \qquad \text{(16)}$$

$$C_1\left(\frac{dr}{dt} + \frac{dr_1}{dt}\right) + F(pq_1 - p_1q) = 0. \qquad \text{(17)}$$

It follows that $dr/dt = 0$, as is otherwise dynamically obvious. Hence

$$r = \text{const.}, \qquad = \omega, \text{ say}, \qquad \text{(18)}$$

and

$$C_1\frac{dr_1}{dt} + F(pq_1 - p_1q) = 0. \qquad \text{(19)}$$

* The free oscillations of an ellipsoid with three unequal axes are discussed by Hough (*l.c.*), the ellipticities being assumed to be small.

In the case of a slight disturbance from a state of steady motion n which the fluid and solid rotate together as one mass about the axis of symmetry, p, q, p_1, q_1 will (initially at all events) be small quantities. If we neglect their products, r_1 will be constant, by (19), and may be taken to be small, since it may be assumed to vanish in the steady motion. With these simplifications the remaining equations of the systems (13) and (14) reduce to

$$\left.\begin{aligned} A\frac{dp}{dt}+F\frac{dp_1}{dt}+(C-A)\,\omega q-F\omega q_1=L,\\ A\frac{dq}{dt}+F\frac{dq_1}{dt}-(C-A)\,\omega p+F\omega p_1=M,\end{aligned}\right\} \qquad (20)$$

$$\left.\begin{aligned} F\frac{dp}{dt}+A_1\frac{dp_1}{dt}-C_1\omega q_1=0,\\ F\frac{dq}{dt}+A_1\frac{dq_1}{dt}+C_1\omega p_1=0.\end{aligned}\right\} \qquad (21)$$

As a typical representation of astronomical disturbing forces we may put

$$L=\kappa\cos\sigma t, \qquad M=\kappa\sin\sigma t. \qquad (22)$$

Hence, writing

$$p+iq=\varpi, \qquad p_1+iq_1=\varpi_1, \qquad (23)$$

we have

$$A\frac{d\varpi}{dt}+F\frac{d\varpi_1}{dt}-i(C-A)\,\omega\varpi+iF\omega\varpi_1=\kappa e^{i\sigma t}, \qquad (24)$$

$$F\frac{d\varpi}{dt}+A_1\frac{d\varpi_1}{dt}+iC_1\omega\varpi_1=0. \qquad (25)$$

Hence, for the forced oscillation,

$$\varpi=-\frac{A_1\sigma+C_1\omega}{\Delta(\sigma)}\,i\kappa e^{i\sigma t}, \qquad (26)$$

$$\varpi_1=\frac{F\sigma}{\Delta(\sigma)}\,i\kappa e^{i\sigma t}, \qquad (27)$$

where

$$\Delta(\sigma)=\begin{vmatrix} A\sigma-(C-A)\,\omega, & F(\sigma+\omega)\\ F\sigma, & A_1\sigma+C_1\omega\end{vmatrix}. \qquad (28)$$

The free oscillations are determined by

$$\Delta(\sigma)=0. \qquad (29)$$

We have chiefly in view the case where the ellipticity of the cavity is slight. If the cavity were exactly spherical we should have, by (4) and (5),

$$A_1=C_1=F, \qquad (30)$$

and therefore

$$\Delta(\sigma)=C_1(\sigma+\omega)\,\{A_0\sigma-(C_0-A_0)\,\omega\}. \qquad (31)$$

Hence for the free oscillations (relative to the rotating axes) we should have

$$\sigma=-\omega, \quad \text{and} \quad \sigma=\frac{C_0-A_0}{A_0}\,\omega. \qquad (32)$$

The former root would make $p=0$, $q=0$, by (25), and corresponds to a slight permanent shift (in space) of the axis of vorticity of the fluid. The second root corresponds to the free 'Eulerian nutation' of the shell, now unaffected by the presence of the fluid. The forced oscillations of the shell would also be independent of the fluid.

In the general case the formula (28) may be put in the form

$$\begin{aligned}\Delta(\sigma)=\;&A_1(\sigma+\omega)\{A_0\sigma-(C-A)\,\omega\}\\ &+(A_1^2-F^2)\,\sigma^2+\{(C_1-A_1)\,A_0+C_1A_1-F^2\}\,\omega\sigma-(C_1-A_1)(C-A)\,\omega^2. \qquad (33)\end{aligned}$$

We write

$$\epsilon=\frac{C_1-A_1}{A_1}=\frac{a^2-c^2}{a^2+c^2}. \qquad (34)$$

This is assumed to be a small quantity, in which case it coincides with the ellipticity of the cavity according to the usual definition. We have also, from (4) and (5),

$$\frac{C_1A_1-F^2}{C_1A_1}=\epsilon, \qquad \frac{A_1^2-F^2}{A_1^2}=\epsilon^2. \qquad (35)$$

As a first approximation to the free oscillations we have from (33)

$$\sigma=-\omega, \quad \text{and} \quad \sigma=\frac{C-A}{A_0}\omega, \qquad (36)$$

the latter root indicating that the period of the free Eulerian nutation is shorter than if the whole mass had been solid, in the ratio A_0/A, as we should expect. The approximation may be continued, but does not present much interest. The effect of a small ellipticity is in any case slight.

The case is different with the forced oscillations, especially those of long period. If the distribution of the disturbing forces were invariable in space, it would have relatively to the moving axes an angular velocity $-\omega$. Putting $\sigma=-\omega$ in (26) and (28), we find

$$p+iq=\varpi=\frac{\kappa}{C\omega}ie^{-i\omega t}. \qquad (37)$$

This may be compared with the formula for the slow precession of a top, which may be regarded as a special case. The result is exactly the same as if the mass had been solid throughout. This conclusion, it is to be noticed, is independent of the smallness of ϵ.

If, however, the distribution of disturbing forces varies slowly in space, the time-factor being e^{int}, we must write $\sigma=-\omega+n$, whence

$$p+iq=-\frac{(C_1-A_1)\omega+A_1 n}{\Delta(-\omega+n)}i\kappa e^{-i(\omega-n)t}. \qquad (38)$$

The denominator may be written

$$\Delta(-\omega+n)=(A_0A_1+A_1^2-F^2)n^2 - \{C_0A_1-A(C_1-A_1)+C_1A_1-F^2\}n\omega-C(C_1-A_1)\omega^2. \qquad (39)$$

It appears, then, on reference to (34) and (35), that if the ratio n/ω be not only small, but small compared with ϵ, the formula (38) reduces to

$$p+iq=\frac{i\kappa}{C\omega}e^{-i(\omega-n)t}, \qquad (40)$$

approximately, the same, again, as if the fluid had been solidified. The assumed condition is that the ratio of the (absolute) period $2\pi/n$ of the disturbing force to the period $2\pi/\omega$ of the rotation should be large compared with $1/\epsilon$.

It follows that a very slight degree of ellipticity of the cavity would suffice to make the forced oscillations of long period practically the same as if the whole mass were rigid. If the earth consisted of a rigid crust surrounding a liquid mass having an ellipticity of the same order ($\frac{1}{300}$) as that of the external surface, the condition would of course be abundantly fulfilled in the case of the luni-solar precession, whose period is 26,000 years. On the other hand, the lunar nineteen-yearly nutation would be appreciably, and the solar and lunar nutations of semi-annual and fortnightly periods (respectively) would be seriously, modified by the internal fluidity*.

It should be added that the results (36) as to the *free* oscillations are based on the assumption that the mass of the envelope is comparable with that of the fluid. In the extreme case where the mass of the shell is negligible we have

$$\Delta(\sigma)=(A_1^2-F^2)\sigma(\sigma+\omega)-C_1(C_1-A_1)\omega^2. \qquad (41)$$

The equation to determine the free periods is therefore

$$(a^2-c^2)\sigma(\sigma+\omega)-2a^2\omega^2=0. \qquad (42)$$

It appears that the periods are real if $c<a$, or $>3a$, but imaginary if $a<c<3a$. This is in accordance with Kelvin's observation† that a liquid gyrostat whose envelope is a slightly prolate ellipsoid of revolution is unstable, whilst the oblate form is stable.

* These propositions were enunciated by Kelvin in 1876 [*Papers*, iii. 322]. The mathematical investigation on which they were based was not published.

† *Papers*, iv. 129, 183. The more precise criterion of stability was given by Greenhill; see also Hough, *l.c.*

385. The precession of a liquid ellipsoid with a *free surface* has also been discussed by Poincaré, who has verified a prevision of Kelvin that if the period of the disturbing forces is sufficiently long the precession will be practically the same as if the mass were solid. The question is more difficult than the former one in that the disturbing forces give rise to tidal oscillations, so that it is necessary to disentangle the precession from the deformation involved in the latter.

Poincaré has recourse to the Lagrangian method of Dirichlet, referred to in Art. 382; but there is some advantage, as well as interest, in pursuing (with the proper modifications) the method of the preceding Art. The procedure is in any case somewhat indirect. We imagine in the first instance that the boundary of the fluid is constrained (if necessary) by suitable pressures to remain ellipsoidal, although its dimension may vary. In the end it appears that the constraining forces are unnecessary (cf. Art. 382).

The equations (1) are now replaced by

$$\left.\begin{aligned} u&=\frac{a}{c}q_1z-\frac{a}{b}r_1y+qz-ry+\frac{\dot a}{a}x,\\ v&=\frac{b}{a}r_1x-\frac{b}{c}p_1z+rx-pz+\frac{\dot b}{b}y,\\ w&=\frac{c}{b}p_1y-\frac{c}{a}q_1x+py-qx+\frac{\dot c}{c}z,\end{aligned}\right\} \qquad (43)$$

in accordance with Art. 110 (5), the variations of the axes being connected by the condition of incompressibility

$$\frac{\dot a}{a}+\frac{\dot b}{b}+\frac{\dot c}{c}=0. \qquad (44)$$

The formula (3) for the kinetic energy is therefore modified by the addition of a term

$$\tfrac{1}{5}\Sigma(m)(\dot a^2+\dot b^2+\dot c^2). \qquad (45)$$

The suffixes in the symbols A_1, B_1, C_1, defined as in (4), may now be omitted, since $A_0, B_0, C_0=0$.

The components of angular momentum are expressed as in (7), and the dynamical equations (13) will accordingly still hold, provided it be remembered that the coefficients A, B, C, F, G, H are no longer constants, since they involve the variables a, b, c. The symbols L, M, N must of course include the moments (if any) of the constraining pressures on the surface.

The components of vorticity being still given by (2), the formula (11) is unaltered; but in place of (10) we have

$$\frac{d\xi}{dt}=\frac{\dot a}{a}\xi+\frac{a}{c}q_1\zeta-\frac{a}{b}r_1\eta, \quad \text{etc., etc.} \qquad (46)$$

Hence, having regard to (44),

$$\frac{d}{dt}(bc\xi)=abq_1\zeta-car_1\eta, \quad \text{etc., etc.} \qquad (47)$$

The Helmholtz equations accordingly retain the form (12), but the coefficients in the form (14) are of course variable.

The component accelerations at any point of the fluid may be derived from the formulae of Art. 12; for example, the acceleration parallel to x is

$$\frac{\partial u}{\partial t}-rv+qw+\frac{\partial u}{\partial x}\frac{Dx}{Dt}+\frac{\partial u}{\partial y}\frac{Dy}{Dt}+\frac{\partial u}{\partial z}\frac{Dz}{Dt}, \qquad (48)$$

where

$$\left.\begin{aligned}\frac{Dx}{Dt}&=\frac{\dot a}{a}x+\frac{a}{c}q_1z-\frac{a}{b}r_1y,\\ \frac{Dy}{Dt}&=\frac{\dot b}{b}y+\frac{b}{a}r_1x-\frac{b}{c}p_1z,\\ \frac{Dz}{Dt}&=\frac{\dot c}{c}z+\frac{c}{b}p_1y-\frac{c}{a}q_1x.\end{aligned}\right\} \qquad (49)$$

The accelerations are therefore linear functions of x, y, z, with coefficients which are functions of t. The conditions of integrability of the hydrodynamical equations shew at once that these functions must reduce to the forms

$$\alpha x+hy+gz, \qquad hx+\beta y+fz, \qquad gx+fy+\gamma z. \qquad (50)$$

This may be verified, with a little trouble, by means of the Helmholtz equations (14), which are in fact the conditions of integrability referred to. The hydrodynamical equations are accordingly of the forms

$$\left.\begin{aligned}-\frac{1}{\rho}\frac{\partial P}{\partial x}&=\alpha x+hy+gz+\frac{\partial\Omega}{\partial x}+\frac{\partial\Omega'}{\partial x},\\ -\frac{1}{\rho}\frac{\partial P}{\partial y}&=hx+\beta y+fz+\frac{\partial\Omega}{\partial y}+\frac{\partial\Omega'}{\partial y},\\ -\frac{1}{\rho}\frac{\partial P}{\partial z}&=gx+fy+\gamma z+\frac{\partial\Omega}{\partial z}+\frac{\partial\Omega'}{\partial z},\end{aligned}\right\} \qquad (51)$$

where P is the pressure, Ω is the potential of the ellipsoidal mass itself, and Ω' is that of disturbing bodies at a distance.

In the notation of Art. 373 we have

$$\Omega=\pi\rho\,(\alpha_0x^2+\beta_0y^2+\gamma_0z^2-\chi_0). \qquad (52)$$

The disturbing potential Ω' can be expanded, for points in the neighbourhood of the origin, in a series of solid spherical harmonics of positive degree. The terms of the first order are without influence on the motion relative to the centre of mass, whilst terms of higher order than the second are usually negligible. We write, therefore,

$$\Omega'=\tfrac{1}{2}\,(A'x^2+B'y^2+C'z^2+2F'yz+2G'zx+2H'xy), \qquad (53)$$

the coefficients, which are known functions of the time, being subject to the relation $A'+B'+C'=0$, in virtue of the equation $\nabla^2\Omega'=0$.

The equations (51) are therefore satisfied by

$$P=\lambda\rho\left(1-\frac{x^2}{a^2}-\frac{y^2}{b^2}-\frac{z^2}{c^2}\right), \qquad (54)$$

provided

$$\alpha+2\pi\rho\alpha_0+A'=\frac{2\lambda}{a^2}, \qquad \beta+2\pi\rho\beta_0+B'=\frac{2\lambda}{b^2}, \qquad \gamma+2\pi\rho\gamma_0+C'=\frac{2\lambda}{c^2}, \qquad (55)$$

and

$$f+F'=0, \qquad g+G'=0, \qquad h+H'=0. \qquad (56)$$

In the equations (14), (44), (55), (56) we have a system of ten equations connecting the ten dependent variables a, b, c, p, q, r, p_1, q_1, r_1, λ with the time.

It is to be noticed that the equations (56) are precisely the equations which would be derived from (51) and (53) by expressing that the rates of increase of the angular momenta with respect to fixed axes coincident with the instantaneous positions of the axes of the ellipsoid are equal to the respective moments of the external forces. They are in fact equivalent to the system (13), where L, M, N may now be taken to refer to the disturbing forces alone, since the pressure-distribution given by (54) has zero moments about the axes. The direct identification of (56) with (13) is also not difficult.

Although it is not essential to our purpose, we may substitute the values of α, β, γ obtained from (48) in (55). Eliminating λ, we get

$$a\ddot{a} - a^2(q^2+r^2+q_1^2+r_1^2) - 2caqq_1 - 2abrr_1 + 2\pi\rho a^2\alpha_0 + A'a^2$$
$$= b\ddot{b} - b^2(r^2+p^2+r_1^2+p_1^2) - 2abrr_1 - 2bcpp_1 + 2\pi\rho b^2\beta_0 + B'b^2$$
$$= c\ddot{c} - c^2(p^2+q^2+p_1^2+q_1^2) - 2bcpp_1 - 2caqq_1 + 2\pi\rho c^2\gamma_0 + C'c^2. \qquad (57)$$

These, together with (13), (14), and (44), may be taken to be our fundamental system of equations.

So far there is no approximation, and the equations would be applicable, for instance, to the finite oscillations of a Jacobian ellipsoid under a disturbing potential of the type (53). In the case, however, of a *slight* disturbance from a state of steady rotation about the axis of z, the quantities p, q, p_1, q_1, r_1 will be small, whilst r will be approximately constant. It follows that, if we neglect small quantities of the second order in the first two of equations (13) and the first two of (14), the coefficients may be treated as constants. The changes in the instantaneous axis are therefore independent of the tidal deformation, and are the same as if the fluid had been enclosed by a rigid envelope of negligible mass.

The tidal oscillations of the free surface, on the other hand, are determined by the equations (57), together with (44) and the third equations of the systems (13) and (14), respectively. These latter, it may be noted, take the forms

$$\frac{d}{dt}(Cr + Hr_1) = N, \qquad \frac{d}{dt}(Hr + Cr_1) = 0. \qquad (58)$$

When the undisturbed ellipsoid is one of revolution about the axis of z, the precessional equations reduce as before to the forms (20) and (21). Moreover, in the astronomical application, that part of the disturbing potential which is effective as regards precession consists of terms of the form

$$\Omega' = -kr^2 \sin\theta \cos\theta \cos(\sigma t + \phi), \qquad (59)$$

where σ is very nearly equal to ω; cf. Art. 219 (1) and p. 361. In Cartesian co-ordinates we have

$$\Omega' = kz(y \sin\sigma t - x\cos\sigma t). \qquad (60)$$

This makes

$$L = -k(C-A)\sin\sigma t, \qquad M = -k(C-A)\cos\sigma t, \qquad N = 0. \qquad (61)$$

Thus

$$L + iM = -ik(C-A)e^{-i\sigma t}. \qquad (62)$$

The argument, leading to the conclusion that the precession is, under a certain condition, the same as if the mass had been solid, then takes the same course as in the preceding Art.

When the disturbing function has the form (59), the oscillations in the semi-axes a and c correspond to diurnal tides in the case of the earth.

LIST OF AUTHORS CITED

The numbers refer to the pages

INDEX

The numbers refer to the pages